Fast Processes in Large-Scale Atmospheric Models

Geophysical Monograph Series

Geophysical Monograph 282

Fast Processes in Large-Scale Atmospheric Models

Progress, Challenges, and Opportunities

Yangang Liu
Pavlos Kollias
Editors

Leo J. Donner
Advisor

This Work is a co-publication of the American Geophysical Union and John Wiley and Sons, Inc.

AGU
ADVANCING EARTH AND SPACE SCIENCES

WILEY

Published under the aegis of the AGU Publications Committee

Matthew Giampoala, Vice President, Publications
Carol Frost, Chair, Publications Committee
For details about the American Geophysical Union visit us at www.agu.org.

Registered Office
John Wiley & Sons, Inc., 111 River Street, Hoboken, NJ 07030, USA

Editorial Office
111 River Street, Hoboken, NJ 07030, USA

For details of our global editorial offices, customer services, and more information about Wiley products visit us at www.wiley.com.

Wiley also publishes its books in a variety of electronic formats and by print-on-demand. Some content that appears in standard print versions of this book may not be available in other formats.

Library of Congress Cataloging-in-Publication Data

Names: Liu, Yangang, editor. | Kollias, Pavlos, editor. | Donner, Leo Joseph, 1956- other.
Title: Fast processes in large scale atmospheric models : progress, challenges, and opportunities / Yangang Liu, Pavlos Kollias, editors ; Leo J. Donner, advisor.
Description: Hoboken, NJ : Wiley-American Geophysical Union, 2024. | Series: Geophysical monograph series | Includes index.
Identifiers: LCCN 2023015913 (print) | LCCN 2023015914 (ebook) | ISBN 9781119528999 (cloth) | ISBN 9781119528968 (adobe pdf) | ISBN 9781119528944 (epub)
Subjects: LCSH: Atmospheric models.
Classification: LCC QC861.3 .F378 2024 (print) | LCC QC861.3 (ebook) | DDC 551.501/13–dc23/eng/20231012
LC record available at https://lccn.loc.gov/2023015913
LC ebook record available at https://lccn.loc.gov/2023015914

Cover image: © Guilherme Almeida da Silva/Getty Images
Cover design: Wiley

Set in 9.5/12pt TimesNewRomanMTStd by Straive, Chennai, India
Printed and bound by CPI Group (UK) Ltd, Croydon, CR0 4YY

C9781119528999_271123

CONTENTS

LIST OF CONTRIBUTORS

M. Joan Alexander
NorthWest Research Associates
Boulder, CO, USA

Vassilis Amiridis
Institute for Astronomy, Astrophysics, Space Applications and Remote Sensing
National Observatory of Athens
Athens, Greece

Julio T. Bacmeister
Climate and Global Dynamics Laboratory
National Center for Atmospheric Research
Boulder, CO, USA

Michael Barlage
National Center for Atmospheric Research
Boulder, CO, USA

Alexei Belochitski(†)
I. M. Systems Group
Environmental Modeling Center
College Park, MD, USA

Kamal Kant Chandrakar
National Center for Atmospheric Research
Boulder, CO, USA

Fei Chen
National Center for Atmospheric Research
Boulder, CO, USA

Anthony B. Davis
NASA Jet Propulsion Laboratory
California Institute of Technology
Pasadena, CA, USA

Xiquan Dong
Department of Hydrology and Atmospheric Sciences
University of Arizona
Tucson, AZ, USA

Leo J. Donner
NOAA Geophysical Fluid Dynamics Laboratory
Princeton University
Princeton, NJ, USA

Graham Feingold
Chemical Sciences Laboratory
NOAA Earth System Research Laboratories
Boulder, CO, USA

Sinan Gao
Collaborative Innovation Center on Forecast and Evaluation of Meteorological Disasters / Key Laboratory for Aerosol-Cloud-Precipitation of China Meteorological Administration
Nanjing University of Information Science and Technology
Nanjing, China

Virendra P. Ghate
Environmental Science Division
Argonne National Laboratory
Lemont, IL, USA

Yu Gu
Joint Institute for Regional Earth System Science and Engineering, and Department of Atmospheric and Oceanic Sciences
University of California Los Angeles
Los Angeles, CA, USA

Christopher Heale
Department of Physical Sciences
Embry–Riddle Aeronautical University
Daytona Beach, FL, USA

Pavlos Kollias
Environmental and Climate Sciences Department
Brookhaven National Laboratory
Upton, NY, USA
and
School of Marine and Atmospheric Science
Stony Brook University
Stony Brook, NY, USA

Ilan Koren
Department of Earth and Planetary Sciences
Weizmann Institute
Rehovot, Israel

Vladimir Krasnopolsky
Environmental Modeling Center/National Centers for Environmental Predictions/National Weather Service
NOAA Center for Weather and Climate Prediction
College Park, MD, USA

Christopher G. Kruse
NorthWest Research Associates
Boulder, CO, USA

Marcin J. Kurowski
NASA Jet Propulsion Laboratory
California Institute of Technology
Pasadena, CA, USA

Katia Lamer
Department of Environmental and Climate Sciences
Brookhaven National Laboratory
Upton, NY, USA

Vincent E. Larson
Department of Mathematical Sciences
University of Wisconsin–Milwaukee
Milwaukee, WI, USA
and
Pacific Northwest National Laboratory
Richland, WA, USA

Wuyin Lin
Department of Environmental and Climate Sciences
Brookhaven National Laboratory
Upton, NY, USA

Alexander Marshak
NASA Goddard Space Flight Center
Greenbelt, MD, USA

Kuo-Nan Liou(†)
Joint Institute for Regional Earth System Science and Engineering, and Department of Atmospheric and Oceanic Sciences
University of California Los Angeles
Los Angeles, CA, USA

Xiaohong Liu
Department of Atmospheric Sciences
Texas A&M University
College Station, TX, USA

Yangang Liu
Environmental and Climate Sciences Department
Brookhaven National Laboratory
Upton, NY, USA

Ulrich Loehnert
Institute of Geophysics and Meteorology
University of Cologne
Cologne, Germany

Chunsong Lu
Collaborative Innovation Center on Forecast and Evaluation of Meteorological Disasters / Key Laboratory for Aerosol-Cloud-Precipitation of China Meteorological Administration
Nanjing University of Information Science and Technology
Nanjing, China

Eleni Marinou
Institute for Astronomy, Astrophysics, Space Applications and Remote Sensing
National Observatory of Athens
Athens, Greece

Allison McComiskey
Department of Environmental and Climate Sciences
Brookhaven National Laboratory
Upton, NY, USA

David B. Mechem
Department of Geography and Atmospheric Science
University of Kansas
Lawrence, KS, USA

Patrick Minnis
Analytical Mechanics Associates, Inc.
Hampton, VA, USA

Jadwiga H. Richter
Climate and Global Dynamics Laboratory
National Center for Atmospheric Research
Boulder, CO, USA

Sabrina Schnitt
Institute of Geophysics and Meteorology
University of Cologne
Cologne, Germany

Raymond A. Shaw
Atmospheric Sciences Program
Michigan Technological University
Houghton, MI, USA

Cheng Sun
Collaborative Innovation Center on Forecast and Evaluation of Meteorological Disasters / Key Laboratory for Aerosol-Cloud-Precipitation of China Meteorological Administration
Nanjing University of Information Science and Technology
Nanjing, China

Kay Suselj
NASA Jet Propulsion Laboratory
California Institute of Technology
Pasadena, CA, USA
and
Joint Institute for Regional Earth System Science and Engineering
University of California Los Angeles
Los Angeles, CA, USA

João Teixeira
NASA Jet Propulsion Laboratory
California Institute of Technology
Pasadena, CA, USA
and
Joint Institute for Regional Earth System Science and Engineering
University of California Los Angeles
Los Angeles, CA, USA

Junhong Wei
School of Atmospheric Sciences and Guangdong Province Key Laboratory for Climate Change and Natural Disaster Studies
Sun Yat-sen University
Guangzhou, China
and
Southern Marine Science and Engineering Guangdong Laboratory
Zhuhai, China

Shaocheng Xie
Atmospheric, Earth and Energy Division
Lawrence Livermore National Laboratory
Livermore, CA, USA

Xiaoqi Xu
Collaborative Innovation Center on Forecast and Evaluation of Meteorological Disasters / Key Laboratory for Aerosol-Cloud-Precipitation of China Meteorological Administration
Nanjing University of Information Science and Technology
Nanjing, China

PREFACE

Computer models are essential tools for understanding atmospheric phenomena and for making accurate predictions of any changes in the Earth's climate, weather, and resources of renewable energy resulting from anthropogenic activities that generate greenhouse gases and particulates into the atmosphere. Many physical processes that influence Earth's climate and weather occur on spatial (temporal) scales smaller (shorter) than typical grid sizes (time steps) of general circulation models, and thus must be parameterized.

This book focuses on the atmospheric subgrid processes—collectively called fast physics—by reviewing and synthesizing relevant physical understanding, parameterization developments, various measurement technologies, and model evaluation framework. The book contains 18 chapters and is divided into three parts to reflect and synthesize the multiple aspects involved.

The first chapter briefly introduces the historical development of fast physics parameterizations and the involved complexities. Part I is devoted to discussing major subgrid processes, with eight chapters (Chapters 2–9) each covering different processes more or less in the conventional compartmentalized format that emphasizes individual processes. Topics covered include, but not limited to, radiative transfer, aerosols, and aerosol direct and indirect effects; entrainment-mixing processes and their microphysical influences; convection and convective clouds; stratiform clouds such as stratus and stratocumulus; planetary boundary layer processes; land surface and its interactions with the atmosphere; and gravity waves. On top of the conventional treatments, some promising ideas/approaches are described that have recently emerged to unify the treatment of individual processes and thus allow for consideration of process interactions.

Part II is devoted to such unifying efforts, with four chapters (Chapters 10–13) covering four different endeavors: the unifying parameterizations based on assumed probability density functions; the EDMF approach that combines the eddy–diffusivity and mass–flux approaches to unify turbulence and convection; application of machine learning techniques; and innovative top-down attempts that consider the involved totality by borrowing ideas from systems theory, statistical physics, and nonlinear sciences.

Part III (Chapters 14–17) is devoted to assessments, model evaluation, and model-measurement integration, with four chapters that focus on satellite and airborne remote-sensing measurements; surface-based remote-sensing measurements; in situ and laboratory measurements; and model evaluation and model-measurement integration, respectively. The final chapter of the book summarizes emerging challenges, new opportunities, and future research directions.

The development of the book happened around two noteworthy events. The first was that the 2021 Nobel Prize in Physics was awarded to three pioneers in modeling climate and weather and studying complex systems (Syukuro Manabe of Princeton University, USA; Klaus Hasselmann of the Max Planck Institute for Meteorology, Germany; and Giorgio Parisi of Sapienza University of Rome, Italy). This exciting choice accentuates not only the critical importance of the subject but also the outstanding challenges of the topics discussed in this book.

The second event was the COVID-19 pandemic, which unfortunately overlapped with the writing of most of the chapters in this book and affected the lives of many of our contributors. We would like to express our special thanks to all the authors and reviewers, as well as to the staff at Wiley and AGU for their hard work and patience as we completed this book under these circumstances.

This book is dedicated to two of our dear colleagues and contributing authors who passed away during this period, Kuo-Nan Liou and Alexei Belochitski. The book is also dedicated to Yangang Liu's mother, Chunlan Sun, who was hospitalized during the pandemic and passed away in China recently without his company.

Yangang Liu
Brookhaven National Laboratory, USA
Pavlos Kollias
Brookhaven National Laboratory, USA
and Stony Brook University, USA

1

Progress in Understanding and Parameterizing Fast Physics in Large-Scale Atmospheric Models

Yangang Liu[1] and Pavlos Kollias[1,2]

ABSTRACT

This introductory chapter discusses the atmospheric subgrid processes – collectively called "fast physics" or "fast processes" – and their parameterizations in large-scale atmospheric models. It presents a brief historical progression of the parameterization of fast processes in numerical models. Despite great efforts and notable advances in understanding, progress in improving fast physics parameterizations has been frustratingly slow, the underlying reasons for which are explored. To guide readers, this chapter describes the main objectives and scope of this book and summarizes each chapter.

1.1. FAST PHYSICS AND PROGRESS OF PARAMETERIZATION DEVELOPMENT

Large-scale atmospheric models are integral components of weather and climate models. Ongoing developments in high-resolution modeling (i.e., global storm-resolving models (GSRMs; Stevens et al., 2019); Energy Exascale Earth System Model (E3SM; Rasch et al., 2019); and large-eddy simulations (LES; Gustafson et al., 2020)) have resulted in ultra high-resolution numerical simulations of atmospheric systems. Despite these advancements, coarser resolution large-scale models remain our main modeling capability for future climate predictions. Many atmospheric processes and phenomena that influence Earth's weather and climate occur at spatiotemporal scales that are too small to be resolved in these large-scale atmospheric models and must be parameterized – approximately represented by the variables that can be resolved by the model grids. In this book, we refer to this array of parameterized subgrid processes and phenomena collectively as "fast physics" or "fast processes" for convenience, including radiative transfer, aerosol/cloud physics, convection, boundary layer processes, gravity wave (GW), and land-atmosphere interactions.

While early parameterizations of fast physics used simple and often empirical or ad hoc relationships (e.g., the Kessler bulk parameterization for representing cloud microphysical processes; Kessler, 1969), later parameterization development was concerned about building conceptual models with increasingly detailed physical processes by leveraging theoretical analysis, observations, and/or detailed process modeling studies.

Furthermore, parallel to the continuing improvement/development of parameterizations for individual fast processes, there has been growing interest in studying and understanding interactions/couplings among different processes. Significant progress has been made and several promising approaches have emerged since late 1900s and early 2000s. Figure 1.1 illustrates the approximate timelines in developing fast physis parameterizations in context of the conventional parameterizations that target individual processes as well as several unifying efforts that addresses multiple processes together.

[1] *Environmental and Climate Sciences Department, Brookhaven National Laboratory, Upton, NY, USA*

[2] *School of Marine and Atmospheric Sciences, Stony Brook University, Stony Brook, NY, USA*

Fast Processes in Large-Scale Atmospheric Models: Progress, Challenges, and Opportunities, Geophysical Monograph 282, First Edition.
Edited by Yangang Liu and Pavlos Kollias.

DOI: 10.1002/9781119529019.ch01

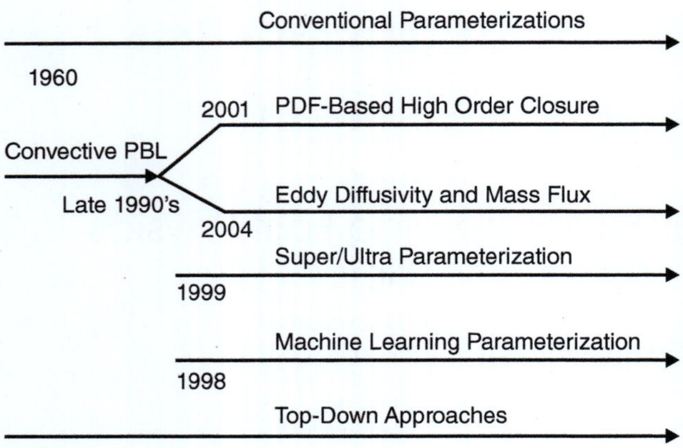

Figure 1.1 Schematic of the approximate timelines of development of fast physics parameterizations. Conventional parameterizations are focused on individual fast processes. The four lines of unifying efforts (PDF-Based High Order Closure, Eddy Diffusivity and Mass Flux, Super/Ultra Parameterization, and Machine Learning Parameterization) aim to unify the representation of more than two physical processes. Top-down approaches borrow holistic ideas that have been scattered in various disciplines (e.g., nonlinear systems dynamics, statistical physics, information theory, self-organization, networks, and pattern formation).

Despite remarkable efforts and increasing recognition of the importance of these fast processes over the past few decades, progress remains frustratingly slow in improving their representation in models. As a result, their impacts on future climate predictions remain poorly understood and highly uncertain. The slow progress is perhaps best attested by the historical lack of change in the ranges of climate sensitivity across models from the celebrated 1979 US National Research Council report (Charney et al., 1979) to the latest (6th) Coupled Model Intercomparison Project (CMIP6) results used in the Intergovernmental Panel on Climate Change (IPCC) report (Figure 1.2). Deficient fast physics parameterizations, especially those related to clouds, have been thought to be primarily responsible for the stubborn large spread of model climate sensitivity (Meehl et al., 2020; Zelinka et al., 2020). Aerosol climate forcing in climate models has been fraught with similarly unchanged uncertainty (for details refer to Chapter 3 of this book).

The slow progress can be attributed to two overarching types of complexities (also see Jakob, 2010; Randall, 2013). The first lies in the "4M-2N complexities" inherently accompanying the atmosphere and associated physical processes (Figure 1.3). Briefly, fast processes and especially those cloud-related ones involve *multibody* (sub)systems with numerous particles of different sizes and shapes, in which multiple physical processes (*multiphysics*) occur over a wide range of spatiotemporal scales (*multiscale*) and interact with one another, and manifest themselves in a variety of cloud types such as cumulus and stratiform clouds (*multitype*). The equations describing these processes are often highly *nonlinear* and exhibit *non-Gaussian* statistics (Lovejoy & Schertzer, 2010).

The other inherent complexity lies with that model development involves an iterative cycle of developing parameterizations, implementing and evaluating parameterizations against observations to identify potential parameterization deficiencies and further improvement. This iterative procedure calls for an organic integration of

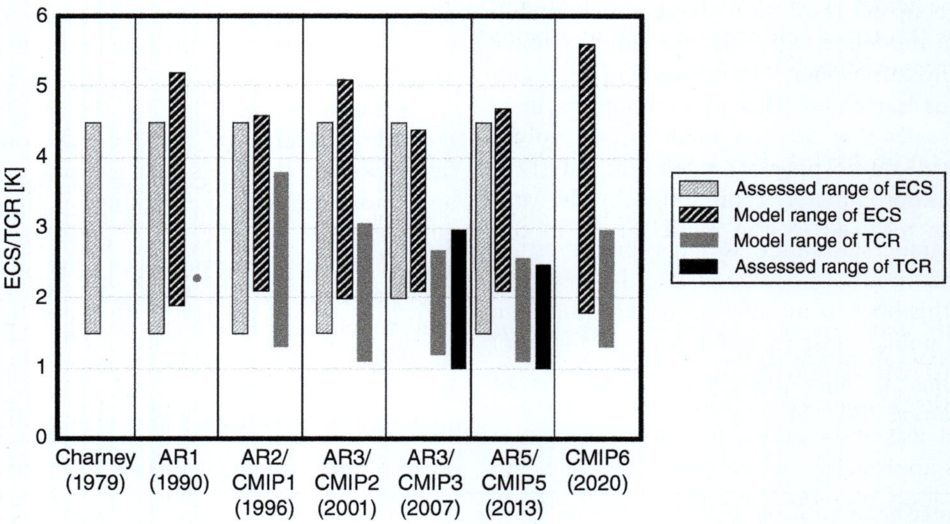

Figure 1.2 Historical values of equilibrium climate sensitivity (ECS) and transient climate response (TCR). Source: Adapted from Meehl et al. (2020), which can be consulted for details on the data sources and definitions.

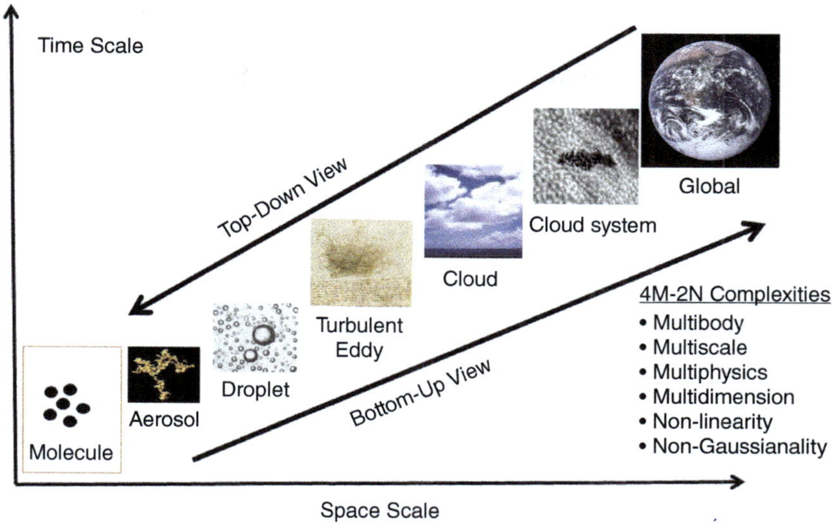

Figure 1.3 Schematic to illustrate the atmospheric scale hierarchy and involved "4M-2N complexities." Together with the "operational complexity" discussed in the text, these science complexities have posed and will continue to pose challenges to model development in general and fast physics parameterizations in particular. Source: Leonardo da Vinci/Wikimedia Commons and NASA/Wikimedia Commons/Public Domain.

the key components involved ranging from modeling to measurements, which in turn demands effective coordination of expertise in distinct areas. However, effective coordination and collaboration across different disciplines and institutions are not trivial, and such an "operational complexity" adds another layer of technological and social challenges in virtually every step of model development. The issue will become more acute as the field is moving toward more emphasis on process interactions with ever-increasing data volumes and model resolutions. To echo Jakob (2010), "... acceleration in model development can only be achieved by significantly strengthening these weak links through additional research and better coordination across existing programs."

1.2. OBJECTIVES AND SCOPE OF THE BOOK

The objectives of this book are threefold. First, to survey advances in understanding of key fast processes and their parameterization developments (Part I). In particular, Part II of this book is uniquely devoted to unifying efforts. Second, unlike most review articles or the book by Stensrud (2007) on fast physics parameterizations, this book includes discussions on measurement techniques and studies that use observations for model evaluation and thus covers approaches to addressing the weak link in the iterative loop of model development. Third, by surveying the recent advances in key areas, we hope to reveal new challenges, opportunities, and directions for future research.

It is worth noting that the related literature is enormous and that the selection of the material in this text is nonexhaustive and likely biased to the authors' own research interests. On the other hand, books focusing on fast physics parameterizations are rare; the only one we are aware of is Stensrud (2007), which is primarily on conventional parameterizations of individual fast processes in numerical weather prediction (NWP) models. Bringing together modeling and measurements with a common goal of parameterization development and evaluation and including multiple unifying efforts are unique to this book.

1.3. BOOK STRUCTURE AND SUMMARY OF CHAPTERS

Fast physics in large-scale atmospheric models involves multiple processes that occur over a wide range of spatiotemporal scales. Progress has been made on many fronts and new promising directions of research are emerging. To reflect and synthesize the multiple facets involved, this book is divided into three parts. Part I deals with the major subgrid processes, with eight chapters (Chapters 2–9) covering different fast processes. Beyond conventional treatments, some promising approaches have recently emerged to unify the treatment of (some) processes and thus allows for consideration of process interactions. Part II is devoted to such unifying efforts, with four chapters (Chapters 10–13) that each cover a different endeavor. Part III is devoted to measurements, model evaluation, and model-measurement integration, with four chapters (Chapters 14–17) that focus on satellite and airborne remote sensing measurements, surface-based remote sensing measurements, in situ and laboratory measurements, and model evaluation and model-measurement integration, respectively.

1.3.1. Process Studies and Parameterizations

Essential to the Earth's climate and weather and understanding climate change is the understanding and representation of the solar (shortwave) and terrestrial (longwave) radiation and of radiative transfer processes such as absorption, and scattering. In Chapter 2, Gu and Liou present the fundamentals of radiative transfer and its interactions with the atmosphere, and summarize the commonly used radiative transfer parameterization schemes in atmospheric models. Also discussed are several more advanced topics in the study of the atmospheric radiation, including cloud vertical overlapping, cloud horizontal inhomogeneity, and 3D radiative transfer in both the cloudy atmosphere and over complex rugged land surfaces such as mountainous terrains. In particular, the chapter highlights that the current commonly used radiation schemes normally represent 1D transport in the vertical direction, although radiative transfer in 3D atmosphere and surfaces could play an important role in determining the radiation budget and radiative heating at the top of the atmosphere, at the surface, and within the atmosphere. Both horizontal and vertical subgrid scale inhomogeneities and 3D radiative transfer may substantially influence the radiative transfer within clouds and cloud-radiation interactions, suggesting the need for further investigation and for improving their representations in models.

Atmospheric aerosols are suspensions of solid particles or liquid droplets in the air. Aerosols contain multiple compositions, exhibit various morphologies, and span a few orders of magnitude in sizes from a few nanometers to tens of micrometers. Aerosol radiative effects constitute one of the largest uncertainties in climate projection, and the large spread of simulated values among general circulation models (GCMs) can be traced to different representations of aerosol processes, including emissions, transport, formation and removal, and aerosol-cloud interactions. In Chapter 3, Liu provides an overview of atmospheric aerosols and their climatic impacts through both aerosol direct effects on radiation (aerosol-radiation interactions) and aerosol indirect effects (aerosol-cloud interactions). The authors focus on addressing topics related to three aerosol-related questions: (1) How are aerosol properties and processes as well as aerosol-cloud interactions represented and compared in current GCMs? (2) What are the major assumptions, simplifications, and weaknesses of the current representations? (3) Why are there large uncertainties in the aerosol climate effects from GCMs? Several future directions are highlighted.

Although entrainment of surrounding dry air into clouds, subsequent turbulent mixing processes, and their microphysical influences have been known to be essential in determining cloud microphysical and related properties for some time, theoretical understanding of these processes is still far from complete, and their parameterizations in atmospheric models are in their infancy. In Chapter 4, Lu, Liu, Xu, Gao, and Sun discuss these issues in shallow clouds (cumulus and stratocumulus clouds), focusing on two critical yet understudied aspects: entrainment-mixing mechanisms and entrainment rate. Different conceptual models of entrainment-mixing mechanisms are reviewed, and latest studies on unifying microphysical measures to quantify different entrainment-mixing mechanisms are presented. Approaches for estimating fractional entrainment rate in cumulus clouds are summarized; relationships of entrainment rate to internal cloud properties (e.g., vertical velocity) or external properties (e.g., relative humidity in environment) are discussed as plausible parameterizations. Three approaches for estimating entrainment velocity in stratocumulus clouds are also discussed. Several topics are highlighted for future research, e.g., the connection between entrainment rate, entrainment-mixing mechanisms, and relationships to other factors (e.g., rain initiation, detrainment, spectral shape of cloud droplet size distributions, entrained aerosols, and environmental relative humidity).

Following the discussion on entrainment in shallow cumulus clouds and its role in shallow convection parameterization, Donner turns to deep convection from the perspective of large-scale flows in Chapter 5. The chapter begins with discussing the effects of convection on large-scale flows in which it is embedded, follows with strategies for solving the problem of cumulus parameterization, and concludes with a brief overview of interactions between convection and momentum, chemistry, tracers, cloud microphysics, and aerosols. Emphasized are the roles of convective vertical velocities in treating aerosol-cloud interactions and cloud microphysics related to cloud feedbacks. Major deficiencies in existing parameterizations are discussed, including interactions between deep convection and aerosols, convection-chemistry interactions, understanding and representation of convective organization, and knowledge of convective-scale pressure-gradient forces in treating effects of convection on momentum fluxes. Limitations of mean-state perspectives and the widely used quasi-equilibrium assumption are discussed. Also touched on are connections with other topics (e.g., scale awareness, higher-order closure, multiscale modeling frameworks and high-resolution models without conventional deep convection parameterizations, shallow convection, boundary-layer processes, and gravity waves) detailed in other chapters.

Besides convective clouds, stratiform clouds including stratus and stratocumulus clouds constitute another critical component of the atmospheric system that significantly affects climate and has long been the subject of active research from many perspectives. In Chapter 6, Dong and Minnus provide an overview of such clouds, with a focus on what we have learned from observational studies in terms of improving their parameterization in atmospheric models. Stratus and stratocumulus cloud properties and their importance are discussed based on measurements from trained surface observers, satellite and ground-based remote sensors, and aircraft field campaigns. The processes that determine the variations in stratocumulus properties and govern where and when they occur are discussed, along with such factors as aerosols, radiation, and humidity. Retrieval methods used for extracting information about stratus and stratocumulus clouds from satellite- and ground-based sensors are also briefly reviewed, with an emphasis on the knowledge learned for improving understanding and parameterizations of such clouds in large-scale models. Unique consistency between the early trained observers and the state-of-the-art technologies is demonstrated; synergy of different observational platforms is highlighted for future investigation. Emerging but understudied phenomena are summarized, including the impact of low-level temperature advection, veil clouds developing at the top of the marine boundary layer in areas of open-cell and unorganized cellular convection, the role of gravity waves in the subtropical jet stream in initiating Pocket of Cells in some closed-cell stratocumulus over the southeast Pacific, and effects of land-sea breezes. Outstanding issues in profiling marine boundary layer cloud and drizzle microphysical properties are highlighted, including the need for incorporating cloud-top entrainment, drizzle, and vertical and horizontal inhomogeneities to address the issue of nonadiabatic multispectral retrievals.

As a layer between the ground surface and the free troposphere, the planetary boundary layer (PBL) is often turbulent and particularly important, because the majority of biota (including humans) and climatically important low clouds like stratocumulus and shallow cumulus reside. Even deep convection is highly related to the properties of the plumes or thermals originating in the PBL. In Chapter 7, Ghate and Mechem introduce the PBL structure and the commonly used theoretical approaches for investigating the PBL. A hierarchy of models for representing the boundary layer is presented, including mixed-layer models, first-order closure, 1.5-order TKE closure, and higher-order closure approaches. Challenges for evaluating the emerging advanced schemes (high order, PDF-based, or EDMF) are also discussed in context of the inherent needs for observations of joint PDFs of vertical air motion and thermodynamic variables. The discussion emphasizes the buoyancy-driven convective boundary layer but briefly mentions impacts of shear and clouds. The chapter concludes with a brief historical context and future outlook for representing the boundary layer in large-scale atmospheric models. To some extent, this chapter can be viewed as an introduction to Chapters 13 and 14 where the PDF-based and EDMF schemes are detailed.

Although the focus of this book is on atmospheric processes, the weather and climate system consists of other subsystems that strongly interact with the atmosphere over a wide range of spatiotemporal scales. In particular, various surface processes are fundamental to the exchange of heat, water, and momentum between the surface and the atmosphere through PBL. Thus, modeling land-surface processes has been an integral component of atmospheric models. In Chapter 8, Barlage and Chen focus on recent progress in understanding and modeling the biophysical effects of the human dimension, especially urbanization and agriculture, on surface water and energy budgets, and their cascading effects on weather and climate including clouds, aerosols, convection, and precipitation. Well-known phenomena are discussed, including the Urban Heat Island and urban impacts on precipitation through both cloud microphysical and/or dynamical effects. Also discussed are the unique roles of rough vegetated or urban canopy in determining turbulent fluxes (e.g., evapotranspiration over vegetated regions can exceed evaporative flux from the oceans because of larger surface roughness and stronger turbulence whereas the moisture flux can be effectively shut off when the land is dry). Land-surface models (parameterizations) of 3D subgrid structures within urban or vegetation canopies are presented, including the most sophisticated multilayer scheme – Building Effect Parameterization. Challenges for evaluating and applying such a comprehensive land-surface model are discussed, including existence and specification of the large number of tunable parameters used in urban canopy models.

Atmospheric GWs have horizontal wavelengths ranging from 1 to thousands of kilometers. Current climate models and even NWP models cannot resolve significant portions of their momentum flux, and parameterizations are necessary to represent their under- or unresolved effects in atmospheric models. In Chapter 9, Kruse, Richter, Alexander, Bacmeister, and Wei discuss GWs that are important at nearly all levels of the atmosphere, especially for the general circulation of the middle and upper atmosphere. GWs in the tropical stratosphere significantly contribute to the driving of the quasi-biennial oscillation and the stratospheric and mesospheric semiannual oscillation, both primary modes of variability up there. In the extratropics, GWs contribute to the driving of the stratospheric Brewer-Dobson circulation and significantly influence the strength of the polar night jet and the corresponding polar temperatures. Additionally, GWs are responsible for the cold summer mesopause and the reversal of extratropical zonal mean winds in the mesosphere. Also discussed are both primary sources of atmospheric GWs (i.e., flows over mountains, moist convection, and imbalances in jets and frontal systems) and secondary GWs generated as a result of dissipation of primary GWs. The basic theory of GW generation, propagation, and dissipation and commonly used GW parameterizations are presented. Uncertainties, parameter tuning, and known missing processes in current parameterizations are explored as well. The importance of gravity wave in shaping clouds (Chapter 7) and in determining cloud microphysical properties (Chapter 3) is gradually recognized as well.

1.3.2. Unifying Efforts

Chapter 10 is the first of the four chapters that introduce the emerging efforts to unify the parameterizations of different processes, with a focus on higher-order equations closed by assuming the shape of the probability density function (PDF) of fields on the subgrid scale (PDF-based method for short). In this chapter, Larson presents the general equations involved. Theoretical analysis of the higher-order equations reveals that they contain the flux-of-flux terms that lead to nonlocal cumulus transport, along with a detailed representation of buoyant generation of turbulence, which is the essential source term of convection and can be closed by a multivariate PDF. Instead, traditional low-order closure omits the flux-of-flux terms that are crucial for representing nonlocal cumulus transport. The popular Cloud Layers Unified By Binormals (CLUBB) is detailed as an example of such PDF-based methods. Other higher-order closure models are also briefly discussed, including the Intermediately Prognostic Higher-Order Closure (IPHOC) parameterization (Cheng et al., 2010) and the Turbulence Kinetic Energy-Scalar Variance (TKESV) parameterization of Mironov and Machulskaya (2017). The IPHOC prognoses all the moments prognosed by CLUBB, plus two additional third-order moments of water vapor and potential temperature; TKESV prognoses TKE and scalar variances, and optionally a third-order moment related to cloud liquid water. The connections of the PDF-based method to the conventional mass-flux method for convection and low-order closure for turbulence are also discussed.

Another approach that seeks to unify the treatment of convection and turbulent processes in PBL is conceptually more direct, combining the widely used eddy diffusivity approach for local turbulent transport with the mass-flux scheme for convection. In Chapter 11, Teixeira, Suselj, and Kurowski discuss the EDMF approach. After briefly reflecting on the early development in the late 1990s, this chapter focuses on the new stochastic multiplume EDMF scheme that can realistically represent the dry boundary layer, stratocumulus, shallow, and even deep cumulus convection within a single framework. The variability of updraft properties is parameterized using joint PDFs of thermodynamic properties to initialize multiple updrafts. Lateral entrainment is parameterized as a stochastic process. Furthermore, the unified EDMF parameterization explicitly considers the horizontal resolution of the model, paving the way to a scale-aware extension of the scheme. Both the fundamentals and latest results of using the new EDMF scheme are introduced. The multiplume framework allows for the coexistence of different convective regimes (i.e., dry plumes, shallow moist convection, and even deep convection) within a single grid box, without any artificial separations between them and with scale-adaptive capabilities for use in next-generation weather and climate models with high and variable horizontal resolutions.

The PDF-based and EDMF approaches both aim to unify the parameterizations of turbulence, PBL, and convection (especially shallow convection). Further coupling with other processes such as cloud microphysics remains an area of active research for both approaches. Around similar times in the late 1990s and early 2000s, ideas of superparameterization – that embed cloud-resolving models (CRM) in climate model grid column – were proposed and developed as a way to replace all the subgrid processes that the embedded CRM model represents, including turbulence, PBL, convection, cloud microphysics, and radiation (Grabowski, 2001; Randall, 2013). Recently, similar ideas were extended to using high-resolution LES models instead of CRMs in so-called ultraparameterization (Parishani et al., 2017). Obviously, the benefits of such multiscale modeling approaches come at high computational cost and call for more computationally effective approaches that can be used as alternative to represent multiple fast processes together. Note that even for LES models, many sub-LES processes such as cloud microphysics and turbulence-microphysics interactions remain to be parameterized (Liu et al., 2023). In Chapter 12, Krasnopolsky and Belochitski describe applications of machine learning (ML) approaches to emulate existing parameterizations and developing new ML surrogate models as new parameterizations. The authors first argue that a parameterization can be formulated as a generic problem of mathematical mapping and then argue that ML tools can be used to emulate and/or approximate the involved mathematical mappings. Four mapping complexities (physical complexity, mathematical complexity, numerical/computational complexity, and functional complexity) are discussed. Further discussed are ML applications to emulate existing parameterizations, to develop new parameterizations, to ensure physical constraints, and to control the accuracy of developed applications. Some ML approaches that allow developers to go beyond the standard parameterization paradigm are discussed as well. Limitations of ML models are also discussed, including inability to provide a meaningful physical interpretation of underlying processes, requirements of large data for training and testing purposes, and their limited generalizability to out-of-sample scenarios. Given that neither an ML-only nor a physically based-only approach can be considered sufficient for complex scientific and engineering applications, the research community has been exploring the hybridization of physically based and ML-based models, where both scientific knowledge and data are integrated in a synergistic manner. It is reasoned that this hybrid paradigm is fundamentally different from the ML mainstream where domain-specific knowledge is often considered secondary, and several differences are discussed. The concept of compound parameterization that combines an ML parameterization, the original physically based parameterization, and a quality control procedure is introduced.

Despite the tremendous advances and different extents in dealing with the number of fast processes and their interactions discussed in the previous chapters, a common theme of those studies is that they are all essentially bottom-up-based and aim to upscale subgrid scale processes to grid variables. However, the climate system, including its atmospheric component, is a multiscale complex system that involves highly nonlinear bottom-up and top-down scale interactions (recall Figure 1.3). Without considering the top-down direction, our understanding would never be complete, and the physical pictures from the unifying efforts could be as murky as understanding the output of a full GCM. As another unique addition of this book compared to existing ones, in Chapter 13, Feingold and Koren summarize innovative ideas that attempt to consider processes holistically but are scattered in various disciplines including nonlinear dynamics (e.g., chaos theory), statistical physics, information theory, self-organization, networks, pattern formation, and general systems theory. The "top-down view" is focused on system-wide behavior and emergent phenomena, distinguishing from the traditional "bottom-up" view that focuses on individual processes. In particular, a new behavior at a larger scale can emerge from interactions/couplings between detailed processes and between the involved subsystems at a finer scale. And this type of order/emergence is not driven by an external force, but instead grows spontaneously from local interactions, or is "self-organized." Spatiotemporal communication between components of a system is key to development of synchronization, patterns, and self-organization. It is argued that the top-down approach can yield simple holistic models that are more amenable to interrogation and digestion than complex, detailed models. Concepts and terminologies that are not familiar to the atmospheric modeling, especially the parameterization and community, are introduced and discussed, including fixed points, attractors, limit cycles, chaotic state, bifurcation points, synchronization, information content, and entropy. In addition to their distinct foci on local and detailed physical processes vs. process interactions and emergence, this chapter also provides some intriguing examples to elucidate the conceptual differences between the bottom-up and top-down approaches from other perspectives: reductionism vs. holism; basic building blocks vs. integrative view; models representing a large vs. a reduced number of degrees of freedom; models rooted in mathematical representation of physical/chemical/biological processes vs. models that are an abstraction of these processes; and complexity vs. simplicity. The authors use aerosol-cloud-precipitation system as a particular example to demonstrate the great potentials of the top-down view and the need to integrate the complementary top-down view and bottom-up thinking in addressing the remaining challenge.

1.3.3. Measurements, Model Evaluation, and Model-Measurement Integration

Reliable observations are always important to improve our understanding of natural phenomena including atmospheric processes and serve as the ground truth to verify and evaluate any theoretical and modeling developments. The synergy between model development and observations is becoming increasingly important as both fields progress. Earth science observations in general and atmospheric observations in particular have unique features, involving different but complementary approaches: surface-based, satellite-based and airborne remote sensing, in situ field measurements, and laboratory studies. This part is devoted to such crucial endeavors, with four chapters focusing on four different topics that are summarized next.

Chapter 14 focuses on surface-based remote sensing for the study of the macrophysical and microphysical structure of clouds, precipitation, aerosols, and PBL. In this chapter, Lamer, Kollias, Amiridis, Arinou, Loehnert, Schnitt, and McComiskey place their emphasis on the unique ability of ground-based system to continuously characterize the atmosphere at high vertical resolution from near the surface to the top of the atmosphere, effectively filling observational gaps left by spaceborne and aircraft platforms. Following an overview of the emergence of ground-based observatories, details about the measurement principles of some cornerstone instruments (e.g., cloud and precipitation radars, lidars, and radiometers) are provided. Modern techniques to retrieve cloud and precipitation location, microphysical and dynamical properties, as well as PBL structure and aerosol properties are discussed, including their underlying assumptions and uncertainties. Several challenges associated with using ground-based observations for model evaluation are discussed. For example, most measurements are related to moments of particle size distributions (e.g., the 6th and 2nd being recorded by radars and lidar, respectively), which differ from those of most interest in models (e.g., the 0th moment for particle number concentration and 3rd moment for mass content). Furthermore, high-resolution observations and large-scale models involve widely different scales; GCMs having grid resolutions larger than 50 km, while radars have a range resolution of ~30 m. The growing role of synthetic Observing Systems Simulation Experiments (OSSEs), instrument simulators, and subcolumn generators in bridging those gaps is emphasized. The chapter concludes with an outlook on the next generation of ground-based observatories that should employ scanning systems and distributed networks.

In Chapter 15, Marshak and Davis cover remote sensing of cloud and aerosol properties from overhead instruments, including satellite-based and airborne sensors with standoff distances ranging from NASA's P-3B aircraft at about 3.5 km above cloud top to the Deep Space Climate Observatory (DSCOVR) platform at the Lagrange-1 point, about 1,500,000 km toward the Sun. The electromagnetic spectrum covered ranges from the ultraviolet to microwaves, and both traditional passive and relatively recent active (e.g., lidar and radar) and modalities are discussed. The emphasis is on the physics behind the sensors as well as on the retrieval algorithms. The chapter starts with remote sensing of cloud properties, introducing the popular Nakajima-King approach widely used in retrievals of cloud optical depth and particle size and the Bréon-Goloub approach that is based on the directional signature of the polarized reflectance. Techniques for retrieving cloud properties (mostly ice) with microwave sensors are also briefly described. The chapter then switches to remote sensing of aerosol properties, describing the main aerosol remote sensing approaches used by the major satellite imagers. In addition to passive remote sensing, the active remote sensing methods for aerosol and cloud profiling are also highlighted, with an emphasis on CALIPSO and CloudSat for lidar and radar, respectively. The chapter explores oxygen A-band and B-band remote sensing of cloud and aerosol layer height, along with other passive methods for estimating cloud top height. A special section is dedicated to cloud remote sensing at very high spatial resolution either from tasked imaging sensors in space or from airborne platforms deployed above the clouds of interest, such as NASA's ER-2. New studies on the transition zone are reviewed. The chapter closes with a brief discussion of retrieval uncertainty quantification for both cloud and aerosol remote sensing.

Laboratory experiments allow more controlled, repeatable measurements of a physical quantity or phenomenon in a well-defined system of interest with known external influences. In Chapter 16, Chandrakar and Shaw describe in situ measurements and laboratory experiments, with a focus on physical processes that are small in spatial or temporal scale such as cloud microphysics and small-scale turbulence. Some illustrative historical examples are given to highlight significant advances and capabilities in three areas of airborne measurement, ground-based measurements, and laboratory experiments, with a focus on cloud studies. The challenges of operating an aircraft and the inherent sampling limitation of high-speed nature and thus low spatial resolution of most measurement platforms are highlighted, and two developments are introduced to address these challenges: the emergence of uncrewed aerial vehicles (drones) for scientific purposes and the HOLODEC instrument based on digital in-line holography. The HOLODEC provides an estimate of the cloud particle size distribution and particle shape from a single sample volume of centimeter scales, providing unique opportunities to measure and study outstanding science questions such as droplet clustering, particle breakup, and particle shattering. It is pointed out that laboratory measurements have become somewhat less common and lagged field measurement capabilities, although their contributions have been profound.

The ultimate test of any models and thus parameterizations is its performance in climate simulations and weather prediction. Efficient and effective evaluation frameworks are needed to test the parameterizations, assess their predictive skills, characterize the model behavior from process level to global scale, and identify sources of potential errors to confidently guide further development. In Chapter 17, Lin and Xie describe the approaches and frameworks used for testing and evaluating fast physics parameterizations in climate and weather models. An integrated yet complementary modeling and evaluation framework is advocated that promotes process-oriented evaluation and effectively bridging parameterization development with observations and modeling, with focus on two exemplary frameworks that have been widely adopted by the modeling centers and the research community. The first is the integrated SCM-CRM-LES framework that has been widely used since it was promoted in the early 1990s. This modeling framework allows for process studies with SCMs,

CRMs, and LES models with the scale ranging from a few hundred kilometers to a few tens of meters. With all the models driven by the same large-scale forcing and initial and boundary conditions, this frame has proven useful to identify strengths and weaknesses of model parameterizations. The second framework is based on the idea of running climate models in "weather forecast mode with initial data from NWP analyses to take advantage of the facts that (1) the large-scale state of the atmosphere in the early periods of a forecast is realistic enough that errors may be ascribed to the parameterizations of the atmospheric physical processes; (2) the atmospheric physical processes (e.g., moist process) are often fast (~hours) and the large-scale state changes slowly (~days); and (3) there is a strong correspondence between the short-term and long-term systematic errors in climate models, particularly for those fields that are related to fast physics (e.g., clouds)." Further integration of the two evaluation frameworks to better capitalize on their respective advantages is also explored. The metrics and diagnostics designed for model evaluation are presented as well, including process-oriented diagnostics and metrics in support of process studies, providing more insights into model errors, and satellite/radar simulator packages that permit direct comparison of model outputs to sensor signals without complicated retrievals.

1.4. HOW TO APPROACH THE CONTENT IN THIS BOOK

The book is targeted at researchers and graduate students working on the relevant areas. Each chapter of this collective volume has its own objectives that are closely related to other chapters and can be read either separately as a stand-alone chapter with its own list of references or together with the other chapters with cross-references as needed.

This book serves a valuable addition to existing literature on fast physics parameterizations in large-scale atmospheric models, with several unique features. It should be better read together with Stensrud (2007), the special fast physics collection in *Journal of Geophysical Research: Atmospheres* (Liu, 2019; https://agupubs.onlinelibrary.wiley.com/doi/toc/10.1002/(ISSN)2169-8996.FASTPHYS1), and various topical review articles (e.g., Morrison et al. (2020) on parameterizations of cloud microphysical processes).

ACKNOWLEDGMENTS

The volume editors thank all the chapter authors for their valuable contributions to this book. Dr. Leo Donner deserves special recognition; he co-convened with the editors various sessions on related scientific topics at the AGU Fall Meetings and helped with the conception of this book. The editors' relevant research has been mainly supported by the US Department of Energy's Office of Science Biological and Environmental Research program as part of the Atmospheric Systems Research (ASR) program and Earth System Modeling (ESM) program, and the Office of Energy Efficiency and Renewable Energy (EERE) under Solar Energy Technologies Office (SETO) award number 33504. Brookhaven National Laboratory is operated by Battelle for the US Department of Energy under contract DE-Sc01112704.

REFERENCES

Charney, J. G., Arakawa, A., Baker, D., Bolin, B., Dickenson, R., Goody, R., et al. (1979). *Carbon Dioxide and Climate: A Scientific Assessment*. Washington, DC: Natl. Acad. Sci.

Cheng, A., Xu, K.-M., & Stevens, B. (2010). Effects of resolution on the simulation of boundary-layer clouds and the partition of kinetic energy to subgrid scales. *Journal of Advances in Modeling Earth Systems*, *2*(3). doi:10.3894/JAMES.2010.2.3

Grabowski, W. (2001). Coupling cloud processes with the large-scale dynamics using the Cloud-Resolving Convection Parameterization (CRCP). *Journal of the Atmospheric Sciences*, *58*, 978–997.

Gustafson, W. I., Vogelmann, A. M., Li, Z., Cheng, X., Dumas, K. K., Endo, S., et al. (2020). The Large-Eddy Simulation (LES) Atmospheric Radiation Measurement (ARM) Symbiotic Simulation and Observation (LASSO) activity for continental shallow convection. *Bulletin of the American Meteorological Society*, *101*, E462–E479. doi:10.1175/bams-d-19-0065.1

Jakob, C. (2010). Accelerating progress in global atmospheric model development through improved parameterizations: Challenges, opportunities, and strategies. *Bulletin of the American Meteorological Society*, *91*(7), 869–876. https://doi.org/10.1175/2009BAMS2898.1

Kessler, E. (1969). On the distribution and continuity of water substance in atmospheric 1966 circulation. *Meteorological Monographs*, No. 32. Amer. Meteor. Soc., 84 pp.

Liu, Y. (2019). Introduction to special section on fast physics in climate models: Parameterization, evaluation and observation. *Journal of Geophysical Research: Atmospheres*, *124*, 8631–8644. https://doi.org/10.1029/2019JD030422

Liu, Y., Yau, M.-K., Shima, S., Lu, C., & Chen, S. (2023). Parameterization and explicit modeling of cloud microphysics: Approaches, challenges, and future directions. *Advances in Atmospheric Sciences*, *40*, 747–790. https://doi.org/10.1007/s00376-022-2077-3

Lovejoy, S., & Schertzer, D. (2010). Towards a new synthesis for atmospheric dynamics: Space–time cascades. *Atmospheric Research*, *96*, 1–52, doi:10.1016/j.atmosres.2010.01.004

Meehl, G. A., Senior, C. A., Eyring, V., Flato, G., Lamarque, J.-F., Stouffer, R. J., et al. (2020). Context for interpreting equilibrium climate sensitivity and transient climate response from the CMIP6 Earth system models. *Science Advances*, *6*, eaba1981. https://doi.org/10.1126/sciadv.aba1981

Mironov, D. V., & Machulskaya, E. E. (2017). A turbulence kinetic energy scalar variance turbulence parameterization scheme (Technical Report 575 No. 30). Consortium for Small-Scale Modeling. (http://www.cosmo576 model.org/content/model/documentation/techReports/docs/techReport30.pdf)

Morrison, H., van Lier-Walqui, M., Fridlind, A. M., Grabowski, W. W., Harrington, J. Y., Hoose, C., et al. (2020). Confronting the challenge of modeling cloud and precipitation microphysics. *Journal of Advances in Modeling Earth Systems*, *12*, e2019MS001689. https://doi.org/10.1029/2019MS001689

Mülmenstädt, J., C. Nam, M. Salzmann, J. Kretzschmar, T. S. L'Ecuyer, U. Lohmann, P.-L., et al. (2020). Reducing the aerosol forcing uncertainty using observational constraints on warm rain processes. *Science Advances*, *6*, eaaz6433.

Parishani, H., M. Pritchard, C. Bretherton, M. Wyant, & M. Khairoutdinov (2017). Toward low-cloud permitting cloud superparameterization with explicit boundary layer turbulence. *Journal of Advances in Modeling Earth Systems*, *9*, 1542–1571. doi: 10.1002/2017MS000968

Rasch, P. J., Xie, S., Ma, P.-L., Lin, W., Wang, H., Tang, Q., et al. (2019). An overview of the atmospheric component of the Energy Exascale Earth System Model. *Journal of Advances in Modeling Earth Systems*, *11*, 2377–2411. doi: https://doi.org/10.1029/2019MS001629

Randall, D. (2013). Beyond deadlock. *Geophysical Research Letters*, *40*, 5970–5976. doi: 10.1002/2013GL057998

Stensrud, D. J. (2007). *Parameterization schemes. Keys to understanding numerical weather prediction models* (459 pp.). Cambridge: Cambridge University Press.

Stevens, B., Satoh, M., Auger, L., Biercamp, J., Bretherton, C. S., Chen, X., et al. (2019). DYAMOND: The Dynamics of the Atmospheric general circulation Modeled On Non-hydrostatic Domains. *Progress in Earth and Planetary Sciences*, *6*, 61. https://doi.org/10.1186/s40645-019-0304-z

Zelinka, M. D., Myers, T. A., McCoy, D. T., Po-Chedley, S., Caldwell, P. M., Ceppi, P., et al. (2020). Causes of higher climate sensitivity in CMIP6 models. *Geophysical Research Letters*, *47*, e2019GL085782. https://doi.org/10.1029/2019GL085782

Part I

Processes and Parameterizations

2

Radiative Transfer and Atmospheric Interactions

Yu Gu and Kuo-Nan Liou†

ABSTRACT

This chapter aims to provide a concise and comprehensive overview of the fundamental principles underlying radiative transfer processes and their interactions with the atmosphere. It encompasses discussions necessary for understanding radiative transfer parameterizations employed in modern climate models. Key concepts covered include emission, absorption, scattering, and the resolution of radiative transfer equations. Specifically, our focus is on the fundamentals of radiative transfer in the plane-parallel atmosphere. We thoroughly examine the intricate interactions between radiation and the Earth's atmosphere, including gaseous absorption and cloud-radiation interaction. The interaction between aerosols and radiation, on the other hand, is discussed in Chapter 3. Furthermore, we delve into the realm of three-dimensional (3D) radiative transfer, exploring its applications to understanding radiative processes in 3D clouds and mountainous regions. Additionally, we provide a brief review of commonly employed radiative transfer schemes in climate and weather research, using the Weather Research and Forecasting (WRF) model as an example.

2.1. BACKGROUND AND INTRODUCTION

Radiative transfer is a subject of study in a variety of fields, including astrophysics, applied physics, optics, planetary sciences, atmospheric sciences, meteorology, and various engineering disciplines. Prior to 1950, radiative transfer was studied principally by astrophysicists, although it was also an important research area in nuclear engineering and applied physics associated with neutron transport. In his groundbreaking book, Chandrasekhar (1950) presented the subject of radiative transfer in plane-parallel (PP; one-dimensional) atmospheres as a branch of mathematical physics and developed numerous solution methods and techniques. The field of atmospheric radiation, which evolved from the study of radiative transfer, is now concerned with the study, understanding, and quantitative analysis of the interactions of solar and terrestrial radiation with molecules, aerosols, and cloud particles in planetary atmospheres as well as the surface on the basis of the theories of radiative transfer and radiometric observations made from the ground, the air, and space (Liou, 1980, 2002). A fundamental understanding of radiative transfer processes is the key to understanding the atmospheric greenhouse effects and global warming, which results from external radiative perturbations of the greenhouse gases and air pollution, and to the development of methods to infer atmospheric and surface parameters through remote sensing.

Almost all the energy that drives the Earth's atmosphere and ocean currents originates from the Sun (Liou, 1992). Therefore, climate of the earth-atmosphere system is mainly determined by the radiation balance at the top of the atmosphere and the surface since radiation is the only

Joint Institute for Regional Earth System Science and Engineering, and Department of Atmospheric and Oceanic Sciences, University of California Los Angeles, Los Angeles, CA, USA

Fast Processes in Large-Scale Atmospheric Models: Progress, Challenges, and Opportunities, Geophysical Monograph 282, First Edition.
Edited by Yangang Liu and Pavlos Kollias.

DOI: 10.1002/9781119529019.ch02

mechanism by which the earth-atmosphere system gains or loses energy. The total radiant energy from the Sun varies only slightly, so we define a term solar constant or total solar irradiance (TSI) to represent the solar energy received per unit time and unit area at the top of the atmosphere when the Sun is at its mean distance from the Earth. On average, the TSI is 1361 W m^{-2} (Coddington et al., 2016) based on NASA's Solar Radiation and Climate Experiment (SORCE) (Rottman, 2005) Total Irradiance Monitor TIM (Kopp et al., 2005) measurements of total solar irradiance (Kopp & Lean, 2011). The Earth's atmosphere contains molecules, gases, aerosols, and clouds, which are significant components that are pertinent to the radiation and atmospheric interactions and determine the energy budget of the earth-atmosphere system. Figure 2.1 provides an updated estimation of the Earth's annual and global mean energy balance for the approximate period 2000–2010 (Stephens et al., 2012). The incoming solar flux for climatological energy balance is about 340 W m^{-2} (round off the decimal point). About 23% of the incoming solar radiation gets reflected by clouds, aerosols, and atmospheric gases and 7% is reflected by surface, providing a planetary albedo of about 30%. About 22% of the incoming solar flux is absorbed by the atmosphere while another half is absorbed by the Earth's surface, resulting from the divergence of net solar fluxes at the top of the atmosphere and the surface. The solar fluxes absorbed by the surface are transferred to the atmosphere by means of the convection and conduction of sensible heat from the surface, by the latent heat of condensation released in precipitation processes, as well as by emission of longwave or thermal infrared (IR) radiation. The global/annual mean upwelling longwave flux of 398 W m^{-2} is equivalent to an effective emission temperature of 289 K. This surface thermal IR radiation is largely absorbed by clouds, aerosols, and greenhouse gases. The atmosphere in turn radiates longwave fluxes downward back to Earth and upward to space. At the top of the atmosphere, the energy is balanced by radiative flux exchanges. At the surface, however, upward sensible and latent heat fluxes must be introduced to maintain radiative flux balance.

The above energy balance of the earth-atmosphere system is obtained when we treat the entire Earth as a single point. Earth's climate is determined by the flows of energy into and out of the planet, and changes to the surface energy balance ultimately also control how the hydrological cycle responds to the small energy imbalances that force climate change. Geographical distributions of these energy flows in the earth-atmosphere system are also particularly important as they drive ocean circulations, fuel the evaporation of water from the Earth's surface, and govern the planetary hydrological cycle (Stephens et al., 2012). Radiative transfer schemes have been used in climate models to compute radiative fluxes and heating rates over regional and global scales in the earth-atmosphere system. In this chapter, we present the fundamentals of radiative transfer in the plane-parallel atmosphere, the interactions between radiation and the Earth's atmosphere (including gaseous absorption and cloud-radiation interaction; aerosol-radiation interaction is covered in Chapter 3), three-dimensional (3D) radiative transfer for 3D clouds and over the mountainous regions,

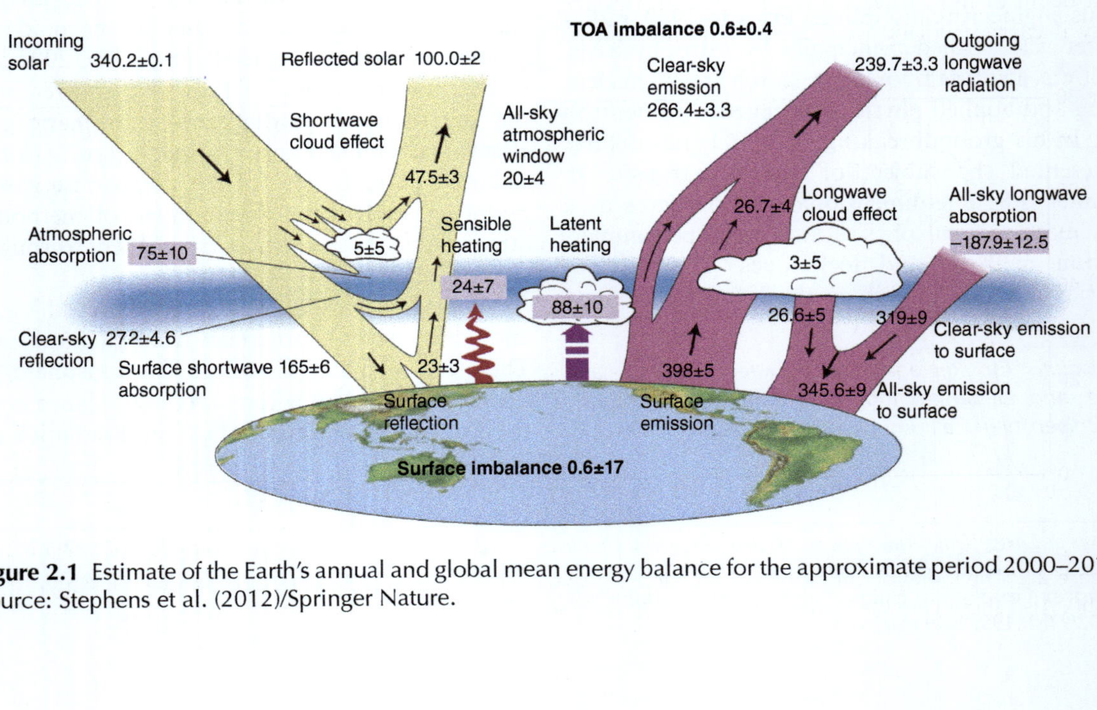

Figure 2.1 Estimate of the Earth's annual and global mean energy balance for the approximate period 2000–2010. Source: Stephens et al. (2012)/Springer Nature.

and a brief review of the commonly used radiative transfer schemes in the climate and weather research using the Weather Research and Forecasting (WRF) model as an example.

2.2. FUNDAMENTALS AND EQUATION GOVERNING RADIATIVE TRANSFER FOR PLANE-PARALLEL ATMOSPHERE

In radiation, we use the Planck function to relate the emitted radiation intensity to the wavelength and the temperature of the emitting substance:

$$B_\lambda(T) = \frac{2hc^2}{\lambda^5(e^{hc/K\lambda T} - 1)}, \qquad (2.1)$$

where h is the Planck's constant, λ is the wavelength, K is Boltzmann's constant, c is the velocity of light, and T is the absolute temperature. Figure 2.2 illustrates the curves of the intensity as a function of wavelength for a star with a temperature range similar to the Sun (~6000 K, on the left) and a planet similar to the Earth (~288 K, on the right). Two facts are evident from the figures and also can be derived from the Planck function: first, the intensity of emitted radiation increases with temperature. The intensity from the Sun is much larger than that from the Earth since the Sun has a much higher temperature. Second, the wavelength of the maximum intensity decreases with increasing temperature, which can be seen from the shift of the location of the maximum values from right (longer wavelengths) to the left (shorter wavelengths) as temperature increases. For the Sun, the maximum intensity is located at about 0.5 μm, while for the Earth, it is located at about 12 μm.

When a beam of light travels through the atmosphere, the energy it carries will be weakened by its interaction with the matter in the atmosphere, such as molecules, gases, cloud particles, and aerosols. Two major processes are considered in the radiative transfer. One is extinction and the other is the source or contribution. Extinction consists of two distinct processes, scattering and absorption. Figure 2.3 illustrates how the particles in the atmosphere scatter and absorb sunlight. Depending on the size and single-scattering properties of the particles, when a beam of sunlight hit on a particle, some energy gets absorbed and transformed to heat. Some light gets scattered and the energy is redirected to different directions. The extinction of light is determined by the so-called extinction coefficient or extinction cross section $k(\nu)$, where ν represents a certain wavenumber. There are two types of scattering: Rayleigh scattering occurs when the particle size is much smaller than the wavelength. This normally happens for molecules in the atmosphere, and the scattered intensity is inversely dependent on the wavelength to the fourth power. Lorenz-Mie scattering is for spherical particles with sizes comparable to or larger than the wavelength, such as aerosols and cloud droplets. The Mie theory is basically a solution to Maxwell's equation that takes the form of analytical infinite series.

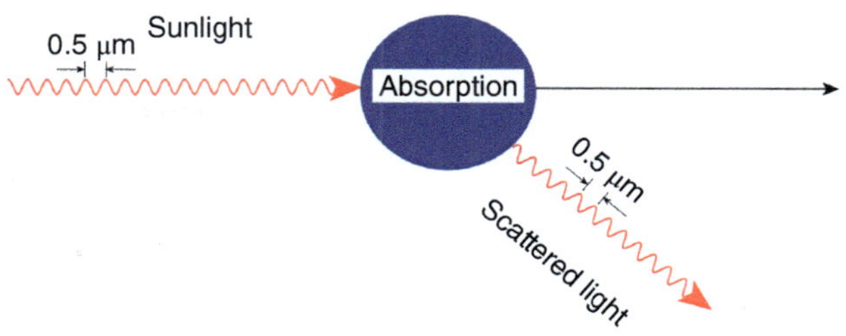

Figure 2.3 Diagram illustrating the scattering and absorption of sunlight by a particle in the atmosphere.

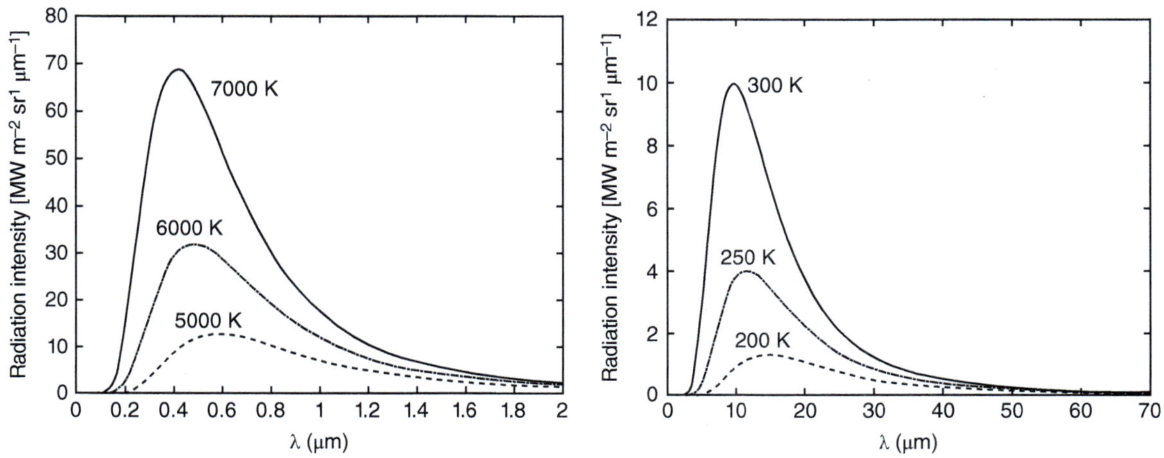

Figure 2.2 The radiation intensity as a function of wavelength for the Sun (~6000 K, left) and the Earth (~300 K, right). Also shown are the curves for a few other temperatures.

On the other hand, the radiation intensity may be strengthened by emission from the materials and scattering from all other directions. We define a source function coefficient j_ν to describe the emission and scattering process. Combining the extinction and source processes, we obtain the basic radiative transfer equation in the form shown below for a given wavenumber ν:

$$dI_\nu = -k(\nu)I_\nu\rho_a ds + j_\nu\rho_a ds, \qquad (2.2)$$

where I_ν is the radiation intensity which is a function of solid angle, $k(\nu)$ is the mass absorption coefficient with units of m^2 kg^{-1}, and ρ_a is the air density. For source or contribution, three factors are considered: (1) emission by the matter, such as from clouds; (2) single scattering of direct solar flux which is associated with the exponential attenuation to the level τ based on extinction law; and (3) the multiple scattering. We further introduce two parameters to represent single scattering and multiple scattering: one is phase function or asymmetry factor, which represents the angular distribution of scattered energy as a function of direction; the other one is single-scattering albedo, defined as the ratio of the scattering cross section to the total extinction (scattering plus absorption) cross section. So, the fundamental parameters in radiative transfer, also called single-scattering properties, are extinction coefficient, single-scattering albedo, and phase function or asymmetry factor. These parameters are functions of the incident wavelength, particle size and shape, and refractive index with respect to wavelength. The refractive index consists of a real part and an imaginary part, corresponding to the scattering and absorption properties, respectively.

In radiative transfer, it is commonly assumed that the atmosphere in a local position is plane-parallel, that is, variations in the radiation intensity and atmospheric parameters, such as temperature and gaseous profiles, are permitted only in the vertical direction. So, changes in distance can be expressed in terms of changes in height together with the cosine of the solar zenith angle μ. The solar zenith angle θ is defined to be the angle between the center of the Sun's disc and the zenith (a line that goes straight up directly above a particular location; Fig. 2.4). Normally, we are interested in the total extinction over a finite distance and hence define a physical parameter called optical depth τ, which is the integration of extinction coefficient over the depth of an atmospheric layer:

$$\tau(\nu) = \int_z^{z_\infty} k_\nu(z')\rho_a(z')dz', \qquad (2.3)$$

so that $d\tau = -k_\nu\rho_a dz$. Then, we obtain the fundamental equation governing the transfer of diffuse solar radiation for the problem of multiple scattering in plane-parallel atmospheres in terms of τ, μ, and azimuthal angle in reference to the x-axis ϕ and by measuring intensity downward

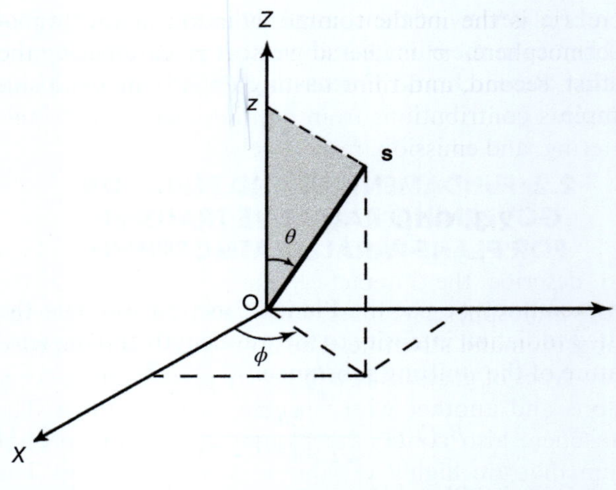

Figure 2.4 Definitions of the zenith angle θ and the azimuthal angle ϕ with reference to the Cartesian coordinate system (x, y, z), where s represents a position vector in space.

from the outer boundary (top of the atmosphere):

$$\mu\frac{dI(\tau,\mu,\phi)}{d\tau} = I(\tau,\mu,\phi) - J(\tau,\mu,\phi). \qquad (2.4)$$

The solar azimuth angle ϕ is the angular distance measured in terms of a 360-degree compass between the x-axis and the projection of the line of sight to the Sun on the ground (Fig. 2.4). Here, we ignore the subscript representing the wavenumber or wavelength, and $J(\tau, \mu, \phi)$ is the source function.

Diffuse solar radiation is associated with multiple scattering processes and is differentiated from direct solar radiation. Considering a vertical layer, the differential change of disuse solar radiation after passing through the layer is due to the following processes: (1) reduction from the extinction attenuation, (2) increase from the single scattering of the unscattered direct solar flux from the original incident direction $(-\mu_0, \phi_0)$ to (μ, ϕ), (3) increase from multiple scattering of the diffuse intensity from directions (μ', ϕ') to (μ, ϕ), and (4) increase from emission within the layer in the direction (μ, ϕ) under the Kirchhoff's law referring to wavelength-specific radiative emission and absorption by a material body in thermal equilibrium condition. Let the phase function corresponding to a volume of particulates be P and $P(\mu, \phi; \mu', \phi')$ denotes the redirection of the incoming intensity defined by (μ', ϕ') to the outgoing intensity defined by (μ, ϕ). The source function $J(\tau, \mu, \phi)$ is then given by

$$J(\tau,\mu,\phi) = \frac{\varpi}{4\pi}\int_0^{2\pi}\int_{-1}^{1} I(\tau,\mu',\phi')P(\mu,\phi;\mu',\phi')d\mu'd\phi'$$
$$+ \frac{\varpi}{4\pi}F_\odot P(\mu,\phi;-\mu_0,\phi_0)e^{-\tau/\mu_0}$$
$$+ (1-\varpi)B[T(\tau)], \qquad (2.5)$$

where $F_{\odot}$ is the incident solar radiation at the top of the atmosphere, ϖ is the single-scattering albedo, and the first, second, and third terms on the right-hand side represents contributions from multiple scattering, single scattering, and emission, respectively.

2.3. GASEOUS ABSORPTION

To describe the interactions of radiation with the Earth's atmosphere, we must understand the atmosphere's composition and structure. The atmosphere is composed of two types of gases, one with nearly permanent concentrations and another with variable concentrations. The atmosphere also contains aerosols, clouds, and precipitation that are highly variable in space and time. The most abundant gas in the atmosphere is nitrogen (~78%), followed by oxygen (~21%) and argon (~1%), which account for more than 99.96% of the atmosphere by volume. The concentrations of these gases, referred to as permanent constituents, are well mixed in the atmosphere with nearly constant volume ratios up to about 60 km in altitude. The amounts of variable gases vary significantly with both space and time and are extremely important in the radiation budget of the atmosphere although they are small. There are a number of trace gases, such as carbon dioxide (CO_2), nitrous oxide (N_2O), methane (CH_4), ozone (O_3), chlorofluorocarbons (CFCs), nitrogen oxides (NO_x), Sulfur oxides (SO_x), and volatile organic carbons (VOCs), which absorb and emit infrared radiation. These are called greenhouse gases and play an essential role in the Earth's energy budget. Besides the trace gases, water vapor is also a major radiative, dynamic, and thermodynamic element in the atmosphere with highly variable mixing ratio. Water vapor is known to be Earth's most abundant greenhouse gas through feedback (Liou, 2002). Increased greenhouse gases in the atmosphere makes the atmosphere more humid, and the increase in humidity amplifies the warming from carbon dioxide. Human activities are most likely to affect the trace gases rather than the permanent gases such as nitrogen and oxygen (Liou, 2002). Among them, carbon dioxide (CO_2) is the most important anthropogenic greenhouse gas with a concentration of ~400 ppm (parts per million in volume) in 2014 and ~407 ppm in 2018, increasing by about 0.4% per year as a result of the combustion of fossil fuels, absorption and release by the oceans, and photosynthesis. Another important variable gas is ozone, which occurs at altitudes between 15 and 30 km with a maximum concentration at ~20–25 km depending on latitude and season. The absorption of solar ultraviolet (UV) radiation by the ozone layer is essential to life on the Earth. Ozone can also be formed in the troposphere by the interaction of ultraviolet light with hydrocarbons and nitrogen oxides, which are emitted by automobile tailpipes and smokestacks. Tropospheric ozone is considered as both air pollutants and greenhouse gases. Nitrogen oxides emitted by transportation and combustion processes at the surface and by aircraft in the upper troposphere and lower stratosphere are important in the determination of ozone concentrations. For methane, a number of measurements indicate that its concentration has increased by 1%–2% per year. The most likely cause is the greater biogenic emissions associated with an increasing human population. There is also some evidence of an increase in N_2O of about 0.2% per year, attributed to the combustion of fossil fuels and, in part, to fertilizer denitrification.

The Earth's atmosphere is divided into different layers—troposphere, stratosphere, mesosphere, and thermosphere—based on vertical temperature profile. The troposphere is characterized by a decrease in temperature with respect to height, up to about 12 km, with a global mean lapse rate of about 6.5 K km^{-1}. The temperature profile in this layer is a result of the radiative balance and transport of energy from surface to the atmosphere. About 80% of the atmosphere is concentrated in this layer, with almost all water vapor, clouds, and precipitation confined below the tropopause—the top of the troposphere. The stratosphere represents an isothermal layer from the tropopause to about 20 km, followed by a layer with increasing temperature with altitude at an average rate of 5 K km^{-1}, largely due to the absorption of Sun's ultraviolet radiation by ozone. The water vapor in the troposphere does not mix much with that above the tropopause due to the temperature inversion in the stratosphere. The temperature in the mesosphere decreases again with altitude from about 50 to 80 km. The air is well mixed from the ground to the top of the mesosphere with composition of gases almost identical except for water vapor and ozone. The thermosphere is the region above the top of mesosphere where the temperature begins to rise again to as high as 2000 K, associated with the absorption of solar radiation by atomic oxygen. This is because the atmosphere there is very thin, so a little heating can significantly raise the temperature.

Figure 2.5 illustrates the results calculated by LINEPAK (Gordley et al., 1994) using the HITRAN2004 (Rothman et al., 2005) spectroscopic database assuming the U.S. 1976 Standard Atmosphere and the solar zenith angle of 0° (Liou & Yang, 2016). The upper panel of Figure 2.5 illustrates solar and thermal IR spectral irradiances at the top of the atmosphere as a function of wavelength and wavenumber. The solar and thermal IR spectra cover wavelength ranges from ~0.1 to 5 and ~5 to 100 μm, respectively, with a slight overlap between ~3 and 5 μm. The middle panel shows absorption by various gases and Rayleigh scattering. The bottom panel illustrates the total atmospheric absorption in terms of the percentage

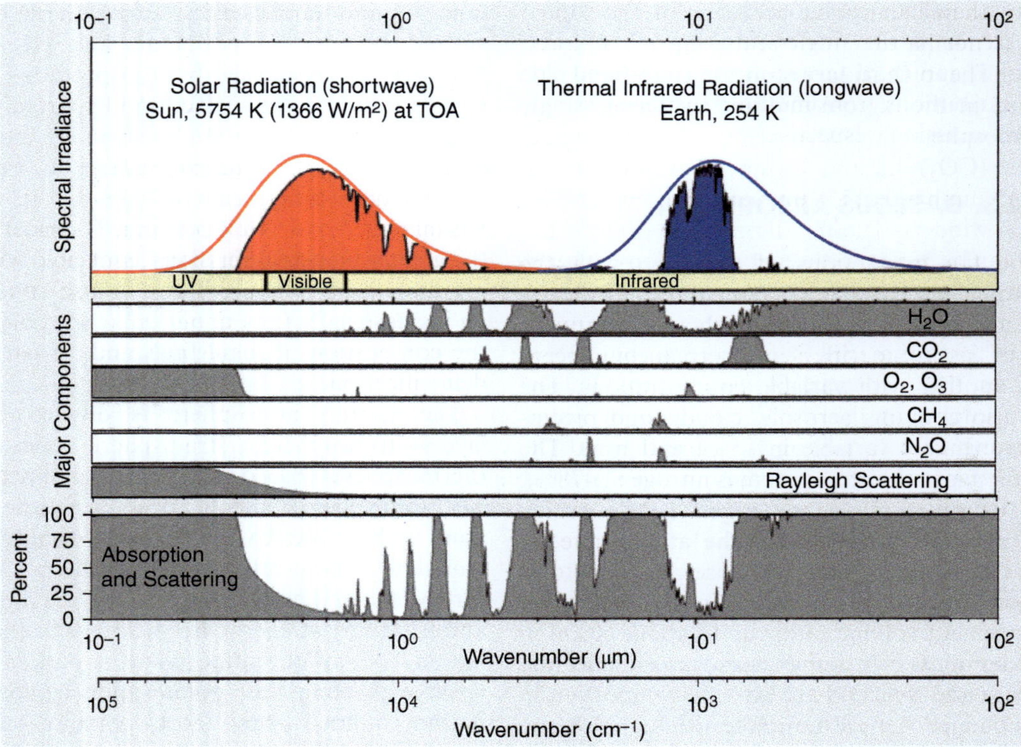

Figure 2.5 Top panel: Spectral irradiances for the solar (red curve) and thermal IR (blue curve, scaled with reference to solar) regions as a function of wavelength and wavenumber. The red area denotes the observed solar fluxes available at the surface; the blue area represents the thermal IR fluxes absorbed in the atmosphere and available at the top of the atmosphere. Middle panel: Absorption by H_2O, CO_2, O_2, O_3, CH_4, and N_2O, and Rayleigh scattering. Bottom panel: Total atmospheric absorption in terms of the percentage of the entire atmosphere corresponding to the U.S. 1976 Standard Atmosphere; 0% and 100% indicate that the atmosphere is clear (total transmission) and opaque to radiation, respectively.

of the entire atmosphere corresponding to the U.S. 1976 Standard Atmosphere.

2.3.1. Absorption in Solar Spectra

The solar spectrum (red curve) is computed from Planck emission at the Sun's photosphere temperature of 5754 K and attenuated to the top of the atmosphere coupled with a corresponding solar constant of 1366 W m^{-2}. The observed solar fluxes available at the surface is represented by the red area (top panel of Fig. 2.5). The depletion of solar flux in the UV region with wavelength shorter than 0.4 μm is mainly due to the absorption of molecular oxygen and ozone together with Rayleigh scattering (middle panel of Fig. 2.5). In the visible region (between 0.4 and 0.7 μm), the depletion is caused by the absorption produced by oxygen, ozone (O_3), and water vapor (H_2O), but main attenuation is associated with Rayleigh scattering. Absorption lines by H_2O produced by vibration-rotation transitions are located at 0.72, 0.82, 0.94, 1.1, 1.38, 1.87, 2.7, and 3.2 μm. In the near thermal IR, the primer absorber is water vapor and carbon dioxide. Absorption due to CO_2 is largely produced by vibrational transitions with spectral bands located at 1.6, 2.0, 2.7, and 4.3 μm in the solar spectrum. H_2O rational lines essentially cover the entire solar spectra. Other minor absorbers in the solar spectra include N_2O (nitrous oxide) and CH_4 (methane).

The bottom panel of Figure 2.5 illustrates the total atmospheric absorption produced by the various molecules discussed above in terms of the percentage of the entire atmosphere corresponding to the U.S. 1976 Standard Atmosphere. Here, 0% and 100% indicate that the gas is clear (total transmission without absorption) and opaque to radiation, respectively. This information is essential to the development of space remote sensing of the atmosphere, oceans, and land surfaces, employing UV, visible, and thermal IR techniques. Note that Rayleigh scattering occurs primarily in the UV and visible regions.

2.3.2. Absorption in Thermal Infrared Spectra

The thermal IR spectrum (blue curve in the top panel of Fig. 2.5) is computed from the global equilibrium temperature of the earth-atmosphere system of 254 K, and its peak is normalized with reference to the solar spectral peak. The blue area denotes the thermal IR fluxes absorbed in the atmosphere and available at the top of the atmosphere. A few important bands include carbon dioxide (CO_2) 4.3 and 15 μm bands, O_3 9.6 μm band, H_2O 6.25 μm band, and 10 μm continuum band in the thermal IR window. The middle panel of Figure 2.5 shows absorption by H_2O produced by vibration-rotation transitions located at 6.3 μm in the thermal IR spectrum. Similar to that in the solar spectra, H_2O rational lines cover the entire thermal IR spectra, but these absorption lines are strongest at wavelengths longer than ~12 μm, known as the rotational band of water vapor. The strong CO_2 15 μm band in the thermal IR spectrum is most pronounced in trapping the emission from the surface and lower atmospheres. This band is the source of about half of the radiative forcing due to anthropogenic greenhouse warming since the era of the Industrial Revolution. Other minor absorbers in the thermal IR spectra include methane (CH_4), N_2O, and CFCs.

Greenhouse gases are those that absorb and emit infrared radiation in the wavelength range emitted by the Earth. The most important greenhouse gases in the Earth's atmosphere include H_2O, CO_2, CH_4, N_2O, O_3, CFCs, and hydrofluorocarbons (including HCFCs and HFCs). These gases mostly transmit visible radiation and partially absorb infrared radiation (between about 4 and 30 μm wavelength) and hence play an important role in the energy budget of the earth-atmosphere system by trapping the IR radiation and warming the Earth's surface temperature by about 33 K.

2.3.3. Absorption Coefficient

The emission spectra of certain gases are composed of a large number of individual spectral lines. However, monochromatic emission is practically never observed. Radiation emitted during repeated energy transitions is non-monochromatic, and spectral lines of finite widths are observed. The broadening of lines is mainly caused by collision, referred to as pressure broadening, and the Doppler effect resulting from the difference in thermal velocities of atoms and molecules.

Absorption line parameters for various gases can be computed from fundamental quantum mechanics theory or measured through lab experiments. Based on theory and measurements, line parameters have been compiled over the range 0–17,900 cm^{-2} (Rothman et al., 1998) with data for more than 1 million lines. The strength of absorption is normally represented by the absorption coefficient that is a function of wavenumber, pressure, and temperature, and can be written in the form

$$k(\nu, p, T) = \sum_{i=1}^{N} S_i(T) f_{\nu,i}(\nu, p, T), \tag{2.6}$$

where S_i is the line strength for the ith absorption line and $f_{\nu,i}$ is the line shape.

Ideally, the absorption coefficients should be computed at wavenumbers, temperatures, and pressures, which are as closely spaced as possible. However, due to computing resource limitations, these spacings would be chosen as large as possible while still retaining the high degree of accuracy needed for climate studies. Here, an example of computing the absorption coefficients for any pressures and temperatures is provided. The absorption coefficients for a given wavenumber and particular gas can be parameterized as a function of pressure and temperature according to Chou and Kouvaris (1986) in the following form:

$$\ln k(\nu, p, T) = \sum_{n=0}^{2} a_n(\nu, p)(T - 250)^n. \tag{2.7}$$

For a given wavenumber and pressure, three temperatures (e.g., 200, 250, and 300 K) can be used to determine the coefficients that can be tabulated at various prescribed pressures for the solar and thermal infrared calculations. The absorption coefficients at other pressures can be linearly interpolated in the pressure coordinate (Fu & Liou, 1992).

2.3.4. Absorption Integration Methods

For solar radiation, if we consider a nonscattering atmosphere and a small spectral interval $\Delta\nu$ such that the spectral solar flux can be taken as constant, the downward solar flux at a given level z can be written in the form

$$F_\nu^{\downarrow}(z) = \mu_0 T_\nu(z/\mu_0) S_\nu \Delta\nu, \tag{2.8}$$

where $S\Delta\nu$ represents the incident solar flux, μ_0 is the cosine of the solar zenith angle, and T_ν is the spectral-mean transmittance over the spectral interval $\Delta\nu$ defined by

$$T_\nu(z/\mu_0) = \int_{\Delta\nu} \exp\left[-\int_z^{\infty} k(\nu, p, T)\rho_a/\mu_0 dz'\right] \frac{d\nu}{\Delta\nu}, \tag{2.9}$$

where ρ_a denotes the density of the absorbing gas.

For thermal infrared radiation, similarly, the spectral-mean transmittance between levels z and z' is given by

$$T_\nu(z, z'; \mu) = \int_{\Delta\nu} \exp\left[-\int_z^{z'} k(\nu, p, T)\rho_a/\mu dz''\right] \frac{d\nu}{\Delta\nu}. \tag{2.10}$$

Transmittance computation can be performed by a line-by-line (LBL) technique or simplified approaches such as the correlated k-distribution (CKD) method.

For a given wavenumber and species, contributions to transmittance arise from the absorption coefficients for N lines as shown in equation (2.6). In order to resolve individual lines in the LBL approach, the absorption coefficient must be computed as wavenumber intervals that are smaller than the line half-width. In the troposphere, absorption due to H_2O generally dominates; however, there are also a number of absorption bands associated with CO_2. Absorption due to CO_2 at the 15 μm vibration and rotation bands is critically important concerning greenhouse warming. H_2O lines are broadened by collision, and their half-widths are >0.01 cm^{-1}. Therefore, computations must be performed at about 1 million points to resolve H_2O lines. In the stratosphere, absorption and emission processes are dominated by CO_2 and O_3. Broadening of the absorption lines is mainly due to the Doppler effect with a half-width of about 0.0005–0.001 cm^{-1}. The computations must be performed at more than half a million points if individual lines are to be resolved. Furthermore, for each point, there are numerous lines and atmospheric conditions that must be considered for applications to atmospheric radiative transfer. The computer time required for line-by-line calculations is very expensive, especially for flux calculations in which an integration over all absorption bands is necessary.

The k-distribution method is based on the grouping of gaseous spectral transmittance according to the absorption coefficient k. Under a homogeneous condition where pressure and temperature are constant, for a given absorption gas and spectral interval $\Delta\nu$, we may introduce the k-distribution function $h(k)$, which is the probability density function such that $h(k)\mathrm{d}k$ is the fraction of $\Delta\nu$ within which the absorption coefficient is between k and $k + \mathrm{d}k$. The spectral-mean transmittance should be dependent on the k-distribution but independent of the ordering of k (Arking & Grossman, 1972). Hence, we may replace the wavenumber integration by an integration over k space. The spectral-mean transmittance may be expressed by

$$T_\nu(u) = \int_{\Delta\nu} \mathrm{e}^{-k(\nu)u} \frac{\mathrm{d}\nu}{\Delta\upsilon} = \int_0^\infty \mathrm{e}^{-ku} h(k)\mathrm{d}k, \qquad (2.11)$$

where $u(z) = \int_z^{z_\infty} \rho_a(z')\mathrm{d}z'$ is the path length (g cm^{-2}) to denote the amount of absorber, and $h(k)$ is normalized to 1 in the domain (0, ∞). This equation defines the k-distribution method for the homogeneous atmosphere.

We may further define a monotonically increasing and smooth cumulative probability function in k space in the form

$$g(k) = \int_0^k h(k)\mathrm{d}k, \qquad (2.12)$$

where $g(0) = 0$, $g(k\to\infty) = 1$, and $\mathrm{d}g(k) = f(k)\mathrm{d}k$. The spectral-mean transmittance can then be expressed in terms of cumulative probability g as

$$T_\nu(u) = \int_{\Delta\nu} \mathrm{e}^{-k(\nu)u} \frac{\mathrm{d}\nu}{\Delta\upsilon} = \int_0^1 \mathrm{e}^{-k(g)u}\mathrm{d}g \cong \sum_{i=1}^{M} \mathrm{e}^{-k(g_i)u}\Delta g_i, \qquad (2.13)$$

where $k(g)$ is the inverse function of $g(k)$ and must also be a smooth function in g space. Here, the integration in g space replaces the tedious wavenumber integration and can be evaluated by a finite sum of exponential terms as shown in equation (2.13).

The k-distribution method can be further extended to nonhomogeneous atmosphere, referred to as the correlated k-distribution method. In this method, the vertical nonhomogeneity of the atmosphere is accounted for by assuming a simple correlation of k-distribution at different temperatures and pressures. The CKD approach allows the use of k-distribution method at each altitude and can be used for absorption bands in both solar and thermal infrared spectra. Proof of the validity of CKD and its accuracy under various atmospheric conditions have been carried out by a number of studies (e.g., Fu & Liou, 1992; Goody et al., 1989). Errors in flux calculations associated with CKD with respects to LBL results are generally on the order of 1%. At the same time, the results from CKD can be directly incorporated into multiple scattering processes associated with cloud and aerosol processes. Thus, the CKD method is a powerful technique in the parameterization of radiative transfer in climate models where cloud and aerosol processes must be accounted for in flux and heating rate calculations.

Domoto (1974) and Wang and Ryan (1983) illustrated the importance of treating overlap absorption in radiative transfer calculations and climate studies. Goody et al. (1989), Lacis and Oinas (1991), Fu and Liou (1992), and Shi (1998) pointed out that overlap absorption by several different gases is an important theoretical and practical problem in CKD since computational speed is essential to radiative transfer modeling, especially when it is applied to the scattering atmosphere. Lacis and Oinas (1991) adopted the multiplication rule for transmittance computations under which the absorption spectra for two gases are assumed to be uncorrelated. Mlawer et al. (1997) developed a method to treat bands containing gases with overlap absorption, in which the key absorbers in each spectral band are treated with high accuracy, whereas a less detailed procedure is employed to compute absorption due to minor gases in the band.

In Fu and Liou (1992), two different approaches were employed to treat overlap absorption in the g space. These approaches have been proven to be both efficient and accurate for treating the overlap problem involving atmospheric radiative transfer. The first approach uses the multiplication assumption in which the absorption of different gases are assumed to be uncorrelated so that the total transmittance can be obtained by the product of the transmittance from each gas. Fu and Liou (1992) concluded that the multiplication approach for overlap gases can achieve excellent accuracy in flux and heating rate calculations over a spectral interval of about 150 cm^{-1}. In the second approach, a new absorption coefficient, which can be considered as the absorption coefficient for a single-mixture gas, is defined. The second approach does not require the assumption that the two absorption spectra are uncorrelated, and it is computationally more efficient.

2.4. COMMONLY USED APPROXIMATIONS OF RADIATIVE TRANSFER

In section 2.2, we introduce the basic radiative transfer equation for the plane-parallel condition (equation (2.5)). For calculations of solar fluxes in the atmosphere, the azimuthal dependence of the intensity expansion can be neglected and we may define the phase function as

$$P(\mu, \mu') = \frac{1}{2\pi}\int_0^{2\pi} P(\mu, \phi; \mu', \phi')\mathrm{d}\phi'. \tag{2.14}$$

The azimuthally independent transfer equation for diffuse radiation can be written as

$$\mu\frac{\mathrm{d}I(\tau,\mu)}{\mathrm{d}\tau} = I(\tau,\mu) - \frac{\varpi}{2}\int_{-1}^{1} I(\tau,\mu')P(\mu,\mu')\mathrm{d}\mu' - \frac{\varpi}{4\pi}F_{\odot}P(\mu,-\mu_0)\mathrm{e}^{-\tau/\mu_0} - (1-\varpi)B[T(\tau)]. \tag{2.15}$$

The above radiation equation is for radiance or intensity. In climate model applications, radiation fluxes are normally needed and can be obtained by the integration of radiation intensity over the hemisphere for upward and downward directions, respectively:

$$F^{\uparrow}_{\text{dif}}(\tau) = \int_0^{2\pi}\int_0^{1} I(\tau,\mu,\phi)\mu\mathrm{d}\mu\mathrm{d}\phi = 2\pi\int_0^{1} I(\tau,\mu)\mu\mathrm{d}\mu, \quad \mu \geq 0, \tag{2.16}$$

$$F^{\downarrow}_{\text{dif}}(\tau) = \int_0^{2\pi}\int_0^{-1} I(\tau,\mu,\phi)\mu\mathrm{d}\mu\mathrm{d}\phi = 2\pi\int_0^{-1} I(\tau,\mu)\mu\mathrm{d}\mu, \quad \mu \leq 0, \tag{2.17}$$

Note that for thermal infrared radiation, the direct solar term involving $F_{\odot}$ (the third term on the right-hand side) does not appear. For solar radiation, the thermal infrared emission term (the fourth term on the right-hand side) will be omitted. In addition, the above equation only describes the diffuse component. We must include the direct component in the calculations of the downward solar radiation. This is given by the simple Beer-Bouguer-Lambert law that describes the extinction of solar radiation in the form

$$F^{\downarrow}_{\text{dir}} = \mu_0 F_{\odot}\mathrm{e}^{-\tau/\mu_0}. \tag{2.18}$$

The total upward and downward flux densities at a given τ are

$$F^{\uparrow}(\tau) = F^{\uparrow}_{\text{dif}}(\tau) = 2\pi\int_0^{1} I(\tau,\mu)\mu\mathrm{d}\mu, \tag{2.19}$$

$$F^{\downarrow}(\tau) = F^{\downarrow}_{\text{dif}}(\tau) + F^{\downarrow}_{\text{dir}}(\tau) = 2\pi\int_0^{-1} I(\tau,\mu)\mu\mathrm{d}\mu + \mu_0 F_{\odot}\mathrm{e}^{-\tau/\mu_0}. \tag{2.20}$$

Thus, the net flux for a given level z is

$$F(z) = F^{\downarrow}(z) - F^{\uparrow}(z). \tag{2.21}$$

For an atmosphere layer with thickness of Δz, the radiative heating in the atmosphere is produced by the divergence of the net flux and is given by

$$\frac{\partial T}{\partial t} = -\frac{1}{\rho C_{\text{p}}}\frac{\Delta F(z)}{\Delta z}, \tag{2.22}$$

where T is the temperature, ρ is air density, C_{p} is the specific heat at constant pressure, and $\Delta F(z) = F(z) - F(z+\Delta z)$.

There are quite a few methods available to provide exact solutions to the radiative transfer equation, including discrete-ordinates method, invariance method, and adding method. The discrete-ordinates method was first developed by Chandrasekhar (1950) and was proved by Liou (1973) to be a useful and powerful method for the computation of radiation in aerosol and cloudy atmospheres. The principle of the adding method was stated by Stokes (1862) when dealing with reflection and transmission by glass plates. van de Hulst (1980) presented a set of adding equations that is now commonly used for multiple scattering. Takano and Liou (1989) modified the adding method for application to randomly oriented ice crystals. The exact adding/doubling method appears to be a powerful tool for multiple scattering calculations, particularly for remote sensing applications from the ground, the air, and space. The discrete-ordinates and adding methods are similar in terms of numerical calculations. The invariance method is, in principle, equivalent to the adding method (Liou, 2002). In this chapter, we focus on the discrete-ordinates method and its approximations.

The discrete-ordinates method involves the discretization of the basic radiative transfer equation and the solution of a set of first-order differential equations.

In order to solve equation (2.15) analytically, the integral must be replaced by a summation over a finite number of quadrature points. The use of Gaussian quadrature is essential because it makes phase function renormalization unnecessary, implying that energy is conserved in the computation (Stamnes et al., 1988). In numerical integrations, double-Gauss's formula has been found to be superior to other formulas for quadratures in the interval (−1, 1). Double-Gauss simply refers to a quadrature rule suggested by Sykes (1951) in which the Gaussian formula is applied separately to the half-ranges $-1 < \mu < 0$ and $0 < \mu < 1$. For any function $f(\mu)$, double-Gauss's formula is expressed by

$$\int_{-1}^{1} f(\mu)d\mu \approx \sum_{j=-n}^{n} w_j f(\mu_j), \tag{2.23}$$

where the weights w_j are

$$w_j = \frac{1}{P'_{2n}(\mu_j)} \int_{-1}^{1} \frac{P_{2n}(\mu)}{\mu - \mu_j} d\mu, \tag{2.24}$$

and μ_j are the zeros of the even-order Legendre polynomials $P_{2n}(\mu)$, and the prime denotes the differentiation with respect to μ_j. Also, the quadrature weights of w and discrete ordinate of μ can be selected to satisfy $w_{-j} = w_j$ $(\sum_{j=-n}^{n} w_j = 2)$ and $\mu_{-j} = -\mu_j$. The Gaussian points and weights for the first two approximations are provided in Table 2.1.

Replacing the integral with a summation and omitting the emission term, the radiative transfer equation can be written as

$$\mu_i \frac{dI(\tau, \mu_i)}{d\tau} = I(\tau, \mu_i) - \frac{\varpi}{2} \sum_{j=-n}^{n} I(\tau, \mu_j) P(\mu_i, \mu_j) w_j - \frac{\varpi}{4\pi} F_{\odot} P(\mu_i, -\mu_0) e^{-\tau/\mu_0}, \qquad i = -n, \ldots, n, \tag{2.25}$$

where the equation is discretized by replacing μ with μ_i $(-n, n)$, which represent the direction of the radiation streams.

The phase function may be numerically expended in Legendre polynomials with a finite number of terms N and can be expressed by

$$P(\mu_i, \mu_j) = \sum_{l=0}^{N} \varpi_l P_l(\mu_i) P_l(\mu_j), \qquad j = -n, \ldots, n. \tag{2.26}$$

Table 2.1 Gaussian points and weights.

n	2n	$\pm\mu_n$	w_n
1	2	$\mu_1 = 0.5773503$	$w_1 = 1$
2	4	$\mu_1 = 0.3399810$	$w_1 = 0.6521452$
		$\mu_2 = 0.8611363$	$w_2 = 0.3478548$

The discretized equation can then be rewritten as

$$\mu_i \frac{dI(\tau, \mu_i)}{d\tau} = I(\tau, \mu_i) - \frac{\varpi}{2} \sum_{l=0}^{N} \varpi_l P_l(\mu_i) \sum_{j=-n}^{n} I(\tau, \mu_j) P_l(\mu_j) w_j - \frac{\varpi}{4\pi} F_{\odot} \left[\sum_{l=0}^{N} (-1)^l \varpi_l P_l(\mu_i) P_l(\mu_0) \right] e^{-\tau/\mu_0}. \tag{2.27}$$

In the discrete-ordinates method, analytical solutions for the diffuse intensity can be explicitly given for any optical depth. Moreover, useful approximations can be developed from this method for flux calculations. The two commonly used approximations of the discrete-ordinates method are the two-stream and four-stream approximations that are introduced in the following sections.

2.4.1. Two-Stream and Eddington's Approximation

Two-stream approximation for radiative transfer based on the discrete-ordinates method has been widely used in radiative flux calculations (Meador & Weaver, 1980) because analytic solutions for upward and downward fluxes can be derived and numerical computations for these fluxes can be efficiently performed.

In the two-stream approximation, two radiation streams, one upward and one downward, are selected, i.e., $N = 1$, and $j = -1$ and 1. A number of two-point quadrature methods have been developed for applications to two-stream radiative transfer (Meador & Weaver, 1980), including the delta-Eddington approximation (Joseph et al., 1976) and the Practical Improved Flux Method (PIFM; Zdunkowski et al., 1980), which used Gaussian quadratures with $\mu_1 = 0.5$ in the shortwave and $\mu_1 = 1/1.66$ in the longwave. Another commonly used Gaussian choice is shown in Table 2.1 (Liou, 1973, 1974) with $\mu_1 = 0.5773503$ and $w_1 = w_{-1} = 1$. Two equations can be obtained as follows after rearranging terms in equation (2.27) and denoting $I^{\uparrow} = I(\tau, \mu_1)$ and $I^{\downarrow} = I(\tau, -\mu_1)$

$$\mu_1 \frac{dI^{\uparrow}}{d\tau} = I^{\uparrow} - \varpi(1-b)I^{\uparrow} - \varpi b I^{\downarrow} - S^{-} e^{-\frac{\tau}{\mu_0}}, \tag{2.28}$$

$$\mu_1 \frac{dI^{\downarrow}}{d\tau} = I^{\downarrow} - \varpi(1-b)I^{\downarrow} - \varpi b I^{\uparrow} - S^{+} e^{-\frac{\tau}{\mu_0}}. \tag{2.29}$$

where

$$g = \frac{\varpi_1}{3} = \frac{1}{2} \int_{-1}^{1} P(\cos\Theta) \cos\Theta d\cos\Theta, \tag{2.30}$$

$$b = \frac{1-g}{2} = \frac{1}{2} \int_{-1}^{1} P(\cos\Theta) \frac{1-\cos\Theta}{2} d\cos\Theta, \tag{2.31}$$

$$S^{\pm} = \frac{F_{\odot}\varpi}{4\pi}(1 \pm 3g\mu_1\mu_0), \tag{2.32}$$

where Θ is the scattering angle, $\cos\Theta = \mu\mu' + (1-\mu^2)^{1/2}(1-\mu'^2)^{1/2}\cos\phi$, and g is the asymmetry factor that is the first moment of the phase function. For isotropic scattering such as Rayleigh scattering, $g = 0$. The asymmetry factor increases as the diffraction peak of the phase function sharpens and more forward scattering occurs. The asymmetry factor may become negative if the phase function is peaked in backward direction (90°–180°). For Lorenz-Mie scattering where the phase function has a sharp peak at 0° scattering angle (forward direction), the asymmetry factor denotes the relative strength of forward scattering. The coefficients b and $1 - b$ represent the integrated fraction of back and forward scattered energy, respectively. Thus, the multiple scattering is represented in the two-stream approximation by the upward and downward intensities weighted by the appropriate fractions of the forward and backward phase functions.

Equations (2.28) and (2.29) are two first-order differential equations that can be solved analytically and the details can be found in Liou (2002). The solutions are given as

$$I^{\uparrow} = I(\tau, \mu_1) = Kve^{k\tau} + Hue^{-k\tau} + \varepsilon e^{-\tau/\mu_0}, \tag{2.33}$$

$$I^{\downarrow} = I(\tau, -\mu_1) = Kue^{k\tau} + Hve^{-k\tau} + \gamma e^{-\tau/\mu_0}, \tag{2.34}$$

where

$$v = (1 + a)/2, \quad u = (1 - a)/2$$

$$a^2 = (1 - \varpi)/(1 + \varpi g),$$

$$\varepsilon = (\alpha + \beta)/2, \quad \gamma = (\alpha - \beta)/2$$

$$\alpha = Z_1 \mu_0^2 / \left(1 - \mu_0^2 k^2\right) \quad \beta = Z_2 \mu_0^2 / \left(1 - \mu_0^2 k^2\right), \tag{2.35}$$

$$Z_1 = \frac{(1 - \varpi g)(S^- + S^+)}{\mu_1^2} + \frac{S^- - S^+}{\mu_1 \mu_0},$$

$$Z_2 = \frac{(1 - \varpi)(S^- - S^+)}{\mu_1^2} + \frac{S^- + S^+}{\mu_1 \mu_0},$$

$$k^2 = (1 - \varpi)(1 - \varpi g)/\mu_1^2,$$

where $\pm k$ are the eigenvalues for the solution of the differential equations, u and v represent the eigenfunctions defined by the similarity parameter a. The constants K and H are to be determined from the diffuse radiation boundary conditions at the top and the bottom of the scattering layer. Assuming no diffuse radiation from the top and the bottom, we have

$$K = -\left(\varepsilon v e^{-\tau/\mu_0} - \gamma u e^{-\tau/\mu_0}\right) / \left(v^2 e^{k\tau_1} - u^2 e^{-k\tau_2}\right), \tag{2.36}$$

$$H = -\left(\varepsilon u e^{-\tau/\mu_0} - \gamma v e^{-\tau/\mu_0}\right) / \left(v^2 e^{k\tau_1} - u^2 e^{-k\tau_2}\right). \tag{2.37}$$

Once the upward and downward intensities have been evaluated, the upward and downward fluxes can be computed as

$$F^{\uparrow}(\tau) = 2\pi \mu_1 I^{\uparrow}, \tag{2.38}$$

$$F^{\downarrow}(\tau) = 2\pi \mu_1 I^{\downarrow}. \tag{2.39}$$

The Eddington's approximation uses an approach similar to that of the two-stream approximation and was originally used for studies of radiative equilibrium in stellar atmosphere (Eddington, 1916). In this approximation, the radiative transfer equation is decomposed using the property of Legendre polynomials. In line with the Legendre polynomial expansion for the phase function denoted in equation (2.26), the diffuse radiative intensity may be expanded in terms of Legendre polynomials such that

$$I(\tau, \mu) = \sum_{l=0}^{N} I_l(\tau) P_l(\mu). \tag{2.40}$$

Letting $N = 1$, the basic radiative transfer equation (2.16) can be reduced to a set of two simultaneous equations similar to equations (2.29) and (2.30), which can be analytically solved and the solutions are similar to those of the two-stream approximation.

Combining two-stream approximation for discrete-ordinates method and the Eddington's approximation, a generalized two-stream approximation may be expressed by

$$\frac{dF^{\uparrow}(\tau)}{d\tau} = \gamma_1 F^{\uparrow}(\tau) - \gamma_2 F^{\downarrow}(\tau) - \gamma_3 \varpi F_{\odot} e^{-\tau/\mu_0}, \tag{2.41}$$

$$\frac{dF^{\downarrow}(\tau)}{d\tau} = \gamma_2 F^{\uparrow}(\tau) - \gamma_1 F^{\downarrow}(\tau) + (1 - \gamma_3) \varpi F_{\odot} e^{-\tau/\mu_0}. \tag{2.42}$$

The coefficients γ_i are provided in Table 2.2. In the two-stream approximation, there are only upward and downward intensities in the directions μ_1 and $-\mu_1$, while the phase function is expanded in two terms in Legendre polynomials. In Eddington's approximation, both the intensity and phase functions are expanded in two polynomials terms.

The solutions for the equations of the generalized two-stream approximation are as follows:

$$F^{\uparrow} = vKe^{k\tau} + uHe^{-k\tau} + \varepsilon e^{-\tau/\mu_0}, \tag{2.43}$$

$$F^{\downarrow} = uKe^{k\tau} + vHe^{-k\tau} + \gamma e^{-\tau/\mu_0}, \tag{2.44}$$

Table 2.2 Coefficients in two-stream approximations.

Method	γ_1	γ_2	γ_3
Two-Stream	$[1 - \varpi(1 + g)/2]/\mu_1$	$\varpi(1 - g)/2\mu_1$	$(1 - 3g\ \mu_1 \mu_0)/2$
Eddington's	$[7 - (4 + 3g)\varpi]/4$	$-[1 - (4 - 3g)\varpi]/4$	$(2 - 3g\ g\mu_0)/4$

where coefficients K and H can be determined from boundary conditions, and

$$k^2 = \gamma_1^2 - \gamma_2^2,$$

$$v = \frac{1}{2}\left[1 + \frac{\gamma_1 - \gamma_2}{k}\right], \qquad u = \frac{1}{2}\left[1 - \frac{\gamma_1 - \gamma_2}{k}\right], \tag{2.45}$$

$$\varepsilon = [\gamma_3(1/\mu_0 - \gamma_1) - \gamma_2(1 - \gamma_3)]\mu_0^2 \varpi F_\odot,$$

$$\gamma = -[(1 - \gamma_3)(1/\mu_0 + \gamma_1) - \gamma_2\gamma_3]\mu_0^2 \varpi F_\odot.$$

2.4.2. Four-Stream Approximation

The four-stream approximation, which is first derived by Liou (1974), is also based on the discrete-ordinates method for radiative transfer. A systematic solution for this approximation has been presented by Liou et al. (1988).

If we consider four radiative streams, with two streams in the upward and downward directions, respectively, i.e., $n=2$, and expand the phase function in four terms, i.e., $N=3$, then we can obtain four first-order differential equations:

$$\frac{\mathrm{d}}{\mathrm{d}\tau}\begin{bmatrix} I_2 \\ I_1 \\ I_{-1} \\ I_{-2} \end{bmatrix} = \begin{bmatrix} b_{2,-2} & b_{2,-1} & b_{2,1} & b_{2,2} \\ b_{1,-2} & b_{1,-1} & b_{1,1} & b_{1,2} \\ -b_{1,2} & -b_{1,1} & -b_{1,-1} & -b_{1,-2} \\ -b_{2,2} & -b_{2,1} & -b_{2,-1} & -b_{2,-2} \end{bmatrix} \times \begin{bmatrix} I_2 \\ I_1 \\ I_{-1} \\ I_{-2} \end{bmatrix} - \begin{bmatrix} b_{2,-0} \\ b_{1,-0} \\ b_{-1,-0} \\ b_{-2,-0} \end{bmatrix}, \tag{2.46}$$

where the terms $b_{i,j}$ ($i=\pm 1, 2; j=-0, \pm 1, 2$) are defined as follows:

$$b_{i,j} = \begin{cases} c_{i,j}/\mu_i \\ (c_{i,j} - 1)/\mu_i \end{cases}$$

$$c_{i,j} = \frac{\varpi}{2} w_j P(\mu_i, \mu_j) = \frac{\varpi}{2} w_j \sum_{l=0}^{N} \varpi_l P_l(\mu_i) P_l(\mu_j),$$

$$j = -n, \ldots, -0, \ldots n, \tag{2.47}$$

$$c_{i,-j} = c_{-i,j}, \qquad c_{-i,-j} = c_{i,j}, \qquad i \neq -0.$$

The Gauss quadrature and weights in the four-stream approximation are $\mu_1 = 0.3399810$, $\mu_2 = 0.8611363$, and $w_1 = 0.6521452$, $w_2 = 0.3478548$, as listed in Table 2.1. The four-by-four matrix represents the contribution of multiple scattering. Thus, the derivative of the diffuse intensity at a specific quadrature angle is the weighted sum of the multiple-scattered intensity from all four quadrature angles. The last term represents the contribution of the unscattered component of the direct flux.

The four equations can be analytically solved with solutions to be

$$\begin{bmatrix} I_1 \\ I_{-1} \\ I_2 \\ I_{-2} \end{bmatrix} = \begin{bmatrix} \Phi_1^+ e_1^- & \Phi_1^- e_1^+ & \Phi_2^+ e_2^- & \Phi_2^- e_2^+ \\ \Phi_1^- e_1^- & \Phi_1^+ e_1^+ & \Phi_2^- e_2^- & \Phi_2^+ e_2^+ \\ \phi_1^+ e_1^- & \phi_1^- e_1^+ & \phi_2^+ e_2^- & \phi_2^- e_2^+ \\ \phi_1^- e_1^- & \phi_1^+ e_1^+ & \phi_2^- e_2^- & \phi_2^+ e_2^+ \end{bmatrix} \begin{bmatrix} G_1 \\ G_{-1} \\ G_2 \\ G_{-2} \end{bmatrix} + \begin{bmatrix} Z_1^+ \\ Z_1^- \\ Z_2^+ \\ Z_1^- \end{bmatrix} \mathrm{e}^{-\tau/\mu_0}, \tag{2.48}$$

where the elements $e_1^- = e^{-k_1\tau}, e_1^+ = e^{k_1\tau}, e_2^- = e^{-k_2\tau}$, and $e_2^+ = e^{k_2\tau}$. The eigenvectors are

$$\phi_{1,2}^{\pm} = \frac{1}{2}\left(1 \pm \frac{b_{11}^- - A_{1,2}b_{21}^-}{a^-}k_{1,2}\right), \tag{2.49a}$$

$$\Phi_{1,2}^{\pm} = \frac{1}{2}\left(A_{1,2} \pm \frac{A_{1,2}b_{22}^- - b_{12}^-}{a^-}k_{1,2}\right). \tag{2.49b}$$

where $b_{ij}^{\pm}$ is defined by

$$b_{22}^{\pm} = b_{2,2} \pm b_{2,-2}, \qquad b_{21}^{\pm} = b_{2,1} \pm b_{2,-1}, \tag{2.50a}$$

$$b_{12}^{\pm} = b_{1,2} \pm b_{1,-2}, \qquad b_{11}^{\pm} = b_{1,1} \pm b_{1,-1}, \tag{2.50b}$$

$k_{1,2}$ are eigenvalues determined by

$$k^2 = [b \pm (b^2 + 4c)^{1/2}]/2, \tag{2.51}$$

where $b = a_{22} + a_{11}, c = a_{21}a_{12} - a_{11}a_{22}, A_{1,2} = (k_{1,2}^2 - a_{22})/a_{21}$, a_{ij} are defined by

$$a_{22} = b_{22}^+ b_{22}^- + b_{12}^+ b_{21}^-, \qquad a_{21} = b_{21}^+ b_{22}^- + b_{11}^+ b_{21}^-, \tag{2.52a}$$

$$a_{12} = b_{22}^+ b_{12}^- + b_{12}^+ b_{11}^-, \qquad a_{11} = b_{21}^+ b_{12}^- + b_{11}^+ b_{11}^-, \tag{2.52b}$$

and $a^- = b_{22}^- b_{11}^- - b_{12}^- b_{21}^-$. The Z functions are defined by

$$Z_{1,2}^{\pm} = \frac{1}{2}(\eta_{1,2} + \eta'_{1,2}), \tag{2.53}$$

where $\eta_{1,2}$ and $\eta'_{1,2}$ are defined by

$$\eta_1 = \frac{d_1/\mu_0^2 + a_{12}d_2 - a_{22}d_1}{f(1/\mu_0)}\frac{F_\odot}{2\pi},$$

$$\eta_2 = \frac{d_2/\mu_0^2 + a_{21}d_1 - a_{11}d_2}{f(1/\mu_0)}\frac{F_\odot}{2\pi}, \tag{2.54}$$

$$\eta'_1 = \frac{d'_1/\mu_0^2 + a'_{12}d'_2 - a'_{22}d'_1}{f(1/\mu_0)}\frac{F_\odot}{2\pi},$$

$$\eta'_2 = \frac{d'_2/\mu_0^2 + a'_{21}d'_1 - a'_{11}d'_2}{f(1/\mu_0)}\frac{F_\odot}{2\pi}.$$

The function f is defined as

$$f(k) = k^4 - bk^2 - c, \tag{2.55}$$

where k is replaced by $1/\mu_0$. d_i are defined by

$$d_1 = b_{12}^- b_2^- + b_{11}^- b_1^- + b_1^+/\mu_0, \tag{2.56a}$$

$$d_2 = b_{22}^- b_2^- + b_{21}^- b_1^- + b_2^+/\mu_0. \tag{2.56b}$$

d_i' and a_{ij}' have the same expressions as those for d_i and a_{ij} except that the superscripts + and – are replaced by – and +, respectively. The coefficients G_j ($j = 1, 2, -1, -2$) can be determined from boundary conditions. Once the intensities for the four streams have been obtained, the upward and downward fluxes at a given τ are given by

$$F^{\uparrow}(\tau) = 2\pi(w_1\mu_1 I_1 + w_2\mu_2), \tag{2.57}$$

$$F^{\downarrow}(\tau) = 2\pi(w_1\mu_1 I_{-1} + w_2\mu_2 I_{-2}) + \mu_0 F_{\odot} e^{-\tau/\mu_0}. \tag{2.58}$$

2.4.3. Delta-Function Adjustment

When calculating the radiative transfer in the atmosphere, the effects of scattering and absorption by clouds and various aerosols must be included. These particles usually have size larger than the incident solar radiation, and in many cases a sharp diffraction occurs near 0° scattering angle. To incorporate the forward peak contribution in multiple scattering, we may consider an adjusted absorption and scattering atmosphere such that the fraction (f) of scattered energy in the forward peak is accounting for. Using delta-Eddington scaling, $f = g^2$ (Joseph et al., 1976). The forward peak can be expressed as $f = \varpi_2/5$ ($l = 1$) for two-stream approximation and $f = \varpi_4/9$ ($l = 4$) for four-stream approximation, to account for strong forward scattering by cloud and aerosol particles. For M-stream, the delta-M method, which is a natural extension of the delta-Eddington approximation developed by Wiscombe (1977), can be applied to all orders M of angular approximation. The delta-M method relies on matching the first $2M$ phase moments and using a Dirac delta-function representation of forward scattering.

We use primes to represent the adjusted single-scattering properties, optical depth τ, single-scattering albedo ϖ, and phase function P or asymmetry factor g. The optical depth is the sum of the scattering (τ_s) and absorption (τ_a) optical depth. Since the forward peak is only associated with scattering without the contribution of absorption, the adjusted scattering and absorption optical depth must be

$$\tau_s' = (1 - f)\tau_s, \tag{2.59}$$

$$\tau_a' = \tau_a. \tag{2.60}$$

The total adjusted optical depth is then

$$\tau' = \tau_s' + \tau_s' = (1 - f)\tau_s + \tau_a = \tau(1 - f\varpi), \tag{2.61}$$

where the single-scattering albedo by definition can be expressed by $\varpi = \tau_s/\tau$. The adjusted single-scattering albedo is defined by

$$\varpi' = \frac{\tau_s'}{\tau'} = \frac{(1 - f)\tau_s}{(1 - f\varpi)\tau} = \frac{(1 - f)\varpi}{1 - f\varpi}. \tag{2.62}$$

Moreover, we may express the normalized phase function expansion by incorporating the δ-forward adjustment in the form

$$P_l(\cos\Theta) = 2f\delta(\cos\Theta - 1) + (1 - f)\sum_{l=0}^{N} \varpi_l' P_l(\cos\Theta), \tag{2.63}$$

where δ denotes the delta-function, and ϖ_l' is the adjusted coefficients in the phase function expansion. The adjusted phase function is then given by

$$P'(\cos\Theta) = \sum_{l=0}^{N} \varpi_l' P_l(\cos\Theta). \tag{2.64}$$

Equations (2.61), (2.62), and (2.64) constitute the generalized similarity principle for radiative transfer. That is, the removal of the forward diffraction peak in scattering processes using adjusted single-scattering parameters is "equivalent" to actual scattering processes.

The similarity principle was first introduced by Sobolev (1975) for isotropic scattering. The inclusion of the asymmetry factor was discussed by van de Hulst (1980). The principle of employing a Dirac delta-function to approximate highly forward peaked scattering in radiative transfer has been presented by Hansen (1969), Potter (1970), and Wiscombe (1977).

2.4.4. Application to Plane-Parallel Atmosphere

The accuracy of the radiative transfer approximations described in the previous sections has been examined by comparing the approximate results with the values computed from the "exact" method, such as the adding method. Figure 2.6 shows the relative accuracy of the delta-two-stream (upper panel) and delta-four-stream (lower panel) approximations with respect to the results derived from the adding method displayed in intervals of 0%, 1%, 2%, 5%, 10%, etc. Here, the relative accuracy is defined by $(\Delta r/r)100\% = [(\hat{r} - r)r]100\%$, where $\hat{r}$ and r represent the reflectance computed from the approximation and the "exact" method, respectively. Likewise, the relative accuracy of the total transmittance is denoted by $(\Delta t/t)100\%$. In general, reflectance and total transmittance values computed from the delta-four-stream approximation are accurate within about 5%, while the delta-two-stream approximation produces errors greater than 5%–10%. In particular, errors of more than 50%

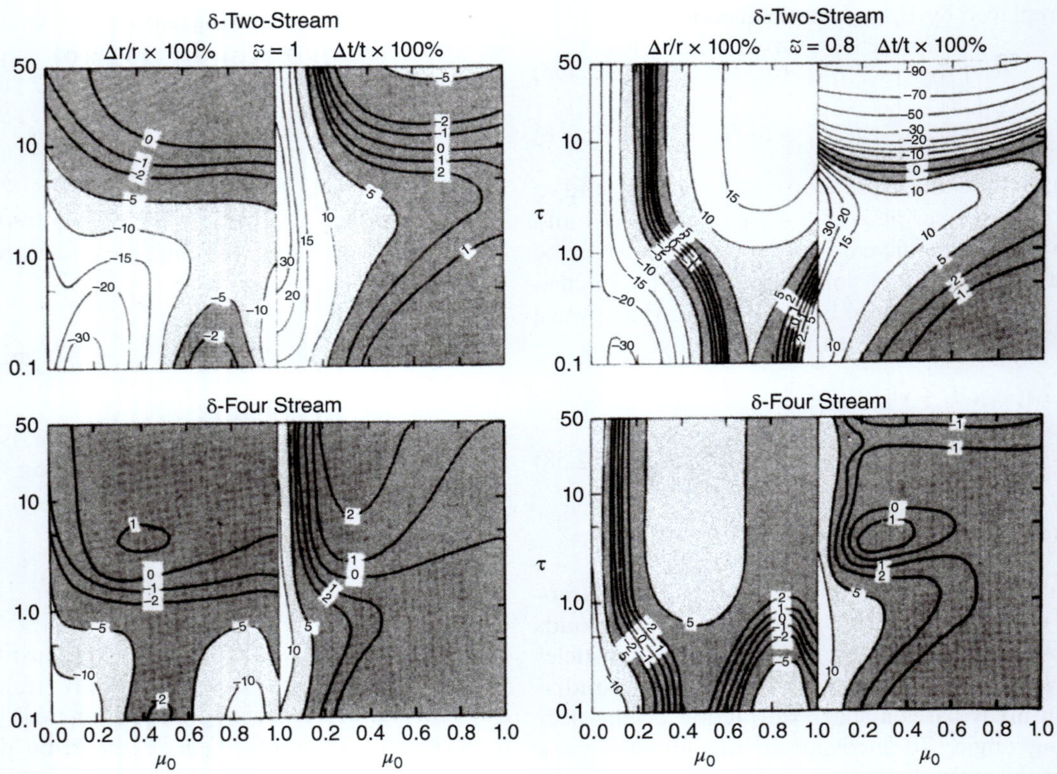

Figure 2.6 Relative accuracy of the reflectance and transmittance computed from the delta-two-stream (upper panel) and delta-four-stream (lower panel) approximations with respect to the results derived from the adding method, for $\varpi = 1$ (conservative scattering, left panel) and $\varpi = 0.8$ (right panel), respectively. The results are shown in the domain of τ (ranging from 0.1 to 50) and μ_0 (0.0–1.0), and expressed in terms of percentage. The heavy and light shadings denote errors with 5% and 5%–10%, respectively, while the white area represent errors larger than 10%. Source: Liou et al. (1988)/American Meteorological Society.

may occur in the case of large optical depth for the delta-two-stream when $\varpi = 0.8$. The accuracy of the delta-two-stream approximation shown in Figure 2.6 is comparable to that of the delta-Eddington approximation reported by King and Harshvardhan (1986).

For application to the solar absorption bands, in which gaseous absorption in scattering atmospheres must be accounted for, the accuracy of the delta-two-stream and delta-four-stream approximations has also been examined using single-scattering albedo of 0.5 and 0.3. For cases involving large absorption, the reflectance values are generally very small. Thus, we investigate the relative accuracy for the absorptance, defined as $(\Delta A/A)100\%$, where $A = 1 - r - t$. Figure 2.7 shows that the delta-four-stream approximation produces better accuracy than the delta-two-stream approximation, with errors for acceptance generally less than 2%. The accuracy increases as ϖ decreases because the effect of multiple scattering on the flux calculations becomes less important. For total transmittance, errors from the delta-four-stream approximation are again within 5%. Large relative errors can be produced by the delta-two-stream approximation when the transmittance values are small.

Stephens et al. (2001) assessed a select number of two-stream models in representing solar and infrared radiative transfer problems. They reported that the delta-Eddington model and the constant-hemispheric two-stream models are shown to be superior to other two-stream methods of solution and also superior to four-stream solutions for the many classes of problems relevant to modeling the global atmosphere. Räisänen (2002) tested four two-stream approximations, including the delta-Eddington approximation (Joseph et al., 1976), the PIFM (Zdunkowski et al., 1980), and two recent modifications of delta-Eddington by Li (1999) and Qiu (1999), with the general circulation model (GCM) data set. It is found that two-stream approximations generally performed fairly well with a negative bias in atmospheric absorption and a positive bias in surface net shortwave flux of about 1 W M^{-2} in the global mean. Furthermore, the absorption errors depended substantially on solar elevation. The tests with GCM data also confirmed

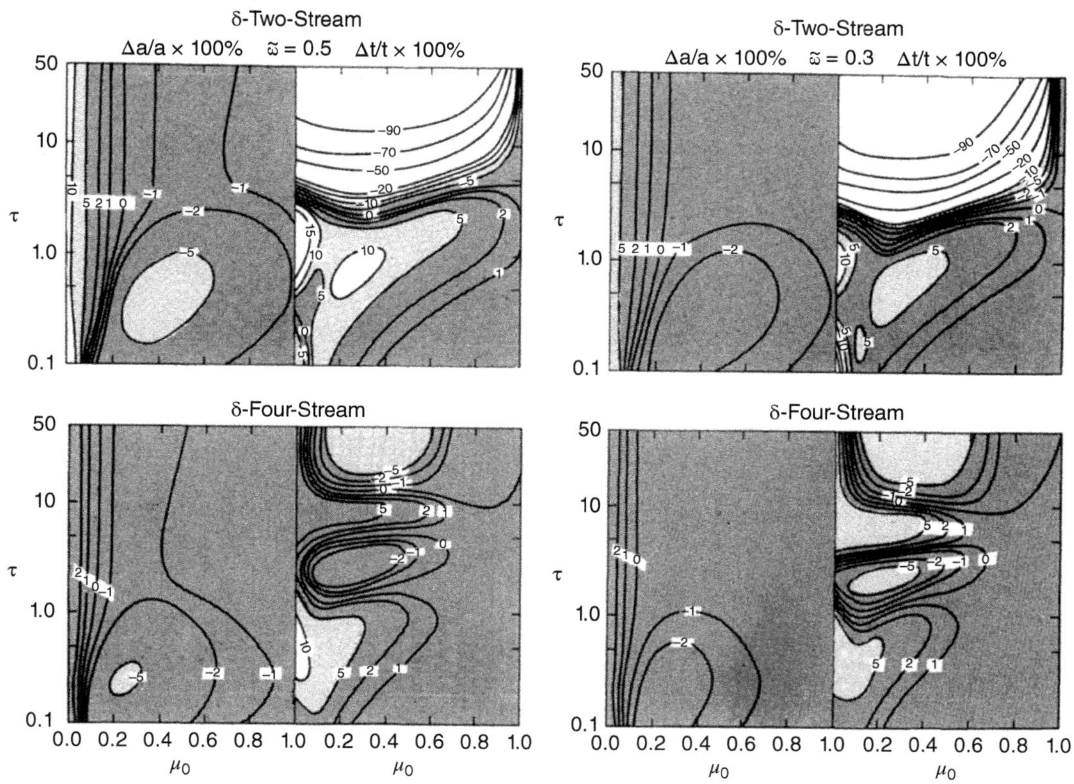

Figure 2.7 Same as Figure 2.6, except for absorptance. The left and right panels are for $\varpi = 0.5$ and $\varpi = 0.3$, respectively. Source: Liou et al. (1988)/American Meteorological Society.

that the four-stream method is considerably more accurate than the two-stream approximations, especially for atmospheric absorption.

While all approximations for radiative transfer calculations have advantages and shortcomings in terms of the computational accuracy, the delta-four-stream approximation can achieve relative accuracy with 5% for all reasonable ranges of single-scattering parameters at a given wavelength. For computations of solar fluxes covering the entire solar spectrum, the averaged accuracy should also be within 5%. The delta-four-stream approximation has all the radiative characteristics inherent in the delta-two-stream approximation, and the solution is also in analytic form so that the computer time involved is reasonable. For radiative transfer parameterizations in numerical models, the delta-four-stream approximation would be an excellent method. Two-stream approximations are currently employed for multiple-scattering calculations in most solar radiation schemes used in climate models. Fu et al. (1997) demonstrated that the delta-two-stream method is the most computationally efficient but produces significant errors in fluxes and heating rates under cloudy conditions. High accuracy can be obtained by using the delta-four-stream method, but substantial computer time is required for the calculation of thermal infrared radiative transfer. The delta-two/four-stream combination method is sufficiently economical for thermal IR calculations; it is 4 times faster than the delta-four-stream method but only 50% slower than the two-stream method and at the same time produces acceptable accuracy under most atmospheric conditions.

2.5. RADIATION-CLOUD INTERACTIONS IN LARGE-SCALE ATMOSPHERIC MODELS

Clouds absorb, reflect, and transmit solar radiation, and the amount is a function of the optical depth, the geometry governing the Sun, and the direction of detection. Clouds can also reflect and transmit the thermal infrared radiation emitted from the surface and the atmosphere and, meanwhile, emit thermal IR radiation according to the temperature structure within the clouds. While considerable advances in the understanding of atmospheric processes and feedbacks in the climate system have led to a better representation of these mechanisms in general circulation models, the greatest uncertainty in predictability of future climate arises from clouds and their interactions with radiation.

Theoretical, observational, and modeling studies have demonstrated the important role of cloud-radiation interactions in climate variability. Role of clouds in climate has twofold: first clouds are generated by large-scale dynamic forcing, radiative cooling in the atmosphere, and turbulent transfer at the surface. On the other hand, clouds provide one of the most important mechanisms for the vertical redistribution of momentum, moisture, and sensible and latent heat. Clouds also influence the coupling between the atmosphere and the surface, as well as the radiative and dynamical and hydrological balance. The atmospheric general circulation model (AGCM) studies of Ramanathan et al. (1983), Slingo and Slingo (1988, 1991), and Randall et al. (1989) have shown that radiation, latent heat release, and small-scale transport are of equal importance in the cloud-climate problem and that many features of the simulated climate are sensitive to the treatment of clouds and radiation in the model. The cloud-radiation interaction and feedback problem have been identified as the highest priority item in climate research nationally and internationally, as illustrated by a number of composite field and satellite observations, including the International Satellite Cloud Climatology Project (ISCCP), the First ISCCP Regional Experiment (FIRE), the Atmospheric Radiation Measurement (ARM) Program, and the Earth Radiation Budget Experiment (ERBE). Therefore, numerical models used for climate studies are required to provide the cloud and radiation fields as accurate as possible. Recent comparisons of feedbacks produced by climate models under climate change show that the current generation of models still exhibits a large spread in cloud feedbacks, which is larger than for other feedbacks (Bony et al., 2006). Webb et al. (2013) diagnosed climate feedback parameters including rapid adjustment in 12 atmosphere/mixed-layer-ocean climate models and found that cloud effects can explain the full range of climate sensitivities, and cloud feedback components contribute 4 times as much as cloud components of CO_2 forcing to the range.

The parameterization of cloud-radiation processes in climate models is a complex task. Radiative transfer in the atmosphere is determined by spectrally dependent optical properties. Calculation of the radiative heating/cooling in clouds is complicated due to difficulties in parameterizing their single-scattering properties, especially those of ice clouds due to complexities in the ice crystal size, shape, and orientation, which cannot be determined from the models (Liou, 1986). Furthermore, clouds have a typical length scale of perhaps several hundred meters and display substantial horizontal variability on scales that are generally smaller than the usual AGCM grid box. A specific challenge in studying the cloud-radiation feedback on climate is to estimate the uncertainty in the vertical profile of cloud cover, the information of which is limited in present satellite and surface observations. In recent years, some models have improved representation of subgrid scale cloud variability, which has important effects on grid-mean radiative fluxes, for example, based on the use of probability density functions of thermodynamic variables (e.g., Watanabe et al., 2009). Stochastic approaches have also been used to efficiently account for this variability (Barker et al., 2008). Improved treatments of overlap have been developed in some models based on new observations that show cloud layers at different levels overlap less often than typically assumed in GCMs (Pincus et al., 2006; Shonk et al., 2012).

2.5.1. Parameterization of Cloud Radiative Properties

Clouds are conventionally classified in terms of their position and appearance in the atmosphere. For radiative transfer calculations, we normally classify clouds as water cloud at temperature above 0 °C, ice below about −38 °C (e.g., Koop et al., 2000), and either or both phases at intermediate temperatures. Low clouds with height <2 km are normally water clouds. High clouds present at height >6 km are normally ice clouds. In between are the middle clouds that could consist of both water and ice clouds. Under the plane-parallel assumption, the radiative transfer methods introduced in section 2.4 can be applied to clouds. We have learned from section 2.2 that the fundamental parameters in radiative transfer are the single-scattering properties, including extinction coefficient, single-scattering albedo, and phase function or asymmetry factor. In the case of clouds, the basic extinction properties are determined by particle size distribution.

The calculations of the single-scattering properties require a detailed light scattering program and particle size distribution and are usually time consuming. If radiation calculations are for an evolving cloud where particle size distribution varies as a function of time and/or space, the computer time needed just for determining the single-scattering properties would be formidable even with a supercomputer. Thus, it is practically required to simplify the computational procedure for the calculations of the single-scattering properties of cloud particles.

Water Clouds

The characteristics of scattered light from clouds depend on the droplet size distribution, $n(r)$. Therefore, the first parameter describing droplet size distribution should be some measure of the mean size. For water clouds, since the spherical droplets scatter an amount of light in proportion to their cross-sectional area, we may define a mean effective radius r_e, which differs from simple mean radius by using droplet cross section as a weight factor:

$$r_e = \frac{\int r \cdot \pi r^2 n(r) \mathrm{d}r}{\int \pi r^2 n(r) \mathrm{d}r}. \tag{2.65}$$

From the definition of liquid water content (LWC), the liquid water path (LWP), which is the vertically integrated LWC over a cloud thickness of Δz, can be expressed in terms of droplet radius as

$$\mathrm{LWP} = \Delta z \cdot \mathrm{LWC} = \Delta z \cdot \rho_l V = \Delta z \cdot \rho_l \cdot \frac{4\pi}{3} \int r^3 n(r) \mathrm{d}r, \tag{2.66}$$

where ρ_l is the density of liquid water and V is the volume of cloud droplets divided by the volume of air. The cloud optical depth for a given droplet size distribution can also be expressed as a function of droplet radius by

$$\tau = \Delta z \int \sigma_e n(r) \mathrm{d}r = \Delta z \int Q_e \pi r^2 n(r) \mathrm{d}r, \tag{2.67}$$

where the extinction cross section $\sigma_e = Q_e \pi r^2$, and Q_e is the efficiency factor that is a function of the droplet radius, wavelength, and refractive index. For visible wavelength, $Q_e \cong 2$ for cloud droplets. Therefore, the mean effective radius r_e (μm) can be related to LWP (g cm^{-2}) and τ in the form

$$r_e \cong \frac{3}{2} \mathrm{LWP}/\tau, \quad \text{or} \quad \tau = \frac{3}{2} \mathrm{LWP}/r_e. \tag{2.68}$$

That is, the cloud optical depth can be obtained once the cloud effective radius is known.

The single-scattering albedo is defined by

$$1 - \varpi = \sigma_a / \sigma_e, \tag{2.69}$$

where σ_a denotes the absorption cross section, which is proportional to the product of the absorption coefficient k and the volume, and it follows that $\sigma_a = k \cdot \frac{4}{3} r_e \int \pi r^2 n(r) \mathrm{d}r$. If an efficiency factor of 2 is used for extinction, then $\sigma_e = 2 \int \pi r^2 n(r) \mathrm{d}r$. The other important factor is asymmetry factor. Based on Mie scattering calculations, the asymmetry factor shows relatively little variation in the solar wavelength.

Parameterization of the solar radiative properties for water clouds has been carried out by Stephens (1978), Liou and Wittman (1979), and Slingo (1989). Based on a number of droplet size distributions, the optical depth, single-scattering albedo, and asymmetry factor, which are all functions of wavelength and position, can be expressed in terms of r_e. Slingo (1989) has developed a solar radiative transfer parameterization scheme based on the two-stream approximation, in which both LWP and r_e are used as the basic parameters for the computation of the spectral solar radiative properties of water clouds. However, this parameterization was designed for a limited cloud droplet size range of 4.2–16.6 μm. Dobbie et al. (1999) have shown that the Slingo (1989) parameterization does not agree with calculations based on Mie theory for large cloud droplets. Nielsen et al. (2014) proposed a new cloud liquid optical property scheme based on the Mie solution to Maxwell's equations (Wiscombe, 1980) in the form

$$\beta(\lambda, x, y, z) = a(\lambda) r_e(x, y, z)^{-b(\lambda)}, \tag{2.70}$$

$$\varpi(\lambda, x, y, z) = c(\lambda) - d(\lambda) r_e(x, y, z), \tag{2.71}$$

$$g(\lambda, x, y, z) = e(\lambda) + f(\lambda) r_e(x, y, z) - h(\lambda) \exp[-j(\lambda) r_e], \tag{2.72}$$

where a–f, h, and j are coefficients that can be determined from numerical fittings based on detailed light scattering and absorption calculations for a range of ice crystal size distributions and shapes.

Ice Clouds

Parameterization for ice clouds can be obtained following the similar procedure. Ice crystals are nonspherical, and ice crystal size distributions are usually expressed in terms of the maximum dimension (or length) L. Representation of the size distribution for ice crystals is much more involved than that for water droplets. To the extent that scattering of light is proportional to the cross-section area of nonspherical particles, we may define a mean effective size D_e, analogous to the mean effective radius r_e, to represent ice crystal size distributions in the form

$$D_e = \frac{\int D \cdot DLn(L) \mathrm{d}l}{\int DLn(L) \mathrm{d}l}, \tag{2.73}$$

where D is the width of the particle, $n(L)$ denotes the ice crystal size distribution, and the numerator and denominator represent the volume-weighted and area-weighted ice crystal size distributions, respectively.

The ice water content (IWC) for a given ice crystal size distribution $n(L)$ is defined by

$$\mathrm{IWC} = \int V_i \rho_i n(L) \mathrm{d}L, \tag{2.74}$$

where ρ_i is the density of ice, and V_i is the volume of an individual ice crystal. Considering a hexagonal ice crystal, the volume is given by

$$V_i = \frac{3\sqrt{3}}{8} LD^2. \tag{2.75}$$

The ice water path (IWP) for a cloud with thickness Δz is then IWP = Δz·IWC.

The optical depth for ice clouds is defined by

$$\tau = \Delta z \int \sigma_e n(L) \mathrm{d}L. \tag{2.76}$$

For randomly oriented hexagonal ice crystals in the limits of geometric optics, the extinction cross section may be expressed by (Takano & Liou, 1989)

$$\sigma_e = \frac{3}{2} D \left(\frac{\sqrt{3}}{4} D + L \right). \tag{2.77}$$

Using all the preceding equations, the optical depth can be determined by two independent cloud physics variables, IWP and D_e, as

$$\tau \simeq \mathrm{IWP} \left(a + \frac{b}{D_e} \right), \tag{2.78}$$

where a and b are certain coefficients.

The absorption cross section is approximately proportional to the product of the absorption coefficient k and the volume for weak absorption. Thus, the single-scattering albedo for randomly oriented hexagonal ice crystals may be parameterized in the form

$$1 - \varpi = \frac{\sigma_a}{\sigma_e} \sim \frac{k\sqrt{3}LD^2}{D\left(\sqrt{3}D + 4L\right)}. \tag{2.79}$$

Based on aircraft observations by Ono (1969) and Auer and Veal (1970), the width D may be related to the length L. Thus, $1 - \varpi$ must be proportional to the mean effective size as

$$1 - \varpi \simeq c + dD_e, \tag{2.80}$$

where c and d are certain coefficients.

In the preceding discussion, the linear relationship derived are based on the geometric optics limit and the assumptions that the ice crystals are randomly oriented in space and their absorption is weak. For general cases, we would expect that higher-order expansion may be needed to define more precisely the single-scattering properties of ice crystals in terms of D_e. Thus, general parameterizations of single-scattering properties involving ice clouds may be written in the forms

$$\tau_i = \left(\sum_{n=0}^{N} \frac{a_{n,i}}{D_e^n} \right) \mathrm{IWP}, \tag{2.81}$$

$$1 - \varpi_i = \sum_{n=0}^{N} b_{n,i} D_e^n, \tag{2.82}$$

where the subscript i denotes the index for the spectral band, a_n and b_n are empirical coefficients that must be determined from numerical fittings with detailed light scattering calculations, and N is the total number of expansion terms to achieve a certain accuracy. The asymmetry factor may also be expressed in a likely expansion as

$$g_i = \sum_{n=0}^{N} c_{n,i} D_e^n, \tag{2.83}$$

where c_n are again empirical coefficients. Based on numerical experiments, it is found that for solar bands the first-order polynomial expansion ($N = 1$) is sufficient to achieve 0.1% accuracy. However, for thermal infrared bands, the second-order polynomial fitting ($N = 2$) is required to achieve this level of accuracy (Fu & Liou, 1993).

Ice clouds consist of ice crystals with complex shapes and sizes. Figure 2.8 shows the observed ice crystal size

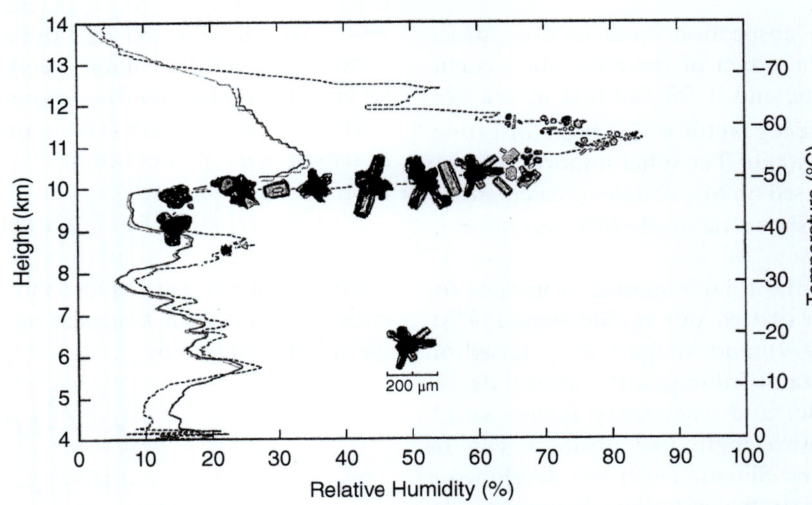

Figure 2.8 Ice crystal size and shape as a function of height and relative humidity captured by a replicator balloon sounding system in Marshall, Colorado on 10 November 1994. The relative humidity was measured by cryogenic hygrometer (dashed line) and Vaisala RS80 instruments (solid line and dots). Also shown is temperature as a function of height. Source: Courtesy of Andrew Heymsfield.

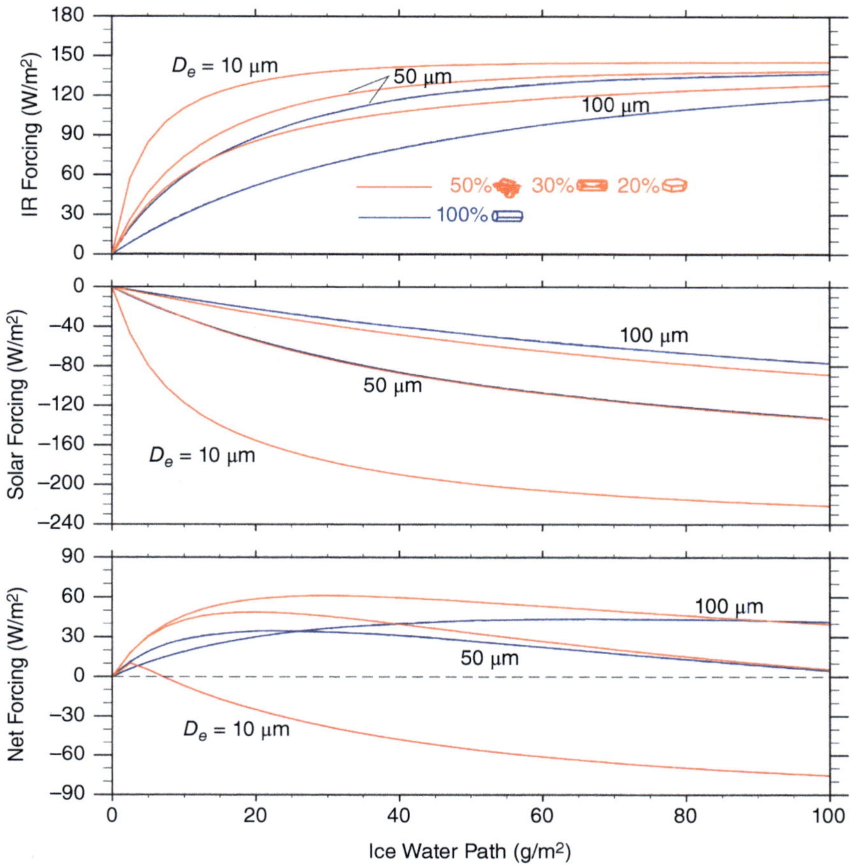

Figure 2.9 Solar, thermal IR, and net radiative forcings for cirrus clouds as functions of mean effective size D_e and ice water path for two ice crystal shape distributions in the standard atmospheric condition. The solar constant, solar zenith angle, surface albedo, cloud base height, and cloud thickness used in the calculations are 1366 W m^{-2}, 60°, 0.1, 9 km, and 2 km, respectively.

and shapes as a function of height and humidity captured by a replicator balloon sounding system. It is found that ice crystals present various shapes in the atmosphere and that that larger ice crystals are normally located in the lower part of cloud while smaller ones are in the upper part. Ice crystal effective size and shape have significant effect on the single-scattering properties of ice clouds and the associated cloud radiative forcing. According to equations (2.82)–(2.83), smaller D_e corresponds to larger extinction coefficient or optical depth, large single-scattering albedo, and smaller asymmetry factor, which would result in more scattering in the backward direction. Figure 2.9 shows the solar, thermal IR, and net radiative forcings for cirrus clouds as functions of IWP for various ice crystal sizes under two ice crystal shape distributions, 100% hexagonal and mixed habit, respectively, in the standard atmospheric condition. It can be seen that for a given IWP, cirrus clouds containing smaller effective sizes reflect more solar radiation and trap more thermal infrared radiation. For the same ice crystal effective size, the radiative forcings under various shapes of ice crystals would also substantially differ, as evidenced in Figure 2.9.

Theoretical and experimental studies also demonstrate the importance of ice particle roughness and irregularity in determining their radiative effect (Baran, 2012). Ulanowski et al. (2006) experimentally measured the scattering pattern of laboratory grown ice analog crystals and suggested that ice crystals with rough surfaces could reflect almost twice as much incident solar radiation than their smooth counterparts. Using a comprehensive in situ data set of ice crystal complexity coupled with measurements of the cloud angular scattering functions collected during a number of observational airborne campaigns at diverse geographical locations, Järvinen et al. (2018) demonstrated that an overwhelming fraction (~61%–81%) of atmospheric ice crystals sampled in the different regions contains mesoscopic deformations. The influence of ice particle surface roughness on the ice cloud radiative effect was estimated through simulations using the Fu-Liou radiation scheme and the GCM version of the Rapid Radiative Transfer Model (RRTMG) and the National Center for Atmospheric Research Community Atmosphere Model (Yi et al., 2013). Their results indicate that ice particle surface roughness may lead to a global-averaged shortwave cloud radiative effect of about 1–2 W m^{-2} and a small but nonnegligible increase in the global longwave cloud radiative effect. By assuming an irregular ice crystal, Zhang et al. (1999) demonstrated that small ice particles could cool the Earth's surface by ~40 W m^{-2}, and larger particles could warm the surface by ~20 W m^{-2}.

Since 1980s, development in cloud modeling has included prognostic equations for the prediction of IWC for high-level clouds formed in GCMs and climate models. This is a milestone accomplishment from the standpoint of incorporating a physically based cloud microphysics scheme in these models, and at the same time, it is also essential from the perspective of studying cloud-radiation interactions. However, cloud particle size is also an important independent parameter that affects radiation transfer. Ice crystal size and shape in the Earth's atmosphere are complex and intricate. After initial homogeneous and/or heterogeneous nucleation involving suitable aerosol particles and atmospheric conditions, ice crystal growth is governed by diffusion processes and subsequent actions by means of collision and coalescence. These physical processes are complicated by the nature of the ice crystal's hexagonal and irregular shape. Incorporating a fully interactive ice microphysics based on the first principle in a GCM appears to be a challenging but an extremely difficult computational task. Innovative D_e parameterization based on theory and observation must be developed for GCM applications.

It has been a common practice to prescribe a mean effective ice crystal size in GCMs (see, e.g., Gu et al., 2003). A number of GCMs have also used temperature to determine D_e (Gu & Liou, 2006; Kristjansson et al., 2005). This approach is rooted in earlier ice microphysics observations from aircraft, and attests to the fact that small and large ice crystals are related to cold and warm temperatures in cirrus cloud layers. Ou and Liou (1995) developed a parameterization equation relating cirrus temperature to a mean effective ice crystal size based on a large number of midlatitude cirrus microphysics data presented by Heymsfield and Platt (1984). Ou et al. (1995) reduced large standard deviations in the size-temperature parameterization by incorporating a dimensional analysis between IWC and D_e. Using CEPEX data, McFarquhar et al. (2003) developed a D_e parameterization as a function of IWC for use in a single column model.

Liou et al. (2008) have developed an ice microphysics parameterization to include interactive mean effective ice crystal size D_e in connection with radiation parameterizations. Correlation analysis between IWC and D_e has been carried out using a large set of observed ice crystal size distributions obtained from a number of cirrus field campaigns in the tropics, midlatitude, and Arctic. It is shown that IWC and D_e are well correlated using this regional division (Fig. 2.10). Including temperature classification in midlatitude cases increases this correlation. Using least-squares fitting to the observed data, the parameterization of mean effective ice crystal size in polynomials in terms of ice water content, with various coefficients for the three different regions, has been obtained. Simulations from the UCLA GCM employing this parameterization showed substantial regional deviations in outgoing longwave radiation (OLR) and precipitation patterns from assuming a prescribed constant D_e. In a more recent study, Guignard et al. (2012) also suggested that the dependence of D_e on temperature was found to be weak and that D_e is better related to IWP as this relationship was found to be more robust.

Recent development in association with cloud process modeling in climate models has been focused on the introduction of more complex representations of microphysical processes, with the dual goals of coupling them better to atmospheric aerosols and linking them more consistently to the subgrid variability assumed by the model for other calculations (IPCC, 2013). For example, more models participating in CMIP5 predict both mass and number mixing ratios for liquid stratiform cloud. Some models explicitly treat subgrid cloud water variability for calculating microphysical process rates (e.g., Morrison & Gettelman, 2008). Ice cloud treatments are similar to those for liquid water but face greater challenges because of the complexity of ice processes. Many CMIP3 models predicted the condensed water amount in just two categories—cloud and precipitation—with a temperature-dependent partitioning between liquid and ice within either category. Although supersaturation with respect to ice is commonly observed at low temperatures, only one CMIP3 GCM (ECHAM) allowed ice supersaturation (Lohmann & Kärcher, 2002).

2.5.2. Cloud Vertical Overlap

As previously mentioned, clouds have a typical length scale of perhaps several hundred meters and display substantial horizontal variability on scales that are generally smaller than the usual AGCM grid box. This introduces the cloud overlap problem for the calculation of radiative transfer under cloudy conditions, which is an important issue in climate model studies. Various parameterizations of cloud overlap effects have been developed, and several AGCM-sensitivity tests have been performed (e.g., Chou

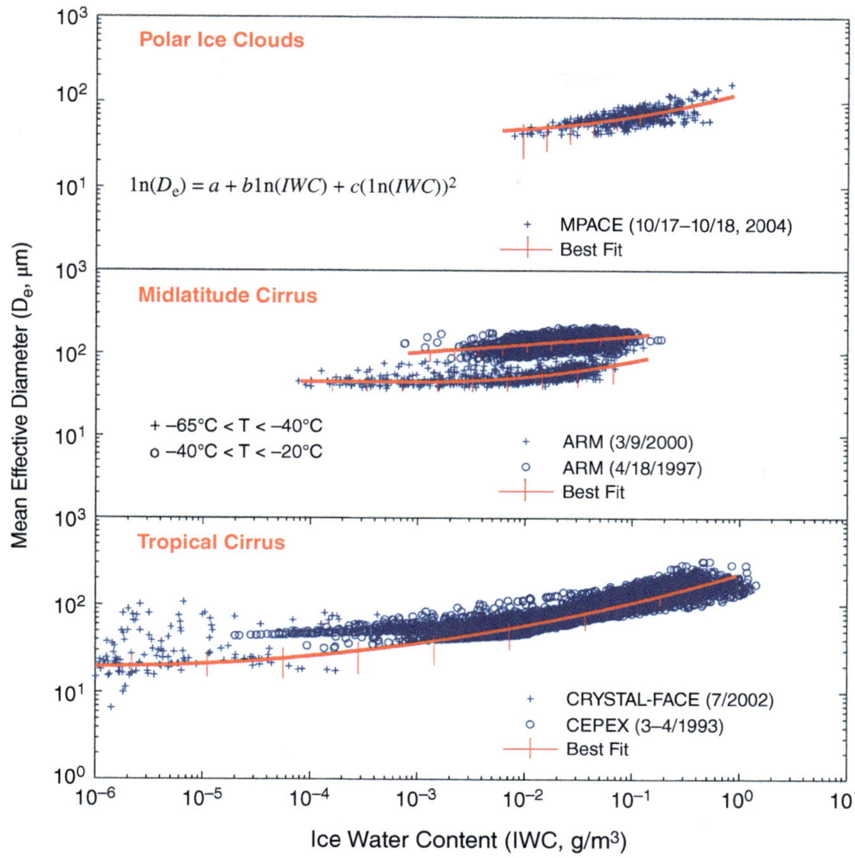

Figure 2.10 Mean correlation curves with standard deviations (vertical bars) for IWC and D_e for the tropics (bottom panel), midlatitude (middle panel), and Arctic region (top panel). Source: Liou et al. (2008)/John Wiley & Sons.

et al., 1998; Collins, 2001; Gu & Liou, 2001; Liang & Wang, 1997).

Previous commonly used approaches to treat cloud vertical overlap include random overlap, maximum overlap, and maximum/random overlap. The most straightforward approach to dealing with fractional cloud cover is the random overlap, in which the sky is divided into sectors within which the cloud amount is either 0 or 1. Radiative fluxes are calculated for each sector and then weighted by the respective cloud amount to obtain grid box fluxes. There could be a lot of possible cloud configurations (2^n possible cloud configurations, where n = total number of cloud layers). Therefore, this method could be computationally very time consuming. It also neglects cloud geometry association and tends to overestimate total cloud cover. In maximum overlap, clouds are closely associated and stack on each other. Thus, there are only two possible cloud configurations and are computationally efficient. However, this method tends to underestimate total cloud cover. In maximum/random overlap, clouds are grouped as regions, for example, by height or by adjacent and discrete clouds. Within each region, maximum overlap is assumed. For clouds of the different regions, random overlap is applied. The most common methods used in contemporary AGCMs are random overlap (Manabe & Strickler, 1964) and maximum/random overlap (Chou et al., 1998; Geleyn & Hollingsworth, 1979). The latter has been shown to be more consistent with the observed cloud distribution (Tian & Curry, 1989).

As an example, Figure 2.11 shows the possible cloud configurations if clouds are grouped as low, middle, and high regions. An atmospheric column, therefore, can be divided into at most eight sectors if clouds are present in all of the three groups. The number of configurations reduces to four if no cloud occurs or if the cloud is overcast in one of the cloud groups. If the cloud amounts in the three groups are $C1$, $C2$, and $C3$, the fractional area for the cloudy case is $C1 \times C2 \times C3$, while for the clear case it is $(1 - C1) \times (1 - C2) \times (1 - C3)$. Radiation calculations can then be performed for each of the cloud configurations, and the all-sky flux can be determined as the weighted sum of the flux computed for each sector. To implement the cloud overlap scheme in the radiation

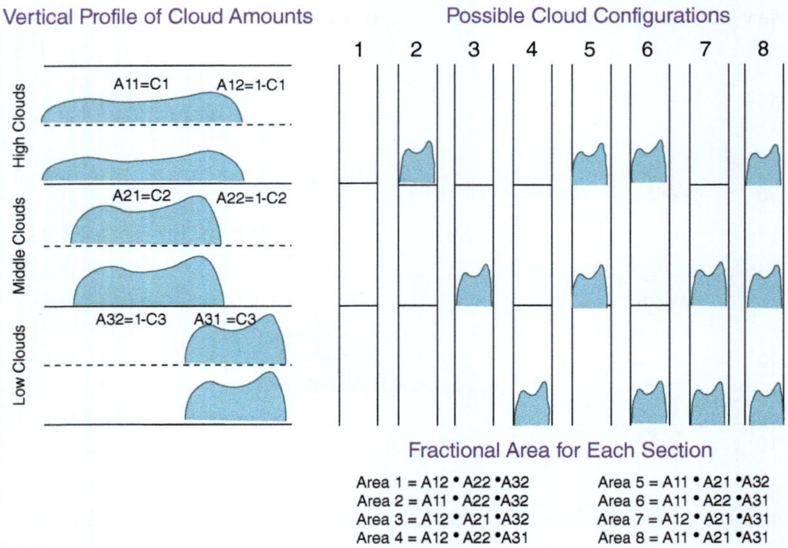

Figure 2.11 An example of possible cloud configurations.

calculation, the optical properties for clear and overcast conditions for each layer are first determined. A combination of the optical properties for clear or overcast conditions corresponding to each layer is then configured throughout the entire atmospheric column. Radiative transfer calculations can subsequently be applied to each configuration to obtain the radiation fluxes at each layer.

While in many models, the maximum/random overlap has been applied in which vertically continuous cloudy layers are assumed to overlap maximally and layers separated by noncloudy layers are assumed to overlap randomly, these assumptions have not been systematically evaluated with a comprehensive data set that can resolve simultaneously occurring cloud layers. Based on high vertical resolution cloud radar data in the United Kingdom, Hogan and Illingworth (2000) examined the overlap characteristics of clouds and found that vertically continuous clouds do not tend to be maximally overlapped, which is contrary to the assumption made in most models. A simple inverse-exponential expression for the degree of overlap as a function of level separation is then proposed, which could be implemented in current GCMs with relatively little difficulty. Using long-term cloud radar data collected by continuously operating millimeter-wavelength instruments deployed at the Atmospheric Radiation Measurement sites in the tropics, middle latitudes, and the Arctic, and the approach proposed by Hogan and Illingworth (2000), Mace and Benson-Troth (2002) also concluded that an assumption of random overlap for layers separated by noncloudy layers is supported by observations. However, the overlap characteristics of vertically continuous layers cannot be considered maximal. Indeed, vertically continuous cloudy layers do not appear to be able to represented by a simple overlap assumption. They also showed that the cloud-layer overlap characteristics in the middle latitudes appear to be a strong function of season, suggesting that an overlap parameterization in terms of cloud system type may be possible.

2.5.3. Cloud Inhomogeneity Effects

Satellite mapping of the optical depth in midlatitude and tropical regions has illustrated that cirrus clouds are frequently finite in nature and display substantial horizontal variabilities (Minnis et al., 1993; Ou et al., 1995). Vertical inhomogeneity of the ice crystal size distribution and ice water content has also been demonstrated in the microphysics balloon sounding observations (Heymsfield and Miloshevich, 1993). Figure 2.12 shows a 3D cirrus from Lidar observations. The data were obtained by the scanning of Lidar across the moving direction of the clouds. The observed cirrus clouds over an area of 24 × 24 km at altitudes from 6 to 12 km were shown to be highly inhomogeneous. Based on 3D radiative transfer calculations, cloud inhomogeneity has been shown to play a significant role in the heating rate profile averaged over mesoscale grids, and the result differs from that computed from the conventional plane-parallel approach (Gu & Liou, 2001). However, creating representations of cloud inhomogeneity in general circulation models for climate study is a difficult task since they generally have spatial scales greater than 100 km primarily due to computational limitations. Moreover, the relationship between the cloud optical properties (in terms of optical depth) and radiative fluxes is nonlinear, leading to

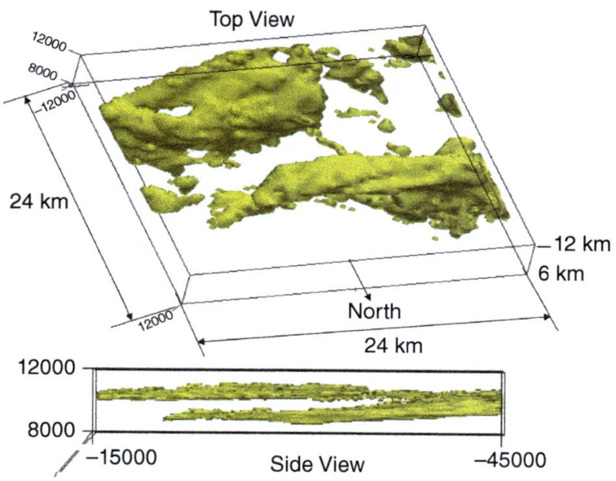

Figure 2.12 A 3D cirrus observation from Lidar (private communication with E. Eloranta).

biases in the model-simulated flux field caused by the use of an average optical depth to represent cloud spatial inhomogeneity.

The cloud horizontal inhomogeneity effect on solar albedo has been examined by Cahalan et al. (1994), Kogan et al. (1995), and more recently by Cairns et al. (2000). Using data obtained from observations, large eddy simulations, and Monte Carlo radiative transfer calculations for stratocumulus and cumulus clouds, these authors found that cloud inhomogeneity leads to a considerably smaller solar albedo as compared to the value computed from the PP radiative transfer model employing a mean liquid water content. As an earlier approach, this effect was taken into account in climate models by reducing the optical depth, e.g., by a factor of 0.7, for all cloud types. Tiedtke (1996) and Rotstayn (1997) tested the assumption of a uniform adjustment in GCMs and pointed out that extension of the uniform reduction factor to all modeled clouds appears unrealistic but might be taken as the lower limit of the adjustment. Rossow et al. (2002) used the International Satellite Cloud Climatology Project data sets to obtain a climatology of the mesoscale variability of cloud optical depth, emissivity, and temperature to test and refine the treatment of cloud-radiation interactions.

A set of inhomogeneity factor for high-level clouds derived from the ISCCP D1 data set averaged over a five-year period (Fig. 2.13) has been incorporated in the UCLA atmospheric GCM to investigate the effect of cirrus cloud inhomogeneity on climate simulation (Gu & Liou, 2006). The inclusion of this inhomogeneous factor improves the global mean planetary albedo by about 4% simulated from the model. It also produces changes in solar fluxes and OLRs associated with changes in cloud fields, revealing that the cloud inhomogeneity not only affects cloud albedo directly but also modifies cloud and radiation fields. The corresponding difference in the geographic distribution of precipitation is as large as 7 mm day^{-1}. Using the climatology cloud inhomogeneity factor also produces a warmer troposphere related to changes in the cloudiness and the corresponding radiative heating, which, to some extent, corrects the cold bias in the UCLA AGCM. Using the MODIS data, Oreopoulos and Cahalan (2005) found that the range of the inhomogeneity parameter is from ~0.65 to 0.8. Shonk et al. (2010) reviewed a wind range of observations of cloud inhomogeneity, including data from radar, microwave radiometer (MWR), aircraft, and other satellite data such as Landsat and FIRE.

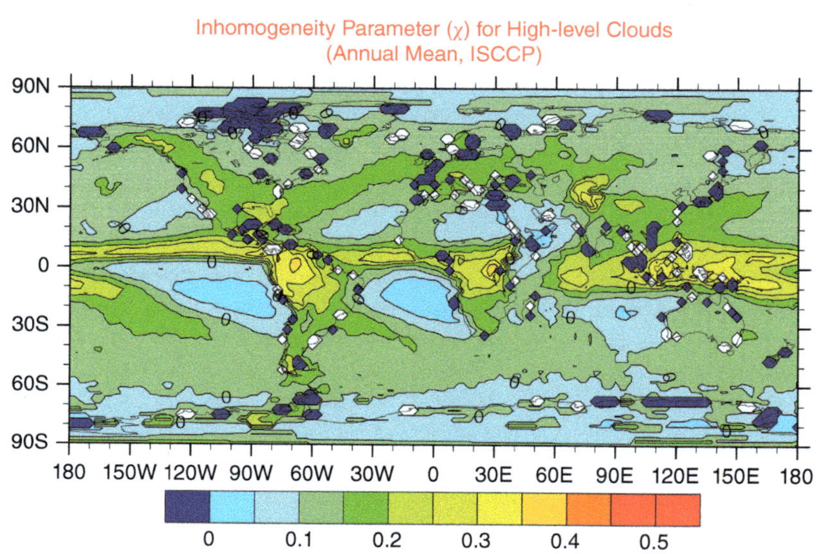

Figure 2.13 Annual mean inhomogeneity factor for high-level clouds derived from the ISCCP D1 data set.

Another suggested approach involves statistically replicating the inhomogeneity by weighting the cloud optical depth by a distribution function, which dramatically reduced the plane-parallel biases (Barker, 1996; Carlin et al., 2002). Pincus et al. (2003) developed a computationally efficient technique based on simultaneously sampling cloud state and spectral interval for computing domain-averaged broadband radiative fluxes in vertically and horizontally variable cloud fields of arbitrary complexity. This technique is referred to as McICA since it is a Monte Carlo integration of the Independent Column Approximation (ICA). Radiative fluxes computed using McICA are unbiased with respect to the benchmark Independent Column Approximation and work well independent of cloud structure. However, they are subject to random, uncorrelated errors depending on the distribution of cloud optical properties. Seasonal forecasts, on the other hand, are not sensitive to this noise, making the method useful in weather and climate prediction models. Another new scheme was presented by Shonk and Hogan (2008), which attempts to account for the cloud inhomogeneity by using two regions of cloud in each vertical level of the model, comprising one optically thinner region and one optically thicker region. Therefore, plus the clear-sky region, the scheme has three regions in each model level and is referred to as Tripleclouds. Furthermore, the scheme has the capability to represent arbitrary vertical overlap between the three regions in pairs of adjacent levels. Compared to the conventional method of accounting for inhomogeneity by adjusting the optical depths, Tripleclouds showed improvements both in terms of top-of-atmosphere radiative flux biases and internal heating rates.

2.6. 3D RADIATIVE TRANSFER

2.6.1. 3D Cloud-Radiation Interaction

As discussed in the previous section, clouds are frequently finite in nature and display substantial horizontal and vertical variabilities. Therefore, the radiation field should be obtained by numerically solving the 3D transport equation (Evans, 1998). Most of the approaches to 3D radiative transfer employ the Monte Carlo method (e.g., Cahalan et al., 1994; O'Hirok & Gautier, 1998). The 3D Monte Carlo method for the transfer of radiation from the Sun as a point source with application to clouds has been presented in some detail by Marshak and Davis (2005). Chen et al. (2006) developed a 3D Monte Carlo radiative transfer model that calculates the flux components exactly given a realistic distribution of scatterers and absorbers in the atmosphere and provided the first definitive assessment of the relative importance of the flux components in a clear-sky atmosphere. Some other methods have also been presented, including the spherical harmonic method employed by Evans (1993) and the Fourier-Riccati approach by Gabriel et al. (1993) for radiative transfer in 2D inhomogeneous clouds. Evans (1998) developed a spherical harmonic discrete-ordinates method (SHDOM) for modeling radiative transfer in inhomogeneous three-dimensional media. Ou and Liou (1982) presented a spherical harmonic method in multiple dimensions, based on which the diffusion approximation for 3D radiative transfer can be developed (Liou, 1992). For application to cirrus clouds, Liou and Rao (1996) used the successive orders of scattering (SOS) approach that can be directly applied to the specific geometry and inhomogeneous structure of a medium.

However, the requirement of computer resources remains the primary obstacle in the modeling of 3D radiative transfer. A lot of efforts on parameterization of 3D cloud-radiation interaction in large-scale models has focused on binary medium or oversimplified closure assumptions (Anisimov & Fukshansky, 1992; Hogan & Shonk, 2013; Kassianov & Veron, 2011; Pomraning, 1996; Stephens, 1988; Titov, 1990; Tompkins & Di Giuseppe, 2007; Vainikko, 1973). Huang and Liu (2014) developed a new statistical-physics-like approach, referred to as statistical radiative transport (SRT), that makes a direct connection between the statistical characterization of cloud structure and the statistical moments of the radiation field by properly averaging the 3D equation, and a spatial autocorrelation function can be used to physically capture the net effects of subgrid cloud interaction with radiation. This approach is able to faithfully reproduce the Monte Carlo 3D simulation results with several orders less computational cost, allowing for more realistic representation of cloud-radiation interactions in large-scale models. It can be seen from that for a checkboard case (Fig. 2.14a), the SRT agrees extremely well with the full 3D calculations (Fig. 2.14b). The difference between 3D and ICA indicates the magnitude of 3D effects. At small solar angles (<18°), the 3D effects reduce the domain-average reflectance by up to 20%, primarily due to photon leaking from clouds to the clear region. For large solar angles, the 3D effects actually enhance the domain-average reflectance by up to 80%. The enhancement of reflectance increases with solar angle and is mainly due to cloud side illumination effects (Hogan & Shonk, 2013).

Gu and Liou (2001) developed a 3D radiative transfer model to simulate the transfer of solar and thermal infrared radiation in inhomogeneous cirrus clouds utilizing a diffusion approximation approach (four-term expansion in the intensity) for application to inhomogeneous media employing Cartesian coordinates. Comparisons of the model results with those from

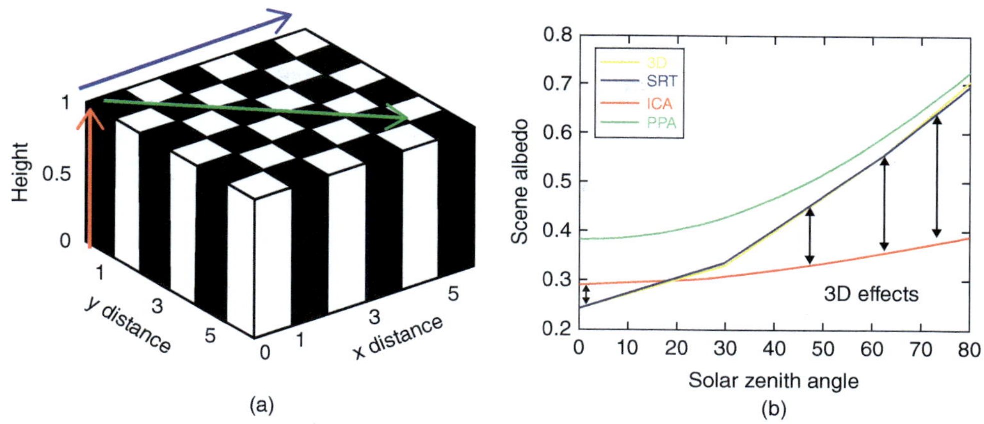

Figure 2.14 (a) Visualization of a checkboard medium and (b) the scene albedo calculated using different approaches as a function of solar zenith angle. The checkboard cloud case with an aspect ratio of 1 is used for the simulations. Source: Huang et al. (2014)/IOP Publishing/CC BY 3.0.

plane-parallel and other 3D models show reasonable agreement for both broadband and monochromatic results. The PP method is shown to be a good approximation under the homogeneous condition when the cloud horizontal dimension is much larger than the cloud thickness. As the horizontal dimension decreases, clouds produce less infrared warming at the bottom as well as less cooling at the top, while more solar heating is generated within the cloud. For inhomogeneous cases, upwelling and downwelling fluxes display patterns corresponding to the extinction coefficient field. Figure 2.15 shows the 3D radiative heating and cooling structure within the clouds computed using a 3D radiative transfer model. The left panel shows a three-dimensional

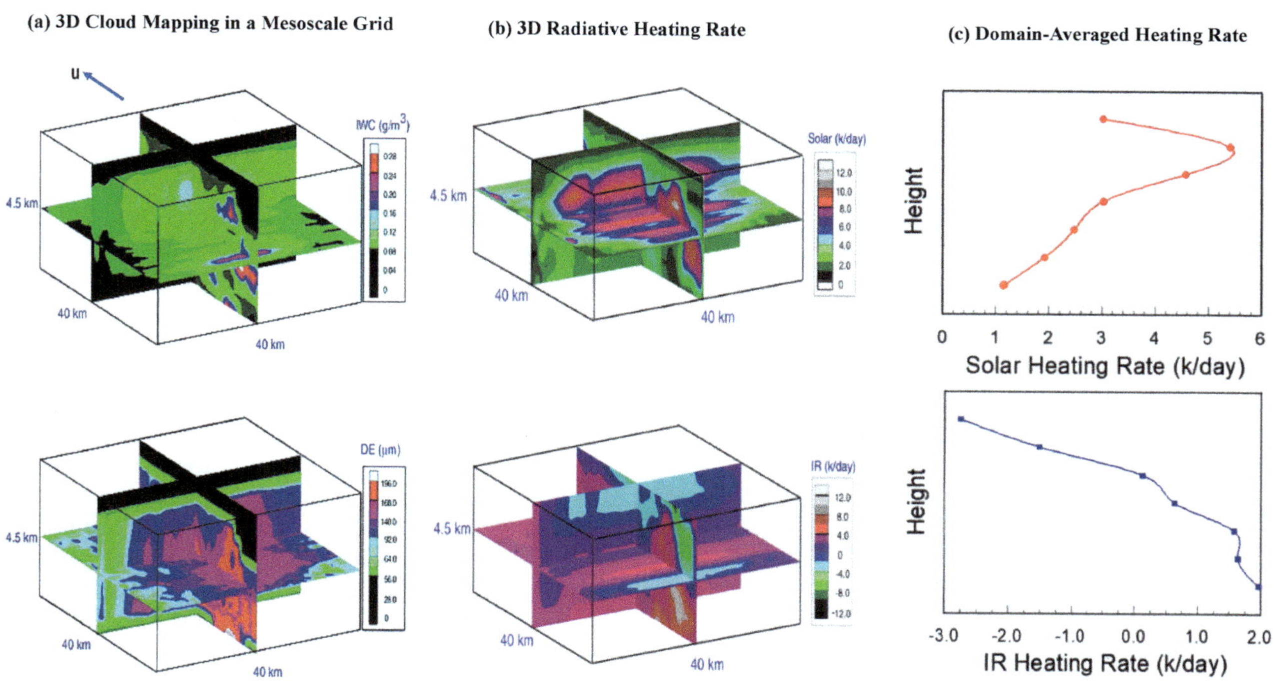

Figure 2.15 (a) 3D ice water content (IWC, g/m^3) and mean effective ice crystal size (D_e, μm) determined from a unification of the optical depth and D_e retrieval from the 0.63 and 3.7 μm AVHRR channels aboard the NOAA-14 satellite and the IWC and D_e retrieved from the 35 GHz cloud radar over the ARM-SGP CART site at 2023 UTC on 18 April 1997. The 3D and D_e results are represented in *xy*, *yz*, and *xz* planes over a 40 km × 40 km × 4.5 km domain. Source: Liou et al. (2002)/with permission of Elsevier. (b) 3D radiative heating rates calculated for the clouds in (a). (c) Domain-averaged heating rate.

ice water content field and mean effective ice crystal size determined from AVHRR data. The 3D solar and thermal IR heating/cooling rates, together with their domain averages, computed for this 3D cloud, are shown in the middle and right panels, respectively. It can be seen that the distribution of solar heating rates closely matches the cloud water distribution. For thermal IR, while cooling generally occurs at the cloud top, warming occurs at the cloud bottom. The overall distribution also follows that of the cloud water.

Recently, a new radiation scheme ecRad has been implemented in the European Centre for Medium-Range Weather Forecasts (ECMWF), which includes McICA, Tripleclouds, as well as the Speedy Algorithm for Radiative Transfer through Cloud Sides (SPARTACUS; Hogan et al., 2016; Schäfer et al., 2016), which makes ECMWF the first global model capable of representing the 3D radiative effects of clouds. It is shown that with SPARTACUS the radiation scheme is 5.8 times slower than with McICA, and 3.6 times slower than the original McRad scheme in the ECMWF. Given the current computing capability, this would be too slow to be employed for operational weather forecasting, but it is fast enough to be used for research.

2.6.2. 3D Radiative Transfer over the Mountainous Regions

The energy emitted by the Sun and received by the Earth's surface is determined by three sets of factors. The first includes the latitude, the solar hour angle, and the Earth's position relative to the Sun, which determine incoming solar radiation at the top of the atmosphere and can be precisely calculated (Liou, 2002). The second involves the attenuation of the solar beam by scattering and absorption caused by atmospheric gases, aerosols, and cloud particles. The last comprises terrain characteristics including elevation, slope, orientation, and surface albedo. On a sloping surface in mountainous terrain, the total solar radiation can be separated into five components according to the Sun-to-surface path (Lee et al., 2011): (1) the direct flux contains the photons arriving on the ground directly in line from the Sun to the surface, unscattered by the atmosphere; (2) the diffuse flux is composed of photons scattered by the atmosphere; (3) the direct-reflected flux is associated with photons traveling directly through the atmosphere unscattered but subsequently reflected to the target surface by neighboring slopes; (4) the diffuse-reflected flux is similar to the direct-reflected irradiance except that it is composed of photons already scattered by the atmosphere and then reflected by neighboring slopes; and (5) the coupling flux is composed of photons that have experienced multiple surface reflections and atmospheric scatterings. The direct, diffuse, and coupling

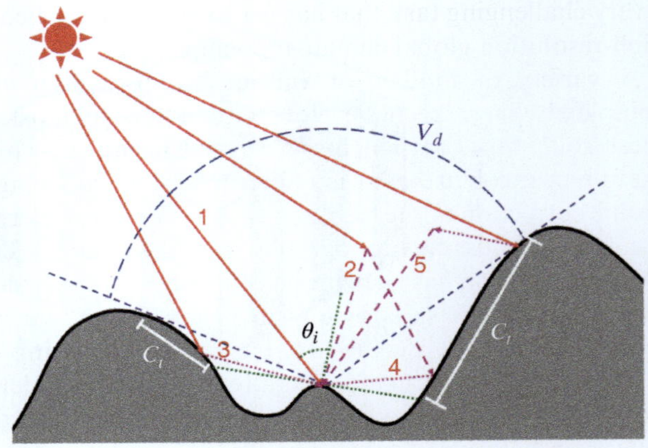

Figure 2.16 A schematic representation of flux components received by the target on an inclined surface in a mountainous area: (1) direct flux, (2) diffuse flux, (3) direct-reflected flux, (4) diffuse-reflected flux, and (5) coupled flux. θ_i is the solar incident angle, V_d is the sky view factor, and terrain C_t is the configuration factor. Source: Lee et al. (2011)/John Wiley & Sons.

components are produced when solar radiation interacts with a flat surface. However, over mountains, additional components are generated by the direct reflected and diffuse-reflected fluxes through interaction with rough topography that are absent for the flat surface (Fig. 2.16).

Topography exerts a major control on land surface processes. Through its influence on atmospheric forcing, most notably temperature, precipitation, and radiation, and river network topology, drainage area, and soil properties, the imprints of topography on the spatial patterns of water storage (e.g. soil moisture and snowpack) and vegetation cover are ubiquitous. The spatial and temporal distributions of surface solar radiation are critical to many climatic processes, including snowmelt, soil moistening, photosynthesis, vegetation, and evapotranspiration (Bonan, 2002; Geiger, 1965; Gu et al., 2002; Müller & Scherer, 2005). This is evident by the prominent vegetation differences often displayed in the north facing versus south facing slopes of mountains. Thus, topographically induced solar insolation and precipitation differences may be regarded as one of the most fundamental variables of land environment, including snow, soil, and vegetation characteristics. Essentially, all modern climate models and ESMs use a plane-parallel radiative transfer schemes although surface topography can cause significant deviations from PP. Because of the intricacy of spatial orientation and inhomogeneous features of mountain topography and the complexity of interactions of direct and diffuse solar beams with mountain surfaces, quantifying these interactions and reliably determining total surface solar fluxes for incorporation in a land-surface energy balance model has been a very challenging task that has yet to be accomplished in high-resolution global climate modeling.

A variety of models of varying sophistication and complexity have been developed to compute most of these solar flux components in rugged terrain in which the direct incident beam is computed by introducing a cosine correction of the local zenith angle and considering the shadow caused by topography (e.g., Olyphant, 1986). Diffuse radiation has been modeled as being proportional to the area of the sky dome visible to the target surface (e.g., Dozier & Frew, 1990). Factors contributing to terrain-reflected flux have generally not been considered explicitly but are approximated by assuming a first-order reflection between surrounding terrain and the target surface. The terrain-reflected radiation is particularly significant over snow surfaces (Dozier, 1980). Miesch et al. (1999) used a Monte Carlo approach to compute flux components at ground level and the radiance reaching the satellite-borne sensor over a very simple 2D rugged terrain with homogeneous atmospheric layers above, and concluded that a 10% error may occur when the terrain-reflected and coupling components may play a dominant role in the signal measured by a satellite sensor over poorly exposed or shadowed areas. As this brief literature survey illustrates, the effect of complex 3D mountainous terrain on the surface flux components of Figure 2.16 has not been systematically and rigorously studied. Chen et al. (2006) also characterized the factors controlling diurnal, seasonal, and geographical variability in the flux components and determined the significance and sources of errors in conventional radiative transfer schemes in areas of intense topography by applying the 3D radiative transfer model to a real 3D midlatitude mountainous surface. Errors in radiative transfer schemes that have been used in GCMs with smoothed topography were assessed by comparing 3D results with the mountainous surface to identical calculations for a flat surface with the same mean elevation. It was shown that the errors are on the order of 5–20 W m^{-2}, depending on the time of the day, and stem from the fact that the atmosphere absorbs a different amount of solar radiation when the topography underneath it is smoothed.

Using a 3D Monte Carlo radiative transfer model, Liou et al. (2007) evaluated the surface solar fluxes over the Tibetan Plateau and illustrated that anomalies of the surface solar flux with reference to a flat surface are significant. It is shown that anomalies of surface solar fluxes with reference to a flat surface can be as large as 600 W m^{-2}, depending on time of day, mountain configuration, and albedo (Fig. 2.17). Surface temperature is the dominating factor in determining anomalies of the surface infrared flux distribution relative to a flat surface with values as high as 70 W m^{-2} at cold mountain surfaces. The average surface solar flux over regional domains of 100 × 100 and 50 × 50 km^2 comprising intense topography can deviate from the smoothed surface conventionally assumed in climate models and GCMs by 10–50 W m^{-2} (Fig. 2.18).

In current climate models, conventional plane-parallel radiative transfer schemes have already been developed to calculate solar fluxes on a flat surface with a known elevation. A parameterization has been developed for radiative transfer over 3D mountainous regions to produce relative deviations of the fluxes from those of a flat surface with the same elevation, given subgrid-scale topographic information (Lee et al., 2011). In addition to the mean values for the sky view factor, the terrain configuration factor, and the cosine of the solar zenith angle (see Fig. 2.16), conventional topographic parameters for each 10 × 10 km^2 domain involving the mean elevation and slope are also employed in multiple linear regression analysis, along with their standard deviation and skewness. Five regression equations have been derived for flux deviations that have the following general form: $F^*_i = a_i + \Sigma\, b_{ij}\, y_j$, i = dir, dif, dir-ref, dif-ref, and coup, where a_i is the intercept, y_j is a specific variable, and b_{ij} are regression coefficients. For example, for the deviation of direct flux, we have $F^*_{\text{dir}} = a_1 + b_{11}\, y_1 + b_{12}\, y_2$, where y_1 is the mean cosine of the solar zenith angle and y_2 is the mean sky view factor.

Figure 2.19 illustrates examples of the regression analysis results for the cosine of the solar zenith angle μ_0 of 0.55. The upper panel displays comparison of the flux components computed from Monte Carlo simulations (real values) and multiple regression analysis (predicted values) for direct (left) and diffuse (right) fluxes, along with multiple determination coefficients R^2. The middle panel is for direct-reflected (left) and diffuse-reflected (right) fluxes. The lower panel shows coupled flux with a surface albedo of 0.1 (left) and 0.7 (right). The most significant term is the direct flux, which has high correlations generally larger than 0.9 with root-mean-square errors less than 3 W m^{-2} (out of 700 W m^{-2}). Deviations from plane-parallel results are on the order of 100 W m^{-2}. The most significant factor determining this flux is the mean solar zenith angle, but the mean sky view factor also has a strong impact because it is an indicator of the shading effect. Deviation of the 3D diffuse flux from plane-parallel results is only about 3 W m^{-2} with elevation, the most significant determining variable. For direct-reflected fluxes, deviation is generally on the order of 3–6 W m^{-2}. For diffuse-reflected and coupled fluxes, deviations are about 0.3 and 0.5 W m^{-2}, respectively. The dominant factor for these reflected-related fluxes is the terrain configuration factor, followed by the skewness of elevation.

The present parameterization can be applied to different atmospheric compositions in view of the fact that variation in the trace gas concentrations (such as

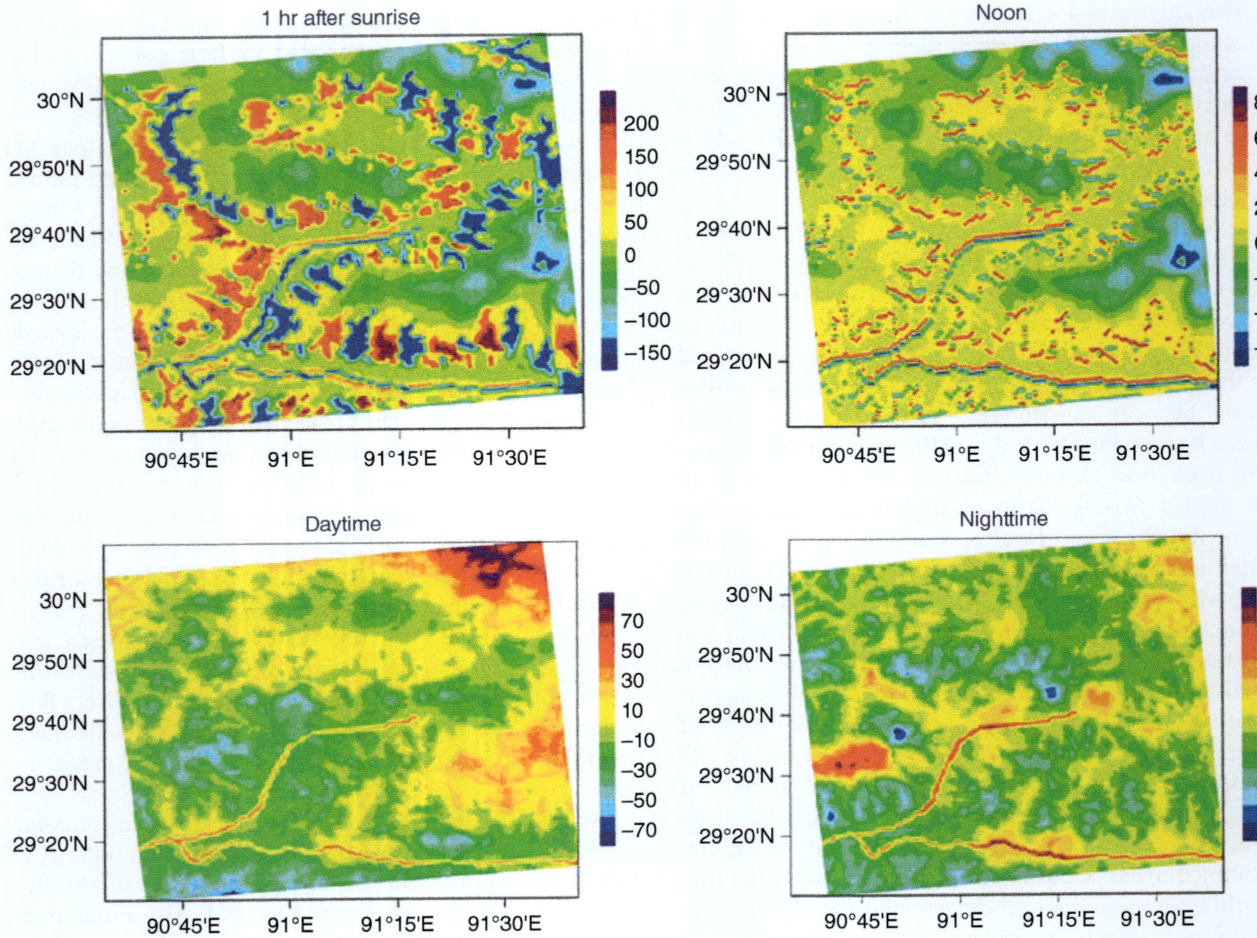

Figure 2.17 The upper panels display the spatial distribution of deviations of the surface net solar flux, defined as difference of the downward and upward fluxes, on mountains from a flat surface (4.37 km with an average albedo of 0.22). The lower panels illustrate the spatial distribution of deviations of the surface net thermal IR flux, defined as difference of the upward and downward fluxes, on mountains from this flat surface during daytime (10.30 am) and nighttime (10.30 pm). Source: Liou et al. (2007)/John Wiley & Sons.

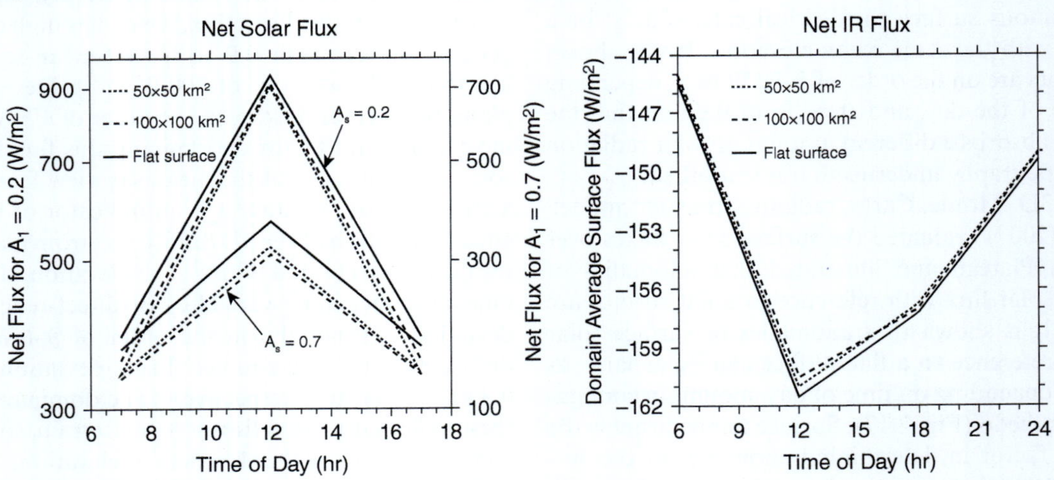

Figure 2.18 Differences between the domain-averaged net radiative flux on mountains and a flat surface as a function of time of day for two domains of 30 × 30 and 50 × 50 km^2 (centered at Lhasa, Tibet) on 21 March (equinox). The left panel shows results for surface albedo values of 0.2 (upper curves) and 0.7 (lower curves), while the right panel illustrates net thermal IR flux differences. Source: Liou et al. (2007)/John Wiley & Sons.

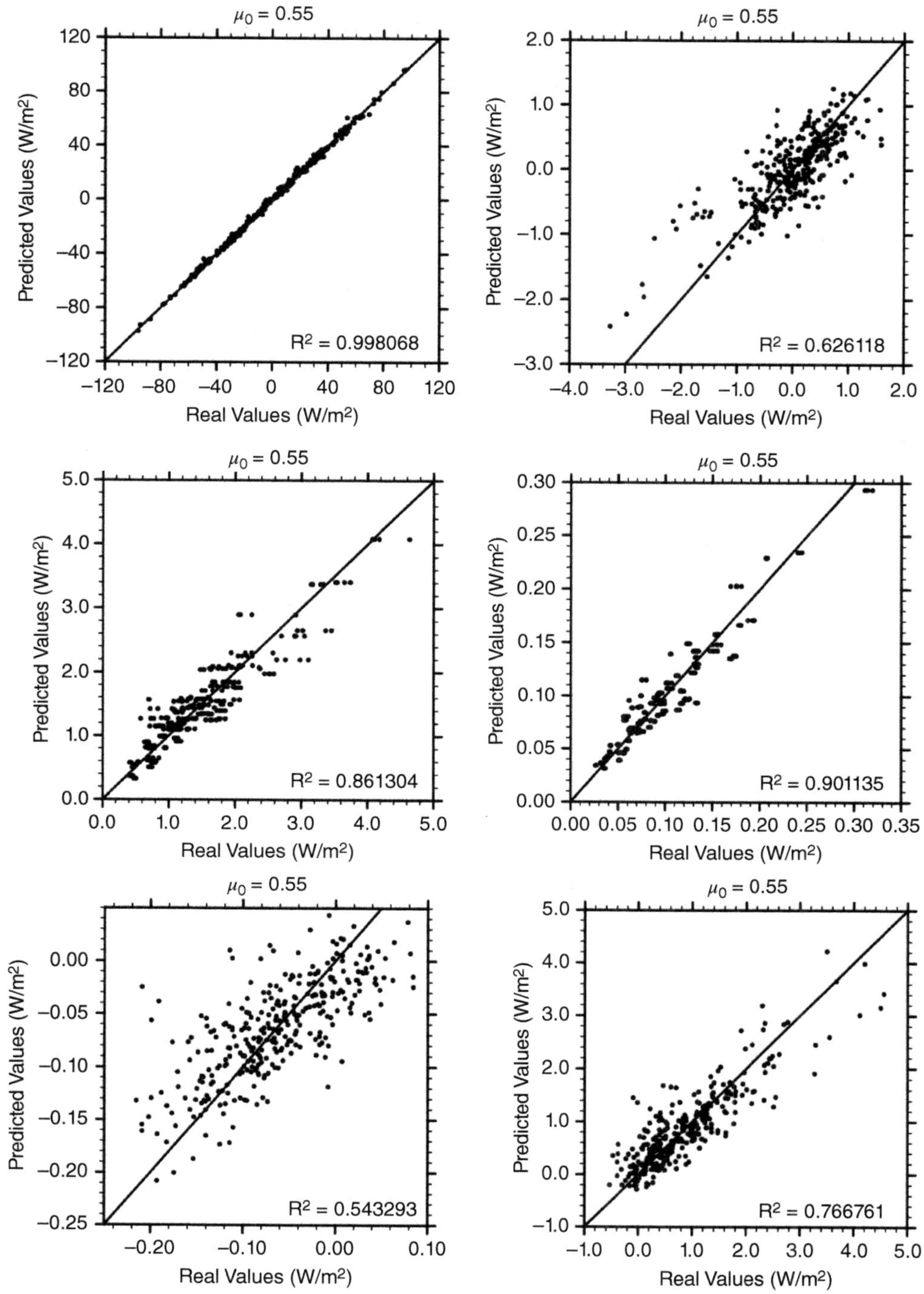

Figure 2.19 Comparison of the deviations of the five flux components computed from Monte Carlo simulations (real values) and multiple regression equations (predicted values) for $\mu_0 = 0.55$. The upper panel is for direct (left) and diffuse (right) fluxes. The middle panel is for direct-reflected (left) and diffuse-reflected (right) fluxes. The lower panel shows the coupled flux with a surface albedo of 0.1 (left) and 0.7 (right). The coefficient of multiple regression R^2 for each case is also depicted in the figures. Source: Lee et al. (2011)/John Wiley & Sons.

water vapor) only alters the absorption coefficient associated with the number of absorbed photons, while the number and direction of the scattered photons remain unchanged. Because the effect due to the change in the atmospheric absorption coefficient has already been accounted for in conventional radiative transfer schemes, the present parameterization in reference to deviations from plane-parallel radiative transfer results can be directly applied to different atmospheric conditions. Surface solar fluxes could be sensitive to terrain spatial resolution and substantial differences could be produced with grid box sizes larger than 200 m (Dubayah et al., 1989; Lee et al., 2011). Several experiments have been carried out using the new SRTM data (Farr et al., 2007) at a 3 arc second (~90 m) resolution over the same area corresponding to 1 km DEM data mentioned above. The results showed that while differences in direct and diffuse fluxes are small, they are much larger for reflect-related fluxes. In Lee et al. (2015), fine resolution SRTM data of 90 m have been used to conduct 3D Monte Carlo calculations to improve the accuracy of parameterization for reflected-related fluxes and applied results to mountain topography over the globe in association with the use of the WRF model and the Community Climate System Model version 4 (CCSM4).

The 3D radiation parameterization has been successfully incorporated into WRF (Gu et al., 2012; Lee et al., 2015; Liou et al., 2013). Figure 2.20 shows the monthly mean snow water equivalent (SWE) values simulated from WRF (30 km spatial resolution) including 3D radiation parameterization for mountains (upper left panel) and estimated from the Northern Hemisphere daily

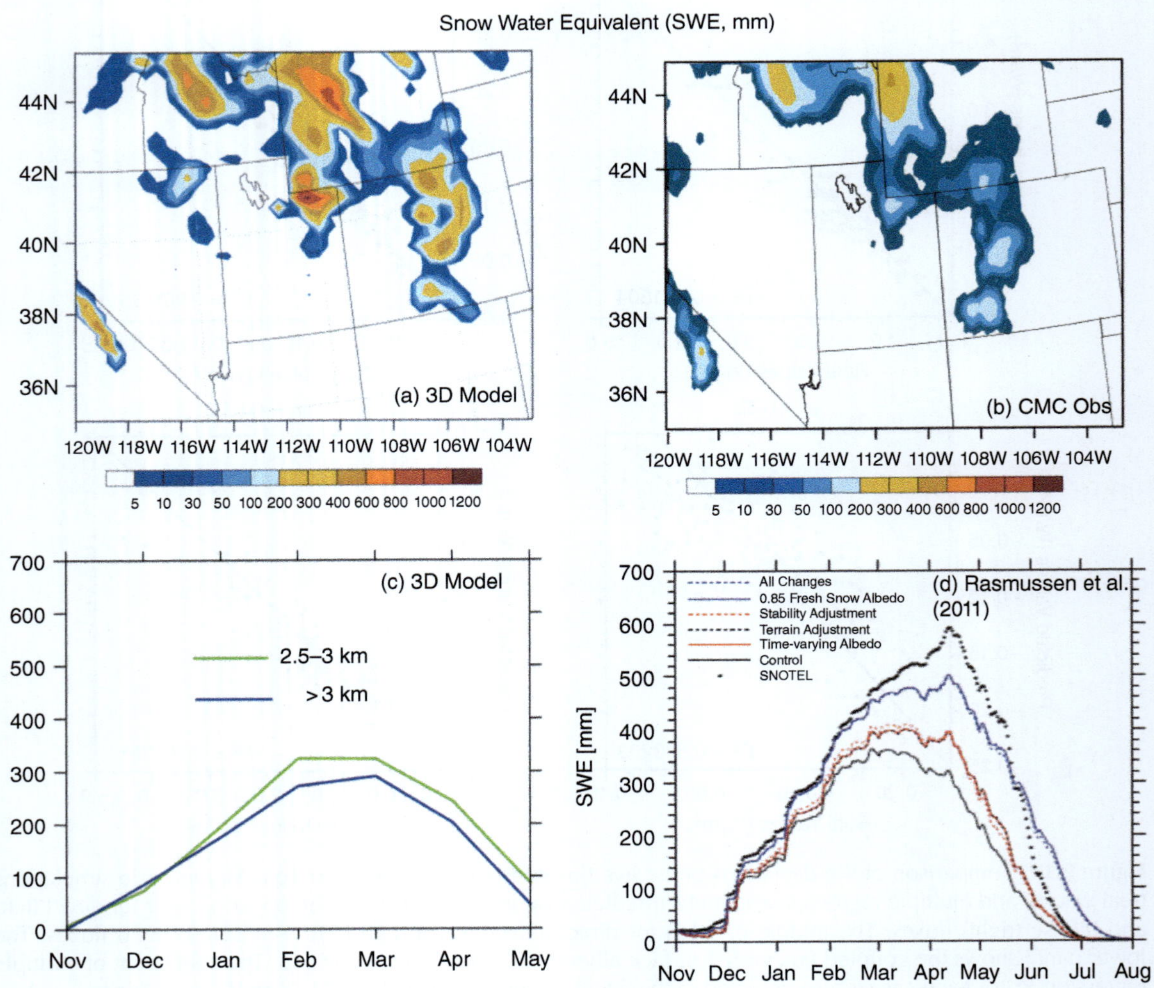

Figure 2.20 The monthly mean SWE map for April 2008: (a) simulated from WRF over the Rocky-Sierra mountains and (b) estimated from the Northern Hemisphere daily snow depth analysis data (CMC). The monthly time series of SWE (c) for the study domain for two elevation zones of 2.5–3 and >3 km and (d) presented in Rasmussen et al. (2011). Source: Liou et al. (2013)/European Geosciences Union/CC BY 3.0.

snow depth analysis data processed by the Canadian Meteorological Centre (CMC; Brown & Brasnett, 2010) at a spatial resolution of 24 km (upper right panel). The CMC data clearly display SWE over the Rocky Mountains and Sierra Nevada regions. The simulated WRF results show a similar spatial pattern to CMC but are larger than their values in terms of both the magnitude and areal coverage. We also compared the monthly time series of SWE for the study domain for two elevation zones of 2.5–3 and >3 km (lower left panel) with the results presented in Rasmussen et al. (2011) (lower right panel) for the same cold season (starting in November 2007). The black dots denote the SWE measurements collected at Snowpack Telemetry (SNOTEL) stations (2.4–3.6 km in Colorado), and the various curves correspond to WRF simulations performed at a 4 km resolution with various adjustments and averaged over all SNOTEL locations. The WRF simulated results at two elevation zones at a 30 km resolution are smaller than the observed SNOTEL data and the results of 4 km resolution WRF simulations. Comparing with CMC observations, WRF model results tend to produce larger SWE values over mountain areas. The incorporation of 3D radiation parameterization has reduced model overestimation to certain degree. We hasten to add, however, that model overestimation bias for SWE can come from a number of sources, including snow parameterization and precipitation bias, and cannot be solely resolved by including 3D radiation effect over mountains.

The monthly mean SWE averaged over the entire domain as a function of elevation is shown in the upper left panel of Figure 2.21, while the corresponding deviations (3D-PP) are displayed in the lower left panel; these deviations show an increase in lower elevations due to the mountain shading effect, with the largest value occurring in March. Positive deviations become smaller after March because the Sun is often overhead during spring, leading to a reduced shading effect. As a result of increased snow accumulation that reduces rainfall and/or snowmelt

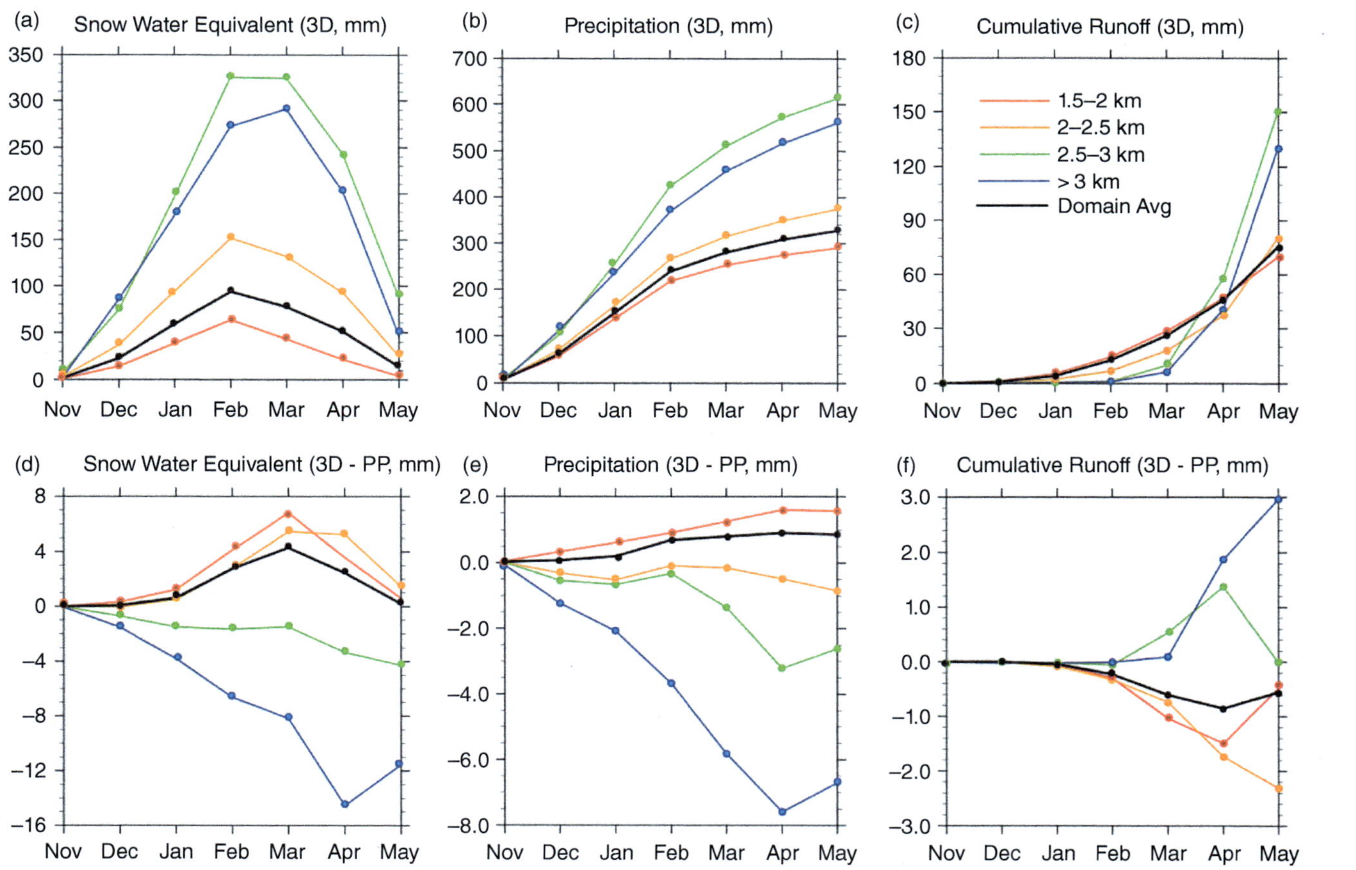

Figure 2.21 The monthly mean (a) SWE, (b) precipitation, and (c) cumulative runoff averaged over the simulation domain. Corresponding deviations (3D-PP) for (d) SWE, (e) precipitation, and (f) runoff as a function of elevations of 1.5–2 km, 2–2.5 km, 2.5–3 m, above 3 km, and whole domain. Source: Liou et al. (2013)/European Geosciences Union/CC BY 3.0.

contributions to runoff, the cumulative runoff deviations (3D-PP) are reduced for lower elevation areas (Fig. 21, lower right panel). Due to the mountain effect, SWE decreases over higher elevation areas in connection with greater solar radiation available at the surface. At the elevation range above 3 km, SWE is reduced by 8 % in April and 24 % in May due to 3D effects. The cumulative runoff shows a maximum increase in April for the elevation range 2.5–3 km and a substantial increase after March for elevations above 3 km because of the increased surface solar flux produced by the 3D mountain effect resulting in increased snowmelt runoff. Precipitation is categorized as snow when the temperature in the lowest atmospheric layer is below 0 °C. The monthly mean precipitation (Fig. 2.21, upper center panel) generally increases with elevation due to orographic forcing—except above 3 km, where moisture is significantly depleted caused by rainout at lower elevations. Precipitation increases from November to May, with substantially larger values for elevation areas higher than 2.5 km. For lower regions, precipitation deviations (Fig. 2.21, lower center panel) increase and result in an increased SWE caused by reduced runoff. Thus, one important impact of the 3D effect is to delay the snowmelt-driven runoff into the warm season for lower elevations and, at the same time, to reduce the SWE in higher elevation regions.

The 3D radiation parameterization has also been into CCSM4 for a 10-year climate run to study the 3D topography effect on variation of solar flux patterns over WUS. The upper panels in Figure 2.22 depict the April mean distributions for precipitation, runoff, and SWE simulated from CCSM4 with the inclusion of 3D radiation parameterization for mountains. Significant precipitation and SWE are mostly seen over the vast Rocky Mountain region as well as over the narrow Sierra Nevada region. Generally, the SWE pattern shows relatively larger values on the west side of the mountains in response to enhanced precipitation on the windward slopes associated with orographic forcing. However, SWE displays smaller values at the highest elevation and on the east side of mountains in response to the reduced precipitation and the largest solar flux available at mountain tops. Corresponding differences (3D-PP) are shown in the lower panels. Due to the 3D mountain effect, SWE generally decreases over mountain tops in the area south of 42°N, in association with reduced precipitation. In Rocky Mountains (~37°N and 107°W), for example, the reduction in SWE is as much as 100 mm or 40%. Decreased/increased SWE patterns correspond closely to increased/decreased net surface solar radiation patterns. Over the northern part of the Sierra Nevada region where precipitation increases, reduced SWE is caused by enhanced runoff over that region.

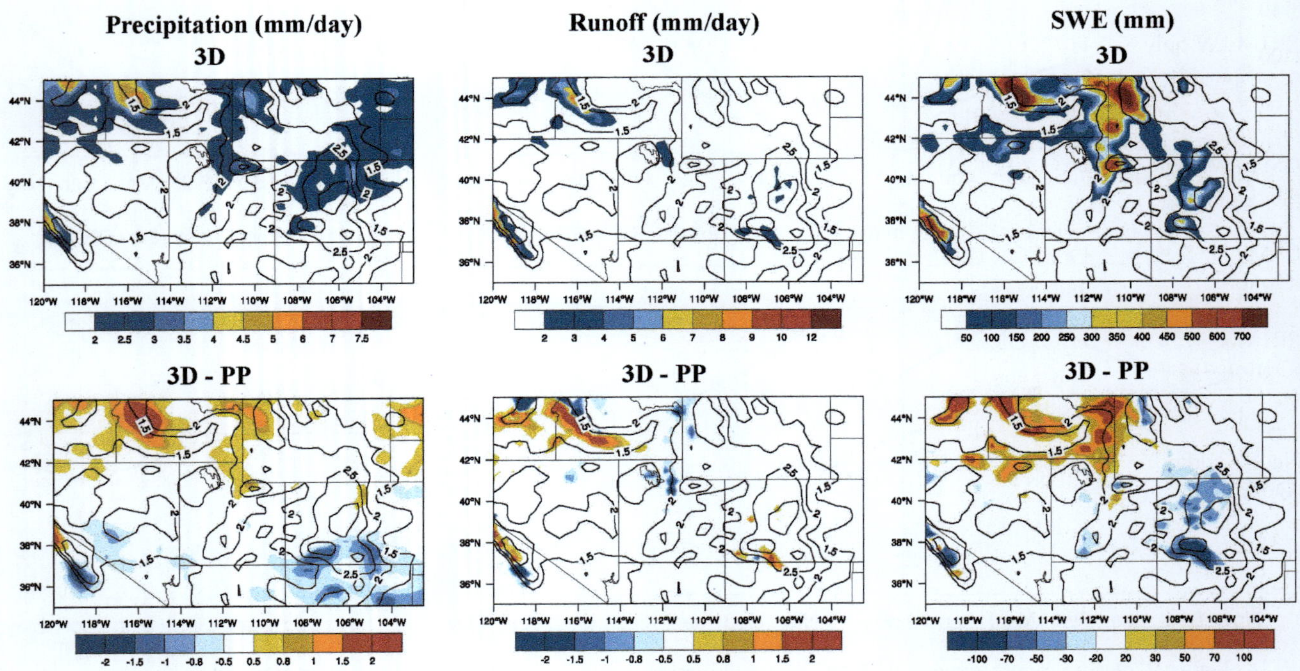

Figure 2.22 The effect of 3D mountains over WUS on the precipitation, runoff, and SWE based on a CCSM4 climate simulation. Upper panels: Incorporation of 3D effects in CCSM4. Lower panels: Deviations (3D-PP) of the simulated results.

2.7. RADIATIVE TRANSFER SCHEMES IN NUMERICAL MODELS

In this section, we review several commonly used radiative transfer schemes in numerical models using WRF as an example. The WRF model (Skamarock et al., 2008) is a fully compressible, nonhydrostatic model (with a hydrostatic option) suitable for a broad spectrum of applications across scales ranging from a few meters to thousands of kilometers. Currently, several physics components have been included in WRF: microphysics (bulk schemes ranging from simplified physics suitable for mesoscale modeling to sophisticated mixed-phase physics suitable for cloud-resolving modeling), cumulus parameterization, longwave and shortwave radiation, planetary boundary layer, surface layer, land-surface parameterization, and subgrid-scale diffusion.

Radiation schemes available in WRF are listed in Table 2.3, which include Dudhia Shortwave Scheme (Dudhia, 1989), RRTM Longwave Scheme (Mlawer et al., 1997), Goddard Shortwave Scheme (Chou & Suarez, 1999; Chou et al., 2001), CAM Shortwave and Longwave Schemes (Collins et al., 2004), RRTMG Shortwave and Longwave Schemes (Iacono et al., 2008), New Goddard Shortwave and Longwave Schemes (Chou & Suarez, 1999), Fu-Liou-Gu (FLG) Shortwave and Longwave Schemes (Fu & Liou, 1992; Gu et al., 2011), and GFDL Shortwave and Longwave Schemes (Fels & Schwarzkopf., 1981). Note that Dudhia and Goddard schemes are for shortwave only, and RRTM is only for longwave radiation. Other schemes include both shortwave and longwave.

For shortwave radiation, the only source is solar insolation that can be scattered, reflected, or absorbed. The WRF longwave radiation schemes include IR radiation emitted and absorbed by the surface and gases. IR emission from the surface is determined by its emissivity, a function of surface temperature and land-use type. Within the atmosphere, the radiation calculation responds to model-predicted cloud and water vapor distributions, as well as specified carbon dioxide, ozone, and other trace gas concentrations. All the schemes use multiple spectral bands corresponding to various absorption gases. Gaseous absorption due to H_2O, O_3, and CO_2 is normally included. Trace gases, such as CH_4, N_2O, O_2, N_2, CFCs, etc., are also considered in many schemes

Table 2.3 Shortwave (SW) and longwave (LW) schemes in the WRF model.

Scheme name	Gases	Clouds	Cloud fraction	Flux solution
Dudhia-SW only (SW option 1)	H_2O (No, O_3, or CO_2)	Q_c, Q_i, Q_s, Q_r, Q_g Look-up tables	1/0	Downward integration
RRTM-LW only (LW option 1)	H_2O, O_3, CO_2 Trace gases *k*-distribution	Q_c, Q_i, Q_s, Q_r, Q_g	1/0	Two-stream
GSFC-SW only (SW option 2)	H_2O, O_3, CO_2	Q_c, Q_i	1/0	Two-stream
CAM (Option 3)	H_2O, O_3, CO_2 Trace gases	Q_c, Q_i, Q_s	Max-ran overlap	Solar: two-stream IR: absorptivity-emissivity
RRTMG (Option 4)	H_2O, O3, CO_2 Trace gases *k*-distribution	Q_c, Q_i, Q_s, Q_r	Max-ran overlap	Two-stream
New Goddard (Option 5)	H_2O, O_3, CO_2 Trace gases *k*-distribution	Q_c, Q_i, Q_s, Q_r, Q_g	1/0	Two-stream
FLG (Option 7)	H_2O, O_3, CO_2, O_2 Trace gases *k*-distribution	Q_c, Q_i, Q_s, Q_r, Q_g	Max-ran overlap	Solar: two-stream or four-stream IR: two-stream or four-stream; two-stream and four-stream combination
GFDL (Option 99)	H_2O, O_3, CO_2	Q_c, Q_i, Q_s, Q_r	Max-ran overlap	Two-stream

(RRTM, CAM, RRTMG, New Goddard, and FLG). Absorption by water vapor continuum and a number of minor absorbers in the solar spectrum, including CH_4, N_2O, NO_2, O_3, CO, SO_2, O_2-O_2, and N_2-O_2, have also been included in FLG scheme, which would lead to an additional absorption of solar flux in a clear atmosphere on the order of 1–3 W m^{-2} depending on the solar zenith angle and the amount of water vapor employed in the calculations (Zhang et al., 2005). The *k*-distribution method is used in RRTM/RRTMG, New Goddard, and FLG schemes. All schemes interact with model-resolved cloud fields, such as water and ice clouds. Some schemes can also consider the radiative effects by rain, snow, and/or graupel (e.g., Dudhia, New Goddard, and FLG). Cloud droplet/particle effective sizes are either prescribed or parameterized in terms of IWC and/or temperature. CAM, RRTMG, FLG, and GFDL schemes allow cloud fraction and maximum-random overlap for subgrid cloud configurations, while other schemes assume either clear condition or overcast in a certain model column.

For the solution of the radiative transfer equation, the Dudhia Shortwave Scheme (Dudhia, 1989) provides a simple downward integration of solar flux that accounts for clear-air scattering, water vapor absorption, and cloud absorption and reflection using look-up tables. Two-stream approximation is commonly used for other schemes, except FLG that allows both two-stream and four-stream approximations and a two-stream and four-stream combination for IR.

The flowchart in Figure 2.23 shows how the radiation scheme is normally used in numerical models. The general input includes solar constant, surface albedo, solar zenith angle, and atmospheric profiles, including pressure, temperature and water vapor, and ozone if available from the model (if ozone profile is not available from the model, some radiation schemes use climatology data prescribed in the radiation scheme). Cloud fields, including liquid and ice water content, mean effective radius, or size if available (if not, radiation schemes may prescribe or parameterize the effective size), are also input to the radiation scheme. Gas concentrations can also be input or given in the radiation schemes. The radiation schemes will then calculate or parameterize the single-scattering properties for the matters in the atmosphere, such as gases, clouds, and aerosols. Finally, depending on the solar zenith angle, the radiative transfer equation will be solved for solar radiation and/or IR radiation, respectively. The output from a radiation scheme normally includes solar and IR fluxes for each atmospheric level, including TOA and surface, and solar and IR heating rates for each atmospheric layer. Note that radiation calculation is too expensive to be called at every time step of the model integration. It is a common practice to use the time interval of 1 minute per km grid size for WRF model.

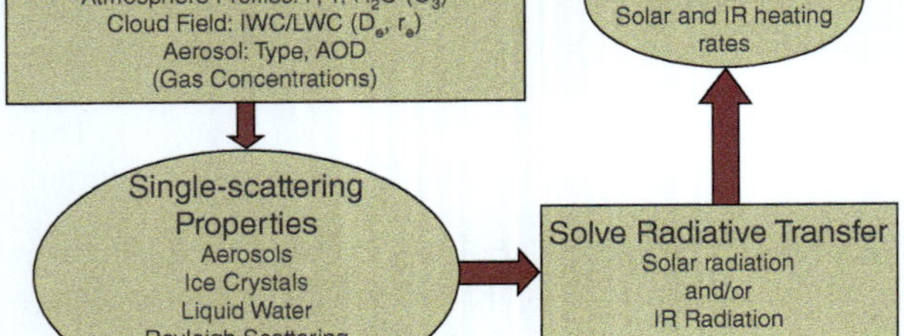

Figure 2.23 Flowchart of the radiation program in a numerical model.

2.8. SUMMARY AND OUTLOOK

An understanding of the Earth's climate and climate change must begin with a comprehensive understanding of the solar and IR radiation emitted from the Sun and the earth-atmosphere system, respectively, and the radiative transfer processes, including absorption, scattering, and transmission of the energy, by the atmosphere and the surface. Physical understanding of climate processes and the prediction of future climate and climate change require numerical atmospheric models that encompass all relevant climate system components. In this chapter, we present the fundamentals and parameterizations of radiative transfer and its interactions with the atmosphere and summarize the commonly used radiative transfer schemes in atmospheric models, using WRF as an example. We also present a few advanced topics in the study of the atmospheric radiation, including the fractional cloud and cloud overlapping, cloud inhomogeneity, and 3D radiative transfer in the atmosphere and over the rugged surfaces.

Given the importance of radiation processes in the earth-atmosphere system, it would be required that an atmospheric model provide radiation computations as accurate as possible. However, the current commonly used radiation schemes are normally 1D programs. Radiative transfer in 3D atmosphere and surface could play an important role in the determination of the radiation budget at the TOA and surface and the fluxes and heating rates within the atmosphere. Therefore, the application of 3D radiative transfer model or parameterizations of 3D radiation effect in climate models are needed. Regarding cloud-radiation interaction, the cloud subgrid-scale structure, including horizontal and vertical inhomogeneities, may substantially influence the radiative transfer within clouds and also need further in-depth research.

REFERENCES

Anisimov, O., & Fukshansky, L. (1992). Stochastic radiation in macroheterogeneous random optical media. *Journal of Quantitative Spectroscopy and Radiative Transfer*, *2*, 169–186.

Arking, A., & Grossman, K. (1972). The influence of line shape and band structure on temperatures in planetary atmospheres. *Journal of the Atmospheric Sciences*, *29*, 937–949.

Auer, A. H. Jr., & Veal, D. L. (1970). The dimension of ice crystals in natural clouds. *Journal of the Atmospheric Sciences*, *27*, 919–926.

Baran, A. J. (2012). From the single-scattering properties of ice crystals to climate prediction: A way forward. *Atmospheric Research*, *112*, 45–69. doi:10.1016/j.atmosres.2012.04.010.

Barker, H. W. (1996). A parametrisation for computing grid-averaged solar fluxes for inhomogeneous marine boundary layer clouds. Part I: Methodology and homogeneous biases. *Journal of the Atmospheric Sciences*, *53*, 2289–2303.

Barker, H. W., Cole, J. N. S., Morcrette, J.-J., Pincus, R., Raisaenen, P., von Salzen, K., & Vaillancourt, P. A. (2008). The Monte Carlo Independent Column Approximation: An assessment using several global atmospheric models. *Quarterly Journal of the Royal Meteorological Society*, *134*, 1463–1478.

Bonan, G. B. (2002). *Ecological Climatology: Concepts and Applications* (678 pp) Cambridge Univ. Press: New York.

Bony, S., Colman, R., Kattsov, V. M., Allan, R. P., Bretherton, C. S., Dufresne, J. -L., et al. (2006). How well do we understand and evaluate climate change feedback processes? *Journal of Climate*, *19*, 3445–3482.

Brown, R. D., & Brasnett, B. (2010). *Canadian Meteorological Centre (CMC) Daily Snow Depth Analysis Data, Environment Canada, 2010*, Boulder, CO, USA, National Snow and Ice Data Center (updated annually).

Cahalan, R., Ridgway, W., & Wiscombe, W. (1994). Independent pixel and Monte Carlo estimates of stratocumulus albedo. *Journal of the Atmospheric Sciences*, *51*, 3776–3790.

Cairns, B., Lacis, A. A., & Carlson, B. E. (2000). Absorption within inhomogeneous clouds and its parameterization in general circulation models. *Journal of the Atmospheric Sciences*, *57*, 700–714.

Carlin, B., Fu, Q., Lohmann, U., Mace, G., Sassen, K., & Comstock, J. M. (2002). High-cloud horizontal inhomogeneity and solar albedo bias. *Journal of Climate*, *15*, 2321–2339.

Chandrasekhar, S. (1950). Radiative Transfer. Dover: New York. Chapters 1, 3, 6, and 7 contain basic materials for the discrete-ordinates method for radiative transfer and the principles of invariance.

Chen, Y., Hall, A., & Liou, K. N. (2006). Application of threedimensional solar radiative transfer to mountains. *Journal of Geophysical Research*, *111*, D21111. doi:10.1029/2006JD007163.

Chou, M. D., & Kouvaris, L. (1986). Monochromatic calculations of atmospheric radiative transfer due to molecular line absorption. *Journal of Geophysical Research*, *91*, 4047–4055.

Chou, M. D., & Suarez, M. J. (1999). A solar radiation parameterization for atmospheric studies. *NASA Tech. Memo.* 104606, 15, 40 pp.

Chou, M. D., Suarez, M. J., Ho, C. H., Yan, M.-H., & Lee, K.-T. (1998). Parameterizations for cloud overlapping and shortwave single-scattering properties for use in general circulation and cloud ensemble models. *Journal of Climate*, *11*, 202–214.

Chou, M. D., Suarez, M. J., Liang, X. Z., & Yan, M. M. H. (2001). A thermal infrared radiation parameterization for atmospheric studies. *NASA Tech. Memo.*, 104606, 19, 68 pp.

Coddington, O., Lean, J. L., Pilewskie, P., Snow, M., & Lindholm, D. (2016). A solar irradiance climate data record. *Bulletin of the American Meteorological Society*, *97* (7): 1265–1282. Bibcode:2016BAMS...97.1265C. doi:10.1175/bams-d-14-00265.1.

Collins, W. D. (2001). Parameterization of generalized cloud overlap for radiative calculations in general circulation models. *Journal of the Atmospheric Sciences*, *58*, 3224–3242.

Collins, W., Rasch, P. J., Boville, B. A., McCaa, J., Williamson, D. L., Kiehl, J. T., et al. (2004). Description of the NCAR Community Atmosphere Model (CAM 3.0). NCAR Tech. Note NCAR/TN–464+STR. 214 pp.

Dobbie, J. S., Li, J., & Chýlek, P. (1999). Two- and four-stream optical properties for water clouds and solar wavelengths. *Journal of Geophysical Research*, *104*, 2067–2079.

Dozier, J. (1980). A clear-sky spectral solar radiation model for snow covered mountainous terrain. *Water Resources Research*, *16*, 709–718.

Dozier, J., & Frew, J. (1990). Rapid calculation of terrain parameters for radiation modeling from digital elevation data. *IEEE Transactions on Geoscience and Remote Sensing*, *28*, 963–969.

Domoto, G. A. (1974). Frequency integration for radiative transfer problems involving homogeneous nongray cases: The inverse transmission function. *Journal of Quantitative Spectroscopy and Radiative Transfer*, *14*, 935–942.

Dubayah, R., Dozier, J., & Davis, F. W. (1989). The distribution of clear- sky radiation over varying terrain. *Proceedings IGARSS '89* (Vol. 2, pp. 885–888). IEEE 89CH2768-0, IEEE, New York.

Dudhia, J. (1989). Numerical study of convection observed during the Winter Monsoon Experiment using a mesoscale two–dimensional model. *Journal of the Atmospheric Sciences*, *46*, 3077–3107.

Eddington, A. S. (1916). On the radiative equilibrium of the stars. *Monthly Notices of the Royal Astronomical Society*, *77*, 16–35.

Evans, K. F. (1993). Two-dimensional radiative transfer in cloudy atmosphere: The spherical harmonic spatial grid method. *Journal of the Atmospheric Sciences*, *50*, 3111–3124.

Evans, K. F. (1998). The spherical harmonics discrete ordinate method for three-dimensional atmospheric radiative transfer. *Journal of the Atmospheric Sciences*, *55*, 429–446.

Farr, T. G., Rosen, P. A., Caro, E., Crippen, R., Duren, R., Hensley, S., et al. (2007). The shuttle radar topography mission. *Reviews of Geophysics*, *45*, RG2004. doi:10.1029/2005RG000183.

Fels, S. B., & Schwarzkopf, M. D. (1981). An efficient, accurate algorithm for calculating CO_2 15 μm band cooling rates. *Journal of Geophysical Research*, *86*, 1205–1232.

Fu, Q., & Liou, K. N. (1992). On the correlated k–distribution method for radiative transfer in nonhomogeneous atmospheres. *Journal of the Atmospheric Sciences*, *49*, 2139–2156.

Fu, Q., & Liou, K. N. (1993). Parameterization of the radiative properties of cirrus clouds. *Journal of the Atmospheric Sciences*, *50*, 2008–2025.

Fu, Q., Liou, K. N., Cribb, M., Charlock, T.P., & Grossman, A. (1997). Multiple scattering parameterization in thermal infrared radiative transfer. *Journal of the Atmospheric Sciences*, *54*, 2799–2812.

Gabriel, P. M., Tsay, S. C., & Stephens, G. L. (1993). A Fourier–Riccati approach to radiative transfer. Part I: Foundations. *Journal of the Atmospheric Sciences*, *50*, 3125–3147.

Geiger, R. (1965). *The Climate Near the Ground (611 pp)*. Harvard Univ. Press: Cambridge.

Geleyn, J.-F., & Hollingsworth, A. (1979). An economical analytical method for the computation of the interaction between scattering and line absorption of radiation. *Beitrage zur Physik der Atmosphare*, *52*, 1–16.

Goody, R. M., West, R., Chen, L., & Crisp, D. (1989). The correlated-k method for radiation calculation in nonhomogeneous atmospheres. *Journal of Quantitative Spectroscopy and Radiative Transfer*, *42*, 539–550.

Gordley, L. L., Marshall, B. T., & Chu, A. D. (1994). LINEPAK: Algorithms for modeling spectral transmittance and radiance. *Journal of Quantitative Spectroscopy & Radiative Transfer*, *52*, 563–580.

Gu, L., Baldocchi, D., Verma, S. B., Black, T. A., Vesala, T., Falge, E. M., & Dowty, P. R. (2002). Advantages of diffuse radiation for terrestrial eco-system productivity. *Journal of Geophysical Research*, *107*(D6), 4050. doi:10.1029/2001JD001242.

Gu, Y., & Liou, K. N. (2001). Radiation parameterization for three dimensional inhomogeneous cirrus clouds: Application to climate models. *Journal of Climate*, *14*, 2443–2457.

Gu, Y., Farrara, J., Liou, K. N., & Mechoso, C. R. (2003). Parameterization of cloud-radiation processes in the UCLA general circulation model. *Journal of Climate*, *16*, 3357–3370.

Gu, Y., & Liou, K. N. (2006). Cirrus cloud horizontal and vertical inhomogeneity effects in a GCM. *Meteorology and Atmospheric Physics*, *91*, 223–235.

Gu, Y., Liou, K. N., Ou, S. C., & Fovell, R. (2011). Cirrus cloud simulations using WRF with improved radiation parameterization and increased vertical resolution. *Journal of Geophysical Research*, *116*, D06119.

Gu, Y., Liou, K. N., Lee, W. L., & Leung, L. R. (2012). Simulating 3-D radiative transfer effects over the Sierra Nevada Mountains using WRF. *Atmospheric Chemistry and Physics*, *12*, 9965–9976. doi:10.5194/acp-12-9965-2012.

Guignard, A., Stubenrauch, C. J., Baran, A. J., & Armante, R. (2012). Bulk microphysical properties of semi-transparent cirrus from AIRS: A six year global climatology and statistical analysis in synergy with geometrical profiling data from CloudSat-CALIPSO. *Atmospheric Chemistry and Physics*, *12*, 503–525. doi:10.5194/acp-12-503-2012.

Hansen, J. E. (1969). Radiative transfer by doubling very thin layers. *Astrophysical Journal*, *155*, 565–573.

Heymsfield, A. J., & Miloshevich, L. M. (1993). Overview of microphysics and state parameter measurements from FIRE-II. *Proceedings of the Conference on FIRE Cirrus Science Results 1993*, *Breckenridge*, CO, NASA, pp 1–4.

Heymsfield, A. J., & Platt, C. M. R. (1984). A parameterization of the particle size spectrum of ice clouds in terms of the ambient temperature and ice water content. *Journal of the Atmospheric Sciences*, *41*, 846–855.

Hogan, R., & Illingworth, A. (2000). Deriving cloud overlap statistics from radar. *Quarterly Journal of the Royal Meteorological Society*, *126*(569), 2903–2909. doi: 10.1002/qj.49712656914.

Hogan, R. & Shonk, J. P. (2013). Incorporating the effects of 3D radiative transfer in the presence of clouds into two-stream multilayer radiation schemes. *Journal of the Atmospheric Sciences*, *70*, 708–724.

Hogan, R., & Bozzo, A. (2018). A flexible and efficient radiation scheme for the ECMWF model. *Journal of Advances in Modeling Earth Systems*, *10*, 1990–2008. https://doi.org/10.1029/2018MS001364.

Hogan, R. J., Schäfer, S. A., Klinger, C., Chiu, J. C., & Mayer, B. (2016). Representing 3-D cloud radiation effects in two-stream schemes: 2. Matrix formulation and broadband evaluation. *Journal of Geophysical Research: Atmospheres*, *121*(14), 8583–8599.

Huang, D., & Liu, Y. (2014). A novel approach for introducing cloud spatial structure into cloud radiative transfer parameterizations. *Environmental Research Letters*, 9124022, 7pp. doi: 10.1000/1748-9326/9/12/124022.

Iacono, M. J., Delamere, J. S., Mlawer, E. J., Shephard, M. W., Clough, S. A., & Collins, W. D. (2008). Radiative forcing by long–lived greenhouse gases: Calculations with the AER radiative transfer models. *Journal of Geophysical Research*, *113*, D13103.

IPCC (2013). Climate Change 2013: The Physical Science Basis. Contribution of Working Group I to the Fifth Assessment Report of the Intergovernmental Panel on Climate Change [Stocker, T. F., Qin, D., Plattner, G.-K., Tignor, M., Allen, S. K., Boschung, J., et al. (Eds.)]. Cambridge University Press: Cambridge, United Kingdom and New York, NY, USA, 1535 pp.

Järvinen, E., Jourdan, O., Neubauer, D., Yao, B., Liu, C., Andreae, M. O., et al. (2018). Additional global climate cooling by clouds due to ice crystal complexity. *Atmospheric Chemistry and Physics*, *18*, 15767–15781. https://doi.org/10.5194/acp-18-15767-2018.

Joseph, J. H., Wiscombe, W. J., & Weinman, J. A. (1976). The delta-Eddington approximation for radiative flux transfer. *Journal of the Atmospheric Sciences*, *33*, 2452–2459.

Kassianov, E., & Veron, D. (2011). Stochastic radiative transfer in Markovian mixtures: Past, present, and future. *Journal of Quantitative Spectroscopy and Radiative Transfer*, *112*, 566–576.

King, M. D., & Harshvardhan (1986). Comparative accuracy of selected multiple scattering approximations. *Journal of the Atmospheric Sciences*, *43*, 784–801.

Kogan, Z. N., Lilly, D. K., Kogan, Y. L., & Filyushkin, V. (1995). Evaluation of radiation parameterizations using an explicit cloud microphysical model. *Atmospheric Research*, *35*, 157–172.

Koop, T., Luo, B. P., Tsias, A., & Peter, T. (2000). Water activity as the determinant for homogeneous ice nucleation in aqueous solutions. *Nature*, *406*, 611–614.

Kopp, G., & Lean, J. (2011). A new, lower value of total solar irradiance: Evidence and climate significance. *Geophysical Research Letters*, *38*, L01706. doi:10.1029/2010GL045777.

Kristjansson, J. E., Iversen, T., Kirkevag, A., Seland, Ø., & Debernard, J. (2005). Response of the climate system to aerosol direct and indirect forcing: Role of cloud feedbacks. *Journal of Geophysical Research*, *110*, D24206. doi:10.1029/2005JD006299.

Lacis, A., & Oinas, V. (1991). A description of the correlated k-distribution method for modeling nongray gaseous absorption, thermal emission, and multiple scattering in vertically inhomogeneous atmospheres. *Journal of Geophysical Research*, *96*, 9027–9063.

Lee, W. L., Liou, K. N., & Hall, A. (2011). Parameterization of solar fluxes over mountain surfaces for application to climate models. *Journal of Geophysical Research*, *116*, D01101. doi:10.1029/2010JD014722.

Lee, W. L., Gu, Y., Liou, K. N., Leung, L. R., & Hsu, H.-H. (2015). A global model simulation for 3-D radiative transfer impact on surface hydrology over Sierra Nevada and Rocky Mountains. *Atmospheric Chemistry and Physics*, *15*, 5405–5413. doi: 10.5194/acp-15-5405-2015.

Li, J., (1999). On the fractional scattering into the forward peak. *Journal of the Atmospheric Sciences*, *56*, 2728–2732.

Liang, X.-Z., & Wang, W.-C. (1997). Cloud overlap effects on general circulation model climate simulations. *Journal of Geophysical Research*, *102*, 11039–11047.

Liou, K. N. (1973). A numerical experiment on Chandrasekhar's discrete-ordinate method for radiative transfer: Applications to cloudy and hazy atmospheres. *Journal of the Atmospheric Sciences*, *30*, 1303–1326.

Liou, K. N. (1974). Analytic two-stream and four-stream solutions for radiative transfer. *Journal of the Atmospheric Sciences*, *31*, 1473–1475.

Liou, K. N., & Wittman, G. D. (1979). Parameterization of the radiative properties of clouds. *Journal of the Atmospheric Sciences*, *36*, 1261–1273.

Liou, K. N. (1980). *An Introduction to Atmospheric Radiation* (392 pp). Academic Press: New York.

Liou, K. N. (1986). Influence of cirrus clouds on weather and climate processes: A global perspective. *Monthly Weather Review*, *114*, 1167–1199.

Liou, K. N. (1992). *Radiation and Cloud Processes in the Atmosphere: Theory, Observation, and Modeling (487 pp)*. Oxford University Press: New York.

Liou, K. N., & Rao, N. (1996). Radiative transfer in cirrus clouds. Part IV: On cloud geometry, inhomogeneity, and absorption. *Journal of the Atmospheric Sciences*, *53*, 3046–3065.

Liou, K. N. (2002). *An Introduction to Atmospheric Radiation* (2nd edition, 583 pp). Academic Press: San Diego.

Liou, K. N., Fu, Q., & Ackerman, T. P. (1988). A simple formulation of the delta-four-stream approximation for radiative transfer parameterizations. *Journal of the Atmospheric Sciences*, *45*, 1940–1947.

Liou, K. N., Lee, W.-L., & Hall, A. (2007). Radiative transfer in mountains: Application to the Tibetan Plateau. *Geophysical Research Letters*, *34*, L23809, doi:10.1029/2007GL031762.

Liou, K. N., Gu, Y., Yue, Q., & McFarguhar, G. (2008). On the correlation between ice water content and ice crystal size and its application to radiative transfer and general circulation models. *Geophysical Research Letters*, *35*, L13805, doi:10.1029/2008GL033918.

Liou, K. N., Gu, Y., Leung, L. R., Lee, W. L., & Fovell, R. G. (2013). A WRF simulation of the impact of 3-D radiative transfer on surface hydrology over the Rocky Mountains and Sierra Nevada. *Atmospheric Chemistry and Physics*, *13*, 11709–11721. doi:10.5194/acp-13-11709-2013.

Liou, K. N., Ou, S. C., Takano, Y., Roskovensky, J., Mace, G. G., Sassen, K., & Poellot, M. (2002). Remote sensing of three-dimensional inhomogeneous cirrus clouds using satellite and mm-wave cloud radar data. *Geophys. Res. Lett.*, 29, doi: 10.1029/2002GL014846.

Liou, K. N., & Yang, Y. (2016). *Light scattering by ice crystals: Fundamentals and applications*. Cambridge, UK: Cambridge University Press. Book Reviews.

Lohmann, U., & Kärcher, B. (2002). First interactive simulations of cirrus clouds formed by homogeneous freezing in the ECHAM general circulation model. *Journal of Geophysical Research*, *107*, 4105.

Mace, G. G., & Benson-Troth, S. (2002). Cloud-layer overlap characteristics derived from long-term cloud radar data. *Journal of Climate*, *15*, 2505–2515. doi: 10.1175/1520-0442(2002)015<2505:CLOCDF>2.0.CO;2.

Manabe, S., & Strickler, R. (1964). Thermal equilibrium of the atmospheres with a convective adjustment. *Journal of the Atmospheric Sciences*, *21*, 361–385.

Marshak, A., & Davis, A. (2005). *3D Radiative Transfer in Cloudy Atmospheres* (701 pp). Berlin/Heidelberg, Germany: Springer Verlag.

McFarquhar, G., Iacobellis, S., & Sommerville, R. (2003). SCM simulations of tropical ice clouds using observationally based parameterizations of microphysics. *Journal of Climate*, *16*, 1643–1664.

Meador, W. E., & Weaver, W. R. (1980). Two stream approximations to radiative transfer in planetary atmospheres: A unified description of existing methods and a new improvement. *Journal of the Atmospheric Sciences*, *37*, 630–643.

Miesch, C., Briottet, X., Herr, Y. H., & Cabot, F. (1999). Monte Carlo approach for solving the radiative transfer equation over mountainous and heterogeneous areas. *Applied Optics*, *38*, 7419–7430.

Minnis, P., Liou, K. N., & Takano, Y. (1993). Inference of cirrus cloud properties using satellite-observed visible and infrared radiances. Part I: Parameterization of radiance field. *Journal of the Atmospheric Sciences*, *50*, 1279–1304.

Mlawer, E. J., Taubman, S. J., Brown, P. D., Iacono, M. J., & Clough, S. A. (1997). Radiative transfer for inhomogeneous atmospheres: RRTM, a validated correlated–k model for the longwave. *Journal of Geophysical Research*, *102*, 16663–16682.

Morrison, H., & Gettelman, A. (2008). A new two-moment bulk stratiform cloud microphysics scheme in the community atmosphere model, version 3 (CAM3). Part I: Description and numerical tests. *Journal of Climate*, *21*, 3642–3659.

Müller, M. D., & Scherer, D. (2005). A grid- and subgrid-scale radiation parameterization of topographic effects for mesoscale weather forecast models. *Monthly Weather Review*, *133*, 1431–1442.

O'Hirok, W., & Gautier, C. (1998). A three-dimensional radiative transfer model to investigate the solar radiation within a cloudy atmosphere. Part I: Spatial effects. *Journal of the Atmospheric Sciences*, *55*, 2162–2179.

Olyphant, G. A. (1986). The components of incoming radiation within a mid-latitude alpine watershed during the snowmelt season. *Arctic, Antarctic, and Alpine Research*, *18*, 163–169.

Ono, A. (1969). The shape and riming properties of ice crystals in natural clouds. *Journal of the Atmospheric Sciences*, *26*, 138–147.

Oreopoulos, L., & Cahalan, R. F. (2005). Cloud inhomogeneity from MODIS. *Journal of Climate*, *18*(23), 5110–5124.

Ou, S. C., & Liou, K. N. (1982). Generalization of the spherical harmonic method to radiative transfer in multi-dimensional space. *Journal of Quantitative Spectroscopy and Radiative Transfer*, *28*, 271–288.

Ou, S. C., & Liou, K. N. (1995). Ice microphysics and climatic temperature perturbations. *Atmospheric Research*, *35*, 127–138.

Ou, S. C., Liou, K. N., Takano, Y., Rao, N. X., Fu, Q., Heymsfield, A. J., et al. (1995). Remote sounding of cirrus cloud optical depths and ice crystal sizes from AVHRR data: Verification using FIRE II IFO measurements. *Journal of the Atmospheric Sciences*, *52*, 4143–4158.

Pincus, R., Barker, H. W., & Morcrette, J.-J. (2003). A fast, flexible, approximate technique for computing radiative transfer in inhomogeneous cloud fields. *Journal of Geophysical Research*, *108*(D13), 4376, doi:10.1029/2002JD003322.

Pincus, R., Hemler, R., & Klein, S. A. (2006). Using stochastically generated subcolumns to represent cloud structure in a large-scale model. *Monthly Weather Review*, *134*, 3644–3656.

Pomraning, G. C. (1996). Transport theory in discrete stochastic mixtures. *Advances in Nuclear Science and Technology*, *24*, 47–93.

Potter, J. F. (1970). The delta-function approximation in radiative transfer theory. *Journal of the Atmospheric Sciences*, *27*, 943–949.

Qiu, J. (1999). Modified delta-Eddington approximation for solar reflection, transmission and absorption calculations. *Journal of the Atmospheric Sciences*, *56*, 2955–2961.

Räisänen, P. (2002). Two-stream approximations revisited: A new improvement and tests with GCM data. *Quarterly Journal of the Royal Meteorological Society*, *128*, 2397–2416. doi: 10.1256/qj.01.161.

Ramanathan, V., Pitcher, E. J., Malone, R. C., & Blackmon, M. L. (1983). The response of a spectral general circulation model to refinements in radiative processes. *Journal of the Atmospheric Sciences*, *40*, 605–630.

Randall, D. A., Harshvardhan, D. A. Dazlich, & Corsetti, T. G. (1989). Interactions among radiation, convection, and large-scale dynamics in a general circulation model. *Journal of the Atmospheric Sciences*, *46*, 1943–1970.

Rasmussen, R. M., Liu, C., Ikeda, K., Gochis, D., Yates, D., Chen, F., et al. (2011). Weather Support to Deicing Decision Making (WSDDM): a winter weather nowcasting system, *Bulletin of the American Meteorological Society*, *82*, 579–595.

Rossow, W. B., Delo, C., & Cairns, B. (2002). Implications of the observed mesoscale variations of clouds for the Earth's radiation budget. *Journal of Climate*, *15*, 557–585.

Rothman, L. S., Rinsland, C. P., Goldman, A., Massie, S. T., Edwards, D. P., Flaud, J.-M., et al. (1998). The HITRAN molecular spectroscopic database and HAWKS (HITRAN Atmospheric Workstation): 1996 Edition. *Journal of Quantitative Spectroscopy and Radiative Transfer*, *60*, 665–710.

Rothman, L. S., Jacquemart, D., Barbe, A., Chris Benner, D., Birk, M., Brown, L. R., et al. (2005). The HITRAN 2004 molecular spectroscopic database. *Journal of Quantitative Spectroscopy & Radiative Transfer*, *96*, 139–204.

Rotstayn, L. D. (1997). A physical based scheme for the treatment of stratiform clouds and precipitation in large-scale models. I: Description and evaluation of the microphysical processes. *Quarterly Journal of the Royal Meteorological Society*, *123*, 1227–1282.

Rottman, G. (2005). The SORCE mission. *Solar Physics*, *230*, 7–25. doi:10.1007/s11207-005-8112-6.

Schäfer, S. A., Hogan, R. J., Klinger, C., Chiu, J. C., & Mayer, B. (2016). Representing 3-D cloud radiation effects in two-stream schemes: 1. Longwave considerations and effective cloud edge length. *Journal of Geophysical Research: Atmospheres*, *121*(14), 8567–8582.

Shi, G. Y. (1998). On the k-distribution and correlated kdistribution models in the atmospheric radiation calculation. *Scientia Atmospherica Sinica*, *22*(4), 659–676.

Skamarock, W. C., Klemp, J. B., Dudhia, J., Gill, D. O., Barker, D. M., Duda, M. G., et al. (2008). *A description of the advanced research WRF version 3. NCAR Technical Note NCAR/TN-475+STR,* 113 pp. doi:10.5065/D68S4MVH.

Shonk, J. K. P., & Hogan, R. J. (2008). Tripleclouds: An efficient method for representing horizontal cloud inhomogeneity in 1D radiation schemes by using three regions at each height. *Journal of Climate*, *21*, 2352–2370. doi: 10.1175/2007JCLI1940.1.

Shonk, J. K. P., Hogan, R. J., Edwards, J. M., & Mace, G. G. (2010). Effect of improving representation of horizontal and vertical cloud structure on the Earth's global radiation budget. Part I: Review and parametrization. *Quarterly Journal of the Royal Meteorological Society*, *136*, 1191–1204. doi:10.1002/qj.647.

Shonk, J. K. P., Hogan, R. J., & Manners, J. (2012). Impact of improved representation of horizontal and vertical cloud structure in a climate model. *Climate Dynamics*, *38*, 2365–2376.

Slingo, A. (1989). A GCM parameterization for the shortwave radiative properties of water clouds. *Journal of the Atmospheric Sciences*, *46*, 1419-1427.

Slingo, A., & Slingo, J. M. (1988). The response of a general circulation model to cloud longwave radiative forcing. Part I: Introduction and initial experiments. *Quarterly Journal of the Royal Meteorological Society*, *114*, 1027–1062.

Slingo, A., & Slingo, J. M. (1991). Response of the National Center for Atmospheric Research community climate model to improvements in the representation of clouds. *Journal of Geophysical Research*, *96*, 15 341–15 347.

Sobolev, V. V. (1975). *Light scattering in Planetary Atmospheres (254 pp)*. Oxford: Pergamon Press.

Stamnes, K., Tsai, S.-C., Wiscombe, W., & Jayaweera, K. (1988). Numerically stable algorithm for discrete-ordinate-method radiative transfer in multiple scattering and emitting layered media. *Applied Optics*, *27*, 2502–2509.

Stephens, G. L. (1978). Radiation profiles in extended water clouds. II. Parameterization schemes. *Journal of the Atmospheric Sciences*, *35*, 2123–2132.

Stephens, G. L. (1988). Radiative transfer through arbitrarily shaped optical media: II. Group theory and simple closures. *Journal of the Atmospheric Sciences*, *45*, 1837–1849.

Stephens, G. L., Gabriel, P. M., & Partain, P. T. (2001). Parameterization of atmospheric radiative transfer. Part I: Validity of simple models. *Journal of the Atmospheric Sciences*, *58*, 3391–3409.

Stephens, G., Li, J., Wild, M., Clayson, C. A., Loeb, N., Kato, S., et al. (2012). An update on Earth's energy balance in light of the latest global observations. *Nature Geoscience*, *5*, 691–696. doi: 10.1038/NGEO1580.

Stokes, G. G. (1862). On the intensity of the light reflected from or transmitted through a pile of plates. *Proceedings of the Royal Society of London*, *11*, 545–556.

Sykes, I. B. (1951). Approximate integration of the equation of transfer. *Monthly Notices of the Royal Astronomical Society*, *11*, 377.

Takano, Y., & Liou, K. N. (1989). Solar radiative transfer in cirrus clouds. Part I. Single-scattering and optical properties of hexagonal ice crystals. *Journal of the Atmospheric Sciences*, *46*, 3–19.

Tian, L., & Curry, J. A. (1989). Cloud overlap statistics. *Journal of Geophysical Research*, *94*, 9925–9935.

Tiedtke, M. (1996). An extension of cloud-radiation parameterization in the ECMWF model. The representation of sub-grid-scale variations of optical depth. *Monthly Weather Review*, *124*, 745–750.

Titov, G. A. (1990). Statistical description of radiation transfer in clouds. *Journal of the Atmospheric Sciences*, *47*, 24–38.

Tompkins, A. M., & Di Giuseppe, F. (2007). Generalizing cloud overlap treatment to include solar zenith angle effects on cloud geometry. *Journal of the Atmospheric Sciences*, *64*, 2116–2125.

Ulanowski, Z., Hesse, E., Kaye, P. H., & Baran, A. J. (2006). Light scattering by complex ice-analogue crystals. *Journal of Quantitative Spectroscopy and Radiative Transfer*, *100*, 382–392.

Vainikko, G. M. (1973). Transfer approach to the mean intensity of radiation in non-continuous clouds Trudy MGK SSSR. *Meteorological Investigations*, *21*, 28–37.

van de Hulst, H. C. (1980). *Multiple Light Scattering. Tables, Formulas, and Applications (Vols. 1 and 2, 739 pp).* Cambridge, MA: Academic Press.

Wang, W. C., & Ryan, P. B. (1983). Overlapping effect of atmospheric H2O, CO2 and O3 on the CO2 radiative effect. *Tellus B*, *35*, 81–91.

Watanabe, M., Emori, S., Satoh, M., & Miura, H. (2009). A PDF-based hybrid prognostic cloud scheme for general circulation models. *Climate Dynamics*, *33*, 795–816.

Webb, M. J., Lambert, F. H., & Gregory, J. M. (2013). Origins of differences in climate sensitivity, forcing and feedback in climate models. *Climate Dynamics*, *40*, 677–707.

Wiscombe, W. (1977). The delta-M method: Rapid yet accurate radiative flux calculations for strongly asymmetric phase functions. *Journal of the Atmospheric Sciences*, *36*, 1408–1422.

Wiscombe, W. J. (1980). Improved Mie scattering algorithms. *Applied Optics*, *19*, 1505–1509.

Yi, B., Yang, P., Baum, B. A., L'ecuyer, T., Oreopoulos, L., Mlawer, E. J., et al. (2013). Influence of ice particle surface roughening on the global cloud radiative effect. *Journal of the Atmospheric Sciences*, *70*, 2794–2807. doi: 10.1175/JAS-D-13-020.1.

Zdunkowski, W. G., Welch, R. M., & Korb, G. (1980). An investigation of the structure of typical two-stream methods for the calculation of solar fluxes and heating rates in clouds. *Beitrage zur Physik der Atmosphare*, *53*, 147–166.

Zhang, Y., Macke, A., & Albers, F. (1999). Effect of crystal size spectrum and crystal shape on stratiform cirrus radiative forcing. *Atmospheric Research*, *52*, 59–75.

Zhang F., Zeng, Q., Gu, Y., & Liou, K. N. (2005). Parameterization of the absorption of the H2O continuum, CO2, O2, and other trace gases in the Fu-Liou solar radiation program. *Advance in Atmospheric Sciences*, *22*, 545–558.

3

Aerosols and Climate Effects

Xiaohong Liu

ABSTRACT

Atmospheric aerosols are suspensions of solid particles or liquid droplets in air, which include sulfate (SO_4), nitrate (NO_3), ammonium (NH_4), organic carbon (OC), black carbon (BC), dust, and sea salt. Aerosol particles exhibit vastly different sizes spanning a few orders of magnitude from 10^{-3} to 10 µm. Aerosol radiative forcing (RF) is one of the largest uncertainties in climate projection. The large uncertainties in aerosol RF can be tracked to diverse representations of aerosol processes and properties including aerosol-radiation and aerosol-cloud interactions in global climate models (GCMs). In this chapter, the processes and properties of atmospheric aerosols are introduced. Particularly, the current status of their representations in GCMs are reviewed. We highlight and advocate more research on the understanding and representation of critical processes and properties. They include emissions of BC and primary organic aerosol (POA) or particulate organic matter (POM) from wildfires and their atmospheric aging; formation and aging of secondary organic aerosol (SOA) and roles of SOA in the nucleation and growth of new particles; convective aerosol processes and interactions with cloud hydrometeors; abundances and optical properties of coarse mode dust and brown carbon (BrC); and ice nucleation and aerosol indirect effects on mixed-phase and ice clouds.

3.1. AEROSOL PROCESSES

Aerosols in the atmosphere experience myriad complicated physical and chemical processes, such as emission, transport and diffusion, dry deposition, gas-phase and aqueous-phase chemical production, aerosol microphysics (nucleation, coagulation, and condensation), water uptake, in-cloud (nucleation) scavenging, and below-cloud wet scavenging. Aerosol processes determine the properties of aerosols (e.g., size distribution, composition, and optics). Figure 3.1 shows a schematic diagram of aerosol processes in clear air and in clouds.

Department of Atmospheric Sciences, Texas A&M University, College Station, TX, USA

3.1.1. Emission

Primary aerosols in the atmosphere are emitted directly from a variety of sources. The rest of aerosols are generated in the atmosphere from chemical transformations involving emitted precursor gases. Efforts have been made to develop data sets for anthropogenic emissions to be used in global climate models (GCMs) (e.g., Dentener et al., 2006; Hoesly et al., 2018; Lamarque et al., 2010). These data sets are prescribed for model simulations, while emissions for natural aerosols (e.g., dust and sea salt) are often calculated interactively in the models as a function of meteorological variables (e.g., wind speeds) and surface properties. Below we separately introduce emissions by different sources (and sectors).

Fast Processes in Large-Scale Atmospheric Models: Progress, Challenges, and Opportunities, Geophysical Monograph 282, First Edition.
Edited by Yangang Liu and Pavlos Kollias.

DOI: 10.1002/9781119529019.ch03

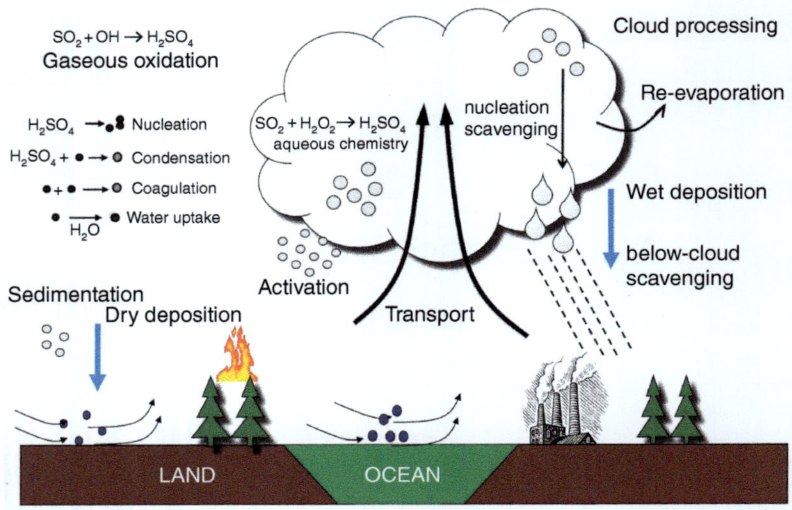

Figure 3.1 Schematic of aerosol processes in the atmosphere.

Anthropogenic Emission

Emissions of anthropogenic aerosols and precursor gases originate from all the aspects of human activities from fossil fuel combustion to manure management. Anthropogenic emission sources are usually aggregated into several major sectors, for example, industrial, energy, transportation, agriculture, residential, solvents production and application, waste treatment, international shipping, and aircraft sectors. Emitted aerosols and gas species include chemically reactive gases (e.g., carbon monoxide (CO), methane (CH_4), ammonia (NH_3), nitrogen oxides (NO_x), sulfur dioxide (SO_2), and nonmethane volatile organic compounds (NMVOCs)), carbonaceous aerosols (BC and OC), and carbon dioxide (CO_2).

Several global and regional emission inventories are widely used for climate modeling, including the Emissions Database for Global Atmospheric Research (EDGAR) (Crippa et al., 2016; Crippa et al., 2018; EC-JRC/PBL, 2012), the Greenhouse gas-Air pollution INteractions and Synergies (GAINS) (Amann et al., 2011), the Hemispheric Transport of Air Pollutants version 2 (HTAP v2) (Janssens-Maenhout et al., 2015), and the **E**valuating the **CL**imate and Air Quality **I**m**P**acts of **S**hort-liv**E**d Pollutants (ECLIPSE) (Klimont et al., 2017). A historical (1850–2000) global emission inventory was developed to provide a consistent data set for the models participating in the Climate Model Intercomparison Program Phase 5 (CMIP5) in support of the Fifth Assessment Report of the Intergovernmental Panel on Climate Change (IPCC AR5) (Lamarque et al., 2010). The data set combined the "best available estimates" from several anthropogenic and biomass burning emission data sources (see the next subsection for biomass burning emission). Data were provided in every 10 years with monthly variations only for biomass burning, and ship and aircraft emissions.

To prepare for the CMIP6, a new global historical (1750–2015) data set was developed on an annual basis with a new methodology that was designed to be more consistent over time and across sectors under the same activity or "underlying driver" (e.g., fuel consumption) (Hoesly et al., 2018). The emissions of a set of species from a sector, a country, and a fuel type (if applicable) under the same underlying driver are estimated as

$$E_i^{c,s,f,t} = A^{c,s,f,t} \times \mathrm{EF}_i^{c,s,f,t} \tag{3.1}$$

where E_i is the emissions for a species i, A is the driver or activity, EF_i is the emission factor for species i, c is the country, s is the sector, f is the fuel type (where applicable), and t is the year. The activity data are the fuel consumption for combustion emissions and population for noncombustion emissions. The emissions were then calibrated to match the existing country-level inventories when available on the sector basis. The emission factors reflect the combined impacts of fuel types, technologies, and abatement measures on emissions and are subject to large uncertainties.

Comparison between main CMIP6 emission sectors shows dramatic increases in anthropogenic emissions since 1750, but the historical trends are featured by different details for different sectors (see Figure 2 in Hoesly et al., 2018). Before 1850, major anthropogenic emissions of OC, BC, NMVOCs, and CO were from residential and agricultural sources. After 1850, emissions of SO_2, NO_x, NMVOCs, and CH_4 in the energy sector increased dramatically, especially after 1950. Emissions of SO_2,

Table 3.1 CMIP6 Emissions of aerosols and chemically reactive gases from different sectors in year 2014 (Tg $year^{-1}$).

	Agriculture	Energy	Industrial	Transportation	RCO[a]	Solvents	Waste	Shipping	Aircraft
SO_2	0.0	47.8	42.9	3.1	8.3	0.0	0.5	8.9	0.3
BC	0.0	1.2	0.8	1.4	3.7	0.0	0.7	0.2	0.0
OC	0.0	2.1	1.5	0.6	11.2	0.0	4.2	0.1	0.0
NO_x	5.9	34.2	19.8	43.0	8.0	0.0	6.0	22.7	3.1
NH_3	45.7	0.9	1.3	0.4	3.5	0.0	9.1	0.0	0.0
CH_4	137.9	154.3	0.9	0.6	12.1	0.0	65.8	0.0	0.0
NMVOC	0.0	53.7	13.5	30.8	33.3	22.9	7.3	3.1	0.1
CO	0.0	48.9	129.3	149.5	245.1	0.0	40.3	0.7	0.6

[a] RCO: Residential, commercial, other.

NO_x, and CO in the industrial sectors also increased with production activities. The SO_2 emission was stabilized in 1970s due to emission control polices in North America and Europe and decreased in 2000s as a result of abatement measures. Transportation emissions have grown steadily and become an important contributor to NO_x, NMVOCs, and CO emissions after 1950. Residential cooking and heating have been important sources of BC and OC since 1750. The use of fertilizers in agricultural productions has brought a fast growth in the NH_3 emission. Table 3.1 compares the global emissions of aerosols and chemically reactive gases from different CMIP6 sectors for the year 2014. We note that due to the subgrid nature of plume-scale emissions, a small fraction of anthropogenic sulfur emissions (e.g., 2.5% by molar) are emitted directly as sulfate aerosols in GCMs (Dentener et al., 2006).

Differences between emission inventories are larger in some countries/regions than the others. Comparison between CMIP5 and CMIP6 emissions was conducted in Hoesly et al. (2018) (see their Figure 7 and supplemental Section K). In Figure 3.2, we further include the comparison of the Multi-resolution Emission Inventory for China (MEIC; Zhang et al., 2009; Li et al., 2017) since anthropogenic emissions in China are subject to large uncertainties due to the complexities of emission factors, fuel properties, and energy use data. The emissions of aerosols and their precursor gases due to the use of fossil fuels have experienced dramatic changes in China after year 1990 (Figure 3.2). CMIP5, CMIP6, and MEIC consistently show that since 1990 emission rates of SO_2, BC, POM, and nitrogen monoxide (NO) gradually rise, slow down in the mid-1990s, increase dramatically after 2000, and stabilize around 2005. However, various inventories show different yearly trends after 2006. The SO_2 and NO emissions in MEIC decrease more dramatically than the CMIP6 inventory after 2006 and 2012, respectively. The emission rates of BC and POM in MEIC slightly decrease after reaching their peaks in 2006, while they keep growing as estimated by the CMIP6 inventory. The discrepancies may be due to the fact that the MEIC emission inventory keeps track of the renewal of energy structure and emission reduction measures in recent years in correspondence to the urgent need of improving the air quality in China. The emission hot spots are located in populated and economically prosperous regions, e.g., the Jing-Jin-Ji Region, Yangtze River Delta, and Pearl River Delta.

Biomass Burning Emission

Biomass burning aerosols and precursor gases are produced from the burning of living and dead vegetation, including human-initiated burning for land clearing and land use change as well as natural, lightning-induced fires. Biomass burning is one of the major sources of many aerosols and tracer gases to the global atmosphere. It accounts for 59% of global BC emissions and 85% of POA emissions (Bond et al., 2013) and one third to one half of global CO and 20% of NO_x emissions (Wiedinmyer et al., 2011). Fires also emit a large amount of greenhouse gases, including CO_2, CH_4, and nitrous oxide (N_2O).

Generally, a biomass burning emission inventory requires two factors: the burned condition and the underlying vegetation distribution. The burned condition describes the burned area and the related fire intensity. The vegetation distribution describes the potential amounts of tracer gases and aerosols that could be released. In GCMs, if the burned condition and the vegetation distribution are based on observations, these inventories are referred as "offline" emission inventories, whereas if the burned condition and vegetation distribution are calculated from an embedded "fire-land model," the emissions are termed as being "online." The offline emission inventories are useful to provide historical and future estimates of biomass burning, while online emissions are helpful for investigating the interactions between climate change

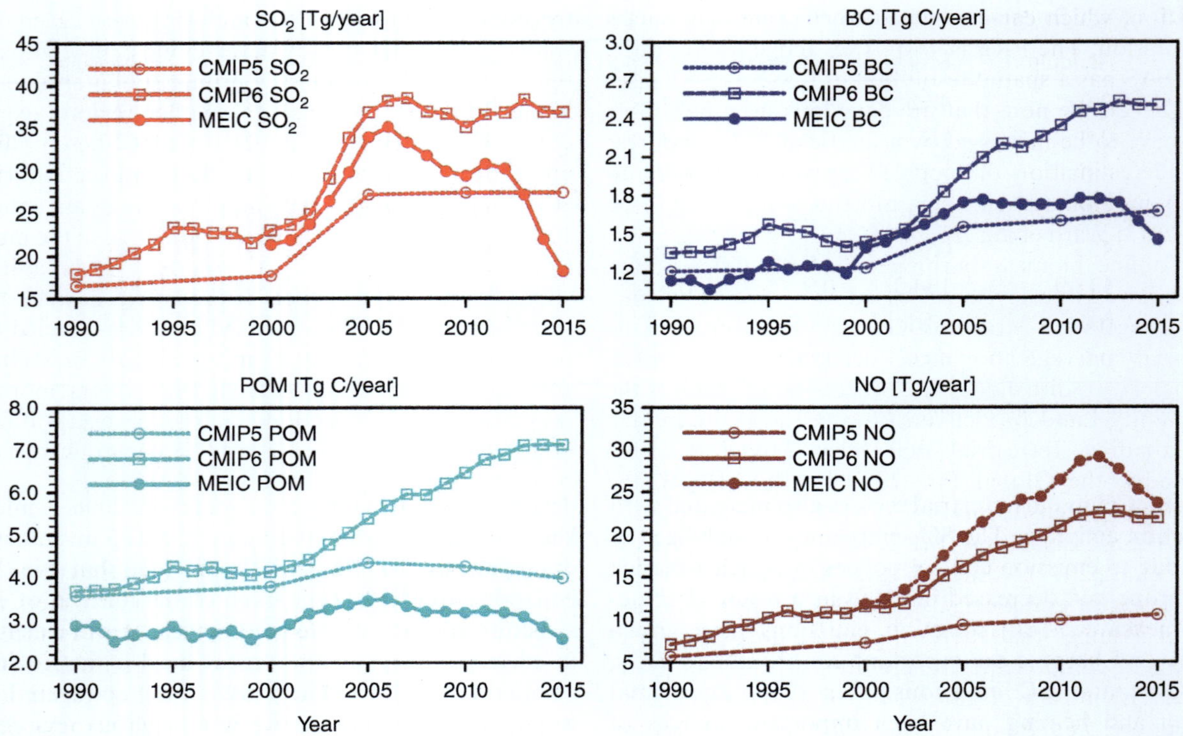

Figure 3.2 Emission rates of SO_2, BC, POM, and NO due to fossil fuel use in China from 1990–2015. Source: Adapted from emission inventories of MEIC, CMIP5 (RCP45 scenario), and CMIP6.

and biomass burning (Lamarque et al., 2010). The offline biomass burning inventories are constrained by satellite observations or by visibility and charcoal records, such as the Global Fire Emission Database (GFED), the Fire INventory from NCAR (FINN), the Global Inventory for Chemistry-Climate studies (GICC), and the Global Fire Assimilation System (GFAS).

The GFED emission inventory is based on the satellite-retrieved burned area product from the Moderate Resolution Imaging Spectroradiometer (MODIS) and the satellite-driven Carnegie-Ames-Stanford Approach (CASA) ecosystem model. The emission for a certain gas is calculated as

$$\text{Emission} = \sum_{i=1}^{n} \text{BA} \times F_i \times \text{CF}_i \times \text{EF}_i, \qquad (3.2)$$

where i denotes a certain landcover type involved in the burning, BA is burned area, F_i is available fuel for the landcover type, CF_i is combustion factor representing the fraction of biomass that is burned (combustion completeness), and EF_i is the emission factor for the landcover type. BA is provided by the MODIS burned area, F_i and CF_i are taken from the CASA model (Van Der Werf et al., 2010), and EF_i in the unit of g [species emitted]/kg [dry matter burnt] is taken from the data set compiled by Andreae and Merlet (2001) with annual updates. The GFED provides emission data from 1997 to present with a 0.25° × 0.25° spatial resolution and 6-hour temporal resolution. The latest GFED4s version that counts in small fires detected by the MODIS active fire product produces higher global biomass burning emissions by 11% than GFED3 and by 30% than GFED4 (i.e., when small fires excluded in GFED4s) (Van Der Werf et al., 2017). GFED4s is the biomass burning emission data set used for CMIP6 present-day simulations. The FINN data set applies a similar approach (Wiedinmyer et al., 2011), but its burned area is mainly derived from the MODIS active fire product, and the fuel loading is based on a study by Hoelzemann et al. (2004). The combustion factor is taken from Wiedinmyer et al. (2006) and the emission factor is taken from Akagi et al. (2011) and Andreae and Merlet (2001). One feature of FINN data set is its high resolution, 1 km × 1 km spatially and 1-day temporally. The GICC emission data set uses the same burned area as GFED4 but uses other parameters from the Global Land Cover 2000 (GLC2000) vegetation map (Mieville et al., 2010). Therefore, the resolution of GICC data is same as GFED4. The GFAS data set is a little different, which does not use a burned area product but rather uses the biomass-burning-induced fire radiative power (FRP) as a proxy of burned area. FRP measures the radiative energy released per time unit by burning vegetation, which can be related to the rate of biomass consumption. The FRP is a MODIS active fire product, and GFAS has a spatial resolution of 0.5° × 0.5° (Kaiser et al., 2012). We note that fire emissions used in GCMs often have to be increased by a factor of 2–8 to address the underestimation of modeled aerosol optical depth (AOD) in biomass burning influenced regions (Tosca et al., 2013; Ward et al., 2012).

For online biomass burning emissions, different fire models have been developed based on various Dynamic Global Vegetation Models (DGVMs), which provide the territorial and vegetational platforms for fires to evolve. The majority of fire models are process-based, such as the Community Land Model version 5 (CLM5) fire model, the Canadian Terrestrial Ecosystem Model (CTEM) fire model, the Global Fire Model (GlobFIRM), the Spread and Intensity of Fire (SPITFIRE) model, and the MAPSS-CENTURY (MC2) Fire model (MC2-Fire) (Rabin et al., 2017). Although these fire models have different levels of complexity, in general the process-based method assumes that the evolution of fires consists of three major components: fire ignition, spread, and termination. The impact of natural factors, such as lighting, temperature, and humidity, and the impact of anthropogenic factors important for the fire suppression, such as population density and gross domestic product (GDP), are characterized through empirical methods, and the burned area is determined by the total fire counts multiplied by the average fire spread rate in each time step. Then, the fire emission is determined by the burned area, combustion completeness, and emission factors. Many fire models do not allow the duration of individual fire events to exceed more than 1 day in natural vegetation regions. Current fire models can capture the pattern and seasonality of global burned area, but most models fail to simulate the interannual variability of global fires (Li et al., 2019). Only a few models can capture the long-term fire trends for some regions, e.g., the increasing trend in North America due to land fragmentation, and the decreasing trend in Northern Africa due to precipitation increase and agriculture activity expansion (Andela et al., 2017; Li et al., 2019; Westerling et al., 2006). Precipitation increase driven by the El Niño/Southern Oscillation (ENSO) substantially contributes to the increasing trend in Southern Africa due to increase in fuel loading. The decreasing trend in Australian fires correlates with precipitation increase, and there is no obvious trend in Indonesian fires (Andela & van der Werf, 2014).

Another important parameter about biomass burning emission is the injection height of fire plumes. Generally, if the injection heights are above the planetary boundary layer (PBL), the emissions in the free troposphere can be transported to locations further away from the burning regions because of higher wind speeds in the free troposphere than the PBL and less wet scavenging of smoke pollutants by low-level clouds; in comparison, the impacts of biomass burning pollutants emitted in the PBL are restricted to smaller regions (e.g., Liu et al., 2014; Paugam et al., 2016) unless these pollutants are transported vertically by deep convection (Fiebig et al., 2009). Observational and modeling data show that 10%–20% plumes could penetrate the PBL in the morning (Val Martin et al., 2012). Currently, GCMs emit these emissions near the surface, at prescribed altitude ranges (Dentener et al., 2006), or based on more detailed plume height parameterizations (Veira et al., 2015). Multiangle Imaging SpectroRadiometer (MISR) offers more than 20 years of plume injection height data (e.g., Val Martin et al., 2018), which can be used to constrain the plume height models.

Dust Emission

Dust emission fluxes are calculated interactively in Earth System Models (ESMs) using parameterizations. In principle, dust emission fluxes depend on surface wind, soil moisture, and surface roughness. Surface roughness elements (SREs) not only consist of vegetation but include other nonerodible elements such as rocks, pebbles, and gravels. SREs suppress dust emission by reducing the erodible surface area and increasing the threshold velocity. The former SRE effect is included in ESMs, while the latter effect is often not represented (Wu et al., 2016). Due to the large uncertainty in dust emissions, practically, a soil erodibility index is introduced to make modeled spatial patterns of dust emissions close to the observations.

Table 3.2 compares the three dust emission parameterizations widely used in ESMs. These parameterizations differ in many aspects, such as soil erodibility, wind velocity, threshold wind velocity, flux equation, and particle size distribution. This may result in considerable differences in the simulated global dust distribution. The dust emission parameterization in the Community Atmosphere Model version 5 (CAM5), the atmosphere component of the Community Earth System Model (CESM), follows the scheme of Zender et al. (2003a). The scheme calculates the vertical dust mass emission flux (F_j for a given size range j) proportional to the horizontal (saltating) sand flux as

$$F_j = TSf_{\text{bare}}\alpha Q_s M_j \tag{3.3}$$

where T is an adjustable tuning factor (e.g., to match observed AOD over dust source regions); S is the soil erodibility index; f_{bare} is the grid cell fraction of exposed bare soil; α is the sandblasting mass efficiency calculated from the mass fraction of clay particles in the soil (Marticorena & Bergametti, 1995); Q_s is the total horizontally saltating sand flux that is calculated as the third

Table 3.2 Comparison of the three dust emission schemes used in GCMs.

Scheme	Zender et al. (2003a)	Ginoux et al. (2001)	Kok et al. (2014a)
Soil erodibility	Static; geomorphic erodibility (Zender et al., 2003b)	Static; topographic erodibility (Ginoux et al., 2001)	Dynamic; dependent on the standardized threshold friction velocity
Threshold wind velocity at dry soils	Marticorena & Bergametti, 1995	Marticorena & Bergametti, 1995	Marticorena & Bergametti, 1995
Threshold wind velocity modified for wet soils	Correction by soil moisture (Fécan et al., 1999)	Correction by soil wetness (Ginoux et al., 2001)	Correction by soil moisture (Fécan et al., 1999)
Wind velocity	Friction velocity	Wind velocity at 10 m	Friction velocity
Equation for vertical flux	Equation 3.3	Equation 3.4	Equation (3.5)
Particle size distribution of dust into atmosphere	Observation near the Saharan source (D' Almeida, 1987)	Simple assumption (Tegen & Fung, 1994)	Theoretical expression (Kok, 2011)
Mass fraction for each size bin (in diameter)[a]	M_j for two modes: 0.032 (0.1–1 μm); 0.968 (1–10 μm);	s_p for five bins: 0.1 (0.2–2 μm); 0.25 (2–3.6, 3.6–6, 6–12, and 12–20 μm)	M_j for two modes: 0.011 (0.1–1 μm); 0.989 (1–10 μm);

[a] Mass fractions of two modes (accumulation and coarse modes) for the three-mode version of the Modal Aerosol Module (MAM3) in CAM5 are shown for the schemes of Zender et al. (2003a) and Kok et al. (2014a). Mass fractions of five bins for the Goddard Chemistry, Aerosol, Radiation, and Transport (GOCART) in GEOS-5 are shown for the scheme of Ginoux et al. (2001).

power of friction velocity and also takes into account the threshold friction velocity (White, 1979); and M_j is the mass fraction in each size range j, which is derived from observed tri-modal lognormal distribution for the background dust aerosol near the Saharan sources (D'Almeida, 1987). The threshold friction velocity is an indicator of surface resistance to wind erosion and can be expressed as

$$u_{*t}(d) = u_{*t0}(d) \times f_z \times f_w \tag{3.4}$$

where $u_{*t0}(d)$ is the threshold friction velocity of particle with size d in the idealized situation when soil is dry and bare following Marticorena and Bergametti (1995), f_z is a correction factor depending on the surface roughness (set to 1 in CAM5), and f_w is a factor depending on soil moisture following Fécan et al. (1999). A geomorphic soil erodibility index based on the upstream runoff collection area is adopted in the emission scheme (Zender et al., 2003b).

Another widely used dust emission scheme is Ginoux et al. (2001), which was implemented into the NASA Goddard Earth Observing System Model, Version 5 (GEOS-5). The scheme calculates the dust emission flux using the following equation:

$$F_p = CSs_p u_{10\text{m}}^2 (u_{10\text{m}} - u_t(d_p))(u_{10\text{m}} > u_t(d_p)) \tag{3.5}$$

where F_p is the dust emission flux for a size range p; C is a dimensional factor for the tuning of dust emission flux; S is the source erodibility index; s_p is the mass fraction of size range p; $u_{10\text{m}}$ is the horizontal wind speed at 10 m; d_p is the effective particle diameter for size range p; and $u_t(d_p)$ is the size-dependent threshold wind friction velocity for size range p following the threshold velocity parameterization of Marticorena and Bergametti (1995). The threshold velocity is modified for real soil conditions in the presence of soil wetness w. Preferably, Ginoux et al. (2001) adopted a topographic erodibility index, which is based on the correlation of potential dust sources with large-scale topographic depressions. Following Tegen and Fung (1994), a simple dust size distribution is assumed, and the four silt ranges (2–3.6, 3.6–6, 6–12, and 12–20 μm in diameter) are each assigned a mass fraction of $s_p = 0.25$. The mass of emitted clay particles (0.2 < diameter < 2 μm) is assumed to be 10% of the total mass of emitted silt, and thus $s_p = 0.1$ for clay particles (Table 3.2).

Kok et al. (2014a) proposed a physically based scheme that can better account for soils' ability to produce dust under different soil and wind conditions. The total dust emission is calculated as

$$F_d = C_d f_{\text{bare}} f_{\text{clay}} \frac{\rho_a (u_*^2 - u_{*t}^2)}{u_{*st}} \left(\frac{u_*}{u_{*t}} \right)^{C_a \frac{u_{*st} - u_{*st0}}{u_{*st0}}} \quad (u_* > u_{*t}) \tag{3.6}$$

where C_d is a dimensionless coefficient that can be regarded as a measure of a soil's ability to produce dust, f_{bare} is the grid cell fraction of exposed bare soil, f_{clay} is the clay fraction in soil from FAO (2012), ρ_a is the air density, u_* is the friction velocity, u_{*t} is the threshold friction velocity, u_{*st} is the standardized threshold friction velocity, u_{*st0} is the minimal value of u_{*st} and $u_{*st0} \approx 0.16\,\mathrm{m\,s^{-1}}$, and C_α is a dimensionless efficient and set to 2.7 ± 1.0 according to measurements (Kok et al., 2014b). C_d is equivalent to the source erodibility in the scheme of Zender et al. (2003a).

The Kok et al. scheme has a better representation of dust emission in the aspect of a soil's increased ability to produce dust under saltation bombardment as it becomes more erodible, and the increased scaling of dust flux with wind speed as a soil becomes less erodible. The total dust emission flux is further partitioned into different size ranges with the mass fraction calculated from the particle size distribution (PSD) of dust given by Kok (2011). The new PSD is derived based on the physics of the scale-invariant fragmentation of brittle materials and shows improved agreement with various dust size distribution measurements (Kok, 2011). This PSD gives much smaller fractions of dust particles with diameter <1 µm compared to those used in the Zender et al. scheme (see also Table 3.2). Nevertheless, this PSD can also be used with other dust emission schemes such as the Zender et al. scheme.

Due to the subgrid variability of surface wind speeds in a GCM grid, the Weibull distribution has been used in GCMs to represent the probability distribution of subgrid wind speeds (Yu et al., 2015; Zhang et al., 2016). The probability density function (PDF) of a Weibull random variable (e.g., surface wind speed U) can be written as

$$p(U,k,\lambda) = \frac{k}{\lambda}\left(\frac{U}{\lambda}\right)^{k-1} \mathrm{e}^{-\left(U/\lambda\right)^{k}}, \tag{3.7}$$

where k is the shape parameter and λ is the scale parameter, both of which can be derived from the mean and standard deviation of wind speeds. Zhang et al. (2016) developed a new method based on model predicted physical quantities to estimate the subgrid standard deviation of surface wind resulting from, e.g., dry convective eddies and small-scale topography. The model-simulated global and annual mean dust emission at 2° horizontal resolution increases by 50% when including the subgrid surface wind variability from the Weibull distribution (Zhang et al., 2016). Even when tuning the dust emission scaling factor (e.g., T in equation 3.3) to match the global mean dust emission and AOD, to the level of the default model (which does not consider the subgrid wind variability), the PDF of dust emission flux changes when considering the subgrid wind variability, with more occurrence of weaker emissions and less occurrence of stronger emissions.

Sea-Spray Aerosol Emission

Sea-spray (SS) aerosols are mainly produced by breaking waves via bubble bursting and by tearing of wave crests (de Leeuw et al., 2011). 90% of the salt dissolved in the oceans is sodium chloride (NaCl), together with smaller amounts of other salts (e.g., SO_4^{2-}, K^+, Mg^{2+}, and Ca^{2+}) (Grythe et al., 2014). Chemical composition of SS aerosols is similar to their composition in bulk seawater, although organic substances at the sea surface can be enriched, resulting in substantial contributions of organic matter to the composition of submicron SS particles (O'Dowd et al., 2004). SS aerosol emission flux depends on surface wind (10-m wind speed, u_{10}), sea surface temperature, wave state, and ocean salinity, but u_{10} dominates (Grythe et al., 2014; Ovadnevaite et al., 2014).

Like dust, SS aerosol emission fluxes are calculated interactively in GCMs. Table 3.3 shows several SS emission parameterizations used in GCMs. Emission fluxes are parameterized as a function of u_{10}, while some parameterizations also include the impacts of sea surface temperature. The Global Model of Aerosol Processes (GLOMAP-mode) aerosol module (Mann et al., 2010) embedded in the UK Earth System Model (Sellar et al., 2019) uses the Gong (2003) parameterization, and the European Centre Hamburg Model-Hamburg Aerosol Module version 2 (ECHAM-HAM2) (Zhang et al., 2012) uses Monahan et al. (1986) for the small particle range (radius ≤ 4 µm) and Smith and Harrison (1998) for the large particle range (radius > 4 µm). The Modal Aerosol Module (MAM) in CESM and DOE's Energy Exascale Earth System Model version 1 (E3SMv1) (Liu et al., 2012a, 2016) uses the Mårtensson et al. (2003) scheme for dry diameter (D_p) <2.8 µm and the Monahan et al. (1986) scheme for D_p >2.8 µm. A sectional aerosol module, the Community Aerosol and Radiation Model for Atmospheres (CARMA) in CESM1 (Yu et al., 2015), also uses a combination of source functions (Clarke et al., 2006; Monahan et al., 1986; Smith et al., 1993) for SS aerosol emission fluxes in different particle size ranges, as introduced in Fan and Toon (2011). In addition, a Weibull wind distribution is used in the SS aerosol source functions in CARMA to account for the subgrid wind variability.

In a few models (Burrows et al., 2022; Yu et al., 2015), a fraction of emitted SS aerosol is assigned to be organics, either internally or externally mixed with sea salt, with the fraction correlated to the chlorophyll concentration (Gantt et al., 2009; O'Dowd et al., 2008). Burrows et al. (2014) developed the mechanistic OCEANFILMS (Organic Compounds from Ecosystems to Aerosols: Natural Films and Interfaces via Langmuir Molecular Surfactants) parameterization that connects emissions of sea-spray organic aerosol to model ocean biogeochemistry. The marine organic species represents five broad classes of macromolecules derived from phytoplankton, zooplankton, and semilabile dissolved organic carbon in ocean surface waters. With the introduction of marine organics (added and internally mixed with sea salt) in E3SM (Burrows et al., 2022), which comprises up to

Table 3.3 Sea-Salt emission parameterizations.

Source Function	Formula†	Size range in radius, *r*
Monahan et al. (1986)	$(dF/dr)_{Monahan} = 1.373u_{10}^{3.41}r^{-3}(1+0.057r^{1.05}) \times 10^{1.19e^{-B^2}}$ * $B = (0.38 - \log r)/0.65$	0.1–10 µm
Gong (2003)	$(dF/dr)_{Gong} = 1.373u_{10}^{3.41}r^{-A}(1+0.057r^{3.45}) \times 10^{1.607e^{-B^2}}$, $A = 4.7(1+\Theta r)^{-0.017r^{-1.44}}$, Θ = adjustable parameter = 30, $B = (0.433 - \log r)/0.433$	0.01–15 µm
Mårtensson et al. (2003)	$(dF/dr)_{Martensson} = 3.84 \times 10^{-4}u_{10}^{3.41}(A_kT_w + B_k)/r$ See Mårtensson et al. (2003) for A_k and B_k. T_w is for water temperature.	0.02–20 µm
Caffrey et al. (2006)	$(dF/dr)_{Caffrey} = (dF/dr)_{Monahan} \times W(r)$ $W(r) = 1.136^{-1.809r^{-0.855}}\left(1 + \frac{0.4}{r}\right)$	0.01–0.3 µm
	$(dF/dr)_{Caffrey} = \begin{cases} u_{10} < 9\ ms^{-1},\ (dF/dr)_{Monahan} \\ u_{10} < 9\ ms^{-1},\ \max((dF/dr)_{Monahan}, (dF/dr)_{Smith}) \end{cases} \times W(r)$ where $(dF/dr)_{Smith} = \sum_{i=1}^{2} A_i exp\left\{-f_i[ln(r/r_i)]^2\right\}$, $f_1 = 3.1$, $f_2 = 3.3$, $r_1 = 2.1$ µm, $r_2 = 9.2$ µm,	0.3–30 µm
Clarke-Monahan-Smith (Fan & Toon, 2011)	$\log(A_1) = 0.0676u_{14} + 2.43$, $\log(A_2) = 0.959u_{14} - 1.476$ $(dF/dr)_{Clarke} = 3.84 \times 10^{-6}{u_{10}}^{3.41}(dF/dr)_{BW}$ where $(dF/dr)_{BW} = 1/r \cdot \sum_{i=1}^{3} A_i$ $A_i = \beta_{0i} + \beta_{1i}r + \beta_{2i}r^2 + \beta_{3i}r^3 + \beta_{4i}r^4 + \beta_{5i}r^5$ See Table 1 in Clarke et al. (2006) for β coefficients.	0.02–1.6 µm
	The same as Caffrey et al. (2006) in size range of 0.3–30 µm	1.6–30 µm

Note. Source functions, dF/dr, unit: particles $m^{-2}\ s^{-1}\ \mu m^{-1}$.
†Here, $r = r_{80}$ is the droplet radius at 80% relative humidity (taken as a typical value in the marine boundary layer). We have converted different notions of particle size in the source functions to a consistent *r* based on the approximation of $2r_{dry} = D_p = r_{80}$, where r_{dry} is the dry particle radius and d_p is the dry particle diameter.
*u_{10} and u_{14} are the wind speed at 10 and 14 m, respectively.

50% of accumulation mode SS aerosol, accumulation mode number emissions are more than doubled over the Southern Ocean and Arctic.

Emissions of Other Natural Aerosols and Precursor Gases

The IPCC emission data sets (Hoesly et al., 2018; Lamarque et al., 2010) do not provide emissions of natural aerosols and precursor gases. Volcanic (sulfur) emissions consider both explosive and continuous degassing volcanos, and are prescribed from different data sets, e.g., the Global Emissions InitiAtive (GEIA) inventory (Andres & Kasgnoc, 1998) and the VolcanEESM database for 1850–2015 (Neely & Schmidt, 2016) for CMIP6. GCMs often prescribe stratospheric aerosol properties (e.g., stratospheric AOD and surface area density) based on satellite observations for explosive volcanic eruptions, while predict sulfate aerosols from diffusive volcanic eruptions. During periods of volcanic quiescence, emissions of carbonyl sulfide (OCS) through biogenic processes are the main source of sulfate aerosols in the stratosphere between the tropopause and about 30 km altitude (the Junge Layer). Some explosive volcanic eruptions can directly inject SO_2 into the stratosphere where it undergoes chemical transformation into sulfate aerosols. For example, the June 1991 eruptions of Mt. Pinatubo in the Philippines injected 12–26 megatons of SO_2 into the stratosphere reaching as high as 30 km (McCormick et al., 1995). The Pinatubo eruptions had near-global effects on weather and climate, reducing the global mean surface temperature by up to 0.5 °C in 1992 (Parker et al., 1996).

Emissions of dimethyl sulfide (DMS) from the marine biosphere are calculated online in the models or simply prescribed such as from the AeroCom protocols (Dentener et al., 2006). Online DMS emission fluxes are calculated using a climatology of seawater DMS concentrations (Kettle & Andreae, 2000; Lana et al., 2011) and an air-sea transfer velocity parameterization (Liss & Merlivat, 1986; Nightingale et al., 2000). We note that Kettle and Andreae (2000) provide a stronger seasonal cycle of DMS seawater concentrations than Lana et al. (2011). The Liss and Merlivat (1986) parameterization predicts much lower DMS gas transfer velocities than Nightingale et al. (2000) under strong winds (u_{10} > $10\,m\,s^{-1}$) and agrees better with direct measurements of DMS air-sea exchange (Yang et al., 2011). DMS emissions have a strong seasonal cycle with highest values in the summer and lowest in the winter. Once emitted into the atmosphere, DMS is oxidized to form dimethyl sulfoxide (DMSO), methanesulfinic acid (MSIA), methanesulfonic acid (MSA), and SO_2. SO_2 is further oxidized to form sulfate aerosols. These sulfate aerosols can act as cloud condensational nuclei (CCN) and increase cloud droplet number concentrations, which in turn increase the liquid water content (LWC) and area coverage of clouds. This acts to increase cloud albedo, leading to a reduction in the amount of solar radiation reaching the ocean surface and thus a reduced growth rate of phytoplankton. This negative feedback loop between ocean ecosystems and the Earth's climate was proposed by Charlson et al. (1987) (the CLAW hypothesis). However, recent studies indicate that the CLAW hypothesis may oversimplify the complicated physical and chemical processes and aerosol-cloud interactions in marine environments (Quinn & Bates, 2011).

Plants emit large amounts of biogenic VOCs (BVOCs). Isoprene and monoterpenes are the two main components of BVOCs. BVOC emissions are controlled by ambient environmental variables, most notably temperature and light. Thus, their emissions are normally higher in the summer than in the winter, and higher during the noontime than during the nighttime. In addition, BVOC emissions may also occur as responses to environmental stressors (e.g., excess light and temperature, air pollutants, and insect infestation). The BVOC emissions in the models are either prescribed (Guenther et al., 1995) or calculated online based on the Model of Emission of Gases and Aerosols from Nature (MEGAN) model (Guenther et al., 2006).

Size Distributions of Emitted Primary Particles and Injection Heights

Size distributions of emitted primary aerosol particles, which are important for their optical properties, are not recommended by the IPCC emission data sets, nor their injection heights. The size information needed for models predicting the aerosol number concentrations and size distributions often follows the AeroCom protocols (Dentener et al., 2006). For instance, emitted biomass/biofuel POM/BC particles have larger sizes (with a geometric mean diameter of 0.15 μm) than fossil fuel sources (with a geometric mean diameter of 0.06 μm). Forest fire and grass fire emissions are injected into six altitude ranges (0–0.1, 0.1–0.5, 0.5–1, 1–2, 2–3, and 3–6 km), with relative fractions in different altitude ranges depending on regions.

3.1.2. Chemical Production

A substantial fraction of aerosols are produced in the atmosphere from the chemical production of emitted precursor gases, the so-called secondary aerosols. For instance, sulfate (SO_4) is formed from the oxidation of SO_2, nitrate (NO_3) from the oxidation of NO_x, and SOA from the oxidation of volatile organic compounds (VOCs). These chemical reactions can take place in the gaseous phase and the aqueous phase (in cloud water and aerosol) and on the surface of aerosols (e.g., mineral dust), called heterogeneous reactions.

Secondary aerosols contribute significantly to regional and global aerosol burdens (Huang et al., 2014a; Textor et al., 2006; Zhang et al., 2007). For instance, organic aerosols (OAs) represent a significant and sometimes a major (20%–90%) mass fraction of submicron aerosols (Kanakidou et al., 2005), while SOA accounts for 64%–95% of the total OA in anthropogenically influenced Northern Hemisphere (Zhang et al., 2007). Secondary aerosols are found to contribute 30%–77% to the mass concentrations of $PM_{2.5}$ (particles with diameter less than 2.5 μm) during severe pollution events in China (Huang et al., 2014a). Because of their abundances in the atmosphere, it is critical to represent the formation of secondary aerosols correctly in GCMs. Almost all GCMs treat the SO_4 formation from the oxidation of SO_2 and DMS based on the gas-phase sulfur chemistry, often with prescribed oxidants such as hydroxyl radical (OH), hydroperoxyl radical (HO_2), and ozone (O_3) taken from simulations by full chemistry models. SO_2 oxidation in bulk water by hydrogen peroxide (H_2O_2) and O_3 is considered, depending on the pH value in the bulk cloud water either with a fixed value or calculated from the electroneutrality equation between cations and anions. This sulfate source in cloud water can contribute about 70% to total sulfate burdens in the atmosphere (Feichter et al., 1996) due to the much faster oxidation rate in aqueous phase than that in the gas phase. However, most of the models do not include the SO_2 aqueous oxidation by oxygen (O_2) catalyzed by transition metals Fe(III) and

Mn(II) in dust, which can contribute about 13% to total sulfate production in China (Huang et al., 2014b). Also, most of the models do not consider the heterogeneous oxidation of SO_2 on dust surface, which has been reported to promote the formation of severe haze pollution in eastern China (He et al., 2014; Y. Wang et al., 2014a; Zheng et al., 2015).

Nitrate is another type of secondary inorganic aerosol, whose concentrations can be comparable to those of sulfate in industrial regions and can become more important in the foreseeing future with the rapid reduction of SO_2 emissions and the steady increase of NO_x emissions in industrial regions (Hoesly et al., 2018). Treating nitrate in GCMs is much more challenging than sulfate due to mainly (1) that more complicated chemical reactions are involved in nitrate formation and (2) that nitric acid gas (HNO_3) is a semivolatile compound and its partitioning between the gas phase and the aerosol phase is difficult to represent well in the models. As a result, a physical treatment of nitrate demands much more computational time. A handful of GCMs and chemistry transport models (CTMs) include HNO_3 gaseous production and its partitioning into the aerosol phase to form nitrate aerosol (Bian et al., 2017). Almost all of these models adopt the thermodynamic equilibrium approach to deal with gas-aerosol partitioning and aqueous-phase and solid-phase reactions. Global nitrate burden differs by a factor of 13 among these models, and nitrate concentrations are also poorly simulated especially over East Asia (Bian et al., 2017). The thermodynamical equilibrium approach is not valid for the partitioning of HNO_3 to coarse mode particles (e.g., coarse dust and sea salt) as the time to reach equilibrium on these particles can be much longer than the model time step. A kinetic approach for these particles is required to accurately treat the partitioning of HNO_3 (Feng & Penner, 2007; Zaveri et al., 2008). Lu et al. (2021) used the Model for Simulating Aerosol Interactions and Chemistry (MOSAIC) (Zaveri et al., 2008) to treat the HNO_3 thermodynamical partitioning in the Community Atmosphere Model version 6 with interactive chemistry (CAM6-Chem). Nitrate aerosol is well simulated by the model when compared with measurements from ground-based networks in North America, Europe, and East Asia.

The formation of SOA involves complicated chemical reactions, and thus it is computationally expensive to treat SOA formation and aging explicitly in GCMs. Many global models apply fixed mass fields to precursor BVOCs (e.g., isoprene and monoterpene) emissions and directly emit this mass as primary organic aerosols (e.g., Bauer et al., 2008; Stier et al., 2005). MAM in CESM adds one more level of complexity by simulating a single lumped semivolatile organic gaseous species (SOAG) (Liu et al., 2012a, 2016). SOAG is emitted at surface with the mass flux calculated by applying fixed mass yields to five primary categories of VOCs from biogenic and anthropogenic sources. SOAG is then partitioned to the aerosol phase based on a thermodynamical approach to form SOA. These simple treatments of SOA bypass the complicated chemical reactions in the atmosphere. The gas-phase oxidation of terpenes is explicitly treated to form a secondary organic species, which is then condensed to form SOA in the GLOMAP mode (Mann et al., 2010). A two-product treatment is used in ECHAM-HAM2 to account for the oxidation of precursors from biogenic source (Zhang et al., 2012). The semivolatile oxidation products are then partitioned between the gas and aerosol phases. More sophisticated approaches like the volatility basis set (VBS) have been implemented in GCMs to represent the SOA formation (Shrivastava et al., 2015; Tilmes et al., 2019), which increase the modeled SOA burdens and generally improve the agreement with observations. The SOA condensation rate in one simulation of Shrivastava et al. (2015) is used for the SOAG emission in the E3SMv1 to replace surface SOAG emission in CESM (Wang et al., 2020). Lin et al. (2012, 2014a) used an explicit scheme in a global CTM (IMPACT) to predict the SOA formation from organic nitrates and peroxides and also from reactions of epoxides, glyoxal, and methylglyoxal on aqueous aerosols and cloud droplets.

3.1.3. Dry Deposition and Gravitational Settling

Dry deposition is an important physical process removing aerosols from the atmosphere. While wet deposition occurs episodically during precipitation events, dry deposition happens all the time and over any surfaces (Zhang & Vet, 2006). The process of dry deposition of aerosol particles is generally represented as consisting of three steps: (1) aerodynamic transport down through the atmospheric surface layer to the quasilaminar sublayer, (2) Brownian transport across the quasilaminar sublayer to the surface, and (3) uptake at the surface (Seinfeld & Pandis, 2016). Dry deposition is a complex process depending on physical and chemical properties of the aerosol, the underlying surface characteristics, and micro-meteorological conditions.

In GCMs, dry deposition is treated with an empirically parameterized dry deposition velocity (V_d). At the lowest vertical layer, V_d is parameterized using the resistance-in-series method and expressed as

$$V_d = V_g + \frac{1}{R_a + R_s} \tag{3.8}$$

where V_g is the gravitational settling velocity, and R_a and R_s are the aerodynamic and surface resistance,

Table 3.4 Aerosol dry deposition scheme in global climate models.

GCM	Aerosol module	Dry deposition scheme
CESM2	MAM4 (Liu et al., 2016)	Zhang et al., 2001
E3SMv1	MAM4 (Liu et al., 2016)	Zhang et al., 2001
AM3	Donner et al., 2011	Gallagher et al., 2002
GISS ModelE2	MATRIX (Bauer et al., 2008)	Wesely & Hicks, 1977
GEOS-5	GOCART (Chin et al., 2002)	Wesely, 1989
ECHAM6	HAM (Stier et al., 2005; Zhang et al., 2012)	Ganzeveld et al., 1998
UK Earth System Model	GLOMAP-mode (Mann et al., 2010)	Slinn, 1982; Zhang et al. 2001

respectively. At other model vertical layers, V_d is equal to V_g. V_g is expressed as

$$V_g = \frac{\rho_p d_p^2 g C_c}{18\mu} \tag{3.9}$$

where ρ_p is the particle density, d_p is the particle diameter, g is the acceleration of gravity, C_c is the Cunningham correction factor as a function of d_p and the mean free path of air molecules (λ), and μ is the viscosity coefficient of air. The large uncertainties in modeling V_d mainly occur from the parameterizations of R_a and R_s.

Table 3.4 lists the aerosol dry deposition schemes used in GCMs, which have not been updated for almost 20 years. Many GCMs adopt the Zhang et al. (2001) (Z01) dry deposition scheme, which tends to overestimate dry deposition of fine particles (with a diameter of 0.1–1.0 μm) as pointed out by Petroff and Zhang (2010). More recently, Petroff and Zhang (2010) (PZ10) developed a dry deposition scheme, which predicts much lower V_d for fine particles, especially over nonvegetated surfaces, than Z01 and is more sensitive to the change of land covers. Zhang and Shao (2014) proposed a new dry deposition scheme considering the effects of gravity and surface roughness elements on particle motion. Khan and Perlinger (2017) evaluated five aerosol dry deposition schemes with field measurements and found that the scheme by Zhang and He (2014) performs the best.

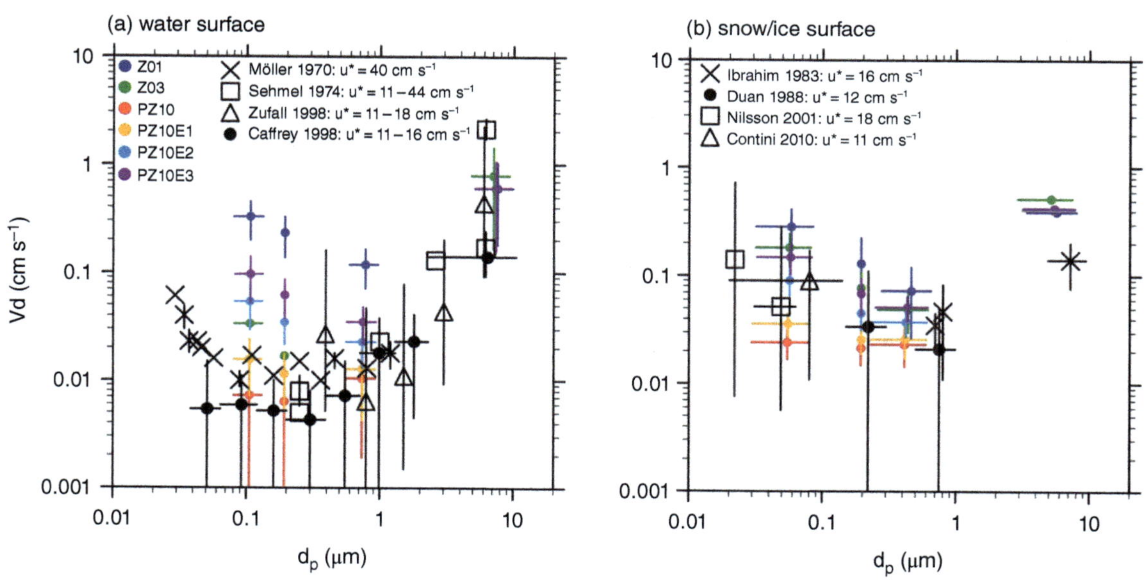

Figure 3.3 Dry deposition velocities on (a) water surfaces and (b) snow and ice surfaces from CESM and observations. Z01: Zhang et al. (2001); Z03: Zender et al. (2003a); PZ10: Petroff and Zhang (2010); PZ10E1: same as PZ10 but replacing $Sc^{-2/3}$ with $Sc^{-1/2}$ in the calculation of collection efficiency due to Brownian diffusion (E_{gb}), where Sc is the Schmidt number; PZ10E2: same as PZ10E1 but additionally applying a factor of 5 to E_{gb}; and PZ10E3: same as PZ10E1 but additionally applying a factor of 10 to E_{gb}. Observation data are from Möller and Schumann (1970), Sehmel and Sutter (1974), Zufall et al. (1998), Caffrey et al. (1998), Ibrahim et al. (1983), Duan et al. (1988), Nilsson and Rannik (2001), and Contini et al. (2010). Source: Wu et al. (2018)/John Wiley & Sons.

Currently, large uncertainties exist in modeled V_d for fine particles, which can vary over 2 orders of magnitude (Figure 3.3). Previous studies based on wind tunnel measurements suggested that fine particles should have V_d around 0.01 cm s^{-1}, whereas studies based on field observations suggested much higher V_d values. It is also worth noting that coefficients that are used to calculate V_d in GCMs, such as canopy height (h), surface roughness (z_0), and leaf area index (LAI), are not well coupled with land models and mostly prescribed. The influence of seasonality of snow/ice cover fraction on aerosol dry deposition is not well represented. Limited studies have been conducted on the evaluation of aerosol dry deposition schemes in GCMs and on the quantification of their contributions to the uncertainties of modeled aerosol spatial distributions. Wu et al. (2018) found that CAM5 with PZ10 improves the simulation of BC concentration at high latitudes and its seasonality in the Arctic due to the reduction of V_d for fine particles.

3.1.4. Wet Scavenging

Droplet activation from CCN is a primary process for cloud formation, which can be represented by the Köhler theory. Likewise, heterogeneous ice nucleation on ice nucleating particles (INPs) is an important source of ice crystals in mixed phase and ice clouds, although ice crystals can also be produced by secondary processes at temperatures warming than −15 °C (Field et al., 2017; Zhao et al., 2021a; Zhao & Liu, 2021). Ice nucleation is much more complicated than droplet activation, and the mechanisms of heterogeneous ice nucleation (e.g., deposition, immersion, condensation, and contact) (Pruppacher & Klett, 1997; Vali, 2014) are far less understood. CCN activity depends in the first order on aerosol size and secondarily on particle chemical composition (Dusek et al., 2006). On the contrary, only special types of aerosols (e.g., dust, biological aerosols, and metallic aerosols) can serve as good INPs, which are only a tiny fraction (10^{-3} to 10^{-6}) of ambient aerosol populations (DeMott et al., 2010; Hoose & Möller, 2012; Murray et al., 2012).

Once aerosol particles are nucleated into cloud hydrometeors, they are subject to wet removal when cloud hydrometeors form rain and ice/snow through cloud microphysical processes, the so-called in-cloud nucleation scavenging. When clouds totally or partially evaporate due to mixing with entrained dry air, or when precipitating hydrometeors fall in the clear air and are totally evaporated, aerosols in cloud and precipitation hydrometeors can be released into the atmosphere with different sizes from initial nucleated particles. Precipitation hydrometeors falling through the clear atmosphere can also remove aerosol particles below cloud base, the so-called below-cloud scavenging. Nucleation scavenging is the most important pathway of removing submicron particles from the atmosphere (critically depending on in-cloud updraft velocity, and thus supersaturation), while supermicron particles are removed predominantly by gravitational settling. Below-cloud scavenging of particles is most efficient for ultrafine particles with diameter (d_p) less than 0.01 μm due to Brownian diffusion and for coarse particles with d_p larger than 3 μm due to inertial impaction. The collection efficiency of accumulation mode particles by rain droplets is much lower (Slinn, 1983; Wang et al., 2010).

It is critically important for large-scale models to represent the in-cloud nucleation scavenging in order to correctly simulate the burden and lifetime of aerosol particles. Table 3.5 lists the nucleation scavenging schemes used in GCMs. Earlier CTMs (e.g., GEOS-Chem, IMPACT, GMI, and GOCART) often used the first-order scheme of Giorgi and Chameides (1986) to treat in-cloud scavenging of aerosols by stratiform and convective precipitation (Chin et al., 2000; H. Liu et al., 2001; Liu et al., 2007b). The fraction of particles removed by in-cloud scavenging depends on the aerosol scavenging efficiency, conversion rate of cloud water to precipitation, and the areal fraction of a grid box experiencing precipitation. The scavenging efficiency of aerosol is assumed to be constant depending on its solubility (e.g., 1.0 for sulfate and sea salt, and fractional values for partially soluble aerosols). The areal fraction of a grid box experiencing precipitation is parameterized as a function of constant values of cloud condensed water content, conversion rate of cloud water to precipitation, and grid-scale precipitation rates prescribed from GCMs. Additionally, aerosol scavenging in subgrid convective updrafts is treated.

In-cloud scavenging of aerosols by stratiform precipitation in NCAR CAM5 and CAM6 (Liu et al., 2012a, 2016) is treated based on Barth et al. (2000) and Rasch et al. (2000). The first-order loss rate of cloud water is calculated from the model-predicted precipitation rate and the cloud water mixing ratio. This loss rate is applied to the stratiform-cloud-borne aerosol particles explicitly treated in the model, assuming the scavenging efficiency of 1.0 for stratiform-cloud-borne aerosols. In-cloud scavenging of aerosols by convective precipitation is treated similarly but applied to interstitial aerosols (i.e., all aerosols except stratiform-cloud-borne aerosol), with the precipitation rate and the cloud water mixing ratio diagnosed from the convection parameterization. Wang et al. (2013, 2020) further improved the in-cloud aerosol scavenging by convective clouds in CAM5 and E3SMv1 by treating the convective aerosol transport and wet scavenging simultaneously and scavenging of convective cloud-borne aerosols separately from that of convective interstitial aerosols. This new treatment was shown to significantly

Table 3.5 In-Cloud aerosol scavenging schemes in global climate models.

GCM	Aerosol module	Scavenging scheme	Reference
CESM2	MAM4 (Liu et al., 2016)	$\Delta q_i = q_i f_i f_{cld} P_r / L \Delta t$, where q_i is the grid-mean mixing ratio of tracer i, f_i is the scavenging efficiency of tracer i, f_{cld} is the cloud fraction, P_r is the precipitation rate, L is the cloud water mixing ratio, and Δt is the model time step. For CESM2, E3SMv1, q_i is for liquid cloud-borne aerosols with $f_i = 1$. In-cloud scavenging of aerosols in ice clouds is not treated.	Barth et al., 2000; Rasch et al., 2000
E3SMv1	MAM4 (Liu et al., 2016)	The same as CESM2 for stratiform clouds. For convective clouds, a unified convective aerosol transport and wet scavenging is used, and scavenging of convective cloud-borne aerosols is separately treated from that of convective interstitial aerosols.	Barth et al., 2000; Rasch et al., 2000; Wang et al., 2013
ECHAM6	HAM (Stier et al., 2005; Zhang et al., 2012)	Similar to CESM2 based on the conversion rate of cloud water to rain.	Stier et al., 2005
UK Earth System Model	GLOMAP-mode (Mann et al., 2010)	Similar to CESM2 based on the conversion rate of cloud water to rain.	Mann et al., 2010; Spracklen et al., 2005
GEOS-5	GOCART (Chin et al., 2002)	$\Delta q_i = \frac{q_i f_i f_{pre} P_r}{L} \Delta t$, where q_i is the grid-mean mixing ratio of tracer i, f_i is the scavenging efficiency of tracer i, P_r is the precipitation rate, L is the cloud water mixing ratio, Δt is the model time step, and f_{pre} is the fraction of grid box experiencing precipitation: $f_{pre} = \frac{f_{pre}^0 P_r \Delta t}{L C_1 T_c}$ $C_1 = C_{1min} + P_r/L$ $C_{1min} = 10^{-4}\ s^{-1}$, and T_c is duration of precipitation over the model time step.	Giorgi & Chameides, 1986
GISS ModelE2	MATRIX (Bauer et al., 2008)	Similar to GEOS-5	Giorgi & Chameides, 1986; Koch et al., 1999
AM3	Donner et al. (2011)	Similar to GEOS-5	Giorgi & Chameides, 1986

reduce the excess BC and sea salt in the upper troposphere compared with in situ observations (Shan et al., 2021; Yu et al., 2019). Aerosol in-cloud wet scavenging closely follows the Giorgi and Chameides (1986) parameterization in GFDL AM3 (Donner et al., 2011) and GISS ModelE (Bauer et al., 2008; Koch et al., 2009). The in-cloud scavenging by stratiform clouds in ECHAM6 and HadGEM3 is more similar to that in CESM2, with the removal rate of activated aerosols proportional to the prescribed scavenging efficiency and the conversion rate of cloud condensate to precipitation (Mann et al., 2010; Stier et al., 2005).

3.1.5. Microphysical Processes

Aerosol microphysics here includes the new particle formation (NPF) due to aerosol nucleation and the particle growth due to condensation and coagulation. NPF is a major source of submicron particles in the atmosphere and thus has important impacts on global CCN number and climate (Kazil et al., 2010; Merikanto et al., 2009; Wang & Penner, 2009). For NPF, clusters and molecules are first formed, then grow in size by the condensation of, e.g., sulfuric acid gas (H_2SO_4), and are stabilized by NH_3, amines, or organic vapors (Kulmala et al., 2013). At the same time, these vapors can condense on preexisting particles, which competes with NPF for the condensable vapors. Thus, NPF may occur preferentially in the free troposphere where temperatures are colder and preexisting aerosol surface areas are lower (Clarke & Kapustin, 2010). However, a recent study found that NPF occurs in a polluted megacity in China despite high preexisting aerosol loadings (Yao et al., 2018). NPF there is a result of strong emissions of precursor gases (e.g., SO_2, NH_3, and VOCs), which favor the nucleation and growth of new particles by H_2SO_4, NH_3, and organic vapors. NPF can also occur purely due to highly oxygenated molecules independent of H_2SO_4 (Bianchi et al., 2016; Kirkby et al., 2016; Tröstl et al., 2016).

Bulk aerosol modules in GCMs do not treat aerosol microphysical processes, prescribe the aerosol size distribution, and assume the external mixing between aerosol species and instantaneous formation of sulfate aerosol from the gas-phase SO_2 oxidation. Many advanced aerosol modules treat aerosol microphysics to explicitly represent aerosol number, size distribution, and mixing state. For instance, ECHAM-HAM1 uses the binary H_2SO_4-H_2O nucleation (Vehkamäki et al., 2002), and ECHAM-HAM2 includes the charged nucleation induced by cosmic rays (Kazil et al., 2010). The GLOMAP model uses the binary H_2SO_4-H_2O nucleation (Kulmala et al., 1998). MATRIX in GISS ModelE has a few options for nucleation parameterizations, including one for the ternary H_2SO_4-NH_3-H_2O nucleation (Napari et al., 2002) and two for binary H_2SO_4-H_2O nucleation (Jaecker-Voirol & Mirabel, 1989; Vehkamäki et al., 2002). MAM includes the ternary H_2SO_4-NH_3-H_2O nucleation for MAM7 (Merikanto et al., 2007), binary H_2SO_4-H_2O nucleation for MAM3 (Vehkamäki et al., 2002), and boundary layer nucleation as the empirical first-order nucleation rate (in H_2SO_4) (Sihto et al., 2006). Until now only a few models consider SOA impacts on NPF (Zhu et al., 2019).

In the aerosol microphysics modules, the condensation of H_2SO_4, NH_3, and semivolatile organics onto preexisting particles is treated dynamically, with the first-order mass transfer coefficients calculated and accommodation coefficients given for these species. In order to account for the competition of H_2SO_4 between aerosol nucleation and condensation, ECHAM-HAM calculates the H_2SO_4 condensation first, and then the remaining H_2SO_4 is used for the nucleation of new clusters (Stier et al., 2005). Wan et al. (2013) investigated the numerical issue due to the sequential operator splitting with a long time step in a GCM and suggested that the source (chemical production) and sinks (condensation and nucleation) of H_2SO_4 should be solved at the same time to avoid the numerical errors. The coagulation between aerosol particles is treated in the aerosol microphysics modules based on the coagulation kernels for self-coagulation (i.e., collisions of particles within the same aerosol mode or bin) and intermodal or interbin coagulations (i.e., collisions of particles between different modes or bins) (Bauer et al., 2008; Liu et al., 2012a; Mann et al., 2010).

3.1.6. Aging

Aerosol particles will undergo physical and chemical processing in the atmosphere. Different aerosol species will mix with each other via the condensation of semivolatile or nonvolatile gases, or via coagulation between particles. These can be referred to as physical aging. Physical aging can alter the physical properties of aerosol particles, e.g., morphology, and solubility, as well as their optical properties, e.g., absorption of BC. Chemical aging refers to as changes in chemical composition of aerosol particles via gas-phase, heterogeneous-phase, and aqueous-phase oxidation.

Many bulk aerosol modules in GCMs without treating aerosol microphysics (e.g., condensation, and coagulation) assume a constant e-folding time (e.g., 1.2 days in Chin et al., 2002) for the aging of emitted hydrophobic carbonaceous aerosols to become hydrophilic. More sophisticated aerosol modules treat the aging process more physically as determined by the coating thickness of condensed soluble species (e.g., sulfate and SOA) on the surface of hydrophobic aerosols (Liu et al., 2005, 2012a, 2016; Mann et al., 2010; Zhang et al., 2012). For instance,

Liu et al. (2012a, 2016) calculated the fraction of carbonaceous aerosols in the primary carbon mode that have been aged based on the sulfate and SOA masses accumulated on the carbonaceous aerosols during a model time step.

Chemical aging of organic aerosols is not well represented in GCMs. Aging of SOA in the atmosphere can reduce its volatility and impact its loadings. Shrivastava et al. (2015) implemented a VBS approach to represent the multigenerational aging by fragmentation and functionalization reactions of gas phase of organics and investigated their impacts on SOA, and found a large sensitivity of SOA loadings to fragmentation reactions.

3.1.7. Water Uptake

Water-soluble or partially soluble particles can uptake water in the atmosphere when relative humidity (RH) is over a threshold depending on chemical composition of particles. This will change the sizes and optical properties of aerosol particles. The water swallowed particles will be more efficient in scattering sunlight than dry particles, significantly reducing the atmospheric visibility under humid haze environments.

Aerosol modules in GCMs treat the water uptake in different ways. Earlier models often used the Geber (1991) scheme to increase the sizes of hygroscopic aerosols as a function of RH (Liu et al., 2005, 2007b) or used look-up tables for the growth factors of particle size (Chin et al., 2002). More recent aerosol modules (e.g., Mann et al., 2010; Stier et al., 2005) treat the water uptake by each compound based on the ZSR relation (Stokes & Robinson, 1966; Zdanovskii, 1948), which calculates the binary electrolyte molalities or calculate aerosol water using the thermodynamical equilibrium schemes (EQSAM or ISORROPIA) (Bauer et al., 2008). MAM (Liu et al., 2012a, 2016) calculates the wet volume-mean radius of each aerosol mode based on the equilibrium Köhler theory (Ghan & Zaveri, 2007), with the hygroscopicity of particles in an aerosol mode calculated from volume-weighted hygroscopicities of each aerosol species in the mode. For treating organic aerosols, models often use the κ-Köhler theory (Petters & Kreidenweis, 2007) by assigning the κ values for POM and SOA derived from observations (Liu & Wang, 2010).

Aerosol water burden in 12 AeroCom II models differs by more than a factor of 10 (Figure 3.4) with mean and standard deviation of 60 ± 45 Tg, compared to 35 ± 28 Tg in the AeroCom I models (Textor et al., 2006). The factors behind these large differences include method to calculate aerosol water, aerosol dry mass, composition, size and hygroscopicity, and RH. Zhang et al. (2012) found a doubling of global aerosol water burden when the RH ceiling for calculating aerosol water is changed from 95% to 100% in ECHAM-HAM2. Therefore, resolving the subgrid variability of RH is important for aerosol water calculated in the models due to the strong nonlinearity of aerosol water uptake with RH at RH higher than 90%.

3.1.8. Aerosol Cycle and Lifetime

In the above sections, we describe the source and sink processes of atmospheric aerosols as illustrated in Figure 3.1. Here, we give the lifecycles of sulfate, nitrate, and ammonium aerosols. The atmospheric sulfur cycle

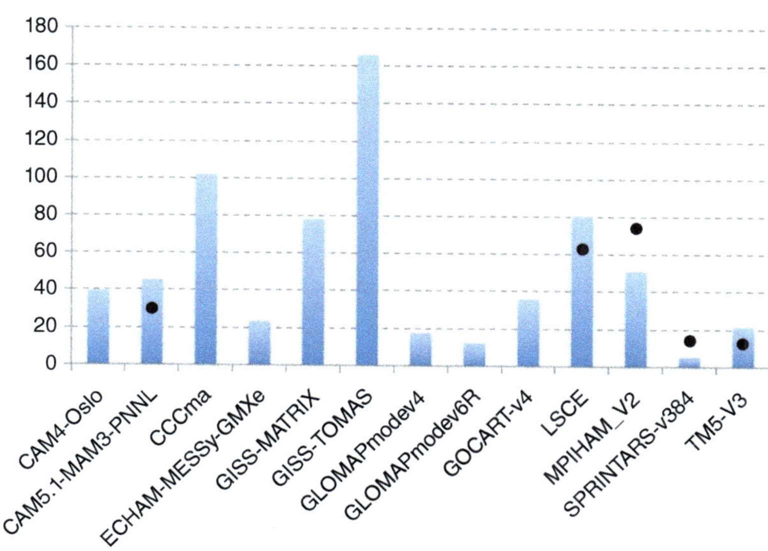

Figure 3.4 Aerosol water burden (Tg) in 12 AeroCom II models. From Kai Zhang (personal communication). The dots are results from the AeroCom I models (Textor et al., 2006).

related to tropospheric sulfate aerosol includes the following processes: (1) nearly all the DMS emitted from oceans is oxidized to form SO_2; (2) most of SO_2 from human and natural emissions (e.g., volcanic eruption, biomass burning) and from DMS oxidation is oxidized to form H_2SO_4 via gas-phase chemical reactions and to form sulfate via aqueous-phase chemical reactions, with a smaller fraction of SO_2 removed from the atmosphere via dry and wet deposition; (3) nucleation and condensation of H_2SO_4 and resuspension of sulfate in cloud water form sulfate aerosol; and (4) majority of sulfate aerosol is removed by wet deposition with a small fraction by dry deposition.

The atmospheric nitrogen cycle related to tropospheric nitrate aerosol includes the following processes: (1) NO_x emitted from sources such as industry, transportation, and energy (Table 3.1) is oxidized in the atmosphere to form HNO_3; (2) a fraction of HNO_3 is partitioned into the aerosol phase to form nitrate aerosol (NO_3) with the remaining fraction removed by dry and wet deposition; and (3) most of NO_3 aerosol is removed from the atmosphere by wet deposition with a small fraction by dry deposition. The atmospheric nitrogen cycle related to tropospheric ammonium aerosol includes the following processes: (1) NH_3 emitted from sources such as agriculture and waste (Table 3.1) is partitioned into the aerosol phase to form ammonium aerosol (NH_4) with the remaining fraction removed by dry and wet deposition; and (2) most of NH_4 aerosol is removed from the atmosphere by wet deposition with a smaller fraction by dry deposition.

Aerosol lifetimes are defined as burdens of aerosols in the atmosphere divided by their total sources or sinks. Aerosol lifetimes are determined to a large extent by aerosol processes, particularly aerosol sinks such as dry deposition/gravitational settling and wet scavenging. Atmospheric lifetimes of tropospheric aerosols are typically only a few days, substantially shorter than major GHGs (e.g., CO_2 and CH_4). Depending on their sizes, lifetimes of coarse particles (e.g., coarse dust and sea salt) are usually less than 3 days due to their efficient removal by dry deposition and gravitational settling (Easter et al., 2004; Ginoux et al., 2001; Liu et al., 2005, 2007b; Luo et al., 2003). Comparatively, fine particles are mainly removed by wet scavenging and have longer lifetimes (3–8 days) (Chin et al., 2002; Easter et al., 2004; Liu et al., 2005, 2007b; Textor et al., 2006). However, depending on their locations relative to clouds and precipitation, fine particles can have much longer lifetimes, e.g., months in the upper troposphere or years in the stratosphere (e.g., stratospheric sulfate formed from major volcanic eruptions).

Atmospheric processes can also alter the lifetimes of aerosols. For example, the hygroscopic growth of coarse sea-salt particles enhances their removal by dry deposition/gravitational settling, leading to a short lifetime (<2 days). Atmospheric aging of insoluble BC particles enables their removal by wet scavenging, thereby reducing their lifetimes in the atmosphere (Liu et al., 2005).

3.2. AEROSOL PROPERTIES

3.2.1. Particle Size Distribution and Mixing State

Comparing with gas species, aerosols are more complicated in terms of physical, chemical and optical properties: aerosol particles can vastly exist in different size (10^{-3} to 10 μm), mixing state (external vs. internal), density, shape (spherical vs. nonspherical), light scattering/absorbing, hygroscopicity, etc. The aerosol mixing state is often defined as the distribution of chemical species across a population of aerosol particles (Riemer et al., 2019), with the same mixture of chemical species in the population as the internal mixing and with all particles consisting of pure but varying chemical species in the population as the external mixing. Aerosol mixture in the real world can be much more complicated than these two extremes. Aerosol properties will determine their effects on radiation and clouds. It is a great challenge to represent the aerosol properties in GCMs due to constraints from computational resources. Table 3.6 summarizes the aerosol treatments in GCMs. The simplest bulk aerosol treatment (e.g., Chin et al., 2000; Donner et al., 2011; Emmons et al., 2010) only predicts mass mixing ratios of different aerosol species, assumes external mixing among aerosol components, and prescribes aerosol size distribution. The bulk models generally neglect the treatment of aerosol microphysics. For sea salt and dust, mass mixing ratios in several size bins are predicted.

More advanced moment-based treatment (e.g., modal and two-moment quadrature method of moments) (Bauer et al., 2008; Grandey et al., 2018; Liu et al., 2012a, 2016; Mann et al., 2010; Stier et al., 2005; Zhang et al., 2012) assumes a functional form of aerosol size distributions (e.g., log-normal). It predicts the evolution of size distributions by predicting mass (third moment) and number (zeroth moment) mixing ratios in each aerosol mode; assumes standard deviations of log-normal distributions, internal mixture within aerosol modes and external mixture between modes; and explicitly treats aerosol microphysical processes. For example, the modal aerosol module developed by Stier et al. (2005) includes seven log-normal modes: one soluble nucleation mode, two Aitken modes (one for soluble/mixed and one for insoluble POM and BC), two accumulation modes (one for soluble/mixed and one for dust), and two coarse modes (one for soluble/mixed and one

Table 3.6 Aerosol treatments in global climate models.

Aerosol treatment	Features	Aerosol module/GCM
Bulk method	Advantage: computational efficient; disadvantage: (1) only mass mixing ratios predicted, (2) no size distribution predicted, and (3) external mixing assumed	GOCART (Chin et al., 2002) in GEOS-5, Donner et al. (2011) in AM3
Modal method (two-moment)	Advantage: (1) both mass and number mixing ratios predicted, (2) both internal and external mixing considered, and (3) medium computational costs; disadvantage: shape of size distribution and modal width assumed for two-moment schemes	MAM3/4 (Liu et al., 2012a, 2016) in CESM1/CESM2/E3SMv1, HAM (Stier et al., 2005; Zhang et al., 2012) in ECHAM, GLOMAP-mode (Mann et al., 2010) in UK Earth System Model, and MATRIX (Bauer et al., 2008) in GISS ModelE
Bin method	Advantage: explicit size distribution and mixing state predicted; disadvantage: computationally expensive	CARMA (Yu et al., 2015) in CESM1 and TOMAS (Adams & Seinfeld, 2002) in GISS ModelE

for dust). Microphysical processes will age/transfer the insoluble modes to soluble/mixed modes. Liu et al. (2012a) developed a seven-mode version of MAM (MAM7), which includes Aitken, accumulation, primary carbon, fine dust, fine sea salt, coarse dust, and coarse sea-salt modes. Particles in the primary carbon mode can be transferred to the accumulation mode due to physical aging. Meanwhile, they developed a simplified three-mode version of MAM (MAM3) by assuming the instantaneous aging of primary carbon particles for the benefit of saving computational costs. Liu et al. (2016) further incorporated the primary carbon mode based on MAM3 to improve the treatment of aging of primary carbon particles. Based on the two-moment quadrature method of moments (QMOM), the MATRIX aerosol module (Bauer et al., 2008) predicts aerosol mass and number mixing ratios and size distributions for up to 16 mixed-mode aerosol populations to simulate the aerosol mixing states comprehensively.

The most accurate yet computationally expensive aerosol treatment is the bin schemes, which split the aerosol size distribution into bins, predict the evolution of size distribution by predicting mass mixing ratios of multiple aerosol species within each bin, and explicitly treat the aerosol microphysics. A sectional aerosol scheme (CARMA) implemented in CESM1 (Yu et al., 2015) separates pure sulfate particles into 20 discrete mass bins from 0.2 to 1.3 μm in radius, and internally mixed aerosols composed of OC, BC, dust, sea salt, and sea-spray sulfate into another set of 20 discrete bins from approximately 0.05–8.7 μm. The TOMAS two-moment sectional scheme (Adams & Seinfeld, 2002) predicts mass mixing ratios of several aerosol species in a varying number of bins as well as the number mixing ratio of aerosol particles in each bin.

3.2.2. Aerosol Optical Properties

Aerosol optical properties (e.g., scattering coefficient, absorption coefficient, and asymmetry parameter) within each spectral interval are calculated from look-up tables (as a function of real and imaginary parts of refractive indices, particle size, and wavelength), derived from the Mie code. Refractive index and its wavelength dependency are prescribed for each aerosol species. Aerosol single scattering albedo (SSA), the ratio of scattering coefficient to extinction coefficient, is a key parameter for assessing the radiative effects (REs) of aerosols. A value of unity for SSA represents purely scattering aerosols while absorbing aerosols (e.g., BC and dust) have SSA values smaller than unity. In bulk aerosol modules, aerosols size distributions are prescribed for each aerosol species. In modal aerosol modules, the volume-averaged refractive indices are derived for each aerosol mode assuming internal mixing of different aerosol species within a mode (including aerosol water). In some aerosol modules (e.g., Yu et al., 2015), aerosol optical properties are calculated based on the Mie core-shell code for aerosol mode (or bin), assuming soluble components (e.g., sulfate, sea salt, and aerosol water) to form the liquid shell and insoluble materials (e.g., BC and dust) to form the core. The aerosol optical properties are then used in the radiation code for the radiative forcing and heating calculations.

3.3. AEROSOL DIRECT EFFECTS

Atmospheric aerosol particles can scatter and absorb solar and terrestrial radiation, perturbing the Earth's radiation balances at top of the atmosphere (TOA), surface and in the atmosphere. These effects are called the aerosol direct radiative effects (or forcings) or radiative effects (or forcings) due to aerosol-radiation interactions (REari or RFari) (Boucher et al., 2013). Here, the direct radiative effect represents the instantaneous radiative impact of aerosols on the energy balance (Heald et al., 2014). The direct radiative forcing represents the change of radiative effect between two different periods, e.g., the preindustrial and the present-day times (Heald et al., 2014; Liu et al., 2007b). Aerosol direct radiative effect increases the diffuse fraction of solar radiation by scattering the sunlight. The increased diffuse radiation from aerosols can impact climate indirectly, e.g., by enhancing the rate of plant photosynthesis, leading to increased land carbon sink (Mercado et al., 2009).

RFari of various aerosol types has been extensively studied as reviewed in Myhre et al. (2013a, 2013b). The uncertainty range of anthropogenic aerosol RFari estimated from the IPCC's Second to the Fifth Assessment Report has been increasing despite the substantially improved understanding of aerosol processes and properties in the past 30 years. This is a result of enhanced complexity of aerosol representations in GCMs with many substantial factors contributing to the uncertainty, such as size distribution, mixing state, hygroscopicity, microphysical processes, and removal processes as reviewed in sections 3.1 and 3.2. For example, anthropogenic aerosol RFari was $-0.50\,W\ m^{-2}$ (-0.90 to -0.10 at the 90% confidence level) in the IPCC's Fourth Assessment Report (AR4). The forcing was shifted toward slightly less negative with a larger uncertainty range of $-0.45\,W\,m^{-2}$ (-0.95 to $+0.05$ at the 90% confidence level) in the IPCC AR5. We note that AR5 used the effective radiative forcing including rapid adjustments. A recent assessment gave more negative forcings (-0.71 to -0.14 W m^{-2} at the 90% confidence level) due to downward revised aerosol absorption related to BC (Bellouin et al., 2020). In addition to anthropogenic aerosol RFari, AR5 also estimated the RFari of natural aerosols (e.g., volcanic aerosols). The radiative effect of volcanic aerosols is highly episodic and is $-0.11\,W\ m^{-2}$ (-0.15 to -0.08) during the period of 2008–2011 as compared to 1750 and $-0.06\,W\,m^{-2}$ (-0.08 to -0.04) as compared to the volcanic quiescent period of 1999–2002 (Myhre et al., 2013a).

Below we only summarize the latest studies related to biomass burning aerosols, dust, and brown carbon since RFari of many other aerosol species (e.g., BC) have been extensively reviewed (Bellouin et al., 2020; Bond et al., 2013; Myhre et al., 2013a, 2013b).

3.3.1. Biomass Burning Aerosols

The TOA REari of biomass burning aerosol was estimated to be $0.18 \pm 0.10\,W\ m^{-2}$ averaged for the period of 1997 to 2009 (Tosca et al., 2013) using CAM4 with the bulk aerosol module. Ward et al. (2012) reported an estimate of 0.13 and $-0.27\,W\,m^{-2}$ for the year 2000 under all-sky and clear-sky conditions, respectively. Jiang et al. (2016) estimated the REari of open fire aerosols to be $0.16 \pm 0.01\,W\ m^{-2}$ using CAM5 with MAM4 and the Ghan (2013) method for diagnosing REari under all-sky conditions. They also estimated the fire BC REari to be $0.25 \pm 0.01\,W\ m^{-2}$ and that of POM to be -0.05 and $0.04 \pm 0.01\,W\ m^{-2}$ based on two different methods. The maximum radiative effect of smoke aerosols is found in S.E. Atlantic west of southern Africa ($\sim 5.0\,W\ m^{-2}$) where dense smokes transported from southern Africa overlap with semipermanent stratocumulus decks over the ocean. Brown et al. (2021) found varying degrees of overestimation in biomass burning aerosol absorption by 9 models participating in the IPCC AR5 compared with a large collection of observational data sets. The REari of the CAM5 model changes from $0.059\,W\,m^{-2}$ to $-0.011\,W\,m^{-2}$ when the model is constrained by observed biomass burning aerosol properties.

When smoke aerosols are transported over the top of bright clouds, they can induce a significant positive (warming) radiative effect at TOA. The relative importance of this effect in comparison with the negative (cooling) effect of smoke aerosols in the clear sky is an important factor for determining the radiative energy budget of the S.E. Atlantic region (Chand et al., 2009). However, despite this qualitative understanding, there is a large spread of cloudy-sky and/or all-sky radiative effects of smoke aerosols in the S.E. Atlantic region reported in the literature. Figure 3.5 shows the TOA all-sky smoke aerosol REari over S.E. Atlantic among eight global models that participated in the AeroCom-Phase II project (Myhre et al., 2013b). Over the Ascension Island region, where the DOE Layered Atlantic Smoke Interactions with Clouds (LASIC) campaign was conducted, the model estimated REari dramatically varies from about -0.6 to $+0.6\,W\ m^{-2}$. Many factors may contribute to these intermodel differences. First of all, there is still a lack of observational constraints due to the large uncertainty in the observational estimates of smoke REari. Moreover, in order to correctly simulate the all-sky radiative effects of smoke aerosols in the S.E. Atlantic region, a model has to properly simulate not only the properties of smoke aerosols, including their absorption and vertical positions of smoke plumes, but also the properties of marine boundary layer (MBL) clouds underneath. However, simulating MBL clouds has proved to be highly challenging, especially for GCMs with 100–200 km horizontal spacings (Kay et al., 2012; Zhang et al., 2005). As shown in Stier et al. (2013), the intermodel difference of cloud fraction is an important factor causing the intermodel difference of the radiative effects of absorbing aerosols, especially over S.E. Atlantic.

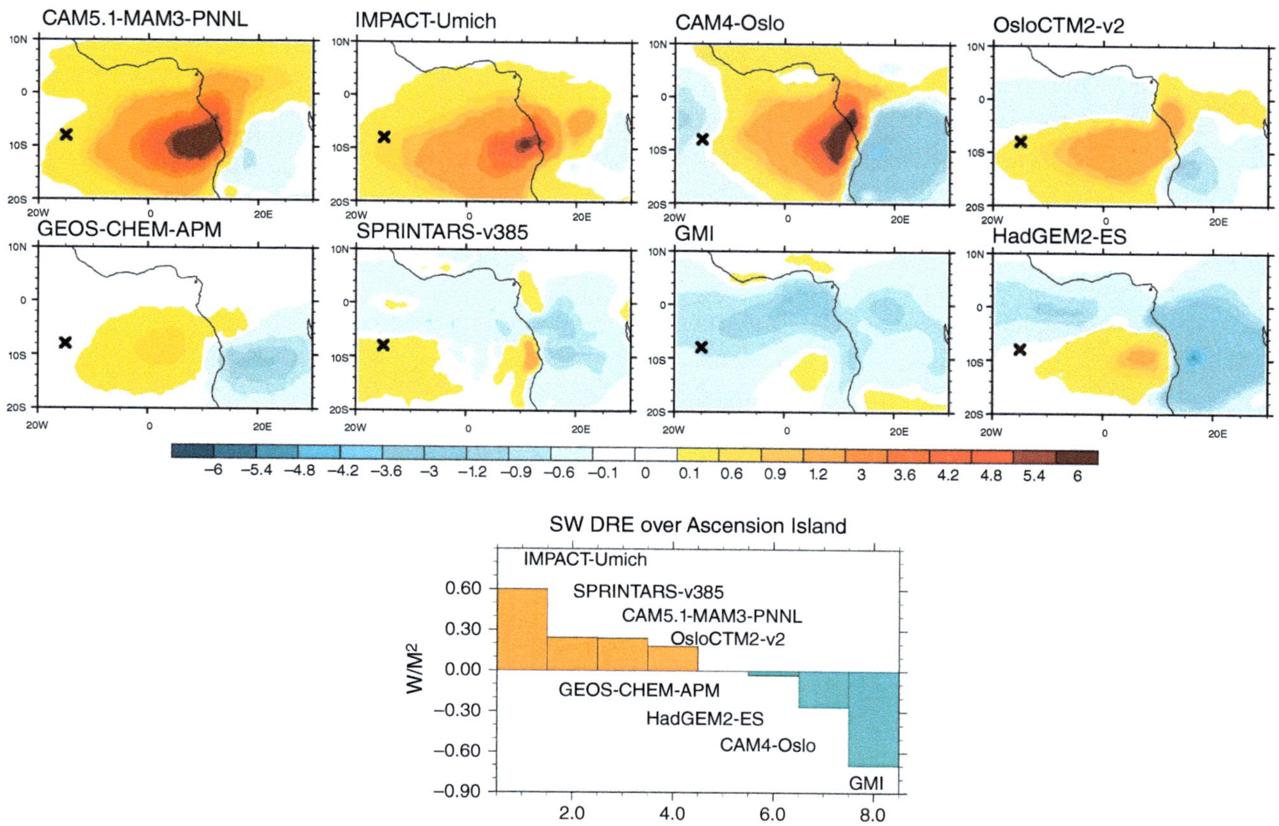

Figure 3.5 Shortwave all-sky DREs of biomass burning aerosols during the fire season (July–October) estimated by eight global models that participated in AeroCom-Phase II project. The right panel shows the seasonal mean of SW all-sky DRE over the Ascension Island (14.5°W, 8°S) ("×" in the left panels) as modeled by these eight models.

3.3.2. Dust

The REari of mineral dust depends on both the optical properties and the size distribution of dust. Large uncertainties exist in the imaginary part of the dust refractive index corresponding to the absorption, which can vary by 1 order of magnitude (Balkanski et al., 2007). Colarco et al. (2014) found that changes in dust refractive index have much stronger impacts on dust REari than the dust particle shape using the NASA GEOS-5. The mineralogical components of dust that absorb solar radiation, such as hematite and goethite, may determine the sign of dust REari. Scanza et al. (2015) speciated dust into eight minerals in CAM5 and found that including dust mineralogy changes the sign of the dust REari from a negative value (-0.17 W m^{-2}) to a small positive value ($+0.05$ W m^{-2}).

The size distribution of emitted mineral dust is poorly constrained. Kok (2011) found that current GCMs produce too much fine dust compared with observed dust size distributions. A theoretical expression of the emitted dust size distribution based on the brittle fragmentation theory was proposed and agrees well with measurements. According to this theory, the mass fraction of emitted fine dust particles (diameter < 1 μm) is 1.1%. However, 3.2% of the dust mass is emitted in the accumulation mode (diameter < 1 μm) in CESM (Zender et al., 2003a). Kok et al. (2017) suggested that the dust REari should have less cooling than -0.46 W m^{-2} (-0.78 to -0.03) estimated by current GCMs and reported an estimate of -0.20 W m^{-2} (-0.48 to $+0.20$), which includes the potential warming effect of dust. Few GCMs consider super-coarse dust particles with diameter larger than 10 μm, which can strongly influence the dust warming effect especially near source regions.

3.3.3. Brown Carbon

Absorbing organic carbon—also known as BrC—has been a new addition to CTMs and GCMs within the past 10 years. BrC is characterized by a spectrally dependent absorption at shorter wavelengths and a stronger contribution of primary particles from biomass burning compared to fossil fuel combustion (Kirchstetter et al., 2004; Saleh et al., 2014). The emission of BrC varies as a function of burn conditions with more flaming biomass fires (i.e., grassfires; higher [BC]/[OA] ratio) contributing to stronger absorbing BrC than more smoldering fires (i.e., boreal forests; lower [BC]/[OA] ratio) (Lu et al., 2015; Saleh et al., 2014). Secondary BrC formation can occur with photooxidation of volatile organic aerosols in addition to contributions from within-cloud, aqueous-phase chemical processing (Ervens et al., 2011; Nakayama et al., 2013). The higher concentration of OA in the atmosphere compared to BC (>3 times) indicates a large potential forcing for this aerosol in the atmosphere. However, this forcing is complicated by the short lifetime of BrC. Field observations of absorption half-life time range from 9 hours to 1 day (Forrister et al., 2015; Wang et al., 2016), with shorter lifetimes—measured in the laboratory—ranging from minutes to hours (Zhao et al., 2015). This aging (i.e., "bleaching") is attributed to photochemical reactions, accelerated by the presence of OH radical and decelerated by the presence of NO_x (Lee et al., 2014; Zhao et al., 2015; Zhong & Jang, 2014).

Early modeling studies of BrC using CTMs did not take into account the bleaching of BrC and assumed moderately and strongly absorbing BrC to take into account the variation in BrC absorption with burn type. Feng et al. (2013) assumed that 66% of biomass burning and biofuel emissions were associated with BrC, while Lin et al. (2014b) assumed that all SOA and biomass burning POA in the model was associated with BrC. The global BrC REari at TOA ranged from +0.04 to +0.11 W m^{-2} in Feng et al. (2013). The RFari (PD-PI) due to BrC absorption in the atmosphere ranged from +0.22 to +0.57 W m^{-2} in Lin et al. (2014b). Saleh et al. (2015) incorporated the first BrC absorption parameterization that was dependent on the emission BC/OA ratio, reporting a BrC REari of +0.12 W m^{-2} (for the case of internal mixture with BC). Wang et al. (2018) included both the aforementioned parameterization as well as a new BrC bleaching parameterization that describes a global half-life time of BrC of around 0.5 days. This aging is dependent on atmospheric [OH] in the model and reduces the global REari of BrC from +0.10 to +0.05 W m^{-2}. All the modeling studies mentioned above have been conducted with CTMs coupled to radiative transfer models. Brown et al. (2018) incorporated both the Saleh et al. (2015) BrC absorption and the Wang et al. (2018) bleaching parameterizations for biomass and biofuel burning aerosols in CESM1, which allows for more complex feedback effects than available in CTMs. This study reported similar global BrC REari as the CTM studies with a range of +0.06 W m^{2} (BrC bleaching) to +0.13 W m^{-2} (no BrC bleaching).

3.4. AEROSOL INDIRECT EFFECTS

In addition to the aerosol effects on climate directly through interactions with radiation, atmospheric aerosols can influence climate indirectly through interactions with cloud microphysics. Aerosol-cloud interactions are two-way processes: on the one hand, aerosols affect the number and size of cloud hydrometeors by acting as CCN and INPs (i.e., first indirect effects), and subsequent changes in cloud water amount, macrophysics, and lifetime of clouds (i.e., second indirect effects); on the other hand, activated aerosols in cloud droplets and ice crystals (i.e., cloud-borne aerosols) will be subject to in-cloud nucleation scavenging when cloud water is converted to precipitation via cloud microphysics. These two-way interactions determine the short lifetime (3–8 days on the global mean) of submicron aerosols in the troposphere. Therefore, it is critically important to treat interactively the fully coupled nature of aerosol-cloud interactions in GCMs. Large uncertainties still persist in the estimate of anthropogenic aerosol radiative forcing due to aerosol-cloud interaction (RFaci). The estimate of anthropogenic aerosol RFaci from the IPCC AR5 ranges from 0 to –1.2 W m^{-2}, together with anthropogenic aerosol RFari of –0.95 to +0.05 W m^{-2}, resulting in a total anthropogenic RF (including GHGs) over the industrial era of 1.1–3.3 W m^{-2} (Boucher et al., 2013; Myhre et al., 2013a).

3.4.1. Aerosol Indirect Effects on Warm Clouds

Cloud droplets form on aerosol particles that serve as CCN under typical supersaturations in the atmosphere. Thus, aerosol particles are a necessary requirement for the formation of clouds. The aerosol activation (or droplet nucleation) can be described by the Köhler theory. The equilibrium Köhler theory determines the equilibrium wet diameter of a particle as a function of its dry diameter for a given RH (Figure 3.6). By solving the maximum of Köhler curves, the critical supersaturation (S_c) can be found, and the particles with the critical supersaturation less than the given supersaturation in a cloud parcel will form cloud droplets. Generally, larger dry particles with higher hygroscopicities will have lower critical supersaturations, and thus they more readily form cloud droplets. The remaining question is then how to decide the history of supersaturation (e.g., the maximum supersaturation) in cloud parcels. The dynamic Köhler theory consists of a set of equations to describe in a cloud parcel the time evolutions of particle size distribution, supersaturation in the parcel due to cooling from upward motions, and the liquid water mixing ratio due to the condensational growth of cloud droplets (Pruppacher & Klett, 1997; Seinfeld & Pandis, 2016).

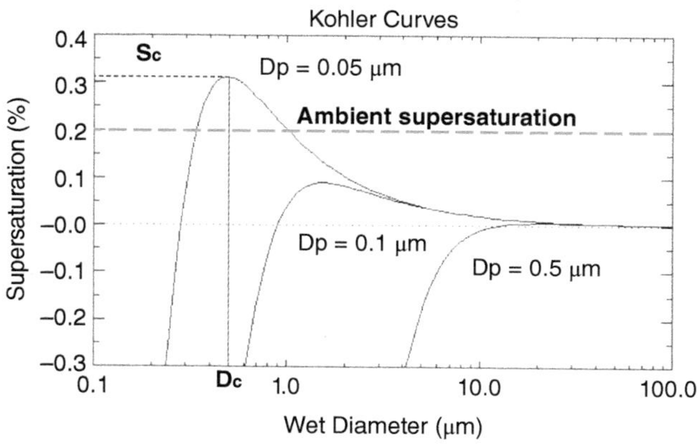

Figure 3.6 Supersaturation as a function of equilibrium wet diameter for particles with three different dry diameters. Black dashed lines correspond to critical supersaturation (S_c) and critical wet diameter (D_c). Green dashed line corresponds to ambient supersaturation (S) in a cloud parcel. Only particles with $S_c \leq S$ are activated into cloud droplets.

Based on the Köhler theory, an increase of CCN number concentration, e.g., due to anthropogenic pollution, can lead to the formation of more cloud droplets at a given updraft velocity. Due to the competition of water vapor in the cloud parcel, these droplets will be smaller. The increase in cloud droplet number concentration (CNDC) and decrease in droplet size will increase the optical depth and albedo of clouds under the same liquid water amount, the so called Twomey effect/first indirect effect/cloud albedo effect of aerosol (Twomey, 1974, 1977). Once cloud droplets are formed, they can grow by the condensation of water vapor and by collision-coalescence, to form rain drops. Autoconversion refers to the collision-coalescence between cloud drops that forms raindrops, while the collection of cloud drops by raindrops is called the accretion process. Cloud droplets with smaller sizes, e.g., due to anthropogenic pollution, take more time to coalescence to raindrops, and thus the autoconversion rate is reduced. This suppresses the precipitation formation and increases the liquid water content, and leads to subsequent changes in cloud macrophysics, e.g., increases of cloud height (Pincus & Baker, 1994) and cloud lifetime (Albrecht, 1989). These climatic effects of aerosol through modifying the cloud properties can be termed as the second indirect effect/cloud lifetime effect/Albrecht effect.

GCMs have generally implemented the treatment of aerosol indirect effects through warm clouds. Many GCMs with one-moment cloud microphysics (i.e., only predicting mass mixing ratios of cloud hydrometeors) use empirical parameterizations (Boucher & Lohmann, 1995; Menon et al., 2002) to diagnose CDNC from model predicted mass concentrations of aerosols (e.g., sulfate, OC, and sea salt). The CDNC is then used in the calculation of cloud droplet effective radius for the first aerosol indirect effect and in the calculation of autoconversion rate for the second aerosol indirect effect (Menon et al., 2002; Rasch & Kristjánsson, 1998). GCMs with two-moment cloud microphysics predict the number mixing ratios of cloud hydrometeors in addition to their mass mixing ratios (Lohmann et al., 2007; Morrison & Gettelman, 2008). Droplet nucleation is a source term of the number mixing ratio of cloud droplets, which is calculated from the physically based parameterization based on the dynamic Köhler theory (e.g., Abdul-Razzak & Ghan, 2000; Ming et al., 2006; Nenes & Seinfeld, 2003). Ghan et al. (2011) compared several droplet nucleation parameterizations widely used in GCMs and found that the Abdul-Razzak and Ghan scheme is more sensitive to increases in aerosol concentration than the Nenes and Seinfeld scheme, which results in a 10% difference (0.2 W m^{-2}) in aerosol indirect forcing. The subgrid updraft velocity driving these parameterizations is often diagnosed from the turbulent kinetic energy (Lohmann, 2002). A lower bound is often applied to this subgrid updraft velocity to account for its uncertainty. Some GCMs prescribe a lower bound to model-predicted CDNC, which can suppress the aerosol indirect effect by influencing the CDNC more in the preindustrial times than in the present-day times (Hoose et al., 2009).

GCMs treat the second aerosol indirect effect by using an autoconversion parameterization, which inversely depends on CDNC (Beheng, 1994; Khairoutdinov & Kogan, 2000; Liu & Daum, 2004). GCMs generally overestimate the response of LWC to aerosol perturbation (Malavelle et al., 2017; Wang et al., 2012). The reasons behind it may be that autoconversion parameterizations used in GCMs are subject to significant uncertainty (Wood, 2005), and GCMs may overestimate the contribution of autoconversion relative to accretion to precipitation formation (Gettelman et al., 2013). Precipitation susceptibility to aerosol (S_{pop}), a proxy for aerosol second indirect forcing (Wang et al., 2012), is much reduced, with the total global annual mean aerosol indirect forcing reduced by 0.26 W m^{-2} (−15%), when the model horizontal resolution increases from 2° to 0.25° (Ma et al., 2015). However, modeled S_{pop} is still higher than observationally constrained values even at 0.25° resolution, indicating the possibility of further reducing aerosol indirect forcing at higher model resolutions.

3.4.2. Aerosol Indirect Effect on Mixed-Phase and Ice Clouds

Aerosol particles are necessary for the formation of ice crystals in cold clouds with temperatures ranging from of 0 to −37 °C. Heterogeneous ice nucleation on INPs can proceed through multiple mechanisms, e.g., deposition, condensation/immersion, and contact freezing (Pruppacher & Klett, 1997). The ice nucleation process is much more complicated and far less understood compared to droplet nucleation. In mixed-phase clouds where cloud droplets and ice crystals coexist, immersion freezing of cloud droplets has been found to be the dominant mechanism for ice crystal formation (de Boer et al., 2011; Westbrook & Illingworth, 2013). At temperatures below −37 °C, cloud droplets freeze spontaneously to form ice crystals. Ice nucleation in cirrus clouds at temperatures below −37 °C can proceed through the heterogeneous nucleation on INPs (e.g., immersion, deposition) and homogeneous freezing of solution droplets. It is still unclear what is the dominant mechanism for ice nucleation in the upper tropospheric cirrus. Cziczo et al. (2013) found the dominant role of heterogeneous nucleation based on the analysis of ice residuals collected in several field campaigns in the tropics and midlatitudes of the Northern Hemisphere.

Mineral dust is generally regarded as the most efficient INPs at temperatures colder than −15 °C (Hoose & Möhler, 2012; Murray et al., 2012). The role of black carbon (soot) is still controversial and may only be important at temperatures colder than −30 °C (Kärcher et al., 2007). Organic aerosols in the glassy state may serve as INPs (Knopf et al., 2018; Murray et al., 2010). Metallic (especially lead-containing) aerosols (Cziczo et al., 2009, 2013) and solid ammonium sulfate (Abbatt et al., 2006) mainly from anthropogenic sources are also suggested to be good INPs in cirrus clouds. B. Zhao et al. (2019) found strong evidence that aerosols from anthropogenic pollution increase the effective radius of ice crystals near the cloud top by serving as INPs in moderate convective clouds based on top-down analyses of satellite observations. Recent studies indicate that marine organic aerosol can act as INPs and may be an important source of INPs in remote marine regions such as the Southern Ocean (Burrows et al., 2013; DeMott et al., 2016; Wilson et al., 2015; Zhao et al., 2021b).

Heterogeneous ice nucleation on INPs displays both random (stochastic or time-dependent) and deterministic (surface-specific or time-independent) features. The dominance of these different factors remains a subject of debate as does the need to parameterize both factors in models to predict ice formation (Murray et al., 2011; Niedermeier et al., 2014; Vali, 2008, 2014). Classical nucleation theory (CNT), which describes ice nucleation rate based on the stochastic approach, requires information on the variability of INP surface properties. Recent efforts to incorporate the probability distribution of INP surface properties in CNT have used measurements for this purpose (Lüönd et al., 2010; Marcolli et al., 2007). Earlier empirical and deterministic INP parameterizations (e.g., Meyers et al., 1992) relate INP number concentrations to temperature and/or supersaturation and have no link to aerosol properties, and may significantly overestimate INP number in clean environments, e.g., the Arctic (Prenni et al., 2007). More recent deterministic parameterizations link INP number concentrations to aerosol composition and surface area (Niemand et al., 2012; Phillips et al., 2008) or to number concentrations of aerosol particles larger than 0.5 μm (DeMott et al., 2010, 2015). Knopf and Alpert (2013) presented a physical model of immersion freezing that depends on water activity (i.e., independent of the nature of solutes) and INP surface area. These parameterizations have been implemented in GCMs to represent the heterogeneous ice nucleation in different modes (immersion/condensation and deposition) (Liu et al., 2007a; Shi & Liu, 2019; Y. Wang et al., 2014b; Wang and Liu, 2014).

Physically based ice nucleation parameterizations with link to aerosol properties have been implemented in GCMs to quantify aerosol effects on cirrus clouds and climate forcing (Barahona & Nenes, 2009; Kärcher et al., 2006; Liu & Penner, 2005). These parameterizations consider both homogeneous and heterogeneous nucleation, and the competition between the two pathways for ice formation in cirrus clouds, and are driven by the vertical updraft or cooling rate. The subgrid updraft velocity is often diagnosed from the turbulent kinetic energy, which should also account for the influence from temperature and water vapor fluctuations (Penner et al., 2018; Wang & Penner, 2010; Y. Wang et al., 2014b) due to gravity waves. Unlike the treatment of droplet nucleation in warm clouds, wet removal of aerosols due to ice nucleation is generally not treated in the models. Much progress has been made in the estimate of aerosol radiative forcing (effect) through cirrus and mixed-phase clouds (e.g., dust INP), indicating appreciable effects on the climate system (Gettelman et al., 2012; Liu et al., 2009; Liu et al., 2012b; Liu & Shi, 2018; Lohmann, 2002; Penner et al., 2018; Shi & Liu, 2019). However, large uncertainties are associated with the estimate call for more research studies on INP abundance and properties, and multiscale wave dynamics.

3.4.3. Aerosol Indirect Effect on Convective Clouds

Aerosol can affect convective cloud microphysics by acting as CCN and INPs, and thus exert a radiative forcing on climate. An invigoration effect is hypothesized by Rosenfeld et al. (2008), in which aerosol pollution inhibits the warm rain formation and thus allows more liquid water to be transported above the freezing level so as to release more latent heat. Fan et al. (2018) found another mechanism of aerosol invigoration effect by which ultrafine pollution aerosols lead to the condensational growth of numerous cloud droplets and thus release more latent heat based on high-resolution modeling and observational data analyses obtained over the Amazon. Profound aerosol effects on climate can also be realized by enlarging sizes of convective anvils (Fan et al., 2013), which may constitute a warming effect on climate. B. Zhao et al. (2019) found that dust and pollution aerosols could increase the effective radius of ice crystals near the cloud top by facilitating the freezing of cloud droplets in convective clouds with moderate strengths.

Only a few GCMs consider the physically based microphysics in convective clouds (Lin et al., 2021; Lohmann, 2008; Song & Zhang, 2011). The detailed convective cloud microphysics allows more physical representations of rain and snow formation and convective detrainment. More observations are needed to constrain the simulated convective cloud microphysics and dynamics (e.g., anvil properties, convective mass flux, and vertical velocity) so as to have an accurate estimate of aerosol effects on convective clouds.

3.5. SUMMARY AND OUTLOOK

GCMs have started the implementation of detailed aerosol and cloud microphysics parameterizations to enable the treatment of aerosol direct and indirect effects on climate. Despite tremendous efforts in studying aerosols and aerosol effects on climate in the past several decades, large uncertainties still persist in the estimate of aerosol radiative forcing. The reasons behind this can be due to limited understandings of some of the key aerosol and cloud processes, as well as their representations in GCMs. Here, we highlight some of them and advocate urgent actions (1) to improve emissions of primary aerosols, e.g., size distributions and injection heights of wildfire smokes, and emissions of primary marine organics; (2) to understand the formation and aging of SOA and roles of SOA in the nucleation and growth of new particles, as well as their parameterizations in GCMs; (3) to transfer knowledge of SOA formation and aging into computationally efficient representation in GCMs; (4) to improve the treatment of aerosol wet scavenging in GCMs, especially transport and wet scavenging of aerosols by convection, which are critical for aerosol vertical profiles and long-range transport; and (5) to understand ice nucleation mechanisms and their relationship to overall aerosol properties and environmental conditions, and to investigate ice nucleation effects on mixed phase and cirrus clouds that may constitute a warming climate effect, and thus partially cancel out some of the cooling from aerosol effects on warm clouds.

We note that many aerosol and aerosol-cloud interaction processes occur at subgrid scales in current GCMs (with a horizontal resolution of 0.5°–1°). Increasing spatial resolutions of GCMs has the potential to better simulate aerosol concentrations and spatial distributions (Ma et al., 2014; Qian et al., 2010), and to reduce aerosol indirect forcing (Ma et al., 2015). Therefore, with the rapid increase of computational power, high-resolution GCMs hold the potential of better simulating aerosol processes and properties and reducing the uncertainties in the estimates of aerosol RFari and RFaci.

Related to anthropogenic aerosol RF, one characteristic of climate system is climate sensitivity, which is defined as the global mean surface air temperature change in response to changes in radiative forcing, for example, due to a doubling of CO_2, the so-called equilibrium climate sensitivity (ECS). Large spread in the climate sensitivity from ESMs has persisted since 1970s. The IPCC AR5 estimates that "the ECS is likely between 1.5 °C and 4.5 °C and very unlikely greater than 6 °C" (Bindoff & Stottt, 2013). Cloud feedback in ESMs is the major contributor to the uncertainty of ECS. With anthropogenic aerosols counteracting a significant fraction of warming effects of GHGs since the industrial era, future reduction of aerosol emissions implies an accelerated warming from GHG increases. However, projections of future climate changes are still highly uncertain due to large uncertainties in climate sensitivity in ESMs (Andreae et al., 2005; A. Zhao et al., 2019).

ACKNOWLEDGMENTS

The author (X.L.) would like to acknowledge the funding supports from the "Enabling Aerosol-cloud interactions at GLobal convection-permitting scalES (EAGLES)" project (74358) funded by the Department of Energy (DOE) Office of Science Biological and Environmental Research, Earth System Model Development (ESMD) program and also the DOE Regional and Global Model Analysis (RGMA) Program (Grant DE-Sc0322065). He also thanks Tianyi Fan, Ziming Ke, Hunter Brown, Zheng Lu, Chenglai Wu, and Minguxuan Wu for their help with writing some of the sections.

REFERENCES

Abbatt, J. P. D., Benz, S., Cziczo, D. J., Kanji, Z., Lohmann, U., & Möhler, O. (2006). Solid ammonium sulfate aerosols as ice nuclei: A pathway for cirrus cloud formation. *Science*, *313*, 1770–1773. doi:10.1126/science.1129726.

Abdul-Razzak, H., & Ghan, S. J. (2000). A parameterization of aerosol activation 2. Multiple aerosol types. *Journal of Geophysical Research*, *105*(D5), 6837–6844. doi: 10.1029/1999JD901161.

Adams, P. J., & Seinfeld, J. H. (2002). Predicting global aerosol size distributions in general circulation models. *Journal of*

Geophysical Research: Atmospheres, *107*, 4370, doi:10.1029/2001JD001010.

Akagi, S. K., Yokelson, R. J., Wiedinmyer, C., Alvarado, M. J., Reid, J. S., Karl, T., et al. (2011). Emission factors for open and domestic biomass burning for use in atmospheric models. *Atmospheric Chemistry and Physics, 11*(9), 4039–4072. doi:10.5194/acp-11-4039-2011.

Albrecht, B. A. (1989). Aerosols, cloud microphysics, and fractional cloudiness. *Science*, *245*(4923), 1227–1230.

Amann, M., Bertok, I., Borken-Kleefeld, J., Cofala, J., Heyes, C., Höglund-Isaksson, L., et al. (2011). Cost-effective control of air quality and greenhouse gases in Europe: Modeling and policy applications. *Environmental Modelling & Software*, *26*, 1489–1501. https://doi.org/10.1016/j.envsoft.2011.07.012.

Andela, N., Morton, D. C., Giglio, L., Chen, Y., van der Werf, G. R., Kasibhatla, P. S., et al. (2017). A human-driven decline in global burned area. *Science*, (80-.), *356*(6345), 1356–1362. doi:10.1126/science.aal4108.

Andela, N., & van der Werf, G. (2014). Recent trends in African fires driven by cropland expansion and El Niño to La Niña transition. *Nature Climate Change*, *4*, 791–795. https://doi.org/10.1038/nclimate2313.

Andreae, M. O., & Merlet, P. (2001). Emission of trace gases and aerosols from biomass burning. *Biogeochemistry*, *15*(4), 955–966.

Andreae, M. O., Jones, C., & Cox, P. (2005). Strong present-day aerosol cooling implies a hot future. *Nature*, *435*, 1187–1190. https://doi.org/10.1038/nature03671.

Andres, R. J., & Kasgnoc, A. D. (1998). A time-averaged inventory of subaerial volcanic sulfur emissions, *Journal of Geophysical Research*, *103*(D19), 25,251–25,261. doi:10.1029/98JD02091.

Balkanski, Y., Schulz, M., Claquin, T., & Guibert, S. (2007). Reevaluation of mineral aerosol radiative forcings suggests a better agreement with satellite and AERONET data. *Atmospheric Chemistry and Physics, 7*, 81–95. https://doi.org/10.5194/acp-7-81-2007.

Barahona, D., & Nenes, A. (2009). Parameterizing the competition between homogeneous and heterogeneous freezing in ice cloud formation—Polydisperse ice nuclei. *Atmospheric Chemistry and Physics, 9*(2), 369–381. https://doi.org/10.5194/acp-9-369-2009

Barth, M. C., Rasch, P. J., Kiehl, J. T., Benkovitz, C. M., & Schwartz, S. E. (2000). Sulfur chemistry in the National Center for Atmospheric Research Community Climate Model: Description, evaluation, features and sensitivity to aqueous chemistry. *Journal of Geophysical Research*, *105*, 1387–1415.

Bauer, S. E., Wright, D. L., Koch, D., Lewis, E. R., McGraw, R., Chang, L.-S., et al. (2008). MATRIX (Multiconfiguration Aerosol TRacker of mIXing state): An aerosol microphysical module for global atmospheric models. *Atmospheric Chemistry and Physics, 8*, 6003–6035, https://doi.org/10.5194/acp-8-6003-2008.

Beheng, K. D. (1994). A parameterization of warm cloud microphysical conversion processes. *Atmospheric Research*, *33*, 193–206.

Bellouin, N., Quaas, J., Gryspeerdt, E., Kinne, S., Stier, P., Watson-Parris, D., et al. (2020). Bounding global aerosol radiative forcing of climate change. *Reviews of Geophysics*, *58*, e2019RG000660. https://doi.org/10.1029/2019RG000660.

Bian, H., Chin, M., Hauglustaine, D. A., Schulz, M., Myhre, G., Bauer, S. E., et al. (2017). Investigation of global particulate nitrate from the AeroCom phase III experiment, *Atmospheric Chemistry and Physics, 17*, 12911–12940. https://doi.org/10.5194/acp-17-12911-2017.

Bianchi, F., Tröstl, J., Junninen, H., Frege, C., Henne, S., Hoyle, C. R., et al. (2016). New particle formation in the free troposphere: A question of chemistry and timing. *Science*, *352*, 1109–1112. 10.1126/science.aad5456.

Bindoff, N. L., & Stott, P. A. (2013). 10.8.2 Constraints on long-term climate change and the equilibrium climate sensitivity. *Climate Change 2013: The Physical Science Basis—IPCC Working Group I Contribution to AR5*. Geneva, Switzerland.

Bond, T. C., Doherty, S. J., Fahey, D. W., Forster, P. M., Berntsen, T., DeAngelo, B. J., et al. (2013). Bounding the role of black carbon in the climate system: A scientific assessment. *Journal of Geophysical Research: Atmospheres*, *118*, 5380–5552.

Boucher, O., & Lohmann, U. (1995). The sulphate-CCN-cloud albedo effect—A sensitivity study with two general circulation models. *Tellus B*, *47*, 281–300.

Boucher, O., Randall, D., Artaxo, P., Bretherton, C., Feingold, G., Forster, P., et al. (2013). Clouds and Aerosols. In: *Climate Change 2013: The Physical Science Basis. Contribution of Working Group I to the Fifth Assessment Report of the Intergovernmental Panel on Climate Change* [Stocker, T. F., Qin, D., Plattner, G.-K., Tignor, M., Allen, S. K., Boschung, J., et al. (Eds.)]. Cambridge University Press: Cambridge, United Kingdom and New York, NY, USA.

Brown, H., Liu, X., Feng, Y., Jiang, Y., Wu, M., Lu, Z., et al. (2018). Radiative effect and climate impacts of brown carbon with the Community Atmosphere Model (CAM5). *Atmospheric Chemistry and Physics, 18*(24), 17745–17768. doi:10.5194/acp-18-17745-2018.

Brown, H., Liu, X., Pokhrel, R., Murphy, S., Zheng Lu, Z., Saleh, R., et al. (2021). Biomass burning aerosols in most climate models are too absorbing. *Nature Communications*, *12*, 277. https://doi.org/10.1038/s41467-020-20482-9.

Burrows, S. M., Hoose, C., Pöschl, U., & Lawrence, M. G. (2013). Ice nuclei in marine air: Biogenic particles or dust?. *Atmospheric Chemistry and Physics, 13*, 245–267. https://doi.org/10.5194/acp-13-245-2013.

Burrows, S. M., Ogunro, O., Frossard, A. A., Russell, L. M., Rasch, P. J., & Elliott, S. (2014). A physically-based framework for modelling the organic fractionation of sea spray aerosol from bubble film Langmuir equilibria. *Atmospheric Chemistry and Physics, 14*, 13601–13629. doi:10.5194/acp-14-13601-2014.

Burrows, S. M., Easter, R. C., Liu, X., Ma, P.-L., Wang, H., Elliott, S. et al. (2022). OCEANFILMS (Organic Compounds from Ecosystems to Aerosols: Natural Films and Interfaces via Langmuir Molecular Surfactants) sea spray organic aerosol emissions—Implementation in a global climate model and impacts on clouds, *Atmospheric Chemistry and Physics, 22*, 5223–5251. https://doi.org/10.5194/acp-22-5223-2022.

Caffrey, P. F., Ondov, J. M., Zufall, M. J., & Davidson, C. I. (1998). Determination of size-dependent dry particle deposition velocities with multiple intrinsic elemental tracers. *Environmental Science & Technology*, *32*, 1615–1622. https://doi.org/10.1021/es970644f.

Caffrey, P. F., Hoppel, W. A., & Shi, J. J. (2006). A one-dimensional sectional aerosol model integrated with mesoscale meteorological data to study marine boundary layer aerosol dynamics. *Journal of Geophysical Research*, *111*, D24201. doi:10.1029/2006jd007237.

Chand, D., Wood, R., Anderson, T. L., Satheesh, S. K., & Charlson, R. J. (2009). Satellite-derived direct radiative effect of aerosols dependent on cloud cover. *Nature Geoscience*, *2*(3), 181–184. doi:10.1038/ngeo437.

Charlson, R. J., Lovelock, J. E., Andreae, M. O., & Warren, S. G. (1987). Oceanic phytoplankton, atmospheric sulphur, cloud albedo and climate. *Nature*, *326*, 655–661. https://doi.org/10.1038/326655a0.

Chin, M., Rood, R. B., Lin, S. J., Müller, J. F., & Thompson, A. M. (2000). Atmospheric sulfur cycle simulated in the global model GOCART: Model description and global properties. *Journal of Geophysical Research: Atmospheres*, *105*(D20), 24671–24687.

Chin, M., Ginoux, P., Kinne, S., Torres, O., Holben, B. N. Duncan, R. V., et al. (2002). Tropospheric aerosol optical thickness from the GOCART model and comparisons with satellite and Sun photometer measurements. *Journal of the Atmospheric Sciences*, *59*(3), 461–483.

Clarke, A., & Kapustin, V. (2010). Hemispheric aerosol vertical profiles: Anthropogenic impacts on optical depth and cloud nuclei. *Science*, *329*(5998), 1488–1492.

Clarke, A. D., Owens, S. R., & Zhou, J. (2006). An ultrafine sea salt flux from breaking waves: Implications for cloud condensation nuclei in the remote marine atmosphere, *Journal of Geophysical Research*, *111*, D06202. doi:10.1029/2005JD006565.

Colarco, P. R., Nowottnick, E. P., Randles, C. A., Yi, B., Yang, P., Kim, K.-M., et al. (2014). Impact of radiatively interactive dust aerosols in the NASA GEOS-5 climate model: Sensitivity to dust particle shape and refractive index. *Journal of Geophysical Research: Atmospheres*, *119*, 753–786. https://doi.org/10.1002/2013JD020046.

Contini, D., Donateo, A., Belosi, F., Grasso, F. M., Santachiara, G., & Prodi, F. (2010). Deposition velocity of ultrafine particles measured with the eddy-correlation method over the Nansen Ice Sheet (Antarctica). *Journal of Geophysical Research*, *115*, D16202. https://doi.org/10.1029/2009JD013600.

Crippa, M., Janssens-Maenhout, G., Dentener, F., Guizzardi, D. Sindelarova, K., Muntean, M., et al. (2016). Forty years of improvements in European air quality: Regional policy-industry interactions with global impacts. *Atmospheric Chemistry and Physics*, *16*, 3825–3841. https://doi.org/10.5194/acp-16-3825-2016.

Crippa, M., Guizzardi, D., Muntean, M., Schaaf, E., Dentener, F., van Aardenne, J. A., et al. (2018). Gridded emissions of air pollutants for the period 1970–2012 within EDGAR v4.3.2. *Earth System Science Data*, *10*, 1987–2013. https://doi.org/10.5194/essd-10-1987-2018.

Cziczo, D. J., Stetzer, O., Worringen, A., Ebert, M., Weinbruch, S., Kamphus, M., et al. (2009). Inadvertent climate modification due to anthropogenic lead. *Nature Geoscience*, *2*, 333–336. doi:10.1038/ngeo499.

Cziczo, D. J., Froyd, K. D., Hoose, C., Jensen, E. J., Diao, M. H., Zondlo, M. A., et al. (2013). Clarifying the dominant sources and mechanisms of cirrus cloud formation. *Science*, *340*, 1320–1324. doi:10.1126/science.1234145.

D'Almeida, G. A. (1987). On the variability of desert aerosol radiative characteristics. *Journal of Geophysical Research*, *92*, 3017–3026.

de Boer, G., Morrison, M. D., Shupe, M. D., & Hildner, R. (2011). Evidence of liquid dependent ice nucleation in high-latitude stratiform clouds from surface remote sensors. *Geophysical Research Letters*, *38*, L01803. https://doi.org/10.1029/2010GL046016.

de Leeuw, G., Andreas, E. L., Anguelova, M. D., Fairall, C. W., Lewis, E. R., O'Dowd, C. et al. (2011). Production flux of sea spray aerosol. *Reviews of Geophysics*, *49*, RG2001. https://doi.org/10.1029/2010RG000349.

DeMott, P. J., Prenni, A. J., Liu, X., Kreidenweis, S. M., Petters, M. D., Twohy, C. H., et al. (2010). Predicting global atmospheric ice nuclei distributions and their impacts on climate. *Proceedings of the National Academy of Sciences of the United States of America*, *107*(25), 11217–11222. 10.1073/pnas.0910818107.

DeMott, P. J., Prenni, A. J., McMeeking, G. R., Sullivan, R. C., Petters, M. D., Tobo, Y., et al. (2015). Integrating laboratory and field data to quantify the immersion freezing ice nucleation activity of mineral dust particles. *Atmospheric Chemistry and Physics*, *15*, 393–409. https://doi.org/10.5194/acp-15-393-2015.

DeMott, P. J., Hill, T. C. J., McCluskey, C. S., Prather, K. A., Collins, D. B., Sullivan, R. C., et al. (2016). Sea spray aerosol as a unique source of ice nucleating particles. *Proceedings of the National Academy of Sciences U.S.A.*, *113*, 5797–5803. doi:10.1073/pnas.1514034112.

Dentener, F., Kinne, S., Bond, T., Boucher, O., Cofala, J., Generoso, S., et al. (2006). Emissions of primary aerosol and precursor gases in the years 2000 and 1750 prescribed data-sets for AeroCom. *Atmospheric Chemistry and Physics*, *6*, 4321–4344. doi:10.5194/acp-6-4321-2006.

Donner, L. J., Wyman, B. L., Hemler, R., Horowitz, L. W., Ming, Y., Zhao, M., et al. (2011). The dynamical core, physical parameterizations, and basic simulation characteristics of the atmospheric component AM3 of the GFDL global coupled model CM3. *Journal of Climate*, *24*, 3484–3519.

Duan, B., Fairall, C. W., & Thomson, D. W. (1988). Eddy correlation measurements of the dry deposition of particle in wintertime. *Journal of Applied Meteorology*, *27*, 642–652. 10.1175/1520-0450(1988)027%3C0642:ECMOTD%3E2.0.CO;2.

Dusek, U., Frank, G. P., Hildebrandt, L., Curtius, J., Schneider, J., Walter, S., et al. (2006). Size matters more than chemistry for cloud-nucleating ability of aerosol particles. *Science*, *312*(5778), 1375–1378.

EC-JRC/PBL (2012). Emission Database for Global Atmospheric Research (EDGAR), release EDGARv4.2 FT2012. http://edgar.jrc.ec.europa.eu.

Easter, R. C., Ghan, S. J., Zhang, Y., Saylor, R. D., Chapman, E. G., Laulainen, N. S., et al. (2004). MIRAGE: Model description and evaluation of aerosols and trace gases. *Journal of Geophysical Research: Atmospheres, 109*, D20210. doi:10.1029/2004JD004571.

Emmons, L. K., Walters, S., Hess, P. G., Lamarque, J.-F., Pfister, G. G., Fillmore, D., et al. (2010). Description and evaluation of the Model for Ozone and Related chemical Tracers, version 4 (MOZART-4). *Geoscientific Model Development, 3*, 43–67. 10.5194/gmd-3-43-2010.

Ervens, B., Turpin, B. J., & Weber, R. J. (2011). Secondary organic aerosol formation in cloud droplets and aqueous particles (aqSOA): A review of laboratory, field and model studies. *Atmospheric Chemistry and Physics, 11*(21), 11069–11102. doi:10.5194/acp-11-11069-2011.

Fan, J., Leung, L. R., Rosenfeld, D., Chen, Q., Li, Z., Zhang, J., & Yan, H. (2013). Microphysical effects determine macrophysical response for aerosol impacts on deep convective clouds. *Proceedings of the National Academy of Sciences U.S.A., 110*, 4581–4590. doi:10.1073/pnas.1316830110.

Fan, J., Rosenfeld, D., Zhang, Y., Giangrande, S., Li, Z., Machado, L., & Martin, S.T. (2018). Substantial convection and precipitation enhancements by ultrafine aerosol particles. *Science, 359*(6374), 411–418. doi:10.1126/science.aan8461.

Fan, T., & Toon, O. B. (2011). Modeling sea-salt aerosol in a coupled climate and sectional microphysical model: Mass, optical depth and number concentration. *Atmospheric Chemistry and Physics, 11*(9), 4587–4610. doi:10.5194/acp-11-4587-2011.

FAO (2012). Harmonized World Soil Database (version 1.2). FAO, Rome, Italy and IIASA, Laxenburg, Austria, 2012.

Fécan, F., Marticorena, B., & Bergametti, G. (1999). Parameterization of the increase of the aeolian erosion threshold wind friction velocity due to soil moisture for arid and semi-arid areas. *Annales Geophysicae—Atmosphere Hydrospheres and Space Sciences, 17*, 149–157.

Feichter, J., Kjellström, E., Rodhe, H., Dentener, F., Lelieveldi, J., & Roelofs, G. J. (1996). Simulation of the tropospheric sulfur cycle in a global climate model. *Atmospheric Environment, 30*(10–11), 1693–1707.

Feng, Y., & Penner, J. E. (2007). Global modeling of nitrate and ammonium: Interaction of aerosols and tropospheric chemistry. *Journal of Geophysical Research: Atmospheres, 112*, D01304. 10.1029/2005jd006404.

Feng, Y., Ramanathan, V., & Kotamarthi, V. R. (2013). Brown carbon: A significant atmospheric absorber of solar radiation?. *Atmospheric Chemistry and Physics, 13*(17), 8607–8621. doi:10.5194/acp-13-8607-2013.

Fiebig, M., Lunder, C. R., & Stohl, A. (2009). Tracing biomass burning aerosol from South America to Troll Research Station, *Antarctica. Geophysical Research Letters, 36*, L14815. 10.1029/2009GL038531.

Field, P., Lawson, P., Brown, G., Lloyd, C., Westbrook, D., Moisseev, A., et al. (2017). Chapter 7: Secondary ice production—current state of the science and recommendations for the future. *Meteorological Monograph. 58*, 7.1–7.20. doi:10.1175/AMSMONOGRAPHS-D-16-0014.1.

Forrister, H., Liu, J., Scheuer, E., Dibb, J., Ziemba, L., Thornhill, K. L., et al. (2015). Evolution of brown carbon in wildfire plumes. *Geophysical Research Letters, 42*(11), 4623–4630. Doi:10.1002/2015GL063897.

Gallagher, M. W., Nemitz, E., Dorsey, J. R., Fowler, D., Sutton, M. A., Flynn, M., et al. (2002). Measurements and parameterizations of small aerosol deposition velocities to grassland, arable crops, and forest: Influence of surface roughness length on deposition. *Journal of Geophysical Research, 107*, 4154. 10.1029/2001JD000817.

Gantt, B., Meskhidze, N., & Kamykowski, D. (2009). A new physically-based quantification of marine isoprene and primary organic aerosol emissions. *Atmospheric Chemistry and Physics, 9*, 4915–4927. doi:10.5194/acp-9-4915-2009.

Ganzeveld, L., Lelieveld, J., & Roelofs, G.-J. (1998). Dry deposition parameterization of sulfur oxides in a chemistry and general circulation model. *Journal of Geophysical Research, 103*, 5679–5694. 10.1029/97JD03077.

Gettelman, A., Liu, X., Barahona, D., Lohmann, U., & Chen, C. (2012). Climate impacts of ice nucleation. *Journal of Geophysical Research, 117*, D20201. Doi:10.1029/2012JD017950.

Gettelman, A., Morrison, H., Terai, C. R., & Wood, R. (2013). Microphysical process rates and global aerosol–cloud interactions. *Atmospheric Chemistry and Physics, 13*, 9855–9867. 10.5194/acp-13-9855-2013.

Ghan, S. J. (2013). Technical Note: Estimating aerosol effects on cloud radiative forcing. *Atmospheric Chemistry and Physics, 13*, 9971–9974, doi:10.5194/acp-13-9971-2013.

Ghan, S. J., & Zaveri, R. A. (2007). Parameterization of optical properties for hydrated internally mixed aerosol. *Journal of Geophysical Research: Atmospheres, 112*, D10201. doi: 10.1029/2006jd007927.

Ghan, S. J., Abdul-Razzak, H., Nenes, A., Ming, Y., Liu, X., Ovchinnikov, M., et al. (2011). Droplet nucleation: Physically-based parameterization and comparative evaluation. *Journal of Advances in Modeling Earth Systems, 3*. doi:10.1029/2011MS000074.

Ginoux, P., Chin, M., Tegen, I., Prospero, J. M., Holben, B., Dubovik, O., & Lin, S.-J. (2001). Sources and distributions of dust aerosols simulated with the GOCART model. *Journal of Geophysical Research, 106*, 20255–20273. 10.1029/2000JD000053.

Giorgi, F., & Chameides, W. L. (1986). Rainout lifetimes of highly soluble aerosols and gases as inferred from simulations with a general circulation model. *Journal of Geophysical Research, 91*, 14,367–14,376.

Gong, S. L. (2003). A parameterization of sea salt aerosol source function for sub and super micron particles. *Global Biogeochemical Cycles, 17*, 1097. doi:10.1029/2003GB002079,.

Grandey, B. S., Rothenberg, D., Avramov, A., Jin, Q., Lee, H.-H., Liu, X., et al. (2018). Effective radiative forcing in the aerosol–climate model CAM5.3-MARC-ARG. *Atmospheric Chemistry and Physics, 18*, 15783–15810. 10.5194/acp-18-15783-2018.

Grythe, H., Ström, J., Krejci, R., Quinn, P., & Stohl, A. (2014). A review of sea-spray aerosol source functions using a large global set of sea salt aerosol concentration measurements. *Atmospheric Chemistry and Physics, 14*, 1277–1297. 10.5194/acp-14-1277-2014.

Guenther, A., Hewitt, N., Erickson, D., Fall, R., Geron, C., Graedel, T., et al. (1995). A global model of natural volatile organic compound emissions. *Journal of Geophysical Research, 100*(D5), 8873–8892.

Guenther, A., Karl, T., Harley, P., Wiedinmyer, C., Palmer, P. I., & Geron, C. (2006). Estimates of global terrestrial isoprene emissions using MEGAN (Model of Emissions of Gases and Aerosols from Nature). *Atmospheric Chemistry and Physics, 6*, 3181–3210.

He, H., Wang, Y. S., Ma, Q., Ma, J., Chu, B., Ji, D., et al. (2014). Mineral dust and NOx promote the conversion of SO_2 to sulfate in heavy pollution days. *Scientific Reports, 4*, 4172. doi:10.1038/srep04172.

Heald, C. L., Ridley, D. A., Kroll, J. H., Barrett, S. R. H., Cady-Pereira, K. E., Alvarado, M. J., & Holmes, C. D. (2014). Contrasting the direct radiative effect and direct radiative forcing of aerosols. *Atmospheric Chemistry and Physics, 14*, 5513–5527. doi:10.5194/acp-14-5513-2014.

Hoelzemann, J. J., Schultz, M. G., Brasseur, G. P., Granier, C., & Simon, M. (2004). Global Wildland Fire Emission Model (GWEM): Evaluating the use of global area burnt satellite data. *Journal of Geophysical Research D: Atmospheres, 109*, D14S04. doi:10.1029/2003JD003666.

Hoesly, R. M., Smith, S. J., Feng, L., Klimont, Z., Janssens-Maenhout, G., Pitkanen, T., et al. (2018). Historical (1750–2014) anthropogenic emissions of reactive gases and aerosols from the Community Emissions Data System (CEDS). *Geoscientific Model Development, 11*, 369–408. 10.5194/gmd-11-369-2018.

Hoose, C., Kristjansson, J. E., Iversen, T., Kirkevag, A., Seland, Ø., & Gettelman, A. (2009). Constraining cloud droplet number concentration in GCMs suppresses the aerosol indirect effect. *Geophysical Research Letters, 36*, L12807. doi:10.1029/2009GL038568.

Hoose, C., & Möhler, O. (2012). Heterogeneous ice nucleation on atmospheric aerosols: A review of results from laboratory experiments. *Atmospheric Chemistry and Physics, 12*, 9817–9854. 10.5194/acp-12-9817-2012

Huang, R. J., Zhang, Y., Bozzetti, C., Ho, K. F., Cao, J. J., Han, Y., et al. (2014a). High secondary aerosol contribution to particulate pollution during haze events in China. *Nature, 514(7521), 218–222*, 2014 Oct 9. doi: 10.1038/nature13774.

Huang, X., Song, Y., Zhao, C., Li, M., Zhu, T., Zhang, Q., & Zhang, X. (2014b). Pathways of sulfate enhancement by natural and anthropogenic mineral aerosols in China. *Journal of Geophysical Research: Atmospheres, 119*, 14,165–14,179. doi:10.1002/2014JD022301.

Ibrahim, M., Barrie, L. A., & Fanaki, F. (1983). An experimental and theoretical investigation of the dry deposition of particles to snow, pine trees and artificial collectors. *Atmospheric Environment, 4*, 781–788. 10.1016/0004-6981(83)90427-4.

Jaecker-Voirol, A., & Mirabel, P. (1989). Heteromolecular nucleation in the sulfuric acid-water system. *Journal of Geophysical Research, 23*, 2053–2057.

Jacobson, M. Z., & R. P. Turco (1995). Simulating condensational growth, evaporation and coagulation of aerosols using a combined moving and stationary size grid. *Aerosol Science and Technology, 22*, 73–92.

Jacobson, M., Turco, R., Jensen, E., & Toon, O. (1994). Modeling coagulation among particles of different composition and size. *Atmospheric Environment, 28*, 1327–1338.

Janssens-Maenhout, G., Crippa, M., Guizzardi, D., Dentener, F., Muntean, M., Pouliot, G., et al. (2015). HTAP_v2.2: A mosaic of regional and global emission grid maps for 2008 and 2010 to study hemispheric transport of air pollution. *Atmospheric Chemistry and Physics, 15*, 11411–11432, 10.5194/acp-15-11411-2015.

Jiang, Y., Lu, Z., Liu, X., Qian, Y., Zhang, K., Wang, Y., & Yang, X.-Q. (2016). Impacts of global open-fire aerosols on direct radiative, cloud and surface-albedo effects simulated with CAM5. *Atmospheric Chemistry and Physics, 16*, 14805–14824, 10.5194/acp-16-14805-2016.

Kaiser, J. W., Heil, A., Andreae, M. O., Benedetti, A., Chubarova, N., Jones, L., et al. (2012). Biomass burning emissions estimated with a global fire assimilation system based on observed fire radiative power. *Biogeosciences, 9*(1), 527–554. doi:10.5194/bg-9-527-2012.

Kanakidou, M., Seinfeld, J. H., Pandis, S. N., Barnes, I., Dentener, F. J., Facchini, M. C., et al. (2005). Organic aerosol and global climate modelling: A review, *Atmospheric Chemistry and Physics, 5*, 1053–1123.

Kärcher, B., Möhler, O., DeMott, P. J., Pechtl, S., & Yu, F. (2007). Insights into the role of soot aerosols in cirrus cloud formation. *Atmospheric Chemistry and Physics, 7*, 4203–4227.

Kärcher, B., Hendricks, J., & Lohmann, U. (2006). Physically based parameterization of cirrus cloud formation for use in global atmospheric models. *Journal of Geophysical Research, 111*, D01205. 10.1029/2005JD006219.

Kay, J. E., Hillman, B. R., Klein, S. A., Zhang, Y., Medeiros, B., Pincus, R., et al. (2012). Exposing global cloud biases in the Community Atmosphere Model (CAM) using satellite observations and their corresponding instrument simulators. *Journal of Climate, 25*(15), 5190–5207. doi:10.1175/JCLI-D-11-00469.1.

Kazil, J., Stier, P., Zhang, K., Quaas, J., Kinne, S., O'Donnell, D., et al. (2010). Aerosol nucleation and its role for clouds and Earth's radiative forcing in the aerosol-climate model ECHAM5-HAM. *Atmospheric Chemistry and Physics, 10*, 10733–10752, doi:10.5194/acp-10-10733-2010.

Kerminen, V. M., & Kulmala, M. (2002). Analytical formulae connecting the real and the apparent nucleation rate and the nuclei number concentration for atmospheric nucleation events. *Journal of Aerosol Science, 33*, 609–622.

Kettle, A., & Andreae, M. (2000). Flux of the dimethylsulfide from the oceans: A comparison of updated data sets and flux models. *Journal of Geophysical Research, 105*, 26,793–26,808.

Khairoutdinov, M., & Kogan, Y. (2000). A new cloud physics parameterization in a large-eddy simulation model of marine stratocumulus. Monthly Weather Review, 128(1), 229–243, 1287. doi:10.1175/1520-0493.

Khan, T. R., & Perlinger, J. A. (2017). Evaluation of five dry particle deposition parameterizations for incorporation into atmospheric transport models. *Geoscientific Model Development, 10*, 3861–3888. 10.5194/gmd-10-3861-2017.

Kirchstetter, T. W., Novakov, T., & Hobbs, P. V. (2004). Evidence that the spectral dependence of light absorption by aerosols is affected by organic carbon. *Journal of Geophysical Research: Atmospheres, 109*, D21208. doi:10.1029/2004JD004999.

Kirkby, J., Duplissy, J., Sengupta, K., Frege, C., Gordon, H., Williamson, C., et al. (2016). Ion-induced nucleation of pure biogenic particles. *Nature*, *533*, 521–524. 10.1038/nature17953.

Klimont, Z., Kupiainen, K., Heyes, C., Purohit, P., Cofala, J., Rafaj, P., et al. (2017). Global anthropogenic emissions of particulate matter including black carbon. *Atmospheric Chemistry and Physics, 17,* 8681–8723. 10.5194/acp-17-8681-2017.

Knopf, D. A., Alpert, P. A., & Wang, B. B. (2018). The role of organic aerosol in atmospheric ice nucleation: A review. *ACS Earth And Space Chemistry*, *2*, 168–202. doi:10.1021/acsearthspacechem.7b00120.

Knopf, D. A., & Alpert, P. A. (2013). Water activity based model of heterogeneous ice nucleation kinetics for freezing of water and aqueous solution droplets. *Faraday Discussions*, *165*, 513–534. doi:10.1039/C3FD00035D.

Koch, D., Jacob, D., Tegen, I., Rind, D., & Chin, M. (1999). Tropospheric sulfur simulation and sulfate direct radiative forcing in the Goddard Institute for Space Studies general circulation model. *Journal of Geophysical Research*, *104*, 23,799–23,822.

Kok, J. F. (2011). A scaling theory for the size distribution of emitted dust aerosols suggests climate models underestimate the size of the global dust cycle. *Proceedings of the National Academy of Sciences of the United States of America*, *108*, 1016–1021. 10.1073/pnas.1014798108.

Kok, J., Albani, S., Mahowald, N., & Ward, D. S. (2014a). An improved dust emission model. Part 2 Evaluation in the Community Earth System Model, with implications for the use of dust source functions. *Atmospheric Chemistry and Physics, 14*, 13043–13061.

Kok, J., Mahowald, N., Fratini, G., Gillies, J., Ishizuka, M., Mikami, M., et al. (2014b). An improved dust emission model. Part 1: Model description and comparison against measurements. *Atmospheric Chemistry and Physics, 14*, 13023–13041.

Kok, J. F., Ridley, D. A., Zhou, Q., Miller, R. L., Zhao, C., Heald, C. L., et al. (2017). Smaller desert dust cooling effect estimated from analysis of dust size and abundance. *Nature Geoscience*, *10*, 274–278. 10.1038/NGEO2912.

Kulmala, M., Laaksonen, A., & Pirjola, L. (1998). Parameterizations for sulfuric acid/water nucleation rates. *Journal of Geophysical Research*, *103*(D7), 8301–8307.

Kulmala, M., Kontkanen, J., Junninen, H., Lehtipalo, K., Manninen, H. E., Nieminen, T., et al. (2013). Direct observations of atmospheric aerosol nucleation. *Science*, *339*, 943–946.

Lamarque, J.-F., Bond, T. C., Eyring, V., Granier, C., Heil, A., Klimont, Z., et al. (2010). Historical (1850–2000) gridded anthropogenic and biomass burning emissions of reactive gases and aerosols: Methodology and application. *Atmospheric Chemistry and Physics, 10*, 7017–7039, 10.5194/acp-10-7017-2010.

Lamarque, J.-F., Emmons, L. K., Hess, P. G., Kinnison, D. E., Tilmes, S., Vitt, F., et al. (2012). CAM-chem: Description and evaluation of interactive atmospheric chemistry in the Community Earth System Model. *Geoscientific Model Development*, *5*, 369–411. 10.5194/gmd-5-369-2012.

Lana, A., Bell, T. G., Simó, R., Vallina, S. M., Ballabrera-Poy, J., Kettle, A. J., et al. (2011). An updated climatology of surface dimethylsulfide concentrations and emission fluxes in the global ocean. *Global Biogeochemical Cycles*, *25*, GB1004. 10.1029/2010GB003850.

Lee, H. J., Aiona, P. K., Laskin, A., Laskin, J., & Nizkorodov, S. A. (2014). Effect of solar radiation on the optical properties and molecular composition of laboratory proxies of atmospheric brown carbon. *Environmental Science & Technology*, *48*(17), 10217–10226. doi:10.1021/es502515r.

Li, F., Martin, M. V., Hantson, S., Andreae, M. O., Arneth, A., Lasslop, G., et al. (2019). Historical (1700–2012) Global Multi-model Estimates of the Fire Emissions from the Fire Modeling Intercomparison Project (FireMIP). *Atmospheric Chemistry and Physics Discussion (March)*.

Li, M., Zhang, Q., Kurokawa, J., Woo, J.-H., He, K., Lu, Z., et al. (2017). MIX: A mosaic Asian anthropogenic emission inventory under the international collaboration framework of the MICS-Asia and HTAP. *Atmospheric Chemistry and Physics*, *17*(2), 935–963. doi:10.5194/acp-17-935-2017.

Lin, G., Penner, J. E., Sillman, S., Taraborrelli, D., & Lelieveld, J. (2012). Global modeling of SOA formation from dicarbonyls, epoxides, organic nitrates and peroxides. *Atmospheric Chemistry and Physics*, *12*(10), 4743–4774.

Lin, G., Sillman, S., Penner, J. E., & Ito, A. (2014a). Global modeling of SOA: The use of different mechanisms for aqueous-phase formation. *Atmospheric Chemistry and Physics*, *14*(11), 5451–5475.

Lin, G., Penner, J. E., Flanner, M. G., Sillman, S., Xu, L., & Zhou, C. (2014b). Radiative forcing of organic aerosol in the atmosphere and on snow: Effects of SOA and brown carbon. *Journal of Geophysical Research: Atmospheres*, *119*(12), 7453–7476. doi:10.1002/2013JD021186.

Lin, L., Fu, Q., Liu, X., Shan, Y., Giangrande, S. E., Elsaesser, G. S., et al. (2021). Improved convective ice microphysics parameterization in the NCAR CAM model. *Journal of Geophysical Research: Atmospheres*, *126*, e2020JD034157. 10.1029/2020JD034157.

Liss, P. S., & Merlivat, L. (1986). Air–sea gas exchange rates: Introduction and synthesis. In P. Buat-Ménard (Ed.), *The Role of Air–Sea Exchange in Geochemical Cycling (pp. 113–127). NATO ASI Series, Vol. 185, Springer. Dordrecht: D.* Riedel. https://doi.org/10.1007/978-94-009-4738-2_5.

Liu, H., Jacob, D. J., Bey, I., & Yantosca, R. M. (2001). Constraints from 210Pb and 7Be on Wet Deposition and Transport in a Global Three-Dimensional Chemical Tracer Model Driven by Assimilated Meteorological Fields. *Journal of Geophysical Research: Atmospheres*, *106*(D11), 12109–12128. doi:10.1029/2000jd900839.

Liu, X., & Penner, J. E. (2005). Ice nucleation parameterization for global models. *Meteorologische Zeitschrift*, *14*, 499–514.

Liu, X., Penner, J. E., and Herzog, M. (2005). Global modeling of aerosol dynamics: Model description, evaluation, and interactions between sulfate and nonsulfate aerosols. *Journal of Geophysical Research*, *110*, D18206, doi:10.1029/2004jd005674.

Liu, X., Penner, J. E., Ghan, S. J., & Wang, M. (2007a). Inclusion of ice microphysics in the NCAR Community Atmospheric Model Version 3 (CAM3). *Journal of Climate*, *20*, 4526–4547.

Liu, X., Penner, J. E., Das, B., Bergmann, D., Rodriguez, J. M., Strahan, S., et al. (2007b). Uncertainties in global aerosol simulations: Assessment using three meteorological datasets. *Journal of Geophysical Research*, *112*, D11212. doi:10.1029/2006JD008216.

Liu, X., Penner, J. E., & Wang, M. (2009). Influence of anthropogenic sulfate and soot on upper tropospheric clouds using CAM3 coupled with an aerosol model. *Journal of Geophysical Research*, *114*, D03204. 10.1029/2008JD010492.

Liu, X., & Wang, J. (2010). How important is organic aerosol hygroscopicity to aerosol indirect forcing? *Environmental Research Letters*, *5*(4), 044010.

Liu, X., Easter, R. C., Ghan, S. J., Zaveri, R., Rasch, P., Shi, X., et al. (2012a). Toward a minimal representation of aerosols in climate models: Description and evaluation in the Community Atmosphere Model CAM5. *Geoscientific Model Development*, *5*, 709–739. doi:10.5194/gmd-5-709-2012.

Liu, X., Shi, X., Zhang, K., Jensen, E. J., Gettelman, A., Barahona, D., et al. (2012b). Sensitivity studies of dust ice nuclei effect on cirrus clouds with the Community Atmosphere Model CAM5. *Atmospheric Chemistry and Physics*, *12*(24), 12,061–12,079. 10.5194/acp-12-12061-2012.

Liu, X., Ma, P.-L., Wang, H., Tilmes, S., Singh, B., Easter, R. C., et al. (2016). Description and evaluation of a new four-mode version of the Modal Aerosol Module (MAM4) within version 5.3 of the Community Atmosphere Model. *Geoscientific Model Development*, *9*, 505–522. doi:10.5194/gmd-9-505-2016.

Liu, X., & Shi, X. (2018). Sensitivity of homogeneous ice nucleation to aerosol perturbations and its implications for aerosol indirect effects through cirrus clouds. *Geophysical Research Letters*, *45*, 1684–1691. 10.1002/2017GL076721.

Liu, Y., & Daum, P. H. (2004). Parameterization of the autoconversion process. Part I: Analytical formulation of the Kessler-type parameterizations. *Journal of the Atmospheric Sciences*, *61*, 1539–1548.

Liu, Y., Goodrick, S., & Heilman, W. (2014). Wildland fire emissions, carbon, and climate: Wildfire-climate interactions. *Forest Ecology and Management*, *317*, 80–96. doi:10.1016/j.foreco.2013.02.020.

Lohmann, U. (2002). Possible aerosol effects on ice clouds via contact nucleation. *Journal of the Atmospheric Sciences*, *59*, 647–656.

Lohmann, U., Stier, P., Hoose, C., Ferrachat, S., Kloster, S., Roeckner, E., & Zhang, J. (2007). Cloud microphysics and aerosol indirect effects in the global climate model ECHAM5-HAM. *Atmospheric Chemistry and Physics*, *7*, 3425–3446, doi:10.5194/acp-7-3425-2007.

Lohmann, U. (2008). Global anthropogenic aerosol effects on convective clouds in ECHAM5-HAM. *Atmospheric Chemistry and Physics*, *8*, 2115–2131. doi:10.5194/acp-8-2115-2008.

Lu, Z., Streets, D. G., Winijkul, E., Yan, F., Chen, Y., Bond, T. C., et al. (2015). Light Absorption properties and radiative effects of primary organic aerosol emissions. *Environmental Science & Technology*, *49*(8), 4868–4877. doi:10.1021/acs.est.5b00211.

Lu, Z., Liu, X., Zaveri, R. A., Easter, R. C., Tilmes, S., Emmons, L. K., et al. (2021). Radiative forcing of nitrate aerosols from 1975 to 2010 as simulated by MOSAIC module in CESM2-MAM4. *Journal of Geophysical Research: Atmospheres*, *126*, e2021JD034809. 10.1029/2021JD034809.

Luo, C., Mahowald, N. M., and del Corral, J. (2003). Sensitivity study of meteorological parameters on mineral aerosol mobilization, transport, and distribution. *Journal of Geophysical Research: Atmospheres*, *108*(D15), 4447, doi:10.1029/2003JD003483.

Lüönd, F., Stetzer, O., Welti, A., & Lohmann, U. (2010). Experimental study on the ice nucleation ability of size-selected kaolinite particles in the immersion mode. *Journal of Geophysical Research*, *115*, D14201.

Ma, P.-L., Rasch, P. J., Fast, J. D., Easter, R. C., Gustafson, W. I., Jr., Liu, X., et al. (2014). Assessing the CAM5 physics suite in the WRF-Chem model: Implementation, resolution sensitivity, and a first evaluation for a regional case study. *Geoscientific Model Development*, *7*, 755–778. doi:10.5194/gmd-7-755-2014.

Ma, P.-L., Rasch, P. J., Wang, M., Wang, H., Ghan, S. J., Easter, R. C., et al. (2015). How does increasing horizontal resolution in a global climate model improve the simulation of aerosol-cloud interactions? *Geophysical Research Letters*, *42*, 5058–5065. doi:10.1002/2015GL064183.

Malavelle, F. F., Haywood, J. M., Jones, A., Gettelman, A., Clarisse, L., Bauduin, S., et al. (2017). Strong constraints on aerosol–cloud interactions from volcanic eruptions. *Nature*, *546*, 485–491. 10.1038/nature22974.

Mann, G. W., Carslaw, K. S., Spracklen, D. V., Ridley, D. A., Manktelow, P. T., Chipperfield, M. P., et al. (2010). Description and evaluation of GLOMAP-mode: A modal global aerosol microphysics model for the UKCA composition-climate model. *Geoscientific Model Development*, *3*, 519–551. 10.5194/gmd-3-519-2010.

Marcolli, C., Gedamke, S., Peter, T., & Zobrist, B. (2007). Efficiency of immersion mode ice nucleation on surrogates of mineral dust. *Atmospheric Chemistry and Physics*, *7*, 5081–5091.

Mårtensson, E. M., Nilsson, E. D., de Leeuw, G., Cohen, L. H., & Hansson, H. C. (2003). Laboratory simulations and parameterization of the primary marine aerosol production. *Journal of Geophysical Research*, *108*, 1–12.

Marticorena, B., & Bergametti, G. (1995). Modeling the atmospheric dust cycle: 1. Design of a soil-derived dust emission scheme. *Journal of Geophysical Research*, *100*(D8), 16,415–416,430.

McCormick, M. P., Thomason, L. W., & Trepte, C. R. (1995). Atmospheric effects of the Mt. *Pinatubo eruption. Nature*, *373*, 399–404.

Meyers, M. P., DeMott, P. J., & Cotton, W. R. (1992). New primary ice-nucleation parameterizations in an explicit cloud model. *Journal of Applied Meteorology*, *31*, 708–721.

Menon, S., Del Genio, A. D., Koch, D., & Tselioudis, G (2002). GCM simulations of the aerosol indirect effect: Sensitivity to cloud parameterization and aerosol burden. *Journal of the Atmospheric Sciences*, *59*, 692–713.

Mercado, L. M., Bellouin, N., Sitch, S., Boucher, O., Huntingford, C., Wild, M., & Cox, P. M. (2009). Impact of changes in diffuse radiation on the global land carbon sink. *Nature*, *458*, 1014–1017.

Merikanto, J., Napari, I., Vehkamaki, H., Anttila, T., & Kulmala, M. (2007). New parameterization of sulfuric acid-ammonia-water ternary nucleation rates at tropospheric conditions. *Journal of Geophysical Research: Atmospheres*, *112*(D15), D15207. doi 10.1029/2006jd007977.

Merikanto, J., Spracklen, D. V., Mann, G. W., Pickering, S. J., & Carslaw, K. S. (2009). Impact of nucleation on global

CCN. *Atmospheric Chemistry and Physics, 9*, 8601. doi: 10.5194/acp-9-8601-2009.

Ming, Y., Ramaswamy, V., Donner, L. J., & Phillips, V. T. J. (2006). A new parameterization of cloud droplet activation applicable to general circulation models. *Journal of the Atmospheric Sciences, 63*(4), 1348–1356. doi: 10.1175/JAS3686.1.

Möller, U., & Schumann, G. (1970). Mechanisms of transport from the atmosphere to the Earth's surface. *Journal of Geophysical Research, 75*, 3013–3019. 10.1029/JC075i015p03013.

Monahan, E. C., Spiel, D. E., & Davidson, K. L. (1986). A model of marine aerosol generation via whitecaps and wave disruption. In E. C. MacNiocaill & M. G. Reidel (Eds.), *Oceanic whitecaps and their role in air-sea exchange processes* (pp. 167–174). Dordrecht: The Netherlands.

Morrison, H., & Gettelman, A. (2008). A new two-moment bulk stratiform cloud microphysics scheme in the Community Atmosphere Model, version 3 (CAM3). Part I: Description and numerical tests. *Journal of Climate, 21*, 3642–3659. doi: 10.1175/2008JCLI2105.1.

Murray, B. J., Wilson, T. W., Dobbie, S., Cui, Z., Jumur, S. M. R. K., Kärcher, B., et al. (2010). Heterogeneous nucleation of ice particles on glassy aerosols under cirrus conditions. *Nature Geoscience, 3*, 233–237.

Murray, B. J., Broadley, S. L., Wilson, T. W., Atkinson, J. D., & Wills, R. H. (2011). Heterogeneous freezing of water droplets containing kaolinite particles. *Atmospheric Chemistry and Physics, 11*, 4191–4207.

Murray, B. J., O'Sullivan, D., Atkinson, J. D., & Webb, M. E. (2012). Ice nucleation by particles immersed in supercooled cloud droplets. *Chemical Society Reviews, 41*, 6519–6554. 10 .1039/C2CS35200A.

Myhre, G., Shindell, D., Bréon, F.-M., Collins, W., Fuglestvedt, J., Huang, J., et al. (2013a). Anthropogenic and natural radiative forcing. In *Climate Change 2013: The Physical Science Basis. Contribution of Working Group I to the Fifth Assessment Report of the Intergovernmental Panel on Climate Change* [Stocker, T. F., Qin, D., Plattner, G.-K., Tignor, M., Allen, S. K., Boschung, J., et al. (Eds.)]. Cambridge University Press: Cambridge, United Kingdom and New York, NY, USA.

Myhre, G., Samset, B. H., Schulz, M., Balkanski, Y., Bauer, S., Berntsen, T. K., et al. (2013b). Radiative forcing of the direct aerosol effect from AeroCom Phase II simulations. *Atmospheric Chemistry and Physics, 13*, 1853–1877. 10.5194/acp-13-1853-2013.

Nakayama, T., Sato, K., Matsumi, Y., Imamura, T., Yamazaki, A., & Uchiyama, A. (2013). Wavelength and NOx dependent complex refractive index of SOAs generated from the photooxidation of toluene. *Atmospheric Chemistry and Physics, 13*(2), 531–545. doi:10.5194/acp-13-531-2013.

Napari, I., Noppel, M., Vehkamaki, H., & Kulmala, M. (2002). Parametrization of ternary nucleation rate for H2SO4-NH3-H2O vapors. *Journal of Geophysical Research, 107*(D19), 4381. doi:10.1029/2002JD002132.

Neely, R. R., III, & Schmidt, A. (2016). VolcanEESM: Global volcanic sulphur dioxide (SO2) emissions database from 1850 to present—Version 1.0. Centre for Environmental Data Analysis, 04 February 2016. doi: 10.5285/76ebdc0b-0eed-4f70- b89e-55e606bcd568.

Nenes, A., & Seinfeld, J. H. (2003). Parameterization of cloud droplet formation in global climate models. *Journal of Geophysical Research, 108*(D14), 4415. doi: 10.1029/2002JD002911.

Niedermeier, D., Ervens, B., Clauss, T., Voigtländer, J., Wex, H., Hartmann, S., & Stratmann, F. (2014). A computationally efficient description of heterogeneous freezing: A simplified version of the Soccer ball model. *Geophysical Research Letters, 41*, 736–741.

Niemand, M., Möhler, O., Vogel, B., Vogel, H., & Hoose, C. (2012). A particle-surface-area based parameterization of immersion freezing on desert dust particles. *Journal of the Atmospheric Sciences, 69*, 3077–3092. https://doi.org/10.1175/JAS-D-11-0249.1

Nightingale, P. D., Malin, G., Law, C. S., Watson, A. J., Liss, P. S., Liddicoat, M. I., et al. (2000). In situ evaluation of air-sea gas exchange parameterizations using novel conservative and volatile tracers. *Global Biogeochemical Cycles, 14*(1), 373–387.

Nilsson, E. D., & Rannik, Ü. (2001). Turbulent aerosol fluxes over the Arctic Ocean: 1. Dry deposition over sea and pack ice. *Journal of Geophysical Research, 106*, 32,125–32,137. 10 .1029/2000JD900605.

O'Dowd, C. D., Facchini, M. C., Cavalli, F., Ceburnis, D., Mircea, M., Decesari, S., et al. (2004). Biogenically driven organic contribution to marine aerosol. *Nature, 431*, 676–680. doi:10.1038/nature02959.

O'Dowd, C. D., Langmann, B., Varghese, S., Scannell, C., Ceburnis, D., & Facchini, M. C. (2008). A combined organic-inorganic sea-spray source function. *Geophysical Research Letters, 35*, L01801. doi:10.1029/2007GL030331.

Ovadnevaite, J., Manders, A., de Leeuw, G., Ceburnis, D., Monahan, C., Partanen, A.-I., et al. (2014). A sea spray aerosol flux parameterization encapsulating wave state. *Atmospheric Chemistry and Physics, 14*, 1837–1852. https://doi.org/10.5194/acp-14-1837-2014.

Parker, D. E., Wilson, H., Jones, P. D., Christy, J. R., & Folland, C. K. (1996). The impact of Mount Pinatubo on world-wide temperatures. *International Journal of Climatology, 16*(5), 487–497.

Paugam, R., Wooster, M., Freitas, S., & Val Martin, M. (2016). A review of approaches to estimate wildfire plume injection height within large-scale atmospheric chemical transport models. *Atmospheric Chemistry and Physics, 16*(2), 907–925. doi:10.5194/acp-16-907-2016.

Penner, J. E., Zhou, C., Garnier, A., & Mitchell, D. L. (2018). Anthropogenic aerosol indirect effects in cirrus clouds. *Journal of Geophysical Research: Atmospheres, 123*(20). doi:10.1029/2018JD029204.

Petroff, A., & Zhang, L. (2010). Development and validation of a size-resolved particle dry deposition scheme for application in aerosol transport models. *Geoscientific Model Development, 3*, 753–769. 10.5194/gmd-3-753-2010.

Petters, M. D., & Kreidenweis, S. M. (2007). A single parameter representation of hygroscopic growth and cloud condensation nucleus activity. *Atmospheric Chemistry and Physics, 7*, 1961–1971. doi:10.5194/acp-7-1961-2007.

Phillips, V. T. J., DeMott, P. J., & Andronache, C. (2008). An empirical parameterization of heterogeneous ice nucleation for multiple chemical species of aerosol. *Journal of the Atmospheric Sciences*, *65*, 2757–2783.

Pincus, R., & Baker, M. B. (1994). Effect of precipitation on the albedo susceptibility of clouds in the marine boundary layer. *Nature*, *372*(6503), 250–252.

Prenni, A. J., Harrington, J. Y., Tjernström, M., DeMott, P. J., Avramov, A., Long, C. N., et al. (2007). Can ice-nucleating aerosols affect Arctic seasonal climate? *Bulletin of the American Meteorological Society, 88*, 541–550.

Pruppacher, H. R., & Klett, J. D. (1997). *Microphysics of Clouds and Precipitation*. Kluwer Acad.: Norwell, Mass.

Qian, Y., Gustafson, W. I., Jr., & Fast, J. D. (2010). An investigation of the sub-grid variability of trace gases and aerosols for global climate modeling. *Atmospheric Chemistry and Physics, 10*, 6917–6946. 10.5194/acp-10-6917-2010.

Quinn, P. & Bates, T. (2011). The case against climate regulation via oceanic phytoplankton sulphur emissions. *Nature, 480*, 51–56 (2011). https://doi.org/10.1038/nature10580

Rabin, S. S., Melton, J. R., Lasslop, G., Bachelet, D., Forrest, M., Hantson, S., et al. (2017). The Fire Modeling Intercomparison Project (FireMIP), phase 1: Experimental and analytical protocols with detailed model descriptions. *Geoscientific Model Development*, *10*(3), 1175–1197. doi:10.5194/gmd-10-1175-2017.

Randerson, J. T., Chen, Y., Van Der Werf, G. R., Rogers, B. M., & Morton, D. C. (2012). Global burned area and biomass burning emissions from small fires. *Journal of Geophysical Research Biogeosciences*, *117*(4). doi:10.1029/2012JG002128.

Rasch, P. J., & Kristjánsson, J. E. (1998). A comparison of the CCM3 model climate using diagnosed and predicted condensate parameterizations. *Journal of Climate*, *11*, 1587–1614.

Rasch, P. J., Barth, M. C., Kiehl, J. T., Schwartz, S. E., & Benkovitz, C. M. (2000). A description of the global sulfur cycle and its controlling processes in the National Center for Atmospheric Research Community Climate Model, Version 3. *Journal of Geophysical Research, 105*, 1367–1385.

Riemer, N., Ault, A. P., West, M., Craig, R. L., & Curtis, J. H. (2019). Aerosol mixing state: Measurements, modeling, and impacts. *Reviews of Geophysics*, *57*, 187–249. 10.1029/2018RG000615.

Rosenfeld, D., Lohmann, U., Raga, G. B., O'Dowd, C. D., Kulmala, M., Fuzzi, S., et al. (2008). Flood or drought: How do aerosols affect precipitation? *Science*, *321*, 1309–1313. doi:10.1126/science.1160606pmid:18772428

Saleh, R., Robinson, E. S., Tkacik, D. S., Ahern, A. T., Liu, S., Aiken, A. C., et al. (2014). Brownness of organics in aerosols from biomass burning linked to their black carbon content. *Nature Geoscience*, *7*(9), 647–650. doi:10.1038/ngeo2220.

Saleh, R., Marks, M., Heo, J., Adams, P. J., Donahue, N. M., & Robinson, A. L. (2015). Contribution of brown carbon and lensing to the direct radiative effect of carbonaceous aerosols from biomass and biofuel burning emissions, *Journal of Geophysical Research: Atmospheres*, *120*(19), 2015JD023697. doi:10.1002/2015JD023697.

Scanza, R. A., Mahowald, N., Ghan, S., Zender, C. S., Kok, J. F., Liu, X., et al. (2015). Modeling dust as component minerals in the Community Atmosphere Model: Development of framework and impact on radiative forcing. *Atmospheric Chemistry and Physics, 15*, 537–561. 10.5194/acp-15-537-2015.

Sehmel, G. A., & Sutter, S. L. (1974). Particle deposition rates on a water surface as a function of particle diameter and air velocity. *Journal De Recherches Atmospheriques*, *8*, 911–920.

Seinfeld, J. H., & Pandis, S. N. (2016). *Atmospheric chemistry and physics: From air pollution to climate change*. New York, NY: John Wiley.

Sellar, A. A., Jones, C. G., Mulcahy, J. P., Tang, Y., Yool, A., Wiltshire, A., et al. (2019). UKESM1: Description and evaluation of the U.K. Earth System Model. *Journal of Advances in Modeling Earth Systems*, *11*, 4513–4558, 10.1029/2019MS001739.

Shan, Y., Liu, X., Lin, L., Ke, Z., & Lu, Z. (2021). An improved representation of aerosol wet removal by deep convection and impacts on simulated aerosol vertical profiles. *Journal of Geophysical Research: Atmospheres*, *126*, e2020JD034173. 10.1029/2020JD034173.

Shi, Y., & X. Liu (2019). Dust radiative effects on climate by glaciating mixed-phase clouds. *Geophysical Research Letters*, *46*, 6128–6137. 10.1029/2019GL082504.

Shrivastava, M., Easter, R. C., Liu, X., Zelenyuk, A., Singh, B., Zhang, K., et al. (2015). Global transformation and fate of SOA: Implications of low-volatility SOA and gas-phase fragmentation reactions. *Journal of Geophysical Research: Atmospheres*, *120*(9), 4169–4195.

Sihto, S. L., Kulmala, M., Kerminen, V.-M., Dal Maso, M., Petäjä, T., Riipinen, I., et al. (2006). Atmospheric sulphuric acid and aerosol formation: Implications from atmospheric measurements for nucleation and early growth mechanisms. *Atmospheric Chemistry and Physics, 6*, 4079–4091.

Slinn, W. G. N. (1982). Predictions for particle deposition to vegetative surfaces. *Atmospheric Environment*, *16*, 1785–1794. 10.1016/0004-6981(82)90271-2.

Slinn, W. G. N (1983). Precipitation scavenging. In *Atmospheric Sciences and Power Production—1979, chap. 11. Division of Biomedical Environmental Research, U.S.* Department of Energy, Washington DC, USA.

Smith, M., & Harrison, N. (1998). The sea spray generation function. *Journal of Aerosol Science*, *29*, 189–190. doi:10.1016/S0021-8502(98)00280-8.

Smith, M. H., Park, P. M., & Consterdine, I. E. (1993). Marine aerosol concentrations and estimated fluxes over the sea. *Quarterly Journal of the Royal Meteorological Society*, *119*, 809–824.

Song, X., & Zhang, G. J. (2011). Microphysics parameterization for convective clouds in a global climate model: Description and single-column model tests. *Journal of Geophysical Research*, *116*, D02201. doi:10.1029/2010JD014833.

Spracklen, D. V., Pringle, K. J., Carslaw, K. S., Chipperfield, M. P., & Mann, G. W. (2005). A global off-line model of size-resolved aerosol microphysics: I. Model development and prediction of aerosol properties. *Atmospheric Chemistry and Physics, 5*, 2227–2252. 10.5194/acp-5-2227-2005.

Stier, P., Feichter, J., Kinne, S., Kloster, S., Vignati, E., Wilson, J., et al. (2005). The aerosol-climate model ECHAM5-HAM. *Atmospheric Chemistry and Physics, 5,* 1125–1156. 10.5194/acp-5-1125-2005.

Stier, P., Schutgens, N. A. J., Bellouin, N., Bian, H., Boucher, O., Chin, M., et al. (2013). Host model uncertainties in aerosol radiative forcing estimates: Results from the AeroCom Prescribed intercomparison study. *Atmospheric Chemistry and Physics*, *13*(6), 3245–3270. doi:10.5194/acp-13-3245-2013.

Stokes, R. H., & Robinson, R. A. (1966). Interactions in aqueous nonelectrolyte solutions. I. *Solute-solvent equilibria. Journal of Physical Chemistry*, *70*, 2126–2130. doi:10.1021/j100879a010.

Tegen, I., & Fung, I. (1994). Modeling of mineral dust in the atmosphere: Sources, transport, and optical thickness. *Journal of Geophysical Research*, *99*(D11), 22897–22914.

Textor, C., Schulz, M., Guibert, S., Kinne, S., Balkanski, Y., Bauer, S., et al. (2006). Analysis and quantification of the diversities of aerosol life cycles within AeroCom. *Atmospheric Chemistry and Physics*, *6*, 1777–1813. 10.5194/acp-6-1777-2006.

Tilmes, S., Hodzic, A., Emmons, L. K., Mills, M. J., Gettelman, A., Kinnison, D. E., et al. (2019). Climate forcing and trends of organic aerosols in the Community Earth System Model (CESM2). *Journal of Advances in Modeling Earth Systems*, *11*, 4323–4351. 10.1029/2019MS001827.

Tosca, M. G., Randerson, J. T., & Zender, C. S. (2013). Global impact of smoke aerosols from landscape fires on climate and the Hadley circulation. *Atmospheric Chemistry and Physics*, *13*, 5227–5241. doi:10.5194/acp-13-5227-2013.

Tröstl, J., Chuang, W. K., Gordon, H., Heinritzi, M., Yan, C., Molteni, U., et al. (2016). The role of low-volatility organic compounds in initial particle growth in the atmosphere. *Nature*, *533*, 527–531.

Twomey, S. (1974). Pollution and the planetary albedo. *Atmospheric Environment, 8*(12), 1251–1256.

Twomey, S. (1977). The influence of pollution on the shortwave albedo of clouds. *Journal of the Atmospheric Sciences*, 34(7), 1149–1152.

Val Martin, M., Kahn, R. A., Logan, J. A., Paugam, R., Wooster, M., & Ichoku, C. (2012). Space-based observational constraints for 1-D fire smoke plume-rise models. *Journal of Geophysical Research: Atmospheres*, *117*(22), 1–18. doi:10.1029/2012JD018370.

Val Martin, M., Kahn, R. A., & Tosca, M. G. (2018). A global analysis of wildfire smoke injection heights derived from space-based multi-angle imaging. *Remote Sensing*, *10*, 1609.

Vali, G. (2008). Repeatability and randomness in heterogeneous freezing nucleation. *Atmospheric Chemistry and Physics, 8*, 5017–5031.

Vali, G. (2014). Interpretation of freezing nucleation experiments: Singular and stochastic; sites and surfaces. *Atmospheric Chemistry and Physics*, *14*, 5271–5294.

Van Der Werf, G. R., Randerson, J. T., Giglio, L., Collatz, G. J., Mu, M., Kasibhatla, P. S., et al. (2010). Global fire emissions and the contribution of deforestation, savanna, forest, agricultural, and peat fires (1997-2009). *Atmospheric Chemistry and Physics*, *10*(23), 11707–11735. doi:10.5194/acp-10-11707-2010.

Van Der Werf, G. R., Randerson, J. T., Giglio, L., van Leeuwen, T. T., Chen, Y., Rogers, B. M., et al. (2017). Global fire emissions estimates during 1997–2016. *Earth System Science Data*, *9*, 697–720. 10.5194/essd-9-697-2017.

van Marle, M. J. E., Kloster, S., Magi, B. I., Marlon, J. R., Daniau, A.-L., Field, R. D., et al. (2017). Historic global biomass burning emissions for CMIP6 (BB4CMIP) based on merging satellite observations with proxies and fire models (1750–2015). *Geoscientific Model Development*, *2017*, 10, 3329–3357. 10.5194/gmd-10-3329-2017.

Vehkamäki, H., Kulmala, M., Napari, I., Lehtinen, K. E. J., Timmreck, C., Noppel, M., & Laaksonen, A. (2002). An improved parameterization for sulfuric acid water nucleation rates for tropospheric and stratospheric conditions. *Journal of Geophysical Research*, *107*, 4622. doi:10.1029/2002JD002184.

Veira, A., Kloster, S., Wilkenskjeld, S., & Remy, S. (2015). Fire emission heights in the climate system—Part 1: Global plume height patterns simulated by ECHAM6-HAM2. *Atmospheric Chemistry and Physics*, *15*, 7155–7171. 10.5194/acp-15-7155-2015.

Wan, H., Rasch, P. J., Zhang, K., Kazil, J., & Leung, L. R. (2013). Numerical issues associated with compensating and competing processes in climate models: An example from ECHAM-HAM. *Geoscientific Model Development*, *6*, 861–874. 10.5194/gmd-6-861-2013.

Wang, H., Easter, R. C., Rasch, P. J., Wang, M., Liu, X., Ghan, S. J., et al. (2013). Sensitivity of remote aerosol distributions to representation of cloud–aerosol interactions in a global climate model. *Geoscientific Model Development*, *6*(3), 765–782. 10.5194/gmd-6-765-2013.

Wang, H., Easter, R. C., Zhang, R., Ma, P. L., Singh, B., Zhang, K., et al. (2020). Aerosols in the E3SM version 1: New developments and their impacts on radiative forcing. *Journal of Advances in Modeling Earth Systems*, *12*(1). 10.1029/2019ms001851.

Wang, M., & Penner, J. E. (2009). Aerosol indirect forcing in a global model with particle nucleation. *Atmospheric Chemistry and Physics*, *9*, 239–260. 10.5194/acp-9-239-2009.

Wang, M., & Penner, J. E. (2010). Cirrus clouds in a global climate model with a statistical cirrus cloud scheme. *Atmospheric Chemistry and Physics*, *10*(12), 5449–5474.

Wang, M., Ghan, S., Liu, X., L'Ecuyer, T. S., Zhang, K., Morrison, H., et al. (2012). Constraining cloud lifetime effects of aerosol using A-Train satellite observations. *Geophysical Research Letters, 39*, L15709. doi:10.1029/2012GL052204.

Wang, M., Liu, X., Zhang, K., & Comstock, J. M. (2014). Aerosol effects on cirrus through ice nucleation in the Community Atmosphere Model CAM5 with a statistical cirrus scheme. *Journal of Advances in Modeling Earth Systems*, *06756–776*. doi:10.1002/2014MS000339.

Wang, X., Zhang, L., & Moran, M. D. (2010). Uncertainty assessment of current size-resolved parameterizations for below-cloud particle scavenging by rain. *Atmospheric Chemistry and Physics*, *10*, 5685–5705. 10.5194/acp-10-5685-2010.

Wang, X., Heald, C. L., Sedlacek, A. J., de Sá, S. S., Martin, S. T., Alexander, M. L., et al. (2016). Deriving brown carbon from multiwavelength absorption measurements: Method and application to AERONET and Aethalometer observations. *Atmospheric Chemistry and Physics*, *16*(19), 12733–12752. doi:10.5194/acp-16-12733-2016.

Wang, X., Heald, C. L., Liu, J., Weber, R. J., Campuzano-Jost, P., Jimenez, J. L., et al. (2018). Exploring the observational constraints on the simulation of brown carbon. *Atmospheric Chemistry and Physics*, *18*(2), 635–653. doi:10.5194/acp-18-635-2018.

Wang, Y., Zhang, Q. Q., Jiang, J., Zhou, W., Wang, B., He, K., et al. (2014a). Enhanced sulfate formation during China's severe winter haze episode in January 2013 missing from current models. *Journal of Geophysical Research*, *119*, 10,425–10,440. doi:10.1002/2013JD021426.

Wang, Y., & Liu, X. (2014). Immersion freezing by natural dust based on a soccer ball model with the Community Atmospheric Model version 5: Climate effects. *Environmental Research Letters*, *9*, 124020. doi:10.1088/1748-9326/9/12/124020.

Wang, Y., Liu, X., Hoose, C., & Wang, B. (2014b). Different contact angle distributions for heterogeneous ice nucleation in the Community Atmospheric Model version 5. *Atmospheric Chemistry and Physics*, *14*, 10411–10430. 10.5194/acp-14-10411-2014.

Ward, D. S., Kloster, S., Mahowald, N. M., Rogers, B. M., Randerson, J. T., & Hess, P. G. (2012). The changing radiative forcing of fires: Global model estimates for past, present and future. *Atmospheric Chemistry and Physics*, *12*, 10857–10886. doi:10.5194/acp-12-10857-2012.

Wesely, M. L. (1989). Parameterization of surface resistance to gaseous dry deposition in regional-scale numerical models. *Atmospheric Environment*, *23*, 1293–1304. 10.1016/0004-6981(89)90153-4.

Wesely, M. L., & Hicks, B. B. (1977). Some factors that affect the deposition rates of sulfur dioxide and similar gases on vegetation. *Journal of the Air Pollution Control Association*, *27*, 1110–1116. 10.1080/00022470.1977.10470534.

Westbrook, C. D., & Illingworth, A. J. (2013). The formation of ice in a long-lived supercooled layer cloud. *Quarterly Journal of the Royal Meteorological Society*, *139*, 2209–2221. 10.1002/qj.2096.

Westerling, A. L., Hidalgo, H. G., Cayan, D. R., & Swetnam, T. W. (2006). Warming and earlier spring increase western U.S. forest wildfire activity. *Science (80-.)*, *313(5789)*, 940–943. doi:10.1126/science.1128834.

White, B. R. (1979). Soil transport by winds on Mars. *Journal of Geophysical Research—Solid Earth, 84*, 4643–4651. 10.1029/JB084iB09p04643.

Wiedinmyer, C., Quayle, B., Geron, C., Belote, A., McKenzie, D., Zhang, X., et al. (2006). Estimating emissions from fires in North America for air quality modeling. *Atmospheric Environment*, *40*(19), 3419–3432. doi:10.1016/j.atmosenv.2006.02.010.

Wiedinmyer, C., Akagi, S. K., Yokelson, R. J., Emmons, L. K., Al-Saadi, J. A., Orlando, J. J., & Soja, A. J. (2011). The Fire INventory from NCAR (FINN): A high resolution global model to estimate the emissions from open burning. *Geoscientific Model Development*, *4*(3), 625–641. doi:10.5194/gmd-4-625-2011.

Wilson, T. W., Ladino, L. A., Alpert, P. A., Breckels, M. N., Brooks, I. M., Browse, J., et al. (2015). A marine biogenic source of atmospheric ice-nucleating particles. *Nature*, *525*(7568), 234–238.

Wood, R. (2005). Drizzle in stratiform boundary layer clouds. Part II: Microphysical aspects. *Journal of the Atmospheric Sciences*, *62*, 3034–3050.

Wu, M., Liu, X., Zhang, L., Wu, C., Lu, Z., Ma, P.-L., et al. (2018). Impacts of aerosol dry deposition on black carbon spatial distributions and radiative effects in the Community Atmosphere Model CAM5. *Journal of Advances in Modeling Earth Systems*, *10*, 1150–1171. 10.1029/2017MS001219.

Wu, C., Lin, Z., He, J., Zhang, M., Liu, X., Zhang, R., & Brown, H. (2016). A process-oriented evaluation of dust emission parameterizations in CESM: Simulation of a typical severe dust storm in East Asia. *Journal of Advances in Modeling Earth Systems*, *8*, 1432–1452. doi:10.1002/2016MS000723.

Yang, M., Blomquist, B. W., Fairall, C. W., Archer, S. D., & Huebert, B. J. (2011). Air-sea exchange of dimethylsulfide in the Southern Ocean: Measurements from SO GasEx compared to temperate and tropical regions. *Journal of Geophysical Research*, *116*, c03F05. 10.1029/2010Jc036526.

Yu, F., & Luo, G. (2009). Simulation of particle size distribution with a global aerosol model: Contribution of nucleation to aerosol and CCN number concentrations. *Atmospheric Chemistry and Physics*, *9*, 7691–7710. 10.5194/acp-9-7691-2009.

Yao, L., Garmash, O., Bianchi, F., Zhen, J., Yan, C., Kontkanen, J., *et al* (2018). Atmospheric new particle formation from sulfuric acid and amines in a Chinese megacity. *Science*, *361*, 278–281.

Yu, P., Toon, O. B., Bardeen, C. G., Mills, M. J., Fan, T., English, J. M., & Neely, R. R. (2015). Evaluations of tropospheric aerosol properties simulated by the community earth system model with a sectional aerosol microphysics scheme. *Journal of Advances in Modeling Earth Systems*, *7*, 865–914. doi:10.1002/2014MS000421.

Yu, P., Froyd, K. D., Portmann, R. W., Toon, O. B., Freitas, S. R., Bardeen, C. G., et al. (2019). Efficient in-cloud removal of aerosols by deep convection. *Geophysical Research Letters*, *46*, 1061–1069. 10.1029/2018GL080544.

Zaveri, R. A., Easter, R. C., Fast, J. D., & Peters, L. K. (2008). Model for Simulating Aerosol Interactions and Chemistry (MOSAIC). *Journal of Geophysical Research*, *113*, D13204. doi:10.1029/2007JD008782.

Zdanovskii, A. B. (1948). New methods for calculating solubilities of electrolytes in multicomponent systems. *Zhurnal fizicheskoi Khimii*, *22*, 1475–1485.

Zender, C., Bian, H., & Newman, D. (2003a). Mineral Dust Entrainment and Deposition (DEAD) model: Description and 1990s dust climatology. *Journal of Geophysical Research*, *108*(D14), 4416. doi:4410.1029/2002JD002775.

Zender, C., Newman, D., & Torres, O. (2003b). Spatial heterogeneity in aerolian erodibility: Uniform, topographic, geomorphic and hydrologic hypotheses. *Journal of Geophysical Research*, *108*(D17), 4543. doi:4510.1029/2002JD003039.

Zhang, J., & Shao, Y. (2014). A new parameterization of particle dry deposition over rough surfaces. *Atmospheric Chemistry and Physics*, *14*, 12429–12440. 10.5194/acp-14-12429-2014

Zhang, K., O'Donnell, D., Kazil, J., Stier, P., Kinne, S., Lohmann, U., et al. (2012). The global aerosol-climate model ECHAM-HAM, version 2: Sensitivity to improvements in process representations. *Atmospheric Chemistry and Physics, 12*, 8911–8949. 10.5194/acp-12-8911-2012, 2012.

Zhang, K., Zhao, C., Wan, H., Qian, Y., Easter, R. C., Ghan, S. J., et al. (2016). Quantifying the impact of sub-grid surface wind variability on sea salt and dust emissions in CAM5. *Geoscientific Model Development*, *9*, 607–632. doi:10.5194/gmd-9-607-2016.

Zhang, L., & He, Z. (2014). Technical Note: An empirical algorithm estimating dry deposition velocity of fine, coarse and giant particles. *Atmospheric Chemistry and Physics*, *14*, 3729–3737. 10.5194/acp-14-3729-2014.

Zhang, L., & Vet, R. (2006). A review of current knowledge concerning size-dependent aerosol removal. *China Particuology*, *4*, 272–282.

Zhang, L., Gong, S., Padro, J., & Barrie, L. (2001). A size-segregated particle dry deposition scheme for an atmospheric aerosol module. *Atmospheric Environment*, *35*, 549–560. 10.1016/S1352-2310(00)00326-5

Zhang, M., Lin, W., Klein, S., Bacmeister, J., Bony, S., Cederwall, R., et al. (2005). Comparing clouds and their seasonal variations in 10 atmospheric general circulation models with satellite measurements. *Journal of Geophysical Research*, *110*(D15S02). doi:10.1029/2004JD005021.

Zhang, Q., Jimenez, J. L., Canagaratna, M. R., Allan, J. D., Coe, H., Ulbrich, I., et al. (2007). Ubiquity and dominance of oxygenated species in organic aerosols in anthropogenically-influenced Northern Hemisphere midlatitudes. *Geophysical Research Letters*, *34*, L13801. doi:10.1029/2007GL029979

Zhang, Q., Streets, D. G., Carmichael, G. R., He, K. B., Huo, H., Kannari, A., et al. (2009). Asian emissions in 2006 for the NASA INTEX-B mission. *Atmospheric Chemistry and Physics*, *9*, 5131–5153. 10.5194/acp-9-5131-2009.

Zhao, A., Bollasina, M. A., & Stevenson, D. S. (2019). Strong influence of aerosol reductions on future heatwaves. *Geophysical Research Letters*, *46*, 4913–4923. 10.1029/2019GL082269.

Zhao, B., Wang, Y., Gu, Y., Liou, K.-N., Jiang, J. H., Fan, J., et al. (2019). Evidence for the ice nucleation ability of anthropogenic aerosols. *Nature Geosciences*, *12*, 602–607. 10.1038/s41561-019-0389-4.

Zhao, R., Lee, A. K. Y., Huang, L., Li, X., Yang, F., & Abbatt, J. P. D. (2015). Photochemical processing of aqueous atmospheric brown carbon. *Atmospheric Chemistry and Physics*, *15*(11), 6087–6100. doi:10.5194/acp-15-6087-2015.

Zhao, X., & Liu, X. (2021). Global importance of secondary ice production. *Geophysical Research Letters*, *48*, e2021GL092581. https://doi.org/10.1029/2021GL092581.

Zhao, X., Liu, X., Phillips, V. T. J., & Patade, S. (2021a). Impacts of secondary ice production on Arctic mixed-phase clouds based on ARM observations and CAM6 single-column model simulations. *Atmospheric Chemistry and Physics*, *21*(7), 5685–5703. https://doi.org/10.5194/acp-2020-1276.

Zhao, X., Liu, X., Burrows, S. M., and Shi, Y. (2021b). Effects of marine organic aerosols as sources of immersion-mode ice-nucleating particles on high-latitude mixed-phase clouds. *Atmospheric Chemistry and Physics*, *21*, 2305–2327. https://doi.org/10.5194/acp-21-2305-2021.

Zheng, B., Zhang, Q., Zhang, Y., He, K. B., Wang, K., Zheng, G. J., et al. (2015). Heterogeneous chemistry: A mechanism missing in current models to explain secondary inorganic aerosol formation during the January 2013 haze episode in North China. *Atmospheric Chemistry and Physics*, *15*, 2031–2049. https://doi.org/10.5194/acp-15-2031-2015.

Zhong, M., & Jang, M. (2011). Light absorption coefficient measurement of SOA using a UV–Visible spectrometer connected with an integrating sphere. *Atmospheric Environment*, *45*(25), 4263–4271. doi:10.1016/j.atmosenv.2011.04.082.

Zhu, J., Penner, J. E., Yu, F., Sillman, S., Andreae, M. O., & Coe, H. (2019). Decrease in radiative forcing by organic aerosol nucleation, climate, and land use change. *Nature Communications*, *10*(1), 423. doi:10.1038/s41467-019-08407-7.

Zufall, M. J., Davidson, C. I., Caffrey, P. F., & Ondov, J. M. (1998). Airborne concentrations and dry deposition fluxes of particulate species to surrogate surfaces deployed in southern Lake Michigan. *Environmental Science & Technology*, *32*, 1623–1628. https://doi.org/10.1021/es9706458.

4

Entrainment, Mixing, and Their Microphysical Influences

Chunsong Lu[1], Yangang Liu[2], Xiaoqi Xu[1], Sinan Gao[1], and Cheng Sun[1]

ABSTRACT

Entrainment, subsequent mixing processes, and their microphysical influences in cumulus and stratocumulus clouds are essential to cloud physics and further affect aerosol-cloud interactions, cloud-climate feedbacks, etc. Theoretical understanding of these processes is still not clear and their parameterizations in atmospheric models are in their infancy. This chapter focuses on two aspects: entrainment-mixing mechanisms (cloud microphysics) and entrainment rate (dynamics). Unifying microphysical measures are defined to quantify different entrainment-mixing mechanisms, and their relationships to the dynamical measures are examined for the purpose of providing physical understanding and developing the parameterization of entrainment-mixing processes in large-scale atmospheric models. For entrainment rate in cumulus clouds, approaches for estimating the fractional entrainment rate are reviewed, including bulk-plume approaches, a mixing fraction approach, and several other approaches. Relationships of entrainment rate to internal cloud properties or external properties are discussed as deterministic parameterizations; entrainment is also parameterized as a stochastic process. Three approaches for estimating entrainment velocity in stratocumulus clouds are also reviewed. Several remaining challenges are explored, including the connection between entrainment-mixing mechanisms and entrainment rate, detailed cloud response to entrainment-mixing in liquid-phase and ice/mixed-phase clouds, secondary entrainment-mixing events, warm rain initiation, detrainment, and effects of initial cloud droplet spectral width and environmental factors.

4.1. INTRODUCTION

Clouds are essential for radiative transfer, weather, and climate (e.g., Betts & Harshvardhan, 1987; Guo et al., 2015a; Li et al., 2017; Li et al., 2021b; Randall et al., 2016; Yan et al., 2016). One reason for the uncertainties in simulations/projections of climate change is cloud-climate feedbacks because many cloud-related physical processes need to be parameterized in large-scale models (Bony & Dufresne, 2005; Cess et al., 1989; Colman, 2003; Guo et al., 2015c; Lin et al., 2020; Li et al., 2021a; Stephens, 2005; Wang et al., 2017; Zhang et al., 2014) and the understanding of these physical processes is far from complete (Donner et al., 2001; Lei et al., 2008; Li et al., 2019; Lin et al., 2015; Zhang et al., 2018). One of such subgrid processes is entrainment between cloud and environmental air, and the subsequent mixing processes (e.g., Liu et al., 2002; von Salzen & McFarlane, 2002).

There are at least two scientific questions on entrainment-mixing processes: (1) What is the response of cloud microphysics to the entrained dry air (entrainment-mixing mechanisms)? (2) How fast is dry air entrained into clouds

[1]*Collaborative Innovation Center on Forecast and Evaluation of Meteorological Disasters / Key Laboratory for Aerosol-Cloud-Precipitation of China Meteorological Administration, Nanjing University of Information Science and Technology, Nanjing, China*

[2]*Environmental and Climate Sciences Department, Brookhaven National Laboratory, Upton, NY, USA*

Fast Processes in Large-Scale Atmospheric Models: Progress, Challenges, and Opportunities, Geophysical Monograph 282, First Edition.
Edited by Yangang Liu and Pavlos Kollias.

DOI: 10.1002/9781119529019.ch04

(entrainment rate)? Entrainment-mixing mechanisms affect many cloud-related issues, including warm-rain initiation, evaluation of aerosol indirect effects, and radar retrieval of cloud microphysics (e.g., Blyth, 1993; Dawe & Austin, 2013; Del Genio & Wu, 2010; Jiang et al., 2006; Kim et al., 2008; Lu et al., 2016; Xue & Feingold, 2006; Yang et al., 2016). Entrainment rate affects cloud lifetimes and the evolution of cloud systems (Duan et al., 2018; Mellado, 2017), and further affect numerical simulations of boundary layer (Stevens, 2002), precipitation (Hirota et al., 2014; Loriaux et al., 2013; Wang et al., 2007; Yang et al., 2015; Yang et al., 2020), lightning (Formenton et al., 2013), Madden-Julian Oscillation (Klingaman & Woolnough, 2014), Asian monsoon (Bush et al., 2015), El Niño-Southern Oscillation (Kim et al., 2011b; Lu & Ren, 2016), tropical cyclone (Lim et al., 2015), and convectively coupled equatorial waves (Peatman et al., 2018). Many efforts have been devoted to the topics, including observations (Beals et al., 2015; Gerber et al., 2005; Liu et al., 2002; Luo et al., 2010; Siebert et al., 2006; Wagner et al., 2013; Wang et al., 2021; Warner, 1973), laboratory experiments (Korczyk et al., 2012; Latham & Reed, 1977), numerical simulations, and theoretical analyses (Andrejczuk et al., 2009; Gregory, 2001; Kumar et al., 2017; Lasher-Trapp et al., 2005; Pinsky et al., 2016b; Simpson, 1971; Wang, 2020). Despite the efforts and progress, the precise nature of these entrainment and subsequent mixing processes is still an active field of research and their parameterizations in various models are still in their infancy (de Rooy et al., 2013; Lu et al., 2018a).

This chapter discusses the topics of entrainment-mixing mechanisms and entrainment rate. For entrainment-mixing mechanisms, we mainly focus on nonprecipitating liquid clouds (shallow cumulus and stratocumulus clouds), with some brief discussion on deep cumulus clouds. Studies of entrainment rate are focused on shallow and deep cumulus clouds; studies on stratocumulus clouds are briefly reviewed.

4.2. IMPORTANCE OF CUMULUS AND STRATOCUMULUS

Low-level clouds (shallow cumulus and stratocumulus) reflect shortwave radiation but have only relatively small effects on outgoing longwave radiation (Jian et al., 2021; Wood, 2012; Zhao & Ishizaka, 2004); therefore, these clouds have strong negative net radiative forcing (Karlsson et al., 2008; Wood, 2012; Zheng et al., 2019) and significant feedbacks to climate change (Jian et al., 2021; Klein et al., 2018; Vial et al., 2017). Deep cumulus clouds are important for precipitation (Liu et al., 2019; Zhao et al., 2018) and play critical roles in transporting water vapor and chemical substances to tropopause or stratosphere (Evan et al., 2020; Holton & Gettelman, 2001). The importance of shallow cumulus clouds is reviewed here. See Chapters 5 and 6 for detailed discussions on deep convective and stratiform clouds, respectively.

Shallow cumulus clouds can occur in large areas on Earth and are one of the most common cloud types in the troposphere (Isotta et al., 2011; Rauber et al., 2007). They have significant impacts on radiative transfer. For example, Berg et al. (2011), based on the measurements of shallow cumulus clouds over eight summers at the U.S. Department of Energy's Atmospheric Radiation Measurement (ARM) Southern Great Plains (SGP) site, found that the average shortwave and longwave cloud radiative forcing at the surface was −45.5 and 15.9 W m^{-2}, respectively. Furthermore, these clouds transport heat and moisture vertically from the boundary layer into the free troposphere and redistribute temperature and moisture (Brown et al., 2002; Hong & Jang, 2018; Raga et al., 1990a; Stevens et al., 1998; Stull, 1985), indirectly affecting cloud cover, depth of the boundary layer, and radiation budget (Bretherton et al., 2004). Demott et al. (2005) found that the vertical transport of water vapor by shallow cumulus clouds paved the way for the development of deep cumulus clouds. Bellenger et al. (2015) directly linked shallow cumulus clouds to moisture variations within the lowest 4 km of the atmosphere with high-frequency observations of humidity and convection. They found that the moistening by shallow cumulus clouds was a steady moisture supply and its importance could increase with the timescale, although large-scale advective tendencies were stronger. Based on long-term observations and reanalysis products, Zermeno-Diaz et al. (2016) estimated daily moistening tendencies of shallow cumulus clouds and found that there were evident increases in low-level moisture before rainfall peaks. Pilon et al. (2016) pointed out that shallow cumulus clouds moistened the lower troposphere, enhanced total precipitation and diabatic heating, and weakened the diurnal cycle in rainfall.

Because of the small spatial and temporal scales, shallow cumulus clouds have to be parameterized in numerical weather prediction and climate models, and the parameterizations are critical for virtually all climate variables. For example, McCaa and Bretherton (2004) found that the transition from stratocumulus to cumulus in the northeast Pacific and the maintenance of cumulus clouds in the southeastern Pacific heavily depended on the shallow cumulus parameterization schemes. Regardless of their detailed differences, most shallow cumulus parameterization schemes are based on the idea of mass flux (Arakawa & Schubert, 1974; Baba, 2020; Freitas et al., 2020; Hack, 1994; Kain, 2004; Tiedtke, 1989; Park & Bretherton, 2009; Zhang & McFarlane, 1995;

Zhao et al., 2016;). Shallow cumulus clouds are usually parameterized separately from deep convection. There have been growing efforts to unify their treatments (Park, 2014; Shin & Park, 2019; see Chapters 10–13 for more discussion on such unifying studies).

As mentioned above, entrainment-mixing in shallow/deep cumulus and stratocumulus is still not clear, preventing accurate description of entrainment-mixing in parameterization schemes and accurate simulation of cloud, precipitation, and radiative transfer. Therefore, this chapter focuses on this topic.

4.3. ENTRAINMENT-MIXING MECHANISMS

4.3.1. Different Conceptual Models of Entrainment-Mixing Mechanisms

The effects of entrainment-mixing processes on cloud microphysics have been extensively studied since the 1960s. The most studied concepts are based on the notion of homogeneous/inhomogeneous entrainment-mixing mechanisms (Baker et al., 1980). Figure 4.1 summarizes the evolution of cloud microphysics in adiabatic clouds and diluted clouds affected by different entrainment-mixing mechanisms (e.g., Liu et al., 2002; Yum, 1998). For the homogeneous entrainment-mixing mechanism, entrained dry air is assumed to mix with cloudy air immediately and all cloud droplets experience the same condition in evaporation, decreasing droplet sizes. Droplet number in the cloudy volume does not change; but droplet number concentration decreases because entrained air increases cloud volume (dilution). Therefore, homogeneous entrainment-mixing leads to a positive correlation between the mean droplet size and droplet number concentration, with cloud droplet size distributions shifting leftward to the end of small droplet sizes. In the extreme inhomogeneous entrainment-mixing process, turbulent mixing is so slow that some droplets completely evaporate to saturate the entrained dry air before mixing occurs. Cloud droplet size distributions shift downward to the end of small number concentration, with mean droplet sizes not changing but droplet number and droplet number concentration decreasing significantly. Fewer droplets mean less competition for water vapor; therefore, the remaining droplets can grow bigger if the cloud parcel is lifted upward again after entrainment-mixing. This process is termed as the inhomogeneous entrainment-mixing with subsequent ascent.

Warner (1973) was the first to study the effects of homogeneous entrainment-mixing mechanism on cloud microphysics. Using a parcel model, Warner (1973) concluded that this mechanism could not reproduce the observed bimodal droplet size distributions reported in Warner (1969). However, the spectra were successfully reproduced by Mason and Jonas (1975), Jonas and Mason (1982), and Lee and Pruppacher (1977), also assuming this mechanism. Latham and Reed (1977) were the first to report the inhomogeneous mechanism based on laboratory experiments.

The homogeneous/inhomogeneous entrainment-mixing mechanisms are often studied in the mixing diagram as shown in Figure 4.2 (Beals et al., 2015; Burnet & Brenguier, 2007; Kumar et al., 2018; Lu et al., 2013a). The states in entrainment-mixing are numbered from 1 to 3. State 1 is an adiabatic cloud with volume-mean radius of r_{va} and number concentration of n_a. State 2 is just diluted due to entrainment of dry air with r_{va} and diluted number concentration of n_h. From States 2 to 3, volume-mean radius changes from r_{va} to r_v, and number concentration changes from n_h to n, due to mixing and evaporation. When a mixing process is homogeneous (State 3″), the number concentration and volume-mean radius are n_h and r_{vh}, respectively. Cubic volume-mean radius normalized by cubic adiabatic volume-mean radius (r_v^3/r_{va}^3) is positively correlated with number concentration normalized by adiabatic number concentration (n/n_a). When a mixing process is extreme inhomogeneous (State 3′), the number concentration and volume-mean radius are n_i and r_{va}, respectively, i.e., droplet size does not change with decreasing number concentration. When inhomogeneous entrainment-mixing with subsequent ascent occurs, State 3‴ is above the horizontal extreme inhomogeneous entrainment-mixing line and r_v is larger than r_{va}. In this case, r_v^3/r_{va}^3 and n/n_a are negatively correlated (Figure 4.1).

Besides the framework of homogeneous/inhomogeneous mechanisms that are based on the idea of isobaric mixing, some ideas based on vertical mixing have been proposed to explain the microphysical variations during entrainment-mixing processes, including the entity-type entrainment-mixing and vertical circulation mixing. The entity-type entrainment-mixing (Telford and Chai, 1980; Telford, 1996) is composed of four stages. In the first stage, the entity forms when cloudy air near cloud top mixes with the dry air above the cloud top. Droplets evaporate completely and the entity becomes saturated. In the second stage, droplets continue to enter the entity without further evaporation because the entity is already saturated. In the third stage, when the entity descends, the droplets evaporate to much smaller sizes than those in the adiabatic clouds and large adiabatic droplets nearby enter the entity. In the fourth stage, the entity could be forced to rise and water vapor mainly condenses on the large droplets, which leads to enhanced growth of big droplets. This stage is similar to the inhomogeneous entrainment-mixing with subsequent ascent. A key property that determines the end result of the entity-type entrainment-mixing is how many times the turbulent entity recycles in clouds.

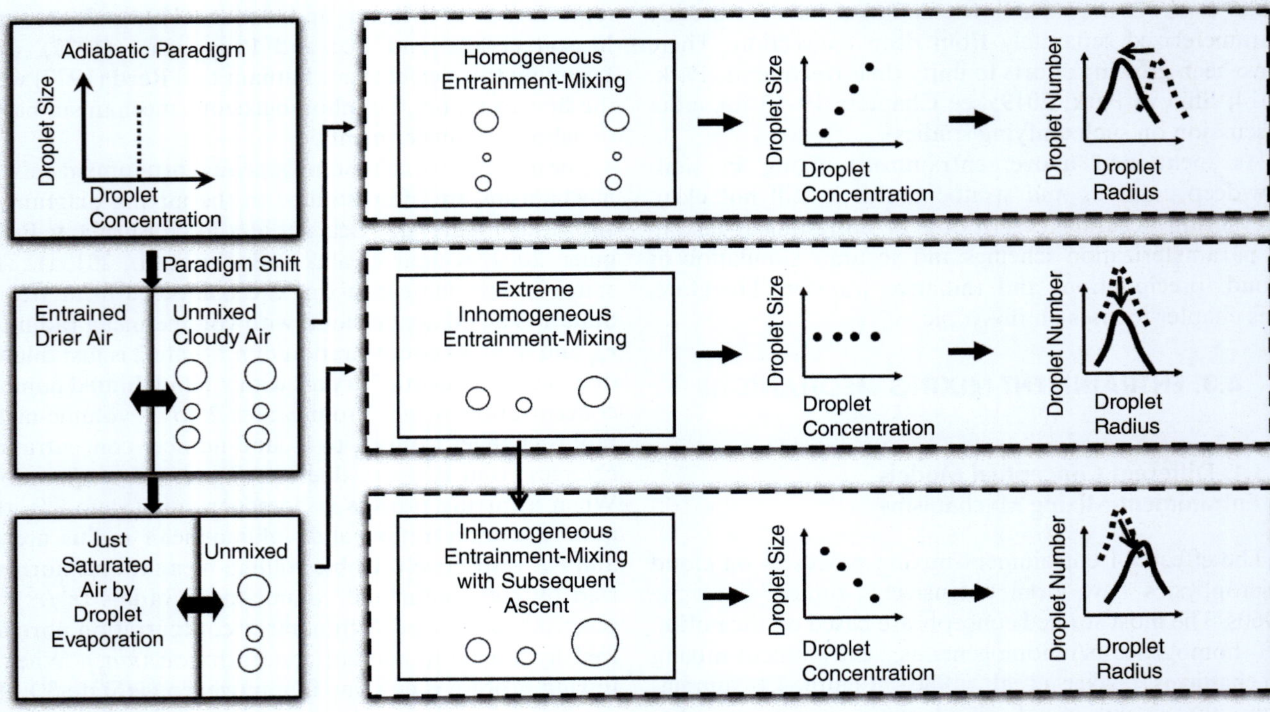

Figure 4.1 Schematic for the evolution of cloud microphysics in adiabatic and diluted clouds affected by homogeneous entrainment-mixing mechanism, extreme inhomogeneous entrainment-mixing mechanism, and inhomogeneous entrainment-mixing with subsequent ascent, respectively (after Yum & Hudson (2001), Liu et al. (2002), and Lu et al. (2018a)).

Similar to the entity-type entrainment-mixing and inhomogeneous entrainment-mixing with subsequent ascent, vertical circulation mixing (Wang et al., 2009; Yeom et al., 2017; Yum et al., 2015) can also produce large droplets in diluted clouds than in adiabatic clouds. Betts (1978, 1983) and Schubert et al. (1979) developed a conceptual model to describe the ascending and descending branches of circulation in the stratocumulus boundary layer. Wang et al. (2009) further related vertical variation of microphysical properties to the mechanism of vertical circulation mixing. In the ascending branch of vertical circulation, a parcel rises from the surface (e.g., ocean) and forms cloud at the height of the cloud condensation level. When the cloud parcel is affected by entrainment at the cloud top, liquid water content is reduced. Due to evaporative and/or radiative cooling, the cloud parcel starts to descend to its cloud base where liquid water content is close to 0. Below the cloud base, further descent of the parcel follows a dry adiabat. When arriving at the surface, the parcel receives water vapor from the surface and starts another circulation. The cloud base of the parcel affected by entrainment is higher than that of the adiabatic cloud parcel due to less water content. Mixtures of cloud parcels with different cloud base heights have different cloud droplet number concentrations. During an ascent, the cloud droplets in the mixture with a lower number concentration can grow larger due to less competition for water vapor. Again, a key property that determines the end result of the vertical circulation mixing is the number of times that the parcel recycles. Based on stratocumulus observations, Yum et al. (2015) speculated that the entrainment-mixing near the cloud top might be inhomogeneous, but due to vertical circulation mixing, the entrainment-mixing mechanism looked like homogeneous entrainment-mixing in the midlevel of clouds. The reason was that, during the descent, cloud droplets evaporated faster in more diluted cloud parcels, leading to a positive correlation between cubic volume-mean radius and liquid water content, regardless of the mixing mechanism near the cloud top. Yeom et al. (2021) developed a new diagram of LWC and phase relaxation time, and found that inhomogeneous entrainment-mixing trait was dominant near the cloud top, but homogeneous entrainment-mixing trait was stronger at lower altitudes. They further pointed out that such a vertical transition was consistent with the vertical circulation mixing. Yeom et al. (2017) also argued that vertical circulation mixing could enhance the signature of homogeneous entrainment-mixing.

Despite the similarity, there are notable differences between the entity-type mixing and the vertical

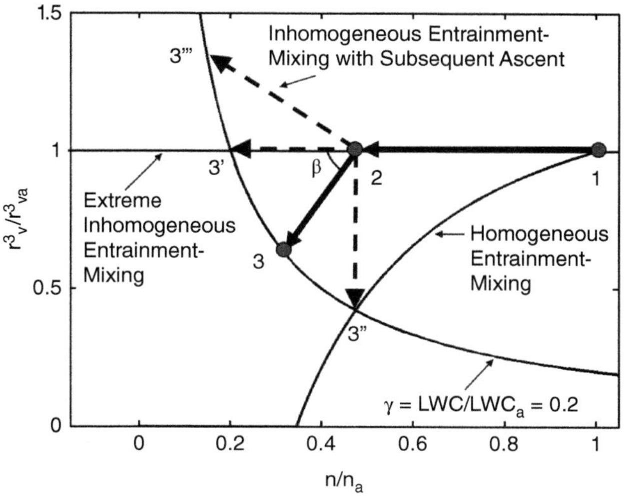

Figure 4.2 Mixing diagram of cubic volume-mean radius over cubic adiabatic volume-mean radius (r_v^3/r_{va}^3) vs. number concentration over adiabatic number concentration (n/n_a), illustrating homogeneous/inhomogeneous entrainment-mixing mechanisms. The black solid lines represent homogeneous entrainment-mixing (relative humidity in environmental air is assumed to be 66% as an example), extreme inhomogeneous entrainment-mixing, and contour of LWC/LWC$_a$ = 0.2; LWC is liquid water content and LWC$_a$ is adiabatic LWC. Inhomogeneous entrainment-mixing with subsequent ascent is also shown. See text for the other symbols' meanings. Source: Adapted from Lu et al. (2013a).

circulation mixing: the recycling in the entity-type entrainment-mixing is within the cloud, and the ascending branch starts at the position where descending branch ends. In the vertical circulation mixing, the recycling is between the surface and cloud top, and ascending and descending branches occur at different positions (Wang et al., 2009). It is also noteworthy that, for entity-type entrainment-mixing and vertical circulation mixing, entrainment is assumed to occur at the cloud top. This is expected to be valid in stratocumulus clouds but may not be true for cumulus clouds. For the concept of homogeneous/inhomogeneous entrainment-mixing mechanisms, entrainment can occur at the cloud top or near the cloud lateral boundary.

4.3.2. Quantification of Entrainment-Mixing Mechanisms

Microphysical Measure for Entrainment-Mixing Mechanisms

In the above-mentioned three main concepts of entrainment-mixing mechanisms, the entity-type entrainment-mixing and vertical circulation mixing are mainly used in qualitative analyses on cloud observations due likely to the difficulty to determine the recycling number of turbulent entities or parcels. To the authors' knowledge, no quantitative measures have been developed for them. Therefore, only the concept of homogeneous/inhomogeneous entrainment-mixing is discussed here.

In the framework of homogeneous/inhomogeneous entrainment-mixing, entrainment-mixing mechanisms often fall between the two extremes—homogeneous entrainment-mixing and extreme inhomogeneous entrainment-mixing (Lehmann et al., 2009; Lu et al., 2011). Besides qualitative description of entrainment-mixing mechanisms, quantitative measures are needed to describe entrainment-mixing mechanisms that encompass all the processes from homogeneous to extreme inhomogeneous; such unifying measures are also desirable in developing parameterizations for various entrainment-mixing processes. It is obvious that the correlations between key microphysical properties (e.g., droplet size, droplet concentration, and liquid water content) or slope of data points in the mixing diagram of droplet size vs. number concentration or droplet size vs. liquid water content can be used to represent different types of entrainment-mixing mechanisms (Andrejczuk et al., 2009; Wang et al., 2009; Yeom et al., 2017; Yum et al., 2015). Morrison and Grabowski (2008) developed an equation for n under arbitrary entrainment-mixing mechanisms:

$$n = n_0\left(\frac{\mathrm{LWC}}{\mathrm{LWC_0}}\right)^{\alpha} \tag{4.1}$$

where LWC$_0$ is the liquid water content after entrainment but before evaporation, and $0 \le \alpha \le 1$ is an empirical parameter determining the type of entrainment-mixing processes. For the homogeneous entrainment-mixing, $\alpha = 0$; $\alpha = 1$ for the extreme inhomogeneous entrainment-mixing. Therefore, α gauges the inhomogeneous entrainment-mixing degree; a larger α indicates more inhomogeneous mixing. Lu et al. (2013a, 2013b) defined three different homogeneous mixing degrees (ψ) based on the mixing diagram shown in Figure 4.2. The first homogeneous mixing degree (ψ_1) is defined as

$$\psi_1 = \frac{\beta}{\pi/2}, \tag{4.2a}$$

$$\beta = \tan^{-1}\left(\frac{\frac{r_v^3}{r_{va}^3} - 1}{\frac{n}{n_a} - \frac{n_h}{n_a}}\right), \tag{4.2b}$$

where β is the angle between the State 2 and State 3 line and the State 2 and State 3′ line. The second measure of homogeneous mixing degree (ψ_2) is based on the fact that n is between n_h and n_i, and r_v is between r_{vh} and r_{va}, such that

$$\psi_2 = \frac{1}{2}\left(\frac{n - n_i}{n_h - n_i} + \frac{r_v^3 - r_{va}^3}{r_{vh}^3 - r_{va}^3}\right), \tag{4.3a}$$

where

$$r_{\mathrm{vh}}^3 = \frac{n}{n_{\mathrm{h}}} r_{\mathrm{v}}^3, \tag{4.3b}$$

$$n_{\mathrm{i}} = \frac{r_{\mathrm{v}}^3}{r_{\mathrm{va}}^3} n. \tag{4.3c}$$

Taking the logarithm of the relationship between $r_{\mathrm{v}}^3/r_{\mathrm{va}}^3$ and n/n_{a}, the third measure of homogeneous mixing degree (ψ_3) is defined

$$\psi_3 = \frac{\ln n - \ln n_{\mathrm{i}}}{\ln n_{\mathrm{h}} - \ln n_{\mathrm{i}}} = \frac{\ln r_{\mathrm{v}}^3 - \ln r_{\mathrm{va}}^3}{\ln r_{\mathrm{vh}}^3 - \ln r_{\mathrm{va}}^3}, \tag{4.4}$$

which can be readily shown to be related to α by

$$\psi_3 = 1 - \alpha. \tag{4.5}$$

See Appendix for the detailed derivation of equation (4.5). Noting that number concentration in Figure 4.2 is sensitive to aerosols and vertical velocity and trying to minimize such effects, Lu et al. (2014b) defined another homogeneous mixing degree (ψ_4) based on the $r_{\mathrm{v}}^3/r_{\mathrm{va}}^3$ vs. LWC/LWC$_{\mathrm{a}}$ mixing diagram:

$$\psi_4 = \frac{1 - \frac{r_{\mathrm{v}}^3}{r_{\mathrm{va}}^3}}{1 - \frac{1}{\chi}\frac{\mathrm{LWC}}{\mathrm{LWC_a}}}, \tag{4.6}$$

where χ is the mixing fraction defined as the ratio of adiabatic cloud mass to total cloud mass when dry air is entrained into the cloud. This definition is similar to the inhomogeneous fraction based on the diagram of effective radius vs. LWC in Gerber et al. (2008).

Gao et al. (2018) compared the various measures of homogeneous mixing degree ($\psi_1 - \psi_4$) with the slope measure in terms of how well they are related to the dynamical measures. They found that the various measures of homogeneous mixing degree performed similarly and had much less uncertainties than the slope measure because they were better constrained by the two extreme mixing types. Accordingly, these homogeneous mixing degrees are used in the following discussions of appropriate timescales used to quantify the dynamical measures, dominant entrainment-mixing mechanisms, development of parameterizations, and dependence on the sampling scales.

Dynamical Measure for Entrainment-Mixing Mechanisms

Two dynamical measures have been used in the study of entrainment-mixing processes. The first one, Damkohler number (*Da*), has been widely used since ~1980, which is defined as the ratio of turbulent mixing timescale to microphysical timescale (Andrejczuk et al., 2009; Baker & Latham, 1979; Baker et al., 1980; Burnet & Brenguier, 2007; Jeffery, 2007; Lehmann et al., 2009; Siebert et al., 2006; Yeom et al., 2017; Yum et al., 2015):

$$Da = \frac{\tau_{\mathrm{mix}}}{\tau_{\mathrm{r}}}, \tag{4.7}$$

where τ_{mix} is the turbulent mixing time and τ_{r} is the microphysical response time of droplet population. The turbulent mixing time can be estimated by

$$\tau_{\mathrm{mix}} = \left(\frac{L^2}{\varepsilon}\right)^{1/3}, \tag{4.8}$$

where ε and L are the dissipation rate and dry air blob size, respectively.

Despite its popular use, *Da* requires knowledge of L used in the calculation of τ_{mix} that is not unique in turbulent clouds and is difficult to know (Lehmann et al., 2009). To overcome this difficulty, Lehmann et al. (2009) proposed use of the transition length (L^*) defined as the eddy size that corresponds to $Da = 1$ and given by

$$L^* = \varepsilon^{1/2}\tau_{\mathrm{r}}^{3/2}. \tag{4.9}$$

Note that a similar transition length was also introduced by Kabanov et al. (1971), Baker et al. (1980), and Jensen and Baker (1989). But Lehmann et al. (2009) conducted a systematic examination of it. They argued that if L^* fell within the turbulent inertial subrange, after a blob of size LE of subsaturated air was initially entrained into a cloud, all eddies of size L in the range of $L^* < L <$ LE experienced inhomogeneous entrainment-mixing processes, whereas eddies smaller than L^* mixed homogeneously. Note that L is the size of dry air that has experienced turbulent mixing after the initial dry air with the size of LE entrained into cloud. Following Lehmann et al. (2009) and recognizing that the lower end of the turbulent inertial subrange is represented by the Kolmogorov length scale (η), Lu et al. (2011) and Kumar et al. (2013) further introduced the dimensionless transition scale number (N_{L}), defined as the ratio of L^* to η:

$$N_L = \frac{L^*}{\eta}, \tag{4.10}$$

where η is given by

$$\eta = \left(\frac{\nu^3}{\varepsilon}\right)^{1/4}, \tag{4.11}$$

where ν is the kinematic viscosity (Wyngaard, 2010). Similar to L^*, N_{L} is a dynamical measure of entrainment-mixing processes but accounts for both L^* and η: a larger N_{L} indicates a more homogenous entrainment-mixing process. The transition scale number has been used in observational studies of entrainment-mixing processes (Lu et al., 2011; Yeom et al., 2017; Yum et al., 2015) and development of entrainment-mixing parameterizations (section 4.3.2.5).

Determination of the Appropriate Microphysical Timescales

Although there is little controversy about τ_{mix} in *Da* or N_L, three microphysical timescales have been used to represent τ_r in the study of entrainment-mixing processes: evaporation time (τ_{evap}) (Andrejczuk et al., 2009; Baker et al., 1980; Burnet & Brenguier, 2007), phase relaxation time (τ_{phase}) (Kumar et al., 2013), and reaction time (τ_{react}) (Lehmann et al., 2009; Lu et al., 2011, 2013a, 2014a). Which microphysical time should be used remains a topic of debate. Here, we first review the use of different τ_r and then discuss which is more appropriate in studying entrainment-mixing processes and under what conditions.

The evaporation time τ_{evap} for a single droplet is defined as follows. In a subsaturated environment, the change of droplet radius (r) during evaporation is described by

$$\frac{\mathrm{d}r}{\mathrm{d}t} = G\frac{S}{r}, \tag{4.12}$$

where S is the saturation ratio deficit, t is the time, and G is a function of temperature and air pressure (Rogers & Yau, 1989). Integration of equation (4.12) from the adiabatic droplet radius (r_a) to 0 yields

$$\tau_{evap} = -\frac{r_a^2}{2GS_0}, \tag{4.13}$$

Baker and Latham (1979) and Baker et al. (1980) compared τ_{evap} with τ_{mix} to discuss spatial scales for different entrainment-mixing mechanisms and found that extreme inhomogeneous entrainment-mixing mechanism was generally dominant for spatial scales over about 1 m. It is noteworthy that the derivation of equation (4.13) assumes constant S (equal to the initial saturation ratio deficit S_0).

Another commonly used microphysical timescale is the phase relaxation time τ_{phase} first introduced by Squires (1952). In an updraft, the variation of S is described by

$$\frac{\mathrm{d}S}{\mathrm{d}t} = Q_1\frac{\mathrm{d}z}{\mathrm{d}t} - Q_2\frac{\mathrm{d}LWC}{\mathrm{d}t}, \tag{4.14}$$

where Q_1 and Q_2 are functions of temperature and air pressure (Rogers & Yau, 1989), and z is the height. Omitting $Q_1\frac{\mathrm{d}z}{\mathrm{d}t}$ and using equation (4.12) and LWC = $\frac{4}{3}\pi\rho_L nr^3$ (ρ_L is liquid water density), equation (4.14) becomes

$$\frac{\mathrm{d}S}{\mathrm{d}t} = -CnrS, \tag{4.15}$$

where C is a function of temperature and air pressure (Kumar et al., 2013; Rogers & Yau, 1989). Integration of equation (4.15) from S_0 to S_0/e yields

$$\tau_{phase} = \frac{1}{Cnr_a}, \tag{4.16}$$

under the assumption of constant n, C, and $r = r_a$ (monodisperse droplet size distribution). Equation (4.16) indicates that τ_{phase} is a function of the product of n and r_a (integral radius) but independent of the initial saturation deficit S_0.

After τ_{phase} was first introduced to cloud physics by Squires (1952), τ_{phase} has been widely used in many aspects of cloud physics. First, quasi steady supersaturation can be calculated using τ_{phase} (Politovich & Cooper, 1988). By comparing τ_{phase} with the characteristic timescale of vertical turbulent fluctuation, Korolev and Mazin (2003) concluded that the quasi steady supersaturation gave a good estimate of supersaturation in liquid-phase and mixed-phase clouds; but in ice clouds, the estimation of such quasi steady supersaturation should be made with caution. Kabanov et al.(1971) compared τ_{phase} with the molecular diffusivity timescale to check if neighboring droplets could grow at different local supersaturation in isolated cells or simply grew at the common quasi steady supersaturation.

Second, fluctuation of supersaturation and cloud droplet spectral broadening in clouds is related to τ_{phase}. Ditas et al. (2012) found that τ_{phase} was a good timescale to explain different fluctuations of supersaturation under different conditions of number concentration and mean radius. Clark and Hall (1979) found that the horizontal variability of all the thermodynamic fields significantly increased with increasing τ_{phase}. Similarly, Korolev (1995) found that an increase in τ_{phase} led to an increase in the fluctuation of liquid water content. Considering the different relationships between τ_{phase} and the characteristic Lagrangian timescale of turbulent fluctuations, Khvorostyanov and Curry (1999) developed a general kinetic equation of stochastic condensation to describe cloud droplet spectral broadening. Austin et al. (1985) used the reciprocal of τ_{phase} to study the broadening of cloud droplet size distributions due to turbulence. Chandrakar et al. (2016) concluded that an increase in the width of the droplet size distribution for low aerosol concentration was consistent with larger variability of supersaturation due to the slow microphysical response, i.e., τ_{phase} larger than turbulence correlation timescale. Chandrakar et al. (2017) further found that the process of cloud cleansing was accelerated by transforming cloudy and polluted air into cloudy and clean air for slow microphysical response.

Third, some studies have used τ_{phase} in the calculation of *Da* in studies of entrainment-mixing processes (Kumar et al., 2013; Pinsky et al., 2016a; Pinsky et al., 2016b; Small et al., 2013). Jensen and Baker (1989) pointed out that when $\tau_{evap} > \tau_{phase}$, entrained air rapidly saturated and the total evaporation of droplets was thus limited, whereas when $\tau_{evap} < \tau_{phase}$, some droplets completely evaporated before the entrained air was saturated.

Based on Da numbers calculated with τ_{evap} and τ_{phase} in six direct numerical simulation runs, Kumar et al. (2014) concluded that the response of the cloud system was governed by the smaller of τ_{evap} and τ_{phase}. Lu et al. (2013a) and Kumar et al. (2013) argued that taking n as n_h ($=n_a * \chi$) worked better than n_a in entrainment-mixing studies. Therefore, the calculations of timescales and then dynamical measures in section 4.3.2.2 are also related to Figure 4.1.

Lehmann et al. (2009) argued that the assumptions involved with the calculation of τ_{evap} and τ_{phase} were unrealistic because both S and r varied with time. They defined the reaction time τ_{react} as the time either for a droplet to completely evaporate or for the dry air to reach a new saturation, which can be calculated by the use of equations (4.12) and (4.15) as a coupled system. Later on, τ_{react} has been used in entrainment-mixing analysis and parameterizations (Lu et al., 2011; Lu et al., 2013a; Lu et al., 2014a).

Which timescale is more appropriate in studying entrainment-mixing processes? Based on the theoretical understanding that a larger ψ corresponds to a smaller Da (e.g., Lehmann et al., 2009), or a larger τ_r given τ_{mix}, a positive correlation between ψ and τ_r should serve as a criterion to select the more appropriate microphysical timescales. By analyzing the measurements collected in the Routine AAF (Atmospheric Radiation Measurement Aerial Facility) Clouds with Low Optical Water Depths (CLOWD) Optical Radiative Observations (RACORO) field campaign in 2009 (Vogelmann et al., 2012) and simulations with the Explicit Mixing Parcel Model (EMPM) (Krueger et al., 1997; Su et al., 1998b), Lu et al. (2018b) compared the use of these different microphysical timescales. For the convenience of discussion, two groups were defined in solving equations (4.12) and (4.15): the "Evaporated" group wherein droplets completely evaporated and the "Saturated" group wherein new saturation was achieved. Lu et al. (2018b) summarized that the more appropriate timescale to use depended on the specific objectives of entrainment-mixing studies (Figure 4.3). First, if the focus was on the variations of LWC and S, then τ_{react} for the "Saturated" group and τ_{phase} were almost equivalently appropriate because they all reflected the rate of dry air reaching saturation or the rate of LWC decrease.

Second, if the focus was on the variations of droplet size and number concentration, neither τ_{react} for the "Saturated" group nor τ_{phase} was appropriate because they only reflected how fast the dry air reached saturation or steady state, and the same variations of S and LWC could be due to complete evaporation of some droplets or partial evaporation of all droplets. Instead, τ_{react} for the "Evaporated" group and τ_{evap} were proper because they characterized how fast droplets evaporated and whether number concentration decreased. For practical use, τ_{evap} is recommended in view of the simplicity of its calculation.

Third, under the condition of constant LWC_0/S_0, all the microphysical timescales were proportional to one another and thus equally appropriate. Baker et al. (1984) defined a similar property to study the influence of mixing

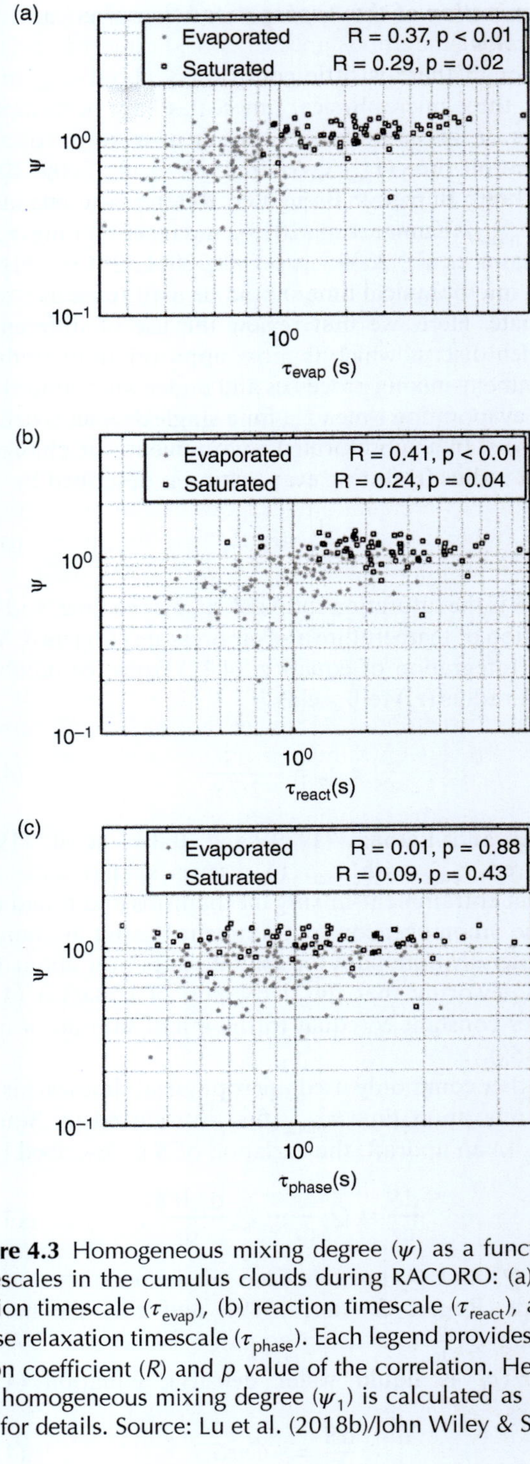

Figure 4.3 Homogeneous mixing degree (ψ) as a function of timescales in the cumulus clouds during RACORO: (a) evaporation timescale (τ_{evap}), (b) reaction timescale (τ_{react}), and (c) phase relaxation timescale (τ_{phase}). Each legend provides correlation coefficient (R) and p value of the correlation. Here, the first homogeneous mixing degree (ψ_1) is calculated as ψ. See text for details. Source: Lu et al. (2018b)/John Wiley & Sons.

on cloud droplet size distributions. Desai et al. (2021) pointed out the importance of considering the variation of S used in the calculation of τ_{evap}.

Dominant Entrainment-Mixing Mechanisms in Observed Clouds

As shown in section 4.3.2.2, entrainment-mixing mechanisms are affected by different factors. Homogeneous entrainment-mixing is found to dominate in some studies, while extreme inhomogeneous entrainment-mixing dominates in other studies, as summarized below.

In stratocumulus clouds, Pawlowska et al. (2000) analyzed microphysical properties along horizontal legs and found that compared to adiabatic values at the same level, subadiabatic regions were characterized by smaller droplet concentrations but similar volume-mean radius, suggesting extreme inhomogeneous entrainment-mixing. By analyzing cloud droplet size distributions measured at 50 Hz, Haman et al. (2007) found that extreme inhomogeneous entrainment-mixing mechanism dominated in view of the variations of droplet number concentration without significant change of the mean droplet diameter and spectral width from the observed cloud droplet spectra. Burnet and Brenguier (2007) found extreme inhomogeneous entrainment-mixing mechanism in one stratocumulus clouds. Differently, Yeom et al. (2017) found that cubic volume-mean radius were positively correlated with liquid water content in most aircraft horizontal penetrations of stratocumulus clouds.

Considering the complicated nature of entrainment-mixing mechanisms, Lu et al. (2011, 2013a) explored and used an integrative approach combining microphysical, dynamical, and thermodynamic relationships to study the dominant entrainment-mixing mechanisms with the help of quantitative measures presented in sections 4.3.2.1–4.3.2.3. They found a mixture of different entrainment-mixing mechanisms (thirteen legs by extreme inhomogeneous entrainment-mixing mechanism, two legs by inhomogeneous entrainment-mixing with subsequent ascent, and one leg by homogeneous entrainment-mixing). This was consistent with the dynamical measure; the only leg dominated by the homogeneous entrainment-mixing process had the largest leg-averaged N_L.

In shallow cumulus clouds, Jensen et al. (1985) applied a mixing diagram of number concentration vs. mixing fraction of adiabatic cloud to in situ observations of a nonprecipitating cumulus cloud and found that the mixing between cloudy and entrained air often appeared to occur without total removal of droplets, supporting homogeneous entrainment-mixing mechanisms. Burnet and Brenguier (2007) found a homogeneous entrainment-mixing mechanism in two cumulus clouds. Lu et al. (2013b, 2014a) found that the main entrainment-mixing mechanism was homogeneous because volume-mean radius and number concentration were positively correlated. Although the homogeneous entrainment-mixing mechanism seems to dominate in shallow cumulus clouds based on the above literature, different parts of cumulus clouds can have different mechanisms. For example, based on aircraft observations of cumulus clouds, Lehmann et al. (2009) concluded that homogeneous entrainment-mixing was more likely to occur in the vicinity of the cloud core, whereas inhomogeneous entrainment-mixing dominated in more diluted cloud regions.

In deep cumulus clouds, Freud et al. (2011) found that the mixing of subsaturated ambient air into the cloud at scales of 100 m and inhomogeneous entrainment-mixing were dominant. Differently, Guo et al. (2015b) found that volume-mean radius and liquid water content were positively correlated, indicating the dominance of the homogeneous entrainment-mixing mechanism. These different opinions need to be reconciled in future studies.

Actually, when relative humidity of entrained air is high, homogeneous and extreme inhomogeneous entrainment-mixing mechanisms cannot be distinguished. For example, Gerber et al. (2008) analyzed shallow cumulus clouds and concluded that because of the cool and humid entrained air, entrainment primarily caused dilution of the droplets without significant size changes and it was not possible to clearly differentiate between the potential mixing mechanisms.

Parameterizations

As discussed above, a variety of entrainment-mixing processes can occur in ambient clouds and lead to different cloud microphysical properties and relationships, other conditions being equal. Thus, accurate representation of these processes in models is expected to affect the model results. Therefore, this subsection focuses on the sensitivity study and parameterizations of entrainment-mixing mechanisms.

Some studies found significant impacts of different entrainment-mixing mechanisms. For example, with a modeling framework including Lagrangian microphysical parcel model, Lasher-Trapp et al. (2005) examined the effects of different inhomogeneous mixing fractions (0%, 5%, 10%, 20%, 50%, and 100%) on cloud droplet size distribution in cumulus clouds. They found that the droplet number concentration decreased as the inhomogeneous mixing fraction increased, allowing the remaining droplets to grow larger. The 100% inhomogeneous mixing simulation produced droplets with a maximum radius of 18 μm and the homogeneous mixing simulation only produced droplets with a maximum radius of 14 μm. Based on large eddy simulation (LES) of stratocumulus clouds, Chosson et al. (2007) revealed that assuming homogeneous and extreme inhomogeneous entrainment-mixing mechanisms at cloud top might substantially bias the simulated cloud albedo. For a fragmented and thin stratocumulus cloud, the albedo bias changed from −3% to −31%, when the assumed mixing changes from homogeneous to inhomogeneous. Grabowski (2006) discussed a cloud-resolving modeling study assuming homogeneous, extreme inhomogeneous entrainment-mixing mechanisms and an intermediate one between the two extremes. It was found that the amount of solar energy reaching the surface could be the same in the pristine case assuming the homogeneous entrainment-mixing scenario and in the polluted case with the extreme inhomogeneous entrainment-mixing. With a semi-Lagrangian-Eulerian model, Slawinska et al. (2008) confirmed the findings in Grabowski (2006).

However, some other studies suggested less important impacts of different entrainment-mixing mechanisms. For example, Morrison and Grabowski (2008) found that changing entrainment-mixing mechanisms from homogeneous to extreme inhomogeneous reduced droplet concentration by 27%–39%, increased effective radius by 0.8–1.5 μm, and reduced cloud optical thickness by about 2%–7%. These impacts were weaker compared with the results in Lasher-Trapp et al. (2005), Chosson et al. (2007), Grabowski (2006), and Slawinska et al. (2008). Similarly, Hill et al. (2009) also found that different entrainment-mixing mechanisms played a weak role. Assuming extreme inhomogeneous rather than homogeneous entrainment-mixing resulted in a small reduction in number concentration, a small increase in droplet effective radius, and about 1% decrease in cloud optical depth.

Slawinska et al. (2012) proposed some explanations for smaller impacts of assuming different entrainment-mixing mechanisms. First, in-cloud aerosol activation replenished cloud droplets more efficiently in assuming extreme inhomogeneous than assuming homogeneous entrainment-mixing, offsetting the effect of reducing number concentration by extreme inhomogeneous entrainment-mixing. Second, RH was diagnosed to be close to 100%, much higher than that in the far environment. With such humid entrained air, differences in cloud microphysics caused by different entrainment-mixing mechanisms were expected to vanish (Burnet & Brenguier, 2007; Gerber et al., 2008). Third, numerical diffusion also played an important role. Different entrainment-mixing mechanisms were assumed only for subgrid-scale processes and the role of the parameterized subgrid-scale mixing was smaller, compared to mixing and evaporation due to numerical diffusion. To reduce numerical diffusion of supersaturation, Hoffmann and Feingold (2019) used an linear eddy model (Kerstein, 1992) to simulate subgrid-scale turbulence fluctuation of supersaturation and found that inhomogeneous entrainment-mixing decreased droplet number concentration and increased droplet growth more significantly, compared with some previous studies (Hill et al., 2009). Finally, different entrainment-mixing processes may occur at different stages of cloud evolution. For example, Morrison and Grabowski (2008) and Hill et al. (2009) argued that changes in the cloud microphysical and optical properties due to different entrainment-mixing mechanisms might be more significant over the entire cloud life cycle, especially during the dissipation stage when entrainment-mixing processes were expected to dominate. This argument was supported by an observational study by Schmeissner et al. (2015); they divided clouds into actively growing, decelerated, and dissolving clouds and observed that there was a significant tendency to more inhomogeneous entrainment-mixing with increasing cloud lifetime.

In addition to the sensitivity studies with prescribed mixing types, people have started to explore the development of entrainment-mixing parameterizations to diagnose homogeneous mixing degree in atmospheric models. Andrejczuk et al. (2009) analyzed dozens of direct numerical simulations with bin microphysics. They used the slope of the mixing line in the mixing diagram of mean radius vs. number concentration to represent homogeneity of mixing and analyzed the relationship between the slope and *Da*. They found a robust relationship for *Da* in the range of 0.5–10. Jarecka et al. (2013) connected the homogeneous mixing degree defined in Morrison and Grabowski (2008) to this slope with an empirical expression as the parameterization in their large eddy simulations. Entrained dry air sizes and subgrid cloud fraction were also predicted from some heuristic equations (Jarecka et al., 2009). They found that coupling homogeneity of mixing with the double-moment microphysical scheme generated spatial/temporal variations of mixing scenarios ranging from homogeneous entrainment-mixing to extreme inhomogeneous entrainment-mixing. However, the impact on mean microphysical characteristics was small, perhaps due to the high humidity of the entrained air involved in the subgrid-scale mixing processes.

Since *Da* needs the entrained dry air size that is not available in main-stream models (e.g., weather and climate models), Lu et al. (2013a, 2014a) and Gao et al. (2018) developed parameterizations of entrainment-mixing mechanisms by relating homogeneous mixing degree to transition scale number, as defined in sections 4.3.2.1 and 4.3.2.2. Positive correlations were found in observations of both stratocumulus (Lu et al., 2013a) and cumulus clouds (Lu et al., 2014a), and also in EMPM simulations under different conditions (i.e., dissipation rate, mixing fraction of entrained air, RH, and adiabatic cloud droplet number concentration) (Lu et al., 2013a; Luo et al., 2020). For example, based on more than

12,000 EMPM simulations, Luo et al. (2020) developed a parameterization (Figure 4.4a):

$$\psi_3 = 113.39 \exp\left(-0.90 N_L^{-0.26}\right), \qquad (4.17)$$

and found that the cloud droplet number concentrations calculated from the parameterization (equation (4.17)) and from cloud droplet size distributions agreed well with each other (Figure 4.4b).

Gao et al. (2018) also investigated the effects of the above factors on entrainment-mixing mechanisms and parameterization using particle-resolved direct numerical simulations. Different measures of microphysical homogeneous mixing degree were systematically examined, modified, and compared in terms of their ability as a unifying metric to represent the effects of various entrainment-mixing mechanisms on cloud microphysics. Both *Da* and N_L were examined and compared as a dynamical measure of different mixing mechanisms. Further analysis of the relationship between the microphysical homogeneous mixing degree and the dynamical measure indicated that the variety of entrainment-mixing mechanisms were reasonably represented with a power-law relationship between one of the microphysical homogeneous mixing degrees and one of the dynamical measures. They also compared the measures of homogeneous mixing degrees with the slope used in Andrejczuk et al. (2009) and Jarecka et al. (2013) as the microphysical measure in the parameterization of entrainment-mixing processes. They found that the various homogeneous mixing degrees perform better generally because they are better constrained by the two limiting mixing scenarios than the slope.

Scale Dependence of Entrainment-Mixing Mechanisms

One of the major reasons for the uncertainties of entrainment-mixing studies is that the related processes occur over a tremendous range of scales (Su et al., 1998a). Therefore, there is a possible dependence of entrainment-mixing mechanism on the sampling/averaging scale. Understanding scale dependence of entrainment-mixing mechanisms is important to developing scale-aware parameterizations needed for improving models at different resolutions (Jeworrek et al., 2019; Xie & Zhang, 2015; Yun et al., 2017).

Burnet and Brenguier (2007) simulated a large number of realizations of entrainment-mixing processes with different mixing fractions between cloud and dry air using a stochastic model. They assumed that each sample corresponded to 0.2 m and that each random series of 50 successive samples represented a 10 m sample. They calculated the mean cloud droplet size distributions for the 10 m samples. Compared with the original sampling scale of 0.2 m, the sampling scale of 10 m significantly affected the appearance of the simulation data in the

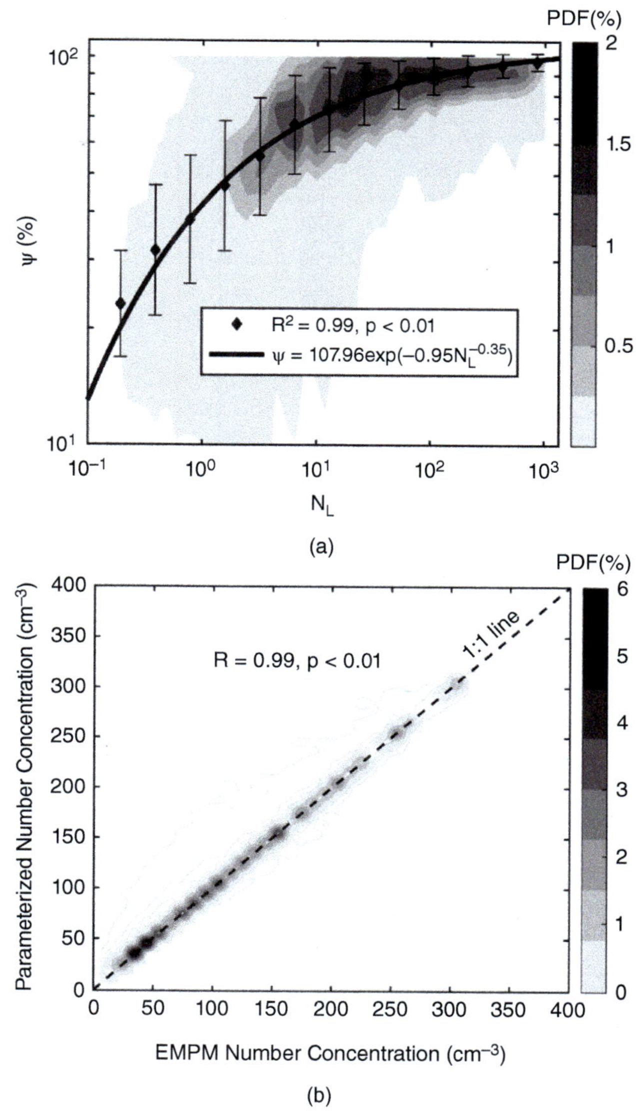

Figure 4.4 (a) Parameterization of cloud entrainment-mixing mechanisms by relating homogeneous mixing degree (ψ) to transition scale number (N_L) based on 12,218 simulations using the Explicit Mixing Parcel Model (EMPM). The contours represent the joint probability density function (PDF) of ψ vs. N_L. The magenta dots and error bars are mean values and standard deviations of ψ in each N_L bin, respectively. The mean values are fitted using a weighted least-squares method with the number of data points in each N_L bin as the weight. The fitting equation, coefficient of determination (R^2), and p value are given. Here, the third homogeneous mixing degree (ψ_3) is calculated as ψ. (b) Joint PDF of the cloud droplet number concentrations calculated from the parameterization in (a) and from cloud droplet size distributions in the EMPM with 12,218 cases. The correlation coefficient (R) and p value are given. Source: Adapted from Luo et al. (2020).

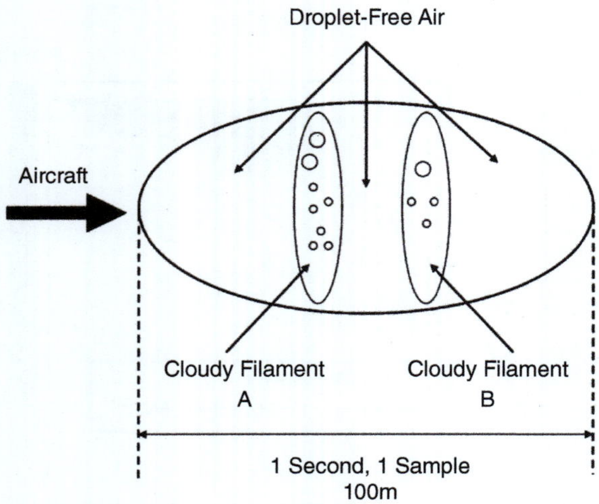

Figure 4.5 Schematic illustration of the filament structure of a 1 Hz sample as collected by an instrumented aircraft at the true air speed of 100 m s^{-1}. The circles in Filaments A and B represent cloud droplets. Source: Lu et al. (2011)/John Wiley & Sons.

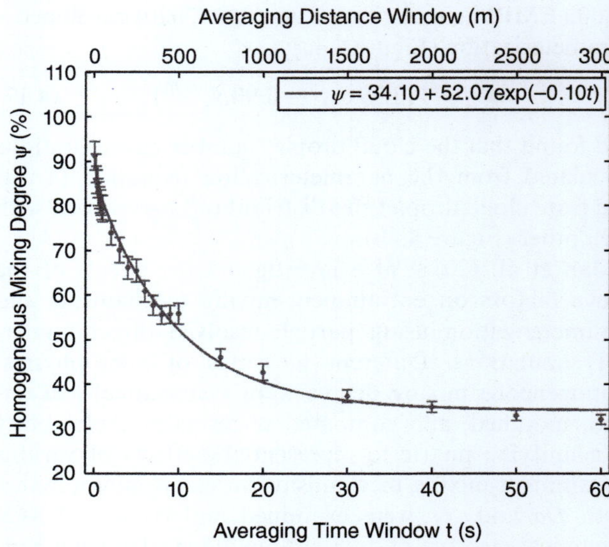

Figure 4.6 Homogeneous mixing degree (ψ) vs. averaging time window (t) or averaging distance window in the growing cumulus clouds during RACORO. The bars represent the standard errors of the mean ψ. Here, the fourth homogeneous mixing degree (ψ_4) is calculated as ψ. Source: Adapted from Lu et al. (2014b).

mixing diagram of r_v^3/r_{va}^3 vs. n/n_a toward the direction of more inhomogeneous entrainment-mixing. This could be caused by the fact that a cloud is a mixture of cloudy air and droplet-free air, as shown in Figure 4.5 from Lu et al. (2011). Lu et al. (2011) pointed out that the dry air filament structure could contribute to the observed small variation in r_v^3/r_{va}^3 but large variation in n/n_a.

Using observational data of cumulus, Lu et al. (2014b) systematically quantified the scale dependence and examined the factors influencing the scale dependence. They found that homogeneous mixing degrees decreased with increasing averaging scales (Figure 4.6), consistent with the early "dilution" argument. Furthermore, they found that the decreasing rate of homogeneous mixing degree with increasing scales became larger with larger droplet-free air fraction, larger microphysical properties (volume-mean radius and liquid water content) adjacent to droplet-free filaments, and higher RH. Differently, based on some observational data collected from stratocumulus clouds during the Aerosol and Cloud Experiments in Eastern North Atlantic (ACE-ENA) and RACORO, Gao et al. (2020) found that entrainment-mixing could become more homogeneous or more inhomogeneous with increasing averaging scales (Figure 4.7). Microphysical properties near and far away from droplet-free air and RH determined both the sign and strength of scale dependence, while droplet-free air fraction only affected the strength. Different from Lu et al. (2014b), Gao et al. (2020) found that microphysical properties far away from droplet-free air were also critical for scale dependence. The highest resolution in the above two studies was 5 or 10 m. Interestingly, Beals et al. (2015) used airborne digital holographic imaging to measure the 3D spatial distribution of droplets and hence droplet size distributions at centimeter scales. Even with this high resolution, the measurements revealed that the entrainment-mixing mechanisms were extreme inhomogeneous. Analyzing similar high-resolution measurements from ACE-ENA, Desai et al. (2021) found existence of both homogeneous and inhomogeneous entrainment-mixing at different altitudes.

By setting different domain sizes of direct numerical simulations with side length from 0.128 to 2 m, Kumar et al. (2018) found that entrainment-mixing mechanisms did not show significant dependence on the domain sizes, but the fluctuation of trajectories in the mixing diagram of volume-mean radius vs. number concentration and relative dispersion of cloud droplet size distribution increased with an increasing domain size.

These few studies are hardly conclusive, instead revealing that the topic of scale-dependence seems more complicated than originally thought and merits further investigation.

4.4. ENTRAINMENT RATE

This section mainly discusses how entrainment rate in cumulus clouds is estimated and parameterized; entrainment rate in stratocumulus clouds is briefly discussed as well.

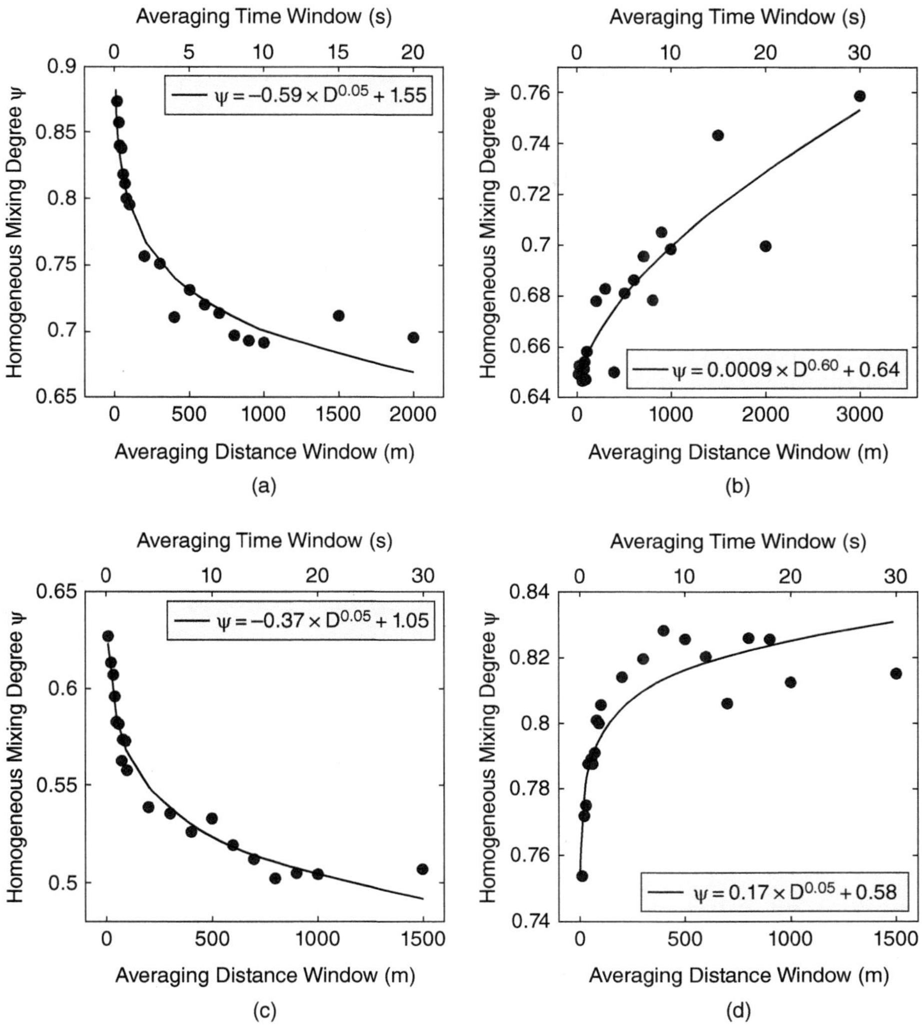

Figure 4.7 Examples of homogeneous mixing degree (ψ) as a function of averaging distance (D)/time (t) on (a) 26 June 2017, (b) 21 June 2017 during ACE-ENA, (c) 6 February 2009, and (d) 20 March 2009 during RACORO. The legends provide the fitting equations. Here, the fourth homogeneous mixing degree (ψ_1) is calculated as ψ. Source: Gao et al. (2020)/John Wiley & Sons.

4.4.1. Approaches for Estimating Entrainment Rate in Cumulus Clouds

This subsection focuses on the approaches for estimating entrainment rate in cumulus clouds. Before presenting entrainment rate estimation, it is necessary to distinguish entrainment and dilution. Hannah (2017) found that entrainment generally corresponded to dilution of convection, but the two properties were not well correlated. Hannah (2017) also pointed out that, strictly, entrainment could only be measured by direct approaches, i.e., the direct calculation/measurement of the flux of mass through the cloud/clear-air interface. For example, Siebesma (1998) defined entrainment rate (E) (in kg m^{-3} s^{-1}):

$$E = -\frac{1}{A}\oint_{\hat{\boldsymbol{m}}\cdot(\boldsymbol{u}-\boldsymbol{u}_i)<0} \rho\hat{\boldsymbol{m}}\cdot(\boldsymbol{u}-\boldsymbol{u}_i)\mathrm{d}l, \tag{4.18}$$

where $\hat{\boldsymbol{m}}$ is a unit vector perpendicular to the cloud surface, $\boldsymbol{u}$ is the air mass flow velocity vector at the surface, $\boldsymbol{u}_i$ is the velocity of the surface, A is the horizontal domain size, ρ is the air density, and $\mathrm{d}l$ represents the path integral around the cloud surface at a certain height. However, this approach needs to know the relative velocity of the air mass with respect to the cloud surface, which is a big challenge in both large eddy simulation models (Dawe & Austin, 2011; Romps, 2010) and measurements. To overcome this problem, Romps (2010) introduced an approach for directly measuring entrainment rate at the grid cell level by analyzing the local sources and sinks of convecting air. Similarly, Dawe and Austin (2011) directly estimated entrainment rate by developing an approach for the subgrid interpolation of a cloud surface on a discrete numerical model grid. Wang (2020) developed an approach to approximately calculate mass fluxes across the cloud surface of cumulus clouds for large eddy simulations. The approach used cloud scalar gradients across the cloud interface without calculating cloud interface velocity. The entrainment rate was calculated as an integration over a small region around the cloud interface with the integrand composed of the total derivative of the cloud scalar.

The concept of fractional entrainment rate focused in this chapter refers to the bulk-plume entrainment rate that measures the dilution of a cloud by environmental air at the same level and has been widely used in cumulus parameterizations in large-scale models. Stommel (1947) was the first to describe the entrainment process in cumulus clouds and develop an approach to evaluate entrainment strength based on temperature and specific humidity inside and outside the cloud. The equation of mass flux including fractional entrainment rate (λ in m^{-1}) is

$$\frac{1}{M_c}\frac{dM_c}{dz} = \lambda - \delta, \tag{4.19}$$

where δ is the detrainment rate, M_c is the cloud mass flux, and z is the height.

Siebesma and Cuijpers (1995) decomposed the prognostic equations of a conserved variable or tracer (ϕ) into active cloudy and environmental parts, and derived the equation for calculating entrainment ($E = \lambda M_c$) together with the continuity equation:

$$E(\phi_e - \phi_c) = M_c\frac{\partial \phi_c}{\partial z} + \frac{\partial \sigma\rho(w'\phi')_c}{\partial z} + \sigma\rho\frac{\partial \phi_c}{\partial t} - \sigma\rho\left(\frac{\partial \bar{\phi}}{\partial t}\right)_{\text{forcing}}, \tag{4.20}$$

where σ is the cloud area, z is the height, t is the time, $(w'\phi')_c$ is the perturbation correlation term of vertical velocity (w) and ϕ, and the subscripts c and e indicate the cloud and environment, respectively. Siebesma and Cuijpers (1995) also pointed out that the forcing term $\left(\sigma\rho\left(\frac{\partial\bar{\phi}}{\partial t}\right)_{\text{forcing}}\right)$ is very small because $\sigma \ll 1$; the time derivative $\left(\sigma\rho\frac{\partial \phi_c}{\partial t}\right)$ is negligible, because of the steady state. Zhang et al. (2016) derived an equation for λ based on moist static energy (h):

$$\lambda = \frac{-\frac{\sigma\rho}{M_c}\frac{\partial h_c}{\partial t} + \delta(h_c - h_d) - \frac{\partial h_c}{\partial z} - \frac{\partial\sigma\rho(w'h')_c}{M_c\partial z} + \frac{\sigma\rho}{M_c}S_c}{h_c - h_e}, \tag{4.21}$$

where h_d is the moist static energy of detrained air and S_c is the source term for moist static energy due to freezing/melting and deposition/sublimation. When the moist static energy in the detrained air is assumed to be the same in the cloud, $\delta(h_c - h_d)$ is equal to 0. Assuming that the time derivative $(-\frac{\sigma\rho}{M_c}\frac{\partial h_c}{\partial t})$ and the perturbation correlation term $[(w'h')_c]$ are negligible, equation (4.21) becomes

$$\lambda = \frac{-\frac{\partial h_c}{\partial z} + \frac{\sigma\rho}{M_c}S_c}{h_c - h_e}. \tag{4.22}$$

Siebesma and Cuijpers (1995) tested the importance of the perturbation correlation term in λ estimation; they found that this term was negligible in the middle cloud layer but was of the same order as the other terms near cloud base and top. For liquid-phase clouds, S_c in equation (4.22) is 0. In this case, equation (4.22) becomes equation (4.23):

$$\lambda = \frac{\partial \phi_c}{\partial z}\frac{1}{\phi_e - \phi_c}, \tag{4.23}$$

where ϕ can be h or other conserved properties (e.g., total water and liquid water potential temperature); this is called the bulk-plume approach (Betts, 1975; Houze, 1993). The approach has been widely used in the estimation of λ from aircraft observations and numerical simulations (de Rooy & Siebesma, 2010; Gerber et al., 2008; Hernandez-Deckers & Sherwood, 2018; Kuang & Bretherton, 2006; Neggers et al., 2003; Raga et al., 1990b; Siebesma and Cuijpers, 1995; Siebesma and Holtslag, 1996; Siebesma et al., 2003).

To retrieve λ based on remote-sensing data, Jensen and Del Genio (2006) and Eissner et al. (2021) rearranged equation (4.23) and used the equivalent potential temperature (θ_e) as the conserved property:

$$\theta_{ec}(z + \Delta z) = \frac{\theta_{ec}(z) + \lambda\Delta z\theta_{ee}}{1 + \lambda\Delta z} \tag{4.24}$$

where Δz is the height interval. Given λ, a new profile of θ_{ec} can be generated, from which the level of neutral buoyancy can be determined and used to match the observed cloud-top height to estimate λ iteratively. The iteration started from λ of $0.001\,km^{-1}$ with the step of $0.001\,km^{-1}$ until the level of neutral buoyancy was close to the observed cloud top height. Instead of θ_e, Luo et al. (2010) and Takahashi and Luo (2012) used moist static energy to estimate λ by matching the estimated and observed moist static energy at the cloud top height. Using the same approach, Stanfield et al. (2019) employed satellite-observed CO concentration as the conserved property to estimate λ.

Besides the bulk-plume approach, Wang et al. (2020) used the difference between the level of neutral buoyancy and the level of maximum detrainment as a proxy for the effective bulk entrainment based on satellite observations. Drueke et al. (2019) presented an approach for inferring

λ based on a turbulent kinetic energy similarity theory. Wagner et al. (2013) developed an algorithm, called the Entrainment Rate In Cumulus Algorithm (ERICA), by applying suite of ground-based instruments as inputs into the EMPM model. An initial guess of λ was iteratively adjusted through the Gauss-Newton optimal estimation scheme until the best match between modeled cloud properties and their observed counterparts was obtained.

Although the above bulk-plume approaches have been widely used to estimate λ, these studies are rarely connected to entrainment-mixing mechanisms. In nature, entrainment and the subsequent mixing are closely related, especially, for the fractional entrainment rate that essentially quantifies the dilution of some variables. To overcome this limitation, Lu et al. (2012b) developed a mixing fraction approach for estimating λ at the height z assuming constant λ from cloud base (z_0) to z:

$$\lambda = -\frac{\ln \chi}{z - z_0}, \tag{4.25}$$

where χ is the mixing fraction of adiabatic cloud estimated from the conservation of total water and energy during the mixing process. The assumption of constant λ from z_0 to z can be relaxed by the simple adjustment:

$$\lambda_{aj} = \frac{\lambda_j z_j - \lambda_{j-1} z_{j-1}}{z_j - z_{j-1}} (j \geq 2), \tag{4.26a}$$

$$\lambda_{aj} = \lambda_j (j = 1), \tag{4.26b}$$

where λ_{aj} is the entrainment rate after adjustment and λ_j represents the entrainment rate from cloud base to the jth height (z_j), calculated in equation (4.25). The adjusted jth height (z_{aj}) is given by

$$z_{aj} = (z_j + z_{j-1})/2. \tag{4.27}$$

Note that λ_{aj} corresponds to z_{aj}. This approach relates λ directly to χ, which is also used in the calculation of homogeneous mixing degrees (section 4.3). Therefore, this approach has the potential to connect the two parts of entrainment-mixing processes. Lu et al. (2012b) evaluated the mixing fraction approach with both observations and numerical simulations, and found that this approach had much smaller uncertainty than the traditional approach (equation (4.23)) in observations. The results from simulations indicated that the traditional approach with total water mixing ratio and liquid water potential temperature as the two conserved properties produced systematically different values of λ, while the λ estimated from the mixing fraction approach ranged between those estimated from the traditional approach, perhaps because χ was estimated from the conservation of both total water and energy. This approach only required liquid water content in cloud, temperature and water vapor mixing ratio in environment and at cloud base height, and did not need temperature and water vapor mixing ratio in cloud, the measurements of which are difficult and problematic. This approach also suggested a possible remote-sensing technique to measure entrainment rate.

The above approaches can be used to estimate λ only. In recent years, studies have appeared to obtain both entrainment and detrainment together based on aircraft observations. Norgren et al. (2016) developed an approach to estimate the gross entrainment and detrainment by optimizing three equations for three conserved properties: cloud mass, total water, and moist static energy, assuming each cloud had a mass determined by the balance of three terms: adiabatic cloudy air, entrained air, and detrained air. The gross entrainment and detrainment are defined as the integrated fractional mass entrained and detrained, respectively, from cloud base to a given height level. Zhu et al. (2021) further developed an approach to estimate fractional entrainment and detrainment rates in cumulus clouds based on aircraft observations, by deriving equations relating the fractional entrainment and detrainment rates to gross entrainment and detrainment in Norgren et al. (2016).

4.4.2. Parameterizations of Entrainment Rate in Cumulus Clouds

The fractional entrainment rate is a key parameter in convection parameterizations in large-scale models; how to accurately represent it cannot be overemphasized (Chikira & Sugiyama, 2010; Del Genio et al., 2012; Kim et al., 2011b; Lu & Ren, 2016; Stanfield et al., 2019; Wu, 2012; Zhao et al., 2012). The common practice is to find out the relationship between λ and some other fundamental properties and then represent it as a function of these properties. The properties commonly used in such relationships can be classified into internal cloud properties (e.g., vertical velocity and buoyancy) (Neggers et al., 2002; Zhang et al., 2016) and external properties (e.g., relative humidity in environment) (Bechtold et al., 2008), which are detailed next in sections 4.4.2.1 and 4.4.2.2, respectively.

Parameterization in Terms of Internal Cloud Properties

Many studies have attempted to relate λ to internal cloud properties, including cloud radius, buoyancy (B), w, and ε. For example, it was proposed that λ was proportional to the reciprocal of cloud radius (Drueke et al., 2021; Eissner et al., 2021; Hernandez-Deckers & Sherwood, 2018; Simpson, 1971; Squires & Turner, 1962) because of smaller surface-to-volume ratio of larger clouds. This conclusion was supported by the analysis based on 5-year CloudSat data in Takahashi et al. (2017). They found that λ was larger, and convective core size was smaller over the Warm Pool than the counterpart over Africa and Amazonia.

Buoyancy is often used to parameterize λ as well. Lin (1999) simulated cumulus ensembles with a cloud-resolving model and found that λ was connected to B with a negative power expression:

$$\lambda = aB^{-b}, \tag{4.28}$$

where a and b are two empirical parameters. Lin (1999) averaged b from different experiments and found that equation (4.28) with average $b = 1.27$ and varying a was able to estimate λ, to a first-order approximation. Lin (1999) argued that B was diluted more through entrainment-mixing if clouds had a larger λ, and clouds with a larger B were likely insulated from environment more and thus have a smaller λ. Bretherton and Smolarkiewicz (1989a, 1989b) simulated convective clouds in a calm atmosphere with a two-dimensional model and found that entrainment was enhanced when buoyancy in clouds increased with the increasing height. Emanuel and Živković-Rothman (1999) fashioned after the modeling work by Bretherton and Smolarkiewicz (1989a) and proposed a parameterization of the mixing fraction of the cloud-base mass flux and environmental air at any level. This parameterization supported the conclusion of λ vs. vertical buoyancy gradient in Bretherton and Smolarkiewicz (1989a). von Salzen and McFarlane (2002) proposed a parameterization of shallow cumulus clouds and linked λ to buoyancy vertical gradient ($\frac{\partial B}{\partial z}$):

$$\lambda = c \max\left(\frac{\partial B}{\partial z}, 0\right), \tag{4.29}$$

where c is a scaling factor that is determined based on closure assumptions.

Besides B, λ is also related to w. Neggers et al. (2002) proposed that λ was inversely proportional to w for a cumulus convection parameterization with an ensemble of rising parcels:

$$\lambda = \frac{d}{\tau_c}\frac{1}{w}, \tag{4.30}$$

where d is a dimensionless constant and τ_c is the turnover timescale for the strongest updraft to rise from cloud base to cloud top. Neggers et al. (2002) claimed that τ_c was approximately constant. A cloud with a larger w took a shorter time to rise to a certain height and had less time to interact with the environment, thus leading to a smaller λ; a cloud parcel with a smaller λ further accelerated and decreased λ again. Zhang et al. (2016) analyzed cloud-resolving model simulation of convective clouds and estimated λ for different cloud types characterized by cloud top heights. They related λ to the inverse of w by a second-order polynomial for different height groups:

$$\lambda = -1.2 + 3.08\frac{1}{w} - 0.36\frac{1}{w^2} \text{ for } z < 5 \text{ km}, \tag{4.31a}$$

$$\lambda = -0.23 - 1.22\frac{1}{w} + 10.0\frac{1}{w^2} \text{ for } z \geq 5\text{km}. \tag{4.31b}$$

Based on ground-based observations, Kirshbaum and Lamer (2021) also found that λ and w were negatively correlated. The relationship between λ and w is further complicated by the balance of forces that produces w in cloud parcels (e.g., Gu et al., 2020). When the role of form drag and wave drag is negligible, the parcels are slippery (Sherwood et al., 2013); otherwise, the parcels are sticky (Romps & Öktem, 2015). Therefore, it is needed to combine entrainment and drag forces to improve understanding of w and the λ-w relationship.

The combination of B and w is also used to parameterize λ. Gregory (2001) suggested an expression that links λ to buoyant production of vertical kinetic energy:

$$\lambda = e\frac{B}{w^2}, \tag{4.32}$$

where e is an empirical parameter. Single-column model studies showed that this parameterization well captured high values of λ in shallow convection. This parameterization is used in the National Aeronautics and Space Administration (NASA) Goddard Institute for Space Studies (GISS) General Circulation Model (Kim et al., 2011a). de Rooy and Siebesma (2010) related λ to w and B through a different expression:

$$\lambda = \frac{\alpha B}{w^2} - \frac{1}{w}\frac{\partial w}{\partial z}, \tag{4.33}$$

where α is a buoyancy reduction factor. Lu et al. (2016) analyzed observations and simulations of cumulus clouds to develop a parameterization of λ in terms of B, w, and ε. They showed that λ was negatively correlated with w, B, and ε, but positively correlated with B/w^2 (Figure 4.8). It is physically consistent that a larger λ leads to smaller B, w, and ε. According to Lin (1999), B is smaller when λ is larger. A smaller B in turn leads to a smaller w and ε because B is the primary driver of w and ε in cumulus clouds. In addition, turbulence near the cloud edge associated with entrainment also creates form drag that acts to decelerate w. A smaller w meant a smaller horizontal shear of vertical velocity in and outside clouds, leading to weaker turbulence, i.e., a smaller ε, in clouds. The mutual correlations between ε, B, and w are also responsible for the positive correlation between λ and B/w^2. Lu et al. (2016) developed several different parameterizations for shallow cumulus clouds in terms of these internal cloud properties by applying the principal component regression:

$$\lambda = 0.11w^{-0.45}B^{-0.50} \text{ (adjusted } R^2 = 0.51;\ p < 0.01), \tag{4.34a}$$

$$\lambda = 0.072w^{-0.31}B^{-0.36}\varepsilon^{-0.15} \text{ (adjusted } R^2 = 0.53;\ p < 0.01), \tag{4.34b}$$

for the observed clouds; and

$$\lambda = 0.13w^{-0.29}B^{-0.31} \text{ (adjusted } R^2 = 0.50;\ p < 0.01), \tag{4.35a}$$

$$\lambda = 0.056w^{-0.18}B^{-0.24}\varepsilon^{-0.16} \text{ (adjusted } R^2 = 0.57;\ p < 0.01), \tag{4.35b}$$

for the simulated clouds, where R^2 and p are the adjusted coefficient of determination and level of significance, respectively. Similar relationships were also found in deep convection, although the coefficients were different (Xu et al., 2021).

Note that these internal cloud properties themselves need to be parameterized as well in large-scale models. The parameterized properties are inherently uncertain. Nevertheless, these relationships enhance our physical understanding of cumulus clouds as an open system with multiple interactions and provide improved constraints over the equations to be solved. Furthermore, there are more mature techniques (e.g., cloud radar or lidar) to measure some cloud properties (e.g., w) than that to measure λ.

Different from the abovementioned deterministic parameterizations, Romps and Kuang (2010) pointed out that entrainment should be parameterized as a stochastic process. They coupled a parcel model with stochastic entrainment and found that it could reproduce cloud properties well, while a parcel model with a constant and continuous λ could not. Böing et al. (2014) also supported the stochastic concept by assuming a gamma distribution of λ. Lu et al. (2012a) and Cheng et al. (2015) examined the use of the gamma, lognormal, or Weibull distributions to represent the probability density function (PDF) of λ based on aircraft observations of shallow cumulus clouds and found that all the three distributions could describe the PDF of λ well, with the Weibull distribution being the best. Nie and Kuang (2012) proposed a framework for diagnosing interaction between cumulus and environment and viewed the vertical evolution of the cloud mass flux distribution as a stochastic mapping process.

Parameterization in Terms of External Properties

External properties, such as vertical wind shear, cloud fraction, cloud depth, aerosol loading, and RH, could play important roles in determining λ. Some researchers parameterized λ with the cloud depth (Eissner et al., 2021) or the reciprocal of height above cloud base (Jakob & Siebesma, 2003; Roode et al., 2000). Small et al. (2009) found that when nonprecipitating cumulus clouds experienced aerosol perturbation gradient, there was stronger vorticity around the cloud core and stronger entrainment (Jiang et al., 2006), constituting an "evaporation-entrainment feedback." The most studied external property used in entrainment rate parameterizations is RH, which is thus the focus of this section.

RH in the lower- and middle-tropospheric moisture and its variability are critical for the formation and development of cumulus clouds (Raymond & Zeng, 2000; Ridout, 2002; Zhang et al., 2003). Derbyshire et al. (2004) emphasized that RH affected cumulus clouds through entrainment-mixing, using a cloud-resolving model and a single-column model with RH of 80% between 1 and 2 km, and 25%, 50%, 70%, and 90% above 2 km in different simulations. Their cloud-resolving model simulations showed strong deep precipitating convection in the moist cases but only shallow convection in the dry cases. However, single-column model simulations had sensitivities of convection to RH but were different in various ways from the cloud-resolving model simulations. Derbyshire et al. (2011) further found that there was no significant correlation between λ and RH above 2 km; in the layer 1–2 km, λ was found to be larger in a drier environment, which could be due to nonlocal impacts, possibly via the overall cloud depths and properties. Bechtold et al. (2008) revised the convection and diffusion parameterizations in the European Center for Medium-Range Weather Forecasts model assuming λ was negatively correlated with RH. Amplitudes of tropical and extratropical mesoscale and synoptic and planetary perturbations were significantly improved. This scheme was later implemented in the National Centers for Environmental Prediction's Global Forecast System (Han & Pan, 2011). Zhao et al. (2018) also assumed a negative correlation between λ and RH in the Global Atmosphere and Land Model AM4.0/LM4.0 of Geophysical Fluid Dynamics Laboratory, to optimize tropical precipitation and top-of-atmosphere radiative fluxes. Based on 90 high-resolution large eddy simulations, Böing et al. (2012) found that λ was rather insensitive to RH; if the values of λ were shown on a much small scale, λ was negatively correlated with RH.

Oppositely, Stirling and Stratton (2012) performed a series of idealized cloud-resolving model simulations for a range of environmental conditions and found a positive correlation between λ and RH; a parameterization relating λ to RH was then developed. Based on the data from the ARM facility at Nauru Island in the tropical western Pacific, Jensen and Del Genio (2006) found positive correlations between λ and RH, and convective available potential energy, which was consistent with the idea that λ was closely related to buoyancy production of turbulent kinetic energy. Kirshbaum and Lamer (2021) also found a robust positive sensitivity of observationally retrieved λ to RH, based on ground-based observations.

Different from the above conclusions, Xu et al. (2021) found a nonmonotonic relationship between λ and RH.

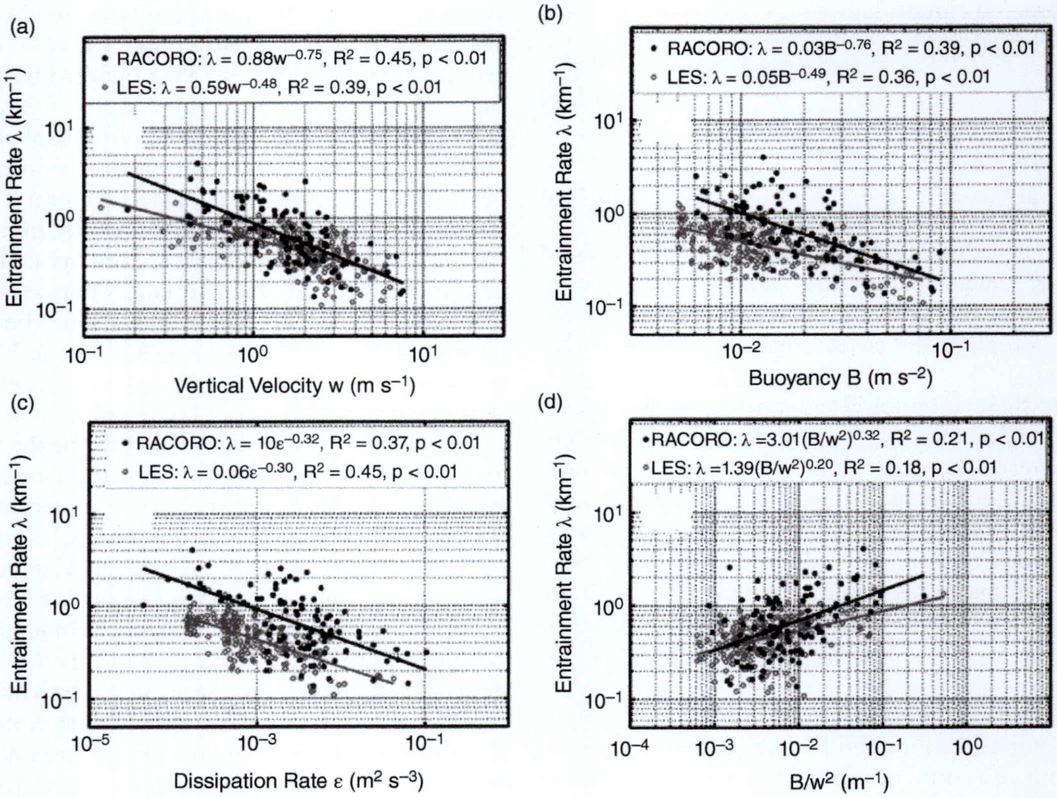

Figure 4.8 Relationships between entrainment rate (λ) and (a) vertical velocity (w), (b) buoyancy (B), (c) dissipation rate (λ), and (d) B/w^2 in the actively growing cumulus clouds during RACORO. Also shown for comparison are the results from actively growing cumulus clouds sampled from the large eddy simulation (LES). The parameter R^2 is the adjusted coefficient of determination and p shows the significant level. Source: Lu et al. (2016)/American Meteorological Society.

The two properties were positively/negatively correlated below/above the 0 °C level. Eissner et al. (2021) found that the relationship between λ and RH was very weak. Stirling and Stratton (2012), Derbyshire et al. (2011), and de Rooy et al. (2016) all claimed that the effects of RH on λ were quite unclear.

The above analyses on the relationship between λ and RH were mainly based on numerical simulations or remote sensing. Differently, Lu et al. (2018c) analyzed aircraft in situ observations of cumulus clouds from two field campaigns and studied λ vs. RH and physical mechanisms underlying the relationships, based on the mixing fraction approach. This approach estimated λ by calculating the mixing fraction between cloudy and environmental air (equations (4.25–4.27) in section 4.4.1). Besides, 11 actively growing clouds observed in the Rain in Cumulus over the Ocean (RICO) field campaign were used to estimate λ using liquid water potential temperature (θ_l) in the bulk-plume approach (Lu et al., 2018c). The positive correlation between λ and RH was found in this observational study of both field campaigns and with the two different approaches.

Furthermore, Lu et al. (2018c) showed that higher RH promoted entrainment by reducing buoyancy in the cloud cores and by weakening downdrafts near the cloud cores. The reduced buoyancy in the cloud cores and weakened downdrafts surrounding the cores further reduced updrafts in the cloud cores; the cloud cores with smaller updrafts were more significantly affected by their environment, resulting in larger λ.

The positive correlation between λ and RH is also supported by the concept of buoyancy sorting, which has been widely used in conceptual cloud models and convective parameterizations (Bretherton et al., 2004; de Rooy & Siebesma, 2008; Emanuel, 1991; Kain & Fritsch, 1990; Raymond & Blyth, 1986). Buoyancy sorting refers to a concept of lateral mixing across the cloud edges. Different mixtures around cloud edges have different proportions of cloudy and dry air. The positively buoyant mixtures with mixing fraction of dry air smaller than its critical value (χ_{dc}) are entrained into cloud cores, and the negatively

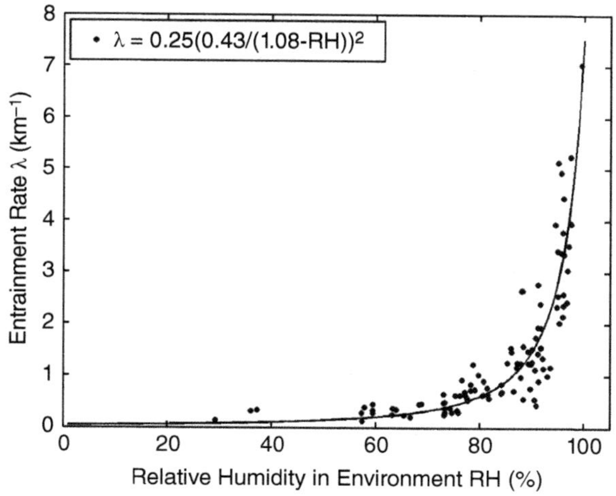

Figure 4.9 Regression based on equation (4.35) between entrainment rate (λ) and relative humidity (RH) in the environment in the actively growing cumulus clouds during RACORO. Source: Adapted from Lu et al. (2018c).

buoyant ones are detrained. Although the approaches to determine χ_{dc} and λ may be different in different studies, the concepts are similar. For example, de Rooy and Siebesma (2008) derived χ_{dc} as a function of RH:

$$\chi_{dc} = \frac{K_1}{K_2 - RH}, \tag{4.36a}$$

where K_1 and K_2 are two empirical parameters. Bretherton et al. (2004) derived

$$\lambda = \varepsilon_0 \chi_{dc}^2, \tag{4.36b}$$

where ε_0 is an empirical parameter related to the height above ground. Both equations indicate that χ_{dc} and λ are larger when RH is higher, i.e., a positive correlation between λ and RH. Figure 4.9 shows that regression of the observational data conforms well with equation (4.36). Using large eddy simulations, Drueke et al. (2020) also found a systematic positive correlation between λ and RH and argued that this correlation was consistent with the buoyancy sorting concept.

Different scientists have tried to explain the different relationships between λ and RH. Derbyshire et al. (2011) suggested two competing mechanisms. The first one was the natural selection of a diverse population of clouds with different λ: in a favorable environment with high environmental RH, clouds had large λ; clouds with too large λ would dissipate and drop out from the ensemble. The remaining clouds in the ensemble tended to have the largest λ they could suffer from. Therefore, the bulk ensemble was constrained to entrain more. The second one was related to cloud size: in a favorable environment, clouds were expected to be bigger; adiabatic cloud cores were more likely to exist and hence λ tended to be smaller. However, Becker et al.'s (2018) results added more complexities. Comparison between two simulations with unaggregated and aggregated convection indicated that λ increased when convection aggregated, which was due to a strong increase of turbulence in the close environment of updrafts. Stirling and Stratton (2012) stated that Bechtold et al.'s (2008) parameterization might be an alternative way of inhibiting convective clouds in low RH environments, and their cloud-resolving model results achieved this through higher detrainment rates.

4.4.3. Entrainment Rate in Stratocumulus Clouds

Studies and approaches for estimating entrainment rate in stratocumulus clouds are briefly discussed here. More studies on entrainment rate in stratocumulus clouds are referred to Stevens (2002), Wood (2012), Mellado (2017) and Schulz and Mellado (2019), and Chapter 6 on stratocumulus clouds.

Briefly, entrainment occurring at the top of stratocumulus clouds is affected by multiple factors (e.g., radiative and evaporative cooling, shear instabilities, and cloud microphysics) and can result in a deepening of the cloud-topped boundary layer and breakup of stratocumulus clouds. Cloud top entrainment instability (CTEI) is an important idea on this topic. As summarized by Wood (2012), buoyancy reversal describes the case where negative buoyant mixtures enhance mixing near cloud top or even in the whole mixed layer, which is a positive feedback that dissipates cloud (Deardorff, 1980; Randall, 1980). However, even when the criteria for CTEI are satisfied, stratocumulus clouds may still exist (Stevens et al., 2003). Buoyancy reversal may not be enough for the rapid breakup of stratocumulus (Mellado, 2010; Wood, 2012).

Unlike in cumulus clouds, entrainment in stratocumulus clouds is often measured by entrainment velocity (w_{entr}) instead of (fractional) entrainment rate. Lilly (1968) showed that entrainment velocity could be estimated by use of the flux-jump approach:

$$w_{entr} = < w \cdot \phi > / \Delta\phi, \tag{4.37}$$

where $< w{\cdot}\phi >$ is the vertical flux of ϕ at cloud top, and $\Delta\phi$ is ϕ's jump across the entrainment interface layer. This approach has been widely used (e.g., Deardorff, 1980; Gerber et al., 2013; Khalsa, 1993). Total water content, ozone, and dimethyl sulfide are often used as ϕ (e.g., Faloona et al., 2005). Gerber et al. (2005, 2013) used the total water mixing ratio (q_t) to calculate w_{entr}:

$$w_{entr} = A_r \left\langle w' \left[q_c' + q_v' \right] \right\rangle / \Delta q_t, \tag{4.38}$$

where w', q_c', and q_v' are the differences of vertical velocity, liquid water mixing ratio, and water vapor mixing ratio between cloud holes and adiabatic clouds, respectively; A_r is the sum of all horizontal hole sizes divided by the flight horizontal distance; and Δq_t is the jump in q_t across the interface between the free atmosphere and cloud top.

Another approach for estimating w_{entr} is through the temporal change of temperature inversion height (z_i) (Ackerman et al., 2009; Carson, 1973; Deardorff, 1974):

$$w_{entr} = \frac{dz_i}{dt} - Dz_i, \tag{4.39}$$

where D is the divergence of the large-scale horizontal winds at z_i. The idea is based on the assumption that the change of z_i is determined by the balance of entrainment and large-scale sedimentation associated with large-scale divergence. A third expression for estimating w_{entr} (Stevens, 2002) is

$$w_{entr} = F\frac{W}{\Delta B} + w_{dir}, \tag{4.40}$$

where W is an appropriate measure of the rate of turbulent production, F is an efficiency factor, ΔB is the buoyancy difference, and w_{dir} represents the effect of nonturbulent processes in deepening the layer. As pointed out by Wood (2012), it is still an open question to determine the appropriate rate of turbulent production W. There are also controversies on F, ΔB, and w_{dir} (Caldwell et al., 2005; Lewellen & Lewellen, 1998; Nicholls & Turton, 1986).

4.5. SUMMARY AND OUTLOOK

This chapter discusses several issues essential to entrainment-mixing processes and entrainment rate in cumulus and stratocumulus clouds. In terms of entrainment-mixing mechanisms, three different frameworks (i.e., homogeneous/inhomogeneous entrainment-mixing mechanisms, entity-type entrainment-mixing, and vertical circulation entrainment-mixing) are discussed, with a focus on the framework of homogeneous/inhomogeneous entrainment-mixing. The major advance is the introduction of continuous microphysical measures—homogeneous mixing degrees that can be used to quantify all the mixing types, instead of qualitative description of different entrainment-mixing mechanisms. The Damkohler number has been used for studying entrainment-mixing mechanisms since around 1980; however, this number needs entrained air blob size that is often unknown; therefore, transition scale number is introduced. The Damkohler and transition scale numbers are discussed as two dynamical measures in understanding entrainment-mixing processes. The quantitative microphysical and dynamical measures provide unifying quantification of entrainment-mixing processes continuously from extreme inhomogeneous to homogeneous entrainment-mixing mechanisms, which is missing in qualitative description. Furthermore, based on the empirical relationships between the homogenous mixing degrees and dynamical measures, some parameterizations of entrainment-mixing mechanism have been developed for use in larger scale models. With the quantitative measures, the appropriate microphysical scales for the studies of entrainment-mixing mechanisms are determined and found to be dependent on scientific objectives; dependence of entrainment-mixing mechanisms on sampling scales or model grid sizes are quantified and the trends can be opposite, depending on cloud microphysics, relative humidity in entrained air, and entrained air size.

In terms of entrainment rate in cumulus, several approaches for estimating entrainment rate are reviewed, including the simplified and full versions of the bulk-plume approach and a mixing fraction approach. In effort of improving cumulus parameterizations, the fractional entrainment rate has been parameterized as a function of internal cloud properties (e.g., vertical velocity and buoyancy), external properties (e.g., relative humidity in environment), or both. A big advance is that relative humidity in environment has been considered in parameterizations since 2004. However, different scientists have different opinions on the relationship between entrainment rate and relative humidity. Besides, entrainment rate can also be parameterized as a stochastic process. Three approaches commonly used to estimate entrainment velocity in stratocumulus clouds are briefly reviewed.

Several challenges remain to be addressed. (1) Entrainment-mixing mechanisms and entrainment rate have been largely studied in separation, although the two parts are closely connected in nature. For example, it has been shown that the two parts can be connected, at least for cumulus clouds, through mixing fraction of the adiabatic cloud during isobaric mixing, which is included in the calculations of homogeneous mixing degree, microphysical timescales (thus transition scale number and Damkohler number), and fractional entrainment rate (Figure 4.10). Besides, both entrainment-mixing mechanisms and entrainment rate are closely related to relative humidity in the entrained dry air (Bechtold et al., 2008; Desai et al., 2021). A higher relative humidity leads to a higher homogeneous mixing degree and a larger entrainment rate, others being the same. In addition, a larger entrainment rate is more likely to produce, or associate with, a larger dry air filament size. The larger filament size and higher relative humidity cause more significant scale dependence of entrainment-mixing mechanisms. Another separation lies in the three frameworks proposed to study entrainment-mixing processes. It is possible that in real clouds the concepts of homogeneous/inhomogeneous entrainment-mixing and vertical mixing work together;

but how to unify/integrate the different concepts remains elusive and merits further study.

(2) The detailed physical processes of entrainment-mixing on cloud droplet size distribution are still not clear. High-resolution laboratory experiments (Chang et al., 2016; Prabhakaran et al., 2020; Shaw et al., 2020), in situ observations (Beals et al., 2015; Gerber et al., 2013; Haman et al., 2007), and direct numerical simulations with large domains (Kumar et al., 2018) are needed.

(3) Due to complexities of both entrainment-mixing and ice/mixed-phase clouds, the effects of entrainment-mixing on cloud microphysics have been mainly studied in warm clouds. Pinsky et al. (2018) explored the process of ice-liquid water interaction in the unsaturated environment both analytically and with the help of a numerical simulation. They found that mixing of a mixed-phase cloud with a dry environment accelerated cloud glaciation. Therefore, extending the concepts of entrainment-mixing in warm clouds to ice/mixed-phase clouds is needed.

(4) In the mixing diagrams of droplet size vs. number concentration or liquid water content, no spectral width of cloud droplet size distribution is considered, although understanding spectral broadening and rain initiation largely prompted the early investigation of entrainment-mixing processes. Luo et al. (2021) attempted to explicitly account for the effect of spectral shape of cloud droplet size distribution by introducing spectral-bin-weighted homogeneous mixing degrees to quantify entrainment-mixing mechanisms. More studies are needed to consider spectral shape in the traditional mixing diagram.

(5) One entrainment-mixing event per cloud life time is often assumed, but in reality secondary mixing may occur (Jensen et al., 1985). With EMPM, Lu et al. (2014a) found that entrainment-mixing is more inhomogeneous considering secondary mixing events than only considering one primary mixing event because the initial droplets in secondary mixing events are smaller than those in the adiabatic cloud. Therefore, how to theoretically quantify the effects of secondary mixing deserves further studies.

(6) The homogeneous mixing degrees need adiabatic values of liquid water content, droplet concentration, or volume mean sizes, which sometimes are hard to be determined. In order to overcome this difficulty, Gao et al. (2021) defined another homogeneous mixing degree as the ratio of relative standard deviation of cubic volume mean radius to relative standard deviation of liquid water content. More studies along this line are needed.

(7) In the current entrainment-mixing concepts, there is no aerosol in the entrained dry air and no secondary aerosol activation. However, as pointed out by Warner (1973) and Krueger (2008), numerical simulations with/without aerosols in the entrained dry air could produce significantly different cloud droplet size distributions and different relationships between microphysical properties if the cloud is lifted upward after entrainment-mixing. The data points in the mixing diagram of droplet size vs. number concentration are quite scattered (e.g., Burnet & Brenguier, 2007), which could be related to the analysis approaches without considering aerosol in the entrained dry air, secondary mixing, and spectral width of cloud droplet size distributions. The same holds true for entrainment parameterization in large-scale models, in which entrainment of aerosols has not been addressed but could be important (Chen et al., 2020).

(8) Whether entrainment-mixing processes promote or inhibit warm rain initiation deserves further examination. Some studies argue that inhomogeneous entrainment-mixing and entity-type entrainment-mixing cause superadiabatic growth of cloud droplets (Baker et al., 1980; Telford, 1996), others argue that rain embryos are more likely to form in the least diluted cloud cores (e.g., Bera et al., 2016).

(9) Studies on the influence of relative humidity in the environment on entrainment rate are far from complete (de Rooy et al., 2016). Aircraft observations, remote sensing, and numerical simulations of shallow and deep cumulus clouds over different underlying surfaces are needed to further explore the relationship between the two properties.

(10) For parameterizations of entrainment rate, more efforts should be devoted to the properties less explored, for example, vertical wind shear, cloud fraction, and aerosol loading. In a two-paper sequence (Drueke et al., 2020, 2021), based on large eddy simulations, the authors found strong sensitivities of entrainment rate to continentality, vertical wind shear, and mean wind speed, but weak sensitivities to cloud-layer and subcloud-layer depths. In an observational study with surface-based retrievals, Kirshbaum and Lamer (2021) evaluated sensitivities of entrainment rate to a wide range of external properties and internal cloud properties. Combination of external properties and internal cloud properties could make parameterizations of entrainment rate more reliable (Xu et al., 2021). In addition, it is hard to tell whether parameterizations developed based on simulations or observations are more reliable. The reason is that both simulations and observations have their own uncertainties. This problem may be solved with more data to obtain better fits. Furthermore, the direct measurement of entrainment rate has been limited to exploration in large eddy simulation studies; further research is needed to develop observational technologies and to build the link to fractional entrainment rate. Entrainment velocity is often used in studying stratocumulus clouds; understanding the

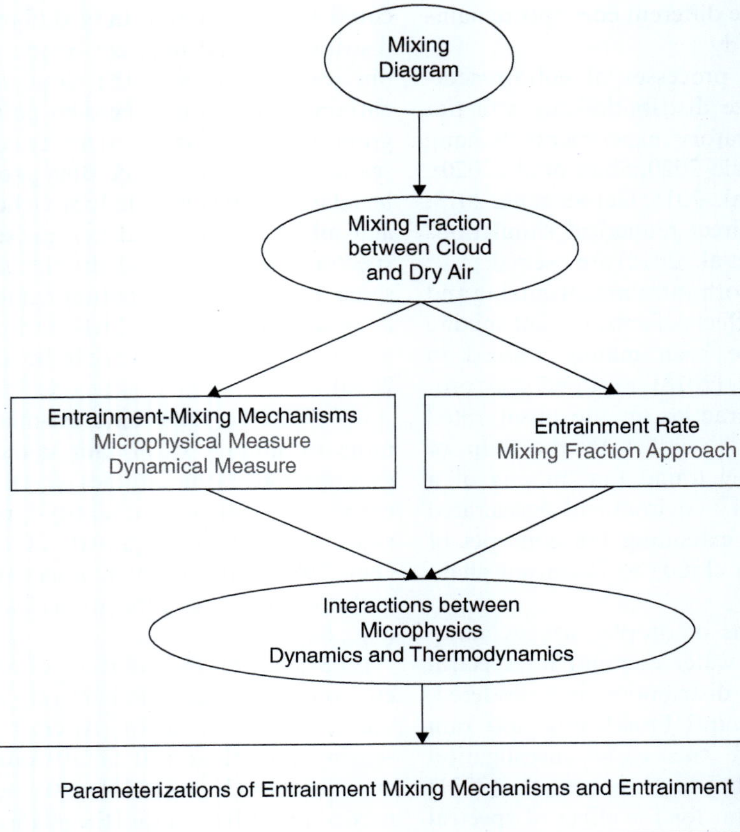

Figure 4.10 Schematic for combined studies on entrainment-mixing mechanisms and entrainment rate in cumulus.

connection between entrainment velocity, direct entrainment rate, and fractional entrainment rate is desirable.

(11) Preferential sorting/concentration due to turbulence can affect different spatial distributions of cloud droplets (Fessler et al., 1994; Morrison et al., 2020; Shaw, 2003). For example, the phenomenon of elongated cloud holes inside cloud is observed and can be explained based on the study of heavy inertial particles sedimenting inside a Burgers vortex (Karpińska et al., 2019). It is challenging to determine whether observed cloud holes are caused by preferential sorting or entrainment-mixing, especially for aircraft observations.

(12) Detrainment is also a key process affecting cumulus clouds. Estimation and parameterization of detrainment rate are mainly based on numerical simulations or theoretical derivations (Böing et al., 2012; Dawe & Austin, 2013; de Rooy et al., 2013; Romps, 2010; Wang, 2020). Norgren et al. (2016) and Zhu et al. (2021) developed an approach for estimating detrainment based on aircraft observations of cumulus clouds. It would be helpful to apply this approach to observations to examine factors affecting detrainment and develop parameterizations of detrainment.

(13) Although it has been suggested since the early 1950s (Houghton & Cramer, 1951) that distinction between dynamical and turbulent entrainment should be made since the process can involve both organized motion and/or small-scale turbulence, the detail as to how to separate and represent them remains elusive and even controversial (Bechtold et al., 2008; de Rooy et al., 2013; Gregory & Rowntree, 1990; Lecoanet & Jeevanjee, 2019; Tiedtke, 1989). This outstanding issue is likely scale-dependent and merits further study.

APPENDIX

Morrison and Grabowski (2008) quantified the effect of entrainment-mixing process by

$$n = n_0\left(\frac{LWC}{LWC_0}\right)^{\alpha}, \quad \text{(A1)}$$

where n and n_0 are the number concentrations of cloud water droplets after and before the evaporation process, respectively; LWC and LWC_0 represent the corresponding liquid water content, respectively. When $\alpha = 0$, homogenous mixing mechanism is assumed, and $n = n_0 = n_h$, where n_h is number concentration assuming homogeneous entrainment-mixing. When $\alpha = 1$, extreme inhomogeneous entrainment-mixing mechanism is assumed, equation (A1) becomes

$$n = n_i = n_h \frac{LWC}{LWC_0}. \quad \text{(A2)}$$

Thus, LWC/LWC$_0$ = n_i/n_h, and equation (A1) can be rewritten as

$$n = n_h \left(\frac{n_i}{n_h} \right)^{\alpha}. \quad \text{(A3)}$$

Therefore,

$$\alpha = \frac{\ln n - \ln n_h}{\ln n_i - \ln n_h}. \quad \text{(A4)}$$

Lu et al. (2013a) defined the third homogeneous mixing degree (ψ_3) as follows:

$$\psi_3 = \frac{\ln n - \ln n_i}{\ln n_h - \ln n_i}. \quad \text{(A5)}$$

Therefore,

$$\psi_3 = 1 - \alpha. \quad \text{(A6)}$$

LIST OF SYMBOLS

a	parameter in equation (4.28)
A	horizontal domain size
A_r	the sum of all horizontal hole sizes divided by the flight horizontal distance
b	empirical parameter in equation (4.28)
B	Buoyancy
c	scaling factor in equation (4.29)
C	function of temperature and air pressure
d	empirical parameter in equation (4.30)
D	divergence of large-scale horizontal winds
Da	Damkohler number
e	empirical parameter in equation (4.32)
E	entrainment rate
F	efficiency factor in equation (4.40)
G	a function of temperature and air pressure
h	moist static energy
h_c	moist static energy in cloud
h_d	moist static energy of detrained air
h_e	moist static energy of entrained air
K_1	an empirical parameter in equation (4.36a)
K_2	an empirical parameter in equation (4.36a)
l	the cloud surface at a certain height
L	dry air blob size
L^*	transition length
LE	blob size of subsaturated air entrained into a cloud
LWC	liquid water content
LWC$_a$	adiabatic liquid water content
LWC$_0$	liquid water content after entrainment but before evaporation
$\hat{\boldsymbol{m}}$	a unit vector perpendicular to the cloud surface
M	cloud mass
M_c	cloud vertical mass flux
n	cloud droplet number concentration
n_a	adiabatic cloud droplet number concentration
n_i	cloud droplet number concentration for extreme inhomogeneous entrainment-mixing in Figure 4.2
n_h	cloud droplet number concentration for homogeneous entrainment-mixing in Figure 4.2
n_0	cloud droplet number concentration after entrainment and before evaporation, which is identical to n_h
N_L	transition scale number
p	significant level
q_c'	difference of liquid water mixing ratio between cloud holes and adiabatic clouds
q_v'	difference of water vapor mixing ratio between cloud holes and adiabatic clouds
q_t	total water mixing ratio
Q_1	function of temperature and air pressure in equation (4.14)
Q_2	function of temperature and air pressure in equation (4.14)
r	droplet radius
r_a	adiabatic droplet radius
r_v	volume-mean radius
r_{va}	adiabatic volume-mean radius
r_{vh}	volume-mean radius for homogeneous entrainment-mixing in Figure 4.2
RH	relative humidity in entrained air
R^2	the adjusted coefficient of determination
S	saturation ratio deficit
S_c	source term for moist static energy due to freezing/melting and deposition/sublimation
S_0	initial saturation ratio deficit
t	time
$\boldsymbol{u}$	air mass flow velocity vector at the cloud surface
$\boldsymbol{u}_i$	velocity of the cloud surface

w	vertical velocity
W	appropriate measure of the rate of turbulent production
w_{dir}	effect of nonturbulent processes in deepening the layer
w_{entr}	entrainment velocity at the top of stratocumulus clouds
X	empirical parameter in equation (4.35a)
Y	empirical parameter in equation (4.35a)
z	Height
z_{aj}	the adjusted *j*th height
z_{i}	temperature inversion height
z_{j}	the *j*th height
z_0	cloud base height
$\frac{\partial B}{\partial z}$	buoyancy vertical gradient
α	parameter used to represent the effect of entrainment-mixing process in Morrison scheme
α_{B}	buoyancy reduction factor
β	angel between two state lines in Figure 4.2
δ	detrainment rate
ε	dissipation rate
ε_0	empirical parameter related to the height above ground
η	Kolmogorov length scale
θ_{e}	equivalent potential temperature
θ_{ec}	equivalent potential temperature of cloud
θ_{ee}	equivalent potential temperature of environment
θ_{l}	liquid water potential temperature
λ	fractional entrainment rate
λ_{aj}	entrainment rate after adjustment
λ_{j}	entrainment rate from cloud base to the *j*th height
ν	kinematic viscosity
ρ	air density
ρ_{L}	liquid water density
σ	cloud area
τ_{c}	turnover timescale
τ_{evap}	evaporation time
τ_{mix}	turbulent mixing time
τ_{phase}	phase relaxation time
τ_{r}	microphysical response time of droplet population
τ_{react}	reaction time
χ	mixing fraction of adiabatic cloud
χ_{dc}	critical value of mixing fraction of dry air
ψ	homogeneous mixing degree
ψ_1	first homogeneous mixing degree
ψ_2	second homogeneous mixing degree
ψ_3	third homogeneous mixing degree
ψ_4	fourth homogeneous mixing degree
ϕ	conserved tracer
ϕ_{c}	conserved tracer of cloud
ϕ_{e}	conserved tracer of environment

REFERENCES

Ackerman, A. S., vanZanten, M. C., Stevens, B., Savic-Jovcic, V., Bretherton, C. S., Chlond, A., et al. (2009). Large-Eddy simulations of a drizzling, stratocumulus-topped marine boundary layer. *Monthly Weather Review*, *137*(3), 1083–1110.

Andrejczuk, M., Grabowski, W. W., Malinowski, S. P., & Smolarkiewicz, P. K. (2009). Numerical simulation of cloud–clear air interfacial mixing: Homogeneous versus inhomogeneous mixing. *Journal of the Atmospheric Sciences*, *66*(8), 2493–2500. doi:10.1175/2009JAS2956.1.

Arakawa, A., & Schubert, W. H. (1974). Interaction of a cumulus cloud ensemble with the large-scale environment, Part I. *Journal of the Atmospheric Sciences*, *31*(3), 674–701. doi:10.1175/1520-0469(1974)031<0674:ioacce>2.0.co;2.

Austin, P. H., Baker, M. B., Blyth, A. M., & Jensen, J. B. (1985). Small-scale variability in warm continental cumulus clouds. *Journal of the Atmospheric Sciences*, *42*(11), 1123–1138. 10.1175/1520-0469(1985)042<1123:ssviwc>2.0.co;2.

Baba, Y. (2020). Shallow convective closure in a spectral cumulus parameterization. *Atmospheric Research*, *233*, 104707.

Baker, M. B., Breidenthal, R. E., Choularton, T. W., & Latham, J. (1984). The effects of turbulent mixing in clouds. *Journal of the Atmospheric Sciences*, *41*(2), 299–304. doi:10.1175/1520-0469(1984)041<0299:TEOTMI>2.0.CO;2.

Baker, M. B., Corbin, R. G., & Latham, J. (1980). The influence of entrainment on the evolution of cloud droplet spectra: I. A model of inhomogeneous mixing. *Quarterly Journal of the Royal Meteorological Society*, *106*(449), 581–598. 10.1002/qj.49710644914.

Baker, M. B., & Latham, J. (1979). The evolution of droplet spectra and the rate of production of embryonic raindrops in small cumulus clouds. *Journal of the Atmospheric Sciences*, *36*(8), 1612–1615. doi:10.1175/1520-0469(1979)036<1612:TEODSA>2.0.CO;2.

Beals, M. J., Fugal, J. P., Shaw, R. A., Lu, J., Spuler, S. M., & Stith, J. L. (2015). Holographic measurements of inhomogeneous cloud mixing at the centimeter scale. *Science*, *350*(6256), 87–90. 10.1126/science.aab0751.

Bechtold, P., Köhler, M., Jung, T., Doblas-Reyes, F., Leutbecher, M., Rodwell, M. J., et al. (2008). Advances in simulating atmospheric variability with the ECMWF model: From synoptic to decadal time-scales. *Quarterly*

Journal of the Royal Meteorological Society, *134*(634), 1337–1351.

Becker, T., Bretherton, C. S., Hohenegger, C., & Stevens, B. (2018). Estimating bulk entrainment with unaggregated and aggregated convection. *Geophysical Research Letters*, *45*(1), 455–462. doi:10.1002/2017GL076640.

Bellenger, H., K. Yoneyama, M. Katsumata, T. Nishizawa, K. Yasunaga, and R. Shirooka (2015). Observation of moisture tendencies related to shallow convection. *Journal of the Atmospheric Sciences*, *72*(2), 305–306.

Bera, S., Prabha, T. V., & Grabowski, W. W. (2016). Observations of monsoon convective cloud microphysics over India and role of entrainment—mixing. *Journal of Geophysical Research*, *121*(16), 9767–9788.

Berg, L. K., Kassianov, E. I., Long, C. N., & Mills, D. L., Jr., (2011). Surface summertime radiative forcing by shallow cumuli at the Atmospheric Radiation Measurement Southern Great Plains site. *Journal of Geophysical Research*, *116*, D01202. https://doi.org/10.1029/2010JD014593.

Betts, A. (1978), Convection in the tropics. *Quarterly Journal of the Royal Meteorological Society, Suppl.*, 105–132.

Betts, A. K. (1975). Parametric interpretation of trade-wind cumulus budget studies. *Journal of the Atmospheric Sciences*, *32*(10), 1934–1945. doi:10.1175/1520-0469(1975)032<1934:piotwc>2.0.co;2.

Betts, A. K. (1983). Thermodynamics of mixed stratocumulus layers: Saturation point budgets. *Journal of the Atmospheric Sciences*, *40*(11), 2655–2670. 10.1175/1520-0469(1983)040<2655:tomsls>2.0.co;2.

Betts, A. K., & Harshvardhan, H. (1987). Thermodynamic constraint on the cloud liquid water feedback in climate models. *Journal of Geophysical Research*, *92*(D7), 8483–8485. 10.1029/JD092iD07p08483.

Blyth, A. M. (1993). Entrainment in cumulus clouds. *Journal of Applied Meteorology*, *32*(4), 626–641. doi: 10.1175/1520-0450(1993)032<0626:eicc>2.0.co;2.

Böing, S. J., Jonker, H. J. J., Nawara, W. A., & Siebesma, A. P. (2014). On the deceiving aspects of mixing diagrams of deep cumulus convection. *Journal of the Atmospheric Sciences*, *71*(1), 56–68. 10.1175/JAS-D-13-0127.1.

Böing, S. J., Siebesma, A. P., Korpershoek, J. D., & Jonker, H. J. J. (2012). Detrainment in deep convection. *Geophysical Research Letters*, *39*, L20816.

Bony, S., & Dufresne, J.-L. (2005). Marine boundary layer clouds at the heart of tropical cloud feedback uncertainties in climate models. *Geophysical Research Letters*, *32*(20), L20806. doi:10.1029/2005gl023851.

Bretherton, C. S., McCaa, J. R., & Grenier, H. (2004). A new parameterization for shallow cumulus convection and its application to marine subtropical cloud-topped boundary layers. Part I: Description and 1D results. *Monthly Weather Review*, *132*(4), 864–882. 10.1175/1520-0493(2004)132<0864:ANPFSC>2.0.CO;2.

Bretherton, C., & Smolarkiewicz, P. K. (1989a). Gravity waves, compensating subsidence and detrainment around cumulus clouds. *Journal of the Atmospheric Sciences*, *46*(6), 740–759.

Bretherton, C. S., & Smolarkiewicz, P. K. (1989b). Gravity waves, compensating subsidence and detrainment around cumulus clouds. *Journal of the Atmospheric Sciences*, *46*(6), 740–759. 10.1175/1520-0469(1989)046<0740:gwcsad>2.0.co;2.

Brown, A. R., Cederwall, R. T., Chlond, A., Duynkerke, P. G., Golaz, J. C., Khairoutdinov, M., et al. (2002). Large-eddy simulation of the diurnal cycle of shallow cumulus convection over land, *Quarterly Journal of the Royal Meteorological Society*, *128*(582), 1075–1093.

Burnet, F., & Brenguier, J. L. (2007). Observational study of the entrainment-mixing process in warm convective clouds. *Journal of the Atmospheric Sciences*, *64*(6), 1995–2011. doi:10.1175/JAS3928.1.

Bush, S. J., Turner, A. G., Woolnough, S. J., Martin, G. M., & Klingaman, N. P. (2015). The effect of increased convective entrainment on Asian monsoon biases in the MetUM general circulation model. *Quarterly Journal of the Royal Meteorological Society*, *141*(686), 311–326. doi:10.1002/qj.2371.

Caldwell, P., Bretherton, C. S., & Wood, R. (2005). Mixed-layer budget analysis of the diurnal cycle of entrainment in southeast pacific stratocumulus. *Journal of the Atmospheric Sciences*, *62*(10), 3775–3791, 10.1175/jas3561.1.

Carson, D. J. (1973). The development of a dry inversion-capped convectively unstable boundary layer. *Quarterly Journal of the Royal Meteorological Society*, *99*(421), 450–467. https://doi.org/10.1002/qj.49709942105.

Cess, R. D., Potter, G. L., Blanchet, J. P., Boer, G. J., Ghan, S. J., Kiehl, J. T., et al. (1989). Interpretation of cloud-climate feedback as produced by 14 atmospheric general circulation models. *Science*, *245*(4917), 513–516. doi:10.1126/science.245.4917.513.

Chandrakar, K. K., Cantrell, W., Chang, K., Ciochetto, D., Niedermeier, D., Ovchinnikov, M., et al. (2016). Aerosol indirect effect from turbulence-induced broadening of cloud-droplet size distributions. *Proceedings of the National Academy of Sciences USA*, *113*(50), 14243–14248.

Chandrakar, K. K., Cantrell, W., Ciochetto, D., Karki, S., Kinney, G., & Shaw, R. (2017). Aerosol removal and cloud collapse accelerated by supersaturation fluctuations in turbulence. *Geophysical Research Letters*, *44*, 4359–4367.

Chang, K., Bench, J., Brege, M., Cantrell, W., Chandrakar, K. K., Ciochetto, D., et al. (2016). A laboratory facility to study gas–aerosol–cloud interactions in a turbulent environment: The Π chamber. *Bulletin of the American Meteorological Society*, *97*(12), 2343–2358. 10.1175/bams-d-15-00203.1.

Chen, J., Liu, Y., & Zhang, M. (2020). Effects of lateral entrainment mixing with entrained aerosols on cloud microphysics. *Geophysical Research Letters*, *47*(13), e2020GL087667. https://doi.org/10.1029/2020GL087667.

Cheng, M., Lu, C., & Liu, Y. (2015). Variation in entrainment rate and relationship with cloud microphysical properties on the scale of 5 m. *Science Bulletin*, *60*, 707–717. doi: 10.1007/s11434-015-0737-8.

Chikira, M., & M. Sugiyama (2010). A cumulus parameterization with state-dependent entrainment rate. Part I: Description and sensitivity to temperature and humidity profiles. *Journal of the Atmospheric Sciences*, *67*(7), 2171–2193. doi:10.1175/2010JAS3316.1.

Chosson, F., Brenguier, J.-L., & Schüller, L. (2007). Entrainment-mixing and radiative transfer simulation in boundary layer clouds. *Journal of the Atmospheric Sciences*, *64*(7), 2670–2682, doi:10.1175/JAS3975.1.

Clark, T. L., & Hall, W. D. (1979). A numerical experiment on stochastic condensation theory. *Journal of the Atmospheric Sciences*, *36*(3), 470–483.

Colman, R. (2003). A comparison of climate feedbacks in general circulation models. *Climate Dynamics*, *20*(7), 865–873.

Dawe, J. T., & Austin, P. H. (2011). Interpolation of LES cloud surfaces for use in direct calculations of entrainment and detrainment. *Monthly Weather Review*, *139*(2), 444–456. doi: 10.1175/2010MWR3473.1.

Dawe, J. T., & Austin, P. H. (2013). Direct entrainment and detrainment rate distributions of individual shallow cumulus clouds in an LES. *Atmospheric Chemistry and Physics*, *13*(15), 7795–7811.

de Rooy, W. C., Bechtold, P., Fröhlich, K., Hohenegger, C., Jonker, H., Mironov, D., et al. (2013). Entrainment and detrainment in cumulus convection: An overview. *Quarterly Journal of the Royal Meteorological Society*, *139*(670), 1–19. 10.1002/qj.1959.

de Rooy, W. C., & Siebesma, A. P. (2008). A simple parameterization for detrainment in shallow cumulus, *Monthly Weather Review*, *136*(2), 560–576. doi:10.1175/2007mwr2201.1.

de Rooy, W. C., & Siebesma, A. P. (2010). Analytical expressions for entrainment and detrainment in cumulus convection. *Quarterly Journal of the Royal Meteorological Society*, *136*(650), 1216–1227. 10.1002/qj.640.

de Rooy, W. C., Yano, J.-I., Bechtold, P., & Böing, S. J. (2016). Entrainment and detrainment formulations for mass-flux parameterization. In R. Plant and J.-I. Yano (Eds.), *Parameterization of atmospheric convection* (pp. 273–323). Imperial College Press: Singapore.

Deardorff, J. W. (1974). Three-dimensional numerical study of the height and mean structure of a heated planetary boundary layer. *Boundary-Layer Meteorology*, *7*(1), 81–106. 10.1007/BF00224974.

Deardorff, J. W. (1980). Cloud Top entrainment instability. *Journal of the Atmospheric Sciences*, *37*(1), 131–147.

Del Genio, A. D., Chen, Y., Kim, D., & Yao, M. S. (2012). The MJO transition from shallow to deep convection inCloudSat/CALIPSO data and GISS GCM simulations. *Journal of Climate*, *25*(11), 3755–3770.

Del Genio, A. D., & Wu, J. (2010). The role of entrainment in the diurnal cycle of continental convection. *Journal of Climate*, *23*(10), 2722–2738. doi: 10.1175/2009JCLI3340.1.

Demott, C. A., Randall, D. A., & Khairoutdinov, M. (2005). Convective precipitation variability as a tool for general circulation model analysis. *Journal of Climate*, *20*(1), 91–112.

Derbyshire, S. H., Beau, I., Bechtold, P., Grandpeix, J. Y., Piriou, J. M., Redelsperger, J. L., & Soares, P. M. M. (2004). Sensitivity of moist convection to environmental humidity, *Quarterly Journal of the Royal Meteorological Society*, *130*(604), 3055–3079. 10.1256/qj.03.130.

Derbyshire, S. H., Maidens, A. V., Milton, S. F., Stratton, R. A., & Willett, M. R. (2011). Adaptive detrainment in a convective parametrization. *Quarterly Journal of the Royal Meteorological Society*, *137*(660), 1856–1871. 10.1002/qj.875.

Desai, N., Liu, Y., Glienke, S., Shaw, R. A., Lu, C., Wang, J., & Gao, S. (2021). Vertical variation of turbulent entrainment mixing processes in marine stratocumulus clouds using high-resolution digital holography. *Journal of Geophysical Research*, *126*(7), e2020JD033527. https://doi.org/10.1029/2020JD033527.

Ditas, F., Shaw, R. A., Siebert, H., Simmel, M., Wehner, B., & Wiedensohler, A. (2012). Aerosols-cloud microphysics-thermodynamics-turbulence: Evaluating supersaturation in a marine stratocumulus cloud. *Atmospheric Chemistry and Physics*, *12*(5), 2459–2468. 10.5194/acp-12-2459-2012.

Donner, L. J., Seman, C. J., Hemler, R. S., & Fan, S. (2001). A cumulus parameterization including mass fluxes, convective vertical velocities, and mesoscale effects: Thermodynamic and hydrological aspects in a General Circulation Model. *Journal of Climate*, *64*(14), 3444–3463.

Drueke, S., Kirshbaum, D. J., & Kollias, P. (2019). Evaluation of Shallow-Cumulus entrainment rate retrievals using Large-Eddy simulation. *Journal of Geophysical Research*, *124*(16), 9624–9643. https://doi.org/10.1029/2019JD030889.

Drueke, S., Kirshbaum, D. J., & Kollias, P. (2020). Environmental sensitivities of shallow-cumulus dilution—Part 1: Selected thermodynamic conditions. *Atmospheric Chemistry and Physics*, *20*(21), 13217–13239. 10.5194/acp-20-13217-2020.

Drueke, S., Kirshbaum, D. J., & Kollias, P. (2021). Environmental sensitivities of shallow-cumulus dilution—Part 2: Vertical wind profile. *Atmospheric Chemistry and Physics*, *21*(18), 14039–14058. 10.5194/acp-21-14039-2021.

Duan, S. Q., Wright, J. S., & Romps, D. M. (2018). On the utility (or futility) of using stable water isotopes to constrain the bulk properties of tropical convection. *Journal of Advances in Modeling Earth Systems*, *10*(2), 516–529. https://doi.org/10.1002/2017MS001074.

Eissner, J. M., Mechem, D. B., Jensen, M. P., & Giangrande, S. E. (2021). Factors governing cloud growth and entrainment rates in shallow cumulus and cumulus congestus during GoAmazon2014/5. *Journal of Geophysical Research*, *126*(12), e2021JD034722. https://doi.org/10.1029/2021JD034722.

Emanuel, K. (1991). A scheme for representing cumulus convection in large-scale models. *Journal of the Atmospheric Sciences*, *48*(21), 2313–2329.

Emanuel, K. A., & Živković-Rothman, M. (1999). Development and evaluation of a convection scheme for use in climate models. *Journal of the Atmospheric Sciences*, *56*(11), 1766–1782, 10.1175/1520-0469(1999)056<1766:daeoac>2.0.co;2.

Evan, S., Brioude, J., Rosenlof, K., Davis, S. M., Vömel, H., Héron, D., et al. (2020). Effect of deep convection on the tropical tropopause layer composition over the southwest Indian Ocean during austral summer. *Atmospheric Chemistry and Physics*, *20*(17), 10565–10586. 10.5194/acp-20-10565-2020.

Faloona, I., Lenschow, D. H., Campos, T., Stevens, B., VanZanten, M., Blomquist, B. W., et al. (2005). Observations of entrainment in eastern Pacific marine stratocumulus using three conserved scalars. *Journal of the Atmospheric Sciences*, *62*(9), 3268–3285.

Fessler, J. R., Kulick, J. D., & Eaton, J. K. (1994). Preferential concentration of heavy particles in a turbulent channel flow. *Physics of Fluids*, *6*(11), 3742–3749. 10.1063/1.868445.

Formenton, M., Panegrossi, G., Casella, D., Dietrich, S., Mugnai, A., Sanò, P., et al. (2013). Using a cloud electrification model to study relationships between lightning activity and cloud microphysical structure. *Natural Hazards and Earth System Sciences*, *13*, 1085–1104.

Freitas, S. R., Grell, G. A., & Li, H. (2020). The GF Convection Parameterization: Recent developments, extensions, and applications. *Geoscientific Model Development Discussions*, *2020*, 1–28. 10.5194/gmd-2020-38.

Freud, E., Rosenfeld, D., & Kulkarni, J. R. (2011). Resolving both entrainment-mixing and number of activated CCN in deep convective clouds. *Atmospheric Chemistry and Physics*, *11*(24), 12887–12900.

Gao, S., Lu, C., Liu, Y., Mei, F., Wang, J., Zhu, L., & Yan, S. (2020). Contrasting Scale Dependence of Entrainment-Mixing Mechanisms in Stratocumulus Clouds, *Geophysical Research Letters*, *47*(9), e2020GL086970. 10.1029/2020gl086970.

Gao, S., Lu, C., Liu, Y., Yum, S. S., Zhu, J., Zhu, L., et al. (2021). Comprehensive quantification of height dependence of entrainment-mixing between stratiform cloud top and environment. *Atmospheric Chemistry and Physics*, Accepted.

Gao, Z., Liu, Y., Li, X., & Lu, C. (2018). Investigation of turbulent entrainment-mixing processes with a new particle-resolved direct numerical simulation model. *Journal of Geophysical Research*, *123*, 2194–2214.

Gerber, H., Frick, G., Malinowski, S. P., Brenguier, J.-L., & Burnet, F. (2005). Holes and entrainment in stratocumulus. *Journal of the Atmospheric Sciences*, *62*(2), 443–459. doi:10.1175/JAS-3399.1.

Gerber, H., Frick, G., Malinowski, S. P., Jonsson, H., Khelif, D., & Krueger, S. K. (2013). Entrainment rates and microphysics in POST stratocumulus. *Journal of Geophysical Research*, *118*(21), 12,094–12,109. 10.1002/jgrd.50878.

Gerber, H. E., Frick, G. M., Jensen, J. B., & Hudson, J. G. (2008). Entrainment, mixing, and microphysics in trade-wind cumulus. *Journal of the Meteorological Society of Japan*, *86A*, 87–106.

Grabowski, W. W. (2006). Indirect impact of atmospheric aerosols in idealized simulations of convective-radiative quasi equilibrium. *Journal of Climate*, *19*(18), 4664–4682. doi:10.1175/JCLI3857.1.

Gregory, D. (2001). Estimation of entrainment rate in simple models of convective clouds. *Quarterly Journal of the Royal Meteorological Society*, *127*(571), 53–72. doi:10.1002/qj.49712757104.

Gregory, D., & Rowntree, P. R. (1990). A mass flux convection scheme with representation of cloud ensemble characteristics and stability-dependent closure. *Monthly Weather Review*, *118*(7), 1483–1506. 10.1175/1520-0493(1990)118 <1483:amfcsw>2.0.co;2.

Gu, J.-F., Stephen Plant, R., Holloway, C. E., & Muetzelfeldt, M. R. (2020). Pressure drag for shallow cumulus clouds: From thermals to the cloud ensemble. *Geophysical Research Letters*, *47*(22), e2020GL090460. https://doi.org/10.1029/2020GL090460.

Guo, X., Fu, D., Li, X., Hu, Z., Lei, H., Xiao, H., & Hong, Y. (2015a). Advances in cloud physics and weather modification in China, *Advances in Atmospheric Sciences*, *32*, 230–249. 10.1007/s00376-014-0006-9.

Guo, X., Lu, C., Zhao, T., Zhang, G., & Liu, Y. (2015b). An observational study of entrainment rate in deep convection. *Atmosphere*, *6*(9), 1362–1376.

Guo, Z., Wang, M., Qian, Y., Larson, V. E., Ghan, S., Ovchinnikov, M., et al. (2015c). Parametric behaviors of CLUBB in simulations of low clouds in the Community Atmosphere Model (CAM). *Journal of Advances in Modeling Earth Systems*, *7*(3), 1005–1025. https://doi.org/10.1002/2014MS000405.

Hack, J. J. (1994). Parameterization of moist convection in the National Center for Atmospheric Research community climate model (CCM2). *Journal of Geophysical Research*, *99*(D3), 5551–5568. https://doi.org/10.1029/93JD03478.

Haman, K. E., Malinowski, S. P., Kurowski, M. J., Gerber, H., & Brenguier, J.-L. (2007). Small scale mixing processes at the top of a marine stratocumulus—a case study. *Quarterly Journal of the Royal Meteorological Society*, *133*(622), 213–226. doi:10.1002/qj.5.

Han, J., & Pan, H.-L. (2011). Revision of convection and vertical diffusion schemes in the NCEP Global Forecast System, *Weather Forecasting*, *26*(4), 520–533. 10.1175/waf-d-10-05038.1.

Hannah, W. M. (2017). Entrainment versus dilution in tropical deep convection. *Journal of the Atmospheric Sciences*, *74*(11), 3725–3747. 10.1175/jas-d-16-0169.1.

Hernandez-Deckers, D., & Sherwood, S. C. (2018). On the role of entrainment in the fate of cumulus thermals. *Journal of the Atmospheric Sciences*, *75*(11), 3911–3924. 10.1175/jas-d-18-0077.1.

Hill, A. A., Feingold, G., & Jiang, H. (2009). The influence of entrainment and mixing assumption on aerosol-cloud interactions in marine stratocumulus. *Journal of the Atmospheric Sciences*, *66*(5), 1450–1464.

Hirota, N., Takayabu, Y. N., Watanabe, M., Kimoto, M., & Chikira, M. (2014). Role of convective entrainment in spatial distributions of and temporal variations in precipitation over tropical oceans. *Journal of Climate*, *27*(23), 8707–8723. 10.1175/jcli-d-13-00701.1.

Hoffmann, F., & Feingold, G. (2019). Entrainment and mixing in stratocumulus: Effects of a new explicit subgrid-scale scheme for large-eddy simulations with particle-based microphysics. *Journal of the Atmospheric Sciences*, *76*(7), 1955–1973. 10.1175/jas-d-18-0318.1.

Holton, J., & Gettelman, A. (2001). Horizontal transport and dehydration in the stratosphere. *Geophysical Research Letters*, *28*, 2799–2802. 10.1029/2001GL013148.

Hong, S.-Y., & Jang, J. (2018). Impacts of shallow convection processes on a simulated boreal summer climatology in a global atmospheric model. *Asia-Pacific Journal of Atmospheric Sciences*, *54*(1), 361–370.

Houghton, H. G., & Cramer, H. E. (1951). A theory of entrainment in convective currents. *Journal of Meteorology*, *8*(2), 95–102. 10.1175/1520-0469(1951)008<0095:atoeic>2.0.co;2.

Houze, R. A. Jr. (1993). *Cloud Dynamics*. Academic Press, Inc.: San Diego, California, US.

Isotta, F. A., Spichtinger, P., Lohmann, U., & von Salzen, K. (2011). Improvement and implementation of a parameterization for shallow cumulus in the global climate model ECHAM5-HAM. *Journal of the Atmospheric Sciences*, *68*(3), 515–532.

Jakob, C., & Siebesma, A. P. (2003). A new subcloud model for mass-flux convection schemes: Influence on triggering, updraft properties, and model climate. *Monthly Weather Review*, *131*(11), 2765–2778. 10.1175/1520-0493(2003)131<2765:ansmfm>2.0.co;2.

Jarecka, D., Grabowski, W. W., Morrison, H., & Pawlowska, H. (2013). Homogeneity of the subgrid-scale turbulent mixing in large-eddy simulation of shallow convection. *Journal of the Atmospheric Sciences*, *70*, 2751–2767.

Jarecka, D., Grabowski, W. W., & Pawlowska, H. (2009). Modeling of subgrid-scale mixing in large-eddy simulation of shallow convection. *Journal of the Atmospheric Sciences*, *66*(7), 2125–2133.

Jeffery, C. A. (2007). Inhomogeneous cloud evaporation, invariance, and Damköhler number. *Journal of Geophysical Research*, *112*, D24S21. doi:10.1029/2007jd008789.

Jensen, J. B., Austin, P. H., Baker, M. B., & Blyth, A. M. (1985). Turbulent mixing, spectral evolution and dynamics in a warm cumulus cloud. *Journal of the Atmospheric Sciences*, *42*(2), 173–192. doi:10.1175/1520-0469(1985)042<0173:TMSEAD>2.0.CO;2.

Jensen, J. B., & Baker, M. B. (1989). A simple model of droplet spectral evolution during turbulent mixing, *Journal of the Atmospheric Sciences*, *46*(18), 2812–2829, doi:10.1175/1520-0469(1989)046<2812:ASMODS>2.0.CO;2.

Jensen, M. P., & Del Genio, A. D. (2006). Factors limiting convective cloud-top height at the ARM Nauru Island Climate Research Facility. *Journal of Climate*, *19*(10), 2105–2117. doi:10.1175/jcli3722.1.

Jeworrek, J., West, G., & Stull, R. (2019). Evaluation of cumulus and microphysics parameterizations in WRF across the Convective Gray Zone. *Weather Forecasting*, *34*(4), 1097–1115. 10.1175/waf-d-18-0178.1.

Jian, B., Li, J., Wang, G., Zhao, Y., Li, Y., Wang, J., et al. (2021). Evaluation of the CMIP6 marine subtropical stratocumulus cloud albedo and its controlling factors. *Atmospheric Chemistry and Physics*, *21*(12), 9809–9828. 10.5194/acp-21-9809-2021.

Jiang, H., Xue, H., Teller, A., Feingold, G., & Levin, Z. (2006). Aerosol effects on the lifetime of shallow cumulus. *Geophysical Research Letters*, *33*(14), L14806. 10.1029/2006gl026024.

Jonas, P. R., & Mason, B. J. (1982). Entrainment and the droplet spectrum in cumulus clouds. *Quarterly Journal of the Royal Meteorological Society*, *108*(458), 857–869. doi:10.1002/qj.49710845808

Kabanov, A. S., Mazin, I. P., & Smirnov V. I. (1971). Comment on "The theory of growth of cloud drops by condensation". *Journal of the Atmospheric Sciences*, *28*(1), 129–130. 10.1175/1520-0469(1971)028<0129:cotogo>2.0.co;2.

Kain, J. S. (2004). The Kain-Fritsch convective parameterization: An update. *Journal of Applied Meteorology 43*, 170–181.

Kain, J. S., & Fritsch, J. M. (1990). A one-dimensional entraining/detraining plume model and its application in convective parameterization, *Journal of the Atmospheric Sciences*, *47*(23), 2784–2802. doi:10.1175/1520-0469(1990)047<2784:AODEPM>2.0.CO;2.

Karlsson, J., Svensson, G., & Rodhe, H. (2008). Cloud radiative forcing of subtropical low level clouds in global models, *Climate Dynamics*, *30*(7), 779–788, 10.1007/s00382-007-0322-1.

Karpińska, K., Bodenschatz, J. F. E., Malinowski, S. P., Nowak, J. L., Risius, S., Schmeissner, T., et al. (2019). Turbulence-induced cloud voids: Observation and interpretation. *Atmospheric Chemistry and Physics*, *19*(7), 4991–5003. 10.5194/acp-19-4991-2019.

Kerstein, A. R. (1992). Linear-eddy modelling of turbulent transport. Part 7. Finite-rate chemistry and multi-stream mixing. *Journal of Fluid Mechanics*, *240*(1), 289–313.

Khalsa, S. J. (1993). Direct sampling of entrainment events in a marine stratocumulus layer. *Journal of the Atmospheric Sciences*, *50:12*(12), 1734–1750.

Khvorostyanov, V. I., & Curry, J. A. (1999). Toward the theory of stochastic condensation in clouds. Part I: A general kinetic equation. *Journal of the Atmospheric Sciences*, *56*(23), 3985–3996. 10.1175/1520-0469(1999)056<3985:tttosc>2.0.co;2.

Kim, B.-G., Miller, M. A., Schwartz, S. E., Liu, Y., & Min, Q. (2008). The role of adiabaticity in the aerosol first indirect effect. *Journal of Geophysical Research*, *113*(D5), D05210. doi:10.1029/2007jd008961.

Kim, D., Del Genio, A. D., & Yao, M.-S. (2011a). Moist convection scheme in Model E2. *NASA GISS Technical Note and Technical Report*.

Kim, D., Jang, Y.-S., Kim, D.-H., Kim, Y.-H., Watanabe, M., Jin, F.-F., & Kug, J.-S. (2011b). El Niño–Southern Oscillation sensitivity to cumulus entrainment in a coupled general circulation model. *Journal of Geophysical Research*, *116*(D22), D22112. 10.1029/2011JD016526.

Kirshbaum, D. J., & Lamer, K. (2021). Climatological sensitivities of Shallow-Cumulus bulk entrainment in continental and oceanic locations. *Journal of the Atmospheric Sciences*, *78*(8), 2429–2443. 10.1175/jas-d-20-0377.1.

Klein, S. A., Hall, A., Norris, J. R., & Pincus, R. (2018). Low-cloud feedbacks from cloud-controlling factors: A review. *Shallow Clouds, Water Vapor, Circulation, and Climate Sensitivity*, *38*, 135–157.

Klingaman, N. P., & Woolnough, S. J. (2014). Using a case-study approach to improve the Madden–Julian oscillation in the Hadley Centre model. *Quarterly Journal of the Royal Meteorological Society*, *140*(685), 2491–2505. doi:10.1002/qj.2314.

Korczyk, P. M., Kowalewski, T. A., & Malinowski, S. P. (2012). Turbulent mixing of clouds with the environment: Small scale two phase evaporating flow investigated in a laboratory by particle image velocimetry. *Physica D: Nonlinear Phenomena*, *241*(3), 288–296. https://doi.org/10.1016/j.physd.2011.11.003.

Korolev, A. V. (1995). The influence of supersaturation fluctuations on droplet size spectra formation. *Journal of the Atmospheric Sciences*, *52*(20), 3620–3634. 10.1175/1520-0469(1995)052<3620:tiosfo>2.0.co;2.

Korolev, A. V., & Mazin, I. P. (2003). Supersaturation of water vapor in clouds. *Journal of the Atmospheric Sciences*, *60*(24), 2957–2974. 10.1175/1520-0469(2003)060<2957:sowvic>2.0.co;2.

Krueger, S., Su, C., & McMurtry, P. (1997). Modeling entrainment and finescale mixing in cumulus clouds. *Journal of the Atmospheric Sciences*, *54*, 2697–2712. doi:10.1175/1520-0469(1997)054<2697:MEAFMI>2.0.CO;2.

Krueger, S. K. (2008). Fine-scale modeling of entrainment and mixing of cloudy and clear air, paper presented at the *15th International Conference on Clouds and Precipitation*, Cancun, Mexico.

Kuang, Z., & Bretherton, C. S. (2006). A mass-flux scheme view of a high-resolution simulation of a transition from shallow to deep cumulus convection. *Journal of the Atmospheric Sciences*, *63*(7), 1895–1909. 10.1175/jas3723.1.

Kumar, B., Schumacher, J., & Shaw, R. (2013). Cloud microphysical effects of turbulent mixing and entrainment. *Theoretical and Computational Fluid Dynamics*, *27*, 361–376.

Kumar, B., Schumacher, J., & Shaw, R. A. (2014). Lagrangian mixing dynamics at the cloudy–clear air interface. *Journal of the Atmospheric Sciences*, *71*(7), 2564–2580. 10.1175/JAS-D-13-0294.1.

Kumar, B., Bera, S., Prabha, T. V., & Grabowski, W. W. (2017). Cloud-edge mixing: Direct numerical simulation and observations in Indian Monsoon clouds. *Journal of Advances in Modeling Earth Systems*, *9*(1), 332–353. doi:10.1002/2016MS000731.

Kumar, B., Götzfried, P., Suresh, N., Schumacher, J., & Shaw, R. A. (2018). Scale dependence of cloud microphysical response to turbulent entrainment and mixing. *Journal of Advances in Modeling Earth Systems*, *10*(11), 2777–2785. 10.1029/2018ms001487.

Lasher-Trapp, S. G., Cooper, W. A., & Blyth, A. M. (2005). Broadening of droplet size distributions from entrainment and mixing in a cumulus cloud. *Quarterly Journal of the Royal Meteorological Society*, *131*, 195–220. doi:10.1256/qj.03.199.

Latham, J., & Reed, R. L. (1977). Laboratory studies of the effects of mixing on the evolution of cloud droplet spectra. *Quarterly Journal of the Royal Meteorological Society*, *103*(436), 297–306. doi: 10.1002/qj.49710343607.

Lecoanet, D., & Jeevanjee, N. (2019). Entrainment in resolved, dry thermals. *Journal of the Atmospheric Sciences*, *76*(12), 3785–3801. 10.1175/jas-d-18-0320.1.

Lee, I.-Y., & Pruppacher, H. R. (1977). A comparative study on the growth of cloud drops by condensation using an air parcel model with and without entrainment. *Pure and Applied Geophysics*, *115*(3), 523–545. doi: 10.1007/BF00876119

Lehmann, K., Siebert, H., & Shaw, R. A. (2009). Homogeneous and inhomogeneous mixing in cumulus clouds: Dependence on local turbulence structure. *Journal of the Atmospheric Sciences*, *66*, 3641–3659. doi:10.1175/2009JAS3012.1.

Lei, H., Hong, Y., Zhao, Z., Xiao, H., & Guo, X. (2008). Advances in cloud and precipitation physics and weather modification in recent years. *Chinese Journal of the Atmospheric Sciences*, *32*, 967–974 (in Chinese).

Lewellen, D. C., & Lewellen, W. S. (1998). Large-Eddy boundary layer entrainment. *Journal of the Atmospheric Sciences*, *55*(16), 2645–2665.

Li, J., Wang, W.-C., Dong, X., & Mao, J. (2017). Cloud-radiation-precipitation associations over the Asian monsoon region: An observational analysis. *Climate Dynamics*, *49*(9), 3237–3255. 10.1007/s00382-016-3509-5.

Li, M., Letu, H., Peng, Y., Ishimoto, H., Lin, Y., Nakajima, T., et al. (2021a). Assessment of ice cloud modeling capabilities for the irregularly shaped Voronoi models in climate simulations with CAM5. *Atmospheric Chemistry and Physics Discussions*, 2021, 1–32. 10.5194/acp-2021-208.

Li, Y., Ge, J., Dong, Z., Hu, X., Yang, X., Wang, M., & Han, Z. (2021b). Pairwise-rotated EOFs of global cloud cover and their linkages to sea surface temperature. *International Journal of Climatology*, *41*(4), 2342–2359. https://doi.org/10.1002/joc.6962.

Li, Z., Y. Wang, J. Guo, C. Zhao, M. Cribb, X. Dong, et al. (2019). East Asian Study of Tropospheric Aerosols and their Impact on Regional Clouds, Precipitation, and Climate (EAST-AIRCPC). *Journal of Geophysical Research*, *124*(23), 13026–13054. https://doi.org/10.1029/2019JD030758.

Lilly, D. K. (1968). Models of cloud topped mixed layers under strong inversion. *Quarterly Journal of the Royal Meteorological Society*, *94*(401), 292–309.

Lim, Y.-K., Schubert, S. D., Reale, O., Lee, M.-I., Molod, A. M., & Suarez, M. J. (2015). Sensitivity of tropical cyclones to parameterized convection in the NASA GEOS-5 Model, *Journal of Climate*, *28*(2), 551–573. 10.1175/jcli-d-14-00104.1.

Lin, C. (1999). Some bulk properties of cumulus ensembles simulated by a cloud-resolving model. Part II: Entrainment profiles. *Journal of the Atmospheric Sciences*, *56*(21), 3736–3748. 10.1175/1520-0469(1999)056<3736:sbpoce>2.0.co;2.

Lin, W., Liu, Y., Vogelmann, A. M., Fridlind, A., Endo, S., Song, H., et al. (2015). RACORO continental boundary layer cloud investigations: 3. Separation of parameterization biases single-column model CAM5 simulations of shallow cumulus. *Journal of Geophysical Research*, *120*(12), 6015–6033. 10.1002/2014JD022524.

Lin, Y., Huang, X., Liang, Y., Qin, Y., Xu, S., Huang, W., et al. (2020). Community Integrated Earth System Model (CIESM): Description and evaluation. *Journal of Advances in Modeling Earth Systems*, *12*(8), e2019MS002036. https://doi.org/10.1029/2019MS002036.

Liu, H., Guo, J., Koren, I., Altaratz, O., Dagan, G., Wang, Y., et al. (2019). Non-monotonic aerosol effect on precipitation in convective clouds over tropical oceans. *Scientific Reports*, *9*(1), 7809. 10.1038/s41598-019-44284-2.

Liu, Y., Daum, P. H., Chai, S. K., & Liu, F. (2002). Cloud parameterizations, cloud physics, and their connections: An overview. *Recent Research Developments in Geophysics*, *4*, 119–142.

Loriaux, J. M., Lenderink, G., Roode, S. R. D., & Siebesma, A. P. (2013). Understanding convective extreme precipitation scaling using observations and an entraining plume model. *Journal of the Atmospheric Sciences*, *70*(11), 3641–3655. 10.1175/jas-d-12-0317.1.

Lu, B., & Ren, H.-L. (2016). Improving ENSO periodicity simulation by adjusting cumulus entrainment in BCC_CSMs. *Dynamics of Atmospheres and Oceans*, *76*, 127–140. http://dx.doi.org/10.1016/j.dynatmoce.2016.10.005.

Lu, C., Liu, Y., & Niu, S. (2011). Examination of turbulent entrainment-mixing mechanisms using a combined approach. *Journal of Geophysical Research*, *116*, D20207. doi:10.1029/2011JD015944.

Lu, C., Liu, Y., & Niu, S. (2014a). Entrainment-mixing parameterization in shallow cumuli and effects of secondary mixing events. *Chinese Science Bulletin*, *59*, 896–903.

Lu, C., Liu, Y., Niu, S., & Endo, S. (2014b). Scale dependence of entrainment-mixing mechanisms in cumulus clouds. *Journal of Geophysical Research*, *119*(24), 13877–13890, 10.1002/2014JD022265.

Lu, C., Liu, Y., Niu, S., & Vogelmann, A. M. (2012a). Lateral entrainment rate in shallow cumuli: Dependence on dry air sources and probability density functions. *Geophysical Research Letters*, *39*(20), L20812. doi:10.1029/2012GL053646.

Lu, C., Liu, Y., Yum, S. S., Niu, S., & Endo, S. (2012b). A new approach for estimating entrainment rate in cumulus clouds. *Geophysical Research Letters*, *39*, L04802. doi:10.1029/2011GL050546.

Lu, C., Liu, Y., Niu, S., Krueger, S. K., & Wagner, T. (2013a). Exploring parameterization for turbulent entrainment-mixing processes in clouds. *Journal of Geophysical Research*, *118*, 185–194. doi:10.1029/2012JD018464.

Lu, C., Niu, S., Liu, Y., & Vogelmann, A. M. (2013b). Empirical relationship between entrainment rate and microphysics in cumulus clouds. *Geophysical Research Letters*, *40*(10), 2333–2338. doi: 10.1002/grl.50445.

Lu, C., Liu, Y., Niu, S., & Xue, Y. (2018a). Broadening of cloud droplet size distributions and warm rain initiation associated with turbulence: An overview. *Atmospheric and Oceanic Science Letters*, *11*(2), 123–135. 10.1080/16742834.2018.1410057.

Lu, C., Liu, Y., Zhang, G. J., Wu, X., Endo, S., Cao, L., et al. (2016). Improving parameterization of entrainment rate for shallow convection with aircraft measurements and large eddy simulation. *Journal of the Atmospheric Sciences*, *73*, 761–773. 10.1175/JAS-D-15-0050.1.

Lu, C., Liu, Y., Zhu, B., Yum, S. S., Krueger, S. K., Qiu, Y., et al. (2018b). On which microphysical time scales to use in studies of entrainment-mixing mechanisms in clouds. *Journal of Geophysical Research*, *123*(7), 3740–3756. doi:10.1002/2017JD027985.

Lu, C., Sun, C., Liu, Y., Zhang, G. J., Lin, Y., Gao, W., et al. (2018c). Observational Relationship Between Entrainment Rate and Environmental Relative Humidity and Implications for Convection Parameterization, *Geophysical Research Letters*, *45*(24), 13,495–413,504, doi:10.1029/2018GL080264.

Luo, S., Lu, C., Liu, Y., Bian, J., Gao, W., Li, J., et al. (2020). Parameterizations of entrainment-mixing mechanisms and their effects on cloud droplet spectral width based on numerical simulations, *Journal of Geophysical Research*, *125(22), e2020JD032972*, https://doi.org/10.1029/2020JD032972.

Luo, S., Lu, C., Liu, Y., Gao, W., Zhu, L., Xu, X., et al. (2021). Consideration of initial cloud droplet size distribution shapes in quantifying different entrainment-mixing mechanisms. *Journal of Geophysical Research*, e2020JD034455. https://doi.org/10.1029/2020JD034455.

Luo, Z. J., Liu, G. Y., & Stephens, G. L. (2010). Use of A-Train data to estimate convective buoyancy and entrainment rate, *Geophysical Research Letters*, *37*(9), L09804, 10.1029/2010gl042904.

Mason, B. J., & Jonas, P. R. (1975). The evolution of droplet spectra and large droplets by condensation in cumulus clouds. *Quarterly Journal of the Royal Meteorological Society*, *101*(428), 397–397. doi: 10.1002/qj.49710142828.

McCaa, J. R., & Bretherton, C. S. (2004). A new parameterization for shallow cumulus convection and its application to marine subtropical cloud-topped boundary layers. Part II: Regional simulations of marine boundary layer clouds. *Monthly Weather Review*, *132*(132), 864.

Mellado, J. P. (2010). The evaporatively driven cloud-top mixing layer. *Journal of Fluid Mechanics*, *660*(4), 5–36.

Mellado, J. P. (2017). Cloud-top entrainment in stratocumulus clouds. *Annual Review of Fluid Mechanics*, *49*(1), 145–169. 10.1146/annurev-fluid-010816-060231.

Morrison, H., & Grabowski, W. W. (2008). Modeling supersaturation and subgrid-scale mixing with two-moment bulk warm microphysics. *Journal of the Atmospheric Sciences*, *65*(3), 792–812. doi:10.1175/2007JAS2374.1.

Morrison, H., Lier-Walqui, M., Fridlind, A. M., Grabowski, W. W., Harrington, J. Y., Hoose, C., et al. (2020). Confronting the challenge of modeling cloud and precipitation microphysics. *Journal of Advances in Modeling Earth Systems*, *12*(8), e2019MS001689. https://doi.org/10.1029/2019MS001689.

Neggers, R. A. J., Duynkerke, P. G., & Rodts, S. M. A. (2003). Shallow cumulus convection: A validation of large-eddy simulation against aircraft and Landsat observations, *Quarterly Journal of the Royal Meteorological Society*, *129*(593), 2671–2696. doi:10.1256/qj.02.93.

Neggers, R. A. J., Siebesma, A. P., & Jonker, H. J. J. (2002). A multiparcel model for shallow cumulus convection. *Journal of the Atmospheric Sciences*, *59*(10), 1655–1668. doi: 10.1175/1520-0469(2002)059<1655:ammfsc>2.0.co;2.

Nicholls, S., & Turton, J. D. N. T. (1986). An observational study of the structure of stratiform cloud sheets: Part II. Entrainment. *Quarterly Journal of the Royal Meteorological Society*, *112*(472), 461–480.

Nie, J., & Z. Kuang (2012). Beyond bulk entrainment and detrainment rates: A new framework for diagnosing mixing in cumulus convection. *Geophysical Research Letters*, *39*(21), L21803.

Norgren, M. S., Small, J. D., Jonsson, H. H., & Chuang, P. Y. (2016). Observational estimates of detrainment and entrainment in non-precipitating shallow cumulus. *Atmospheric Chemistry and Physics*, *16*(1), 21–33. 10.5194/acp-16-21-2016.

Park, S. (2014). A Unified Convection Scheme (UNICON). Part I: Formulation. *Journal of the Atmospheric Sciences*, *71*, 3902–3930.

Park, S., & Bretherton, C. S. (2009). The University of Washington shallow convection and moist turbulence schemes and their impact on climate simulations with the Community Atmosphere Model. *Journal of Climate*, *22*(12), 3449–3469.

Pawlowska, H., Brenguier, J. L., & Burnet, F. (2000). Microphysical properties of stratocumulus clouds. *Atmospheric Research*, *55*(1), 15–33.

Peatman, S. C., Methven, J., & Woolnough, S. J. (2018). Isolating the effects of moisture entrainment on convectively coupled equatorial waves in an Aquaplanet GCM. *Journal of the Atmospheric Sciences*, *75*(9), 3139–3157. 10.1175/jas-d-18-0098.1.

Pilon, R., Zhang, C., & Dudhia, J. (2016). Roles of deep and shallow convection and microphysics in the MJO simulated by the model for prediction across scales. *Journal of Geophysical Research*, *121*(18), 10,575–10,600.

Pinsky, M., Khain, A., & Korolev, A. (2016a). Theoretical analysis of mixing in liquid clouds—Part 3: Inhomogeneous mixing. *Atmospheric Chemistry and Physics*, *16*, 9273–9297. 10.5194/acpd-15-30321-2015.

Pinsky, M., Khain, A., & Korolev, A. (2018). Theoretical analysis of liquid–ice interaction in the unsaturated environment with application to the problem of homogeneous mixing. *Journal of the Atmospheric Sciences*, *75*(4), 1045–1062. 10.1175/jas-d-17-0228.1.

Pinsky, M., Khain, A., Korolev, A., & Magaritz-Ronen, L. (2016b). Theoretical investigation of mixing in warm clouds—Part 2: Homogeneous mixing. *Atmospheric Chemistry and Physics*, *16*(14), 9255–9272. 10.5194/acp-16-9255-2016.

Politovich, M. K., & Cooper, W. A. (1988). Variability of the supersaturation in cumulus clouds., *Journal of the Atmospheric Sciences*, *45*(11), 1651–1664. 10.1175/1520-0469(1988)045<1651:votsic>2.0.co;2.

Prabhakaran, P., Shawon, A. S. M., Kinney, G., Thomas, S., Cantrell, W., & Shaw, R. A. (2020). The role of turbulent fluctuations in aerosol activation and cloud formation. *Proceedings of the National Academy of Sciences*, *117*(29), 16831–16838. 10.1073/pnas.2006426117.

Raga, G. B., Jensen, J. B., & Baker, M. B. (1990a). Characteristics of cumulus band clouds off the coast of Hawaii. *Journal of the Atmospheric Sciences*, *47*(3), 338–356. doi:10.1175/1520-0469(1990)047<0338:COCBCO>2.0.CO;2.

Raga, G. B., Jensen, J. B., & Baker, M. B. (1990b). Characteristics of cumulus band clouds off the coast of Hawaii. *Journal of Atmospheric Sciences*, *47*(3), 338–356. 10.1175/1520-0469(1990)047<0338:cocbco>2.0.co;2.

Randall, D. A. (1980). Conditional instability of the first kind upside-down. *Journal of the Atmospheric Sciences*, *37*(1), 125–130.

Randall, D. A., Genio, A. D. D., Donner, L. J., Collins, W. D., & Klein, S. A. (2016). The impact of ARM on Climate Modeling. *Meteorological Monographs*, *57*, 26.21–26.16. 10.1175/amsmonographs-d-15-0050.1.

Rauber, R. M., Stevens, B., Ochs, H. T. III, Knight, C., Albrecht, B. A., Blyth, A. M., et al. (2007). Rain in shallow cumulus over the ocean: The RICO campaign, *Bulletin of the American Meteorological Society*, *88*(12), 1912–1928. doi: 10.1175/BAMS-88-12-1912.

Raymond, D. J., & Blyth, A. M. (1986). A stochastic mixing model for nonprecipitating cumulus clouds. *Journal of the Atmospheric Sciences*, *43*(22), 2708–2718. 10.1175/1520-0469(1986)043<2708:asmmfn>2.0.co;2.

Raymond, D. J., & Zeng, X. (2000). Instability and large-scale circulations in a two-column model of the tropical troposphere. *Quarterly Journal of the Royal Meteorological Society*, *126*, 3117–3135.

Ridout, J. A. (2002). Sensitivity of tropical pacific convection to dry layers at mid- to upper levels: Simulation and parameterization tests. *Journal of the Atmospheric Sciences*, *59*(23), 3362–3381. 10.1175/1520-0469(2002)059<3362:sotpct>2.0.co;2.

Rogers, R. R., & Yau, M. K. (1989). *A short course in cloud physics*. Butterworth Heinemann: Burlington, MA, USA.

Romps, D. M. (2010). A direct measure of entrainment. *Journal of the Atmospheric Sciences*, *67*(6), 1908–1927. doi: 10.1175/2010JAS3371.1.

Romps, D. M., & Kuang, Z. (2010). Nature versus nurture in shallow convection, *Journal of the Atmospheric Sciences*, *67*(5), 1655–1666. doi: 10.1175/2009JAS3307.1.

Romps, D. M., & Öktem, R. (2015). Stereo photogrammetry reveals substantial drag on cloud thermals. *Geophysical Research Letters*, *72*(42), 5051–5057.

Roode, S. R. d., Duynkerke, P. G., & Siebesma, A. P. (2000). Analogies between Mass-Flux and Reynolds-Averaged Equations. *Journal of the Atmospheric Sciences*, *57*(10), 1585–1598. 10.1175/1520-0469(2000)057<1585:abmfar>2.0.co;2.

Schmeissner, T., Shaw, R. A., Ditas, J., Stratmann, F., Wendisch, M., & Siebert, H. (2015). Turbulent mixing in shallow trade wind cumuli: Dependence on cloud life cycle. *Journal of the Atmospheric Sciences*, *72*(4), 1447–1465.

Schubert, W. H., Wakefield, J. S., Steiner, E. J., & Cox, S. K. (1979). Marine Stratocumulus Convection. Part I: Governing Equations and Horizontally Homogeneous Solutions, *Journal of the Atmospheric Sciences*, *36*(7), 1286–1307. doi:10.1175/1520-0469(1979)036<1286:MSCPIG>2.0.CO;2.

Schulz, B., & Mellado, J. P. (2019). Competing effects of droplet sedimentation and wind shear on entrainment in Stratocumulus. *Journal of Advances in Modeling Earth Systems*, *11*(6), 1830–1846. https://doi.org/10.1029/2019MS001617.

Shaw, R. A. (2003). Particle-turbulence interactions in atmospheric clouds. *Annual Review of Fluid Mechanics*, *35*, 183–227, doi: 10.1146/annurev.fluid.35.101101.161125.

Shaw, R. A., Cantrell, W., Chen, S., Chuang, P., & Xue, L. (2020). Cloud–Aerosol–Turbulence interactions: Science priorities and concepts for a large-scale laboratory facility. *Bulletin of the American Meteorological Society*, *101*(7), E1026–E1035.

Sherwood, S. C., Hernández-Deckers, D., Colin, M., & Robinson, F. (2013). Slippery thermals and the cumulus entrainment paradox. *Journal of the Atmospheric Sciences*, *70*(8), 2426–2442.

Shin, J., & Park, S. (2019). A Stochastic Unified Convection Scheme (UNICON). Part I: Formulation and single-column simulation for shallow convection. *Journal of the Atmospheric Sciences*, *77*(2), 583–610.

Siebert, H., Franke, H., Lehmann, K., Maser, R., Wei Saw, E., Schell, D., et al. (2006). Probing finescale dynamics and microphysics of clouds with helicopter-borne measurements. *Bulletin of the American Meteorological Society*, *87*(12), 1727–1738. doi:10.1175/BAMS-87-12-1727.

Siebesma, A. P., & Cuijpers, J. W. M. (1995). Evaluation of parametric assumptions for shallow cumulus convection. *Journal of the Atmospheric Sciences*, *52*(6), 650–666. 10.1175/1520-0469(1995)052<0650:eopafs>2.0.co;2.

Siebesma, A. P., & Holtslag, A. A. M. (1996). Model impacts of entrainment and detrainment rates in shallow cumulus convection. *Journal of the Atmospheric Sciences*, *53*(16), 2354–2364. doi:10.1175/1520-0469(1996)053 <2354:MIOEAD>2.0.CO;2.

Siebesma, A. P. (1998). Shallow cumulus convection. In E. J. Plate, et al. (Eds.), *Buoyant Convection in Geophysical Flows* (pp. 441–486). Springer: Netherlands, Dordrecht.

Siebesma, A. P., Bretherton, C. S., Brown, A., Chlond, A., Cuxart, J., Duynkerke, P. G., et al. (2003). A Large Eddy simulation intercomparison study of shallow cumulus convection. *Journal of the Atmospheric Sciences*, *60*(10), 1201–1219. 10.1175/1520-0469(2003)60<1201:alesis>2.0.co;2.

Simpson, J. (1971). On cumulus entrainment and one-dimensional models. *Journal of the Atmospheric Sciences*, *28*(3), 449–455. 10.1175/1520-0469(1971)028<0449:oceaod> 2.0.co;2.

Slawinska, J., Grabowski, W. W., Pawlowska, H., & Morrison, H. (2012). Droplet activation and mixing in Large-Eddy simulation of a shallow cumulus field. *Journal of the Atmospheric Sciences*, *69*(2), 444–462.

Slawinska, J., Grabowski, W. W., Pawlowska, H., & Wyszogrodzki, A. A. (2008). Optical properties of shallow convective clouds diagnosed from a bulk-microphysics large-eddy simulation. *Journal of Climate*, *21*(7), 1639–1647.

Small, J. D., Chuang, P. Y., Feingold, G., & Jiang, H. (2009). Can aerosol decrease cloud lifetime? *Geophysical Research Letters*, *36*(16), L16806. 10.1029/2009gl038888.

Small, J. D., Chuang, P., & Jonsson, H. (2013). Microphysical imprint of entrainment in warm cumulus. *Tellus B*, *65*, 6647–6662.

Squires, P. (1952). The growth of cloud drops by condensation. I. General characteristics. *Australian Journal of Scientific Research*, *5*(1), 59–86.

Squires, P., & Turner, J. S. (1962). An entraining jet model for cumulonimbus updraughts. *Tellus*, *14*(4), 422–434. doi:10.1111/j.2153-3490.1962.tb01355.x.

Stanfield, R. E., Su, H., Jiang, J. H., Freitas, S. R., Molod, A. M., Luo, Z. J., et al. (2019). Convective entrainment rates estimated from Aura CO and CloudSat/CALIPSO observations and comparison with GEOS-5. *Journal of Geophysical Research*, *124*(17–18), 9796–9807, 10.1029/2019jd030846.

Stephens, G. L. (2005). Cloud feedbacks in the climate system: A critical review. *Journal of Climate*, *18*(2), 237–273. doi:10.1175/JCLI-3243.1.

Stevens, B., Cotton, W. R., Feingold, G., & Moeng, C.-H. (1998). Large-Eddy simulations of strongly precipitating, shallow, stratocumulus-topped boundary layers. *Journal of the Atmospheric Sciences*, *55*(24), 3616–3638.

Stevens, B. (2002). Entrainment in stratocumulus-topped mixed layers. *Quarterly Journal of the Royal Meteorological Society*, *128*(586), 2663–2690. doi:10.1256/qj.01.202.

Stevens, B., Lenschow, D. H., Faloona, I., Moeng, C.-H., Lilly, D. K., Blomquist, B., et al. (2003). On entrainment rates in nocturnal marine stratocumulus. *Quarterly Journal of the Royal Meteorological Society*, *129*(595), 3469–3493. https://doi.org/10.1256/qj.02.202.

Stirling, A. J., & Stratton, R. A. (2012). Entrainment processes in the diurnal cycle of deep convection over land. *Quarterly Journal of the Royal Meteorological Society*, *138*(666), 1135–1149.

Stommel, H. (1947). Entrainment of air into a cumulus cloud. *Journal of Meteorology*, *4*(3), 91–94. doi:10.1175/1520-0469(1947)004<0091:EOAIAC>2.0.CO;2.

Stull, R. B. (1985). A fair-weather cumulus cloud classification scheme for mixed-layer studies. *Journal of Climate and Applied Meteorology*, *24*(1), 49–56.

Su, C.-W., Krueger, S. K., McMurtry, P. A., & Austin, P. H. (1998a). Linear eddy modeling of droplet spectral evolution during entrainment and mixing in cumulus clouds. *Atmospheric research*, *47*, 41–58. 10.1016/S0169-8095(98)00039-8.

Su, C.-W., Krueger, S. K., McMurtry, P. A., & Austin, P. H. (1998b). Linear eddy modeling of droplet spectral evolution during entrainment and mixing in cumulus clouds. *Atmospheric Research*, *47–48*, 41–58.

Takahashi, H., & Luo, Z. (2012). Where is the level of neutral buoyancy for deep convection?. *Geophysical Research Letters*, *39*(15), L15809. https://doi.org/10.1029/2012GL052638.

Takahashi, H., Luo, Z. J., & Stephens, G. L. (2017). Level of neutral buoyancy, deep convective outflow, and convective core: New perspectives based on 5 years of CloudSat data. *Journal of Geophysical Research*, *122*(5), 2958–2969. https://doi.org/10.1002/2016JD025969.

Telford, J. W., & Chai, S. K. (1980). A new aspect of condensation theory. *Pure and Applied Geophysics*, *118*(2), 720–742.

Telford, J. W. (1996). Clouds with turbulence; the role of entrainment. *Atmospheric Research*, *40*(2–4), 261–282.

Tiedtke, M. (1989). A comprehensive mass flux scheme for cumulus parameterization in large- scale models. *Monthly Weather Review*, *117*(8), 1779–1800.

Vial, J., Bony, S., Stevens, B., & Vogel, R. (2017). Mechanisms and model diversity of trade-wind shallow cumulus cloud feedbacks: A review. *Shallow Clouds, Water Vapor, Circulation, and Climate Sensitivity*, 159–181.

Vogelmann, A. M., McFarquhar, G., Ogren, J. A., Turner, D. D., Comstock, J. M., Feingold, G., et al. (2012). RACORO extended-term aircraft observations of boundary layer clouds. *Bulletin of the American Meteorological Society*, *93*(6), 861–878. doi: 10.1175/bams-d-11-00189.1.

von Salzen, K., & McFarlane, N. A. (2002). Parameterization of the bulk effects of lateral and cloud-top entrainment in transient shallow cumulus clouds. *Journal of the Atmospheric Sciences*, *59*(8), 1405–1430. doi:10.1175/1520-0469(2002)059<1405:POTBEO>2.0.CO;2.

Wagner, T. J., Turner, D. D., Berg, L. K., & Krueger, S. K. (2013). Ground-based remote retrievals of cumulus entrainment rates. *Journal of Atmospheric and Oceanic Technology*, *30*(7), 1460–1471. 10.1175/JTECH-D-12-00187.1.

Wang, D., Jensen, M. P., D'Iorio, J. A., Jozef, G., Giangrande, S. E., Johnson, K. L., et al. (2020). An observational comparison of level of neutral buoyancy and level of maximum detrainment in tropical deep convective clouds. *Journal of Geophysical Research*, *125*(16), e2020JD032637. https://doi.org/10.1029/2020JD032637.

Wang, J., Daum, P. H., Yum, S. S., Liu, Y., Senum, G. I., Lu, M.-L. et al. (2009). Observations of marine stratocumulus microphysics and implications for processes controlling droplet spectra: Results from the Marine Stratus/Stratocumulus Experiment. *Journal of Geophysical Research*, *114*, D18210. doi:10.1029/2008jd011035.

Wang, Y., Zhang, G. J., & He, Y. J. (2017). Simulation of precipitation extremes using a stochastic convective parameterization in the NCAR CAM5 under different resolutions. *Journal of Geophysical Research Atmospheres*, *122*(23), 12,875–12,891.

Wang, Y., Zhao, C., McFarquhar, G. M., Wu, W., Reeves, M., & Li, J. (2021). Dispersion of droplet size distributions in supercooled non-precipitating stratocumulus from aircraft observations obtained during the Southern Ocean Cloud Radiation Aerosol Transport Experimental Study. *Journal of Geophysical Research*, *126*(6), e2020JD033720. https://doi.org/10.1029/2020JD033720.

Wang, Y., Zhou, L., & Hamilton, K. (2007). Effect of convective entrainment/detrainment on the simulation of the tropical precipitation diurnal cycle. *Monthly Weather Review*, *135*(2), 567–585.

Wang, Z. (2020). A method for a direct measure of entrainment and detrainment. *Monthly Weather Review*, *148*(8), 3329–3340. 10.1175/mwr-d-20-0046.1.

Warner, J. (1969). The microstructure of cumulus cloud. Part I. General features of the droplet spectrum. *Journal of the Atmospheric Sciences*, *26*(5), 1049–1059. doi:10.1175/1520-0469(1969)026<1049:TMOCCP>2.0.CO;2.

Warner, J. (1973). The microstructure of cumulus cloud: Part IV: The effect on the droplet spectrum of mixing between cloud and environment. *Journal of the Atmospheric Sciences*, *30*(2), 256–261. doi:10.1175/1520-0469(1973)030<0256:TMOCCP>2.0.CO;2.

Wood, R. (2012). Review: Stratocumulus clouds. *Monthly Weather Review*, *140*(8), 2373–2423.

Wu, T. (2012). A mass-flux cumulus parameterization scheme for large-scale models: Description and test with observations. *Climate Dynamics*, *38*(3–4), 725–744. 10.1007/s00382-011-0995-3.

Wyngaard, J. C. (2010). *Turbulence in the Atmosphere*. Cambridge University Press: New York.

Xie, X., & Zhang, M. (2015). Scale-aware parameterization of liquid cloud inhomogeneity and its impact on simulated climate in CESM. *Journal of Geophysical Research*, *120*(16), 8359–8371, https://doi.org/10.1002/2015JD023565.

Xu, X., Sun, C., Lu, C., Liu, Y., Zhang, G. J., & Chen, Q. (2021). Factors affecting entrainment rate in deep convective clouds and parameterizations. *Journal of Geophysical Research*, *126*(15), e2021JD034881. https://doi.org/10.1029/2021JD034881.

Xue, H., & G. Feingold (2006). Large-eddy simulations of trade wind cumuli: Investigation of aerosol indirect effects. *Journal of the Atmospheric Sciences*, *63*(6), 1605–1622. 10.1175/jas3706.1.

Yan, Y., Liu, Y., & Lu, J. (2016). Cloud vertical structure, precipitation, and cloud radiative effects over Tibetan Plateau and its neighboring regions. *Journal of Geophysical Research*, *121*(10), 5864–5877. 10.1002/2015JD024591.

Yang, B., Zhang, Y., Qian, Y., Huang, A., & Yan, H. (2015). Calibration of a convective parameterization scheme in the WRF model and its impact on the simulation of East Asian summer monsoon precipitation. *Climate Dynamics*, *44*(5), 1661–1684. 10.1007/s00382-014-2118-4.

Yang, B., Wang, M., Zhang, G. J., Guo, Z., Qian, Y., Huang, A., & Zhang, Y. (2020). Simulated precipitation diurnal variation with a deep convective closure subject to shallow convection in Community Atmosphere Model Version 5 coupled with CLUBB. *Journal of Advances in Modeling Earth Systems*, *12*(7), e2020MS002050. https://doi.org/10.1029/2020MS002050.

Yang, J., Wang, Z., Heymsfield, A. J., & French, J. R. (2016). Characteristics of vertical air motion in isolated convective clouds. *Atmospheric Chemistry and Physics*, *16*(4), 10159–10173.

Yeom, J. M., Yum, S. S., Liu, Y., & Lu, C. (2017). A study on the entrainment and mixing process in the continental stratocumulus clouds measured during the RACORO campaign. *Atmospheric Research*, *194*, 89–99. https://doi.org/10.1016/j.atmosres.2017.04.028.

Yeom, J. M., Yum, S., Shaw, R. A., La, I., Wang, J., Lu, C., et al. (2021). Vertical variations of cloud microphysical relationships in marine stratocumulus clouds observed during the ACE-ENA campaign. *Journal of Geophysical Research*, *126*(24), e2021JD034700. https://doi.org/10.1029/2021JD034700.

Yum, S. (1998). Cloud droplet spectral broadening in warm clouds: An observational and model study, Dissertation for the Doctoral Degree thesis, University of Nevada, Reno, Nevada, USA.

Yum, S. S., & Hudson, J. G. (2001). Microphysical relationships in warm clouds. *Atmospheric Research*, *57*(2), 81–104.

Yum, S. S., Wang, J., Liu, Y., Senum, G., Springston, S., McGraw, R., & Yeom, J. M. (2015). Cloud microphysical relationships and their implication on entrainment and mixing mechanism for the stratocumulus clouds measured during the VOCALS project. *Journal of Geophysical Research*, *120*(10), 5047–5069. 10.1002/2014JD022802.

Yun, Y., Fan, J., Xiao, H., Zhang, G. J., Ghan, S. J., Xu, K.-M., et al. (2017). Assessing the resolution adaptability of the Zhang-McFarlane cumulus parameterization with spatial and temporal averaging. *Journal of Advances in Modeling Earth Systems*, *9*(7), 2753–2770. https://doi.org/10.1002/2017MS001035.

Zermeno-Diaz, D., Zhang, C., Kollias, P., & Kalesse, H. (2016). The role of shallow cloud moistening in MJO and non-MJO convective events over the ARM Manus site. *Journal of the Atmospheric Sciences*, 150831134708004.

Zhang, C., Mapes, B. E., & Soden, B. J. (2003). Bimodality in tropical water vapour. *Quarterly Journal of the Royal Meteorological Society*, *129*, 2847–2866.

Zhang, F., Yan, J.-R., Li, J., Wu, K., Iwabuchi, H., & Shi, Y.-N. (2018). A new radiative transfer method for solar radiation in a vertically internally inhomogeneous medium, *Journal of the Atmospheric Sciences*, *75*(1), 41–55. 10.1175/jas-d-17-0104.1.

Zhang, G., Wu, X., Zeng, X., & Mitovski, T. (2016). Estimation of convective entrainment properties from a cloud-resolving model simulation during TWP-ICE. *Climate Dynamics*, *47*(7), 2177–2192.

Zhang, G. J., & McFarlane, N. A. (1995). Sensitivity of climate simulations to the parameterization of cumulus convection in the Canadian Climate Centre general circulation model. *Atmosphere-Ocean*, *33*, 407–446.

Zhang, H., Jing, X., & Li, J. (2014). Application and evaluation of a new radiation code under McICA scheme in BCC_AGCM2.0.1. *Geoscientific Model Development*, *7*, 737–754.

Zhao, C., & Ishizaka, Y. (2004). Modeling marine stratocumulus with a detailed microphysical scheme. *Advances in Atmospheric Sciences*, *21*, 61–74.

Zhao, M., Golaz, J.-C., Held, I. M., Guo, H., Balaji, V., Benson, R., et al. (2018). The GFDL Global Atmosphere and Land Model AM4.0/LM4.0: 2. Model description, sensitivity studies, and tuning strategies. *Journal of Advances in Modeling Earth Systems*, *10*(3), 735–769. doi:10.1002/2017MS001209.

Zhao, M., Golaz, J. C., Held, I. M., Ramaswamy, V., Lin, S. J., Ming, Y., et al. (2016). Uncertainty in model climate sensitivity traced to representations of cumulus precipitation microphysics. *Journal of Climate*, *29*(2), 543–560.

Zhao, M., Held, I. M., & Lin, S.-J. (2012). Some counterintuitive dependencies of tropical cyclone frequency on parameters in a GCM. *Journal of the Atmospheric Sciences*, *69*(7), 2272–2283. 10.1175/jas-d-11-0238.1.

Zheng, Y., Rosenfeld, D., Zhu, Y., & Li, Z. (2019). Satellite-Based Estimation of Cloud Top Radiative Cooling Rate for Marine Stratocumulus, *Geophysical Research Letters*, *46*(8), 4485–4494. https://doi.org/10.1029/2019GL082094.

Zhu, L., Lu, C., Yan, S., Liu, Y., Zhang, G. J., Mei, F., et al. (2021). A new approach for simultaneous estimation of entrainment and detrainment rates in nonprecipitating shallow cumulus, *Geophysical Research Letters, Accepted*.

5

Deep Convection and Convective Clouds

Leo J. Donner

ABSTRACT

Deep convection is a "fast process" important broadly for weather and climate with myriad, interacting impacts on the Earth system through its dynamic, thermodynamic, and transport properties. Its interactions with and behaviors as a function of large-scale flows are a central research problem in atmospheric science and pose a major challenge for numerical models. Large-scale models typically are unable to fully resolve deep convection, with the resolution requirements to explicitly simulate deep convection in its full microphysical and turbulent complexity comprising an active research question. Parameterizations for effects of deep convection on large-scale flows have been developed based on (1) adjusting large-scale flows to account for convection, (2) aggregating the statistical impacts of idealized convective structures embedded in large-scale flows, subject to closure assumptions, and (3) invoking higher-order turbulence theory with assumed distributions and closures. This chapter reviews these methods and places them in the context of multiscale methods (super-parameterization) and the evolution of global atmospheric models toward horizontal resolutions of several kilometers. Impacts of convection on the thermal, dynamic, moisture, and chemical structures of the atmosphere will be noted, as will be microphysical and aerosol interactions with deep convection.

5.1. INTRODUCTION

Deep convection and convective clouds are among the most striking atmospheric phenomena, with important manifestations on the full range of space and timescales in weather and climate. Heat, moisture, momentum, and chemical tracers are all transported vertically, in some cases over a substantial fraction of the depth of the troposphere and even overshooting into the stratosphere at times. In-cloud processes can modify properties at cloud base during transport with entrainment-mediating and detrainment-mediating interactions between clouds and their surroundings. In general circulation models (GCMs) devoted to climate applications, resolved horizontal scales have typically been significantly larger than convective scales, and the aggregate effects of ensembles of convective clouds have required parameterization. Models for weather prediction resolve scales closer to those of convection, and this may become the case for climate models within a decade or so. Many aspects of climate and climate sensitivity have been traced to the representation of convection in climate models (Rougier et al., 2009; Sanderson et al., 2010; Stainforth et al., 2005; Zhao, 2014). Extensive uncertainty characterizes interactions between deep convection and aerosols (Donner et al., 2016; Fan et al., 2012; Fan et al., 2016; Rosenfeld et al., 2013; Rosenfeld et al., 2014; Varble, 2018), related to processes ranging from dependence of aerosol activation on updraft speeds (Feingold, 2003; McFiggans et al., 2006) to aerosol-driven changes in ice microphysics (Cui et al., 2011), convective intensity, and organization (Lee et al., 2010). The simultaneous limitations convection has placed on narrowing uncertainty in both climate forcing and sensitivity

NOAA Geophysical Fluid Dynamics Laboratory, Princeton University, Princeton, NJ, USA

Fast Processes in Large-Scale Atmospheric Models: Progress, Challenges, and Opportunities, Geophysical Monograph 282, First Edition.
Edited by Yangang Liu and Pavlos Kollias.

DOI: 10.1002/9781119529019.ch05

pose a major challenge to understanding fully the cause of anthropogenic climate change from preindustrial to present times. Deep convection also remains a key challenge for weather prediction, with initiation of convection in space and time, diurnal cycle, organization, and intensity, including association with extreme events such as heavy rain, high winds, hail, and dangerous lightning, among high-impact forecast challenges.

This chapter discusses deep convection from the perspective of larger-scale flows, which, as a practical matter, comprises the problem of cumulus parameterization for numerical weather prediction and Earth system modeling. The chapter begins by discussing the effects of convection on larger-scale flows in which it is embedded. Deep convection is closely related to shallow convection (Chapter 7) and boundary-layer processes (Chapter 8). Next, strategies for solving the problem of cumulus parameterization are discussed, followed by a discussion of the relationship of deep convection to microphysics, aerosols, tracers, and chemistry, and their treatment in numerical weather prediction and Earth system models.

5.2. BASIC STRUCTURE OF PROBLEM AND RELATIONSHIP TO OTHER PROCESSES

Generally, convection is considered to be a buoyantly driven circulation, whose characteristic horizontal scales are of the order of a kilometer and smaller. The largest vertical scales are of the order of the depth of the troposphere but can be much smaller. The scales of deep convection are toward the larger ends of these ranges, with the scales of convective circulation extending down into shallow convection and turbulence. The presence of liquid and ice distinguishes convective systems from their surrounding environments, but these systems can consist of multiple updrafts with closely associated downdrafts and mesoscale circulations (Leary & Houze, 1980). To the extent that deep convection is vertically coherent, its base is often considered to be related to near-surface properties, with a height approximately at the lifting condensation level, but elevated convection also occurs. Connections between the planetary boundary layer (PBL) and deep convection are important in many cases.

Given the small scale of convection relative to scales resolved in models for climate and weather prediction, convection is often cast as consisting of unresolved variations in a mean flow, having a turbulent character relative to the mean flow. For the mean potential temperature $\overline{\theta}$, water vapor mixing ratio $\overline{q}$, and general tracer $\overline{X}$, the governing equations are

$$\frac{\mathrm{d}\overline{\theta}}{\mathrm{d}t} - \frac{\overline{Q_r^{\mathrm{e}}}}{\pi c_p} - \frac{\sum_{k=1}^{6} L_k \overline{\gamma_k^{\mathrm{e}}}}{\pi c_p} = \frac{\overline{Q_r^*}}{\pi c_p} + \frac{\sum_{k=1}^{6} L_k \overline{\gamma_k^*}}{\pi c_p} - \frac{\partial \overline{\omega'\theta'}}{\partial p} - \nabla \cdot \overline{\mathrm{v}'\theta'} = Q_\theta \quad (5.1)$$

$$\frac{\mathrm{d}\overline{q}}{\mathrm{d}t} + \sum_{k=1}^{4} \mathrm{sgn}(L_k)\overline{\gamma_k^{\mathrm{e}}} = -\sum_{k=1}^{4} \mathrm{sgn}(L_k)\overline{\gamma_k^*} - \frac{\partial \overline{\omega' q'}}{\partial p} - \nabla \cdot \overline{\mathrm{v}' q'} = -Q_q \quad (5.2)$$

$$\frac{\mathrm{d}\overline{X}}{\mathrm{d}t} - \overline{S_X^{\ \mathrm{e}}} = \overline{S_X^{\ *}} - \frac{\partial \overline{w' X'}}{\partial p} - \nabla \cdot \overline{\mathrm{v}' X'} = Q_X. \quad (5.3)$$

Here, Q_r denotes radiative heating, c_p is the specific heat at constant pressure, $\pi = \left(\frac{p}{p_0}\right)^{\frac{R_\mathrm{d}}{c_p}}$ is the Exner function, and R_d is the gas constant for dry air. The summations represent phase changes, with L_1, L_3, and L_5 the latent heats of vaporization, sublimation, and freezing, respectively. The phase transformation rates are denoted by γ_1, condensation; γ_2, evaporation; γ_3, deposition (from vapor to ice); γ_4, sublimation, γ_5, freezing, and γ_6, melting. The phase transformation rates are defined to be positive, so L_2, L_4, and L_6 for evaporation, sublimation, and melting are assigned to be negative. In these equations, the flow is idealized as consisting of convective components (including updrafts and downdrafts of various scales), denoted by asterisks, and a nonconvective environment, denoted by a superscript e. In forming the large-scale horizontal average, denoted by overbars, there are zero contributions from the convective and environmental variables outside the relevant portions of the flow. The variable X denotes a generic tracer, and S_X a source of X. Convective microphysical and aerosol transports, sources, and sinks are represented by equation (3). The components of the horizontal momentum vector are among the tracers whose evolution is determined in part by convection, with in-cloud pressure-gradient forces included in the source terms. Convective potential temperature, water vapor mixing ratio, tracer mixing ratios, and vertical velocities all differ characteristically from the surrounding environment, with strong correlations between vertical velocity on the one hand and convective thermodynamics, moisture, and tracers on the other hand. Thus, vertical convergences of eddy fluxes associated with convection can be large. Scale analysis suggests that horizontal convergences of eddy fluxes associated with deep convection are relatively small for sufficiently large horizontal grids but possibly not so for grid sizes much less than 50 km (Anthes, 1977). Cho and Cheng (1980) parameterize horizontal transport of vorticity by convection, finding this transport important in studies of observed tropical vorticity budgets. Although not discussed further in this chapter, radiative heating and cooling associated with convective systems, $\overline{Q_r^*}$, can be important (Ackerman et al., 1988). Although the areas associated with deep convective updrafts may not be large, the associated mesoscale updrafts can be optically thick and spatially extensive (Leary & Houze, 1980). The spatial extents and microphysical properties of convective systems can be combined with parameterizations for radiative transfer (Chapter 2) to calculate $\overline{Q_r^*}$ in equation (1).

Equations (1)–(3) decompose the flow into large-scale, resolved means (related to, but generally somewhat coarser than, the model grid size) and smaller-scale, unresolved fluctuations about these means. These fluctuations arise not only from deep convection but also from shallow convection (Chapter 7), turbulence in PBLs (Chapter 10) and elsewhere, and gravity waves (Chapter 12). These processes are closely related, and the scale separations are not sharp. The distinctions drawn among them by model parameterizations represent idealizations that have facilitated parameterization development. Efforts to tie these processes together (Park, 2014) characterize eddy diffusivity and mass flux parameterizations (Chapter 15) and representations of convection as turbulence with assumed probability density functions (PDFs) for subgrid variability (Chapter 14).

5.3. STRATEGIES FOR SOLUTION

5.3.1. Scale Awareness

An essential characteristic of the decomposition between the mean flow and fluctuations in equations (1)–(3) is that the physical character of the submean processes strongly depends on the scales over which the means are taken. For example, in a GCM whose horizontal resolution is around 200 km or so, the mesoscale components associated with many convective systems require parameterization (Donner, 1993), but this becomes less necessary as model resolutions approach tens of kilometers. Model resolutions of a kilometer or so (with nonhydrostatic dynamics) are comparable to the diameters of large observed convective updrafts (LeMone & Zipser, 1980). Arakawa and Wu (2013) show that the contributions of subgrid vertical eddy fluxes to total vertical fluxes decrease sharply as horizontal averaging scales decrease from around 50 to 2 km. Model resolutions near or somewhat coarser than the observed scales of particular convective phenomena are often referred to as in the gray zone for these phenomena. For example, a model whose horizontal resolution is around 50 km would be in the gray zone for mesoscale parts of convective systems, while a model whose resolution is around 3–5 km would be in the gray zone for deep convective updrafts.

Many parameterizations do not by design allow for their application over a wide range of resolutions. Robust solutions to the parameterization problem should do so. Such parameterizations are variously referred to as scale-aware, scale-sensitive, scale-accommodating, or scale-adaptive. A range of issues are involved, including the appropriateness of assumptions regarding the physical nature of unresolved flow structures or PDFs for unresolved variables and the extent to which a parameterization by construction transitions between resolved and parameterized flows. An important element of scale awareness is likely to be the characteristic scale of the convective flow structure relative to the model resolution (Arakawa & Wu, 2013), a complex issue as the size of convective systems can vary. Although the issue of scale awareness has a long history in the parameterization of turbulence, including formulations that explicitly incorporate grid size (Smagorinsky, 1963), the topic has received limited attention in cumulus parameterizations.

5.3.2. Turbulence with Higher-Order Closure

Especially if the thermodynamic and moisture variables in equations (1)–(2) are expressed in terms of variables conserved for adiabatic, nonprecipitating processes (e.g., liquid water potential temperature and total water mixing ratio), the right-hand sides of equations (1)–(2) appear formally as turbulence. By assuming forms for the PDF for unresolved variations in a set of variables that characterize the thermodynamics, moisture, and dynamics of convection, higher-order closure methods can be used to obtain the convective contributions to the evolution of large-scale flows (Lappen & Randall, 2001). Golaz, Larson, and Cotton (2002) and Bogenschutz and Krueger (2013) assume double Gaussian PDFs for the triple joint PDF among vertical velocity, a conserved temperature variable (liquid water potential temperature and liquid water/ice moist static energy, respectively), and total water mixing ratio. The assumed double Gaussian forms are supported by aircraft observations and large-eddy simulation (LES) of stratocumulus and shallow cumulus clouds (Larson et al., 2002) and also for deep convection by LES (Bogenschutz et al., 2010). Using the assumed PDFs in prognostic higher-order closure (ADHOC, for assumed-distribution higher-order closure) simulates well cumulus and stratocumulus clouds in single-column tests with prescribed large-scale advection (Guo et al., 2010), at least with sufficiently fine vertical and temporal resolution. Guo et al. (2015) used ADHOC to parameterize both shallow and deep convection in a unified framework in a GCM. The realism of the resulting simulation was within the range of state-of-the-science models.

The unification of deep and shallow convection is an attractive aspect of ADHOC methods, but challenges include enforcing vertical coherence in unresolved cumulus structures for microphysical processes and treatment of the full suite of liquid and ice microphysical processes. The former may be amenable to solution by statistical sampling of the PDFs and using the samples to construct vertical columns (Larson & Schanen, 2013).

ADHOC methods can work at any resolution for which their assumed distributions provide reasonable characterizations of the unresolved flow, with the resolved flow "taking over" from the parameterized flow as the resolution increases. As resolutions become increasingly coarse, the assumed distributions are likely less realistic. Guo et al. (2015) show that cloud simulations using ADHOC change with resolution. The use of a single variable for total water mixing ratio in current ADHOC methods poses challenges for tracking ice and liquid contributions to the mixing ratio in the vertical and time.

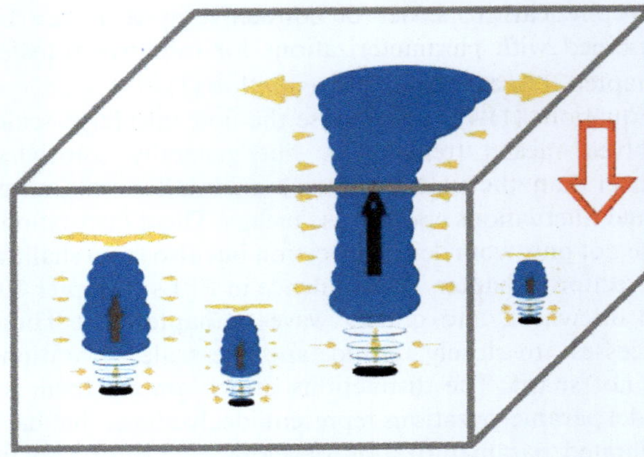

Figure 5.1 A cumulus ensemble idealized as an ensemble of laterally entraining plumes embedded in a uniform mean state, from Kipling et al. (2017). *Source:* Reproduced with permission from https://creativecommons.org/licenses/by/4.0.

5.3.3. Plumes and Mass Fluxes

An alternative to using assumed PDFs to represent convection is to exploit well-established views of convection as buoyantly driven thermals or plumes (Turner, 1973). These approaches, by treating convection as vertically coherent structures superposed or embedded in a large-scale flow, offer a means of incorporating cloud and aerosol microphysics, as well as including unresolved fluctuations in equations (1)–(3) through physically consistent equation sets. Most deep cumulus parameterizations have used steady-state plumes rather than rising thermals produced by buoyant pulses, but these perspectives may be related (Morrison, 2017).

In fact, the physics of plumes and thermals is not entirely settled. Only recently have fundamental issues on the influence of vertical wind shear on thermals been addressed (Peters et al., 2019), and there are contrasting views on the critical question of mixing between plumes and their surroundings. Some parameterizations are built around an ensemble of laterally entraining updrafts, detraining at plume top (Figure 5.1) (Anthes, 1977; Arakawa & Schubert, 1974; Chikira & Sugiyama, 2010; Donner, 1993; Moorthi & Suarez, 1992; Wagner & Graf, 2010; Zhang & McFarlane, 1995). Another approach is to conceptualize plumes as both entraining and detraining throughout their vertical extents (Figure 5.2) (Gregory & Rowntree, 1990; Grell & Freitas, 2014; Tiedtke, 1989). These plumes are often presented as bulk representations of ensembles of entraining plumes, thereby justifying the presence and extent of detrainment throughout the depth of the plume. Buoyancy and microphysical processes are generally functions of the plume thermodynamic and moisture properties. The bulk plume approach simplifies parameterizations computationally, enabling multiple plumes to be replaced by a single plume but can ignore nonlinear behaviors that could be captured by multiple plumes. Constraints from theory and observations can be difficult to impose on bulk plumes because they entangle properties of individual plumes amenable to theoretical and observational analysis.

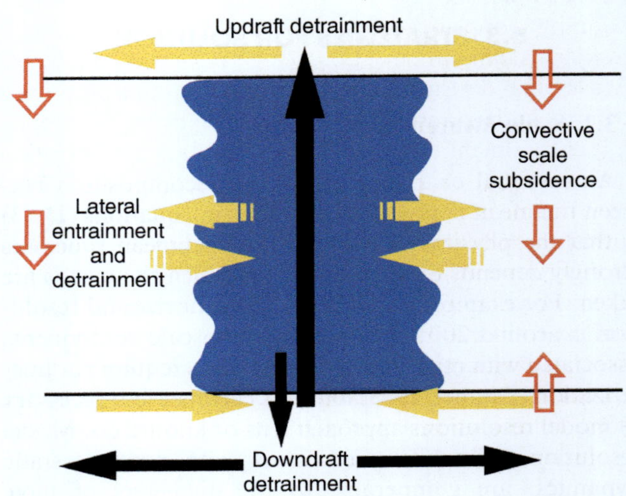

Figure 5.2 A plume with both entrainment and detrainment, from Grell and Freitas (2014). Bulk plumes typically both entrain and detrain in this manner. *Source:* Reproduced with permission from https://creativecommons.org/licenses/by/4.0.

Limited understanding of the environmental response to rising plumes, which effectively reduces buoyancy through nonhydrostatic vertical perturbation pressure-gradient forces, and entrainment has introduced significant uncertainty into cumulus parameterizations that use them. Recent analytic studies, supported by LES, have narrowed the uncertainty in formulations for effective buoyancy (Jeevanjee & Romps, 2016; Morrison, 2016; Peters, 2016; Tarshish et al., 2018). Cloud-resolving models (CRMs), LES, and direct numerical simulation have been used to estimate entrainment into convective plumes and thermals (Anber et al., 2019;

Dawe & Austin, 2011, 2013; de Rooy et al., 2013; Hannah, 2017; Hernandez-Deckers & Sherwood, 2018; Lecoanet & Jeevanjee, 2019; Lu et al., 2016; Moser & Lasher-Trapp, 2017; Romps, 2010; Zhang et al., 2016). Convective entrainment appears to be a particularly important property for critical emergent properties of the climate models in which plumes are used. In particular, multiple parameter-sensitivity studies suggest that the entrainment rate in parameterized convection is a key control on modeled climate sensitivity (Rougier et al., 2009; Sanderson et al., 2010; Stainforth et al., 2005; Zhao, 2014). Despite this, attempts to constrain entrainment in convective parameterizations using observations or models have been limited. Although characterizing and measuring entrainment pose challenges, inferences can be drawn from observations of related cloud properties. As examples, Donner (1993) uses observations of convective vertical velocities to constrain entrainment. Donner et al. (2016) and Labbouz, Kipling, Stier, and Protat (2018) emphasize the potential of new observations from scanning Doppler radars (Collis et al., 2013), radar wind profilers (Giangrande et al., 2016), and satellite retrievals (Masunaga & Luo, 2016) to build on this approach toward understanding and more realistically modeling entrainment and other internal parameterization variables.

Although studies with CRMs suggest that lateral entrainment explains mixing between deep cumulus and its environment, at least well above cloud base (C. Lin & Arakawa, 1997), numerous field studies, especially of shallow convection, suggest alternatives to lateral entrainment as explanations for cumulus mixing (LaMontagne & Telford, 1983; Paluch, 1979; Raymond & Blyth, 1986). These studies suggest that cumulus clouds are dominated by mixtures of environmental air from near the base and top of the clouds. Simple cumulus models based on buoyancy sorting, often with stochastic mixing, are broadly consistent with these observations. Parameterizations for shallow (Bretherton et al., 2004), deep (Emanuel, 1991; Zhao et al., 2016), and unified (Park, 2014) convection are based on cumulus models including buoyancy sorting.

Observations (Leary & Houze, 1980) show that important contributions to the right-hand sides of (1)–(3) are associated with components of convective systems other than convective-scale updrafts. Many parameterizations include convective downdrafts (Del Genio & Yao, 1993; Donner, 1993; Emanuel, 1991; Grell & Freitas, 2014; Schmidt et al., 2014; Tiedtke, 1989; Zhang & McFarlane, 1995). Depending on the grid size of the large-scale model, mesoscale updrafts and downdrafts associated with deep convection may be unresolved. Donner's (1993) parameterization includes representations of these mesoscale circulations tied closely to the properties of convective-scale updrafts (Figure 5.3). Radar retrievals of vertical velocities in convective systems exhibit structures similar to those of the parameterization, including multiple updrafts originating in the lower troposphere and a mesoscale updraft in the upper troposphere (Figure 5.4). While the parameterized and retrieved PDFs of the vertical velocities exhibit qualitative agreement, the parameterization does not capture the full extent of the retrieved variations in the PDFs (Donner et al., 2016).

Satellite observations (Machado & Rossow, 1993; Mapes & Houze, 1993) show that around 60% of tropical cloud shields are larger than 10^4 km, consistent with frequent occurrence of convective organization at a range of scales. Donner et al. (2001) show (in model with 3° resolution) that cumulus parameterizations including mesoscale circulations can simulate distributions of cloud sizes consistent with these satellite observations. Other multiscale phenomena associated with convective organization, notably convective momentum transport and coherent propagation of orogenic precipitation, have proved more difficult to capture without higher resolutions. Miyakawa et al. (2012) analyze convective momentum transport in a model with 7-km resolution. Moncrieff and Liu (2006) develop a hybrid approach to representing convective organization in models with resolutions around 10 km by enhancing resolved downdrafts in the presence of convective heating. Pritchard et al. (2011) show that propagating orogenic precipitation is possible in a climate model in which deep convection is explicitly represented using superparameterization (described later in this chapter).

The terms on the right-hand sides of (1)–(3) can be evaluated if the rates of phase changes, sources and sinks, and properties are known for each component comprising the convective systems. For example, the grid-average eddy flux of X, $\overline{\omega' X'}$ is

$$\overline{\omega' X'} = a_e \omega_e X_e + \sum_{i=1}^{N} a_i \omega_i X_i - \overline{\omega}\overline{X} \quad (5.4)$$

where a denotes fractional area, the subscript e refers to the part of the grid not occupied by the convective system, and the subscript i refers to the ith of the N components comprising the convective system, i.e., convective-scale updrafts, convective-scale downdrafts, mesoscale updrafts, and mesoscale downdrafts. There may be distinct, multiple realizations of each component. Solving for the parameterized terms in equations (1)–(3) then requires a submodel for the thermodynamics, moisture, phase changes, and vertical velocity of each component of the convective system, as well as the generic tracer X and its sources and sinks. The properties of these submodel components distinguish the convective components from the part of the grid not occupied by the

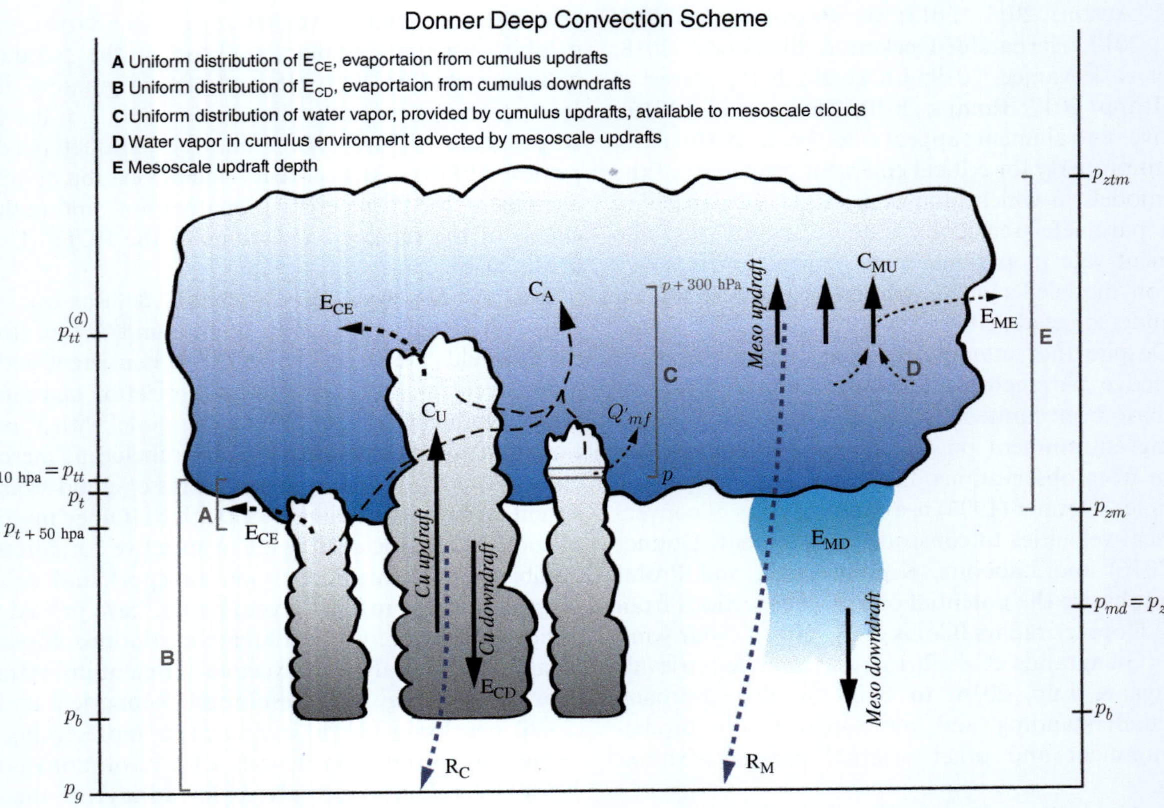

Figure 5.3 Parameterized convective system including mesoscale processes (Benedict et al., 2013; Donner, 1993). Convective-scale updrafts and downdrafts are shaded gray. Mesoscale updrafts and downdrafts are shaded blue. Condensate formed in convective updrafts (C_U) can evaporate directly into the environment within the cloud-top zone (E_{CE}), evaporate in convective-scale downdrafts (E_{CD}), precipitate at convective scale (R_C), or be transferred to mesoscale updrafts (C_A), along with water vapor (Q'_{mf}). Condensate formed in the mesoscale updraft (C_{MU}) can precipitate (R_M), evaporate in mesoscale downdrafts (E_{md}), or either evaporate into the non-convective part of the grid box or be transferred into large-scale stratiform clouds (E_{ME}). Ice processes are also treated. The subensembles for convective-scale updrafts can have distinct detrainment zones around p_t, the last level at which the ensemble member has a nonzero vertical velocity, with the deepest ensemble member top at $p_{tt}^{(d)}$. The base of the mesoscale updraft (p_{zm}) is where the ensemble of convective-scale updrafts begins to detrain vapor, with a top (p_{ztm}) near the level of zero buoyancy (p_{LZB}). The base of the mesoscale downdraft is at p_{md}. *Source:* Reproduced with permission from © American Meteorological Society.

convective system ("environment"). Also required are the values of the component fractional areas a_i. These are provided by a closure, as discussed in the next section.

A powerful simplification links the sum of the parameterization terms on the right-hand sides of equations (1)–(3) closely to the total convective mass flux, M_c:

$$M_c = \sum_{i=1}^{N} a_i \rho_i w_i, \tag{5.5}$$

where a_i, ρ_i, and w_i are the fractional area, density, and vertical velocity (relative to height coordinates) of the ith component of the convective system. This approximation is made evident by examining the local time tendency of X_e (or the other large-scale variables influenced by convection), cf., Arakawa and Schubert (1974):

$$\frac{\partial X_e}{\partial t} = \frac{1}{\rho a_e}\sum_{dc} D_i(X_i - X_e) + \frac{1}{a_e}\left(\frac{M_c}{\rho} - \overline{w}\right)\frac{\partial X_e}{\partial z} - \frac{1}{\rho a_e}[\overline{\nabla \cdot \boldsymbol{v}\rho X} - X_e \overline{\nabla \cdot \boldsymbol{v}\rho}] \tag{5.6}$$

The horizontal velocity is $\mathbf{v}$, and use has been made that the total mass flux in the grid box, $\rho\overline{w}$, is the sum of the mass fluxes associated with the convective system and its environment. The density $\rho(z)$ has been assumed not to vary over the grid, and D_i denotes the detrainment rate, summed over clouds detraining at height z.

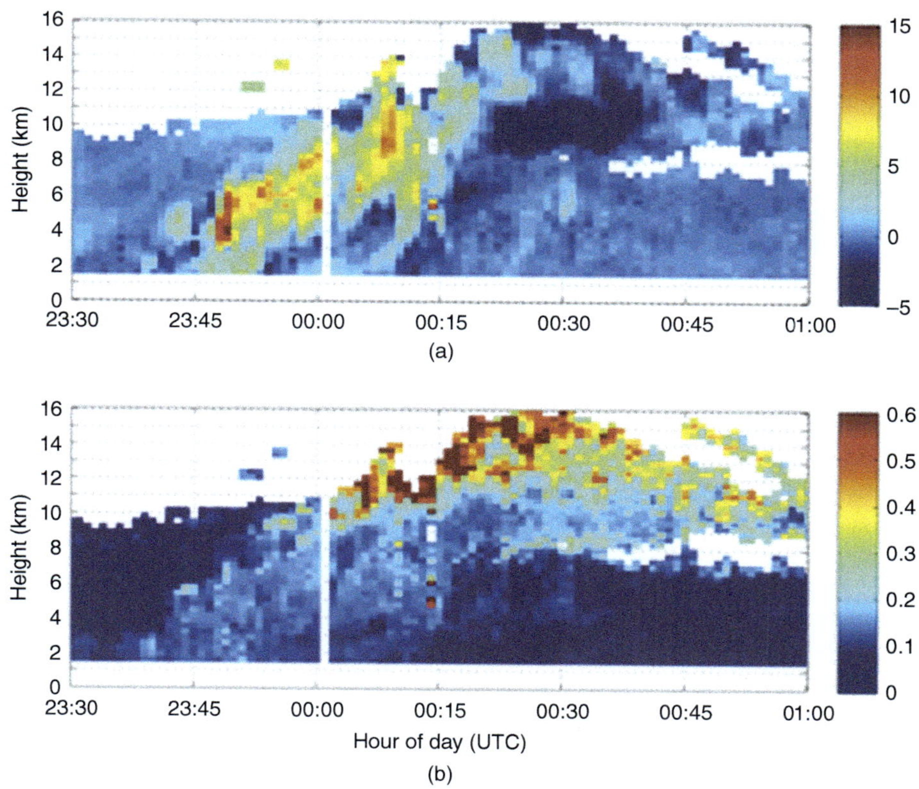

Figure 5.4 Vertical velocities (top) with retrieval uncertainties (bottom) (m s^{-1}) observed in a convective system (Collis et al., 2013). *Source:* Reproduced with permission from © American Meteorological Society.

With the assumptions that the fractional area of the convective system is small ($a_e \approx 1$); and, consistently, $X_e \approx \overline{X}$; and, plausibly, net lateral transports across the boundary by convection are much smaller than transports across the boundary by the grid-resolved flow, i.e., $\overline{\nabla \cdot (\rho \mathbf{v} X)} \approx \nabla \cdot (\rho \overline{\mathbf{v}} \overline{X})$, equation (3) (with analogs for (1) and (2)) takes the form

$$Q_x = \frac{1}{\rho} \sum_{dc} D_i (X_i - \overline{X}) + \frac{M_c}{\rho} \frac{\partial \overline{X}}{\partial z}. \tag{5.7}$$

Comparing the right-hand sides of (3) and (7) shows that the latter requires only a representation of X_i (and potential temperature and mixing ratio for (1) and (2)) and mass fluxes in the convective updrafts. Separate evaluations of a_i and w_i are not necessary. Further, since $\theta_i \approx \overline{\theta}$ as updrafts become neutrally buoyant as they detrain, the total convective updraft mass flux is seen to be a dominant control on the large-scale tendency of potential temperature by convection. Many cumulus parameterizations are based on mass fluxes. Simple plume models can provide estimates for the vertical dependence of M_c, but its magnitude requires a closure. At model resolutions approaching the convective scale, the assumptions $a_e \approx 1$, $X_e \approx \overline{X}$, and $\overline{\nabla \cdot (\rho \mathbf{v} X)} \approx \nabla \cdot (\rho \overline{\mathbf{v}} \overline{X})$ break down, apparently imposing a limit on the scale awareness of the approximation (7).

Vertical velocities are an important control on cloud-aerosol interaction and microphysics (Sullivan et al., 2016). Convective mass fluxes (as products of density, vertical velocity, and area) alone provide insufficient information for coupling physically based treatments of these processes into cumulus parameterizations. Constraining precipitation efficiency (the fraction of condensate removed as precipitation) in convection is therefore difficult with cumulus parameterizations not treating convective vertical velocities, limiting the utility of the mass-flux simplification. Mass fluxes simulated by climate and global transport models vary widely, by factors of 3–4 through much of the middle and upper troposphere (Figure 5.5) (Folkins et al., 2006). GCMs with highly discrepant convective mass fluxes can have fairly similar patterns for precipitation, which vertically integrates Q_q and the environmental phase changes on the left in equation (2) (Donner et al., 2007). The decoupling of mass fluxes (and associated transport of tracers without significant in-convection sources and sinks) and precipitation can be traced to the inability of mass fluxes alone to constrain precipitation efficiency. Relatively

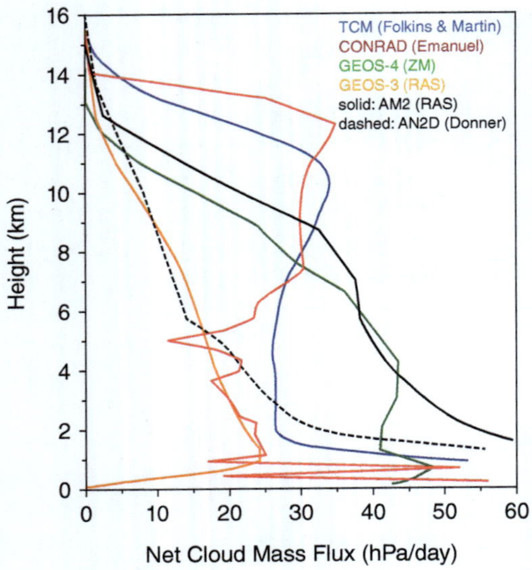

Figure 5.5 Mass fluxes for a number of deep convective parameterizations, described in detail in Folkins et al. (2006). AM2(RAS) and AM2D(Donner) have similar precipitation patterns as described in Donner et al. (2007).

large magnitudes of M_c can be associated with convective systems whose precipitation efficiency is relatively low. The relatively large amounts of detrained condensate, implied by low precipitation efficiency, are then subject to evaporation (or sublimation). The moistening and cooling associated with the evaporation offsets some of the drying and heating associated with the M_c term on the right of equation (7). Similar precipitation is possible with smaller mass fluxes but higher precipitation efficiencies.

5.3.4. Closures and Triggers

Solutions for parameterizations based on plumes and mass fluxes require vertical profiles for thermodynamics, moisture, tracers, and associated sources and sinks for the components of convective systems. Plume models, as described in the preceding section, provide this information, once lower boundary conditions are specified. For updrafts, the top can be defined as the height where buoyancy or vertical velocity vanishes. Corresponding approaches can be used for downdrafts based on negative buoyancy.

For mass-flux parameterizations that do not treat the vertical velocities of the convective components, the solution is then closed by specifying the mass flux at the plume bases. For multiple-plume models, the partitioning of the mass flux among the plumes must also be specified. Many closures have been proposed, differing greatly in details. A common underlying concept is quasi-equilibrium (Arakawa & Schubert, 1974), which asserts that total changes in the cloud work function are small relative to changes in the cloud work function produced by nonconvective processes. The cloud work function $A_i(\lambda)$ for the ith plume in a convective system measures the vertically integrated Archimedean buoyancy of a plume whose fractional entrainment rate is λ:

$$A_i(\lambda) = \int_{z_B}^{z_D(\lambda)} \frac{g\eta}{\overline{T}_v}(T_{v,i} - \overline{T}_v)dz. \qquad (5.8)$$

Here, z_B is the height of the plume base, $z_D(\lambda)$ is the height at the top of a plume whose entrainment rate is λ, T_v denotes virtual temperature, g is the gravity constant, and η is the convective mass flux $M(\lambda, z)$ (cf., equation (5)) normalized by its value at cloud base $M(\lambda, z_B)$. The fractional entrainment rate is defined as

$$\lambda = \frac{1}{M}\frac{dM}{dz} \qquad (5.9)$$

In the limit of a nonentraining plume ($\lambda = 0$), the cloud work function becomes the convective available potential energy (CAPE). In Arakawa and Schubert (1974) and Wagner and Graf (2010), the convective plumes interact with each other and nonconvective changes in the large-scale flow to determine $M(\lambda, z_B)$, the plume-base mass flux for each fractional entrainment rate. Computationally simpler approaches are often employed. For example, Moorthi and Suarez (1992) relax the cloud work functions for individual members of the ensemble toward equilibrium without explicit cloud-cloud interactions. Anderson et al. (2004) report improved tropical precipitation, eddy, and storm activity when members of the ensemble with low rates of entrainment (deeply penetrative) are eliminated, following Tokioka et al. (1988). Focusing on deep convection, Zhang and McFarlane (1995) eliminate plumes not penetrating above the minimum in saturated moist static energy. These deep plumes are then assumed to have identical values of $M(\lambda, z_B)$ for all values of λ. Bulk plumes and ensembles of plumes with fixed partitioning of the base mass flux among plumes require a closure for the total mass flux, M_B, integrated over all values of λ. A range of closure assumptions are used, including relating M_B to moisture convergence (Tiedtke, 1989), atmospheric stability near cloud base (Gregory & Rowntree, 1990), available lifting power produced by processes like density currents (Rio et al., 2013), and CAPE or the cloud work function (Donner et al., 2011; Zhang & McFarlane, 1995; Zhao et al., 2018). Relaxation timescales, the choice of which can be used to tune emergent properties of the model in which the parameterization is used, are often imposed, especially with closures based on CAPE and the cloud work function.

For parameterizations that include convective vertical velocities as well as mass fluxes, vertical velocities must be specified at the base of the plumes, along with information on the fractional areas of the plume bases. Among the approaches to specifying base vertical velocities are fixing them at typical values (Chikira & Sugiyama, 2010; Donner, 1993) and relating them to PBL turbulent kinetic energy (Wagner & Graf, 2010; Zhao et al., 2018). In multiple-plume parameterizations, the relative frequencies of ensemble members must also be specified. Donner (1993) does so by using observations of convective updraft speeds to infer a PDF of entrainment coefficients by inverting a convective plume model constrained by these observations. Wagner and Graf (2010) calculate the populations of plumes of various sizes by placing them in competition with each other as a Lotka-Volterra system. Chikira and Sugiyama's (2010) ensemble is distributed uniformly across a specified range of plume-base vertical velocities.

Observational support for deep-convection closures is not strong. Quasi-equilibrium requires that total changes in the cloud work function or CAPE be small relative to changes produced by nonconvective processes. J. Wang and Randall (1994), Zhang (2002, 2003), and Donner and Phillips (2003) use observations from various field campaigns to investigate the behavior of CAPE and related quantities. Quasi-equilibrium holds when changes to CAPE produced by changes in the PBL are small, for example, during field campaigns over ocean surfaces as opposed to over land surfaces with a strong diurnal cycle. Quasi- equilibrium also holds when large, high-frequency changes in the PBL are averaged away, for example, in diurnal averages (Figure 5.6). Based on these results, Zhang (2002) suggests that convection is in quasi-equilibrium only with CAPE changes produced by nonconvective processes above the PBL, which is called free-troposphere quasi-equilibrium. Free-troposphere quasi-equilibrium is problematic if the equilibrium is viewed as buoyancy based since the source of buoyancy should not matter. Zhang (2009) points out that free-troposphere quasi-equilibrium is approximately equivalent to a balance between the vertical integrals of advection of dry static energy and convective heating, which holds widely in the tropics and maintains

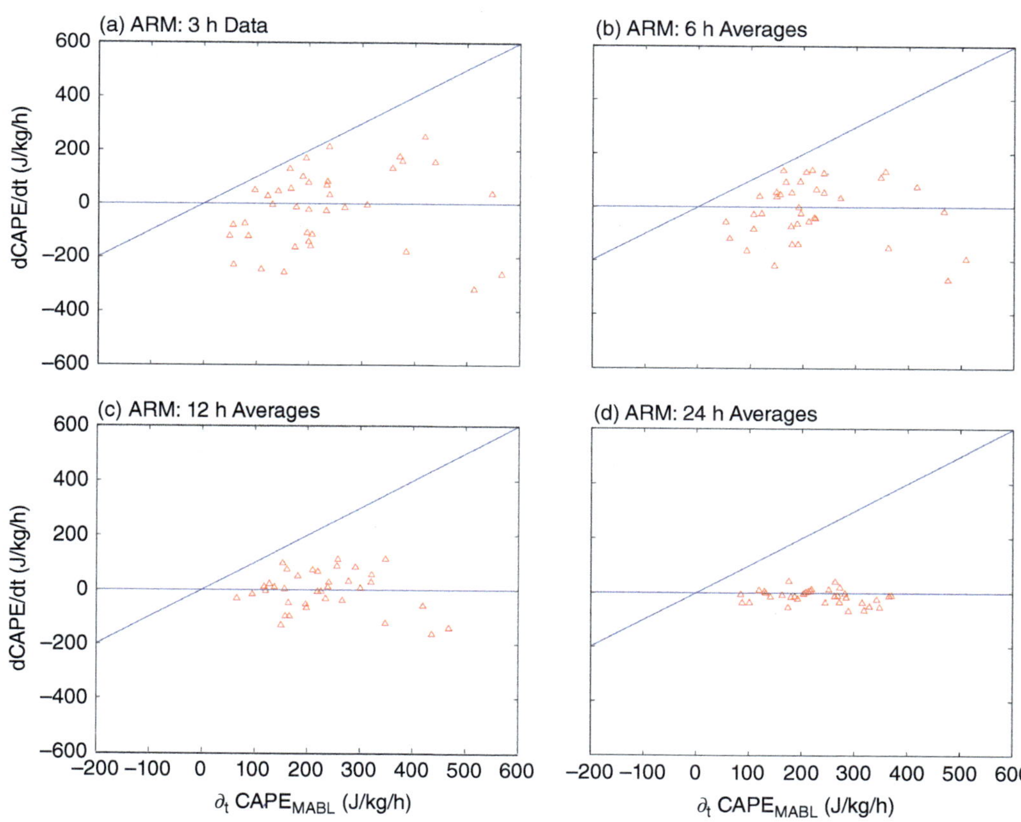

Figure 5.6 Change of CAPE by mean advection and PBL processes (horizontal axis) plotted against total CAPE change (vertical axis), using observations from central Oklahoma from the Atmospheric Radiation Measurements (ARM) program, averaged over indicated time intervals (Donner and Phillips, 2003). Quasi equilibrium predicts the total CAPE change should be much smaller than the change due to mean advection and PBL processes.

observed weak temperature gradients there. Bechtold et al. (2014) construct a closure based on free-troposphere quasi-equilibrium modified by the PBL to varying degrees over ocean and land surfaces.

The closures discussed so far are primarily relevant for deep convection. The mass fluxes at the base of shallow convection are generally tied to the properties of the PBL where the convection is rooted. These properties include surface heat fluxes, PBL turbulent velocities or kinetic energy, mixed layer depth, and convective inhibition (CIN) (Bretherton et al., 2004; Sŭselj et al., 2013; Zhao et al., 2018).

In the foregoing, closures are assumed to be diagnostic, i.e., the base mass fluxes do not explicitly depend on their prior values. By noting that the kinetic energy in a cumulus ensemble member increases as the product of its base mass flux and the cloud work function, allowing for dissipation of kinetic energy, and asserting a proportionality between kinetic energy and the square of the base mass flux, Pan and Randall (2006) develop a prognostic closure for base mass fluxes.

In ADHOC parameterizations, the assumed distribution itself is a closure. In addition, many closure assumptions typical of third-order turbulence closure are required, as discussed in Golaz et al. (2002). As is the case with closure assumptions for mass-flux parameterizations, the characteristics of simulations using ADHOC parameterizations are sensitive to the choices of parameters in these closure assumptions (Bogenschutz et al., 2013; Guo et al., 2014).

In addition to the closures required for mass-flux parameterizations, the conditions under which convection occurs, often referred to as a trigger, must also be specified. Typically, these are tied to atmospheric structure and usually require some combination of positive CAPE, positive cloud work function, or cloud-base instability; CIN below a threshold, possibly dependent on PBL turbulence; positive CAPE or cloud work function tendencies due to nonconvective processes (or their free-troposphere portion (Xie & Zhang, 2000)); and satisfaction of criteria on levels of zero (or neutral) buoyancy and free convection. An advantage of ADHOC methods is their avoidance of triggers; when convection does not occur with these methods, that is signified by the absence of upward motion or condensed water phases in the relevant subgrid PDFs.

5.3.5. Multiscale Modeling Frameworks and High-Resolution Models without Deep-Convection Parameterizations

The use of closure assumptions and simplifications, whose impact on the realism of simulations using parameterizations can be difficult to assess, can be overcome by embedding process models into a GCM (Grabowski, 2001). By restricting the dimensionality and domain of these models, the computational cost is reduced considerably relative to a full model of comparable resolution, though typically still far more expensive than parameterizations of the character described in the foregoing sections. Nonconvective tendencies of temperature and moisture provided by the GCM are applied to the process models, which in turn provide Q_θ, Q_q, and, in some applications, Q_X. When CRMs are used as process models, this approach is often described as superparameterization, and its application has yielded improvements in some convection-related fields, for example, the diurnal cycle and Madden-Julian Oscillation, that have been difficult to obtain with other parameterization approaches (Randall, 2013). CRMs themselves require parameterizations for subgrid turbulence, a key process for low clouds and the PBL. Cheng and Xu (2011, 2015) show that the treatment of parameterized processes in CRMs used in superparameterization is critical to simulating cloud systems whose scales remain unresolved in these CRMs. Even with advanced treatments of subgrid turbulence in CRMs used with superparameterization, errors in superparameterized simulations of key fields like cloud forcing and precipitation are similar to those in GCMs using other parameterization methods (Guo et al., 2015). Parishani et al. (2017) have embedded process models whose vertical scales approach LES (ultraparameterization) to extend the multiscale modeling framework to explicitly include PBL clouds. Although this simulates vertical structures of marine stratocumulus more realistically, climatological errors are consistent with those of superparameterization.

Increased realism in simulations using superparameterization and ultraparameterization, and evident limits in these increases, suggests that a fundamental limitation remains in all parameterization approaches discussed thus far. In all of these parameterizations, grid-scale, nonconvective tendencies are applied uniformly to the parameterized processes, whether represented as coherent structures with a closure, through assumed PDFs, or by higher-resolution process models. A common thread is the implicit assumption of scale separation, which may well be problematic, including in the multiscale modeling framework. Models with sufficiently high resolution to explicitly simulate deep convection and eliminate the need to parameterize it would be a path forward, if computationally feasible.

Even for deep convection, the resolution required to do this successfully poses large computational demands. Bryan and Morrison (2012); Bryan et al. (2003); Jeevanjee (2017); Lebo and Morrison (2015); Petch et al. (2002); and Varble et al. (2020) explore dependence of simulated convection on resolutions ranging from 20 m to 2 km in settings from squall lines to radiative-convective equilibrium. Solutions generally fail to converge until resolutions approach 100–200 m; even then, as well as at coarser resolutions, simulated vertical velocities do not match observations (Nicol et al., 2015; Varble et al., 2014). Generally, Craig and Dörnbrack (2008) find that simulations of a thermal converge at the smaller of the initial thermal diameter and the buoyancy length scale. Very short (12 hours) integrations with a global model of resolution 0.87 km (Miyamoto et al., 2013) have been reported, and a range of coarser resolutions and longer simulations have been studied (Sato et al., 2019). Stevens et al. (2019) compare 40-day simulations using models with resolutions from 2.5 to 4.8 km. As with superparameterization using CRMs without advanced representations of turbulence, simulations of low clouds in models at these resolutions are problematic. The superparameterization results of Cheng and Xu (2011, 2015) suggest that global-scale models of kilometer-scale resolution in conjunction with ADHOC approaches for turbulence could be a promising path forward.

Given the high computational cost of multiscale modeling frameworks and global models with grids approaching kilometer scales in the horizontal, machine learning has been proposed as a strategy by which computationally less expensive models can "learn" from these models. Gentine et al. (2018) use neural nets on a superparameterized global model to parameterize Q_θ and Q_q for lower-resolution models. Brenowitz and Bretherton (2019) also use neural nets for this purpose but train on a global atmospheric model with a horizontal grid around 4 km. Machine learning has proven problematic when applied outside the parameter space in which training takes place. In advanced models of the Earth system, cumulus parameterizations provide not only Q_θ and Q_q but also sources and sinks for horizontal momentum, microphysical properties, aerosol properties, and large numbers of tracers. These are among the challenges for machine learning, but early results encourage further investment in this research strategy. Hybrid approaches combining physically formulated cumulus parameterizations, especially the unified framework for thermodynamics, moisture, tracers, aerosols, and microphysics these can provide, with machine learning, especially its potential to elicit relationships between large-scale flows and convection for parameterization elements like closure, could yield major advances.

5.3.6. Adjustment Approaches

Adiabatic, isentropic processes are characterized by well-defined lapse rates. If a GCM grid cell becomes convectively unstable, its lapse rate should not exceed dry or saturated adiabatic, depending on whether or not it is saturated. Manabe et al. (1965) use this principle to adjust grid boxes whose lapse rates exceed dry or saturated (if relevant) lapse rates back to dry or saturated adiabats, conserving energy while doing so with requisite phase changes. On horizontal scales of a hundred kilometers or so, the atmosphere frequently becomes convectively active before lapse rates become unstable. Kuo (1965, 1974) adjusts temperature profiles toward a saturated adiabat for a parcel undergoing deep ascent from the PBL, doing so at a rate proportional to the sum of surface moisture fluxes and large-scale moisture convergence. Betts (1986) and Betts and Miller (1986) relax toward thermodynamic structures as empirically observed in large-scale flows.

Empirically based adjustments, like neural nets trained on limited sets of multiscale or high-resolution model integrations, are subject to error that is difficult to assess when they are applied outside the observational or model parameter space on which they are based. This can be especially problematic in applications related to climate change or prediction of extreme or unusual weather events. For example, Keil et al. (2014) note that the timescale for convective adjustment distinguishes between strong and weak synoptically forced precipitation. In large-scale models, treating effects of convection on atmospheric chemistry requires knowledge of mass fluxes and the chemical properties of convective plumes. Treating aerosol-cloud interactions and microphysics related to cloud feedbacks requires knowledge of convective vertical velocities (Sullivan et al., 2016). Treating effects of convection on horizontal momentum requires mass fluxes and knowledge of convective-scale pressure-gradient forces. Adjustments of temperature and moisture fields, or their tendencies inferred from machine learning, do not provide this information. If adjustments or tendencies inferred from machine learning were viewed as sufficiently robust, the mass-flux formulations discussed in preceding sections could be used to consistently infer other convection-related processes from adjustments or tendencies of temperature and moisture. In principle, machine learning could be extended to additional convective effects directly, though the "domain knowledge" captured by theories of parameterization could prove beneficial in structuring solutions from machine learning in a physically consistent manner.

5.3.7. Limitations of Mean-State Perspective

Although parameterizations using machine learning on high-resolution models and adjustment methods do not, many of the strategies discussed in this chapter rely on a conceptual frame in which convective plumes or turbulence are embedded on a uniform mean state (Figure 5.1). A multiscale modeling framework overcomes this to some extent by generating internal variability in its CRM or

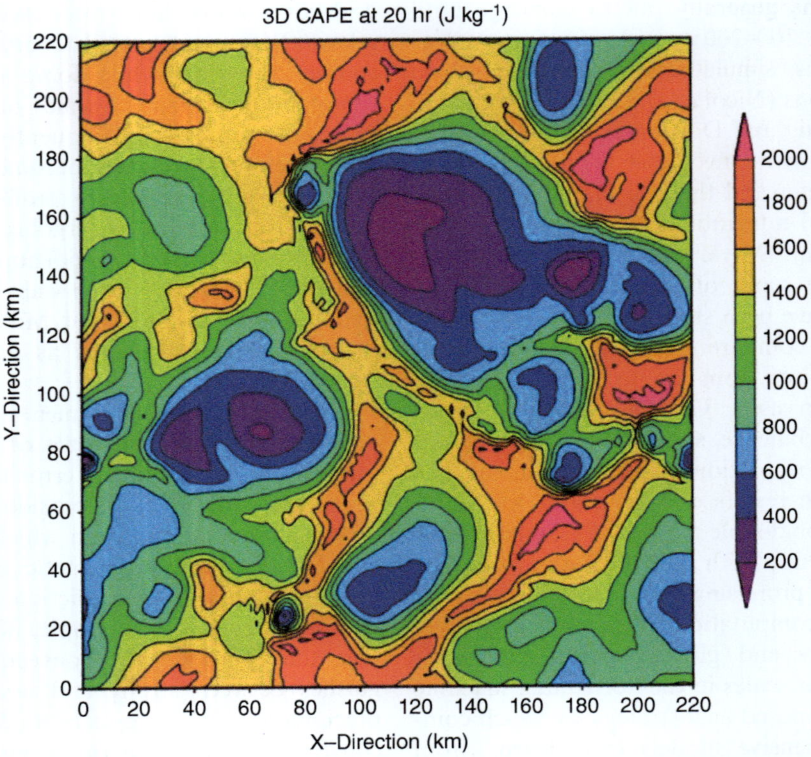

Figure 5.7 CAPE for tropical convection simulated by a CRM (Donner et al., 1999). The mean CAPE for the entire domain is around 1,100 J kg^{-1}. Note that only a small fraction of the domain has this CAPE.

LES, but these are forced uniformly by the mean state. The limitations of this perspective are especially severe for methods idealizing convective plumes with well-defined surroundings, as evident in the Donner et al. (1999) CRM analysis of thermodynamic variability on the kilometer scale (Figure 5.7). These results show that CAPE varies widely about its value averaged over horizontal scales of around 200 km. Brast et al. (2016) show that variations of termination heights for shallow cumulus plumes similarly are mostly controlled by heterogeneities in the subcloud layer, with details of entrainment secondary. In both of these studies, embedded convective plumes would generally form in environments differing substantially from the mean. Convection can be organized (Moncrieff, 1992) such that plumes interact with environments depending on how the convection is organized. The implication is that even if plume dynamics and microphysics could be realistically represented (noting progress along these lines discussed above), the parameterization would still fall short as these plumes would not interact with realistic environments.

Rather than envisioning convection as plumes on a base state, the parameterization problem could be posed in terms of generating a set of representative substates, which would, in turn, provide plume environments. Rather than multiple plumes forming on a single, mean state, fewer plumes would form in multiple environments. It is possible that research on the fundamental dynamics of plumes will limit current wide ranges in entrainment assumptions. If so, the observed variations in plume behaviors, e.g., vertical velocities (Collis et al., 2013; Donner et al., 2016; Giangrande et al., 2016), could be attributed not to variations in entrainment but rather to variations in plume environments.

The implications of viewing a key element of convective parameterization as the generation of thermodynamic substates has implications for both closures and plume properties. Wide variations in CAPE about its mean value can be generated by cold pools and density currents produced by deep convection. In general, convective organization will be associated with variations in thermodynamic structures that provide local environments for convective plumes or thermals. These structures evolve over time, and generating associated variations in thermodynamic profiles almost certainly will require prognostic closures. Parameterizations in which closures are tied to subgrid thermodynamic variability rather than mean-state equilibrium have been developed and implemented in GCMs (Grandpeix & Lafore, 2010; Rio et al., 2013).

5.3.8. Stochastic Convection

The possibility that uncertainties associated with the various aspects of convective parameterization discussed in preceding sections may not be entirely eliminated, even in principle, suggests the utility of introducing stochastic perturbations to the parameterizations (Berner et al., 2017). Stochastic perturbations have been introduced into many elements of convective parameterizations, including relaxation toward quasi-equilibrium (Grell & Dévényi, 2002; Khouider & Leclerc, 2019), convective organization (Cardoso-Bihlo et al., 2019; Goswami et al., 2017), cloud models (Bergemann et al., 2017), mass-flux closure (Dorrestijn et al., 2016; Plant & Craig, 2008), large-scale forcing (Grell & Dévényi, 2002), moisture-convergence closure (Grell & Dévényi, 2002), entrainment (Grell & Dévényi, 2002; Romps, 2016; Sŭselj et al., 2013), triggering (Rochetin et al., 2014a,b), and CAPE (J. W.-B. Lin & Neelin, 2000). The near inevitability of stochasticity in the interplay between convection and large-scale flows renders this research important. A promising strategy is to carefully bound stochastic elements within limits imposed by the best physical understanding that can be obtained.

5.4. MOMENTUM, CHEMISTRY, MICROPHYSICS, AND AEROSOL INTERACTIONS

Historically, convective parameterizations have focused on the effects of convection on large-scale thermal and moisture fields, Q_θ and Q_q. In addition to interacting with large-scale dynamics through its thermodynamic effects, convection directly impacts dynamics at multiple scales. For example, Helfand (1979) shows that vertical mixing of horizontal momentum by convection ("cumulus friction") strengthens the winter Hadley circulation. Although vertical mixing of horizontal momentum by convection has been parameterized as diffusive in proportion to M_c (Anderson et al., 2004), these diffusive approaches do not capture cloud-scale pressure gradients that modify horizontal momentum. Zhang and Cho (1991a) and Zhang and Cho (1991b) include cloud-scale pressure fields using specified cloud dynamic fields in a simple cloud model, which produces cloud pressure-gradient forces generally in the direction of the wind shear. Kershaw and Gregory (1997) and Gregory et al. (1997) parameterize cloud pressure gradients based on CRM simulations. Alternatively, this problem can be approached from the perspective of vorticity transport by convection (Cho et al., 1979). Convective organization is fundamental to the interactions between convective systems and large-scale flows (Moncrieff & Liu, 2006; Moncrieff et al., 2017), posing a major parameterization challenge and limiting the validity of approaches based on isolated convective plumes.

Deep convection also plays important roles in atmospheric chemistry, both through tracer transport and reactions in the presence of liquid and ice. The large range of mass fluxes consistent with much smaller variations in Q_θ and Q_q, discussed in section 3, is problematic for simulating aerosol and chemical distributions in the atmosphere. Field campaigns to measure Q_θ and Q_q have been useful in evaluating convective parameterizations (Donner & Phillips, 2003). Some inferences on convective mass fluxes can be drawn from observed tracer profiles (Folkins et al., 2006). Even stronger conclusions could probably be drawn from a closed chemistry-convection campaign in which all large-scale terms and tendencies in the budget for $\overline{X}$ would be measured, enabling Q_X to be inferred. The demands upon such a chemistry campaign would be large, requiring three-dimensional budgets of dynamics, thermodynamics, moisture, and chemical tracers for an ensemble of convective clouds. Campaigns such as DC3 (Barth et al., 2015) have characterized the chemistry of thunderstorm inflow and outflow, along with storm structure and chemical transformations in the convective plume. Extracting Q_X with confidence from a field campaign requires even more comprehensive observations, similar to those in the pioneering Global Atmospheric Research Program Atlantic Tropical Experiment (GATE) (Thompson et al., 1979) but augmented by simultaneous observations of chemical tracers. As implied in section 3, constraining mass fluxes would, in conjunction with Q_θ and Q_q, also constrain convective entrainment and precipitation efficiency, possibly in turn narrowing the range of estimates in climate sensitivity (Zhao et al., 2016), as discussed in preceding sections.

Given the importance of convective microphysics for both climate sensitivity and cloud-aerosol interactions with associated aerosol climate forcing, the current treatments of convective microphysics, which mostly assume precipitation is a fraction of cloud condensate, possibly in conjunction with or replaced by a threshold in cloud condensate for precipitation formation (Zhang & McFarlane, 1995; Zhao et al., 2016), offer major opportunities to increase process realism. These treatments of convective microphysics stand in contrast to treatments of stratiform microphysics in GCMs, which in some cases include two moments for key microphysical species (Morrison & Gettelman, 2008) and invariably at least include key processes such as autoconversion, collection, and particle terminal speeds, even if only a single moment is included (Rotstayn, 1997). Single-moment microphysics treating key interactions among microphysical species are treated in the convective plumes in the parameterizations of Anthes (1977), Donner (1993), and Kipling et al. (2017). Convective microphysics treating the major interactions between liquid and ice species with two moments have been developed (Song & Zhang, 2011) and produced profiles of convective liquid and ice in better agreement with satellite and field observations than treating precipitation as a fraction of cloud liquid. At best, microphysics in parameterized convection remains limited by fundamental uncertainties in microphysics apparent in cloud-resolving models, e.g., biases in ice sizes (Stanford et al., 2017). Bryan and Morrison (2012) and Fan et al. (2015) discuss the dependencies of key properties of deep convection on the treatment of microphysics.

Climate forcing by aerosol-cloud interactions remains highly uncertain (Stocker et al., 2013). Deep convective clouds and their associated stratiform systems may provide significant positive offsets for negative forcing associated with low clouds (Rosenfeld et al., 2013, 2014). Aerosol-cloud interactions are tightly coupled to microphysics in deep convection and can be simulated using CRMs (Fan et al., 2012; Heikenfeld et al., 2019; Tao et al., 2012), albeit with many uncertainties. However, the highly simplified treatments of convective microphysics in many GCMs preclude treatment of aerosol-cloud interactions associated with deep convection, possibly leading to significant errors in estimates of aerosol climate forcing. M. Wang et al. (2011) compare the change in shortwave cloud radiative effect due to aerosols in a GCM whose deep convective parameterization does not include aerosol interactions with the corresponding change in a superparameterized version of the model, which does include aerosol interactions in its CRM. The magnitude of the reduction in shortwave cloud radiative effect in the superparameterized GCM due to anthropogenic aerosols is less than half that in the version of the GCM lacking aerosol interactions with deep convection. The high computational expense of superparameterized GCMs renders them unsuitable for integrations long enough to explore aerosol-cloud interactions on the timescales of climate change. The latter problem requires including more advanced treatments of aerosols and associated convective microphysics in more computationally efficient parameterizations.

5.5. CONCLUSION

Understanding the role of deep convection in weather and climate, and its parameterization and modeling, remains a major scientific challenge. Multiple, viable frameworks for attacking the problem have been developed. All have associated with them interlocking, multiscale problems that would require solution for the merits and weaknesses of these frameworks to be fully apparent. From a parameterization perspective, it is important to realize that completely capturing the effects of unresolved phenomena on the resolved scale may not be possible, if resolved properties do not uniquely constrain the aggregate effects of the unresolved phenomena. Conceivably, the problem of deep convective parameterization cannot be cast any more robustly than as a range of possible scale interactions, in turn propagating into a range of outcomes for emergent behaviors in weather, climate, and Earth system models.

While the foregoing considerations are daunting, understanding deep convection and its interactions with other scales summons cross-disciplinary and cross-methodological approaches that evoke the best of scientific practice. Modeling across space and timescales, incisive analysis of emergent behaviors with their process dependencies in complex models, theoretical and "big data" analysis, and observations from satellites and field campaigns are all intrinsic to the problem. Promising research is underway on scale awareness (Chapter 16) and integrating both eddy diffusivity and convective mass fluxes more consistently, referred to as eddy-diffusivity/mass-flux parameterizations (Chapter 15). Surface-based (Chapter 19), satellite-based (Chapter 20), laboratory, and in situ (Chapter 21) measurements of key cloud properties are providing multiple perspectives on deep convection. As extensive as observational analyses of convection have been, vertical velocities, which until recently have been a technical challenge to obtain, provide novel and alternative constraints on parameterizations that may be highly relevant for climate forcing and sensitivity (Donner et al., 2016). Emerging computational capabilities will likely enable kilometer-scale coupled global models within the next decade. Although their success as climate models will depend on successful parameterization of cloud macrophysics and microphysics at even smaller scales, the prospect of resolved modeling of deep convection and its interactions with larger-scale flows in coupled, global models is widely anticipated as among the most important of the next steps forward.

ACKNOWLEDGMENTS

Assistance with figure preparation from Charles Seman and Cathy Raphael is appreciated.

Data availability for the papers in this review is discussed in those respective papers. Discussions with many colleagues over several decades have played important roles in shaping the perspectives presented here.

Manuscript reviews by Charles Seman, Ming Zhao, and three anonymous reviewers for AGU Books are appreciated.

REFERENCES

Ackerman, T., Liou, K.-N., Valero, F., & Pfister, L. (1988). Heating rates in tropical anvils. *Journal of the Atmospheric Sciences*, *45*, 1606–1623.

Anber, U., Giangrande, S., Donner, L., & Jensen, M. (2019). Updraft constraints on entrainment: Insights from Amazonian deep convection. *Journal of the Atmospheric Sciences*, *76*, 2429–2442. doi: 10.1175/JAS-D-18-0234.1.

Anderson, J., Balaji, V., Broccoli, A. J., Cooke, W. F., Delworth, T. L., Dixon, K. W., et al. (2004). The new GFDL global atmosphere and land model AM2-LM2: Evaluation with prescribed SST simulations. *Journal of Climate*, *17*, 4641–4673.

Anthes, R. (1977). A cumulus parameterization scheme using a one-dimensional cloud model. *Monthly Weather Review*, *105*, 270–286.

Arakawa, A., & Schubert, W. (1974). Interaction of a cumulus cloud ensemble with the large-scale environment. *Journal of the Atmospheric Sciences*, *31*, 674–701.

Arakawa, A., & Wu, C.-M. (2013). A unified representation of deep moist convection in numerical modeling of the atmosphere. Part I. *Journal of the Atmospheric Sciences*, *70*, 1977–1992. doi: 10.1175/JAS-D-12-0330.1.

Barth, M., Cantrell, C. A., Brune, W. H., Rutledge, S. A., Crawford, J. H., Huntrieser, H., et al. (2015). The Deep Convective Clouds and Chemistry (DC3) field campaign. *Bulletin of the American Meteorological Society*, *96*, 1281–1309. doi: 10.1175/BAMS-D-13-00290.1.

Bechtold, P., Semane, N., & Lopez, P. (2014). Representing equilibrium and nonequilibrium convection in large-scale models. *Journal of the Atmospheric Sciences*, *71*, 734-753. doi: 10.1175/JAS-D-13-0163.1.

Bergemann, M., Khouider, B., & Jakob, C. (2017). Coastal tropical convection in a stochastic modeling framework. *Journal of Advances in Modeling Earth Systems*, *9*, 2561–2582. doi: 10.1002/2017MS001048.

Berner, J., Achatz, U., Batté, L., Bengtsson, L., de la Cámara, A., Christensen, H. M., et al. (2017). Stochastic parameterization: Toward a new view of weather and climate models. *Bulletin of the American Meteorological Society*, *98*, 565–588. doi: 10.1175/BAMS-D-15-00268.1.

Betts, A. (1986). A new convective adjustment scheme. Part I: Observational and theoretical basis. *Quarterly Journal of the Royal Meteorological Society*, *112*, 677–691.

Betts, A., & Miller, M. (1986). A new convective adjustment scheme. Part II: Single column tests using GATE wave, BOMEX, ATEX, and arctic air-mass data sets. *Quarterly Journal of the Royal Meteorological Society*, *112*, 693–709.

Bogenschutz, P., Gettelman, A., Morrison, H., Larson, V., Craig, C., & Schanen, D. (2013). Higher-order turbulence closure and its impact on climate simulations in the Community Atmosphere Model. *Journal of Climate*, *26*, 9655–9676. doi:10.1175/JCLI-D-13-00075.1.

Bogenschutz, P., & Krueger, S. (2013). A simplified PDF parameterization of subgrid scale clouds and turbulence for cloud-resolving models. *Journal of Advances in Modeling Earth Systems*, *5*, 195–211. doi: 10.1002/jame.20018.

Bogenschutz, P., Krueger, S., & Khairoutdinov, M. (2010). Assumed probability density functions for shallow and deep convection. *Journal of Advances in Modeling Earth Systems*, *2*. doi: 10.3894/JAMES.2010.2.10.

Brast, M., Neggers, R., & Heus, T. (2016). What determines the fate of rising parcels in a heterogeneous environment? *Journal of Advances in Modeling Earth Systems*, *8*, 1674–1690. doi: 10.1002/2016MS000750.

Brenowitz, N., & Bretherton, C. (2019). Spatially extended tests of a neural network parameterization trained by coarse-graining. *Journal of Advances in Modeling Earth Systems*, *11*, 2728–2744. doi: 10.1029/2019MS001711.

Bretherton, C., McCaa, J., & Grenier, H. (2004). A new parameterization for shallow cumulus parameterization and its application to marine subtropical cloud-topped boundary layers. Part I: Description and 1D results. *Monthly Weather Review*, *132*, 864–882. doi: 10.1175/1520-0493(2004):ANPFSC>2.0CO;2.

Bryan, G., & Morrison, H. (2012). Sensitivity of a simulated squall line to horizontal resolution and parameterization of microphysics. *Monthly Weather Review*, *140*, 202–225. doi: 10.1175/MWR-D-11-00046.1.

Bryan, G., Wyngaard, J., & Fritsch, J. (2003). Resolution requirements for the simulation of deep moist convection. *Monthly Weather Review*, *131*, 2394–2416. doi: 10.1175/1520-0493(2003)131<2394:RRFTSO>2.0CO;2.

Cardoso-Bihlo, E., Khouider, B., Schumacher, C., & Chevrotière, M. D. L. (2019). Using radar data to calibrate a stochastic parametrization of organized convection. *Journal of Advances in Modeling Earth Systems*, *11*, 1655–1684. doi: 10.1029/2018MS001537.

Cheng, A., & Xu, K. (2011). Improved low-cloud simulation from a multiscale modeling framework with a third-order turbulence closure in its cloud-resolving model component. *Journal of Geophysical Research*, *116*, D14101. doi: 10.1029/2010JD015362.

Cheng, A., & Xu, K. (2015). Improved low-cloud simulation from the Community Atmosphere Model with an advanced third-order turbulence closure. *Journal of Climate*, *28*, 5737–5762. doi: 10.1175/JCLI-D-14-00776.1.

Chikira, M., & Sugiyama, M. (2010). A cumulus parameterization with state, dependent entrainment rate. Part I: Description and sensitivity to temperature and humidity profiles. *Journal of the Atmospheric Sciences*, *67*, 2171–2193. doi: 10.1175/2010JAS3316.1.

Cho, H.-R., & Cheng, L. (1980). Parameterization of horizontal transport of vorticity by cumulus convection. *Journal of the Atmospheric Sciences*, *37*, 812–826. doi: 10.1175/15200469(1980)037<0812:POHTOV>2.0CO;2.

Cho, H.-R., Cheng, L., & Bloxam, R. (1979). The representation of cumulus cloud effects in the large-scale vorticity equation. *Journal of the Atmospheric Sciences*, *36*, 127-139, doi: 10.1175/1520-0469(1979)036<0127:TROCCE>2.0CO;2.

Collis, S., Protat, A., May, P., & Williams, C. (2013). Statistics of storm updraft velocities from TWP-ICE including verification with profiling measurements. *Journal of Applied Meteorology and Climatology*, *52*, 1909–1922. doi: 10.1175/JAMC-D-12-0230.1.

Craig, G., & Dörnbrack, A. (2008). Entrainment in cumulus clouds: What resolution is cloud-resolving? *Journal of the Atmospheric Sciences*, *65*, 3978–3988. doi: 10.1175/2008JAS2613.1.

Cui, Z., Davies, S., Carslaw, K., & Blythe, A. (2011). The response of precipitation to aerosol through riming and melting in deep convective clouds. *Atmospheric Chemistry and Physics*, *11*, 3495–3510. doi: 10.5194/acp-11-3495-2011.

Dawe, J., & Austin, P. (2011). Interpolation of LES cloud surfaces for use in direct calculations of entrainment and detrainment. *Monthly Weather Review*, *139*, 444–456. doi: 10.1175/2010MWR3473.1.

Dawe, J., & Austin, P. (2013). Direct entrainment and detrainment rate distributions of individual shallow cumulus clouds in an LES. *Atmospheric Chemistry and Physics*, *13*, 7795–7811. doi: 10.5194/acp-13-7795-2013.

de Rooy, W., Bechtold, P., Fröhlich, K., Hohenegger, C., Jonker, H., Mironov, D., & Yano, J.-I. (2013). Entrainment and detrainment in cumulus convection: an overview. *Quarterly Journal of the Royal Meteorological Society*, *139*, 1–19.

Del Genio, A., & Yao, M.-S. (1993). Efficient cumulus parameterization for long term climate studies: The GISS scheme. In K. Emanuel & D. Raymond (Eds.), Meteorological Monographs: The Representation of Cumulus Convection in Numerical Models (Vol. 24, pp. 181–184).

Donner, L. (1993). A cumulus parameterization including mass fluxes, vertical momentum dynamics, and mesoscale effects. *Journal of the Atmospheric Sciences*, *50*, 889–906.

Donner, L., Wyman, B. L., Hemler, R. S., Horowitz, L. W., Ming, Y., Zhao, M., et al. (2011). The dynamical core, physical parameterizations, and basic simulation characteristics of the atmospheric component of the GFDL global coupled model CM3. *Journal of Climate*, *24*, 3484–3519. doi: 10.1175/2011JCLI3955.1.

Donner, L., Horowitz, L., Fiore, A., Seman, C., Blake, D., & Blake, N. (2007). Transport of radon-222 and methyl iodide by deep convection in the GFDL Global Atmospheric Model AM2. *Journal of Geophysical Research: Atmospheres*, *112*. doi: 10.1029/2006JD007548.

Donner, L., O'Brien, T., Rieger, D., Vogel, B., & Cooke, W. (2016). Are atmospheric updrafts a key to unlocking climate forcing and sensitivity? *Atmospheric Chemistry and Physics*, *16*, 12,983–12,292. doi: 10.5194/acp-16-12983-2016.

Donner, L., & Phillips, V. (2003). Boundary layer control on convective available potential energy: Implications for cumulus parameterization. *Journal of Geophysical Research*, *108*, 4701. 10.1029/2003JD003773.

Donner, L., Seman, C., & Hemler, R. (1999). Three-dimensional cloud-system modeling of GATE convection. *Journal of the Atmospheric Sciences*, *56*, 1885–1912.

Donner, L., Seman, C., & Hemler, R. (2001). A cumulus parameterization including mass fluxes, convective vertical velocities, and mesoscale effects: Thermodynamic and hydrological aspects in a general circulation model. *Journal of Climate*, *14*, 3444–3463.

Dorrestijn, J., Crommelin, D., Siebesma, A., Jonker, H., & Selten, F. (2016). Stochastic convection parameterization with Markov chains in an intermediate complexity GCM. *Journal of the Atmospheric Sciences*, *16*, 1367–1382. doi: 10.1175/JAS-D-15-0244.1.

Emanuel, K. (1991). A scheme for representing cumulus convection in large-scale models. *Journal of the Atmospheric Sciences*, *48*, 2313–2335.

Fan, J., Liu, Y.-C., Xu, K.-M., North, K., Collis, S., Dong, X., et al. (2015). Improving representation of convective transport for scale-aware parameterization: 1. convection and cloud properties simulated with spectral bin and bulk microphysics. *Journal of Geophysical Research: Atmospheres*, *120*, 3485–3509. doi: 10.1002/2014JD022142.

Fan, J., Rosenfeld, D., Ding, Y., Leung, L. R., & Li, Z. (2012). Potential aerosol indirect effects on atmospheric circulation and radiative forcing through deep convection. *Geophysical Research Letters*, *39*, L09806. doi: 10.1029/2012GL051851.

Fan, J., Wang, Y., Rosenfeld, D., & Liu, X. (2016). Review of aerosol-cloud interactions: Mechanisms, significance, and challenges. *Journal of the Atmospheric Sciences*, *73*, 4221-4252. doi: 10.1175/JAS-D-16-0037.1.

Feingold, G. (2003). Modeling of the first indirect effect: Analysis of measurement requirements. *Geophysical Research Letters*, *30*, 1997. doi: 10.1029/2003GL017967.

Folkins, I., Bernath, P., Boone, C., Eldering, L. D. A., Lesins, G., Martin, R. V. et al. (2006). Testing convective parameterizations with tropical measurements of HNO_3, CO, H_2O, and O_3: Implications for the water vapor budget. *Journal of Geophysical Research*, *111*. doi: 10.1029/2006JD007325.

Gentine, M. P., P., Rasp, S., Reinaudi, G., & Yacalis, G. (2018). Could machine learning break the convection parameterization deadlock? *Geophysical Research Letters*, *45*, 5742–5751. doi: 10.1029/2018GL078202.

Giangrande, S., Toto, T., Jensen, M. P., Bartholomew, M., Feng, Z., Protat, A., et al. (2016). Convective cloud vertical velocity and mass-flux characteristics from radar wind profiler observations during GoAmazon2014/5. *Journal of Geophysical Research: Atmospheres*, *121*, 12,891–12,913. doi: 10.1002/2016JD025303.

Golaz, J.-C., Larson, V., & Cotton, W. (2002). A PDF based model for boundary layer clouds. Part I: Method and model description. *Journal of the Atmospheric Sciences*, *59*, 3540–3551.

Goswami, B., Khouider, B., Phani, R., Mukhopadhyay, P., & Majda, A. (2017). Implementation and calibration of a stochastic multicloud convective parameterization in the NCEP Climate Forecast System (CFSv2). *Journal of Advances in Modeling Earth Systems*, *8*, 1721–1739. doi: 10.1002/2017MS001014.

Grabowski, W. (2001). Coupling cloud processes with the large-scale dynamics using the Cloud-Resolving Convection Parameterization (CRCP). *Journal of the Atmospheric Sciences*, *58*, 978–997.

Grandpeix, J.-Y., & Lafore, J.-P. (2010). A density current parameterization coupled with Emanuel's convection scheme. Part I: The models. *Journal of the Atmospheric Sciences*, *67*, 881–897. doi: 10.1175/2009JAS3044.1.

Gregory, D., Kershaw, R., & Inness, P. (1997). Parameterization of momentum transport by convection. II. Tests in single-column and general circulation models. *Quarterly Journal of the Royal Meteorological Society*, *123*, 1153–1183. doi: 10.1002/qj.49712354103.

Gregory, D., & Rowntree, P. (1990). A mass flux convection scheme with representation of cloud ensemble characteristics and stability-dependent closure. *Monthly Weather Review*, *118*, 1483–1506.

Grell, G., & Dévényi, D. (2002). A generalized approach to parameterizing convection containing ensemble and data assimilation techniques. *Geophysical Research Letters, 14,* 1693. doi: 10.1029/2002GL015311.

Grell, G., & Freitas, S. (2014). A scale and aerosol aware stochastic convective parameterization for weather and air quality modeling. *Atmospheric Chemistry and Physics, 14,* 5233–5250. doi: 10.5194/acp-14-5233-2014.

Guo, H., Golaz, J.-C., Donner, L. J., Ginoux, P., Hemler, R., & Schanen, D. (2014). Multi-variate probability density functions with dynamics in the GFDL atmospheric general circulation model: Global tests. *Journal of Climate, 27,* 2087–2108. doi: 10.1175/JCLI-D-13-00347.1.

Guo, H., Golaz, J.-C., Donner, L. J., Larson, V., Schanen, D., & Griffin, B. (2010). Multi-variate probability density functions for cloud droplet activation in large-scale models: Single column tests. *Geoscientific Model Development, 3,* 475–486. doi: 10.5194/gmd-3-475-2010.

Guo, H., Golaz, J.-C., Donner, L. J., Wyman, B., Zhao, M., & Ginoux, P. (2015). CLUBB as a unified cloud parameterization: Opportunities and challenges. *Geophysical Research Letters, 42,* 4540–4547. doi: 10.1002/2015GL063672.

Hannah, W. (2017). Entrainment versus dilution in tropical deep convection. *Journal of the Atmospheric Sciences, 74,* 3725–3747. doi: 10.1175/JAS-D-16.0619.1.

Heikenfeld, M., While, B., Labbouz, L., & Stier, P. (2019). Aerosol effects on deep convection: The propagation of aerosol perturbations through convective cloud microphysics. *Atmospheric Chemistry and Physics, 19,* 2601–2627. doi: 10.5194/acp-19-2601-2019.

Helfand, H. (1979). The effect of cumulus friction on the simulation of the January Hadley circulation by the GLAS model of the general circulation. *Journal of the Atmospheric Sciences, 1827–1843.* doi: 10.1175/1520-0469(1979)036<1827:TEOCFO>2.0CO;2.

Hernandez-Deckers, D., & Sherwood, S. (2018). On the role of entrainment in the fate of cumulus thermals. *Journal of the Atmospheric Sciences, 75,* 3911–3924. doi: 10.1175/JAS-D-18.0077.1.

Jeevanjee, N. (2017). Vertical velocity in the gray zone. *Journal of Advances in Modeling Earth Systems, 9.* doi: 10.1002/2017MS001059.

Jeevanjee, N., & Romps, D. (2016). Effective buoyancy at the surface and aloft. *Quarterly Journal of the Royal Meteorological Society, 142,* 811–820. doi: 10.1002/qj.2683.

Keil, C., Heinlein, F., & Craig, G. (2014). The convective adjustment time-scale as indicator of predictability of precipitation. *Quarterly Journal of the Royal Meteorological Society, 140,* 480–490. doi: 10.1002/qj.2143.

Kershaw, R., & Gregory, D. (1997). Parameterization of momentum transport by convection. I: Theory and cloud modeling results. *Quarterly Journal of the Royal Meteorological Society, 123,* 1133–1151. doi: 10.1002/qj.49712354102.

Khouider, B., & Leclerc, E. (2019). Toward a stochastic relaxation for the quasi-equilibrium theory of cumulus parameterization: Multicloud instability, multiple equilibria, and chaotic dynamics. *Journal of Advances in Modeling Earth Systems, 11,* 2474–2502. doi: 10.1029/2019MS001627.

Kipling, Z., Stier, P., Labbouz, L., & Wagner, T. (2017). Dynamic subgrid heterogeneity of convective cloud in a global model: Description and evaluation of the Convective Cloud Field Model (CCFM) in ECHAM6-HAM2. *Atmospheric Chemistry and Physics, 17,* 327–342. doi: 10.5194/acp-17-327-2017.

Kuo, H.-L. (1965). On formation and intensification of tropical cyclones through latent heat release by cumulus convection. *Journal of the Atmospheric Sciences, 22,* 40–63.

Kuo, H.-L. (1974). Further studies of the parameterization of the influence of cumulus convection on large-scale flow. *Journal of the Atmospheric Sciences, 31,* 1232–1240.

Labbouz, L., Kipling, Z., Stier, P., & Protat, A. (2018). How well can we represent the spectrum of convective clouds in a climate model? Comparisons between internal parameterization variables and radar observations. *Journal of the Atmospheric Sciences, 75,* 1509–1524. doi: 10.1175/JAS-D-17-0191.1.

LaMontagne, R., & Telford, J. (1983). Cloud top mixing in small cumuli. *Journal of the Atmospheric Sciences, 40,* 2148–2156.

Lappen, C., & Randall, D. (2001). Toward a unified parameterization of the boundary layer and moist convection. *Journal of the Atmospheric Sciences, 58,* 2021–2036. doi: 10.1175/1520-0469(2001)058<2021:TAUPOT>2.0.CO;2.

Larson, V., Golaz, J.-C., & Cotton, W. (2002). Small-scale and mesoscale variability in cloud boundary layers: Joint probability density functions. *Journal of the Atmospheric Sciences, 59,* 3519–3539.

Larson, V., & Schanen, D. (2013). The subgrid importance Latin hypercube sampler (SILHS): A multivariate subcolumn generator. *Geoscientific Model Development, 6,* 1813–1829. doi: 10.5194/gmd-6-1813-2013.

Leary, C., & Houze, R. (1980). The contribution of mesoscale motions to the mass and heat fluxes of an intense tropical convective system. *Journal of the Atmospheric Sciences, 37,* 784–796.

Lebo, Z., & Morrison, H. (2015). Effects of horizontal and vertical grid spacing on mixing in simulated squall lines and implications for convective strength and structure. *Monthly Weather Review, 143,* 4355-4375. doi: 10.1175/MWR-D-15-0154.1.

Lecoanet, D., & Jeevanjee, N. (2019). Entrainment in resolved, dry thermals. *Journal of the Atmospheric Sciences, 76,* 3785–3801. doi: 10.1175/JAS-D-18-0320.1.

Lee, S., Donner, L. J., & Penner, J. E. (2010). Thunderstorm and stratocumulus: How does their contrasting morphology affect their interactions with aerosols? *Atmospheric Chemistry and Physics, 10,* 6819–6837.

LeMone, M., & Zipser, E. (1980). Cumulonimbus vertical velocity events in GATE. Part 1: Diameter, intensity, and mass flux. *Journal of the Atmospheric Sciences, 37,* 2444–2457. doi: 10.1175/1520-0469(1980)037< 2444:CVVEIG>2.0CO;2.

Lin, C., & Arakawa, A. (1997). The macroscopic entrainment processes of simulated cumulus ensemble. Part II: Testing the entraining-plume model. *Journal of the Atmospheric Sciences, 54,* 1044–1053.

Lin, J. W.-B., & Neelin, J. (2000). Influence of a stochastic moist convective parameterization on tropical climate variability. *Geophysical Research Letters, 27,* 3691–3964.

Lu, C., Liu, Y., Zhang, G. J., Wu, X., Endo, S., Cao, L., et al. (2016). Improving parameterization of entrainment rate for shallow convection with aircraft measurements and large-eddy simulation. *Journal of the Atmospheric Sciences, 73*, 761–773. doi: 10.1175/JAS-D-15-0050.1.

Machado, L., & Rossow, W. (1993). Structural characteristics and radiative properties of tropical cloud clusters. *Monthly Weather Review, 121*, 3234–3260.

Manabe, S., Smagorinsky, J., & Strickler, R. (1965). Simulated climatology of a general circulation model with a hydrologic cycle. *Monthly Weather Review, 93*, 769–798.

Mapes, B., & Houze, R. (1993). Cloud clusters and super-clusters over the oceanic warm pool. *Monthly Weather Review, 121*, 1398–1415.

Masunaga, H., & Luo, Z. (2016). Convective and large-scale mass flux profiles over tropical oceans determined from synergistic analysis of a suite of satellite observations. *Journal of Geophysical Research: Atmospheres, 121,* 7958–7974. doi: 10.1002/2016JD024753.

McFiggans, G., Artaxo, P., Baltensperger, U., Coe, H., Facchini, M., Feingold, G., et al. (2006). The effect of physical and chemical aerosol properties on warm cloud droplet activation. *Atmospheric Chemistry and Physics, 6*, 2593–2649. doi: 10.5194/acp-6-2593-2006.

Miyakawa, T., Takayabu, Y., Nasuno, T., Miura, H., Satoh, M., & Moncrieff, M. (2012). Convective momentum transport by rainbands within a Madden-Julian Oscillation in a global nonhydrostatic model with explicit deep convective processes. Part I: Methodology and general results. *Journal of the Atmospheric Sciences, 69*, 1317–1338. doi: 10.1175/JAS-D-11-024.1.

Miyamoto, Y., Kajikawa, Y., Yoshida, R., Yamaura, T., Yashiro, H., & Tomita, H. (2013). Deep moist atmospheric convection in a subkilometer global simulation. *Geophysical Research Letters, 40*, 4922–4926. doi: 10.1002/grl.50944.

Moncrieff, M. (1992). Organized convective systems: Archetypal dynamical models, mass and momentum flux theory, and parameterization. *Quarterly Journal of the Royal Meteorological Society, 118*, 819–850. doi: 10.1002/qj.49711850703.

Moncrieff, M., & Liu, C. (2006). Representing convective organization in prediction models by a hybrid strategy. *Journal of the Atmospheric Sciences, 63*, 3403–3420. doi: 10.1175/JAS3812.1.

Moncrieff, M., Liu, C., & Bogenschutz, P. (2017). Simulation, modeling, and dynamically based parameterization of organized tropical convection for global climate models. *Journal of the Atmospheric Sciences, 74*, 1363–1380. doi: 10.1175/JAS-D-16.0166.1.

Moorthi, S., & Suarez, M. (1992). Relaxed Arakawa-Schubert: A parameterization of moist convection for general circulation models. *Monthly Weather Review, 120*, 978–1002.

Morrison, H. (2016). Impacts of updraft size and dimensionality on the perturbation pressure and vertical velocity in cumulus convection: Part II: Comparison of theoretical and numerical solutions and fully dynamical simulations. *Journal of the Atmospheric Sciences, 73*, 1455–1480. doi: 10.1175/JAS-D-15-0041.1.

Morrison, H. (2017). An analytic description of the structure and evolution of growing deep cumulus updrafts. *Journal of the Atmospheric Sciences, 74*, 809–834. doi: 10.1175/JAS-D-16-0234.1.

Morrison, H., & Gettelman, A. (2008). A new two-moment stratiform cloud microphysics scheme in Community Atmosphere Model, Version 3 (CAM3). Part I: Description and numerical tests. *Journal of Climate, 21*, 3642–3659. doi: 10.1175/2008JCLI2105.1.

Moser, D., & Lasher-Trapp, S. (2017). The influence of successive thermals on entrainment and detrainment in a simulated cumulus congestus. *Journal of the Atmospheric Sciences, 74*, 375–392. doi: 10.1175/JAS-D-16-0144.1.

Nicol, J., Hogan, R., Stein, T., Hanley, K., Clark, P., Halliwell, C., et al. (2015). Convective updraught evaluation in high-resolution NWP simulations using single-Doppler radar measurements. *Quarterly Journal of the Royal Meteorological Society, 3177–3189*. doi: 10.1002/qj.2602.

Paluch, I. (1979). The entrainment mechanism in Colorado cumuli. *Journal of the Atmospheric Sciences, 36*, 2467–2478.

Pan, D.-M., & Randall, D. (2006). A cumulus parametrization with a prognostic closure. *Quarterly Journal of the Royal Meteorological Society, 124*, 949–981. doi: 10.1002/qj.49712454714.

Parishani, H., Pritchard, M., Bretherton, C., Wyant, M., & Khairoutdinov, M. (2017). Toward low-cloud permitting cloud superparameterization with explicit boundary layer turbulence. *Journal of Advances in Modeling Earth Systems, 9*, 1542–1571. doi: 10.1002/2017MS000968.

Park, S. (2014). A unified convection scheme (UNICON). Part I: Formulation. *Journal of the Atmospheric Sciences, 71*, 3902–3930. doi: 10.1175/JAS-D-13-0233.1.

Petch, J., Brown, A., & Gray, M. (2002). The impact of horizontal resolution on the simulation of convective development over land. *Quarterly Journal of the Royal Meteorological Society, 128*, 2031–2044.

Peters, J. (2016). The impact of effective buoyancy and dynamic pressure forcing on vertical velocities within two-dimensional updrafts. *Journal of the Atmospheric Sciences, 73*, 4531–4551. doi: 10.1175/JAS-D-16-0016.1.

Peters, J., Hannah, W., & Morrison, H. (2019). The influence of vertical wind shear on moist thermals. *Journal of the Atmospheric Sciences, 76*, 1645–1659. doi: 10.1175-JAS-D-18.0296.1.

Plant, R., & Craig, G. (2008). A stochastic parameterization for deep convection based on equilibrium statistics. *Journal of the Atmospheric Sciences, 65*, 87–105. doi: 10.1175/2007JAS2263.1.

Pritchard, M., Moncrieff, M., & Somerville, R. (2011). Orogenic propagating precipitation systems over the United States in a global climate model with embedded explicit convection. *Journal of the Atmospheric Sciences, 68*, 1821–1840. doi: 10.1175/2011JAS3699.1.

Randall, D. (2013). Beyond deadlock. *Geophysical Research Letters, 40*, 5970–5976. doi: 10.1002/2013GL057998.

Raymond, D., & Blyth, A. (1986). A stochastic mixing model for nonprecipitating cumulus clouds. *Journal of the Atmospheric Sciences, 43*, 2708–2718.

Rio, C., Grandpeix, J.-Y., Hourdin, F., Guichard, F., Couvreux, F., Lafore, J.-P. et al. (2013). Control of deep convection by sub-cloud lifting processes: The ALP closure in the LMDZ5B general circulation model. *Climate Dynamics, 40*, 2271–2292. doi: 10.1007/s00382-012-1506-x.

Rochetin, N., Couvreux, F., Grandpeix, J.-Y., & Rio, C. C. (2014a). Deep convection triggering by boundary layer thermals. Part I. LES analysis and stochastic triggering formulation. *Journal of the Atmospheric Sciences, 71*, 496–514. doi: 10.1175/JAS-D-12-0336.1.

Rochetin, N., Grandpeix, J.-Y., Rio, C., & Couvreux, F. (2014b). Deep convection triggering by boundary layer thermals. Part II. Stochastic triggering parameterization for the LMDZ GCM. *Journal of the Atmospheric Sciences, 71*, 515–538. doi: 10.1175/JAS-D-12-0337.1.

Romps, D. (2010). A direct measure of entrainment. *Journal of the Atmospheric Sciences, 67*, 1908–1927. doi: 10.1175/2010JAS3371.1.

Romps, D. (2016). The stochastic parcel model: A deterministic parameterization of stochastically entraining convection. *Journal of Advances in Modeling Earth Systems, 8*, 319–344. doi: 10.1002/2015MS000537.

Rosenfeld, D., Sherwood, S., Wood, R., & Donner, L. (2014). Climate effects of aerosol-cloud interactions. *Science, 343*, 379–380. doi: 10.1126/science.1247490.

Rosenfeld, D., Wood, R., Donner, L., & Sherwood, S. (2013). Aerosol cloud-mediated radiative forcing: Highly uncertain and opposite effects from shallow and deep clouds. In *Climate science for serving society: Research, modeling, and prediction priorities*. Dordrecht: Springer.

Rotstayn, L. (1997). A physically based scheme for the treatment of stratiform clouds and precipitation in large-scale models. I. Description and evaluation of microphysical processes. *Quarterly Journal of the Royal Meteorological Society, 123*, 1227–1282.

Rougier, J., Sexton, D., Murphy, J., & Stainforth, D. (2009). Analyzing the climate sensitivity of the HadSM3 climate model using ensembles from different but related experiments. *Journal of Climate, 22*, 3540–3557. doi: 10.1175/2008JCLI2533.1.

Sanderson, B., Shell, K., & Ingram, W. (2010). Climate feedbacks determined using radiative kernels in a multi-thousand member ensemble of AOGCMs. *Climate Dynamics, 35*, 1219–1236.

Sato, M., Stevens, B., Judt, F., & Coauthors. (2019). Global cloud-resolving models. *Current Climate Change Reports, 5*, 172–184. doi: 10.1007/s40641-019-00131-0.

Schmidt, G., Kelley, M., Nazarenko, L., Ruedy, R., Russell, G. L., Aleinov, I., et al. (2014). Configuration and assessment of the GISS ModelE2 contributions to the CMIP5 archive. *Journal of Advances in Modeling Earth Systems, 6*, 141–184. doi: 10.1002/2013MS000265.

Smagorinsky, J. (1963). General circulation experiments with the primitive equations. I. *The basic experiment. Monthly Weather Review, 91*, 99–164.

Song, X., & Zhang, G. (2011). Microphysics parameterization for convective clouds in a global climate model: Description and single-column model tests. *Journal of Geophysical Research, 116*. doi: 10.1029/2010JD014833.

Stainforth, D., Alna, T., Christensen, C., Collins, M., Faull, N., Frame, D., et al. (2005). Uncertainty in predictions of the climate response to rising levels of greenhouse gases. *Nature, 433*, 403–406.

Stanford, M., Varble, A., Zipser, E., Strapp, J., Leroy, D., Schwarzenboeck, A., et al. (2017). A ubiquitous ice size bias in simulations of tropical deep convection. *Atmospheric Chemistry and Physics, 9599–9621*. doi: 10.5194/acp-17-9599-2017.

Stevens, B., Satoh, M., Auger, L., Biercamp, J., Bretherton, C., Chen, X., et al. (2019). DYAMOND: The DYnamics of the atmospheric general circulation modeled on non-hydrostatic domains. *Progress in Earth and Planetary Science, 6*, 61, doi: 10.1186/s40645-019-0304-z.

Stocker, T., Qin, D., Plattner, G.-K., Tignor, M., Allen, S., Boschung, J., et al. (2013). Summary for policymakers. *Climate Change 2013: The Physical Science Basis. Contribution of Working Group I to the Fifth Assessment Report of the Intergovernmental Panel on Climate Change*. Cambridge/New York: Cambridge University Press.

Sullivan, S., Lee, D., Oreopoulos, L., & Nenes, A. (2016). Role of updraft velocity in temporal variability of global cloud hydrometeor number. *Proceedings of the National Academy of Sciences U. S. A.*, 113, 5791–5796. doi:10.1073/pnas.1514039113.

Sŭselj, K., Teixeira, J., & Chung, D. (2013). A unified model for moist convective boundary layers based on a stochastic eddy-diffusivity/mass-flux parameterization. *Journal of the Atmospheric Sciences, 70*, 1929–1953. doi: 10.1175/JAS-D-12-0106.1.

Tao, W., Chen, J.-P., Li, Z., Wang, C., & Zhang, C. (2012). Impact of aerosols on convective clouds and precipitation. *Reviews of Geophysics, 50*. doi: 10.1029/2011RG000369.

Tarshish, N., Jeevanjee, N., & Lecoanet, D. (2018). Buoyant motion of a turbulent thermal. *Journal of the Atmospheric Sciences, 75*, 3233–3244. doi: 10.1175/JAS-D-17-0371.1.

Thompson, R., Payne, S., Recker, E., & Reed, R. (1979). Structure and properties of synoptic-scale wave disturbances in the intertropical convergence zone of the eastern Atlantic. *Journal of the Atmospheric Sciences, 36*, 53–72. doi: 10.1175/1520-0469(1979)036<0053:SAPOSS>2.0CO;2.

Tiedtke, M. (1989). A comprehensive mass flux scheme for cumulus parameterization in large-scale models. *Monthly Weather Review, 117*, 1779–1800.

Tokioka, T., Yamazaki, K., Kitoh, A., & Ose, T. (1988). The equatorial 30–60 day oscillation and the Arakawa-Schubert penetrative cumulus parameterization. *Journal of the Meteorological Society of Japan, 66*, 883–901.

Turner, J. (1973). Buoyant convection from isolated sources. In *Buoyancy effects in fluids* (chap. 6, 367 pp). Cambridge: Cambridge University Press.

Varble, A. (2018). Erroneous attribution of deep convective invigoration to aerosol concentration. *Journal of the Atmospheric Sciences, 75*, 1351–1368. doi: 10.1175/JAS-D-17-0217.1.

Varble, A., Morrison, H., & Zipser, E. (2020). Effects of under-resolved convective dynamics on the evolution of a squall line. *Monthly Weather Review, 148*, 289–311. doi: 10.1175/MWR-D-19-0187.1.

Varble, A., Zipser, E., Fridland, A., Zhu, P., Ackerman, A., Chaboureau, J.-P., et al. (2014). Evaluation of cloud-resolving and limited area model intercomparison simulations using TWP-ICE observations. 1. Deep convective updraft properties. *Journal of Geophysical Research: Atmospheres, 119*, 13,891–13,918. doi: 10.1002/2013JD021371.

Wagner, T., & Graf, H.-F. (2010). An ensemble cumulus convection parameterization with explicit cloud treatment. *Journal of the Atmospheric Sciences, 67*, 3854–3869. doi: 10.1175/2010JAS3485.1.

Wang, J., & Randall, D. (1994). The moist available energy of a conditionally unstable atmosphere: Part II: Further analysis of GATE data. *Journal of the Atmospheric Sciences, 51*, 703–710.

Wang, M., Ghan, S., Ovchinnikov, M., Liu, X., Easter, R., Kassianov, E., & Morrison, H. (2011). Aerosol indirect effects in a multi-scale aerosol climate model PNNL-MMF. *Atmospheric Chemistry and Physics, 11*, 5431–5455. doi: 10.5194/acp-11-5431-2011.

Xie, S., & Zhang, M. (2000). Impact of the convection triggering function on single column model simulations. *Journal of Geophysical Research, 105*, 14,983–14,996.

Zhang, G. (2002). Convective quasi-equilibrium in midlatitude continental environment and its effect on convective parameterization. *Journal of Geophysical Research, 107(*D14*)*, 4220. doi: 10.1029/2001JD001005.

Zhang, G. (2003). Convective quasi-equilibrium in the tropical western Pacific: Comparison with midlatitude continental environment. *Journal of Geophysical Research, 108(*D19*)*, 4592. doi: 10.1029/2003JD003520.

Zhang, G. (2009). Effects of entrainment on convective available potential energy and closure assumptions in convection parameterization. *Journal of Geophysical Research, 114(*D7*)*. doi: 10.1029/2008JD010976.

Zhang, G., & Cho, H.-R. (1991a). Parameterization of the vertical transport of momentum by cumulus clouds. Part II: Application. *Journal of the Atmospheric Sciences, 48*, 2448–2457. doi: 10.1175/1520-0469(1991)048<2448:POTVTO>2.0.CO;2.

Zhang, G., & Cho, H.-R. (1991b). Parameterization of the vertical transport of momentum by cumulus clouds. Part I: Theory. *Journal of the Atmospheric Sciences, 48*, 1483–1492. doi: 10.1175/1520-0469(1991)048<1483:POTVTO>2.0.CO;2.

Zhang, G., & McFarlane, N. (1995). Sensitivity of climate simulations to the parameterization of cumulus convection in the Canadian Climate Centre general circulation model. *Atmosphere-Ocean, 33*, 407–446.

Zhang, G., Wu, X., Zeng, X., & Mitovski, T. (2016). Estimation of convective entrainment properties from a cloud-resolving model simulation during TWP-ICE. *Climate Dynamics, 47*, 2177–2192. doi: 10.1007/s00382-015-2957-7.

Zhao, M. (2014). An investigation of the connections among convection, clouds, and climate sensitivity in a global climate model. *Journal of Climate, 27*, 1845–1862. doi: 10.1175/JCLI-D-13-00145.1.

Zhao, M., Golaz, J.-C., Held, I. M., Ramaswamy, V., Lin, S.-J., Ming Y., et al. (2016). Uncertainty in model climate sensitivity traced to representations of cumulus precipitation microphysics. *Journal of Climate, 29,* 543–560. doi: 10.1175/JCLI-D-15-0191.1.

Zhao, M., Golaz, J.-C., Held, I. M., Guo, H., Balaji, V., Benson, R., et al. (2018). The GFDL global atmosphere and land model AM4.0/LM4.0: 2. Model Description, Sensitivity Studies, and Tuning Strategies. *Journal of Advances in Modeling Earth Systems, 10,* 735–769. doi: 10.1002/2017MS001209.

6

Stratus, Stratocumulus, and Remote Sensing

Xiquan Dong[1] and Patrick Minnis[2]

ABSTRACT

Stratus and stratocumulus clouds comprise a critical component of the atmospheric system and have long been the subject of research from many perspectives. Observations from the surface, air, and space have provided the data necessary to quantify their occurrence, properties, and interactions with radiation, dynamics, and other atmospheric components. Such data have served to develop and improve parameterizations for characterizing these clouds in climate and weather prediction models. This chapter provides an overview of this cloud type and its role in climate with a focus upon what we have learned from ground-based and satellite remote sensing. The processes that determine the variations in these low-level cloud properties and govern where and when they occur are discussed along with factors, such as aerosols, radiation, and humidity, that influence their characteristics. Retrieval methods used to infer the properties of these stratiform clouds from satellite-based and ground-based sensors are also reviewed. Remote sensing is a critical means to further our understanding of these clouds and the processes involving them, as well as, for monitoring and predicting their role in a changing climate.

6.1. INTRODUCTION

A key, radiatively sensitive component of the climate that must be addressed accurately in general circulation models (GCMs) comprises the low-level stratiform clouds: stratus and stratocumulus, as well as fog. Widespread around the globe, they are among the most studied cloud types. Based on the Latin noun, stratum, their names indicate that they are confined in relatively well-defined layers, so their vertical extent is generally quite small compared to their horizontal dimensions. Except for ground fog, the layer containing the cloud can usually be defined by the cloud base near the lifting condensation level (LCL) and by the cloud top underneath a capping temperature inversion. The term, stratus, refers to a continuous, often featureless cloud sheet that is gray when viewed from below (Figure 6.1a). Stratus often forms by slow uplift caused by orography (upslope), rising air produced near a low-pressure system, or moist air rising over a stable atmospheric layer near a frontal boundary. Stratocumulus clouds can be produced in a similar manner but have the added component of small-scale thermal differentials and turbulence of convection, the source of cumulus (Latin for "heap") clouds. Or, they can form directly from convective processes as cumulus clouds that are flattened out by the temperature inversion (Figure 6.1b and c). Stratocumulus clouds can have sky coverage ranging from partly cloudy to overcast with highly variable geometrical structures. They can be composed of water droplets, supercooled water droplets, ice crystals, or mixtures of droplets and crystals depending on the ambient temperatures and aerosols.

Because they are low in the troposphere, stratus clouds radiate longwave (LW) radiation out to space at temperatures only a few degrees less than those at the surface. Yet they are often sufficiently thick to have relatively

[1]*Department of Hydrology and Atmospheric Sciences, University of Arizona, Tucson, AZ, USA*

[2]*Analytical Mechanics Associates, Inc., Hampton, VA, USA*

Fast Processes in Large-Scale Atmospheric Models: Progress, Challenges, and Opportunities, Geophysical Monograph 282, First Edition.
Edited by Yangang Liu and Pavlos Kollias.

DOI: 10.1002/9781119529019.ch06

Figure 6.1 Examples of (a) stratus generated around a low-pressure system, (b) stratocumulus undulatus ahead of cold front, and (c) stratocumulus opacus formed after passage of cloud front.

high albedos and therefore reflect a significant amount of shortwave (SW) solar radiation back to space that would otherwise have heated the surface. Thus, except for some notable exceptions, these clouds generally produce a net cooling at the surface (e.g., Randall et al., 1984; Slingo, 1990) and exert complex feedbacks on the climate system (Hang et al., 2019; L'Ecuyer et al., 2019; Stephens, 2005; Wood, 2012). The radiative effect of these clouds contributes to one of the largest uncertainties in climate modeling documented in the IPCC Fifth and Sixth Assessment Reports (IPCC, 2013, 2021). Due to their relative ubiquity, shading effect, and other roles in the general circulation, it is critical that these stratiform clouds and the processes that form, maintain, and dissipate them be observed, understood, and modeled as accurately as possible.

This chapter provides a general review of this cloud type and its role in climate with a focus upon what we have learned from current ground-based and satellite remote sensing and models informed by the observations. For the sake of brevity, unless otherwise indicated, stratus will hereafter be used to refer to both stratus and stratocumulus clouds. An overview of these clouds, their distributions, radiative effects, and the means for observing them is given followed by a technical summary of how these clouds form and dissipate and by a delineation of the structural characteristics that define stratus. Observations of their properties at various spatial and temporal scales are then presented followed by the outlook for improved remote sensing and other means to further our understanding and modeling of this critical cloud type.

6.2. OVERVIEW

According to the World Meteorology Organization (WMO) International Cloud Atlas, which was developed for surface observers, both stratus (St) and stratocumulus (Sc) are defined as having a cloud-base height no greater than 2 km above the surface. Thus, their base altitudes would be <2 km over the oceans but could reach much higher altitudes over land surfaces. That base altitude defines all low clouds. The more quantitative and extensive objective observations of clouds from satellites directly infer the cloud-top altitude or pressure, not the cloud-base height, from observations of infrared brightness temperatures. In an attempt to be relatively consistent with the WMO definition, while keeping the classification as simple as possible for global application, the International Satellite Cloud Climatology Project (ISCCP; Rossow & Schiffer, 1991) defined low clouds as those having cloud-top heights (CTHs) at pressures above 680 hPa or altitudes below ~3.3-km MSL. This definition should include most St and Sc clouds because they form under temperature inversions that prevent vertical development but excludes cumulonimbus clouds and other vertically developed cumulus clouds, which are classified as low clouds by the WMO. It also misses the low clouds that form over high terrain, e.g., Tibetan plateau. Despite the notable differences between the WMO and satellite definitions, there should be general consistency between the satellite and surface climatologies.

6.2.1. Surface Observers

Prior to the satellite era, cloud-type occurrence and fractional sky coverage were entirely based on a trained weather observer looking at the sky and making estimates based on internationally sanctioned methods. Although prone to some subjective bias and illumination restrictions, such measurements provided valuable information for understanding the distribution of clouds around the globe. Temporally regular observations were taken from weather stations over land and a few dedicated weather ships over the oceans, the distribution of the stations is anything but homogeneous and few observations were made in sparsely populated areas. Over the open oceans, commercial and military ships have provided the bulk of the cloud observations. Thus, knowledge of clouds over many marine areas was extremely limited.

Figure 6.2 highlights this constraint in the distribution of mean stratus coverage from surface observations

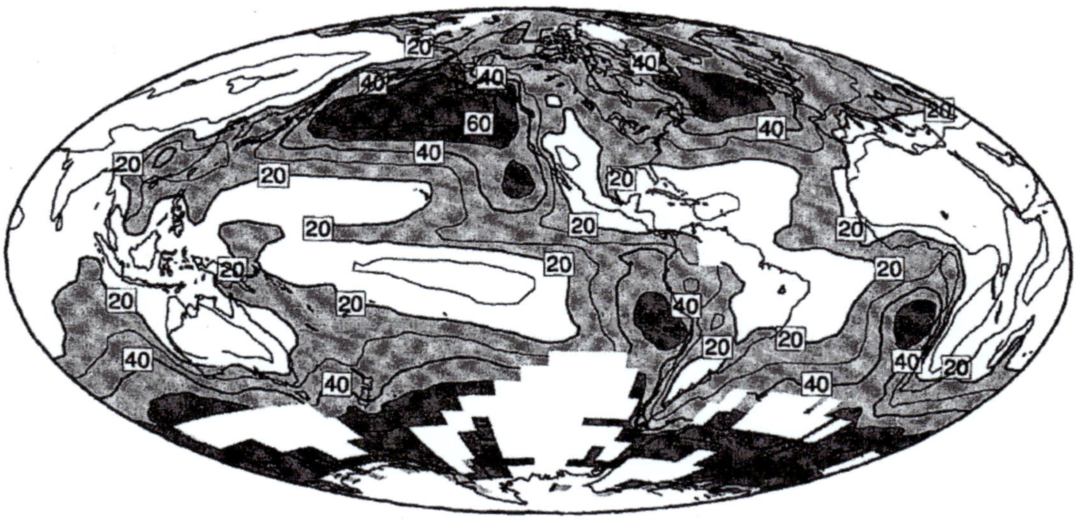

Figure 6.2 Combined annual mean coverage of stratus and stratocumulus clouds in percent derived from surface observations, 1971–1981. Source: Klein and Hartmann (1993)/American Meteorological Society.

for the period 1971–1981. Extensive gaps exist over the ocean south of 30°S latitude. Nevertheless, the surface observations uniquely show the major areas of stratus, the most widespread being over the midlatitude oceans, where mean coverage exceeds 40% in most areas. Other, noteworthy systems are evident over the subtropical oceans off the west coasts of the continents, over the Arctic, and over land in southeastern China and eastern Canada. Minima occur over tropical and subtropical land areas, the western sides of the tropical and subtropical oceans, Antarctica, and the interior of Asia. Using the compilations of Warren et al. (1986, 1988) for the same period, Klein and Hartmann (1993) determined that the mean fractional sky covered by stratus is 0.29 with 0.34 and 0.17 over marine and land areas, respectively, exclusive of the areas without observations.

The subtropical marine stratus cloud systems form over the cold currents that flow equatorward on the eastern sides of subtropical oceans, namely the California, Peru, Canary, Benguela (south Atlantic), and West Australia Currents. Above them and centered to their west, the subtropical highs that constitute the descending branches of the Hadley circulation maintain the subsidence capping the marine boundary layer (MBL). This sinking air minimizes the occurrence of clouds above the stratus. The contours in Figure 6.2 show a negative east-to-west gradient indicating the transition of the stratocumulus regime to that of trade cumulus (Cu) clouds. These trade Cu are scattered clouds that arise from a combination of factors, such as warmer sea surface temperature, weaker inversion, and boundary-layer deepening. Notable exceptions to the confinement of subtropical stratus to the eastern oceans are the relative maxima extending into the central equatorial Pacific and slightly into the mideequatorial Atlantic, east and northeast of Australia, and west of China. The equatorial stratus forms over the cooler water upwelling between the North and South Equatorial Currents in the two oceans. East of Australia, the stratus forms as part of the South Pacific Convergence Zone, which produces clouds at all levels. Over eastern China, the relative peak in stratus, which primarily occurs during winter, results from complex processes forced by orography and the seasonal circulation patterns (Y. Zhang et al., 2013). The cold season peak over China contrasts with that of the northeastern ocean stratus that tends to have maximum coverage during the summer. Seasonal variations of stratus coverage follow the boundary-layer static stability, which is defined as the difference in potential temperatures at 700 hPa and sea level (Klein & Hartmann, 1993).

In the midlatitudes, stratus clouds form mostly as a result of cyclonic and frontal activity. A snapshot of clouds covering the Earth on 22 September 2020 taken by the Visible and Infrared Imager/Radiometer Suite (VIIRS; see Hillger et al., 2013) on the Suomi National Polar-orbiting Partnership (S-NPP) shows the east-to-west succession of cyclones and associated frontal clouds in the midlatitudes in both hemispheres (Figure 6.3). In many cases, fields of stratocumulus (mottled areas) can be seen west of the frontal clouds. These types of systems give rise to the maxima seen in Figure 6.2. The stratus associated with the cyclone and ahead of the front are often coincident with clouds at higher levels in areas of rising motion, while the subsidence behind

Figure 6.3 Merged true color corrected reflectance images from 22 September 2020 taken 1330 Local Time by the Suomi National Polar-orbiting Partnership (S-NPP) Visible and Infrared Imager/Radiometer Suite (VIIRS) SNPP-VIIRS. The nearly vertical gray strips over the tropics in the image are areas of sunglint viewed from successive orbits. Source: NASA/https://worldview.earthdata.nasa.gov/last accessed 16, March 2023.

the front prevents cloud generation above the boundary layer. Passage of cold air behind the fronts over the warmer waters, particularly the Kuroshio and Gulf Stream Currents, can produce spectacular outbreaks of stratocumulus. While those outbreaks primarily occur during the winter months (Wood, 2012), summer is the season of maximum stratus in the midlatitudes (Klein & Hartmann, 1993).

Summer is also the season of peak stratus coverage over the Arctic Ocean. From 37 years of surface observations, Eastmann and Warren (2010) found that the stratus coverage rises from ~28% during the winter to around 61% during the summer accounting for ~3/4 of the total cloud cover. Based on the lower tropospheric stability, the maximum stratus coverage should occur during the winter, but the extremely low temperatures and lack of low-level humidity during winter limit low cloud formation (Klein & Hartmann, 1993). Minimum stability occurs during the summer, but its magnitude is similar to that under subtropical highs. Thus, during the summer the presence of low-level moisture advected from the south and evaporated from ice-free areas and melt ponds readily generates stratus clouds over the region, often in various layers, leading to the peak in low cloudiness (e.g., Huang et al., 2021). Due to the high elevations and low temperatures, the Arctic stratus phenomenon is not repeated over Antarctica, even though the continent is ringed by stratus with mean coverage that can reach as high as 60% (e.g., Stanfield et al., 2014).

6.2.2. Satellite Observations

Surface observer reports have been a valuable resource for estimating stratus coverage since low clouds are not obscured by other types of clouds and because they provide a relatively long-term record. Satellites, however, can provide a wealth of more quantitative information about stratus systems than somewhat subjective surface cloud observations. Most satellite information about clouds has come from analyses of radiances measured by passive visible and infrared imagers onboard sun-synchronous and geostationary orbiters. The presence of high and midlevel clouds between stratus and these satellite instruments still limits the conditions when stratus can be directly viewed. For example, for satellite sensors, high-level clouds are more prevalent (40%–50%) than low clouds, which account for only about 40%–45% of the total cloud cover, including cumulus. Microwave passive radiometers measure at wavelengths that are mostly transparent to cloud ice and sensitive to liquid water. They, therefore, provide information about liquid water clouds, specifically, liquid water path (LWP; e.g., O'Dell et al., 2008), the depth of water that would result from compressing all of the cloud droplets in the atmospheric column into a single, air-free layer. Cloud information from microwave sensors, though, is mostly limited to ocean regions because of the large variability of land surface emissivities. Since 2006, it has been possible to study clouds globally throughout the atmospheric column using active sensors on orbiting

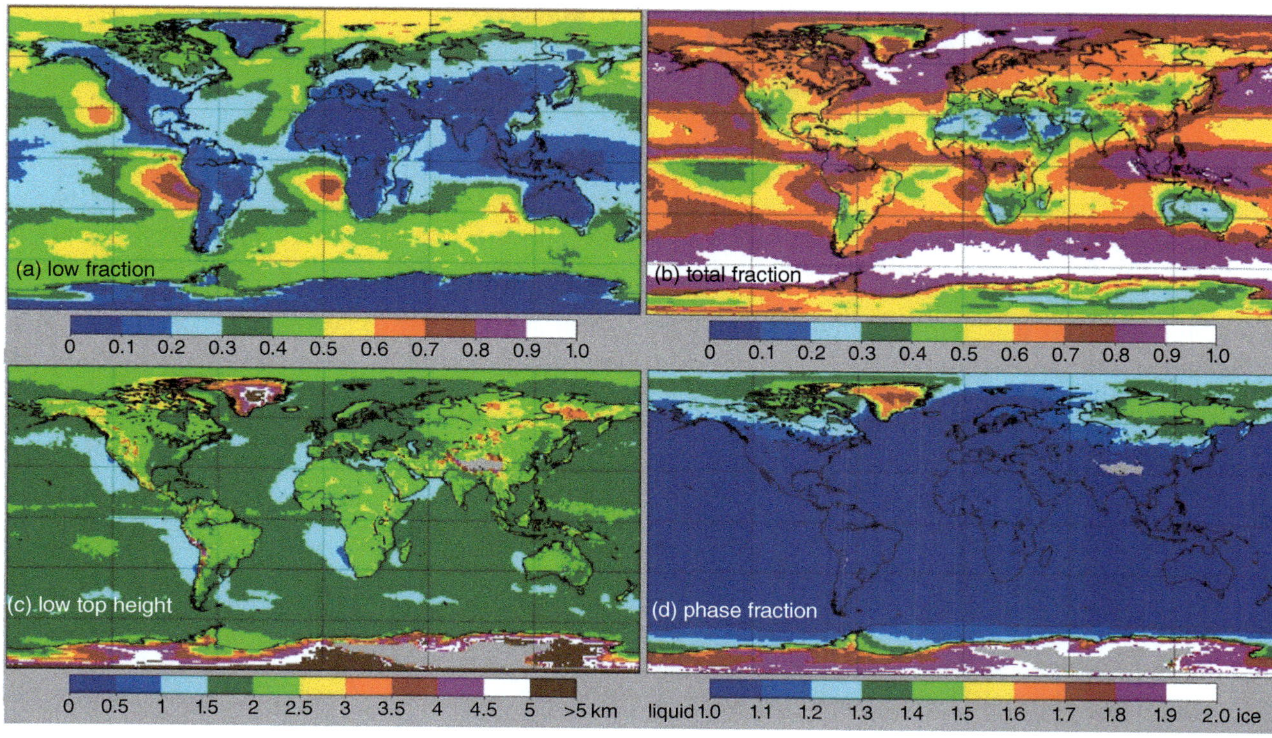

Figure 6.4 Mean 2018 cloud properties from CERES Aqua MODIS retrievals. (a) Low cloud fraction CF_L, (b) total cloud fraction CF_T, (c) low cloud-top height CTH above MSL, and (d) low-cloud phase retrieval fraction, where a value of 1.0 is all liquid and 2.0 is entirely ice.

platforms. At that time, the Cloud-Aerosol Lidar and Infrared Pathfinder Satellite Observations (CALIPSO; Winker et al., 2009) and the CloudSat (Stephens et al., 2008) satellites were launched into orbits in formation with Aqua and other satellites in the sun-synchronous Afternoon Constellation or A-Train. CALIPSO carries the Cloud-Aerosol Lidar with Orthogonal Polarization and the CloudSat instrument is the Cloud Profiling Radar (CPR). While limited to very narrow nadir views at 0130 and 1330 LT, the combined CALIOP and CPR cloud profiles have been extremely valuable alone and even more so when merged with the Aqua imager data. This synergism has advanced our understanding of clouds, the quantification of errors in the passive retrievals from imagers, and the amount of information that can be inferred from the imager data. Because they provide more extensive temporal and spatial coverage and have a longer record than other satellite data, passive imager cloud retrievals (see Stubenrauch et al. 2013, and references therein) are still the most valuable for studying clouds of all types on a global scale.

Some examples of the information that can be gleaned from the passive retrievals is presented in Figure 6.4 using retrievals from Aqua 1-km MODerate-resolution Imaging Spectroradiometer (MODIS) data for the Clouds and Earth's Radiant Energy System (CERES) project (Wielicki et al., 1996). The CERES Edition 4 (Ed4) cloud retrievals were performed as in Trepte et al. (2019) and Minnis et al. (2021). Using the ISCCP definition above, the distribution of mean low cloud fraction (CF_L) for 2018 (Figure 6.4) is similar in many respects to the surface observations (Figure 6.1). The subtropical maxima are clearly defined; the CERES-MODIS peak values are slightly larger, exceeding 0.7 off the Chile-Peru and Namibian coasts, compared to maxima less than 0.6 in those areas. The equatorial Pacific tongue and relative maximum over China are also present. Mean CF_L in the surface-data gaps over the southern oceans is between 0.4 and 0.6. Over the northern oceans, the low cloud coverage is somewhat less than the 50%–60% amounts seen in the surface observations. This is not surprising, given that some of the stratus observed are associated with low-pressure systems. Over the Arctic Ocean, the mean CF_L is between 0.4 and 0.6, comparable to that reported from the surface. On average, the global mean CF_L from Aqua is 0.30 compared to 0.29 from the surface-based climatology noted above.

Aside from the positive values, the differences between the CERES-MODIS and surface estimates can be mostly explained by the obscuration effect of multilayer (ML) clouds. Total cloud fraction (CF_T, Figure 6.4b), 0.66 on average, is significantly greater than the low cloud fraction (CF) in areas where the surface-observed stratus exceeds the low cloud cover in Figure 6.4a. The positive differences can be due to several factors. These include limited sampling over marine areas, inclusion of cumulus clouds in the low-cloud category, and limitations in accurately estimating low cloud cover at night from the surface. The 1-year sample in Figure 6.4 may also cause some differences, but it is typical for the CERES-MODIS record.

Cloud-top height is also retrieved from the satellite imager data. In Figure 6.4c, the mean CTH is lowest over the eastern oceans and over relatively small regions in the midlatitude seas. The mean minima, <1 km, occur near the coasts of California, Chile, and Namibia with rising heights to the west and the Equator, where average CTH exceeds 2.0 km. The lower values over the northern midlatitudes are probably the result of detecting the low clouds under the strong postfrontal inversions. The average CTH over the Arctic Ocean is between 1.5 and 2.5 km. Continental CTH exceeds its marine counterpart and, in many instances, follows the terrain. Overall, the mean low CTH is 1.91 km. As shown in section 6.4, CTH and other cloud properties are subject to significant seasonal and diurnal variations.

The great majority of low clouds (i.e., stratus) are composed only of liquid droplets as shown in Figure 6.4d, which maps the mean 2018 low-cloud phase fraction. In this instance, the average phase fraction index is at or close to unity for much of the Earth. Phase fraction indices greater than 1.0 indicate that some clouds are identified as ice up to a value of 2.0, which denotes that all of the cloudy pixels were found to be composed of ice. Most of the low clouds identified over Greenland and Antarctica are ice clouds, while less than 50% of the low clouds over the rest of the Arctic are ice. If only daytime data are considered, then the fraction of ice clouds dramatically drops over the Arctic. Liquid phase is dominant with some mixed-phase clouds in the midlatitudes. The determination of the cloud as ice or liquid in the Arctic depends on the proportion of cloud ice to liquid and whether the ice is on or below the cloud top. The formation of ice in these clouds affects the energy budget through release of latent heat, precipitation, and changes in the radiative properties.

Other cloud properties that affect the radiation budget are the cloud-top temperature (CTT), the cloud optical depth (COD), and the cloud-droplet effective radius (CER). The product of CER and COD is proportional to the liquid water path for liquid clouds and, likewise, IWP for ice clouds. CTT, which corresponds to CTH, along with COD determines the longwave flux emitted and transmitted upward by the cloud. COD and CER together determine that the amount of shortwave solar radiation is reflected by the cloud. For a given COD, smaller droplets will reflect more radiation than their larger counterparts. Figure 6.5 plots the distribution of CERES-MODIS daytime means for these properties from Aqua for 2018. Average CTT (Figure 6.5a) mainly follows zonal variation with values ranging from less than 250 K over Greenland to 295 K in the Tropics. Notable breaks from the zonal pattern are evident in lower means over the subtropical cool currents and warmer averages penetrating poleward over the Gulf Stream, and the North Atlantic and North Pacific Drifts. The global mean CTT is 280.0 K, while the polar (|latitude| > 60°) and nonpolar means are 258.7 and 283.2 K, respectively.

Between 30°N and 30°S, the mean low-cloud COD (Figure 6.5b) is generally less than 6, with larger values over the primary stratus areas. The maximum over southeastern China, where average COD exceeds 24, is presumably due to very high concentrations of low-level pollutant aerosols that limit precipitation (Qian et al., 2009), thus permitting the development of thick stratus with relatively small values of CER (e.g., Li et al., 2020). Low-cloud COD increases poleward from the tropics with mean values generally between 12 and 18; maxima are seen over northeastern Europe, the southern Pacific, and the Arctic Ocean above Asia. Globally, COD has an average value of 8.3. In nonpolar areas, the mean is 7.3, which is less than half the 15.2 average over the polar regions. Part of this dramatic meridional difference may be due to biases in the polar retrievals because of the reduced contrast between the surface and cloud reflectances, and because of a tendency to overestimate COD with satellite retrievals for solar zenith angles (SZAs) above 60° (e.g., Dong et al., 2016; McHardy et al., 2018).

The distribution of CER (Figure 6.5c) differs from the COD pattern. Subtropical stratus regimes can be characterized by CER mean minima of 9–11 μm near the coast that increase westward and equatorward into the open ocean where the maxima between 16 and 18 μm are found. Over land, the lowest means, between 6 and 10 μm, are found over areas where low clouds are sparse, except over southeastern China, where boundary-layer aerosol concentrations suppress droplet growth and enhance droplet concentrations. Over the globe, the CER average is 13.5 μm for low clouds. Variations in mean CER are due to a number of factors. Foremost are the cloud vertical depth and the cloud condensation nucleus (CCN) density. Areas with more aerosols in the boundary layer are likely to have more CCN and hence will generally have a greater number of small cloud droplets (higher N_c and lower CER), assuming that all other factors are equal. Less vertical development will also minimize the droplet size because of reduced growth time as a nascent droplet ascends through the cloud. The mean CER of 11.0 μm for low clouds over land, where aerosols are generally more

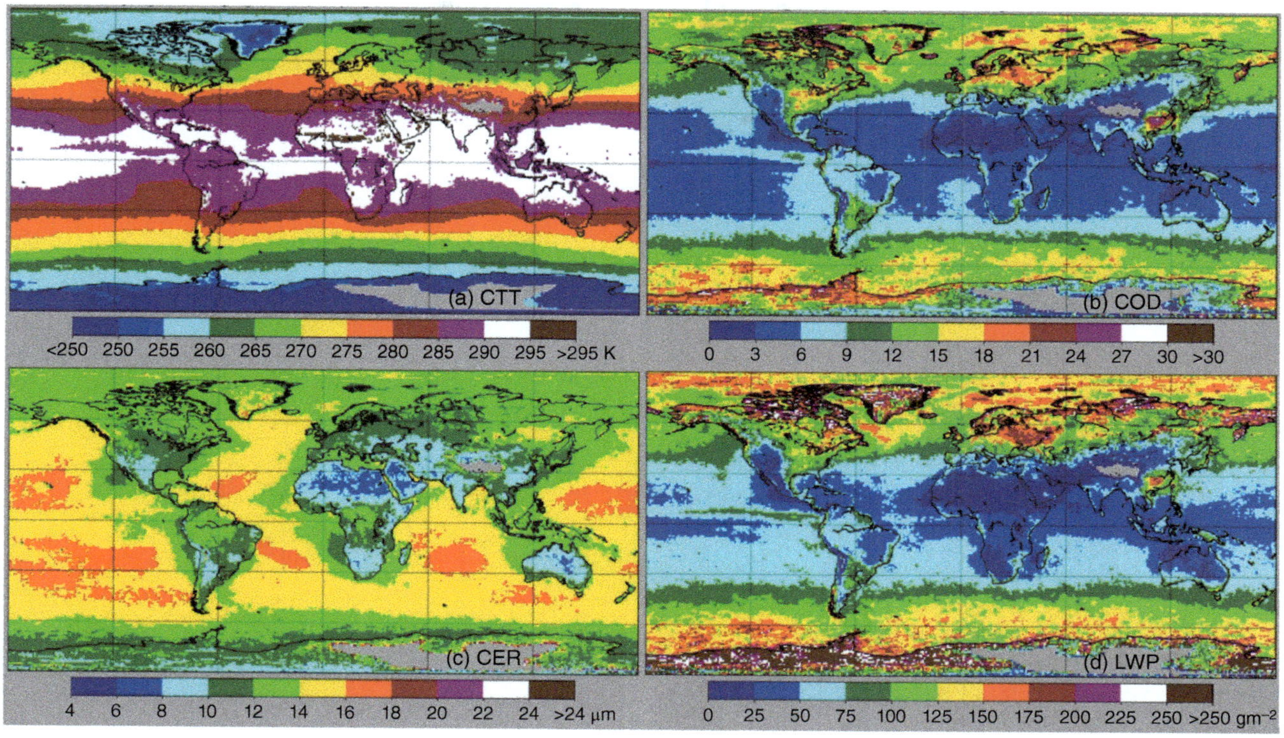

Figure 6.5 Same as Figure 6.4, but for daytime only and for low (a) cloud-top temperature (CTT), (b) cloud optical depth (COD), (c) cloud-droplet effective radius (CER), and (d) cloud fraction weighted liquid water path (LWP). Source: Courtesy of S. Sun-Mack.

abundant, is 3.5 μm less than that over ocean. Thus, it would appear that aerosols are a major factor governing CER. This topic is addressed in later sections.

Combining COD and CER together, assuming that the cloud-droplet size distribution is uniform, yields the LWP distributions as shown in Figure 6.5d. The equatorial Pacific maximum of ~85 g m^{-2} is readily apparent, while much of the subtropical stratus LWP is between 50 and 75 g m^{-2}. Over China, the COD peak translates to a standout LWP maximum exceeding 150 g m^{-2}. Elsewhere, the variations in LWP tend to follow those of COD. If it is assumed that the clouds develop adiabatically, the LWP estimates can be reduced by 17%. The adiabatic assumption is more accurate in many stratus areas (e.g., Minnis et al., 2021; Seethala & Horvath, 2010).

The modulation of the radiant energy in the earth-atmosphere system by these clouds, which affects the climate, can most easily be quantified in the parameter, cloud radiative effect, CRE defined as the total amount of radiation leaving a given area F minus that for cloud-free skies only F_c (e.g., Minnis & Harrison, 1984b; Ramanathan et al., 1989). The net radiative heating of the earth-atmosphere system is

$$F_{NET} = S(1 - a_{SW}) - F_{LW}, \tag{6.1}$$

where S is the incoming solar irradiance, a_{SW} is the albedo, and F_{LW} is outgoing longwave radiation at the top of the atmosphere (TOA). The CRE can be estimated for the net radiation,

$$CRE_{NET} = F_{NET} - F_{cNET}, \tag{6.2}$$

where F_{NET} represents all-sky net radiation and F_{cNET} represents clear-sky one. CRE_{SW} and CRE_{LW} represent the SW and LW cloud radiative effects, respectively. Negative CREs indicate that clouds have a cooling effect and, conversely, a warming effect for positive numbers. Values for the CRE components are found by first measuring the Earth's radiation budget for all conditions and then using the cloudless subset of those measurements to compute the cloud-free radiation on the desired spatial scales.

Global distributions of the average SW and LW CREs in Figure 6.6 are based on the CERES Level 4 Energy Balance and Filled at the top of the atmosphere (EBAF-TOA) Ed2.6r products (Loeb et al., 2009) for the period, 2000–2010, reported by Calisto et al. (2014). In the CRE_{SW} means (Figure 6.6a), the stratus areas are immediately noticeable in the darker green and blue shades that denote averages ranging from −50 to −150 W m^{-2}, although values of −65 W m^{-2} or so occur in some regions not dominated by stratus, such as the maritime

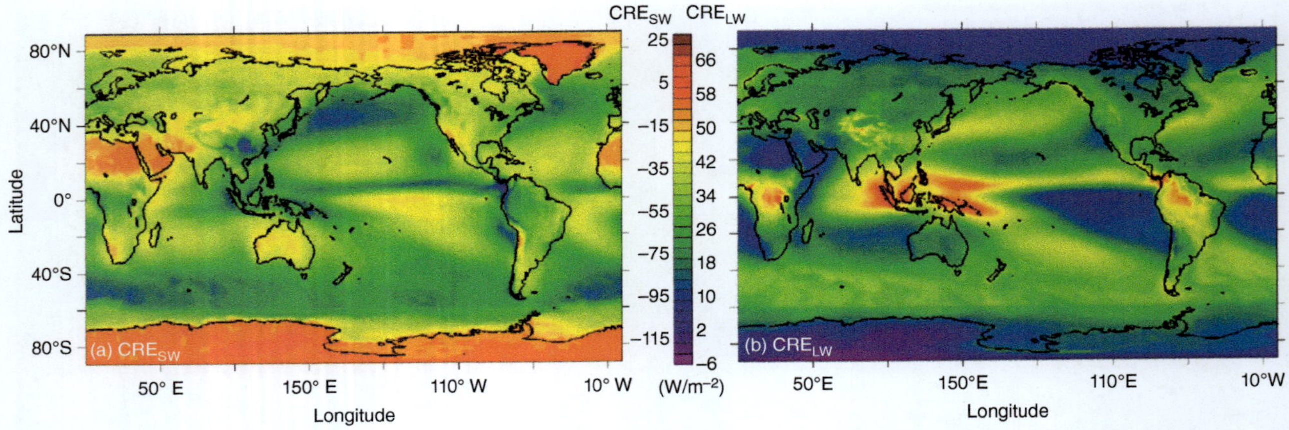

Figure 6.6 Mean cloud radiative effects from CERES measurements (2000–2010) for (a) shortwave and (b) longwave radiation. Source: Calisto et al. (2014)/European Geosciences Union/CC BY 3.0.

continent and other tropical convective areas. The maximum SW cooling of approximately -120 W m^{-2} is seen over Panama and surrounding areas. In the tropical convective areas, CRE_{LW} (Figure 6.6b) is also strongly positive with values up ~65 W m^{-2}. Those large CRE_{LW} values contrast with the averages over the stratus regimes that range from ~10 W m^{-2} over the subtropics to 40 W m^{-2} over the equatorial Pacific. The lower values in tropics reflect the absence of higher-level clouds. Subtracting CRE_{LW} from CRE_{SW} yields CRE_{NET}, which averages -18 W m^{-2} over the globe (Calisto et al., 2014), a value that was found by Harrison et al. (1990) who used Earth Radiation Budget Experiment (ERBE) data and by Loeb et al. (2018) employing CERES Ed4.0 EBAF.

To illustrate the net CRE on a projection similar to that employed in Figure 6.2, Figure 6.7 presents the CRE_{NET} image from the NASA Hyperwall (https://svs.gsfc.nasa.gov/30603), which is based on the 2000–2015 average from the CERES 4.0 EBAF dataset (Loeb et al., 2018). It clearly shows the influence of the stratus regimes; the most negative values occur over the subtropical stratus areas, southeast China, and the midlatitude oceans. Cloud warming is mainly found over deserts and the high-altitude polar regions, while most of the continental areas have a neutral or slightly negative mean CRE_{NET}. The one stratus-dominant area with a nearly neutral average cloud effect is the Arctic Ocean, which is confirmed by Dong et al. (2010) who used 10 years of ground-based observations taken near Barrow, Alaska, at ~70°N. According to the Hartmann et al. (1992) study using ERBE, the magnitude of low-cloud CRE_{NET} is double that of any other cloud type. Thus, as the atmosphere warms in response to increasing greenhouse gases, it is critical for predicting the future climate to understand how stratus and its characteristics will change. At present, stratus clouds are the most effective cooling agents in the atmosphere. Will the stratus cloud impact remain static, enhance warming, or act as a buffer and minimize feedbacks?

6.2.3. Surface and Airborne Observations

That question and others related to this important cloud type are most directly answered through modeling studies, specifically those using GCMs for climate studies because they attempt to simultaneously account for all of the processes that occur as a reaction to a given change in the earth-atmosphere system. Since the models cannot hope to explicitly account for every process in detail, they employ generalized parameterizations that attempt to simulate the static and time-dependent relationships among the many variables that define the climate on many scales. These parameterizations heavily rely on observations. But, while the complete climate models attempt to match the large-scale data taken by satellites, the parameterizations have to link smaller scale processes that produce the observed large-scale fields. Those small-scale processes are best understood and modeled with the aid of both in situ and remotely sensed observations taken from aircraft and surface instrumentation. Those same observations can in many cases be used to ensure through direct comparisons that the interpretation of large-scale satellite measurements is accurate or, at least, provide the basis for improved retrievals.

Aircraft have been employed to study marine stratus for more than nine decades. For example, Dean (1928) used in situ temperature and humidity measurements to define the marine inversion that accompanies the "fog" below. Early surface-based remote sensors, such as the ceilometer, were valuable for monitoring stratus cloud-base height (Vernon, 1936). With the advent of the radiosonde in the 1930s, vertical profiles of temperature and humidity

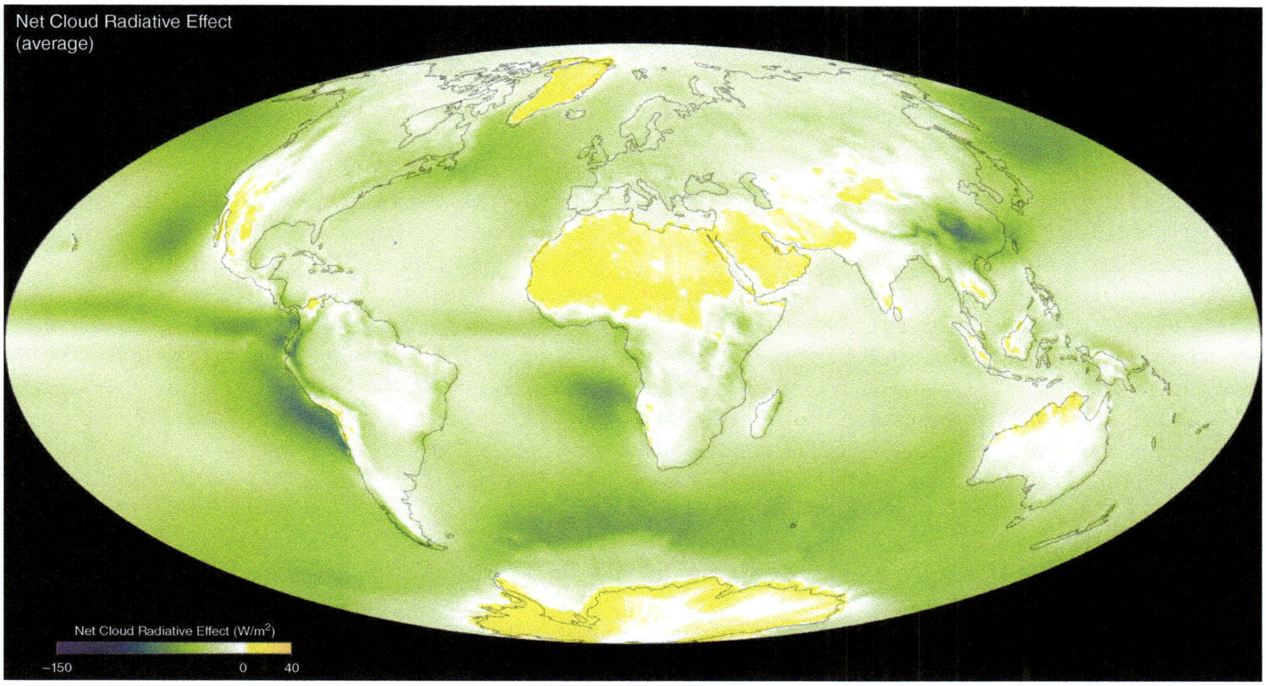

Figure 6.7 Net radiative cloud effect from CERES, 2000–2015, from the NASA Hyperwall. Source: https://svs.gsfc.nasa.gov/30603.

could be taken regularly to understand how MBL inversion heights varied as the California coastal stratus formed and dissipated (e.g., Neiburger, 1944). These and other observations along with theoretical considerations gradually shed light on the formation and maintenance of low stratiform clouds. As knowledge of these clouds and instrumentation sophistication increased, so did the questions about the stratus role in climate, leading to coordinated aircraft, satellite, and surface-based expeditions beginning in the late twentieth century, as well as isolated single-aircraft or surface-based studies.

Examples of the early comprehensive expeditions are the 1987 First ISCCP Experiment (FIRE-87; Albrecht et al., 1988), the 1992 Atlantic Stratocumulus Transition Experiment (ASTEX; Albrecht et al., 1995), and the 1998 FIRE Arctic Clouds Experiment (FIRE-ACE; Curry et al., 2000). FIRE-87 was anchored to San Nicolas Island west of Los Angeles, CA, where a surface site deployed standard meteorological and research instruments including microwave radiometers (for LWP), acoustic sounders, and a UHF wind profiler (Albrecht et al., 1988). Five aircraft were employed to measure standard meteorological parameters as well as spectral, SW, and LW radiation, droplet distributions, chemistry, aerosols, and turbulence using a variety of instruments including an airborne lidar. While the study of the east-west transition of stratus to stratocumulus to cumulus between California and Hawaii corresponding to the CF_L gradient in Figure 6.4a was the initial primary focus, only a few cases of cumulus were encountered. Yet, many other stratocumulus processes were examined. A better opportunity to study the transition was found during ASTEX, which was centered on the surface sites at Santo Porto and Santa Maria in the Azores Islands. The array of instruments was similar to those used in FIRE 87, but 8.8-mm and 3-mm radars were also deployed to remotely profile the clouds. ASTEX also had dedicated ship platforms for several instruments. As summarized by Randall et al. (1996a), both of these subtropical stratus experiments led to the development of new parameterizations that improve the representation of stratus in GCMs, new theories about the role of cloud-top entrainment instability, improved stratus remote sensing, and many advances in our understanding of marine stratocumulus.

Those two experiments also influenced the development of long-term comprehensive surface-based measurement systems by the Atmospheric Radiation Measurement (ARM) Program (Ackerman & Stokes, 2003; Stokes & Schwartz, 1994). ARM was developed to continuously monitor the radiation and atmospheric components at different climate regimes with a great variety of instruments based at semipermanent sites. These sites and their surface measurements were to be complemented by focused field programs and deployment of similar instruments at temporary sites to focus on a different climate area. Thus, stratus clouds around the globe have been monitored for several decades with a host of surface, airborne, and satellite measurement systems. For example, Dong et al. (2006) found that low cloud cover measured with uplooking cloud radar and lidar at the Southern Great Plains (SGP) ARM site in Oklahoma is consistent with the climatology of the same cloud type based on earlier surface observer data from Warren et al. (1986). However, if only single-layer low clouds were considered, the Dong et al. (2006) climatology is consistent with the satellite-retrieved low clouds. This complementarity between the objective surface and satellite measurements has been exploited in numerous studies of stratiform and other cloud types.

Among other locales, semipermanent ARM sites were established on the north slope of Alaska (NSA) in 1997 at Barrow, AK, to study Arctic clouds (Stamnes et al., 1999) and in the eastern North Atlantic (ENA; Wood et al., 2015) at Graciosa Island in the Azores to study marine stratocumulus clouds. Field campaigns employing aircraft have been conducted over the semipermanent ARM sites to gain additional insight into the clouds of the region. Over the Arctic Ocean, these include ships embedded in the polar ice cap during the 1997–1998 Surface Heat Budget of the Arctic Ocean project (SHEBA; Intrieri et al., 2002) and the 2019–2020 Multidisciplinary Drifting Observatory for the Study of Arctic Climate (MOSAIC; Shupe et al., 2018). FIRE ACE was conducted to complement the SHEBA surface data with aircraft and satellite measurements. The 2004 Mixed-Phase Arctic Cloud Experiment (MPACE; Verlinde et al., 2007) used in situ measurements over the NSA to examine Arctic stratus clouds occurring during autumn. From the ground-based observations, Shupe et al. (2008b) and Qiu et al. (2015) found that those clouds have a relatively deep layer of liquid water on top and mainly precipitate in ice form although liquid water is the dominant component of the cloud. It was suspected that the concentration of ice nuclei was low because of the limited ice fraction in the clouds. Over the ENA, the 2009–2010 Clouds, Aerosols, and Precipitation in the Marine Boundary Layer (CAP-MBL; Wood et al., 2015) deployed aircraft over the surface site to study the interactions of aerosols and clouds to improve model parameterizations (e.g., Liu et al., 2016). Additional aircraft measurements were taken during the 2017–2018 Aerosols and Clouds Experiment over the ENA (ACE-ENA) to further study cloud microphysical properties and validate surface remote sensing of those properties (e.g., Wu et al., 2020a).

The single-layered low cloud macrophysical and microphysical properties at the ARM ENA and SGP sites are plotted in Figures 6.8 and 6.9, respectively. The macrophysical cloud properties include cloud-base height (CBH), cloud-top height, cloud thickness (ΔH), and cloud-top and cloud-base temperatures (CTT and CBT). The monthly means of MBL CBH, CTH, and ΔH are plotted in Figure 6.8 with the annual means of 0.98, 1.57, and 0.59 km, respectively. There are essentially no strong seasonal variations in ΔH, but CBH and CTH vary slightly along with changes in their seasonal synoptic patterns. During summer, CBH and CTH are lower than during other months, which is consistent with the persistent high pressure and subsidence conditions at that time of year. The CBH and CTH during other seasons, especially during winter, are slightly higher than the annual averages due to the relatively frequent passage of low-pressure systems that make the MBL clouds deeper and thicker (Dong et al., 2014a). Unlike the MBL cloud properties, the cloud layers at the ARM SGP site are deeper and thicker during warm months (April–September) than in cold months (October–March) due to stronger solar radiation (Dong et al., 2005 and 2006). On average, the annual mean CBH at the ENA is only 50 m greater than its counterpart at the SGP, while the SGP CTH and ΔH are 250 m deeper and 300 m thicker than those at the ENA. Both CBT and CTT at the two sites increase from winter to summer. However, there are stronger seasonal variations at the SGP than those at the ENA due to the lower heat capacity of the land surface. The temperature difference between summer and winter at the SGP (~20 K) is almost double that at the ENA (~10 K). The CTH and CTT means at the ENA and SGP sites are consistent with the satellite retrievals as shown in Figures 6.4c and 6.5a.

Figure 6.9 shows the seasonal variation of MBL cloud microphysical properties. The MBL cloud LWP, LWC, and COD/τ from April to July are slightly higher than during other months, and monthly mean CER/r_e values are nearly constant and fluctuate within 1 μm around the annual mean of 12.0 μm. Monthly mean N_c values also fluctuate slightly around the annual mean of 80 cm^{-3}. These annual and monthly means represent typical MBL cloud CER and N_c values, and are close to the aircraft in situ measurements summarized in Miles et al. (2000) and the results derived from the recent field campaign during the Aerosol and Cloud Experiments in the Eastern North Atlantic (ACE-ENA) (Wu et al., 2020a).

The monthly mean values of LWP, LWC, N_c, and COD at the ARM SGP site during warmer months (May–August) are less than those during cooler months, unlike the more invariant ENA properties. However, there is no apparent monthly pattern in the mean CER. The annual mean CER and N_c of 8.9 μm and 244 cm^{-3} at the SGP site represent typical continental stratus cloud microphysical properties. Compared to the ENA cloud properties, the SGP CER is about 3 μm smaller, but N_c is three times greater than the ENA N_c. With nearly the same cloud LWC and LWP and atmospheric precipitable water vapor (2.47 cm at SGP vs. 2.33 cm at ENA), the

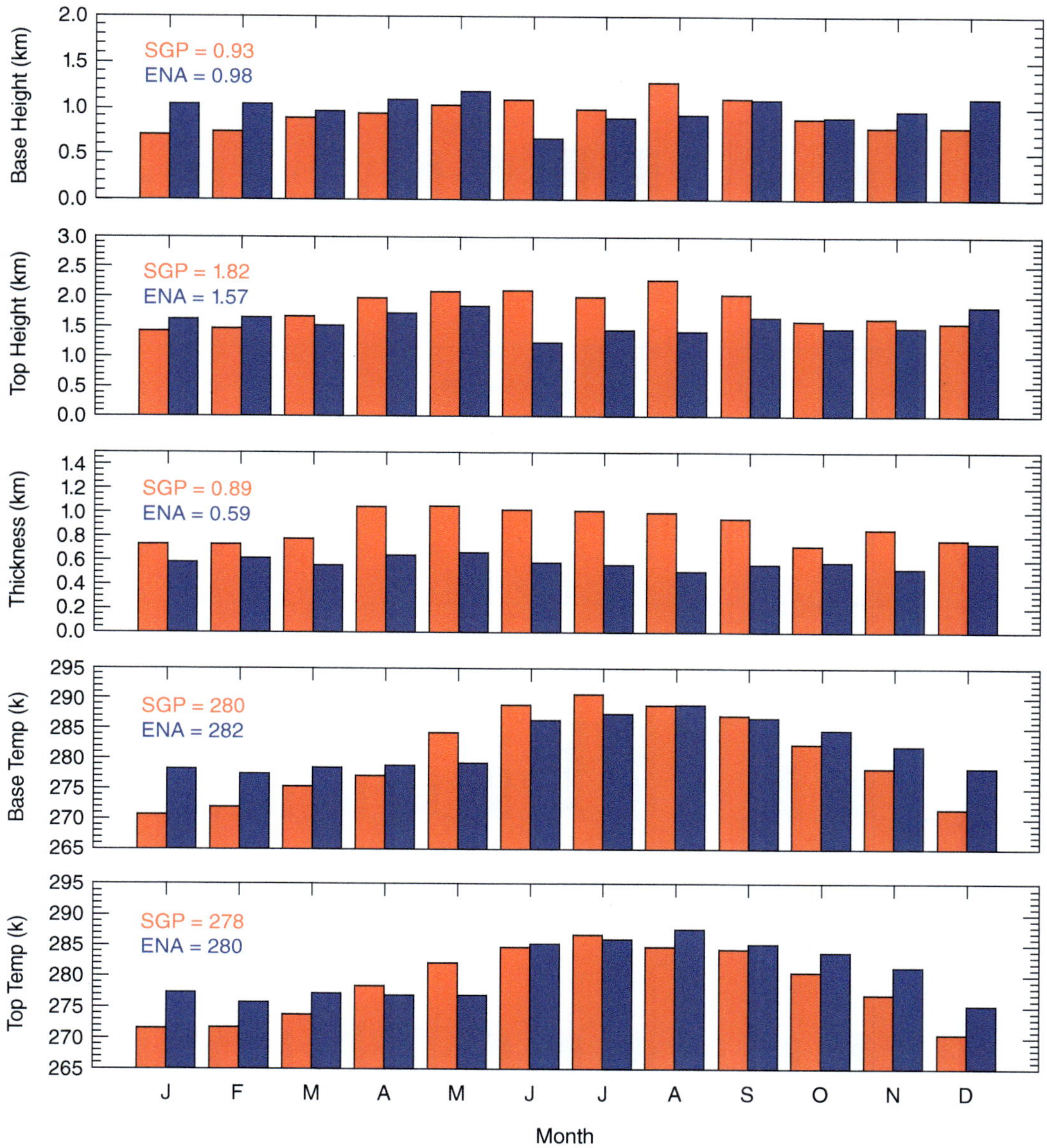

Figure 6.8 Monthly means of single-layered stratus cloud (a) base height CBH, (b) top height CTH, (c) thickness ΔH, (d) base temperature CBT, and (e) top temperature CTT at the ARM SGP site during the period 1997–2006 and the ENA site during the periods June 2009 to December 2010 and October 2015 to February 2017. Source: Adapted from Dong et al. (2005, 2014a).

significant differences in r_e and N_c at two sites primarily arise from different cloud condensation nuclei concentrations (N_{CCN}, 608 cm^{-3} cm at SGP vs. 212 cm^{-3} at ENA). The CER values at both the ENA and SGP sites are about 2–3 μm smaller than the satellite retrievals (Figure 6.5c), the LWP and COD at SGP are close to, but those at ENA are larger than the satellite retrievals (Figure 6.5b and d). More comprehensive comparisons and detailed discussions of stratus cloud macrophysical and microphysical properties, their corresponding SW and LW fluxes and CREs at the ARM ENA and SGP sites are found in Dong et al. (2023).

Suites of ARM instruments (ARM Mobile Facilities or AMF and ARM Aircraft Facilities or AAF) have also been deployed in other areas for limited time periods often in conjunction with field programs conducted by the ARM Program or under other auspices. Several deployments addressed low stratiform clouds specifically. During the multiagency sponsored 2008 VAMOS Ocean-Cloud-Atmosphere-Land Study Regional Experiment (VOCALS-Rex; Wood et al.,

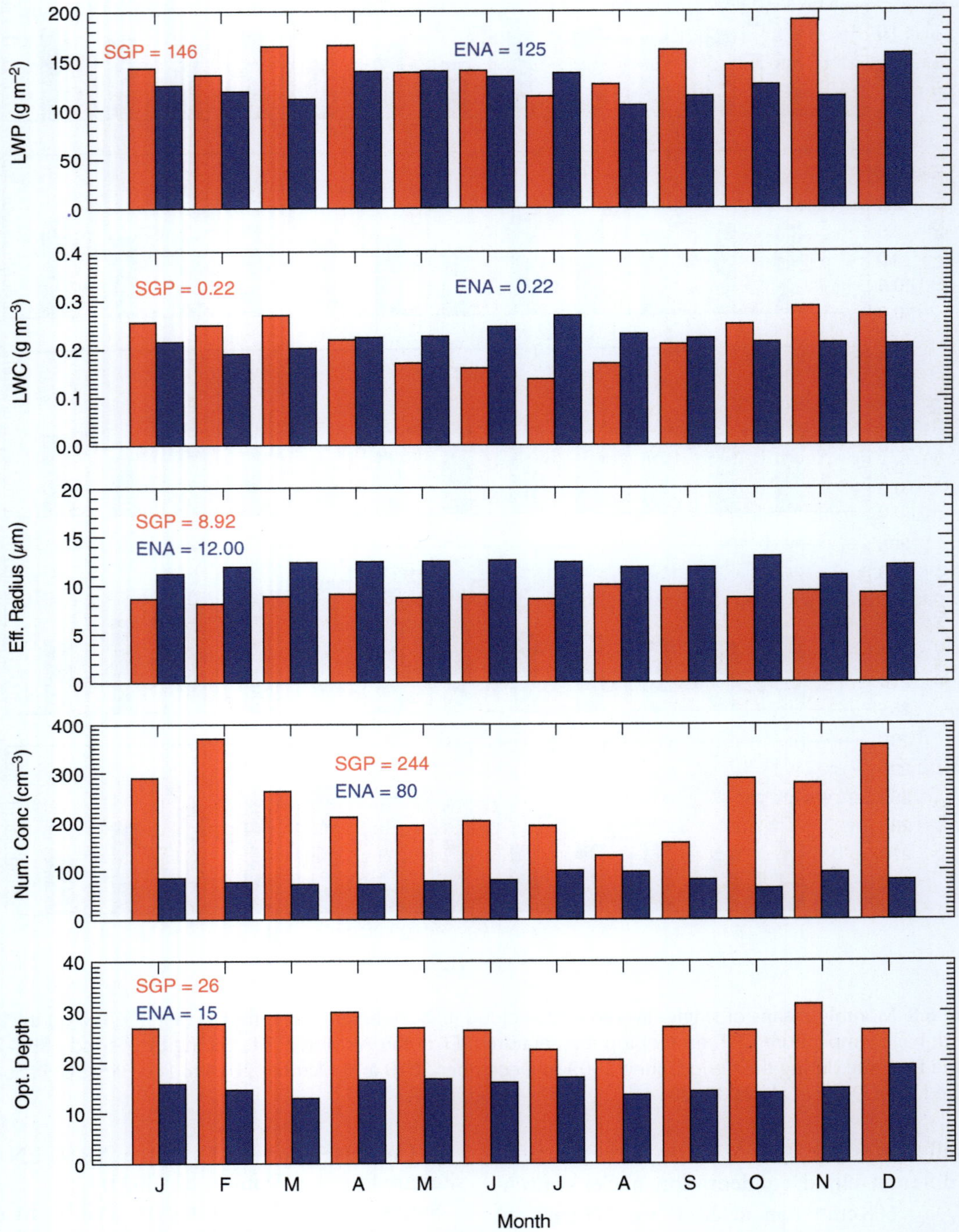

Figure 6.9 Same as Figure 6.8, except for single-layered stratus clouds (a) LWP, (b) LWC, (c) layer-mean cloud-droplet effective radius CER/r_e, (d) number concentration N_c, and (f) optical depth COD/τ. The monthly means include both day and night observations and retrievals at the ARM SGP site during the period 1997–2006 and the ENA site during the periods June 2009 to December 2010 and October 2015 to February 2017. Source: Adapted from Dong et al. (2005, 2014a).

2011a) in the southeastern Pacific, the AAF provided in situ cloud microphysical data in marine stratocumulus. Two other aircraft carried several remote-sensing instruments and took additional in situ data. They overflew a stationary instrumented buoy and the Research Vessel Ronald Brown.

Other marine stratus regions have also been intensively studied. The 2012–2013 Marine ARM GPCI Investigation of Clouds (MAGIC; Lewis & Teixeira, 2015) sited the AMF on a commercial ship that provided a wealth of information looking at the stratocumulus to cumulus transition (SCT) between California and Hawaii. The 2015 National Science Foundation Cloud System Evolution in the Trades (CSET; Albrecht et al., 2019) experiment returned to the northeast Pacific using aircraft flying between California and Hawaii to further study the transition area. The 2016–2017 Layered Atlantic Smoke Interactions with Clouds (LASIC; Zuidema et al., 2016) centered on Ascension Island focused on the impact of biomass burning in southern Africa on the southeastern Atlantic stratus systems. Midlatitude stratus systems were examined in detail with the AMF on a research ship sailing between Hobart, Tasmania, and Antarctica and Macquarie Island. The MBL aerosol, radiation, and clouds over the Southern Ocean have been measured by a few more recent ship-based campaigns (The 2017–2018 Measurements of Aerosols, Radiation, and Clouds over the Southern Ocean (MARCUS) and Clouds Aerosols Precipitation Radiation and atmospheric Composition over the Southern Ocean (CAPRICORN) as well as an aircraft-based field campaign called the Southern Ocean Clouds Radiation Aerosol Transport Experimental Study (SOCRATES; McFarquhar et al., 2021). To study the stratocumulus forming as cold continental air pours over much warmer waters, the 2019–2020 Cold-air Outbreaks in the Marine Boundary Layer Experiment (COMBLE; Geerts, 2019) deployed the AMF to a site on Bear Island in the Norwegian Sea, and aircraft conducted a field campaign off the US east coast (Aerosol Cloud meTeorology Interactions oVer the western ATlantic Experiment; Sorooshian et al., 2019).

The measurement systems and deployments summarized above represent just a sampling of the many experiments conducted in the field to better understand boundary-layer stratiform cloudiness. Nearly all of the major stratus regimes have been investigated to some degree, although empirical data over southeastern China appear to be somewhat limited. Nevertheless, these field measurements, both remote and in situ, have markedly advanced the efforts to model the boundary layer and thereby improve stratus representation in climate models. While it is not possible to review the results from all of these experiments, some findings will be included in the following sections to further illuminate the complex nature of this critical cloud type. Moreover, many of the research results reported in the following section are based on the many remote and in situ observations discussed above.

6.3. FORMATION AND MAINTENANCE, AND CLOUD MORPHOLOGY

Development and dissipation of boundary-layer stratus clouds are governed by many variables that ultimately determine their climatic impact (e.g., Albrecht et al., 1995; Dong et al., 2005, 2014a; Lilly, 1968; Rémillard et al.,

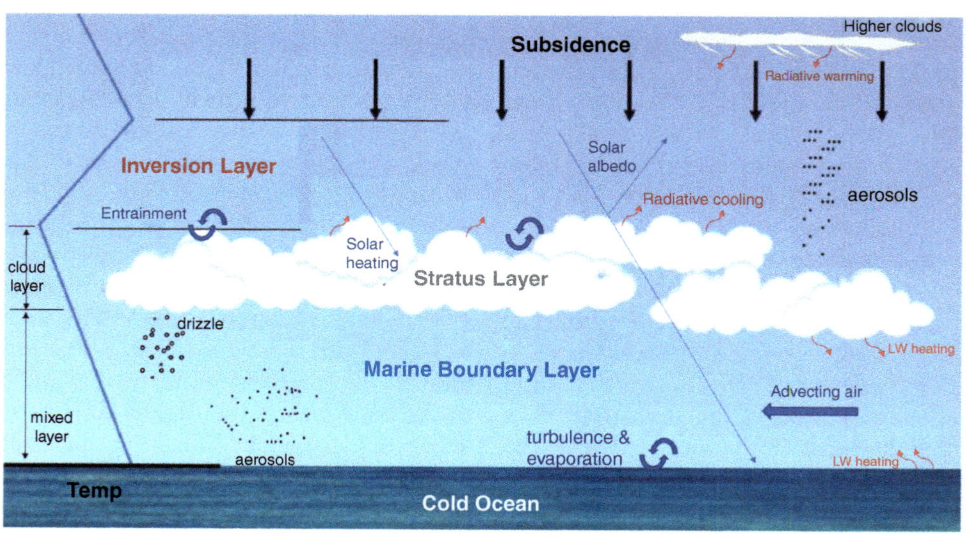

Figure 6.10 Schematic diagram of variables affecting marine boundary-layer (MBL) clouds and their properties. Source: Courtesy of Francine Dong.

2012; Wood, 2012; Wood et al., 2015). Many of those variables are schematically represented in Figure 6.10 for MBL clouds, but they also apply to stratus over land and ice. The roles of these parameters and the processes governing stratus and stratocumulus clouds are briefly discussed in this section. Further details and discussion of the processes and parameterizations used to represent them in models can be found in other chapters (e.g., Ghate, 2023; Larson, 2023) and in the references herein.

6.3.1. Formation and Maintenance

As noted earlier, a strong temperature inversion above the planetary boundary layer (PBL) is essential for capping cloud vertical development to produce a stratiform cloud. Sources of moisture, either advected horizontally, produced locally by evaporation or both, and a lapse rate conducive to upward mixing are essential to cloud formation. Turbulence and evaporation at the surface can initiate upward mixing of the moist air. Among the aerosols in the boundary layer, cloud condensation nuclei (CCN) are needed to form cloud droplets as the moist parcels reach the vicinity of the lifting condensation level. Release of latent heat increases the buoyancy adding more momentum to the upward rising parcel. In such conditions, the cloud continues growing until it reaches the inversion, where its buoyancy is matched by the subsiding air. Depending on the strength of the inversion and the vigor of the rising air, some parcels can briefly penetrate the inversion forcing or entraining some warmer, drier air down into the PBL. Together with the degree of supersaturation and updraft speed, the CCN concentration will govern the number of droplets formed near cloud base, which ultimately determines the cloud-droplet number concentration (N_c) and effective radius (CER). N_c and CER not only affect the radiation budget but also impact the rate of drizzle (e.g., Wood, 2005b). Virga, drizzle drops that evaporate after falling out of the cloud base, provide another water vapor source for MBL clouds (Wu et al., 2020b). Virga can destabilize the subcloud layer and promote penetrating cumulus development that can sustain the stratocumulus layer (e.g., Wang & Lenschow, 1995). Conversely, drizzle drops reaching the surface can cool the entire subcloud column resulting in stabilization of the boundary layer, cutting off the supply of moisture to the cloud layer (e.g., Paluch & Lenschow, 1991).

Once a cloud layer forms, turbulence generated by the strong LW radiative cooling at the cloud top plays a key role in maintaining the cloud layer. The MBL stratocumulus clouds are turbulently mixed downward from the top due to negative buoyancy through a combination of LW radiative cooling and evaporative cooling at the cloud top (Wood, 2012). The net cooling at the top depends on the humidity of the overlying atmosphere and the presence of overlying clouds. Increases in either, but especially higher clouds, reduce the net cooling at cloud top and, hence, the turbulent mixing (e.g., Shupe et al., 2013). Free tropospheric air entrained into the PBL can incorporate additional aerosols, if present, into the boundary layer, further affecting the droplet distribution by reducing CER (e.g., Brioude et al., 2009; Painemal et al., 2014). Some of the cloud-top air is also detrained into the overlying atmosphere. Entrainment rates tend to increase with decreasing inversion strength (e.g., Gerber et al., 2013). During the day, solar radiation is absorbed in and below the cloud layer increasing the turbulence within the cloud layer itself while counteracting the sinking due to LW cooling. Thus, mixing in the lower part of the MBL is reduced, in effect, cutting off the cloud from the mixing at the surface and decoupling the cloud layer and the moisture-laden lower boundary layer (e.g., Randall et al., 1996a). Any small amount of cloud-top entrainment will then dry the layer initiating thinning and breakup of the clouds, if no additional moisture is provided from below. When solar radiation diminishes during the late afternoon and evening, the cloud layer can gradually regenerate as LW cooling again predominates. Thus, both drizzle and solar heating can contribute to the decoupling of the boundary layer, albeit the heating only occurs during daytime.

In more general terms, if the stratocumulus topped boundary layer (STBL) is well mixed with moisture transported from the surface to the top of the cloud layer and capped by a strong temperature and humidity inversion just above the cloud layer, this layer is normally defined as a coupled boundary layer. When the STBL deepens, it begins to separate into two layers with the upper layer becoming decoupled from the surface moisture supply (Wood, 2012). The potential temperature method has been widely used to differentiate between coupled and decoupled MBL stratocumulus (Jones et al., 2011). Figure 6.11 shows two coupled cases and one decoupled case. Typically, for the coupled cases, the turbulence is strong enough to mix the boundary layer so that liquid water potential temperature and total water mixing ratio are relatively constant vertically from the surface to the MBL cloud base. For the decoupled case, both potential temperature and water vapor mixing ratios vary with height. Jones et al. (2011) found that decoupling, as indicated by both drizzle intensity and humidity profiles, is strongly correlated with the theoretical mixed layer cloud thickness, which is the difference between the inversion height and the LCL. Thus, the thicker the mixed cloud layer, the more likely the MBL will be decoupled.

The planetary boundary layer, the general form of MBL, deepens westward as the surface temperature rises and the subsidence ebbs. Decoupling generally occurs in areas 400 or 500 km from the coast, and cumuli develop

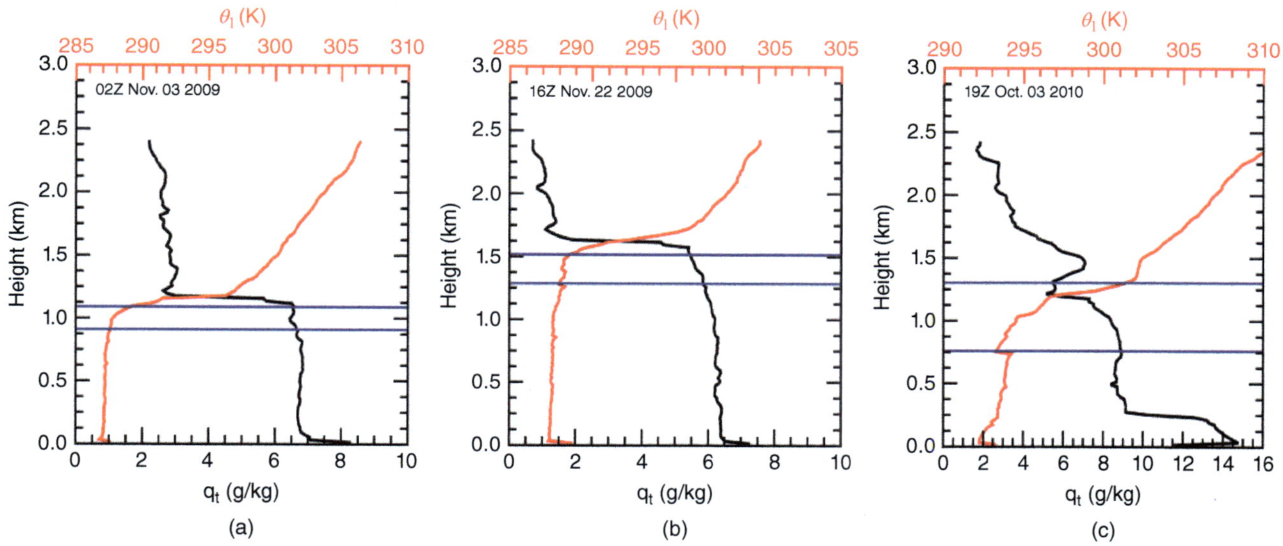

Figure 6.11 Liquid water potential temperature (θ_L, red lines) and total water mixing ratio (q_t, black lines) profiles from the ARM merged soundings for well-mixed boundary layers on (a) 02Z 3 November 2009 and (b) 16Z 22 November 2009, and (c) a decoupled case on 19Z 3 October 2010 at the ARM Eastern North Atlantic (ENA) site. Source: Dong et al. (2015)/John Wiley & Sons.

beneath the Sc and form mesoscale cellular cloud complexes as the PBL deepens (Brendeche et al., 2021). Deeper cumulus clouds farther from the coast leave distinctive thin veil clouds at the inversion base (e.g., Wood et al., 2018). Approaching Hawaii, stratiform clouds are absent, replaced with less vigorous trade cumuli that do not reach the top of the PBL. Among the other factors that can affect the subtropical marine stratus formation and dissipation are the orography of the surrounding continents (Richter & Mechoso, 2006) and gravity waves generated by disturbances far from the Sc decks (Allen et al., 2012).

The cold tongue stratus in the equatorial convergence zones appears to form in a slightly different manner than the subtropical marine Sc discussed above. As shown in Figure 6.4a, the Pacific equatorial Sc is a separate feature resulting from inflow to the intertropical convergence zone (ITCZ). The air moving from the south flows from the subtropical marine stratus transition zone over an east-to-west tongue of cold water, where the cloud top (e.g., Figure 6.4c) and base (Pyatt et al., 2005) heights decrease. The cloud cover over the cold tongue is highly variable but generally increases to the north along with the cloud height as the air passes over warmer waters before entering into the deep convection parts of the east Pacific ITCZ. The ~6 °C surface temperature gradient occurs over a distance of ~50 km compared to ~1800 km as indicated in Figure 6.12 resulting in rapid destabilization and increased wind speeds (Pyatt et al., 2005). Deepening of the low clouds over the warmer water results from an increase of moisture throughout the boundary layer. In contrast to the subtropical stratus, the MBL is often decoupled over the cold tongue and coupled over the warmer water. For the latter, this produces thicker clouds with higher tops and bases and greater LWP than the surrounding low clouds (e.g., Figure 6.5d).

Another variant of this decoupling in warm air advection occurs in the warm sectors of marine cyclones or ahead of cold fronts. Data taken during MARCUS over the Southern Ocean revealed that air flowing from warm to colder water produced a highly stratified layer with a relatively thick cloud layer on top below the STBL inversion (Y. Zheng et al., 2020). Unlike the typical subtropical Sc generated by low-level cold air advection, the warm air advection suppresses the development of cumulus under the stratiform deck. Utilizing various campaign results, Y. Zheng et al. (2020) provide an excellent summary of the impact of low-level temperature advection on the STBL.

In a study utilizing a numerical weather prediction model and observations from the ARM ENA site, where midlatitude fronts meet subtropical stratocumulus, kezemirad and Miller (2020) concluded that the post-frontal stratocumulus developed in a manner consistent with the deepening-warming process observed in the subtropics. The overall correlation is best for wintertime and the Southern Hemisphere (SH). During summer, the correlation is very weak over the Northern Hemisphere (NH), consistent with the results of kezemirad and Miller (2020), who found an alternative parameter to predict LCF: the vertical gradient of the difference between the saturated equivalent potential temperature and the equivalent potential temperature.

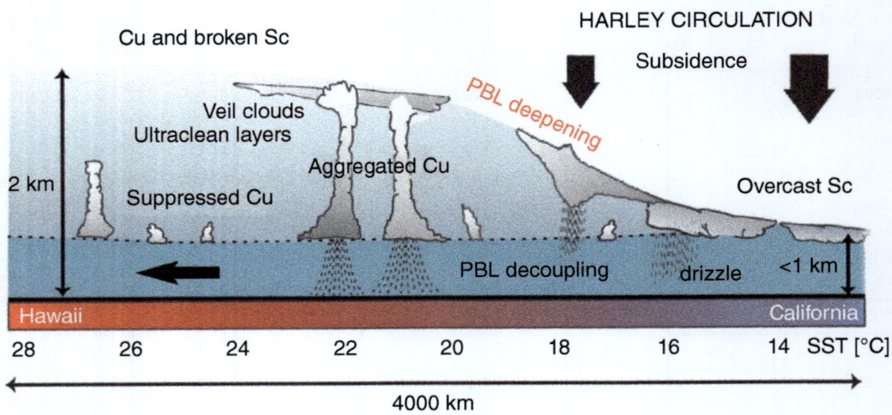

Figure 6.12 Schematic showing cloud and boundary-layer evolution from California to Hawaii from the CSET perspective. Source: Albrecht et al. (2019)/American Meteorological Society.

Although MBL stratus is usually a very long-lived, but variable feature of the subtropics, the formation and dissipation processes of MBL clouds are also mostly applicable to more short-lived continental low clouds with some notable differences (e.g., Del Genio & Wolf, 2000; Mechem et al., 2010). For MBL clouds, the water vapor directly comes from the ocean surface, whereas the water vapor over land is typically supplied by advection of a moist air mass into the region or less by local evaporation when the surface is wet. In that case, solar absorption at the land surface, especially for moist surfaces produced by preceding frontal precipitation, can easily be converted into sensible and latent heat providing moisture directly to the cloud layer; thus, continental low stratiform clouds are often surface-driven and mixed from the bottom-up (Dong et al., 2005). At night, wind shear near the land surface is critical for providing the turbulence necessary to initiate cloud formation, while the LW cooling maintains the cloud after formation (Zhu et al., 2001). During the day, the conversion of absorbed solar radiation to latent and sensible heat is much faster over land than over water. More sunlight is absorbed by a water body because of its lower albedo, but the heating occurs through a much greater depth than over land. Thus, both the cloud and surface absorb sunlight and generate local turbulence altering both the cloud and mixed layer. Opposite to more stratus during night and less during afternoon over the sea (e.g., Dong et al., 2014a, 2014b), stratus over the ARM SGP site increases monotonically from morning to local noon and then declines until sunset, primarily following the absorbed solar radiation at the surface. This sequence is more obvious during summer than during winter (e.g., Dong et al., 2005). Therefore, a stronger diurnal variation in stratus over land is found than over ocean, which is at least true at the ARM SGP site. Continental stratus clouds are also subject to dynamic effects not found over ocean. Stratus can develop over land due to orographic uplift that causes upslope winds. An extreme case is the area of stratus over China noted earlier, where the Tibetan Plateau forces large-scale lifting at low levels, while the Siberian high feeds low-level cold air converging with slightly warmer moist air from the Bay of Bengal that results in stability and upslope cloud formation (Y. Zhang et al., 2013).

These formation and maintenance mechanisms are mainly suitable for explaining the low clouds over the subtropical and midlatitude regions, and possibly for summertime Arctic mixed-phase clouds (AMCs) over open water. However, most of the low clouds over the Arctic regions are formed over snow or ice surfaces where there is minimal surface moisture contribution to the Arctic atmosphere. Leads in the ice surface and meltwater pools on the ice during late summer provide the only local sources of humidity. AMCs are the dominant low clouds and can persist for hours to days, and sometimes even weeks (Qiu et al., 2015, 2018; Shupe et al., 2006, 2008b, 2011). It is apparent that the formation, maintenance, and dissipation processes of Arctic low clouds significantly differ from what typically occurs in the midlatitudes. Tsay and Jayaweera (1984) found that Arctic low clouds are normally formed near the sea-ice surface if warm moist air advects over the cold Arctic Ocean or by a convective-type process if cold polar air flows over a warmer sea-ice surface. Once Arctic low clouds form, they persist (Herman & Goody, 1976) because the dissipative mechanisms found in the midlatitudes such as precipitation, convective heating from the surface, absorption of solar radiation, and destruction by synoptic activity are either nonexistent or relatively weak (Dong & Mace, 2003a). This is the significant difference between the Arctic and midlatitude low clouds. Arctic low clouds persist in a near steady-state situation where dissipative processes are balanced by advection of air masses into a region suitable for cloud formation and maintenance.

Using ship-based remote-sensing data taken during 22 days in late summer at 87°N, Shupe et al. (2013) found that 75% of the observed stratus clouds were decoupled from the surface. The decoupling did not result from the surface driven convection but rather from warm moist layers advecting over the surface air. Aerosols necessary for droplet and ice crystal initiation were also brought in with those layers, whereas CCNs can develop locally from turbulence at the surface of open water or lifted from a land surface. When the clouds were coupled with the surface, the average LWP was generally higher; the median LWP increased from ~55 to ~75 g m^{-2}. The average IWP was around 5–7 g m^{-2} for both coupled and decoupled cases. Similar to the Shupe et al. (2013) observations, but based on ARM NSA measurements, Qiu et al. (2015) found that thin layers of increased humidity above cloud top provide moisture for forming and maintaining AMCs. This source of moisture balances the loss of water from ice precipitation, helping to reconcile the persistence of AMCs even when the surface is covered by snow and ice and decoupled from the AMC layer (Sedlar & Tjernström, 2009; Solomon et al., 2011). Morrison et al. (2012) summarized the physical processes related to the resilience of persistent AMCs and explained the counterintuitive persistence of AMCs despite microphysical instability arising from the Wegener-Bergeron-Findeisen mechanism. The liquid layer at the AMC top generates strong LW radiative cooling that can decrease the static stability inside the cloud and generate turbulence, facilitating droplet condensational growth. These processes consist of a self-maintaining feedback pathway between liquid phase, radiation, and turbulence.

6.3.2. Cloud Morphology: MCC, Closed and Open Cells

From the above discussion, it is clear that convective processes play a signature role in stratocumulus formation. Because of the capping inversion, vertical development is constrained to a relatively small depth (0.5–3.0 km) of the boundary layer. As turbulence tends to exhibit horizontal structure on roughly the same scale as the vertical, most of the vertical motion energy or cellularity occurs at that scale. Despite this characteristic scale, the stratocumulus convection frequently, especially for marine Sc, organizes at the mesoscale (a few kilometers up to tens of kilometers). This often results in the 3D mesoscale cellular convection (MCC; e.g., Wood, 2012) noted earlier. Three-dimensional cellular convection generally occurs in light wind conditions. As wind speed in the PBL increases (e.g., >5 m s^{-1} throughout the PBL), roll vortices often develop across the wind direction, producing cloud streets when the heat flux is sufficient (e.g., Weckwerth et al., 1997). Figure 6.13 shows examples of these two types of organizations. Different types of 3D cellular organization are prominent over the subtropical SE Pacific and in the postfrontal Sc in the GOES-16 image (Figure 6.13a). Cloud streets are readily apparent off the east coast of the USA in the S-NPP VIIRS image in Figure 6.13b. These streets form when the cold postfrontal air passes over the warm waters. Naud et al. (2020) showed that a cold-air outbreak is a strong predictor of cloud cover and height in the extra tropics. This relationship is also confirmed for winter stratiform clouds over the Gulf Stream by Painemal et al. (2021).

Wood and Hartmann (2006) used a neural network technique based on visual inspection to classify marine stratocumulus clouds into four general mesoscale morphological types: (i) no cellularity at the mesoscale (rare), (ii) organized closed mesoscale cellular convection (MCC) (common), (iii) organized open MCC (common), and (iv) unorganized mesoscale cells (rare). These types are illustrated with the scenes in Figure 6.14, which were selected from NOAA-20 VIIRS imagery to match the blues boxes in Figure 6.13a. For the most part, these four types represent different stages of an air mass transition from shallow coastal marine stratus to trade cumulus over the subtropical-tropical eastern oceans. Wood and Hartmann (2006) concluded that MCCs are extremely common over the eastern subtropical oceans, where an MCC usually transitions from closed cells (near-coastal regions) to open cells (far offshore). This transition is evinced by the presence of large areas of open cells interspersed with closed MCCs in the upper left portion of Fig. 6.13a. The open-cell area in Figure 6.14c, occurring in the midst of a large area of closed cells, is representative of pockets of open cells (POCs; Stevens et al., 2005). POCs are notable for having dramatically different microphysical characteristics (low aerosol and droplet concentrations) from the surrounding closed cells and for having much heavier drizzle rates (e.g., Sharon et al., 2006). Drizzle intensity is greatest at the open-closed cell interface and is credited for removing both aerosol and liquid water from the MBL (e.g., Comstock et al., 2007), resulting in ultraclean layers (UCLs) (e.g., Wood et al., 2011b, 2018).

Veil clouds often develop at the top of the MBL in areas of open-cell and unorganized cellular convection (Wood et al., 2018). These clouds, as shown schematically in Figure 6.12, are typically 200-m thick, composed of relatively large droplets and can last for several hours. Hypotheses regarding their formation include detrainment of saturated air with low droplet concentrations due to collision-coalescence in the strong updrafts of the drizzling cells and in situ formation from slowly rising

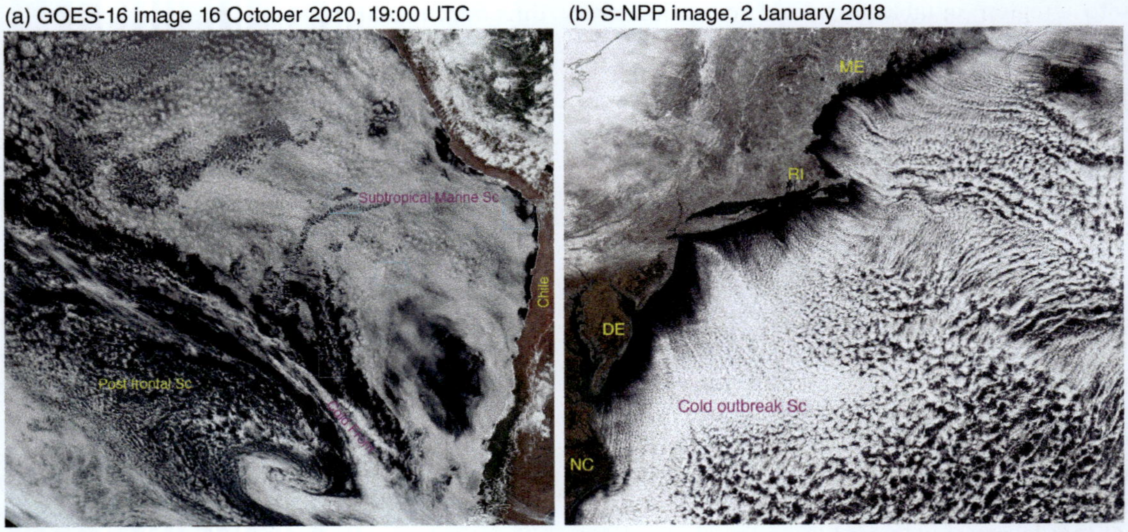

Figure 6.13 Visible satellite imagery showing stratocumulus clouds in the (a) midlatitudes and subtropics and (b) associated with a cold-air outbreak. Blue boxes in (a) refer to images in Figure 6.14. Images are from (a) the Sixteenth Geostationary Operational Environmental Satellites (GOES-16) and (b) S-NPP VIIRS: NC, North Carolina; DE, Delaware; RI, Rhode Island; ME, Maine. The scale is different in each image. Sources: (a) National Oceanic and Atmospheric Administration (NOAA); (b) NASA/https://worldview.earthdata.nasa.gov/last accessed 16, March 2023.

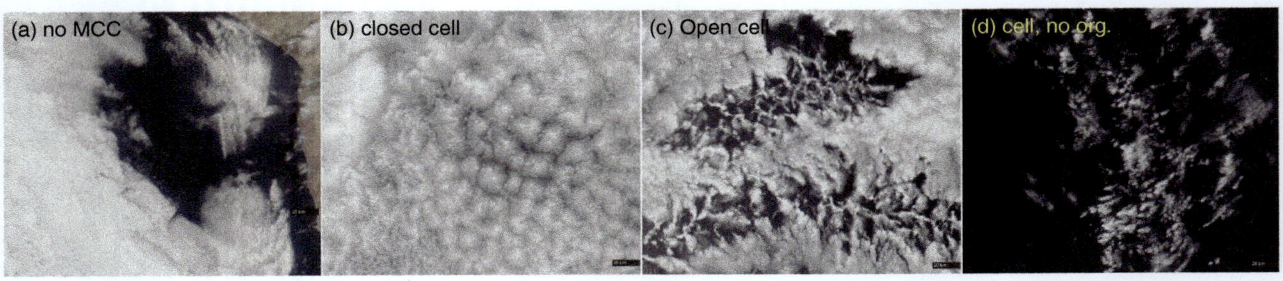

Figure 6.14 Examples of four mesoscale structure types occurring in marine stratocumulus from 16 October 2020, 375-m NOAA-20 VIIRS true color imagery corresponding to blue boxes in Figure 6.13. (a) No MCC, (b) closed MCC, (c) open MCC, and (d) cellular but disorganized scene types. Each image is at the same scale, shown by small black box indicating 20-km distance in each image. Source: NASA/https://worldview.earthdata.nasa.gov//last accessed 16, March 2023.

air in the UCLs. Figure 6.15 shows an example of veil clouds observed during CSET. The GOES-15 image (Figure 6.15a) shows the GV aircraft flight track over the open and unorganized cellular convection, while the photograph taken from the aircraft window shows veil clouds among the larger cells and above lower-level Cu (Figure 6.15c). The structure of the clouds is revealed in the backscatter cross-section plot from the nadir-view High Spatial Resolution Lidar (HSRL). Cumulus clouds are in the lower boundary layer with bases near 400 m. Laminar, optically thin veil clouds are located between about 1100 and 1900 m. The Sc top at the right is at ~1900 m and is ~400-m thick. Clear UCLs below the inversion are evident in the dark blue shade. Optically thick clouds are evinced by white shadows below cloud base. The nadir-viewing HIAPER Cloud Radar (HCR) in Figure 6.15d shows the drizzling cloud structure to the right of the location corresponding to minute 3.1 in Figure 6.15b. Scattered Cu and a few veil clouds are present between the two main drizzling elements at 30 and 90 km. The latter cell has separate outflow near cloud top and base. These heavily drizzling clouds are typical of the deeper elements of the open-cell convection.

While the Sc cloud-top heights in Figure 6.13a follow the patterns typified in Figure 6.4c, the microphysical property distribution is more interesting, especially that for the cloud-droplet effective radius CER. Figure 6.16 shows the CER and LWP distributions derived from the GOES-16 imagery (Minnis et al., 2008) corresponding to Figure 6.13a. The stratus near the Chilean coast has

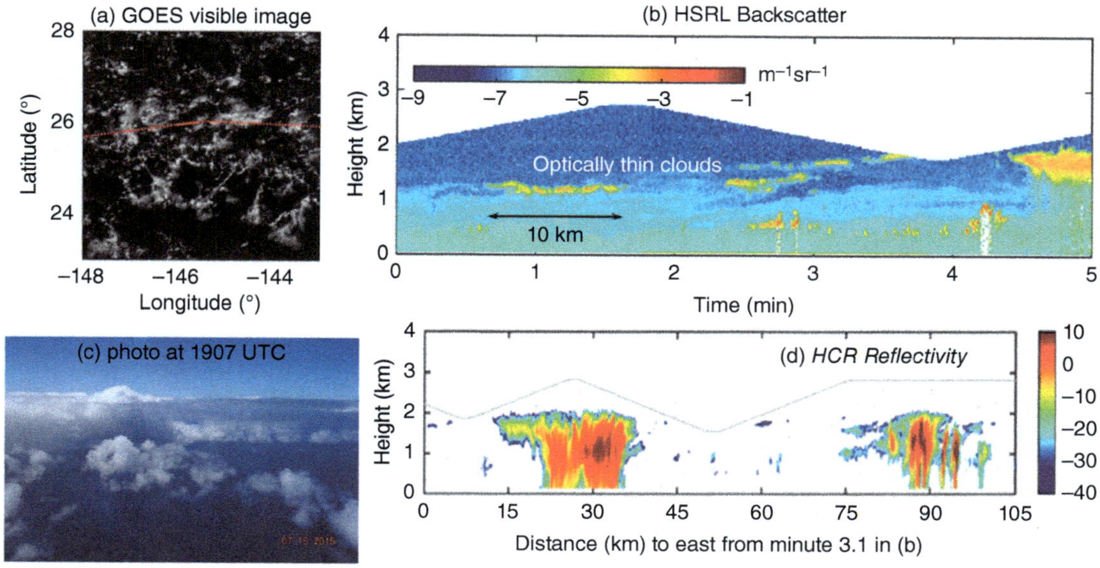

Figure 6.15 Optically thin veil clouds, ultraclean layers, and drizzling Sc sampled from the Gulfstream V, 1900–1915 UTC, 19 July 2015 during CSET. (a) GV flight track in red on 1900 UTC GOES-15 1-km visible image, (b) nadir-viewing HSRL backscatter coefficient image, (c) photo taken at 1907 UTC from the GV. (d) HIAPER Cloud Radar (HCR) reflectivity. Source: Albrecht et al. (2019)/American Meteorological Society.

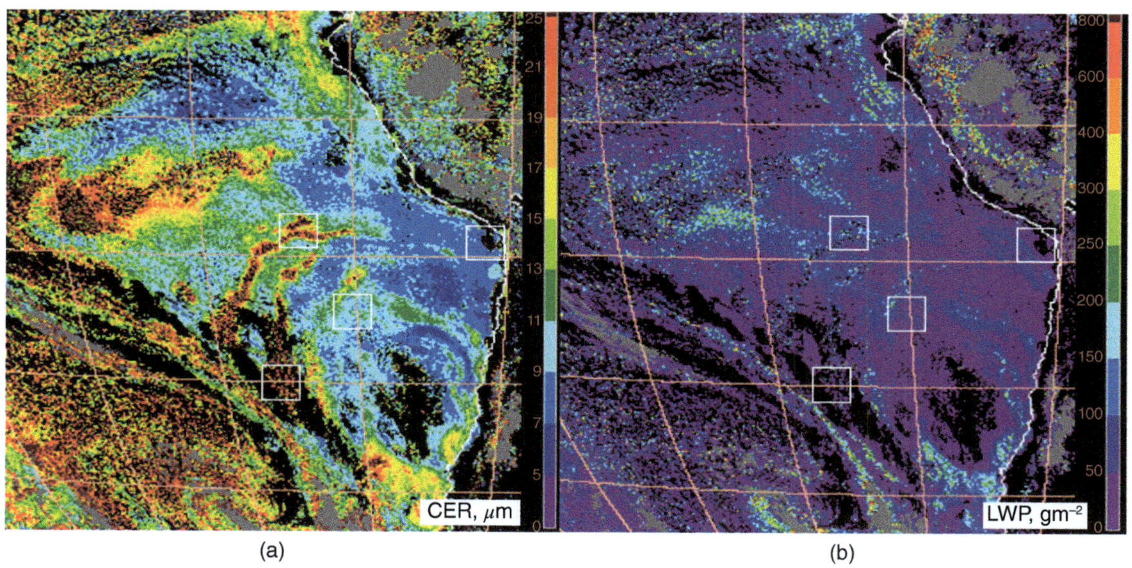

Figure 6.16 Cloud properties retrieved from GOES-16 imagery corresponding to Figure 6.13a. Taken from NASA Langley Satellite ClOud and Radiation Property retrieval System (SatCORPS). Source: NASA/https://satcorps.larc.nasa.gov//last accessed 16, March 2023.

some of smallest values of CER (<7 μm; Figure 6.16a), while LWP is mostly <50 g m^{-2}. In general, overcast areas without organized cellular structure tend to have CER < 10 μm and relatively small LWP values (Figure 6.16b). The closed cell MCCs generally have CER values of 6–12 μm, while the open and unorganized cells have large water droplets, CER > 15 μm, and low droplet number concentrations N_c, as well as drizzle. It is notable that LWP is mostly between 25 and 100 g m^{-2} over much of the area, but in many POC and POC-adjacent areas LWP rises up to 200 g m^{-2} or more. These relative maxima are due to the larger droplets and drizzle, corresponding to the presumably more intense convection in these areas.

A final example illustrating the transition from stratus to unorganized cellular cumulus is seen in the data taken

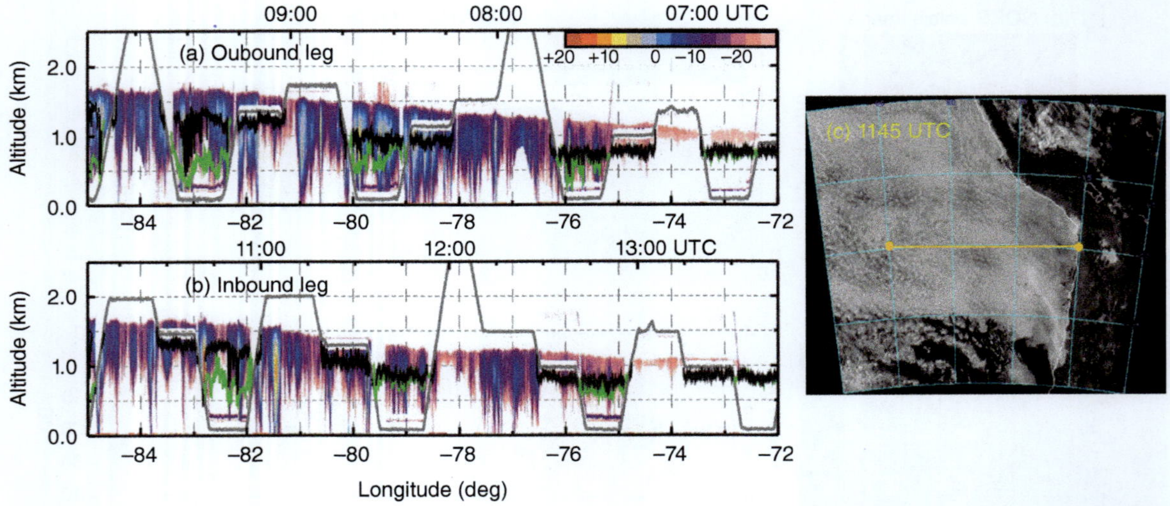

Figure 6.17 Plot of 94-GHz cloud radar reflectivity (both downward and upward-pointing radar means) along 20°S for the (a) outbound and (b) return portions of NSF C130 aircraft flight (RF03; 21 October 2008). During subcloud legs, the in situ LCL (green) and the lidar-detected cloud base (black) are superimposed. During cloud legs, the black line shows the cloud base adiabatically derived from in situ LWC_c. Gray line traces the aircraft flight track. Source: Bretherton et al. (2010)/European Geosciences Union. Large-scale cloud structure at 1145 UTC is seen in (c) GOES-10 visible image with overlaid flight track in gold. Source: NASA/https://satcorps.larc.nasa.gov//last accessed 16, March 2023.

by the University of Wyoming 94-GHz cloud radar and upward-pointing lidar aboard the NSF C130 aircraft during a flight over the southeastern Pacific during VoCALS-Rex (Figure 6.17). Those instruments measured cloud boundaries, and cloud and precipitation properties between 73°W and 85°W during an early-morning C130 mission at 20°S (indicated by the overlaid line in the GOES-10 image of Figure 6.17c). The radar reflectivity plots in Figure 6.15a and b show that the MBL deepened from ~1 km in the coastal zone to ~1.7-km far offshore. Radar reflectivity increased from roughly −15 dBZ near the shore to 0–10 dBZ ~1500-km offshore. West of 75°W, the clouds are more cellular and drizzle is common (e.g., Figure 6.17c), though much of it appears to evaporate before reaching the surface. The cloud tops are slightly higher and the drizzle more frequent and intense during the predawn outbound flight (Figure 6.17a) compared to the inbound leg that included the first two daylight hours (Figure 6.17b). Cloud-base heights appear to be higher during the inbound flight in many areas.

The postfrontal Sc in Figure 6.13b shows similar, but mostly open-celled organization, which is common in polar and midlatitude regions. Cloud streets produced by marine cold-air outbreaks (MCAOs), exemplified in Figure 6.13b, constitute a special form of postfrontal Sc because they mostly occur in air flowing off a continent or ice sheet. MCAOs are most common east of continental coasts and over open water near ice sheets. MCAOs are defined objectively as air masses having positive values of the MCAO index, which is the difference between the potential skin temperature of the water body and the potential temperature at 800 hPa. Figure 6.18 shows the occurrence frequency of positive index values in the Northern Hemisphere during winter (Figure 6.18a) and in the Southern Hemisphere during boreal summer (Figure 6.18b) based on 35 years of reanalysis data (Fletcher et al., 2016). MCAOs are most frequent over the Pacific east of Japan, where they often extend nearly 4000 km over the ocean. Other areas of frequent occurrence are east of North America over the Gulf Stream and north of Europe over the Norwegian Current. The cold Labrador Current probably reduces the frequency off Nova Scotia. During SH winter, the occurrence maxima north of Antarctica and south of Africa are less than half the magnitudes in the NH. MCAOs are negligible during the NH summer. On average, MCAOs occur more often in the SH because of the greater ocean area. More intense MCAOs account for more events in the NH than in the SH.

Although the outbreaks often cover great distances over the ocean, the cloud streets typically extend into the ocean only several hundred kilometers before transitioning to closed and/or open-celled convection as seen in the lower right of Figure 6.13b. Using the SatCORPS analysis, it was found that the values of CER (not shown) for Figure 6.13b are small (<9 μm) near the coast and increase up 11–14 μm before reaching 20 μm or greater in the open-celled area. The CTH (not shown) rose from

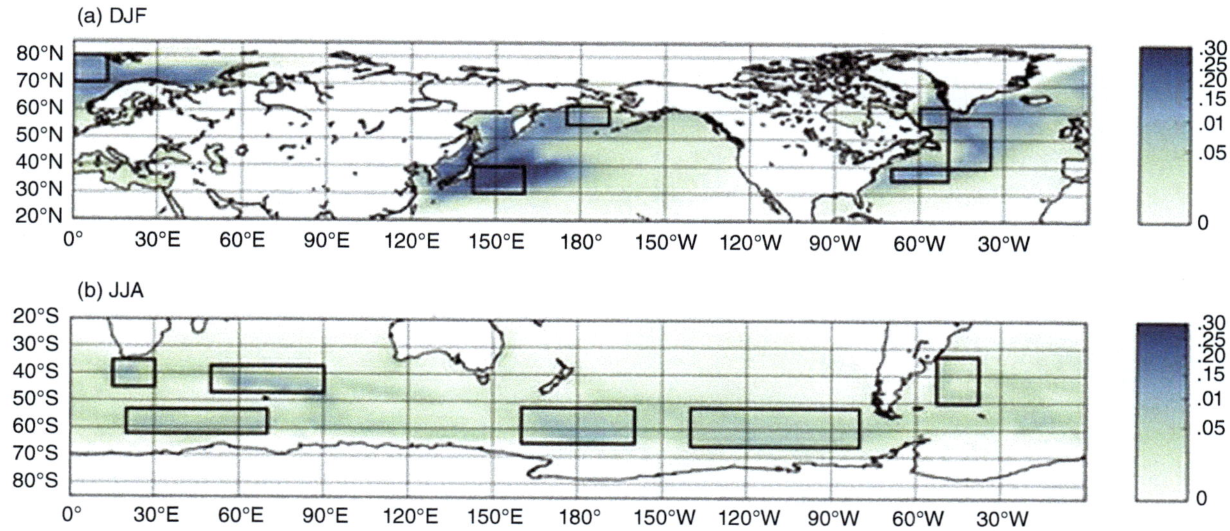

Figure 6.18 Occurrence frequency of positive MCAO index in (a) Northern Hemisphere during boreal winter and in (b) Southern Hemisphere during boreal summer. Black boxes denote areas of special interest in the original study. Source: Fletcher et al. (2016)/American Meteorological Society.

less than 1 km up to nearly 3.4 km in the closed-cell area for this case before decreasing slightly in the open-celled area. This behavior seems to be quite typical.

In a study of cold-air outbreaks off the Arctic ice near Svalbard, Hartmann et al. (1997) found that the inversion height is very low near the surface and rapidly deepens offshore, driven by the extreme air-surface temperature difference. A composite plot demonstrating cold-air outbreak (Figure 6.19) shows an example of thickening and moistening of the MBL as cold air spreads over the warm ocean. In this case, the air at the surface leaving the ice was −17 °C compared to ~3 °C at the sea surface (not shown). The yellow circles in Figure 6.19a estimate where the profiles in Figure 6.19b and c were taken. As indicated by potential temperature profiles, the MBL 3-km offshore is at ~120 m and rises to ~450 and ~ 600 m at 40- and 73-km offshore, respectively (Figure 6.19b). At 3-km offshore, the specific humidity (Figure 6.19c) is relatively low and uniform up to 1100 m. It jumps by ~60% over the ensuing 70-km fetch. The inversion gradually rose to ~1250 m at 225-km offshore, where the cloud thickness peaked at ~800 m and the roll convection appears to have begun breaking down. This change in inversion and cloud depth is consistent with the CTH variation occurring within many midlatitude cold-air outbreaks. The smaller CER near the coast would naturally arise from the reduced thickness of cloud, which limits the growth time for a given droplet. While roll vortices are common for MCAOs, they also occur over land when relatively cool moist air passes over a warmer surface (e.g., Weckwerth et al., 1997).

6.4. DIURNAL AND SEASONAL VARIATIONS

One of the most notable features of stratus and stratocumulus clouds is their tendency to vary semiregularly over the diurnal cycle. This characteristic can often be found in a given series of geostationary satellite images over subtropical Sc areas. For example, Figure 6.20 shows multichannel, pseudocolor GOES-16 images for the area between 10°S and 30°S and 70°W and 100°W. Night and day images use different combinations of channels. Early in the morning (10 UTC), the cloud deck is mostly unbroken east of 90°W, with a mix of open and closed cells to the west. As the day progresses, thin areas near the coast begin to open up (12–14 UTC). The clouds in the west start shrinking after 14 UTC, while near-coastal holes continue expanding through 16 UTC. By 18 UTC, a large POC opens near 25°S and 80°W and clouds near the coast reach their minimum coverage for the day. At 20 UTC, clear areas off the coast have begun shrinking, while open areas west of 75°W reach peak coverage before ebbing by 22 UTC. Cloud horizontal growth continues at least through 08 UTC in nearly all parts of the scene. As observations taken around the world have shown, this type of cloud cover change, though highly variable, is relatively common.

Surface observations (e.g., Sherr et al., 1968) and early satellite studies (e.g., Short & Wallace, 1980) found systematic diurnal changes in cloud cover over many parts of the globe. Simon (1977) visually analyzed three-hourly geostationary satellite imagery to estimate the average variation of stratus near California. The diurnal variability in cloud amount, heights, and top temperatures for

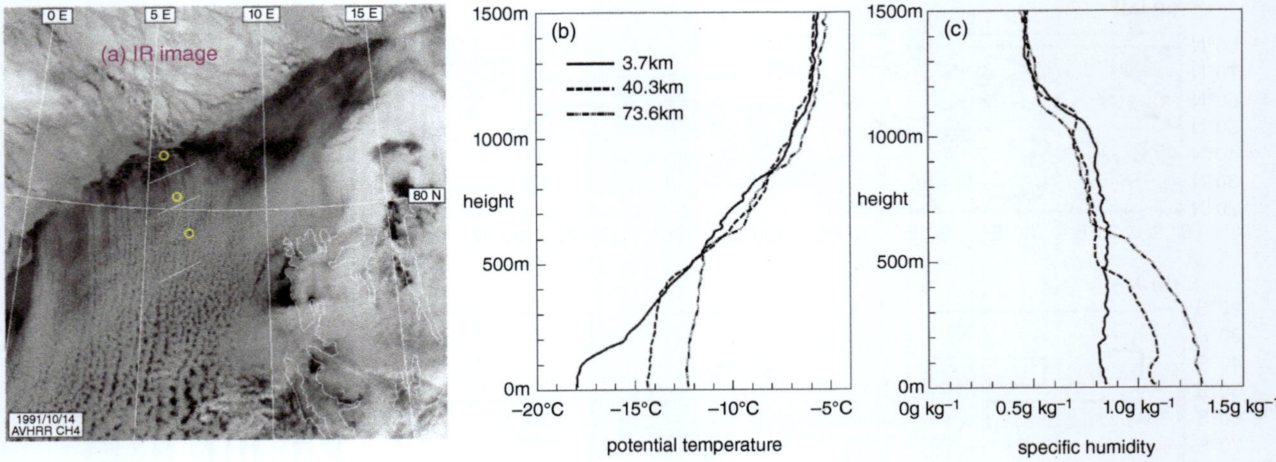

Figure 6.19 Horizontal roll convection example in a cold-air outbreak. (a) NOAA-11 AVHRR infrared (IR) image of clouds over the Fram Strait between the polar ice cap and Svalbard at 1100 UTC, 14 October 1991. Yellow circles indicate approximate locations of soundings, (b) potential temperature profiles, and (c) specific humidity profiles. Source: Hartmann et al. (1997)/Springer Nature. Distances are measured from the coast. Source: Reproduced from Jörg Hartmann et al. (1977)/Springer Nature.

much of the western hemisphere was objectively quantified by Minnis and Harrison (1984a) for one month using GOES-2 hourly infrared and visible imagery. They found that over the SE Pacific, low cloud fraction (CF_L) decreased, on average, by more than 0.30 from early morning to midafternoon before gradually increasing into the night. Mean CTT reached a minimum early in the morning (~0600 LT) and increased by ~2.5 K by ~1500 LT before diminishing again, suggesting a deepening (thinning) of the MBL during the night (day). This variation is exemplified in Figure 6.21, which shows November 1978 mean hourly CF_L over a region in the SE Pacific based on GOES-2 data. Similar, though less intense, diurnal variations were found for the Pacific equatorial Sc using GOES-6 data (Minnis et al., 1987). Over some areas, the hourly variability was found to have both diurnal and semidiurnal components with phases at ~0600 and ~2100 LT, respectively.

Development and destruction of cloud cover also imply changes in other parameters such as COD and LWP. Variations in any or all of the cloud properties can have significant impacts on the radiation budget because they alter the SW albedo and outgoing LW radiation (OLR). For example, Minnis and Harrison (1984b) found that the drop of 0.3 in CF_L from early morning to midafternoon resulted in ~14 W m^{-2} more OLR, and an ~0.17 decrease in SW albedo over the SE Pacific. The latter translates to a morning-afternoon increase of ~43 W m^{-2} in SW radiation absorbed at the surface. Consideration of cloud properties at 1 or 2 times of the day or as a daily average, without accounting for systematic changes during the day, can make a significant difference in the energy budget and how the clouds are modeled. Therefore, it is critical to both measure and model clouds and their properties over the entire daily cycle to accurately determine their role in and response to climate change.

While that early research established some basic characteristics of the stratus diurnal cycle, subsequent investigations have expanded and refined our knowledge of this important feature of stratocumulus. Using 7 years of global three-hourly ISCCP cloud data, Rozendaal et al. (1995) found that the magnitude of the diurnal cycle in marine CF_L is greatest along the littoral and in the transition zones of the subtropical stratus regimes during summer. Over portions of the SE Atlantic and SE Pacific, the mean summertime range (amplitude) approached 0.26 (0.13). Over the NE Atlantic, the summertime range was ~0.15, which is similar to that (0.17) found by Dong et al. (2014a) from active remote-sensing instruments at the ARM ENA site. Diurnal variability was smaller in amplitude but still significant over midlatitudes and even polar waters. In nearly all cases, the phase over ocean regions was between 0300 and 1000 LT. Even greater amplitudes, up to 0.20, were found by Burleyson et al. (2013) in the southern subtropics.

The macrophysical and microphysical properties associated with these CF_L diurnal variations are also subject to systematic day-night changes. For example, Minnis et al. (1992) found, using satellite and SODAR data taken off the California coast during FIRE-87, that CTH, LWP, CER, and cloud thickness systematically dropped by 200 m, 105 g m^{-2}, 6 μm, and 150 m, respectively, during the day along with the decreasing top temperature before increasing again after 1600 LT. Similar systematic

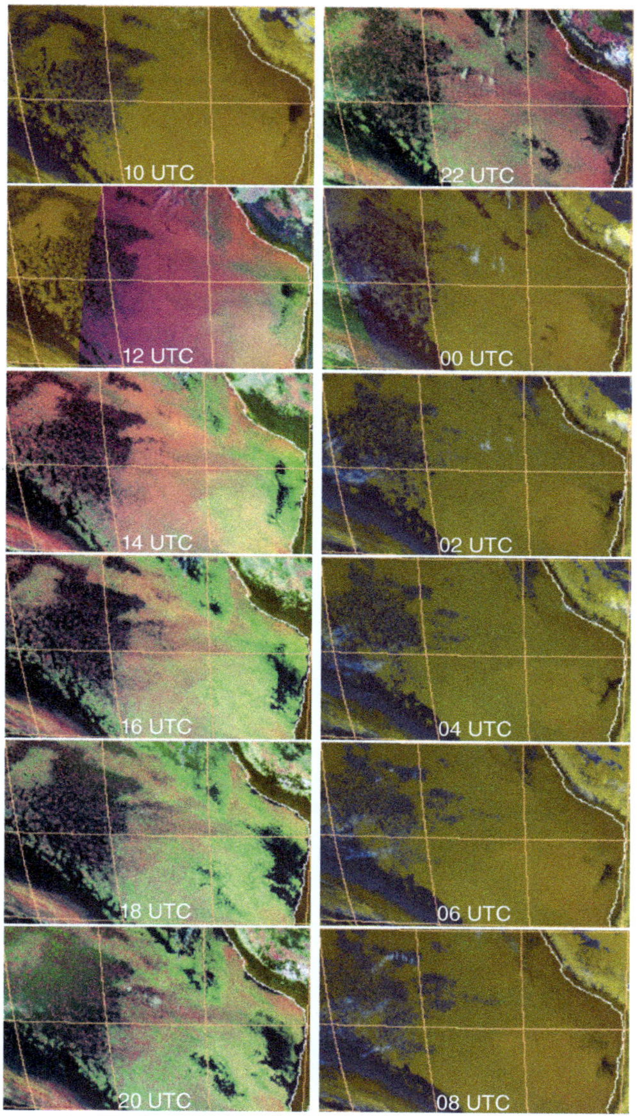

Figure 6.20 Multichannel pseudocolor two-hourly GOES-16 images of SE Pacific, 1–2 November 2020. Nighttime images in blue and gold tones, and daytime in multicolor tones. Sunglint affects northern parts of the images at 18 and 20 UTC.

changes were also observed over a larger portion of the NE Pacific by Greenwald and Christopher (1999) and over the SE Pacific by Bretherton et al. (2004) and by Painemal et al. (2012, 2013a). Figure 6.22 shows the daytime hourly mean CF_L, LWP, COD or τ, and CER (or r_e) retrieved from GOES-10 data along 21°S at 72°W near the coast, 76°W, and 80°W near the transition to open cells. For these three regions, CF_L (Figure 6.22a) drops precipitously, near the coast, after 0800 LT to a minimum at 1400 LT. In the closed (open) cell region near 76°W (80°W), the cloud breakup is delayed until 1000 LT, dropping to minima around 1500 LT. COD (Figure 6.22c) plummets by a factor of 2 or more from 0700 to ~1330 LT before recovering. Similarly, CER (Figure 6.22d) decreases after 0900 LT until ~1330 LT for all three regions. The diurnal range in mean CER rises from 1.5 μm at 72°W to 3.0 μm at 80°W. As a result of the CER and COD changes, the LWP (Figure 6.22b) also varies by a factor of ~3 for all three areas. CTH (not shown) increased by ~200 m during the day over the area (Painemal et al., 2013a). The average decreases in CF_L, COD, and LWP are all consistent with the cloud thinning and diminished drizzle observed in the flight discussed earlier. These cases are also typical of the SE Pacific domain during VOCALS-REx but not of all Sc regimes. For example, based on active remote sensing at the ARM ENA site, Dong et al. (2014b) found a diurnal range in LWP of ~55 g m^{-2} and in COD of ~36% using averages over all seasons. No diurnal cycle was found in CER in contrast to observations elsewhere. That difference may owe, in part, to differences in the averaging periods (seasons vs. annual) used in other studies.

Currently, geostationary satellite retrievals of COD, CER, and therefore LWP are limited to daytime observations because they depend on reflected solar radiation. Additionally, the retrievals may suffer from some solar and viewing angle biases and are impacted by the presence of ice clouds in multilayer situations. Microwave radiometers on lower Earth-orbiting satellites can, at least, provide LWP over the entire diurnal cycle over water surfaces and are mostly unaffected by ice clouds, so they can provide complete coverage. O'Dell et al. (2008) combined 18 years of microwave radiometer data taken by multiple satellites in various orbits to determine the complete diurnal cycle of mean LWP over oceans. Figure 6.23 shows their results for January and July. The 24-hour mean LWP for January (Figure 6.23a) and July (Figure 6.23d) peaks at more than 175 g m^{-2} along the ITCZ at all longitudes. During January, the MCAOs east of Japan and North America are likely responsible for the maxima seen in those areas. In the north Pacific, LWP has a July maximum, while in the SH midlatitudes, the maxima of ~100 g m^{-2} are much broader during January than in July, probably as a result of the MCAOs. Yet, the LWP is greater in July than during January over all of the subtropical marine status regions.

For most areas, the microwave LWP means are noticeably larger than the MODIS averages in Figure 6.5d. This difference is due to, at least, two main factors. The MODIS retrievals cannot see through relatively thick ice clouds to retrieve CER and COD in areas such as the ITCZ. More importantly, the Aqua retrievals are taken at 1330 LT, near the minimum in the diurnal cycle. The diurnal cycles in LWP characterized by the relative amplitude (in percent) and the phase (maximum in LT) are significant over large portions of the oceans. Amplitudes

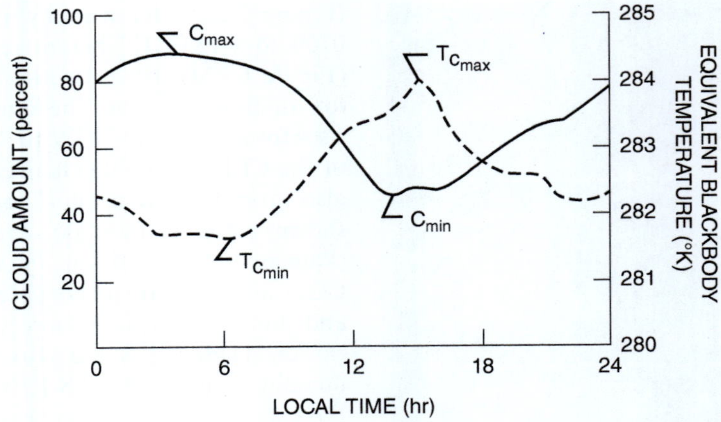

Figure 6.21 Mean November 1978 low cloud fraction in percent from GOES-2 visible and infrared radiances for a 250 km × 250 km region centered at 21.4°S, 86.3°W. Source: Minnis et al. (1984a)/American Meteorological Society.

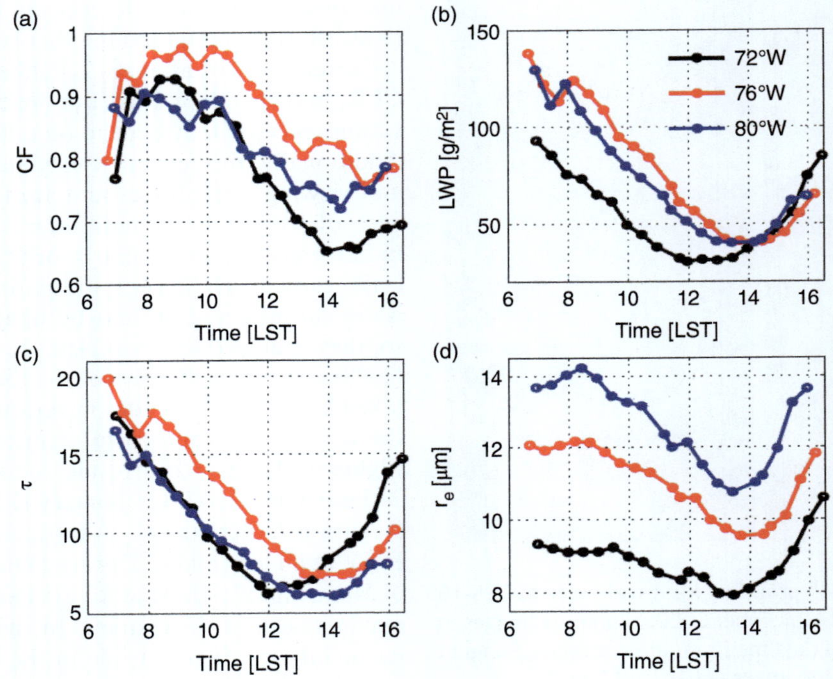

Figure 6.22 Example of daytime variations in cloud microphysics along 21°S latitude at 72°W (black, near the coast), 76°W (red, offshore), and 80°W (blue, far offshore) longitude. (a) CF, (b) LWP, (c) COD, and (d) CER. Source: Adapted from Painemal et al. (2012).

exceed 50% in the parts of the southern subtropics during January. Many southern areas and the subtropical NE Atlantic have amplitudes of ~20% (Figure 6.23b). The diurnal range over the ITCZ is insignificant.

During July (Figure 6.23e), the diurnal range is relatively large over the northern and SE Pacific and Atlantic subtropical regimes, and even into the northern midlatitudes in agreement with Rozendaal et al. (1995). The large diurnal ranges off the eastern coasts are likely due to sea breeze and deep convective processes over warm water. For most regions, the diurnal phase (Figures 6.23c and f) is between 0400 and 0800 LT, although some Sc areas have maxima near 0200 LT. Midday and late afternoon phases are associated with the strong diurnal cycles off the west coasts. These results are consistent with the GOES-10 retrievals in Figure 6.22. O'Dell et al. (2008) also report that some areas have significant semidiurnal components similar to those reported by Minnis et al. (1984b).

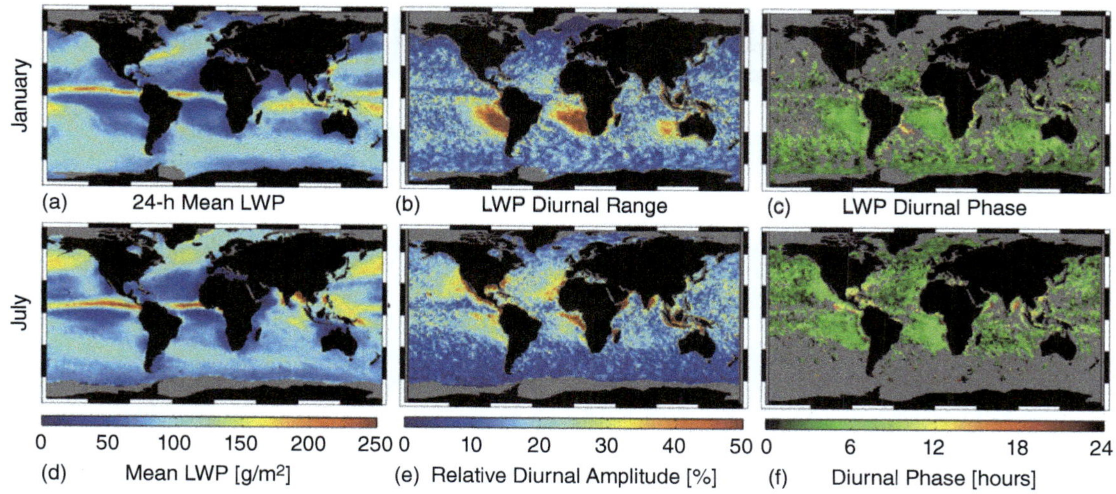

Figure 6.23 Daily average (left) and relative amplitude (middle) and phase (right) of the mean January (top) and July LWP diurnal cycle from 18 years of satellite microwave radiometer data. Black pixels denote land and gray pixels denote either missing data or the absence of a well-defined diurnal cycle. Source: Adapted from O'Dell et al. (2008).

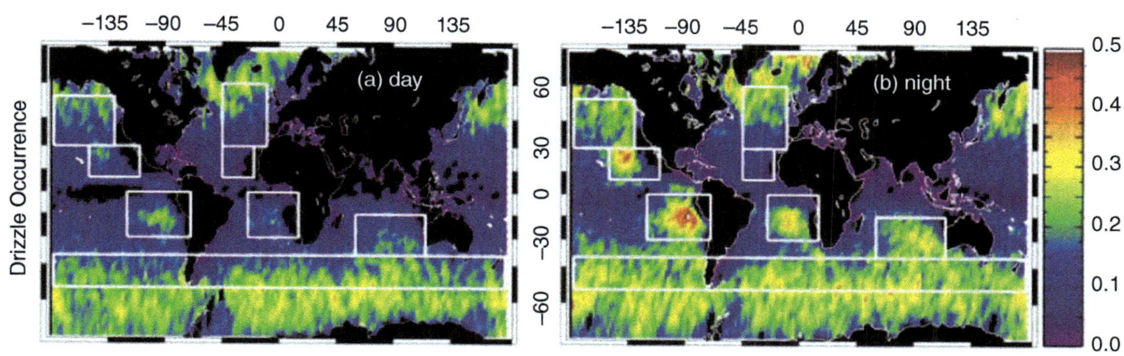

Figure 6.24 Mean drizzle occurrence during July 2006–June 2007 determined from CloudSat CPR data for (a) day and (b) night. White boxes denote special areas of interest in the original study. Source: Adapted from Leon et al. (2008).

These diurnal cycles of CF_L and LWP, as well as the convective nature of the marine Sc suggest that drizzle is also subject to diurnal variations. This is implied to some extent by the hourly change in CER (e.g., Figure 6.22) and the radar images in Figure 6.17. Leon et al. (2008) determined the frequency of drizzle occurrence for CTH <4 km using CloudSat CPR and CALIPSO lidar data for the period July 2006 to June 2007. These results, based on measurements at 0130 LT (night) and 1330 LT (day), were matched with LWP and CER from Aqua MODIS (Platnick et al., 2003). Those image times roughly correspond to the CF_L maxima and minima. The 1-year means (Figure 6.24) demonstrate that drizzle occurs significantly more often at night (Figure 6.24b) than during the day (Figure 6.24a) in nearly all oceanic locations. Over the SE Pacific, the day and night peaks are centered at 20°S and 90°W, far offshore in the transition region. As in the other subtropical Sc areas, the littoral minimum is more extensive during the day than at night. Rémillard et al. (2012) found from surface radar measurements that drizzle was most frequent over the ARM ENA at night, near the center of the NE Atlantic box in Figure 6.24.

The seasonal variation in drizzle occurrence (not shown) is significant over many areas. For example, drizzle events are most frequent at 0.35 during August and sparsest during April at 0.23 in the NE Pacific according to Leon et al. (2008). The seasonal difference is comparable to the diurnal difference during the same months. Seasonal maxima occur during local winter over the North Atlantic, North Pacific, and Southern Ocean. Over those areas, the drizzle peaks coincide with the maxima in average CER as determined from satellite retrievals (Minnis et al., 2011). The CloudSat results are confirmed over the NE Atlantic by surface radar results from Rémillard et al. (2012). The diurnal range is greatest during summer for those same areas. Drizzle occurs most often during the August–October period over the Australian, SE Atlantic, and SE Pacific Sc regions, when the diurnal range is greatest. The Canaries region (small box west of Africa) has the least drizzle and the weakest diurnal cycle of the subtropical domains.

Drizzle seasonal and diurnal variabilities appear to be in sync with LWP. Rémillard et al. (2012) found that, on average, drizzling Sc layers are physically thicker and have greater LWP than nondrizzling layers. A threshold of ~75 g m^{-2} appears to identify most drizzling layers, although nondrizzling MBL clouds can have LWP up to 150 g m^{-2} or more, and drizzling clouds with LWP ~20 g m^{-2} are not uncommon. The overlap range of drizzling and nondrizzling clouds (roughly 75–150 g m^{-2}) is likely due to variable CER in the clouds. Drizzle is more likely to be present in clouds having large values of CER (e.g., Leon et al., 2008).

Based on more than 4 years of ground-based observations and retrievals at the ARM ENA site, Wu et al. (2020b) found that the drizzling frequency was highest in winter (>70%) and lowest in summer (~45%) with an annual mean of ~55%. The annual mean CF_L of all single-layered MBL clouds is 0.259, which includes 0.115 for nondrizzle clouds, 0.084 for virga (drizzle did not reach the surface), and 0.06 for rain clouds (drizzle reached the surface). No strong day-night differences were found in drizzle properties. Drizzle particle size (r_d) increased from the cloud top down to the cloud base from collision-coalescence process, where r_d reached a maximum and then decreased below the cloud base due to evaporation. Drizzle drop number concentration (N_d) kept decreasing from the cloud top to below cloud base due to self-collection. Drizzle liquid water content (LWC_d) peaked in the middle of the cloud and decreased toward the cloud boundaries. However, the fraction of LWC_d in total water content increased monotonically from cloud top to base. The adiabaticity (f_{ad}) was higher in nondrizzling clouds than those in drizzling cloud, and drizzle could decrease f_{ad} by 0.25–0.4.

The CALIPSO data in the study of Leon et al. (2008) reveal an average day-to-night increase of 100–200 m in CTH over the major subtropical Sc regimes, consistent with the more limited studies noted previously. Maximum CTH occurs at night in all cases, while the diurnal range is greatest in the Canaries and smallest over the northern Pacific, Southern Ocean, and the NE Atlantic. Dong et al. (2014a) also found minimal day-night changes in the average CTH over the ARM ENA. Mean CTH is highest over the NE (SE) Pacific during October (February) and over the Australian domain during winter. The mean CER has also been shown to follow a seasonal cycle over the oceans (Minnis et al., 2011), with the greatest zonal range of up to 4 μm over the southern midlatitudes, where the minimum occurs during austral summer and the maximum during the winter. This seasonal cycle appears to hold for the latitudes south of 15°S. A similar seasonal variation occurs in the NH between 40° and 60°N, but the zonal range maximizes at ~2 μm. The CER seasonal cycle is consistent with that for drizzle occurrence. Dong et al. (2014a,2014b) found no seasonal variation in CER over the ARM ENA, and Wu et al. (2020b) detected no strong seasonal variations of MBL cloud and drizzle microphysical properties.

Daytime breakup of marine stratus in the subtropics and midlatitudes due to absorption of SW radiation was discussed in section 6.3.1. This dissolution of the clouds is apparently the main driving force for the observed diurnal cycles. Other factors include drizzle, which may help diminish the cloud-layer moisture and cool the surface. However, the fact that drizzle increases during the night everywhere, on average, indicates that it is not a large factor in reducing the cloud population. Changes in subsidence can also contribute to the Sc dissipation (e.g., Brill & Albrecht, 1982). Allen et al. (2012) demonstrated that gravity waves generated by disturbances in the subtropical jet stream were responsible for initiating POCs in some closed-cell Sc over the SE Pacific. In a modeling study, Garreaud and Munoz (2004) found that deep (up to 5 km) warming of the air over land during the morning along the west coast of Chile produces an "upsidence" wave that propagates westward lifting the MBL briefly as it passes causing additional entrainment that further dries the cloud layer. This has been supported by both surface-based (Rahn & Garreaud, 2010) and satellite observations (e.g., Painemal et al., 2013a). Daily land-sea breezes could also contribute to decreasing cloudiness during the day and possibly to semidiurnal variations where the clouds are near a coast.

Since boundary-layer cumulus clouds are relatively frequent over land, studies of the diurnal characteristics of continental stratus and stratocumulus per se are less common than for their marine counterparts. Eastmann and Warren (2014) tackled this issue using visual surface observations from 1971 to 1996. They found that both St and Sc peak in the morning between 0400 and 0700 LT and reach a minimum during midafternoon. The greatest amplitudes for St and Sc (~0.09 zonal mean) are found over tropical land, while the mean zonal range over the midlatitudes is ~0.03 during the hemispheric summer. During boreal winter, the strongest St and Sc diurnal cycles are found over North America, Europe, western Asia, and southeast China. During boreal spring, they occur over the central plains of the USA, eastern China, and the northern taiga areas. St and Sc diurnal variations are seen over Argentina, Pakistan, and SE Asia during the NH summer and again over SE China, the Great Plains, central Canada, and NW Europe during NH autumn.

Stratus/stratocumulus cloud-base heights also rose systematically during the day peaking at midafternoon. These results reinforce the conclusions of Del Genio and Wolf (2000) that the same mechanism is responsible for maintenance and breakup of stratus clouds for both land and ocean areas.

6.5. AEROSOL IMPACTS ON STRATUS AND STRATOCUMULUS

Aerosols can directly affect the radiation budget and atmospheric heating rates by absorbing and reflecting SW radiation and absorbing and emitting LW radiation. When present in a cloud-free atmospheric column, aerosols increase the column albedo and absorption causing a decrease in sunlight reaching the surface. The relative amount of absorption (aerosol layer warms) to albedo (surface cools) depends on the spectral scattering properties of the particular particle, especially its single-scattering albedo. The overall effect also depends on the surface albedo and aerosol concentration. Larger aerosols can also affect LW radiation, especially those comprising mineral dust (e.g., Dufresne et al., 2002); however, other aerosols are generally too small to interact with radiation at longer wavelengths. In addition to their direct radiative effects, aerosols indirectly impact the radiation field and atmospheric thermodynamics by altering cloud properties. The first aerosol indirect effect is the change in cloud albedo by aerosols impacting the cloud-droplet size (Twomey, 1977). Essentially, for a cloud with a fixed LWP, the droplet size will be determined by N_c. Increasing N_c is accompanied by decreasing CER. Smaller droplets absorb less solar radiation and a large N_c causes more scattering so that together they raise the cloud albedo due to increased scattering area. Cloud amount and LWP changes due to aerosols constitute the second indirect effect (Albrecht, 1989), while heating by aerosol absorption, the semidirect effect, decreases cloud coverage. These are just a few of the indirect aerosol radiative effects. Others include glaciating of supercooled water clouds by the presence of ice nuclei. All of the effects are sometimes lumped together as aerosol-cloud interactions (ACI). Fully understanding these processes and reducing the uncertainties in GCMs is essential to estimating changes in the future as natural or anthropogenic aerosol generation changes with the climate.

Both modeling and empirical studies have been used to estimate ACI effects, particularly the SW effective radiative forcing (ERF). Zelinka et al. (2014) employed a variety of coupled climate models to estimate the global ERF_{ACI}. The nine-model ensemble mean SW ERF_{ACI} is -1.04 ± 0.67 W m^{-2}, which greatly exceeds their estimated LW ERF_{ACI} of 0.13 W m^{-2}. Roughly 60% of the SW ERF_{ACI} is attributed to sulfate aerosols, which are

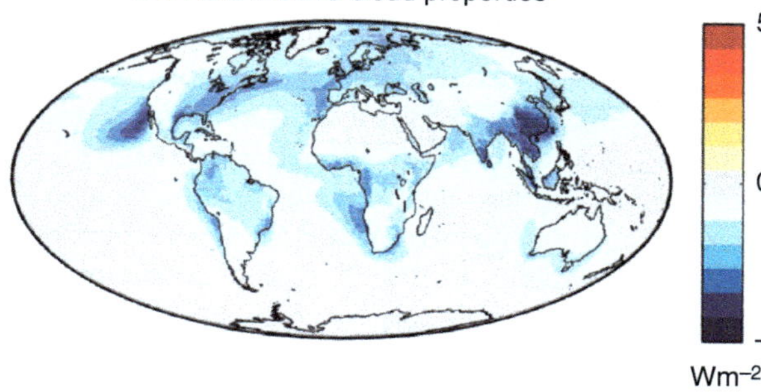

Figure 6.25 ERF_{ACI} during industrial era computed using MODIS COD and CF, CERES SW fluxes, and AeroCom sulfate concentrations. Source: McCoy et al. (2017)/John Wiley & Sons.

effective droplet nuclei. McCoy et al. (2017) used N_c inferred from MODIS-retrieved CER and COD under overcast conditions with aerosol analyses to determine that N_c is highly dependent on sulfate concentrations. With that dependency, they estimated that relative to the preindustrial period, the SW ERF_{ACI} due to anthropogenic sulfates is -0.97 ± 0.23 W m^{-2} in the subtropical Sc regimes. That estimate falls within the range of -2.0 to -0.3 W m^{-2} from other studies (Bellouin et al., 2013) that include both models and observations. Thus, a great deal of uncertainty remains about their impacts and how they will change.

As noted by the range in ERF_{ACI} estimates, there is still much uncertainty in the aerosol distributions and methods for estimating the aerosol impacts. Figure 6.25 shows annual mean ERF_{ACI} from McCoy et al. (2017) based on industrial era sulfate aerosol optical thickness (AOT) from AeroCom (Mhyre et al., 2013). The distributions in this figure are very similar to the annual mean CER in Figure 6.5c, suggesting that CER is highly susceptible to SO_4 concentrations in line with previous studies.

In addition to relating aerosols and CER horizontally, aerosols need to be incorporated into the clouds. Simply knowing the distributions of the aerosols is insufficient to explain their impact on clouds. The boundary-layer conditions are very important for investigating the aerosol-cloud interactions. Dong et al. (2015) found that CER and N_c under coupled conditions strongly depend on surface CCN concentration (N_{CCN}) and have higher correlations (-0.56 and 0.59) with N_{CCN} than decoupled results (-0.14 and 0.25). Similarly, Liu et al. (2016) found that the first indirect effect of susceptibility of CER to AOT is stronger for clouds with lower base heights. When aerosols are above the boundary layer, they also need to contact the clouds. Wang et al. (2020) used WRF-LES simulations and found that the aerosol plume aloft cannot affect underlying MBL cloud properties over the

ARM ENA in the first scenario. Even when the aerosols were close to the cloud top in the second scenario, the deepening of cloud is critical for producing significant aerosol-cloud interactions. Painemal et al. (2014) analyzed CALIPSO, MODIS, and CERES data to find that widespread smoke over the SE Atlantic from biomass burning over southern Africa only had noticeable ACI when the aerosol layer base was near to or abutting the Sc cloud tops. Susceptibility of N_c to aerosols is greater for clouds with relatively low N_c compared to those with high N_c. Thus, clouds over land should generally be less affected by aerosols because the sulfate AOT is generally much greater over land (not shown). Open ocean clouds are more prone to be altered by the introduction of additional CCNs.

In order to investigate the ACI*s* over the ARM SGP site, X. Zheng et al. (2020) examined the physicochemical properties of aerosols and their impacts on cloud microphysical properties using 16 selected stratus cases under daytime coupled boundary-layer conditions. They used the ACI index (ACI_r, representing the relative change in layer mean CER with respect to the relative change in aerosol loading) to quantify the aerosol impacts with respect to CER and found that the magnitude of ACI_r decreased with increasing LWP, which suggests a diminished cloud microphysical response to aerosol loading. Furthermore, they found that in the presence of weak absorbing aerosols, the stratus clouds feature a higher N_{CCN} and smaller CER, and the measured ACI_r values are relatively higher, indicating that clouds have greater microphysical responses to aerosols, owing to the favorable thermodynamic conditions. For strong absorbing aerosols, on the other hand, the stratus clouds feature a lower N_{CCN} and larger CER, and the reduced ACI_r values are due to the cloud-layer heating effect induced by strongly absorbing aerosols.

For the three major subtropical marine Sc domains, analyses of CERES cloud and radiation data showed that near the coast the susceptibility of SW albedo to the natural logarithm of either N_c or COD is relatively high (~0.06) for clouds with LWP of ~20 g m^{-3} (Painemal & Minnis, 2012) and rises to 0.08 for 100 g m^{-3}. It does not increase for clouds with greater LWP, which is consistent with the results over the ARM SGP site (X. Zheng et al., 2020). Offshore, the susceptibility is only 0.04 at 20 g m^{-2} but rises steadily up to ~0.09 for LWP values around 100 g m^{-2}. Thus, the susceptibility of a cloud albedo to changes in CCN must account for the LWP or COD of a given cloud.

By reducing the cloud-droplet size, aerosols can decrease the amount of drizzle because fewer larger drizzle drops are available for collision and coalescence (e.g., Ferek et al., 2000; Wood et al., 2005a). In some cases, the lack of drizzle could extend cloud lifetime and, therefore, the cloud's radiative impact (e.g., Christensen et al., 2020). As seen above, it is anthropogenic aerosols that seem to have the most impact on low-level clouds, thus changes in emissions could have dramatic impacts on these clouds in the future. Continued monitoring of all of the relevant parameters with ever-increasing precision and accuracy is essential. For additional information on these parameters and aerosols, in general, see Chapter 3 (Liu et al., 2023).

6.6. REMOTE SENSING

Monitoring stratus cloud parameters on large time and space scales requires remote sensing of macrophysical and microphysical properties of low clouds. The former include cloud-base and cloud-top heights CBH, CTH, cloud effective height CEH, cloud thickness (ΔH = CTH − CBH), cloud-top and cloud-base temperatures (CTT and CBT), cloud effective temperature CET, cloud layering, and LWP, which includes both cloud LWP_c and drizzle LWP_d. CET is the 11-μm brightness temperature corresponding to the radiance leaving the top of the cloud. Typically, it corresponds to some altitude ~25–100 m below the cloud top due to the diffuse nature of the clouds. CEH is the height corresponding to CET in the profile. The height parameters can also be expressed as pressures. Cloud heterogeneity provides information on the spatial structure of the cloud field. A first-order macrophysical parameter is cloud fraction, CF. CF_L is determined based on CTH for those pixels determined to be cloudy. Microphysical properties include cloud phase, COD, CER (may combine cloud and drizzle particle sizes, CER_c and CER_d), N (includes both cloud N_c and drizzle N_d), and liquid water content LWC (both cloud and drizzle components LWC_c and LWC_d may be retrieved). For glaciated supercooled clouds, both LWC and ice water content IWC are necessary. Since LWC and CER vary with height in the cloud, vertical profiles of those two parameters are valuable information.

Many of these properties are retrieved using data from remote sensors on satellites and aircraft and from instruments on the surface. Chapter 18 (Marshak & Davis, 2023) provides a general overview of satellite and aircraft remote sensing of clouds and aerosols, while surface-based remote sensing of the atmosphere is reviewed in Chapter 17 (Lamer et al., 2023). This section primarily focuses on boundary-layer cloud remote sensing. Details concerning retrievals of aerosols, midlevel and high clouds, and other atmospheric components, as well as additional information applicable to low clouds can be found in the cited chapters.

Differences in the instruments, their sensitivities, and viewing perspectives can lead to discrepancies between the nadir- and zenith-viewing remote sensors that could

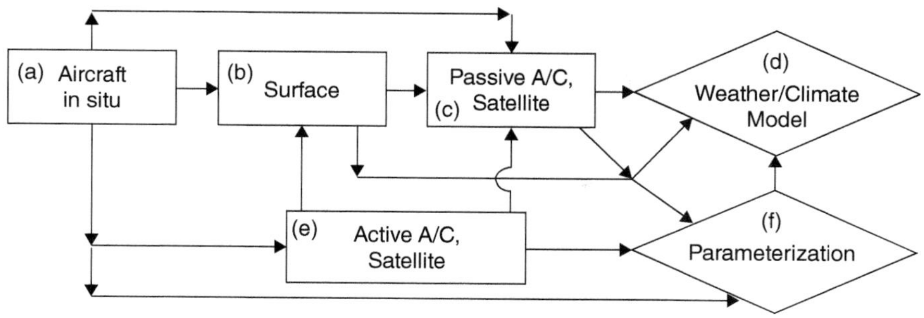

Figure 6.26 Schematic illustration of using observations to validate and improve weather and climate models.

be large due to spatial and temporal differences among the three platforms. Passive satellite retrievals typically do not distinguish between cloud and precipitation, so CER, COD, and LWP (e.g., Figure 6.5) are the retrieved parameters. These quantities can include both drizzle and cloud water to the extent that each contributes to the spectral radiances used in a given retrieval. Verifying a given retrieval and quantifying its uncertainty is critical for using it in a given application.

In addition to expanding our knowledge of stratus clouds, the measurements are also used to validate and improve weather and climate model estimates of the same variables. A simplified overview of the complex approach to validate model results from the observations is illustrated in Figure 6.26. First, surface and passive satellite remote-sensing observations, which are most applicable for comparisons with the models, must be validated as indicated by the arrows from boxes (a) and (e) going directly to boxes (b) and (c). Aircraft in situ measurements are often treated as a cloud truth to verify certain parameters retrieved remotely using various algorithms applied to both ground-based observations (e.g., Dong et al., 1998; Dong & Mace, 2003b; Lin et al., 2001; Wu et al., 2020a) and to satellite or airborne remote measurements (e.g., Painemal et al., 2012; Painemal & Zuidema, 2011; Platnick & Valero, 1995; Zheng et al., 2011). As indicated in the arrow from box (b) to box (c), validated ground-based observations are then used to assess satellite retrievals (e.g., Dong et al., 2002, 2008, 2016; Painemal et al., 2016; Xi et al., 2010; Zhang et al., 2017). Additionally, certain parameters such as cloud fraction, LWP, cloud heights, and cloud layering are assumed to be accurate enough from surface instruments that they provide cloud truth for evaluating satellite retrievals (e.g., Ackerman et al., 2008; Chang et al., 2010; Huang et al., 2005; Marchand, 2016; Pavolonis et al., 2005). Similarly, satellite and airborne active remote sensors (box (e)) can provide cloud truth for many macrophysical cloud properties (e.g., Minnis et al., 2021; Noh et al., 2017; Yost et al., 2021, 2023), although some parameters require validation with aircraft (e.g., McGill et al., 2007) and surface observations (e.g., Kim et al., 2011). After their uncertainties are evaluated, the observations can be more confidently compared with the models.

The weather and climate models represented by box (d) cover a wide range of spatial (several meters to >100 km) and temporal (minutes to several hours or a day) scales. To facilitate computation and simplify formulation, a wide variety of parameterizations (box (f)) are employed in the models (arrow from box (f) to box (d)) to replace the explicit physical equations. As indicated by the arrows directed at box (f), observations of all types are used to inform and assess these parameterizations (e.g., Albrecht et al., 2016; Del Genio et al., 1996; Kogan & Kogan, 2001; Yoo et al., 2013; Zhu & Zhao, 2008). Large-scale satellite data and sometimes surface data are used directly to evaluate weather and climate model results (e.g., Del Genio et al., 2005; Guo et al., 2020; Kennedy et al., 2010; Randall et al., 1996b; Stanfield et al., 2014, 2015). In many field programs (e.g., Albrecht et al., 2019), the model forecasts are used to aid the flight mission planning (indicated by the dashed line from (d) to (a) in Figure 6.26) to collect the proper data for parameterization assessment. In all cases, the validations, which are basically comparisons, are sensitive to temporal and spatial mismatches between any pair of observations or retrievals. For example, in situ measurements by an aircraft must account for the vertical location of the plane and the fact that it samples only a tiny pencil of air at the given altitude, while the surface or satellite instruments cover 100 m to 30 km or more. Thus, temporal averaging and spatial averaging are employed to mitigate the differences (e.g., Dong et al., 2008). Similar strategies are employed when comparing model and observational results.

6.6.1. Satellite Remote Sensing

Cloud remote sensing is possible using data from many different satellites that carry a multitude of passive imaging and sounding radiometers as well as active sensors such as those on CALIPSO and CloudSat. It is not possible in this chapter to summarize every type of sensor and technique, so only those most relevant to and most frequently employed for low-cloud monitoring are discussed here. Despite providing a wealth of information about the vertical profiles of clouds and aerosols in the atmosphere, satellite-borne cloud radars and lidars are generally confined to a nadir view and have very narrow footprints (<2 km) that severely constrain their spatial coverage. Nevertheless, they provide unique information, especially when combined with passive imagery, and serve as sources for validating passive retrievals. Typically, sounders such as the TIROS Operational Vertical Sounder (TOVS) including the High-Resolution Infrared Sounder (HIRS) or the Atmospheric Infrared Sounder (AIRS) have many channels in the infrared and are often combined with a microwave sounder. Visible wavelength channels may also be included for better cloud detection on those instruments. They have relatively large footprints, however, such that a given pixel corresponds to a resolution of 14 km or more. Such sensors can be used to accurately determine cloud fraction and some cloud properties but are limited in their ability to resolve many of the features of Sc that have been discussed. Multispectral microwave scanning radiometers that are used for retrieving LWP have been carried on many satellites (e.g., O'Dell et al., 2010). Although they typically have large footprints, the retrieval of LWP, which is confined to marine areas only, often serves as the standard for that particular parameter.

More relevant for comprehensive study of stratus are multispectral imagers, such as MODIS, the Advanced Baseline Imager (ABI) on GOES-16/17, and VIIRS, which have resolutions ranging from 250 m up to 2 km for certain wavelengths. Remote sensing of cloud properties from current imagers makes use of channels in the visible (e.g., 0.55 and 0.65 μm), near-infrared (0.87 μm), SW infrared (e.g., 1.24, 1.64, 2.13, and 2.25 μm), midwave infrared (3.70–4.00 μm), and LW infrared (8.5, 10.8, 11.0, and 12.0 μm), and, sometimes, in the CO_2 absorption band (13.3–15.0 μm). Sun-synchronous satellites sample the entire globe twice per day nominally at the same local times each day, while geostationary satellites image a particular portion of the Earth at high temporal resolution. For example, GOES-16 and the Himawari-8, which carries an imager similar to the ABI, routinely record multispectral full disk images every 10 min. Certain sectors of the viewed disk can also be imaged at 1-minute resolution for special applications, such as field campaigns (e.g., Bedka et al., 2018). These newer satellites represent significant advances as older imagers, particularly those on geostationary satellites, had fewer channels than current models and lower temporal (15 min to 1 h) and spatial resolutions (3–8 km). Both types of satellites are required for monitoring clouds. Geostationary satellites have fixed viewing angles for a given region and have oblique views for a large portion of the full disk, which does not include areas poleward of ~72° latitude. Sun-synchronous satellites view the entire globe and view a given area at different viewing angles each overpass, thus minimizing any view angle bias for purposes of averaging. Moreover, sampling frequency considerably increases near the poles and the lower orbits enable higher spatial resolution for the imagers. Together, the current satellites and their imagers are providing information at temporal and spatial scales that should facilitate continuing advances in stratus and stratocumulus research.

Retrieving stratus cloud properties from radiances measured by a satellite sensor first requires determination if a cloud is responsible for the observed radiances in a given pixel. A great variety of cloud detection algorithms have been developed for passive imagers, while several have been developed for active sensors such as CALIPSO and CloudSat. Stubenrauch et al. (2013) provide a good summary of algorithms as of 2012. Many of those algorithms have since been updated (e.g., Baum et al., 2012; Heidinger et al., 2016; Trepte et al., 2019) and others have become operational (e.g., Karlsson et al., 2017, Minnis et al., 2021). Trepte et al. (2019) compared CF averages from seven different algorithms and found that all six of the passive methods detected fewer clouds than CALIPSO in all seasons and that the means ranged from ~0.62 to ~0.69. Although differences among algorithms are found at all latitudes, the greatest discrepancies are in the tropics, where small cumulus and thin cirrus are difficult to detect, and during polar night when thermal contrast between clouds and the surface is very small. It should be noted that nearly all of these cloud detection techniques assume that the pixel is completely overcast. A few other methods attempt to determine pixel fractional cloudiness using the assumption that within a given cluster of pixels, CTT is constant (e.g., Coakley & Baldwin, 1984; Coakley et al., 2005). This assumption is most appropriate for stratocumulus or other layer clouds.

After declaring that a pixel is cloudy, the cloud altitude (pressure) has to be estimated. As noted earlier, a maximum (minimum) CTH (CTP) of ~3.3 km (680 hPa) above the surface is used to define low clouds. Here, low clouds are the focus, so other cloud types are of concern only if they are concurrent with the low clouds. While some techniques can directly estimate CTP from pertinent reflected spectral radiances (e.g., Yang et al., 2019), most methods retrieve CET and relate that temperature to an altitude in a vertical temperature profile. Those profiles typically are taken from analyses or reanalyses by numerical weather prediction models. Stratus often occurs under very sharp inversions that are difficult to resolve in the limited vertical resolution of numerical models. Furthermore, CET is often a degree or two less than the air temperature measured at cloud top with aircraft (e.g.,

Painemal et al., 2013a) and radiosondes (e.g., Garreaud et al., 2001). It is suspected that this discrepancy is due to evaporative and radiative cooling at the droplet surface at cloud top, although further study of the phenomenon is required. The net result is that St and Sc heights are often overestimated if a model or even a radiosonde profile is used to relate CET to CTT. Such errors complicate the use of satellites for monitoring.

One means to mitigate this problem is to alter the profile in the lower atmosphere by inserting a fixed lapse rate in the lower part of the profile. Minnis and Harrison (1984a, 1984b) and Minnis et al. (1992) used the lapse rates of −6.5 and −7.1 K km^{-1}, respectively, to estimate CTHs at all levels. For CERES Ed2, Minnis et al. (2011) used the lapse rate of −7.1 K km^{-1} only for the lower part of the profile around the globe. The MODIS Atmosphere Science Team (MAST) used a zonal average Lapse rate for low clouds (Baum et al., 2012), while a diurnally and seasonally dependent regional lapse rate (Sun-Mack et al., 2014) was employed for the SatCORPS analysis of the AVHRR climate record (Minnis et al., 2016a) and for CERES Ed4 [Minnis et al., 2021). The NOAA NESDIS cloud height product (Heidinger, 2012) uses the lapse rate of −9.8 K km^{-1} whenever an inversion is detected in the sounding. Presumably, that approach is also used by the Pathfinder Atmospheres-Extended (PATMOS-X, Heidinger et al., 2014) algorithm. Lutz et al. (2011) use the inversion height and temperature in the sounding, when a low-level inversion is detected in the profile.

The impact of using these various approaches can be seen in Figure 6.27, which plots the distribution of mean CTP for October 2008 derived using six different satellite retrievals. The CTP means for SatCORPS AVHRR (Minnis et al., 2016a) in Figure 6.27a, CERES Ed4 Aqua MODIS (Figure 6.27b), and PATMOS-X AVHRR (Figure 6.27d) have very similar patterns in the subtropical Sc regimes; CTP decreases westward from the coasts toward the open ocean. The MAST Aqua MODIS product (Figure 6.27c) also shows some of the same trends except it is spotty west of Peru and the gradients are not as clear. The ISCCP multiple satellite product (Figure 6.27e) and CM-SAF AVHRR cLoud, Albedo and Radiation product (CLARA-A1; Karlsson et al., 2013) in Figure 6.27f both use the unaltered model reanalysis vertical profiles and the reverse trend results, and CTP is least near the coast and increases downwind. These examples are fairly typical of the results for the different approaches.

Sun-Mack et al. (2014) present a more detailed analysis using low clouds only. An alternative lapse rate from Zuidema et al. (2009) is also discussed therein. It appears to be as accurate as or more so than the regional lapse rate, at least, for marine Sc, but it tends to underestimate CTH for other low clouds and does not apply to land surfaces. Overall, Sun-Mack et al. (2014) found that the mean day (night) low CTH differences with CALIPSO over water are 0.04 ± 0.61 km (0.10 ± 0.62 km) for CERES Ed4 and −0.20 ± 0.64 km (−0.07 ± 0.66 km) for the Zuidema et al. (2009) approach. Using an unaltered reanalysis profile produced differences of 0.24 ± 0.98 km (0.64 ± 0.96 km). It is clear that the use of low-level lapse rate is considerably more accurate than entirely relying on a reanalysis temperature profile. Active sensor retrievals are the most accurate; they are the standard. Because of their sparse temporal and spatial sampling, however, they are not normally used for monitoring purposes but serve to verify the passive retrievals.

Knowing CTT, it is possible to retrieve the phase, COD, and CER for a cloudy pixel. For imagers, the bispectral method of Nakajima and King (1990) is the typical approach for estimating COD and CER simultaneously. Retrieval of COD is usually performed based on visible (VIS, e.g., 0.65 μm) or near-infrared (NIR; e.g., 0.87 μm) reflectance because droplets absorb little radiation at those wavelengths and there is only a small dependency on droplet size. A SW-Infrared channel (SWIR; e.g., 1.6 or 2.1 μm) or midwave infrared channel (MWIR; e.g., 3.75 μm) is used to estimate CER because liquid and ice water absorb radiation at these wavelengths, while being relatively unaffected by atmospheric attenuation. This is illustrated in Figure 6.28, which shows the reflectance for five of the bands commonly employed for the bispectral method. There is little separation between the reflectance curves at a wavelength λ of 0.63 μm (Figure 6.28a), but the usual pattern is evident, increasing reflectance with rising COD. As λ increases, the separation of the curves grows up to 2.1 μm (Figure 6.28d) before dropping again at 3.75 μm (Figure 6.28e). The curves flatten at different CODs for wavelength. At low CODs, the greatest separation occurs for 3.75 μm and the curves become asymptotic for COD <12. The 1.24-μm reflectance curves (Figure 6.28b) plateau at higher CODs suggesting that it could be used for retrieving COD for a greater range than for 1.65 μm (Figure 6.28c) and 2.15 μm. All of those SWIR and MWIR channels are useful for retrieving CER.

This potential is seen in Figure 6.29, which shows an SWIR (2.11 μm) band reflectance as a function of the nonabsorbing 0.86-μm band reflectance at various CODs for a range of CER at a viewing zenith angle (VZA) of 5° and relative azimuth angle (RAA) of 144.5° at a solar zenith angle of 55.6°. As CER decreases, the 2.1-μm reflectance increases monotonically, except for CER = 2 μm. Its reflectance falls below those for larger CER at smaller COD values (<4). It is especially evident for a cloud over the low-albedo surface (Figure 6.29a). For a majority of viewing and illumination angles, there is negligible nonmonotonic variation of reflectance with CER. On the other hand, the 0.86-μm reflectance is

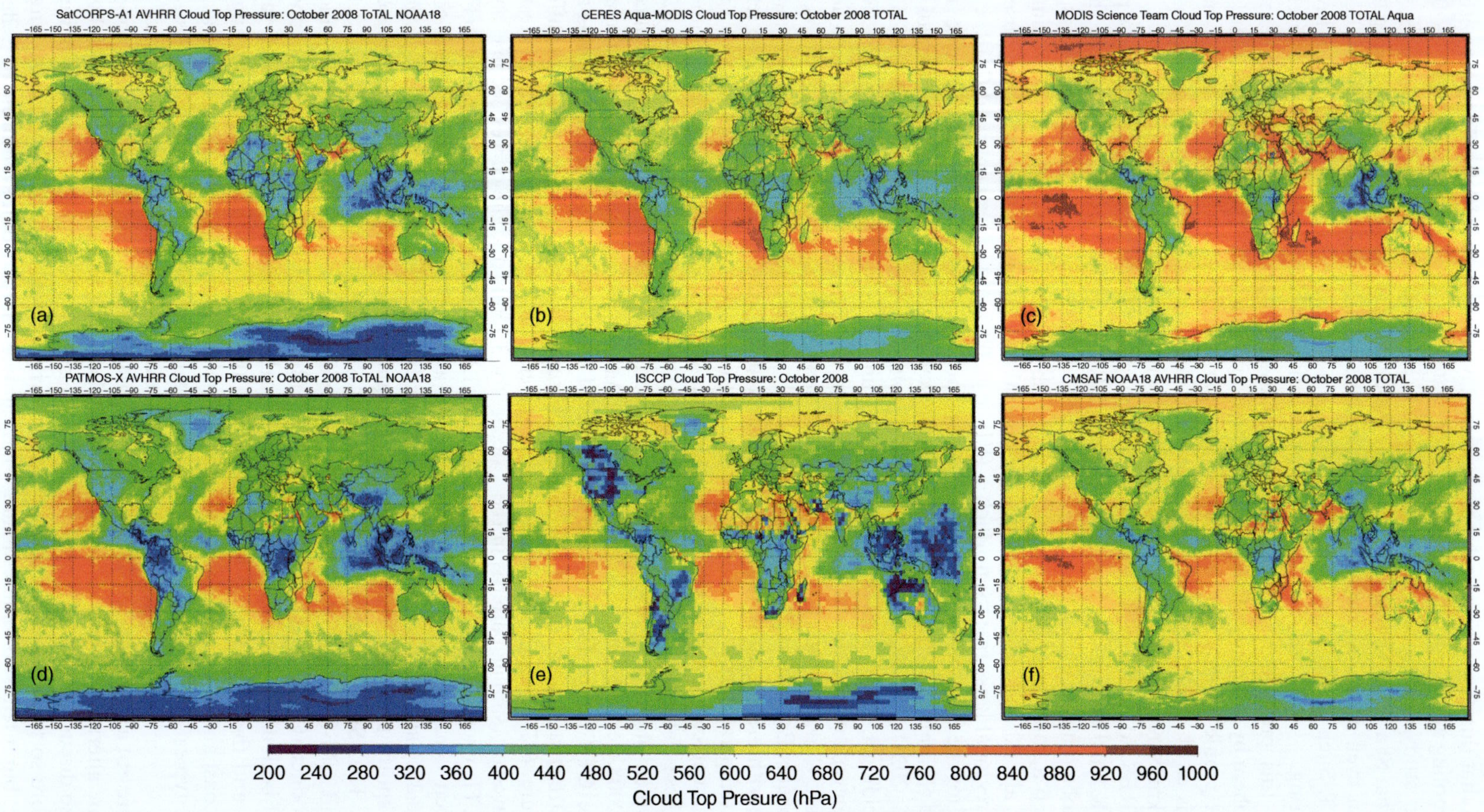

Figure 6.27 October 2008 day + night cloud-top pressure. Regional means from (a) SatCORPS-A1 NOAA18 AVHRR, (b) CERES Ed4 Aqua MODIS, (c) MAST Collection 6 Aqua MODIS, (d) PATMOS-X NOAA18 AVHRR, (e) ISSCP D2 multiple satellites, and (f) CLARA-A1 NOAA18 AVHRR. Resolution is 1°, except for ISCCP (2.5°). Source: Adapted from Minnis et al. (2016a).

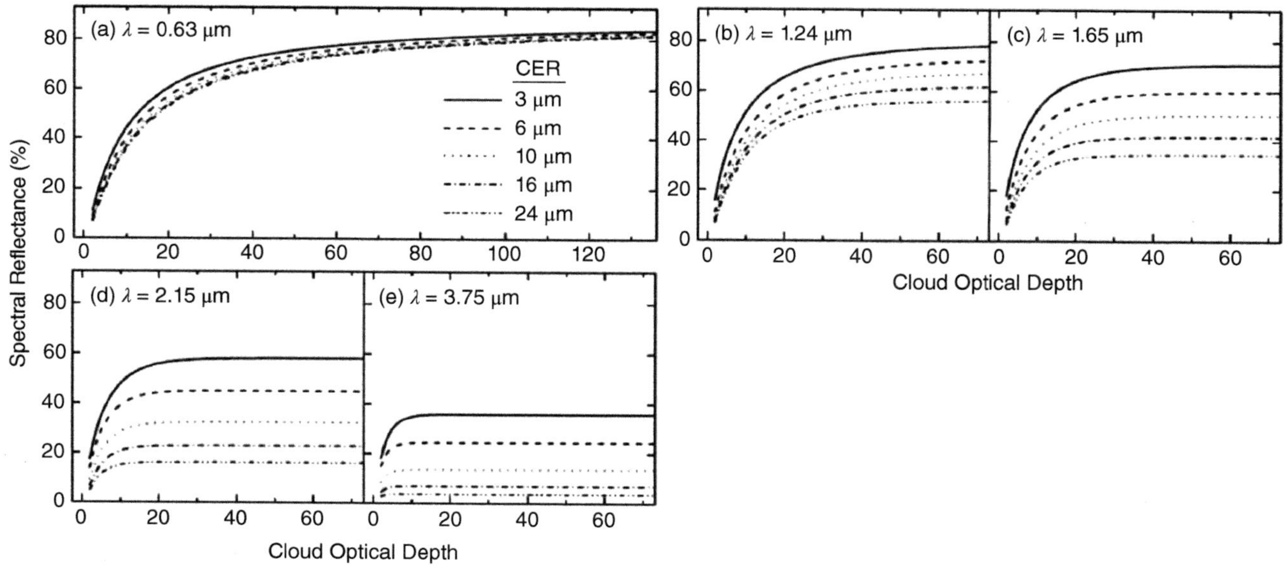

Figure 6.28 Dependence of spectral reflectance on cloud optical depth and droplet effective radius for wavelengths (a) 0.63 μm, (b) 1.24 μm, (c) 1.65 μm, (d) 2.15 μm, and (e) 3.75 μm. Reflectances computed for SZA = 60°, VZA = 7.2°, and relative azimuth angle = 0°. Source: Chang and Li (2002)/John Wiley & Sons.

nearly constant for a given COD regardless of CER with a slight tendency toward greater reflectance for a given COD as CER decreases, similar to the results for 0.63 μm in Figure 6.28a. For large COD (>4), the 0.86- and 2.1-μm reflectances are nearly orthogonal, whereas the differences in reflectance between larger and smaller droplets for small CODs become ambiguous, leading to sizeable retrieval uncertainties. When the 0.86-μm surface albedo is large (Figure 6.29b), the reflectance ambiguities extend to even greater values of COD (<8), limiting the usefulness of this approach over bright surfaces. In practice, different wavelength pairs, such as 1.24 μm and an MWIR or another SWIR channel, are employed over surfaces, such as snow, that are highly reflective in the visible and NIR ranges (e.g., Minnis et al., 2021; Platnick, 2000; Platnick et al., 2003). Since snow and ice albedos are relatively small for longer solar wavelengths, their reflectances provide contrast between clouds and underlying snow.

Retrieving COD from a pair of reflectances is accomplished by matching the reflectances to the computed values on this type of diagram in the form of a lookup table (LUT), after accounting for any contribution and/or attenuation by the atmosphere. A similar daytime approach solves for CTT, COD, and CER simultaneously using a VIS or NIR reflectance and 11-μm and MWIR temperatures. CTT, derived mainly with the 11-μm radiance, depends on COD when COD is less than ~5, while the MWIR brightness temperature depends on CTT, COD, and primarily CER. Han et al. (1994) and Minnis et al. (2011) solve this three-channel approach iteratively, while optimal estimation solutions are used by Heidinger et al. (2014) and others.

Without solar reflectance at night, only infrared channels are available for retrieving COD and CER. Generally, three channels are used: 11 and 12 μm, and either 3.8 μm (e.g., Minnis et al., 2011) or 8.5 μm (e.g., Lin & Coakley, 1993). Microphysical property retrievals using infrared measurements, however, are challenging during nighttime and prone to large errors, particularly for low clouds, and should be used carefully for any stratus analyses. Much additional research is needed to monitor low cloud properties at night.

To determine the cloud-top phase, a variety of approaches are taken that can be as simple as using a temperature threshold of 253 K to divide liquid and ice clouds (e.g., Austin et al., 2009) or can be highly complex. For example, the CERES Ed2 and Ed4 algorithms and the MAST Collection 6 phase (e.g., Platnick et al., 2017) algorithms use a variety of tests to determine phase beginning with two model solutions. The iterative analysis is run twice, once with water droplet reflectance LUTs and again with ice crystal reflectance LUTs. If there is no result for the ice model run and a result for water, and CTT > 233 K, the cloud is liquid and vice versa for deciding ice. If there is a solution for both, then a series of tests are applied using a variety of channels. Phase selection for the ABI and presumably the PATMOS-X algorithms uses emissivity ratios that employ four infrared channels (Pavolonis, 2010). That approach tends to detect more ice clouds than other methods (Minnis et al., 2016b) because it often identifies thin ice clouds above lower water clouds,

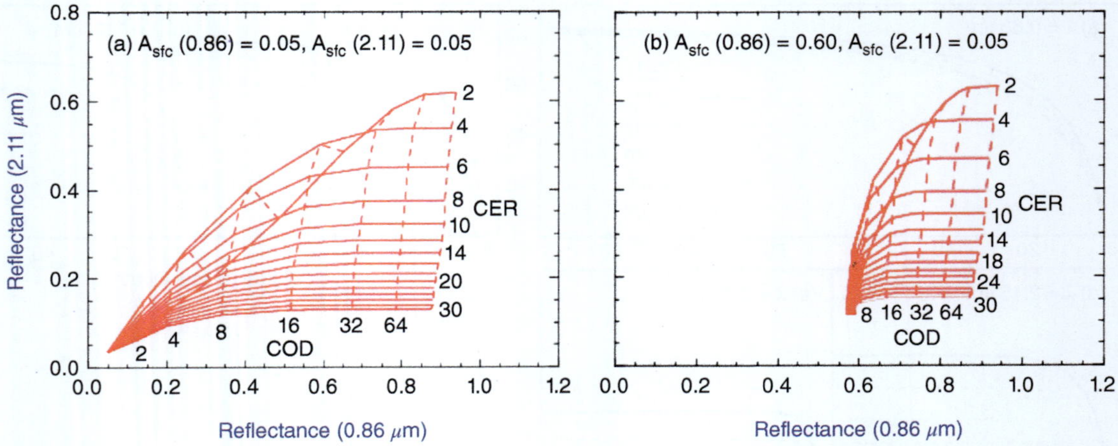

Figure 6.29 Simulated reflectances at 0.86 and 2.11 μm for a range COD and CER values over surfaces with 0.86-μm albedo A_{sfc} of (a) 0.05 and (b) 0.60. Solar zenith angle = 55.6°, viewing zenith angle = 5°, and relative azimuth angle = 144.5°. Source: Adapted from Platnick et al. (2003).

but the COD is then applied to the ice cloud. Other methods tend to classify such overlapped clouds as liquid when the upper-level COD is small. The trispectral infrared technique of Baum et al. (2000) employs the 8.5-, 11-, and 12-μm channels to discriminate between ice and liquid clouds and is applicable both day and night. During the daytime, it is enhanced with reflectances in the visible and SWIR bands.

When several SWIR and MWIR channels are available, CER is estimated at more than one wavelength (Minnis et al., 2021; Platnick et al., 2017). Since absorption rises with increasing wavelength, the retrieved radius for each channel corresponds to a layer between the cloud top and a particular depth. Theoretical computations for an adiabatic cloud (CER increases with height within the cloud layer) indicate that as the absorbing channel wavelength decreases the physical thickness and optical depth of the cloud layer increase such that the retrieved CER corresponds to the layer thickness for the given SWIR or MWIR wavelength. For example, Figure 6.30 plots CER as function of COD as measured from the top of cloud (Figure 6.30a) and as a function of height (Figure 6.30b). For the adiabatic cloud, both CER and liquid water content (LWC_c) increase with height in the cloud. When the CER is retrieved with the 3.7-μm channel, the value corresponds to an optical depth layer of ~2. For the 2.1- and 1.6-μm channels, that depth increases to ~3 and 4, respectively. Theoretically, then CER(3.7) > CER(2.1) > CER(1.6). For the theoretical cloud in Figure 6.30a, CER(3.7) = 11 μm, CER(2.1) = 10.5 μm, and CER(1.6) = 10 μm. If 1.24-μm reflectances were used, then CER(1.24) would be even smaller. Similar behavior would be seen for the nearly adiabatic cloud CER profile measured in situ (Figure 6.30b) within a cloud containing drizzle, except the values would be slightly larger. Ideally, retrievals of CER using two or more of these channels should provide information about the vertical profile of CER. Attempts to retrieve vertical profiles have been met with limited success (Chang & Li, 2002; King & Vaughan, 2012) and only for nonprecipitating clouds.

Assuming adiabatic growth, MBL CER and LWC values should increase from the cloud base to cloud top. This is observed from aircraft in situ measurements (Miles et al., 2000; Wood, 2005; Wu et al., 2020a). Satellite studies often fail to record this adiabatic growth, however. CER(3.75) is generally smaller than CER(SWIR) as seen in various studies (e.g., Zhang et al., 2017; Xi et al., 2014) and in Figure 6.31, which shows MODIS (Platnick et al., 2003) CER differences. Nearly everywhere in the figure, CER(1.6) exceeds CER(3.7). The only exceptions are found off the western coasts in the three of the subtropical Sc domains. As these are nonprecipitating clouds, the differences cannot be explained by drizzle and are likely due to either nonadiabatic cloud-droplet distributions or cloud heterogeneity effects, which are discussed below.

Multiple hypotheses on the nonadiabaticity in the satellite data have been proposed. Nakajima et al. (2010a) and Nagao et al. (2013) suggested that the CER difference derived from different bands is due to vertical inhomogeneity of cloud-droplet size; specifically, drizzle-sized drops in the lower portion of the cloud layer potentially increase CER (SWIR) retrievals and small cloud droplets at the cloud top because evaporation from entrainment could yield smaller CER(3.7) retrievals. Drizzle in MBL clouds causes CER to increase toward cloud base, which could yield larger CER(SWIR) retrievals (e.g., Chang and Li, 2002; Nakajima et al., 2010b). Studies from both observations and model simulations have produced mixed conclusions about the impact of drizzle (vertical inhomogeneity) on CER retrievals (Painemal & Zuidema, 2011;

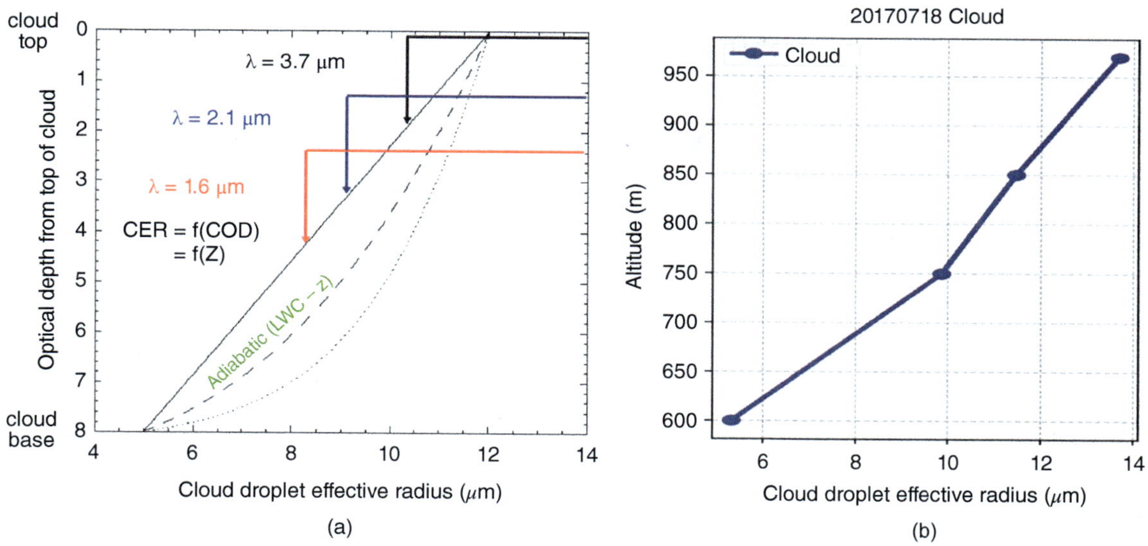

Figure 6.30 (a) Theoretically retrieved CER values from MODIS 3.7-, 2.1-, and 1.6-μm wavelengths for adiabatic clouds. Source: Adapted from Platnick, (2000). (b) MBL CER profile measured by G-1 FCDP sensor (diameter < 50 μm) on 18 July 2017 during ACE-ENA.

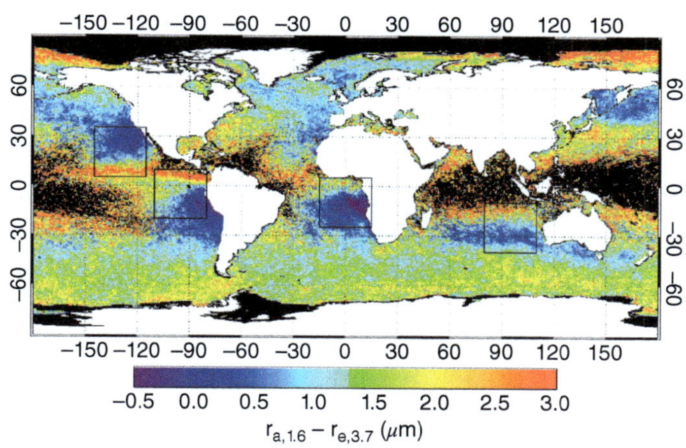

Figure 6.31 Annual mean difference between MODIS 1.6- and 3.7-μm CER retrievals for nonprecipitating marine overcast clouds. Source: Seethala and Horváth (2010)/John Wiley & Sons.

Xi et al., 2014; Z. Zhang et al., 2013; Zinner et al., 2010) despite mostly adiabatic profiles from aircraft in situ measurements. From retrievals of CER and LWC profiles using surface-based radar in the ARM ENA transition area, Wu et al. (2020a) found that CER and LWC profiles peak at a depth ~25% from the top of the cloud and gradually decrease toward the cloud base. This peak is most prominent for precipitating clouds. Such profiles could explain the relative values of CER in Figure 6.31 in the transition areas but not necessarily near the coasts.

Zhang and Platnick (2011) demonstrated that horizontal inhomogeneity in the cloud field can enhance SWIR absorption, while having little impact on CER(3.7). Thus, the multispectral retrievals should be in closer agreement for St rather than for Sc or Cu cloud fields. The negligible heterogeneity effect at λ = 3.7 μm, confirmed by other studies (e.g., Painemal et al., 2013b), is likely due to the small range of sensitivity to COD for CER(3.7) noted above. That is, 3D scattering in inhomogeneous conditions increases the photon path length through the cloud and, thus, the absorption. At λ = 3.7 μm, the extra scattering should have little influence on the absorption when COD exceeds ~2–4, depending on the actual CER. The SWIR and MWIR retrievals are more likely to agree when CER(SWIR) is less than 15 μm (Zhang & Platnick, 2011). For larger droplets, a given increase in path length has more impact on the retrieved CER than for smaller droplets. Thus, over open ocean where trade Cu and more open-celled Sc predominate, CER(1.6) is very likely to be impacted by such 3D effects and is greater than CER from other channels (e.g., Figure 6.31). It is apparent that a combination of cloud-top entrainment, drizzle, and vertical and horizontal inhomogeneities is probably responsible for the relatively rare adiabatic multispectral CER retrievals. Regardless of the cause, such uncertainties in the vertical profiles of CER can complicate the estimation of N_c, which relies on the adiabatic assumption (e.g., Bennartz, 2007; Painemal & Zuidema, 2011), and LWP, which requires certain assumptions. Additionally, it exacerbates the retrieval of vertical cloud-droplet profiles.

Cloud LWP can be determined from imager retrievals or from microwave radiometer measurements (e.g., Wentz, 1997). For the true effective radius CER(col), which represents the entire cloud column,

$$\text{LWP} = 2/3 * \text{CER(col)} * \text{COD}. \tag{6.3}$$

As discussed above, most imager retrievals yield a value of CER that is more representative of the top of cloud. Typically, the SWIR or MWIR retrieval has been assumed to represent the entire column of cloud and LWP has been computed with (6.3). For adiabatic clouds, however, that assumption should represent an overestimate of CER(col). For an adiabatic cloud,

$$\text{LWP} = 5/9 * \text{CER(top)} * \text{COD}. \tag{6.4}$$

These two values differ by a factor of 0.83 if treating CER(col) same as CER(top), making it simple to compute one from the other.

Since LWP can be retrieved directly from microwave radiometer (MWR) measurements and COD is relatively insensitive to CER, CER(col) can be retrieved from equation (6.3) using matched imager and microwave measurements/retrievals (e.g., Bennartz, 2007; Minnis et al., 1992). With data from matched MODIS and Advanced Scanning Microwave Radiometer (AMSR) on Aqua, Bennartz (2007) demonstrated that equation (6.4) is appropriate for Sc clouds near the coast in the subtropical regimes. This was confirmed by Seethala and Horvath (2010) and Painemal et al. (2016). Better agreement was also found with equation (6.4) for midlatitude and polar oceanic regions, but LWP from MODIS was still too large compared to that from AMSR. Part or all of that difference is likely due to the high SZAs at the time of measurements. 3D effects become more pronounced at SZA > 60° and yield an overestimate of the imager COD (e.g., Loeb & Coakley, 1998; McHardy et al. 2018) that will similarly impact LWP. In the open oceans and transition areas with warm broken clouds, Seethala and Horvath (2010) found that the MODIS LWP is lower than the MWR LWP with equation (6.4) and less so with equation (6.3). Conversely, the MODIS overestimates the MWR LWP for overcast warm clouds using equation (6.3) but matches it, on average, using equation (6.4). A dependence of the bias on cloud fraction (Horvath and Davies, 2007) is due to overestimation of LWP by the MWR retrieval in partly cloudy scenes as seen in the comparisons with surface MWR retrievals plotted in Figure 6.32a (Painemal et al., 2016). Furthermore, the use of CER(3.75) with equation (6.4) in those cases yields a more accurate estimate of LWP than using CER(SWIR) (e.g., Figure 6.32b).

As the preceding discussion indicates, the MWR data should be used carefully when validating optical retrievals and when used for LWP monitoring. Fractional cloud cover, a clear-sky bias, and sensitivity to atmospheric water vapor are potential sources of error in the MWR retrievals (Painemal et al., 2016; Seethala & Horvath, 2010). And even though such retrievals are only accurate over water surfaces, there are many more satellites carrying MWRs now as part of the Global Precipitation Measurement (GPM; see Hou et al., 2014). Thus, it should be possible to monitor marine stratus clouds more thoroughly around the globe. Most retrievals of LWP from the optical imager data have been compared with MWR estimates of LWP only for warm (CTT > 273 K), nonprecipitating clouds. Additional uncertainties in the MWR LWP are introduced when considering supercooled or mixed-phase clouds (e.g., Horvath and Davies, 2007). Nevertheless, MWR and optical retrievals are critical resources for understanding and monitoring Sc properties.

Cloud-base height has been estimated using a variety of methods that either directly estimate the base height or estimate cloud geometric thickness and subtract it from CTH. The former category includes stereoscopic methods that require nearly simultaneous measurements from at least two angles (e.g., Böhm et al., 2019; Marshak & Davis, 2021) and estimating CBH as being the same as the LCL (e.g., Wang & Rossow, 1995). The latter include both theoretical (Bennartz, 2007; Hutchison et al., 2006; Meerkötter & Zinner, 2007) and empirical parameterizations based on retrieved properties from matched imager and active sensor datasets (e.g., Minnis et al., 1992, 2021; Noh et al., 2017). Key among those properties are COD and LWP. For the thickness subtraction approach, the CBH accuracy depends on the uncertainties in both thickness and CTH. Overall, for low water clouds, this translates to accuracies in CBH of 0.0 ± 0.5 km for single-layer warm clouds (Minnis et al., 2021) and to 0.4 ± 1.1 km for a class of supercooled clouds (Noh et al., 2017). While the stereoscopic methods are relatively accurate, they require multiangle views and space between clouds, thus limiting the sensors and

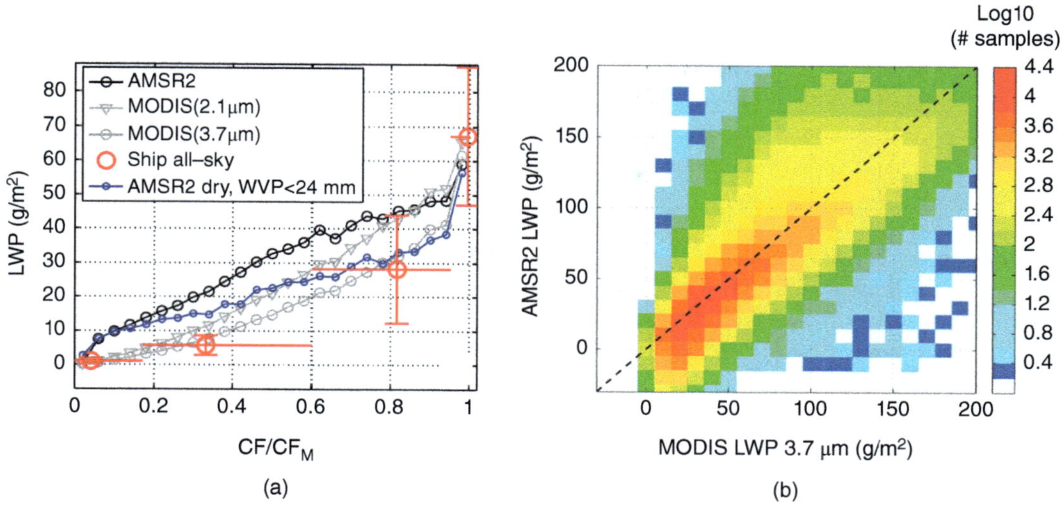

Figure 6.32 (a) LWP binned as a function of ceilometer CF and CERES MODIS CF, CF_M. AMSR2 (black), MODIS 2.1 μm (gray triangles), and 3.7 μm (gray circles) are spatially collocated with ship transect. Ship-based all-sky LWP (red circles) is a data subset taken between 11:00 and 15:00 LT. Blue lines denote AMSR2 LWP for samples with precipitable water vapor, WVP < 24 mm. Horizontal bars are CF bin range and vertical bars denote standard deviation. (b) Bivariate histogram of AMSR2 LWP and CERES MODIS adiabatic 3.7-μm LWP over the MAGIC domain (160–120°W and 19–35°N) for overcast scenes. Source: Painemal et al. (2016)/John Wiley & Sons.

sampling. Thus, the route through cloud thickness may be a more promising approach.

Reliably detecting stratus and retrieving low cloud properties underneath higher clouds is a continuing challenge because of the wide variation in both low and high cloud properties that do not necessarily produce unique spectral signatures. A thin veil of cirrus over a bright stratus has negligible impact on visible reflection and a noisy signal in infrared channels because of its low COD. Likewise, stratus below a thick ice cloud may contribute to the reflectance, but nothing to the infrared radiation, so that the source of additional reflectivity in the cloud could just as well be more ice in the cloud. Still, efforts to unscramble the spectral signals emanating from multilayer systems have been underway for several decades and are continuing.

Typically, consistency between retrievals using different wavelengths is used to detect multiple-layer (ML) clouds. The current methods that have either been demonstrated or are being applied routinely include two groups: multi-instrument and single instrument with multichannels. Examples of the former are Lin et al. (1998), Minnis et al. (2007), and Joiner et al. (2010). The other class of methods typically examines discrepancies between the visible channel COD and that determined from another channel, often infrared, that allows the detection of optically thin cirrus clouds only. Thus, they only detect ML clouds for thin cirrus over a lower cloud. Techniques that employ this approach include those of Pavolonis and Heidinger (2004), Chang and Li (2005), and Chang et al. (2010). Another strategy employs channels sensitive to clouds at different heights (e.g., Wind et al., 2010) along with appropriate spectral tests to isolate ML cloud pixels. To further improve ML detection, one or more methods have been applied together (e.g., Marchant et al., 2020; Platnick et al., 2017). To identify ML clouds, Desmons et al. (2017) used multiangle polarized spectral reflectances and two different retrievals of CTP from the Polarization and Directionality of the Earth's Reflectance (POLDER) on a satellite in the A-Train. Success rates for some of these ML detection methods have been evaluated using CloudSat and/or CALIPSO profiles. For example, Desmons et al. (2017) evaluated their method and that of MODIS Collection 6 (Platnick et al., 2017) for COD > 5 and CF > 0.95. They found that the Collection 6 and POLDER hit rates for ML clouds are 56% and 58%, respectively, while the corresponding risks of misclassifying the pixel as single layer or ML are 31% and 33%.

Once ML clouds are detected, it is helpful to retrieve their properties. Minnis et al. (2007) used a two-instrument approach to estimate low cloud properties underneath higher clouds, but spatial resolution, temporal sampling, and confinement to water surfaces limit the method's application. Multispectral imager data can be used alone with the proper assumptions and channel selection (e.g., Chang et al., 2010; Li & Chang, 2005; Minnis et al., 2021). Their methods tend to underestimate the lower and overestimate the upper cloud CTPs. Watts et al. (2011) constructed an optimal estimation method that detected ML clouds and solved for the upper and lower cloud properties at the same time. Several of the newer methods appear promising, but additional evaluation and improvements will be necessary before ML retrievals become a routine product.

Drizzle detection and intensity are also needed for comprehensive Sc system monitoring. Although the CloudSat CPR has proven to be a reference resource for these quantities (e.g., Leon et al., 2008) it is restricted in sampling, as noted earlier. The GPM core satellite carries a dual-frequency precipitation radar (DPR) that is less sensitive to light precipitation than the CPR. Attempts to use the DPR for quantifying drizzle have improved the detection capability in marine stratus, but the product is less than ideal for subtropical Sc (Hamadai & Takayuba, 2016). Precipitation can also be detected using MWR data, but the MWR retrieval is highly uncertain in marine Sc areas (e.g., Adhikari et al., 2019).

Although several approaches have been proposed, drizzle detection and quantification from imager data have not yet become standard products. Most methods rely on estimates of CER, since the autoconversion rate, or conversion of cloud droplets to drizzle drops (radius > 20 μm), is inversely proportional to N_c (e.g., Wood, 2005b). Additionally, drizzle droplets are most abundant in the upper portion of the cloud, and then their number decreases below because of collision and coalescence as the drops fall and grow (e.g., Wood, 2005b; Wu et al., 2020a). Gerber (1996) found that for clouds having CER > 15 μm, drizzle is quite likely. The cloud must also be thick enough for droplets to grow to drizzle size. Drizzling cloud-layer thickness tends to be almost twice that of nonprecipitating cloud layers (e.g., Wu et al., 2020b). Although there are exceptions, aircraft measurements taken in the SE Pacific indicate that the likelihood of drizzle occurring is nearly zero until the cloud is 200 m thick, increasing rapidly to nearly 100% for $\Delta Z = 500$ m (Serpetzoglou et al., 2008). Thus, drizzle is most likely for thick clouds in a nearly pristine boundary layer. Instead of using a single-channel CER threshold, Chen et al. (2008) used multispectral retrievals in the linear method of Chang and Li (2002) to estimate CER at cloud top and base. They found relatively good correlation between rain rates and cloud CER(top), and better correlation for CER(base) when CER > 15 μm. Additionally, they determined when CER(top) < CER(base) and LWP > 100 g m^{-2}, the drizzle rate exceeds 0.1 mm h^{-1}. Shao and Liu (2004) combined MWR LWP and retrievals of CER(1.6) and COD to construct a drizzle index that can be used to estimate separate drizzling and nondrizzling clouds. In a different approach using MWR attenuation and adiabatic estimates of LWP from CER and COD, Bennartz et al. (2010) estimated rainwater path in stratus systems. That initial study shows some promise for quantifying drizzle amounts over the open oceans. While the initial research into passive satellite drizzle detection and retrieval is encouraging, there remains much more investigation before drizzle from stratus can be reliably monitored globally on a routine basis.

Many of the low-cloud products discussed above rely on CER and COD. There are numerous limitations and sources of uncertainty in the optical retrievals of these parameters that should be mitigated in future retrievals. For example, the artificial increase in COD and CER with SZA (e.g., Ahn et al., 2018) due to 3D effects hampers the ability to accurately determine their diurnal cycle and averages, especially in higher latitudes. It might be possible to account for this effect by developing either theoretical or empirical models based on imager retrievals matched with surface or satellite radar-lidar and MWR data with contextual information from the imager, such as visible and/or infrared spatial heterogeneity. The effects of viewing and illumination angles must be taken into account in any such approach (e.g., Grosvenor & Wood, 2014). Nocturnal retrievals of these parameters are quite limited due to saturation of infrared channels. Efforts to utilize reflected moonlight in highly responsive visible channel sensors like that on VIIRS show some skill but are limited by the phase of the moon and surface light sources (e.g., Walther et al., 2013). Further research is required to fully exploit the potential of moonlight reflectance. Another factor confusing the retrievals is the presence of ice and water in an observed low cloud. Minimal research has been conducted on detecting mixed-phase clouds and partitioning the ice and liquid amounts, although the use of the multispectral retrievals using the VIIRS imager show some skill (Miller et al., 2014).

Low cloud detection is difficult whenever the cloud element is small or the background provides little contrast to the cloud. The former is common over the marine trade wind regions and over land, while the latter is most prominent at night, especially in polar regions where the clouds and surface can be quite cold and of similar temperatures. Higher resolution sensors such as VIIRS are helpful in detecting smaller clouds, but the potential of correctly identifying clouds at night over polar ice caps remains around 80%, compared to ~93% for the rest of the globe (e.g., Trepte et al., 2019). Training cloud masks with active sensors have not yet pushed the accuracies past that limit either (e.g., Heidinger et al., 2012). Thus, proper assessment of cloud properties over the poles, where cloud impacts can be sensitive to climate change, will necessitate additional innovations.

Traditionally, most retrievals of cloud parameters have been based on physical retrievals. That is, the atmosphere, surface, and cloud properties are modeled to produce radiances at the top of the atmosphere and matched with observations. Over the last decade, machine learning techniques have been explored to further improve estimates of those properties, particularly in situations where the radiance signals are ambiguous or the information content is small compared to the noise. These methods are especially suited to some of the difficulties noted above, and with the advent of the surface-based and orbiting active sensors providing reference information around the globe, it is possible to train the methods with some confidence. Kox et al. (2014) used an artificial neural network (ANN) approach to improve cirrus CTH and COD, while Minnis et al. (2016b) estimated COD for thick ice clouds with some skill at night using a similar scheme. Cloud-top height estimates at all levels are achieved with higher accuracy than in the past using ANNs (Håkansson et al., 2018). Multilayer cloud detection and cloud thickness can also be improved for both day and night using ANNs (Minnis et al., 2019; Sun-Mack et al., 2017). Although much of the machine learning methods so far have focused on high clouds, it is expected that similar results could be obtained for low cloud thickness, COD, and LWP, as well as for other properties such as drizzle, mixed phase, or cloud vertical profiles.

This section has provided an overview of the state of mainly passive low-cloud remote sensing. It is by no means exhaustive. Many scientists around the globe have been researching and refining methods for monitoring low clouds and their properties for decades and their efforts continue. Innovative measures are still being developed to address the many shortcomings in accurately characterizing stratus clouds everywhere, at all times of the day. It is hoped that with the aid of specialized surface, aircraft, and satellite-based instruments and new approaches, the needed advances will be realized.

6.6.2. Ground-Based Remote Sensing

As mentioned above, passive satellite sensors, such as MODIS and GOES, can provide large-scale, reliable low cloud properties and diurnal and seasonal variations, with certain limitations, especially in defining their vertical distributions. Active satellite remote sensors, such as radar and lidar from CloudSat and CALIPSO, can observe the majority of clouds and yield a nearly complete vertical cloud profile, but they cannot provide longtime, continuous observations over a specific location. Observations taken from surface sites, however, can produce much valuable information for a given location. There have been a great number of studies of low clouds using surface observations that complement the satellite and aircraft research. It is not possible to consider all of them here, so they will be briefly reviewed with some relevant examples and literature to complement those presented in the previous sections.

In addition to surface observers, many modern weather stations, particularly in North America and Europe, measure a variety of parameters, including low-cloud CBH and CF automatically and continuously using ceilometers (Illingworth et al., 2019; NOAA, 1998). Most ceilometers use a laser with the lidar method to detect clouds and measure their base height above ground level. Rain and snow are detected by the Light-Emitting Diode Weather Identifier (LEDWI), an instrument in the Automated Surface Observing System (ASOS) complement; it also has potential for detecting drizzle (e.g., Wade, 2003). Additional ASOS instruments provide visibility estimates and icing detection. To date, these instruments have not been fully exploited for their potential applications for studying and monitoring Sc properties.

Cloud fraction and cloud radiative effects at the surface can also be determined from radiative flux measurements taken by surface-based SW and LW broadband radiometers (e.g., Dong et al., 2006, 2010; Liu et al., 2011; Long & Ackerman, 2000; Long et al., 2006; Xie et al., 2014). These radiometers routinely measure radiation at many sites around the world as part of systems such as the international Baseline Surface Radiation Network (BSRN; Driemel et al., 2018). Total column aerosol optical depth is also derived operationally from surface spectral radiometers at a variety of locations around the globe (e.g., Giles et al., 2019; Holben et al., 1998), but those measurements require cloud-free skies. Additionally, they provide no information about the vertical location of the aerosols, which is necessary to determine their impact on Sc clouds. To partially address those shortcomings, a network of lidars has been employed in Europe to continuously monitor vertical profiles of aerosol concentrations (Bösenberg et al., 2004). While the aerosol and flux measurements have potential for enhanced monitoring of low stratiform clouds, it is not clear that any concerted effort has been made to exploit those data for Sc research. That undertaking has mainly been left to scientists analyzing data from heavily instrumented research sites.

Since the early 1990s, the DOE ARM program has pioneered the designing and continuous operation of comprehensive ground-based atmospheric observations, especially for observing and retrieving the profiles of low cloud properties (Ackerman et al., 2016; Shupe et al., 2016). The high-frequency, nearly continuous observations by ARM cloud radar-lidar systems can provide more accurate cloud vertical distributions. Such observing systems are limited in that they view only a small column of the atmosphere above the instruments, providing only a pencil beam measurement. Nevertheless, they sample all of the weather that passes over the site. In the case of the shipborne ARM AMF, the location changes at the ship speed, so the observed area is broader than at the fixed sites, but at the expense of not sampling all of the clouds over a given location.

Ackerman et al. (2016) briefly describe the history of the ARM program, surface sites, and key science goals, and summarize the organized effort to establish continuously operating ground-based observatories for monitoring cloud properties and atmospheric states. Mather and Voyles (2013) summarize the available instruments, data available, and potential for the ARM sites. Mather et al. (2016) point out that improving cloud parameterizations in GCMs through greater understanding of cloud and radiation processes obtained from a combination of modeling and data analysis is the fundamental goal of the ARM program. The ARM sites serve as the model for developing similar observatories in other nations. In Europe, the Cloudnet program (Illingworth et al., 2007) was established to utilize the measurements of cloud properties from ARM-like instrument complements in Chilbolton, UK; Palaiseau, France (Haeffelin et al., 2005); and Cabauw, Netherlands. Haeffelin et al. (2018) describe the sites having a complement of instruments similar to those used by ARM. In addition to the original Cloudnet sites, cloud profiling systems are deployed at Lindenberg, Germany, Potenza, Italy, and other European locales (Lamer et al., 2021). The products generated from those measurements and the algorithms used to retrieve the cloud properties, as well as some highlights of the analyses are also summarized by Haeffelin et al. (2018). Collaborative efforts between ARM and European researchers have been underway for several years and have led to some fruitful advances in retrieval algorithms and cloud parameterizations (Haeffelin et al., 2018). A similarly equipped observatory was established in Lanzhou, China, to address many of the ARM goals and those of most interest to weather in China (Ge et al., 2018; Huang et al., 2008). Together these sites should provide a good characterization of continental stratus and stratocumulus, while the coastal sites and ship deployments noted earlier are valuable for marine Sc studies. Examples of studies from these other sites are given in the references. Below, a few examples from ARM are presented, which, together with highlights given in previous sections, should be representative of the instrument complements and the types of studies that have been conducted with similar measurements.

The primary ARM aerosol, CCN and cloud observations and retrievals, as well as their uncertainties and references used in this chapter are listed in Table 6.1. The centerpiece of the ARM cloud instrument array comprises the 35- and 95-GHz cloud radars, which operate at wavelengths of 8.6 and 3.2 mm, respectively, in a vertically pointing mode. Following the pioneer work of Lhermitte (1987) and development by NOAA and Penn State groups (e.g., Frisch et al., 1995; Miller & Albrecht, 1995), ARM researchers modified the cloud radars for the ARM community. Those modifications permitted operational use of the radars, which provide continuous profiles of radar reflectivity from hydrometeors in the radar field of view, allowing the identification of clear and cloudy conditions (Moran et al., 1998; Widener et al., 2012). Kollias et al. (2020) introduced the history of ARM cloud radar development from the first generation of millimeter-wavelength cloud radars (MMCRs; Moran et al., 1998) to the Ka-band ARM zenith radars (KAZRs) and second-generation KZAR (KAZR2). In addition to these profiling radars (zenith view), they also briefly introduced and discussed the newly developed scanning radars in last decade to improve our understanding of cloud and precipitation processes. These scanning radar measurements can provide spatial variability of cloud system evolution and more importantly can evaluate the representativeness of the profiling radar column measurements of regional processes. The radar data, when combined with ceilometer, lidar, and other instrument data at the ARM sites, can be used to retrieve column and vertical profiles for a wide variety of cloud parameters.

Cloud fraction derived from ground-based remote sensors (vertical pointing) is simply the percentage of radar-lidar returns that are cloudy within a specified sampling time period (e.g., an hour or a month). The CF is given by the ratio of the number of hours when both the radar and lidar/ceilometer detected clouds to the total number of hours when all measurements (radar/lidar/ceilometer) were available. Total CF is the fraction of time when a cloud is detected anywhere in the vertical column, and low CF (CF_L) is the fraction of time when clouds occur below an altitude of 3.3 km. Since CF is defined differently for satellite imagers, surface observers, and ground-based remote sensors (Dong et al., 2006; Xi et al., 2010), special attention must be paid when comparing the CFs derived from these three sources due to their large temporal and spatial average differences. This is usually accomplished by temporally averaging the smaller field of view measurements to accommodate the greater spatial coverage of the satellite or other surface observation.

Cloud-base height (CBH) is derived from a composite of Belfort laser ceilometer, micropulse lidar (MPL), and cloud radar data (CloudBaseBestEstimate, Clothiaux et al., 2000), while CTH is usually derived from the cloud radar reflectivity profile. CBH values derived from ceilometer and lidar measurements are more accurate than those derived from cloud radar reflectivity because the laser ceilometer and lidar are sensitive to the second moment of the particle distribution (or the cross-sectional area of the particle), whereas cloud radar reflectivity is sensitive to the sixth moment and can detect precipitation-sized drops (such as drizzle) below cloud base. Ceilometer and lidar signals can be severely attenuated due to absorption by optically thick liquid cloud

Table 6.1 ARM instruments and their corresponding measurements and uncertainties, as well as the low stratiform cloud retrievals.

Parameter	Instruments/methods	Uncertainty	References
Cloud-base height CBH	Ceilometer Micropulse lidar	15 m	Clothiaux et al. (2000)
Cloud-top height CTH	35/95-GHz cloud radar	30/43 m	Clothiaux et al. (2000); Kollias et al. (2016)
Cloud-base and cloud-top temperatures CBT/CTT	Radiosonde sounding	0.2 °C	ARM website: www.arm.gov
Profiles of radar reflectivity, velocity, and spectra	K-band ARM Zenith Radar (KAZR), 35 GHz W-band ARM Cloud Radar (WACR), 95 GHz	Sensitivity: −50 dBZ at 2 km	Clothiaux et al. (2000); Kollias et al. (2016)
Vertical air velocity	Doppler Lidar	Vary with time and height	ARM Doppler Lidar Handbook. DOE/SC-ARM-TR-101
Cloud LWP	Microwave radiometer	~20 g m^{-2} for LWP < 200; ~10% for LWP > 200	Liljegren et al. (2001) Cadeddu et al. (2013)
Layer mean of cloud LWC/r_e/N_C	Retrieval method	~10% for LWC and r_e ~30% for N_C	Dong et al. (1998, 2014a, 2014b); Dong and Mace (2003b); Wu et al. (2020a)
Profile of cloud LWC_c, r_c	Retrieval method	~30% for LWC_c, ~15% for r_c	Wu et al. (2020a)
Cloud optical depth, τ	Shortwave radiometer Retrieval method	~10%	Dong et al. (1997, 1998); Dong and Mace (2003a)
Profile of drizzle LWC_d/r_d/N_d	Retrieval method	~30% for r_d, and ~50% for LWC_d and N_d	Wu et al. (2020a)
CCN and aerosol properties	Aerosol Observing System	1-min resolution; uncertainties <10%	ARM website: www.arm.gov

layers, and these signals can only penetrate about 200 m through the cloud base (Sassen, 1991). Therefore, the ceilometer/lidar derived CBH is used as the lowest CBH. For optically thin clouds that do not completely attenuate the MPL beam, the highest lidar cloud detection is used for CTH because the radar is less sensitive to smaller hydrometeors.

The development of active cloud remote-sensing technology in the last three decades has provided new opportunities to improve our understanding of cloud physical processes. Several radar-radiometer-based retrieval methods have been developed to retrieve low cloud microphysical properties (Dong et al., 1997, 2014b; Frisch et al., 1995; Huang et al., 2012; Zhao et al., 2012). These algorithms can be classified into two types: (a) using the solar transmission as additional independent information (e.g., Dong et al., 1997, 1998, 2014a,2014b; Fielding et al., 2014, 2015) and (b) applying additional assumptions without using solar transmission (e.g., Frisch et al., 1995, 1998). Each type of algorithm has its strengths and weaknesses. For example, the solar transmission-based algorithms have been demonstrated to be more accurate because they use additional constraints, but they can only be applied in a narrow range of conditions (daytime and overcast). In contrast, the algorithms using additional assumptions can be implemented more generally but tend to yield larger uncertainties. To take the advantage of both types of retrieval algorithms, regression-based power-law relationships were developed to retrieve cloud microphysical properties (e.g., Dong & Mace, 2003b; Dong et al., 2014b; Fox & Illingworth, 2002). These methods retrieve cloud microphysics that agree statistically well with aircraft in situ measurements or other retrieval techniques but may not perform well in a case-by-case manner.

In a historical overview of ARM contributions to cloud property retrievals, Shupe et al. (2016) describe advances in using specific passive and active remote sensors or multiple remote sensors. While previous studies and routine ARM ground-based observations and retrievals have significantly advanced our knowledge and understanding of maritime and continental low cloud properties, a quantitative characterization of their cloud properties has not yet been fully realized. Some of the processes (transition from cloud droplets to drizzle drops) are still poorly understood and outstanding issues remain. For example, what are the roles of autoconversion and accretion processes in warm rain processes? And what is the representativeness of surface-measured CCN at the cloud base? Most previous studies have assumed that the surface-measured CCN and aerosol properties represent those at cloud base. Vertical velocity can be estimated in clear-sky conditions and below cloud base using Doppler lidar measurements, which can potentially provide the relationship between the surface-measured CCN and aerosol properties with those at cloud base. Furthermore, Zheng et al. (2016) found a statistical relationship between the updrafts at cloud-base and cloud-top radiative cooling rates based on MAGIC ship-based measurements. As for boundary-layer dynamics, the higher-order moments of vertical velocity are widely used in different model parameterizations, such as higher-order turbulence closure and probability density function methods (Ghate et al., 2010), while the vertical velocity variance can be used to represent turbulence intensity in the subcloud boundary layer (Feingold et al., 1999).

Another outstanding issue is the profiling of MBL cloud and drizzle microphysical properties. Theoretically, MBL cloud r_e and LWC values should increase from cloud base to cloud top assuming adiabatic growth. This is observed from aircraft in situ measurements (Miles et al., 2000; Wood, 2005a,2005b; Wu et al., 2020a), whereas this does not always occur. Sometimes, the opposite results have been retrieved from the ground-based and satellite observations due to frequent drizzling in MBL clouds. Two of the great challenges for both ground-based and satellite remote sensing of low clouds are the retrieval of MBL cloud and drizzle properties and the retrieval of mixed-phase clouds in the polar regions. As the latter is even more difficult (Shupe et al., 2008a,2008b), the former will serve as the focus for further discussion.

The greatest challenge to simultaneously retrieve both cloud and drizzle properties within an MBL cloud layer is the differential response of the radar pulse to various hydrometeors. Radar reflectivity depends on the sixth power of the particle size and, thus, can be highly weighted by a few large drizzle drops in a drizzling cloud layer. Very often the reflectivity peaks near cloud base, especially for MBL clouds. One of the early retrieval algorithms using the ground-based radar-lidar observations was developed by Frisch et al. (1995, 1998) who suggested a threshold of −15 dBZ for separately identifying cloud droplets and drizzle drops. They also estimated the cloud LWC_c profile by distributing the observed LWP according to the shape of radar reflectivity. Recently, Küchler et al. (2017) modified the method of Frisch et al. (1998) to retrieve cloud and drizzle liquid water content (LWC_c and LWC_d), and found the height of zero skewness, where cloud and drizzle reflectivity values are the same. Their retrieval uncertainty in drizzling clouds is below 20%, which is much smaller than the original Frisch et al. (1998) method. However, the zero-skewness height may not always exist in radar measurements or there may exist multiple zero-skewness heights across the cloud layer (Luke & Kollias, 2013).

Another representative algorithm uses radar-lidar observations to retrieve the profiles of drizzle properties below the cloud base because of the strong attenuation of lidar signals by cloud droplets. O'Connor et al. (2005) used the Doppler spectrum width and the ratio of radar reflectivity and attenuated lidar backscatter to retrieve the drizzle median diameter (D_0), shape parameter (μ), and number concentration (N_d) below cloud base. Westbrook et al. (2010) used the differences in lidar backscatter at two wavelengths to retrieve drizzle drop size. Different to those two methods, Fielding et al. (2015) developed an ensemble-based optimal estimation method to retrieve cloud and drizzle microphysical properties in which the algorithm converges fast and provides full assessment of the error statistics given the observational uncertainties. The retrieval method uses surface-measured shortwave zenith radiance as one of the observations to constrain the forward models, which limits its application to daytime only.

Building on previous studies, such as Dong and Mace (2003b), Frisch et al. (1995), and O'Connor et al. (2005), Wu et al. (2020a) developed a new method to simultaneously retrieve profiles of MBL cloud and drizzle microphysical properties using ground-based observations. This technique was validated by comparisons with aircraft in situ measurements taken during ACE-ENA. The first step in the retrieval is to decompose cloud and drizzle reflectivity within an MBL cloud layer from ARM cloud radar reflectivity measurements. It then finds the height where cloud droplets grow to their maximum size and assume that drizzle drops start to form below this height. All of the in situ cloud and drizzle PSDs measured during ACE-ENA were used to calculate their reflectivities at 35 GHz. The resulting PDFs of cloud and drizzle reflectivity are plotted in Figure 6.33a. The cloud-droplet reflectivity values cover a range of −40 to −12 dBZ, with

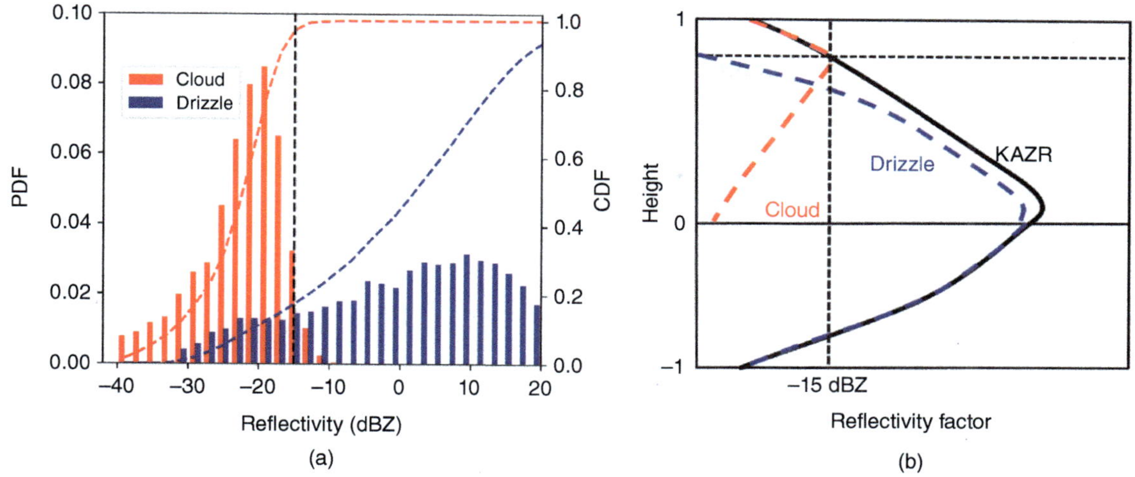

Figure 6.33 (a) Probability density functions (PDFs, solid lines) and cumulative density functions (CDFs, dashed lines) of cloud (red) and drizzle (blue) reflectivity calculated from aircraft in situ measurements during ACE-ENA. Black dashed line marks −15 dBZ. (b) Schematic diagram showing the separation of cloud (red) and drizzle (blue) reflectivity within a cloud layer. Black curve shows Ka-band ARM zenith radar (KAZR) measured reflectivity. −1, 0, and 1 represent drizzle-base, cloud-base, and cloud-top heights, respectively. Vertical and horizontal dashed lines show the location and height of −15 dBZ reflectivity factor. Source: Wu et al. (2020a)/John Wiley & Sons.

less than 1% of the samples having reflectivity greater than −15 dBZ (vertical dashed line in Figure 6.33a), while more than 80% of the drizzle samples have reflectivities exceeding this value.

From condensational growth, cloud droplets reach their maximum size near cloud top as does the cloud reflectivity where the maximum reflectivity of cloud droplets is assumed to be −15 dBZ (vertical dashed lines in Figure 6.33b) based on the results in Figure 6.33a and previous studies. The height of −15 dBZ near the cloud top (horizontal dashed line in Figure 6.33b) is defined as drizzle initiation height above which only cloud droplets contribute to the radar reflectivity. The KAZR reflectivity is due to both cloud droplets and drizzle drops above cloud base, while below only drizzle drops contribute to the reflectivity. Cloud reflectivity at the first range gate above cloud base is obtained by subtracting KAZR reflectivity at one range gate below the cloud base, which is shown as the first point in the red dashed line at the cloud base in Figure 6.33b. A cloud reflectivity profile is constructed starting at the cloud base up to the height of −15 dBZ using the adiabatic assumption of linearly increasing LWCc. If the cloud-droplet number concentration *N*c is constant in the cloud, then this assumption is equivalent to a linear increase of $\sqrt{Z_c}$. After obtaining the cloud reflectivity (Z_c, red dashed line in Figure 6.33b), the drizzle reflectivity (Z_d, blue dashed line in Figure 6.33b) is simply the difference between the KAZR observed reflectivity (Z_{KAZR}, black line in Figure 6.33b) and Z_c.

After decomposing the cloud and drizzle reflectivities within the MBL cloud layer, the drizzle properties below the cloud base were retrieved using the method of O'Connor et al. (2005) with a Gamma distribution assumed for the DSD. The cloud DSD is represented by a lognormal distribution (Dong et al., 1997, 2014a; Frisch et al., 1995; Miles et al., 2000). Details of the retrieval methods are described in Wu et al. (2020a). Figure 6.34 shows the decomposed cloud and drizzle reflectivities and retrieved cloud and drizzle microphysical properties for the case of 18 July 2017 during ACE-ENA.

The retrieved cloud and drizzle property profiles in Figure 6.34 were matched and plotted with the aircraft in situ measurements taken during the same period (Figure 6.35). As shown in Figure 6.35a and c, both the retrieved and the in situ measured r_c and LWC_c basically follow adiabatic growth, that is, increasing with height due to condensational growth. The retrieved N_c values were nearly constant with height while the in situ measured N_c values reached the maximum in the middle of the cloud (Figure 6.35b). For drizzle microphysical properties, both retrieved and in situ measured $r_{m,d}$ values rose (Figure 6.35d) from cloud top toward cloud base, whereas both N_d values decreased (Figure 6.35e). This makes physical sense because drizzle drops form near the cloud top, where they have smallest sizes but largest number densities as shown in Figure 6.35d and e. As they fall, drizzle drops grow by collecting cloud droplets and smaller drizzle drops through the collision-coalescence process, so drizzle number density decreases toward the cloud base. The in situ measured LWC_d values (Figure 6.35f) drops from cloud top toward cloud base, whereas the retrieved LWC_d values are slightly smaller in

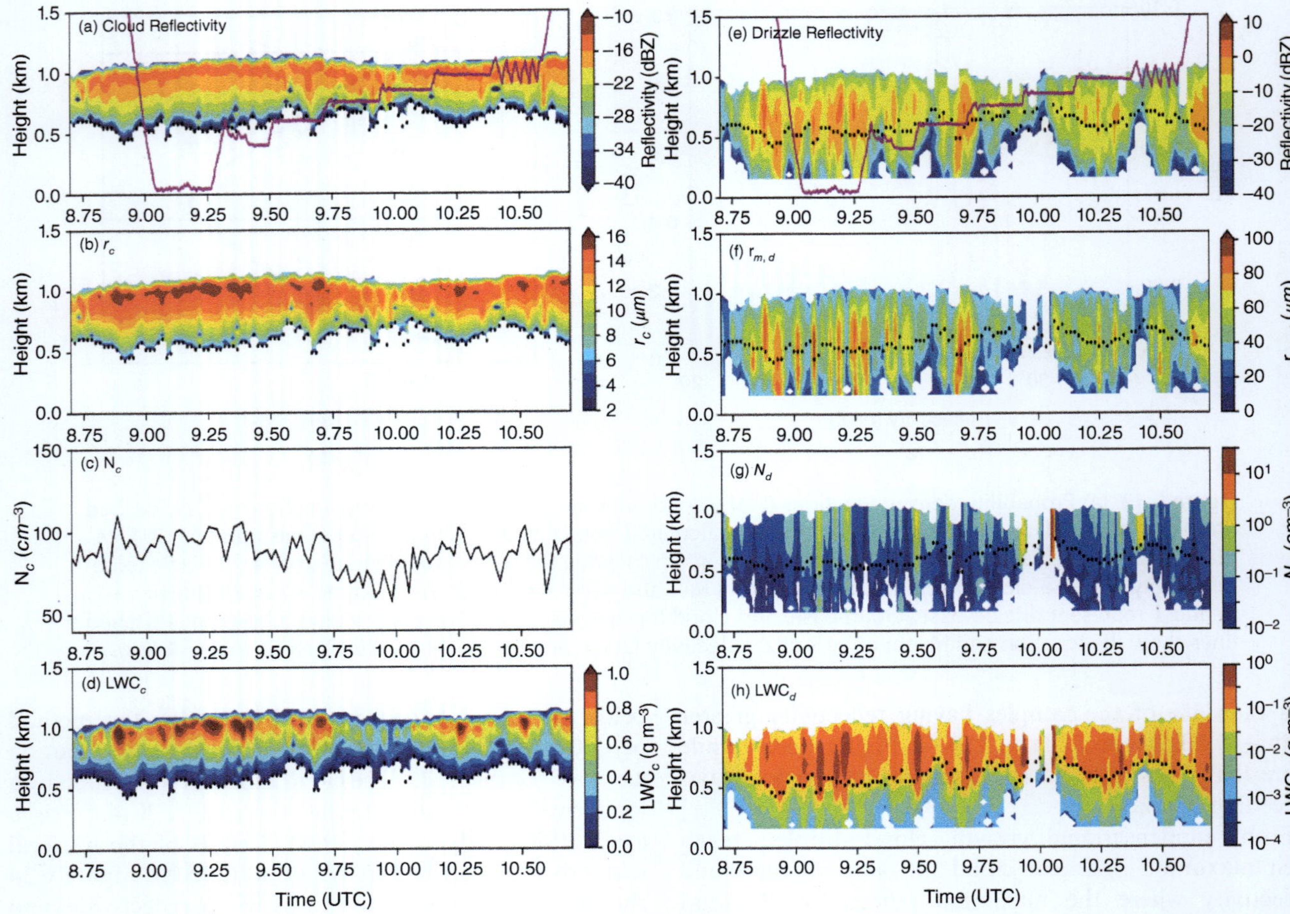

Figure 6.34 (a) Temporal variation of the profiles of decomposed cloud reflectivity from the DOE ARM KAZR reflectivity superimposed with aircraft flight track (magenta), (b) retrieved cloud-droplet median radius r_c, (c) cloud-droplet number concentration (N_c, layer mean), and (d) liquid water content LWC_c from ground-based observations on 18 July 2017 during ACE-ENA over the ARM ENA site. (e–h) Profiles of decomposed drizzle reflectivity from DOE ARM KAZR reflectivity with ceilometer measured cloud-base heights in black dots, retrieved drizzle drop median radius r_d, number concentration N_d, and liquid water content LWC_d. Source: Wu et al. (2020a)/John Wiley & Sons.

the upper level and larger near the bottom level than the in situ data. Using all matched aircraft and surface cloud samples and treating the aircraft in situ measurements as cloud truth during ACE-ENA, the estimated median retrieval errors are ~15% for r_c, ~35% for N_c, ~30% for LWC_c and r_d, and ~50% for N_d and LWC_d.

While these retrieval methods are designed for single-layered MBL clouds, they can be extended to shallow cumulus clouds to retrieve their cloud and drizzle microphysical properties as long as there are no ice particles existing in the clouds. That is, these retrieval methods cannot be used for mixed-phase clouds. However, they could be applied in multilayer conditions when low clouds occur underneath upper-level cirrus clouds. Such an application would assume that the MWR-retrieved LWP is only in low clouds. This is a big advantage compared to those using solar radiation as a constraint during the retrievals.

Wu et al. (2020b) used more than 4 years of ground-based observations and retrievals at the ARM ENA site to investigate the seasonal and diurnal variations of single-layered MBL (with three subsets: nondrizzling, virga, and rain) cloud and drizzle properties, as well as their vertical and horizontal variations. The mean drizzle frequencies are ~70% in winter and ~45% in summer with an annual mean drizzle frequency of ~55%. Average cloud thickness is deepest (0.8 km) for rain clouds, which is 4 (2) times that of nondrizzling (virga) clouds.

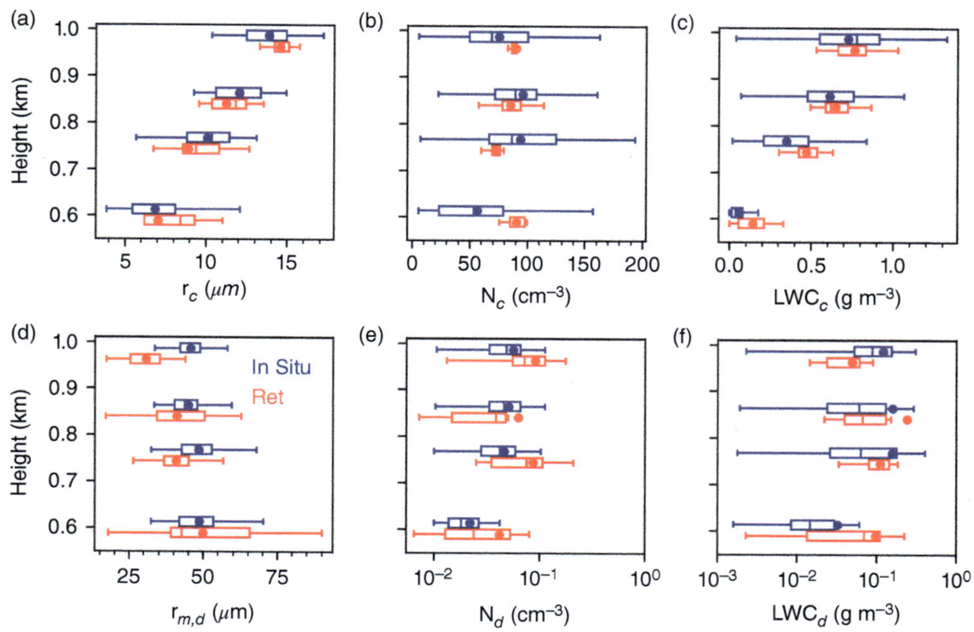

Figure 6.35 Boxplot of 1-s aircraft in situ measured (blue) and 1-min surface retrieved (red) cloud (top row) and drizzle (bottom row) properties for the four horizontal flight legs between 9:30 to 10:21 UTC in Fig. 6.34. Retrieved values in lower three levels are extracted from three range gates centered at the flight heights. Top-level retrieved values are extracted from the range gate at the flight level and the one below. Dots show the mean values at each level, and the vertical bars from left to right represent 10%, 25%, 50%, 75%, and 90% values. The aircraft data were confined to be within 25 km of the site. Source: Wu et al. (2020a)/John Wiley & Sons.

Profiles of retrieved cloud and drizzle microphysical properties during the 4-year period are illustrated in Figure 6.36. The retrieved CER (Figure 6.36a) is greatest (smallest) for rain (nondrizzling) clouds, and nighttime values exceed those during daytime (not shown). The drizzle drop radius, r_d, is up to 50% larger than its virga counterpart (Figure 6.36d). Drizzle number concentration (N_d, Figure 6.36e) and liquid water content (LWC$_d$, Figure 6.36f) are 3 and 1 order(s) lower than their cloud counterparts (Figures 6.36b and c), respectively.

Nearly all algorithms for retrieving cloud properties and investigating cloud processes are based on vertically pointing MMCR and KAZR observations. In the past decade, the DOE ARM facility not only upgraded and expanded its profiling millimeter radars but also developed six scanning cloud radars (SACRs) using W-band, Ka-band, X-band, and C-band wavelengths (Ka-W SCAR(2), Ka-X SCAR, XSCAR(2), and CSCAR(2)). For more details, see Table 6.1 in Kollias et al. (2020). These SCARs can provide large spatial coverage of different cloud and precipitation systems, such as Sc, shallow cumulus, and even deep convective clouds, by taking advantage of both millimeter and centimeter bands. These SCARs have been deployed at the ARM permanent sites and in field campaigns.

6.7. CONCLUDING REMARKS

Stratus and stratocumulus clouds, the most common type on the Earth, are a critical component of the weather and climate system that needs to be quantitatively understood and parameterized accurately in GCMs and weather prediction models to determine how they will impact a changing climate. While it is not possible to cover this topic in every detail, the discussions and examples as well as the studies cited here should facilitate more in-depth examination of this familiar but complex cloud type.

Based on our current knowledge, this chapter has provided a brief summary of where, why, and how they form and how they interact with radiation and the local environment. That knowledge, which was won from a myriad of studies using in situ and remotely sensed data from sea, land, air, and space, is continually growing as the instrumentation, computing systems, and retrieval technologies advance. Remote sensing is critical to furthering understanding of low stratiform cloud properties and processes, and for monitoring and predicting the occurrence and nature of these clouds around the globe. As the references noted herein have reported, the practical results of the various remote-sensing efforts have been used to verify and constrain both weather and climate

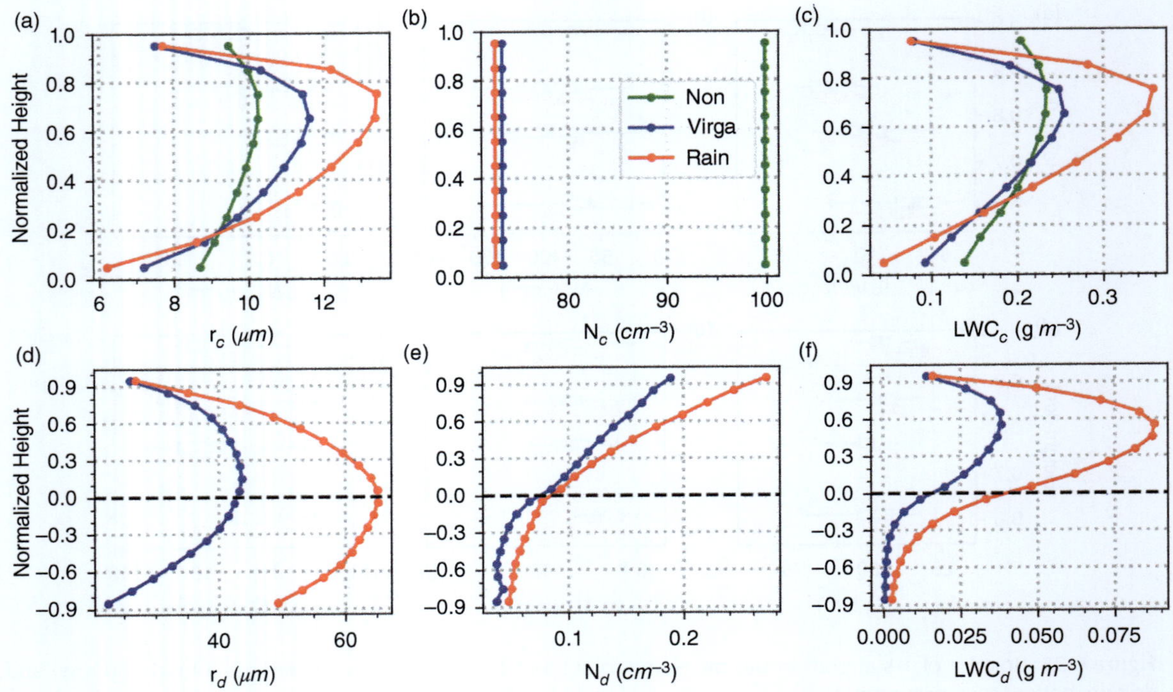

Figure 6.36 Profiles of cloud (top row) and drizzle (bottom row) microphysical properties in normalized height ($z_i = \frac{z - z_{base}}{z_{top} - z_{base}}$), with drizzle base as −1, cloud base as 0, and cloud/drizzle top as 1. Wu et al. (2020b)/American Meteorological Society.

models. Together with in situ measurements, they also have provided the bases for improved and new parameterizations such as those summarized in other chapters of this book. Additionally, remotely sensed stratus cloud properties are being assimilated into numerical weather prediction models to improve short-term forecasts (e.g., Benjamin et al., 2021). Even with a number of shortcomings and limitations, such data have proven extremely valuable.

Some of the outstanding problems remaining in surface and satellite remote sensing of stratus have been identified, and some possible avenues of addressing them have been discussed. In addition to the challenges of retrieving multilayer cloud properties, mixed-phase clouds, and separating MBL drizzle and cloud properties, more research is needed to improve the passive satellite remote sensing of cloud properties at night, particularly over the poles. Mixed-phase and drizzling low cloud retrievals from surface instruments also need much additional development. Investigations into retrievals of other parameters relevant to boundary-layer clouds, such as updraft speeds at cloud base (Zheng et al., 2016) and detection of decoupling (Zheng et al., 2018), should be pursued to complement the cloud retrievals. New satellite programs will facilitate the measurement of new parameters. These include, among others, the Aeolux (Reitebuch et al., 2012), a Doppler wind lidar that provides continuous retrievals of wind profiles around the globe, and the upcoming Earth Clouds, Aerosols and Radiation Explorer (EarthCARE; Illingworth et al., 2015), which combines a multispectral imager and broadband radiometers with a new lidar and cloud profiling radar.

Subtropical marine stratocumulus has garnered the focus of much of the experimental and modeling work on low stratiform clouds, but these clouds are important in many other areas (e.g., Mace, 2010). Some of the recent field campaigns noted herein have diffused that focus somewhat by gathering in situ and remote-sensing data on stratus clouds in the midlatitudes and polar regions. Analyses of these forthcoming datasets are already enhancing our understanding and characterization of these clouds poleward of 30°. Hopefully, future field programs and ARM-like deployments will continue to fill in the gaps from stratocumulus over oceans.

Overcoming the remaining obstacles for both active and passive remote sensing should lead to better modeling of low clouds that eventually will reduce the uncertainties in both weather prognoses and longer-term forecasts of climatic conditions. Although this chapter has mainly focused on ground-based and satellite remote sensing of stratus clouds, the important role of modeling cannot be ignored. An integrative analysis of model simulations and observations can provide a greater understanding of cloud and radiation processes than that using observations alone. It can also better inform cloud parameterizations developed for GCMs. Providing detailed model information is beyond the scope of this chapter; however, discussions of models relevant to stratocumulus can be found elsewhere in this book.

ACKNOWLEDGMENTS

Dr. Xiquan Dong's work was supported by the NSF grant AGS-2031750 and the NASA CERES project under grant 80NSSC19K0172 at the University of Arizona. Dr. Patrick Minnis's contributions were fully supported by NASA CERES project. The authors want to thank two reviewers providing constructive suggestions to significantly improve this manuscript. Special thanks to Dr. Sunny Sun-Mack, Ms. Rita Smith, Dr. Baike Xi, and Ms. Francine Dong, who provided plots for this manuscript.

LIST OF ACRONYMS

Acronym	Meaning
ABI	Advanced Baseline Imager
ACE-ENA	Aerosols and Clouds Experiment over the Eastern North Atlantic
ACI	Aerosol-Cloud interactions
AIR	Atmospheric Infrared Sounder
AMC	Arctic Mixed-phase Cloud
AMSR	Advanced Scanning Microwave Radiometer
ANN	Artificial Neural Network
AOT	Aerosol Optical Thickness
ARM/AMF	Atmospheric Radiation Measurement/ARM Mobile Facilities
AVHRR	Advanced Very High Resolution Radiometer
BTD	Brightness Temperature Difference
CACTI	Cloud, Aerosol, and Complex Terrain Interactions
CALIPSO	Cloud-Aerosol Lidar and Infrared Pathfinder Satellite Observations
CAP-MBL	Clouds, Aerosols, and Precipitation in the Marine Boundary Layer
CBH/CBT	Cloud-Base Height/ Cloud-Base Temperature
CCN	Cloud Condensation Nuclei
CER/CEH/CET	Cloud Effective Radius/Height/Temperature
CERES	Clouds and Earth's Radiant Energy System
CF_L/CF_T	Low and Total Cloud Fraction
COD	Cloud Optical Depth
COMBLE	Cold-air Outbreaks in the Marine Boundary Layer Experiment
CPR	Cloud Profiling Radar on CloudSat
CRE	Cloud Radiative Effect (CRE_{SW}, CRE_{LW}, and CRE_{net} for SW, LW, and Net, respectively)
CSET	Cloud System Evolution in the Trades
CTEI	Cloud-Top Entrainment Instability
CTH /CTT/CTP	Cloud-Top Height/Cloud-Top Temperature/Cloud-Top Pressure
Cu	Cumulus
EBAF	Energy Balance and Filed
ENA	Eastern North Atlantic
ERBE	Earth Radiation Budget Experiment
ERF	Effective Radiative Forcing
FCDP	Fast Cloud Droplet Probe
FIRE-ACE	First ISCCP Experiment Arctic Clouds Experiment
GCM/SCM	General Circulation Model/Single Column Model
GOES	Geostationary Operational Environmental Satellite
GPM	Global Precipitation Measurement
HCR	HIAPER Cloud Radar
HIRS	High-Resolution Infrared Sounder
HSRHI	Hemispherical Sky Range-Height Indicator
IPCC	Intergovernment Panel on Climate Change
ISCCP	International Satellite Cloud Climatology Project
ITCZ	Intertropical Convergence Zone
KAZR	Ka-band ARM Zenith Radar
LACIS	Layered Atlantic Smoke Interactions with Clouds
LCL	Lifting Condensation Level
LCF/TCF	Low and Total Cloud Fraction
LW	Longwave
LWP/LWC	Liquid Water Path/Content (LWC_c for cloud and LWC_d for drizzle)
MACC	Monitoring Atmospheric Composition and Climate Reanalysis
MAGIC	Marine ARM GPCI Investigation of Clouds

MARCUS	Measurements of Aerosols, Radiation, and Clouds over the Southern Ocean
MAST	MODIS Atmosphere Science Team
MBL	Marine Boundary Layer
MCC	Mesoscale Cellular Convection
MCAO	Marine Cold-Air Outbreak
MMCR	Millimeter-wavelength Cloud Radar
MODIS	MODerate-resolution Imaging Spectroradiometer
MPL	Micropluse Lidar
MSL/ML	Mean Sea Level/Multilayer
MWIR /MWR	Midwave Infrared (e.g., 3.75 μm)/Microwave Radiometer
N_c/N_d	Cloud-Droplet Number Concentration/Drizzle Number Concentration
NH/SH	Northern/Southern Hemisphere
NIR	Near-Infrared (e.g., 0.87 μm)
NSA	North Slope of Alaska
OLR	Outgoing LW Radiation
PBL	Planetary Boundary Layer
POC	Pockets of Open Cell
POLDER	Polarization and Directionality of the Earth's Reflectance
PSD /DSD	Particle/Drop Size Distribution
RAA	Relative Azimuth Angle
r_c/r_d	Cloud-Droplet Effective Radius (same as CER) and Drizzle Drop Radius
SCAR	SCanning Cloud Radar
SCM	Single Column Model
SGP	Southern Great Plains
SHEBA	Surface Heat Budget of the Arctic Ocean
S-NPP	Suomi National Polar-orbiting Partnership
St/Sc	Stratus/Stratocumulus
SW/SWIR	Shortwave (0.3–5 μm); SW-Infrared (e.g., 1.6 or 2.1 μm)
SZA	Solar Zenith Angle
TOA	Top of Atmosphere
VIIRS	Visible and Infrared Imager/Radiometer Suite
VOCALS-Rex	VAMOS Ocean-Cloud-Atmosphere-Land Study Regional Experiment
VZA	View Zenith Angle
WMO	World Meteorology Organization

REFERENCES

Ackerman, T. P., & Stokes, G. M. (2003). The atmospheric radiation measurement program. *Physics Today*, *56*, 38–44.

Ackerman, S. A., Holz, R. E., Frey, R., Eloranta, E. W., Maddux, B. C., & McGill, M. (2008). Cloud detection with MODIS. Part II: Validation. *Journal of Atmospheric and Oceanic Technology*, *25*, 1073–1086. doi:10.1175/2007JTECHA1053.1.

Ackerman, T. P., Cress, T. S., Ferrell, W. R., Mather, J. H., & Turner, D. D. (2016). The programmatic maturation of the ARM program. The Atmospheric Radiation Measurement (ARM) Program: The First 20 Years. Meteorological Monographs, No. 57, Amer. Meteor. Soc., https://doi.org/10.1175/AMSMONOGRAPHS-D-15-0054.1.

Adhikari, A., Liu, C., & Hayden, L. (2019). Uncertainties of GPM Microwave Imager precipitation estimates related to precipitation system size and intensity. *Journal of Hydrometeorology, 20*(9), 1907–1923. doi:10.1175/JHM-D-19-0038.1

Ahn, E., Huang, Y., Siems, S. T., & Manton, M. J. (2018). A comparison of cloud microphysical properties derived from MODIS and CALIPSO with in situ measurements over the wintertime Southern Ocean. *Journal of Geophysical Research: Atmospheres*, *123*, 11120–11140. doi:10.1029/2018JD028535.

Albrecht, B. A., Randall, D. A., & Nicholls, S. (1988). Observations of marine stratocumulus during FIRE. *Bulletin of the American Meteorological Society*, *69*, 618–626.

Albrecht, B. A. (1989). Aerosols, cloud microphysics, and fractional cloudiness. *Science*, *245*, 1227–1230.

Albrecht, B. A., Bretherton, C. S., Johnson, D., Schubert, W. H., & Frisch, A. S. (1995). The Atlantic Stratocumulus Transition Experiment—ASTEX. *Bulletin of the American Meteorological Society*, *76*, 889–904.

Albrecht, B., Fang, M., & Ghate, V. (2016). Exploring stratocumulus cloud-top entrainment processes and parameterization using Doppler cloud radar observations. *Journal of the Atmospheric Sciences*, *73*, 729–742. doi:10.1175/JAS-D-15-0147.1.

Albrecht, B., Ghate, V. P., Mohrmann, J., Wood, R., Zuidema, P., Bretherton, C., et al. (2019). Cloud System Evolution in the Trades—CSET, Following the evolution of boundary layer cloud systems with the NSF/NCAR GV. *Bulletin of the American Meteorological Society*, *100*, 93–121. doi:10.1175/BAMS-D-17-0180.1.

Allen, G., Vaughan, G., Tonlazzo, T., Coe, H., Connolly, P., Yuter, S., et al. (2012). Gravity wave-induced perturbations in marine stratocumulus. *Quarterly Journal of the Royal Meteorological Society*, *139*, 32–45. doi: 10.1002/qj.1947.

Austin, R. T., Heymsfield, A. J., & Stephens, G. L. (2009). Retrieval of ice cloud microphysical parameters using the CloudSat millimeter-wave radar and temperature. *Journal of Geophysical Research*, *114*, D00A23. doi:10.1029/2008JD010049.

Baum, B. A., Soulen, P. F., Strabala, K. I., King, M. D., Ackerman, S. A., Menzel, W. P., & Yang, P. (2000). Remote sensing of cloud properties using MODIS airborne simulator imagery during SUCCESS 2. Thermodynamic phase. *Journal of Geophysical Research*, *105*, 11781–11792.

Baum, B. A., Menzel, P., Frey, R. A., Tobin, D. C., Holz, R. E., & Ackerman, S. A. (2012). MODIS cloud-top property refinements for Collection 6. *Journal of Applied Meteorology and Climatology*, *51*, 1145–1163.

Bedka, K., Murillo, E. M., Homeyer, C. R., Scarino, B., & Mersiovsky, H. (2018). The above-anvil cirrus plume: An important severe weather indicator in visible and infrared satellite imagery. *Weather Forecasting, 33*, 1159–1181. doi:10.1175/WAF-D-18-0040.1.

Bellouin, N., Quass, J., Morcrette, J.-J., & Boucher, O. (2013). Estimates of aerosol radiative forcing from the MACC re-analysis. *Atmospheric Chemistry and Physics*, *13*, 2045–2062. doi:10.5194/acp-13-2045-2013.

Benjamin, S. G., James, E. P., Hu, M., Alexander, C. R., Ludwig, T. T., Brown, J. M., et al. (2021). Stratiform cloud hydro-meteor assimilation for HRRR and RAP model short-range weather prediction. *Monthly Weather Review*, *149*, accepted.

Bennartz, R. (2007). Global assessment of marine boundary layer cloud droplet number concentration from satellite. *Journal of Geophysical Research*, *112*, D02201. doi:10.1029/2006JD007547.

Bennartz, R., Watts, P., Meirink, J. F., & Roebeling, R. (2010). Rainwater path in warm clouds derived from combined visible, near-infrared and microwave satellite observations, *Journal of Geophysical Research*, *115*, D19120. doi:10.1029/2009JD013679.

Böhm, C., Sourdeval, O., Mülmenstädt, J., Quass, J., & Crewell, S. (2019). Cloud base height from multi-angle satellite data. *Atmospheric Measurement Techniques*, *12*, 1841–1860. doi:10.5194/amt-12-1841-2019.

Bösenberg, J., Matthias, V., Amodeo, A., Amoiridis, V., Ansmann, A., Baldasano, J. M., et al. (2003). *EARLINET: A European Aerosol Research Lidar Network to establish an aerosol climatology. MPI Report 343*, Max–Planck–Inst. für Meteorol., Hamburg, Germany.

Brendecke, J., Dong, X., Xi, B., & Wu, P. (2021). Maritime cloud and drizzlemicrophysical properties retrievedfrom ship-based observations during MAGIC. *Earth and Space Science*, *8*, e2020EA001588. https://doi.org/10.1029/2020EA001588.

Bretherton, C. S., Uttal, T., Fairall, C. W., Yuter, S. E., Weller, R. A., Baumgardner, D., et al. (2004). The EPIC 2001 stratocumulus study. *Bulletin of the American Meteorological Society*, *84*, 967–977.

Bretherton, C. S., Wood, R., George, R. C., Leon, D., Allen, G., & Zheng, X. (2010). Southeast Pacific stratocumulus clouds, precipitation and boundary layer structure sampled along 20°S during VOCALS-Rex. *Atmospheric Chemistry and Physics*, *10*, 10639–10654.

Brill, K., & Albrecht, B. (1982). Diurnal variation of the trade-wind boundary layer. *Monthly Weather Review*, *110*, 601–613.

Brioude, J., Cooper, O. R., Feingold, G., Trainer, M., Freitas, S. R., Kowal, D., et al. (2009). Effect of biomass burning on marine stratocumulus clouds off the California coast. *Atmospheric Chemistry and Physics*, *9*, 8841–8856. doi:10.5613.194/acp-9-8841-2009.

Burleyson, C. D., De Szoeke, S. P., Yuter, S. E., Wilbanks, M., & Brewer, W. A. (2013). Ship-based observations of the diurnal cycle of southeast Pacific marine stratocumulus clouds and precipitation. *Journal of the Atmospheric Sciences*, *70*, 3876–3894.

Cadeddu, M. P., Liljegren, J. C., & Turner, D. D. (2013). The Atmospheric Radiation Measurement (ARM) program network of microwave radiometers: Instrumentation, data, and retrievals. *Atmospheric Measurement Techniques*, *6*, 2359–2372. https://doi.org/10.5194/amt-6-2359-2013.

Calisto, M., Folini, D., Wild, M., & Bengtsson, L. (2014). Cloud radiative forcing intercomparison between fully coupled CMIP5 models and CERES satellite data. *Annales Geophysicae*, *32*, 793–807. doi:10.5194/angeo-32-793-2014.

Chang, F. L., & Li, Z. (2002). Estimating the vertical variation of cloud droplet effective radius using multispectral near-infrared satellite measurements. *Journal of Geophysical Research*, *107*, D15. 10.1029/2001JD000766

Chang, F. L., & Li, Z. (2005). A new method for detection of cirrus overlapping water clouds and determination of their optical properties. *Journal of the Atmospheric Sciences*, *62*, 3993–4009. doi:10.1175/JAS3578.1.

Chang, F.-L., Minnis, P., Lin, B., Khaiyer, M., Palikonda, R., & Spangenberg, D. (2010). A modified method for inferring cloud top height using GOES-12 imager 10.7- and 13.3-μm data. *Journal of Geophysical Research*, *115*, D06208. doi:10.1029/2009JD012304.

Chen, R., Wood, R., Li, Z., Ferraro, R., & Chang, F.-L. (2008). Studying the vertical variation of cloud droplet effective radius using ship and space-borne remote sensing data. *Journal of Geophysical Research*, *113*, D00A02. doi:10.1029/2007JD009596.

Christensen, M. W., Jones, W. K., & Stier, P. (2020). Aerosols enhance cloud lifetime and brightness along the stratus-to-cumulus transitions. *Proceedings of the National Academy of Sciences*, *117*, 17591–17598. doi:10.1073/pnas.1921231117.

Clothiaux, E. E., Ackerman, T. P., Mace, G. G., Moran, K. P., Marchand, R. T., Miller, M. A., & Martner, B. E. (2000). Objective determination of cloud heights and radar reflectivities using a combination of active remote sensors at the Atmospheric Radiation Measurement Program Cloud and Radiation Test Bed (ARM CART) sites. *Journal of Applied Meteorology*, *39*, 645–665.

Coakley, J. A. Jr., & Baldwin, D. G. (1984). Towards the objective analysis of clouds from satellite imagery data. *Journal of Climate and Applied Meteorology*, *23*, 1065–1099.

Coakley, J. A. Jr., Friedman, M. A., & Tahnk, W. R. (2005). Retrieval of cloud properties for partly cloudy imager pixels. *Journal of Atmospheric and Oceanic Technology*, *22*, 3–17.

Comstock, K., S. E. Yuter, R. Wood, and C. S. Bretherton (2007). The three-dimensional structure and kinematics of drizzling stratocumulus. *Monthly Weather Review*, *135*, 3767–3784.

Curry, J. A., Hobbs, P. V., KJing, M. D., Randall, D. A., Minnis, P., Isaac, G. A., et al. (2000). FIRE Arctic clouds experiment. *Bulletin of the American Meteorological Society*, *81*, 5–29.

Dean, B. (1928). Temperature inversions at San Diego from aerographic observations by airplane. *Monthly Weather Review*, *56*, 221–224.

Del Genio, A. D., & Wolf, A. B. (2000). The temperature dependence of the liquid water path of low clouds in the Southern Great Plains. *Journal of Climate, 13*, 3465–3486.

Del Genio, A. D., Wolf, A., & Yao, M.-S. (2005). Evaluation of regional cloud feedbacks using single-column models. *Journal of Geophysical Research, 110*, D15S13. doi:10.1029/2004JD005011.

Del Genio, A. D., Yao, M.-S., Kovari, W., & Lo, K. K-W. (1996). A prognostic cloud water parameterization for global climate models. *Journal of Climate, 9*, 270–304. doi:10.1175/1520-0442(1996)009<0270:APCWPF>2.0.CO;2.

Desmons, M., Ferlay, N., Parol, F., & Thieuleux, F. (2017). A global multilayer cloud identification with POLDER/PARASOL. *Journal of Applied Meteorology and Climatology, 56*, 1121–1139. doi:10.1175/JAMC-D-16-0159.1.

Dong, X., Ackerman, T. P., Clothiaux, E. E., Pilewskie, P., & Han, Y. (1997). Microphysical and Radiative properties of stratiform clouds deduced from ground-based measurements. *Journal of Geophysical Research, 102*, 23,829–23,843.

Dong, X., Ackerman, T. P., & Clothiaux, E. E. (1998). Parameterizations of the microphysical and shortwave radiative properties of boundary layer stratus from ground-based measurements. *Journal of Geophysical Research, 103*(D24), 31681–31693, doi:10.1029/1998JD200047.

Dong, X., Minnis, P., Mace, G. G., Smith, W. L. Jr., Poellot, M., Marchand, R., & Rapp, A. (2002). Comparison of stratus cloud properties deduced from surface, GOES, and aircraft data during the March 2000 ARM Cloud IOP. *Journal of the Atmospheric Sciences, 59*, 3265–3284.

Dong, X., & Mace, G. G. (2003a). Arctic stratus cloud properties and radiative forcing derived from ground-based data collected at Barrow, Alaska. *Journal of Climate, 16*, 445–461.

Dong, X., & Mace, G. G. (2003b). Profiles of low-level stratus cloud microphysics deduced from ground-based measurements. *Journal of Atmospheric and Oceanic Technology, 20*, 42–53.

Dong, X., Minnis, P., & Xi, B. (2005). A climatology of midlatitude continental clouds from ARM SGP site. Part I: Low-level cloud macrophysical, microphysical and radiative properties. *Journal of Climate, 18*, 1391–1410.

Dong, X., Xi, B., & Minnis, P. (2006). A climatology of midlatitude continental clouds from ARM SGP site. Part II: Cloud fraction and surface radiative forcing. *Journal of Climate, 19*, 1765–1783.

Dong, X., Minnis, P., Xi, B., Sun-Mack, S., & Chen, Y. (2008). Validation of CERES-MODIS Stratus cloud properties using ground-based measurements at the DOE ARM SGP site. *Journal of Geophysical Research, 113*, D03204. doi:10.1029/2007JD008438.

Dong, X., Xi, B., Crosby, K., Long, C. N., Stone, R., & Shupe, M. (2010). A 10-yr climatology of Arctic cloud fraction and radiative forcing at Barrow, *Alaska. Journal of Geophysical Research, 115*, D12124. doi:10.1029/2009JD013489.

Dong, X., Xi, B., Kennedy, A., Minnis, P., & Wood, R. (2014a). A 19-month record of marine aerosol-cloud-radiation derived from the DOE ARM AMF deployment at the Azores: Part I: Cloud fraction and single-layered MBL cloud properties. *Journal of Climate, 27*, 3665–3682, doi:10.1175/JCLI-D-1300553.1.

Dong, X., Xi, B., & Wu, P. (2014b). Investigation of the diurnal variation of marine boundary layer cloud microphysical properties at the Azores. *Journal of Climate, 27*, 8827–8835. doi: 10.1175/JCLI-D-14-00434.1.

Dong, X., Schwantes, A. C., Xi, B., & Wu, P. (2015). Investigation of the marine boundary layer cloud properties under the coupled and decoupled conditions over the Azores, *Journal of Geophysical Research, 120*, 6179–6191. doi: 10.1002/2014JD022939.

Dong, X., Xi, B., Qiu, S., Minnis, P., Sun-Mack, S., & Rose, F. (2016). A radiation closure study of Arctic cloud microphysical properties using the collocated satellite-surface data and Fu-Liou radiative transfer model. *Journal of Geophysical Research, 121*, 10,175–10,198. doi:10.1002/2016JD025255.

Dong, X., Zheng, X., Xi, B., & Xie, X. (2023). A climatology of mid-latitude maritime cloud fraction and radiative effect derived from the ARM ENA ground-based observations. *Journal of Climate, 36*, 536–546, doi:10.1175/JCLI-D-22-0290.1.

Driemel, A., Augustine, J., Behrens, K., Colle, S., Cox, C., Cuevas-Agulló, E., et al. (2018). Baseline Surface Radiation Network (BSRN): Structure and data description (1992–2017). *Earth System Science Data, 10*, 1491–1501. doi:10.5194/essd-10-1491-2018.

Dufresne, J.-L., Gautier, C., Ricchiazzi, P., & Fouquart, Y. (2002). Longwave scattering effects of mineral aerosols. *Journal of the Atmospheric Sciences, 59*, 1959–1966.

Eastmann, R., & Warren, S. G. (2010). Interannual variations of Arctic cloud types in relation to sea ice. *Journal of Climate, 23*, 4216–4232. doi:10.1175/2010JCLI3492.1.

Eastmann, R., & Warren, S. G. (2014). Diurnal cycles of cumulus, stratus, and stratocumulus, and fog from surface observations over land and ocean. *Journal of Climate, 23*, 2386–2404. doi:10.1175/JCLI-D-13-00352.1.

Feingold, G., Frisch, A. S., Stevens, B., & Cotton W. R. (1999). On the relationship among cloud turbulence, droplet formation and drizzle as viewed by Doppler radar, microwave radiometer and lidar. *Journal of Geophysical Research: Atmospheres, 104*, 22195–22203, doi:10.1029/1999JD900482.

Ferek, R. J., Garrett, T., Hobbs, P. V., Strader, S., Johnson, D., Taylor, J. P., et al. (2000). Drizzle suppression in ship tracks. *Journal of the Atmospheric Sciences, 57*, 2707–2728.

Fielding, M. D., Chiu, J. C., Hogan, R. J., & Feingold, G. (2014). A novel ensemble method for retrieving properties of warm cloud in 3-D using ground-based scanning radar and zenith radiances. *Journal of Geophysical Research, 119*, 10912–10930. doi:10.1002/2014jd021742.

Fielding, M. D., Chiu, J. C., Hogan, R. J., Feingold, G., Eloranta, E., O'Connor, E. J., & Cadeddu, M. P. (2015). Joint retrievals of cloud and drizzle in marine boundary layer clouds using ground-based radar, lidar and zenith radiances. *Atmospheric Measurement Techniques Discussions, 8*, 1833–1889.

Fletcher, J., Mason, S., & Jakob, C. (2016). The climatology, meteorology, and boundary layer structure of marine cold air outbreaks in both hemispheres. *Journal of Climate, 29*, 1999–2014. doi://10.1175/JCLI-D-15-0268.1.

Fox, N. I., & Illingworth, A. (2002). The retrieval of stratocumulus cloud properties by ground-based cloud radar. *Journal of Applied Meteorology*, *36*(5), 485–492. https://doi.org/10.1175/1520-0450(1997)036<0485:troscp>2.0.co;2

Frisch, A., Fairall, C. W., & Snider, J. B. (1995). Measurement of stratus cloud and drizzle parameters in ASTEX with a K-band Doppler radar and a microwave radiometer. *Journal of the Atmospheric Sciences, 52*, 2788–2799.

Frisch, A., Feingold, G., Fairall, C. W., & Uttal, T. (1998). On cloud radar and microwave radiometer measurements of stratus cloud liquid water profiles. *Journal of Geophysical Research, 103,* 23 195–23 197.

Garreaud, R. D., Ruttlant, J., Quintana, J., Carrasco, J., & Minnis, P. (2001). CIMAR-5: A snapshot of the lower troposphere over the subtropical southeast Pacific. *Bulletin of the American Meteorological Society*, *82*, 2193–2207.

Garreaud, R. D., & Muñoz, R. (2004). The diurnal cycle in circulation and cloudiness over the subtropical southeast Pacific: A modeling study. *Journal of Climate, 17*, 1699–1710. doi: 10.1175/1520-0442(2004)017<1699:TDCICA>2.0.CO;2

Ge, J., Zheng, C., Xie, H., Xin, Y., Huang, J., & Fu, Q. (2018). Midlatitude cirrus clouds at the SACOL site: Macrophysical properties and large-scale atmospheric states. *Journal of Geophysical Research: Atmospheres*, *123*(4), 2256–2271. doi:10.1002/2017JD027724.

Geerts, B. (2019). Cold-air outbreaks in the Marine Boundary Layer Experiment (COMBLE) Science and Implementation Plan, DOE/SC-ARM-19-002, 24 pp.

Gerber, H. (1996). Microphysics of marine stratocumulus with two drizzle modes. *Journal of the Atmospheric Sciences*, *53*, 1649–1662.

Gerber, H., Frick, G., Malinkowski, S. P., Jonsson, H., Khelif, D., & Krueger, S. K. (2013). Entrainment rates and microphysics in POST stratocumulus. *Journal of Geophysical Research*, *118*, 12094–12109, doi:10.1002/jgrd.50878.

Ghate, V. P., Albrecht, B. A., & Kollias, P. (2010). Vertical velocity structure of nonprecipitating continental boundary layer stratocumulus clouds. *Journal of Geophysical Research: Atmospheres*, *115*, D13204, doi:10.1029/2009JD013091.

Ghate, V. (2021). Chapter 9. Planetary boundary layer and processes. In Y. Liu, P. Kollias, & L. Donner (Eds.), *Fast Physics in Large Scale Atmospheric Models: Parameterization, Evaluation, and Observations*. AGU-Wiley Publication.

Giles, D. M., Sinyuk, A., Sorokin, M. S., Schafer, J. S., Smirnov, A., Slutsker, I., et al. (2019). Advancements in the Aerosol Robotic Network (AERONET) Version 3 database—automated near-real-time quality control algorithm with improved cloud screening for Sun photometer aerosol optical depth (AOD) measurements. *Atmospheric Measurement Techniques*, *12*, 169–209. doi:10.5194/amt-12-169-2019.

Greenwald, T. J., & Christopher, S. A. (1999). Daytime variation of marine stratocumulus microphysical properties as observed from geostationary satellite. *Geophysical Research Letters*, *26*, 1723–1726.

Grosvenor, D. P., & Wood, R. (2014). The effect of solar zenith angle on MODIS cloud optical and microphysical retrievals within marine liquid water clouds, *Atmospheric Chemistry and Physics*, *14*, 7291–7321. doi:10.5194/acp-14-7291-2014.

Guo, Z., Wang, M., Peng, Y., & Luo, Y. (2020). Evaluation on the vertical distribution of liquid and ice phase cloud fraction in Community Atmosphere Model version 5.3 using spaceborne lidar observations. *Earth and Space Science*, *7*, e2019EA001029. doi:10.1029/2019EA001029.

Håkansson, N., Adok, C., Thoss, A., Scheirer, R., & Hörnquist, S. (2018). Neural network cloud top pressure and height for MODIS. *Atmospheric Measurement Techniques*, *11*, 3177–3196. doi:10.5194/amt-11-3177-2018.

Hamadai, A., & Takayabu, Y. N. (2016). Improvements in detection of light precipitation with the Global Precipitation Measurement Dual-frequency Precipitation Radar (GPM DPR). *Journal of Atmospheric and Oceanic Technology*, 33,653–33,667. doi:10.1175/JTECH-D-15-0097.1.

Han, Q., Rossow, W. B., & Lacis, A. A. (1994). Near-global survey of effective droplet radii in liquid water clouds using ISCCP data. *Journal of Climate*, *7*, 465–497. doi:10.1175/1520-0442(1994)007<0465:NGSOED>2.0.CO;2.

Hang, Y., L'Ecuyer, T. S., Henderson, D. S., Matus, A. V., & Wang, Z. (2019). Reassessing the effect of cloud type on earth's energy balance in the age of active spaceborne observations. Part II: Atmospheric heating. *Journal of Climate, 32*, 6219–6236.

Harrison, E. F., Minnis, P., Barkstorm, B. R., Ramanathan, V., Cess, R. D., & Gibson, G. G. (1990). Seasonal variation of cloud radiative forcing derived from the Earth Radiation Budget Experiments, *Journal of Geophysical Research*, *95*, 18,687–18,703.

Hartmann, D. L., Ockert-Bell, L. M. E., & Michelsen, M. L. (1992). The effect of cloud type on earth energy's balance: Global analysis. *Journal of Climate*, *5*, 1281–1304.

Hartmann, J., Kottmeier, C., & Raasch, S. (1997). Roll vortices and boundary-layer development during a cold air outbreak. *Boundary-Layer Meteorology*, *84*, 45–65.

Haeffelin, M., Crewell, S., Illingworth, A. J., Pappalardo, G., Russchenburg, H., Chiriaco, M., et al. (2018). Parallel developments and formal collaboration between European atmospheric profiling observatories and the U. S. ARM research program. *Meteorological Monographs*, *57*, 29.1–29.34. doi:10.1175/AMSMONOGRAPHS-D-15-0045.1.

Haeffelin, M., Barthès, L., Bock, O., Boitel, C., Bony, S., Bouniol, D., et al. (2005). SIRTA, a ground-based atmospheric observatory for cloud and aerosol research. *Annales Geophysicae*, *23*, 253–275. doi:10.5194/angeo-23-253-2005.

Heidinger, A. (2012). ABI Cloud Height, *Algorithm Theoretical Basis Document Version 3, NOAA NESDIS Center for Satellite Applications and Research*, 79 pp. Available at https://www.star.nesdis.noaa.gov/goesr/documents/ATBDs/Baseline/ATBD_GOES-R_Cloud%20Height_v3.0_July%202012.pdf.

Heidinger, A. K., Evan, A. T., Foster, M. J., & Walther, A. (2012). A naïve Bayesian cloud-detection scheme derived from CALIPSO and applied within PATMOS-x. *Journal of Applied Meteorology and Climatology*, *51*, 1129–1144. doi: 10.1175/JAMC-D-11-02.1.

Heidinger, A., Foster, M., Botambekov, D., Hiley, M., Walther, A., & Li, Y. (2016). Using the NASA EOS A-Train to probe the performance of the NOAA PATMOS-X cloud fraction CDR. *Remote Sensing*, *8, 511*, 18 pp. doi:10.3390/rs8060511.

Heidinger, A., Foster, M., Walther, A., & Zhao, X. (2014). The Pathfinder Atmospheres—Extended AVHRR climate dataset. *Bulletin of the American Meteorological Society*, *94*, 909–922. doi:10.1175/BAMS-D-12-00246.1.

Herman, G. F., & Goody, R. (1976). Formation and persistence of summertime Arctic stratus clouds. *Journal of the Atmospheric Sciences*, *33*, 1537–1553.

Hillger, D., Kopp, T., Lee, T., Lindsey, D., Seaman, C., Miller, S., et al. (2013). First-light imagery from Suomi NPP VIIRS. *Bulletin of the American Meteorological Society*, *93*, 1019–1029. doi:10.1175/BAMS-D-12-00097.1.

Holben, B., Eck, T. F., Slutsker, I., Tanré, D., Buis, J. P., Setzer, A., et al. (1998). AERONET—A federated instrument network and data archive for aerosol characterization. *Remote Sensing of Environment*, *66*, 1–16. doi:10.1016/S0034-4257(98)00031-5

Horváth, Á., & Davies, R. (2007). Comparison of microwave and optical cloud water path estimates from TMI, MODIS, and MISR. *Journal of Geophysical Research*, *112*, D01202. doi:10.1029/2006JD007101.

Hou, A. Y., Kakar R. K., Neeck S., Azarbarzin A. A., Kummerow C. D., Kojima M., et al. (2014). The Global Precipitation Measurement mission. *Bulletin of the American Meteorological Society*, *95*, 701–722. doi:10.1175/BAMS-D-13-00164.1.

Huang, D., Zhao, C., Dunn, M., Dong, X., Mace, G. G., Jensen, M. P., et al. (2012). An Intercomparison of radar-based liquid cloud microphysics retrievals and implication for model evaluation studies. *Atmospheric Measurement Techniques, 5*, 1409–1424. doi:10.5194/amt-5-1409-2012.

Huang, J., Minnis, P., Lin, B., Yi, Y., Khaiyer, M. M., Arduini, R. F., & Mace, G. G. (2005). Advanced retrievals of multilayered cloud properties using multi-sensor and multi-spectral measurements. *Journal of Geophysical Research*, *110*, D15S18. 10.1029/2004JD005101.

Huang, J. P., Zhang, W., Zuo, J., Bi, J., Shi, J., Wang, X., et al. (2008). An overview of the semi-arid climate and environment research observatory over the Loess Plateau. *Advances in Atmospheric Sciences*, *25*(6), 906–921. https://doi.org/10.1007/s00376-008-0906-7.

Huang, Y., Ding, Q., Dong, X., Xi, B., & Baxter, I. (2021). Summertime low clouds mediate the impact of the large-scale circulation on Arctic Sea ice. *Communications Earth and Environment*, *2(*1), 38. doi:10.1038/s43247-021-00114-w.

Hutchison, K. D., Wong, E., & Ou, S. C. (2006). Cloud base height retrieval during nighttime conditions with MODIS data. *International Journal of Remote Sensing*, *27*, 2847–2862. doi:10.1080/01431160500296800.

Illingworth, A. J., Barker, H. W., Beljaars, A., Ceccaldi, M., Chepfer, H., Clerbaux, N., et al. (2015). The EarthCARE satellite: The next step forward in global measurements of clouds, aerosols, precipitation, and radiation. *Bulletin of the American Meteorological Society*, *96*, 1311–1332. doi:10.1175/BAMS-D-12-00227.1.

Illingworth, A. J., Hogan, R. J., O'Connor, E. J., Bouniol, D., Brooks, M. E., Delanoë, J., et al. (2007). Cloudnet: Continuous evaluation of cloud profiles in seven operational models using ground-based observations. *Bulletin of the American Meteorological Society*, *88*, 883–898. doi:10.1175/BAMS-88-6-883.

Illingworth, A. J., Cimini, D., Haefele, A., Haeffelin, M., Hervo, M., Kotthaus, S., et al. (2019). How can existing ground-based profiling instruments improve European weather forecasts? *Bulletin of the American Meteorological Society*, *100*, 605–619. doi:10.1175/BAMS-D-17-0231.1

IPCC (2013). In Stocker, T. F., Qin, D., Plattner, G.-K., Tignor, M., Allen, S. K., Boschung, J., et al. (Eds.), Climate Change 2013: The physical science basis. Contribution of working group I to the fifth assessment report of the intergovernmental panel on climate change. Cambridge, United Kingdom and New York, NY, USA: Cambridge University Press, 1535 pp. https://doi.org/10.1017/CBO9781107415324

IPCC (2021). In V. Masson-Delmotte et al. (Eds.), *Climate Change 2021: The physical science basis. Contribution of working group I to the sixth assessment report of the intergovernmental panel on climate change*. Cambridge, UK: Cambridge University Press, 2391 pp.

Intrieri, J. M., Shupe, M. D., Uttal, T., & McCarty, B. J. (2002). An annual cycle of Arctic cloud characteristics observed by radar and lidar at SHEBA. *Journal of Geophysical Research, 107,* 8039. doi:10.1029/2000Jc060439.

Joiner, J., Vasilkov, A. P., Bhartia, P. K., Wind, G., Platnick, S., & Menzel, W. P. (2010). Detection of multi-layer and vertically-extended clouds using A-Train sensors. *Atmospheric Measurement Techniques*, *3*, 233–247.

Jones, C., Bretherton, C., & Leon, D. (2011). Coupled vs. decoupled boundary layers in VOCALS-Rex. *Atmospheric Chemistry and Physics*, *11*, 7143–7153.

Karlsson, K.-G., Riihelä, A., Müller, R., Meirink, J. F., Sedlar, J., Stengel, M., et al. (2013). CLARA-A1: A cloud, albedo, and radiation data record from 28 years of global AVHRR data. *Atmospheric Chemistry and Physics*, *13*, 5351–5367. doi:10.5194/acp-12-5351-2013.

Karlsson, K.-G., Anttila, K., Trentmann, J., Stengel, M., Meirink, J. F., Devasthale, A., et al. (2017). CLARA-A2: The second edition of the CM SAF cloud and radiation data record from 34 years of global AVHRR data. *Atmospheric Chemistry and Physics*, *17*, 5809–5828. doi:10.5194/acp-17-5809-2017.

kezemirad, M., & Miller, M. (2020). Summertime post-cold-frontal marine stratocumulus transition processes over the eastern North Atlantic. *Journal of the Atmospheric Sciences*, *77*. 10.1175/JAS-D-19-0167.1.

Kennedy, A., Dong, X., Xi, B., Minnis, P., Del Genio, A., Wolf, A., & Khaiver, M. (2010). Evaluation of the NASA GISS Single Column Model simulated clouds using combined surface and satellite observations. *Journal of Climate, 23,* 5175–5192. doi: 10.1175/2010JCLI3353.1

Kim, S. W., Chung, E.-S., Yoon, S.-C., Sohn, B.-J., & Sugimoto, N. (2011). Intercomparisons of cloud-top and cloud-base heights from ground-based lidar, CloudSat and CALIPSO measurements. *International Journal of Remote Sensing*, *32*, 1179–1197, doi:10.1080/01431160903527439

King, N. J., & Vaughan, G. (2012). Using passive remote sensing to retrieve the vertical variation of cloud droplet size in marine stratocumulus: An assessment of information content and the potential for improved retrievals from hyperspectral measurements. *Journal of Geophysical Research*, *117*(D15).

Klein, S. A., & Hartmann, D. L. (1993). The seasonal cycle of low stratiform clouds. *Journal of Climate*, *6*, 1587–1606.

Kogan, Z. N., & Kogan, Y. L. (2001). Parameterization of drop effective radius for drizzling marine stratocumulus. *Journal of Geophysical Research*, *106*, 9757–9764. doi:10.1029/2001JD900041.

Kollias, P., Clothiaux, E. E., Ackerman, T. P., Albrecht, B. A., Widener, K. B., Moran, K. P., et al. (2016). Development and applications of ARM Millimeter-Wavelength Cloud Radars, *Meteorological Monographs*, *57*, 17.1–17.19. https://doi.org/10.1175/AMSMONOGRAPHS-D-15-0037.1

Kollias, P., Bharadwaj, N., Clothiaux, E. E., Lamer, K., Oue, M., Hardin, J., et al. (2020). The ARM Radar Network. *Bulletin of the American Meteorological Society*, E588–E607. doi:10.1175/BAMS-D-18-0288.1

Kox, S., Bugliaro, L., & Ostler, A. (2014). Retrieval of cloud optical thickness and top altitude from geostationary remote sensing. *Atmospheric Measurement Techniques*, *7*, 3233–3246. doi:10.5194/amt-7-3233-2014.

Küchler, N., Kneifel, S., Kollias, P., & Löhnert, U. (2018). Revisiting liquid water content retrievals in warm stratified clouds: The modified Frisch. *Geophysical Research Letters, 45*(17), 9323–9330. doi:10.1029/2018GL079845.

Lamer, K., Kollias, P., Amiridis, V., Marinou, E., Loenhert, U., Schmitt, S., & McComiskey, A. (2023). Chapter 18. Surface-based remote sensing of key properties. In Y. Liu, P. Kollias, & L. Donner (Eds.), *Fast Physics in large scale atmospheric models: Parameterization, evaluation, and observations*. AGU-Wiley Publications.

Larson, V. (2023). Chapter 13. Parameterizations based on assumed probability distribution function. In Y. Liu, P. Kollias, & L. Donner (Eds.), *Fast Physics in large scale atmospheric models: Parameterization, evaluation, and observations*. AGU-Wiley Publications.

L'Ecuyer, T. S., Hang, Y., Matus, A. V., & Wang, Z. (2019). Reassessing the role of cloud type in Earth's radiation budget after a decade of active spaceborne observations. Part I: Top of atmosphere and surface. *Journal of Climate*, *32*, 6197–6217.

Leon, D. C., Wang, Z., & Liu, D. (2008). Climatology of drizzle in marine boundary layer clouds based on 1 year of data from CloudSat and Cloud-Aerosol Lidar and Infrared Pathfinder Satellite Observations (CALIPSO). *Journal of Geophysical Research*, *113*, D00A14. doi:10.1029/2008JD009835.

Lewis, E., & Teixeira, J. (2015). Dispelling clouds of uncertainty. *Eos.*, *96*(12). doi:10.1029/2015EO031303.

Lhermitte, R. M. (1987). A 94-GHz doppler radar for cloud observations. *Journal of Atmospheric and Oceanic Technology*, *4*, 36–48.

Li, Z., Wang, Y., Guo, J., Zhao, C., Cribb, M. C., Dong, X., et al. (2020). East Asian Study of Tropospheric Aerosols and Impact on Regional Cloud, Precipitation, and Climate (EAST-AIR$_{CPC}$). *Journal of Geophysical Research*, *124*, 13,026–13,054. https://doi.org/10.1029/2019JD030758

Lilly, D. K. (1968). Models of cloud-topped mixed layers under a strong inversion. *Quarterly Journal of the Royal Meteorological Society, 94*, 292–309.

Liljegren, J. C., Clothiaux, E. E., Mace, G. G., Kato, S., & Dong, X. (2001). A new retrieval for cloud liquid water path using a ground-based microwave radiometer measurements of cloud temperature. *Journal of Geophysical Research*, *106*, 14485–14500.

Lin, X., & Coakley, J. A. Jr. (1993). Retrieval of properties for semitransparent clouds from multispectral infrared imagery data. *Journal of Geophysical Research*, *98*, 18501–18514.

Lin, B., Minnis, P., Wielicki, B., Doelling, D. R., Palikonda, R., Young, D. F., & Uttal, T. (1998). Estimation of water cloud properties from satellite microwave, infrared, and visible measurements in oceanic environments 2. *Results. Journal of Geophysical Research*, *103*, 3887–3905.

Lin, B., Minnis, P., Fan, A., Curry, J. A., & Gerber, H. (2001). Comparison of cloud liquid water paths derived from in situ and microwave radiometer data taken during the SHEBA/FIREACE. *Geophysical Research Letters*, *28*, 975–978.

Liu, Y. G., Wu, W., Jensen, M., & Toto, T. (2011). Relationship between cloud radiative forcing cloud fraction and cloud albedo, and new surface-based approach for determining cloud albedo. *Atmospheric Chemistry and Physics*, *11*, 7155–7170.

Liu, J., Li, Z., & Cribb, M. (2016). Response of marine boundary layer clouds properties to aerosol perturbations associated with meteorological conditions from the 19-month AMF-Azores Campaign. *Journal of the Atmospheric Sciences*, *73*, 4253–. doi:10.1175/JAS-D-15-0364.1.

Loeb, N. G., & Coakley, J. A. Jr. (1998). Inference of marine stratus cloud optical depths from satellite measurements: Does 1D theory apply? *Journal of Climate*, *11*, 215–233.

Loeb, N. G., Wielicki, B. A., Doelling, D. R., Smith, G. L., Keyes, D. F., Kato, S., et al. (2009). Toward optimal closure of the Earth's top-of-atmosphere radiation budget. *Journal of Climate, 22*, 748–766. doi:10.1175/2008JCLI2637.1.

Loeb, N. G., Doelling, D., Wang, H., Su, W., & Kato, S. (2018). Clouds and the Earth's Radiant Energy System (CERES) Energy Balanced and Filled (EBAF) Top-of-Atmosphere (TOA) Edition-4.0 Data Product. *Journal of Climate*, *31*, 895–918. doi: 10.1175/JCLI-D-17-0208.1

Long, C., & Ackerman, T. P. (2000). Identification of clear skies from broadband pyranometer measurements and calculation of downwelling shortwave cloud effects. *Journal of Geophysical Research, 105*, 15609–15626. doi:10.1029/2000JD900077.

Long, C., Ackerman, T. P., Gaustad, K. L., & Cole, J. N. S. (2006). Estimation of fractional sky cover from broadband shortwave radiometer measurements. *Journal of Geophysical Research*, *111*, D11204. doi:10.1029/2005JD0066475.

Luke, E. P., & Kollias, P. (2013). Separating cloud and drizzle radar moments during precipitation onset using doppler spectra. *Journal of Atmospheric and Oceanic Technology*, *30*(8), 1656–1671. https://doi.org/10.1175/JTECH-D-11-00195.1

Lutz, H.-J., Wilson, H. K., Heinemann, T., Samain, O., Gustafsson, J., Holmlund, K., & de Smet, A. (2011). MSG Meteorological Products Extraction Facility Algorithm Specification Document, *EUMETSAT Document Nr. EUM/MSG/SPE/022*, 259 pp.

Mace, G. G. (2010). Cloud properties and radiative forcing over the maritime storm tracks of the Southern Ocean and

North Atlantic derived from A-Train. *Journal of Geophysical Research*, *115*, D10201, doi:10.1029/2009JD012517.

Marchand, R. (2016). ARM and satellite cloud validation. *Meteorological Monographs*, *57*, 30.1–30.11. doi:10.1175/AMSMONOGRAPHS-D-15-0038.1.

Marchant, B., Platnick, S., Meyer, K., & Wind, G. (2020). Evaluation of the MODIS Collection 6 multilayer cloud detection algorithm through comparisons with CloudSat Cloud Profiling Radar and CALIPSO CALIOP products. *Atmospheric Measurement Techniques*, *13*, 3263–3275. doi:10.5194/amt-13-3263-2020.

Marshak, A., & Davis, A. B. (2021). Chapter 19. Satellite and airborne remote sensing of clouds and aerosols. In Y. Liu, P. Kollias, and L. Donner (Eds.), *Fast Physics in Large Scale Atmospheric Models: Parameterization, Evaluation, and Observations*. AGU-Wiley Publications.

Mather, J. H., & Voyles, J. W. (2013). The ARM climate research facility: A review of structure and capabilities, *Bulletin of the American Meteorological Society*, *94*, 377–392. doi:10.1175/BAMS-D-11-00218.1.

Mather, J. H., Turner, D. D., & Ackerman, T. P. (2016). Scientific maturation of the ARM Program. The Atmospheric Radiation Measurement (ARM) Program: The First 20 Years. Meteorological Monographs, No. 57, Amer. Meteor. Soc., doi:10.1175/AMSMONOGRAPHS-D-15-0053.1.

McCoy, D. T., Bender, F. A.-M., Mohrmann, J. K. C., Hartmnn, D. L., Wood, R., & Grosvenor, D. P. (2017). The global aerosol-cloud first indirect effect estimated using MODIS, MERRA, and AeroCOm. *Journal of Geophysical Research*, *122*, 1779–1796. doi:10.1002/2016JD026141.

McFarquhar, G. M., Bretherton, C. S., Marchand, R., Protat, A., DeMott, P. J., Alexander, S. P., et al. (2021). Observations of clouds, aerosols, precipitation and surface radiation over the Southern Ocean: An overview of CAPRICORN, MARCUS, MICRE, and SOCRATES. *Bulletin of the American Meteorological Society*, *102*, E894–E928. doi 10.11175/BAMS-D-20-0132.1.

McGill, M. J., Vaughan, M. A., Trepte, C. R., Hart, W. D., Hlavka, D. L., Winker, D. M., & Kuehn, R. (2007). Airborne validation of spatial properties measured by the CALIPSO lidar. *Journal of Geophysical Research*, *11*, D20201. doi:10.1029/2007JD008768.

McHardy, T. M., Dong, X., Xi, B., Thieman, M. M., & Minnis, P. (2018). Comparison of daytime low- level cloud properties derived from GOES and ARM SGP measurements. *Journal of Geophysical Research*, *123*, 8221–8237. https://doi.org/10.1029/2018JD028911.

Mechem, D. B., Kogan, Y. L., & Schultz, D. M. (2010). Large-eddy observation of post-cold frontal continental stratocumulus. *Journal of the Atmospheric Sciences*, *67*, 3368–3383. doi:10.1175/2010JAS3389.1.

Meerkötter, R., & Zinner, T. (2007). Satellite remote sensing of cloud base height for convective cloud fields: A case study. *Geophysical Research Letters*, *34*, 117805. doi:10.1029/2007GL030347.

Mhyre, G. D., et al. (2013). Anthropogenic and natural radiative forcing. In T. F. Stocker, D. Qin, G.-K. Plattner, M. Tignor, S. K. Allen, J. Doschung, et al. (Eds.), *Climate change 2013: The physical science basis. Contribution of working group I to the fifth assessment report of the intergovernmental panel on climate change* (pp. 659–740). Cambridge: Cambridge University Press. doi:10.1017/CBO9781107415324.018.

Miles, N. L., Verlinde, J., & Clothiaux, E. E. (2000). Cloud-droplet size distributions in low level stratiform clouds. *Journal of the Atmospheric Sciences*, *57*, 295–311.

Miller, M. A., & Albrecht, B. A. (1995). Surface-based observations of mesoscale cumulus–stratocumulus interaction during ASTEX. *Journal of the Atmospheric Sciences*, *52*, 2809–2826. https://doi.org/10.1175/1520-0469(1995)052,2809:SBOOMC.2.0.CO;2.

Miller, S. D., Noh, Y.-J., & Heidinger, A. K. (2014). Liquid-top mixed phase cloud detection from shortwave-infrared satellite radiometer observations: A physical basis. *Journal of Geophysical Research: Atmospheres*, *119*, 8245–8267. doi:10.1002/2013JD021262.

Minnis, P., Bedka, K., Trepte, Q., Yost, C. R., Bedka, S. T., Scarino, B., et al. (2016a). A consistent long-term cloud and clear-sky radiation property dataset from the Advanced Very High Resolution Radiometer (AVHRR). *Climate Algorithm Theoretical Basis Document (C-ATBD), CDRP-ATBD-0826 AVHRR Cloud Properties—NASA*, NOAA CDR Program, 19 September, 159 pp., DOI:10.7289/V5HT2M8T. Available at https://www1.ncdc.noaa.gov/pub/data/sds/cdr/CDRs/AVHRR_Cloud_Properties_NASA/AlgorithmDescription_01B-30b.pdf

Minnis, P., & Harrison, E. F. (1984a). Diurnal variability of regional cloud and clear-sky radiative parameters derived from GOES data, Part II: November 1978 cloud distributions. *Journal of Climate and Applied Meteorology*, *23*, 1012–1031.

Minnis, P., & Harrison, E. F. (1984b). Diurnal variability of regional cloud and clear-sky radiative parameters derived from GOES data, Part III: November 1978 radiative parameters. *Journal of Climate and Applied Meteorology*, *23*, 1032–1052.

Minnis, P., Harrison, E. F., & Gibson, G. G. (1987). Cloud cover over the equatorial eastern Pacific derived from July 1983 ISCCP data using a hybrid bispectral threshold method. *Journal of Geophysical Research*, *92*, 4051–4073.

Minnis, P., Heck, P. W., Young, D. F., Fairall, C. W., & Snider, J. B. (1992). Stratocumulus cloud properties derived from simultaneous satellite and island-based instrumentation during FIRE. *Journal of Applied Meteorology*, *31*, 317–339.

Minnis, P., Hong, G., Sun-Mack, S., Smith, W. L. Jr., Chen, Y., & Miller, S. (2016b). Estimating nocturnal opaque ice cloud optical depth from MODIS multispectral infrared radiances using a neural network method. *Journal of Geophysical Research*, *121*, 4907–4932. doi:10.1002/2015JD024456.

Minnis, P., Huang, J., Lin, B., Yi, Y., Arduini, R. F., Fan, T.-F., et al. (2007). Ice cloud properties in ice-over-water cloud systems using TRMM VIRS and TMI data. *Journal of Geophysical Research*, *112*, D06206. doi:10.1029/2006JD007626.

Minnis, P., Nguyen, L., Palikonda, R., Heck, P. W. Spangenberg, D. A., Doelling, D. R., et al. (2008). Near-real time cloud retrievals from operational and research meteorological satellites, *Proceedings of SPIE 7107, Remote Sensing of Clouds and the Atmosphere XIII*, Cardiff, Wales, UK, 15–18 September, 710703-710703-8, https://doi.org/10.1117/12.800344.

Minnis, P., Sun-Mack, S., Smith, W. L. Jr., Hong, G., & Chen, Y. (2019). Advances in neural network detection and retrieval of multilayer clouds for CERES using multispectral satellite data. Proceedings of SPIE Conference, Remote Sensing Clouds and the Atmosphere XXIV, Strasbourg, France, Sept. 9–12, 11152, 1–12 pp., doi: 10.1117/12.2532931.

Minnis, P., Sun-Mack, S., Young, D. F., Heck, P. W., Garber, D. P., Chen, Y., et al. (2011). CERES Edition-2 cloud property retrievals using TRMM VIRS and Terra and Aqua MODIS data, Part II: Examples of average results and comparisons with other data. *IEEE Transactions on Geoscience and Remote Sensing*, *49*, 11, 4401–4430. doi:10.1109/TGRS.2011.2144602.

Minnis, P., Sun-Mack, S., Chen, Y., Chang, F. L., Yost, C. R., Smith, W. L., et al. (2021). CERES MODIS cloud product retrievals for Edition 4—Part I: Algorithm changes, *IEEE Transactions on Geoscience and Remote Sensing*, *59*, 2744–2780. doi:10.1109/TGRS.2020.3008866.

Moran, K. P., Martner, B. E., Post, M. J., Kropfli, R. A., Welsh, D. C., & Widener, K. B. (1998). An unattended cloud profiling radar for use in climate research. *Bulletin of the American Meteorological Society, 79*, 443–455.

Morrison, H., de Boer, G., Feingold, G., Harrington, J., Shupe, M. D., & Sulia, K. (2012). Resilience of persistent Arctic mixed-phase clouds. *Nature Geoscience, 5*, 11–17.

Nagao, T. M., Suzuki, K., & Nakajima, T. Y. (2013). Interpretation of multiwavelength-retrieved droplet effective radii for warm water clouds in terms of in-cloud vertical inhomogeneity by using a spectral bin microphysics cloud model. *Journal of the Atmospheric Sciences*, *70*(8), 2376–2392.

Nakajima, T., & King, M. D. (1990). Determination of the optical thickness and effective radius of clouds from reflected solar radiation measurements. Part I: Theory. *Journal of the Atmospheric Sciences*, *47*(15), 1878–1893.

Nakajima, T. Y., Suzuki, K., & Stephens, G. L. (2010a). Droplet growth in warm water clouds observed by the A-Train. Part I: Sensitivity analysis of the MODIS-derived cloud droplet sizes. *Journal of the Atmospheric Sciences*, *67*(6), 1884–1896.

Nakajima, T. Y., Suzuki, K., Stephens, G. L., Nakajima, T. Y., Suzuki, K., & Stephens, G. L. (2010b), Droplet growth in warm water clouds observed by the A-Train. Part II: A multisensor view. *Journal of the Atmospheric Sciences*, *67*(6), 1897–1907. https://doi.org/10.1175/2010JAS3276.1

Naud, C. M., Booth, J. F., Lamer, K., Marchand, R., Protat, A., & McFarquhar, G. M. (2020). On the relationship between the marine cold air outbreak M parameter and low-level cloud heights in the midlatitudes. *Journal of Geophysical Research: Atmospheres*, *125*(13), e2020JD032465. doi:10.1029/2020JD032465.

Neiburger, M. (1944). Temperature changes during formation and dissipation of west coast stratus. *Journal of Meteorology*, *1*, 29–41.

NOAA (1998). Automated Surface Observing System (ASOS) user's guide. *National Weather Service Doc.*, 61 pp., http://www.nws.noaa.gov/asos/pdfs/aum-toc.pdf.

Noh, Y.-J., Forsythe, J. M., Miller, S. D., Seaman, C. J., Li, Y., Heidinger, A. K., et al. (2017). Cloud base height estimation from VIIRS. Part II: A statistical algorithm based on A-Train satellite data. *Journal of Atmospheric and Oceanic Technology*, *34*, 585–598. doi:10.1175/JTECH-D-16-0110.1.

O'Connor, E. J., Hogan, R. J., & Illingworth, A. J. (2005). Retrieving stratocumulus drizzle parameters using Doppler radar and lidar. *Journal of Applied Meteorology*, *44*(1), 14–27. https://doi.org/10.1175/jam-2181.1

O'Dell, C. W., Wentz, F. J., & Bennartz, R. (2008). Cloud liquid water path from satellite-based passive microwave observations: A new climatology over the global oceans. *Journal of Climate*, *21*, 1721–1739. doi:10.1175/2007JCLI1958.1.

Painemal, D., Greenwald, T., Cadeddu, M., & Minnis, P. (2016). First extended validation of satellite microwave liquid water path with ship-based observations of marine low clouds. *Geophysical Research Letters*, *43*, 6563–6570. doi:10.1002/2016GL069061.

Painemal, D., Kato, S., and Minnis, P. (2014). Boundary-layer regulation in the southeast Atlantic cloud microphysics during the biomass burning season as seen by the A-Train satellite constellation. *Journal of Geophysical Research*, *119*, 11288–11302. doi:10.1002/2014JD022182.

Painemal, D., & Minnis, P. (2012). On the dependence of albedo on cloud microphysics over marine stratocumulus cloud regimes determined from Clouds and the Earth's Radiant Energy System (CERES) data. *Journal of Geophysical Research, 117*, D06203. doi:10.1029/2011JD017120.

Painemal, D., Minnis, P., Ayers, J. K., & O'Neill, L. (2012). GOES-10 microphysical retrievals in marine warm clouds: Multi-instrument validation and daytime cycle over the southeast Pacific. *Journal of Geophysical Research*, *17*, D19212. doi:10.1029/2012JD017822.

Painemal, D., Minnis, P., & O'Neill, L. (2013a). Diurnal cycle of cloud-top height and cloud cover over the southeastern Pacific as observed by GOES-10. *Journal of the Atmospheric Sciences*, *70*, 2393–2408. doi:10.1175/JAS-D-12-0325.1.

Painemal, D., Minnis, P., & Sun-Mack, S. (2013b). The impact of horizontal heterogeneities, cloud fraction, and cloud dynamics on warm cloud effective radii and liquid water path from CERES-like Aqua MODIS retrievals. *Atmospheric Chemistry and Physics*, *13*, 9997–10003. doi:10.5194/acp-13-9997-2013.

Painemal, D., & Zuidema, P. (2011). Assessment of MODIS cloud effective radius and optical thickness retrievals over the Southeast Pacific with VOCALS-REx in situ measurements. *Journal of Geophysical Research, 116*(D24), D24206. Dot:10.1029/2011JD016155.

Painemal, D., Corral, A. F., Sorooshian, A., Brunke, M. A., Chellappan, S., Afzali Go rooh, V., et al. (2021). An overview of atmospheric features over the Western North Atlantic Ocean and North American East Coast—Part 2: Circulation, boundary layer, and clouds. *Journal of Geophysical Research: Atmospheres*, *126*, e2020JD033423. doi:10.1029/2020JD033423.

Paluch, I. R., & Lenschow, D. H. (1991). Stratiform cloud formation in the marine boundary layer. *Journal of the Atmospheric Sciences*, *48*, 2141–2158.

Pavolonis, M. (2010). ABI cloud type and cloud phase. *Algorithm Theoretical Basis Document Version 2,* NOAA NESDIS *Center for Satellite Applications and Research*, 86 pp., Available at https://www.star.nesdis.noaa.gov/goesr/documents/ATBDs/Baseline/ATBD_GOES-R_Cloud_Phase_Type_v2.0_Sep2010.pdf

Pavolonis, M. J., & Heidinger, A. K. (2004). Daytime cloud overlap detection from AVHRR and VIIRS. *Journal of Applied Meteorology*, *43*(5), 762–778. doi:10.1175/2099.1.

Pavolonis, M. J., Heidninger, A. K., & Uttal, T. (2005). Global cloud typing from AVHRR and VIIRS: Algorithm description, validation, and comparisons. *Journal of Applied Meteorology*, *44*, 805–826. doi:10.1175/JAM2236.1.

Platnick, S., & Valero, F. P. J. (1995). A validation of a satellite cloud retrievals during ASTEX. *Journal of the Atmospheric Sciences*, *52*, 2985–3001.

Platnick, S. (2000). Vertical photon transport in cloud remote sensing problems. *Journal of Geophysical Research*, *105*(D18), 22,919–22,935. doi:10.1029/2000JD900333.

Platnick, S., King, M. D., Ackerman, S. A., Menzel, W. P., Baum, B. A., Riedi, J. C., & Frey, R. A. (2003). The MODIS cloud products: Algorithms and examples from Terra. *IEEE Transactions on Geoscience and Remote Sensing, 41*(2), 459–473. doi:10.1109/TGRS.2002.808301.

Platnick, S., Meyer, K., King, M., Wind, G., Amarasinghe, N., Marchant, B., et al. (2017). The MODIS cloud optical and microphysical products: Collection 6 updates and examples from Terra and Aqua. *IEEE Transactions on Geoscience and Remote Sensing, 55*, 502–525. https://doi.org/10.1109/TGRS.2016.2610522.

Pyatt, H., Albrecht, B. A., Fairall, C., Hare, J. E., Bond, N., Minnis, P., & Ayers, J. K. (2005). Evolution of marine boundary layer structure across the cold tongue ITCZ complex. *Journal of Climate*, *15*, 737–753.

Qian, Y., Gong, D., Fan, J., Leung, L. R., Bennartz, R., Chen, D., & Wang, W. (2009). Heavy pollution suppresses light rain in China: Observations and modeling. *Journal of Geophysical Research*, *114*, D00K02. doi:10.1029/2008JD011575.

Qiu, S., Dong, X., Xi. B., & Li, J. (2015). Characterizing Arctic mixed-phase cloud structure and its relationship with humidity and temperature inversion using ARM NSA observations. *Journal of Geophysical Research*, *120*, 7737–7746. doi:10.1002/2014JD023022.

Qiu, S., Xi, B., & Dong, X. (2018). Influence of wind directions on thermodynamic properties and Arctic mixed-phase clouds at Barrow, Alaska in autumn season. *Journal of Geophysical Research*, *123*, 9589–9603. https://doi.org/10.1029/2018JD028631

Rahn, D. A., & Garreaud, R. (2010). Marine boundary layer over the subtropical southeast Pacific during VOCALS-Rex—Part I: Mean structure and diurnal cycle. *Atmospheric Chemistry and Physics*, *10*, 4491–4506. doi:10.5194/acp-10-4491-2010.

Ramanathan, V., Cess, R. D., Harrison, E. F., Minnis, P., Barkstorm, B. R., Ahmad, E., & Hartmann, D. (1989). Cloud-radiative forcings and climate: Results from the Earth Radiation Budget Experiment, *Science*, *243*, 57–63.

Randall, D. A., Coakley, J. A., Lenschow, D. H., Fairall, C. W., & Kropfli, R. A. (1984). Outlook for research on subtropical marine stratification clouds. *Bulletin of the American Meteorological Society*, *65*(12), 1290–1301. https://doi.org/10.1175/1520-0477(1984)065<1290:ofrosm>2.0.co;2

Randall, D., Albrecht, B., Cox, S., Johnson, D., Minnis, P., Rossow, W., & Starr, D. (1996a). On FIRE at ten. *Advances in Geophysics, 38*, 37–177.

Randall, D., Xu, K.-M., Somerville, R. J. C., & Iacobellis, S. (1996b). Single-column models and cloud ensemble models as links between observations and climate models, *Journal of Climate*, *9*, 1683–1697.

Reitebuch, O. (2012). The Spaceborne Wind Lidar Mission ADM-Aeolus. In U. Schumann (Ed.), *Atmospheric physics: Background, methods, trends* (pp. 815–827). Research Topics in Aerospace. Berlin, London: Springer.

Rémillard, J., Kollias, P., Luke, E., & Wood, R. (2012). Marine boundary layer cloud observations in the Azores. *Journal of Climate, 25*, 7381–7398. doi:10.1175/JCLI-D-11-00610.1.

Richter, I., & Mechoso, C. R. (2006). Orogaphic influences on subtropical stratocumulus. *Journal of the Atmospheric Sciences*, *63*, 2585–2601. doi:10.1175/JAS3756.1.

Rossow, W. B., & Schiffer, R. A. (1991). *International Satellite Cloud Climatology Project* (ISCCP) cloud data products. *Bulletin of the American Meteorological Society*, *72*, 2–20.

Rozendaal, M. A., Leovy, C. B., & Klein, S. A. (1995). An observational study of the diurnal cycle of marine stratiform cloud. *Journal of Climate*, *8*, 1795–1809.

Sassen, K. (1991). The polarization lidar technique for cloud research: A review and current assessment. *Bulletin of the American Meteorological Society, 72*, 1848–1866.

Sedlar, J., & Tjernström, M. (2009). Stratiform cloud–inversion characterization during the Arctic melt season. *Boundary Layer Meteorology*, *132*, 455–474.

Seethala, C., & Horváth, Á. (2010). Global assessment of AMSR-E and MODIS cloud liquid water path retrievals in warm oceanic clouds. *Journal of Geophysical Research*, *115*, D13202. doi:10.1029/2009JD012662.

Serpetzoglou, E., Albrecht, B. A., Kollias, P., & Bretherton, C. W. (2008). Boundary layer, cloud, and drizzle variability in the southeast Pacific stratocumulus regime. *Journal of Climate*, *21*, 6191–6214. doi:10.1175/2008JCLI2186.1.

Sherr, P. E., Glaser, A. H., Barnes, J. C., & Willand, J. H. (1968). *World-wide cloud cover distributions for use in computer simulations. NASA CR-61226,* 272 pp. [NTIS-N68-32282].

Shao, H., & Liu, G. (2004). Detecting drizzle in marine warm clouds using combined visible, infrared, and microwave satellite data. *Journal of Geophysical Research*, *109*, D07205. doi:10.1029/2003JD004286.

Sharon, T. M., Albrecht, B. A., Jonsson, H. H., Minnis, P., Khaiyer, M. M., Van Reken, T. M., et al. (2006). Aerosol and cloud microphysical characteristics of rifts and gradients in maritime stratocumulus clouds. *Journal of the Atmospheric Sciences*, *63*, 983–997.

Short, D. A., & Wallace, J. M. (1980). Satellite-inferred morning to evening cloudiness changes. *Monthly Weather Review*, *108*, 1160–1169.

Shupe, M. D., Matrosov, S. Y., & Uttal, T. (2006). Arctic mixed-phase cloud properties derived from surface-based sensors at SHEBA. *Journal of the Atmospheric Sciences, 63,* 697–711.

Shupe, M. D., Comstock, J. M., Turner, D. D., & Mace, G. G. (2016). Cloud property retrievals in the ARM Program, The Atmospheric Radiation Measurement (ARM) Program: The first 20 years. Meteorological Monographs, No. 57, Amer. Meteor. Soc. doi:10.1175/AMSMONOGRAPHS-D-15-0030.1.

Shupe M., de Boer, G., Dethloff, K., Hunke, E., Maslowski, W., McComiskey, A., et al. (2018). The Multidisciplinary Drifting Observatory for the Study of Arctic Climate (MOSAIC) Atmosphere Science Plan. *Ed. by Robert Stafford, ARM Climate Research Facility, DOE/SC-ARM-18-005*, 23 pp.

Shupe, M. D., Daniel, J. S., De Boer, G., Eloranta, E. W., Kollias, P., Long, C. N., et al. (2008a). A focus on mixed-phase clouds: The status of ground-based observational methods. *Bulletin of the American Meteorological Society, 89*, 1549–1562.

Shupe, M. D., Kollias, P., Persson, P. O., & McFarquhar, G. M. (2008b). Vertical motions in mixed-phase stratiform clouds. *Journal of the Atmospheric Sciences, 65*, 1304–1322. doi:10.1175/2007JAS2479.1.

Shupe, M. D., Persson, P. O. G., Brooks, I. M., Tjernström, M., Sedlar, J., Mauritsen, T., et al. (2013). Cloud and boundary layer interactions over the Arctic sea ice in late summer. *Atmospheric Chemistry and Physics, 13*, 9379–9400. doi:10.5194/acp-13-9379-2013.

Shupe, M. D., Valden, V. P., Eloranta, E., Uttal, T., Campbell, J. R., Starkweather, S. M., & Masataka, M. (2011). Clouds at Arctic atmospheric observatories, Part I: Occurrence and macrophysical properties. *Journal of Applied Meteorology and Climatology, 50*, 626–644.

Simon, R. L. (1977). The summertime stratus over the offshore waters of California. *Monthly Weather Review, 105*, 1310–1314.

Slingo, A. (1990). Sensitivity of the Earth's radiation budget to changes in low clouds. *Nature, 343*, 49–51.

Solomon, A., Shupe, M. D., Persson, P. O. G., & Morrison, H. (2011). Moisture and dynamical interactions maintaining decoupled Arctic mixed-phase stratocumulus in the presence of a humidity inversion. *Atmospheric Chemistry and Physics, 11*, 10127–10148.

Sorooshian, A., Anderson, B., Bauer, S. E., Braun, R. A., Cairns, B., Crosbie, E., et al. (2019). Aerosol–Cloud–Meteorology Interaction Airborne Field Investigations: Using lessons learned from the U.S. West Coast in the design of ACTIVATE off the U.S. East Coast. *Bulletin of the American Meteorological Society*, 1511–1528. doi: https://doi.org/10.1175/BAMS-D-18-0100.1

Stamnes, K., Ellingson, R. G., Curry, J. A., Walsh, J. E., & Zak, B. D. (1999). Review of science issues, deployment strategy, and status for the ARM North Slope of Alaska–Adjacent Arctic Ocean climate research site. *Journal of Climate, 12*(1), 43–63. 10.1175/1520-0442-12.1.46.

Stanfield, R. E., Dong, X., Xi, B., Kennedy, A., Del Genio, A. D., Minnis, P., and Jiang, J. H. (2014). Assessment of NASA GISS CMIP5 and post-CMIP5 simulated clouds and TOA radiation budgets using satellite observations: Part I: Cloud fraction and properties. *Journal of Climate, 27*, 4189–4208. doi:10.1175/JCLI-D-13-00558.1.

Stanfield, R. E., Dong, X., Xi, B., Del Genio, A. D., Minnis, P., Doelling, D., & Loeb, N. (2015). Assessment of NASA GISS CMIP5 and Post-CMIP5 simulated clouds and TOA radiation budgets using satellite observations. Part II: TOA radiation budget and CREs. *Journal of Climate, 28*, 1842–1864. doi:10.1175/JCLI-D-14-00249.1

Stephens, G. L. (2005). Cloud feedbacks in the climate system: A critical review. *Journal of Climate, 18*(2), 237–273. https://doi.org/10.1175/JCLI-3243.1

Stephens, G. L., Vane, D. G., Tanelli, S., Im, E., Durden, S., Rokey, M., et al. (2008). CloudSat mission: Performance and early science after the first year of operation, *Journal of Geophysical Research, 113*, D00A18. doi:10.1029/2008JD009982.

Stevens, B., Moeng, C. -H., Ackerman, A. S., Bretherton, C. S., Chlond, A., De Roode, S., et al. (2005). Evaluation of large-eddy simulations via observations of nocturnal marine stratocumulus. *Monthly Weather Review, 133*, 1443–1462. doi:10.1175/MWR2930.1.

Stokes, G. M., & Schwarz, S. E. (1994). The Atmospheric Radiation Measurement (ARM) Program: Programmatic background and design of the Cloud and Radiation Test Bed. *Bulletin of the American Meteorological Society, 75*, 1201–1221.

Stubenrauch, C., Rossow, W. B., Kinne, S., Ackerman, S., Cesana, G., Chepfer, H., et al. (2013). Assessment of global cloud datasets from satellites: Project and database initiated by the GEWEX Radiation Panel. *Bulletin of the American Meteorological Society, 94*, 1031–1049. doi:10.1175/BAMS-D-12-00117.

Sun-Mack, S., Minnis, P., Chen, Y., Kato, S., Yi, Y., Gibson, S., et al. (2014). Regional apparent boundary layer lapse rates determined from CALIPSO and MODIS data for cloud height determination. *Journal of Applied Meteorology and Climatology, 53*, 990–1011, doi:10.1175/JAMC-D-13-081.1.

Sun-Mack, S., Minnis, P., Smith, W. L., Hong, G., & Chen, Y. (2017). Detection of single and multilayer clouds in an artificial neural network approach. *Proceedings of SPIE Conference, Remote Sensing Clouds and the Atmosphere XXII*, Warsaw, Poland, *Sept. 11–14*, 10424-7, 12 pp., doi: 10.1117/12.2277397.

Trepte, Q. Z., Bedka, K. M., Chee, T. L., Minnis, P., Sun-Mack, S., Yost, C. R., et al. (2019). Global cloud detection for CERES edition 4 using terra and aqua MODIS data. *IEEE Transactions on Geoscience and Remote Sensing, 57*, 9410–9449. doi: 10.1109/TGRS.2019.2926620

Tsay, S.-C., & Jayaweera, K. (1984). Physical characteristics of Arctic stratus clouds. *Journal of Applied Meteorology, 23*, 584–596.

Twomey, S. (1977). The influence of pollution on the shortwave albedo of clouds. *Journal of the Atmospheric Sciences, 34*, 1149–1152.

Verlinde, J., Harrington, J. Y., McFarquhar, G. M., Yanuzzi, V. T., Avramov, A., Greenberg, S., et al. (2007). The mixed-phase arctic cloud experiment. *Bulletin of the American Meteorological Society, 88*, 205–221.

Vernon, E. M. (1936). The diurnal variation in ceiling height beneath stratus clouds. *Monthly Weather Review, 64*, 14–16.

Wade, C. G. (2003). A multisensor approach to detecting drizzle on ASOS. *Journal of Atmospheric and Oceanic Technology, 20*, 820–832. https://doi.org/10.1175/1520-0426(2003)020\LY1\textbackslash0820:AMATDD[2.0.CO;2.

Walther, A., Heidinger, A. K., & Miller, S. D. (2013). The expected performance of cloud optical and microphysical properties derived from Suomi NPP VIIRS day/night band lunar reflectance. *Journal of Geophysical Research: Atmospheres, 118*, 13,230–13,240. doi:10.1002/2013JD020478.

Wang, Q., & Lenschow, D. H. (1995). Observational study of the role of penetrating cumulus in a marine stratocumulus-topped boundary layer. *Journal of the Atmospheric Sciences*, *52*, 2778–2787.

Wang, J., & Rossow, W. B. (1995). Determination of cloud vertical structure from upper-air observations. *Journal of Applied Meteorology*, *34*, 2243–2258. doi:10.1175/15200450(1995)034<2243: DOCVSF>2.0.CO;2.

Wang, Y., Zheng, X., Dong, X., Xi, B., Wu, P., Logan, T., & Yung, Y. L. (2020). Long-range transport of aerosols on marine boundary cloud properties. *Atmospheric Chemistry and Physics*, *20*, 14741–14755. https://doi.org/10.5194/acp-20-14741-2020.

Warren, S. G., Hahn, C. J., London, J., Chervin, R. M., & Jenne, R. L. (1986). *Global distribution of total cloud cover and cloud type amounts over land, NCAR Tech.* Note, *NCAR/TN-273+STR,* 229 pp., Natl. Cent. for Atmos. Res., Boulder, Colo.

Warren, S. G., Hahn, C. J., London, J., Chervin, R. M., & Jenne, R. L. (1988). *Global distribution of total cloud cover and cloud type amounts over ocean, NCAR Tech.* Note, *NCAR/TN-317 +STR,* 170 pp., Natl. Cent. for Atmos. Res., Boulder, Colo.

Watts, P. D., Bennartz, R., & Fell, F. (2011). Retrieval of two-layer cloud properties from multispectral observations using optimal estimation. *Journal of Geophysical Research*, *116*, D16203. doi:10.1029/2011JD015883.

Weckwerth, T. M., Wilson, J. W., Wakimoto, R. M., & Crook, N. A. (1997). Horizontal roll convection: Determining the environmental conditions supporting their existence and characteristics. *Monthly Weather Review*, *123*, 505–526.

Wentz, F. J. (1997). A well-calibrated ocean algorithm for special sensor microwave/imager, *Journal of Geophysical Research*, *102*, 8703–8718. doi:10.1029/96JC01751.

Westbrook, C. D., Hogan, R. J., O'Connor, E. J., & Illingworth, A. J. (2010). Estimating drizzle drop size and precipitation rate using two-colour lidar measurements. *Atmospheric Measurement Techniques*, *3*(3), 671–681. https://doi.org/10.5194/amt-3-671-2010

Wielicki, B. A., Barkstrom, B. R., Harrison, E. F., Lee, R. B. III, Smith, G. L., & Cooper, J. E. (1996). Clouds and the Earth's Radiant Energy System (CERES): An Earth Observing System experiment. *Bulletin of the American Meteorological Society*, *77*, 853–868.

Widener, K., Bharadwaj, N., & Johnson, K. (2012). Ka-band ARM Zenith Radar (KAZR) instrument handbook, *Rep. DOE/SC-ARM/TR-106,* DOE Office of *Science Atmospheric Radiation Measurement (ARM) Program (United States)*.

Wind, G., Platnick, S., King, M. D., Hubanks, P. A., Pavolonis, M. J., Heidinger, A. K., et al. (2010). Multilayer cloud detection with the MODIS near-infrared water vapor absorption band. *Journal of Applied Meteorology and Climatology*, *49*, 2315–2333. doi:10.1175/2010JAMC2364.1.

Winker, D. M., Vaughan, M. A., Omar, A., Hu, Y., Powell, K. A., Liu, Z., et al. (2009). Overview of the CALIPSO mission and CALIOP data processing algorithms. *Journal of Atmospheric and Oceanic Technology*, *26*, 2310–2323. doi:10.1175/2009JTECHA1281.1.

Wood, R. (2005a). Drizzle in stratiform boundary layer clouds: Part I: Vertical and horizontal structure. *Journal of the Atmospheric Sciences*, *62*, 3011–3033.

Wood, R. (2005b). Drizzle in stratiform boundary layer clouds: Part II: Microphysical aspects. *Journal of the Atmospheric Sciences*, *62*, 3034–3050.

Wood, R., & Hartmann, D. L. (2006). Spatial variability of liquid water path in marine low cloud: The importance of mesoscale cellular convection. *Journal of Climate*, *19*. doi:10.1175/JCLI3702.1.

Wood, R., Mechoso, C. R., Weller, R. A., Huebert, B., Straneo, F., Albrecht, B. A., et al. (2011a). The VAMOS Ocean-Cloud-Atmosphere-Land Study Regional Experiment (VOCALS-REx): Goals, platforms, and field operations, *Atmospheric Chemistry and Physics*, *11*, 627–654.

Wood, R., Bretherton, C. S., Leon, D., Clarke, A. D., Zuidema, P., Allen, G., & Coe, H. (2011b). An aircraft case study of the spatial transition from closed to open mesoscale cellular convection over the Southeast Pacific. *Atmospheric Chemistry and Physics*, *11*, 2341–2370.

Wood, R. (2012). Review: Stratocumulus clouds. *Monthly Weather Review*, *140*, 2373–2423.

Wood, R., Kuan-Ting, O., Bretherton, C. S., Mohrmann, J., Albrecht, B., Zuidema, P., et al. (2018). Ultraclean layers and optically thin clouds in the stratocumulus transition: Part I: Observations. *Journal of the Atmospheric Sciences*, *75*, 1631–1652. doi:10.1175/JAS-D-17-0213.1.

Wood, R., Wyant, M., Bretherton, C. S., Rémillard, J., Kollias, P., Fletcher, J., et al. (2015). Clouds, aerosol, and precipitation in the marine boundary layer: An ARM Mobile Facility deployment. *Bulletin of the American Meteorological Society*, *96*, 419–439.

Wu, P., Dong, X., Xi, B., & Ward, D. M. (2020a). Profiles of MBL cloud and drizzle microphysical properties retrieved from ground-based observations and validated by aircraft in-situ measurements over the Azores. *Journal of Geophysical Research*, *125*, e2019JD032205. doi:10.1029/2019JD032205.

Wu, P., Dong, X., & Xi, B. (2020b). A climatology of marine boundary layer cloud and drizzle properties derived from ground-based observations over the Azores. *Journal of Climate*, *30*, 10133–10148. doi:10.1175/JCLI-D-20-0272.1.

Xi, B., Dong, X., Minnis, P., & Khaiyer, M. M. (2010). A 10-year climatology of cloud cover and vertical distribution derived from both surface and GOES observations over the DOE ARM SGP site. *Journal of Geophysical Research*, *115*, D12124. doi:10.1029/2009JD012800.

Xi, B., Dong, X., Minnis, P., & Sun-Mack, S. (2014). Comparison of marine boundary layer cloud properties from CERES-MODIS Edition 4 and DOE ARM AMF Measurements at the Azores. *Journal of Geophysical Research*, *119*, 9509–9529. doi:10.1002/2014JD021813.

Xie, Y., Liu, Y., Ling, C. N., & Min, Q. (2014). Retrievals of cloud fraction and cloud albedo from surface-based shortwave radiation measurements: A comparison of 16 year measurements, *Journal of Geophysical Research: Atmospheres*, *119*, 8925–8940. doi:10.1002/2014JD021705.

Yang, Y., Meyer, K., Wind, G., Zhou, Y., Marshak, A., Platnick, S., et al. (2019). Cloud products from the Earth Polychromatic Imaging Camera (EPIC): Algorithms and initial evaluation. *Atmospheric Measurement Techniques*, *12*, 2019–2031. doi:10.5194/amt-12-2019-2019.

Yoo, H., Li, Z., Hou, Y.-T., Lord, S., Weng, F., & Barker, H. W. (2013)., Diagnosis and testing of low-level cloud parameterizations for the NCEP/GFS model using satellite and ground-based measurements. *Climate Dynamics*, *41*, 1595–1613. doi:10.1007/s00382-013-1884-8.

Yost, C. R., Minnis, P., Sun-Mack, S., Chen, Y., & Smith, W. L. Jr. (2021). CERES MODIS cloud product retrievals for Edition 4—Part II: Comparisons to CloudSat and CALIPSO. *IEEE Transactions on Geoscience and Remote Sensing*, *59*, 3695–3724. doi:10.1109/TGRS.2020.3015155.

Yost, C. R., Minnis, P., Sun-Mack, S., Smith, W. L. Jr., Trepte, Q. Z., & Chen, Y. (2023). VIIRS Edition 1 cloud properties for CERES. Part 2: Evaluation with CALIPSO. *Remote Sensing*, *15*, 1349. doi:10.3390/rs15051349.

Zelinka, M. D., Andrews, T., Forster, P. M., & Taylor, K. E. (2014). Quantifying components of aerosol-cloud-radiation interactions in climate models. *Journal of Geophysical Research*, *119*, 7599–7615. doi:10.1002/2014JD021710.

Zhang, Y., Yu, R., Li, J., Yuan, W., & Zhang, M. (2013). Dynamic and thermodynamic relations of distinctive stratus clouds on the lee side of the Tibetan Plateau in the cold season. *Journal of Climate*, *26*, 8378–8391. doi:10.1175/JCLI-D-13-00009.1.

Zhang, Z., & Platnick, S. (2011). An assessment of differences between cloud effective particle adius for marine water clouds from three MODIS spectral bands. *Journal of Geophysical Research*, *116*, D20215. doi:10.1029/2011JD016216.

Zhang, Z., Ackerman, A. S., Feingold, G., Platnick, S., Pincus, R., & Xue, H. (2013). Effects of cloud horizontal inhomogeneity and drizzle on remote sensing of cloud droplet effective radius: Case studies based on large-eddy simulations. *Journal of Geophysical Research*, *117*, D19208. doi:10.1029/2012JD017655.

Zhang, Z., Dong, X., Xi, B., Song, H., Ma, P.-L., Ghan, S., et al. (2017). Inter-comparisons of marine boundary layer cloud properties from two MODIS products and ground-based retrievals over the ARM Azores site. *Journal of Geophysical Research*, *122*, 2351–2369. doi:10.1002/2016JD025763.

Zhao, C., Xie, S., Klein, S., Protat, A., Shupe, M., McFarlane, S., et al. (2012). Toward understanding of differences in current cloud retrievals of ARM ground-based measurements. *Journal of Geophysical Research*, *117*, D10206. doi:10.1029/2011JD016792.

Zheng, X., Albrecht, B., Jonsson, H. H., Khelif, D., Feingold, G., Minnis, P., et al. (2011). Observations of the boundary layer, cloud, and aerosol variability in the southeast Pacific near-coastal marine stratocumulus during VOCALS-Rex, *Atmospheric Chemistry and Physics*, *11*, 9943–9959. doi:10.5194/acp-11-9943-2011.

Zheng, X., Xi, B., Dong, X., Logan, T., Wang, Y., & Peng, W. (2020). Investigation of aerosol and cloud interactions under different absorptive aerosol regimes using ARM SGP ground-based measurements. *Atmospheric Chemistry and Physics*, *20*, 3483–3501. doi:10.5194/acp-20-3483-2020.

Zheng, Y., Rosenfeld, D., & Li, Z. (2016). Quantifying cloud base updraft speeds of marine stratocumulus from cloud top radiative cooling. *Geophysical Research Letters*, *43*, 11,407–11,413. doi:10.1002/2016GL071185.

Zheng, Y., Rosenfeld, D., & Li, Z. (2018). Estimating the decoupling degree of subtropical stratocumulus decks from satellite. *Geophysical Research Letters*, *45*, 12560–12568. doi:10.1002/2018GL078382.

Zheng, Y., Rosenfeld, D., & Li, Z. (2020). A more general paradigm for understanding the decoupling of stratocumulus-topped boundary layers: The importance of horizontal temperature advection. *Geophysical Research Letters*, *47*, e2020GL087697. doi:10.1002/2020GL087697.

Zhu, P., & Zhao, W. (2008). Parameterization of continental boundary layer clouds. *Journal of Geophysical Research*, *113*, D10201. doi:10.1029/2007JD009315.

Zhu, P., Albrecht, B., & Gottschalk, J. (2001). Formation and dissipation of nocturnal boundary layer clouds over the Southern Great Plains. *Journal of the Atmospheric Sciences*, *58*, 1409–1426.

Zinner, T., Wind, G., Platnick, S., & Ackerman, A. S. (2010). Testing remote sensing on artificial observations: Impact of drizzle and 3-D cloud structure on effective radius retrievals. *Atmospheric Chemistry and Physics*, *10*(19), 9535–9549.

Zuidema, P., Painemal, D., de Szoeke, S., & Fairall, C. (2009). Stratocumulus cloud-top height estimates and their climatic implications. *Journal of Climate*, *22*, 4652–4666.

Zuidema, P., Redemann, J., Haywood, J., Wood, R., Piketh, S., Hipondoka, M., & Formenti, P. (2016). Smoke and clouds above the southeast Atlantic: Upcoming field campaigns probe absorbing aerosol's impact on climate. *Bulletin of the American Meteorological Society*, *97*, 1131–1135. doi:10.1175/BAMS-D-15-00082.1.

7

Planetary Boundary Layer and Processes

Virendra P. Ghate[1] and David B. Mechem[2]

ABSTRACT

The planetary boundary layer (PBL) is a linchpin of many tropospheric meteorological phenomena and serves as an interface of exchange between the surface and the free troposphere. The behavior of deep convection, for example, is largely determined by the properties of the boundary-layer air feeding the convective updrafts. Climatically important low clouds such as stratocumulus and shallow cumulus reside in the boundary layer, and the PBL is particularly important because the majority of biota (including humans) reside there. This chapter introduces the PBL structure and provides an in-depth look at a specific case study of the development of a summertime convective boundary layer. The chapter then summarizes the most common theoretical approaches for addressing the boundary layer, including Reynolds averaging, the turbulence kinetic energy equation, and the surface layer. A hierarchy of models for representing the boundary layer is then presented, including mixed-layer models, first-order closure, 1.5-order TKE closure, and higher-order closure approaches. The discussion throughout emphasizes the buoyancy-driven convective boundary layer but briefly mentions impacts of shear and clouds. The chapter concludes with a brief historical context and future outlook of representing the boundary layer in large-scale atmospheric models.

7.1. BACKGROUND AND INTRODUCTION

The planetary (or atmospheric) boundary layer (PBL) can be defined as the layer of the atmosphere directly influenced by the Earth's surface through processes that typically occur on the timescale of an hour (Stull, 1988). The boundary layer is the linchpin that ties together land-atmosphere or ocean-atmosphere interactions via fluxes of heat, moisture, carbon, and momentum from the surface with properties of the free troposphere. The boundary layer may be thought of as the portion of the troposphere above which the atmosphere, on short timescales, is not influenced by the surface.

Over land, the boundary-layer depth greatly varies both temporally (over few hours) and spatially (over few kilometers), with deepest boundary layers over the arid deserts and shallowest boundary layers over the frozen polar ice sheets (Figure 7.1). Due to the relative homogeneity of surface temperatures and the surface fluxes over the oceans, the PBL depth over the ocean demonstrates little variability diurnally and over short distances (<20 km), although it does exhibit changes on seasonal and annual timescales (not shown).

The land or ocean surface defines the lower boundary of the PBL, whereas the upper boundary is often marked by one or more of the following: (i) base of a temperature and/or moisture inversion, (ii) changes in the lapse rate of a conserved variable, or (iii) changes in the wind speed and/or direction. A well-mixed PBL itself can be further

[1] *Environmental Science Division, Argonne National Laboratory, Lemont, IL, USA*

[2] *Department of Geography and Atmospheric Science, University of Kansas, Lawrence, KS, USA*

Fast Processes in Large-Scale Atmospheric Models: Progress, Challenges, and Opportunities, Geophysical Monograph 282, First Edition.
Edited by Yangang Liu and Pavlos Kollias.

DOI: 10.1002/9781119529019.ch07

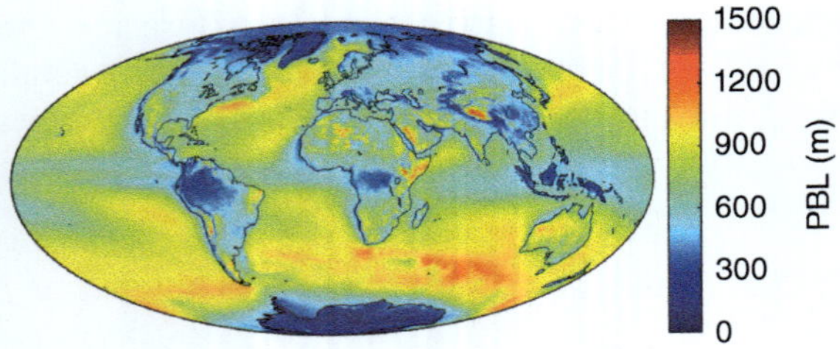

Figure 7.1 Average planetary boundary layer (PBL) depths as reported by the ERA5 reanalysis model for the years 1979–2019 (Hersbach et al., 2020).

subdivided vertically into a (i) surface layer, (ii) mixed layer, and (iii) entrainment layer, where air from the free troposphere is mixed into the turbulent boundary layer (described in greater detail in the next section). The thicknesses of the surface layer and the entrainment layer are often less than 10% of the boundary-layer depth and are characterized by the presence of large gradients (and hence fluxes) of the thermodynamic properties and wind fields. In the clear convective boundary layer, the conserved thermodynamic properties and winds are nearly constant within the mixed layer due to the turbulence within the boundary layer that is maintained and modulated by the surface buoyancy fluxes, entrainment, and wind shear. If the surface lifting condensation level (LCL, the height at which air ascending from the surface reaches condensation) lies below the PBL inversion, clouds can form within the boundary layer, becoming an important component of boundary-layer dynamics. The focus of this chapter is on the cloud-free boundary layer, with much of the discussed theory applicable to both cloudy and clear-air boundary layers.

Below we first describe the diurnal cycle of the summertime planetary boundary layer over land, followed by the theoretical foundation of the PBL parameterizations and different approaches used for representing the boundary layer in large-scale atmospheric models. The chapter is concluded with brief summary of the historical research on PBL and discussion on the similarities and differences in PBL parameterizations.

7.2. DIURNAL CYCLE OF SUMMERTIME BOUNDARY LAYER OVER LAND

An example of the surface radiation budget during summertime conditions over land is shown in Figure 7.2. At the surface, the upwelling longwave radiation is greater than the downwelling longwave radiation, resulting in cooling of the surface in the longwave spectrum. During the daytime, the surface absorbs most of the incident solar radiation (>80%), resulting in substantial heating of the surface that increases the surface skin temperature by tens of degrees. The net radiation absorbed by the surface is greatly determined by the surface albedo that in turn depends on the land use and vegetation. Collectively, this

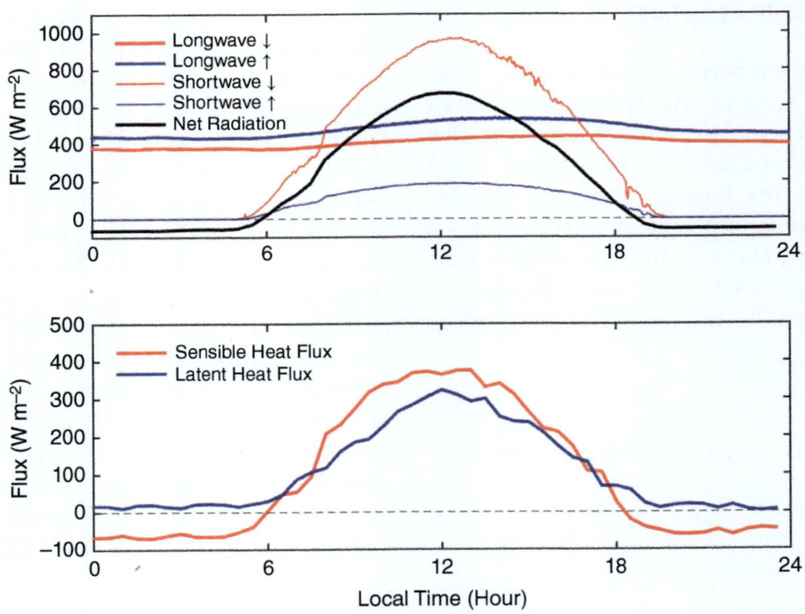

Figure 7.2 Top: Diurnal cycle of downwelling and upwelling shortwave and longwave radiation at the surface. Bottom: Diurnal cycle of sensible heat flux and latent heat flux at the surface. The observations are from 12 June 2013 at the Atmospheric Radiation Measurement (ARM) Southern Great Plains (SGP) site near Lamont Oklahoma.

results in heating of the surface during the daytime and cooling during the nighttime, making the surface skin temperature higher than the surface air temperature during the daytime and lower during the nighttime.

Net radiation at the surface is partitioned into surface-sensible and latent heat fluxes (SHF and LHF), which represent the transfer of heat and moisture from the surface to the air. The surface-sensible heat and moisture fluxes increase the temperature and moisture content of air parcels at the surface, destabilizing them and providing the buoyancy necessary to drive turbulent boundary-layer eddies. These eddies originating from the surface heating have higher temperature and moisture compared to the mean air properties, and hence positive buoyancy and vertical velocity. The eddies ascend to increasingly higher levels over the course of the day consistent with the buoyancy forcing. The strength of the turbulence may be enhanced by wind shear, especially near the surface. These turbulent fluxes homogenize the boundary layer over its depth, meaning that properties of certain conserved variables such as potential temperature and total water mixing ratio remain constant throughout this layer (Figure 7.3).

The boundary layer deepens by mixing-in free-tropospheric air into the turbulent boundary layer in a process called entrainment. Entrainment typically mixes air from the free troposphere that is warmer, drier, and of greater horizontal momentum into the boundary layer. Free-tropospheric properties are mixed downward via entrainment of the laminar free-tropospheric flow into the turbulent boundary layer. The nature, magnitude, and scales of entrainment remain one of the least understood and most active areas of research.

When sufficient low-level moisture is available, the LCL lies below the boundary-layer inversion, leading to the formation of shallow cumulus clouds that then penetrate into and sometimes through the inversion. Depending on the thermodynamic stability, deeper clouds such as cumulus congestus and cumulonimbus can result. These deeper clouds distribute heat, moisture, and momentum upward from the boundary layer into the troposphere, thereby also venting gas and aerosol constituents (either natural or anthropogenic), specifically those originating at the surface or in the boundary layer, into the free troposphere. The boundary-layer inversion can be viewed as an interface between the boundary layer and free troposphere, but also as an energy barrier since energy is required in the entrainment process to mix the warm, dry free-tropospheric air downward. The depth and strength of the boundary layer is also modulated by synoptic-scale vertical air motion, as well as the stability of the inversion and overlying free troposphere. The stability of the

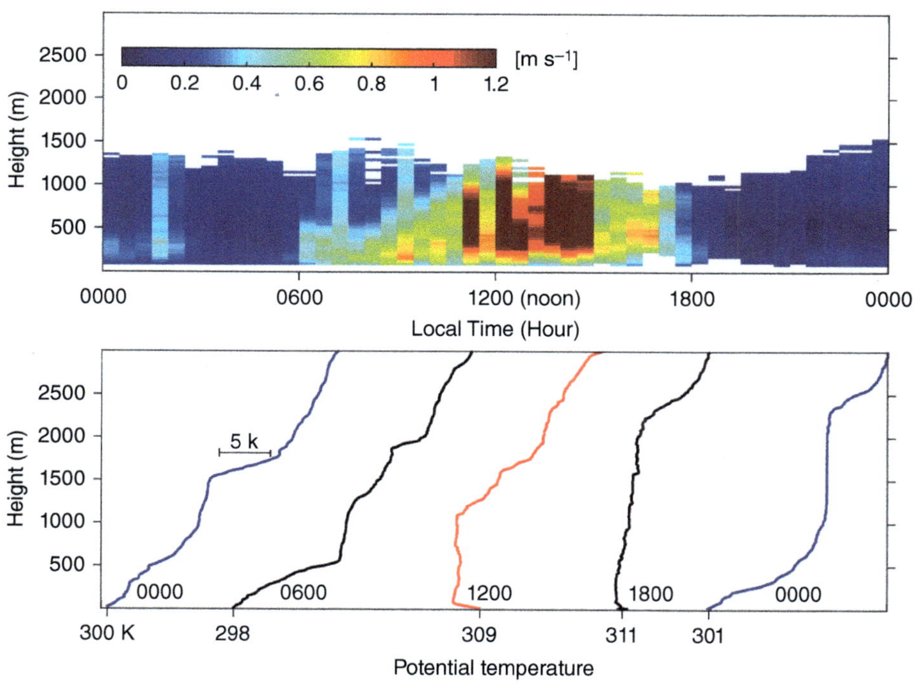

Figure 7.3 An example of the diurnal cycle of updraft velocity within the boundary layer (top) and profiles of potential temperature (bottom) for a clear-sky summertime boundary layer over land. The data were collected on 12 June 2013 at the ARM SGP site (36°N, 97°W, 316 m).

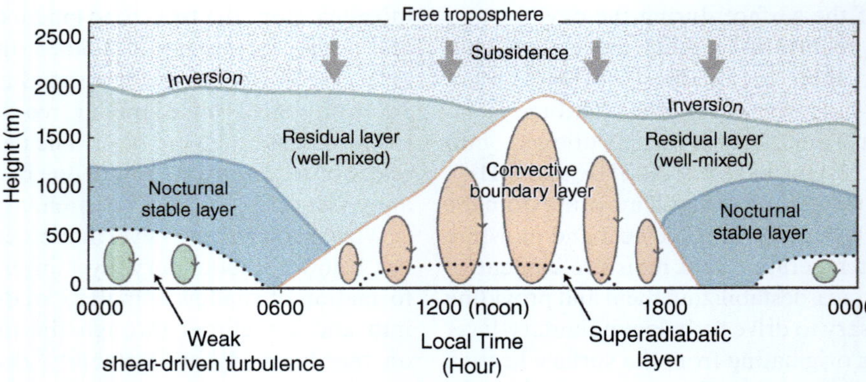

Figure 7.4 Conceptual schematic of the evolution of summertime planetary boundary layer over land.

boundary layer is typically cast in terms of unsaturated static stability where the stability is proportional to the vertical gradient of potential (or virtual potential) temperature (i.e., $\frac{d\theta_v}{dz} > 0$ for stable conditions).

After sunset, solar heating of the land heating ceases and the main mechanism generating turbulence over the depth of the boundary layer is removed (Figure 7.3), although shear may continue to drive turbulence. In the absence of solar heating, the boundary layer that has deepened over the course of the day becomes a so-called "residual" layer, which can remain well mixed and is characterized by weak turbulence. Strong surface cooling results in a stable layer that decouples a shallow layer near the surface (appropriately called the surface layer) with the residual layer above. This nocturnal decoupling is particularly important in understanding the structure and diurnal behavior of low-level jets such as the Great Plains Low-Level Jet in the central United States. Heating of the surface from the rising Sun rapidly heats this shallow layer near the surface, and the boundary layer begins to grow again.

A conceptual schematic of a typical diurnal cycle of the summertime boundary layer over land is shown in Figure 7.4. The primary mechanism producing turbulence during the nighttime is wind shear, while during the daytime it is the surface heating. Although less important than surface heating and wind shear, the radiative cooling of the atmosphere due to changes in temperature and gases (water vapor) and mixing of different air masses through entrainment also generates turbulence. The interaction between these processes collectively determines the strength of the turbulence. Boundary-layer depth is set by the balance between large-scale vertical motion (typically subsidence, though not always) and entrainment. The near-dawn hours and the early evening hours often exhibit large and rapid changes in the depths of these layers, presenting a particular challenge in simulating them in large-scale atmospheric models.

7.3. THEORETICAL REPRESENTATIONS OF THE BOUNDARY LAYER

This section summarizes the principal analysis frameworks used for exploring boundary-layer dynamics and the hierarchy of models used to represent boundary layers.

7.3.1. Reynolds Averaging

One defining characteristic of boundary layers is that the flow is turbulent. Any discussion of boundary-layer dynamics is intrinsically tied to theories of turbulent flow. To characterize the turbulent fluctuations of the flow, the technique of Reynolds averaging is employed to partition the flow into a slowly varying mean component and the turbulent fluctuations around it. In the case of temperature (T), it is separated as $T = \overline{T} + T'$, where $\overline{T}$ is the temperature averaged over sufficient length of time that the high-frequency fluctuations are averaged out. (Note here that any long-term or "secular" trend in the variables must also be removed.) In this framework, the mean of the perturbation quantity is then by definition zero, i.e., $\overline{T'} = 0$. The Reynolds-averaging framework separates contributions of transport from the mean and turbulent components of the flow. In the case of vertical transport of temperature,

$$\overline{wT} = \overline{(\overline{w} + w')(\overline{T} + T')} = \overline{w}\,\overline{T} + \overline{w'T'} \tag{7.1}$$

Note that the mean of covariance terms like $\overline{w'T'}$ are generally not zero, while the terms $\overline{\overline{w}T'}$ and $\overline{w'\overline{T}}$ are by definition zero. A case of $\overline{w'T'} > 0$ represents warm, upward motion or cool, downward motion, which corresponds to a thermodynamically direct circulation associated with convection and the conversion from potential energy to kinetic.

In studying the boundary layer, we typically apply Reynolds averaging to the conservation equations to draw out the explicit effects of turbulence on the mean fields. As a simple example, consider a thermodynamic energy equation for unsaturated adiabatic flow,

$$\frac{D\theta}{Dt} = 0 \tag{7.2}$$

After separating the potential temperature into mean and fluctuating components, and then taking the Reynolds average, we can derive an equation for the time tendency of Reynolds-averaged potential temperature (note here that strictly speaking $\overline{\theta}$ is the Reynolds average of the potential temperature departure from a height-dependent base state θ_0):

$$\frac{\overline{D}\overline{\theta}}{Dt} = -\overline{w}\frac{d\theta_0}{dz} - \left[\frac{\partial\, \overline{u'\theta'}}{\partial x} + \frac{\partial\, \overline{v'\theta'}}{\partial y} + \frac{\partial\, \overline{w'\theta'}}{\partial z}\right] \tag{7.3}$$

where

$$\frac{\overline{D}}{Dt} = \frac{\partial}{\partial t} + \overline{u}\frac{\partial}{\partial x} + \overline{v}\frac{\partial}{\partial y} + \overline{w}\frac{\partial}{\partial z}$$

The covariance terms represent the influences of the turbulent fluxes on the mean flow. Equation (7.3) can be used to prognose the mean potential temperature given the knowledge of the mean vertical air motion ($\overline{w}$), the lapse rate of potential temperature at the given time, and the horizontal gradient of the covariance between the turbulence fluctuation of wind vectors and the turbulence fluctuation of potential temperature. The terms in the square bracket in equation (7.3) are sometimes referred to as the spatial gradient of fluxes.

In the Reynolds-averaging framework, surface fluxes correspond directly to the covariance of vertical velocity and enthalpy, and the covariance of vertical velocity and water vapor mixing ratio, using the following equations:

$$\text{SHF} = \overline{\rho} c_p \,\overline{w'T'} \tag{7.4}$$

$$\text{LHF} = \overline{\rho}\, L_v \,\overline{w'q'} \tag{7.5}$$

where $\overline{\rho}$ is the mean air density at the surface, c_p is the specific heat of air at constant pressure, and L_v is the latent heat of vaporization.

The latent heat flux is usually positive during the entire day denoting transfer of moisture from surface to the air (which cools the surface), with higher values at the daytime and lower values during the nighttime. The sensible heat flux largely follows the difference between the surface skin temperature and air temperature, with positive values during the daytime and negative values during the nighttime. Similar to (7.4) and (7.5), equations can also be written to quantify the transport of tracers like carbon dioxide from the surface to the atmosphere.

Power spectra of meteorological variables demonstrate aspects of the nature of the turbulent boundary layer (Figure 7.5). Turbulent spectra are continuous, reflecting an input of kinetic energy at large scales, a continuous cascade of energy to smaller scales, and finally dissipation into heat energy from molecular diffusion at small scales (which typically occurs at scales smaller than most spectra can demonstrate). Details of the spectra are governed by the forcing mechanisms, boundary-layer depth, and stability. A peak in the spectra lies at a scale of typical anisotropic eddies that are of order boundary-layer depth

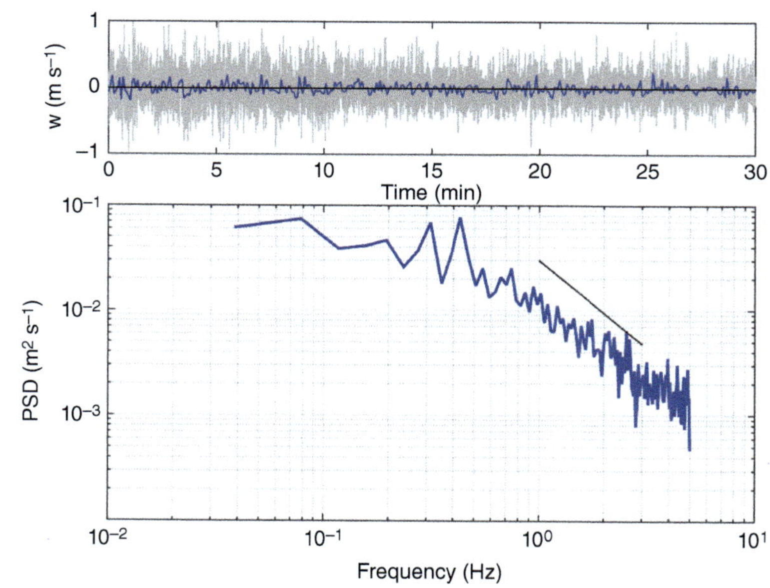

Figure 7.5 Top: High-resolution (10 Hz) vertical velocity (w) at 2 m (gray) and running 5 s average (blue). Bottom: Power spectrum density of the w along with −5/3rd slope line.

(H), with a timescale governed by the depth and characteristic vertical velocity W ($\sim H/W$). At scales lying between the energy-containing eddies and scales governed by viscous dissipation is the inertial subrange, where an equilibrium is present between energy coming from large eddies and the cascade of energy to smaller and smaller eddies. The slope of the energy spectrum is proportional to $f^{-\frac{5}{3}}$, which can be derived from dimensional analysis and can also be observationally verified. Even over the same analysis period, momentum, scalars, and covariances between different variables often exhibit somewhat different spectral characteristics. This deviation from the −5/3rd power spectrum scaling could be due to intermittency, which complicates representing PBL processes using mean values of variables.

The eddies that carry most of the turbulent energy in the boundary layer are appropriately referred to as "large eddies" in the context of the large-eddy simulation (LES), a class of fluid-dynamics models aimed at explicitly representing turbulent flows. These models have horizontal and vertical resolutions that allow them to resolve most of the turbulence energy in the boundary layer. However, as evident from the spectra, the LES models also cannot fully resolve the boundary-layer turbulence, necessitating a need to parameterize the subgrid scale (SGS) turbulence.

7.3.2. Mechanisms Governing Boundary-Layer Turbulence

The mean turbulence kinetic energy normalized by mass ($\overline{e}$) can be derived by applying the definition of kinetic energy to the turbulent (perturbation) portions of the wind vectors and then taking Reynolds average:

$$\overline{e} = \frac{1}{2}(\overline{u'^2} + \overline{v'^2} + \overline{w'^2}) \tag{7.6}$$

The Reynolds-averaging approach can be applied to the three-dimensional momentum equations to derive a conservation equation for the TKE associated with the turbulent contributions to the flow. Given a slightly simplified framework that assumes a coordinate system aligned with the mean flow, horizontal homogeneity in order to neglect horizontal transports, and neglecting large-scale vertical motion, the TKE equation can be written as

$$\frac{\partial \overline{e}}{\partial t} = \frac{g}{\overline{\theta}_v}\left(\overline{w'\theta_v'}\right) - \overline{u'w'}\frac{\partial \overline{U}}{\partial z} - \overline{v'w'}\frac{\partial \overline{V}}{\partial z} - \frac{\partial(\overline{w'e})}{\partial z} - \frac{1}{\overline{\rho}}\frac{\partial(\overline{w'p'})}{\partial z} - \varepsilon \tag{7.7}$$

The terms on the right-hand side of the above equation from left to right represent generation of TKE by buoyancy, mechanical generation due to vertical shear of the zonal wind, mechanical generation due to vertical shear of the meridional wind, transport of TKE by the turbulent eddies themselves, pressure correlation, and dissipation of the turbulence via molecular diffusion. The buoyancy term can be positive or negative, denoting that buoyancy can either generate or dissipate turbulence. The momentum flux $\overline{u'w'}$ is usually opposite in sign of the wind shear, and hence shear term is always positive, indicating that any form of wind shear always increases turbulence. The turbulent redistribution term simply redistributes TKE within different eddies in the boundary layer and could be positive or negative. The pressure correlation term is associated with changes in pressure due to phenomena like gravity waves. The last term is TKE dissipation, which represents one of the fundamental characteristics of turbulent flows. Although a steady-state balance could be envisioned for TKE, it is not a conserved quantity.

To explore the salient points of the mechanisms governing boundary-layer TKE, we note that the pressure correlation term is typically quite small and empirically challenging to determine. Additionally, when considering the mean TKE budget of the entire boundary layer, the turbulent transport term is near zero. Dissipation is always positive, and the term acts as a sink for TKE to represent the cascade of energy to smaller and smaller scales; otherwise, energy would unrealistically pile up at any given scale. The terms that remain are the mechanical/shear and buoyancy terms. The mechanical/shear term is always positive, while buoyancy can either generate or suppress turbulence, dependent upon the static stability.

When conditions are unstable or neutrally stratified, the buoyancy term is positive $\overline{w'\theta_v'} > 0$ and acts to increase the TKE. This behavior describes convection—warmer updrafts and colder downdrafts—which energetically is a thermally direct circulation and a conversion of potential energy into kinetic energy. The stable state where typically $\overline{w'\theta_v'} < 0$ (colder updrafts and/or warmer downdrafts) corresponds to a conversion of kinetic energy into potential energy where the turbulence must do work against the stratification to rearrange the vertical distribution of mass into a higher potential energy state. The most obvious example of turbulence acting in a stable state is the entrainment of warm, dry free-tropospheric air from above the inversion into the turbulent boundary layer, which requires a substantial amount of work to be done by the turbulence.

The growth or suppression of turbulence in a layer is predominantly governed by the relative measure of the shear production and buoyancy production/loss. When conditions are unstable or neutrally stratified, both terms are usually positive and the TKE increases. In stable conditions, when the buoyancy term is negative, turbulence can still be maintained but only if the shear-generation term substantially exceeds it. The relative importance of these two terms can be quantified using the flux Richardson number,

$$R_f = \frac{\left(\frac{g}{\theta_v}\right)\left(\overline{w'\theta'_v}\right)}{\overline{u'w'}\frac{\partial \overline{u}}{\partial z} + \overline{v'w'}\frac{\partial \overline{v}}{\partial z}}$$

where the barred quantities represent the Reynolds-averaged (mean) fields. In unstable conditions, $R_f < 0$; for neutral conditions, $R_f = 0$; and for stable conditions, $R_f > 0$. Even in stable conditions, observations demonstrate that turbulence can be maintained if the shear-generation term is ~4 times greater than the buoyancy damping, i.e., $R_f < 0.25$. Thus, turbulence can be maintained in a stable layer given strong enough shear. It should be noted that values different than 0.25 have been used in large-scale atmospheric models (Zhang et al., 2015).

7.4. IDEALIZED MODELS OF THE BOUNDARY LAYER

Using the theoretical framework described above, several approaches have been proposed to represent the boundary layer in large-scale models. These approaches utilize the mean profiles of thermodynamic and dynamic variables at a particular model grid box along with the surface fluxes. Traditionally (before 2000), different schemes were used to represent boundary layer, clouds, and the surface. However, because these different physical components interact with each other in a nonlinear fashion, there have been several attempts to represent them in a unified framework, especially as pertained to accurately representing the boundary-layer clouds (Arakawa, 2004). Below we describe the traditional boundary-layer frameworks and also some of the more recent approaches.

7.4.1. Mixed-Layer Model

During strong surface turbulent flux conditions, the boundary layer can be idealized as being well mixed in the vertical. The boundary layer over the open oceans is also often well mixed primarily due to uniformity in the surface conditions and forcing. In the case of a well-mixed boundary layer, the temperature and moisture profiles can be distilled into a single temperature value and a single moisture value, provided care is taken to use conservative thermodynamic variables (e.g., the various flavors of potential temperature and mixing ratio). This assumption results in a low-dimensional representation of the turbulent boundary layer called a mixed-layer model (MLM). MLMs trace their history to Ball (1960), who formulated the MLM approach to address the regulation of inversion height from surface heating and large-scale subsidence.

Assuming a mixed layer purely governed by turbulent fluxes, a conservation equation for potential temperature as a function of time and height can be developed:

$$\frac{\partial \theta}{\partial t} = -\frac{\partial \left(\overline{w'\theta'}\right)}{\partial z} \tag{7.8}$$

It is straightforward to demonstrate that to maintain a profile of constant potential temperature (i.e., a mixed layer), the fluxes must be linear. These can be integrated from the surface ($z = 0$) to the boundary-layer height h to yield an equation for mean boundary-layer potential temperature:

$$h \times \frac{d\overline{\theta}}{dt} = -\left(\overline{w'\theta'}\Big|_h - \overline{w'\theta'}\Big|_{sfc}\right) \tag{7.9}$$

Note here that the $\overline{\theta}$ signifies a quantity that is averaged in the vertical over the mixed layer and Reynolds-averaged in time. The quantity $\overline{w'\theta'}\big|_h$ is referred to as the entrainment flux, and $\overline{w'\theta'}\big|_{sfc}$ is the surface-sensible flux of θ. These turbulent fluxes are both unknown quantities and require closure assumptions (parameterizations), formulated as a function of the mean variables ($\overline{\theta}$ in this case) or the boundary conditions. Over land, the surface flux is driven by shortwave heating of the surface and is often parameterized using a bulk aerodynamic approach:

$$\overline{w'\theta'}\Big|_{sfc} = -C_d \times V \times (\overline{\theta} - \theta_{sfc}) \tag{7.10}$$

where V is the wind speed, C_d is an aerodynamic exchange coefficient, and θ_{sfc} is the surface skin temperature. The entrainment flux $\overline{w'\theta'}\big|_h$ also requires a closure assumption and in MLMs is typically represented as

$$\overline{w'\theta'}\Big|_h = -w_e \times \Delta\theta \tag{7.11}$$

where $\Delta\theta$ is the difference in potential temperature (the "jump") between the mixed layer and the free troposphere just above the inversion. The quantity w_e represents the entrainment rate at which the boundary layer rises in the absence of any large-scale vertical motion; w_e must be parameterized. In dry convective boundary layers, entrainment flux is typically assumed to be related to a fraction A of the surface-sensible flux, so

$$w_e = \frac{A \times \overline{w'\theta'}\Big|_{sfc}}{h} \tag{7.12}$$

where A is dimensionless parameter, typically taken to be 0.2 for cloud-free, buoyancy-driven boundary layers but can be much more variable in, for example, stratocumulus-topped boundary layers. These assumptions produce an equation for the mixed-layer potential temperature,

$$\frac{d\overline{\theta}}{dt} = \frac{\overline{w'\theta'}\big|_{\mathrm{sfc}}}{h} + \frac{A \times \overline{w'\theta'}\big|_{\mathrm{sfc}}}{h} \tag{7.13}$$

with the terms representing the change of mixed-layer potential temperature, surface-sensible heat flux, and entrainment flux, respectively:

$$\frac{d\overline{\theta}}{dt} = (1 + A)\frac{\overline{w'\theta'}\big|_{\mathrm{sfc}}}{h} \tag{7.14}$$

The above set of equations present a framework to calculate the average value of mixed-layer potential temperature at a future time, given the surface flux and mixed-layer depth at the current time. In a similar vein, by assuming horizontal homogeneity (i.e., neglecting the possibility of a sloped boundary layer), the boundary-layer depth h evolves according to

$$\frac{dh}{dt} = w_{\mathrm{e}} + w_{\mathrm{ls}} \tag{7.15}$$

where w_{ls} is the large-scale vertical air motion. On average, the boundary-layer depth is in equilibrium with the vertical motion, and the large-scale vertical motion helps determine the equilibrium boundary-layer depth. Equations (7.14) and (7.15) together with closure assumptions (7.12) constitute a closed system for a mixed-layer model. A typical MLM model also includes equations for moisture and momentum, which we omit for brevity.

Lilly (1968) extended the MLM approach to stratocumulus-topped boundary layers. These boundary layers are also well mixed but include mechanisms beyond simple surface heating that may enhance or suppress turbulence (e.g., cloud-top radiative cooling, evaporative cooling of precipitation, and wind shear). Assumption that the boundary layer be well mixed looms large in PBLs with drizzle, shortwave heating, and other mechanisms that can often violate the mixed-layer assumption. MLMs received a second life in the early 2000s as a framework to analyze entrainment closures (Lilly, 2002a,b; Stevens, 2002) and the behavior of heat and moisture budgets governing low-cloud behavior (e.g. Caldwell et al., 2005; Ghate et al., 2019; McMichael et al., 2019; Van der Dussen, 2014; Wood, 2007).

7.4.2. First-Order Closure

Stable boundary layers are by definition not well mixed and require an approach that includes height dependence. The most common approach is to employ the concept of molecular viscosity (diffusion) as an analog for turbulent fluxes and assume the turbulent fluxes to be proportional to the gradients of the mean variables. Considering here just the vertical variation,

$$\overline{w'\theta'} = -K_{\mathrm{h}}\frac{\partial\overline{\theta}}{\partial z} \tag{7.16}$$

$$\overline{w'u'} = -K_{\mathrm{m}}\frac{\partial\overline{u}}{\partial z} \tag{7.17}$$

Here K_{h} is the eddy viscosity and K_{m} is the eddy diffusivity (m^2 s^{-1}). This parameterization for the turbulent fluxes using this flux-gradient or K-theory approach is termed first-order closure. Unlike for molecular viscosity and diffusivity coefficients, K_{m} and K_{h} values for turbulent flows vary substantially based on characteristics of the flow.

Formulating realistic values of these eddy coefficients can be highly challenging. The most common approach, based on the work of Prandtl (1925) and summarized in Stull (1988), is to consider vertical finite displacements in an environment characterized by linear profiles in mean scalar and momentum quantities. The difference between those displaced values and the environment are then

$$\theta' = \frac{\partial\overline{\theta}}{\partial z} \times z' \tag{7.18}$$

$$u' = \frac{\partial\overline{u}}{\partial z} \times z' \tag{7.19}$$

Now assuming that the magnitude of the vertical motion of those displaced parcels scales with the horizontal motion, i.e., $w' \propto u'$,then

$$w' = c\frac{\partial\overline{u}}{\partial z} \times z' \tag{7.20}$$

where c is a constant of proportionality.

Multiplying θ' and w' and taking Reynolds averaging give the turbulent flux of potential temperature

$$\overline{w'\theta'} = -c \times (z')^2 \times \frac{\partial\overline{u}}{\partial z} \times \frac{\partial\overline{\theta}}{\partial z} \tag{7.21}$$

We can then define a mixing length as

$$l^2 = c \times (z')^2$$

and

$$K_{\mathrm{h}} = l^2 \times \frac{\partial\overline{u}}{\partial z}$$

This above expression illustrates that K_{h} increases as the shear of the mean environment increases, and increases proportionally with the square of the turbulent displacements.

Not all meteorological situations are amenable to the K theory. For example, K theory does not apply well in the mixed layer since the mean gradient of wind and conserved variables is small (almost negligible), yet turbulent transport can be significant. Furthermore, for $K_{\mathrm{h}} > 0$

first-order closure also assumes that all turbulent fluxes are downgradient (diffusive), whereas some atmospheric phenomena exhibit countergradient transports (e.g., momentum transports in convective squall lines). Even with these drawbacks, however, first-order closure is very commonly employed.

One of the most elegant examples is the boundary-layer model of Ekman (1925), who used one-dimensional, steady-state, Reynolds-averaged momentum equations, together with K theory, appropriate boundary conditions, and a novel solution approach, to obtain the classic Ekman spiral profile of flow in the ocean as a function of depth. A similar approach can be applied in the atmosphere (see Holton, 2004 for a concise summary).

7.4.3. 1.5-Order (TKE) Closure

Higher-order closures add prognostic equations for the higher-order moments and covariances. One natural step beyond first-order closure that is commonly employed is a 1.5-order closure, so called because it includes a prognostic equation for subgrid-scale TKE (and sometimes potential temperature variance $\overline{\theta'^2}$ (Stensrud, 2007)) but not for the full set of variances and covariances. We note here for clarity that the papers of Mellor and Yamada (1974, 1982) term this approach as a "level-2.5" closure. The classic paper by Deardorff (1980) describing a large-eddy simulation treatment of stratocumulus contains an especially clear summary of 1.5-order turbulence closure approach. The prognostic TKE equation in 1.5-order closure is similar in form to equation (7.7), with one very important crucial and often underappreciated difference. Because it represents the subgrid TKE, the fluxes on the right-hand side of the TKE equation cannot be computed from grid-mean calculations but rather must be approximated using closure assumptions, typically K theory. Upon having an expression for subgrid-scale TKE, the 1.5-order closure then assumes that the eddy viscosity for momentum is proportional to mixing length (as in first-order closure) and to the square root of the TKE (Deardorff, 1980):

$$K_m = 0.1 \times l \times \overline{e}^{0.5}$$

Furthermore, it is assumed that the mixing length l itself scales to the square root of the subgrid scale TKE and is inversely proportional to the stability. This is especially necessary to represent much smaller mixing lengths in strongly stable regions. For the classical 1.5-order closure with only the SGS TKE equation, it is noteworthy that as long as the eddy viscosity and diffusivity coefficients are positive, the turbulent fluxes will be downgradient just as they are in first-order closure. Including an equation for SGS variance of perturbation potential temperature allows for countergradient SGS transport (Stensrud, 2007), but most 1.5-order schemes include only the TKE equation.

7.4.4. Higher-Order Closure

Second- or third-order closures are referred to as higher-order closures and require prognostic equations for variances of (potentially) all of the model prognostic variables, as well as prognostic equations for the covariances between the different variables. Typically, second-order closure schemes only use a selected subset of all possible equations. Nevertheless, second-order closure schemes are quite complex and include third-order terms on the right-hand side of the equation, which are typically challenging to represent and physically interpret. Classical second-order approaches are summarized in Mellor and Yamada (1974, 1982) and Stull (1988) and are not repeated here. These schemes also have been successfully applied to simulate cloudy boundary layers (Wang & Wang, 1994).

A recent approach to higher-order closure uses assumed-PDF (distributions) among different variables. Currently, the Cloud Layers Unified by Binormals (CLUBB) parameterization (Golaz et al., 2002; Larson & Golaz, 2005; Larson et al., 2012) represents the state of the art of this approach. The most commonly employed version of CLUBB assumes joint double-Gaussian distributions of liquid water potential temperature (θ_l), total water (q_t), and vertical motion (w). These assumed PDFs, along with appropriate closure assumptions, are applied to prognostic equations for variances and covariances (plus an equation for the third-order quantity $\overline{w'w'w'}$, proportional to the skewness of the vertical velocity field). One of the theoretical advantages of this approach is that it reduces the requirements for separate, regime-based parameterizations such as schemes for stratocumulus entrainment, shallow cumulus, and deep convection. Instead, the higher-order CLUBB framework can represent all three in an internally consistent manner that negates the need for artificial boundaries between the schemes.

7.5. SURFACE LAYER

It is typically assumed that fluid-flow velocity at a boundary is zero (the so-called no-slip boundary condition), which necessitates additional assumptions to blend the no-slip boundary condition with the boundary-layer parameterizations discussed above. Scale analysis shows that the change of momentum flux over the lowest portion of the boundary layer—termed the "surface layer"—is only ~10% of the surface flux itself, so it is reasonable to assume that the flux is constant over this layer.

Surface-layer parameterizations vary greatly, with Monin-Obukhov Similarity Theory (MOST) forming the main basis of most of them.

We define a term called friction velocity u^* (ustar), which is the square root of the surface momentum flux:

$$\overline{w'u'}_{\text{sfc}} = u^{*2} \tag{7.22}$$

Here, the horizontal wind has been rotated to lie entirely in the u' component. Now assuming that fluxes over the depth of the surface layer are equal to the surface value, we employ K theory on the LHS to obtain

$$K_{\text{m}} \left| \frac{\partial \overline{u}}{\partial z} \right| = u^{*2} \tag{7.23}$$

Representing K_{m} using mixing-length theory as $K_{\text{m}} = l^2 \frac{\partial \overline{u}}{\partial z}$ and $l = k \times z$, where k is von Karman's constant, experimentally determined to be ~0.4, we can rewrite equation (7.22) as

$$\frac{\partial \overline{u}}{\partial z} = \frac{u^*}{k \times z} \tag{7.24}$$

Note here that we assume a positive $\overline{u}$, implying a positive shear, which permits neglecting the absolute value operator. Now we integrate the above equation in height, using the realization that the mean wind $\overline{u} = 0$ at $z = z_0$. The quantity z_0 is termed the roughness length and depends upon the physical characteristics of the surface. For grassy fields, it ranges from 1–4 cm and much smaller for a smooth ocean surface. This procedure yields the classic surface-layer logarithmic wind profile

$$\overline{u} = \left(\frac{u^*}{k}\right) \ln\left(\frac{z}{z_0}\right) \tag{7.25}$$

Although many assumptions are made to obtain the log wind profile, observations over the depth of the surface layer demonstrate that it is a reasonable approximation to the distribution of wind speed over the surface layer.

As an important aside, equation (7.24) pertains only to neutrally stable conditions:

$$\frac{\partial \overline{\theta}}{\partial z} = 0$$

In non-neutral conditions, stability adjustments pertaining to the universal functions derived from MOST must be imposed. See Stull (1988) for details about MOST and on how to represent the surface layer in different stability regimes.

7.6. DISCUSSION AND FUTURE OUTLOOK

The boundary layer is the lowest part of the atmosphere that is affected by the surface and is the part of the atmosphere where humans reside. Hence, the boundary layer has received special attention by atmospheric scientists, with first theoretical attempts to represent the surface layer and boundary-layer turbulence proposed in the mid-twentieth century (Kolmogorov, 1941; Monin & Obukhov, 1954), followed by field experiments focused on validating those theories (e.g. Ball, 1960; Kaimal et al., 1968; Kaimal & Wyngaard, 1990). The understanding from these and other field experiments underpinned the representation of the boundary layer in early large-scale models (e.g. Clarke, 1970; Deardorff, 1972; Mellor & Yamada, 1974). Over the years, the PBL schemes in large-scale models have been subsequently refined with inclusion of higher-order closure schemes. This overlap between the parameterizations of the PBL, subgrid-scale fluxes, and cloud cover and microphysical processes became more evident in the 1980s and 1990s with fundamental development of large-eddy simulation models to understand interaction between boundary-layer turbulence and shallow clouds (e.g. Moeng, 1984). It is no coincidence that first-order and 1.5-order (TKE closure) closures are regularly used to represent not only the turbulent boundary layer but, more broadly, the subgrid-scale fluxes in fluid-flow models. Further studies included the development of schemes that could simulate cloudy and clear boundary layers (e.g., Lock et al., 2000; Park & Bretherton, 2009), and schemes that combined the higher-order closures and mass-flux approaches (e.g. Golaz et al., 2002; Lappen & Randall, 2001).

Due to the nonlinear interactions between the intensity of PBL turbulence and cloud variables, it is desirable to have schemes that simulate both cloudy and noncloudy boundary layers. Although useful, these schemes are inherently difficult to evaluate through observations as there are rare observations of joint PDFs of vertical air motion and thermodynamic variables. A more simplistic approach like application of different turbulence closures based on the mean thermodynamic profile (Lock et al., 2000) is inherently easier to evaluate through observations. However, such an approach will fail to accurately simulate boundary-layer processes in grid boxes containing heterogeneous surfaces (coast or mountains) or significant large-scale advection.

With the improvements in computing infrastructure and techniques used for running large-scale models, it will be feasible in the near future to make large-scale model simulations at spatial and temporal resolutions that resolve boundary-layer eddies (~100 m). Such model simulations will crudely resolve the larger eddies and might resolve most of the cloud elements. At smaller scales, direct numerical simulation (DNS) is an additional tool that can play a role in exploring fundamental processes like entrainment with more fidelity than the LES (Haghshenas & Mellado, 2019; Mellado et al., 2018). The high-resolution simulations of large-scale models will nevertheless require turbulence closure schemes to represent subgrid-scale turbulence. Hence, the schemes representing the turbulence processes in the planetary boundary layer will be an integral part of the large-scale atmospheric models for a foreseeable future.

ACKNOWLEDGMENTS

This research was supported by the National Science Foundation (NSF) Grant AGS-1445831 awarded to the University of Chicago and the U.S. Department of Energy's (DOE) Atmospheric System Research (ASR), an Office of Science, Office of Biological and Environmental Research (BER) program, under Contract DE-AC02-06CH11357 awarded to Argonne National Laboratory. Most of the data used in this study were obtained from the Atmospheric Radiation Measurement (ARM) Program sponsored by the U.S. Department of Energy, Office of Science, Office of Biological and Environmental Research, Climate and Environmental Sciences Division (CESD). David Mechem acknowledges support from the Department of Energy Office of Science grant DE-Sc0716522, DE-Sc0723083, and a subaward from University of Maryland Baltimore County to the University of Kansas from DE-Sc0720057.

REFERENCES

Arakawa, A. (2004). The cumulus parameterization problem: Past, present, and future. *Journal of Climate*, *17*, 2493–2525.

Ball, F. K. (1960). Control of inversion height by surface heating. *Quarterly Journal of the Royal Meteorological Society*, *86*, 483–494.

Caldwell, P., Bretherton, C. S., & Wood, R. (2005). Mixed-layer budget analysis of the diurnal cycle of entrainment in SE Pacific stratocumulus. *Journal of the Atmospheric Sciences*, *62*, 3775–3791.

Clarke, R. H. (1970). Recommended methods for the treatment of the boundary layer in numerical models. Australian Meteorological Magazine, 18(2), 51–71. Commonwealth Bureau of Meteorology, Melbourne, June 1970.

Deardorff, J. W. (1972). Parameterization of the planetary boundary layer for use in general circulation models. *Monthly Weather Review*, *100,* 93–106. 10.1175/1520-0493(1972)100 <0093:POTPBL>2.3.CO;2

Deardorff, J. W. (1980). Stratocumulus-capped mixed layers derived from a three-dimensional model. *Boundary-Layer Meteorology*, *18*, 495–527.

Ekman, V. W. (1905). On the influence of the Earth's rotation on ocean currents. *Archives for Mathematics Astronomical Physics*, *2*, 1–52

Ghate, V. P., Mechem, D. B., Cadeddu, M. P., Eloranta, E. W., Jensen, M. P., Nordeen, M. L., & Smith, W. L., Jr., (2019). Estimates of entrainment in closed cellular marine stratocumulus clouds from the MAGIC field campaign. *Quarterly Journal of the Royal Meteorological Society*, in press.

Golaz, J.-C., Larson, V. E., & Cotton, W. R. (2002). A PDF-based model for boundary layer clouds. Part I: Method and model description. *Journal of the Atmospheric Sciences*, *59,* 3540–3551.

Haghshenas, A., & Mellado, J. (2019). Characterization of wind-shear effects on entrainment in a convective boundary layer. *Journal of Fluid Mechanics*, *858*, 145–183. doi:10.1017/jfm.2018.761

Hersbach, H., Bell, B., Berrisford, P., Hirahara, S., Horányi, A., Muñoz-Sabater, J., Nicolas, C., Radu, R, et al. (2020). The ERA5 global reanalysis. *Quarterly Journal of the Royal Meteorological Society*, *146(730)*, 1999–2049.

Holton, J. R. (2004). Chapter 5—The planetary boundary layer. *Dynamic Meteorology. International Geophysics Series. 88* (4th edition). Burlington, MA: Elsevier Academic Press.

Kaimal, J. C., Wyngaard, J. C., & Haugen, D. A. (1968). Deriving power spectra from a three component sonic anemometer. *Journal of Applied Meteorology*, *7*, 827–837.

Kaimal, J. C., & Wyngaard, J. C. (1990). The Kansas and Minnesota experiments. *Boundary-Layer Meteorology*, *50*, 31–47.

Kolmogorov, A. N. (1941). Energy dissipation and locally isotropic turbulence. *Doklady Akademii nauk SSSR*, *32*(1), 19–21.

Lappen, C., & Randall, D. A. (2001). Toward a unified parameterization of the boundary layer and moist convection. Part I: A new type of mass-flux model. *Journal of the Atmospheric Sciences*, *58*, 2021–2036. 10.1175/1520-0469(2001)058<2021:TAUPOT>2.0.CO;2

Larson, V. E., & Golaz, J.-C. (2005). Using probability density functions to derive consistent closure relationships among higher-order moments. *Monthly Weather Review*, *133*, 1023–1042.

Larson, V. E., Schanen, D. P., Wang, M., Ovchinnikov, M., & Ghan, S. (2012). PDF parameterization of boundary layer clouds in models with horizontal grid spacings from 2 to 16 km. *Monthly Weather Review*, *140*, 285–306.

Lilly, D. K. (1968). Models of cloud-topped mixed layers under a strong inversion. *Quarterly Journal of the Royal Meteorological Society*, *94*, 292–309. doi:10.1002/qj.49709440106

Lilly, D. K. (2002a). Entrainment into mixed layers. Part I: Sharp-edged and smooth tops. *Journal of the Atmospheric Sciences*, *59*, 3340–3352.

Lilly, D. K. (2002b). Entrainment into mixed layers. Part II: A new closure. *Journal of the Atmospheric Sciences*, *59*, 3353–3361.

Lock, A. P., Brown, A. R., Bush, M. R., Martin, G. M., & Smith, R. N. (2000). A New Boundary Layer Mixing Scheme. Part I: Scheme description and single-column model tests. *Monthly Weather Review*, *128*, 3187–3199.

McMichael, L. A., Mechem, D. B., Wang, S., Wang, Q., Kogan, L., & Teixeira, J. (2019). Assessing the mechanisms governing the daytime evolution of marine stratocumulus using large-eddy simulation. *Quarterly Journal of the Royal Meteorological Society*, *145*, 845–866.

Mellado, J. P., Bretherton, C. S., Stevens, B., & Wyant, M. C. (2018). DNS and LES for simulating stratocumulus: Better together. *Journal of Advances in Modeling Earth Systems*, *10*, 1421–1438. 10.1029/2018MS001312

Mellor, G. L., & Yamada, T. (1974). A hierarchy of turbulence closure models for planetary boundary layers. *Journal of the Atmospheric Sciences*, *13*, 1791–1806.

Mellor, G. L., & Yamada, T. (1982). Development of a turbulence closure model for geophysical fluid problems. *Reviews of Geophysics and Space Physics*, *20*, 851–875.

Moeng, C. (1984). A Large-Eddy-Simulation model for the study of planetary boundary-layer turbulence. *Journal of the Atmospheric Sciences*, *41*, 2052–2062.

Monin, A. S., & Obukhov, A. M. (1954). Basic laws of turbulent mixing in the atmosphere near the ground. *Transactions Geophysics Institute Akademii Nauk USSR, No. 24*(151), 1963–1987.

Park, S., & Bretherton, C. S. (2009). The University of Washington shallow convection and moist turbulence schemes and their impact on climate simulations with the Community Atmosphere Model. *Journal of Climate*, *22*, 3449–3469.

Prandtl, L. (1925). Bericht über Untersuchungen zur ausgebildeten Turbulenz. *Zeitschrift für Angewandte Mathematik und Mechanik*, *5*, 136–139.

Stensrud, D. (2007). *Parameterization schemes: Keys to understanding numerical Weather Prediction Models*. Cambridge: Cambridge University Press. doi:10.1017/CBO97805118 12590

Stevens, B. (2002). Entrainment in stratocumulus-topped mixed layers. *Quarterly Journal of the Royal Meteorological Society*, *128*, 2663–2690.

Stull, R. B. (1988). *An introduction to boundary layer meteorology (671 pp)*. Dordrecht: Kluwer Academic Publishers.

Van der Dussen, J. J., de Roode, S. R., & Siebesma, A. P. (2014). Factors controlling rapid stratocumulus cloud thinning. *Journal of the Atmospheric Sciences*, *71*, 655–664.

Wood, R. (2007). Cancellation of aerosol indirect effects in marine stratocumulus through cloud thinning. *Journal of the Atmospheric Sciences*, *64*, 2657–2669. 10.1175/JAS3942.1

Zhang, N., Gao, Z., Liu, Y., & Li, D. (2015). Sensitivity of a global climate model to the critical Richardson number in the boundary layer parameterization. *Journal of Geophysical Research: Atmospheres*, *120*, 3310–3328. doi: 10.1002/2014JD022015.

8

Human Impacts on Land Surface-Atmosphere Interactions

Michael Barlage and Fei Chen

ABSTRACT

The land surface plays a vital role in the Earth system, with exchanges of energy and water between the land and atmosphere significantly influencing local and regional weather and climate. This chapter presents a review of recent advancements in comprehending and modeling the effects of human activities, particularly urbanization and agriculture management, on land-atmosphere interactions. It summarizes how these processes modify land surface physical properties, impacting the exchange of water, heat, and momentum between the land and atmosphere. Additionally, the chapter explores the influence of urbanization and agriculture on atmosphere boundary layer structures and convection activities. These processes collectively modify cross-scale precipitation patterns, weather phenomena, and climate dynamics. Understanding the complexities of these interactions is crucial for effective land and water resource management, urban planning, and strategies for mitigating climate change impacts.

8.1. INTRODUCTION

The Earth system is comprised of several subsystems, including the atmosphere, biosphere, hydrosphere, cryosphere, and lithosphere. These are strongly coupled subsystems that interact on a broad spectrum of spatial and temporal scales and can be greatly affected by exchange processes between components, especially in regions near the subsystem interfaces. In the full Earth system, exchanges of energy and water from the surface to the atmosphere are of primary importance. Land subsystems are much more variable, on both temporal and spatial scales, than the oceans. Due to its relatively low heat capacity, the land surface rapidly responds to changes in local thermal forcing. These changes result from the diurnal cycle of solar radiation as well as local forcing due to clouds. When land is wet, evapotranspiration can exceed evaporative flux from the oceans. This is possible because larger surface roughness over land results in more turbulence in the boundary layer and vegetated regions have large surface area for evaporation and transpiration from leaves. When land is dry, this moisture flux can be effectively shut off. The heterogeneous nature of vegetation and soil results in large variability in surface-atmosphere exchange of energy and water vapor over land, which drives the diurnal variations of the planetary boundary layer (PBL), creates mesoscale circulations, and affects the formation and evolution of summer convective precipitation.

Therefore, modeling land-surface processes has been an integral component in the weather and climate models. In today's integrated earth system models, land-surface models (LSMs) include sophisticated representation of global carbon, nitrogen, and ecosystem processes (Lawrence et al., 2012); hydrological cycle (Gochis et al., 2013; Voisin et al., 2013); and cryosphere. There exists rich literature regarding the general structures and physics parametrization of LSMs that readers can reference (e.g., Bonan, 1998; Chen & Dudhia, 2001; Dai et al.,

National Center for Atmospheric Research, Boulder, CO, USA

Fast Processes in Large-Scale Atmospheric Models: Progress, Challenges, and Opportunities, Geophysical Monograph 282, First Edition.
Edited by Yangang Liu and Pavlos Kollias.

DOI: 10.1002/9781119529019.ch08

2003; Dickinson et al., 1993; Ek et al., 2003; Lawrence et al., 2012; Niu et al., 2011; Sellers et al., 1986; etc.). Accompanying these new exciting advancements in LSM development, uncertainties in those models emerged as equally challenging due to incomplete knowledge of the physical processes, the inherent assumptions in model structures, parameterization schemes, and model parameters. The land modeling community has been developing methods of assessing and mitigating these uncertainties (e.g., Best et al., 2015; Clark et al., 2015; Newman et al., 2017; Niu et al., 2011). An in-depth review of broader impacts of land-surface biogeochemistry and ecology is beyond the scope of this chapter. Instead, we focus on recent progress in understanding and modeling the biophysical aspects of the human dimension, especially urbanization and agriculture, and its interactions with the atmosphere in this chapter.

Let us start with the surface-energy budgets for a natural surface (e.g., see Oke, 1987):

$$\mathrm{SW}\downarrow(1-\alpha)+\mathrm{LW}\downarrow(1-\varepsilon)+\varepsilon\sigma T_s^4 = H+\mathrm{LE}+G \tag{8.1}$$

where SW↓ is the downward shortwave radiation (W m^{-2}) at the surface depending on the solar constant, seasonal and diurnal geometry, atmospheric attenuation, clouds, and aerosols; α is the surface albedo; LW↓ is the downward longwave radiation (W m^{-2}) depending on atmospheric temperature profile, water vapor, and clouds; ε is the surface emissivity; σ is the Stefan-Boltzmann constant (5.67 × 10^{-8} W m^{-2} K^{-4}); H is the turbulent flux of sensible heat; LE the turbulent flux of latent heat; and G is the ground heat fluxes. The physical properties of land surfaces determine the values of α and ε and hence the surface radiation budgets, and, along with the soil water and ice content, play central role in partitioning the net radiation (Rn; left terms in equation (8.1)) into H, LE, and G.

Similarly, the water budget for a natural column or a drainage basin can be written as

$$P-E-R=\Delta S \tag{8.2}$$

where P is the precipitation (including snow) amount at the surface; E is the sum of plant evapotranspiration, evaporation from bare ground, and sublimation from snowpack; R is the runoff (surface and baseflow) or streamflow; and ΔS is the change in soil moisture (in liquid and solid phases) storage. Land vegetation and forest structures intercept precipitation and modify the amount of precipitation reaching the surface, and together with soil properties play central role in partitioning the net radiation (left terms in equation (8.1)) into H, LE, and G.

With human intervention, the surface-energy budget equation, for instance for a volume of air, buildings, and the interfacial surface layer with the ground, equation (8.1), can be rewritten as (Oke et al., 2017)

$$\mathrm{SW}\downarrow(1-\alpha)+\mathrm{LW}\downarrow(1-\varepsilon)+\varepsilon\sigma T_s^4+\mathrm{Ah} = H+\mathrm{LE}+G+\Delta Q_s \tag{8.3}$$

Compared to equation (8.1), the new terms are Ah representing anthropogenic sensible and latent heat release, and ΔQ_s representing the net change in the heat storage of building materials and enclosed air volume. Moreover, the man-made surface substantially modifies α and ε.

Considering human impact, equation (8.2) can be rewritten as

$$P+\Delta P-E-R=\Delta S \tag{8.4}$$

where the new term ΔP represents the irrigated water from agriculture and city green space added to the water system, as an example. Water released to the atmosphere by combustion, and built surfaces, agriculture (e.g., tile drainage), and water management modify the amount of E, R, and ΔS.

The following sections further discuss human impacts on surface energy and water budgets and their cascading effects on weather and climate.

8.2. CROSS-SCALE INTERACTIONS OF CITIES WITH THE ATMOSPHERE

8.2.1. Urban Impact on Surface Energy and Water Budgets

In 2007, the world reached a historical milestone of half of its people living in cities, and that trend is projected to accelerate eventually climbing to 60% in 2030 (Chen et al., 2012; United Nations, 2007). Such rapid urbanization in the past century inevitably alters regional weather, climate, and hydrology. When replacing the natural surfaces with built-up environments, various components are significantly modified in equation (8.3). Using remotely sensed data, Brest (1987) showed lower values of α in downtown areas than those in suburban and natural areas, as a combination of lower albedo of building materials and enhanced solar radiation absorption due to multiple reflections between urban structures (Oke, 1982). Similarly, emissivity of building roofs, or for urbanized surfaces, is generally lower than that of natural surfaces (Artis & Carnahan, 1982; Stull, 2012).

The amount of anthropogenic sensible heat (part of the *Ah* term in equation (8.3), released by heating, air conditioning (AC), and other electricity consumption, cooking, and vehicle combustion, etc.) depends on population density, energy consumption, and numbers and types of vehicles, ranging from a few W m^{-2} to hundreds W m^{-2} (Oke, 1987; Sailor & Lu, 2004). Yang and

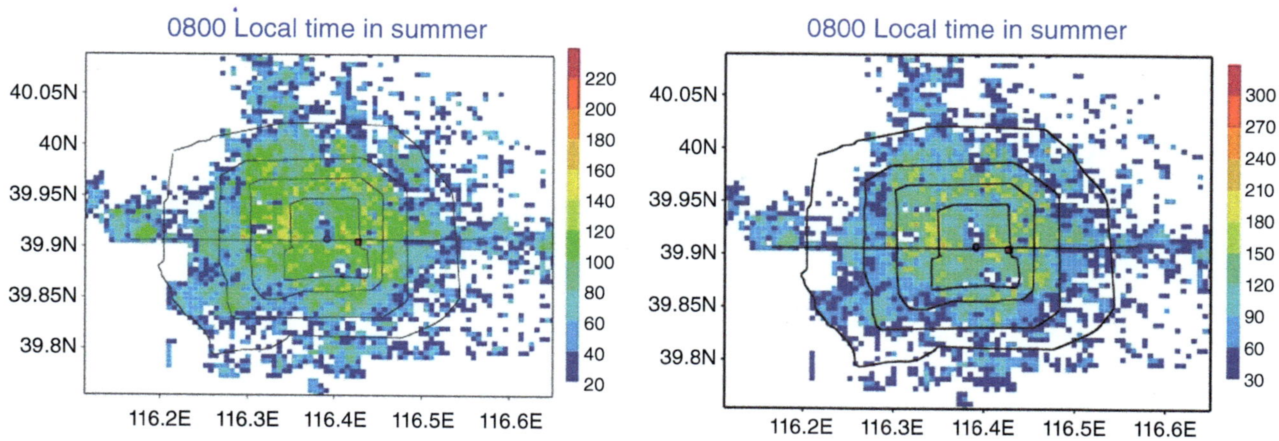

Figure 8.1 Anthropogenic sensible heat fluxes (W m^{-2}) in the Beijing Metro for summer and winter seasons. Black lines delineate, from inner to outer, the second, fourth, and fifth beltways in Beijing. Source: Adapted from Miao et al. (2009).

Bou-Zeid (2018) showed that *Ah* released from buildings alone contributed to 30% of the urban heat island (UHI) during cold periods in the United States. As an example, Figure 8.1 shows the heterogeneous distributions of Ah in the Beijing Metro area. Note that the peak noontime winter $SW\downarrow$ in Beijing is about 500 W m^{-2}, so the maximum *Ah* value (~250 W m^{-2}) is nearly half of incoming solar radiation, representing an important source of heating to the overlying atmosphere in winter and must not be excluded in weather forecast models.

While the urban LE is usually lower than over rural areas from the lack of vegetated surfaces, it is somewhat compensated by the anthropogenic latent heat release, which is less studied than waste heat release. Fuel combustion and the use of air conditioning in cities produce water vapor (Pigeon et al., 2007). Narumi et al. (2009) estimated that the amount of summer anthropogenic latent heat fluxes (maximum value ~50 W m^{-2}) is usually half of the anthropogenic sensible heat fluxes for a Japanese megacity (Keihanshin district).

Similar to Ah, ΔQ_s in equation (8.3) is an important term of energy source and sink for urban environments, and quantifying ΔQ_s is a challenging and an actively pursued research area in urban climate studies. Due to lack of direct observations, ΔQ_s is usually determined as the energy balance residual from direct observations of other terms in equation (8.3). Grimmond and Oke (1999) showed that for seven cities in North America, the values of summer ΔQ_s are generally lower than H but close to LE. Other studies (e.g., Kawai & Kanda, 2010; Miao et al., 2012; Moriwaki & Kanda, 2004) revealed that the winter daytime ΔQ_s is higher than H for downtown and suburban areas in Beijing and Tokyo. Using a similar approach and combining Terra ASTER remote-sensing data, Kato and Yamaguchi (2007) estimated ΔQ_s being ~250 W m^{-2}in the central urban area, which was much higher than ~200 W m^{-2}in residential areas.

Regarding the water balance equation (8.4), ΔP (water supply piped in from rivers and reservoirs, plus city irrigation) significantly increased water input to the urban hydrologic system. Urban evaporation (E) is expected to decrease when compared to rural areas due to reduced vegetated surfaces, but anthropogenic water-vapor release from air conditioning systems and vehicle combustion would partly compensate such loss in evaporation. ΔS is also expected to decrease because of dominant impermeable surfaces in cities. More water input, but less capacity to store water, results in more urban runoff and discharge (e.g., Cuo et al., 2008; Hollis, 1975), being one leading cause for severe urban flooding. Taylor (1977) showed that the most (least) significant increase in urban runoff occurs during spring snowmelt and rain-on-snow events (summer rainstorms).

8.2.2. Urban Impacts on the Atmosphere

With lower albedo (i.e., more absorbed solar radiation), greater capacity to store and release heat from buildings, and the addition of anthropogenic sensible heat, the air in cities generally trends warmer than the air in rural areas. This phenomenon is called urban heat island. UHI is perhaps the most known example of how human activity inadvertently modifies weather and climate. The existence of UHI was first discovered probably by Luke Howard (1833) pointing out that the temperature recorded by one weather station within London is higher than those recorded by three stations outside London. It was later confirmed by numerous observations taken in different cities in the world (e.g., Arnfield 2003; Chandler, 1970; Landsberg, 1981; Oke, 1982). Oke (1995) reviewed observational studies on the formation and effects of UHI.

Urban impacts are felt by the overlying atmosphere beyond the UHI in urban canopy layers. Buildings can effectively influence airflow in cities, PBL thermodynamic structures, and transport and dispersion of pollutants. Bornstein (1968) observed these effects, in New York City, occurring mostly under nighttime clear skies with weak wind conditions at night. Miao et al. (2009) noted an increase of air temperature over Beijing Metro up to 1.8 km above ground during the day and to 0.2 km at night, and the airflow over Beijing is dominated by mountain-valley flows that are modified by urban-rural circulations.

As summarized in Grimmond (2006), various field campaigns were conducted in different cities to understand the urban impacts on UHI and PBL evolution. A recent SURF (Study of Urban-Impacts on Rainfall and Fog/Haze; Liang et al., 2018) field campaign was designed to gather multiscale observations to understand and better predict urban effects on heavy convective precipitation and air-pollutant episodes in the Greater Beijing Metro Area. Those field campaigns also embraced the use of new observation technologies. For instance, observations from Doppler LiDAR were used to derive the PBL depth (e.g., Barlow et al., 2011), a key variable used in weather and air-quality forecasts. Data from the Doppler LiDAR were used by Huang et al. (2017) to study PBL evolution for 6 days during the 2015 summer SURF campaign. Convective boundary layer (CBL) depth was estimated by a vertical velocity variance threshold, which is more accurate than using the maximum wind-shear method. This method, however, had problems detecting nocturnal boundary layers (NBLs). Huang et al. (2017) combined it with a background variance (LeMone et al., 2014) to determine the PBL depth. Their results showed that average PBL depths in downtown Beijing varied from 270 to 1500 m over the 6-day observation period (Figure 8.2).

Modified air flow and PBL structures of urban canopies inevitably affect air quality in cities (e.g., Banta et al., 2005; Taha et al., 1998). As illustrated in Figure 8.3, the very existence of Houston's paved surfaces produces prominent nighttime UHI, and therefore reduces the nighttime thermal contrast between water in the Galveston Bay and the Houston Metro areas (Chen et al., 2011a). This reduces the strength of regional nighttime land breezes and causes wind stagnation leading to the accumulation of air pollutants. Wang et al. (2009) explored (using preurbanization and current urban land use) the influence of rapid urban growth in southern China, and showed that urbanization increases daily mean 2-m air temperature and increases daily PBL depth. Changes in those meteorological conditions resulted in an increase of surface ozone concentrations by about 4.7%–8.5% for nighttime and about 2.9%–4.2% for daytime. Using the WRF-Chem model coupled to UCM, Jiang et al. (2008) explored impacts of future urbanization on surface ozone in Houston under future A1B scenario for 2051–2053. Their study revealed a 2 °C increase in surface air temperature, due to the combined effects of climate warming and future urbanization, which led to an increase of surface ozone by 6.2 ppbv.

Anthropogenic aerosol emissions from cities not only affect air quality but also the radiation budgets and clouds

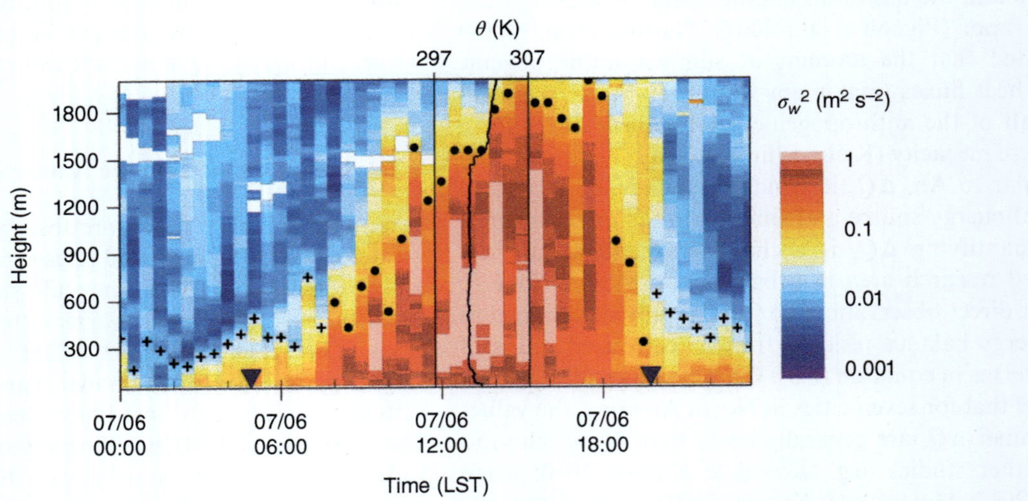

Figure 8.2 Doppler LiDAR Beam Swinging (DBS) scan 30-min mean vertical-velocity variances σ_w^2 (colors, $m^2\ s^{-2}$) on 6 July 2015, with CBL depths from the threshold (•) and NBL depths from the fractional (+) methods; potential temperature θ (K) profile (solid line) from the 1315 local time radiosonde; triangles indicate sunrise and sunset. Source: Adapted from Huang et al. (2017).

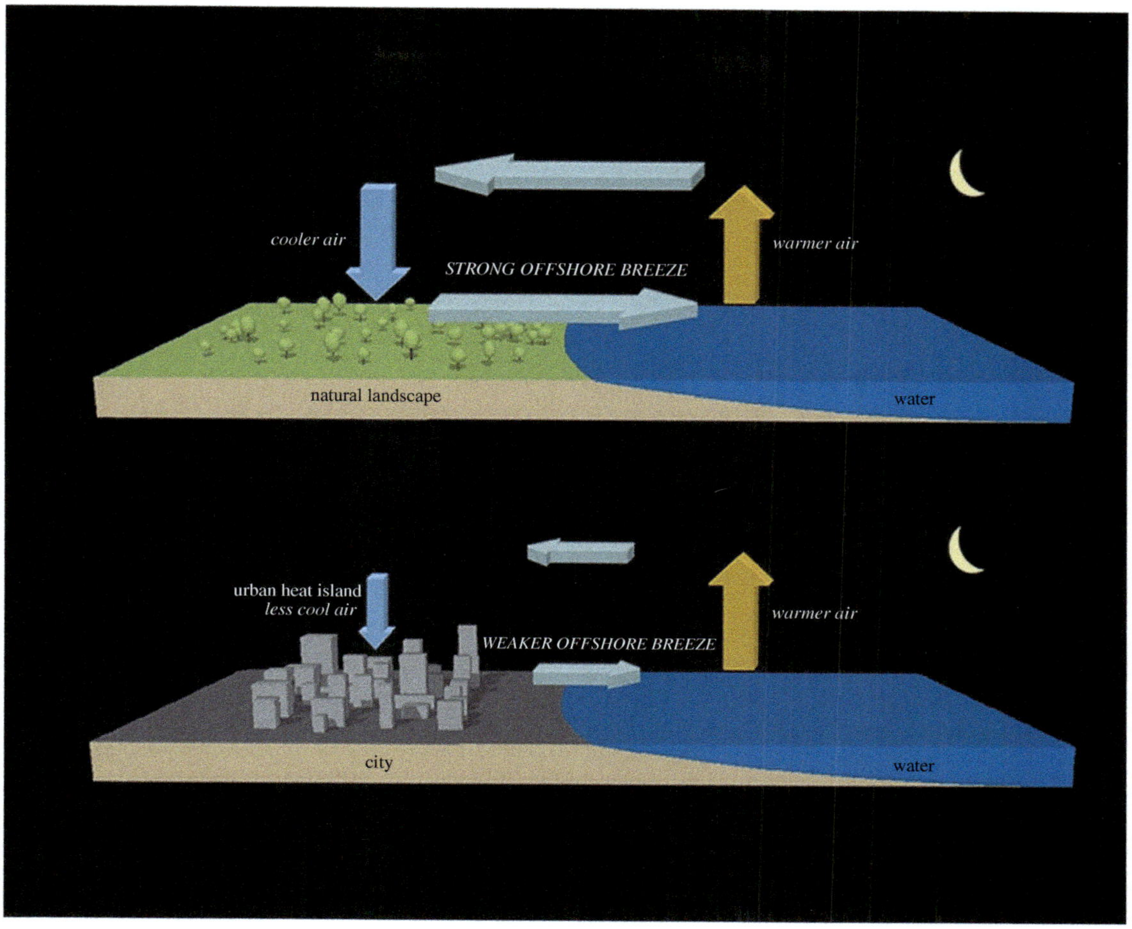

Figure 8.3 Paved surfaces in the Greater Houston Metro area keep the nighttime city warmer than surrounding natural surfaces and weakens the offshore land breeze. Source: Courtesy of University Corporation for Atmospheric Research.

in the PBL (Andreae et al., 2005), PBL thermodynamics (Zhao et al., 2006), and summer precipitation (Rosenfeld, 2000). Shepherd (2005) provided an excellent review of studies on the urban precipitation. Urbanization affects summer precipitation through thermodynamic effects, dynamic effects of precipitation bifurcation (Bornstein & Lin, 2000; Dou et al., 2015), and aerosol-induced microphysics (Rozenfeld et al., 2008). Earlier field experiments such as METROMEX (Huff & Changnon, 1973) showed downwind urban enhancement, and Niyogi et al.'s (2011) results confirmed the downwind summer storm enhancement. On the other hand, long-term (1981–2005) observation showed the rapid urban expansion in Beijing since 1981 is statistically correlated to summer rainfall reduction in the northeast areas of Beijing (Zhang et al., 2009) due to reduction of regional evaporation leading to less convective available potential energy (CAPE). The same study also argued that planting more trees in Beijing would enhance summer precipitation. A modeling study by Miao et al. (2011) showed that urban effects on summer precipitation in Beijing depend on the degrees of urbanization.

8.2.3. Urbanizing Large-Scale Models

Given these significant influences of urbanization on weather and climate, it is imperative to represent urban surface-atmosphere interactions in atmospheric models. The last two decades have witnessed remarkable progress in that direction. Urbanization schemes were implemented in the Fifth-generation Pennsylvania State University (PSU)-NCAR Mesoscale Model (MM5) (Dupont et al., 2004; Liu et al., 2006; Taha, 1999), the Weather Research and Forecast (WRF) model (Chen et al., 2011b), the UK Met Office operational mesoscale model (Best, 2005), the French Meso-NH model (Lemonsu and Masson 2002), the NCAR global climate model (Oleson et al., 2008), and the GFDL's

ESM (D. Li et al., 2016). The urban parameterization schemes incorporated in these weather and climate models were evaluated against site observations, notably through the International Urban Energy Balance Models Comparison Project (Grimmond et al., 2009).

Among those urban-modeling efforts, the WRF-Urban model perhaps incorporated the most comprehensive suite of urban parameterization schemes with varying degrees of complexities, satisfying the need for short-term weather forecast and long-term regional climate simulations (Chen et al., 2011b). The simplest scheme is bulk urban parameterization to modify the following parameters in the Noah land model (Liu et al., 2006): higher surface roughness length, lower surface albedo, higher heat capacity of soils, higher soil thermal conductivity, and a lower green-vegetation fraction over urban areas. The intermediate scheme is the single-layer urban canopy model (SLUCM) based on Kusaka et al. (2001), which assumes infinitely long street canyons to represent the urban geometry but recognizes the three-dimensional nature of the urban surfaces. This SLUCM accounts for the shadowing, reflections, and trapping of radiation, and temperature variations within roofs, walls, and roads. As shown in Figure 8.4, surface-sensible heat fluxes from each facet are calculated using the Monin-Obukhov similarity theory and the Jurges formula, and the anthropogenic heating and its diurnal variation are considered by adding them to the sensible heat flux from the urban canopy layer.

The most sophisticated urban scheme in WRF is the multilayer UCM named building effect parameterization (BEP) developed by Martilli et al. (2002), representing the three-dimensional nature of the urban canopies so the model vertically distributes sources and sinks of heat, moisture, and momentum through the whole urban canopy layer (Figure 8.4). It also considers the effect of the vertical walls, horizontal streets, and roofs on momentum transfer and modifies the turbulent kinetic energy (TKE) terms in the WRF TKE-based PBL schemes. As illustrated in Figure 8.4, unlike the SLUCM, the BEP scheme intersects with several layers above the surface and therefore allows a direct interaction with the PBL. This also adds the capability to have higher vertical resolution close to the ground surface. The BEP model includes a simple building energy model (BEM; Salamanca & Martilli, 2010) accounting for indoor-outdoor radiation and heat transfer, the waste heat generated by building occupants and equipment, and from air conditioning.

One challenge when applying coupled urban and atmospheric models is the inability to specify a vast number of parameters used in urban canopy models. One approach is to use high-resolution remotely sensed (e.g., 30-m Landsat) data to determine the spatial distribution of urban land use type and relate that to urban morphological and thermal properties that can be prescribed in a parameter table. Another approach is to directly incorporate gridded urban canopy parameters. One example of such data is the National Urban Database and Access Portal Tool (NUDAPT) (Ching et al., 2009), including the aggregated urban parameters derived from individual buildings, clusters of buildings, and vegetation in urban areas that can be obtained from airborne lidar systems or photogrammetric techniques. The NUDPAT is currently expanded to World Urban Database and Access Portal Tools (WUDAPT; Ching et al., 2018). WUDAPT contains a basic level of information based on local climate zone (LCZ; Stewart & Oke, 2012) maps of cities, related to surface parameters (roughness, impermeable surface cover, roof area, building height, etc.). It also contains next levels of data that add more information about building morphological forms, material composition, and energy usage. As a new attempt to improve WRF-Urban urban-model-parameter specification, Y. Li et al. (2016) developed a method of using Landsat images to calculate a fractal dimension (FD), which was then used to define new categories of building heights in Beijing.

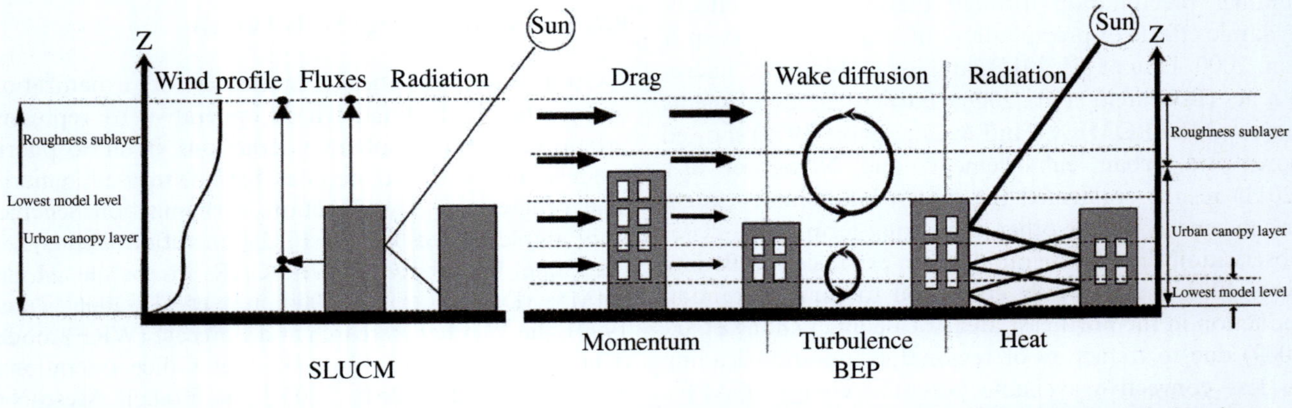

Figure 8.4 A schematic of the SLUCM (on the left-hand side) and the multilayer BEP models (on the right-hand side) in WRF. Source: Adapted from Chen et al. (2011b).

To improve the hydrologic aspects of urban modeling, Miao and Chen (2014) implemented urban irrigation, oasis effect, and anthropogenic latent heat in the WRF-Urban model, which was later expanded by including green-roof parameterization by Yang et al. (2015). This enhanced urban-modeling capability was evaluated against data collected in Beijing, Phoenix, Vancouver, and Montréal. A more recent effort by Salamanca et al. (2018) coupled a more advanced Noah multiple physics (Noah-MP) land model with WRF-Urban. Evaluated against data obtained in Phoenix and Tucson metropolitan areas, this Noah-MP-based urban model appears to perform slightly better than the Noah-based with respect to the daily evolution of surface skin temperature and near-surface air temperature (especially nighttime temperature) and wind speed.

Given the significant impacts of urban flooding and air pollution, numerical weather prediction models continue to improve their forecast skills for urban regions by increasing their spatial resolutions, assimilating emerging urban data, and enhancing urban models and parameter specification. For example, Barlage et al. (2016), through a comprehensive evaluation of various parameterization schemes of PBL, radiation, and UCMs in the WRF-based operational model used at the Institute of Urban Meteorology (IUM, Beijing), highlighted the equal or more important contributions of urban process treatment to forecasted biases in near-surface meteorological variables when compared to PBL and radiation schemes. They found that refining parameters in BEP/BEM schemes to include local urban morphology and building materials in Beijing was able to effectively reduce biases in forecasted temperature and wind speed.

8.3. CROSS-SCALE INTERACTIONS OF AGRICULTURE WITH THE ATMOSPHERE

The representation of agriculture, including cropland and rangeland with and without irrigation, in weather and climate models is becoming increasingly important both due to the large footprint that agriculture has made on the natural landscape and the stark contrast of biophysical properties agriculture introduces. According to the Food and Agriculture Organization of the United Nations (FAO, 2018), 37% of global land area is dedicated to some form of agriculture. This percentage has remained relatively constant over the past few decades. However, how the land is used and whether it is irrigated has changed substantially in several regions. Worldwide the amount of harvested hectares has increased by 18% from 1995 to 2016 even though agricultural land has remained constant. This is due to increased cropping intensity (multiple crops planted per year) and conversion of rangeland to cropland. Spatially, harvested area trends over the past two decades are highly variable with Europe showing almost zero change and Africa showing a nearly 50% increase.

The use of water for irrigation is also becoming increasingly important since most of the optimal rain-fed arable land is already in production. Worldwide there has been a 23% increase in area equipped for irrigation in the past two decades. Irrigated agriculture accounts for about 20% of global croplands (Siebert & Döll, 2010) and consumes about 70% of global total freshwater withdrawals (FAO (Food and Agriculture Organization of the United Nations), 2010). Over this time period, Africa and Asia have increased irrigation-equipped area by nearly 30% each. At the country level, even though China and India are equipped with 2–3 times more acres for irrigation than the next highest country (United States), both countries have expanded their irrigation-equipped hectares by 30%–40% in the past two decades. Given this clear ongoing expansion in both agriculture and irrigation, accurate representation of the biophysical processes and physical landscape changes are vital in land-atmosphere interactions.

8.3.1. Agriculture Impacts on Surface Water and Energy Budgets

The Earth's surface has been and is continuing to be greatly modified by natural and human activities. Land use practices, such as agriculture and intensive grazing, have had considerable influence on climate. These land use modifications alter the albedo, roughness, and evaporative potential of the surface. Land conversion from grassland/forest to agriculture affects the regional climate through magnitude modulation or distribution modification of surface fluxes. The extent of these changes depends on the location and severity of the land use alteration.

Barlage (2001) showed that replacing native grassland and forest with agriculture in the northern United States decreased monthly-averaged summer sensible heat flux by about 20 W m^{-2} with a similar magnitude increase in latent heat flux primarily due to the decrease of stomatal resistance in agricultural vegetation types. Adegoke et al. (2007) also showed that replacing potential natural vegetation with corn in the U.S. state of Nebraska increased latent heat flux by 42%. Due to increased agricultural intensity in India over the past several decades (the so-called Green Revolution), many studies have looked at the effect agriculture has had on local and regional surface energy and water budgets. Mean annual vapor fluxes from the surface due to agricultural land cover conversion in India were simulated to increase by 17% with dry season increases of 55%, though two-thirds of the increase was attributed to irrigation (Douglas et al., 2006).

Southwestern Australia has also been classified as a natural laboratory for land cover change analysis due to a stark contrast between regions of native vegetation and agriculture. Convective heat flux over the native vegetation can be 2–3 times higher compared to agriculture primarily due to lower albedo (Lyons, 2002). In this region, latent heat fluxes over the native vegetation do not vary substantially between the winter and summer seasons, indicating that the transpiration from the native vegetation is less sensitive to soil moisture availability (Nair et al., 2011). The deeper roots of native vegetation have access to the underground aquifer whereas the shallow rooted agricultural crops rely on surface soil moisture. Therefore, replacing native vegetation with agricultural crops can break the link between the underground aquifer and the atmosphere.

In an observational study, Hatfield et al. (2007) showed that, in the U.S. state of Iowa, as corn and soybeans crops grow through the season the heat flux into the soil decreases from 100 to 70 W m^{-2} for corn and from 150 to 100 W m^{-2} for soybeans due to the relative canopy cover between crops and from the beginning to end of the phenology cycle. For sensible heat flux, daytime maximum values reached 50 W m^{-2} for corn while for soybeans it was 200 W m^{-2}. During the peak growing season, total accumulated corn evapotranspiration is 10%– 20% higher than the total accumulated for soybeans.

Irrigated landscapes can enhance regional surface energy modifications in addition to those induced by agriculture. The surface energy changes due to irrigation have been shown to be more than those due to land cover conversion from native to agriculture in simulations over India (Douglas et al., 2009). In sensitivity studies, the effect of irrigation was shown to increase latent heat flux by 36% and decrease sensible heat flux by 15% in the U.S. state of Nebraska (Adegoke et al., 2003). Area-averaged latent heat flux in India due to irrigation has been simulated to increase by 9 W m^{-2} though this amount is an average of locations where agriculture increased by up to 40 W m^{-2} latent heat flux (grassland to cropland) and decreased by 20 W m^{-2} latent heat flux (tropical forest to cropland). Indian subcontinent latent heat increased by 10–20 W m^{-2} per decade during intensive irrigation growth in the 1980s and 1990s, but shows more spatial variability for later periods (late 1990s to early 2000s) when irrigation growth slowed (Lee et al., 2009). Over northern India, the daytime moisture flux is about 100% higher when the surface cropland is irrigated (Roy & Avissar, 2002).

Observational evidence also exists for irrigation to play a role in influencing surface fluxes. Using long-term data collected from two contrasting (irrigated and rainfed) corn-soybean rotation fields in Nebraska, Chen et al. (2018) looked at the effects of irrigation memory on local hydroclimate and noticed a reduction in the transport of heat (i.e., lower sensible heat flux by ~25 W m^{-2}) for irrigated corn but for irrigated soybean there was a negligible change in sensible heat fluxes. Irrigated corn and soybeans were associated with the similar magnitude in increased latent heat fluxes (~20 W m^{-2}). The measured increase in cumulative evaporation (usually less than 50 mm during the growing season) was only 15%–30% of the applied irrigation amount signifying that a significant fraction of applied irrigation was not transported to the atmosphere.

8.3.2. Agriculture Impacts on the Atmosphere

Several modeling and observational studies have quantified the impact of agriculture on the atmosphere through the communication of surface energy alterations to PBL temperature and water-vapor content, cloud cover, and precipitation. However, studies focusing on agriculture effects on the near-surface atmosphere show widely varying results. Barlage (2001) showed that replacing native grassland and forest with agriculture in the northern United States decreased near-surface temperatures about 0.5–1.0 °C in early summer. In July and August, the surface temperature decrease was about 1.5 °C with decreases of 0.5 °C in the lower atmosphere (850 mb). Mixing ratio increased by 0.5–1.0 g kg^{-1} near the surface but decreased by 0.1 to 0.2 g kg^{-1} at 850 mb, indicating that reduced mixing over the agriculture land did not transport the increased surface moisture upward. Bonan (2001) also showed in climate model simulations that conversion of natural forest to cropland lowered maximum temperature by 1.5 °C over the central United States and decreased the diurnal temperature range. In temperate climates, conversion of grassland and forest into cropland cooled surface air by 0.7 °C in summer due to increases in latent heat flux and by 1.1 °C in winter due to increased albedo. In the tropics and subtropics, the transformation of vegetation by agriculture increased temperatures by about 0.8 °C throughout the year by reducing surface roughness and thereby reducing surface-atmosphere exchange, and replacing native vegetation with crops, which both reduce the latent heat flux (Bounoua et al., 2002).

Bentley and Stallins (2008) showed that evapotranspiration from the large agricultural region in the central United States significantly contributed to the PBL moisture and to extreme dew-point temperature events. Restricted low-level mixing, likely due to the presence of crops, was also instrumental to allow moisture trapping (Bentley & Stallins, 2008), similar to the work by Barlage (2001). A potential reason for the greater increases in dew-point temperature over the central United States, a primary corn and soybean region, may be enhanced evapotranspiration from crops. Given that the maximum daily evapotranspiration rates for corn and soybeans occur in July and August, this suggests that even minor regional land use changes, such as changing farming methods and intensity, can have implications on the regional climate. (Sandstrom et al., 2004). Observational studies have also highlighted the contributions of agriculture to enhancing heat stress during heat waves. Changnon et al. (2003) analyzed the temporal distribution of dew-point temperature during extreme heat events at Chicago using a 75-year record and showed an increase of about 3 °C in dew-point temperature in mid-1990s compared to that before mid-1980s. One primary reason is the intensification of agriculture in this region leading to enhanced evapotranspiration in the Midwest.

Irrigated agriculture landscapes can further alter the regional surface energy balance and its associated temperature, humidity, and climate features. An analysis of the monthly climatological surface data sets at the regional level over India showed that agriculture and irrigation substantially reduced the air temperature over different regions during the growing season. Typically, the irrigated crop landscape was about 2–3 °C cooler during the day and about 5 °C cooler at night. Irrigated crop led to about 3 g kg^{-1} or a 20% increase in the boundary-layer-specific humidity. For surface radiative temperature, the difference between the irrigated crops versus nonirrigated was about 5 °C during day and about 2 °C during night. (Roy & Avissar, 2002).

Irrigation can amplify the effect of agriculture on the atmosphere. Adegoke et al. (2003, 2007) found that in sensitivity simulations the near-surface temperature decreased by 1.2 °C and dew-point temperature increased by 2.6 °C with a 15% increase in water-vapor flux 500 m above the surface with the application of irrigated water in Nebraska. But depending on the intensity of irrigation, the temperature decrease can be up to 3.5 °C and water-vapor flux at 500 m could increase by 38%. In the long-term temperature trend, irrigated sites in Nebraska show a decreasing temperature trend in contrast to increasing trends at nonirrigated sites.

For a 12-year average based on observations in Nebraska, Chen et al. (2018) showed that crop irrigation decreases summer surface-air temperature by less than 1 °C and increases surface humidity by 0.5 g kg^{-1} relative to nonirrigated crops. The irrigation cooling effect is more pronounced and longer lasting for corn than for soybeans. Irrigation reduced maximum, minimum, and averaged temperature over corn by more than 0.5 °C for the first 6 days after irrigation, but its temperature effect over soybean was mixed and negligible 2 or 3 days after irrigation. Irrigation increased near-surface humidity over maize by about 1 g kg^{-1} up to 10 days and increased surface humidity over soybeans (~0.8 g kg^{-1}) with a similar memory. These differing effects of irrigation memory on temperature and humidity were associated with respective changes in the surface-sensible and latent heat fluxes for maize and soybean.

Barlage (2001) also showed that mesoscale circulations could be induced through large-scale subsidence over the center of the agriculture with enhanced upward motion generated near the boundary of modified land cover. With weak winds, peak cloud development occurred almost 2 hours earlier over cropland than over forest and the clouds had deeper development (Adegoke et al., 2007). In an observation study, cumulus clouds were found to initiate first and persist longer over a dry surface with an initially dry atmosphere relative to a moist surface. However, the impact of irrigation on convection could depend more on the overlying atmosphere since in a moist environment cumulus development may be enhanced over relatively moist cropland, especially if irrigated (Allard & Carleton, 2010). Background synoptic circulation also modulates how the atmosphere responds to surface forcing changes induced by agriculture. Carleton et al. (2001) showed that on "weak flow" days, when the inferred rate of mid-tropospheric subsidence is reduced relative to that on very weak flow days, cloud enhancement occurs near the cropland-natural boundary suggesting a role for agriculture-induced mesoscale circulations, similar to Barlage (2001).

In southwestern Australia, conversion of large areas of forests into agricultural land began more than a century ago. Changes in albedo, surface roughness, the soil water budget, and the planetary boundary layer evolved over decades. Besides different meteorology, Junkermann et al. (2009) found a significant source of aerosol over the agriculture area. The enhanced number of cloud condensation nuclei greatly affected cloud droplet distribution and is coupled through the hydrological groundwater cycle with agricultural conversion.

Conversion of land to agriculture induces changes in the atmospheric boundary layer. From an analysis of case studies and climatological records, it was shown that in western Australia this change in surface characteristics, and in particular the albedo (changed from 0.09 for native vegetation to 0.13/0.25 for agriculture), has reduced the occurrence of convective cloud formation through limiting the vertical development of the boundary layer. Thus, although there is greater moisture in the atmosphere over the agricultural areas, greater convective mixing over the native vegetation is the driving mechanism for cloud formation. This suggests that the clearing of land in the southwestern Australia for agricultural production has reduced the incidence of cloud formation, particularly during the spring months. Furthermore, the enhanced PBL heights over the native vegetation increase the probability of cloud formation by allowing surface air to reach the lifting condensation level. Rainfall has decreased in this region, likely due to changes in low-level convergence, caused by alteration of both mesoscale circulations and aerodynamic roughness as a result of conversion to agriculture (Lyons, 2002; Nair et al., 2011).

Over harvested wheat in U.S. state of Oklahoma, clouds are observed to form first over a mesoscale-size area (100 × 300 km) where the ground temperature is warmer than adjacent areas dominated by growing vegetation, similar to western Australia. Clouds form earliest over regions characterized by high sensible heat flux and are suppressed over regions characterized by high latent heat flux. This observation supports the hypothesis that clouds form first over areas with higher sensible heat flux during relatively dry atmospheric conditions (Rabin et al., 1990)

A likely candidate for enhancing low-level moisture convergence and promoting deep convection are the vegetation boundaries between the two major land cover types of crop and forest since the aerodynamic contrasts that exist along these boundaries allow air to be lofted above the level of free convection when the lower atmosphere is moist (Carleton et al., 2008). In early summer when cropland-natural boundaries are most evident, precipitation on weaker flow days is significantly greater near the vegetation boundary, but there is no statistical difference on stronger flow days. These changes appear related, primarily, to contrasting soil moisture and implied evapotranspiration differences (Carleton et al., 2008).

Irrigation has caused a disruption of the local hydrological cycle near the Ogallala aquifer in the west-central United States and has enhanced the regional precipitation. Using station and gridded precipitation observations for the warm season months over and downwind of the aquifer, DeAngelis et al. (2010) detected increased precipitation of 15%–30% during July from the eastern edge of the aquifer to as far downwind as 1000 km. Irrigation was also shown to increase the regional moisture flux and modify the convective available potential energy, causing a reduction in the surface temperature and leading to a modified regional circulation pattern and changes in mesoscale precipitation (Douglas et al., 2009).

Rosenan (1963) and Barnston and Schickedanz (1984) provided observational evidences revealing increased precipitation as a result of irrigation in the Great Plains and in Israel. Pielke and Zeng (1989) found that, relative to nonirrigated land, the equivalent potential temperature increased up to 10 °C at 150 m above irrigated land and up to 5 °C at 440 m. Therefore, it appears there were two impacts on severe thunderstorm development when irrigated land is adjacent to native land cover in regions where natural thunderstorm development already occurs: elevation of CAPE occurred over the irrigated land and even further enhanced CAPE in convergence zones that develop as a result of sensible heat gradients between the irrigated and dryland areas.

Significant increase in agricultural activity during recent years in coastal and inland regions of India supports a decreasing tendency in rainfall. Irrigation hinders development and intensity of sea breeze circulation reducing low-level moisture supply. Over the Indian subcontinent vegetation anomalies during premonsoon season due to increased agriculture and increases in the irrigated area significantly decreased July rainfall in central and southern India. Irrigation affects the local temperature, which in turn effects large-scale circulations and precipitation. Decreased July surface temperature in the Indian subcontinent from increased evapotranspiration due to irrigation and increased vegetation leads to a reduced land-sea thermal contrast and therefore weakens the monsoon circulation. A weak early monsoon appears to be at least partially a result of irrigation and the resultant increased vegetation and crop activity prior to the monsoon (Lee et al., 2009; Lohar & Pal, 1995). Regions of agricultural intensive land use suggest that premonsoon vegetation decreases July month precipitation over peninsular India. The results are more robust over peninsular and northern India indicating that an increase in vegetation activity has possibly weakened the early monsoon rainfall in this region. This suggests that land use change associated with agricultural intensification could be reducing the summer monsoon rainfall over certain regions of India (Niyogi et al., 2010; Saeed et al., 2009).

During the Indian monsoon season, atmospheric conditions where soil moisture positively influences precipitation are present about 25% of the time. During onset and retreat of the monsoon, the south and east of India show more potential for feedbacks than the north. These feedbacks suggest that large-scale irrigation in the south and east may increase local precipitation in contrast to, but not contradictory to, other above studies showing delayed onset of the monsoon. A positive trend in the precipitation just before the monsoon onset is found for irrigated stations. For nonirrigated stations, there is no such upward trend in precipitation. The precipitation trend for irrigated areas might be due to a positive trend in the extent of irrigated areas, with land-atmosphere feedbacks inducing increased precipitation (Tuinenburg et al., 2011).

8.3.3. Representing Agriculture in Large-Scale Models

Many modeling studies used a simple approach to incorporate agriculture in large-scale atmospheric models (see e.g., Chen & Dudhia, 2001; Skamarock et al., 2008). In these simple methods, land surface models use vegetation-specific physical and biological parameters, such as roughness length, albedo, leaf area index, and stomatal resistance, to define how different land covers are treated in models, while using the same parameterization equations for physical processes, such as transpiration, surface conductance, and radiative transfer. Using this approach, relative to grassland, crops have higher roughness (causing increased mixing), higher leaf area and lower stomatal resistance (causing increased latent heat flux), and similar albedo (depending on time of year) and rooting depth. Relative to forest, crops have lower roughness (causing reduced mixing), similar summer leaf area, much lower winter leaf area (causing exposed soil or snow), lower stomatal resistance, shallower rooting depth (allowing forest to access deeper soil moisture), and higher albedo (lowering net surface radiation). Tsvetsinskaya et al. (2001) incorporated daily crop-growth development based on a physiological crop-growth CERES-Maize model into a regional climate simulation model. They showed that such representation produced 20%–35% changes in sensible heat fluxes and 30%–45% changes in latent heat fluxes, and consequently 2–4 °C differences in air temperatures.

In recent years, more complex models have been coupled to earth system models to represent dynamic crop-growth and associated biogeophysical and biogeochemical processes (e.g., Levis et al., 2012; Liu et al., 2016; Lokupitiya et al., 2009, etc.). See McDermid et al.'s (2017) review about some of those efforts in modeling agriculture management. One example of these complex models is to simulate the seasonal evolution of various biomass carbon pools (e.g., root mass, leaf mass, stem mass, grain mass, and soil carbon) of different crop types (Liu et al., 2016). In this approach, the land surface model calculates photosynthesis by absorbing radiation energy and using it to convert atmospheric carbon dioxide into carbohydrate in leaves. This total assimilated carbon is then allocated to different carbon pools (leaf, stem, root, grain, and soil). The fraction of assimilated carbon that flows to different pools is dependent on agronomic crop-growth stages: initial vegetative, normal vegetative, initial reproductive, and physiological maturity (see Figure 8.5). Agronomical growth stages advance through the simulated seasons by the accumulating growing degree days (GDD) as a function of air temperature and crop types.

In different crop-growth stages, the allocation of carbohydrate fluxes and respiration rates varies, as does the crop turnover and senescence rate. During vegetative periods, the carbohydrate fluxes are mainly allocated to leaf, stem, and root, and during the reproductive period, the carbohydrate fluxes are mainly allocated to grain. At the mature reproduction stage, carbohydrate fluxes are only allocated to grain and root, and leaves are continually senescent and the carbohydrate in stems and roots will be partially translocated to grain. In addition, considering field management can introduce additional stages beyond the growing season: before planting, from maturity to harvesting, and after harvesting. Management factors were incorporated into the crop model because it is anticipated that they can contribute to improved performance when the model is used in different locations where management practices or climate regimes may change. Incorporating these agriculture management data in the coupled atmosphere-crop modeling system is critical to improving the realism of crop modeling.

Parameterization schemes of agriculture irrigation with varying degrees of complexity have been developed in the past decades (McDermid et al., 2017; Nazemi & Wheater, 2015; Pokhrel et al., 2016) for earth system models. Commonly, these irrigation models use the "soil

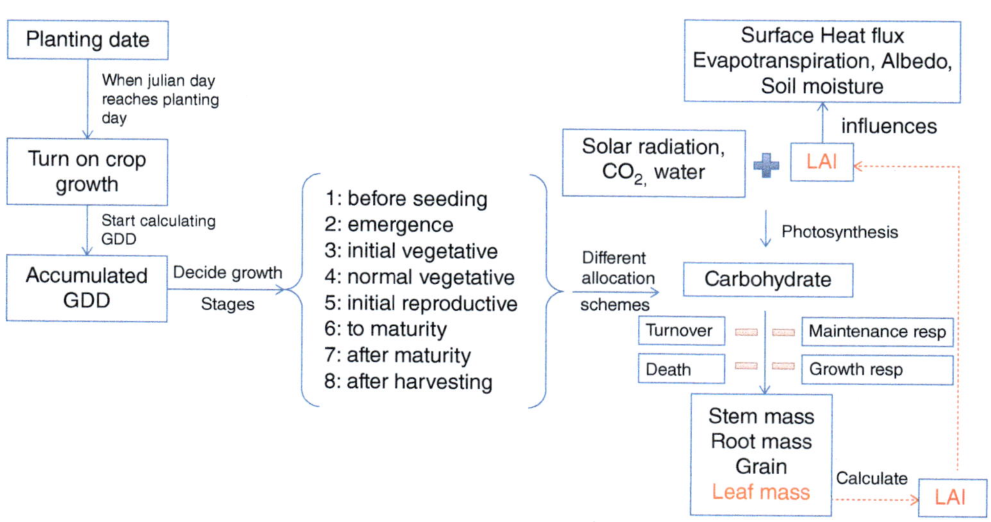

Figure 8.5 A schematic of modeling crop-growth processes in the Noah-MP-Crop model. Source: Liu et al. (2016)/John Wiley & Sons.

moisture deficit" concept to estimate irrigation demand during crop-growing season and then calculate the irrigation amount based on the departure of the root-zone soil moisture from the water-stress-free field capacity. Remote-sensing data such as the 500-m Moderate Resolution Imaging Spectroradiometer (MODIS)-based irrigated fractional area map (Ozdogan & Gutman, 2008) are used to define the spatial extent of irrigated regions (Leng et al., 2013; Xu et al., 2019).

Leaf growth in the crop model is a primary link to hydrology/atmosphere/cropland interactions since these dynamic crop-growth models linked to an irrigation model calculate time-varying leaf area index, which in turn directly impacts the surface radiation, photosynthesis, soil moisture, and surface energy fluxes. As demonstrated earlier, it is these changes in the partitioning of sensible heat flux and latent heat flux that will determine how agriculture and irrigation will affect weather and climate when coupled with atmospheric models.

ACKNOWLEDGMENTS

This review work was supported by NCAR Water System, USDA NIFA Grants 2015-67003-23460, NSF INFEWS Grant #1739705, and NOAA OAR Grant NA18OAR4590381.

REFERENCES

Adegoke, J. O., Pielke, R. A., Sr, & Carleton, A. M. (2007). Observational and modeling studies of the impact of agriculture-related land use change on climate in the central U.S. *Agricultural and Forest Meteorology, 142*, 203–215.

Adegoke, J. O., Pielke, R. A., Sr, Eastman, J., Mahmood, R., & Hubbard, K. G. (2003). Impact of irrigation on midsummer surface fluxes and temperature under dry synoptic conditions: A regional atmospheric model study of the U.S. *High Plains. Monthly Weather Review*, *131*, 556–564.

Allard, J., & Carleton, A. M. (2010). Mesoscale associations between Midwest land surface properties and convective cloud development in the warm season. *Physical Geography, 31*, 107–136.

Andreae, M. O., Jones, C. D., & Cox, C. P. (2005). Strong present-day aerosol cooling implies a hot future. *Nature*, *435*, 1187–1190.

Arnfield, A. J. (2003). Two decades of urban climate research: A review of turbulence, exchanges of energy and water, and the urban heat island. *International Journal of Climatology*, *23*(1), 1–26.

Artis, D. A., & Carnahan, W. H. (1982). Survey of emissivity variability in thermography of urban areas. *Remote Sensing of Environment*, *12*(4), 313–329.

Banta, R. M., Senff, C. J., Nielsen-Gammon, J., Darby, L. S., Ryerson, T. B., Alvarez, R. J. et al. (2005). A bad air day in Houston. *Bulletin of the American Meteorological Society*, *86*, 657–669. doi:10.1175/BAMS-86-5-657.

Barlage, M. (2001). The effects of land use change on the numerical modeling of regional climate and watershed runoff in the Great Lakes region, Ph.D. Dissertation, University of Michigan, 188pp.

Barlage, M., Miao, S., & Chen, F. (2016). Impact of physics parameterizations on high-resolution weather prediction over two Chinese megacities. *Journal of Geophysical Research: Atmospheres*, *121*(9), 4487–4498.

Barlow, J. F., Dunbar, T. M., Nemitz, E. G., Wood, C. R., Gallagher, M. W., Davies, F., et al. (2011). Boundary layer dynamics over London, UK, as observed using Doppler lidar during REPARTEE-II. *Atmospheric Chemistry and Physics*, *11*(5), 2111–2125.

Barnston, A. G., & Schickedanz, P. T. (1984). The effect of irrigation on warm season precipitation in the Southern Great Plains. *Journal of Climate and Applied Meteorology*, *23*, 865–888.

Bentley, M. L., & Stallins, J. A. (2008). Synoptic evolution of Midwestern US extreme dew point events. *International Journal of Climatology, 28*, 1213–1225.

Best, M. J. (2005). Representing urban areas within operational numerical weather prediction models. *Boundary-Layer Meteorology*, *114*, 91–109.

Best, M. J., Abramowitz, G., Johnson, H. R., Pitman, A. J., Balsamo, G., Boone, A., et al. (2015). The plumbing of land surface models: Benchmarking model performance. *Journal of Hydrometeorology*, *16*(3), 1425–1442.

Bonan, G. B. (1998). The land surface climatology of the NCAR Land Surface Model coupled to the NCAR Community Climate Model. *Journal of Climate*, *11*(6), 1307–1326.

Bonan, G. B. (2001). Observational evidence for reduction of daily maximum temperature by croplands in the Midwest United States. *Journal of Climate*, *14*, 2430–2442.

Bornstein, R. D. (1968). Observations of the urban heat island effects in New York City. *Journal of Applied Meteorology*, *7*, 575–582.

Bornstein, R. D., & Lin, Q. (2000). Urban heat islands and summertime convective thunderstorms in Atlanta: Three case studies. *Atmospheric Environment*, *34*, 507–516.

Bounoua, L., DeFries, R., Collatz, G.J., Sellers, P., & Khan, H. (2002). Effects of land cover conversion on surface climate. *Climatic Change*, *52*, 29–64. https://doi.org/10.1023/A:1013051420309

Brest, C. L. (1987). Seasonal albedo of an urban/rural landscape from satellite observations. *Journal of Climate and Applied Meteorology*, *26*(9), 1169–1187.

Carleton, A. M., Adegoke, J. O., Allard, J., Arnold, D. L., & Travis, D. J. (2001). Summer season land cover-convective cloud associations for the Midwest U.S. *Corn Belt. Geophysical Research Letters, 28*, 1679–1682.

Carleton, A. M., Travis, D. J., Adegoke, J. O., Arnold, D. L., & Curran, S. (2008). Synoptic circulation and land surface influences on convection in the Midwest U.S. Corn Belt during the summers of 1999 and 2000. Part II: Role of vegetation boundaries. *Journal of Climate, 21*, 3617–3641.

Chandler, T. J. (1970). Selected bibliography on urban climate. Tech Note (No. 155).

Changnon, D., Sandstrom, M., & Schaffer, C. (2003). Relating changes in agricultural practices to increasing dew points in extreme Chicago heat waves. *Climate Research*, *24*(3), 243–254.

Chen, F., Bornstein, R., Grimmond, S., Li, J., Liang, X., Martilli, A., et al. (2012). Research priorities in observing and modeling urban weather and climate. *Bulletin of the American Meteorological Society*, *93*, 1725–1728.

Chen, F., & Dudhia, J. (2001). Coupling an advanced land surface–hydrology model with the Penn State–NCAR MM5 modeling system. Part I: Model implementation and sensitivity. *Monthly Weather Review*, *129*(4), 569–585.

Chen, F., Kusaka, H., Bornstein, R., Ching, J., Grimmond, C. S. B., Grossman-Clarke, S., et al. (2011b). The integrated WRF/urban modeling system: Development, evaluation, and applications to urban environmental problems. *International Journal of Climatology*, *31*, 273–288. doi: 10.1002/joc.2158.

Chen, F., Miao, S., Tewari, M., Bao, J.-W., & Kusaka, H. (2011a). A numerical study of interactions between surface forcing and sea-breeze circulations and their effects on stagnant winds in the Greater Houston Area. *Journal of Geophysical Research, 116*(D12). doi:10.1029/2010JD015533.

Ching, J., Brown, M., McPherson, T., Burian, S., Chen, F., Cionco, R., et al. (2009). National Urban Database and Access Portal Tool, NUDAPT. *Bulletin of the American Meteorological Society*, *90*(8), 1157–1168.

Ching, J., Mills, G., Bechtel, B., See, L., Feddema, J., Wang, X., et al. (2018). WUDAPT: An urban weather, climate, and environmental modeling infrastructure for the anthropocene. *Bulletin of the American Meteorological Society*, *99*(9), 1907–1924.

Clark, M. P., Nijssen, B., Lundquist, J. D., Kavetski, D., Rupp, D. E., Woods, R. A., et al. (2015). A unified approach to process-based hydrologic modeling: 1. *Modeling concept. Water Resources Research*, *51*. doi:10.1002/2015WR017198.

Cuo, L., Lettenmaier, D. P., Mattheussen, B. V., Storck, P., & Wiley, M. (2008). Hydrologic prediction for urban watersheds with the Distributed Hydrology–Soil–Vegetation Model. *Hydrological Processes*, *22*(21), 4205–4213.

Dai, Y., Zeng, X., Dickinson, R. E., Baker, I., Bonan, G. B., Bosilovich, M. G., et al. (2003). The common land model. *Bulletin of the American Meteorological Society*, *84*(8), 1013–1023.

DeAngelis, A., Dominguez, F., Fan, Y., Robock, A., Kustu, M. D., & Robinson, D. (2010). Evidence of enhanced precipitation due to irrigation over the Great Plains of the United States. *Journal of Geophysical Research, 115*, D15115. doi: 10.1029/2010JD013892

Dickinson, R. E., Henderson-Sellers, A., & Kennedy, P. J. (1993). Biosphere-atmosphere Transfer Scheme (BATS) Version 1e as Coupled to the NCAR Community Climate Model. NCAR Technical Note NCAR/TN-387+STR. doi:10.5065/D67W6959.

Dou, J. J., Wang, Y., Bornstein, R., & Miao, S. (2015). Observed spatial characteristics of Beijing urban climate impacts on summer thunderstorms. *Journal of Applied Meteorology and Climatology*, *54*, 94–105.

Douglas, E. M., Beltran-Przekurat, A., Niyogi, D., Pielke, R. A., Sr, & Vorosmarty, C. J. (2009). The impact of agricultural intensification and irrigation on land–atmosphere interactions and Indian monsoon precipitation—a mesoscale modeling perspective. *Global and Planetary Change, 67*, 117–128.

Douglas, E. M., Niyogi, D., Frolking, S., Yeluripati, J. B., Pielke, R. A. Sr, Vorosmarty, C. J., et al. (2006). Changes in moisture and energy fluxes due to agricultural land use and irrigation in the Indian Monsoon Belt. *Geophysical Research Letters, 33*. doi:10.1029/ 2006GL026550

Dupont, S., Otte, T. L., & Ching, J. K. S. (2004). Simulation of meteorological fields within and above urban and rural canopies with a mesoscale model (MM5). *Boundary-Layer Meteorology*, *113*, 111–158.

Ek, M. B., Mitchell, K. E., Lin, Y., Rogers, E., Grunmann, P., Koren, V., et al. (2003). Implementation of Noah land surface model advances in the National Centers for Environmental Prediction operational mesoscale Eta model. *Journal of Geophysical Research: Atmospheres*, *108*(D22) (2003). https://doi.org/10.1029/2002JD003296.

FAO (2010). AQUASTAT—FAO's global information system on water and agriculture, FAO, http://www.fao.org/nr/aquastat, last access: 16 March 2010, Rome, Italy.

FAO (2018). World Food and Agriculture Statistical Pocketbook 2018, Food and Agriculture Organization of the United Nations, Rome, 254 pp. License: CC BY-NC-SA 3.0 IGO.

Grimmond, C. S. B. (2006). Progress in measuring and observing the urban atmosphere. *Theoretical and Applied Climatology*, *84*, 3–22.

Grimmond, C. S. B., Blackett, M., Best, M. J., Barlow, J., Baik, J. J., Belcher, S. E., et al. (2009). The international urban energy balance models comparison project: First results from Phase 1. *Journal of Applied Meteorology and Climatology*, accepted subject to revision.

Grimmond, C. S. B., & Oke, T. R. (1999). Heat storage in urban areas: Local-scale observations and evaluation of a simple model. *Journal of Applied Meteorology*, *38*(7), 922–940.

Gochis, D. J., Yu, W., & Yates, D. N. (2013). The WRF-Hydro Model Technical Description and User's Guide, Version 1.0, NCAR Technical Document, 120 pp., NCAR, Boulder, Colo. [Available at http://www.ral.ucar.edu/projects/wrf_hydro/.]

Hatfield, J. L., Prueger, J. H., & Kustas, W. P. (2007). Spatial and temporal variation of energy and carbon dioxide fluxes in corn and soybean fields in central Iowa. *Agronomy Journal 99*, 285–296.

Howard, L. (1833). *The climate of London: Deduced from meteorological observations made in the metropolis and at various places around it*. London: Harvey and Darton.

Hollis, G. E. (1975). The effect of urbanization on floods of different recurrence interval. *Water Resources Research*, *11*(3), 431–435.

Huang, M., Gao, Z., Miao, S., Chen, F., LeMone, M. A., Li, J., & Hu, F. (2017). Estimate of boundary-layer depths over Beijing, China, using Doppler Lidar data during SURF-2015. *Boundary-Layer Meteorology*, *162*, 503–522.

Huff, F. A., & Changnon, S. A. (1972). Climatological assessment of urban effects on precipitation at St. *Louis. Journal of Applied Meteorology*, *11*, 823–842.

Jiang, X. Y., Wiedinmyer, C., Chen, F., Yang, Z. L., & Lo, J. C. F. (2008). Predicted impacts of climate and land-use change on surface ozone in the Houston, Texas, *Area. Journal of Geophysical Research*, *113*, D20312. doi:10.1029/2008JD009820.

Junkermann, W., Hacker, J., Lyons, T., & Nair, U. S. (2009). Land use change suppresses precipitation. *Atmospheric Chemistry and Physics, 9*, 11481–11500.

Kawai, T., & Kanda, M. (2010). Urban energy balance obtained from the comprehensive outdoor scale model experiment. Part I: Basic features of the surface energy balance. *Journal of Applied Meteorology and Climatology*, *49*(7), 1341–1359.

Kato, S., & Yamaguchi, Y. (2007). Estimation of storage heat flux in an urban area using ASTER data. *Remote Sensing of Environment*, *110*(1), 1–17.

Kusaka, H., Kondo, H., Kikegawa, Y., & Kimura, F. (2001). A simple single-layer urban canopy model for atmospheric models: Comparison with multi-layer and slab models. *Boundary-Layer Meteorology*, *101*, 329–358.

Landsberg, H. E. (1981). *The urban climate* (Vol. *28*). Academic Press.

Lawrence, D. M., Oleson, K., Flanner, M. G., Fletcher, C. G., Lawrence, P., Levis, S., et al. (2012). The CCSM4 land simulation, 1850–2005: Assessment of surface climate and new capabilities. *Journal of Climate*, *25*, 2240–2260.

Lee, E., Chase, T. N., & Rajagopalan, B. (2008). Highly improved predictive skill in the forecasting of the East Asian summer monsoon. *Water Resources Research, 44*, W10422. doi: 10.1029/2007WR0 06514

Lee, E., Chase, T. N., Rajagopalan, B., Barry, R. G., Biggs, T. W., & Lawrence, P. J. (2009). Effects of irrigation and vegetation activity on early Indian summer monsoon variability. *International Journal of Climatology*, *29*, 573–581.

LeMone, M. A., Tewari, M., Chen, F., & Dudhia, J. (2014). Objectively determined fair-weather NBL features in ARW-WRF and their comparison to CASES-97 observations. *Monthly Weather Review*, *142*, 2709–2732.

Leng, G., Leung, L. R., & Huang, M. (2017). Significant impacts of irrigation water sources and methods on modeling irrigation effects in the ACME Land Model. *Journal of Advances in Modeling Earth Systems*, *9*, 1665–1683. doi:10.1002/2016MS000885

Levis, S., Gordon, B. B., Kluzek, E., Thornton, P. E., Jones, A., Sacks, W. J., & Kucharik, C. J. (2012). Interactive crop management in the community earth system model (CESM1): Seasonal influences on land-atmosphere fluxes. *Journal of Climate*, *25*(14), 4839–4859.

Li, D., Malyshev, S., & Shevliakova, E. (2016a). Exploring historical and future urban climate in the Earth System Modeling framework: 1. Model development and evaluation. *Journal of Advances in Modeling Earth Systems*, *8*(2), 917–935.

Li, Y., Miao, S., Chen, F., & Liu, Y. (2016b). Introducing and evaluating a new building-height categorization based on the fractal dimension into the Coupled WRF/urban Model. *International Journal of Climatology*, doi:10.1002/joc.4903

Liang, X., Miao, S., Li, J., Bornstein, R., Zhang, X., Gao, Y., et al. (2018). SURF: Understanding and predicting urban convection and haze. *Bulletin of the American Meteorological Society, 99*(7), 1391–1413. https://doi.org/10.1175/BAMS-D-16-0178.1

Liu, X., Chen, F., Barlage, M., Zhou, G., & Niyogi, D. (2016). Noah-MP-Crop: Introducing dynamic crop growth in the Noah-MP Land-Surface Model. *Journal of Geophysical Research, 121*(23), 13,953–13,972. doi:10.1002/2016JD025597.

Liu, Y., Chen, F., Warner, T., & Basara, J. (2006). Verification of a mesoscale data-assimilation and forecasting system for the Oklahoma City area during the Joint Urban 2003 Field Project. *Journal of Applied Meteorology*, *45*, 912–929.

Lohar, D., & Pal, B. (1995). The effect of irrigation on pre-monsoon season over southwest Bengal, India. *Journal of Climate, 8*, 2567–2570.

Lokupitiya, E., Denning, S., Paustian, K., Baker, I., Schaefer, K., Verma, S., et al. (2009). Incorporation of crop phenology in Simple Biosphere Model (SiBcrop) to improve land-atmosphere carbon exchanges from croplands. *Biogeosciences*, *6*(1), 969–986.

Lyons, T. J. (2002). Clouds prefer native vegetation. *Meteorology and Atmospheric Physics, 80*, 131–140.

McDermid, S. S., Mearns, L. O., & Ruane, A. C. (2017). Representing agriculture in Earth System Models: Approaches and priorities for development. *Journal of Advances in Modeling Earth Systems*, *9*, 2230–2265. doi:10.1002/2016MS000749.

Miao, S., & Chen, F. (2014). Enhanced modeling of latent heat flux from urban surfaces in the Noah/single-layer urban canopy coupled model. *Science China Earth Sciences*, *57*, 2408–2416. doi: 10.1007/s11430-014-4829-0

Miao, S., Chen, F., LeMone, M. A., Tewari, M., Li, Q., & Wang, Y. (2009). an observational and modeling study of characteristics of urban heat island and boundary layer structures in Beijing. *Journal of Applied Meteorology and Climatology*, *48*, 484–501.

Miao, S., Chen, F., Li, Q., & Fan, S. (2011). Impacts of urban processes and urbanization on summer precipitation: A case study of heavy rainfall in Beijing on 1 August 2006. *Journal of Applied Meteorology and Climatology*, *50*, 806–825.

Miao, S., Dou, J., Chen, F., Li, J., & Li, A. (2012). Analysis of observations on the urban surface energy balance in Beijing. *Science China Earth Sciences*, *55*, 1881–1890. doi: 10.1007/s11430-012-4411-6

Moriwaki, R., & Kanda, M. (2004). Seasonal and diurnal fluxes of radiation, heat, water vapor, and carbon dioxide over a suburban area. *Journal of Applied Meteorology*, *43*(11), 1700–1710.

Nair, U. S., Wu, Y., Kala, J., Lyons, T. J., Pielke, R. A., Sr, & Hacker, J. M. (2011). The role of land use change on the development and evolution of the west coast trough, convective clouds, and precipitation in southwest Australia. *Journal of Geophysical Research*, *116*, D07103. doi: 10.1029/2010JD014950

Narumi, D., Kondo, A., & Shimoda, Y. (2009). Effects of anthropogenic heat release upon the urban climate in a Japanese megacity. *Environmental Research*, *109*(4), 421–431.

Nazemi, A., & Wheater, H. S. (2015). On inclusion of water resource management in Earth system models—Part 1: Problem definition and representation of water demand. *Hydrology and Earth System Sciences, 19*(1), 33–61. https://doi.org/10.5194/hess-19-33-2015

Newman, A. J., Mizukami, N., Clark, M. P., Wood, A. W., Nijssen, B., & Nearing, G. (2017). Benchmarking of a physically based hydrologic model. *Journal of Hydrometeorology*, *18*, 2215–2225.

Niu, G. Y., Yang, Z. -L., Mitchell, K. E., Chen, F., Ek, M. B., Barlage, M. J., et al. (2011). The community Noah land surface model with multiparameterization options (Noah-MP): 1. Model description and evaluation with local-scale measurements. *Journal of Geophysical Research: Atmospheres*, *116*(D12). doi:10.1029/2010JD015140.

Niyogi, D., Kishtawal, C. M., Tripathi, S., & Govindaraju, R. S. (2010). Observational evidence that agricultural intensification and land use change may be reducing the Indian summer monsoon rainfall. *Water Resources Research*, *46*, W03533. doi: 10.1029/2008WR007082

Niyogi, D., Pyle, P., Lei, M., Aalarya, S., Kishtawal, M., Shepherd, M., et al. (2011). Urban modification of thunderstorms: An observational storm climatology and model case study for the Indianapolis urban region. *Journal of Applied Meteorology and Climatology*, *50*, 129–144.

Oke, T. R. (1982). The energetic basis of the urban heat island. *Quarterly Journal of the Royal Meteorological Society, 108, 1–24.*

Oke, T. R. (1987). *Boundary layer climates* (2nd edition). Methuen: London.

Oke, T. R. (1995). The heat island of the urban boundary layer: Characteristics, causes and effects. In J. E. Cermak et al. (Eds.), *Wind climate in cities* (Vol. *227*, 81–107). NATO ASI Series E, Kluwer Academic.

Oke, T. R., Mills, G., Christen, A., & Voogt, J. A. (2017). *Urban climates*. Cambridge University Press.

Oleson, K. W., Bonan, G. B., Feddema, J., Vertenstein, M., & Grimmond, C. S. B. (2008). An urban parameterization for a global climate model: 1. Formulation & evaluation for two cities. *Journal of Applied Meteorology and Climatology*, *47*, 1038–1060. doi: 10.1175/2007JAMC1597.1

Ozdogan, M., & Gutman, G. (2008). A new methodology to map irrigated areas using multi-temporal MODIS and ancillary data: An application example in the continental US. *Remote Sensing of Environment, 112*(9), 3520–3537. https://doi.org/10.1016/j.rse.2008.04.010

Ozdogan, M., Rodell, M., Beaudoing, H. K., & Toll, D. L. (2010). Simulating the effects of irrigation over the United States in a land surface model based on satellite-derived agricultural data. *Journal of Hydrometeorology*, *11*(1), 171–184. https://doi.org/10.1175/2009JHM1116.1

Pielke, R. A., & Zeng, X. (1989). Influence on severe storm development of irrigated land. *National Weather Digest*, *14*, 16–17.

Pigeon, G., Legain, D., Durand, P., & Masson, V. (2007). Anthropogenic heat release in an old European agglomeration (Toulouse, France). *International Journal of Climatology*, *27*(14), 1969–1981.

Pokhrel, Y. N., Hanasaki, N., Wada, Y., & Kim, H. (2016). Recent progresses in incorporating human land–water management into global land surface models toward their integration into Earth system models. *Wiley Interdisciplinary Reviews: Water, 3*(4), 548–574. https://doi.org/10.1002/wat2.1150

Rabin, R. M., Stadler, S., Wetzel, P. J., Stensrud, D. J., & Gregory, M. (1990). Observed effects of landscape variability on convective clouds. *Bulletin of the American Meteorological Society*, *71*, 272–280.

Rosenan, N. (1963). The map of evaporation from water surface in Israel. *Atlas of Israel: Tel Aviv, Survey of Israel*.

Rosenfeld, D. (2000). Suppression of rain and snow by urban and industrial air pollution. *Science*, *287*(5459), 1793–1796.

Rozenfeld, H. D., Rybski, D., Andrade, J. S., Jr., Batty, M., Stanley, H. E., & Makse, H. A. (2008). Laws of population growth. *Proceedings of the National Academy of Sciences of the United States of America*, *105*, 18702–18707, https://doi.org/10.1073/pnas.0807435105.

Roy, S. B., & Avissar, R. (2002). Impact of land use/land cover change on regional hydrometeorology in Amazonia. *Journal of Geophysical Research*, *107*, 1–12.

Saeed, F., Hagemann, S., & Jacob, D. (2009). Impact of irrigation on the South Asian summer monsoon. *Geophysical Research Letters*, *36*, L20711. doi: 10.1029/2009GL040625

Sailor, D. J., & Lu, L. (2004). A top–down methodology for developing diurnal and seasonal anthropogenic heating profiles for urban areas. *Atmospheric Environment*, *38*(17), 2737–2748.

Salamanca, F., & Martilli, A. (2010). A new building energy modelcoupled with an urban canopy parameterization for urban climatesimulations – part II. Validation with one dimension offlinesimulations. *Theoretical and Applied Climatology*, *99*, 345–356, doi: 10.1007/s00704-009-0143-8.

Salamanca, F., Zhang, Y., Barlage, M., Chen, F., Mahalov, A., & Miao, S. (2018). Evaluation of the Noah-MP land surface model coupled to WRF in a semiarid urban environment. *Journal of Geophysical Research.* doi: 10.1002/2018JD028377

Sandstrom, M. A., Lauritsen, R. G., & Changnon, D. (2004). A central U. S. summer extreme dew point climatology (1949–2000). *Physical Geography*, *25*, 191–207.

Sellers, P. J., Mintz, Y., Sud, Y. E. A., & Dalcher, A. (1986). A simple biosphere model (SiB) for use within general circulation models. *Journal of the Atmospheric Sciences*, *43*(6), 505–531.

Shepherd, J. M. (2005). A review of current investigations of urban induced rainfall and recommendations for the future. *Earth Interactions*, *9*, 1–27.

Siebert, S., & Döll, P. (2010). Quantifying blue and green water uses and virtual water contents in global crop production as well as potential production losses without irrigation. *Journal of Hydrology*, *384*(3–4), 198–217. https://doi.org/10.1016/j.jhydrol.2009.07.031

Skamarock, W. C., Klemp, J. B., Dudhia, J., Gill, D. O., Barker, D., Duda, M. G., et al. (2008). A Description of the Advanced Research WRF Version 3. NCAR Technical Note NCAR/TN-475+STR, doi:10.5065/D68S4MVH.

Stewart, I., & Oke, T. (2012). Local climate zones for urban temperature studies. *Bulletin of the American Meteorological Society*, *93*, 1879–1900. https://doi-org.cuucar.idm.oclc.org/10.1175/BAMS-D-11-00019.1.

Stull, R. B. (2012). *An introduction to boundary layer meteorology* (Vol. *13*). Springer Science & Business Media.

Taha, H. (1999). Modifying a mesoscale meteorological model to better incorporate urban heat storage: A bulk-parameterization approach. *Journal of Applied Meteorology, 38*, 466–473.

Taha, H., Konopacki, S., & Akbari, H. (1998). Impacts of lowered urban air temperatures on precursor emission and ozone air quality. *Journal of the Air & Waste Management Association, 48*(9), 860–865.

Taylor, C. H. (1977). Seasonal variations in the impact of suburban development on runoff response: Peterborough, *Ontario. Water Resources Research, 13*(2), 464–467.

Tsvetsinskaya, E. A., Mearns, L. O., & Easterling, W. E. (2001). Investigating the effect of seasonal plant growth and development in three-dimensional atmospheric simulations. Part II: Atmospheric response to crop growth and development. *Journal of Climate, 14*(5), 711–729.

Tuinenburg, O. A., Hutjes, R., Jacobs, C., & Kabat, P. (2011). Diagnosis of local land-atmosphere feedbacks in India. *Journal of Climate, 24*, 251–266.

United Nations (2007). World Urbanization Prospects: The 2007 Revision. http://esa.un.org/unup.

Voisin, N., Li, H., Ward, D. L., Huan g, M., Wigmosta, M. S., & Leung, L. R. (2013). On an improved sub-regional water resources management representation for integration into earth system models. *Hydrology and Earth System Sciences, 17*(9), 3605–3622. PNNL-SA-93513. doi:10.5194/hess-17-3605-2013.

Wang, X. M., Chen, F., Wu, Z., Zhang, M., Tewari, M., Guenther, A., & Wiedinmyer, C. (2009). Impacts of weather conditions modified by urban expansion on surface ozone: Comparison between the Pearl River Delta and Yangtze River Delta regions. *Advances in Atmospheric Sciences, 2009b, 26* 962–972.

Xu, X., Chen, F., Barlage, M., Gochis, D., Miao, S., & Shen, S. (2019). Lessons learned from modeling irrigation from field to regional scales. *Journal of Advances in Modeling Earth Systems*, in revision.

Yang, J., & Bou-Zeid, E. (2018). Should cities embrace their heat islands as shields from extreme cold? *Journal of Applied Meteorology and Climatology, 57*, 1309–1320. doi: 10.1175/JAMC-D-17-0265.1.

Yang, J., Wang, Z.-H., Chen, F., Miao, S., Tewari, M., Voogt, J. A., & Myint, S. (2015). Enhancing hydrologic modeling in the coupled WRF-urban modeling system. *Boundary-Layer Meteorology, 155*, 87–109. doi 10.1007/s10546-014-9991-6.

Zhang, C. L., Chen, F., Miao, S. G., Li, Q. C., Xia, X. A., & Xuan, C. Y. (2009). Impacts of urban expansion and future green planting on summer precipitation in the Beijing metropolitan area. *Journal of Geophysical Research, 114*, D02116. doi:10.1029/2008JD010328.

Zhao, C., Tie, X., & Lin, Y. (2006). A possible positive feedback of reduction of precipitation and increase in aerosols over eastern central China. *Geophysical Research Letters, 33*, L11814. doi:10.1029/2006GL025959.

9

Gravity Wave Drag Parameterizations for Earth's Atmosphere

Christopher G. Kruse[1], Jadwiga H. Richter[2], M. Joan Alexander[1], Julio T. Bacmeister[2], Christopher Heale[3], and Junhong Wei[4]

ABSTRACT

Atmospheric gravity waves (GWs), or buoyancy waves, transport momentum and energy through Earth's atmosphere. GWs are important at nearly all levels of the atmosphere, though the momentum they transport is particularly important in general circulation of the middle and upper atmosphere. Primary sources of atmospheric GWs are flow over mountains, moist convection, and imbalances in jet/frontal systems. Secondary GWs can also be generated as a result of dissipation of a primary GWs. Gravity waves typically have horizontal wavelengths of tens to hundreds of kilometers, though they can have scales of one to thousands of kilometers as well. Current effective resolutions of climate models, and even numerical weather prediction models, do not resolve significant portions of the momentum-carrying and energy-flux-carrying GW spectrum, and so parameterizations are necessary to represent under-resolved and unresolved GWs in most current models. Here, an overview of GWs generated by orography, convection, jet/front systems, primary wave breaking, and secondary wave generation is provided. The basic theory of GW generation, propagation, and dissipation relevant to parameterization is presented. Conventionally used GW parameterizations are then reviewed. Lastly, we describe uncertainties and parameter tuning in current parameterizations and discuss known processes that are currently missing.

9.1. INTRODUCTION AND BASIC EQUATIONS

Gravity waves (GWs), or buoyancy waves, are waves in Earth's atmosphere for which buoyancy is the restoring force. Most are familiar with the concentric waves that form on the surface of the water and emanate away from the perturbation from a falling stone: such waves exist on the boundary of a denser fluid (water) and less dense air above. Atmospheric GWs are analogous to surface waves, with the same buoyancy restoring force. However, the atmosphere is continuously stratified, with a more-or-less smooth decrease in potential density with height. In the continuously stratified atmosphere, lower-level atmospheric GW perturbations displace the stably stratified atmospheric flows above, allowing propagation in the vertical as well as the horizontal (i.e., in all three dimensions).

Sources of atmospheric GWs are numerous. Essentially any process, often an instability, that produces transient perturbations of air parcels (e.g., in their altitude or potential density) can be a GW source. However, three primary sources are flow over mountains, moist convection, and imbalances associated with jets and fronts.

[1] *NorthWest Research Associates, Boulder, CO, USA*

[2] *Climate and Global Dynamics Laboratory, National Center for Atmospheric Research, Boulder, CO, USA*

[3] *Department of Physical Sciences, Embry–Riddle Aeronautical University, Daytona Beach, FL, USA*

[4] *School of Atmospheric Sciences, and Guangdong Province Key Laboratory for Climate Change and Natural Disaster Studies, Sun Yat-sen University, Guangzhou, China; and Southern Marine Science and Engineering Guangdong Laboratory, Zhuhai, China*

Fast Processes in Large-Scale Atmospheric Models: Progress, Challenges, and Opportunities, Geophysical Monograph 282, First Edition.
Edited by Yangang Liu and Pavlos Kollias.

DOI: 10.1002/9781119529019.ch09

Recent studies have focused on secondary GWs generated as a result of localized breaking/dissipation of a primary GW. Gravity waves typically have horizontal wavelengths of tens to hundreds of kilometers, and vertical scales of 3–30 km, with periods ranging from about 10 min to several hours. Still, GWs, often orographically forced, with horizontal scales of a few kilometers are not uncommon. Some sources (e.g., jet/front imbalances) can generate waves with larger horizontal wavelengths, too, with scales of ≈500 km and periods between 6 and 24 hours. These waves are typically classified as inertia-gravity waves, where the Coriolis effect cannot be neglected. The properties of GWs at various levels of the atmosphere vary depending on the GW source, altitude, and atmospheric properties that they have propagated through.

Numerous review articles (e.g., Fritts & Alexander, 2003; Smith, 1979; Teixiera, 2014) and textbooks (e.g., Holton, 2004) present the basic, and not so basic, theory of atmospheric GWs. For full derivations and rigorous mathematical treatment, the reader is directed to these references. Here, basic relations from linear GW theory are presented to demonstrate fundamental ideas relevant to GW parameterizations.

Characteristics of GWs are governed by a dispersion relation that gives the relationship between the wave frequency and its horizontal and vertical wavenumbers. The following dispersion relation for two-dimensional (i.e., x and z) gravity waves in a nonrotating atmosphere can be derived from linearized Boussinesq equations of motions (e.g., Holton, 2004):

$$\omega^{*2} = (\omega - Uk)^2 = \frac{N^2k^2}{(k^2+m^2)}. \tag{9.1}$$

Here, ω^{*2} is the intrinsic frequency of the wave, which is the oscillation frequency an air parcel experiences as it is advected through the wave, ω is the wave, or parcel oscillation, frequency relative to the ground $k = 2\pi/\lambda_x$ and $m = 2\pi/\lambda_z$ are the horizontal and vertical wavenumbers, respectively, U is the background wind, and $N^2 = \frac{g}{\theta}\frac{\partial\theta}{\partial z}$ is the background atmospheric buoyancy frequency, where θ is the background potential temperature.

Equation (9.1) can be solved for the vertical wavenumber as follows:

$$m^2 = \frac{k^2(N^2 - \omega^{*2})}{\omega^{*2}} \tag{9.2}$$

Another key property of gravity waves is the horizontal wave phase speed, c_{px}:

$$c_{px} = \frac{\omega}{k} \tag{9.3}$$

The above is the wave phase speed relative to the ground. The intrinsic wave phase speed, or phase speed relative to the mean wind, is defined as

$$c_{px}^* = \frac{\omega^*}{k} = c_{px} - U = \pm\frac{N}{(k^2+m^2)^{\frac{1}{2}}} \tag{9.4}$$

The wave phase speed, as will be described in later sections, largely determines whether waves are able to propagate upward and where they will break and deposit momentum to the mean flow. Horizontal group velocity, c_{gx}, or the speed of energy propagation, is related to the intrinsic phase speed through

$$c_{gx} = \frac{\partial\omega}{\partial k} = U + c_{gx}^* \tag{9.5}$$

where

$$c_{gx}^* = \frac{\partial\omega^*}{\partial k} = c_{px}^*\left(1 - \frac{k^2}{k^2+m^2}\right). \tag{9.6}$$

For stationary orographic GWs (OGWs), or mountain waves (MWs), the ground-relative phase speed is zero, so the horizontal intrinsic phase speed exactly opposes the background wind (i.e., $c_{px}^* = -U$ via (9.4)). In the hydrostatic limit ($k \ll m$), the horizontal group velocity is the same as the horizontal phase speed. This means the horizontal group velocity is zero for hydrostatic OGWs; therefore, these OGWs primarily stay over the orography that generated them. However, for $k \sim m$, the horizontal group velocity becomes nonzero in the direction of the background flow for nonhydrostatic OGWs, leading to their energy propagation and presence downstream. The speed at which the GW energy propagates in the vertical is described by the vertical group velocity:

$$c_{gz} = \frac{\partial\omega}{\partial m} = \frac{\omega^*}{m}\left(\frac{m^2}{m^2+k^2}\right) \tag{9.7}$$

In the hydrostatic limit and using (9.1) and (9.3), the vertical group velocity becomes

$$c_{gz} \approx \pm\frac{(c_{px} - U)^2k}{N}. \tag{9.8}$$

According to (9.8), the speed at which hydrostatic GWs propagate upward is proportional to the squared intrinsic phase speed and inversely proportional to horizontal wavelength and stability. For a given environment, GWs with shorter horizontal scales propagate upward more quickly, at least until nonhydrostatic effects become important (i.e., when k becomes comparable to m).

9.1.1. Mountain Waves

MWs are GWs generated by stably stratified flow over mountains. MWs are perhaps the most well-known type of atmospheric GWs, being quite visible, quasi-stationary, and the subject of research for at least a century (e.g., Lyra, 1940; Smith, 2019). MWs typically form above and downwind of topographic features. Their presence is often indicated by quasi-stationary lenticular clouds visible from the ground and in satellite images. The properties of mountain waves are determined by the size and shape of the topography as well as by the vertical profiles of wind, temperature, and moisture in the surrounding flow. Linear theory can predict the general features of MWs when the mountain height is small in comparison to the vertical wavelength of the wave.

Conventional linear MW theory assumes MWs are stationary, with zero ground-relative horizontal phase speed or frequency ($c_{px}, \omega = 0$). Such MWs have phase lines (e.g., wave crests) that tilt upstream with height. The dispersion relation (9.2) can be written for stationary waves as

$$m^2 = k^2 \left(\frac{N^2 - (Uk)^2}{(Uk)^2} \right). \quad (9.9)$$

Solutions of flow over small-scale, sinusoidal ridges, where the intrinsic frequency of these parcels is higher than the buoyancy frequency (i.e., $Uk > N$ and hence m in (9.9) is imaginary), decay exponentially with height. Solutions of flow over wider ridges (for which $f < Uk < N$) generate MWs that propagate vertically (Durran, 1986b; Holton, 1992; Smith, 1979).

Linear theory has also been used successfully to describe the basic properties of mountain waves generated by compact, or spatially limited, obstacles (e.g., Durran, 2003; Lyra, 1940, 1943; Queney, 1947; Scorer, 1949; Smith, 1979). Some nonlinear solutions to flow over compact obstacles exist as well (e.g., Huppert & Miles, 1969), though here linear theory is the focus. Flow over an isolated ridge will generate a horizontal spectrum of waves, and each wave component will either propagate upward or decay depending on its m derived from (9.9) for its k. Equation (9.6) shows that for a 2D flow over a wider ridge the (hydrostatic) horizontal group velocity is close to the horizontal phase speed so wave energy is contained above the mountain only, with little propagation downstream of the mountain. However, if the mountain ridge is sufficiently narrow, nonhydrostatic waves with downstream group velocity can be generated. If atmospheric conditions are such that $m^2 > 0$ near the surface, but becomes negative further aloft, MWs are reflected and a train of trapped lee waves may be generated downstream of the forcing (Scorer, 1949). This can happen, for example, when there is strong positive vertical shear of the wind, $U_z > 0$, or a sharp reduction in N due to an inversion, over narrow ridges. In three-dimensional flows over obstacles with finite scales in both horizontal dimensions, oblique modes (those at an angle between 0° and 90° to the incident wind) can also propagate downstream, even in the hydrostatic limit (Jiang et al., 2019; Smith, 1980; Sato et al., 2012). This contributes to the appearance of broad, large-amplitude perturbations often seen in satellite data significantly downwind of large orographic features such as the Antarctic Peninsula and the terminus of the Andes (e.g., Alexander et al., 2009; Hoffmann et al., 2013; Hoffmann et al., 2016). High-resolution, three-dimensional simulations have been critical in developing realistic representations of MW generation and have shown that three-dimensional dispersion spreads the gravity wave energy, reducing the amplitude with increasing height relative to only vertically propagating MWs (Eckermann et al., 2015a).

In addition to upward propagating mountain waves, flow in a mountainous region can produce low-level blocking upstream and downslope windstorms downstream. Low-level blocking occurs when the mean flow does not have enough kinetic energy to traverse an obstacle and either stops upstream or diverges around the obstacle (Hughes et al., 2009; Lin & Wang, 1996; Pierrehumbert & Wyman, 1985; Smolarkiewicz & Rotunno, 1990). Blocking can significantly impact the waves generated, the orographic drag on the flow, and precipitation patterns. The Froude number, $F_r = \frac{U}{Nh_m}$, where h_m is the height of the mountain, in large part determines whether blocking will occur (Lin & Wang, 1996; Pierrehumbert & Wyman, 1985). When $F_r > > 1$, the flow is said to be linear. A flow in this regime has enough kinetic energy to easily traverse the mountain, and dynamic wave amplitudes are small (i.e., $u' < < U$). When $F_r \sim 1$, the validity of linear theory is questionable and nonlinear terms become important. When $F_r < 1$, the flow is strongly nonlinear and low-level flow may not traverse the mountain, becoming blocked upstream or diverting around the obstacles (Drazin, 1961; Leo et al., 2016; Sheppard, 1956). The evolution of nonlinear low-level flows depends on obstacle aspect ratios (e.g., Miranda & James, 1992; Olafsson and Bougeault, 1996), orientation, and shape (Phillips, 1984). In this situation, the pressure force by the atmosphere on the orography and the corresponding drag by the orography back on the low-level flow in part due to the nonlinear blocked flow and in part due to a vertically propagating MW response in the atmosphere a bit further aloft. When orographic blocking occurs, the momentum flux of the vertically propagating MWs is reduced relative to what linear theory alone predicts.

Downslope windstorms can exist on a downstream side of a mountain, in particular ones with steep lee-side slopes, with wind gusts that can exceed 50 m s^{-1} (Bacmeister & Pierrehumbert, 1988; Clark & Peltier, 1977; Durran, 1990; Durran, 1986a; Lilly, 1978; Lilly & Klemp, 1979; Peltier & Clark, 1979; Smith, 1985). In addition, flow configurations analogous to hydraulic trans-critical flows (Durran, 1986a; Smith, 1985) can form leading to significantly enhanced drag on the low-level flow. These flows are responsible for many of the named downslope winds that occur around large mountain ranges, e.g., Chinooks, Santa Anas, Föhns, and Zondas. Nonlinear two-dimensional flow over obstacles has been investigated extensively over several decades (e.g., Bacmeister & Schoeberl, 1989; Clark & Peltier, 1977; Clark & Peltier, 1984; Durran & Klemp, 1983; Durran, 1986a; Durran & Klemp, 1987; Farmer & Armi, 1999; Lott, 1998; Winters & Armi, 2014). These studies have led to important insights about stability, hydraulic analogs, downslope winds, and time dependence in mountain wave flows. Results from 3D numerical simulations were explicitly used in developing more complete orographic gravity wave drag parameterization schemes for global models (e.g., Lott & Miller, 1997; Scinocca & McFarlane, 2000; Webster et al., 2003).

In addition to drag from mesoscale OGWs, another important source of low-level orographic drag in the atmosphere is turbulent orographic form drag (TOFD). This is drag produced in the boundary layer by small obstacles (<5 km). In contrast to OGW, this drag is produced in all stratification regimes but does not carry momentum flux out of the boundary layer (e.g., Beljaars et al., 2004).

Figure 9.1a and b shows the vertical and horizontal properties of mountain waves generated over an isolated obstacle. In this case, the MWs forced by the topography are linear and approximately hydrostatic. These hydrostatic waves primarily stay over the idealized mountain below (Figure 9.1b). Wave amplitudes grow with height, resulting in breaking in the upper atmosphere above $z \sim 60$ km. Figure 9.1a shows MW perturbations in the horizontal plane at the height of 65 km. Strongest perturbations are right over the obstacle and show a bow-shaped structure with decaying amplitudes away from the obstacle in the direction perpendicular to the mean flow. At 65 km in this particular simulation, gravity wave breaking has already begun.

Supplementary Animation 1 (http://www.cgd.ucar.edu/staff/jrichter/animations.html) shows the time evolution of MW generation shown in Figure 9.1a and b. Gravity waves form almost immediately over the obstacle and then propagate upward. Smaller-scale MWs appear first, within ~5 hours, at mesospheric altitudes, consistent with (9.8). The MW at $z = 65$ km grows in amplitude and scale with time, as longer waves build in having had enough time to reach this altitude. By ~15 hours into the simulation, the MW field begins to break, in this case via wave overturning and static instability as no turbulence parameterization was used. The entire wave field quickly dissipates after ~24 hours into the simulation, in part due to wave breaking and also in part due to cessation of cross-barrier flow at $t = 24$ hours. The complexity of the wave breaking is well visualized by the right-hand panel of Supplementary Animation 1 that shows a horizontal cross section through the wind field at 65 km. Up to 15 hours into the visualization, coherent wave crest/troughs are present throughout the model domain. After breaking begins, turbulent features at the grid scale become apparent and secondary waves are generated,

Figure 9.1 Depiction of three dominant sources of GWs: (a, b) orographic, (c, d) convective, and (e, f) frontal. Source: Corresponding animations can be found at UCAR/http://www.cgd.ucar.edu/staff/jrichter/animations.html/ last accessed March 18, 2022. (a, b) 3D simulation of orographic gravity waves by a 200-km-wide, 1000-m-high isotropic compact-cosine mountain using the WRF model. Bottom: Zonal winds (gray shaded) and isentropes (contoured) are shown in the x–z slice through the middle of the 3D domain. Top: Zonal winds (gray shaded) are shown in an x–y slice through $z = 65$ km. The single contour shows the spatial extent of the mountain. Both panels are from 18.33 hours after the start of cross-mountain flow, just after wave breaking has begun. Buoyancy frequency, N, is 0.02 s^{-1}, and the constant environmental zonal mean wind of 30 m s^{-1} was specified. A Rayleigh damping layer starts at $z = 70$ km. The WRF setup here is a 3D extension of that described by Kruse and Smith (2018). (c, d) 3D simulation of convectively generated gravity waves using the Complex Geometry Compressible Atmospheric Model (CGCAM) (Felten & Lund, 2006). Latent heating is used as a proxy for convection. Top panel shows a cross section at $z = 85$ km and 1.755 hours into the simulation, whereas the bottom panel shows a vertical cross section through the center of the domain at the same time. Shading indicates vertical velocity perturbations. The solid thin line depicts the background zonal mean wind profile. *Source:* Adapted from Heale et al. (2020). (e, f) Gravity waves simulated by a high-resolution idealized weak moist baroclinic wave simulation using the WRF model in Wei and Zhang (2014). (e) The horizontal view of the simulated 1-km temperature (yellow lines; contour interval is 5 K), 7-km dynamic tropopause where potential vorticity equals 1.5 PVU (turquoise lines), 8-km horizontal wind (black lines; contours at 40, 45, 50, and 55 m s^{-1}), and 12-km horizontal divergence (blue lines, positive; red lines, negative; contour interval is 2.0×10^{-6} s^{-1}; range is between -1.2×10^{-5} and 1.2×10^{-5} s^{-1}; zero value omitted). (f) The vertical cross section along the green line in (e) for the simulated potential temperature (yellow lines; contour interval is 5 K), dynamic tropopause where potential vorticity equals 1.5 PVU (turquoise lines), horizontal wind (black lines; contours at 30, 35, 40, 45, 50, 55, 60, and 65 m s^{-1}), and horizontal divergence (blue lines, positive; red lines, negative; contour interval is 2.0×10^{-6} s^{-1}; range is between -1.2×10^{-5} s^{-1}and 1.2×10^{-5} s^{-1}; zero value omitted). Source: Courtesy of J. Wei.

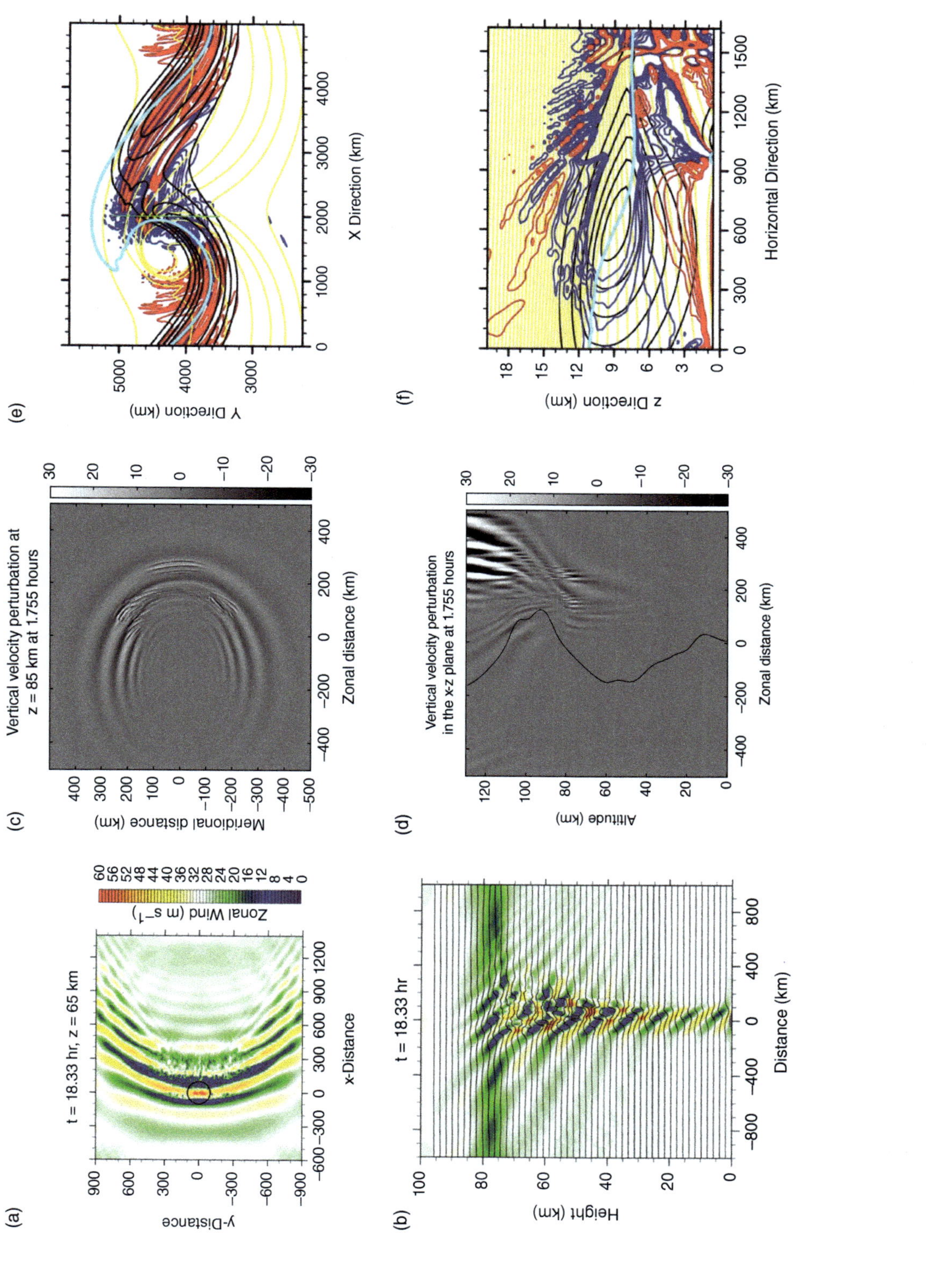
(a)
t = 18.33 hr, z = 65 km
y-Distance
x-Distance
Zonal Wind (m s−1)
(b)
t = 18.33 hr
Height (km)
Distance (km)
(c)
Vertical velocity perturbation at
z = 85 km at 1.755 hours
Meridional distance (km)
Zonal distance (km)
(d)
Vertical velocity perturbation
in the x-z plane at 1.755 hours
Altitude (km)
Zonal distance (km)
(e)
Y Direction (km)
X Direction (km)
(f)
z Direction (km)
Horizontal Direction (km)

which appear throughout the doubly periodic domain. Supplementary Animation 2 shows the characteristics of the gravity wave field at various altitudes at 18.33 hours into the simulation. The visualization shows largest wave amplitudes within 200 km from the center of the obstacle in the cross-wind direction, with waves extending in a quasi V shape away from the obstacle, if looking in the direction of the mean wind at the wave field. Hence, most of the perturbations associated with the flow over the obstacle are directly over the obstacle and in the bow-shaped region downwind and away from the obstacle. Wave breaking begins at 47 km in this particular simulation, intensifying with altitude.

9.1.2. Convectively Generated Gravity Waves

Moist convection is another prominent source of GWs and the dominant GW source in the tropics. Two-dimensional and three-dimensional numerical simulations of convection reveal that convection excites a broad spectrum of waves with horizontal wavelengths between tens and hundreds of kilometers, and periods from a few minutes to several hours (e.g., Alexander et al., 1995; Alexander & Holton, 1997; Piani et al., 2000; Piani & Durran, 2001). The vertical wavelengths of convectively generated gravity waves can vary between several kilometers up to 40 km (e.g., Alexander & Holton, 2004). Convectively generated gravity waves are rarely symmetric in the horizontal plane, and typically have a preferred propagation direction, which is determined by the vertical structure of the horizontal wind within which the convection and GW generation occur (Beres et al., 2002).

Generation of gravity waves by convection is a complex nonlinear process but has been described via three linear mechanisms, all of which are ultimately a response to latent heating:

(1) *Thermal or diabatic forcing*: In this mechanism, temporal and spatial variations of convective heating produce perturbations in potential density that force a spectrum of GWs. This mechanism was found to be of primary importance by Bretherton (1988), Lin et al. (1998), Chun and Baik (1998), and Pandya and Alexander (1999). Based on 2D linearized governing equations, the dominant characteristics of the spectrum of convectively generated GWs by thermal forcing are determined by the vertical scale of the heating region and the horizontal wavenumber and frequency distribution of the heating (Holton et al., 2002). Beres (2004) has shown that the three-dimensional wave forcing problem can be treated as a multiple two-dimensional problem, and the gravity wave spectrum in a given azimuthal direction depends on the heating and mean wind projection in that direction.

(2) *Mechanical oscillator*: Oscillating updrafts and downdrafts about a level of neutral buoyancy (LNB) (e.g., the top of the boundary layer and tropopause) perturb the stably stratified atmosphere at and above the top of the convective motions. For deep, moist convection, updrafts often overshoot their LNBs (i.e., the tropopause), potentially forcing vertically propagating GWs. Transience in convective updraft strength can perpetuate this forcing. These oscillations can produce upward propagating waves in a manner similar to a mechanical oscillator in a stratified fluid (Clark et al., 1986; Fovell et al., 1992). Some studies have found this oscillator mechanism to dominate GW generation (e.g., Lane et al., 2001), whereas others found thermal and mechanical forcing terms are both equally important (Song et al., 2003). In the set of nonlinear equations, flux terms couple the thermodynamic and momentum equations, and the nonlinear momentum term can oppose the heat source giving some apparent cancellation (Chun et al., 2008; Pandya & Alexander, 1999).

(3) *Obstacle effect or moving mountain*: In this mechanism, the top of a convective element acts as a barrier to the background mean flow, producing upstream propagating waves in a manner similar to flow over a mountain (Clark et al., 1986; Pfister et al., 1993). Vertical wind shear, at least near the tops of the convection, is required for this mechanism. The nonzero net vertically propagating GW momentum flux is fundamentally derived from the convective momentum fluxes, as convective updrafts vertically transport air parcels with low (high) momentum into regions of high (low) above. Waves generated by the obstacle effect are not stationary relative to the ground as in the case of mountain waves but rather have horizontal phase speeds similar to the horizontal speed of the convection.

Gravity waves generated by a thermal forcing (proxy for convection) are shown in Figure 9.1c and d and Supplementary Animation 3. Figure 9.1c and the left panel of Animation 3 show that gravity waves propagate away from the convective source in nearly concentric circles spreading away from the source as they propagate upward. In an x–z cross section (Figure 9.1d and the right panel of Animation 3), convectively generated gravity waves form a fan-like structure with a broad spectrum of wave phase speeds and horizontal and vertical wavelengths. In this particular simulation, westward propagating waves are not as apparent at $z = 85$ km due to much slower vertical group velocity (9.8) and dissipation in the stratosphere, resulting in asymmetry in Figure 9.1c. The eastward propagating waves have higher intrinsic phase speeds in the stratosphere and lower mesosphere, propagate upward more quickly, and break in the mesopause region where the mesospheric wind shear brings the environmental winds close to the phase speeds of these eastward propagating waves (Figure 9.1d).

9.1.3. Gravity Waves Generated by Fronts and Jets

Fronts and jets are another major source of GWs in the atmosphere and a significant source of GWs in midlatitudes. GWs from front/jet systems have been observed on numerous occasions (e.g., Plougonven et al., 2003; Sato 1994; Uccellini & Koch, 1987; Wang & Geller, 2003; Zhang & Yi, 2005, 2008) and simulated with numerical models (Jin, 1997; Kaplan et al., 1997; Powers & Reed, 1993; Schmidt & Cotton, 1990; Zhang & Fritsch, 1988; Zhang & Koch, 2000; Zhang et al., 2001, 2003). Jets/fronts generate mesoscale and low-frequency gravity waves (inertia-gravity waves), for which the rotation of the Earth has an influence and the frequency is of the order of the Coriolis parameter, f. Wei and Zhang (2014), for example, examined gravity waves in moist baroclinic jet-front systems with varying degrees of moisture and found the gravity waves to have horizontal scales between 50 and 500 km, vertical scales between 1 and 6 km, and frequencies of $(1–15) \times 10^{-4}\ s^{-1}$ (periods of 1–17 hours). Most intense gravity wave activity has been observed and modeled near the vicinity of the maximum of the jet velocity (strong curvature of the jet) (Plougonven et al., 2003) and in the exit region of upper-tropospheric jet streaks (e.g., Guest et al., 2000; Wu & Zhang, 2004; Zhang et al., 2001, 2003; Zhang, 2004).

The generation mechanisms of gravity waves generated by jet/front systems are still not well understood. Plougonven and Zhang (2014) present a complete review of possible generation mechanisms by such systems. Below, three dry-idealized mechanisms that are thought to be most important are described, two of which form the basis of gravity wave source parameterizations described in subsequent sections:

(1) *Geostrophic adjustment* (see section 3.1 of Plougonven & Zhang, 2014): In this mechanism, GWs are generated and radiated away as some initially imbalanced flow comes back into balance (Rossby, 1938). This concept does not address the cause of the initial imbalance, but only considers the GW emission during the evolution back toward geostrophic balance. "Spontaneous balance adjustment" framework (Zhang, 2004) is further developed to generalize Rossby's classical geostrophic adjustment theory. This mechanism is responsible for emission of large-amplitude inertia-gravity waves in regions of strong horizontal curvature of the wind where the flow becomes unbalanced (Fritts & Luo, 1992; Luo & Fritts, 1993). The residual of the nonlinear balance equation is found to be a useful quantity in diagnosing regions of flow imbalance and predicting regions of wave generation (Zhang et al., 2000, 2001; Zhang, 2004). Readers are also referred to the review by Ruppert et al. (2022) on this topic.

(2) *Spontaneous adjustment emission* (see sections 3.2–3.4 of Plougonven & Zhang, 2014): In this mechanism, well-balanced flow more continuously radiates GWs during the course of its near-balanced evolution. An early example of such physics was presented by Lighthill (1952), where acoustic waves are generated within fluids by turbulent motions. Spontaneous adjustment emission has successfully replicated the salient characteristics of gravity waves emitted from vortices and jets in the shallow water model (e.g., Ford, 1994a; Ford, 1994b; Sugimoto et al., 2008) and in a stratified fluid (e.g., Plougonven & Zeitlin, 2002; Schecter, 2008). Transient generation in sheared disturbances describes how the evolution of potential vorticity anomalies in a sheared flow leads to a transient generation of gravity waves, which has been discussed in horizontal (Vanneste & Yavneh, 2004) and vertical shear (Lott et al., 2010).

(3) *Shear instability*: In this mechanism, gravity wave emission occurs via nonlinear interaction between Kelvin-Helmholtz (K-H) instability and propagating modes (Bühler et al., 1999; Scinocca & Ford, 2000). Shear instability is usually considered by neglecting the Coriolis effect. This mechanism of wave generation can occur in very intense shear layers near the surface or at upper levels, above tropopause jets. Kelvin-Helmholtz instability occurs on very small scales (hundreds of meters in the vertical and tens of kilometers in the horizontal), hence in order for gravity waves to propagate upward, nonlinear emission on the envelope scale (i.e., scale characterizing the extent of K-H instabilities and mean-flow influence) must occur (Chimonas & Grant, 1984; Fritts, 1984).

Convection often occurs in association with a frontal system and can provide an additional source of GWs and/or influence the generation of GWs by frontal/jet system. Many earlier studies of GWs generated by fronts focused on dry idealized baroclinic wave simulations (e.g., Zhang, 2004); however, the role of moisture can potentially be very important (Lane & Reeder, 2001; Powers, 1997; Zhang et al., 2001). Complementary to the work of Zhang (2004), Wei and Zhang (2014), using cloud-permitting mesoscale baroclinic system simulations, showed that moisture enhances GW amplitudes and generates additional wave modes in comparison with a dry simulation. Furthermore, based on the study of GW spectral characteristics using multidimensional discrete Fourier transforms, Wei et al. (2016) further demonstrated that the dry jet/front GW source generates a relatively narrow and less symmetric power spectrum centered around lower phase speeds and horizontal wavenumbers, whereas the moist gravity wave source generates a broader and more symmetrical power spectrum, with a broader range of phase speeds and horizontal wavenumbers. Generation of GWs in frontal systems with a lot of moisture is still a subject of recent research, and diabatic forcing could be a more important generation mechanism. The role of moisture in producing significant momentum fluxes from front/jet systems has been emphasized by many studies (e.g., Holt et al., 2017; Plougonven et al., 2015; Wei et al., 2016).

GWs generated by idealized weak moist baroclinic jet-front systems are illustrated in Figure 9.1e and f and Supplementary Animation 4. As the upper level baroclinic jet develops, with an accompanying deepening surface cyclone, the lower-stratospheric (12-km altitude) flow shows a well-recognized pattern of convergence upstream of the trough and divergence downstream of the trough (Figure 9.1e and the left panel of Animation 4). GW generation begins first with weak amplitude mesoscale GWs appearing in the jet entrance region upstream of the upper level trough, along the surface warm front. As the system matures, mesoscale GW generation primarily occurs in the jet exit region, downstream of the trough and above the surface frontal system (Figure 9.1f and the right panel of Animation 4). In addition, based on a series of four-dimensional ray-tracing experiments, Wei and Zhang (2015) investigated the propagating wave characteristics and the potential source mechanisms of several identified lower-stratospheric GWs in this particular idealized simulation. It was further demonstrated that moist convection may force new wave modes, modify/enhance the existing dry jet/front wave modes through latent heat release, and/or modify the new/existing waves through modification of large-scale flow.

9.1.4. Secondary Wave Generation

An understudied aspect of the GW lifecycle and a likely underestimated GW source is secondary wave generation (e.g., Bacmeister & Schoeberl, 1989), which is generation of GWs by the momentum deposition of a dissipating primary GW. While the research on secondary GW generation, propagation, dissipation, and impact is relatively nascent, there is a growing body of literature on the topic. Currently, at least two secondary GW generations mechanisms have been explored:

(1) Large-scale secondary GW generation by localized momentum deposition, or body forces, resulting from a primary GW. The body forcing and secondary GW generation occur at scales larger than the horizontal scales of primary GW activity. The body forcing from these primary GWs can be both dissipative (e.g., where small-scale instabilities dissipate the wave) and nondissipative (e.g., sometimes reversible forcing on a layer as a GW transiently propagates into and out of it). The latter effect is sometimes referred to as "self-acceleration" (e.g., Fritts & Dunkerton, 1984). Dissipative secondary GW generation is the focus of Vadas et al. (2003, 2018). Wilhelm et al. (2018) have investigated the nondissipative, resonant radiation of mesoscale inertia-gravity waves by a horizontally as well as vertically confined submesoscale gravity wave packet that propagates vertically. It is known from long-short-wave interaction theory (Tabaei & Akylas, 2007; Van den Bremer & Sutherland, 2014) that such a packet of small-scale waves is able to generate a mean flow consisting of mesoscale wave structures connected to a resonance mechanism, wherein the vertical phase velocity of the emitted long waves match the vertical group velocity of the small-scale gravity wave packet, which acts as a traveling wave source.

(2) Secondary GWs are also generated on scales smaller than the horizontal scales of the primary GW activity. Numerous small-scale instabilities can take place during GW breaking and dissipation. Nonlinear dynamics can transfer energy from these small-scale instabilities to larger scales that can force a propagating GW (Franke & Robinson, 1999). Instabilities can occur inhomogenously within a wave field (e.g., in particular phases where stratification is reduced, or unstable, where wind shear is increased), also forcing waves at scales smaller than wave field (Bossert et al., 2017, Bossert et al., 2018; Chun & Kim, 2008; Fritts et al., 2006, Heale et al., 2017; Holton & Alexander, 1999; Lane & Sharman, 2006; Satomura & Sato, 1999).

At this time, the understanding of secondary GWs is limited; their contribution to the energy and momentum budget of the atmosphere is not well understood, nor represented in weather and climate models that do not resolve the relevant mechanisms.

9.1.5. Gravity Wave Propagation and Dissipation

GW characteristics change as they propagate through an atmosphere with changing buoyancy frequency and mean wind, as suggested by (2). As a GW propagates upward through wind shear that is opposite to the wave's propagation direction, the vertical wavelength of the wave will increase. Similarly, the vertical wavelength of the wave will decrease when the wave propagates through wind shear in the same direction as the wave propagation. If wind in the direction of the wave propagation changes so that $\omega^* = \omega - Uk$ reaches zero, a "critical level" is reached. This is the level at which the horizontal phase speed, c_x, matches the background wind. As a GW approaches a critical level, linear theory predicts the wave vertical wavelength goes to zero, zonal wind perturbations go to infinity, and the wave steepens and overturns. Prior to reaching the critical level, however, instabilities (e.g., K-H and static instability) are typically triggered that dissipate the wave, ending its vertical propagation. This is illustrated in the top panel of Figure 9.2, which shows a wave packet propagating upward and in the positive x direction into a region of increasing zonal mean wind

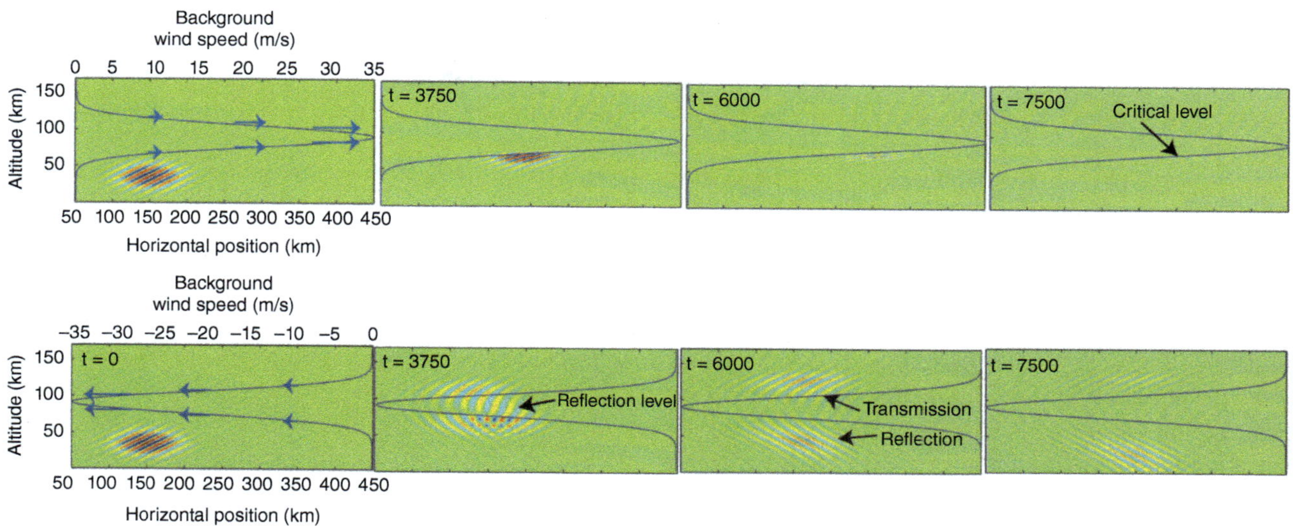

Figure 9.2 Simulated propagation of a gravity wave packet through two different vertically varying background winds (top and bottom) at four different times (from left to right) using the MAGIC model (Snively & Pasko, 2008). The top panel shows a packet propagating in the same direction as the background wind flow (indicated by the black line), which approaches a critical level at ~80 km and is then absorbed into the mean flow. The bottom panel shows the same packet propagating against the background wind flow. In this case, the wave encounters a turning point (reflection) level and partial transmission and reflection occur. The gray shading indicates the perturbation horizontal wind associated with the gravity wave packet. Times of the simulation in seconds is depicted in top left corner of the panels. Source: Reproduced from Heale and Snively (2015)/John Wiley & Sons.

(in the same direction as the wave propagation). As the wave packet propagates upward, its vertical wavelength decreases as it approaches the critical level near $t = 7500$ and altitude of ~75 km, and the GW packet dissipates.

When such a wave packet propagates into a region where the wind speed through the wave increases, the intrinsic phase speed and frequency both increase. According to (2), the wave packet can reach a level where the intrinsic frequency matches the environmental buoyancy frequency and the vertical wavelength approaches infinity. This level is referred to as a turning or reflection level as at least partial GW reflection occurs here. Above the reflection level, m is imaginary and wave evanesces, or decays, exponentially with height. This situation is illustrated in the bottom row of Figure 9.2. Where a wave packet is propagating upward and in the positive x direction with shear against the direction of propagation above the wave. As the wave packet approaches the reflection level, the phase lines become more vertically oriented. At t = 3750 s, both upward and downward portions of the wave packet superpose to produce the wave field below the reflection level. Above the reflection level, nearly vertical phase lines are seen, with a portion of the wave perturbations decaying with height. The wave does not decay above the reflection level immediately. In this case, the wave maintains enough amplitude to "tunnel" through this evanescent layer into the slower winds aloft that allow vertical propagation once again (e.g., Mixa et al., 2021). In this case, part of the wave packet was reflected at the reflection level, while part was able to tunnel through the evanescent layer and continue propagating upward.

Another important characteristic of GWs is how their amplitudes grow with height, even in an environment that is constant in height. Linear, 2D GWs in a horizontally uniform background have a constant vertical flux of horizontal momentum (e.g., $\overline{\rho u' w'}$) with height (e.g., Eliassen & Palm, 1960). As density decreases exponentially with height, this requires the GW perturbations to grow exponentially with height. Similar to encountering a critical level, eventually, wave amplitudes become large and trigger instabilities that dissipate the wave.

9.1.6. Gravity Wave Impacts

Both energy and momentum are extracted from the mean flow, or via interaction with topography, when GWs are generated. Propagating GWs transport this energy and momentum, depositing them wherever the waves are dissipated. While both are important, the energy extracted, transported, and deposited is typically neglected. The momentum, however, is not neglected and represents an important forcing on the background horizontal flow.

MWs attain their momentum flux and drag through interactions with orography. Vertically propagating MWs induce positive pressure perturbations upstream of the mountains and negative perturbations downstream during generation, resulting in a net horizontal pressure force by the atmosphere on the mountains. The mountains exert an equal and opposite force on a lowest layer of the atmosphere. This low-level layer than perturbs a layer immediately above, and a pressure drag by the layer above is exerted on this low-level layer. If no MW dissipation occurs, the forces by the mountain and the layer above on the low-level layer cancel, resulting in no net horizontal force on this layer. Layers increasingly perturb layers above, propagating the MW upward and fluxing atmospheric momentum downward. When the MW dissipates, the top of the dissipation layer is no longer displaced by the primary MW. Atmospheric momentum above the dissipation layer no longer balances the momentum fluxed out the bottom of this layer, and a net force is exerted on the flow. In short, MW drag is exerted wherever the MW is dissipated is fundamentally a reaction to pressure forces by the atmosphere on mountains.

While the forcing exerted by nonorographic GW dissipation technically comes out of thin air, it does come from somewhere. In general, nonorographic GWs attain the momentum and energy they flux from the flow in which they were generated and then spatially redistribute the energy and momentum within the atmosphere. Wave fields generated by nonorographic GW fields tend to be more complex than MWs, having both a spectrum of spatial scales and a spectrum of phase speeds and intrinsic frequencies. For example, in the absence of environmental shear, GWs radiate away from moist convection symmetrically in all directions. While the eastward-propagating, westward-propagating, northward-propagating, and southward-propagating GWs all flux momentum, the net vertical flux of zonal and meridional momentum by these GW beams cancel, resulting in net zero momentum flux. While the net momentum fluxes are zero at the source level, wind shear aloft forces the different beams of GWs to encounter critical levels at different altitudes and exert forces on flows in different directions at different altitudes. Such a phase speed spectrum is critically important in forcing the quasi-biennial oscillation (QBO) and semiannual oscillation (SAO) in the tropical stratosphere.

While the previous discussion suggests convective GWs are symmetric, they rarely are in reality, having a preferred propagation direction and a net momentum flux due to the fact that convection often occurs in environments with shear. Wind shear both allows there to be flow relative to the tops of the convection and allows convection to transport low-momentum air from low lower levels and act as a barrier to the flow across the top of convection. The convective momentum fluxes are likely important in determining a portion of the convective GW spectrum.

GWs generated by jets and fronts derive the energy and momentum they flux from the various imbalances that are produced as the systems evolve in time. The GWs are generated by these imbalances and adjust the flow in the direction of balance.

Quantitatively, the influence by GWs on the zonal momentum appears when deriving an equation for a background or, traditionally, a Reynolds-averaged horizontal wind. These equations that govern a larger-scale horizontal flow contain the following term, representing the influence of the vertical convergence of the vertical flux of horizontal momentum on the background horizontal flow:

$$\text{GWD} = -\frac{1}{\overline{\rho}}\frac{\partial}{\partial z}(\overline{\rho u'w'}, \overline{\rho v'w'}) \tag{9.10}$$

where $\overline{\rho}$ is the background atmospheric density, u', v', and w' are the horizontal and vertical perturbation velocities, and the overline operator, $\overline{(.)}$, is a linear background operator of some kind (e.g., a low-pass filter) used to define the background flow. This term is defined as GW drag (GWD) and has units of acceleration or drag force per unit mass. While momentum fluxes and their divergences are obviously relevant to forcing of the background momentum, pseudomomentum is more relevant to GW dynamics in many current GW parameterizations, and its divergence is sometimes used to quantify GW influences on the mean flow (see Wei et al. (2019) for a comparison of these approaches).

The momentum fluxed and deposited by GWs is significant in Earth's general circulation, especially in the stratosphere, mesosphere, and lower thermosphere. Since GWs grow in amplitude with height, and background horizontal momentum (e.g., $\overline{\rho u}$) decreases with height, the importance of GWs in the general circulation tends to increase with height. In the lower troposphere to midtroposphere, GWs have less influence on large-scale momentum but are still important in initiating convection and convective organization (Bretherton & Smolarkiewicz, 1989; Fovell et al., 2006; Lac et al., 2002; Lane & Zhang, 2011; Pandya & Durran, 1996; Ruppert et al., 2022; Shige & Satomura, 2000; Su & Zhai, 2017). In the upper troposphere and lower stratosphere, GWD does become important in the zonal mean climate (Bacmeister, 1993; Butchart et al., 1998; Palmer et al., 1986). In the tropical stratosphere, gravity waves contribute 50%–90% of the forcing of the quasi-biennial oscillation (Alexander & Ortland, 2010; Bushell et al., 2020; Geller et al., 2016; Kawatani et al., 2010; Richter et al., 2014). In climate models, GW parameterizations largely determine the period and frequency of the QBO, and most models without a gravity wave parameterization will not be able to produce a QBO (Butchart et al., 2018; Geller et al., 2016; Giorgetta et al., 2006;

Richter et al., 2014). The representation of the QBO in climate models is becoming more important as impacts of the QBO on the tropospheric variability are becoming clearer (Giorgetta et al., 1999; Wang et al., 2018a, b; Yoo & Son, 2016). GWs also contribute to the driving of the semiannual oscillation (Hitchman & Leovy, 1988; Ray et al., 1998; Richter & Garcia, 2006) and the mesospheric semiannual oscillation (MSAO) (Dunkerton, 1982).

In the extratropical stratosphere, GWs provide a portion of the driving of the Brewer Dobson circulation, especially during the spring-to-summer transition season in each hemisphere (Alexander & Rosenlof, 1996; Alexander & Rosenlof, 2003; de la Camara, et al., 2016; Okamoto et al., 2011; Rosenlof, 1996). MW drag peaks in the upper stratosphere and mesosphere, (e.g., Kruse, 2020), and together with other extratropical GWs, largely from fronts and jets, control the state of the polar stratospheric temperatures and strength of stratospheric polar night jet (Boville 1991; Garcia & Boville, 1994). Inadequate representation of GW drag in general circulation models (GCMs) can lead to a cold bias of southern winter stratosphere temperature in the polar region (e.g., Austin et al., 2003; Eyring et al., 2007; McLandress et al., 2012). One of the largest effects that GWs have on the atmosphere occurs in the mesosphere. Nonorographic GWs dominate GW drag here and deposit net westward momentum in the winter mesosphere and net eastward momentum in the summer hemisphere causing the reversal of the zonal mean jets and driving a mean transport circulation from the summer to winter hemisphere, leading to a warm winter and cold summer mesopause (Garcia & Solomon, 1985; Holton 1982, 1983; Lindzen, 1981).

Waves that are not filtered by critical levels, reflected by turning levels, and do not break in the middle atmosphere will be dissipated in the thermosphere by increasing molecular viscosity and thermal conductivity or, in polar regions, ion drag. The damping rate is inversely proportional to the vertical wavelength (Walterscheid & Hickey, 2011) so waves with larger vertical wavelengths (and phase speeds) can propagate higher into the thermosphere before dissipating (Heale et al., 2014, 2018; Vadas & Fritts, 2005; Vadas, 2007; Vadas & Nicholls, 2012). However, spectra of wave packets that propagate into the thermosphere from below often evolve from longer to shorter vertical wavelengths in time as a result of dispersion. This occurs because the longer, faster vertical wavelength components reach the thermosphere and are dissipated first, while the shorter, slower components arrive later (Heale et al., 2014, 2018). The dissipation of these waves produces local body forcing and heating/cooling of the thermosphere (Hickey et al.; 2011; Miyoshi et al., 2014; Vadas et al., 2014; Yiğit & Medvedev 2009; Yiğit et al., 2009) and will also generate secondary waves (Vadas et al., 2018). It is suggested that wave dissipation in the thermosphere leads to a drag that opposes the mean zonal winds and is stronger at high latitudes (Miyoshi et al., 2014) and in the winter hemisphere (Yiğit et al., 2009). Thermospheric dissipation of waves from deep convective sources can also lead to in situ generation of planetary-scale diurnal and semidiurnal tides (Vadas et al., 2014). Waves that reach the thermosphere can also couple to the ionosphere, producing traveling ionospheric disturbances (e.g., Azeem et al., 2017; Liu & Vadas, 2013; Yu et al., 2017) and ion outflow (Burleigh et al., 2018). Compared to other regions of the atmosphere, the impacts of waves in the thermosphere are still unknown and require further investigation.

9.2. REPRESENTATION IN LARGE-SCALE MODELS

9.2.1. History and Basic Components

The need for representing GWs in general circulation models began with the recognition of "missing drag" in such models. Without explicit drag in the middle atmosphere, if the atmosphere was in radiative equilibrium, the polar night jet in the stratosphere in models would be much stronger than observed and the winter (summer) mesopause would not be warm (cold). Early GCMs that extended to the stratosphere and mesosphere often used Rayleigh friction to provide a crude parameterization of the effect of breaking gravity waves in the mesosphere and were able to reproduce the observed features of the zonal mean wind and temperature in the stratosphere and mesosphere (e.g., Boville, 1986). First implementations of GW parameterizations in GCMs focused on representing GWs generated by orography (Boer et al., 1984; McFarlane, 1987; Palmer et al., 1986). Subsequently, parameterizations in GCMs were extended to include representation of nonorographic GWs (Alexander & Dunkerton, 1999; Fritts & Lu, 1993; Hines, 1997a, 1997b; Medvedev & Klaasen, 1995; Rind et al., 1988; Warner & McIntyre, 2001).

GW parameterizations in GCMs have three basics components: (1) specification of waves at the source levels, (2) wave propagation with height, and (3) wave dissipation, from which momentum deposition to the mean flow is estimated. Parameterization of MWs is traditionally distinguished from nonorographic GWs, with the horizontal phase speed of orographic MWs assumed to be zero, while nonorographic GWs have a spectrum of phase speeds. Hence, MW drag has historically been treated separately, and this distinction generally remains today. The source spectra of nonorographic waves initially were specified to be uniform in space and time in the first implementations of parameterizations in GCMs. Spatially uniform sources continue to be standard practice in many

GCMs (e.g., Adachi et al., 2013, Davini et al., 2018; Scinocca et al., 2008). However, in recent years, separate source spectrum parameterizations have been developed for waves generated by convection and fronts, and these are described in section 12.2.3. These source spectra parameterizations replace the arbitrary/globally defined nonorographic source spectra in GW parameterizations even though they still carry large uncertainties.

9.2.2. Description of GW Parameterizations

The primary function of GW parameterizations as currently applied in global models is to compute the wave-driven force on the mean flow. The mean flow in this context is the grid-box mean, and the waves are meant to represent subgrid, or otherwise unresolved, GW anomalies. The essential ingredients of GW parameterizations include specification of input parameters describing the gravity wave sources, estimation of the wave dissipation as a function of height, and output vertical profile of the vector momentum forcing via (9.10). The wave dissipation profile implies an energy dissipation rate profile and, in some models (e.g., the Whole Atmosphere Community Climate Model; Gettelman et al. 2019), this energy dissipation is tied to vertical mixing of trace gases. Conversion of energy dissipation to vertical mixing is not direct, however, because the mixing may be more or less perpendicular to isentrope and tracer gradients (Lelong & Dunkerton, 1998) and so must be scaled by an uncertain Prandtl number, with values $O \sim 1$–100 (Smith & Brasseur, 1991).

GW parameterizations typically start with some specification of the wave stress or momentum flux along with the wave propagation properties (wavenumbers, phase speeds, and propagation directions) at a source level, which is selected to be somewhere between the surface and the 90 hPa (e.g., near the terrain for MWs and upper troposphere and lower stratosphere for nonorographic GWs). The vast majority of GW parameterizations assume GWs propagate only vertically and instantaneously through the column of atmosphere above the source level. Two notable exceptions are Amemiya and Sato (2016), where 3D GW propagation was accounted for, and Eckermann et al. (2015b), which accounted for lateral spreading of MW activity and how this influenced MW amplitude and breaking levels. Wave dissipation is estimated with a variety of techniques depending on the parameterization. The plane wave assumption is always made, so the flux and force both lie along a specified direction of wave propagation. This direction is specified at the wave source level and is assumed to remain constant through the column until the wave is completely absorbed. Thus, gravity wave parameterizations are one dimensional (vertical), utilizing parameters and model fields that are projected along the direction of wave propagation. The output force is applied to the vector momentum equations by projection onto zonal and meridional directions.

Differences among parameterizations include (a) specification of the sources and (b) assumptions that control the wave dissipation with height. For dissipation, Lindzen's (1981) saturation theory, with modifications formulated by Holton (1982), forms a starting point for most parameterizations currently in use. Here, parameterized waves are treated as individual steady hydrostatic monochromatic plane waves. Using (2), the continuity polarization relation ($k\hat{u} = -m\hat{w}$), and the fact that momentum flux for such idealized plane waves is constant in height, the following nondimensional wave amplitude, which quantifies both nonlinearity and wave steepness, can be derived:

$$\frac{\hat{u}}{|U-c|} = \left(\frac{2\,MF_{\rm src}\,N}{\overline{\rho}|U-c_{px}|^3 k}\right)^{\frac{1}{2}} \tag{9.11}$$

where hatted quantities are the sinusoidal amplitudes. Here $MF_{\rm src}$, k, and c_{px} are constant in height. Changes to this nondimensional GW amplitude in height result from changes in $\overline{\rho}$, N, and $|c^*_{px}| = |U - c_{px}|$. When this nondimensional amplitude is small, the GW is linear and the steepness of the wave, as predicted by this linear theory, is low. As a GW propagates upward, density decreases exponentially, and so wave amplitude increases exponentially. Increased stratification and environmental wind shear that brings the wind closer to the phase speed can both force GW nonlinearity.

When the nondimensional amplitude (9.11) exceeds unity, linear theory predicts the wave will loft dense fluid over light fluid and induce static instability. The saturation hypothesis (Lindzen, 1981) assumes that instabilities (e.g., static, Kelvin-Helmholtz instabilities) continuously and instantaneously prevent the GW from exceeding some nondimensional wave amplitude (e.g., unity for static instability). For example, if a GW is propagating upward in shear that brings the environment closer to its phase speed (i.e., as it approaches a critical level), the wave amplitude grows. Eventually, this critical wave amplitude is reached. Then, the momentum fluxed by the GW is reduced such that (9.11) gives the critical nondimensional wave amplitude. The momentum flux is reduced with height when the wave is saturated, the vertical derivative of which gives the forcing to the mean flow. If the wind shear reverses so that the intrinsic frequency increases and the nondimensional GW amplitude is reduced below the critical amplitude, then the GW once again propagates upward conserving its momentum flux and exerting no force on the mean flow.

Other parameterizations in use in global climate models today make different wave dissipation assumptions.

Alexander and Dunkerton (1999) represent a finely resolved spectrum of discrete, independently treated, monochromatic waves and assumes complete annihilation of an individual wave at its breaking level (Alexander & Dunkerton, 1999). Hines (1997a, 1997b) proposed a "Doppler Spread" mechanism, assuming that nonlinear interactions among waves in the spectrum reshape the spectrum with altitude. (See McLandress (1998) for a concise summary of the application of the Hines (1997a, 1997b) parameterization.) Warner and McIntyre (2001) assume a similar reshaping of the spectrum with altitude but based on the empirical observations of the shape, coupled to Lindzen's wave saturation concept. Both the Hines and the Warner and McIntyre approaches assume a particular vertical wavenumber spectrum shape at the source level. A more in-depth overview of such spectral parameterizations not using the conventional saturation concept above is provided in Medvedev and Yiğit (2019).

9.2.3. GW Source Parameterizations

Orography

Specification of MW sources is rather simple in most GCMs. The first formulations prescribed only a single, monochromatic vertically propagating wave with zero horizontal phase speed (Boer et al., 1984; Palmer et al., 1986; and McFarlane, 1987), and these formulations are still being used in most GCMs. The GW source specifications are based on 2D theory assuming hydrostatic, steady, horizontally uniform flow over an obstacle. Wave amplitudes at the source levels are defined based on a measure of subgrid-scale orographic variance. In early orographic GW parameterizations, the surface stress vector is parallel to and opposite of the mean flow at the lowest level of the model, assuming isotropic unresolved topography.

The first orographic gravity wave drag schemes (e.g., McFarlane, 1987; Palmer et al., 1986) neglected the impact of topographic anisotropy and did not consider drag produced by nonlinear dynamics near the surface (e.g., downslope winds, blocking, and flow splitting), which can lead to large amplification of surface stress and hence be important to general circulation (e.g., Baines & Palmer, 1990; Pierrehumbert & Wyman, 1985; Sandu et al., 2019). Lott and Miller (1997) described an orographic drag parameterization that incorporated the impact of near-surface nonlinearities due to flow diversion around obstacles (also referred to as blocking). The key elements in their approach are schematically illustrated in Figure 9.3. When mountain heights h_m exceed a critical value $F_1\left(\frac{U}{N}\right)$, a portion of the flow is

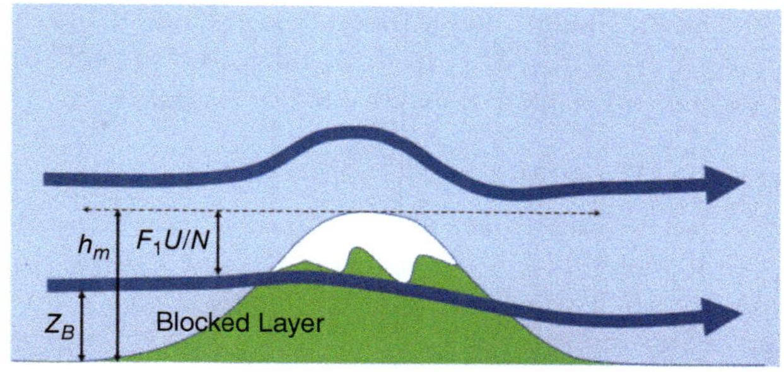

Figure 9.3 Schematic depiction of flow over and around a large mountain. Source: Adapted from Lott and Miller (1997). h_m is the height of the mountain and Z_B is the height of the blocked layer. Blocked flow can develop if the mountain height exceeds $F_1(U/N)$ (see text).

assumed to be diverted or blocked. The depth of this layer is given by

$$Z_B = \text{Max}\left(h_m - F_1\frac{U}{N}, 0\right) \tag{9.12}$$

The parameter F_1, a critical nondimensional mountain height or vertical displacement, is known from numerical and tank experiments to be of order 1 but may depend on obstacle shape. The forcing amplitude for vertically propagating gravity waves in Lott and Miller (1997) is taken to be the full mountain height; however, they suggest that this should be reduced to something approaching $F_1(U/N)$ in three-dimensional flows.

Within the blocked layer, Lott and Miller (1997) assume the drag follows a bluff-body law:

$$\frac{\partial U}{\partial t} \propto -\frac{1}{\bar{\rho}}\frac{\partial(\bar{\rho}\,\overline{u'w'})}{\partial z} = 0.5\,\bar{\rho}\,C_d\,l(z)U|U| \tag{9.13}$$

where C_d is a nondimensional drag coefficient close to 1 and $l(z)$ is the cross-stream length presented by the obstacle. Lott and Miller's (1997) formulation still employs a single wave.

The Scinocca and McFarlane (2000) parameterization employs two, instead of one, vertically propagating waves in order to provide a representation of the azimuthal distribution of momentum flux in the parameterized gravity wave field launched by a "best-fit" elliptical barrier, similarly to Lott and Miller (1997), but with a new way of defining unresolved topography. Scinocca and McFarlane (2000) also extended the representation of low-level drag to include enhancement by the downslope windstorm regimes. These flows are analogous to hydraulic supercritical flow (Smith, 1989) and are thought to be related to near-surface wave breaking (e.g., Bacmeister & Pierrehumbert, 1988; Clark & Peltier, 1977). Three-dimensional numerical studies cited by Scinocca and McFarlane (e.g., Miranda & James, 1992; Olaffson & Bougeault, 1996) suggest that downslope winds with enhanced drag appear for a limited range of mountain heights, roughly for $F_1 < Nh_m/U < 3F_1$. In these flows, the surface drag applied in the scheme is determined by

$$\tau_{\text{sfc}} = (1 + \beta(F))\tau_{\text{lin}} - \tau_c \quad (9.14)$$

where τ_c is the moment flux carried by the freely propagating waves and τ_{lin} is the nominal linear wave drag $\sim\rho NUh_m^2/L$. This enhanced drag is applied via a linearly decreasing momentum flux profile in a layer from the surface to the first breaking level above the mountain. The height of this breaking level is approximated using linear theory. The enhancement factor $\beta(F)$ peaks around $F = Nh_m/U = 1.5F_1$ and has values between 2 and 4 depending on obstacle geometry. This parameterization of form drag can change the direction of the low-level flow to be more parallel to unresolved topographic ridges.

A new approach to orographic GW source parameterization is presented by van Niekerk and Vosper (2021), where they use the spectral, hydrostatic linear MW theory of Garner (2005) and Smith and Kruse (2018). Here, a two-by-two matrix of orographic GW drag coefficients is computed from the 2D Fourier transform of the subgrid-scale terrain. Multiplication of the source-level wind vector with this drag matrix produces a source MW momentum flux vector that takes into account all subgrid scales and orographic anisotropy, eliminating the monochromatic assumption common to all previous MW drag parameterizations, at least at the source. van Niekerk and Vosper (2021) further develop a parameterization for how the elements of the drag coefficient matrix depend on low-level blocking. Initial implementation in the Met Office Unified Model demonstrates this approach does a much better job at keeping total GW drag (resolved + parameterized) constant as grid resolution is changed relative to previous monochromatic parameterizations and improved weather prediction performance.

Convectively Generated Gravity Waves

Source spectra parameterizations for convectively generated gravity waves were developed on the basis of the three dominant generation mechanisms: thermal forcing, mechanical oscillator, and obstacle effect (described in detail in section 9.1.2). Rind et al. (1988) was first to implement a parameterization for nonstationary GWs linked to wave sources. The parameterization included waves generated by convection and wind shear based on theoretical assumptions. GW momentum flux of convectively generated GWs was related to the convective mass flux generated by the model. The phase speed of gravity waves was set to the mean wind over the convective region $\pm 10\,\text{m s}^{-1}$, and for deeper convection additional waves with phase speeds equal to the mean wind ± 20 and $\pm 40\,\text{m s}^{-1}$ were added. Kershaw (1995) developed a convective GW source parameterization that can be viewed as a parametrization of the obstacle effect as it focuses on parameterizing the effect of wave generation by flow over heating. A similar parameterization was developed by Chun and Baik (1998) parameterizing the effect of wave generation by mean flow over steady heating, representing only gravity waves that are stationary relative to the heat source. This parameterization was further extended to include effects of vertical wind shear by Chun and Baik (2002).

Beres et al. (2004) and Beres (2004) developed a parameterization of convectively generated GWs assuming that thermal forcing is the dominant GW generation mechanism. This method is based on linear theory, and both steady and oscillatory components of the heating are considered; hence, stationary and nonstationary GWs, relative to the heating, are represented. The dominant spectral properties of the GWs depend on the horizontal and vertical scales of the heating. The dominant GW phase speed is primarily determined by the convective heating depth, h, leading to a dominant wave phase speed of $\pm 15\,\text{m s}^{-1}$ for $h = 5$ km and a dominant wave phase speed of $25\,\text{m s}^{-1}$ for $h = 10\,\text{km}$ (assuming horizontal scale of heating of 2.5 km). The horizontal scale of the heating primarily changes the amplitude and not the characteristics of the wave spectrum. The momentum flux of convectively generated GWs is proportional to the square of the heating. The effects of environmental wind in and above the convective region are also incorporated into the parameterization as wind shear can create a large asymmetry in the GW spectrum (Beres et al., 2002). A similar parameterization was developed by Song and Chun (2005) based on a more complex vertical structure of the zonal mean wind and stability.

The convective source parameterizations by Beres et al. (2004) and Song and Chun (2005) based on thermal forcing are a large improvement over fixed source representations as they provide physically based connections between GWs and their evolving tropospheric sources, and hence respond to changes in convection on all timescales including changes resulting from climate change. However, both of these parameterizations omit the effects of the nonlinear forcing (Chun et al. 2005). Chun et al. (2008) proposed a method of including the effects of nonlinear forcing effect on a spectrum of convectively generated GWs and showed that this inclusion reduced cloud-top momentum flux by about 10%, except for middle latitude storm-track regions where the cloud-top momentum flux was amplified. Choi and Chun (2011) updated the Song and Chun (2005) parameterization by determining two free parameters of that parameterization: the moving speed of the convective source and the wave propagation direction.

Taking a slightly different approach, Lott and Guez (2013) developed a source spectrum parameterization for convectively generated GWs based on a stochastic approach presented in Eckermann (2011). In this approach, a few monochromatic waves chosen randomly from a probability distribution are launched at each time step. The amplitudes of the waves are proportional to the square of the diabatic heating derived from the precipitation field. This approach leads to a wider range of wave amplitudes, leading to a lower level of momentum deposition as compared to uniform sources. In addition to the complex convective source parameterization described above, Bushell et al. (2015) implemented a relatively simple convective GW source representation by linking GW momentum flux amplitude to the square root of total precipitation. The introduction of the amplitude dependence generated launch-level flux amplitudes with greater spatial and temporal variability, increasing realism of parameterized convectively generated GWs.

The inclusion of all of the above-described convective source parameterizations had a positive impact on simulations of climate in several GCMs. Parameterizations of Chun and Baik (1998) and Chun and Baik (2002) improved the representation of the middle atmosphere in the Yonsei University atmospheric GCM (YONU AGCM; Chun et al. 2001), the National Center for Atmospheric Research (NCAR) Community Climate Model (CCM) version 3 (Chun et al., 2004), and the National Centers for Environmental Prediction (NCEP) global spectral model (GSM) (Jeon et al 2010). The Beres et al. (2005) parameterization was implemented in the Whole Atmosphere Community Climate Model, version 2 (WACCM2), instead of the arbitrarily specified source spectra only in the tropics and resulted in an improved representation of the stratospheric semiannual oscillation (Beres et al., 2005). The Song and Chun (2005) parameterization was implemented in the Whole Atmosphere Community Climate Model, version 1b, resulting in an alleviation of the model's zonal mean wind biases, primarily in the low latitudes to midlatitudes of the upper stratosphere and the mesosphere, and improving the structure and the magnitude of the SAO (Song et al., 2007). Choi and Chun (2013) demonstrated a reduction in wind biases and alleviation of cold temperature biases in the winter polar stratosphere using the Choi and Chun (2011) spectrum and the ray-based parameterization of Song and Chun (2008). Unfortunately, neither one of these parameterizations remedied the lack of an internally generated QBO in WACCM. However, a decade later, an internally generated QBO was developed with the Beres et al. (2005) parameterization in the Community Atmosphere Model, version 5 (Richter et al., 2014), and subsequently in WACCM (Garcia & Richter, 2019), but only after the vertical resolution of these models was doubled to ~500 m in the free troposphere and lower stratosphere. The implementation of the Beres et al. (2004) parameterization in the ECHAM6 model led to improvements of several aspects of the QBO. With the Lott and Guez (2013) parameterization, the LMDz model was able to obtain a QBO with vertical resolution in the stratosphere of ~500 m. The modifications to the source-level amplitudes by Bushell et al. (2015) led to the improved representation of the QBO in the UK Met Office global model.

Gravity Waves Generated by Jets/Fronts

Parameterizations of GWs generated by jets/fronts are less developed and less common in GCMs as compared to parameterizations of orographic and convective gravity wave sources. As mentioned in the previous section, Rind et al. (1988) parameterized nonorographic GW sources by convection and shear. Shear-generated GWs were launched at jet stream level and assigned a single wavenumber and phase speed in each GCM grid box dependent on the direction of the shear and wind velocity in the shear layers. Wave momentum flux magnitude was set to be proportional to the square of the wind shear between two successive layers. Subsequently, Charron and Manzini (2002) parameterized frontally generated gravity waves. They used the frontogenesis function (Hoskins, 1982; Miller, 1948) to diagnose the location of fronts. In this approach, GWs were launched at a fixed level of 600 hPa, and if the frontogenesis function exceeded a critical threshold, a relatively high gravity wave variance was imposed in two cross-front directions. At grid points where the frontogenesis function was not exceeded, GWs were launched with a much lower variance representing other possible GW sources. The implementation of this parameterization in the MAECHAM4 model produced a reasonable representation of the stratospheric and mesospheric dynamics. Richter et al. (2010) used a modified version of the Charron and Manzini (2002) approach to represent frontally generated GWs. They also used the frontogenesis function and a launching level of 600 hPa, and only launched frontally generated gravity waves when the frontogenesis threshold was exceeded, and no small amplitude spectrum was employed. In order to obtain enough drag in the stratosphere/mesosphere via this approach, the frontogenesis threshold used was ~half of that used by Charron and Manzini (2002); however, no additional background GW spectrum was used. Richter et al. (2010) used the Beres et al. (2004) convective GW parameterization as well; hence, GWs in the tropics were primarily generated from the convective scheme and in the extratropics from the frontal scheme. When the frontogenesis threshold was exceeded, a Gaussian GW momentum flux phase speed spectrum of constant value was launched. Hence, the frontogenesis function was

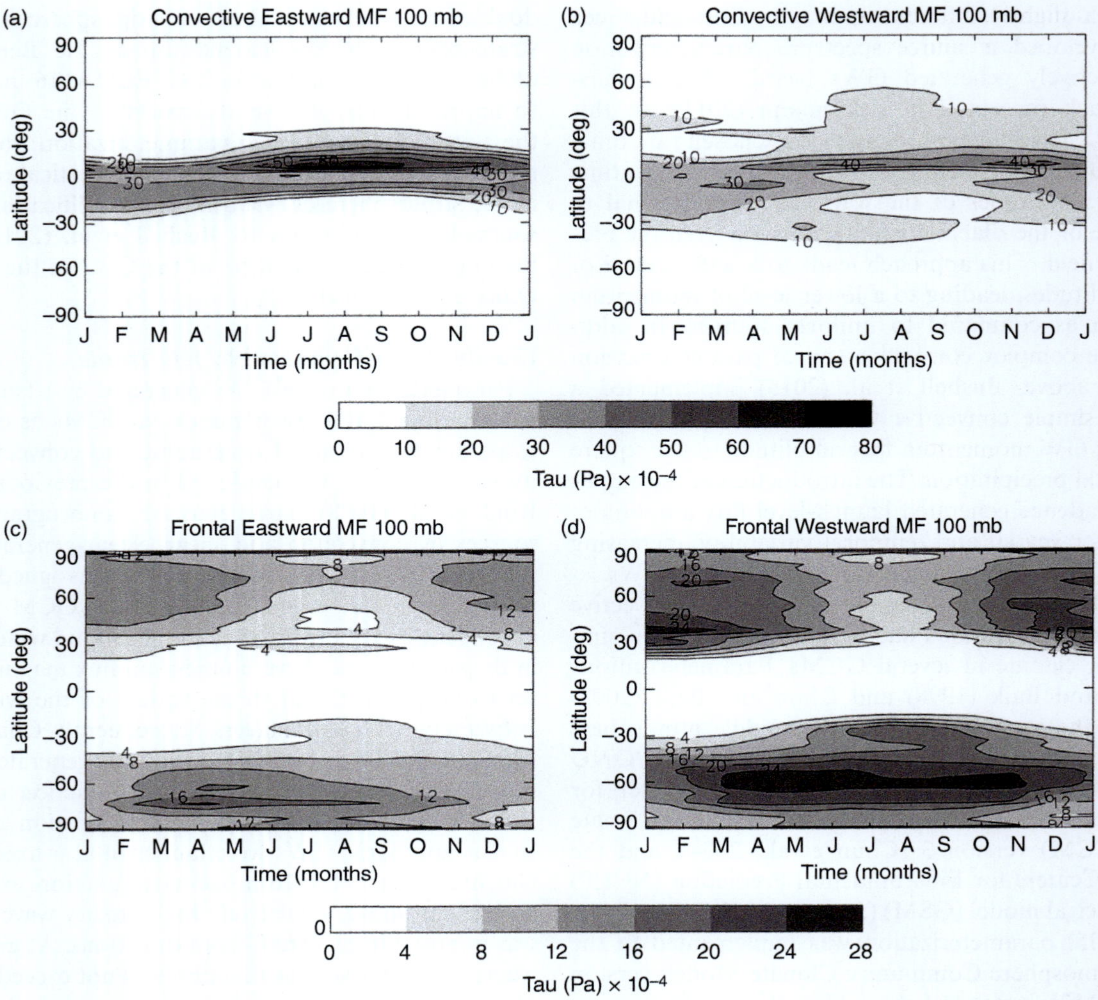

Figure 9.4 Total eastward (left panels) and westward (right panels) momentum flux (in Pa) for convectively (top panels) and frontally (bottom panels) generated gravity waves as a function of latitude and time of year derived from WACCM3.5. Source: Adapted from Richter et al. (2010).

used to produce realistic spatial and temporal variability of frontally generated GWs, including seasonality. However, there was no relation between the properties of fronts and the properties of the GWs generated by them. Figure 9.4 shows the convective (top panels) and frontal (bottom panels) eastward and westward momentum fluxes at 100 hPa in the Whole Atmosphere Community Climate Model, version 3.5 (WACCM3.5), used in Richter et al. (2010). The figure clearly shows the dominance of convectively generated gravity waves in the tropics and of frontally generated gravity waves in the extratropics. Convectively generated GWs follow the seasonal cycle of tropical convection with highest values of GW momentum flux in the NH winter (summer) primarily south (north) of the equator. Although the relationship between the spectrum of waves launched from fronts is not linked to their properties, the 100 hPa momentum flux reflects the seasonal cycle of frontal systems with a maximum in the winter season, and an asymmetry between eastward and westward propagating waves resulting from the strong filtering of eastward propagating GWs by strong tropospheric westerlies between 600 and 100 hPa.

Camara and Lott (2015) parameterized GW generation via spontaneous adjustment mechanism (as described in section 9.1.3) in combination with stochastic approach used in Lott and Guez (2013) to parameterize convective gravity wave sources. In this parameterization GW Eliassen-Palm flux is related to the amplitude and depth of potential vorticity anomalies, hence determining the location and amplitude of GWs. Even though spontaneous adjustment theory predicts exponentially small GW perturbations, the implementation of the scheme in LMDz provided enough extratropical wave drag to obtain reasonable circulations in the stratosphere and mesosphere (Camara and Lott, 2015).

9.2.4. Uncertainties and Parameter Tuning

As GW drag parameterizations are designed, constants are inevitably involved in the derived relations. These constants are not always well constrained and, in practice, are often tuned (more below). Examples include horizontal wavenumber (estimated from unresolved orography in OGW parameterizations and a constant in nonorographic GW parameterizations) and a critical nondimensional wave amplitude, defining a nondimensional GW amplitude at which instabilities will begin dissipation. An additional "efficiency factor," e, is typically introduced into GW parameterizations, having values between zero and unity. This term was introduced into early OGW parameterizations in order to reduce excessive MW drag in the lower stratosphere (e.g., Klinker & Sardeshmukh, 1992). An efficiency factor is a common part of most GW parameterizations, both orographic and nonorographic. This factor is typically applied in one of two ways: (1) e is multiplied with the drag profile, reducing drag but not influencing the levels where the parameterized wave breaks, or (2) just applied to the source-level momentum flux. The latter option reduces the wave amplitude, influencing initial breaking levels and where GW drag is exerted.

A variety of physical justifications for an efficiency factor have been put forward. GWs may occur intermittently within spatial and/or temporal grid scales. There can also be spectral intermittency, where the spectrum that is specified at the source may not be fully represented at all points and times. Mountains are 3D and not 2D, as treated by the parameterizations, which can reduce source-level momentum fluxes, as ridges are not necessarily perpendicular to the flow or the mountains represented could be more isolated. These real effects may result in reducing the momentum fluxes and drags below that predicted by the simple 2D, steady, hydrostatic, instantaneous, Boussinesq, linear, monochromatic, only vertically propagating, Wentzel-Kramer-Boussinesq theory upon which most GW parameterizations are based.

Ideally, the tuning constants involved should be constrained by observations. However, current observation platforms (aircraft, radiosondes, superpressure balloons, satellite-borne nadir, and limb sounders) have limited capability for quantitatively constraining 3D GW characteristics and momentum flux globally due to low frequency in space and time and/or lack of sensitivity to the entire spectrum of GWs. Still, progress toward verifying GW parameterizations has been made in recent years with derivations of global GW momentum fluxes (e.g., Alexander et al., 2008; Ern et al., 2004; Ern et al., 2017; Hertzog et al., 2008; Hindley et al., 2020; Vincent et al., 1997). Geller et al. (2013) made a first attempt at comparing parameterized gravity wave momentum fluxes in climate models to gravity wave momentum fluxes derived from observations. This work focused on absolute momentum fluxes (sum over all directions) and showed a general good spatial agreement between models and global satellite estimates, but the observations were not yet able to provide meaningful constraints. Recent measurements from superpressure balloon campaigns are able to derive GW momentum fluxes with higher accuracy (Jewtoukoff et al., 2013, 2015), but these are limited to a single level in the lower stratosphere, limited in latitude, and limited to campaign periods. Despite new sophisticated methods of analyzing satellite observations and estimating vector momentum fluxes, GW drag on the mean flow cannot be directly estimated from observations without orders of magnitude uncertainty (Alexander & Sato, 2015).

Without a direct way of measuring key quantities in GW drag parameterizations (i.e., momentum flux as a function of wave direction and phase speed at the source level) globally, tuning parameters are estimated indirectly. The free parameters are chosen to be physically reasonable and then the mean wind and temperature of the middle atmosphere, which are relatively well observed, provide an indirect verification measure for gravity wave parameterizations. In other words, during the development of new versions of GCMs, gravity wave parameterizations are "tuned" in order to arrive at a reasonable representation of the observed zonal mean climate. Such an exercise assumes that errors from other parts of the model (e.g., the dynamical core and other parameterizations) are small, that the form of the GW drag parameterizations are correct, and that errors that remain are attributable to the GW drag parameterizations and ultimately incorrectly specified tuning parameters. These tuning parameters are adjusted in order to best represent the observed mean climate. These assumptions are, of course, not valid; however, such an exercise is common practice in developing climate models. Tuning of GW drag parameterizations likely results in the GW parameterizations compensating for other model errors (e.g., horizontal and vertical discretization errors, numerical diffusion, errors in other parameterizations and errors in GW parameterization structure).

Gravity wave tuning is an iterative process that consists of changing the unconstrained parameterization parameters, running the climate model, and assessing model climatology and biases. The process is repeated typically several to several dozen times until an acceptable modeled climate state is achieved. Models with fixed GW sources typically have one set of tunable parameters and with those they need to arrive at reasonable tropical and extratropical mean wind and temperature in the stratosphere, and also in the mesosphere (if the model extends this high). Despite many GW tuning efforts, many GCMs end up with a "cold-pole bias" in the Southern Hemisphere winter polar stratosphere (e.g., Austin et al., 2003; Eyring et al., 2006), which can be improved by additional GW drag in the Southern Hemisphere (Garcia et al., 2017). In addition, many recent climate models have internally generated QBOs that are largely driven by parameterized gravity waves (e.g., Giorgetta et al., 2006; Richter et al., 2014; Geller et al., 2016), and hence GW parameterization parameters must be just right to get a reasonable period of the QBO close to observed in addition to the mean wind and temperature in the middle atmosphere. Source-oriented GW parameterizations allow more flexibility while tuning GWs, as typically in models such as WACCM (Richter et al., 2010) and LMDz (Lott et al., 2012a, b; Lott & Guez, 2013), and convectively generated GWs control the tropics, QBO and SAO, whereas the orographic and frontal waves affect the extratropical mean state. More complexity and options in GW parameterization parameters, however, also means endless combinations of poorly constrained parameters, which is also difficult to deal with.

9.2.5. Missing Processes

Again, GWs are treated as linear, hydrostatic, only propagating vertically (in a slowly varying background in z, no less), propagating upward instantaneously, and propagating through a steady ambient environment with Boussinesq governing equations. All these simplifications lurking in the underlying physics prevent representation of many GW characteristics that are well observed and physically understood.

For example, nonorographic GW observations show strong amplitude intermittency, where infrequently observed waves with very large amplitudes can represent very high fractions of time-averaged momentum flux. Indeed, from observations in the lower stratosphere (Hertzog et al., 2012), distributions of individual event wave momentum fluxes have shown a log-normal distribution of amplitudes, with the largest amplitude waves occurring <10% of the time but carrying ~60% of the average flux. Only recently has intermittency been represented in some nonorographic GW parameterizations (see Plougonven et al., 2020), having important implications for the total momentum fluxed and levels where it gets deposited. GW parameterizations that have GW sources tied to particular mechanisms (e.g., convective parameterizations, frontogenesis metrics, MW parameterizations; see following sections) do have some representation of intermittency, tied to the intermittency of the sources represented by the host model. Still, the infrequent, but very large amplitude GWs remain underrepresented (e.g., Stephan et al. (2016)).

Stochasticity in parameterized gravity wave sources is another way to account for wave intermittency. Eckermann (2011) demonstrated that stochastic representation of the nonorographic wave spectrum in different grid points gave similar middle atmospheric circulation changes as including the full spectrum uniformly, but at much lower computational cost. These ideas have found use in modern parameterization schemes (de la Camara et al., 2014; Serva et al., 2018) with some demonstrated improvements in model performance.

Numerous other characteristics of GWs are unrepresented as a result of the many conventional, and sometimes pragmatic, assumptions and simplifications made. Nonhydrostatic influences occur when the GW intrinsic frequency becomes close to the buoyancy frequency and can reduce orographic drag (e.g., Smith & Kruse, 2017; see section 8 of Tiexeira 2014 for an overview). Neglecting lateral propagation results in significant overestimates of MW amplitudes, breaking and drag at altitudes that are too low, and drag that is too spatially confined (e.g., Eckermann et al., 2015b). Trailing MWs launched terrain orientations oblique to the source-level flow can propagate horizontally for $O(1000)$ kilometers (e.g., Amemiya & Sato, 2016; Jiang et al., 2019; Sato et al., 2012), and neglect of such long-distance propagation is likely in part responsible for an artificial gap in GW drag near 60°S in climate models that significantly influence Southern Hemisphere polar night jet strength (Kruse et al., 2022; McLandress et al., 2012).

Transience of both the GWs and their background can also be important. Transient forcing results in vertical dispersion and spreading of GWs due to the spectrum of vertical group velocities of the generated spectrum of GWs (e.g., Chen et al., 2007; Kruse & Smith, 2018), which influence wave amplitudes and breaking levels. Increasing and decreasing flow over terrain result in MWs with positive and negative phase speeds and propagation downstream and upstream of the mountains, respectively. Transient forcing also induces transient wave packet propagation, allowing interaction of the GW packet with the environment it propagates into and out of even without breaking (Bölöni et al., 2016; Bühler & McIntyre, 1999, 2003, 2005; Dosser & Sutherland, 2011; Fritts & Dunkerton, 1984; Kruse & Smith, 2018). Ray-tracing methods have been applied in research models to account for horizontal propagation and group velocity effects (Bölöni et al., 2016; Song & Chun, 2008), but these have not found wider application.

To summarize, GWs are very important in Earth's atmosphere, and their representation in all weather and climate models that do not fully resolve the entire GW spectrum is essential. The various parameterizations implemented do improve weather and climate model skill. Still, there is much potential to improve parameterizations via both better observational constraints and improving the underlying physics.

ACKNOWLEDGMENTS

C.G.K. was supported by the National Science Foundation (NSF, grant #2004512) and in part by an Advanced Study Program fellowship at the National Center for Atmospheric Research, which is also supported by the NSF under Cooperative Agreement 1852977. J.H.R. and J.T.B. were supported by NCAR as well. M.J.A. was supported by both the NSF (grant #1829373) and NASA (grant #80NSSC17K0169). J.W. was supported by the National Natural Science Foundation of China (Grant 42075005 and Grant 42375003) and Guangdong Province Key Laboratory for Climate Change and Natural Disaster Studies (Grant 2020B1212060025).

AVAILABILITY STATEMENT

The mountain wave simulation in Figure 9.1 and Supplemental Animation 1 were performed using high-performance computing support from Cheyenne (doi:10.5065/D6RX99HX) provided by NCAR's Computational and Information Systems Laboratory, sponsored by the National Science Foundation.

REFERENCES

Adachi, Y., Yukimoto, S., Deushi, M., Obata, A., Nakano, H., Tanaka, T. Y., et al. (2013). Basic performance of a new earth system model of the Meteorological Research Institute (MRI-ESM1). *Papers in Meteorological and Geophysics*, *64*, 1–19.

Alexander, M. J., & Dunkerton, T. J. (1999). A spectral parameterization of mean-flow forcing due to breaking gravity waves. *Journal of the Atmospheric Sciences*, *56*, 4167–4182,.

Alexander, M. J., Eckermann, S. D., Broutman, D., & Ma, J. (2009). Momentum flux estimates for South Georgia Island mountain waves in the stratosphere observed via satellite. *Geophysical Research Letters*, *36*, L12816, doi:10.1029/2009GL038587.

Alexander, M. J., Gille, J., Cavanaugh, C., Coffey, M., Craig, C., Eden, T., et al. (2008). Global estimates of gravity wave momentum flux from High Resolution Dynamics Limb Sounder observations. *Journal of Geophysical Research*, *113, D15S18*, doi:10.1029/ 2007JD008807.

Alexander M. J., & Holton, J. (1997). A model study of zonal forcing in the equatorial stratosphere by convectively induced gravity waves. *Journal of the Atmospheric Sciences*, *54*, 408–419.

Alexander M. J., & Holton, J. R. (2004). On the spectrum of vertically propagating gravity waves generated by a transient heat source. *Atmospheric Chemistry and Physics Discussions, European Geosciences Union*, *4*(1), 1063–1090.

Alexander, M. J., Holton, J. R., & Durran, D. R. (1995). The gravity wave response above deep convection in a squall line simulation. *Journal of the Atmospheric Sciences, 52*, 2212–2226.

Alexander, M. J., & Ortland, D. A. (2010). Equatorial waves in High Resolution Dynamics Limb Sounder (HIRDLS) data. *Journal of Geophysical Research*, *115*, D24111. doi:10.1029/ 2010JD014782.

Alexander, M., & Rosenlof, K. (1996). Nonstationary gravity wave forcing of the stratospheric zonal mean wind. *Journal of Geophysical Research*, *101*, 23,465–23,474.

Alexander, M. J., & Rosenlof, K. H. (2003). Gravity wave forcing in the stratosphere: Observational constraints from UARS and implications for parameterization in global models. *Journal of Geophysical Research*, *108*(D19). https://doi .org/10.1029/2003JD003373.

Alexander, M. J., & Sato, K. (2015). SPARC Newsletter No. 44, p. 9: Gravity Wave Dynamics and Climate: An Update from the SPARC Gravity Wave Activity.

Amemiya, A., & Sato, K. (2016). A new gravity wave parameterization including three-dimensional propagation. *Journal of the Meteorological Society of Japan*, *94*(3), 237–256. doi:10.2151/jmsj.2016-013

Austin, J., Shindell, D., Beagley, S. R., Brühl, C., Dameris, M., Manzini, E., et al. (2003). Uncertainties and assessments of chemistry-climate models of the stratosphere. *Atmospheric Chemistry and Physics*, *3*, 1–27.

Azeem, I., Vadas, S. L., Crowley, G., & Makela, J. J. (2017). Traveling ionospheric disturbances over the United States induced by gravity waves from the 2011 Tohoku tsunami and comparison with gravity wave dissipative theory. *Journal of Geophysical Research: Space Physics*, *122*, 3430–3447. doi:10.1002/2016JA023659.

Bacmeister, J. T. (1993). Mountain-wave drag in the stratosphere and mesosphere inferred from observed winds and a simple mountain-wave parameterization scheme. *Journal of the Atmospheric Sciences*, *50*, 377–399.

Bacmeister, J. T., & Pierrehumbert, R. T. (1988). On high-drag states of nonlinear stratified flow over an obstacle. *Journal of the Atmospheric Sciences, 45*, 63–80. https://doi.org/10.1175/ 1520-0469(1988)045<0063:OHDSON>2.0.CO;2

Bacmeister, J. T., & Schoeberl, M. R. (1989). Breakdown of vertically propagating two-dimensional gravity waves forced by orography. *Journal of the Atmospheric Sciences, 46*, 2109–2134.

Baines, P. G., & Palmer, T. N. (1990). Rationale for a new physically-based parameterisation of sub-grid scale orographic effects., Tech. Rep. 169, ECMWF.

Beljaars, A. C. M., Brown, A. R. & Wood, N. (2004). A new parametrization of turbulent orographic form drag. *Quarterly Journal of the Royal Meteorological Society*, *130*, 1327–1347. doi:10.1256/qj.03.73.

Beres, J. H., Alexander, M. J., & Holton, J. R. (2002). Effects of tropospheric wind shear on the spectrum of convectively generated gravity waves. *Journal of the Atmospheric Sciences, 59*, 1805–1824.

Beres, J. H., Alexander, M. J., & Holton, J. R. (2004). A method of specifying the gravity wave spectrum above convection based on latent heating properties and background wind. *Journal of the Atmospheric Sciences, 61*, 324–337.

Beres, J. H. (2004). Gravity wave generation by a three-dimensional thermal forcing. *Journal of the Atmospheric Sciences, 61*(14), 1805–1815.

Beres, J. H., Garcia, R. R., Boville, B. A., & Sassi, F. (2005). Implementation of a gravity wave source spectrum parameterization dependent on the properties of convection in the Whole Atmosphere Community Climate Model (WACCM). *Journal of Geophysical Research, 110*, D10108. doi:10.1029/2004JD005504.

Boer, G. J., McFarlane, N. A., Laprise, R., Henderson, J. D., & Blanchet, J.-P. (1984). The Canadian Climate Centre spectral atmospheric general circulation model. *Atmosphere-Ocean, 22*, 397–429.

Bossert, K., Fritts, D. C., Heale, C. J., Eckermann, S. D., Plane, J. M. C., Snively, J. B., et al. (2018). Momentum flux spectra of a mountain wave event over New Zealand. *Journal of Geophysical Research: Atmospheres, 123*, 9980–9991. https://doi-org.cuucar.idm.oclc.org/10.1029/2018JD028319

Bossert, K., Kruse, C., Heale, C., Fritts, D. C., Snively, J., Pautet, P.-D., et al. (2017). Secondary gravity wave generation over New Zealand during the DEEPWAVE campaign. *Journal of Geophysical Research: Atmospheres, 122*(15), 7834–7850. doi:10.1002/2016JD026079.

Bölöni, G., Ribstein, B., Muraschko, J., Sgoff, C., Wei, J., & Achatz, U. (2016). The interaction between atmospheric gravity waves and large-scale flows: An efficient description beyond the nonacceleration paradigm. *Journal of the Atmospheric Sciences, 73*, 4833–4852. doi:10.1175/JAS-D-16-0069.1.

Boville, B. (1986). Wave-mean flow interactions in a general circulation model of the troposphere and stratosphere. *Journal of the Atmospheric Sciences, 43*, 1711–1725.

Boville, B. A. (1991). Sensitivity of simulated climate to model resolution. *Journal of Climate, 4*, 469–485.

Bretherton, C. (1988). Group velocity and the linear response of stratified fluids to internal heat or mass sources. *Journal of the Atmospheric Sciences, 45*, 81–93.

Bretherton, C. S., & Smolarkiewicz, P. K. (1989). Gravity waves, compensating subsidence and detrainment around cumulus clouds. *Journal of the Atmospheric Sciences, 46*, 740–759.

Bühler, O., & McIntyre, M. E. (1999). On shear-generated gravity waves that reach the mesosphere. part II: Wave propagation. *Journal of the Atmospheric Sciences, 56*, 3764–3773.

Bühler, O., & McIntyre, M. E. (2003). Remote recoil: A new wave-mean interaction effect *Journal of Fluid Mechanics, 492*, 207–230.

Bühler, O., & McIntyre, M. E. (2005). Wave capture and wave-vortex duality. *Journal of Fluid Mechanics, 534*, 67–95. doi:10.1017/S0022112005004374.

Bühler, O., McIntyre, M., & Scinocca, J. (1999). On shear-generated gravity waves that reach the mesosphere. Part I: Wave generation. *Journal of the Atmospheric Sciences, 56*, 3749–3763.

Burleigh, M. R., Heale, C. J., Zettergren, M. D., & Snively, J. B. (2018). Modulation of low-altitude ionospheric upflow by linear and nonlinear atmospheric gravity waves. *Journal of Geophysical Research: Space Physics, 123*. https://doi.org/10.1029/2018JA025721.

Bushell, A. C., Anstey, J. A., Butchart, N., Kawatani, Y., Osprey, S. M., Richter, J. H., et al. (2020). Evaluation of the Quasi-Biennial Oscillation in global climate models for the SPARC QBO-initiative. *Quarterly Journal of The Royal Meteorological Society, 1–31*. https://doi.org/10.1002/qj.3765

Butchart, N., Anstey, J., Hamilton, K., Osprey, S., McLandress, C., Bushell, A., et al. (2018). Overview of experiment design and comparison of models participating in phase 1 of the SPARC Quasi-Biennial Oscillation initiative (QBOi). *Geoscientific Model Development, 11*, 1009–1032.

Butchart, N., & Austin, J. (1998). Middle atmosphere climatologies from the troposphere-stratosphere configuration of the UKMO's unified model. *Journal of the Atmospheric Sciences, 55*, 2782–2809.

Bushell, A. C., Butchart, N., Derbyshire, S. H., Jackson, D. R., Shutts, G. J., Vosper, S. B., & Webster, S. (2015). Parameterized gravity wave momentum fluxes from sources related to convection and large-scale precipitation processes in a global atmosphere model. *Journal of the Atmospheric Sciences, 72*(11), 4349–4371.

de la Cámara, A., Lott, F., & Hertzog, A. (2014). Intermittency in a stochastic parameterization of nonorographic gravity waves. *Journal of Geophysical Research: Atmospheres, 119* (21), 11,905–11,919.

Cámara, A., & Lott, F. (2015). A parameterization of gravity waves emitted by fronts and jets. *Geophysical Research Letters, 42*(6), 2071–2078.

Cámara, A. D. L., Lott, F., Jewtoukoff, V., Plougonven, R., & Hertzog, A. (2016). On the gravity wave forcing during the southern stratospheric final warming in LMDZ. *Journal of the Atmospheric Sciences, 73*(8), 3213–3226.

Charron, M., & Manzini, E. (2002). Gravity waves from fronts: Parameterization and middle atmosphere response in a general circulation model. *Journal of the Atmospheric Sciences, 59*, 923–941. https://doi.org/10.1175/1520-0469(2002)059<0923:GWFFPA>2.0.CO;2

Chen, C., Hakim, G. J., & Durran, D. R. (2007). Transient mountain waves and their interaction with large scales. *Journal of the Atmospheric Sciences, 64*(7), 2378–2400.

Chimonas, G., & Grant, J. (1984). Shear excitation of gravity waves. Part II: Upscale scattering from Kelvin-Helmholtz waves. *Journal of the Atmospheric Sciences, 41*, 2278–2288.

Choi, H.-J., & Chun, H.-Y. (2011). Momentum flux spectrum of convective gravity waves. Part I: An update of a parameterization using mesoscale simulations. *Journal of the Atmospheric Sciences, 68*(4), 739–759.

Choi, H.-J., & Chun, H.-Y. (2013). Effects of convective gravity wave drag in the southern hemisphere winter stratosphere. *Journal of the Atmospheric Sciences, 70*(7), 2120–2136.

Chun, H.-Y., & Baik, J.-J. (1998). Momentum flux by thermally induced internal gravity waves and its approximation for large-scale models. *Journal of the Atmospheric Sciences, 55*, 3299–3310.

Chun, H.-Y., & Baik, J.-J. (2002). An updated parameterization of convectively forced gravity wave drag for use in large-scale models. *Journal of the Atmospheric Sciences, 59*, 1006–1017.

Chun, H.-Y., Baik, J.-J., & Horinouchi, T. (2005). Momentum flux spectrum of convectively forced gravity waves: Can diabatic forcing be a proxy for convective forcing? *Journal of the Atmospheric Sciences*, *62*, 4113–4120.

Chun, H.-Y., Choi, H.-J., & Song, I.-S. (2008). Effects of nonlinearity on convectively forced internal gravity waves: Application to a gravity wave drag parameterization. *Journal of the Atmospheric Sciences*, *65*, 557–575. doi:https://doi.org/10.1175/2007JAS2255.1.

Chun, H.-Y., & Kim, Y.-H. (2008). Secondary waves generated by breaking of convective gravity waves in the mesosphere and their influence in the wave momentum flux, *Journal of Geophysical Research*, *113*, D23107. doi:10.1029/2008JD009792

Chun, H.-Y., Song, M.-D., Kim, J.-W., & Baik, J.-J. (2001a). Effects of gravity wave drag induced by cumulus convection on the atmospheric general circulation. *Journal of the Atmospheric Sciences*, *58*, 302–319.

Chun, H.-Y., Song, I.-S., Baik, J.-J., & Kim, Y.-J. (2004). Impact of a convectively forced gravity wave drag parameterization in NCAR CCM3. *Journal of Climate*, *17*(18), 3530–3547.

Clark, T. L., Hauf, T., & Kuettner, J. P. (1986). Convectively forced internal gravity waves: Results from two-dimensional numerical experiments. *Quarterly Journal of the Royal Meteorological Society*, *112*, 899–925.

Clark, T. L., & Peltier, W. R. (1977). On the evolution and stability of finite-amplitude mountain waves. *Journal of the Atmospheric Sciences*, *34*, 1715–1730. https://doi.org/10.1175/1520-0469(1977)034<1715:OTEASO>2.0.CO;2

Clark, T. L., & Peltier, W. R. (1984). Critical level reflection and the resonant growth of nonlinear mountain waves. *Journal of the Atmospheric Sciences*, *41*, 3122–3134. https://doi.org/10.1175/1520-0469(1984)041<3122:CLRATR>2.0.CO;2

Davini, P., von Hardenberg, J., Corti, S., Christiansen, H. M., Juricke, S., Subramanian, A., et al. (2017). Climate SPHINX: evaluating the impact of resolution and stochastic physics parameterisations in the EC-EARTH global climate model. *Geoscientific Model Development*, *10*, 1383–1402. doi:10.5194/gmd-10-1383-2017,.

Dosser, H. V., & Sutherland, B. R. (2011). Anelastic internal wave packet evolution and stability. *Journal of the Atmospheric Sciences*, *68*, 2844–2859. doi:10.1175/JAS-D-11-097.1.

Drazin, P. G. (1961). On the steady flow of a fluid of variable density past an obstacle. *Tellus*, *13*, 239–251.

Dunkerton, T. J. (1982). Theory of the mesopause semiannual oscillation. *Journal of the Atmospheric Sciences*, *39*, 2681–2690.

Durran, D. R. (1986a). Another look at downslope windstorms. Part I: On the development of analogs to supercritical flow in an infinitely deep continuously stratified fluid. *Journal of the Atmospheric Sciences*, *93*, 2527–2543.

Durran, D. R. (1986b). *Mountain waves (Mesoscale Meteorology and Forecasting)* (pp. 472–492). American Meteorological Society: Boston.

Durran, D. R. (1990). Mountain waves and downslope winds. In W. Blumen, (Ed.), *Atmospheric Processes over Complex Terrain. Meteorological Monographs, American Meteorological Society*, *23*(45), 59–81.

Durran, D. R. (2003). Lee waves and mountain waves. In J. R. Holton, J. Pyle, & J. A. Curry (Eds.), *Encyclopedia of Atmospheric* Sciences (pp. 1161–1170). Elsevier Science Ltd.: London.

Durran, D. R. & Klemp, J. B. (1983). A compressible model for the simulation of moist mountain waves. *Monthly Weather Review*, *111*, 2341–2361. https://doi.org/10.1175/1520-0493(1983)111<2341:ACMFTS>2.0.CO;2

Durran, D. R., & Klemp, J. B. (1987). another look at downslope winds. Part II: Nonlinear Amplification beneath wave-overturning layers. *Journal of the Atmospheric Sciences*, *44*, 3402–3412. https://doi.org/10.1175/1520-0469(1987)044<3402:ALADWP>2.0.CO;2

Eckermann, S. D. (2011). Explicitly stochastic parameterization of nonorographic gravity wave drag. *Journal of the Atmospheric Sciences*, *68*, 1749–1765.

Eckermann, S. D., Broutman, D., & Knight, H. (2015a). Effects of horizontal geometrical spreading on the parameterization of orographic gravity wave drag. Part II: Analytical solutions. *Journal of the Atmospheric Sciences*, *72*(6), 2348–2365. Retrieved Oct 7, 2021, from https://journals.ametsoc.org/view/journals/atsc/72/6/jas-d-14-0148.1.xml

Eckermann, S. D., Ma, J., & Broutman, D. (2015b). Effects of horizontal geometrical spreading on the parameterization of orographic gravity wave drag. Part I: Numerical transform solutions. *Journal of the Atmospheric Sciences*, *72*, 2330–2347. https://doi.org/10.1175/JAS-D-14-0147.1.

Eliassen, A., & Palm, E. (1960). On the transfer of energy in stationary mountain waves. Geofysiske Publikasjoner, *22*(3), 1–23.

Ern, M., Hoffmann, L., & Preusse, P. (2017). Directional gravity wave momentum fluxes in the stratosphere derived from high-resolution AIRS temperature data. *Geophysical Research Letters*, *44*(1), 475–485. https://doi.org/10.1002/2016GL072007.

Ern, M., Preusse, P., Alexander, M. J., & Warner, C. D. (2004). Absolute values of gravity wave momentum flux derived from satellite data. *Journal of Geophysical Research*, *109*, D20103. doi:10.1029/2004JD004752.

Eyring, V., Butchart, N., Waugh, D. W., Akiyoshi, H., Austin, J., Bekki, S., et al. (2006). Assessment of temperature, trace spices, and ozone in chemistry-climate model simulations of the recent past. *Journal of Geophysical Research*, *111*, D22308. doi:10.1029/2006JD007327.

Eyring, V., Waugh, D. W., Bodeker, G. E., Cordero, E., Akiyoshi, H., Austin, J., et al. (2007). Multimodel projections of stratospheric ozone in the 21st Century. *Journal of Geophysical Research*, 112(D16303). https://doi.org/10.1029/2006JD008332.

Farmer, D. M. & Armi, L. (1999). The generation and trapping of solitary waves over topography. *Science*, *283*, 188–190.

Felten, F. N., & Lund, T. S. (2006). Kinetic energy conservation issues associated with the collocated mesh scheme for incompressible flow. *Journal of Computational Physics*, *215*, 465–484.

Ford, R. (1994a). Gravity wave radiation from vortex trains in rotating shallow water. *Journal of Fluid Mechanics*, *281*, 81–118.

Ford, R. (1994b). The response of a rotating ellipse of uniform potential vorticity to gravity wave radiation. *Physics of Fluids*, *6*(11), 3694–3704.

Fovell, R., Durran, D., & Holton, J. R. (1992). Numerical simulations of convectively generated stratospheric gravity waves. *Journal of the Atmospheric Sciences*, *49*, 1427–1442.

Fovell, R. G., Mullendore, G. L., & Kim, S. (2006). Discrete propagation in numerically simulated nocturnal squall lines, *Monthly Weather Review*, *134*(12), 3735–3752.

Franke, P. M., & Robinson, W. A. (1999). Nonlinear behavior in the propagation of atmospheric gravity waves. *Journal of the Atmospheric Sciences*, *56*, 3010–3027.

Fritts, D. (1984). Shear excitation of atmospheric gravity waves. 2: Nonlinear radiation from a free shear layer. *Journal of the Atmospheric Sciences*, *41*, 524–537.

Fritts, D. C., & Alexander, M. J. (2003). Gravity wave dynamics and effects in the middle atmosphere. *Reviews of Geophysics*, *41*, 1003. doi:10.1029/2001RG000106, 1.

Fritts, D. C., & Dunkerton, T. J. (1984). A quasi-linear study of gravity-wave saturation and self-acceleration. *Journal of Atmospheric Sciences*, *41*(22), 3272–3289.

Fritts, D. C., & Lu, W. (1993). Spectral estimates of gravity wave energy and momentum fluxes, II, Parameterization of wave forcing and variability. *Journal of the Atmospheric Sciences*, *50*, 3695–3713.

Fritts, D. C., & Luo, Z. (1992). Gravity wave excitation by geostrophic adjustment of the jet stream. Part I: Two-dimensional forcing. *Journal of the Atmospheric Sciences*, *49*(8), 681–697.

Fritts, D. C., Vadas, S. L., Wan, K., and Werne, J. A. (2006). Mean and variable forcing of the middle atmosphere by gravity waves. *J. Atmos. Sol. Terr. Phys.*, *68*, 247–265. doi: https://doi.org/10.1016/j.jastp.2005.04.010.

Garcia, R. R., & Boville, B. A. (1994). "Downward control" of the mean meridional circulation and temperature distribution of the polar winter stratosphere. *Journal of the Atmospheric Sciences*, *51*, 2238–2245.

Garcia, R. R., & Richter, J. H. (2019). On the momentum budget of the quasi-biennial oscillation in the whole atmosphere community climate model. *Journal of the Atmospheric Sciences*, *76*, 69–87. https://doi.org/10.1175/JAS-D-18-0088.1

Garcia, R. R., Smith, A. K., Kinnison, D. E., de la Cámara, Á., & Murphy, D. J. (2017). Modification of the gravity wave parameterization in the Whole Atmosphere Community Climate Model: Motivation and results. *Journal of the Atmospheric Sciences*, *74*, 275–291. doi:10.1175/JAS-D-16-0104.1.

Garcia, R. R., & Solomon, S. (1985). The effect of breaking gravity waves on the dynamics and chemical composition of the mesosphere and lower thermosphere, *Journal of Geophysical Research*, *90*, 3850–3868.

Garner, S. T. (2005). A topographic drag closure built on an analytical base flux. *Journal of the Atmospheric Sciences*, *62*, 2302–2315. https://doi.org/10.1175/JAS3496.1.

Geller, M. A., Alexander, M. J., Love, P. T., Bacmeister, J., Ern, M., Hertzog, A., et al. (2013). A comparison between gravity wave momentum fluxes in observations and climate models. *Journal of Climate*, *26*, 6383–6405. https://doi.org/10.1175/JCLI-D-12-00545.1

Geller, M. A., Zhou, T. H., Shindell, D., Ruedy, R., Aleinov, I., Nazarenko, L., et al. (2016). Modeling the QBO—Improvements resulting from higher model vertical resolution. *Journal of Advances in Modeling Earth Systems*, *8*(3), 1092–1105. doi:10.1002/2016MS000699.

Gettelman, A., Mills, M. J., Kinnison, D. E., Garcia, R. R., Smith, A. K., Marsh, D. R., et al. (2019). The whole atmosphere community climate model version 6 (WACCM6). *Journal of Geophysical Research: Atmospheres*, *124*, 12380–12403. https://doi.org/10.1029/2019JD030943

Giorgetta, M. A., Bengtsson, L., & Arpe, K. (1999). An investigation of QBO signals in the east Asian and Indian monsoon in GCM experiments. *Climate Dynamics*, *15*(6), 435–450. doi:10.1007/s003820050292.

Giorgetta, M. A., Manzini, E., Roeckner, E., Esch, M., & Bengtsson, L. (2006). Climatology and forcing of the quasi-biennial oscillation in the MACHAM5 model. *Journal of Climate*, *19*, 3882–3901.

Guest, F. M., Reeder, M. J., Marks, C. J., & Karoly, D. J. (2000). Inertia-gravity waves observed in the lower stratosphere over Macquarie Island. *Journal of the Atmospheric Sciences*, *57*, 737–752.

Heale, C. J., Bossert, K., Snively, J. B., Fritts, D. C., Pautet, P.-D., & Taylor, M. J. (2017). Numerical modeling of a multiscale gravity wave event and its airglow signatures over Mount Cook, New Zealand, during the DEEPWAVE campaign. *Journal of Geophysical Research Atmospheres*, *122*, 846–860. doi: 10.1002/2016JD025700.

Heale, C. J., Lund, T., & Fritts, D. C. (2020). Convectively generated gravity waves during solstice and equinox conditions. *Journal of Geophysical Research: Atmospheres*, *125*, e2019JD031582. https://doi.org/10.1029/2019JD031582

Heale, C. J., Snively, J. B., Hickey, M. P., & Ali, C. J. (2014). Thermospheric dissipation of upward propagating gravity wave packets. *Journal of Geophysical Research: Space Physics*, *119*, 3,857–3,872. https://doi.org/10.1002/2013JA019387

Heale, C. J., & Snively, J. B. (2015). Gravity wave propagation through a vertically and horizontally inhomogeneous background wind. *Journal of Geophysical Research Atmos.*, *120*, 5931–5950, doi:10.1002/2015JD023505.

Heale, C. J., Walterscheid, R. L., & Snively, J. B. (2018). Localization effects on the dissipation of gravity wave packets in the upper mesosphere and lower thermosphere. *Journal of Geophysical Research: Atmospheres*, *123*, 8915–8935. https://doi.org/10.1029/2017JD027617

Hertzog, A., Alexander, M. J., and Plougonven, R. (2012). On the Intermittency of Gravity Wave Momentum Flux in the Stratosphere. *J. Atmos. Sci.*, *69*, 3433–3448. https://doi.org/10.1175/JASD-12-09.1

Hertzog, A., Boccara, G., Vincent, R. A., Vial, F., & Cocquerez, P. (2008). Estimation of gravity-wave momentum fluxes and phase speeds from quasi-Lagrangian stratospheric balloon flights. 2: Results from the Vorcore campaign in Antarctica. *Journal of the Atmospheric Sciences*, *65*, 3056–3070.

Hindley, N. P., Wright, C. J., Hoffmann, L., Moffat-Griffin, T., & Mitchell, N. J. (2020). An 18-year climatology of directional stratospheric gravity wave momentum flux from 3-d satellite observations. *Geophysical Research Letters*, *47*(22), e2020GL089557. https://doi.org/https://doi.org/10.1029/2020GL089557.

Hitchman, M. H., & Leovy, C. B. (1988). Estimation of the Kelvin wave contribution to the semiannual oscillation. *Journal of the Atmospheric Sciences*, *45*, 1462–1475.

Hickey, M. P., Walterscheid, R. L., & Schubert, G. (2011). Gravity wave heating and cooling of the thermosphere: Sensible heat flux and viscous flux of kinetic energy. *Journal of Geophysical Research*, *116*, A12326. doi: https://doi.org/10.1029/2011JA016792

Hines, C. O. (1997a). Doppler-spread parameterization of gravity-wave momentum deposition in the middle atmosphere. Part 1: Basic formulation. *Journal of Atmospheric and Solar-Terrestrial Physics*, *59*(4), 371–386. https://doi.org/10.1016/S1364-6826(96)00079-X

Hines, C. O. (1997b). Doppler-spread parameterization of gravity-wave momentum deposition in the middle atmosphere. Part 2: Broad and quasi monochromatic spectra, and implementation. *Journal of Atmospheric and Solar-Terrestrial Physics*, *59*(4), 387–400.

Hoffmann, L., Grimsdell, A. W., & Alexander, M. J. (2016). Stratospheric gravity waves at Southern Hemisphere orographic hotspots: 2003–2014 AIRS/Aqua observations. *Atmospheric Chemistry and Physics*, *16*, 9381–9397. https://doi.org/10.5194/acp-16-9381-2016.

Hoffmann, L., Xue, X., & Alexander, M. J. (2013). A global view of stratospheric gravity wave hotspots located with Atmospheric Infrared Sounder observations. *Journal of Geophysical Research Atmospheres*, 118, 416–434. doi:10.1029/2012JD018658.

Holt, L. A., Alexander, M. J., Coy, L., Liu, C., Molod, A., Putman, W., & Pawson, S. (2017). An evaluation of gravity waves and their sources in the Southern Hemisphere in a 7-km global climate simulation. *Quarterly Journal of the Royal Meteorological Society*, *143*(707), 2481–2495. doi:10.1002/qj.3101.

Holton, J. R. (1982). The role of gravity wave induced drag and diffusion in the momentum budget of the mesosphere. *Journal of the Atmospheric Sciences*, *39*, 791–799.

Holton, J. R. (1983). The influence of gravity wave breaking on the general circulation of the middle atmosphere. *Journal of the Atmospheric Sciences*,*40*, 2497–2507.

Holton, J. R. (2004). *An introduction to dynamic meteorology* (4th edition). Volume 88 of International Geophysics Series. Elsevier Inc.

Holton, J. R., & Alexander, M. J. (1999). Gravity waves in the mesosphere generated by tropospheric convection. *Tellus*, *51A-B*, 45–58.

Holton, J. R., Beres, J. H., & Zhou, X. (2002). On the vertical scale of gravity waves excited by localized thermal forcing. *Journal of the Atmospheric Sciences*, *59*, 2019–2023. doi:https://doi.org/10.1175/1520-0469(2002)059<2019:OTVSOG>2.0.CO;2.

Hoskins, B. J. (1982). The mathematical theory of frontogenesis. *Annual Review of Fluid Mechanics*, *14*, 131–151.

Hughes, M., Hall, A., & Kim, J. (2009). Anthropogenic reductions of Santa Ana winds. California Climate Change Center Rep. CEC-500-2009-015-D, 19 pp. www.energy.ca.gov/2009publications/CEC-500-2009-015/CEC-500-2009-015-D.pdf.

Huppert, H., & Miles, J. (1969). Lee waves in a stratified flow Part 3. *Semi-elliptical obstacle. Journal of Fluid Mechanics*, *35*(3), 481–496. doi:10.1017/S0022112069001236

Jiang, Q., Doyle, J. D., Eckermann, S. D., & Williams, B. P. (2019). Stratospheric trailing gravity waves from New Zealand. *Journal of the Atmospheric Sciences*, in press.

Jewtoukoff, V., Hertzog, A., Plougonven, R., Cámara, A. D., & Lott, F. (2015). Comparison of gravity waves in the southern Hemisphere derived from balloon observations and the ECMWF analyses. *Journal of the Atmospheric Sciences*, *72*, 3449–3468. https://doi.org/10.1175/JAS-D-14-0324.1

Jewtoukoff, V., Plougonven, R., & Hertzog, A. (2013). Gravity waves generated by deep tropical convection: Estimates from balloon observations and mesoscale simulations, *Journal of Geophysical Research Atmospheres*, 118, 9690–9707. doi:10.1002/jgrd.50781.

Jeon, J., Hong, S.-Y., Chun, H.-Y., & Song, I.-S. (2010). Test of a convectively forced gravity wave drag parameterization in a general circulation model. *Asia-Pacific Journal of Atmospheric Sciences*, *46*(1), 1–10.

Jiang, Q., Doyle, J. D., Eckermann, S., & Williams, B. P. (2019). Stratospheric trailing gravity waves from New Zealand. *Journal of the Atmospheric Sciences*, under review.

Jin, Y. (1997). A numerical model study of the role of mesoscale gravity waves in rainband dynamics in the central United States during STORM-FEST, PhD dissertation. North Carolina State University.

Kaplan, M., Koch, S., Lin, Y.-L., Weglarz, R., & Rozumalski, R. (1997). Numerical simulations of a gravity wave event over CCOPE. Part I: The role of geostrophic adjustment in Mesoscale Jetlet Formation. *Monthly Weather Review*, *125*, 1185–1211.

Kawatani, Y., Watanabe, S., Sato, K., Dunkerton, T. J., Miyahara, S. & Takahashi, M. (2010) The Roles of Equatorial Trapped Waves and Internal Inertia–Gravity Waves in Driving the Quasi-Biennial Oscillation. Part II: Three-Dimensional Distribution of Wave Forcing. *Journal of the Atmospheric Sciences*, *67*, 981–997.

Kershaw, R. (1995). Parameterization of momentum transport by convectively generated gravity waves. *Quarterly Journal of the Royal Meteorological Society*, *121*, 1023–1040.

Klinker, E., & Sardeshmukh, P. D. (1992). The diagnosis of mechanical dissipation in the atmosphere from large-scale balance requirements. *Journal of Atmospheric Sciences*, *49*(7), 608–627. Retrieved Oct 8, 2021, from https://journals.ametsoc.org/view/journals/atsc/49/7/1520-0469_1992_049_0608_tdomdi_2_0_co_2.xml

Kruse, C. G. (2020). Regional to Global Evolution of Impacts of Parameterized Mountain-Wave Drag in the Lower Stratosphere. *J. Climate*, *33*, 3093–3106. https://doi.org/10.1175/JCLID-19-0076.1.

Kruse, C. G., Alexander, M. J., Hoffmann, L., van Niekerk, A., Polichtchouk, I., Bacmeister, J. T., et al. (2022). Observed and modeled mountain waves from the surface to the mesosphere near the Drake Passage. *Journal of the Atmospheric Sciences*, *79*, 909–932. https://doi.org/10.1175/JASD-21-0252.1.

Kruse, C. G., & Smith, R. B. (2018). Nondissipative and dissipative momentum deposition by mountain wave events in sheared environments. *Journal of the Atmospheric Sciences, 75*, 2721–2740. https://doi.org/10.1175/JAS-D-17-0350.1

Lac, C., Lafore, J. P., & Redelsperger, J. L. (2002). Role of gravity waves in triggering deep convection during TOGA COARE. *Journal of the Atmospheric Sciences*, *59*, 1293–1316. doi:https://doi.org/10.1175/1520-0469(2002)059<1293:ROG WIT>2.0.CO;2.

Lane, T. P., & Reeder, M. J. (2001). Convectively generated gravity waves and their effect on the cloud environment. *Journal of the Atmospheric Sciences*, *58*, 2427–2440.

Lane, T. P., Reeder, M. J., & Clark, T. L. (2001). Numerical modeling of gravity wave generation by deep tropical convection. *Journal of the Atmospheric Sciences*, *58*, 1249–1274. doi:https://doi.org/10.1175/1520-0469(2001)058<1249:NMO GWG>2.0.CO;2.

Lane, T. P., & Sharman, R. D. (2006). Gravity wave breaking, secondary wave generation, and mixing above deep convection in a three-dimensional cloud model. *Geophysical Research Letters*, *33*, L23813. doi:10.1029/2006GL027988.

Lane, T. P., & Zhang, F. (2011). Coupling between gravity waves and tropical convection at mesoscales. *Journal of the Atmospheric Sciences*, *68*, 2582–2598. doi:10.1175/2011JAS3577.1.

Leo, L. S., Thompson, M. Y., Di Sabatino, S., & Fernando, H. J. S. (2016). Stratified flow past a hill: dividing streamline concept revisited. *Boundary-Layer Meteorology*, *159*(3), 611–634.

Lelong, M., & Dunkerton, T. J. (1998). Inertia–gravity wave breaking in three dimensions. Part II: Convectively unstable waves. *Journal of the Atmospheric Sciences*, *55*, 2489–2501. https://doi.org/10.1175/1520-0469(1998)055<2489:IGW BIT>2.0.CO;2

Lighthill, J. M. (1952). On sound generated aerodynamically, I. General theory. *Proceedings of the Royal Society of London*, *211(A)*, 564–587.

Lilly, D. K. (1978). A severe downslope windstorm and aircraft turbulence event induced by a mountain wave. *Journal of the Atmospheric Sciences, 35*, 59–77. https://doi.org/10.1175/1520-0469(1978)035<0059:ASD WAA>2.0.CO;2

Lilly, D., & Klemp, J. (1979). The effects of terrain shape on nonlinear hydrostatic mountain waves. *Journal of Fluid Mechanics, 95*(2), 241–261. doi:10.1017/S0022112079001452

Lin, Y.-L., Deal, R. L., & Kulie, M. S. (1998). Mechanisms of cell regeneration, development, and propagation within a two-dimensional multicell storm. *Journal of the Atmospheric Sciences*, *55*, 1867–1886.

Lin, Y-L., & Wang, T. A. (1996). Flow regimes and transient dynamics of two-dimensional stratified flow over an isolated mountain ridge. *Journal of the Atmospheric Sciences, 53*, 139–158.

Lindzen, R. S. (1981). Turbulence and stress owing to gravity wave and tidal breakdown. *Journal of Geophysical Research*, *86*, 9707–9714.

Liu, H.-L., & Vadas, S. L. (2013). Large-scale ionospheric disturbances due to the dissipation of convectively-generated gravity waves over Brazil. *Journal of Geophysical Research*, *118*, 2419–2427. doi:10.1002/jgra.50244.

Lott, F. (1997). The transient emission of propagating gravity waves by a stably stratified shear layer, *Quarterly Journal of the Royal Meteorological Society*, *123*, 1603–1619.

Lott, F. (1998). Linear mountain drag and averaged pseudo-momentum flux profiles in the presence of trapped lee waves. *Tellus*, *50A*, 12–25.

Lott, F., & Guez, L. (2013). A stochastic parameterization of the gravity waves due to convection and its impact on the equatorial stratosphere. *Journal of Geophysical Research: Atmospheres*, *118*(16), 8897–8909.

Lott, F., Guez, L., & Maury, P. (2012b). A stochastic parameterization of non-orographic gravity waves: Formalism and impact on the equatorial stratosphere. *Geophysical Research Letters*, *39*, L06807. doi:10.1029/2012GL051001.

Lott, F., & Miller, M. (1997). A new sub-grid orographic drag parametrization: Its formulation and testing. *Quarterly Journal of the Royal Meteorological Society, 123*, 101–127.

Lott, F., Plougonven, R., & Vanneste, J. (2010). Gravity waves generated by sheared potential vorticity anomalies. *Journal of the Atmospheric Sciences*, *67*, 157–170. doi:10.1175/2009JAS3134.1.

Lott, F., Plougonven, R., & Vanneste, J. (2012a). Gravity waves generated by sheared three-dimensional potential vorticity anomalies. *Journal of the Atmospheric Sciences*, *69*, 2134–2151.

Luo, Z., & Fritts, D. (1993). Gravity wave excitation by geostrophic adjustment of the jet stream. Part II: Three dimensional forcing. *Journal of the Atmospheric Sciences*, *50*(1), 104–115.

Lyra, G. (1940). Über den Einfluß von Bodenerhebungen auf die Str ¨omung einer stabil geschichteten Atmospha ¨re. *Beitrage zur Physik der freien Atmosphare*, *26*, 197–206.

Lyra, G. (1943). Theorie der stationa ¨ren Leewellen- str ¨omung in freir Atmospha ¨re. *Zeitschrift für Angewandte Mathematik und Mechanik*, *23*, 1–28.

McFarlane, N. A. (1987). The effect of orographically excited gravity-wave drag on the general circulation of the lower stratosphere and troposphere. *Journal of the Atmospheric Sciences*, *44*, 1775–1800.

McLandress, C. (1998). On the importance of gravity waves in the middle atmosphere and their parameterization in general circulation models. *Journal of Atmospheric and Solar-Terrestrial Physics*, *60*, 1357–1383.

Medvedev, A. S., & Klaassen, G. P. (1995). Vertical evolution of gravity wave spectra and the parameterization of associated wave drag. *Journal of Geophysical Research*, *100*, 25,841–25,853.

Medvedev, A. S., & Yiğit, E. (2019). Gravity waves in planetary atmospheres: Their effects and parameterization in Global Circulation Models. *Atmosphere*, *10*(9), 531. https://doi.org/10.3390/atmos10090531

Miller, J. E. (1948). On the concept of frontogenesis. *Journal of Meteorology, 5*, 169–171.

Miranda, P. M. A., & James, I. N. (1992). Non-linear three-dimensional effects on gravity wave drag: splitting flow and breaking waves. *Quarterly Journal of the Royal Meteorological Society*, *118*, 1057–1081.

Miyoshi, Y., Fujiwara, H., Jin, H., & Shinagawa, H. (2014). A global view of gravity waves in the thermosphere simulated by a general circulation model. *Journal of Geophysical Research: Space Physics*, *119*, 5807–5820. doi: 10.1002/2014JA019848.

McFarlane, N. A. (1987). The effect of orographically excited wave drag on the general circulation of the lower stratosphere and troposphere. *Journal of the Atmospheric Sciences*, *44*, 1775–1800.

McLandress, C., Shepherd, T. G., Polavarapu, S., & Beagley, S. R. (2012). Is missing orographic gravity wave drag near 60°S the cause of the stratospheric zonal wind biases in chemistry–climate models?. *Journal of the Atmospheric Sciences*, *69*, 802–818. https://doi.org/10.1175/JAS-D-11-0159.1

Mixa, T., Dörnbrack, A., & Rapp, M. (2021). Nonlinear simulations of gravity wave tunneling and breaking over Auckland Island. *Journal of the Atmospheric Sciences*, *78*(5), 1567–1582. Retrieved Sep 30, 2021, from https://journals.ametsoc.org/view/journals/atsc/78/5/JAS-D-20-0230.1.xml

van Niekerk, A., & Vosper, S. (2021). Towards a more "scale-aware" orographic gravity wave drag parametrization: Description and initial testing. *Quarterly Journal of the Royal Meteorological Society*, *147*(739), 3243– 3262. Available from: https://doi.org/10.1002/qj.4126

Okamoto, K., Sato, K., & Akiyoshi, H. (2011). A study on the formation and trend of the Brewer-Dobson circulation. *Journal of Geophysical Research, 116*, D10117. doi:10.1029/2010JD014953.

Ólafsson, H., & Bougeault, P. (1996). Nonlinear flow past an elliptic mountain ridge. *Journal of the Atmospheric Sciences*, *53*, 2465–2489.

Palmer, T. N., Shutts, G. J., & Swinbank, R. (1986). Alleviation of a systematic westerly bias in general circulation and numerical weather prediction models through an orographic gravity wave drag parameterization. *Quarterly Journal of the Royal Meteorological Society*, *112*, 1001–1039.

Pandya, R. & Alexander, M. J. (1999). Linear stratospheric gravity waves above convective thermal forcing. *Journal of the Atmospheric Sciences*, *56*, 2434–2446.

Pandya, R. E., & Durran, D. R. (1996). The influence of convectively generated thermal forcing on the mesoscale circulation around squall lines. *Journal of the Atmospheric Sciences*, *53*, 2924–2951. doi:10.1175/1520-0469(1996)053, 2924:TIOCGT.2.0.CO;2.

Peltier, W. R., & Clark, T. L. (1979). The evolution and stability of finite-amplitude mountain waves. Part II: Surface wave drag and severe downslope windstorms. *Journal of the Atmospheric Sciences*, *36*, 1498–1529. https://doi.org/10.1175/1520-0469(1979)036<1498:TEA SOF>2.0.CO;2

Pfister, L., Scott, S., & Loewenstein, M. (1993). Mesoscale disturbances in the tropical stratosphere excited by convection: Observations and effects on the stratospheric momentum budget. *Journal of the Atmospheric Sciences*, *50*, 1058–1075.

Phillips, D. S. (1984). Analytical surface pressure and drag for linear hydrostatic flow over three-dimensional elliptical mountains. *Journal of the Atmospheric Sciences*, *41*, 1073–1084. https://doi.org/ 10.1175/1520-0469(1984)041,1073:ASPADF. 2.0.CO;2.

Piani, C., & Durran, D. (2001). A numerical study of stratospheric gravity waves triggered by squall lines observed during the TOGA COARE and COPT-81 experiments. *Journal of the Atmospheric Sciences*, *58*, 3702–3723.

Piani, C., Durran, D., Alexander, M., & Holton, J. (2000). A numerical study of three-dimensional gravity waves triggered by deep tropical convection and their role in the dynamics of the QBO. *Journal of the Atmospheric Sciences*, *57*, 3689–3702. doi:10.1175/1520-0469(2000)057,3689: ANSOTD.2.0.CO;2.

Pierrehumbert, R. T., & Wyman, B. (1985). Upstream effects of mesoscale mountains. *Journal of the Atmospheric Sciences*, *42*, 977–1003.

Plougonven, R., de la Cámara, A., Hertzog, A., & Lott, F. (2020). How does knowledge of atmospheric gravity waves guide their parameterizations?. *Quarterly Journal of the Royal Meteorological Society*, *146*, 1529–1543. https://doi.org/10.1002/qj.3732

Plougonven, R., Hertzog, A., & Alexander, M. J. (2015). Case studies of nonorographic gravity waves over the Southern Ocean emphasize the role of moisture. *Journal of Geophysical Research: Atmospheres*, *120*, 1278–1299. doi:10.1002/2014JD022332.

Plougonven, R., Teitelbaum, H., & Zeitlin, V. (2003). Inertia-gravity wave generation by the tropospheric mid-latitude jet as given by the FASTEX radiosoundings, *Journal of Geophysical Research*, *108*(D21), 4686. doi:10.1029/2003JD003535.

Plougonven, R., & Zeitlin, V. (2002). Internal gravity wave emission from a pancake vortex: an example of wave-vortex interaction in strongly stratified flows. *Physics of Fluids*, *14*(3), 1259–1268.

Plougonven, R., & Zhang, F. (2014). Internal gravity waves from atmospheric jets and fronts. *Reviews of Geophysics*, *52*, 33–76. doi:10.1002/2012RG000419.

Powers, J. (1997). Numerical model simulation of a mesoscale gravity-wave event: Sensitivity tests and spectral analyses, *Monthly Weather Review*, *125*, 1838–1869.

Powers, J., & Reed, R. (1993). Numerical simulation of the large-amplitude mesoscale gravity wave event of 15 December 1987 in the Central United States, *Monthly Weather Review*, *121*, 2285–2308.

Queney, P. (1947). Theory of perturbations in stratified currents with applications to airflow over mountain barriers. *Dept. of Meteorology. Univ. of Chicago, Misc. Report No. 23*.

Ray, E. A., Alexander, M. J., & Holton, J. R. (1998). An analysis of structure and forcing of the equatorial semiannual oscillation in zonal wind. *Journal of Geophysical Research*, *103*(D2), 1759–1774.

Richter, J., & Garcia, R. R. (2006). On the forcing of the mesospheric semi-annual oscillation in the Whole Atmosphere Community Climate Model. *Geophysical Research Letters*, *33*, L01806. doi:10.1029/2005GL024378.

Richter, J. H., Sassi, F., & Garcia, R. R. (2010). Towards a physically based gravity wave source parameterization in a general circulation model. *Journal of the Atmospheric Sciences*, *67*, 136–156.

Richter, J. H., Solomon, A., & Bacmeister, J. T. (2014). On the simulation of the quasi-biennial oscillation in the

Community Atmosphere Model, version 5. *Journal of Geophysical Research Atmospheres*, *119*, 3045–3062. doi:10.1002/2013JD021122.

Rind, D., Suozzo, R., Balachandran, N. K., Lacis, A., & Russell, G. (1988). The GISS global climate-middle atmosphere model. Part I: Model structure and climatology. *Journal of the Atmospheric Sciences, 45*, 329–370.

Rosenlof, K. (1996). Summer hemisphere differences in temperature and transport in the lower stratosphere. *Journal of Geophysical Research, 101*, 19,129–19,136.

Rossby, C.-G. (1938). On the mutual adjustment of pressure and velocity distributions in certain simple current systems, II. *Journal of Marine Research*, *1*, 239–263. https://doi.org/10.1357/002224038806440520.

Ruppert, J. H., Koch, S. E., Chen, X., Du, Y., Seimon, A., Sun, Y. Q., et al. (2022). Mesoscale gravity waves and midlatitude weather: A tribute to Fuqing Zhang. *Bulletin of the American Meteorological Society*, *103*, E129–E156. https://doi.org/10.1175/BAMS-D-20-0005.1.

Sandu, I., van Niekerk, A., Shepherd, T. G., Vosper, S. B., Zadra, A., Bacmeister, J., et al. (2019). Impacts of orography on large-scale atmospheric circulation. *npj Climate and Atmospheric Science*, *2*, 10. https://doi.org/10.1038/s41612-019-0065-9

Sato, K. (1994). A statistical study of the structure, saturation and sources of inertio-gravity waves in the lower stratosphere observed with the MU radar. *Journal of Atmospheric and Terrestrial Physics*, *56*(6), 755–774.

Sato, K., Tateno, S., Watanabe, S., & Kawatani, Y. (2012). Gravity wave characteristics in the southern hemisphere revealed by a high-resolution middle-atmosphere general circulation model. *Journal of the Atmospheric Sciences*, *69*(4), 1378–1396. https://journals.ametsoc.org/view/journals/atsc/69/4/jas-d-11-0101.1.xml

Satomura, T., & Sato, K. (1999). Secondary generation of gravity waves associated with the breaking of mountain waves. *Journal of the Atmospheric Sciences*, *56*, 3847–3858. https://doi.org/10.1175/1520-0469(1999)056<3847:SGOGWA>2.0.CO;2

Schecter, D. (2008). The spontaneous imbalance of an atmospheric vortex at high Rossby number. *Journal of the Atmospheric Sciences*, *65*, 2498–2521.

Schmidt, J., & Cotton, W. (1990). Interactions between upper and lower tropospheric gravity waves on squall line structure and maintenance. *Journal of the Atmospheric Sciences*, *47*, 1205–1222.

Scinocca, J., & Ford, R. (2000). The nonlinear forcing of large-scale internal gravity waves by stratified shear instability. *Journal of the Atmospheric Sciences*, *57*, 653–672.

Scinocca, J. F., & McFarlane, N. A. (2000). The parametrization of drag induced by stratified flow over anisotropic orography. *Quarterly Journal of the Royal Meteorological Society*, *126*(568), 2353–2393.

Scinocca, J. F., McFarlane, N. A. Lazare, M., Li, J., & Plummer, D. (2008). Technical Note: The CCCma third generation AGCM and its extension into the middle atmosphere. *Atmospheric Chemistry and Physics*, *8*, 7055–7074. doi:10.5194/acp-8-7055-2008.

Scorer, R. S. (1949). Theory of waves in the lee of mountains. *Quarterly Journal of the Royal Meteorological Society*, *75*, 41–56.

Serva, F., Cagnazzo, C., Riccio, A., & Manzini, E. (2018). Impact of a stochastic nonorographic gravity wave parameterization on the stratospheric dynamics of a general circulation model. *Journal of Advances in Modeling Earth Systems*, *10*(9), 2147–2162.

Sheppard, P. A. (1956). Airflow over mountains. *Quarterly Journal of the Royal Meteorological Society*, *82*, 528–529.

Shige, S., & Satomura, T. (2000). The gravity wave response in the troposphere around deep convection. *Journal of the Meteorological Society of Japan, 78*(6), 789–801. https://doi.org/10.2151/jmsj1965.78.6_789

Smith R. B. (1979). The influence of mountains on the atmosphere. In B. Saltzman (Ed.), *Advances in Geophysics* (Vol. 21, pp. 87–230). Academic Press. https://doi.org/10.1016/S0065-2687(08)60262-9

Smith, R. B. (1980). Linear theory of stratified hydrostatic flow past an isolated mountain. *Tellus, 32*, 348–364.

Smith, R. B. (1985). On severe downslope winds. *Journal of the Atmospheric Sciences, 42*, 2597–2603. https://doi.org/10.1175/1520-0469(1985)042<2597:OSDW>2.0.CO;2

Smith, R. B. (1989). Hydrostatic airflow over mountains. *Advances in Geophysics*, *31*, 1–41

Smith, R. B. (2019). *100 years of progress on mountain meteorology research.* Book chapter under review.

Smith, A. K., & Brasseur, G. P. (1991). Numerical simulation of the seasonal variation of mesospheric water vapor. *Journal of Geophysical Research*, *96*, 7553–7563. doi:10.1029/91JD00226.

Smith R. B., & Kruse, C. G. (2017). Broad-spectrum mountain waves. *Journal of the Atmospheric Sciences*, *74*, 1381–1402. https://doi.org/10.1175/JAS-D-16-0297.1.

Smith, R. B., & Kruse, C. G. (2018). A gravity wave drag matrix for complex terrain. *Journal of the Atmospheric Sciences, 75*, 2599–2613. https://doi.org/10.1175/JAS-D-17-0380.1

Smolarkiewicz, P. K., & Rotunno, R. (1990). Low Froude number flow past three-dimensional obstacles. Part II: Upwind flow reversal zone. *Journal of the Atmospheric Sciences*, *47*, 1498–1511.

Snively, J. B., & Pasko, V. P. (2008). Excitation of ducted gravity waves in the lower thermosphere by tropospheric sources. *Journal of Geophysical Research*, *113*, A06303. doi:10.1029/2007JA012693.

Song, I.-S. & Chun, H.-Y. (2005). Momentum flux spectrum of convectively forced internal gravity waves and its application to gravity wave drag parameterization. Part I: Theory. *Journal of the Atmospheric Sciences*, *62*(1), 107–124.

Song, I.-S., & Chun, H.-Y. (2008). A Lagrangian spectral parameterization of gravity wave drag induced by cumulus convection. *Journal of the Atmospheric Sciences*, *65*, 1204–1224.

Song, I.-S., Chun, H.-Y., & Lane, T. P. (2003). Generation mechanisms of convectively forced internal gravity waves and their propagation to the stratosphere. *Journal of the Atmospheric Sciences*, *60*, 1960–1980. doi:https://doi.org/10.1175/1520-0469(2003)060<1960:GMOCFI>2.0.CO;2.

Stephan, C., Alexander, M. J., & Richter, J. H. (2016). Characteristics of gravity waves from convection and implications for their parameterization in global circulation models. *Journal of the Atmospheric Sciences, 73*, 2729–2742. https://doi.org/10.1175/JAS-D-15-0303.1

Su, T., & Zhai, G. (2017). The role of convectively generated gravity waves on convective initiation: A case study. *Monthly Weather Review,145*, 335–359.

Sugimoto, N., Ishioka, K., & Ishii, K. (2008). Parameter sweep experiments on spontaneous gravity wave radiation from unsteady rotational flow in an f-plane shallow water system. *Journal of the Atmospheric Sciences, 65*, 235–249.

Tabaei, A., & Akylas, T. R. (2007). Resonant long-short wave interactions in an unbounded rotating stratified fluid. *Studies in Applied Mathematics, 119*, 271–296.

Teixiera, M. A. C. (2014). The physics of orographic gravity wave drag. *Frontiers in Physics, 2*, 43. https://doi.org/10.3389/fphy.2014.00043

Uccellini, L., & Koch, S. (1987). The synoptic setting and possible energy sources for mesoscale wave disturbances. *Monthly Weather Review, 115*, 721–729.

Vadas, S. L. (2007). Horizontal and vertical propagation, and dissipation of gravity waves in the thermosphere from lower atmospheric and thermospheric sources. *Journal of Geophysical Research, 112*, A06305. doi:10.1029/2006JA011845.

Vadas, S. L., & Fritts, D. C. (2005). Thermospheric responses to gravity waves: Influences of increasing viscosity and thermal diffusivity. *Journal of Geophysical Research, 110*, D15103. doi:10.1029/2004JD005574.

Vadas, S. L., Fritts, D. C., & Alexander, M. J. (2003). Mechanism for the generation of secondary waves in wave breaking regions. *Journal of the Atmospheric Sciences, 60*, 194–214.

Vadas, S. L., Liu, H.-L., & Lieberman, R. S. (2014). Numerical modeling of the global changes to the thermosphere and ionosphere from the dissipation of gravity waves from deep convection. *Journal of Geophysical Research: Space Physics, 119*. doi:10.1002/2014JA020280

Vadas, S. L. & Nicolls, M. J. (2012). The phases and amplitudes of gravity waves propagating and dissipating in the thermosphere: Theory. *Journal of Geophysical Research, 117*, A05322. doi:10.1029/2011JA017426.

Vadas, S. L., Zhao, J., Chu, X., & Becker, E. (2018). The Excitation of secondary gravity waves from local body forces: Theory and observation. *Journal of Geophysical Research Atmospheres, 123*, 9296–9325. https://doi.org/10.1029/2017JD027970

Van den Bremer, T. S., & Sutherland, B. R. (2014). The mean flow and long waves induced by two-dimensional internal gravity wavepackets. *Physics of Fluids, 26*(10), 106601. doi:10.1063/1. 4899262.

Vanneste, J., & Yavneh, I. (2004). Exponentially small inertia-gravity waves and the breakdown of quasi-geostrophic balance. *Journal of the Atmospheric Sciences, 61*, 211–223.

Vincent, R. A., Allen, S. J., & Eckermann, S. D. (1997). Gravity wave parameters in the lower stratosphere. In K. Hamilton (Ed.), *Gravity wave processes: Their parameterization in global climate models* (pp. 7–25). Springer.

Walterscheid, R. L., & Hickey, M. P. (2011). Group velocity and energy flux in the thermosphere: Limits on the validity of group velocity in a viscous atmosphere. *Journal of Geophysical Research, 116*, D12101. doi: 10.1029/2010JD014987.

Wang, J., Kim, H.-M., & Chang, E. K. M. (2018a). Interannual modulation of Northern Hemisphere winter storm tracks by the QBO. *Geophysical Research Letters, 45*. https://doi.org/10.1002/2017GL076929

Wang, J., Kim, H.-M., Chang, E. K. M., & Son, S.-W. (2018b). Modulation of the MJO and North Pacific storm track relationship by the QBO. *Journal of Geophysical Research Atmospheres, 123*, 3976–3992. https://doi.org/10.1029/2017JD027977.

Wang, L., & Geller, M. (2003). Morphology of gravity-wave energy as observed from 4 years (1998–2001) of high vertical resolution U.S. radiosonde data. *Journal of Geophysical Research, 108*, 4489. doi:10.1029/2002JD002786.

Warner, C. D., & McIntyre, M. E. (2001). An ultra-simple spectral parameterization for non-orographic gravity waves. *Journal of the Atmospheric Sciences, 58*, 1837–1857.

Webster, S., Brown, A. R., Cameron, D. R., & Jones, C. P. (2003). Improvements to the representation of orography in the Met Office Unified Model. *Quarterly Journal of the Royal Meteorological Society, 129*, 1989–2010. https://doi.org/10.1256/qj.02.133.

Wei, J., Bölöni, G., & Achatz, U. (2019). Efficient modelling of the interaction of mesoscale gravity waves with unbalanced large-scale flows: Pseudomomentum-flux convergence versus direct approach. *Journal of the Atmospheric Sciences, 76*, 2715–2738. doi: https://doi.org/10.1175/JAS-D-18-0337.1

Wei, J., & Zhang, F. (2014). Mesoscale gravity waves in moist baroclinic jet–front systems. *Journal of the Atmospheric Sciences, 71*, 929–952. doi:10.1175/JAS-D-13-0171.1.

Wei, J., & Zhang, F. (2015). Tracking gravity waves in moist baroclinic jet-front systems. *Journal of Advances in Modeling Earth Systems, 7*, 67–91. doi: 10.1002/2014MS000395

Wei, J., Zhang, F., & Richter, J. H. (2016). An analysis of gravity wave spectral characteristics in moist baroclinic jet-front systems. *Journal of the Atmospheric Sciences, 73*, 3133–3155. doi:10.1175/JAS-D-15-0316.1.

Wilhelm, J., Akylas, T. R., Bölöni, G., Wei, J., Ribstein, B., Klein, R., & Achatz, U. (2018). Interactions between mesoscale and submesoscale gravity waves and their efficient representation in mesoscale-resolving models. *Journal of the Atmospheric Sciences, 75*, 2257–2280. doi: https://doi.org/10.1175/JAS-D-17-0289.1

Winters, K., & Armi, L. (2014). Topographic control of stratified flows: Upstream jets, blocking and isolating layers. *Journal of Fluid Mechanics, 753*, 80–103. doi:10.1017/jfm.2014.363

Wu, D. L., & Zhang, F. (2004). A study of mesoscale gravity waves over the North Atlantic with satellite observations and a mesoscale model. *Journal of Geophysical Research, 109*, D22104. doi:10.1029/ 2004JD005090.

Yiğit, E., & Medvedev, A. S. (2009). Heating and cooling of the thermosphere by internal gravity waves. *Geophysical Research Letters, 36*, L14807. doi: 10.1029/2009GL038507.

Yiğit, E., Medvedev, A. S., Aylward, A. D., Hartogh, P., & Harris, M. J. (2009). Modeling the effects of gravity wave momentum deposition on the general circulation above the turbopause. *Journal of Geophysical Research, 114*, D07101. doi: 10.1029/2008JD011132.

Yoo, C., & Son, S.-W. (2016). Modulation of the boreal wintertime Madden-Julian oscillation by the stratospheric quasi-biennial oscillation. *Geophysical Research Letters*, *43*, 1392–1398. doi:10.1002/2016GL067762.

Yu, Y., Wang, W., & Hickey, M. P. (2017). Ionospheric signatures of gravity waves produced by the 2004 Sumatra and 2011 Tohoku tsunamis: A modeling study. *Journal of Geophysical Research: Space Physics*, *121*, 1146–1162. doi:10.1002/2016JA023116.

Zhang, D.-L., & Fritsch, J. (1988). Numerical simulation of the meso-b scale structure and evolution of the 1977 Johnstown flood. Part III: Internal gravity waves and the squall line. *Journal of the Atmospheric Sciences*, *45*, 1252–1268.

Zhang, F. (2004). Generation of mesoscale gravity waves in the upper-tropospheric jet-front systems. *Journal of the Atmospheric Sciences*, *61*, 440–457.

Zhang, F., & Koch, S. (2000). Numerical simulations of a gravity wave event over CCOPE. Part II: Waves generated by an orographic density current. *Monthly Weather Review*, *128*(8), 2777–2796.

Zhang, F., Koch, S. E., Davis, C. A., & Kaplan, M. L. (2000). A survey of unbalanced flow diagnostics and their application. *Advances in Atmospheric Sciences*, *17*, 165–183.

Zhang, F., Koch, S., Davis, C., & Kaplan, M. (2001). Wavelet analysis and the governing dynamics of a large amplitude mesoscale gravity wave event along the east coast of the United States. *Quarterly Journal of the Royal Meteorological Society*, *127*, 2209–2245.

Zhang, F., Koch, S., & Kaplan, M. (2003). Numerical simulations of a large-amplitude gravity wave event. *Meteorology and Atmospheric Physics*, *84*, 199–216.

Zhang, S., & Yi, F. (2005). A statistical study of gravity waves from radiosonde observations at Wuhan (30 degrees N, 114 degrees E) China. *Annales Geophysicae*, *23*, 665–673.

Zhang, S., & Yi, F. (2008). Intensive radiosonde observations of gravity waves in the lower atmosphere over Yichang (111 degrees 18' E, 30 degrees 42' N), *China. Annales Geophysicae*, *26*(7), 2005–2018.

Part II
Unifying Efforts

10

Higher-Order Equations Closed by the Assumed PDF Method: Suitability for Parameterizing Cumulus Convection

Vincent E. Larson

ABSTRACT

One method for parameterizing turbulence and clouds in the Earth's atmosphere is higher-order closure. In this method, equations are formed for relevant subgrid-scale moments, such as vertical turbulent fluxes and vertical velocity variance. These equations contain unclosed terms. Some of these terms may be closed via an assumption about the shape of the probability density function (PDF) of fields on the subgrid scale.

This chapter briefly summarizes the method of higher-order PDF closure. This method is compared to two other methods: the mass-flux method and low-order closure. It is argued that higher-order closure contains sufficient physics to parameterize cumulus convection.

10.1. INTRODUCTION

It is often believed that atmospheric boundary-layer turbulence can be adequately parameterized by turbulence closure methods, but that cumulus convection is better left to mass-flux schemes. This chapter expresses a different belief, the belief that higher-order closure models can also serve as viable convection parameterizations. It argues that accurately parameterizing cumulus convection requires in part the representation of key vertical turbulent advection terms. These terms are important for representing nonlocal transport. They are neglected by low order turbulence closure but are indeed retained by high-order turbulence closure. (For the purpose of this chapter, a closure model will be deemed "high order" if it keeps the key turbulent advection terms and "low order" if it omits those terms and also prognoses no more than one turbulent moment, e.g., turbulence kinetic energy.)

Department of Mathematical Sciences, University of Wisconsin-Milwaukee, Milwaukee, WI, USA; and Pacific Northwest National Laboratory, Richland, WA, USA

In addition, closing the equations by use of the assumed probability density function (PDF) method permits a detailed representation of buoyancy. Hence, this chapter argues that PDF-based higher-order closure is a suitable method for parameterizing cumulus convection because its equation set retains the necessary physics to do so.

The broader purpose of this chapter is twofold: first, it gives a brief overview of the methodology of PDF-based higher-order closure; and second, it compares higher-order closure to the perhaps more familiar mass flux and low-order approaches. The three methods of high-order closure, low-order closure, and mass flux may be seen as competing visions of how to treat nonlocal transport and buoyancy. Our discussion draws in part from the discussion in Mironov (2008).

For specificity, this chapter focuses on a particular higher-order PDF closure model, namely, the Cloud Layers Unified By Binormals (CLUBB) parameterization (Golaz et al., 2002; Larson, 2017). We focus on that model because it is the one that we are most familiar with. We do not attempt to write a broad overview of higher-order PDF closure models. The chapter concludes by displaying sample output from a global simulation that includes CLUBB acting as a deep convective parameterization.

Fast Processes in Large-Scale Atmospheric Models: Progress, Challenges, and Opportunities, Geophysical Monograph 282, First Edition.
Edited by Yangang Liu and Pavlos Kollias.

DOI: 10.1002/9781119529019.ch10

10.2. HISTORICAL REVIEW

Single-column turbulence closure modeling of cumulus clouds has a long history. We now mention a few landmark papers in order to give a flavor of the early promise and impediments.

Second-order closure models evolved early. Such models predict one or more second-order moments, such as the variance of the vertical component of velocity $\overline{w'^2}$. For instance, Rotta (1951) introduced a parameterization for pressure covariances of the form $\overline{p'\partial w'/\partial z}$. Such early closure models ignored clouds and were often targeted at idealized flows or engineering applications. By the 1970s, they had reached an advanced state (see, e.g., the review of Mellor and Herring (1973)).

Mellor and Yamada (1982) present a mature hierarchy of closure models, including a "Level $2\frac{1}{2}$" model that prognoses the turbulence kinetic energy but no other higher-order moments. It became widely applied as a boundary-layer parameterization in large-scale models (e.g., Janjic, 2001). Mellor and Yamada (1982) presented a simulation of a trade-wind cumulus layer in Figure 17 of their paper. That figure shows that although most simulated features are plausible, the simulated cloud fraction profile is "upside down," with its maximum near cloud top, rather than near cloud base as subsequently observed in simulations of trade-wind cumulus layers by fine-scale, three-dimensional, large-eddy models (e.g., Siebesma and Co-authors (2003)). Possibly this failure discouraged further modeling of cumulus layers using closure models. However, we argue that some of the physics needed to simulate cumuli accurately with closure models is contained only in models of higher order than Level $2\frac{1}{2}$.

Work on third-order closures which predict one or more third-order moments such as $\overline{w'^3}$ had reached an advanced state by the publication of André, de Moor, Lacarrère, and du Vachat (1978), which simulated the diurnal evolution of a dry atmospheric boundary layer. This model was extended to handle trade-wind cumulus clouds by Bougeault (1981). The results were quite encouraging, but there was little follow-up work for a number of years, perhaps because of the model's complexity and computational cost.

Meanwhile, a different approach, the mass-flux method, was being developed (see, e.g., Arakawa and Schubert (1974)). The mass-flux approach has become the dominant method to parameterize cumulus clouds in large-scale models.

10.3. OVERVIEW OF HIGHER-ORDER CLOSURE

The goal of higher-order closure is the same as the goal of any parameterization of atmospheric turbulence, namely, the parameterization of the subgrid vertical turbulent fluxes of heat, moisture, and momentum. In addition, higher-order closure shares a goal of macrophysical parameterizations, namely, the estimation of information about subgrid clouds, such as cloud fraction or liquid water content, which can then be fed into a microphysical or radiative parameterization.

Turbulence closure modeling begins with analysis rather than phenomenology. In particular, closure modeling starts with the advection-diffusion and Navier-Stokes equations. The equations are manipulated to form prognostic equations for horizontally averaged moments, in a manner akin to Reynolds averaging (Stull, 1988). Let $\overline{()}$ denote a grid box average and $()'$ denote a perturbation from that average. Then, for instance, to form an equation for the vertical turbulent flux of total water mixing ratio, $\overline{w'r'_t}$, one first writes down equations for perturbation vertical velocity w' and total water mixing ratio (cloud liquid plus vapor) r'_t:

$$\frac{\partial w'}{\partial t} = \cdots, \qquad \frac{\partial r'_t}{\partial t} = \cdots, \tag{10.1}$$

where t denotes time, and the right-hand sides have been omitted but come directly from the Navier-Stokes and advection-diffusion equation, respectively. Then, the following linear combination of these equations is added together and the product rule is used:

$$r'_t \frac{\partial w'}{\partial t} + w' \frac{\partial r'_t}{\partial t} = \frac{\partial w' r'_t}{\partial t} = \cdots \tag{10.2}$$

The resulting equation is horizontally averaged (more precisely, subjected to a running mean spatial filter, as in Germano (1992)). The process is repeated to form other moments of interest, such as the vertical turbulent flux of liquid water potential temperature $\overline{w'\theta'_l}$ or the vertical velocity variance $\overline{w'^2}$.

At this point, we have built a set of equations that forms a framework that is largely free of phenomenological assumptions. Unfortunately, it contains unclosed terms on the right-hand sides. These terms are not prognosed or diagnosed by any equation in the set. They must be parameterized in terms of available moments, and here phenomenology must be introduced. As an example, a listing of CLUBB's prognostic equation set is provided in Appendix A.

One strategy for closing some kinds of unclosed terms is the assumed probability density function method (Golaz et al., 2002; Lewellen & Yoh, 1993; Lappen & Randall, 2001; Mellor, 1977; Sommeria & Deardorff, 1977). In this method, the subgrid variability within a model grid box is assumed to be distributed according to a simple PDF shape. This is an effective way of parameterizing the buoyancy terms that are so central to cumulus convection, if (and only if) the PDF is multivariate. Therefore, among higher-order closure methods, this chapter focuses on those closed by the assumed PDF method. A synopsis of CLUBB's PDF closure method is provided in Appendix B.

In the end, the closed equation set contains many parameterized terms, but the turbulent production terms, which are typically a major source, are exact in functional form. (However, the moments that compose the production terms must be prognosed or diagnosed and hence may contain large errors in practice.)

The solution procedure for a higher-order model that is closed by the assumed PDF method, such as CLUBB, may be summarized in three steps:

1. *Advance higher-order moment equations one time step*: In addition to the grid-box-mean equations for wind, temperature, and moisture that are prognosed by a host model, CLUBB prognoses a number of second- and third-order moments. These include turbulent vertical fluxes ($\overline{w'r'_t}$, $\overline{w'\theta'_l}$), variances of vertical and horizontal velocity components ($\overline{w'^2}$, $\overline{u'^2}$, $\overline{v'^2}$), scalar variances and covariances ($\overline{r_t'^2}$, $\overline{\theta_l'^2}$, and $\overline{r'_t\theta'_l}$), and the third-order moment of vertical velocity ($\overline{w'^3}$). These prognostic equations contain terms that must be closed by use of modeling assumptions.

2. *Close pressure and dissipation terms*: Any higher-order equation that prognoses a moment of the form $\overline{w'x'}$, where x' is a scalar or velocity component, includes an unclosed factor of the form $\overline{x'\partial p'/\partial z}$, where p is pressure and z is the height coordinate. In addition, any equation that prognoses a variance contains an unclosed molecular dissipation term. In CLUBB, these terms are closed using classical closures that do not make use of an assumption about the shape of the subgrid PDF. Such terms cannot be closed using CLUBB's PDF because it does not include gradients of fields.

3. *Close remaining terms with an assumption about the subgrid PDF*: Aside from pressure and dissipation terms, all other terms in CLUBB are closed by integrating them over a subgrid PDF. This includes buoyancy-related moments that contain the perturbation virtual potential temperature, θ'_v. This variable indicates the buoyancy of an air parcel, and hence accurate parameterization of it is essential for accurate parameterization of convective flows. The other class of terms that may be closed by integration over the PDF are third-order terms, such as turbulent transport terms.

CLUBB's PDF shape is assumed to be a multivariate double Gaussian, i.e., a normal mixture:

$$P(w', r'_t, \theta'_l) = aG_1(w', r'_t, \theta'_l) + (1-a)G_2(w', r'_t, \theta'_l), \quad (10.3)$$

where G_1 and G_2 are multivariate *single* Gaussians, and $0 < a < 1$ is the mixture fraction, i.e., the fraction of probability mass associated with Gaussian component 1. To close a third-order moment, such as $\overline{w'^2r'_t}$, one needs to integrate the moment over the assumed PDF:

$$\overline{w'^2r'_t} = \int_{-\infty}^{-\infty}\int_{-\infty}^{-\infty}\int_{-\infty}^{-\infty} w'^2r'_tP(w', r'_t, \theta'_l)\mathrm{d}w'\,\mathrm{d}r'_t\,\mathrm{d}\theta'_l. \quad (10.4)$$

Besides CLUBB, there are other higher-order closure models in current use. One example of a higher-order closure parameterization that closes via a subgrid PDF and that is currently implemented in a large-scale atmospheric model is the Intermediately Prognostic Higher-Order Closure (IPHOC) parameterization (Cheng & Xu, 2006, 2008, 2009; Cheng et al., 2010), which prognoses all the moments prognosed by CLUBB, plus two additional third-order moments, $\overline{r_t'^3}$ and $\overline{\theta_l'^3}$.

An intermediate-order ("Level 3") closure parameterization is the Turbulence Kinetic Energy-Scalar Variance (TKESV) parameterization of Mironov and Machulskaya (2017), which prognoses TKE and scalar variances, and optionally a third-order moment related to cloud liquid water. It does not make explicit use of a multivariate PDF for closing the buoyancy terms. It also does not retain the turbulent advection terms in turbulent flux equations such as equation (10.5). A similar Level-3 model has been developed by Nakanishi and Niino (Nakanishi & Niino, 2004, 2006).

10.4. EXAMPLE: PROGNOSIS OF VERTICAL TURBULENT FLUX OF TOTAL WATER

To impart a flavor of higher-order equations, we now present CLUBB's prognostic equation for $\overline{w'r'_t}$, with some minor terms omitted:

$$\frac{\partial\overline{w'r'_t}}{\partial t} = \underbrace{-\frac{1}{\rho}\frac{\partial\rho\overline{w'^2r'_t}}{\partial z}}_{\text{turb advection}} \underbrace{-\overline{w'^2}\frac{\partial\overline{r_t}}{\partial z}}_{\text{turb production}} \underbrace{+\frac{g}{\theta_{vs}}\overline{r'_t\theta'_v}}_{\text{buoy production}}$$
$$\underbrace{-C_7\frac{g}{\theta_{vs}}\overline{r'_t\theta'_v}}_{\text{buoy pressure}} \underbrace{-\frac{C_6}{\tau}\overline{w'r'_t}}_{\text{ret isotropy}} + \underbrace{\left.\frac{\partial\overline{w'r'_t}}{\partial t}\right|_{\text{mc}}}_{\text{microphysics}}. \quad (10.5)$$

Here g is the acceleration due to gravity, θ_{vs} is a base-state virtual potential temperature, and C_7 and C_6 are constants. Each term can be given a physical interpretation.

The (vertical) *turbulent advection term* is a flux-of-flux term. In particular, it is the vertical advection of the flux, $\overline{w'r'_t}$, by the turbulent velocity, w'. Because this term contains a vertical derivative, $\partial/\partial z$, its vertical integral is zero, aside from boundary contributions. That is, it is neither a net sink nor a net source. In typical atmospheric cumulus layer, it takes variability from near cloud base and deposits it near cloud top.

The *turbulent production term* (usually) generates $w'r'_t$, that is, it creates a correlation between w' and r'_t, when there is turbulence ($\overline{w'^2}$) in the presence of a vertical gradient of moisture ($\partial\overline{r}_t/\partial z$). Typically, moisture mixing ratio drops off with increasing altitude. When turbulence is present in such a moisture profile, the upward-moving parcels carry up moister air, leading to moist updrafts and drier downdrafts. Hence, the turbulent production term is typically a source. Occasionally, however, it is a sink, and hence some authors refer to it as the "mean-gradient term."

The *buoyancy production/destruction term* indicates the covariance between moisture (r'_t) and buoyancy $(g/\theta_{vs})\theta'_v$. It can have either sign, but it is often negative in cumulus clouds (Heinze et al., 2015), indicating destruction rather than production.

The *return-to-isotropy term* is a parameterized pressure term. It damps $\overline{w'r'_t}$ to zero with a time scale τ. In this way, it mimics the tendency of non hydrostatic pressure fluctuations to drive the turbulence toward an isotropic state in which w' and r'_t are uncorrelated.

The *buoyancy-pressure term* opposes the buoyancy production/destruction term. It represents the fact that a rising parcel pushes overlying air out of its way and experiences a high-pressure region above the parcel and a low-pressure region underneath (section 7.2 of Houze (1993)).

The *microphysics term* accounts for the effects on $w'r'_t$ by microphysical processes, e.g., r'_t, which includes vapor and cloud liquid water but not rain water, may be diminished by formation of rain or increased by evaporation of rain.

10.5. HOW DO MASS-FLUX SCHEMES, HIGHER-ORDER CLOSURES, AND LOW-ORDER CLOSURES REPRESENT NONLOCAL VERTICAL TRANSPORT OF MOISTURE?

"Nonlocal" vertical transport means transport across multiple vertical grid levels within one model time step. Such transport is caused in nature by buoyant, convective updrafts that penetrate upward. Nonlocal transport is thought to be needed in models with long time steps in order to represent transport by cumulus clouds.

We see that penetrative convection is represented by mass-flux and higher-order closure parameterizations by analogous vertical turbulent advection terms (de Roode et al., 2000; Mironov, 2008). Low-order closure, however, omits the relevant term.

10.5.1. Nonlocal Transport in Mass-Flux Schemes

Whereas higher-order equations are phrased in terms of a single grid-layer averaging operator, $\overline{()}$, including all clear and cloudy areas, mass-flux equations are phrased in terms of a conditional average over a cloudy plume, $\overline{()}^{\text{cld}}$, plus a separate conditional average over the environment. Because of the conditional averaging, mass-flux equations must consider transport through the lateral boundary of the cloudy plume, that is, entrainment (E) and detrainment (D). A prototype equation for transport of total water (vapor plus cloud liquid), r_t, within a cloudy plume by a simple bulk mass-flux scheme may be written as (Arakawa & Schubert, 1974; Plant, 2010)

$$\frac{\partial \rho\sigma\overline{r_t}^{\text{cld}}}{\partial t} = -\frac{\partial M\overline{r_t}^{\text{cld}}}{\partial z} + Er_{t,\text{env}} - D\overline{r_t}^{\text{cld}} - \overline{\text{Precip}}^{\text{cld}}, \quad (10.6)$$

where the mass flux, M, is given by

$$M = \rho\sigma\overline{w}^{\text{cld}}. \quad (10.7)$$

Here, ρ denotes air density and σ denotes plume fractional area.

The role of transport by mass flux is to restore balance when there is an imbalance between sources and sinks at a particular altitude. One could conceive of a flow in which the source of plume moisture (entrainment) exactly balances the sinks (detrainment and precipitation) at each altitude. In this flow, there would be a local balance at each altitude, and there would be no need for either the mass flux M or the time tendency to be non-zero. In nature, however, the entrainment near cloud base is often large, creating a net source of moisture there, and the detrainment near cloud top is also large, creating a net sink of moisture aloft. To restore a quasi steady-state equilibrium at each altitude, the mass-flux term transports excess moisture from the lower-level source region to the upper-level sink region.

To solve equation (10.6) numerically, the time tendency term is typically dropped (Bretherton et al., 2004; Zhang & McFarlane, 1995) and the resulting steady-state equation is integrated in the vertical. The mass flux vertically transports moisture "nonlocally" with the updraft speed $\overline{w}^{\text{cld}}$. That is, moisture is transported across several vertical grid levels within one time step. This "nonlocal" behavior originates from the vertical derivative and the assumption of a steady state. Although a vertical derivative may appear to be a local operator because it only affects adjacent grid levels, it does connect different levels, and in that respect differs from the other "local" terms in the equation.

A mass-flux model typically contains similar, additional equations to transport heat content and possibly cloud liquid water. All quantities are transported by the same updraft speed, $\overline{w}^{\text{cld}}$.

10.5.2. Nonlocal Transport in Higher-Order Closure Models

One key similarity between the mass flux and higher-order closure equations is that they both contain a term with a vertical derivative. The mass-flux equation (10.6) contains $\partial(M\overline{r_t}^{\text{cld}})/\partial z$, which transports the within-plume r_t by the mass flux, M. Here, the quantity in parentheses is given by

$$M\overline{r_t}^{\text{cld}} = \rho\sigma\overline{w}^{\text{cld}}\overline{r_t}^{\text{cld}}. \quad (10.8)$$

It is this term that gives rise to "nonlocal" vertical transport by convection. Analogously, the higher-order closure equation (10.5) contains $-(1/\rho)\partial(\rho\overline{w'^2r_t'})/\partial z$, which transports the flux, $\overline{w'r_t'}$, by the turbulent perturbation, w'. How closely related are these two terms? To see the relationship, we approximate the flux-of-flux term.

The flux of flux, $\overline{w'^2r_t'}$, can be expressed in terms of simpler moments (Larson & Golaz, 2005) if the subgrid PDF is assumed to have a particular double Gaussian shape (namely, the Analytic Double Gaussian 1 (ADG1) shape of Larson, Golaz, and Cotton (2002)):

$$\rho\overline{w'^2r_t'} = \rho a_1 \frac{\overline{w'^3}}{\overline{w'^2}}\,\overline{w'r_t'}. \quad (10.9)$$

where the coefficient a_1 is

$$a_1 = \frac{1}{(1-\tilde{\sigma}_w^2)} \quad (10.10)$$

and $0 < \tilde{\sigma}_w^2 < 1$ determines the width of each Gaussian component in w versus the width of the overall PDF, including both components (Larson, 2017). A similar expression was proposed earlier by Abdella and McFarlane (1997) and Zilitinkevich, Gryanik, Lykossov, and Mironov (1999). (See also the review by Mironov (2008).)

The flux-of-flux expression (10.9) resembles the mass-flux expression (10.8). Both appear under the $\partial/\partial z$ operator. Here, the mass-flux velocity $\overline{w}^{\text{cld}}$ is the average velocity within the cloudy plume. It is related to convective strength and is less relevant in nonconvective (stratiform) regimes, where the mass-flux contribution is weak or zero. The higher-order closure analog to $\overline{w}^{\text{cld}}$ is the factor $\overline{w'^3}/\overline{w'^2}$, which has units of velocity. This "convective velocity" determines how fast the flux-of-flux term transports $\overline{w'r_t'}$ vertically (Zilitinkevich et al., 1999). The velocity $\overline{w'^3}/\overline{w'^2}$ is large and positive in cumulus cases, for which $\overline{w'^3}$ is large. It is small in stratocumulus cases, for which $\overline{w'^3}$ is small.

We have seen that the third-order term in the $\overline{w'r_t'}$ equation vertically transports $\overline{w'r_t'}$ by the "convective velocity" $\overline{w'^3}/\overline{w'^2}$. An analogous situation holds for the scalar variances, such as $\overline{r_t'^2}$. The corresponding third-order term, $\overline{w'r_t'^2}$, in CLUBB's $\overline{r_t'^2}$ equation is derived by integrating over CLUBB's PDF. We find (see equation (45) of Larson and Golaz (2005)):

$$\rho\overline{w'r_t'^2} = \rho a_1 \frac{\overline{w'^3}}{\overline{w'^2}}\overline{r_t'^2}\left(\frac{1}{3}\beta + \left(1-\frac{1}{3}\beta\right)a_1 \text{corr}^2_{w',r_t'}\right), \quad (10.11)$$

where $0 < \beta < 3$ is a tunable parameter and $-1 < \text{corr}_{w',r_t'} < 1$ is the linear correlation between w' and r_t'. Once again, the third-order term lofts the moment of interest (here, $\overline{r_t'^2}$) by the same convective velocity that lofted the flux $\overline{w'r_t'}$. Expression (10.11) extends that of Abdella and McFarlane (1999), Abdella and Petersen (2000), and Mironov, Gryanik, Lykossov, and Zilitinkevich (1999) by assuming that the skewnesses of r_t and w are proportional to each other (see equation (33) of Larson and Golaz (2005)).

In principle, one could imagine numerically solving the prognostic equations for $\overline{w'r_t'}$, (10.9), and $\overline{r_t'^2}$, (10.11), analogously to how the mass-flux equation is solved: the time tendency term could be dropped, and the $\partial/\partial z$ term could be integrated vertically, forming a kind of steady-state "plume" equation. Instead, however, CLUBB retains the time tendency term and discretizes the equation using a short time step (~5 min in global climate simulations, such as Bogenschutz et al. (2013)). Nevertheless, the higher-order equations contain the essential advective physics, i.e., a vertical derivative, that leads to nonlocal transport in mass-flux schemes.

10.5.3. Nonlocal Transport in Low-Order Closure Models

Low-order closure models include the widely used Mellor-Yamada 2.5-level model (Mellor & Yamada, 1982) and the more recent Simplified Higher-Order Closure (SHOC) model of Bogenschutz and Krueger (2013). Both these models prognose a single moment, the turbulence kinetic energy (TKE), but SHOC uses a double-Gaussian PDF to close some of the buoyancy terms.

A low-order closure approach would simplify equation (10.5) by dropping the time tendency and turbulent advection terms:

$$0 = \underbrace{-\overline{w'^2}\frac{\partial \overline{r_t}}{\partial z}}_{\text{turb production}} \underbrace{+\frac{g}{\theta_{vs}}\overline{r_t'\theta_v'}}_{\text{buoy production}} \underbrace{-C_7\frac{g}{\theta_{vs}}\overline{r_t'\theta_v'}}_{\text{buoy pressure}} \underbrace{-\frac{C_6}{\tau}\overline{w'r_t'}}_{\text{ret isotropy}} \quad (10.12)$$

Unlike the mass-flux model, which contains a vertical advection term, low-order closure contains only local source and sink terms. If we were to drop the buoyancy/pressure term, pull the return-to-isotropy term to the left-hand side, and solve for $\overline{w'r'_t}$, then we would find

$$\overline{w'r'_t} = -K\frac{\partial \overline{r_t}}{\partial z}, \tag{10.13}$$

where K is an eddy diffusivity defined as

$$K \equiv \frac{\tau}{C_6}\overline{w'^2}. \tag{10.14}$$

A closure approach in the spirit of Mellor and Yamada (1982) would retain the buoyancy-related terms and effectively close them by downgradient diffusion, thereby replacing K in (10.13) with a modified eddy diffusivity K_H.

Nevertheless, because the vertical turbulent advection term has been dropped, the nonlocal character of the mass-flux scheme has disappeared. What remains is a local, downgradient scheme that is missing the cumulus transport physics of the turbulent advection term.

There have been efforts to restore nonlocal transport to low-order schemes (Deardorff, 1972; Holtslag & Moeng, 1991), but these simple approaches do not retain the vertical derivative that is a key to advective vertical transport. Instead, they use quite different functional forms.

10.6. EXAMPLE: PROGNOSIS OF THE VERTICAL COMPONENT OF TURBULENCE KINETIC ENERGY, $\overline{W'^2}$

Whereas the equation for $\overline{w'r'_t}$ governs transport of moisture, the equation for $\overline{w'^2}$ governs the generation of turbulent updrafts and downdrafts, which, in convective flows, is mostly by buoyancy:

$$\frac{\partial \overline{w'^2}}{\partial t} = \underbrace{-\frac{1}{\rho}\frac{\partial \rho \overline{w'^3}}{\partial z}}_{\text{turb advection}} \underbrace{+\frac{2g}{\theta_{vs}}\overline{w'\theta'_v}}_{\text{buoy production}} \underbrace{+\frac{2}{3}C_5\left(-\frac{2g}{\theta_{vs}}\overline{w'\theta'_v}\right)}_{\text{buoy pressure}}$$
$$\underbrace{-\frac{C_4}{\tau}\left(\overline{w'^2}-\frac{2}{3}\overline{e}\right)}_{\text{ret isotropy}} \underbrace{-\frac{C_1}{\tau}\overline{w'^2}}_{\text{dissipation}}. \tag{10.15}$$

Here, C_5, C_4, and C_1 are constants, τ serves as both a return-to-isotropy time scale and a dissipation time scale, and $\overline{e}$ is turbulence kinetic energy.

The prognostic equation for $\overline{w'^2}$ has similarities and differences to the prognostic equation for $\overline{w'r'_t}$ (10.5). Like the $\overline{w'r'_t}$ equation, the $\overline{w'^2}$ equation contains a turbulent advection term that integrates vertically to zero. Unlike the $\overline{w'r'_t}$ equation, the $\overline{w'^2}$ equation contains a return-to-isotropy term that drives the value of $\overline{w'^2}$ toward the values of $\overline{u'^2}$ and $\overline{v'^2}$. This behavior results from the action of nonhydrostatic pressure fluctuations on turbulent eddies. The dissipation term parameterizes viscous dissipation under the assumption that the dissipation rate is controlled by the large eddies.

In the $\overline{w'^2}$ equation, the buoyancy production/destruction is the dominant source term in convective flows. A key point is that buoyancy production/destruction depends on the *covariance* of w' and θ'_v: turbulence is produced if updrafts are buoyant. However, if updrafts overshoot their level of neutral buoyancy, then w' and θ'_v are anticorrelated, and turbulence kinetic energy is consumed.

10.7. HOW DO MASS-FLUX SCHEMES, HIGHER-ORDER CLOSURE, AND LOW-ORDER CLOSURE REPRESENT BUOYANCY?

Because convective flows are driven by buoyancy, parameterization of convection requires first and foremost an accurate estimate of buoyant dynamics. For instance, as shown in equation (10.15), the main source term in the $\overline{w'^2}$ equation is the buoyant production term, $\frac{2g}{\theta_{vs}}\overline{w'\theta'_v}$. Buoyancy is represented differently in mass-flux, higher-order closure, and low-order closure models, as we discuss presently.

10.7.1. Buoyancy in Mass-Flux Schemes

Most mass-flux equations are advection-diffusion equations with the form of (10.6). Such equations do not explicitly exhibit the acceleration due to gravity, g, and hence it may not be obvious at first where the physics of buoyancy is parameterized. For many standard mass-flux schemes, this physics is contained in the closure assumption for the mass-flux, M. This closure is often written in terms of the convective available potential energy, CAPE, and a time scale over which CAPE is consumed, $\tau_{\text{convection}}$ (see, e.g., Zhang and McFarlane (1995), Yano and Plant (2012), or Section 11 of Chapter 11 of Plant and Yano (2016)):

$$M \propto \text{CAPE}/\tau_{\text{convection}} \tag{10.16}$$

where

$$\text{CAPE} = g\int_{z_{\text{cloud base}}}^{z_{\text{cloud top}}} \frac{\theta_{v,\text{parcel}} - \theta_{v,\text{environment}}}{\theta_{v,\text{environment}}}\, dz, \tag{10.17}$$

which has units of velocity squared. Note that the desired quantity is the *rate* of conversion from potential energy to kinetic energy, but that CAPE itself is only the *total amount* of potential energy available for conversion. To form an expression for a *rate*, CAPE is divided by $\tau_{\text{convection}}$, but this places a large burden on $\tau_{\text{convection}}$ because the rate is greatly influenced by factors such as the strength of the capping inversion above the planetary boundary layer. Because of such concerns, there has been considerable effort to formulate improved closure assumptions (Plant & Yano, 2016; Yano, 2014), but these are not discussed here.

The correlation between updraft speed and buoyancy is essential to convection. Only a positive correlation leads to generation of turbulence kinetic energy by buoyancy. The correlation is not treated explicitly in mass-flux formulations but instead is implicitly built into the closure and the calculation of CAPE.

10.7.2. Buoyancy in Higher-Order Closure

In higher-order closure, the buoyancy term, $\frac{2g}{\theta_{vs}}\overline{w'\theta'_v}$, appears explicitly in the $\overline{w'^2}$ equation (10.15). The buoyancy term is not merely a measure of available potential energy such as CAPE, but rather it is the rate of buoyant generation of turbulence kinetic energy (TKE). It generates TKE in convective flows but can consume TKE in stably stratified layers.

The quantity $\overline{w'\theta'_v}$ is computed in CLUBB by integration over the subgrid PDF. For this to be feasible, the PDF must be multivariate. That is, it must capture the relationship (i.e., covariance) between updraft speed w' and parcel buoyancy θ'_v. Predicting separate univariate PDFs for w' and θ'_v is insufficient. The accuracy of this integral depends on both the realism of the PDF shape and also the accuracy with which the moments are prognosed.

10.7.3. Buoyancy in Low-Order Closure

Low-order closure also contains the buoyancy term, $\frac{2g}{\theta_{vs}}\overline{w'\theta'_v}$, in the $\overline{w'^2}$ equation. In this respect, it is similar to higher-order closure.

10.8. MUST HIGHER-ORDER CLOSURE MODELS OF CUMULUS LAYERS INCLUDE VERTICAL COHERENCE AND ICE?

One may wonder whether higher-order closure models can accurately simulate cumulus layers without an explicit representation of vertical coherence. Although "vertical coherence" is informally understood to mean the degree to which a buoyant plume can rise while maintaining its identity, the term needs to be defined mathematically if it is to be modeled. Suppose that one defines "vertical coherence" as a large positive two-point correlation between vertical velocity measured at two points separated in the vertical (see section 6.3 of Pope (2000)). Then, vertical coherence need not be treated explicitly in higher-order closure models because two-point correlations are not required for closure of higher-order closure models (or for closure of mass-flux models). What is required for closure of higher-order models and is of distinctive importance in cumulus layers, is (1) the third-order turbulent advection terms, which tend to deepen the layers, and (2) the buoyancy terms, which depend sensitively on cloud fraction. Parameterizing these terms with adequate generality is a challenge, of course, but we argue that the higher-order framework itself is sufficiently comprehensive even if two-point correlations are ignored and vertical coherence exists only implicitly.

One may also wonder whether an accurate simulation of cumulus convection can be achieved by a higher-order closure model that omits ice. Liquid can be readily included if a model assumes "saturation adjustment," that is, assumes that condensation occurs as soon as air becomes supersaturated. In contrast, ice does not behave this way (Kärcher & Burkhardt, 2008). Hence, ice effects should not be embedded in a turbulence model via saturation adjustment. Rather, they ought to be calculated by a separate prognostic microphysics scheme. It is true that the effects of ice ought to be applied not only to the horizontal mean fields but also to the turbulent fluctuations. Methods for doing so are described in, e.g., Griffin and Larson (2016).

10.9. GLOBAL SIMULATIONS USING HIGHER-ORDER CLOSURE TO REPRESENT DEEP CONVECTION

Higher-order closure has been used to represent deep atmospheric convection in single-column simulations (Storer et al., 2013) and in two global simulations (Guo et al., 2015; Thayer-Calder et al., 2015). Both global studies found that use of higher-order closure performed comparably to a mass-flux parameterization for several metrics related to clouds and precipitation.

The global studies employed version 1 of the Morrison-Gettelman (MG1) microphysics (Morrison & Gettelman, 2008), which diagnoses rain and snow mixing ratios. Here, we show a couple sample plots from a configuration of the community atmosphere model (CAM; Neale et al. (2012)) that is very similar to Thayer-Calder et al. (2015), but that uses version 2 of the Morrison-Gettelman (MG2) microphysics, which prognoses rain and snow (Gettelman & Morrison, 2015). MG2 microphysics does omit graupel and hail but is otherwise a fairly comprehensive double-moment microphysics scheme.

The simulation includes a Monte Carlo sampler in order to interface cloud fields to the microphysics (Larson & Schanen, 2013). The simulation has a horizontal resolution of 2 degrees (approximately 200 km × 200 km) and a duration of 5 years. Sea-surface temperatures are prescribed. The Zhang-McFarlane deep convective mass-flux scheme (Zhang & McFarlane, 1995) is shut off, and all turbulence and cloud processes are simulated in a unified way by CLUBB.

Figure 10.1 compares the simulated surface precipitation rate versus satellite observations. There is excessive

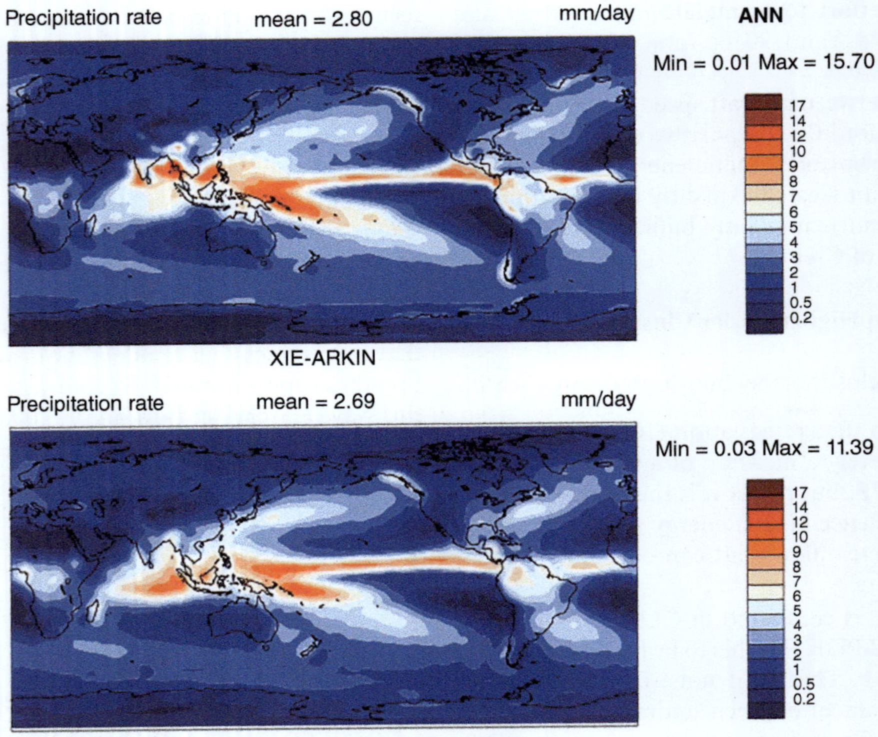

Figure 10.1 Simulation of surface precipitation rate by CAM-CLUBB averaged over 5 years (upper panel) and observations of surface precipitation (lower panel *Source*: Adapted from Xie and Arkin (1997)).

rainfall over the Indian subcontinent and in the South Pacific Convergence Zone, but there little hint of a spurious double Inter Tropical Convergence Zone, as desired, and generally the precipitation climatology is credible.

Figure 10.2 shows the zonally averaged atmospheric temperature and errors versus re-analysis. The simulated polar tropopause is too cold, and the tropical upper troposphere is too warm. Nevertheless, the temperature is consistent with a tropical atmosphere that is experiencing latent heating by deep convection.

In summary, these plots, in conjunction with the studies of Guo et al. (2015) and Thayer-Calder et al. (2015), suggest that higher-order closure is capable of parameterizing deep convection.

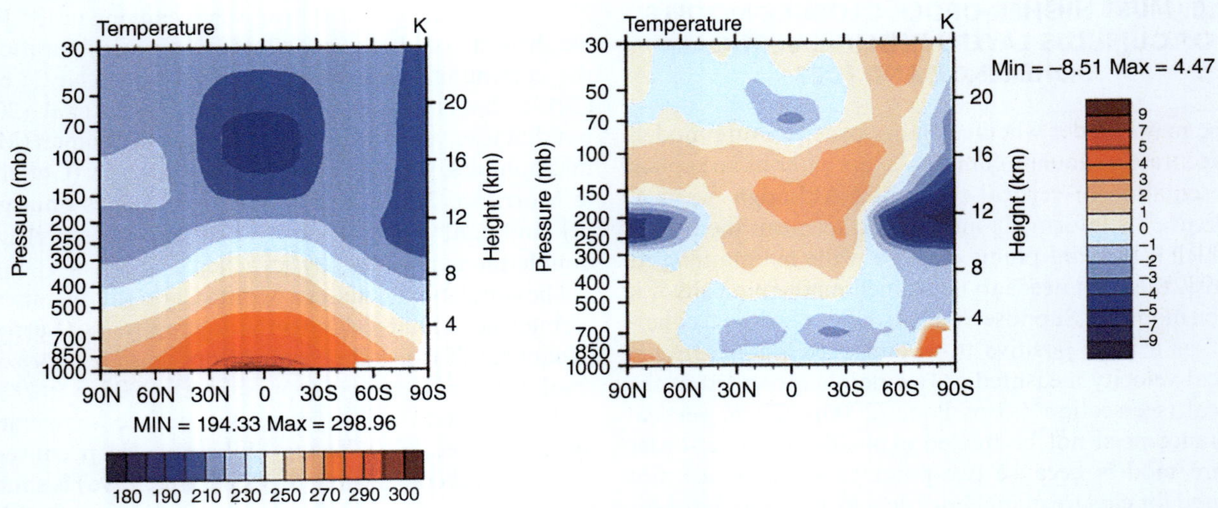

Figure 10.2 Simulation of zonally averaged absolute temperature by CAM-CLUBB averaged over 5 years (left panel) and errors with respect to the Japanese 25-year reanalysis (simulation reanalysis, right panel). *Source*: Adapted from Onogi et al. (2007).

10.10. CONCLUSIONS

There has been skepticism that higher-order closure is well suited to parameterizing the effects of cumulus convection. But an inspection of the higher-order equations shows that they contain the flux-of-flux terms that give rise to nonlocal cumulus transport. In addition, the higher-order equations contain a detailed representation of buoyant generation of turbulence, which is the essential source term of convection. That representation can be closed by a multivariate PDF.

Perhaps the reason for skepticism of high-order closure is that *low-order* closure does omit the crucial flux-of-flux terms. These terms are important for representing non-local cumulus transport. But these omissions are not flaws of turbulence closure methods *per se*. If the relevant terms are retained, accurate simulation of cumulus clouds becomes feasible.

APPENDIX A
PROGNOSTIC EQUATIONS IN CLUBB

For reference, the prognostic equations in CLUBB are listed here.

We assume that grid-mean equations for heat content, moisture, and momentum are provided by the host model, with the turbulent vertical flux terms in those equations supplied by CLUBB. We do not list those equations here.

We further assume that the equation for $\overline{w'r'_t}$ is given by (10.5), and the equation for $\overline{w'^2}$ is given by (10.15).

As before, for clarity we omit minor terms, such as terms that include the grid-mean vertical wind, $\overline{w}$, which is small for large grid boxes, and "background" diffusion terms. See Larson (2017) for more information. In the following equations, all variables denoted with a capital C are constants.

$$\frac{\partial \overline{r_t'^2}}{\partial t} = \underbrace{-\frac{1}{\rho_s}\frac{\partial \rho_s \overline{w' r_t'^2}}{\partial z}}_{\text{turb advection}} \underbrace{-2\overline{w'r'_t}\frac{\partial \overline{r_t}}{\partial z}}_{\text{turb production}} \underbrace{-\frac{C_2}{\tau}\overline{r_t'^2}}_{\text{dissipation}} + \underbrace{\left.\overline{\frac{\partial r_t'^2}{\partial t}}\right|_{\text{mc}}}_{\text{microphysics}} \tag{A.1}$$

CLUBB also contains equations for $\overline{w'\theta'_l}$ and $\overline{\theta_l'^2}$ that are analogous to (10.5) and (A.1) (not shown).

$$\frac{\partial \overline{r'_t\theta'_l}}{\partial t} = \underbrace{-\frac{1}{\rho_s}\frac{\partial \rho_s \overline{w' r'_t \theta'_l}}{\partial z}}_{\text{turb advection}} \underbrace{-\overline{w'r'_t}\frac{\partial \overline{\theta_l}}{\partial z} - \overline{w'\theta'_l}\frac{\partial \overline{r_t}}{\partial z}}_{\text{turb production}} \underbrace{-\frac{C_2}{\tau}\overline{r'_t\theta'_l}}_{\text{dissipation}} + \underbrace{\left.\overline{\frac{\partial r'_t\theta'_l}{\partial t}}\right|_{\text{mc}}}_{\text{microphysics}} \tag{A.2}$$

Table A.1 List of Prognosed Moments in CLUBB.

Prognosed moment	Type of moment	Equation
$\overline{w'r'_t}$	Turbulent vertical flux	(10.5)
$\overline{w'\theta'_l}$	Turbulent vertical flux	analogous to (10.5)
$\overline{w'^2}$	Vertical velocity variance	(10.5)
$\overline{u'^2}$	Eastward velocity variance	(A.3)
$\overline{v'^2}$	Northward velocity variance	Analogous to (A.3)
$\overline{r_t'^2}$	Scalar variance	(A.1)
$\overline{\theta_l'^2}$	Scalar variance	Analogous to (A.1)
$\overline{r'_t\theta'_l}$	Scalar covariance	(A.2)
$\overline{w'^3}$	Third-order moment of vertical velocity	(A.4)

$$\frac{\partial \overline{u'^2}}{\partial t} = \underbrace{-\frac{1}{\rho_s}\frac{\partial \rho_s \overline{w'u'^2}}{\partial z}}_{\text{turb advection}} \underbrace{-(1-C_5)\,2\overline{u'w'}\frac{\partial \overline{u}}{\partial z}}_{\text{turb production + press}} \underbrace{-\frac{2}{3}C_{14}\frac{\overline{e}}{\tau}}_{\text{dissipation}} \underbrace{+\frac{2}{3}C_5\left(\frac{g}{\theta_{vs}}\overline{w'\theta'_v} - \overline{u'w'}\frac{\partial \overline{u}}{\partial z} - \overline{v'w'}\frac{\partial \overline{v}}{\partial z}\right)}_{\text{pressure}} \underbrace{-\frac{C_4}{\tau}\left(\overline{u'^2} - \frac{2}{3}\overline{e}\right)}_{\text{ret isotropy}} \tag{A.3}$$

An analogous equation to (A.3) but for $\overline{v'^2}$ is also contained in CLUBB (not shown).

$$\frac{\partial \overline{w'^3}}{\partial t} = \underbrace{-\frac{1}{\rho_s}\frac{\partial \rho_s \overline{w'^4}}{\partial z}}_{\text{turb advection}} \underbrace{+3\frac{\overline{w'^2}}{\rho_s}\frac{\partial \rho_s \overline{w'^2}}{\partial z}}_{\text{turb production}} \underbrace{+\frac{3g}{\theta_{vs}}\overline{w'^2\theta'_v}}_{\text{buoy production}} \underbrace{-\frac{C_8}{\tau}\overline{w'^3}}_{\text{ret isotropy}} \underbrace{-C_{11}\left(+\frac{3g}{\theta_{vs}}\overline{w'^2\theta'_v}\right)}_{\text{pressure}} \tag{A.4}$$

In the current version of CLUBB, $\overline{u'w'}$ and $\overline{v'w'}$ are diagnosed using eddy diffusivity rather than prognosed.

A summary of the prognostic equations in CLUBB is provided in Table A.1.

APPENDIX B
ASSUMED PDF CLOSURE IN CLUBB

Closure by integration over a PDF can be applied to any moments that are functions of variables contained within the PDF. The PDF variables are, in the case of CLUBB, w', θ_l', and r_t', if we ignore hydrometeors. A multivariate PDF like this can be used to close nonlocal transport terms, such as $\overline{w'^2 r_t'}$, and the buoyancy production moment, $\overline{w'\theta_v'}$, assuming that θ_v' can be written as a function of θ_l' and r_t'.

Each moment is closed by calculating an integral over a PDF, P, such as

$$\overline{w'^2 r_t'} = \int_{\theta_l'=-\infty}^{\theta_l'=\infty} \int_{r_t'=-\infty}^{r_t'=\infty} \int_{w'=-\infty}^{w'=\infty} w'^2 r_t' P(w', \theta_l', r_t') \, \mathrm{d}w' \, \mathrm{d}r_t' \, \mathrm{d}\theta_l'. \tag{B.1}$$

Such integrals can be calculated analytically, yielding relatively simple formulas if the PDF shape is simple enough.

If hydrometeors are ignored, then CLUBB's PDF shape is a double Gaussian (i.e., mixture of normals), namely, the Analytic Double Gaussian 1 (ADG1) PDF of Larson et al. (2002). A double Gaussian is the sum of two Gaussians, denoted here by P_1 and P_2, weighted by a mixture fraction, a:

$$P_{\mathrm{ADG1}}(w', \theta_l', r_t') = aP_1(w', \theta_l', r_t') + (1-a)P_2(w', \theta_l', r_t') \tag{B.2}$$

Each component of the ADG1 PDF does not retain the complexity of a general multivariate Gaussian PDF but instead assumes that w is uncorrelated with r_t and θ_l within a generic component i:

$$\begin{aligned} P_i(w', \theta_l', r_t') &= \frac{1}{(2\pi)^{3/2}\sigma_{wi}\sigma_{r_t i}\sigma_{\theta_l i}(1 - r_{r_t\theta_l i}^2)^{1/2}} \\ &\times \exp\left\{-\frac{1}{2}\left(\frac{w' - (w_i - \overline{w})}{\sigma_{wi}}\right)^2\right\} \\ &\times \exp\left\{-\frac{1}{2(1 - r_{r_t\theta_l i}^2)}\left[\left(\frac{r_t' - (r_{ti} - \overline{r}_t)}{\sigma_{r_t i}}\right)^2\right.\right. \\ &+ \left(\frac{\theta_l' - (\theta_{li} - \overline{\theta}_l)}{\sigma_{\theta_l i}}\right)^2 \\ &\left.\left. - 2r_{r_t\theta_l i}\left(\frac{r_t' - (r_{ti} - \overline{r}_t)}{\sigma_{r_t i}}\right)\left(\frac{\theta_l' - (\theta_{li} - \overline{\theta}_l)}{\sigma_{\theta_l i}}\right)\right]\right\}. \end{aligned} \tag{B.3}$$

Here, σ is the standard deviation of w, r_t, or θ_l, and $r_{r_t\theta_l i}$ is the correlation between r_t and θ_l. Also, w_i is the component mean whereas $\overline{w}$ is the grid mean, averaged over both components. A similar convention holds for r_t and θ_l. While this form simplifies calculations, it could be made more realistic (Naumann et al., 2013; Fitch, 2019).

ACKNOWLEDGMENTS

I am grateful for helpful comments by Jun-Ichi Yano, Dmitrii Mironov, and an anonymous reviewer.

REFERENCES

Abdella, K., & McFarlane, N. (1997). A new second-order turbulence closure scheme for the planetary boundary layer. *Journal of the Atmospheric Sciences, 54*, 1850–1867.

Abdella, K., & McFarlane, N. (1999). *Journal of the Atmospheric Sciences, 56*, 3482–3483.

Abdella, K., & Petersen, A. C. (2000). Third-order moment closure through a mass-flux approach. *Boundary-Layer Meteorology, 95*, 303–318.

André, J. C., de Moor, G., Lacarrère, P., & du Vachat, R. (1978). Modeling the 24-hour evolution of the mean and turbulent structures of the planetary boundary layer. *Journal of the Atmospheric Sciences, 35*, 1861–1883.

Arakawa, A., & Schubert, W. H. (1974). Interaction of a cumulus cloud ensemble with the large-scale environment, part I. *Journal of the Atmospheric Sciences, 31*(3), 674–701.

Bogenschutz, P. A., Gettelman, A., Morrison, H., Larson, V. E., Craig, C., & Schanen, D. P. (2013). Higher-order turbulence closure and its impact on climate simulations in the Community Atmosphere Model. *Journal of Climate, 26*, 9655–9676. doi: 10.1175/JCLID-13-00075.1.

Bogenschutz, P. A., & Krueger, S. K. (2013). A simplified PDF parameterization of subgrid-scale clouds and turbulence for cloud-resolving models. *Journal of Advances in Modeling Earth Systems, 5*, doi:10.1002/jame.20018.

Bougeault, P. (1981). Modeling the trade-wind cumulus boundary layer. Part II: A higher-order one-dimensional model. *Journal of the Atmospheric Sciences, 38*, 2429–2439.

Bretherton, C. S., McCaa, J. R., & Grenier, H. (2004). A new parameterization for shallow cumulus convection and its application to marine subtropical cloud-topped boundary layers. Part I: Description and 1D results. *Monthly Weather Review, 132*, 864–882.

Cheng, A., & Xu, K.-M. (2006). Simulation of shallow cumuli and their transition to deep convective clouds by cloud-resolving models with different third-order turbulence closures. *Quarterly Journal of the Royal Meteorological Society, 132*, 359–382.

Cheng, A., & Xu, K.-M. (2008). Simulation of boundary-layer cumulus and stratocumulus clouds using a cloud-resolving model with low- and third-order turbulence closures. *Journal of the Meteorological Society of Japan, 86A*, 67–86.

Cheng, A., & Xu, K.-M. (2009). A pdf-based microphysics parameterization for simulation of drizzling boundary layer clouds. *Journal of the Atmospheric Sciences, 66*, 2317–2334.

Cheng, A., Xu, K.-M., & Stevens, B. (2010). Effects of resolution on the simulation of boundary-layer clouds and the transfer of kinetic energy to subgrid scales. *Journal of Advances in Modeling Earth Systems.* DOI:10.3894/JAMES.2010.2.3.

de Roode, S. R., Duynkerke, P. G., & Siebesma, A. P. (2000). Analogies between mass-flux and Reynolds-averaged equations. *Journal of the Atmospheric Sciences, 57*, 1585–1598.

Deardorff, J. (1972). Theoretical expression for the counter gradient vertical heat flux. *Journal of Geophysical Research, 77*(30), 5900–5904.

Fitch, A. C. (2019). An improved double-gaussian closure for the subgrid vertical velocity probability distribution function. *Journal of the Atmospheric Sciences, 76*(1), 285–304.

Germano, M. (1992). Turbulence: The filtering approach. *Journal of Fluid Mechanics, 238*, 325–336.

Gettelman, A., & Morrison, H. (2015). Advanced two-moment bulk microphysics for global models. Part I: Off-line tests and comparison with other schemes. *Journal of Climate, 28*, 1268–1287. doi: http://dx.doi.org/10.1175/JCLI-D-14-00102.1.

Golaz, J.-C., Larson, V. E., & Cotton, W. R. (2002). A PDF-based model for boundary layer clouds. Part I: Method and model description. *Journal of the Atmospheric Sciences, 59*, 3540–3551.

Griffin, B. M., & Larson, V. E. (2016). Parameterizing microphysical effects on variances and covariances of moisture and heat content using a multivariate probability density function: A study with CLUBB (tag MCVS). *Geoscientific Model Development, 9*(11), 4273–4295. doi: 10.5194/gmd-9-4273-2016.

Guo, H., Golaz, J.-C., Donner, L., Wyman, B., Zhao, M., & Ginoux, P. (2015). CLUBB as a unified cloud parameterization: opportunities and challenges. *Geophysical Research Letters, 42*, 4540–4547. doi:10.1002/2015GL063672).

Heinze, R., Mironov, D., & Raasch, S. (2015). Second-moment budgets in cloud topped boundary layers: A large-eddy simulation study. *Journal of Advances in Modeling Earth Systems, 7*(2), 510–536.

Holtslag, A., & Moeng, C.-H. (1991). Eddy diffusivity and countergradient transport in the convective atmospheric boundary layer. *Journal of the Atmospheric Sciences, 48*(14), 1690–1698.

Houze, R. A. (1993). *Cloud dynamics*. Academic Press.

Janjic, Z. (2001). *Nonsingular implementation of the Mellor-Yamada level 2.5 scheme in the NCEP meso model* (Tech. Rep.). *NOAA/NWS/NCEP Office Note No. 436*. (http://www.emc.ncep.noaa.gov/officenotes/newernotes/on437.pdf)

Kärcher, B., & Burkhardt, U. (2008). A cirrus cloud scheme for general circulation models. *Quarterly Journal of the Royal Meteorological Society, 134*, 1439–1461. doi: 10.1002/qj.301.

Lappen, C.-L., & Randall, D. A. (2001). Towards a unified parameterization of the boundary layer and moist convection. Part I: A new type of mass-flux model. *Journal of the Atmospheric Sciences, 58*, 2021–2036.

Larson, V. E.(2017). *CLUBB-SILHS: A parameterization of subgrid variability in the atmosphere.* (arXiv preprint arXiv:1711.03675)

Larson, V. E., & Golaz, J.-C. (2005). Using probability density functions to derive consistent closure relationships among higher-order moments. *Monthly Weather Review, 133*, 1023–1042.

Larson, V. E., Golaz, J.-C., & Cotton, W. R. (2002). Small-scale and mesoscale variability in cloudy boundary layers: Joint probability density functions. *Journal of the Atmospheric Sciences, 59*, 3519–3539.

Larson, V. E., & Schanen, D. P. (2013). The Subgrid Importance Latin Hypercube Sampler (SILHS): A multivariate subcolumn generator. *Geoscientific Model Development, 6*, 1813–1829. doi:10.5194/gmdd-6-1813-2013.

Lewellen, W. S., & Yoh, S. (1993). Binormal model of ensemble partial cloudiness. *Journal of the Atmospheric Sciences, 50*(9), 1228–1237.

Mellor, G. L. (1977). The Gaussian cloud model relations. *Journal of the Atmospheric Sciences, 34*(2), 356–358.

Mellor, G. L., & Herring, H. J. (1973). A survey of the mean turbulent field closure models. *AIAA Journal, 11*(5), 590–599.

Mellor, G. L., & Yamada, T. (1982). Development of a turbulence closure model for geophysical fluid problems. *Reviews of Geophysics, 20*, 851–875.

Mironov, D. V. (2008). Turbulence in the lower troposphere: Second-order closure and mass–flux modelling frameworks. In *Interdisciplinary aspects of turbulence* (pp. 1–61). Springer.

Mironov, D. V., Gryanik, V. M., Lykossov, V. N., & Zilitinkevich, S. S. (1999). Comments on "A new second-order turbulence closure scheme for the planetary boundary layer". *Journal of the Atmospheric Sciences, 56*, 3478–3481.

Mironov, D. V., & Machulskaya, E. E. (2017). *A turbulence kinetic energy—scalar variance turbulence parameterization scheme* (Technical Report No. 30). Consortium for Small-Scale Modeling. (http://www.cosmo-model.org/content/model/documentation/techReports/docs/techReport30.pdf)

Morrison, H., & Gettelman, A. (2008). A new two-moment bulk stratiform cloud microphysics scheme in the Community Atmosphere Model, Version 3 (CAM3). Part I: Description and numerical tests. *Journal of Climate., 21*, 3642–3659.

Nakanishi, M., & Niino, H. (2004). An improved Mellor–Yamada level-3 model with condensation physics: Its design and verification. *Boundary-Layer Meteorology, 112*(1), 1–31.

Nakanishi, M., & Niino, H. (2006). An improved Mellor–Yamada level-3 model: Its numerical stability and application to a regional prediction of advection fog. *Boundary-Layer Meteorology, 119*(2), 397–407.

Naumann, A. K., Seifert, A., & Mellado, J. P. (2013). A refined statistical cloud closure using double-Gaussian probability density functions. *Geoscientific Model Development, 6*, 1641–1657. doi: 10.5194/gmd-6-1641-2013.

Neale, R. B., Gettelman, A., Park, S., Chen, C.-C., Lauritzen, P. H., Williamson, D. L., et al. (2012). *Description of the NCAR Community Atmosphere Model (CAM5.0)* (Tech. Rep. Nos. NCAR/TN-486+STR). Boulder, CO, USA: National Center for Atmospheric Research.

Onogi, K., Tsutsui, J., Koide, H., Sakamoto, M., Kobayashi, S., Hatsushika, H., et al. (2007). The JRA-25 reanalysis. *Journal of the Meteorological Society of Japan. Ser. II, 85*(3), 369–432.

Plant, R. S. (2010). A review of the theoretical basis for bulk mass flux convective parameterization. *Atmospheric Chemistry and Physics, 10*(8), 3529–3544. Retrieved from http://www.atmos-chem-phys.net/10/3529/2010/ doi: 10.5194/acp-10-3529-2010.

Plant, R. S., & Yano, J.-I. (Eds.). (2016). *Parameterization of atmospheric convection: Volume 1: Theoretical background and formulation.* Imperial College Press.

Pope, S. B. (2000). *Turbulent flows*. Cambridge University Press.
Rotta, J. C. (1951). Statistische theorie nichthomogener turbulenz. *Zeitschrift für Physik*, *129*(6), 547–572.
Siebesma, A. P., Bretherton, C. S., Brown, A., Chlond, A., Cuxart, J., Duynkerke, P. G., et al. (2003). A large Eddy simulation A intercomparison study of shallow cumulus convection. *Journal of the Atmospheric Sciences*, *60*(10), 1201–1219.
Sommeria, G., & Deardorff, J. W. (1977). Subgrid-scale condensation in models of nonprecipitating clouds. *Journal of the Atmospheric Sciences*, *34*(2), 344–355.
Storer, R., Larson, V. E., Wang, M., Ghan, S., Ovchinnikov, M., & Rasch, P. (2013). *Single-column simulations of deep convection with a PDF parameterization*. (Talk, All Hands Multiscale Science Team Meeting, Denver, CO).
Stull, R. B. (1988). *An introduction to boundary layer meteorology*. Kluwer Academic Publishers.
Thayer-Calder, K., Gettelman, A., Craig, C., Goldhaber, S., Bogenschutz, P. A., Chen, C.-C., et al. (2015). A unified parameterization of clouds and turbulence using CLUBB and subcolumns in the Community Atmosphere Model. *Geoscientific Model Development*, *8*(12), 3801–3821. Retrieved from http://www.geosci-model-dev.net/8/3801/2015/ doi: 10.5194/gmd-8-3801-2015.
Xie, P., & Arkin, P. A. (1997). Global precipitation: A 17-year monthly analysis based on gauge observations, satellite estimates, and numerical model outputs. *Bulletin of the American Meteorological Society*, *78*(11), 2539–2558.
Yano, J.-I. (2014). Formulation structure of the mass-flux convection parameterization. *Dynamics of Atmospheres and Oceans*, *67*, 1–28.
Yano, J.-I., & Plant, R. (2012). Convective quasi-equilibrium. *Reviews of Geophysics*, *50*(4). https://doi.org/10.1029/2011RG000378.
Zhang, G. J., & McFarlane, N. A. (1995). Sensitivity of climate simulations to the parameterization of cumulus convection in the Canadian Climate Centre general circulation model. *Atmosphere Ocean*, *33*, 407–446.
Zilitinkevich, S. S., Gryanik, V. M., Lykossov, V. N., & Mironov, D. V. (1999). Third-order transport and nonlocal turbulence closures for convective boundary layers. *Journal of the Atmospheric Sciences*, *56*, 3463–3477.

11

An Introduction to the Eddy-Diffusivity/Mass-Flux (EDMF) Approach: A Unified Turbulence and Convection Parameterization

João Teixeira[1,2], Kay Suselj[1,2], and Marcin J. Kurowski[1]

ABSTRACT

The eddy–diffusivity/mass–flux (EDMF) approach is based on the unification of concepts typically used for the parameterization of turbulence in the boundary layer (eddy–diffusivity) and of moist convection (mass–flux). The EDMF approach was first proposed and tested in the European Centre for Medium-Range Weather Forecasts model in the late 1990s. Further studies have shown the potential of EDMF to represent dry and moist convective boundary layers. In this chapter, we focus on the stochastic multiplume EDMF scheme that can realistically represent the dry boundary layer, stratocumulus, shallow and even deep cumulus convection within a single framework. In this context, this EDMF scheme can be considered a fully unified turbulence and convection parameterization, spanning from the dry boundary layer to deep convection. This version uses a multiplume approach where surface variability of updraft properties is parameterized using joint probability density function of thermodynamic properties to initialize multiple updrafts. Lateral entrainment of the plumes is parameterized as a stochastic process. This unified EDMF parameterization explicitly considers the horizontal resolution of the model, paving the way to a scale-adaptive extension of the scheme. In this chapter, we introduce the fundamentals of the EDMF approach and discuss some of the latest results of the new multiplume EDMF version.

11.1. INTRODUCTION AND MOTIVATION

Turbulence, convection, and clouds are at the core of key weather and climate prediction problems. For example, (i) to reduce uncertainties in climate projections, it is critical to improve predictions of cloud-climate feedbacks (i.e., how clouds respond to, and influence, climate change), which are controlled by interactions between turbulence, water phase transitions, and radiation (e.g., Bony & Dufresne, 2005), mostly in the boundary layer (e.g., Solomon et al., 2007); (ii) to improve extreme weather prediction as climate changes, it is crucial to improve our understanding of how deep convection responds to a warmer world (e.g., Allen & Ingram, 2002).

The fundamental problem in the representation of turbulence and convection in weather and climate models is that numerical discretization implies a limit to the temporal/spatial scales below which the flow cannot be resolved by the models. Due to the nonlinear nature of the atmosphere, unresolved scales can significantly influence the resolved larger scales. Since explicitly characterizing the unresolved processes is not feasible, the statistical properties of the flow at these scales need to be parameterized as a function of the resolved flow. This is known as the turbulence and convection parameterization problem. Parameterizations have, historically, been modular with a variety of different algorithms (and subroutines) representing particular aspects of turbulence and convection, with artificial separations (and interactions) between these modules.

[1] *NASA Jet Propulsion Laboratory, California Institute of Technology, Pasadena, CA, USA*

[2] *Joint Institute for Regional Earth System Science and Engineering, University of California Los Angeles, Los Angeles, CA, USA*

Fast Processes in Large-Scale Atmospheric Models: Progress, Challenges, and Opportunities, Geophysical Monograph 282, First Edition.
Edited by Yangang Liu and Pavlos Kollias.

DOI: 10.1002/9781119529019.ch11

It is increasingly clear that to realistically represent the different manifestations of turbulence and convection in the atmosphere, new unified parameterizations that consider all types of subgrid flow in one single scheme are needed (e.g., Teixeira et al., 2008). In this context, parameterizations based on the eddy–diffusivity/mass–flux (EDMF) approach that unifies the boundary layer with moist convection (both shallow and deep) have been demonstrated to successfully represent these essential components of the climate system. Furthermore, the fully unified multiplume EDMF parameterization described in this chapter has the potential to be made consistently scale-adaptive in order to take advantage of high and variable horizontal resolutions of the next-generation weather and climate models.

Although the EDMF approach is described in detail below, it is important to highlight from the start, in a simple yet physically insightful manner, why EDMF is able to represent boundary layer and moist convection mixing in a unified way. The first important aspect, as depicted in Figure 11.1, is that EDMF combines the parameterization of small-scale mixing (using the eddy–diffusivity; ED closure) and the parameterization of larger plumes (using the mass–flux; MF approach). A key aspect of the EDMF scheme described in this chapter is the multiplume nature of its MF approach. The multiplume framework allows for the coexistence of different convective regimes (i.e., dry plumes, shallow moist convection, and even deep convection) within a single grid box, without any artificial separations between them. In particular, these EDMF plumes are able to realistically represent the skewed part of the subgrid scale distribution of thermodynamic properties at each model grid box, as further shown in Figure 11.4.

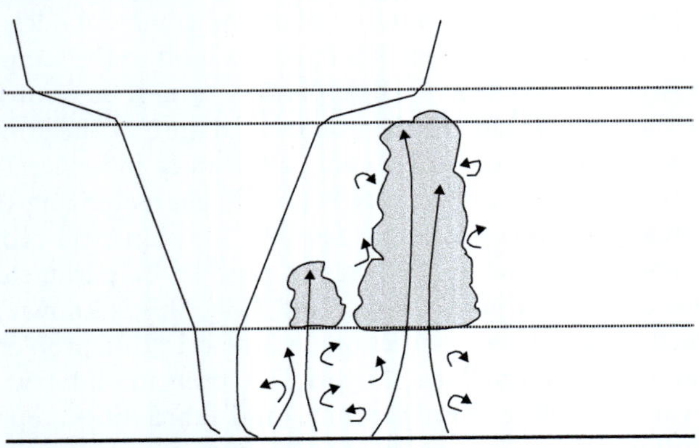

Figure 11.1 Schematic illustrating the thermodynamic structure of a typical shallow cumulus boundary layer, with representative profiles of total water and liquid water potential temperature, as well as a depiction of cumulus clouds and different turbulent eddies. The EDMF approach is able to represent the different scales of turbulent and convective mixing, with ED representing the smaller eddies and MF representing larger (cumulus) scale plumes. Adapted from Soares et al. (2004).

11.2. EDDY-DIFFUSIVITY/MASS-FLUX APPROACH: INTRODUCTION

The EDMF approach is based on the unification of concepts typically used for the parameterization of boundary layer turbulence (eddy–diffusivity) and moist convection (mass–flux). The EDMF approach was first proposed, implemented, and tested in the late 1990s as a unified parameterization of convection and turbulence in global atmospheric models (Siebesma & Teixeira, 2000). This early EDMF version was implemented and evaluated in the European Centre for Medium-range Weather Forecasts (ECMWF) model (Teixeira & Siebesma, 2000). A variety of studies have shown the potential of EDMF to represent dry convection (Siebesma et al., 2007; Witek et al., 2011a, 2011b), moist convection (Neggers 2009, 2015; Soares et al., 2004; Suselj et al., 2012, 2013, 2022; Smalley et al., 2022), and in the context of air quality studies (Angevine, 2005). While most versions are based on steady-state plumes, some developments also include nonstationary plumes (e.g., Tan et al., 2018). The EDMF parameterization is an ideal candidate approach for a full unification of the parameterization of all turbulent and convective processes that occur in the Earth's atmosphere, including deep moist convection, since MF is often the method of choice for deep convection.

In the past few years, a version of EDMF has been developed that is able to represent in a realistic manner dry convection, stratocumulus, shallow convection, and deep convection with one single scheme. This version (Suselj et al., 2013, 2019a, 2019b) uses a multiplume approach where surface variability of thermodynamic properties is parameterized using a diagnostic variance equation. This joint normal probability density function (PDF) can be sampled in a Monte Carlo way to initiate multiple plumes (updrafts) at the surface. Their lateral entrainment is parameterized as a stochastic process, inspired by Romps and Kuang (2010). We assume that the area occupied by each of the plumes remains constant in height as long as the plume is active—the total updraft area in a grid box (at a certain height) is simply the sum of the areas of the convectively active plumes at that height. LES results for key cases indicate that the fixed plume area approximation is a reasonable one at least to first order (e.g., Neggers, 2015; Wu et al., 2020). If the individual plumes terminate at different heights, as it is often the case for moist shallow convection, the total updraft area gradually decreases with height. In this framework, detrainment does not need to be explicitly parameterized.

The EDMF scheme has been implemented in a variety of global weather prediction models. A version of EDMF has been implemented in the operational ECMWF model and shown to significantly improve the simulation of stratocumulus (Koehler, 2005). More recently, EDMF versions have been implemented and evaluated in the U.S. Navy Global Environmental Model (NAVGEM), the National Centers for Environmental Prediction (NCEP) Global Forecasting System (GFS) model, and the National Aeronautics and Space Administration (NASA) Goddard Earth Observing System (GEOS) model (Suselj et al., 2021). NCEP has transitioned a dry EDMF GFS version into operations in January 2015 (Han et al., 2016). A moist multiplume EDMF version, similar to the one described in this chapter, has been implemented operationally in NAVGEM. The EDMF NAVGEM version has been operational since November 2013 as the global weather prediction model of the U.S. Navy (Suselj et al., 2014). Additional recent implementations include Olson et al. (2019) and Hourdin et al. (2019).

The multiplume EDMF parameterization of Suselj et al. (2013, 2019a) has recently been extended to deep convection (Suselj et al., 2019b), allowing for a full unification of the parameterization of all turbulent and convective processes that occur in the Earth's atmosphere within a single framework. A necessary step to extend EDMF to deep convection is to include cloud microphysics in the plume model. A parameterization of downdrafts driven by rain evaporation and the associated cold pool effects are also added.

This new EDMF parameterization is able to explicitly consider horizontal resolution, which means that it can be made scale-adaptive. A necessary step is to retain the updraft area term in the full EDMF decomposition of the subgrid vertical fluxes (see equation (11.1)). Until this new version, EDMF versions (and most MF parameterizations) typically assumed that the updraft area is significantly smaller than the horizontal grid size and can be neglected in some terms. This is not the case any longer with the latest EDMF version. In addition, this multiplume EDMF scheme offers an extra avenue to deal with the issue of making the parameterization scale-adaptive: the number of plumes can be drawn in a stochastic manner and be made dependent on the grid size by considering the typical convective cell size versus grid size. If the horizontal area of plumes and the model grid area are of a comparable magnitude, the stochastic occurrence of plumes in a particular grid becomes dominant. In addition, as recently shown in Kurowski and Teixeira (2018), the ED component can also be made scale-adaptive by judiciously choosing a mixing length formulation that explicitly includes both the grid size as well as physically relevant length scales (e.g., the boundary layer height). Currently, the scale-adaptive nature of parameterizations of turbulence and convection is a critical topic of research (e.g., Honnert et al., 2020; Neggers et al., 2019).

11.3. EDMF FORMULATION

To parameterize turbulent and convective subgrid vertical fluxes, a common practice has been to use the ED approach for the stable boundary layer, the dry convective boundary layer, the subcloud layer, and in the stratocumulus cloud layer. The MF approach is often used in the cumulus layer, where strong mixing occurs in and around cumulus plumes (note that although they have different definitions, we are, as is often the case in the published literature, using the terms plume, thermal, and updraft more or less interchangeably).

This separation is problematic and often results in discontinuities associated with interactions and transitions between different regimes. By combining the ED and MF approaches, EDMF can circumvent many shortcomings of current schemes and help improve the representation of processes such as boundary layer or cloud-top entrainment, countergradient fluxes, and the interaction between the cloud and the subcloud layers. EDMF allows for smoother and more realistic transitions between different cloud regimes (e.g., Suselj et al., 2013) because the ED and MF contributions are treated within a single algorithm and numerical solver, without any trigger functions or additional closure techniques. The MF cumulus updrafts originate from the subcloud layer as physically expected from a variety of previous studies (e.g., Lareau et al., 2018; Nicholls & LeMone, 1980; Tian et al., 2019).

The EDMF approach is based on a scale decomposition where local mixing by small eddies is parameterized by the ED term while nonlocal mixing due to cumulus and planetary boundary layer (PBL) scale plumes is represented by the MF term (see Figure 11.1). In both the subcloud and cloud layers, the MF term represents the effect of strong updrafts (dry and cloudy). These updrafts are represented by one or more laterally-entraining rising plumes, which help determine the PBL height and, if present, the lifting condensation level and cloud-top height. The MF profile is a function of the vertical velocity of the plumes and their cumulative area. The governing steady-state equation for the plume velocity includes all of the relevant processes including buoyancy, advection, entrainment, and pressure drag (see de Roode et al., 2012 for details). In the dry convective PBL, the MF term represents the countergradient transport. Because the MF term is active in the dry convective and subcloud layers, one avoids discontinuities associated with switching between the PBL and moist convection parameterizations, which is a typical issue for models treating the PBL, shallow convection, and deep convection as separate parameterizations/modules (based on different assumptions and applying different numerical algorithms). It also avoids the need to make ad hoc assumptions about the cloud-base MF, removing the need for using a cloud-base closure.

Formally, a model grid box is divided into the updraft area fraction a_u containing the strongest updrafts and the remaining environmental region. Within each region, a prime denotes perturbations about the horizontal mean for that region. After applying Reynolds decomposition and spatially averaging the equations, the subgrid vertical flux of a variable φ is represented by three terms (e.g., Siebesma & Teixeira, 2000, Siebesma et al., 2007):

$$\overline{w'\varphi'} = a_u\overline{w'\varphi'}_u + \left(1 - a_u\right)\overline{w'\varphi'}_e + a_u\left(1 - a_u\right)(w_u - w_e)(\varphi_u - \varphi_e) \qquad (11.1)$$

where the u and e subscripts refer to the updraft and the environmental regions, respectively, and the overbar represents the mean value. For deep convection, this decomposition also needs to account for downdrafts.

In the EDMF approach, the following approximations have often been made: (1) the fractional area covered by strong vertical motion is small, $a_u \ll 1$, so the first term on the r.h.s. of (11.1) can be neglected and $1 - a_u \approx 1$ in the last two terms (this is not the case for our latest EDMF version as described in Suselj et al. (2019a)); (2) $\varphi_e \approx \overline{\varphi}$; (3) the magnitude of the vertical velocity in an updraft is much larger than in the environment, such that $w_u - w_e \approx w_u$; and (4) the environmental turbulent flux (second term on the r.h.s.) is parameterized by the ED approach. With these assumptions, the EDMF vertical subgrid flux approximation becomes a linear combination of ED and MF terms:

$$\overline{w'\varphi'} = -K_H\frac{\partial\overline{\varphi}}{\partial z} + M(\varphi_u - \overline{\varphi}) \qquad (11.2)$$

where $M = a_u w_u$ is the MF associated with the updrafts, which have the fractional area coverage $a_u(z)$ and vertical velocity $w_u(z)$, and K_H is the ED coefficient associated with the downgradient transport and mixing by small-scale turbulence. To close the EDMF parameterization, the updraft vertical velocity $w_u(z)$ and the updraft properties $a_u(z)$ are determined based on the ascent of one or more steady-state entraining plumes initiated just above the surface, with their initial conditions based on the subgrid variability of thermodynamic variables and vertical velocity in the surface layer. For the ED part, a turbulent kinetic energy (TKE) closure is often used.

The grid-mean EDMF tendencies are then calculated as

$$\frac{\partial\overline{\varphi}}{\partial t} = \frac{\partial}{\partial z}\left[K_H\frac{\partial\overline{\varphi}}{\partial z} - M(\varphi_u - \overline{\varphi})\right]. \qquad (11.3)$$

The profiles of vertical velocity and moist conserved variables in the updraft are modeled as

$$\frac{1}{2}\frac{\partial w_u^2}{\partial z} = aB - (b + c\varepsilon)w_u^2 \qquad (11.4)$$

$$\frac{\partial\varphi_u}{\partial z} = -\varepsilon(\varphi_u - \varphi) \text{ with } \varphi = \{\theta_l, q_t\} \qquad (11.5)$$

where $B = g\left(\frac{\theta_{vu}}{\theta_v} - 1\right)$ represents the buoyancy of the plume with respect to the environment, θ_v is virtual potential temperature, and the a, b, and c coefficients are consistent with previous studies (e.g., de Roode et al., 2012; Rio et al., 2010). In equation (11.5), variable φ represents two moist-conserved thermodynamic variables: liquid-water potential temperature, θ_l, and total-water mixing ratio, q_t. Parameter ε represents the lateral entrainment rate. In equations (11.4) and (11.5), it is assumed that (i) the updrafts interact with the environment only through lateral entrainment, (ii) the entrained air has the same properties as the mean environmental air at the level of entrainment (Kurowski et al., 2019), and (iii) the entrained air mixes instantaneously with the air in the updraft. This rules out possible interactions between different updrafts, which may turn out to be relevant in the context of the multiplume approach to deep convection.

11.4. EDMF AND THE DRY CONVECTIVE BOUNDARY LAYER

The EDMF approach is not only particularly well suited to represent the skewed nature of shallow moist convection (see below), but it also plays a key role in the simpler dry convective PBL, where there is no condensation and the convection is essentially driven by surface sensible heat fluxes. For the dry convective PBL, the EDMF approach produces two significant improvements compared to the ED approach: (a) the countergradient flow is now represented in a physically based manner by the MF term and (b) the top entrainment is more realistically represented because of the role played by the larger plumes parameterized by the MF term.

To illustrate the above, Figure 11.2 compares the hourly mean potential temperature profiles for a typical dry convective PBL from a large-eddy simulation (LES) model with two different parameterizations, implemented in a single column model (SCM), at 4 and 8 hours of the simulation; "BL89" refers to an ED (TKE-based) approach and "new" is the EDMF approach. The agreement between the LES profiles and EDMF is clearly better than with the ED approach. In particular, EDMF captures much better the well-mixed nature of the entire PBL with a slightly stable stratification in the upper half

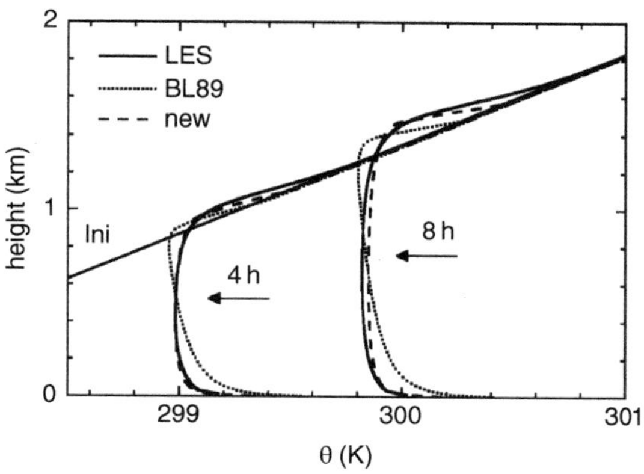

Figure 11.2 Hourly mean potential temperature profiles (4 and 8 hours of simulation) from an LES of a dry convective boundary layer, the "BL89" ED approach, and the "new" EDMF approach. Source: From Soares et al. (2007)/Física de la Tierra.

of the convective boundary layer and a more accurate top entrainment.

11.5. EDMF AND SHALLOW MOIST CONVECTION

Different extensions of EDMF to shallow moist convection have been proposed (e.g., Angevine et al., 2010; Neggers, 2009; Soares et al., 2004). A key advantage of any of these EDMF approaches is that they unify the treatment of unsaturated (dry) convection and the development of overlying cumulus updrafts that originate within the subcloud layer. This is because the strong updrafts within the subcloud layer, represented by the MF component of EDMF, are also the ones that form shallow cumulus clouds. The moist EDMF basically unifies the conventional PBL and shallow convection schemes in a single framework and replaces the typical MF closure as well as the specification of cloud-base properties.

The EDMF version that is discussed below in more detail parameterizes the MF component using multiple plumes, each with its own updraft properties. At the surface, the PDF of thermodynamic properties is sampled by generating plumes with different properties that are integrated in the vertical independently of each other. The lateral entrainment is parameterized as a stochastic process following Romps and Kuang (2010).

11.6. THE MULTIPLUME EDMF PARAMETERIZATION

For the Suselj et al. (2012, 2013, 2019a, 2019b) multiplume moist EDMF parameterization, the subgrid fluxes of moist conserved thermodynamic variables and virtual potential temperature (used in the TKE equation) are written as

$$\overline{w'\varphi'} = -K_{\rm H}\frac{\partial\overline{\varphi}}{\partial z} + \sum_i M_i(\varphi_{{\rm u}i} - \overline{\varphi}), \text{ where } \varphi = \{\theta_{\rm L}, q_{\rm t}, \theta_{\rm v}\} \tag{11.6}$$

This EDMF version partitions the horizontal grid area into multiple updrafts and the remaining environment. Note that in this approximation the different plumes interact only with the surrounding environment and not directly with each other. As mentioned earlier, the ED part of the flux is often parameterized with a TKE approach (e.g., Teixeira & Cheinet, 2004) and is not described here in detail. The MF term is the product of the vertical velocity and area fraction of each plume, while $\varphi_{{\rm u}i}$ is the value of φ in the ith plume. Vertical fluxes of horizontal momentum are often, but not in the latest EDMF version of Suselj (2019a), parameterized using only the ED approach. Note that a multiplume MF formulation was proposed for boundary layer convection by Cheinet (2003, 2004). Neggers et al. (2009) applied a multiplume approach as part of an EDMF framework but only for two plumes/updrafts (dry and moist).

A fixed number of updrafts are initialized in the surface layer when the surface buoyancy flux is positive. The properties of each of the updrafts are drawn from the tail of a joint Gaussian PDF of moist conserved variables and vertical velocity (e.g., Suselj et al., 2013, 2019a), assuming that the updrafts represent the most buoyant and energetic portions of the atmosphere. Each updraft equation is independently integrated until the vertical velocity drops to zero.

The interaction between the plumes and the environment, which is parameterized by lateral entrainment, is a key process determining the fate of a plume. It is not yet completely clear which processes control lateral entrainment and how to parameterize them. Although a variety of entrainment formulations have been proposed (e.g., Neggers et al., 2002; Siebesma et al., 2003; de Rooy et al., 2013; Tian & Kuang, 2016), this is still very much an open topic of research.

Besides the sampling of updraft properties at the surface, lateral entrainment is the process that, in this EDMF version, accounts for the variability of properties among different updrafts in order to allow them to terminate at different levels, as suggested by observations and LES studies. We follow the approach of Romps and Kuang (2010) to represent lateral entrainment as a stochastic process, adjusted to steady-state plumes. In this model, it is assumed that lateral entrainment happens as discrete events and that for each event the entrainment is proportional to the mass flux as is typically assumed. Along a finite grid length Δz, the number of entrainment events

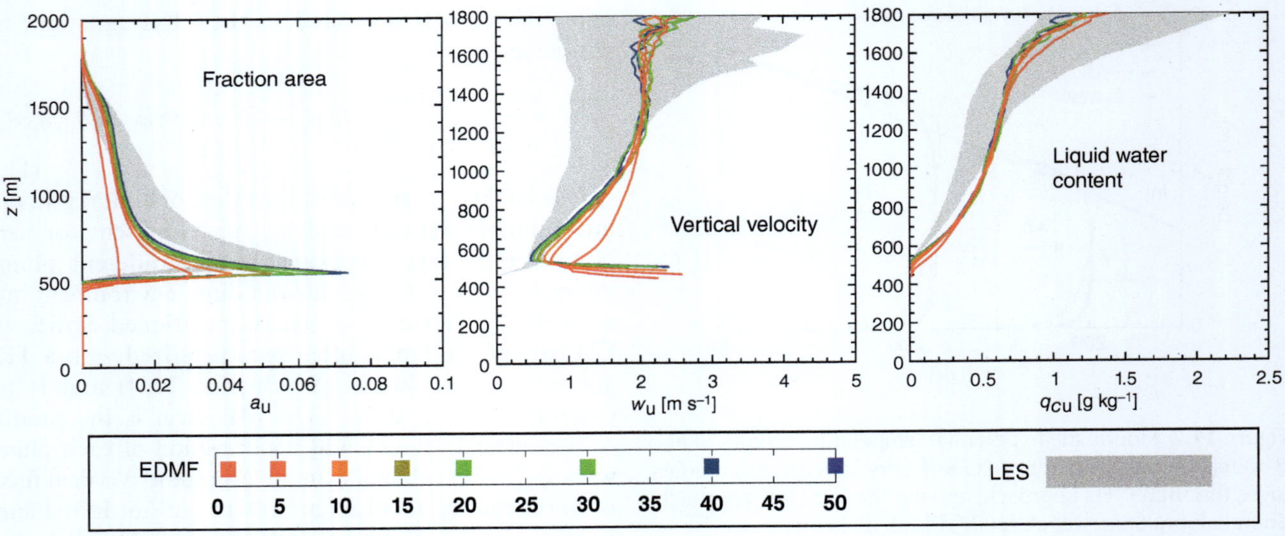

Figure 11.3 BOMEX cumulus case: profiles of updraft fraction area, updraft vertical velocity, and updraft liquid water content from LES and from the multiplume EDMF for a variety of surface updraft fraction areas. Source: Adapted from Suselj et al. (2019a).

is assumed to follow a Poisson distribution (see Suselj et al., 2013).

The multiplume approach assumes that individual plumes have no internal variability although each one of them has different properties contributing to the plume-ensemble variability that is explicitly represented by the multiple plumes. This is an important advantage of multi-plume schemes that is especially useful for simulating the interactions between plume dynamics and microphysics (e.g., Suselj et al., 2022).

The main novelties of this multiplume EDMF approach are (i) parameterization of cumulus properties using multiple plumes, (ii) sampling of the PDF of updraft properties in the surface layer, and (iii) stochastic parameterization of lateral entrainment.

11.7. MULTIPLUME MOIST EDMF RESULTS

The ability of the multiplume EDMF to represent moist convective boundary layers has been investigated by simulating several cases, including the shallow cumulus Barbados Oceanographic and Meteorological Experiment BOMEX (Siebesma et al., 2003), and several stratocumulus to cumulus transition cases from Chung et al. (2012). EDMF is implemented in a standard SCM developed at the Jet Propulsion Laboratory (e.g., Suselj et al., 2013), and the results are evaluated using LES.

This version of EDMF well represents the mean and flux profiles of moist-conserved variables for all cases. A key aspect for realistic simulations of stratocumulus is a consistent coupling between the turbulence, condensation, and radiation, which is achieved with the EDMF parameterization. A key aspect in simulating cumulus cases is a proper representation of moist updraft properties such as updraft area, vertical velocity, and the excess of moist-conserved variables. Figure 11.3 shows that properties such as updraft area, updraft vertical velocity, and updraft liquid water content are well represented when compared to LES results.

For shallow cumulus regimes, LES shows that the joint PDFs of moist-conserved variables and vertical velocity significantly deviate from normal distributions. The EDMF approach is well suited to represent these deviations. Figure 11.4 shows total water and liquid

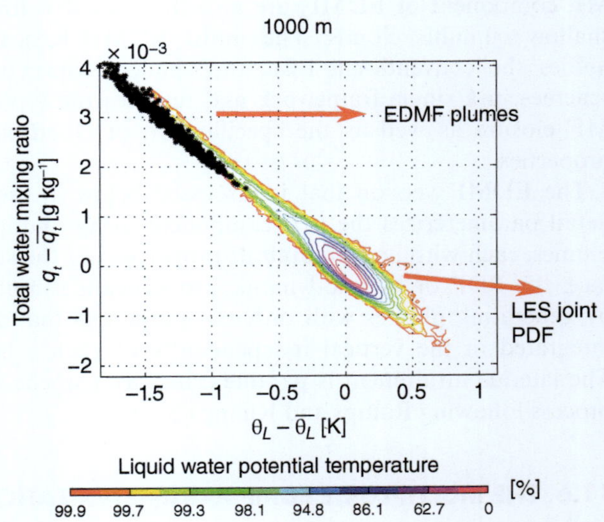

Figure 11.4 Joint distribution of total-water mixing ratio and liquid-water potential temperature sampled in the BOMEX cumulus layer at 1000 m from LES output (color contours) and the corresponding values from the EDMF parameterized plumes over 60 min of integration (black stars). Source: Adapted from Suselj et al. (2013).

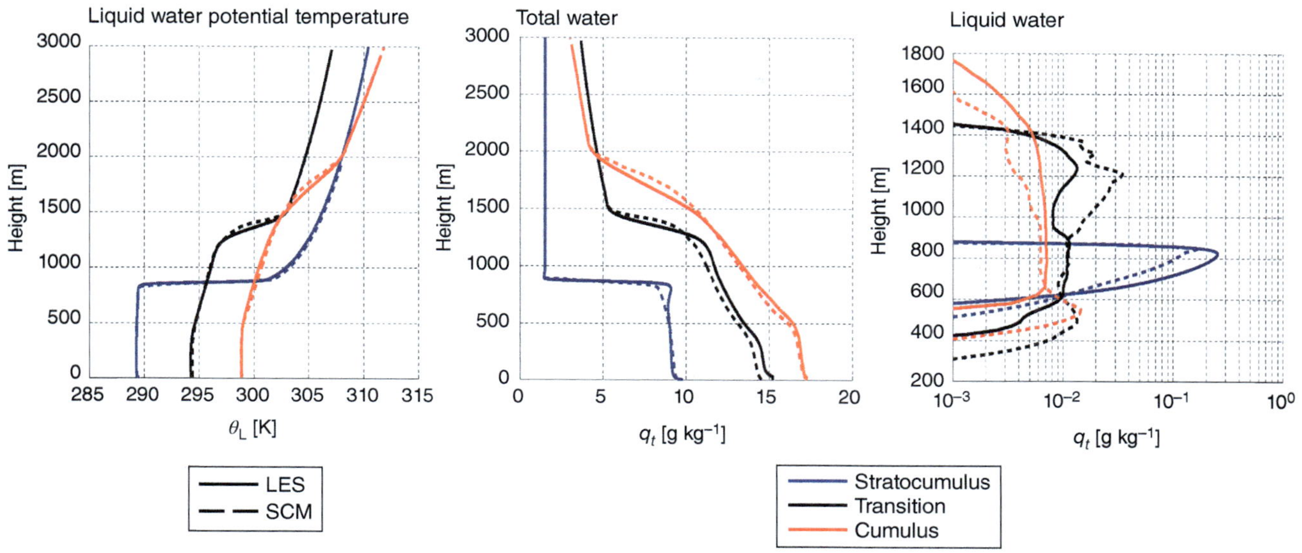

Figure 11.5 Mean liquid water potential temperature, total water, and liquid water content from the LES and EDMF (SCM in caption) models, for stratocumulus, transition, and cumulus cases.

water potential temperature distributions sampled in the BOMEX cumulus layer at 1000 m height from LES output (color contours) and the corresponding values from multiple EDMF plumes over 60 min of integration (black crosses). It is clear that the thermodynamic properties from the EDMF updrafts resemble the LES results in this highly skewed region of the distribution.

A variety of LES cases that represent the stratocumulus to cumulus transition regimes (Chung et al., 2012) were analyzed as well. As shown in Figure 11.5, the multiplume version of EDMF is able to represent a realistic transition from well-mixed stratocumulus to a conditionally unstable cumulus PBL, not only in terms of moist conserved variables but also liquid water content.

As part of the evaluation process, the sensitivity of the multiplume EDMF version to vertical resolution and to key model parameters has also been studied in detail. A vertical resolution that is sufficient to resolve inversions, such as in stratocumulus cases, is important. For cumulus cases, the sensitivity of EDMF to vertical resolution is much less pronounced. An encouraging result is that EDMF is not overly sensitive to reasonable variations of key model parameters that are often hard to characterize precisely (e.g., Suselj et al., 2013). The version described above is not yet fully unified since it is not able to represent deep convection. A fully unified version is described in detail in the next chapters.

11.8. THE FULLY UNIFIED (SHALLOW AND DEEP) EDMF PARAMETERIZATION

The fully unified (in the sense of being able to represent PBL, shallow and deep moist convection with one single parameterization) version of EDMF builds on the previous version of the multiplume EDMF described above and allows for a comprehensive representation of atmospheric convection. Since the model employs a multiplume approach, all these types of convection can coexist within a grid box. The type of convection represented by each plume depends on its vertical development, which in turn is controlled by its initial surface properties and lateral mixing with the environment. The following parameterizations have been added to the previous version to simulate deep convection: (i) cloud ice and mixed-phase physics, (ii) precipitation formation within convective updrafts, (iii) downdrafts driven by evaporative cooling, and (iv) the effects of organized convection and cold pools. The definitions of the thermodynamic variables are generalized to account for both liquid and ice water phases.

In this new multiplume EDMF version, we decompose the horizontal grid box into three separate elements: convective updrafts, convective downdrafts, and the remaining environment. Note that the equations for the vertical subgrid fluxes are similar to the ones discussed above, but they (i) include the additional effects of downdrafts and (ii) the total area of updrafts and downdrafts is not considered negligible as compared to the horizontal grid-box size.

The mean cloud cover and cloud water of the environment (the cloud macrophysics) are computed using a classic PDF-based cloud scheme (e.g., Sommeria & Deardorff, 1977). For simplicity, we assume that the phase of cloud water (either liquid, mixed, or ice) is only a function of temperature (e.g., Grabowski, 1998). Below the freezing temperature, the fraction of ice water linearly increases as the temperature decreases until it entirely turns into ice at 243 K.

Convective updrafts are represented by laterally entraining steady-state plumes (similar to equation (11.5)), with added source terms due to cloud microphysics. Multiple updraft equations are initialized at the surface and are integrated upward as long as the updraft vertical velocity remains positive. The entrainment parameterization follows the stochastic approach described above. Updraft vertical velocity is calculated with an equation similar to equation (11.4).

As described previously, to define the surface conditions for the updrafts, we prescribe a joint normal PDF of the near-surface vertical velocity and moist thermodynamic variables. The tail of this PDF is used to determine the updraft properties since these plumes represent the most energetic convective events. In the case of nonprecipitating and weakly precipitating convection, the second moments used to define this joint PDF are assumed to be a function of surface fluxes as in the previous version. For strongly precipitating convection, they are also a function of parameterized cold pools (see below).

For the thermodynamic variables, an extra source term due to precipitation formation is included in the updraft equations. The rate of precipitation formation follows a simple and traditional approach (e.g., D'Andrea et al., 2014), which resembles the autoconversion approach of Kessler (1969). This formulation essentially assumes that the excess of cloud water above a threshold is converted into rain with a prescribed timescale. For each of the updrafts, the timescale depends on the plume depth.

For each precipitating updraft, we create a complementary downdraft that is void of cloud water but usually contains precipitating water. Evaporative cooling of precipitation is the source of negative buoyancy driving the downdraft. For each variable, downdraft equations are essentially the same as the updraft equations except for the opposite signs due to the negative vertical velocity. The source terms for thermodynamic variables in the downdrafts are limited to evaporation, and their formulation follows Boville et al. (2006).

One important effect of precipitating convection is cold pool formation, which can affect near-surface conditions and stimulate convection organization (e.g., Boeing et al., 2012; Jeevanjee & Romps, 2013; Khairoutdinov & Randall, 2006; Kurowski et al., 2018; Torri et al., 2015). Two mechanisms associated with this process are considered in our parameterization. First, the development of cold pools induces low-level air gusts that can increase the variability of near-surface properties and induce or strengthen mechanical lifting (Kurowski et al., 2018; Schlemmer & Hohenegger, 2014; Torri et al., 2015). Second, as convective clouds aggregate, they become less exposed to the environment, which can suppress their lateral entrainment (Kuang & Bretherton, 2006; Khairoutdinov & Randall, 2006; Kurowski et al., 2018).

To account for the above effects, in Suselj et al. (2019b) we define a new vertical velocity scale, which is meant to represent the effects of cold pools on convection. This velocity scale, which is only nonzero when sufficiently strong precipitating convection develops, depends on the mass flux from the downdrafts in the subcloud layer averaged over a certain "memory" timescale associated with convection. The effects of cold pools on convection development are parameterized by the modification of updraft surface conditions (their variability increases) and lateral entrainment rate (its value decreases) in order to represent, although in a simplified manner, the processes mentioned above.

11.9. RESULTS FROM THE FULLY UNIFIED MULTIPLUME EDMF

The fully unified EDMF is evaluated by analyzing two well-known precipitating convection cases: shallow convection and shallow-to-deep convection transition. The first case is based on measurements collected during an undisturbed period of the Rain in Cumulus over the Ocean (RICO) field campaign (Rauber et al., 2007). It represents slowly deepening and (warm) precipitating convection. The initial and boundary conditions are from the study by van Zanten et al. (2011). We simulate the development of convection during a 24-hour time period and validate the new EDMF results against the LES ensemble results from van Zanten et al. (2011).

To increase the statistical significance of the comparison, we create an ensemble of 10 different EDMF simulations and then analyze their median and interquartile results. The ensemble members only differ in the entrainment rate. For each model run, the entrainment rate is one possible realization of the stochastic entrainment rate.

We compare the EDMF and LES profiles of rain rate averaged over the last simulation hour (not shown). For both models, a gradual increase in rain rate from near the cloud top down to around 1.5 km is simulated. This maximum is above the mean cloud base, and in EDMF it is largely controlled by the evaporation of rain in the downdrafts. The EDMF updraft rain profiles remain constant below about 1.5 km. The maximum rain rate in the downdrafts is also at similar height but decreases due to evaporation of rain as it approaches the surface. Overall, the EDMF precipitation profile is realistically close to the LES in the sense that all of the EDMF ensemble members are clearly within the LES uncertainty range. In particular, the EDMF and LES mean profiles are fairly close to each other and almost indistinguishable between 1.5 and 3 km (see Suselj et al., 2019b for more details).

To investigate the representation of convective updrafts, Figure 11.6 compares the updraft properties averaged over the last simulation hour against values from the

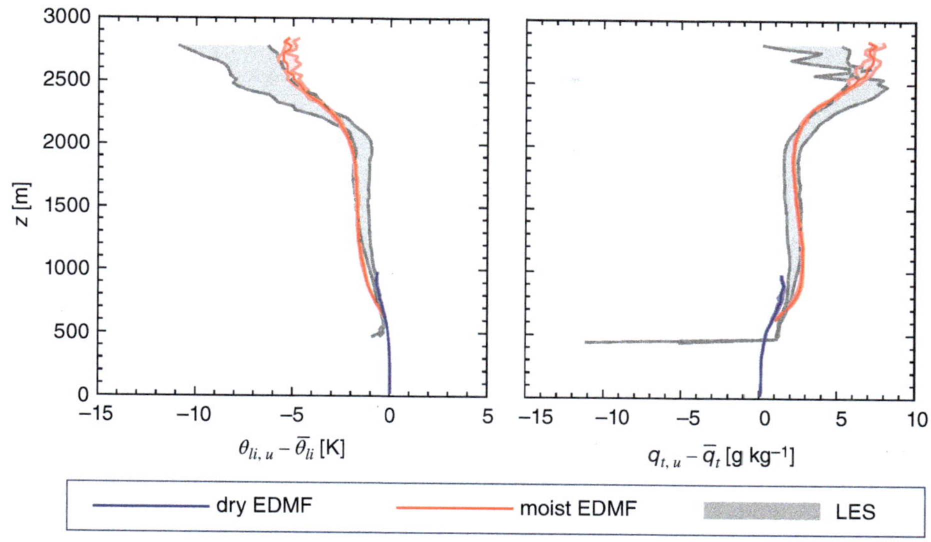

Figure 11.6 Profiles of the difference (excess) of moist conserved variables in moist updrafts with respect to the mean moist conserved variables averaged during the last simulation hour of the RICO case. Gray lines represent the LES results (uncertainty in gray shading); the red line and red area are the mean and the interquartile range from the EDMF moist updrafts; and the blue line represents the dry EDMF updrafts. Source: Adapted from Suselj et al. (2019b).

LES ensemble. Figure 11.6 shows that EDMF updraft properties, such as the excess of liquid water potential temperature and total water content, agree well with the LES results for the RICO case.

The case representing the diurnal cycle of continental convection that we analyze is based on measurements performed during the Large-scale Biosphere-Atmosphere (LBA) experiment in Amazonia. In this case, the early-morning dry boundary layer transitions first to a nonprecipitating shallow cumulus boundary layer. As the layer deepens with time, surface precipitation starts to occur around noon. Further development allows the convective plumes to reach as high as 15 km by early afternoon. The LBA simulation setup closely follows Grabowski et al. (2006). It is initialized at 7:30 a.m. (local time) and it is mostly driven by prescribed time-dependent surface fluxes of heat and moisture, which peak around local noon. Additionally, a relatively weak mean radiative forcing is prescribed. We integrate the model in time for 10 hours, until 5:30 p.m. (local time), to investigate the skill of the fully unified EDMF in simulating the transition from shallow-to-deep convection.

Figure 11.7 compares mean cloud-base and cloud-top heights, and cumulative surface rain rates from the new EDMF versus LES results. In the first 2 hours, a dry boundary layer is simulated by both LES and EDMF. For the following 2 hours, a shallow cumulus (non-precipitating) boundary layer develops. After 4 hours of simulation, convective plumes start growing deeper in both models leading to the onset of surface precipitation.

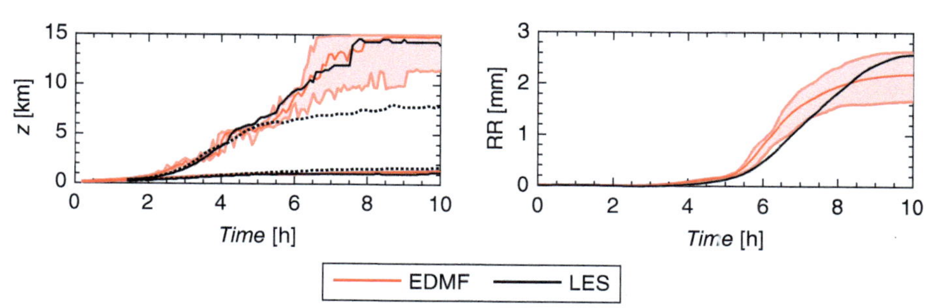

Figure 11.7 EDMF and the LBA case: (left) cloud base and top height, (right) cumulative surface rain rate. Black lines are for LES and red lines for EDMF (the red area represents the interquartile range). The dashed lines are the LES cloud-base and cloud-top heights from the same case but with precipitation switched off. Source: Adapted from Suselj et al. (2019b).

In the next 4 hours (i.e., from the fourth to the eighth hour), a significant growth of convective plumes takes place, with plume depth reaching about 15 km in both models. This time period is also characterized by strong surface precipitation. During the past 2 hours, the depth of the convective layer remains relatively constant and the surface precipitation decreases substantially. Overall, the EDMF and LES representations of the development of deep convection agree well with each other. The EDMF cumulative surface precipitation slightly underestimates the LES rain rate but is pretty realistic. Note that, in general, LES precipitation results can have significant uncertainties.

In Suselj et al. (2019b), we study the impact of key processes, such as the formation of precipitation and cold pools, on the evolution of convection. Our findings show that precipitation formation has only a small direct impact on convection evolution, but the enhancement of precipitation through cold-pool-driven convection invigoration is crucial. We also demonstrate that the impacts of those processes agree well with LES simulations.

As with any parameterization, uncertainty quantification of the EDMF approach is a critical topic. A key question is: what is the impact of individual uncertainties associated with specific physical processes in the overall model results? In the context of shallow convection, Suselj et al. (2020) developed a method for identifying the specific physical processes, and corresponding parameterizations, responsible for the largest model uncertainties. They found that, perhaps unsurprisingly, the uncertainties in the formulations of the MF lateral entrainment (for the convective plumes) and of the ED mixing length (for turbulent diffusion) are the largest contributors to overall model uncertainty. Ongoing work includes the investigation of key parameterization uncertainties in the context of ocean-atmosphere-coupled simulations using an atmospheric model with the multiplume EDMF parameterization.

11.10. SUMMARY

The EDMF approach was devised to unify in one single algorithm and numerical solver the parameterization of boundary layer mixing and moist convection. The initial studies focused on the dry convective boundary layer and showed that the combination of ED to represent small-scale mixing and MF to represent PBL-scale plumes, responsible for the countergradient flow, is ideally suited to model this particular type of boundary layer. Subsequent studies have successfully extended EDMF to stratocumulus and shallow cumulus boundary layers highlighting the ability of EDMF to represent all types of convective boundary layers in a unified manner.

In this chapter, three topics are discussed in some detail: (i) the basics of the overall EDMF formulation, (ii) a recent stochastic multiplume EDMF version for boundary layer and shallow convection mixing, and (iii) its extension to deep convection. It is shown that this multiplume EDMF version is capable of representing realistically unique aspects of moist convective mixing, such as the highly skewed nature of the subgrid distributions of thermodynamic variables within the convective layers of the atmosphere.

Of particular significance is the fact that this multiplume EDMF parameterization is able to represent deep and precipitating convection in a realistic manner, as well as transitions from shallow-to-deep convection, and as such is a fully unified turbulence and convection parameterization capable of representing all types of subgrid vertical mixing in atmospheric models with one single parameterization. EDMF has been implemented operationally in weather prediction models and is currently being implemented and evaluated in global climate models.

ACKNOWLEDGMENTS

This research was carried out at the Jet Propulsion Laboratory, California Institute of Technology, under a contract with the National Aeronautics and Space Administration. Parts of this research were supported by the NASA Modeling, Analysis and Prediction Program; the NOAA/CPO MAPP Program; the U.S. Department of Energy, Office of Biological and Environmental Research, Earth System Modeling (DESc1119242); the ONR Marine Meteorology Program (N000141812432); and by the National Science Foundation Climate Process Team (CPT) Program (AGS 1916619). We thank the reviewers for the detailed comments and helpful suggestions, and M. Chinita for discussions and support.

REFERENCES

Allen, M. R., & Ingram, W. J. (2002). Constraints on the future changes in climate and the hydrological cycle. *Nature*, *419*, 224–232.

Angevine, W. M. (2005). An integrated turbulence scheme for boundary layers with shallow cumulus applied to pollutant transport. *Journal of Applied Meteorology*, *44*, 1436–1452.

Angevine, W. M., Jiang, H., & Mauritsen, T. (2010). Performance of an Eddy diffusivity–mass flux scheme for shallow cumulus boundary layers. *Monthly Weather Review*, *138*, 2895–2912.

Boeing, S. J., Jonker, H. J. J., Siebesma, A. P., & Grabowski, W. (2012). Influence of the subcloud layer on the development of a deep convective ensemble. *Journal of the Atmospheric Sciences*, *69*, 2682–2698.

Bony, S., & Dufresne, J.-L. (2005). Marine boundary layer clouds at the heart of tropical cloud feedback uncertainties in climate models. *Geophysical Research Letters*, *32*, L20 806.

Boville, B. A., Rasch, P. J., Hack, J. J., & McCaa, J. R. (2006). Representation of clouds and precipitation processes in the Community Atmospheric Model (CAM3). *Journal of Climate*, *19*, 2184–2198.

Cheinet, S. (2003). A multiple mass-flux parameterization for the surface-generated convection. Part I: Dry plumes. *Journal of the Atmospheric Sciences*, *60*, 2313–2327.

Cheinet, S. (2004). A multiple mass flux parameterization for the surface generated convection. Part II: Cloudy cores. *Journal of the Atmospheric Sciences*, *61*, 1093–1113.

Chung, D., Matheou, G., & Teixeira, J. (2012). Steady-state Large-Eddy simulations to study the stratocumulus to shallow-cumulus cloud transition. *Journal of the Atmospheric Sciences*, *69*, 3264–3276.

D'Andrea, F., Gentine, P., Betts, A. K., & Lintner, B. R. (2014). Triggering deep convection with a probabilistic plume model. *Journal of the Atmospheric Sciences*, *71*, 3881–3901.

de Roode, S. R., Siebesma, A. P., Jonker, H. J. J., & de Voogd, Y. (2012). Parameterization of the vertical velocity equation for shallow cumulus clouds. *Monthly Weather Review*, *140*, 2424–2436.

de Rooy, W. C., Bechtold, P., Frohlich, K., Hohenegger, C., Jonker, H., Mironov, D., et al. (2013). Review article, entrainment and detrainment in cumulus convection: An overview. *Quarterly Journal of the Royal Meteorological Society*, *139*, 1–19. doi:10.1002/qj.1959.

Grabowski, W. W. (1998). Toward cloud resolving modeling of large-scale tropical circulations: A simple cloud microphysics parameterization. *Journal of the Atmospheric Sciences*, *55*, 3282–3298.

Grabowski, W., Bechtold, P., Cheng, A., Forbes, R., Halliwell, C., Khairoutdinov, M., et al. (2006). Daytime convective development over land: A model intercomparison based on LBA observations. *Quarterly Journal of the Royal Meteorological Society*, *132*, 317–344.

Han, J., Witek, M. L., Teixeira, J., Sun, R., Pan, H.-L., Fletcher, J. K., & Bretherton, C. S. (2016). Implementation in the NCEP GFS of a hybrid Eddy-Diffusivity Mass-Flux (EDMF) boundary layer parameterization with dissipative heating and modified stable boundary layer mixing. *Weather and Forecasting*, *31*, 341–352.

Honnert, R., Efstathiou, G. A., Beare, R. J., Ito, J., Lock, A., Neggers, R., et al. (2020). The atmospheric boundary layer and the "gray zone" of turbulence: A critical review. *Journal of Geophysical Research: Atmospheres*, *125*, e2019JD030317.

Hourdin, F., Jam, A., Rio, C., Couvreux, F., Sandu, I., Lefebvre, M.-P., et al. (2019). Unified parameterization of convective boundary layer transport and clouds with the thermal plume model. *Journal of Advances in Modeling Earth Systems*, *11*, 2910–2933.https://doi.org/10.1029/2019MS001666.

Jeevanjee, N., & Romps, D. M. (2013). Convective self-aggregation, cold pools, and domain size. *Geophysical Research Letters*, *40*(5), 994–998. doi:10.1002/grl.50204.

Kessler, E. (1969). On the distribution and continuity of water substance in atmospheric circulations. *Meteorological Monographs* (Vol. 32). American Meteorological Society.

Khairoutdinov, M., & Randall, D. (2006). High-resolution simulation of shallow to deep convection transition over land. *Journal of the Atmospheric Sciences*, *63*, 3421–3436.

Koehler, M. (2005). Improved prediction of boundary layer clouds. *ECMWF Newsletter*, *104*, 18–22. ECMWF, Reading, United Kingdom.

Kuang, Z., & Bretherton, C. S. (2006). A mass-flux scheme view of a high-resolution simulation of a transition from shallow to deep cumulus convection. *Journal of the Atmospheric Sciences*, *63*, 1895–1909.

Kurowski, M., & Teixeira, J. (2018). A scale-adaptive turbulent kinetic energy closure for the dry convective boundary layer. *Journal of the Atmospheric Sciences*, *75*, 675–690.

Kurowski, M., Suselj, K., Grabowski, W., & Teixeira, J. (2018). Shallow-to-deep transition of continental moist convection: Cold pools, surface fluxes, and mesoscale organization. *Journal of the Atmospheric Sciences*, *75*, 4071–4090.

Kurowski, M. J., Suselj, K., & Grabowski, W. W. (2019). Is shallow convection sensitive to environmental heterogeneities? *Geophysical Research Letters*, *46*, 1785–1793.

Lareau, N. P., Zhang, Y., & Klein, S. A. (2018). Observed boundary layer controls on shallow cumulus at the ARM southern great plains site. *Journal of the Atmospheric Sciences*, *75*, 2235–2255.

Neggers, R. (2009). A dual mass flux framework for boundary layer convection. Part II: Clouds. *Journal of the Atmospheric Sciences*, *66*, 1489–1506.

Neggers, R. A. J. (2015). Exploring bin-macrophysics models for moist convective transport and clouds. *Journal of Advances in Modeling Earth Systems*, *7*, 2079–2104.

Neggers, R. A. J., Griewank, P. J., & Heus, T. (2019). Power-law scaling in the internal variability of cumulus cloud size distributions due to subsampling and spatial organization. *Journal of the Atmospheric Sciences*, *76*, 1489–1503.

Neggers, R. A. J., Köhler, M., & Beljaars, A. C. M. (2009). A dual mass flux framework for boundary layer convection. Part I: Transport, *Journal of the Atmospheric Sciences. 66*, 1465–1487.

Neggers, R. A. J., Siebesma, A. P., & Jonker, H. J. J. (2002). A multiparcel model for shallow cumulus convection. *Journal of the Atmospheric Sciences*, *59*, 1655–1668.

Nicholls, S., & Lemone, M. A. (1980). The fair weather boundary layer in GATE: The relationship of subcloud fluxes and structure to the distribution and enhancement of cumulus clouds. *Journal of Atmospheric Sciences*, *37*, 2051–2067.

Olson, J. B., Kenyon, J. S., Angevine, W. M., Brown, J. M., Pagowski, M., & Sušelj, K. (2019). A description of the MYNN-EDMF scheme and the coupling to other components in WRF-ARW. *NOAA Technical Memorandum OAR GSD*, *61*, 37. https://doi.org/10.25923/n9wm-be49.

Rauber, R. M., Stevens, B., Ochs, H. T., Knight, C., Albrecht, B. A., Blyth, A. M., et al. (2007). Rain in shallow cumulus over the ocean: The RICO Campaign. *Bulletin of the American Meteorological Society*, *88*, 1912.

Rio, C., Hourdin, F., Couvreux, F., & Jam, A. (2010). Resolved versus parametrized boundary-layer plumes. part II: Continuous formulations of mixing rates for mass-flux schemes. *Boundary-Layer Meteorology*, *135*, 469–483.

Romps, D. M., & Kuang, Z. (2010). Nature versus nurture in shallow convection. *Journal of the Atmospheric Sciences, 67*(5), 1655–1666.

Schlemmer, L., & Hohenegger, C. (2014). The formation of wider and deeper clouds as a result of cold-pool dynamics. *Journal of the Atmospheric Sciences, 140421141808009*.

Siebesma, A. P., Bretherton, C. S., Sanchez, E., Stevens, B., Stevens, D. E., Brown, A., et al. (2003). A large eddy simulation intercomparison study of shallow cumulus convection. *Journal of the Atmospheric Sciences, 60*, 1201–1219.

Siebesma, A. P., Soares, P. M. M., & Teixeira, J. (2007). A combined eddy-diffusivity mass-flux approach for the convective boundary layer. *Journal of the Atmospheric Sciences, 64*, 1230–1248.

Siebesma, A. P., & Teixeira, J. (2000). An advection–diffusion scheme for the convective boundary layer: Description and 1D results. *Proceedings of AMS 1^{4t}h Symposium on Boundary Layer and Turbulence*, 133–136.

Smalley, M. A., Suselj, K., Lebsock, M. D., & Witte, M. K. (2022). Coupling warm rain with an eddy diffusivity/mass flux parameterization: 2. Sensitivities and comparison to observations. *Journal of Advances in Modeling Earth Systems, 14, e2021MS002729*.

Soares, P., Miranda, P., Siebesma, A. P., & Teixeira, J. (2004). An eddy-diffusivity/mass-flux parametrization for dry and shallow cumulus convection. *Quarterly Journal of the Royal Meteorological Society, 130*, 3365–3384.

Soares, P. M. M., Miranda, P. M. A., Teixeira, J., & Siebesma, A. P. (2007). An eddy-diffusivity/mass-flux boundary layer parameterization based on the turbulent kinetic energy equation. *Special Issue on Micrometeorology, Fisica de la Tierra, 19*, 147–161.

Solomon, S., Qin, D., Manning, M., Chen, Z., Marquis, M., Averyt, K. B., et al. (2007). *The Physical Science Basis. Contribution of Working Group I to the Fourth Assessment Report of the Intergovernmental Panel on Climate Change*. Cambridge University Press: Cambridge, United Kingdom and New York, NY, USA.

Sommeria, G., & Deardorff, J. W. (1977). Subgrid-scale condensation in models of non-precipitating clouds. *Journal of the Atmospheric Sciences, 34*, 344–355.

Suselj, K., Hogan, T. F., & Teixeira, J. (2014). Implementation of a stochastic Eddy-Diffusivity/Mass-Flux parameterization into the Navy Global Environmental Model. *Weather and Forecasting, 29*, 1374–1390.

Suselj, K., Kurowski, M., & Teixeira, J. (2019a). On the factors controlling the development of shallow convection in Eddy-Diffusivity/Mass-Flux Models. *Journal of the Atmospheric Sciences, 76*, 433–456.

Suselj, K., Kurowski, M., & Teixeira, J. (2019b). A unified Eddy-Diffusivity/Mass-Flux approach for modeling atmospheric convection. *Journal of the Atmospheric Sciences, 76*, 2505–2537.

Suselj, K., Posselt, D., Smalley, M., Lebsock, M. D., & Teixeira, J. (2020). A new methodology for observation-based parameterization development. *Monthly Weather Review, 148*, 4159–4184.

Suselj, K., Smalley, M., Lebsock, M., Kurowski, M., Witte, M., & Teixeira, J. (2022). Coupling warm rain with an Eddy Diffusivity/Mass Flux parameterization. Part 1: Model description and validation. *Journal of Advances in Modeling Earth Systems, 14,* e2021MS002736.

Suselj, K., Teixeira, J., & Chung, D. (2013). A unified model for moist convective boundary layers based on a stochastic Eddy-Diffusivity/Mass-Flux parameterization. *Journal of the Atmospheric Sciences, 70*, 1929–1953.

Suselj, K., Teixeira, J., Kurowski, M., & Molod, A. (2021). Improving the representation of subtropical boundary layer clouds in the NASA GEOS Model with the Eddy-Diffusivity/Mass-Flux parameterization. *Monthly Weather Review, 149*, 793–809.

Suselj, K., Teixeira, J., & Matheou, G. (2012). EDMF and shallow cumulus boundary layer: An updraft pdf multiple mass-flux scheme. *Journal of the Atmospheric Sciences, 69*, 1513–1533.

Tan, Z., Kaul, C. M., Pressel, K. G., Cohen, Y., Schneider, T., & Teixeira, J. (2018). An extended eddy-diffusivity mass-flux scheme for unified representation of subgrid-scale turbulence and convection. *Journal of Advances in Modeling Earth Systems, 10*, 770–800.

Teixeira, J., & Cheinet, S. (2004). A simple mixing length formulation for the Eddy-Diffusivity parameterization of dry convection. *Boundary-Layer Meteorology, 110*, 435–453.

Teixeira, J., & Siebesma, A. P. (2000). A mass-flux/k-diffusion approach to the parameterization of the convective boundary layer: Global model results. *Proceedings of AMS 14th Symposium on Boundary Layers and Turbulence*, 231–234.

Teixeira, J., Stevens, B., Bretherton, C. S., Cederwall, R., Doyle, J. D., Golaz, J. C., et al. (2008). The parameterization of the atmospheric boundary layer: a view from just above the inversion. *Bulletin of the American Meteorological Society, 89*, 453–458.

Tian, Y. & Kuang, Z. (2016). Dependence of entrainment in shallow cumulus convection on vertical velocity and distance to cloud edge. *Geophysical Research Letters, 43*, 4056–4065.

Tian, Y., Kuang, Z., Singh, M., & Nie, J. (2019). The vertical momentum budget of shallow cumulus convection: Insights from a Lagrangian perspective. *Journal of Advances in Modeling Earth Systems, 11*, 113–126.

Torri, G., Kuang, Z., & Tian, Y. (2015). Mechanisms for convection triggering by cold pools. *Geophysical Research Letters, 42*, 1943– 1950.

van Zanten, M. C., Stevens, B., Nuijens, L., Siebesma, A. P., Ackerman, A. S., Burnet, F., et al. (2011). Controls on precipitation and cloudiness in simulations of trade-wind cumulus as observed during RICO. *Journal of Advances in Modeling Earth Systems, 3*(6).

Witek, M., Teixeira, J., & Matheou, G. (2011a). An integrated TKE based eddy-diffusivity/mass-flux boundary layer scheme for the dry convective boundary layer. *Journal of the Atmospheric Sciences, 68*, 1526–1540.

Witek, M. L., Teixeira, J., & Matheou, G. (2011b). An eddy-diffusivity/mass-flux approach to the vertical transport of turbulent kinetic energy in convective boundary layers. *Journal of the Atmospheric Sciences, 68*, 2385–2394.

Wu, E., Yang, H., Kleissl, J., Suselj, K., Kurowski, M. J., & Teixeira, J. (2020). On the parameterization of convective downdrafts for marine stratocumulus clouds. *Monthly Weather Review, 148*, 1931–1950.

12

Application of Machine Learning to Parameterization Emulation and Development

Vladimir Krasnopolsky[1] and Alexei Belochitski[2(†)]

ABSTRACT

In this chapter, a generic mathematical object (mapping) is introduced, and its relation to model physics parameterization is explained. Machine learning (ML) tools that can be used to emulate and/or approximate mappings are introduced. Applications of ML to emulate existing parameterizations, to develop new parameterizations, to ensure physical constraints, and to control the accuracy and stability of developed applications are described. Some ML approaches that allow developers to go beyond the standard parameterization paradigm are discussed.

12.1. INTRODUCTION

The scientific and practical significance of interdisciplinary complex numerical models and prediction systems has increased tremendously during the past few decades due to improvements in their quality via better understanding of the basic processes and their relationships (see section II of this book), as well as progress in numerical methods and computational techniques (see section III of this book), and in Earth-observing system (see section IV of this book). A well-pronounced trend has been revealed in weather and climate numerical modeling. This trend marks a transition from investigating simpler linear or weakly nonlinear single-disciplinary systems, such as simplified atmospheric or oceanic systems that include a limited description of the physical processes, to studying complex nonlinear systems, such as coupled atmospheric-oceanic-land-ice systems that take into account atmospheric physics, chemistry, land-surface interactions, etc. (see Chapter 9 of this book).

[1]*Environmental Modeling Center/National Centers for Environmental Predictions/National Weather Service, NOAA Center for Weather and Climate Prediction, College Park, MD, USA*
[2]*I. M. Systems Group, Environmental Modeling Center, College Park, MD, USA*

Due to their growing complexity (e.g., increase of horizontal and vertical resolutions and increase in sophistication of model physics and chemistry), global and regional modeling activities consume a tremendous amount of computational resources, which presents a significant challenge despite growing computing capabilities. Model data assimilation and ensemble systems have already faced the computational resources problem that limits the resolution and/or the number of the ensemble members in these systems. It is noteworthy that model physics is one of the most computationally demanding parts of weather and climate models. Various approaches are being considered to alleviate this problem. For example, in the European Centre for Medium-Range Weather Forecasts (ECMWF), the modelers consider transitioning from double-precision to single-precision calculation to improve the performance of Integrated Forecasting System and to allow further increase in vertical resolution (Lang et al., 2021). Therefore, flexible and powerful numerical techniques are required to reduce growing demands for computer resources that outrun the growth of computer resources.

Despite significant progress in understanding of physical phenomena and improvement of observing systems, substantial uncertainties remain in representation of many processes in numerical models, e.g., effects of clouds

Fast Processes in Large-Scale Atmospheric Models: Progress, Challenges, and Opportunities, Geophysical Monograph 282, First Edition.
Edited by Yangang Liu and Pavlos Kollias.

DOI: 10.1002/9781119529019.ch12

in general circulation models of the atmosphere (e.g., Ceppi et al., 2017). New flexible and powerful numerical techniques are required to learn underlying physics from data that are available (Christensen & Zanna, 2022).

During the past decade, machine learning (ML) began to play an important role in advancing scientific discovery in domains traditionally dominated by physically based (first-principles) models: in climate, weather, and earth system sciences (Hsieh, 2009, 2023; Krasnopolsky & Fox-Rabinovitz, 2006a; Krasnopolsky, 2013; Reichstein et al., 2019; Wang et al., 2022; Bochenek and Ustrnul, 2022), in oceanography in prediction of ocean weather and climate, habitat modeling and distribution, species identification, coastal water monitoring, marine resources management, detection of oil spill, and pollution and wave modeling (Ahmad, 2019); in solid Earth geoscience (Bergen et al., 2019), in astronomy (Izi'c et al., 2019), in fluid dynamics (Kutz, 2017); in image processing (Ebert-Uphoff & Hilburn, 2020; Wang et al., 2018); in biomedical sciences (Hassoun et al., 2021); in atomic physics (Brockherde et al., 2017); in high energy and fusion physics (Kasim et al., 2020); in engineering (Alizadeh et al., 2020), etc. Such a broad application of ML is possible because problems encountered in these different fields are similar from the mathematical point of view. Many of them belong to various aspects of mapping problems (see the next section for description of mapping). The use of ML models is particularly promising in scientific problems involving processes that are not completely understood, or where it is computationally infeasible to run physically based models at desired resolutions in space and time. Numerical Weather/Climate Prediction Systems (NWPS) represent one of the most complex systems that face such kinds of problems.

Attempts to completely substitute physically based models with the state-of-the-art black box ML models have often met limited success in scientific domains due to inability to provide a meaningful physical understanding and interpretation of underlying processes, their large data requirements, and their limited generalizability to out-of-sample scenarios. Given that neither an ML-only nor a physically based only approach can be considered sufficient for complex scientific and engineering applications, the research community explores the continuum of hybrids of physically based and ML models, where both scientific knowledge and data are integrated in a synergistic manner (Krasnopolsky & Fox-Rabinovitz, 2006a; Rai & Sahu, 2020). This paradigm is fundamentally different from mainstream practices in the ML community for making use of domain-specific knowledge, albeit in subservient roles, e.g., feature engineering or postprocessing. In contrast to these practices that can only work with simpler forms of heuristics and constraints, this presentation is focused on hybrid NWPSs incorporating a deeper coupling of ML methods with scientific knowledge embedded into physically based models.

ML is increasingly being applied to reduce demands for computing resources (e.g., Alizadeh et al., 2020; Kasim et al., 2020) and to accelerate calculations in NWPSs in particular (Chevallier et al., 2000; Gentine et al., 2018; Krasnopolsky et al., 2002; Krasnopolsky et al., 2005; Krasnopolsky, 2013; Lagerquist et al., 2021; Wang et al., 2019). Another venue for application of ML in NWPSs is the development of new ML-based parameterizations derived from observed and/or simulated data (Beucler et al., 2020; Brenowitz & Bretherton, 2018; Krasnopolsky et al., 2013; O'Gorman & Dwyer, 2018; Schneider et al., 2017; Watt-Meyer et al., 2021).

In section 12.2 of this overview, a generic mathematical object (mapping) is introduced, and its relation to model physics parameterization is explained. In section 12.3, ML tools that can be used to emulate and/or approximate mappings are introduced. Section 12.4 describes the use of ML to emulate existing parameterizations. Section 12.5 discusses using ML to develop new parameterizations, section 12.6 introduces the ML ensembles, section 12.7 is devoted to describing methods to ensure physical constraints, section 12.8 introduces some ML tools that allow developers to go beyond the standard parameterization paradigm, and section 12.9 summarizes the discussion.

12.2. PARAMETERIZATIONS AS MAPPINGS

A mapping, M, is a relationship between two vectors X (input vector) and Y (output vector) that can be symbolically written as

$$Y = M(X); X \in R^N \text{ and } Y \in R^M \qquad (12.1)$$

where N and M are the dimensionalities of the input and output spaces correspondingly. Any parameterization of model physics, and even the entire suite of physically based parameterizations as well as the model itself, is a mapping between a vector of input parameters (e.g., profiles of atmospheric state variables) and a vector of output parameters (e.g., a profile of heating rates in radiation parameterization). In terms of Y vs. X dependencies, parameterization mappings may be continuous or almost continuous (in other words, containing only a finite number of step function like discontinuities). Usually, physically based parameterizations of physics do not contain singularities.

Another very important property of the mapping (12.1), relevant for parameterization mappings, is mapping complexity (Krasnopolsky, 2013; Chapter 2.2.2). There are many types of complexity that are discussed in the literature (e.g., Alizadeh et al., 2020). At least four different types of the parameterization mapping complexity can be suggested as relevant to our topic. The parameterization mapping, M (12.1), is usually a symbolic representation of a mathematical formalism based on first principles and describing a chain of physical processes or hierarchy of interacting physical processes (e.g., atmospheric radiation). Therefore, we can talk about the *physical complexity* of the mapping (12.1) that corresponds to the complexity of the hierarchy of physical processes described by the mapping. We can also introduce quantitative or semiquantitative characteristics of physical complexity: the number of physical processes involved and the number of levels (depth of the hierarchy) of the processes involved.

Also, we can consider *mathematical complexity* of the mapping (12.1) and introduce quantitative or semiquantitative measures: the number of equations, the type of equations (e.g., linear vs. nonlinear, ODE vs. PDE vs. integro-differential equations), and the dimensionality of the equations. It is noteworthy that the measures of physical and mathematical complexity are not unambiguous. For a particular physical process, alternative mathematical formalisms based on first principles often exist that lead to different types and numbers of equations for the description of the same physical system. As a result, several different estimates of physical and/or mathematical complexity may be obtained for the same mapping (12.1). Euler vs. Lagrange formulations of the equations of geophysical fluid dynamics and the Schrödinger vs. Heisenberg formulations of quantum mechanics are examples.

The third measure of complexity that can be introduced is *numerical/computational complexity* of a mapping, e.g., the number of elementary numerical operations required for calculating Y given X. Numerical/computational complexity of a parameterization is closely related to the amount of computational resources required for its application. However, this measure is also ambiguous because different numerical schemes applied to the same set of equations (e.g., finite differences vs. variational methods for solving PDEs) may lead to dramatically different counts of elementary numerical operations. Here again, several different estimates of numerical complexity may be obtained for the same mapping (12.1).

The fourth type of mapping complexity is called *functional complexity*. It describes the complexity of the functional dependency of the outputs, Y, versus the inputs, X, or the "smoothness" of this dependency. If the three previous definitions in some respects depend on, or are conditioned by, our knowledge of the internal structure of the target mapping (12.1), this fourth definition characterizes the complexity of the mapping, as a whole, as a single, elementary object that transforms the input vector into the output vector. It is intuitively clear that the functional complexity of the mapping (12.1) can, in principle, be measured unambiguously. Unfortunately, it does not mean that there exists a satisfactory measure of functional complexity for a multidimensional mapping (12.1). Approximating a parameterization mapping using ML tools (e.g., NN) can provide a quantitative measure of the mapping functional complexity. For example, the complexity of NN emulation (see equations (12.7) and (12.8)) may be used as a measure of the functional complexity of the parameterization mapping.

The functional complexity is the most important one when the complexity of an ML tool is evaluated. A parameterization of very high physical and/or numerical complexity may have low functional complexity. For example, parameterization composed of many complex equations may produce a simple output vs. input (Y vs. X) dependence. A significant increase in computational performance achieved when physically based parameterizations are replaced with their ML-based emulators suggests that the functional complexity of the physically based schemes is lower than their physical and/or computational complexities. This situation is optimal when the goal of the ML application is to accelerate calculations because the computational burden in terms of required computer resources of an ML-based emulator is proportional to the functional, not physical or computational complexity, of the parameterization being emulated.

In some cases (e.g., stochastic physics or superparameterization) the parameterization mapping contains an internal source of stochasticity. It may be due to several reasons: a stochastic process that the mapping describes, a stochastic method (e.g., Monte Carlo methods) implemented in mathematical formulation of the mapping, contribution of subgrid processes, or uncertainties in the data that are used to define the mapping. In this case, we can modify the symbolic representation of the mapping (12.1) as

$$Y = M(X, \varepsilon); X \in R^N \text{ and } Y \in R^M \qquad (12.2)$$

where ε is a vector stochastic variable explicitly reflecting a stochastic nature of the mapping—the mapping uncertainty. A stochastic mapping symbolically represented by (12.2) is a family of mappings sampled from a distribution of mappings $Y = M(X, \varepsilon)$. The uncertainty ε is an inherent informative part of the stochastic mapping, which contains important statistical information about the mapping.

The parameterization mapping may be defined explicitly or implicitly. It can be defined explicitly as a set of equations based on first principles and/or empirical dependencies (e.g., radiative transfer or heat transfer equations) or as a computer code. A collection of data records (e.g., observations, measurements, and computer simulations) implicitly represents the parameterization mapping.

In what follows, an ML technique is used to approximate (or emulate) existing physically based parameterization and develop emulating ML parameterizations (EMLPs) or ML emulation of existing parameterization. Both of these terms are used as synonyms. Also, ML can be used to develop new ML parameterizations (MLPs) based on learning from the observed or/and simulated data. Both applications are possible because ML techniques (e.g., neural networks (NNs), support vector machines (SVMs), or decision trees) are also mappings, which can be used to approximate mappings that are given by parameterizations or collections of data records. A more detailed description of mappings and their properties can be found in Krasnopolsky (2013; Chapter 2.2).

12.3. MACHINE LEARNING TOOLS TO APPROXIMATE PARAMETERIZATION MAPPINGS

ML is a subfield of artificial intelligence that uses statistical techniques to give computers the ability to "learn" (i.e., progressively improve performance on a specific task) from data, without being explicitly programmed (Bishop, 2006). This definition explains why ML is sometimes also called statistical learning or learning from data (Cherkassky & Muller, 2007). The set of ML tools includes a large variety of different algorithms such as NNs, different kinds of decision trees (like random forest (RF) algorithm), kernel methods (like SVM and principal component analysis), Bayesian algorithms, etc. (see Figure 12.1). Some of these algorithms are more universal (like generic multilayer perceptron NNs), and some of them are more focused on a specific class of problems (e.g., convolutional NNs show an impressive performance as image/pattern recognition algorithms (Ebert-Uphoff & Hilburn, 2020)).

To the best of our knowledge, two major types of ML tools that have been applied to model physics parameterization mappings are NNs (e.g., Beucler et al., 2020; Brenowitz & Bretherton, 2018; Chevallier et al., 2000; Geer, 2021; Krasnopolsky et al., 2002; Rasp et al., 2018; Rasp, 2019; Xu et al., 2021) and tree algorithms (e.g., Belochitski et al., 2011; O'Gorman & Dwyer, 2018). There are many different types of NNs: shallow, deep, convolutional, recurrent, etc., as well as many types of tree algorithms. Belochitski et al. (2011) showed that, when using a shallow (see the definition below) NN (SNN) to emulate an existing radiation parameterization, the NN performs better than tree algorithms in terms of accuracy and speedup. O'Gorman and Dwyer (2018) stated that when attempting to derive a new parameterization of subgrid processes based on an SNN, NN did not conserve energy and was slightly less accurate than the tree algorithm used. Also, NN was not stable in prognostic single-column integrations. In their work, they experimented with using the simplest shallow, single hidden layer, NNs with a limited number of neurons (60 neurons). Krasnopolsky et al (2008a, 2010) and Belochitski and Krasnopolsky (2021) showed that, when the accuracy of approximation is sufficient, the NN emulation preserves invariants (e.g., energy or moist enthalpy) with an accuracy, sufficient for long-term integration of the NN in the model without accumulation of errors and instability. The stability in the process of integrating NN in the model could also be achieved by using a loss function that involves many time steps (Brenowitz & Bretherton, 2018). Using multiple time steps helps NN learn temporal continuity during the training that is important when the trained NN is integrated in the model. Introducing physical constraints into the loss function also results in stable and accurate integrations in the hybrid model (Beucler et al., 2020).

The vast majority of published work on applying ML to model parameterization mappings used various types of NNs. The few works that used tree algorithms did not show significant advantages as compared to NNs. It is noteworthy that mostly successful tree algorithm – RF algorithm requires more memory than other algorithms because this algorithm stores multiple trees. This can be a problem if the dataset is large. To apply a trained random forest algorithm, the entire training set must be kept in memory. Also, it will not be able to predict any value outside the available training set values since averaging various trees, each of which is built upon the training set, is a big part of random forest models. Thus, we cannot expect reliable predictions/extrapolations/generalizations when using the random trees algorithm. Thus, in the following, we focus on properties of NNs and NN applications to model parameterization mappings. A limited discussion of other ML algorithms is given in the summary.

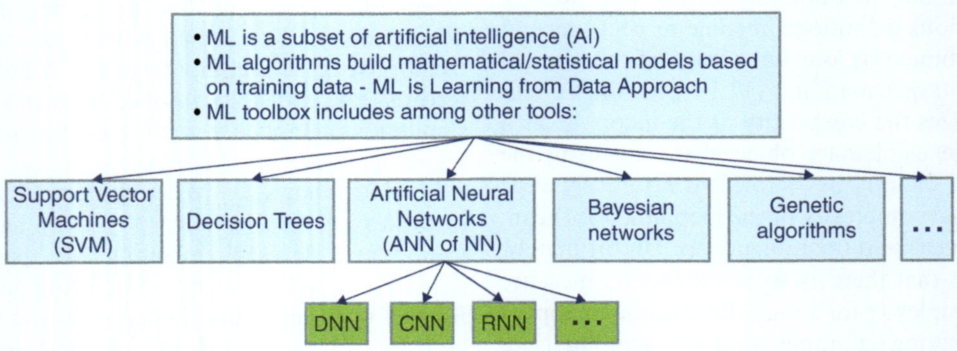

Figure 12.1 Various types of ML tools. New ML tools emerge very often.

12.3.1. Neural Networks

A generic NN that is used for modeling/approximating nonlinear multidimensional mappings is called the multilayer perceptron. It is composed of "neurons" that are situated in layers. A generic neuron can be expressed as

$$x_j^{n+1} = \phi\left(b_j^n + \sum_{i=1}^{k_n} a_{ji}^n \cdot x_i^n\right) \tag{12.3}$$

Equation (12.3) represents a neuron number j in the layer number n. x_j^{n+1} is output of the neuron that, at the same time, is an input to neurons of the layer number $n + 1$. Here, x_j^n are inputs to neurons of the layer number n (outputs of neurons of the layer number $n - 1$, the input layer corresponds to $n = 0$), a and b are fitting parameters or NN weights and biases, and ϕ is a so-called activation function (AF), and k_n is the number of neurons in the layer number n. The entire layer number n can be represented by a matrix equation:

$$X^{n+1} = \phi(B^n + A^n \cdot X^n) \tag{12.4}$$

where for $n = 0$, $X^n = X$, a vector of the NN inputs. If the layer number $n + 1$ is the output layer, where linear neurons are often used,

$$y_j = x_j^{n+1} = b_l^n + \sum_{i=1}^{k_n} a_{ji}^n \cdot x_i^n$$

or for the entire output layer,

$$Y = X^{n+1} = B^n + A^n \cdot X^n \tag{12.5}$$

where Y is a vector of outputs.

The activation function ϕ is a nonlinear function, often specified as the hyperbolic tangent; however, rectangular linear unit, softmax, leaky rectangular linear unit, Gaussian, trigonometric functions, etc., are also used (e.g., Hsieh, 2023).

All layers of multilayer perceptron NN between input and output layers are called "hidden layers." NNs with multiple hidden layers are called "deep neural networks" (DNN). The simplest multilayer perceptron NN with one hidden layer is called a "shallow" NN (SNN),

$$Y = B^1 + A^1 \cdot \phi(B^0 + A^0 \cdot X)$$

or

$$y_j = b_j^1 + \sum_{i=1}^{k} a_{ji}^1 \cdot t_i, \quad j = 1, \ldots, M$$
$$t_i = \phi\left(b_i^0 + \sum_{s=1}^{N} a_{is}^0 \cdot x_s\right) \tag{12.6}$$

The SNN is a generic analytical nonlinear approximation or model for the parameterization mapping (12.1) and a mathematical solution of the machine learning problem (Vapnik, 2019).

The expansion (12.6) is a linear combination of neurons or basis function t_j, where the coefficients a_{ji}^1 ($j = 1, \ldots, M$ and $i = 1, \ldots, k$) are the linear coefficients of this expansion. It is essential that the basis functions t_j (12.6) are nonlinear with respect to inputs x_s ($s = 1, \ldots, N$) and to the fitting parameters or coefficients a_{is}^0 and b_i^0 ($i = 1, \ldots, k$). Due to the nonlinear dependence of the basis functions on multiple fitting parameters a_{is}^0 and b_i^0, the basis $\{t_i\}_{i=1,\ldots,k}$ becomes a very flexible set of nonorthogonal basis functions that adjust to the functional complexity of the approximated mapping (12.1). It has been shown by many authors in a variety of contexts that the family of functions (12.6) can approximate any continuous or almost continuous (with a finite number of finite discontinuities, like step functions) mapping (12.1) (Cybenko, 1989; Funahashi, 1989; Hornik, 1991; Chen & Chen, 1995). Vapnik (2019) proved that "for wide sets of functions of Hilbert space the solutions that minimize Empirical Risk functional in a set of functions with bounded norm defines 'shallow network' (with just one hidden layer)." The accuracy of the SNN approximation or the ability of the NN to resolve details of the parameterization mapping (12.1) is proportional to the number of basis functions, k (Attali & Pagès, 1997).

The NN (12.3 and 12.4) itself belongs to the class of mappings given by (12.1). Both the *computational and functional complexity* of an DNN can be defined as the number of fitting parameters a and b in (12.3) and (12.4) (Krasnopolsky, 2013). Therefore, for a deep neural network, the complexity is given by

$$\mathbb{C} = \sum_{i=0}^{n} k_{i+1}(k_i + 1) \tag{12.7}$$

where n is the number of hidden layers in the DNN, k_i is the number of neurons in the hidden layer number i, $k_0 = N$ is the number of inputs, and $k_n = M$ is the number of outputs.

For the shallow NN, this equation can be simplified as

$$\mathbb{C} = k \cdot (N + M + 1) + M \tag{12.8}$$

Different SNNs approximating the same target mapping (12.1) with a fixed number of inputs N and outputs M differ in complexity proportionally to the number of basis functions (neurons), k, used in individual NNs. The NN complexity linearly grows with the growth of the dimensionalities of both the input space (the number of inputs), N, and the output space (the number of outputs), M. The number of NN weights per output,

$$c = \mathbb{C}/M, \tag{12.9}$$

reflects the computational and functional complexity of the dependency of each NN output vs. NN inputs. This measure of the NN complexity is useful when NNs with different numbers of outputs are compared.

A pictographic language reminiscent of a data flowchart is traditionally used in the NN field. In this language, a basis function t_j (12.4), or neuron, is represented as shown in Figure 12.2a and b. The neurons are situated into *layers* inside the NN (Figure 12.2c shows a shallow NN). The input layer is in a sense a symbolic layer. Input neurons do not perform any numerical function; they simply distribute inputs to neurons in the following hidden layer.

Hidden layers are usually composed of nonlinear neurons (equations (12.4) and Figure 12.2a). The neurons in the output layer are usually linear (Figure 12.2b). The connections (arrows) in Figure 12.2 correspond to the NN weights, the name used for the fitting parameters a and b in NN jargon.

For the simplest shallow NN shown in Figure 12.2c, there is a one-to-one correspondence between equations (12.6) and Figure 12.2c. However, in general (e.g., for more complex types of NNs), the pictographic language (Figure 12.2) is not redundant. This language can suggest NN architectures that probably cannot be represented analytically in terms of a closed set of equations (e.g., some convolutional and recurrent NNs, NNs with missed connections, and complex NNs that are composed of several NNs of different types). The pictograms that represent the design of such NNs cannot be described by a closed set of equations; however, these pictograms can be translated into computer codes.

As we mentioned above, the SNN architecture is sufficient for the approximation of any continuous (or almost continuous) mapping. Additional hidden layers (DNN) and/or nonlinear neurons in the output layer can be introduced to solve more complex and special mapping problems and problems of different nature, for example, principal component analysis (Hsieh, 2001), extreme event prediction (e.g., forecasting large hail; Gagne et al., 2019), and pattern recognition (e.g., detecting synoptic-scale fronts; Lagerquist et al., 2019). DNNs have been very successful in many areas such as pattern recognition, classifications, image processing, etc. However, as pointed out by Vapnik (2019), from the standpoint of statistical learning theory, only the SNN has been formally shown to be a solution to the mapping approximation problem. Successful approximation of the mapping (12.1) by a DNN cannot be guaranteed, and this specific application of DNNs should be considered a heuristic approach until otherwise is proven. Both SNNs and DNNs have been applied to parameterization mappings by different authors (also see the discussion in section 12.9.3).

12.3.2. Training Set

In a practical application, a parameterization mapping (12.1) is usually represented and presented to the NN by a data or training set, C_T, that consists of P pairs of input and output vectors X and Y,

$$C_T = \{X_p, Y_p\}_{p=1,\ldots,P} \tag{12.10}$$

where $Y_p = M(X_p) + \xi_p$, $X_p \in D$, and $Y_p \in R$, ξ_p represents any errors associated with the observations or calculation with a probability density function $\rho(\xi)$, and D and R are the mapping domain and range, respectively. The set C_T can also be considered as a combination of two rectangular matrices,

$$C_T = \{C_X, C_Y\}, \tag{12.11}$$

where C_X is a matrix of dimensionality $N \times P$ composed of all input vectors X, and C_Y is a matrix of dimensionality $M \times P$ composed of all output vectors Y. The training set is all that the NN "knows" about the parameterization mapping that it is expected to approximate. This is the reason why NN belongs to a class of *data driven* or *learning from data* methods (Cherkassky & Muller, 2007).

The training set represents the mapping (12.1) for the NN and, therefore, it needs to be *representative*. It means that the training set must have a sufficient complexity to represent the complexity of the target mapping (12.1), allowing the NN to achieve the desired accuracy of the approximation of the target mapping. The set should have a sufficient size, P, of properly distributed data points that adequately resolve the functional complexity of the parameterization mapping (12.1). The training set should have finer resolution where the target mapping is not smooth and coarser resolution where it is smoother, namely, the domain D should be properly sampled. If the domain is undersampled, the training set is not representative. There are no generic recipes of how to determine if the domain is undersampled. In each particular case, this problem should be solved taking into account spatial and temporal scales of underlying physical processes, the model resolution, and the frequency and spatial and temporal distributions of extreme events.

Understanding the configuration of the mapping domain and its properties is essential for any application of the mapping (12.1) and for proper NN training and application (Bishop, 1995). If all components of the input vector X are scaled to the range [−1., 1.], the volume of the input space R^N is equal to 2^N and, therefore, exponentially grows with N. In a numerical model, once the space is discretized on a grid by K values per dimension, then the problem grows even faster, as K^N. It means that in the input space we have K^N grid cells, and to represent our mapping, we need an exponentially large training set in order to ensure that each grid cell contains at least one

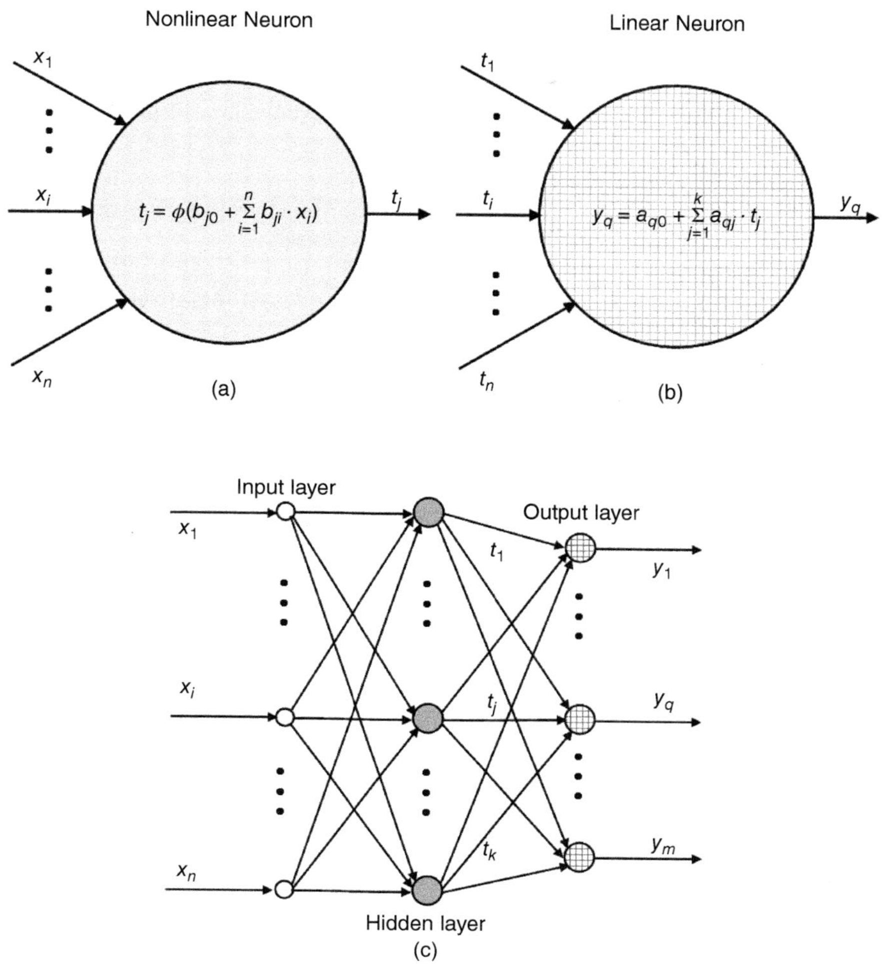

Figure 12.2 (a) Nonlinear neuron, (b) linear neuron (equation (12.4)), and (c) the simplest shallow NN with one hidden layer and linear neurons in the output layer (equation (12.6).

data point. This exponential growth in the amount of data with the increase of the input space dimensionality is often called the *curse of dimensionality* (Bishop, 1995; Vapnik & Kotz, 2006).

Fortunately, for parameterization mappings, the input vector X consists of atmospheric state variables that are related and correlated due to multiple physical and chemical processes in the atmosphere. The interrelations and correlations between inputs reduce the size and the effective dimensionality of the domain. Thus, the mapping domain, D, spanned by the input vectors X occupies only a part of the input space R^N. As a result, for parameterization mappings, the mapping domain, D, may be significantly smaller than an N-dimensional cube in the input space R^N. This may significantly simplify (in many cases make possible) the sampling task, especially for cases of high input dimensionality, helping to avoid the curse of dimensionality. In general, it is often very difficult, if possible at all, to determine the actual shape, location, and effective dimensionality of D, which makes it difficult to adequately sample the mapping domain D. However, for parameterization mappings, when observed data or data simulated by an atmospheric model are used, the nature or model automatically, by default, place the data inside the domain D.

Also, understanding relationships between components of the input vectors X allows to reduce dimensionality of the input vector, removing redundant components (Krasnopolsky et al., 2009; Krasnopolsky et al., 2010). For example, for the short-wave radiation parameterization in NCEP Climate Forecast System (CFS), the input vector X has more than three thousand components. By removing the redundant components, the dimensionality of X was reduced to $N = 562$. Even after such a tremendous reduction of dimensionality, generally speaking, $P > 2^{562}$ data points are necessary to sample the input space and create a training set. However, due to the use of data simulated by CFS, which samples only the parameterization domain D, the size of the training set could be reduced to a reasonable size, $P = 2 \times 10^5$ (Krasnopolsky et al., 2010). When MLPs are developed, simulated data or observations fused with the simulated data (analysis) are used. Usually, there is no shortage of the training data. The only problem here may be the big size of the training set (especially if DNNs are used) that may increase training time and memory requirements even beyond capacities of modern HPC.

Another important consideration is a relationship between the size of the training set, the complexity of the NN architecture, and concepts of underfitting and overfitting that are well known in statistics. Figure 12.3 demonstrates some important issues in this regard. The orange squares represent the data set that was generated using a parabola ($y = a \cdot (x + b)^2$) with some noise added. Figure 12.3a shows an approximation of the data using a minimal complexity linear model. The linear approximation completely misses the major feature of the data—the y quadratic dependence on x. Thus, the data are underfitted; the model is too simple; and the complexity of the approximation model is lower than the complexity of the data set.

Figure 12.3c shows the opposite situation. The complexity of the selected approximation model is significantly higher than the complexity of data; the model overfits the data (fits the noise). Figure 12.3b shows the case when the complexity of the model corresponds to the complexity of the data set.

Figure 12.3d illustrates cases when the data set is incomplete, and a generalization is required. In the first case (left panel), successful generalization (interpolation) is performed. In the second case, generalization fails because the data set is not representative (entire left corner is missed) and the model is forced to extrapolate; however, the nonlinear extrapolation is an ill-posed problem and, in this case, extrapolation is erroneous.

12.3.3. ML Modeling Parameterizations and Special Requirements

ML can be applied to parameterization at least in two different ways (Krasnopolsky, 2013): (1) as an *emulation technique* for *accelerating calculation* of parameterizations previously developed based on approximate description of underlying physical processes (e.g., NN emulation of radiation parameterizations; Krasnopolsky et al., 2010) and (2) for *developing new* "empirical" *parameterizations* based on observed data or data simulated by high-resolution models in situations when underlying physical processes are very complicated and not very well understood (e.g., cloud physics; Beucler et al., 2020; Brenowitz & Bretherton, 2018; Krasnopolsky et al., 2013). The great flexibility of ML tools also allows a combination of these two approaches. Thus, the terms an "emulating ML parameterization (EMLP)" or an "ML parameterization (MLP)" are used here to distinguish between these two problems.

It is noteworthy that both aforementioned tasks are different from a typical ML problem, where minimizing the error function for the process in consideration is the end goal. The ML tool is trained on a training set and validated on a test set. If approximation errors on both sets are sufficiently small, the problem is solved and an ML tool is

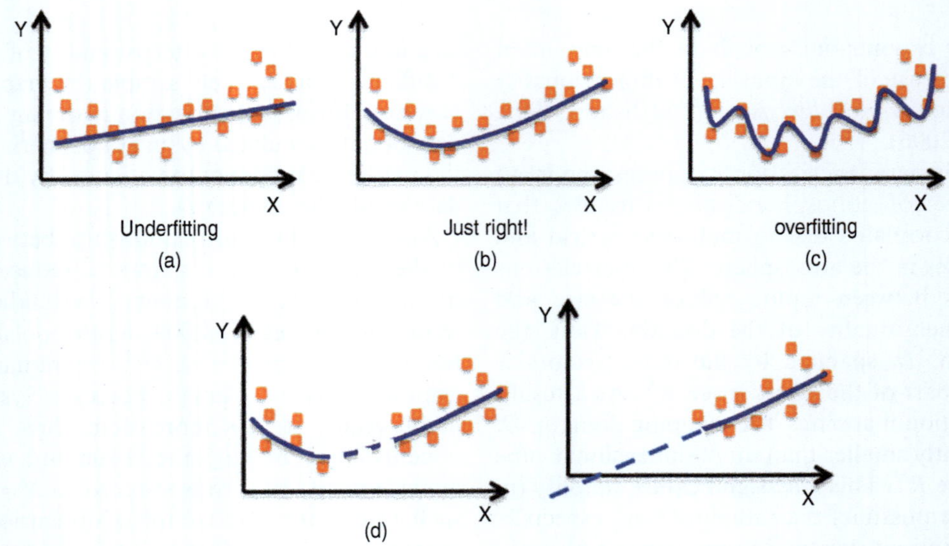

Figure 12.3 Relationship between complexity of approximating function and underfitting and overfitting (see the text).

ready to use. In the case of applying ML to model parameterization mappings, validation on an independent data set is a necessary but not sufficient step. The developed MLP or EMLP (in the rest of this section we use acronym MLP for the both of them) is a subsystem/component of larger systems (atmospheric, oceanic, or coupled models). After introducing MLP in the larger system, the larger system becomes a hybrid model, which is a combination of physically based components (e.g., model dynamical core and other parameterizations) with the MLP. In the hybrid model, MLP can behave differently than on the test set, even when test set approximation errors are small, because of two reasons: (i) the larger host systems (e.g., GCM) are nonlinear complex systems; in such systems, small systematic, and even random, approximation errors can accumulate over time and produce a significant impact on the quality of the model results, and (ii) erroneous behaviors and instability can arise if the system travels to the boundaries of the training set domain. At the same time, Pincus et al. (2003) have demonstrated that unbiased, random, uncorrelated errors in radiative forcing do not statistically affect forecast skill in a state-of-the-art weather forecasting model. Pincus and Stevens (2013) explain this result in the following way: random small local heating rate errors in the bulk of the atmosphere lead to local small-scale instabilities that are mixed away by the flow. They caution, however, that there is no such mechanism for the surface variables, such as skin temperature, and errors in surface fluxes can be more consequential. Therefore, on the one hand, the development and application framework for MLPs should be focused on obtaining both the highest possible approximation accuracy and the highest possible accuracy and stability of MLP in hybrid model simulations. On the other hand, in some applications, the hybrid model can tolerate approximation errors in the MLP component if they remain random, uncorrelated, and unbiased.

Thus, when an MLP is developed, in addition to the criterion of small approximation errors, at least three other criteria are used: (i) the NN complexity, $\mathbb{C}$, (12.7 and 12.8) is controlled and restricted to a minimal level sufficient for good approximation and interpolation; (ii) independent training, validation, and test data sets are used; the validation set is used in the process of training to control overfitting and the test set is used after the training to evaluate interpolation accuracy; redundant training set (additional redundant data points added "in-between" training data points sufficient for a good approximation) may be used for improving the NN interpolation abilities; and (iii) the most important validation criteria for MLP are obtaining the highest possible approximation accuracy, the highest possible accuracy of MLP in GCM simulations, and stability of these simulations.

12.4. ML EMULATIONS TO SPEED UP CALCULATION OF PARAMETERIZATIONS

Here, we discuss the development of NN emulations for the model physics components. Thus, the terms an "emulating ML" or an "ML emulation" of the parameterization mapping (12.1) are used here to emphasize the differences between developing an ML emulation of an existing physically based parameterization and developing a new MLP based on learning from data simulated by finer resolution models and/or observations. In the case of a complex high-resolution multicomponent modern model, calculations of model physics usually take a significant part of the total model computation time, especially if physics or a particular parameterization is calculated at each grid point and time step. For example, radiative transfer parameterizations supply their host model with broadband fluxes and heating rates, which are obtained by integration over time, space, and frequency. Therefore, a trade-off between accuracy and computational expense can be found in how finely these dimensions are discretized (Hogan et al., 2017).

- *Discretization in time*: All GCMs update their radiative heating/cooling rates less frequently than the rest of the model fields. For example, National Centers for Environmental Prediction (NCEP) Global Forecast System (GFS) v16 general circulation model (GCM) in its operational configuration updates its radiative fields once per model hour, while updates to temperature, moisture, most cloud properties, etc., happen every 150 model seconds or 24 times per single radiation call. Updates due to dynamical processes happen even more frequently, every 12.5 s (Kain et al., 2020). This approximation is good for slowly changing fields of certain radiatively active gases but is less justified for small-scale clouds with lifetimes of an hour or less.
- *Discretization in space*: Some GCMs calculate radiative fields on a coarser spatial grid and interpolate them onto a finer grid used for the rest of the model variables. For example, the radiation grid in the European Centre for Medium-Range Weather Forecasts (ECMWF) Integrated Forecast System (IFS) v43R3 in the ensemble mode is 6.25 times coarser than the physics grid (Hogan et al., 2017). This may cause 2 meter temperature errors in areas of surface heterogeneity, e.g., coasts (Hogan & Bozzo, 2015).
- *Discretization/sampling in frequency space*: The Rapid Radiative Transfer Model (RRTMG), a parameterization of radiative transfer for GCMs used in NCEP GFS and ECMWF IFS, utilizes 14 bands in the short wave (Mlawer et al., 1997), while the parameterization used at United Kingdom Met Office Unified Model utilizes 6 bands (Edwards & Slingo, 1996). Monte Carlo Spectral Integration (Pincus & Stevens,

2009) performs integration over only a part of the radiative spectrum, randomly chosen in each point in time and space, allowing to increase temporal/spatial resolution of radiation calculations. Monte Carlo Integration of the Independent Column Approximation (McICA) (Pincus et al., 2003) integrates over the entire spectrum, but samples subgrid-scale (SGS) cloud properties in a random, unbiased manner, in each grid column in time and space, instead of integrating over them.

These techniques reduce horizontal, or vertical, or the temporal variability of the radiation fields, reducing the accuracy of a model's radiation transfer calculations and spatial or/and temporal consistency of radiative fields with other parts of the model physics and dynamics, which may, in turn, impact the accuracy of climate simulations and weather predictions. More frequent calculations of model physics, and future introduction of more sophisticated parameterizations, will result in further increase in computational burden imposed by model physics.

Another approach currently used to reduce the cost of model physics calculations is the use of simplifying physical and mathematical approximations when developing parameterizations. For example,

- In the wind wave model (Tolman, 2002), the calculation of nonlinear wave-wave interactions (Hasselmann & Hasselmann, 1985; Hasselmann et al., 1985) requires roughly 10^3 to 10^4 times more computational effort than all other aspects of the wave model calculations. Present operational constraints require that the computational effort to estimate nonlinear wave-wave interactions should be of the same order of magnitude as for the rest of the wave model. This requirement was met with the development of the discrete interaction approximation (DIA; Hasselmann et al., 1985). Two decades of experience with the DIA in wave models has identified DIA's significant shortcomings (Tolman et al., 2005).
- The approach, originally labeled as a "superparameterization" (SP) (see Chapter 1 of this book), but later as the Multiscale Modeling Framework (MMF), couples cloud-scale and large-scale dynamics within a single modeling framework (Khairoutdinov & Randall, 2001). The MMF replaces cloud parameterizations with a cloud resolving model (CRM). It is noteworthy that, even though 3D cloud dynamics is not considered, MMF still increases the run time of the host global model by a factor of 200–250, which severely limits its applicability (Wang et al., 2011)

State-of-the-art modeling of microphysical (MP) cloud processes is tremendously computationally intensive and most atmospheric models include it in parameterized form (e.g., Morrison et al., 2020). Microphysical parameterizations introduce substantial simplifications and are sometimes derived for specific atmospheric scenarios while being applied globally. Even in parameterized form, microphysics calculations are computer resources and time consuming (see Chapter 4 of this book).

The aforementioned situation provides the motivation to look for alternative, faster, and accurate ways of calculating model physics, chemistry, hydrology, and other processes. During the past decade, an ML approach based on NN approximations or emulations was applied for the accurate and fast calculation of atmospheric radiative processes (Chevallier et al., 1998; Krasnopolsky, 1996, 1997) and for emulations of model physics parameterizations in atmospheric numerical models (Krasnopolsky et al., 2002, 2005, 2008a, 2010). In these works, the calculation of the model physics components has been accelerated by factors of 10 to 10^5 as compared to the corresponding original parameterizations.

12.4.1. NNs Emulating Radiative Parameterizations

Radiation parameterizations (see Chapter 2 of this book) are one of the most computationally demanding parts of model physics. Approaches formulated by Chevallier et al. (1998, 2000), Krasnopolsky (1996), and Krasnopolsky et al. (2002, 2005, 2010) represent two different ways of hybridization of first principles and NN components in the physics parameterizations as well as in complex climate and NWP models. These two approaches lead to the concepts of a hybrid parameterization (HP) (Chevallier et al., 1998, 2000) and a hybrid model or hybrid GCM (HGCM) (Krasnopolsky et al., 2002, 2005; Krasnopolsky and Fox-Rabinovitz, 2006a,b; Goldstein & Coco, 2015).

Chevallier et al. (1998, 2000), following the physical structure of the long-wave radiation (LWR) parameterization, a component of the ECMWF global atmospheric model, considered it as a combination of fluxes at vertical levels, F_i:

$$F(S, T, V, C) = \sum_i \alpha_i(C) \cdot F_i(S, T, V) \tag{12.12}$$

where i is an index for the vertical level, the vector S represents surface variables, T is a vector (profile) of atmospheric temperatures, C is a profile of cloud variables, and the vector V includes all other variables (humidity profile, different gas mixing ratio profiles, etc.). Each partial or individual flux $F_i(S,T,V)$ is a continuous mapping and all discontinuities related to the cloudiness are included in $\alpha_i(C)$. In their hybrid parameterization, referred to as the "NeuroFlux," Chevallier et al. (1998, 2000) combined first-principles-based calculations of cloudiness functions $\alpha_i(C)$ with NN approximations for a partial or individual

Table 12.1 Statistics for estimating the accuracy of the heating rate calculations (in K day^{-1}) and the computational performance (speedup) of the NN LWR emulation vs. the original parameterization for NCAR CAM (T42L26) and NCEP CFS (T126L64) LWR and SWR parameterizations, L is the number of the vertical layers.

		NCAR CAM ($L = 26$)		NCEP CFS ($L = 64$)	
	Statistics	LWR	SWR	RRTMG LWR	RRTMG SWR
Total	Bias	$3. \times 10^{-4}$	$6. \times 10^{-4}$	$2. \times 10^{-3}$	$5. \times 10^{-3}$
Error statistics	RMSE	0.34	0.19	0.49	0.2
NN complexity	c	490	810	520	706
Speedup, η	*Times*	150	20	16 (20)	60 (88)

Note. Total statistics show the bias, RMSE, the NN complexity c (12.7), and average speedup η, where η shows an averaged (over a global data set) speedup or how many times NN emulation is faster than the original parameterization in a sequential single processor code-by-code comparison. RRTMG is the Rapid Radiative Transfer Model for GCM.

flux $F_i(S, T, V)$. As a result, the "NeuroFlux" hybrid LWR parameterization developed by Chevallier et al. (1998, 2000) is an array or battery of about 40 NNs. The number of required NNs (and speedup of calculations) depends on the number of vertical layers in the GCM, and at a vertical resolution of 60 layers or more, both sufficient accuracy and speedup of NeuroFlux cannot be achieved simultaneously (Morcrette et al., 2008).

Krasnopolsky et al. (2002, 2005) developed an emulation approach that considers the entire parameterization as a mapping (12.1), which is emulated using either a single SNN or an ensemble of SNNs. This ML emulation approach is based on the fact that any parameterization of model physics can be considered as a continuous or almost continuous mapping. For this approach, the desired accuracy of the NN emulation is achieved simultaneously with a significant speedup (see Table 12.1). Evaluation of accuracy of the NN emulation, as the first step, includes the calculation of approximation errors by applying the NN emulation to an independent test data set. This step is necessary to identify NNs that are not sufficiently complex (approximation errors are significantly larger than errors on training set). However, it is practically impossible to determine if the approximation errors are "sufficiently small" without a second crucially necessary step: interactive coupling of the NN emulation to its host model. Only evaluation of NN emulations in parallel model runs allows us to conclude that the approximation errors are sufficiently small and that the errors do not accumulate during the hybrid model run and have almost negligible impacts on the hybrid model behavior. It was demonstrated by Krasnopolsky et al. (2008a, 2010, 2012) for NCAR CAM and NCEP CFS and GFS that approximation errors presented in Table 12.1 are sufficiently small.

In terms of the accuracy statistics presented in Table 12.1, there are practically no differences between NCAR CAM with 26 vertical layers and NCEP CFS with 64 vertical layers. Thus, the accuracy of the NN emulation approach does not significantly depend on vertical resolution of the model. This fact illustrates the robustness of the NN emulation approach with respect to the changes in the model vertical resolution.

The NN complexity c (12.9) and average speedup η (how many times NN emulations are faster than the original parameterization) are also shown in Table 12.1. The NN complexity per output, c, is used because NNs with different numbers of outputs are compared here. In this case, c provides a more adequate metric for comparisons. For the LWR parameterization, we see a significant decrease in the speedup for NCEP CFS with 64 vertical layers vs. NCAR CAM with 26 vertical layers although the LWR NN emulation for NCEP CFS is still 16 times faster than the original parameterization. For the SWR parameterization, the opposite tendency is observed; that is, the NCEP CFS SWR NN is more than 3 times faster than the NCAR CAM SWR NN.

These seemingly contradictory speedups for LWR and SWR emulations can be explained by the interplay of the two main contributing factors: the physical and mathematical complexities of the radiation calculation itself (the number of treated species, spectral bands, etc.), and the dependence of the particular numerical scheme implemented in the radiative transfer parameterization on the number of vertical layers in the model (for a detailed discussion of this topic see Krasnopolsky et al., 2010). The results presented in Table 12.1 reflect the fact that the numerical scheme implemented in the NCEP CFS RRTMG-LW parameterization (Clough et al., 2005) is significantly more efficient (linear with respect to the number of vertical levels L) than that of the NCAR CAM LWR (Collins et al., 2002) parameterization (quadratic with respect to L). Thus, a smaller speedup factor is produced by the NN emulation for NCEP CFS LWR than that of NCAR CAM LWR. The NCEP CFS's RRTMG-SW includes more spectral bands and uses more complex treatment for a larger variety of absorbing/scattering species than NCAR CAM SWR; thus, NN emulation shows a larger speedup value, η, for NCEP CFS SWR than that of NCAR CAM SWR and NCEP CFS LWR. It was shown recently (Belochitski & Krasnopolsky, 2021) that shallow-neural-network-based emulators of radiative transfer parameterizations developed almost a decade ago for a state-of-the-art general circulation model NCEP CFS are robust with respect to the substantial structural and parametric change in the host model: when used in two seven month-long experiments with the current significantly changed CFS, they remain stable and generate realistic output (see, for example, Figure 12.4).

The radiative transfer calculations take a different amount of time under different cloud conditions because of the varying complexity of cloud-radiation interaction. More detailed estimations of speedup have been separately performed for three different types of cloudiness: clear sky, three cloud layers, and a more complex cloud condition where deep convection occurs. Calculation speedup for a deep convection case is shown in Table 12.1 in parentheses. For a more complex cloud-radiation interaction in the deep convection case, the calculation of the original LWR and SWR parameterizations takes ~22% and ~57% more time, respectively, than for clear sky conditions. The time required for the NN radiation calculations does not depend on the cloud conditions; as a result, the speedup achieved by the NN emulation is significantly higher for more complex cloud-radiation interactions. Additionally, using NN emulations in place of original physically based parameterization allows for a better load balancing in a parallel computational environment due to reduced idle processor time: calculations over clear sky atmospheric columns do not have to wait for cloudy sky columns calculations to complete.

Using the approach developed at NCEP, an emulator of RRTMG consisting of a single shallow NN that replaces both LW and SW parameterizations at once was developed at Korean Meteorological Agency for the short-range weather forecast model Korea Local Analysis and Prediction System in an idealized configuration with 39 vertical layers (Roh & Song, 2020). Inputs and outputs to RRTMG were saved on each 3 s time step of a 6-hour long simulation of a squall line, and about 270,000 input/output pairs were randomly chosen from this data set to create training, validation, and testing sets. Dimensionality reduction was performed by removing constant inputs. Several activation functions were tested (tanh, sigmoid, softsign, arctan, and linear) with hyperbolic tangent providing best overall accuracy of approximation. The emulator was 2 orders of magnitude faster than the original parameterization and was stable in a 6-hour long simulation.

Two dense, fully connected, feed-forward deep-NN-based emulators with three hidden layers, one emulator per parameterization, were developed for LW and SW components of RRTMG-P for the Department of Energy's Super-Parametrized Energy Exascale Earth System Model (SP-E3SM) (Pal et al., 2019). In SP-E3SM, radiative transfer parameterizations act in individual columns of a 2D cloud-resolving model with 31 vertical levels embedded into columns of the host GCM. Calculations of cloud and aerosol optical properties were not emulated; instead, original RRTMG-P subroutines were used. Inputs and outputs of radiative parameterizations were saved on every time step of a year-long model run, with 9% of these data randomly chosen to form a data set of 12,000,000 input/output records for LW and of 6,000,000 input/output records for SW emulator training and validation. 90% of the data in these sets were used for training and 10% for validation and testing. No additional dimensionality reduction was performed. Sigmoid AF was chosen as it was found to provide slightly better training convergence than the hyperbolic tangent. The emulator was an order of magnitude faster than the original parameterization and was stable in a year-long run.

A number of ML-based radiative transfer parameterizations or their components have been developed, but, to our knowledge, have not yet been tested in an online setting, that is in interactive coupling to an atmospheric model. Among them are deep-NN-based parameterizations of gas optical properties for RTTMG-P (Ukkonen et al., 2020; Veerman et al., 2021) and an SW radiative transfer parameterization based on convolutional deep neural networks (Lagerquist et al., 2021).

A more detailed description of the NN emulation approach can be found in Chapter 4.2 of Krasnopolsky (2013) and in Pal et al. (2019).

12.4.2. NNs Emulating Super Parameterizations

The super-parameterization (SP) in an MMF (see Chapter 1 of this book) is conceptually similar to a traditional parameterization of model physics (Krasnopolsky et al., 2014). On each time step, the embedded CRM receives a vector of input parameters, X, from the host GCM, which describes a state of the atmosphere in a column in terms of the GCM variables. After integration of the CRM in the column of the GCM, the SP returns to the GCM a vector of output parameters Y, which describes the physical forcing for this column in terms/variables of the GCM. As a result, the entire SP, from a mathematical point of view, can be considered as a mapping. Given the physical and mathematical properties of the CRM, this mapping is continuous or almost continuous and can be emulated by an ML tool with a desired accuracy.

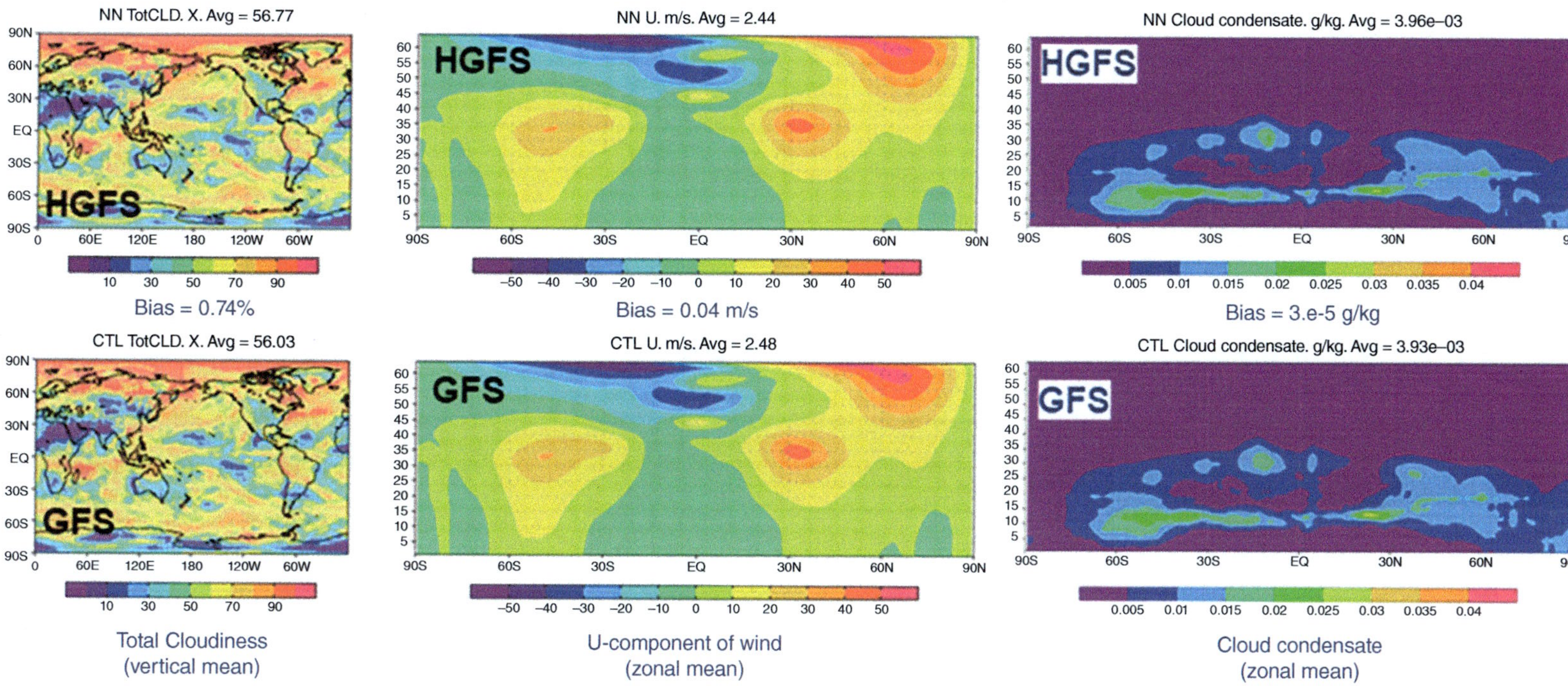

Figure 12.4 The 10-day average of the total cloudiness (left), the U-component of wind (center), and the cloud condensate (right) over a forecast initialized at 0Z on 1 January 2018 produced by HGFS (upper panels) and the control GFS (lower panels).

However, when it comes to emulating SP, stochastic behavior of the parameterization has to be taken into account. This behavior emerges due to variability associated with internal states of the CRM, ξ, that are stored between the GCM time steps and serve as initial conditions for the CRM integration. The internal state is "hidden" from the host GCM environment. Because of this, given the same inputs from the GCM, the SP may generate different outputs, depending on the stored initial conditions, ξ. Hence, the SP is not completely deterministic, the SP outputs are not completely determined by the SP inputs, and the SP should be considered a stochastic mapping (12.2), $Y = F(X, \xi)$. As a result, uncertainty will emerge in the simulated data, which, if the uncertainty is significant, cannot be accounted for by a single emulating NN. Thus, an ensemble of NNs is, probably, better suited for emulating the SP than a single emulating NN (Krasnopolsky et al., 2014). As it could be expected, not all SP outputs demonstrate stochastic behavior to the same degree. For example, while temperature tendencies and cloud fractions are almost deterministic, water vapor and ice water tendencies clearly demonstrate a significant stochastic behavior. As a result, using an ensemble of NNs reduces approximation errors for water vapor and ice water tendencies but has almost no effect on the accuracy of temperature tendencies and cloud fractions (Krasnopolsky et al., 2014). Jones et al. (2019a, 2019b) averaged several SP models to create a "deterministic SP." They obtained some significant differences in the climate and variability, but overall the effect of stochasticity was limited. Rasp et al. (2018) emulated the SP using a single deep NN. Their approach does not take into account the uncertainty discussed above. They showed that using an NN speeds up calculation 20-fold. Such performance improvement opens the possibility to make experimentation with MMF substantially less computationally demanding. Also, the development of SP including 3D CRM, after emulation of this SP with NN, may become feasible.

Recently, Wang et al. (2022) used a group of deep residual multilayer perceptrons with strong nonlinear fitting ability to emulate embedded CRM in an MMF configuration with realistic orography. The resulting hybrid MMF is 20–30 times as fast as the original, reproduces reasonable climatology and climate variability, and remains stable in multiyear experiments.

12.4.3. NN Emulating Microphysics

As mentioned above (see also Chapter 4 of this book), parameterizations necessarily simplify representation of microphysical processes; as a result, the applicability of a given MP parameterization may be limited to a subset of possible atmospheric scenarios. For example, Gilmore et al (2004) have found pronounced differences in the amount of predicted rainfall in numerical simulations of a deep convective storm using liquid-only vs. liquid-ice MP schemes. Moreover, even when MP parameterizations of comparable complexity are used, fundamental lack of sufficient understanding of many individual processes results in differences in simulations of atmospheric phenomena, e.g., supercell thunderstorms (Morrison & Milbrandt, 2011), atmospheric rivers (Jankov et al., 2009), or tropical mesoscale convective systems (van Weverberg et al., 2014). Thus, a particular MP scheme performs well in certain atmospheric situations and performs not so well in others. When and why one scheme outperforms others is often not well understood. It appears that none of the existing MP parameterizations may offer comprehensive treatment of the natural processes involved. Also, even in parameterized form, MP calculations are time consuming.

Thus, ML tools can perform two different but related tasks when applied to MP parameterizations:

1. Create fast ML MPs by emulating various MP parameterizations; for example, the Thompson MP scheme (Thompson et al., 2008) was emulated with an ensemble of NNs (Krasnopolsky et al., 2017), also Brenowitz et al. (2020) built an emulator of an entire cloud microphysical parameterization, including fast phase changes. The emulator performs well in offline and online tests (i.e., when coupled to the rest of the atmospheric model). The FV3GFS weather model with the simplified Zhao and Carr (1997) microphysics was used to generate training sets.
2. Integrate existing MP parameterizations in a more comprehensive scheme that is able to offer better treatment of subgrid processes involved, cover greater variety of subgrid scenarios, and stochastically represent uncertainty in MP schemes.

12.4.4. NN Emulating Entire Model Physics and GCM

Building ML emulation of the entire block of atmospheric physics parameterizations in a state-of-the-art GCM is an attractive task: if successful, it could contribute to development of methodology for constructing new ML-based "empirical" parameterizations in addition to the speedup of model calculations. This task may be aided by the fact that the full diabatic forcing profiles in atmospheric models are generally smoother than forcing profiles from individual processes because the latter often balance each other. Krasnopolsky et al. (2009) discussed problems arising when emulating full physics using NNs and developed methods to solve some of these problems.

Several attempts have been made to create ML emulations of regional and simplified GCMs (Dueben & Bauer,

2018; Scher, 2018; Scher & Messori, 2019; Van der Merwe et al., 2007; Willard et al., 2021). It was shown by Van der Merwe et al. (2007) that NN can provide very fast (1,000 times faster) and accurate emulation of a coupled large-scale circulation model for the Columbia River, its estuary, and near ocean regions. The circulation model is highly nonlinear and is driven by ocean, atmospheric, and river influences at its boundaries. The NN provides an accurate emulation of the entire model. Scher (2018) used a convolutional NN to emulate a dry hydrostatic aquaplanet at ~500 km horizontal resolution with 10 vertical levels, without either diurnal or seasonal cycles and with idealized physics. This NN used the global state of winds, temperature, and geopotential height as inputs, and predicted the global state of the same variables on the next time step. It runs stably and with reasonable realism as compared to the control for 1,000 model years. In the follow-up work, Scher and Messori (2019) assessed how the complexity of a climate model affects the forecast skill of the emulating NN, and how dependent the skill was on the length of the training period. They found that using neural networks to emulate the climate of a general circulation model with a seasonal cycle remained a challenge—in contrast to earlier promising results on a model without a seasonal cycle. Dueben and Bauer (2018) used a toy model for global weather prediction to identify challenges and fundamental design choices for a forecast system based on neural networks. Such fast and accurate emulations (if successful) will enable significant advances in the development of geophysical modeling systems. They may serve as an improved incarnation of statistical models widely used before the numerical weather prediction era for short-term and local forecasting.

12.5. USING ML TO DEVELOP NEW PARAMETERIZATIONS: TRAINING COARSE-GRID PARAMETERIZATION USING DATA FROM FINE-GRID SIMULATIONS

The ML techniques can also be used to improve model physics. Substantial uncertainties remain in representation of cloud properties in GCMs (e.g., Brenowitz & Bretherton, 2018; Ceppi et al., 2017; Rasp et al., 2018). Cloud-resolving models resolve many of the phenomena that lower-resolution global and regional models do not (e.g., higher resolution fluid dynamic motions supporting updrafts and downdrafts, convective organization, mesoscale circulations, and stratiform and convective components that interact with each other, etc.; see also Chapters 5–12 of this book).

ML techniques can be used to build a bridge or interface between CRMs and GCMs, for example, to develop an ML parameterization of moist processes, which can be used as a parameterization in GCMs and can effectively account for major subgrid scale effects taken into account by other approaches (e.g., MMF approach; see section 12.4.2). The idea is to develop an MLP, which emulates the behavior of a CRM or large-eddy simulation (LES) model and can be run at larger scales (e.g., GCM scales) in a variety of regimes and initial conditions. The resulting emulation can be used as a novel and computationally viable parameterization of subgrid processes in an HGCM (e.g., Brenowitz & Bretherton, 2018; Gentine et al., 2018; Krasnopolsky et al., 2011, 2013; O'Gorman & Dwyer, 2018; Schneider et al., 2017), planetary boundary layer processes (Wang et al., 2019), etc. It may produce a parameterization of similar or better quality compared to the SP (that uses a simplified 2D CRM), effectively taking into account subgrid scale effects (with respect to the scales of a GCM) at a fraction of the computational cost. Bolton and Zanna (2019) trained convolutional NNs on data from a high-resolution quasi-geostrophic ocean model. They demonstrated that convolutional NNs successfully replicated the spatiotemporal variability of the subgrid eddy momentum forcing. Also, they could generalize to a range of dynamical behaviors and could be forced to respect global momentum conservation.

Figure 12.5 summarizes the process of developing such an ML parameterization. The CRM uses data, for example, from the Tropical Ocean Global Atmosphere Coupled Ocean-Atmosphere Response Experiment (TOGA-COARE), Department of Energy Atmospheric Radiation Measurement Southern Great Plains observatory, or other observational companies for initialization and forcing. The development of an MLP is a multi-step process. These steps are as follows (Krasnopolsky et al., 2011):

1. *Generating CRM data*: The model is run for period of time necessary to capture phenomena of interest, which

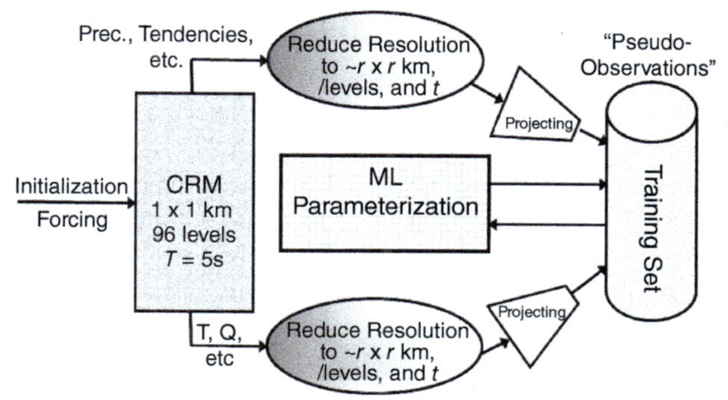

Figure 12.5 The process of preparation for training of an ML convection parameterization includes several steps: (1) simulating CRM data, (2) reducing the resolution of simulated data, (3) projecting a CRM vector of atmospheric states onto a GCM vector of atmospheric states, and (4) creating training set of "pseudo-observations."

may be limited by the data available for initialization and forcing, and the high-resolution output of the model is archived.

2. *Reducing the resolution of simulated data*: The high-resolution CRM simulated data are averaged in space and time. The data are averaged to a reduced horizontal resolution of $\rho < r \leq R$, where ρ and R are the CRM and GCM resolutions, correspondingly, and are interpolated/averaged to the number of vertical layers $l = L$, where L is the number of vertical layers in the GCM.

3. *Projecting a CRM vector of atmospheric states onto a GCM vector of atmospheric states*: From the reduced resolution CRM data created on the previous steps, the subset of variables is selected. We refer to this subset as the NN development set. Only variables that have corresponding GCM variables or can be calculated from or converted to prognostic or diagnostic variables available in the GCM are included in the development set (they are called "pseudo-observations" in Figure 12.5). Only these variables can be used as inputs and outputs in the NN parameterization. In general, CRM models utilize a different set of prognostic variables than a GCM that may include, for example, turbulent kinetic energy, sophisticated treatment of precipitation species, etc. If these variables are not available in the GCM, they cannot be included as MLP inputs and/or outputs. From the point of view of GCM "model reality," these variables are "hidden" variables responsible for subgrid scale variability. The acknowledgement of this challenge requires the introduction of the concept of uncertainty and "stochasticity." The development set of pseudo-observations implicitly represents a stochastic convection parameterization (i.e., a stochastic mapping) with an uncertainty that is an inherent feature of such a parameterization.

4. *Adjusting the differences*: The pseudo-observations used for the development of the MLP are not real observations. They represent averaged CRM simulated data. Pseudo-observations are used for development of the ML parameterization that is introduced into a GCM. The GCM has its own simulated atmosphere, which may not be in complete agreement with the averaged CRM simulated data and, therefore, with the MLP trained on pseudo-observations derived from the averaged CRM simulated data. Thus, special effort may be required to synchronize or make consistent the atmospheric realities of the GCM and the averaged CRM simulated data. CRM and the GCM mean differences for all variables selected as MLP inputs and outputs must be determined and compensated for (Krasnopolsky et al., 2011). These differences are the result of GCM and CRM being two different models with different temporal and spatial scales and resolutions, with different dynamics and physics; they also have different boundary and initial conditions and different forcing.

5. *Creating data sets*: The development set of pseudo-observations is separated into the independent training and test/validation data sets. Then, the MLP is trained using the training set. Due to the inherent uncertainty of pseudo-observations, the MLP derived from these data is a stochastic parameterization. One of the ways to take the stochasticity into account is to implement the MLP as an ensemble of NNs (Krasnopolsky et al., 2011, 2013; see also the next section).

6. The validation procedure for the MLP consists of two steps. First, the trained ensemble stochastic MLP is applied to the test set and error statistics are calculated. Second, the tested MLP is introduced into the host GCM to validate its behavior in the model simulations. *This last step is the most important step of the validation process.*
The same procedures can be applied for developing parameterization of subgrid processes for any coarse resolution model using a fine resolution model. For example, an LES model can be used to generate data for the following development of an ML tool.

Another approach that was introduced recently proposes to use the existing DA system to learn aspects of the model by doing parameter estimation or by learning new functional representations using neural networks or another ML approach (Bocquet et al., 2019). Such an approach of learning within an existing data assimilation system is feasible for a weather forecasting center that is already operating a data assimilation system that encodes a wealth of weather forecasting knowledge accumulated during decades (Geer, 2021).

ML can also be used for improving existing parameterization without following hybridization of the model. For example, genetic algorithms were used to optimize parameters of generalized multiple discrete interaction approximation for wind wave model (Tolman, 2013; Tolman & Grambine, 2013). The free parameters of the parameterization are optimized holistically by optimizing full model behavior in the WAVEWATCH wave model (Tolman, 2002). Then, this optimized parameterization was used in the WAVEWATCH model, and it improved model forecasts especially under extreme weather conditions.

12.6. USING ENSEMBLE OF ML TOOLS

An ensemble of statistical models is usually used when a single model cannot unambiguously account for the variability of the data. Such a situation may occur when, for example, (1) the data contain a significant amount of noise, and deriving an ensemble of models with the following averaging of the ensemble reduces the influence of the noise; (2) the data contain an inherent stochastic component and can be considered as representing a family of functions, as in the case of a stochastic function or mapping; (3) the problem is ill posed and a single statistical model does not give stable results, as in the case of nonlinear extrapolation or calculating derivatives of statistical model.

12.6.1. Stochastic Parameterizations and Stochastization of Deterministic Parameterizations

As mentioned above, the stochastic parameterization mapping (12.2) may be considered as a family of mappings distributed with a probability density function. Each mapping of the family may be considered to be one possible realization of subgrid processes. The range and shape of the distribution function are determined by the uncertainty vector ε. An average over subgrid scenarios represented by the stochastic parameterization can be approximated by a single ML tool, e.g., a single NN. However, an ensemble of ML tools can give a better and more adequate representation of a stochastic parameterization, provided each member of the ML ensemble emulates one or a few functions of the family (one or a few subgrid scenarios).

This ML ensemble emulation can also be used for stochastization of deterministic parameterizations to create perturbed physics ensembles (Krasnopolsky et al., 2008b). Usually, perturbed physics (or parameterization) is created by adding a small random value to a deterministic physics. When ensembles of ML emulations are used, a particular emulation (an ensemble member number j) may be considered to be the jth perturbed version of the unperturbed model physics (or parameterization), P, and can be written as

$$P_j^{\mathrm{ML}} = P + \varepsilon_j \tag{12.13}$$

where P_j^{ML} is an ML emulation (ensemble member) number j of the original model physics, P, and ε_j is an emulation error for the ML emulation number j. If an SNN is used for ML emulation, both the mean value and statistical properties of ε_j can be controlled and altered significantly by varying the number of hidden neurons k in equation (12.6) (Krasnopolsky & Fox-Rabinovitz, 2006a). For example, the systematic component of the approximation error (bias) can be made negligible and, therefore, ε_j almost purely random. Thus, ε_j can be made the same order of magnitude as the natural uncertainty of the model physics (or of a particular parameterization) due to the unaccounted variability of subgrid processes. A single ML emulation can be considered a stochastic version of the original deterministic scheme.

12.6.2. Calculating Jacobian of a Parameterization

In some applications, derivatives of developed MLP are required, e.g., for the calculation of an adjoint of an ML parameterization in a 4D variational data assimilation system (Chantry et al., 2021) or for error and sensitivity analysis. It means that the ML emulation Jacobian, J, that is a matrix of the first derivatives of the outputs of the ML emulation (e.g., 12.6) over its inputs has to be calculated,

$$J = \left[\frac{\partial y_q}{\partial x_i}\right]_{i=1,\ldots,N}^{q=1,\ldots,M} \tag{12.14}$$

More specifically, for SNN,

$$\frac{\partial y_j}{\partial x_p} = \sum_{i=1}^{k} a_{ji}^1 \cdot \left(1 - t_i^2\right) \cdot a_{ip}^0$$

After substituting (12.6) into the above expression, a simple equation for elements of the Jacobian matrix (12.14) for an SNN can be obtained,

$$\frac{\partial y_j}{\partial x_p} = \sum_{i=1}^{k} a_{ji}^1 \cdot \left\{1 - \left[\tanh\left(b_i^0 + \sum_{s=1}^{N} a_{is}^0 \cdot x_s\right)\right]^2\right\} \cdot a_{ip}^0$$

However, from a theoretical point of view, inference of the NN Jacobian is an ill-posed problem (Vapnik, 1995), which leads in practice to significant uncertainties in NN Jacobians calculated in a manner described above (Aires et al., 1999; Aires et al., 2004; Chevallier & Mahfouf, 2001). Several solutions have been proposed to reduce these uncertainties. First, if a data set for a Jacobian is available, the Jacobian can be trained as a separate additional NN (Krasnopolsky et al., 2002). Second, the mean Jacobian (over time or space) can be calculated and used (Chevallier & Mahfouf, 2001). Third, regularization techniques like "weight smoothing" (Aires et al., 1999) or a technique based on a principal component decomposition (Aires et al., 2004) can be used to stabilize the Jacobian. Fourth, if a data set for the Jacobian is available, the Jacobian can be trained, i.e., included as actual additional outputs in the NN and in the training data set. To do this, the error (or cost) function, which is minimized in the process of NN training, should be modified to accommodate the Jacobian; in other words, the Euclidean norm, which is usually used for calculating the error function, should be changed to the first-order Sobolev norm. Actually, it was shown that the function of the Sobolev space can be approximated by an NN with all their derivatives (Hornik et al., 1990).

Finally, an MLP Jacobian can be calculated using an NN ensemble approach (Krasnopolsky, 2007a), which allows one to calculate the ML emulation Jacobian with sufficient accuracy to be used in applications. The ensemble approach (a) significantly reduces the systematic and random error in ML emulation Jacobian, (b) significantly reduces the magnitudes of the extreme outliers, and (c), in general, significantly reduces the number of larger errors. In the NN ensemble approach, an ensemble of q ML emulations of a parameterization is trained. Then, for each particular ML emulation, a partial Jacobian, J_i, is calculated (if shallow NNs are used, by an analytical differentiation of equation (12.6)). Eventually, the Jacobian, J, is calculated as an average of partial Jacobians:

$$J = \frac{1}{q}\sum_{i=1}^{q} J_i \tag{12.15}$$

12.7. ENSURING PHYSICAL CONSTRAINTS

Physical constraints/relationships, such as energy conservation, are embedded in the model. The physically based parameterizations usually preserve physical invariants (energy, momentum, etc.) with high accuracy if these relationships are explicitly (or implicitly) built into the parameterizations. MLPs usually preserve physical invariants only approximately but can approximate physical constraints with high accuracy (Krasnopolsky et al., 2008c; O'Gorman & Dwyer, 2018). For example, for the random forest convection parameterization developed by O'Gorman and Dwyer (2018), the root-mean-squared error (RMSE) in conservation of column-integrated moist enthalpy in the control climate is very small at 0.2 W m^{-2} for both the training data set and the RF predictions on the test data set. For the NN emulations of radiation parameterizations developed by Krasnopolsky et al. (2010), the mean error in the balance equation (12.15) does not exceed 6.5×10^{-4} K day^{-1}. Both these errors do not cause any significant deviation from the control run in parallel run validations. In NCEP CFS parallel runs with NN LWR and NN SWR, which were run for 17 years, the deviation from the balance was corrected using the following approach (Krasnopolsky et al., 2010). The integral relationship for the imbalance, ε, that relates pressure, heating rates, and fluxes was used,

$$\varepsilon = \frac{\sum_{k=1}^{L}\alpha_k \cdot h_k}{\sum_{k=1}^{L}\alpha_k} + \frac{\Phi}{\sum_{k=1}^{L}\alpha_k} = 0.$$
$$\Phi = \begin{cases} F_{\text{tup}} - F_{\text{sup}} + F_{\text{sdn}} & \text{for LWR} \\ F_{\text{tup}} - F_{\text{tdn}} - F_{\text{sup}} + F_{\text{sdn}} & \text{for SWR} \end{cases} \tag{12.16}$$

where $\alpha_k = \frac{p_k - p_{k-1}}{G}$, p_k and h_k are pressure and heating rates at a vertical level k, G is a constant, F_{tup} is the total sky outgoing LWR or SWR flux at the top of the atmosphere, F_{tdn} is the total sky downward SWR flux at the top of the atmosphere, F_{sup} is the total sky upward LWR or SWR flux at the surface, and F_{sdn} is the total sky downward LWR or SWR flux at the surface.

The outputs of the original radiation parameterizations satisfy the relationship (12.16) with high accuracy because these relationships are explicitly (or implicitly) included into the parameterizations. The outputs of the NN emulations obviously satisfy (12.16) only approximately, i.e., in this case, the imbalance $\varepsilon \neq 0$ ε however, is small. For example, for the RRTMG-NN LWR emulation, the mean value for ε is 6.5×10^{-4} K day^{-1}.

A correction can be introduced for the heating rates, h_k. The correction makes the corrected or balanced heating rates, $\sim h_k = h_k + \varepsilon$, to satisfy the relationship (12.16) exactly. This correction is very small and, as a result, this balancing procedure does not practically affect the overall accuracy of LWR NN, marginally improves the overall accuracy of SWR NN, and does not change the results of parallel runs (Krasnopolsky et al., 2010).

In general, ML emulations that are trained, using data simulated by a model, and simply mimic the existing parameterization may not require additional physical constraints as these constraints are implicitly incorporated in the simulated data that have very low levels of noise. If the training set is representative enough, which is comparatively easy to achieve with simulated data, ML emulations learn physical constraints from the data with sufficient accuracy, as illustrated by the results of Krasnopolsky et al. (2010) outlined above. On the other hand, when a new ML parameterization is developed using observed data or data simulated by finer resolution models, the fact that many sources of noise and uncertainties contribute to the training data has to be taken into account. In this case, applying physical constraints during the training may be the only way to avoid instabilities after the trained ML parameterization is incorporated into a hybrid model.

Several approaches could be used to ensure physical constraints. For example, physical constraints can be included as penalty terms in the loss function, which is minimized during the training process. ML approaches (like trees and SVM) that explicitly use training set values to calculate predicted values for the new inputs will preserve physical constraints with the same degree of accuracy as the data comprising the training set (O'Gorman & Dwyer, 2018). However, these techniques are known to be less accurate with generalizing to conditions outside of the training set domain. Beucler et al. (2019) proposed two methods to enforce energy conservation laws in neural-network emulators of physical models: (1) modifying the loss function to include energy conservation laws and (2) constraining the architecture of the network itself. They showed that when applied to the emulation of explicitly resolved cloud processes in an idealized multiscale climate model, the architecture constraints can enforce conservation laws to satisfactory numerical precision, while adding constraints on the loss function helps the NN better generalize to conditions outside of its training set domain.

12.8. GOING BEYOND THE TRADITIONAL PARADIGM

12.8.1. Compound Parameterization to Improve Robustness of MLPs

The accuracy of MLPs significantly depends on our ability to generate a representative training set and to avoid as much as possible using MLPs for extrapolation beyond the domain covered by the training data. Because of high dimensionality of the input domain (i.e., dimensionality of the MLP input vector X), which is usually of the order of several hundred or more, it is difficult if not impossible to cover the entire domain, which may have a very complex shape, even when we use model simulated data for the MLP training. Also, the domain may change with the structural and parametric modifications to the host model, e.g., change or replacement of the dynamical core or other model components, change in horizontal resolution, or chemical composition of the atmosphere due to climate change. In these situations, the MLP may be forced to extrapolate beyond its generalization ability, which may lead to larger errors in MLP outputs and, correspondingly, to instability of the numerical model simulations employing the MLP.

NN emulations of model radiation (Krasnopolsky et al., 2010) are very accurate. Larger errors and outliers (a few extreme errors) in NN emulation outputs occur only when NN emulations are exposed to inputs not represented sufficiently in the training set. These errors have a very low probability and are randomly distributed in space and time. However, when long multidecadal climate simulations are performed, and NN emulations are used in a very complex and nonlinear climate model over long integration times, the probability of larger errors and their undesirable impact on the model simulations increase. As demonstrated in Krasnopolsky et al. (2008a, 2010), a GCM is robust enough to overcome rare large randomly distributed errors without their accumulating over time. In another application of an NN approximation for nonlinear interactions in a wave model, the model did not prove sufficiently robust to retain stability for a sufficient integration time (Krasnopolsky et al., 2008c). Therefore, for some applications of MLPs, it is essential to introduce a quality control (QC) procedure, which can predict and eliminate larger errors from MLP during the integration of highly nonlinear numerical models.

Krasnopolsky et al. (2008c) introduced the concept of a compound parameterization (CP) that combines an MLP, an original physically based parameterization, and a QC procedure. CP makes the MLP emulation approach even more reliable, robust, and generic. It also provides a tool for developing MLP emulations adjustable to the structural and parametric changes in the host model and in the external earth system. An effective QC design is based on training an additional NN to specifically predict the errors of the MLP outputs for a given input. The "error" NN has the same inputs as the MLP and one or several outputs, which predict errors of outputs generated by the MLP for given inputs. The original parameterization, its NN emulation (MLP), the error NN, and the QC block constitute the CP, the design of which is shown in Figure 12.6.

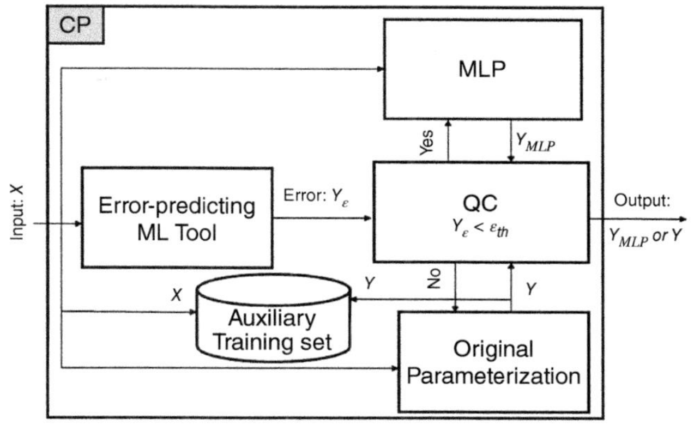

Figure 12.6 A compound parameterization design.

During the model integration, CP works in the following way: if the error predicted by the error-predicting ML tool (error NN) does not exceed a predefined threshold value, ε_{th}, the MLP is used; otherwise, the original parameterization is called instead of the MLP. The auxiliary training set (ATS) is updated each time when QC invokes the original parameterization instead of the MLP. ATS is used for the following adjustments of the MLP. CP can be used for an online adjustment of MLP. After each update of the ATS, the new record can be used to update the coefficients of MLP, using a sequential training algorithm.

As it was shown in Krasnopolsky et al. (2008c), errors predicted by the error NN are close to and highly correlated with the actual errors of the MLP calculated for the same input vector. The use of a CP (a) does not increase the systematic MLP error (bias) that is almost zero and (b) significantly reduces the MLP random error. Especially significant is the reduction of extreme errors or outliers. It is noteworthy that, for example, for the SWR parameterization only less than 1% of the MLP outputs are rejected by QC and calculated using the original parameterization. Other designs of CP can be developed and used. For example, Song et al. (2021) used the cloud fraction to estimate the uncertainty of the NN emulator. Also, a design based on the combination of forward and inverse NNs was successfully applied for the nonlinear wave-wave interaction parameterization in a wind wave model (Krasnopolsky et al., 2008c).

12.8.2. Adjusting MLP to Changes in the Environment

After the MLP is trained and introduced into the host model, it can successfully function in the online mode, replacing the original parameterization, as long as the input domain (domain spanned by MLP inputs provided by the model) does not change significantly. The domain may change with the evolution of the environment, both internal model environment (change of dynamical core, model horizontal resolution, other parameterizations, etc.) and external environment (the system that this model is simulating, e.g., due to climate changes). With domain changes the accuracy of MLP can deteriorate when it encounters input vectors, X, underrepresented or not represented at all in the training set. To mitigate this problem, a concept of dynamical adjustment of MLP (Krasnopolsky & Fox-Rabinovitz, 2006a,b) was introduced. When MLP outputs are rejected by the QC block in the CP described in the previous subsection (Figure 12.6), the original parameterization inputs and outputs are accumulated in the auxiliary training set, augmenting the data underrepresented in the originally generated training set. The auxiliary training set can be used to further adjust the NN emulation. After accumulating a sufficient number of records, an adjustment of the MLP emulation is performed by a short MLP retraining, using the accumulated input/output records.

The MLP can be updated directly in the model with addition of each new record to the auxiliary training set, using a one iteration of sequential training algorithm (see Krasnopolsky, 2013, Chapter 2). As a result, the MLP is dynamically adjusted online to the changes and/or new events/states produced by the host model. Obviously, DA can also alleviate the problem of extreme or rare events. Of course, after some period of such adjustments, the MLP will require full retraining with new data. Dynamically adjusted MLP can be used during transitional period from old MLP to a new completely retrained MLP, allowing enough time to collect a new training set. Rasp (2020) developed a similar concept of "coupled online learning," using the high-resolution simulation that is kept in sync with the neural network-driven lower-resolution model through continuous nudging.

12.9. SUMMARY AND DISCUSSION

In this chapter, we discussed using ML approaches in weather and climate numerical modeling systems. We mentioned here three generic types of ML applications to model physics:

1. ML emulations of existing parameterizations for speeding up the model integration process.

2. Development of new MLP to improve the accuracy of representing atmospheric processes in model physics.

3. Using ML to integrate several parameterizations, each of which works under limited atmospheric conditions, in one MLP working in broader domain.

A new type of model based on a synergetic combination of deterministic modeling and ML within a hybrid model is introduced. This approach uses ML tools (like NNs) to develop new MLPs that are fast and more accurate model components, and/or highly accurate and fast emulations of existing model physics. Many advantages and limitations of ML technique (and NN technique in particular) are discussed in detail in Chapters 2 and 4 of Krasnopolsky (2013). Here, we mainly mention those that were understood after publishing this book and were relevant for the development of MLPs.

12.9.1. Possible Advantages of Using MLPs

The results presented in this chapter and cited papers show the following facts:

(i) There exists the conceptual and practical possibility of developing hybrid models with accurate and fast MLPs as model components, which preserve the integrity and all the detailed features in the original model (e.g., GCM). ML components can be successfully combined with deterministic host model components within the hybrid model so their synergy can be efficiently used for environmental, weather, and climate modeling without any negative impacts on the forecast and simulation quality.

(ii) ML emulations of existing model physics parameterizations are functionally identical to the original physical parameterization due to the capability of ML techniques to accurately approximate complex mappings like parameterizations of the model physics. This capability allows the integrity and level of functional complexity of the parameterizations of the model physics to be preserved. As a result, a hybrid model, using these MLPs, produces climate simulations and weather forecasts that are practically identical to those of the original GCM. It is noteworthy that the MLPs have the same inputs and outputs as the original parameterizations and are used precisely as their functional substitutes within the model.

(iii) Accurate MLPs are robust and very fast (10 to 10^5 times faster than the original parameterization) so the significant speed-up of the model calculations can be achieved without compromising accuracy.

(iv) New computationally efficient MLPs can be developed based on learning from observed data, analysis, and/or data simulated by higher resolution models.

(v) The productive synergy of state-of-the-art deterministic and ML approaches leads to new opportunities of using hybrid models for environmental and climate simulations and weather prediction. For example, new more sophisticated parameterizations, or even "super-parameterizations" using CRM or LES, that are extremely time consuming or even computationally prohibitive, if used in their original form, will become computationally "affordable" in hybrid models when using their accurate and computationally much more efficient ML emulations.

(vi) The stochastic nature of some of the components in the model physics can be adequately represented using ensembles of ML tools to represent or emulate the stochastic component of the model physics.

(vii) ML techniques are very flexible; they could provide an opportunity to include in consideration important variables that have not been included in physically based parameterization. In MLPs, these variables can be included as additional inputs.

(viii) In principle, MLPs can go beyond the single-column physics paradigm and provide a useful tool to include in model physics interaction between neighboring vertical columns, producing ML 3D model physics.

12.9.2. Stability of MLPs, Limitations of the Current Hybrid Modeling Framework, and Possible Solutions

Development of MLPs significantly depends on our ability to generate a representative training set to avoid using NNs for extrapolation far beyond the domain covered by the training set. Because of the high dimensionality of the input domain that is often several hundred or more, it is rather difficult to cover the entire domain, especially the "far corners" associated with rare events, even when we use simulated data for MLP training.

Representativeness of a training set is so important because, as is the case with any nonlinear statistical tool, ML techniques are not very good at generalization, especially extrapolation; nonlinear extrapolation is an ill-posed problem that requires a regularization to provide meaningful results. Poor generalization may lead to instability of the emulation in the host hybrid model. The ML ensemble approach could be of significant help here. Using an ensemble of ML tools can help to regularize the extrapolation and result in MLPs that are more stable when the inputs approach or cross the boundary of the training domain.

Developing a stable and robust NN-based emulator is a multifaceted problem that requires deep understanding of multiple technical aspects of the training process and details of NN architecture. Many techniques for stabilization of hybrid statistical-deterministic models have been developed. Compound parameterization has been proposed for climate and weather modeling applications where an additional NN is trained to predict errors of the NN emulator, and, if the predicted error is above a certain threshold, compound parameterization falls back to calling the original physically based scheme (Krasnopolsky et al, 2008c). Stability theory was used to identify the causes and conditions for instabilities in ML parameterizations of moist convection when coupled to an idealized linear model of atmospheric dynamics (Brenowitz et al., 2020). An NN optimization via random search over hyperparameter space resulted in considerable improvements in the stability of subgrid physics emulators in the Super-parameterized Community Atmospheric Model version 3.0 (Ott et al., 2020). A coupled online learning approach was proposed where a high-resolution simulation is nudged to the output of a parallel lower-resolution hybrid model run, and the ML component of the latter is retrained to emulate tendencies of the former, helping to eliminate biases and unstable feedback loops (Rasp, 2020). Random forests approach was successfully used to build a stable ML parameterization of convection (Yuval & O'Gorman, 2020). Physical constraints were used to achieve stability of hybrid models (e.g., Kashinath et al., 2021; Yuval et al., 2021).

Reusing methodologies from one ML field in another is often suggested as a particularly powerful tool (e.g., Boukabara et al., 2019), allowing to leverage existing knowledge developed in other ML applications. One example of such an approach is transfer learning. However, this powerful tool should be used carefully. ML has been applied to many different problems such as image processing, classification, pattern recognition, asynchronous signal processing, feature detection, data fusion, merging, morphing, mapping, etc. From the mathematical standpoint, some of these problems belong to different mathematical classes. Applying a methodology developed for a problem belonging to a mathematical class different from the class of the problem at hand may not necessarily be justified. Therefore, caution should be exercised when applying techniques designed for problems other than mapping approximation to development of model physics emulators or new ML-based parameterizations. One of such very popular techniques is deep learning.

Application of shallow NNs to the problem of mapping approximation has thorough theoretical support. The universal approximation theorem proves that an SNN is a generic and universal tool for approximating any continuous and almost continuous mappings under very broad assumptions and for a wide class of activation functions (e.g., Hornik et al., 1990; Hornik, 1991). Similarly broad results for deep NNs (DNNs) do not exist as of yet (Vapnik, 2019); however, specific combinations of DNN architectures and activation functions have theoretical support (e.g., Elbrachter et al., 2021; Leshno et al., 1993; Lu et al., 2017). Until there is a universal theory, it has been suggested to consider DNN a heuristic approach since, in general, "from the theoretical point of view, deep networks cannot guarantee a solution of any selection problem that constitutes a complete learning problem" (Vapnik, 2019). These considerations are important to keep in mind when selecting NN architecture for the emulation of model physics or its components.

Next, we compare some properties of DNNs and SNNs to further emphasize their differences and to point out some properties of DNNs that may lead to instabilities in deterministic models coupled to DNN-based model components.

To avoid overfitting and instability, complexity and nonlinearity of approximating/emulating NN should not exceed complexity and nonlinearity of the mapping to be approximated. The complexity of the SNN (12.8) linearly increases with the number of neurons in the hidden layer, k. For given numbers of inputs and outputs, there is only one SNN architecture/configuration with a specified complexity C. The complexity of the DNN (12.7) geometrically increases with the increasing number of layers, n.

Both (12.8) and (12.7) are simply numbers of parameters of the NN that are trained or fit during SNN/DNN training. While there exists a one-to-one correspondence between the SNN complexity and the SNN architecture, given the fixed number of neurons in the input and output layers, correspondence between the DNN complexity and the DNN architecture is multivalued: many different DNN architectures/configurations have the same complexity C given the same size of input and output layers. Overall, controlling the complexity of DNNs is more difficult than controlling the complexity of SNN.

For an SNN given by the expression (12.6) nonlinearity increases arithmetically or linearly with addition of each new hidden neuron, $t_i = \phi\left(b_i^0 + \sum_{s=1}^{n} a_{is}^0 \cdot x_s\right)$, to the single hidden layer of the NN.

For a DNN, symbolically written as

$$Y = X^{n+1} = B^n + A^n \cdot \phi(B^{n-1} + A^{n-1} \cdot \phi(B^{n-2} + A^{n-2} \cdot \phi(B^{n-3} + \cdots \phi(B^0 + A^0 \cdot X)))),$$

each new hidden layer/neuron introduces additional nonlinearity on top of nonlinearities of the previous hidden layers; thus, the nonlinearity of the DNN geometrically increases with the addition of new hidden layers, much quicker than the nonlinearity of the SNN. Thus, controlling nonlinearity of DNNs is more difficult than controlling nonlinearity of SNNs. The higher the nonlinearity of the model the more unstable and unpredictable generalization is (especially nonlinear extrapolation that is an ill-posed problem).

DNNs are a very powerful and flexible technique that is extensively used for the emulation of model physics and its components (Kasim et al., 2020). Discussion of its limitations can be found in Thompson et al. (2020). The arguments listed here are intended to point out possible sources of instability of DNNs in the models and the need for careful handling of this very sensitive tool. Some other factors that can influence the stability of hybrid models are briefly discussed below.

Activation function: Universal approximation theorem for SNNs is satisfied for a wide class of bounded, nonlinear AFs. Note that many popular AFs used in DNN applications, e.g., variants of ReLU, do not belong to this class. However, for a specific problem of mapping approximation, it may be useful to consider additional restrictions on AFs.

If the AF is almost continuous, or, in other words, has only finite discontinuities (e.g., step function), the first derivative (Jacobian) of the NN using this AF will be singular. If the AF is not continuously differentiable (e.g., ReLU), its first derivative will not be continuous (will have finite discontinuities) and so will be the NN Jacobian. Using a noncontinuously differentiable NN as a model component may lead to instability, especially if the Jacobian of this component is calculated in the model. Using gradient-based optimization algorithms for training such NNs may be challenging due to discontinuities in gradients.

If the AF is monotonic, the error surface associated with a single-layer model is guaranteed to be convex, simplifying the training process (Wu, 2009). When AF approximates identity function near the origin (i.e., $\phi(0) = 0$, $\phi'(0) = 1$, and ϕ' is continuous at 0), the neural network will learn efficiently when its weights are initialized with small random values. When the activation function does not approximate identity near the origin, special care must be used when initializing the weights (Sussillo & Abbott, 2014).

It is noteworthy that the sigmoid and hyperbolic tangent AFs, popular in SNN applications, meet all aforementioned criteria. Additionally, in the context of emulation of model physics parameterizations, these AFs provide one of the lowest training losses as compared to other AFs (Chantry et al., 2021).

Introducing physical constraints also helps to regularize the generalization and develop hybrid models that can be stably integrated for a long time.

Another related problem arises when ML emulations are developed for a nonstationary environment or weather/climate system that changes with time. This means that the domain configuration for a climate simulation may evolve when using, for example, a future climate change scenario. Thus, the MLP may be forced to extrapolate beyond its generalization ability leading to errors in MLP outputs and result in simulation errors and instability of the corresponding hybrid model. In this situation, CP and dynamical adjustment as well as using the ML ensemble approach could be helpful.

It is noteworthy that ML still requires human expertise to succeed. As mentioned above, the development of MLPs is not a standard ML problem. While MLPs can, in principle, be used as a black box, the development of an ML physics will require domain knowledge about the Earth system. Close collaborations between computer scientists, atmospheric physicists, and modelers will be essential even if petabytes of training data and GPU supercomputers are available. A deep understanding of how to use physical knowledge of the Earth system and the connectivity between degrees of freedom to improve the development of network architectures and network training and how to preserve conservation properties will be required. There are a lot of decisions that must be made in the process of developing MLPs that cannot be made automatically. Selecting and normalizing of inputs and outputs, preparing training sets, selecting a training algorithm and its parameters, making decisions about sufficient approximation accuracy, etc., require active human participation. This creates some problems with using commercial ML software packages for the MLP development. For example, for the MLP development, a very specific normalization of outputs or rearrangement of the loss function (e.g., in the case of missed outputs or physical constraints) is sometimes required, which is not always possible with a commercial ML software. Commercial ML packages are usually written not in Fortran (usually in Python) and run on GPUs, while NWCS are written in Fortran/C/C++ and usually run on CPUs. Additional efforts are thus needed to connect these compiled language codebases to rapidly evolving ML tools (Partee et al., 2022). Also, confusion may arise when estimating the speedup of ML emulations when comparing the speedup estimated in development environment (Python and GPUs) and in the host NWP model. The former estimate may be significantly larger than the latter one simply because the NWP model has been already parallelized and parallelizations of ML tool performed in the development environment are not possible in the host model.

12.9.3. Concluding Remarks

NNs are the ML tool that is used in the majority of works for developing MLPs. As mentioned in section 12.3.1, SNNs are backed by well-established mathematical theory and have a long history of successful applications. DNNs are a very powerful and flexible technique as they have already demonstrated their power in many applications (Kasim et al., 2020), including developments of MLPs (Rasp et al., 2018). However, when it comes to approximation of mappings, deep learning should currently be considered as a heuristic approach (Vapnik, 2019), as only specific combinations of DNN architectures and activation functions have been demonstrated to be solutions to the approximation problem. Discussion of limitations of the DNN technique can be found in Thompson et al. (2020). More than 1,000 papers and other sources were analyzed by Thompson and coauthors to understand how deep learning performance depends on computational power in different application domains. They showed that computational requirements have escalated rapidly in each of these domains and that these increases in computing power requirements have been central to performance improvements. They conclude that if progress continues along current lines, these computational requirements will rapidly become technically and economically prohibitive. Thus, their analysis suggests that deep learning progress will be constrained by its computational requirements and that the machine learning community will be pushed to either dramatically increase the efficiency of deep learning or to move to more computationally efficient machine learning techniques.

Deep learning is computationally expensive not by accident, but by design. The same flexibility that makes it excellent at modeling diverse phenomena and outperforming expert models also makes it dramatically more complex and computationally expensive. Equation (12.7) shows how quickly the complexity of DNN (the number of DNN weights and the dimensionality of DNN training space) increases with the increase in the number of hidden layers. This rapidly increasing number of weights has to be matched by the corresponding increase in size of the training set. Additionally, increased complexity of DNN leads to increase in nonlinearity of the DNN model. With each additional hidden layer, DNN becomes more and more nonlinear, and prediction and generalization (extrapolation as well as interpolation) using this DNN may become erroneous and unstable (e.g., Rasp, 2019). Understanding of these limitations is necessary when DNN is used for developing MLPs. Functional complexity of the ML tool should correspond to the functional complexity of the modeled physical process. The parsimony principle is valid in application to ML.

In a few papers mentioned in this review, various kinds of tree algorithms were explored for MLP development. Belochitski et al. (2011) compared an SNN with decision tree algorithms (nearest neighbors, regression trees, and sparse occupancy trees) in application to emulate an existing radiation parameterization and have shown that the NN algorithm performs better than tree-based algorithms in terms of accuracy and speedup. In another study, O'Gorman and Dwyer (2018) used the random forest algorithm that consists of an ensemble of decision trees. Each tree makes predictions, using a subset of the training data. The final prediction from the RF is the average over all trees. RF has the advantage that it automatically ensures conservation of energy and positive definiteness of surface precipitation because it always produces the output that is an average of the training data that it uses. On the other hand, this property of the tree (including RT) algorithms leads to the inability of tree algorithms to generalize, to produce an output outside the domain covered by the training data, whereas NN is capable of limited generalization.

However, the same property may make tree algorithms more stable than NNs when integrated in the model because to any unexpected input they respond with the output inside the training domain. O'Gorman and Dwyer (2018) showed that when attempting to derive a new parameterization of subgrid processes based on an SNN, NN did not conserve energy and was slightly less accurate than the tree algorithm. An advantage of NN as compared to tree algorithms is that the training set is used only during the training process, which is a time-consuming and memory-consuming operation. After an NN is trained, its application is fast and does not require the use of training or test data sets. The only data that are required are NN weights. With tree algorithms, each application of the trained algorithm involves using the training set, which is a time-consuming and memory-consuming operation, significantly reducing speedup that this MLP provides. A comparatively small number of publications where other than NN ML tools are used in geoscientific modeling is not sufficient to make well-supported comparisons of NNs with other ML tools and to derive any conclusions about their respective comparative benefits.

It is noteworthy that, in addition to developing MLPs, ML tools can be successfully used in many other parts of the numerical weather prediction and climate projection systems. In data assimilation systems, fast ML emulations of forward models can be used for direct assimilation of satellite measurements (Krasnopolsky, 2007b). ML observation operators were used to propagate surface observations vertically (Krasnopolsky, 2013). Also, MLPs are well suited for calculating adjoints and tangent-linear models for 4D-Var data assimilation (Chantry et al., 2021). In satellite remote sensing, ML retrieval algorithms that simultaneously retrieve several geophysical parameters from satellite data, fast ML forward models for using in retrieval algorithms and data assimilation systems (Krasnopolsky, 2008), and an ML algorithm for filling gaps in satellite data (Krasnopolsky et al., 2016) have been developed. For postprocessing model outputs (Haupt et al., 2021), ML tools were used to average a multimodel ensemble for precipitation (Krasnopolsky & Lin, 2012) and for averaging wind wave model ensembles (Campos et al., 2019). Also, an ML-based biological model for ocean color has been developed to enable the feedback between physical and biological processes in the upper ocean (Krasnopolsky et al., 2018).

REFERENCES

Ahmad, H. (2019). Machine learning applications in oceanography. *Aquatic Research, 2*(3), 161–169. https://doi.org/10.3153/AR19014

Aires, F., Prigent, C., & Rossow, W. B. (2004). Neural network uncertainty assessment using Bayesian statistics with application to remote sensing: 3. Network Jacobians. *Journal of Geophysical Research, 109*, D10305.

Aires, F., Schmitt, M., Chedin, A., & Scott, N. (1999). The "weight smoothing" regularization of MLP for Jacobian stabilization. *IEEE Transactions on Neural Networks, 10*, 1502–1510.

Alizadeh, R., Allen, J. K., & Mistree, F. (2020). Managing computational complexity using surrogate models: Critical review. *Research in Engineering Design.* https://doi.org/10.1007/s00163-020-00336-7

Attali, J.-G., & Pagès, G. (1997). Approximations of functions by a multilayer perceptron: A new approach. *Neural Networks, 6*, 1069–1081

Belochitski, A., Binev, P., DeVore, R., Fox-Rabinovitz, M., Krasnopolsky, V., & Lamby, P. (2011). Tree approximation of the long wave radiation parameterization in the NCAR CAM Global Climate Model, *Journal of Computational and Applied Mathematics, 236*, 447–460. Doi: 10.1016/j.cam.2011.07.013

Belochitski, A., & Krasnopolsky, V. (2021). Robustness of neural network emulations of radiative transfer parameterizations in a state-of-the-art general circulation model. *Geoscientific Model Development, 14*, 7425–7437. https://doi.org/10.5194/gmd-14-7425-2021

Bergen, K. J., Johnson, P. A., Maarten, V., & Beroza, G. C. (2019). Machine learning for data-driven discovery in solid Earth geoscience. *Science, 363*, 6433, eaau0323.

Beucler, T., Pritchard, M., Gentine, P., & Rasp S. (2020). Towards physically-consistent, Data-driven models of convection. IGARSS 2020 — 2020 IEEE International Geoscience and Remote Sensing Symposium (2020): 3987–3990.

Beucler, T., Rasp, S., Pritchard, M., & Gentine, P. (2019). Achieving conservation of energy in neural network emulators for climate modeling. arXiv:1906.06622 [physics.ao-ph]

Bishop, C. M. (1995). *Neural networks for pattern recognition.* Oxford, UK: Oxford University Press.

Bishop, C. M. (2006). *Pattern recognition and machine learning.* New York, NY: Springer. ISBN 0-387-31073-8

Bochenek, B. & Ustrnul, Z. (2022). Machine learning in weather prediction and climate analyses — Applications and perspectives. *Atmosphere, 13*, 180-196. https://doi.org/10.3390/atmos13020180

Bocquet, M., Brajard, J., Carrassi, A., & Bertino L. (2019). Data assimilation as a learning tool to infer ordinary differential equation representations of dynamical models. *Nonlinear Processes in Geophysics, 26*(3), 143–162.

Bolton, T., & Zanna, L. (2019). Applications of deep learning to ocean data inference and subgrid parameterization. *Journal of Advances in Modeling Earth Systems*, *11*, 376–399. https://doi.org/10.1029/2018MS001472

Boukabara, S.-A., Krasnopolsky, V., Stewart, J. Q., Maddy, E. S., Shahroudi, N., & Hoffman, R. N. (2019). Leveraging modern artificial intelligence for remote sensing and NWP: Benefits and challenges. *Bulletin of the American Meteorological Society*, *100*(12), ES473–ES491.

Brenowitz, N. D., Beucler, T., Pritchard, M., & Bretherton, C. S. (2020). Interpreting and stabilizing machine-learning parametrizations of convection. *Journal of the Atmospheric Sciences*, *77*(12), 4357–4375. https://doi.org/10.1175/JAS-D-20-0082.1

Brenowitz, N. D., & Bretherton, C. S. (2018). Prognostic validation of a neural network unified physics parameterization. *Geophysical Research Letters*, *45*, 6289–6298. https://doi.org/10.1029/2018GL078510

Brockherde, F., Vogt, L., Li, L., Tuckerman, M. E., Burke, K., & Müller, K.-R. (2017). Bypassing the Kohn-Sham equations with machine learning. *Nature Communications*, *8*, 872. https://doi.org/10.1038/s41467-017-00839-3

Campos, R., Krasnopolsky, V., Alves, J. H., & Penny, S. (2019). Nonlinear wave ensemble averaging in the Gulf of Mexico using neural networks. *Journal of Atmospheric and Oceanic Technology*, *36*, 113–127. https://doi.org/10.1175/JTECH-D-18-0099.1

Ceppi, P., Brient, F., Zelinka, M. D., & Hartmann, D. L. (2017). Cloud feedback mechanisms and their representation in global climate models. *WIREs Climate Change*, *8*, e465. https://doi.org/10.1002/wcc.465

Chantry, M., Hatfield, S., Dueben, P., Polichtchouk, I., & Palmer, T. (2021). Machine learning emulation of gravity wave drag in numerical weather forecasting *Journal of Advances in Modeling Earth Systems*, *13*, e2021MS002477. https://doi.Org/10.1029/2021MS002477.

Chen, T., & Chen, H. (1995). Universal approximation to nonlinear operators by neural networks with arbitrary activation function and its application to dynamical systems. *Neural Networks*, *6*, 911–917.

Cherkassky, V., & Muller, F. (2007). *Learning from data: Concepts, theory, and methods*. Hoboken, NJ: Wiley. https://onlinelibrary.wiley.com/doi/pdf/10.1002/9780470140529.fmatter

Chevallier, F., Chéruy, F., Scott, N. A., & Chedin, A. (1998). A neural network approach for a fast and accurate computation of longwave radiative budget. *Journal of Applied Meteorology*, *37*, 1385–1397.

Chevallier, F., & Mahfouf, J.-F. (2001). Evaluation of the Jacobians of infrared radiation models for variational data assimilation. *Journal of Applied Meteorology*, *40*, 1445–1461.

Chevallier, F., Morcrette, J.-J., Chéruy, F., & Scott, N. A. (2000). Use of a neural-network-based longwave radiative transfer scheme in the ECMWF atmospheric model. *Quarterly Journal of the Royal Meteorological Society*, *126*, 761–776.

Christensen, H., & Zanna, L. (2022). Parametrization in Weather and Climate Models. *Oxford Research Encyclopedias*. Climate Science. Parametrization in Weather and Climate Models. https://doi.org/10.1093/acrefore/9780190228620.013.826

Clough, S. A., Shephard, M. W., Mlawer, E. J., Delamere, J. S., Iacono, M.J., Cady-Pereira, K., et al. (2005). Atmospheric radiative transfer modeling: A summary of the AER codes. *Journal of Quantitative Spectroscopy & Radiative Transfer*, *91*, 233–244. Doi:10.1016/j.jqsrt.2004.05.058

Collins, W. D., Hackney, J. K., & Edwards, D. P. (2002). A new parameterization for infrared emission and absorption by water vapor in the National Center for Atmospheric Research Community Atmosphere Model. *Journal of Geophysical Research*, *107*, 1–20

Côté, J., Desmarais, J.-G., Gravel, S., Me'thot, A, Patoine, A, Roch, M, & Staniforth, A. (1998b). The operational CMC-MRB global environmental multiscale (GEM) model Part II: Mesoscale results. *Monthly Weather Review*, *126*, 1397–1418.

Côté, J., Gravel, S., Me'thot, A., Patoine, A., Roch, M., & Staniforth, A. (1998a). The operational CMC-MRB global environmental multiscale (GEM) model Part I: Design considerations and formulation. *Monthly Weather Review*, *126*, 1373–1395.

Cybenko, G. (1989). Approximation by superposition of sigmoidal functions. *Mathematics of Control, Signals, and Systems*, *2*, 303–314.

Dueben, P. D., & Bauer, P. (2018). Challenges and design choices for global weather and climate models based on machine learning, *Geoscientific Model Development*, *11*, 3999–4009. https://doi.org/10.5194/gmd-11-3999-2018.

Ebert-Uphoff, I., & Hilburn, K. (2020). Evaluation, tuning, and interpretation of neural networks for working with images in meteorological applications. *Bulletin of the American Meteorological Society*, *101*, E2149–E2170. doi: https://doi.org/10.1175/BAMS-D-20-0097.1

Edwards, J. M., & Slingo, A. (1996). Studies with a flexible new radiation code: 1. Choosing a configuration for a large-scale model. *Quarterly Journal of the Royal Meteorological Society*, *122*, 689–719. https://doi.org/10.1002/qj.49712253107

Elbrachter D., Perekrestenko D., Grohs P., & Bölcskei H. (2021). Deep neural network approximation theory. *IEEE Transactions on Information Theory*, *67*(5), 2581–2623. DOI: 10.1109/TIT.2021.3062161

Funahashi, K. (1989). On the approximate realization of continuous mappings by neural networks. *Neural Networks*, *2*, 183–192.

Gagne, D. J., II, Haupt, S. E., Nychka, D. W., & Thompson, G. (2019). Interpretable deep learning for spatial analysis of severe hailstorms. *Monthly Weather Review*, *147*(8), 2827–2845.

Geer, A. J. (2021). Learning Earth System Models from Observations: Machine Learning or Data Assimilation. *Philosophical Transactions of the Royal Society A*, *379*, 20200089. https://doi.org/10.1098/rsta.2020.0089

Gentine, P., Pritchard, M., Rasp, S., Reinaudi, G., & Yacalis, G. (2018). Could machine learning break the convection parameterization deadlock? *Geophysical Research Letters*, *45*, 5742–5751. https://doi.org/10.1029/ 2018GL078202

Gilmore, M. S., Straka, J. M., & Rasmussen, E. N. (2004). Precipitation and evolution sensitivity in simulated deep convective storms: comparisons between liquid-only and simple ice and liquid phase microphysics. *Monthly Weather Review*, *132*, 1897–1916.

Goldstein, E. B., & Coco, G. (2015). Machine learning components in deterministic models: hybrid synergy in the age of data. *Frontiers in Environmental Science, 3,* article 33. https://doi.org/10.3389/fenvs.2015.00033

Hasselmann, S., & Hasselmann, K. (1985). Computations and parameterizations of the nonlinear energy transfer in a gravity wave spectrum Part I: a new method for efficient computations of the exact nonlinear transfer integral. *Journal of Physical Oceanography*, *15*, 1369–1377.

Hasselmann, S., Hasselmann, K., Allender, J. H., & Barnett, T. P. (1985). Computations and parameterizations of the nonlinear energy transfer in a gravity wave spectrum Part II: parameterization of the nonlinear transfer for application in wave models. *Journal of Physical Oceanography*, *15*, 1378–1391.

Hassoun, S., Jefferson, F., Shi, X., Stucky, B., Wang, J., & Rosa, E. (2021). Artificial Intelligence for Biology. *Integrative and Comparative Biology*, *61*, 2267–2275. https://doi.org/10.1093/icb/ica188

Haupt S. E., Chapman, W., Adams, S. V., Kirkwood, C., Hosking, J. S., Robinson, N. H., et al. (2021). Towards implementing artificial intelligence post-processing in weather and climate: proposed actions from the Oxford 2019 workshop. *Philosophical Transactions of the Royal Society A*, *379*, 20200091. https://doi.org/10.1098/rsta.2020.0091

Hogan, R., Ahlgrimm, M., Balsamo, G., Beljaars, A., Berrisford, P., Bozzo, A., et al. (2017). Radiation in numerical weather prediction. *ECMWF Technical Memorandum.* doi: 10.21957/2bd5dkj8x.

Hogan, R., & Bozzo A. (2015). Mitigating errors in surface temperature forecasts using approximate radiation updates. *Journal of Advances in Modeling Earth Systems, 7*, 836–853. https://doi.org/10.1002/2015MS000455,

Hornik, K. (1991). Approximation capabilities of multilayer feedforward network. *Neural Networks, 4*, 251–257.

Hornik, K., Stinchcombe, M., & White, H. (1990). Universal approximation of an unknown mapping and its derivatives using multilayer feedforward network. *Neural Networks, 3*, 551–560.

Hsieh, W. W. (2001). Nonlinear principal component analysis by neural networks. *Tellus, 53A*, 599–615.

Hsieh, W. W. (2009). *Machine learning methods in the environmental sciences: Neural networks and kernels.* Cambridge: Cambridge University Press. https://doi.org/10.1017/CBO9780511627217

Hsieh W.W. (2023). *Introduction to Environmental Data Science.* Cambridge: Cambridge University Press; ISBN: 9781107065550

Ivezi'c, Ž., Connolly, A. J., VanderPlas, J. T., & Gray, A. (2019). *Statistics, data mining, and machine learning in astronomy: A practical Python guide for the analysis of survey data*. Princeton, NJ: Princeton University Press.

Jankov, I., Bao, J.-W., Neiman, P. J., Schultz, P. J., Yuan, H., & White, A. B. (2009). Evaluation and comparison of microphysical algorithms in ARW-WRF model simulations of atmospheric river events affecting the California coast. *Journal of Hydrometerology, 10*, 847–870.

Jones, T. R., Randall, D. A., & Branson, M. D. (2019a). Multiple-Instance Superparameterization: 1. Concept, and Predictability of Precipitation. *Journal of Advances in Modeling Earth Systems, 11*, 3497–3520.

Jones, T. R., Randall, D. A., & Branson, M. D. (2019b). Multiple-Instance Superparameterization: 2. The Effects of Stochastic Convection on the Simulated Climate. *Journal of Advances in Modeling Earth Systems, 11*, 3521–3544.

Kain, J. S., Moorthi, S., Yang, F., Yang, R., Wei, H., Wu, Y., et al. (2020). Advances in model physics for the next implementation of the GFS (GFSv16). AMS Annual Meeting, Boston, MA, 6A.3., 2020.

Kashinath, K., Mustafa, M., Albert, A., Wu, J. L., Jiang, C., Esmaeilzadeh, S., et al. (2021). Physics-informed machine learning: case studies for weather and climate modelling. *Philosophical Transactions of the Royal Society A, 379*, 20200093. https://doi.org/10.1098/rsta.2020.0093

Kasim, M. F., Watson-Parris, D., Deaconu, L., Oliver, S., Hatfield, P., Froula, D. H., et al. (2020). Up to two billion times acceleration of scientific simulations with deep neural architecture search. https://arxiv.org/pdf/2001.08055.pdf

Khain, A., Rosenfeld, D., & Pokrovsky, A. (2000). Aerosol impact on the dynamics and microphysics of deep convective clouds. *Quarterly Journal of the Royal Meteorological Society: A Journal of the Atmospheric Sciences, Applied Meteorology and Physical Oceanography*, *131*, 2639–2663.

Khairoutdinov, M. F., & Randall, D. A. (2001). A cloud resolving model as a cloud parameterization in the NCAR Community Climate System Model: Preliminary results. *Geophysical Research Letters*, *28*, 3617–3620.

Krasnopolsky, V. (1996). A neural network forward model for direct assimilation of SSM/I brightness temperatures into atmospheric models. Working Group on Numerical Experimentation, Blue Book. 1.29–1.30. http://polar.ncep.noaa.gov/mmab/papers/tn134/OMB134.pdf

Krasnopolsky, V. (1997). A neural network based forward model for direct assimilation of SSM/I brightness temperatures. Technical Note, Ocean Modeling Branch Contribution No 140, NCEP/NOAA. http://polar.ncep.noaa.gov/mmab/papers/tn140/OMB140.pdf

Krasnopolsky, V. (2008). Neural network applications to solving forward and inverse problems in atmospheric and oceanic satellite remote sensing. In S. E. Haupt, A. Pasini, and C. Marzban (Eds.), *Artificial intelligence in environmental sciences* (pp. 191–205). Dordrecht: Springer.

Krasnopolsky, V. (2013). The application of neural networks in the earth system sciences. Neural network emulations for complex multidimensional mappings. *Atmospheric and Oceanic Science Library* (Vol. 46). Dordrecht, Heidelberg, New York, London: Springer. Doi 10.1007/978-94-007-6073-8

Krasnopolsky, V., Belochitski, A., Hou, Y.-T., Lord, S., & Yang, F. (2012). Accurate and fast neural network emulations of long and short wave radiation for the NCEP Global Forecast System Model. *NCEP Office Note*, 471. http://www.lib.ncep.noaa.gov/ncepofficenotes/files/on471.pdf.

Krasnopolsky, V., Middlecoff, J., Beck, J, Geresdi, I., & Toth, Z. (2017). *A neural network emulator for microphysics schemes*. Paper presented at AMS annual meeting. https://ams.confex.com/ams/97Annual/webprogram/Paper310969.html

Krasnopolsky, V., Nadiga, S., Mehra A., & Bayler, E. (2018), Adjusting neural network to a particular problem: Neural network-based empirical biological model for chlorophyll concentration in the upper ocean. *Applied Computational Intelligence and Soft Computing*, 2018, *Article ID 7057363*, 10 pages. https://doi.org/10.1155/2018/7057363

Krasnopolsky, V., Nadiga, S., Mehra, A., Bayler, E., & Behringer, D. (2016). Neural networks technique for filling gaps in satellite measurements: Application to ocean color observations. *Computational Intelligence and Neuroscience Research, 2016,* Article ID 6156513, 9 pages. http://dx.doi.org/10.1155/2016/6156513

Krasnopolsky, V. M. (2007a). Reducing uncertainties in neural network Jacobians and improving accuracy of neural network emulations with NN ensemble approaches. *Neural Networks, 20*, 454–461.

Krasnopolsky, V. M. (2007b). Neural network emulations for complex multidimensional geophysical mappings: Applications of neural network techniques to atmospheric and oceanic satellite retrievals and numerical modeling, *Reviews of Geophysics*, *45*, RG3009. doi:10.1029/2006RG000200.

Krasnopolsky, V. M., Chalikov, D., & Tolman, H. L. (2002). A neural network technique to improve computational efficiency of numerical oceanic models. *Ocean Modelling, 4*, 363–383.

Krasnopolsky, V. M., & Fox-Rabinovitz, M. S. (2006a). Complex hybrid models combining deterministic and machine learning components for numerical climate modeling and weather prediction. *Neural Networks*, *19*, 122–134.

Krasnopolsky, V. M., & Fox-Rabinovitz, M. S. (2006b). A new synergetic paradigm in environmental numerical modeling: hybrid models combining deterministic and machine learning components. *Ecological Modelling*, *191*, 5–18.

Krasnopolsky, V. M., Fox-Rabinovitz, M. S., & Belochitski, A. (2008a). Decadal climate simulations using accurate and fast neural network emulation of full, long- and short-wave radiation. *Monthly Weather Review*, *136*, 3683–3695. Doi:10.1175/2008MWR2385.1

Krasnopolsky, V. M., Fox-Rabinovitz, M. S., & Belochitski, A. (2008b). *Using neural network emulations of model physics in numerical model ensembles.* Paper presented at 2008 IEEE International Joint Conference on Neural Networks, Hong Kong, China (2008 IEEE World Congress on Computational Intelligence). https://doi.org/10.1109/IJCNN.2008.4633998

Krasnopolsky, V. M., Fox-Rabinovitz, M. S., & Belochitski, A. A. (2013). Using ensemble of neural networks to learn stochastic convection parameterization for climate and numerical weather prediction models from data simulated by cloud resolving model. *Advances in Artificial Neural Systems*, Article ID 485913, 13 pages. Doi:10.1155/2013/485913

Krasnopolsky, V. M., Fox-Rabinovitz, M. S., Belochitski, A., Rasch, J. P., Blossey, P. & Kogan, Y. (2011). Development of neural network convection parameterizations for climate and NWP models using Cloud Resolving Model simulations, *NCEP Office Note*, 469. http://www.lib.ncep.noaa.gov/ncepofficenotes/files/on469.pdf

Krasnopolsky, V. M., Fox-Rabinovitz, M. S., & Chalikov, D. V. (2005). New approach to calculation of atmospheric model physics: accurate and fast neural network emulation of long wave radiation in a climate model. *Monthly Weather Review, 133*, 1370–1383.

Krasnopolsky, V. M., Fox-Rabinovitz, M. S., Hou, Y. T., Lord, S. J., & Belochitski, A. A. (2010). Accurate and fast neural network emulations of model radiation for the NCEP coupled climate forecast system: Climate simulations and seasonal predictions. *Monthly Weather Review, 138*, 1822–1842. Doi: 10.1175/2009MWR3149.1

Krasnopolsky, V. M., Fox-Rabinovitz, M. S., Rasch, P. J., & Wang, M. (2014). Fast NN Emulation of the super-parameterization in the multi-scale modeling framework, *NCEP Office Note 476*. https://www.lib.ncep.noaa.gov/ncepofficenotes/files/on476.pdf

Krasnopolsky, V. M., Fox-Rabinovitz, M. S., Tolman, H. L., & Belochitski, A. (2008c). Neural network approach for robust and fast calculation of physical processes in numerical environmental models: Compound parameterization with a quality control of larger errors. *Neural Networks*, *21*, 535–543.

Krasnopolsky V. M., & Lin, Y. (2012). A neural network nonlinear multimodel ensemble to improve precipitation forecasts over continental US. *Advances in Meteorology*, *2012, Article ID* 649450, 11 pages. http://dx.doi.org/10.1155/2012/649450

Krasnopolsky, V. M., Lord, S. J., Moorthi, S., & Spindler, T. (2009). *How to deal with inhomogeneous outputs and high dimensionality of neural network emulations of model physics in numerical climate and weather prediction models*. Paper presented at International Joint Conference on Neural Networks, Atlanta, Georgia, USA, June 14–19, 2009, pp.1668–1673. https://ieeexplore.ieee.org/document/5178898

Kutz, J. N. (2017). Deep learning in fluid dynamics. *Journal of Fluid Mechanics, 814*, 1–4.

Lagerquist, R., McGovern, A., & Gagne D. J., II (2019). Deep learning for spatially explicit prediction of synoptic-scale fronts. *Weather and Forecasting*, *34*(4), 1137–1160.

Lagerquist, R., Turner, D., Ebert-Uphoff, I., Stewart, J., & Hagerty, V. (2021). Using deep learning to emulate and accelerate a radiative-transfer model. *Journal of Atmospheric and Oceanic Technology*, *38,* 1673–1696. https://doi.org/10.1175/JTECH-D-21-0007.1

Lang, S. T. K., Dawson, A., Diamantakis, M., Dueben, P., Hatfield, S., Leutbecher, M., et al. (2021). More accuracy with less precision. *Quarterly Journal of the Royal Meteorological Society, 147,* 4358–4370. https://doi.org/10.1002/qj.4181

Leshno, M., Lin, V. Ya., Pinkus, A., & Schocken, S. (1993). Multilayer feedforward networks with a nonpolynomial activation function can approximate any function. *Neural Networks, 6*, 861–867. https://doi.org/10.1016/S0893-6080(05)80131-5

Lu, Z., Pu, H., Wang, F., Hu Z., & Wang, L. (2017). The expressive power of neural networks: A view from the width. *Advances in Neural Information Processing Systems, 30*, 6231–6239. Curran Associates, Inc.: arXiv:1709.02540

Mlawer, E. J., Taubman S. J., Brown P. D., Iacono M. J., & Clough S. A. (1997). Radiative transfer for inhomogeneous atmospheres: RRTM, a validated correlated-k model for the longwave. *Journal of Geophysical Research: Atmospheres, 102*, 16663–16682. https://doi.org/10.1029/97JD00237

Morcrette, J.-J., Mozdzynski, G., & Leutbecher, M. (2008). A reduced radiation grid for the ECMWF Integrated Forecasting System. *Monthly Weather Review*, *136*, 4760–4772. Doi:10.1175/2008MWR2590.1

Morrison, H., & Milbrandt, J. (2011). Comparison of two-moment bulk microphysics schemes in idealized supercell thunderstorm simulations. *Monthly Weather Review, 139*, 1103–1130.

Morrison, H., van Lier-Walqui, M., Fridlind, A. M., Grabowski, W. W., Harrington, J. Y., Hoose, C., et al. (2020). Confronting the challenge of modeling cloud and precipitation microphysics. *Journal of Advances in Modeling Earth Systems*, *12*, e2019MS00168. https://doi.org/10.1029/2019MS001689

O'Gorman, P. A., & Dwyer, J. G. (2018). Using machine learning to parameterize moist convection: potential for modeling of climate, climate change and extreme events. *Journal of Advances in Modeling Earth Systems 10*. 2548–2563. https://doi.org/10.1029/2018MS001351

Ott, J., Pritchard, M., Best, N., Linstead, E., Curcic, M., & Baldi, P. (2020). A Fortran-Keras deep learning bridge for scientific computing. *Scientific Programming*, 2020, 13 pp., Article ID 8888811. https://doi.org/10.1155/2020/8888811

Pal, A., Mahajan, S., & Norman, M. R. (2019). Using deep neural networks as cost-effective surrogate models for Super-Parameterized E3SM radiative transfer. *Geophysical Research Letters*, *46*, 6069–6079. https://doi.org/10.1029/2018GL081646

Partee, S., Ellis, M., Rigazzi A., et al. (2022). Using Machine Learning at scale in numerical simulations with SmartSim: An application to ocean climate modeling, *Journal of Computational Science*, Article 101707, doi: https://doi.org/10.1016/j.jocs.2022.101707.

Pincus, R., Barker, H. W., & Morcrette, J.-J. (2003). A fast, flexible, approximate technique for computing radiative transfer in inhomogeneous cloud fields. *Journal of Geophysical Research*, *108*, 4376. doi:10.1029/2002JD003322, D13.

Pincus, R., & Stevens, B. (2009). Monte Carlo Spectral Integration: A consistent approximation for radiative transfer in large Eddy simulations. *Journal of Advances in Modeling Earth Systems*, *1*, 1. doi:10.3894/JAMES.2009.1.1.

Pincus, R., & Stevens, B. (2013). Paths to accuracy for radiation parameterizations in atmospheric models. *Journal of Advances in Modeling Earth Systems*, *5*, 225–233. doi:10.1002/jame.20027

Rai, R., & Sahu, C. K. (2020). Driven by data or derived through physics? A review of hybrid physics guided machine learning techniques with Cyber-Physical System (CPS) focus. In *IEEE Access*, *8*, 71050–71073, 2020. doi: 10.1109/ACCESS.2020.2987324.

Rasp, S. (2019). Online learning as a way to tackle instabilities and biases in neural network parameterizations. arXiv preprint arXiv:1907.01351, 2019 - arxiv.org

Rasp, S. (2020). Coupled online learning as a way to tackle instabilities and biases in neural network parameterizations: general algorithms and Lorenz 96 case study (v1.0). - *Geosci. Model Dev., 13*, 2185–2196. https://doi.org/10.5194/gmd-13-2185-2020

Rasp, S., Pritchard, M. S., & Gentine, P. (2018). Deep learning to represent subgrid processes in climate models. *PNAS Latest Articles*.

Reichstein, M., Camps-Valls, G., Stevens, B., Jung, M., Denzler, J., & Carvalhais, N. (2019). Deep learning and process understanding for data-driven Earth system science. *Nature, 566*(7743), 195–204.

Roh, S., & Song, H.-J. (2020). Evaluation of neural network emulations for radiation parameterization in cloud resolving model. *Geophysical Research Letters*, *47*, e2020GL089444. https://doi.org/10.1029/2020GL089444

Scher, S. (2018). Toward data-driven weather and climate forecasting: Approximating a simple general circulation model with deep learning. *Geophysical Research Letters*, 45, 12,616–12,622. https://doi.org/10.1029/2018GL080704

Scher, S., & Messori, G. (2019). Weather and climate forecasting with neural networks: using GCMs with different complexity as study-ground. *Geoscientific Model Development Discussions.* https://doi.org/10.5194/gmd-2019-53

Schneider, T., Lan, S., Stuart, A., & Teixeira, J. (2017). Earth System Modeling 2.0: A blueprint for models that learn from observations and targeted high-resolution simulations. *Geophysical Research Letters*, *44*, 12,396–12,417. https://doi.org/10.1002/2017GL076101

Song, H.-J., Rho, S., & Park, H. (2021). Compound parameterization to improve the accuracy of radiation emulator in a numerical weather prediction model. *Geophysical Research Letters*, *48*, 1–9, *e2021GL095043*. https://doi.org/10.1029/2021GL095043

Sussillo, D., & Abbott, L. F. (2014). Random walk initialization for training very deep feedforward networks. arXiv:1412.6558

Thompson, G., Field, P. R., Rasmussen, R. M., & Hall, W. D. (2008). Explicit forecasts of winter precipitation using an improved bulk microphysics scheme. Part II: Implementation of a new snow parameterization. *Monthly Weather Review*, *138*, 5095–5115.

Thompson, N. C., Greenewald, K., Lee, K., & Manso, G. F. (2020). *The computational limits of deep learning*. https://arxiv.org/abs/2007.05558

Tolman, H. L. (2002). User manual and system documentation of WAVEWATCH III version 2.22. Technical Note, 222, NOAA/NWS/NCEP/MMAB. http://polar.ncep.noaa.gov/mmab/papers/tn222/MMAB_222.pdf

Tolman, H. L. (2013). A Generalized Multiple Discrete Interaction Approximation for resonant four-wave interactions in wind wave models. *Ocean Modelling*, *70*, 11–24. https://doi.org/10.1016/j.ocemod.2013.02.005

Tolman, H. L., & Grumbine, R. W. (2013). Holistic genetic optimization of a Generalized Multiple Discrete Interaction Approximation for wind waves. *Ocean Modelling*, *70*, 25–37. https://doi.org/10.1016/j.ocemod.2012.12.008

Tolman, H. L., Krasnopolsky, V. M, & Chalikov, D. V. (2005). Neural network approximations for nonlinear interactions in wind wave spectra: direct mapping for wind seas in deep water. *Ocean Modelling*, *8*, 253–278.

Ukkonen, P., Pincus, R., Hogan, R. J., Nielsen, K. P., & Kaas, E. (2020). Accelerating radiation computations for dynamical models with targeted machine learning and code optimization. *Journal of Advances in Modeling Earth*

Systems, 12, e2020MS002226. https://doi.org/10.1029/2020MS002226.

Van der Merwe, R., Lee, T. K., Lu, Z., Frolov, S., & Baptista, A. M. (2007). Fast neural network surrogates for very high dimensional physics-based models in computational oceanography. *Neural Networks, 20*(4), 462–478.

Van Weverberg, K., Goudenhoofdt, E., Blahak, U., Brisson, E., Demuzere, M., Marbaix, P., & van Ypersele, J.-P. (2014). Comparison of one-moment and two-moment bulk microphysics for high-resolution climate simulations of intense precipitation. *Atmospheric Research, 147*, 145–161.

Vapnik, V. N. (1995). *The nature of statistical learning theory* (p. 189). New York: Springer

Vapnik, V. N. (2019). Complete statistical theory of learning. *Automation and Remote Control, 80*(11), 1949–1975.

Vapnik, V. N., & Kotz, S. (2006). *Estimation of dependences based on empirical data (Information Science and Statistics)*. New York: Springer.

Veerman, M. A., Pincus, R., Stoffer, R., van Leeuwen, C. M., Podareanu, D. & van Heerwaarden, C. C. (2021). Predicting atmospheric optical properties for radiative transfer computations using neural networks. *Philosophical Transactions of the Royal Society A, 379*, 20200095. https://doi.org/10.1098/rsta.2020.0095

Wang, J., Balaprakash, P., & Kotamarthi, R. (2019). Fast domain-aware neural network emulation of a planetary boundary layer parameterization in a numerical weather forecast model. *Geoscientific Model Development, 12*, 4261–4274, https://doi.org/10.5194/gmd-12-4261-2019

Wang, M., Ghan, S., Easter, R., Ovchinnikov, M., Liu, X., Kassianov, E., et al. (2011). The multi-scale aerosol-climate model PNNL-MMF: Model description and evaluation. *Geoscientific Model Development, 4*(1), 137–168.

Wang, X., Han, Y., Xue, W., Yang, G., & Zhang G. J. (2022). Stable climate simulations using a realistic GCM with neural network parameterizations for atmospheric moist physics and radiation processes. *Geoscientific Model Development, 15*, 3923–3940. https://doi.org/10.5194/gmd-15-3923-2022

Wang, Z., Di, H., Amir Shafiq, M., Alaudah, Y., & AlRegib G. (2018). Successful leveraging of image processing and machine learning in seismic structural interpretation: A review. *The Leading Edge, 37*(6), 451–461.

Watt-Meyer, O., Brenowitz, N. D., Clark, S. K., Henn, B., Kwa, A., McGibbon, J., et al. (2021). Correcting weather and climate models by machine learning nudged historical simulations. *Geophysical Research Letters, 48*. https://doi.org/10.1029/2021GL092555

Willard, J., Jia, X., Xu, Sh., Steinbach, M., & Kumar, V. (2021, July). Integrating physics-based modeling with machine learning: A survey. *1*(1), 34 pages. https://doi.org/10.1145/1122445

Wu, H. (2009). Global stability analysis of a general class of discontinuous neural networks with linear growth activation functions, *Information Sciences, 179*(19), 3432–3441. doi:10.1016/j.ins.2009.06.006

Xu, F., Shi, W., Du, Y., Chen, Z., & Lu, Y. (2021). DeepPE: Emulating Parameterization in Numerical Weather Forecast Model through Bidirectional Network. https://2021.ecmlpkdd.org/wp-content/uploads/2021/07/sub_39.pdf

Yuval, J., & O'Gorman, P. (2020). Stable machine-learning parameterization of subgrid processes for climate modeling at a range of resolutions. *Nature Communications, 11*, 3295. doi:10.1038/s41467-020-17142-3

Yuval, J., O'Gorman, P. A., & Hill, C. N. (2021). Use of neural networks for stable, accurate and physically consistent parameterization of subgrid atmospheric processes with good performance at reduced precision. *Geophysical Research Letters, 48*, e2020GL091363. https://doi.org/10.1029/2020GL091363

Zhao, Q., & Carr, F.H. (1997). A prognostic cloud scheme for operational NWP models. *Monthly Weather Review, 125*(8), 1931–1953. doi: 10.1175/1520-0493(1997)1252.0.co;2.

13

Top-Down Approaches to the Study of Cloud Systems

Graham Feingold[1] and Ilan Koren[2]

ABSTRACT

We approach the problem of convection and clouds using a "top-down view" that focuses on system-wide behavior and emergent phenomena. We distinguish this from the traditional "bottom-up," or reductionist approach, in which the focus is on individual interacting processes, from which one attempts to advance understanding of the system as a whole. With its focus on the overall system behavior, the top-down methodology applies a different conceptual paradigm to the problem of complex systems and can be viewed as complementary to bottom-up reductionist thinking. We discuss the advantages of merging bottom-up and top-down approaches for maximum benefit: reductionism elucidates detailed physical processes and tests ideas about their interplay, while the broader view of atmospheric system behavior uncovers the outcome of interactions between local processes, revealing preferred states and bifurcation behaviors of the system, and responses to perturbations that might change the preferred states.

13.1. INTRODUCTION

Complex systems abound in the natural world and the atmosphere is no exception. An example illustrating such complexity in the case of a marine boundary layer capped by clouds is shown in Figure 13.1a. Through rigorous analytical methods, detailed laboratory and field observation, and modeling at a range of scales, atmospheric systems such as this one have been probed and prodded to gain insights into their inner workings and to produce models with predictive capability. Given the complexity of the system, the methods often focus on details of components such as cloud microphysics, precipitation, radiative properties, or convection, with full knowledge that the couplings between such components ultimately determine the behavior. This is termed the bottom-up approach, and most of the individual chapters in the current volume are documentations thereof.

Closer inspection of the cloud fields in Figure 13.1a reveals system-wide structure that encapsulates the net result of the myriad boundary layer dynamical and microphysical interactions. The eye is drawn to mostly cloudy, cellular structures alongside open, mostly cloud-free regions that contain a honeycomb-like cloudy structure. More complex open patterns that take on structures defined as *actiniae* (Agee, 1984), from the Greek "actino" or "ray," have led to a broad definition of actinoform clouds (Garay et al., 2004). What is clear from this satellite image is that cloud systems often exhibit system-wide "emergence"—or a behavior at a large scale that arises from detailed processes and interactions between parts of the system at a finer scale. It is properties like emergence that characterize the top-down view of systems. Importantly, this type of order is not driven by a central force, but instead grows spontaneously from local interactions or is "self-organized." One might contrast this with

[1]*Chemical Sciences Laboratory, NOAA Earth System Research Laboratories, Boulder, CO, USA*

[2]*Department of Earth and Planetary Sciences, Weizmann Institute, Rehovot, Israel*

Fast Processes in Large-Scale Atmospheric Models: Progress, Challenges, and Opportunities, Geophysical Monograph 282, First Edition.
Edited by Yangang Liu and Pavlos Kollias.

DOI: 10.1002/9781119529019.ch13

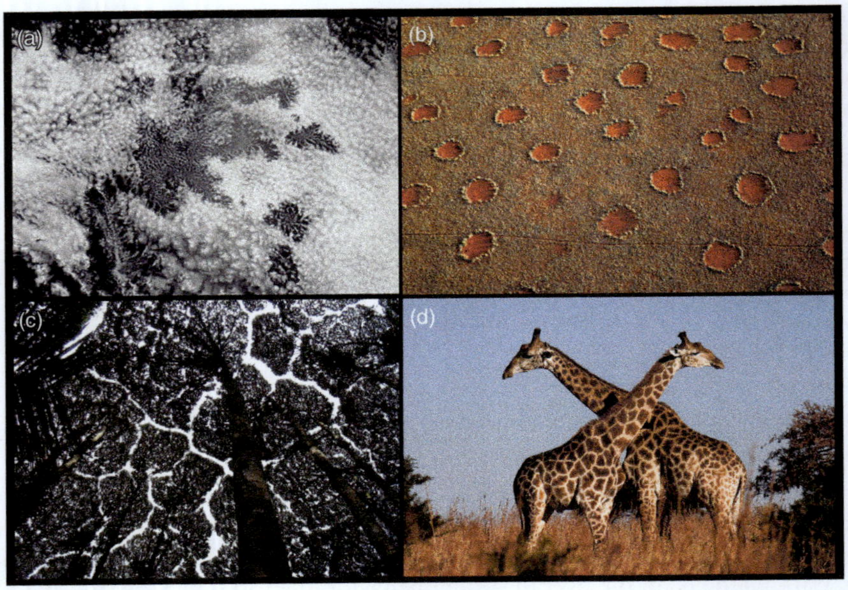

Figure 13.1 From top left in a clockwise direction: Examples of pattern and emergence in (a) a complex atmospheric system (closed-cell and open-cell patterns in a stratocumulus deck off the coast of Peru). *Source:* Courtesy of NASA. Open-cell regions are caused by precipitation. They are characterized by low cloud fraction that reveal the dark ocean surface where solar energy is absorbed. In contrast, closed cells are bright, nonprecipitating regions that reflect much more energy to space. (b) A natural landscape: "Fairy Circles" in the Namibian desert—barren, dry spots, often encircled by healthier vegetation. The underlying mechanisms are still disputed with some arguing that the pattern results from plants competing for scarce water resources and others that underground termite colonies, which consume moisture and destroy root systems, are responsible. (c) The canopy structure of the Dryobalanops Aromatica trees in Borneo is organized such that trees do not touch each other and still cover most of the canopy area, creating relatively ordered polygons (also called Crown shyness). (d) Giraffe markings showing similar partitioning to Crown shyness although the mechanism for this pattern formation can be fully explained by the reaction-diffusion equation (Murray, 1988). Image (b) is from "In African 'Fairy Circles', a template for nature's many patterns (23 January 2017)." *Sources:* (b) Princeton University/Phys.org; (c) Patrice78500/Wikimedia Commons/Public Domain; and (d) Luca Galuzzi/Wikimedia Commons/CC BY-SA 2.5.

conditions in which order *is* imposed by larger scale forcing, e.g., weather patterns associated with synoptic waves.

Because natural systems of different kinds tend to self-organize, the top-down approach has produced a commonality of concepts, tools, and analyses that cut across disciplines.

All panels in Figure 13.1 show mechanisms by which natural systems organize themselves. Figure 13.1b shows "Fairy Circles" in the Namibian desert—barren, dry spots, often encircled by healthier vegetation, which for many years have defied explanation by the ecological community. Self-organization by biomass-water-driven feedbacks is one of the plausible explanations (Getzin et al., 2016). The canopy structure of the Dryobalanops Aromatica trees (panel (c)) in Borneo is organized such that trees do not touch each other and still cover most of the canopy area creating relatively ordered polygons. The giraffe markings in panel (d) show similar partitioning. Other examples in the natural world are patterned mud cracks (Goehring & Morris, 2014) and self-organized mounds generated by earthworms in Colombia and Venezuela (Zangerlé et al., 2016). The field of economics and sociology abounds with analogous examples of self-organization based on human behavior and market competition. For example, the supply of goods and services distributes in distinct patterns through cities based on local needs, without any centralized decision-making regarding their placement. Networks often emerge from these interactions, and the underlying theory has been applied to problems as diverse as social interactions, the flow of information in computer networks, or connections between airline hubs. The examples are seemingly endless. The notion that different phenomena yield similar patterns, drawn from a limited pattern bank, has been explored by Feigenbaum (1983).

While bottom-up approaches tend to focus on local and detailed physical processes, and top-down approaches on emergence and pattern formation, there are other distinct conceptual differences between the two. The differences can be subsumed into the following contrasting approaches to the study of complex systems: Reductionism vs. Holism[1]; Basic building blocks vs. an Integrative view; models representing a Large vs. a Reduced number of degrees of freedom; models rooted in mathematical representation of physical/chemical/biological processes vs. models that are an abstraction of these processes; the researcher's inherent emphasis on Complexity vs. Simplicity.

The distinctions between the approaches are sometimes blurrier than implied by "Approach A" vs. "Approach B": even detailed reductionist models rely on statistical descriptions at the molecular level as in the case of the description of temperature in Maxwell-Boltzmann theory. In addition, "scale" isn't necessarily a good determinant of bottom-up approaches vs. top-down approaches since both can be applied at a range of scales. But a fundamental difference in the two is that top-down approaches favor the deliberate use of simpler, more universal and interdisciplinary ideas and methods as a guiding principle for addressing complex systems.

The advantages of blending these approaches has been discussed elsewhere (e.g., Harte, 2002; Feingold et al., 2016), and the balance in emphasis is often driven by the choice of problem and the conflicting needs to capture the essence of a system while simultaneously representing nuance[2]. A source of tension is often that in a complex system one might not grasp the extent to which seemingly unimportant detailed processes might be important, both alone and in combination with others.

This chapter encourages a more generous serving of the top-down approach in our more routine bottom-up fare. The top-down approach can yield simple models that are more amenable to interrogation and digestion than complex, detailed models. Ultimately all models have their place at the table, and the challenge is to use them in an integrative manner for maximum benefit.

13.2. A BRIEF HISTORICAL PERSPECTIVE

We attempt a brief recapitulation of some of the key investigators who have dedicated efforts to top-down approaches. This will serve as a backdrop to the atmospheric phenomena to be described in section 13.4. Broader and more rigorous perspectives can be found in, e.g., Prigogine and Stengers (1984) and Heylighen (2002). Modern, documented writings on self-organization include W. R. Ashby (1947) who, from the perspective of a psychiatrist and mathematician, showed that the human nervous system can be considered both a deterministic physicochemical system, but also undergo "self-induced internal reorganizations resulting in changes of behavior." This viewpoint captures in a nutshell the essence of emergence; it allows for physical determinism at the process level, while simultaneously allowing for changes in overall system behavior based on changes in local processes. These ideas were later extended to define self-organization in terms of the tendency of dynamical systems to evolve toward equilibrium (the steady state), a state in which parts of the system are mutually adapted.

Because self-organization represents a decrease in entropy, an early concern was its apparent contradiction with the second law of thermodynamics, namely, the tendency of a system to a state of increased entropy. von Foerster (1960) resolved this by stressing the importance of defining the *environment* of self-organization; for entropy to increase in systems characterized by structures having low entropy, the entropy of the environment has to increase so as not to violate the second law. Prigogine introduced rigorous thermodynamic considerations to describe internal self-organization in open systems. He called systems that continuously export entropy in order to maintain their organization "dissipative structures." These dissipative structures constantly generate entropy, which is dissipated out of the system and to the surroundings (e.g., Prigogine & Stengers, 1984). To illustrate this, we consider the case of living organisms, which are thermodynamically open and far-from-equilibrium systems. These highly ordered entities maintain their high metabolic energy/low entropy state by exporting entropy through dissipation of heat and waste products to the environment (e.g., Davies et al., 2013; Schrödinger, 1967). Similarly, phase transitions such as gas to liquid involve maintenance of a more ordered state through the export of latent heat to the environment.

Top-down approaches also include the significant contributions of von Bertalanffy (1950), who studied open systems with his general systems theory (GST) as an alternative to conventional models of organization.

Mathematical rigor has accompanied the development of top-down approaches. Alan Turing was one of a number of scientists who explored equations like the reaction-diffusion equation to describe the emergence of self-organized patterns. The reaction component has many forms that describe chemical, biological, and physical processes, e.g., Rayleigh-Bénard convection, which is

1 In holism, a system and its properties are viewed as a whole, not just as a collection of its parts. Von Bertalanffy emphasized holism in his General Systems Theory. The term was coined by South African statesman Jan Smuts.

2 The more complex a model, the larger the number of degrees of freedom, and the more the potential for being unconstrained and nonphysical. The simpler a model, the more constrained, the easier to understand system behavior, with the risk of a lack of nuance, and unrepresented modes of behavior.

often used to describe cloud formation and patterns such as in Figure 13.1a or giraffe patterns (Figure 13.1d).

A different class of self-organization emerges in systems that comprise components like coupled oscillators that "communicate" with one another. The classic example is the case of pendulum clocks oscillating in unison discovered by the Dutch scientist Huygens, who identified the cause of this synchronization as the weak coupling occurring through the wall on which both clocks were hanging. Modern demonstrations consider a large number of metronomes that experience coupling through small vibrations in the platform on which they all rest. The solution space for a system of coupled oscillators will likely have a dominant frequency, and therefore a higher likelihood of self-organization. Pioneering work by Wiener, Winfree, and Kuramoto (1975) showed that when a population of coupled components (e.g., the above-mentioned metronomes) is sufficiently homogeneous, the system itself can manifest coupled oscillations. Intermediate states can also occur—namely when the population is somewhat more diverse, in which case the less diverse population oscillates in synch but other populations are not synchronized. When the components are too diverse, oscillations are not sustainable. Strogatz and many other colleagues have introduced mathematical rigor and application of synchronization to dynamical systems of various kinds and made these ideas accessible to a broad audience (Strogatz, 2003).

Finally, Barabási and Albert (1999) built on earlier work by graph theorists such as Erdős and Renyi (1959) and applied scale-free network theory (i.e., networks with power-law degree distributions) to demonstrate that large-scale-free networks are robust, self-organizing phenomena that are common to a multitude of systems. They suggested that large networks (such as the internet) tend to self-organize such that they can be characterized as scale-free, with a few main nodes that serve as "hubs" with many links, and allowing redundancy of connections that make the network relatively immune to failure. Together with colleagues they have applied this approach to a host of problems ranging from physical to social (e.g., Barabási, 2002).

13.3. ORGANIZING PRINCIPLES AND TERMINOLOGY

Figure 13.2 lays out the conceptual study of Complex Systems from a top-down view, broken down into seven subdisciplines. The top-down view merges ideas from (mostly) Systems Theory, Pattern Formation, Nonlinear Dynamics, Collective Behavior, and Networks. A given complex system can often be addressed using approaches that derive from multiple subdisciplines. Note that Chapter 17 (Krasnapolsky) deals with machine learning (Evolution and Adaptation) and is discussed only briefly here. We note that the concept of "parametrization" does not appear explicitly in Figure 13.2, but by abstracting the essence of an aspect of system behavior, parameterization can in general be categorized as a top-down approach (regardless of scale, e.g., the Maxwell-Boltzmann description of temperature). Parameterizations come in many forms—ranging from the analytical to the statistical to the highly empirical—but all aim to represent key aspects of a complex system in a reduced form.

Self-organizing systems from different fields tend to yield common patterns. This leads to concepts that cross disciplinary boundaries, with unique language and terminology, some of which is highlighted here. Nonlinear dynamical systems are often controlled by *feedbacks*. Self-organization is typically related to positive feedback and *fixed points* or *attractors* in which local organization that might start by chance drives global organization, followed by an attractor state. Some of these systems have *bifurcation points* in which small changes in the system parameters can shift the system into a different fixed point or attractor. Linear stability analysis allows one to explore the system behavior near the fixed points and as the system parameters approach bifurcation points. Fixed points can be either *stable* (when the system's time derivative vanishes), in which case any small perturbation will be damped to equilibrium, or *unstable*, in which case perturbations can shift the system to another fixed point, *limit cycle* or *chaotic* state.

We use the example of the aerosol-cloud-precipitation system (Figure 13.1a) to illustrate dynamical systems thinking and to reinforce terminology. Aerosol particles can act as nuclei upon which droplets form, thereby controlling drop concentration, which in turn regulates rain formation (clouds generate rain more readily when aerosol and drop concentrations are low). The system can reach a steady state in which the rain depletes the cloud at the same rate at which the cloud is created (e.g., the closed cell state in Figure 13.1a). Cloud systems can go through dramatic changes in their properties once a significant amount of rain falls below cloud base (Figure 13.3). Rain generates evaporative cooling of the air below cloud base, negative buoyancy, and downdrafts that cause cold pools. When these cold pools converge, updrafts generate clouds, which are again a source of cloud water, and ultimately rain. The open (mostly dark) areas in Figure 13.1a are an example of the raining, open cell state that experiences charge/discharge (oscillatory) cycles. The oscillation cycle starts as the cloud develops and thickens until significant rain is produced. The rain

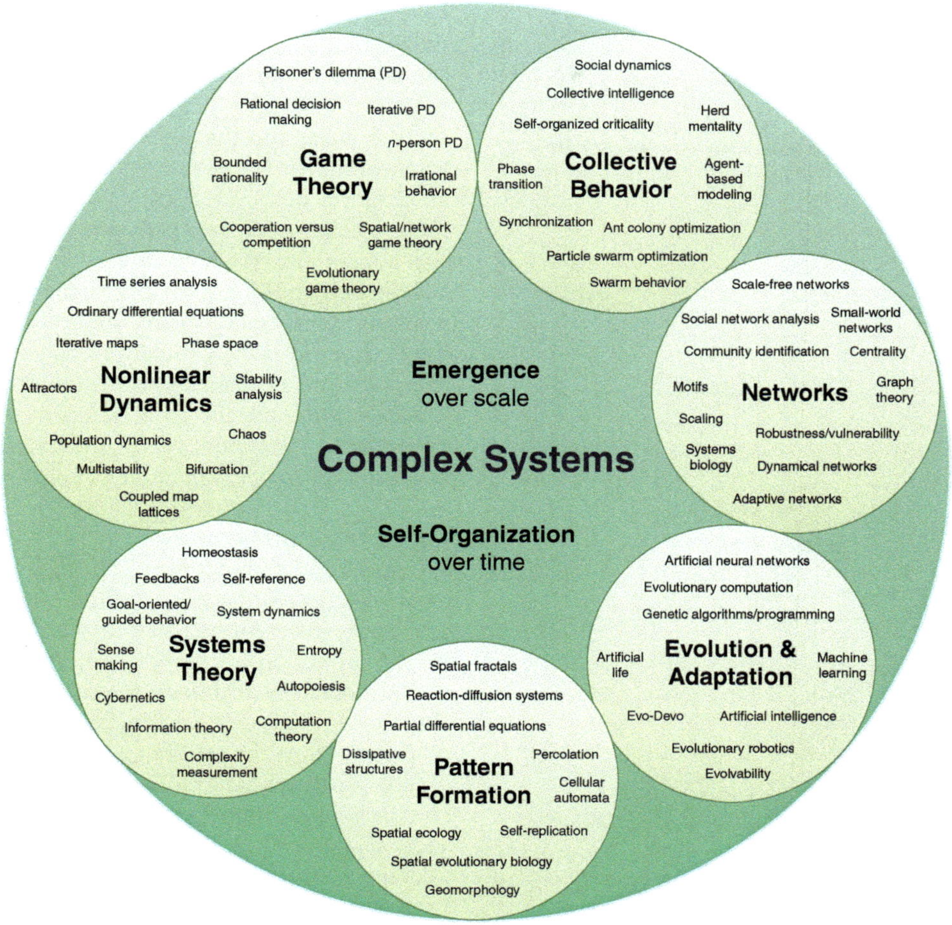

Figure 13.2 Overview of subdisciplines within the study of complex systems. This chapter primarily draws on nonlinear dynamics, systems theory, pattern formation, and networks. *Source:* Courtesy Hiroki Sayama, Collective Dynamics of Complex Systems (CoCo) Research Group at Binghamton University, State University of New York and Wikipedia.

depletes the cloud significantly and therefore weakens, giving way to the next cycle (Figure 13.3). Thus, for a given set of environmental conditions, the aerosol determines whether the system will be steady (a fixed point) or oscillating (approximating a limit cycle). The aerosol concentration that marks the transition between the two states is a bifurcation point (Baker & Charlson, 1990).

Spatiotemporal *communication* between components of a system is key to development of synchronization, patterns, and self-organization. In the aerosol-cloud-rain system described above, communication occurs via interacting cold-pool outflows. In the case of the Fairy Circles (Figure 13.1b), communication is either a result of plants competing for soil moisture (Meron, 2019) or competition between subterranean termite colonies (e.g., Tarnita et al., 2017). Communication and its information content can be quantified as shown by luminaries such as A. Turing and C. Shannon; the more information required to describe the system, the less organized it is. For example, describing the spatial properties of closed cells in Figure 13.1a requires less information than describing a field of randomly distributed trade-wind cumulus clouds. "Noise" or perturbations applied to self-organizing systems can, perhaps counterintuitively, facilitate order by allowing the system to explore its state space and find basins of attraction more readily (von Foerster, 1960). Noise can also be disruptive and dispatch conditionally unstable states into different attractors.

Having introduced these concepts, the following section delves into some examples of dynamical systems encountered in atmospheric physics. These examples also serve to illustrate a multiplicity of tools and preferences, which we hope will serve to encourage creativity in the application of top-down methods.

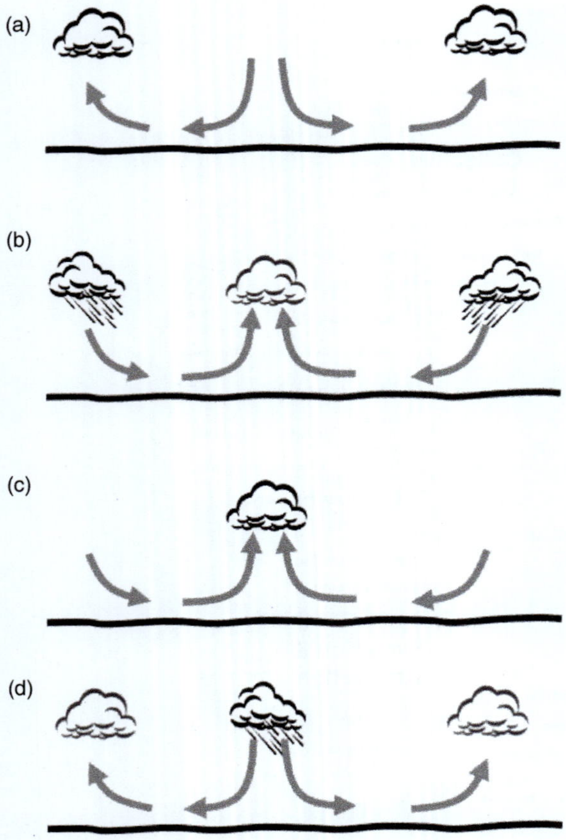

Figure 13.3 Marine stratocumulus clouds undergoing alternating precipitating and nonprecipitating phases (see also Smart, 2011). (a) Subsiding air generates surface divergence that feeds the growth of adjacent cells. (b) Adjacent cells deepen and begin to precipitate, creating negative buoyancy and driving downdrafts that converge in the region of the previously cloud-free sky. (c) As precipitating cells dissipate, cloud formation in the center is strengthened. (d) Rain falls from the newly formed cloud and generates negative buoyancy and downdrafts that converge to feed adjacent, growing clouds. The cycle then repeats.

13.4. EXAMPLES OF TOP-DOWN APPROACHES IN ATMOSPHERIC PHYSICS

13.4.1. Atmospheric Dynamics

Dynamics Systems Approach

The fundamental work of Lorenz (1963) is the cornerstone of the top-down approach in dynamical meteorology. Lorenz's dynamical system, a set of three coupled equations representing a truncated form of the governing equations of the flow, led to an understanding of the inherent sensitivity of atmospheric flow to initial conditions and the development of chaos theory. Lorenz demonstrated how small perturbations, essentially truncation errors associated with a model restart, can shift the system to another attractor. Lorenz's work together with early discoveries of Poincaré and others set the stage for the field of chaos theory. While the implications of Lorenz's work may not always be obvious at the level of physical cloud and convective processes, the concepts and approach are universal. Interestingly, the intervening years have seen far more emphasis on building bottom-up atmospheric models that incorporate a seemingly ever-increasing multitude of processes and level of complexity. Climate models represent one such example, and problems such as cloud-climate feedbacks might greatly benefit from simpler models with their potential to generate a better understanding of preferred states and possible system bifurcations.

Convective Parameterizations

Convective parameterizations have received attention from the top-down approach through a number of studies that apply different methodologies. A sample is included here. Convection is one of the more challenging aspects to parameterize within large-scale models (Emanuel & Raymond, 1993) and yet constitutes a critical component of the relationship between atmospheric circulation and clouds, and attendant problems such as precipitation, and climate forcing via aerosol-cloud interactions. Deep convection is traditionally described as a quasi-equilibrium system, which forms the basis of most parameterizations (e.g., Arakawa & Schubert, 1974). Nober and Graf (2005) use the quasi-equilibrium closure of Arakawa and Schubert but pose the convective system as an analog to the predator-prey problem where convection represents the predator, and convective available potential energy represents the prey. By running a one-dimensional cloud model, they simulate a temporally evolving size distribution of clouds that is self-consistent with the convective environment. The method has the salutary effect of providing information on the probability distribution of updraft velocities, cloud depth, and cloud fraction. The cloud size distribution is an additional useful test of the realism of the approach since cloud size distributions can be approximated by power laws (see also section 13.3.2). Using this approach, Wagner and Graf (2010) demonstrate clear improvement in single-column model simulations of the spatiotemporal distribution of convection relative to other standard approaches.

Craig and Cohen (2006) provide a theoretical basis for stochastic parameterization of cumulus convection by deriving the equilibrium fluctuations of a field of noninteracting cumulus clouds based on the Gibbs canonical ensemble from statistical mechanics. More recently, Sakradzija et al. (2015) take this further: they represent shallow cumulus convection by combining a random

Poisson sampling of convective elements with the cloud mass flux sampled from a Weibull distribution.

Yano and Plant (2012) describe convective quasi-equilibrium using a simple dynamical system analog of a nonlinear discharge-recharge model. Similar to Randall and Pan (1993), the system comprises two prognostic equations for the cloud-base mass flux and the cloud work function (a measure of the rate of generation of convective kinetic energy), and produces periodic cycles, regardless of the initial conditions. Reminiscent of the convective parameterization of Graf and colleagues, the cloud work function first has to be fully recharged before enhanced convective activity is triggered. The enhanced convective activity results in a rapid discharge of convective energy (a rapid reduction in both mass flux and work function) which then leads to the next cycle of recharge. The associated delay might be thought of as representing the delay in the onset of convection over the course of the diurnal cycle through, e.g., the development of shallow convection, which plays an important role in triggering deep convection.

In an extensive review, Yano and Plant (2012) interpret convective quasi-equilibrium as, *inter alia*, a thermodynamic balance, in which the system is considered to be in homeostasis, a concept typically applied to self-regulating biological systems. In this view, convection is uniquely defined by its environment and self-regulation maintains the stability of that state.

13.4.2. Cloud Microphysics

Wacker (1995) was the first to apply a dynamical system, top-down approach to a microphysical scheme, incorporating themes of self-organization and an analog of the predator-prey model to different populations of cloud particles either consuming water vapor through condensation, or one another through collection processes, or the Wegener-Bergeron-Findeisen process (Findeisen, 1938) in which ice grows by water vapor diffusion at the expense of liquid water. Through parameter selection in a multilayered cloud system, she shows the presence of single-point attractors and periodic attractors and multistability. In the latter case, the choice of initial conditions "selects" the attractor. Selection of different source rates is also shown to determine how one of the microphysical species will eventually dominate the system through consumption of the others. Wacker (2006) extends this approach to a two-layer cloud model to account for vertical transport between layers via an updraft or drop sedimentation. In the absence of an updraft, microphysical processes in the upper layer drive those in the lower layer, while in the presence of an updraft, solutions include bifurcations that depend on the magnitude of the updraft and the microphysical conditions.

Arabas and Shima (2017) study the aerosol activation equations and identify a saddle node bifurcation at the peak of the Köhler curve describing equilibrium water vapor pressure over a droplet and a cusp catastrophe—essentially a form of bifurcation when the system transitions from one state to another. The authors show that the activation/deactivation process exhibits hysteretic behavior associated with kinetic limitations to drop growth when the system is far from equilibrium and with the cusp catastrophe when the system is close to equilibrium.

Spreitzer et al. (2017) apply a predator-prey model to subvisible cirrus clouds—i.e., ice clouds with very low optical depth—formed via homogeneous freezing. Based on first principles, they formulate the problem as a set of ordinary differential equation representing ice nucleation, diffusional growth, and sedimentation. Their results show two solutions: damped oscillations to a stable fixed point and an oscillatory limit cycle. Transitions between the states are Hopf bifurcations determined by updraft velocity and temperature. Solutions are shown to mimic a much more complex (bottom-up) microphysical model.

Application of microphysical top-down models such as those highlighted above readily highlights key model variables and parameters that are much harder to discern in bottom-up models. Standard microphysical parameterizations (e.g., Berry, 1968) can be effective at representing processes, although often less elegantly, especially when curve fitting is applied, and without the insights offered by dynamical systems models.

13.4.3. Cloud Systems

A broad view of the dynamics systems approach is provided by Garrett (2012), who generalizes complex systems to a few modes that depend on the time integral of past "material flows" and the current availability of resources. The modes include a law of diminishing returns (early rapid growth followed by saturation), logistic behavior (i.e., a sigmoidal increase in growth to saturation), and superexponential growth. The theme of the study is one of generality of these concepts in terms of systems as widespread as cloud physics and fish populations. We focus here on examples of the top-down approach to shallow and deep cloud systems.

Shallow Cloud Systems

While Wacker (1995) used a predator-prey analog to microphysical processes, Koren and Feingold (2011) took a similar approach but for a cloud *system*. Earlier work with detailed large eddy simulation simulated the drizzle-induced transition from closed-cellular to open-cellular convection as illustrated in Figure 13.1a and described in section 13.3 (e.g., Feingold et al., 2010).

The Koren and Feingold (2011) *cloud-rain model* considers rain as the predator of cloud water; it comprises two coupled, time-dependent delayed differential equations in cloud depth and drop concentration, and one diagnostic equation for rain. The equations include time delays associated with the time required for cloud water to be converted to rain water. The simple equation set can successfully simulate both the steady-state, closed-cell state where drop concentration, cloud water, and rain are in balance, and limit cycles associated with open cells that strongly precipitate and require a recharge of drop concentration and cloud water before rain can develop again (Chekroun et al., 2020). An interesting aspect of the cloud-rain model is its more rapid evolution to its attractors when exposed to "noise," as discussed by von Foerster (1960).

Extending the cloud-rain model developed for a cloud system to a spatially interacting population of clouds, Feingold & Koren (2013) use the analog of coupled cloudy oscillators, each going through cycles of cloud deepening and rain formation. Coupling of the cloudy elements is achieved through a coupling term related to the rate of discharge of rain, which is intended to mimic the development of cold pools. This study identifies complex behavior that includes various degrees of homogeneity within the system (similar to Kuramoto, 1975). By varying the degree of coupling strength, the authors show that in the case of weak coupling, the clouds behave as independent oscillators. As the local coupling strength increases, the system reaches a point at which it becomes highly synchronized via the interacting outflows (as in Sakaguchi et al., 1987), but if the coupling strength exceeds some magnitude it can result in (nonphysical) unstable growth.

An even simpler version of the cloud-rain model is analyzed in Koren et al. (2017), where analytical and numerical solutions are obtained for a single delayed differential equation treating the drop concentration as a free parameter. Using linear stability analysis, the first bifurcation at which the system transitions from steady state to a limit cycle is calculated analytically in terms of normalized system parameters, expressed as a nondimensional parameter that includes drop concentration, cloud depth "carrying capacity" (the maximum depth achievable in the absence of sinks), and the time to achieve this carrying capacity. Solutions are again shown to be consistent with knowledge gained from bottom-up, detailed modeling (Berner et al., 2013; Feingold et al., 2010; Kazil et al., 2011). By revealing preferred states and bifurcations, this back-and-forth between detailed large eddy simulation and a simple dynamic systems analog presents rich opportunity for improving both bottom-up and top-down models. Equally importantly, it allows for new templates for observation-based analyses. A recent application in this direction, using model output as a proxy for observations, is a fitting of the Koren et al. (2017) model parameters to large eddy simulation output via Bayesian inversion (Lunderman et al., 2020).

Finally, the complex cloudy boundary layer system includes cloud and convective processes that act at a range of scales leading Bretherton et al. (2010) to define the system as a "slow manifold" that evolves on a timescale associated with the multiday variability in inversion strength. Fast thermodynamic (order 1 day) and microphysical (order hours) are shown to slave themselves to the slow manifold. In essence, the slowly changing inversion is a slowly varying boundary condition, which allows the faster thermodynamic response time to adjust to the instantaneous boundary conditions. The ideas are investigated with large eddy simulations that exhibit multiple equilibria and bifurcations that depend on initial boundary layer depth and precipitation (as in other work). Morrison et al. (2012) discuss the Arctic mixed-phase cloud system in similar terms: the slow manifold is well represented by surface pressure, which determines whether the system is cloud-free or cloudy; microphysical processes slave to the slow manifold, but may be influential enough to affect a switch in state. In the above examples, we speculate that the slaving of the fast to the slow manifold is a result of the downscale transfer of energy to the smaller scales.

Deep Cloud Systems

A somewhat different class of top-down modeling of organized cloud systems focusing more on convection has been presented by Böing (2016). The model comprises two layers—a lower layer in which instability can accumulate and an upper layer where instability is consumed. "Particles" are used to represent instability and convection on a lattice grid. Coupling between these layers is achieved via the effects of rain and cold pools. Similar work by Härter et al. (2019) expands this approach to convective aggregation, showing how local scale updrafts driven by cold pools can drive convective self-organization at much larger scales (order 1000 km).

Drawing on themes from convective parameterization (section 13.1.2), Garrett et al. (2018) describe the evolution of a tropical cloud system in terms of the net entrainment around cloud boundaries as a top-down constraint on the cloud size (perimeter) distribution. Competition for the sum total of this net entrainment among all clouds in a field leads to a power law for cloud perimeter within moist isentropic layers (or a negative exponential relationship across moist isentropic layers). The interactions are mediated by the thermodynamic state of the atmosphere—particularly stability—which affects the rate of interactions among clouds through the frequency of buoyant circulations around cloud edge.

These circulations in turn can be related to rates of exchange of air between one cloud and the others, which cascade from larger to smaller perimeter clouds. Thus, within a population, any given cloud necessarily grows at the expense of others, all constrained by cloud perimeter entrainment. The authors are also able to show close statistical agreement between tropical cloud ensembles produced with this top-down model and those produced by detailed deterministic models.

Synchronization within deep convective cloud systems has been observed to manifest as a synchronous discharge of lightning (Yair et al., 2009) across a mesoscale system. Based on these observations, the authors describe the system as a coupled network, where coupling occurs through a network-wide sharing of the change in the electric field strength E. The network coupling occurs as a result of a lightning strike at a given location in the system: an individual cloud element in the evolution of its life cycle of E experiences a contribution ΔE based on its distance to the cloud experiencing discharge. When the group coupling strength in the network is higher than a threshold value, the clouds discharge in unison. The coupling is global, as in Kuramoto (1975). Moreover, the delay time for communication of changes in E is extremely short (seconds), thus facilitating the communication across large distances (order 100 km). In this sense, it differs significantly from the stratocumulus system where communication between cells occurs via outflows that have to travel tens of kilometers at speeds on the order of $10\,\mathrm{m\,s^{-1}}$.

13.4.4. Network Approaches

Complex Networks

The burgeoning field of network theory, so widely applied to natural systems and social sciences, is increasingly applied in the atmospheric sciences. A "network" comprises a system of "nodes" interconnected by "edges." In the social sciences, the nodes may be individuals or organizations, and the edges, the existence of flow of information, or contact between individuals or organizations. In climate science, nodes are defined by geographical locations on a climate model's gridded mesh, having physical properties such as temperature or pressure. Whether two nodes are connected by an edge might depend, for example, on the statistical similarity between time series of these physical properties over some period of time. The similarity might be expressed as a correlation coefficient above some statistically significant threshold. If one regards the nodes as a system of coupled oscillators, then the coupling strength is a measure of the interconnectedness of the edges. An early example of the application of network approaches to climate science can be found in Tsonis et al. (2006). Yuan (2011) has described shallow cumulus cloud fields as a network of clouds that randomly clump and merge. An example describing South American extreme rainfall in terms of network connectivity with synchronization can be found in Boers et al. (2014).

Cellular Networks Applied to Cloud Fields

Cellular networks describe networks in which spatial structure is key to the nature and evolution of the network, i.e., when nodes and edges are constrained by geometry (usually) in a two-dimensional plane. The contiguous cellular patterns observed in stratocumulus cloud fields (Figure 13.1a) are reminiscent of natural cellular networks observed in films, emulsions, biological tissue, grains, and other media. Glassmeier and Feingold (2017) use metrics commonly applied to cellular networks in these physical and biological systems to quantify and characterize cloud cellular structures. They show that the network structure produced by detailed numerical modeling of both closed-cell and open-cell stratocumulus can be replicated using cloud system analogs of cell division and cell merging. The authors build a heuristic model of network dynamics that incorporates our physical understanding of cloud processes. The heuristic model is shown to replicate key properties of the detailed model-generated system.

The cellular network analysis sheds light on the role of synchronization of outflows within the open-cell cloud system; it shows that the open-cell pattern can be reproduced by local random transformations that are constrained by overall system topology and does not require close timing of interacting outflows from rainy patches within the system, in agreement with conjecture by Feingold and Koren (2013). Moreover, the authors show that because the nodes of the stratocumulus network have a narrow degree distribution (i.e., most cells are surrounded by six cells, with small spread), there are no cells that take on the role of central hubs, as speculated by Feingold et al. (2010) based on detailed large eddy simulation. Only at the local level do the parent nodes drive the network rearrangement. Studies such as these are an example of the benefits of combining bottom-up approaches alongside of top-down approaches to illuminate the study of a complex system.

13.4.5. Statistical Mechanics Analogs

Drop size distributions are a basic fingerprint of all the physical processes that shape cloud drops and are usually parameterized by lognormal or gamma functions based on least squares or moment-conserving fits. Instead, Liu et al. (1995) use a more general approach that connects between statistical mechanics of molecular systems and particle systems (where particles are on the order of drop sizes). They define the maximum likelihood for the drop size distribution in terms of the Shannon entropy,

$H(y) = -c \int \rho(y) \ln \rho(y) dy$, where $\rho(y)$ represents the maximum likelihood of the drop size distribution and c is a constant (Shannon, 1948). By applying a generalized partition function in the form of an exponential function, and using a power law to relate drop diameter to variables that define physical processes ("restriction variables"), they arrive at a maximum likelihood of a Weibull distribution. An example is given for a simple system in which drop concentration and mass are conserved, yielding $n(D) = N_0 D^2 \exp(-\lambda D^3)$, where D is drop diameter, and N_0 and λ are the remaining free parameters. (We note that most measurements follow a D^1 rather than a D^3 dependence in the exponent.) Follow-on work by Liu and Hallett (1998) extends these ideas to a bound between the maximum likelihood (Weibull) and a minimum likelihood (delta function). They suggest that the existence of multiple size distributions requires a probabilistic description for complete characterization of drop size distributions. More recent studies that consider maximum entropy principles include Yano et al. (2016) and Wu and Mc Farquhar (2018).

Garrett (2019) takes a different approach by treating the size distribution as an open system at steady state, defined by a continual flux of liquid water through size classes. This flux or "cascade" generates a characteristic gamma distribution $n(D) = N_0 D^{\mu} \exp(-\lambda D)$, consistent with the large body of observations. The shape parameter μ is controlled by cloud liquid water path and cloud updraft speed, and can be both positive, as in the case of a concave function, and negative, as in the case of a convex function (observational evidence exists for both).

13.4.6. Statistical Fitting

Various statistical approaches relevant to the top-down view are increasingly considered in the atmospheric sciences. The key approaches include elements such as emulation, machine learning, and parameter fitting. Only emulation is discussed here. A full treatment of machine learning can be found in Chapter 17 (Krasnapolsky).

Bayesian emulators are essentially proxies for detailed models that are designed to mimic the complex system as simulated by a detailed model. As a form of machine learning, they conceptually and organizationally belong to the "Evolution and Adaptation" approach to the study of complex systems (Figure 13.2; Krasnapolsky, Chapter 17). They have been used for some years in the atmospheric sciences to assess parametric uncertainty in models—namely the relationship between settings in parameterizations of physical processes and model output (e.g., Carslaw et al., 2013). Typically, emulators consider the relationship between a number of model inputs and their impact on selected model outputs. To build the emulator, a set of training simulations using parameter settings based on expert solicitation is performed (order 10 simulations per input variable or parameter). Gaussian fitting to this n-dimensional surface (n = number of inputs) yields an emulator, which is then validated using input settings from previously unsampled variable or parameter space. The process is repeated until the emulator is deemed sufficiently precise for the problem under consideration. Once the (inexpensive) emulator has been constructed, it can be run for many different scenarios covering variable or parameter setting combinations that are computationally unfeasible with the complex model.

The approach has been applied to the question of aerosol-cloud interactions and their influence on the radiative properties of stratocumulus clouds by Glassmeier et al. (2019). Starting with six key meteorological and aerosol input variables to a detailed large eddy simulation (bottom-up) model, they simulate hundreds of cases covering a wide range of atmospheric conditions and simulated output states. To reduce the dimensionality of the problem, the six input variables are mapped to a liquid water path, $\mathcal{L}$, and drop concentration, N state space (which is itself a bridge between the bottom-up approaches and the top-down approaches). Then, emulators of the detailed model outputs of the stratocumulus system are constructed and validated against the detailed large eddy simulation. The authors demonstrate successful emulation of cloud fraction, albedo, and radiative effect of the cloud system in terms of $\mathcal{L}$ and N, as well as the bifurcation in the system in $\mathcal{L}; N$ space induced by precipitation. Thus, the emulator synthesizes a complex model into a simple predictive framework in $\mathcal{L}, N$ state space, and facilitates assessment of system behavior and the sensitivities of cloud field properties over a wide range of states.

13.5. INTERPRETATIVE AND DIAGNOSTIC METHODS IN THE WORLD OF THE TOP-DOWN APPROACH TO ATMOSPHERIC SYSTEMS

In addition to the mathematical analyses of dynamical systems in terms of stability, attractors, fixed points, limit cycles, and bifurcations (section 13.3), a number of other diagnostic and interpretative methods are pertinent to the discussion.

13.5.1. Equifinality

Equifinality is a concept that dates back to the work of the developmental biologist, Hans Driesch, in the early twentieth century and popularized by von Bertalanffy

(1950) in his general systems theory. It refers to the fact that in open systems a system end state can evolve from many different pathways. For example, models with different but equally plausible representations of processes may match observations of the global-mean temperature record or produce the same amount of rainfall (Mülmenstädt & Feingold, 2018). Of course, not all open systems exhibit equifinality, e.g., an oscillating system that does not reach a limit cycle or a system that is highly sensitive to initial conditions (Lorenz, 1963). Because they achieve the same result along a different pathway, open systems that do manifest equifinality pose limitations on our ability to untangle causal pathways—particularly through observations—and to definitively exclude particular detailed numerical models of a system. Furthermore, equifinality limits the extent to which understanding of complex dynamical systems can be advanced by the back-and-forth between bottom-up process models and top-down models.

13.5.2. Fingerprints

In the atmospheric sciences there exist a number of different meta-analyses of complex systems that we loosely refer to as "fingerprints." These are typically state space diagrams that encapsulate system behavior but that can be linked to processes. The fingerprints should ideally be focused on key geophysical and readily observable parameters, i.e., on manifestations of the system in an important state space, rather than fussing with unnecessary detail. Examples include the cloud optical depth vs. radar reflectivity state space (Suzuki et al., 2010) that lends itself to interpretation of dominant cloud microphysical processes; the albedo vs. cloud fraction state space (Bender et al., 2011) that can be tied to cloud brightening/cloud deepening (e.g., Feingold et al., 2017); and the cloud depth/liquid water vs. drop number state space that encapsulates the effect of microphysical processes on the system state (e.g., Bretherton et al., 2010; Glassmeier et al., 2019; Koren & Feingold, 2011, Koren et al., 2017). Fluxes of physical properties such as liquid water and drop concentration can provide valuable insights into system behavior, as shown by Hoffmann et al. (2020).

13.5.3. Emergent Constraints

Emergent constraints are model-derived, physically plausible relationships between a characteristic of the predicted state and of the current state of the atmosphere. The constraint enters when the current state of the system can be *observed*, i.e., if a robust relationship between predicted and current state emerges, observations can serve as a constraint on the predicted state variable (e.g., Klein & Hall, 2015). The success of this approach depends on the plausibility of the emergent constraint, which in turn depends on the ability of the model(s) to resolve with fidelity the detailed processes that generate the emergent relationship. Emergent constraints have been widely used to constrain topics as diverse as equilibrium climate sensitivity, cloud feedbacks, aerosol-related cloud forcing (Quaas et al., 2009), and aerosol effects on cloud lifetime (Lebo & Feingold, 2014; Wang et al., 2012). For aerosol-cloud physics, emergent constraints may provide an alternative approach to derive the state of a system or its susceptibility, provided the models that generate the emergent relationship are adequate and not scale-dependent, and the constraining observations precise enough to be useful.

13.6. GUIDELINES FOR TOP-DOWN APPROACHES AND CONCLUDING REMARKS

This chapter has endeavored to provide examples of top-down approaches as well as ways in which they can be applied. The broad scope of problems in atmospheric science makes it unwise to be prescriptive, both in terms of the top-down methods that might be considered and the balance between bottom-up and top-down emphasis. Each problem has its own knowledge base and uncertainties, and these should ultimately guide choices. Nevertheless, nonlinear problems in which system stability is poorly established may benefit most significantly from top-down approaches. While not discussed here, climate feedbacks and tipping points may be one such application deserving of more attention. Bethiany et al. (2016) uncover some of the questions and challenges that emerge when top-down and bottom-up approaches are confronted on the topic of tipping points in the climate system.

We conclude with some guiding principles that might be beneficial when applying top-down approaches:

(i) *Carefully pose the scientific questions* to be addressed by the model. A dynamical system model cannot be expected to represent all detail.
(ii) *Simplicity*: Include no more detail in the top-down model than necessary (for the particular question(s) being posed).
(iii) *Incorporate key physical processes*: Stay as true to physical processes as possible.
Incorporate in the top-down model new developments from the detailed process model world as they become available; they may change understanding of system state.
(iv) Incorporate *physical and energetic constraints* when possible. This will ensure physically realistic model solutions.
(v) Study *evolving systems*; snapshots of a system do not necessarily yield robust information on processes

(Mülmenstädt & Feingold, 2018). Tracking evolving boundary layer systems has been shown to be very powerful (e.g., Glassmeier et al., 2019; Goren & Rosenfeld, 2015). The synergy between new geostationary satellites and models should be exploited. Deriving the microscopic dynamics of a complex system from observations of its evolving response to external perturbations also seems like a fruitful approach (Barzel et al., 2015). A caveat to this recommendation is that when a system exhibits equifinality its evolution will need to be examined in a state space that allows one to discern the microscopic dynamics. And perhaps most importantly:

(vi) *Integrate top-down and bottom-up approaches* for maximum benefit (e.g., section 13.4.2 and Harte, 2002). An illustrative example dwelled upon here is the combination of the bottom-up approach in the form of detailed process understanding/modeling of the aerosol-cloud-rain system through large eddy simulation alongside the top-down approach that invokes a rain-cloud (predator-prey) model generated by a few simple equations describing either the entire system or a system of coupled cloud oscillators. This integration of heuristic and detailed modeling—and, equally importantly, observations—presents opportunity for improving understanding of preferred system states.

In conclusion, we have endeavored to elucidate the benefits of an appropriate balance between the top-down approach and the more traditional bottom-up approach for addressing complex atmospheric systems—a merging of the philosophy of holism, emergence, and pattern formation via simple equation sets that describe system-wide behavior, with the bottom-up detailed physical and numerical descriptions of the system. The appropriate balance will very likely be driven by the nature of the problem, particularly the level of understanding of the system dynamics and its preferred states. The to-and-fro between these two approaches will lead to a firmer basis of understanding and predictability of these systems.

ACKNOWLEDGMENTS

The authors acknowledge Timothy J. Garrett and a second reviewer for their insightful comments. We thank Wikipedia for Figures 13.1c and d, and Prof. Hiroki Sayama and Wikipedia for Figure 13.2.

REFERENCES

Agee, E. M. (1984). Observations from space and thermal convection: A historical perspective. *Bulletin of the American Meteorological Society, 65*, 938–949.

Arabas, S., & Shima, S. (2017). On the CCN (de)activation nonlinearities. *Nonlinear Processes in Geophysics, 24*, 535–542. https://doi.org/10.5194/npg-24-535-2017.

Arakawa, A., & Schubert, W. H. (1974). Interaction of a cumulus cloud ensemble with the large-scale environment, Part 1. *Journal of the Atmospheric Sciences, 31*, 674–701.

Ashby, W. R. (1947). Principles of the self-organizing dynamic system. *Journal of General Psychology, 37*, 125–128.

Baker, M., & Charlson, R. (1990). Bistability of CCN concentrations and thermodynamics in the cloud-topped boundary-layer. *Nature, 345*(6271), 142–145.

Barabási, A.-L., & Albert, R. (1999). Emergence of scaling in random networks. *Science, 286*(5439), 509. doi:10.1126/science.286.5439.509

Barabási, A. (2002). *Linked: The new science of networks*. Perseus Books Group. ISBN 9780738206677.

Barzel, B., Liu, Y.-Y., & Barabási, A. (2015). Constructing minimal models for complex system dynamics. *Nature Communications, 6*, 1–8. doi: 10.1038/ncomms8186.

Bender, F. A., Charlson, R. J., Ekman, A. M. L., & Leahy, L. V. (2011). Quantification of monthly mean regional-scale albedo of marine stratiform clouds in satellite observations and GCMs. *Journal of Applied Meteorology and Climatology, 50*, 2139–2148. https://doi.org/10.1175/JAMC-D-11-049.1.

Berner, A. H., Bretherton, C. S., Wood, R., & Muhlbauer, A. (2013). Marine boundary layer cloud regimes and POC formation in a CRM coupled to a bulk aerosol scheme. *Atmospheric Chemistry and Physics, 13*, 12549–12572. https://doi.org/10.5194/acp-13-12549-2013.

Berry, E. X. (1968). Modification of the warm rain process. In *Proceedings of the 1st Conference on Weather Modification*, April 28–May 1, 81–85 (American Meteorological Society, Albany, 1968).

Bethiany, S., Dijkstra, H., Crucifix, M., Dakos, V., Brovkin, V., Williamson, M. S., et al. (2016). Beyond bifurcation: using complex models to understand and predict abrupt climate. *Dynamics and Statistics of the Climate System, 1*(1), 2016, dzw004. https://doi.org/10.1093/climsys/dzw004.

Boers, N., Rheinwalt, A., Bookhagen, B., Barbosa, H. M. J., Marwan, N., Marengo, J. & Kurths, J. (2014). The South American rainfall dipole: A complex network analysis of extreme events. *Geophysical Research Letters, 41*, 7397–7405. https://doi.org/10.1002/2014GL061829.

Böing, S. (2016). An object-based model for convective cold pool dynamics. *Mathematics of Climate and Weather Forecasting, 2*, 43–60. https://doi.org/10.1515/mcwf-2016-0003.

Bretherton, C. S., Uchida, J., & Blossey, P. N. (2010). Slow manifolds and multiple equilibria in stratocumulus-capped boundary layers. *Journal of Advances in Modeling Earth Systems, 2*, Article 14. https://doi.org/10.3894/JAMES.2010.2.14.

Carslaw, K. S., Lee, L. A., Reddington, C. L., Pringle, K. J., Rap, A., Forster, P. M., et al. (2013). Large contribution of natural aerosols to uncertainty in indirect forcing. *Nature, 503*, 67–71.

Chekroun, M. D., Koren, I., & Liu, H. (2020). Efficient reduction for diagnosing Hopf bifurcation in delay differential systems: Application to cloud-rain models. *Chaos, An Interdisciplinary Journal of Nonlinear Science, 30*(5), 053130, 1–30. https://doi.org/10.1063/5.0004697.

Craig, G. C., & Cohen, B. G. (2006). Fluctuations in an equilibrium convective ensemble, Part I: Theoretical formulation. *Journal of the Atmospheric Sciences*, *63*, 1996–2004.

Davies, P. C. W., Rieper, E., & Tuszynskic, J. A. (2013). Self-organization and entropy reduction in a living cell. *Biosystems*, *111*, 1–10. doi: 10.1016/j.biosystems.2012.10.005.

Emanuel, K. A., & Raymond, D. J. (1993). The representation of cumulus convection in numerical models. *Meteorological Monographs, 46*, Amer. Meteor. Soc., 246 pp.

Erdős, P., & Rényi, A. (1959). *On Random Graphs. I, Publicationes Mathematicae*, *6*, 290–297.

Feigenbaum, M. J. (1983). Universal behavior in nonlinear systems. *Physica D: Non linear Phenomena*, *7*, 16–39. https://doi.org/10.1016/0167-2789(83)90112-4.

Feingold, G., Koren, I., Wang, H., Xue, H., & Brewer, W. A. (2010). Precipitation-generated oscillations in open cellular cloud fields. *Nature*, *466*(7308), 849–852. https://doi.org/10.1038/nature 09314.

Feingold, G., & Koren, I. (2013). A model of coupled oscillators applied to the aerosol-cloud-precipitation system. *Nonlinear Processes in Geophysics*, *20*, 1011–1021. doi:10.5194/npg-20-1011-2013.

Feingold, G., McComiskey, A. C., Yamaguchi, T., Johnson, J. S., Carslaw, K., & Schmidt, K. S. (2016). New approaches to quantifying aerosol influence on the cloud radiative effect. *Proceedings of the National Academy of Sciences*, *13*, 5812–5819. https://www.pnas.org/cgi/doi/10.1073/pnas.1514035112.

Feingold, G., Balsells, J., Glassmeier, F., Yamaguchi, T., Kazil, J., & McComiskey, A. C. (2017). Analysis of albedo vs. cloud fraction relationships using heuristic models and large eddy simulation. *Journal of Geophysical Research*, *122*, 7086–7102. doi:10.1002/2017JD026467.

Findeisen, W. (1938). *Kolloid-meteorologische Vorgänge bei Neiderschlags-bildung. Meteorologiche Zeitschrift*, *55*, 121–133.

Garay, M. J., Davies, R., Averill, C., & Westphal, J. A. (2004). Actinoform clouds: Overlooked examples of cloud self-organization at the mesoscale. *Bulletin of the American Meteorological Society*, *85*, 1585–1594.

Garrett, T. J. (2012). Modes of growth in dynamic systems. *Proceedings of the Royal Society A*, *468*, 2532–2549. https://doi.org/10.1098/rspa.2012.0039

Garrett, T. J. (2019). Analytical solutions for precipitation size distributions at steady state. *Journal of the Atmospheric Sciences*, *76*, 1031–1037. https://doi.org/10.1175/JAS-D-18-0309.1.

Garrett, T. J., Glenn, I. B., & Krueger, S. K. (2018). Thermodynamic constraints on the size distributions of tropical clouds. *Journal of Geophysical Research: Atmospheres*, *123*, 8832–8849. https://doi.org/10.1029/2018JD028803

Getzin, S. H., Yizhaq, B., Bell, T. E., Erickson, A. C., Postle, I., Katra, O., et al. (2016). Discovery of fairy circles in Australia supports self-organization theory. *Proceedings of the National Academy of Sciences*, *113*, 3551–3556. https://doi.org/10.1073/pnas.1522130113

Glassmeier, F., & Feingold, G. (2017). Network approach to patterns in stratocumulus clouds. *Proceedings of the National Academy of Sciences*, *114*, 10,578–10,583. https://doi.org/10.1073/pnas.1706495114.

Glassmeier, F., Hoffmann, F., Yamaguchi, Y., Johnson, J. S., Carslaw, K. S., & Feingold, G. (2019). An emulator approach to stratocumulus susceptibility. *Atmospheric Chemistry and Physics*, 10191–10203. https://doi.org/10.5194/acp-19-10191-2019.

Goehring, L., & Morris, S. W. (2014). Cracking mud, freezing dirt, and breaking rocks. *Physics Today*, *67*, 39–44.

Goren, T., & Rosenfeld, D. (2015). Extensive closed-cell marine stratocumulus downwind of Europe—A large cloud mediated radiative effect or forcing? *Journal of Geophysical Research: Atmospheres*, *120*(12), 6098–6116.

Härter, J. O., Böing, S. J., Henneberg, O., & Nissen, S. B. (2019). Circling in on convective organization. *Geophysical Research Letters*, *46*, 7024–7034. https://doi.org/10.1029/2019GL082092.

Harte, J. (2002). Toward a synthesis of the Newtonian and Darwinian worldviews. *Physics Today*, *55*, 29–34. https://doi.org/10. 1063/1.1522164.

Heylighen, F. (2002). *The science of self-organization and adaptivity*. Knowledge Management, Organizational Intelligence and Learning, and Complexity: v. 3. editor/L. D. Kiel. EOLSS Publishers Co Ltd.

Hoffmann, F., Glassmeier, F., Yamaguchi, T., & G. Feingold (2020). Liquid water path steady states in stratocumulus: insights from process-level emulation and mixed-layer theory. *Journal of the Atmospheric Sciences*, *77*, 2203–2215. https://doi.org/10.1175/JAS-D-19-0241.1.

Kazil, J., Wang, H., Feingold, G., Clarke, A. D., Snider, J. R., & Bandy, A. R. (2011). Chemical and aerosol processes in the transition from closed to open cells during VOCALS-REx. *Atmospheric Chemistry and Physics*, *11*, 7491–7514.

Klein, S. A., & Hall, A. (2015). Emergent constraints for cloud feedbacks. *Current Climate Change Reports*, *4*, 276–287. https://doi. org/10.1007/s40641-015-0027-1.

Koren, I., & Feingold, G. (2011). Aerosol-cloud-precipitation system as a predator-prey problem. *Proceedings of the National Academy of Sciences*, *108*(30), 12,227–12,232. https://doi.org/10.1073/pnas.1101777108.

Koren, I., Tziperman, E., & Feingold, G. (2017). Exploring the nonlinear cloud and rain equation. *Chaos*, *27*, 013107. https://doi. org/10.1063/1.4973593.

Kuramoto, Y. (1975). International Symposium on Mathematical Problems in Theoretical Physics. In Araki, H. (Eds.), *Lecture Notes Physics, 39*, 420–422. New York: Springer.

Lebo, Z. J., & Feingold, G. (2014). On the relationship between responses in cloud water and precipitation to changes in aerosol. *Atmospheric Chemistry and Physics*, *14*(21), 11,817–11,831. https://doi.org/10.5194/acp- 14-11817-2014.

Liu, Y., You, L., Yang, W., & Liu, F. (1995). On the size distribution of cloud droplets. *Atmospheric Research*, *35*, 201–216.

Liu, Y., & Hallett, J. (1998). On size distributions of cloud droplets growing by condensation: A new conceptual model. *Journal of the Atmospheric Sciences*, *55*, 527–536. https://doi.org/10.1175/1520-0469(1998)055<0527:OSDOCD>2.0.CO;2

Lorenz, E. (1963). Deterministic nonperiodic flow. *Journal of the Atmospheric Sciences*, *20*(2), 130–141. https://doi.org/10.1175/1520-0469(1963)020<0130:DNF>2.0.CO;2.

Lunderman, S., Morzfeld, M., Glassmeier, F. & Feingold, G. (2020). Estimating parameters of the nonlinear cloud and rain equation from large-eddy simulations. *Physica D: Nonlinear Phenomena, 410*. https://doi.org/10.1016/j.physd.2020.132500.

Meron, E. (2019). Vegetation pattern formation: The mechanisms behind the forms. *Physics Today, 72*. 10.1063/PT.3.4340.

Morrison, H., de Boer, G., Feingold, G., Harrington, J. Y., Shupe, M. D., & Sulia, K. (2012) Resilience of persistent mixed-phase clouds in the Arctic. *Nature Geoscience*, *5*, 11–17. doi:10.1038/ngeo1332.

Mülmenstädt, J., & Feingold, G. (2018). The radiative forcing of aerosol-cloud interactions in liquid clouds: Wrestling and embracing uncertainty. *Current Climate Change Reports*. Springer Verlag. doi.org/10.1007/s40641-018-0089-y.

Murray, J. D. (1988). How the leopard gets its spots. *Scientific American*, *258*, 80–87.

Nober, F. J., & Graf, H. F. (2005). A new convective cloud field model based on principles of self-organisation. *Atmospheric Chemistry and Physics*, *5*, 2749–2759.

Prigogine, I. & Stengers, I. (1984). *Order out of Chaos*. New York: Bantam Books.

Quaas, J., Ming, Y., Menon, S., Takemura, T., Wang, M. Penner, J. E., et al. (2009). Aerosol indirect effects—general circulation model intercomparison and evaluation with satellite data. *Atmospheric Chemistry and Physics, 22*, 8697–8717. https://doi.org/10.5194/acp-9-8697-2009.

Randall, D. A., & Pan, D-M. (1993). Implementation of the Arakawa—Schubert cumulus parametrization with a prognostic closure. In K. A. Emanuel & D. J. Raymond (Eds.), *The Representation of Cumulus Convection in Numerical Models, Meteorological Monographs*, *46*, 137–144. Boston, MA: Amer. Meteorol. Soc.

Sakaguchi, H., Shinomoto, S., & Kuramoto, Y. (1987). Local and global self-entrainments in oscillator lattices. *Progress in Theoretical Physics*, *77*, 1005–1010, doi:10.1143/PTP.77.1005.

Sakradzija, M., Seifert, A. & Heus, T. (2015). Fluctuations in a quasi-stationary shallow cumulus cloud ensemble. *Nonlinear Processes in Geophysics*, *22*, 65–85, 2015. https://www.nonlin-processes-geophys.net/22/65/2015/ doi:10.5194/npg-22-65-2015.

Schrödinger, E. (1967). *What is life? The physical aspect of the living cell*. Cambridge: Cambridge University Press.

Shannon, C. E. (1948). A mathematical theory of communication. *The Bell System Technical Journal*, *27*, 379–423.

Spreitzer, E. J., Marschalik, M. P., & Spichtinger, P. (2017). Subvisible cirrus clouds—A dynamical system approach. *Nonlinear Processes in Geophysics*, *24*, 307–328. https://doi.org/10.5194/npg-24-307-2017.

Smart, A. G. (2011). Minimalist model captures water-cycle complexities. *Physics Today*, *64*, 9, 19. https://doi.org/10.1063/PT.3.1242

Strogatz, S. (2003). *Sync, The emerging science of spontaneous order* (352 pp.). New York: Hachette Books. ISBN 0786868449.

Suzuki, K., Nakajima, T. Y., & Stephens, G. L. (2010). Particle growth and drop collection efficiency of warm clouds as inferred from joint CloudSat and MODIS Observations. *Journal of the Atmospheric Sciences*, *67*, 3019–3032. https://doi.org/10.1175/2010JAS3463.1.

Tarnita, C. E., Bonachela, J. A., Sheffer, E., J., Guyton, A., Coverdale, T. C., Long, R. A. & Pringle, R. M. (2017). A theoretical foundation for multi-scale regular vegetation patterns. *Nature*, *541*, 398–401, https://doi.org/10.1038/nature20801.

Tsonis, A. A., Swanson, K. L., & Roebber, P. J. (2006). What do networks have to do with climate? *Bulletin of the American Meteorological Society*, *87*, 585–596. https://doi.org/10.1175/BAMS-87-5-585.

von Bertalanffy, L. (1950). The theory of open systems in physics and biology. *Science, 111*(2872), 23–29. https://doi.org/10.1126/ science.111.2872.23.

von Foerster, H. (1960). On self-organising systems and their environments. In M. C. Yovits and S. Cameron (Eds.), *Self-organising systems* (pp. 30–50). London: Pergamon Press.

Wacker, U. (1995). Competition of precipitation particles in a model with parameterized cloud microphysics. *Journal of the Atmospheric Sciences*, *52*(2), 2577–2589. https://doi.org/10.1175/1520-0469(1995)052<2577:COPPIA>2.0.CO;2.

Wacker, U. (2006). Nonlinear effects in a conceptual multi-layer cloud model. *Nonlinear Processes in Geophysics*, *13*(1), 99–107.

Wagner, T. M., & Graf, H.-F. (2010). An ensemble cumulus convection parameterization with explicit cloud treatment. *Journal of the Atmospheric Sciences*, *67*, 3854–3869. https://doi.org/10.1175/2010JAS3485.1.

Wang, M., Ghan, S., Liu, X., L'Ecuyer, T. S., Zhang, K., Morrison, H., et al. (2012). Constraining cloud lifetime effects of aerosols using A-Train satellite measurements. *Geophysical Research Letters*, *39*. doi:10.1029/2012GL052204.

Wu, W., & McFarquhar, G. M. (2018). Statistical theory on the functional form of cloud particle size distributions. *Journal of the Atmospheric Sciences, 75*, 2801–2814. https://doi.org/10.1175/JAS-D-17-0164.1

Yair, Y., Aviv, R., & Ravid, G. (2009). Clustering and synchronization of lightning flashes in adjacent thunderstorm cells from lightning network data. *Journal of Geophysical Research*, *114*, D09210. doi:10.1029/2008JD010738.

Yano, J.-I., & Plant, R. (2012). Finite departure from convective quasi-equilibrium: periodic cycle and discharge-recharge mechanism. *Quarterly Journal of the Royal Meteorological Society*, *138*, 626–637. doi: 10.1002/qj.957.

Yano, J.-I., Heymsfield, A. J., & Phillips, V. T. (2016). Size distributions of hydrometeors: Analysis with the maximum entropy principle. *Journal of the Atmospheric Sciences, 73*, 95–108. https://doi.org/10.1175/JAS-D-15-0097.1

Yuan, T. (2011). Cloud macroscopic organization: Order emerging from randomness. *Atmospheric Chemistry and Physics*, *11*, 7483–7490. doi:10.5194/acp-11-7483-2011.

Zangerlé, A., Renard, D., Iriarte, J., Suarez Jimenez, L. E., Adame Montoya, K. L., Juilleret, J., et al. (2016). The Surales, self-organized earth-mound landscapes made by earthworms in a seasonal tropical wetland. *PLoS ONE*, *1*, e0154269. doi:10.1371/journal.pone.0154269.

Part III

Measurements, Model Evaluation, and Model-Measurement Integration

14

Ground-Based Remote-Sensing of Key Properties

Katia Lamer[1], Pavlos Kollias[1,2], Vassilis Amiridis[3], Eleni Marinou[3], Ulrich Loehnert[4], Sabrina Schnitt[4], and Allison McComiskey[1]

ABSTRACT

Over the past two decades, the sophistication, and number of ground-based networks that have zenith-pointing millimeter radars, lidars, and radiometers as their cornerstones have increased significantly. Their measurements capture the structure of clouds, and precipitation, as well as the thermodynamical and dynamical state of the clear boundary layer. The vertically comprehensive and temporally continuous measurements that ground-based networks provide have filled a gap that was left by spaceborne and aircraft platforms. Toward improving our understanding of fast physics and its representation in numerical models, sophisticated retrieval algorithms have been developed to provide estimates of geophysical quantities. In parallel, forward simulators have been used to convert model output to measurements for apples-to-apples model–experiment comparisons, to quantify retrieval uncertainty, and to optimize experimental design. The multiparametric observations captured by ground-based observatories have supported numerical modeling in many ways, such as parameterization development, forcing datasets, and benchmarking. That said, the scale gap between high-resolution observations and large-scale models particularly complicates their intercomparison. Although methods exist to help bridge this gap, issues related to the representativeness of point observations across scales remain. While scanning sensors and spatially distributed instrument networks may help bridge this gap, further improvements in observing systems and model computational capabilities are desirable.

14.1. THE EMERGENCE OF GROUND-BASED OBSERVATORIES

Advancements achieved over the years have enabled scientists to deploy in situ and remote-sensing instruments on spaceborne, airborne, and ground-based platforms capable of covering scales ranging from several kilometers to only a few millimeters. The observations collected by these instruments, with their unique strengths and limitations, have contributed to improving our fundamental understanding of fast physics and its representation in numerical models.

For instance, passive sensors operating from spaceborne platforms offer a global view of cloud properties such as size, organization, cover, and liquid water path (LWP); however, these sensors are unable to provide information about the vertical distribution of such properties (Karlsson & Devasthale, 2018). Information about the vertical location of clouds has been commonly provided by active sensors such as the CloudSat-Cloud Profiling

[1]*Department of Environmental and Climate Sciences, Brookhaven National Laboratory, Upton, NY, USA*
[2]*School of Marine and Atmospheric Science, Stony Brook University, Stony Brook, NY, USA*
[3]*Institute for Astronomy, Astrophysics, Space Applications and Remote Sensing, National Observatory of Athens, Athens, Greece*
[4]*Institute of Geophysics and Meteorology, University of Cologne, Cologne, Germany*

Fast Processes in Large-Scale Atmospheric Models: Progress, Challenges, and Opportunities, Geophysical Monograph 282, First Edition.
Edited by Yangang Liu and Pavlos Kollias.

DOI: 10.1002/9781119529019.ch14

Radar (Stephens et al., 2018) and the CALIPSO lidar (Winker et al., 2003). That being said, even state-of-the-art spaceborne radar systems suffer from surface clutter, which limits their ability to detect clouds and precipitation in the lowest kilometer of the atmosphere (Burns et al., 2016; Huang et al., 2012a; Huang et al., 2012b); their coarse vertical resolution (>250 m) is also often insufficient to accurately characterize shallow clouds leaving a gap in our understanding of low-level clouds and precipitation (Battaglia et al., 2020; Lamer et al., 2020b).

Although unable to provide a global perspective, airborne sensors have provided much needed fine resolution information about the vertical structure of shallow clouds (e.g., Damiani et al., 2006; Hobbs et al., 2001; Vali et al., 1995). Their capability to collect one-dimensional in situ observations (e.g., Bailey & Hallett, 2009; Heymsfield & Miloshevich, 1993; Lawson et al., 2008; Rangno & Hobbs, 2001) with collocated active and passive measurements has enabled breakthroughs in our understanding of cloud microphysics (e.g., Wang et al., 2012). However, the temporal sampling of airborne observations is limited, and key properties in the temporal domain including diurnal variations generally elude these platforms.

Fixed ground-based remote sensors can capture high-resolution (<30 m) diurnal information about the vertical structure of clouds, precipitation, aerosols, and clear air and can collect complementary multiparametric measurements not available from spaceborne platforms including velocity and polarimetric information. Measurements collected by ground-based sensors have been used for a range of application including the derivation of large-scale forcing to drive large-scale model simulations and in data assimilation to improve numerical weather forecasts. That being said, their comprehensive nature makes them especially well suited for model evaluation and fast physics parametrization improvement because of the intricacies and intertwined nature of fast physics process.

Over the past half century, suites of ground-based sensors have been deployed to support hundreds of large-scale field experiments in various climatological regimes. For example, the Surface Heat Budget of the Arctic Ocean (SHEBA) experiment took place in the Artic region (Uttal et al., 2002); the Rain in Shallow Cumulus Over the Ocean (RICO) experiment was hosted in the Eastern Atlantic (Rauber et al., 2007); the Midlatitude Continental Convective Clouds experiment like the name suggests occurred in the midlatitude region [MC3E (Jensen et al., 2016)]; the Marine ARM GCSS Pacific Cross-Section Intercomparison Investigation of Clouds (MAGIC) ship-based experiment covered the cumulus-stratocumulus transition region (Zhou et al., 2015); the ARM Madden-Julian Oscillation Investigation Experiment/Dynamics of the MJO (AMIE/DYNAMO) and the Cirrus Regional Study of Tropical Anvils and Cirrus Layers Florida Area Cirrus Experiment (CRYSTAL-FACE) experiments both took place in tropical regions (Evans et al., 2005; Feng et al., 2015); and the Southern Ocean Clouds, Radiation, Aerosol Transport Experimental Study (SOCRATES) experiment explored the Southern ocean region (McFarquhar et al., 2014). One drawback of such large multiagency, international field experiments is that they tend to be short term (1–3 months) and intermittent (a large international experiment happens every 3–5 years). While we recognize the contribution of short campaigns, describing their setup and achievements is beyond the scope of this chapter and the reader is directed toward the references listed above for additional information.

This chapter instead emphasizes the science enabled by permanent, ground-based observatories that collect continuous long-term observations. The U.S. Department of Energy (DoE) Atmospheric Radiation Measurement (ARM) program represents the first large organized effort to establish continuously operating ground-based observatories (Mather & Voyles, 2013; Stokes & Schwartz, 1994). The ARM concept was conceived in the late-1980s early-1990s when mounting evidence pointed to cloud radiative feedbacks as the single most important factor affecting the future climate response to human activity. ARM's first well-instrumented observatory was established in 1992 in the Southern Great Plains (SGP) of Oklahoma, U.S.A. At first, its instrumentation focused more on radiation and cloud measurements, but today ARM collects comprehensive measurements of radiation, aerosol, clouds, precipitation, and clear air properties (macrophysical, dynamical, and thermodynamical). ARM's requirement to operate year round, 7 days a week and 24 hours a day, led to technological breakthroughs in ground-based remote-sensing technology such as the development of unattended millimeter-wavelength radars (e.g., Moran et al., 1998). Today, the ARM program has become a national science user facility and operates several fixed and relocatable observatories including the SGP observatory in Oklahoma, the North Slope of Alaska (NSA) observatory in Barrow, and the Eastern North Atlantic (ENA) observatory on Graciosa Island (Miller et al., 2016).

In the 2000s, a similar ground-based multisensors observatory concept also emerged in Europe (the CloudNet project; Illingworth et al., 2007). CloudNet initially formed with three observatories: Chilbolton in the United Kingdom, Palaiseau in France, and Cabauw in the Netherlands, and focused on cloud and precipitation measurements. Despite its later start, the CloudNet project quickly led to the development of robust postprocessing algorithms and geophysical retrievals (the CloudNet suite of algorithms) enabling long-term comparisons between ground-based measurements and several numerical weather prediction (NWP) models (Illingworth et al., 2007). Today, the number of European ground-based observatories has significantly increased with the noticeable additions of supersites in Cologne, Germany (Löhnert et al., 2015); Hyytiälä, Finland (Hirsikko et al., 2014); Barbados, Caribbean (Stevens et al., 2016); and relocatable observing capabilities such as the Karlsruhe Institute of Technology mobile KITcube (Kalthoff et al., 2013) and the Leipzig Institute for Tropospheric Research, TROPOS (https://www.tropos.de/en/). It is however important to note that, unlike ARM, the CloudNet's instrument network remains nonstandardized across its various observatories.

14.2. CORNERSTONE GROUND-BASED REMOTE SENSORS

Gathering comprehensive and robust measurements of the properties of cloud, precipitation, aerosol, clear air, and radiation requires the operation of multiple collocated sensors. From the early beginning a handful of instruments were identified as key components for any site seeking the designation of "cloud observatory": a zenith-pointing millimeter-wavelength "cloud-sensitive" radar, a zenith-pointing "aerosol-sensitive" backscatter lidar, and a "liquid-water-sensitive" microwave radiometer (Illingworth et al., 2007). Since then, the number of instruments operated by ground-based cloud observatories and their sophistication have expanded considerably. A typical modern-day ARM observatory operates over 100 different sensors collecting hundreds of different measurements. Nevertheless, the three aforementioned sensors remain critical for characterizing the atmospheric state, and technological advancements affecting these sensors are the main focus of this section.

14.2.1. Cloud and Precipitation Radars

Generalities

The configuration of modern-day radars is fairly universal; its main components are a transmitter (Tx), a receiver (Rx), and an antenna (Rinehart, 2004). In a nutshell, the radar transmitter generates pulses of electromagnetic waves at a specific frequency/wavelength (depicted by the waves in Figure 14.1a); the antenna focuses these pulses into precise observation volumes (depicted by the cylinders in Figure 14.1a and b) that contain a finite number of atmospheric targets (on the order of 10^6 to 10^{10} depicted as black dots in Figure 14.1b). Each target within the observation volume scatters portion of the incident radar signal, and their backscatter contributions are summed incoherently to provide the radar return signal that is measured by the receiver. Since pulses travel at the known speed of light (c), the time interval (t) between pulse transmission, backscattering, and subsequent reception can be related to the distance of targets along the radar beam path (r) as follows: $r = ct/2$. The 1/2 factor arises because the energy must make a round trip from transmission by the radar to interaction with the target to reception by the radar.

The radar continuously emits pulses at intervals dictated by its pulse repetition time (PRT). Through the averaging of a large number of received pulses (~5,000–10,000 for zenith pointing cloud radars), the uncertainty introduced by the scattering process and from instrument noise, formally referred to as white noise, can be reduced for the estimation of radar reflectivity. Radar reflectivity (η in mm^2 m^{-3}) depends on the radar wavelength. That being said, by taking into consideration the wavelength-dependent response of liquid water (through the K factor), it can be converted to the more common equivalent reflectivity factor $Z_e = (\lambda^4 \cdot \eta)/(\pi^5 \cdot K^2)$ (Z_e in mm^6 m^{-3}). The equivalent reflectivity factor is usually expressed in logarithmic scale: $dBZ_e = 10 \cdot \log_{10}(Z_e)$ and is commonly referred to as radar reflectivity in everyday discourse and will be for the remainder of this chapter. In the Rayleigh scattering regime, where transmitted wavelength is much larger than the size of the targets, Z_e also becomes independent of the radar wavelength and is equal to the sixth moment of the hydrometeor size distribution $Z_e = \int_0^\infty N(D)D^6 dD$, where D is the hydrometeor diameter (in units of m) and $N(D)$ is the hydrometeor size distribution (in units of mm^{-1} m^{-3}). This implies that larger hydrometeors more readily backscatter (or influence) radars signals than smaller hydrometeors, which makes it challenging to observe the properties of, for example, cloud droplets coexisting with drizzle or rain.

Doppler radars are able to exploit the phase shift from pulse to pulse to extract information about the motion of the hydrometeors within the radar sampling volume relative to the radar. Figure 14.1b depicts factors contributing to the observe motion, which include the hydrometeor fall speed (small arrows), turbulence at scales smaller than the radar sampling volume (dashed oval and arrows), and turbulence at scales larger than the radar sampling volume (dotted wave and arrow). Pulse-pairing techniques can be used to estimate the reflectivity-weighted motion of the particles present within the observation volume. Alternatively, a radar Doppler spectrum can be estimated by performing a Fourier decomposition of a large number of received pulses (~5,000–10,000 for zenith pointing cloud radars) (Figure 14.1c). The radar Doppler spectrum describes the distribution of received power as a

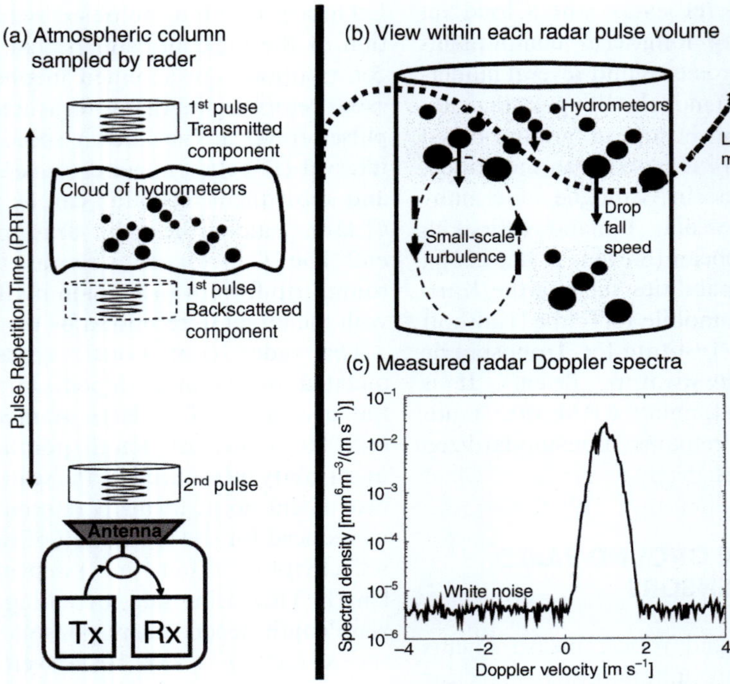

Figure 14.1 (a) Radars transmit pulses of electromagnetic waves using a transmitter (Tx) at intervals dictated by the pulse repetition time (PRT). These pulses are focused by the radar antenna creating distinct observation volumes (solid cylinders) for accurate pointing. The pulses travel across the atmospheric column and interact with hydrometeors (black dots) that backscatter portions of the transmitted pulse (dashed cylinder) back toward the radar. The radar receiver (Rx) measures the backscattered waves amplitude and phase. (b) The amplitude and phase of the backscattered wave are influence by the size, number density, shape, and motion of the hydrometeors (black dots) present within the radar pulse volume (solid cylinder). Factors influencing hydrometeor motion include their fall speed (down-pointing solid arrow), turbulence within the radar pulse volume (dashed oval and arrows), and turbulence occurring at larger scales (dotted wave and arrow). (c) Fourier decomposition can be used to generate a radar Doppler spectrum from a collection of several received pulses; albeit noisy, it provides information about the distribution of received power as a function of velocity relative to the radar (i.e., Doppler velocity).

function of velocity across the range of observable velocities bound by what is known as the Nyquist velocity. A radar's Nyquist velocity, or largest observable velocity, is proportional to its transmitted wavelength and four times its PRT. It is interesting to point out that following white noise removal, the integral of the radar Doppler spectrum gives the previously defined equivalent radar reflectivity factor, the first moment of the radar Doppler spectrum is equal to the mean Doppler velocity, and the second moment to the radar Doppler spectrum width. Until recently, storage limitations only enabled the recording of these three first moments of the radar Doppler spectrum. Recent advancements lead to the storage of the entire Doppler spectrum enabling the estimation of higher moments such as skewness or the application of novel spectral decomposition techniques opening new cloud and precipitation retrieval research avenues (additional information about retrieval techniques is provided in section 14.3).

Particularities

Radars that operate at wavelengths much longer than the size of atmospheric targets can capitalize on phenomena specific to the Rayleigh scattering regime. In the Rayleigh regime, backscatter cross section is not directly equal to geometrical size but rather it is inversely proportional to the fourth power of the transmitted wavelength. This implies that shorter wavelength radars generate higher hydrometeor backscattering cross section and as such require less power to generate the same backscattering signal. For example, the backscattering cross section of a hydrometeor would be a factor of $10^{4.2}$ (42 dB) greater at 8.6-mm wavelength and 10^6 greater (60 dB) at 3.2-mm wavelength than it would be at a 10-cm wavelength (Kollias et al., 2007). For this reason, millimeter wavelength radars can afford to use low power transmitters while longer wavelength radars require powerful transmitters for the observation of clouds and light precipitation, hence the designation of millimeter-wavelength radars as cloud radars. On the flip side, the fact that millimeter wavelengths are highly backscattered by cloud and precipitation also implies that they are more quickly attenuated by cloud and precipitation such that longer wavelength radars can more readily penetrate deep precipitation systems, hence the designation of centimeter wavelength radar as weather radars. It is important to consider that absorption by atmospheric gases also limits the useful operational range of cloud radars. To be exact, a humid boundary layer with a water vapor mixing ratio of 20–25 g kg^{-1} can cause a 7.5% loss of radar signal power per kilometer at 8.6-mm wavelengths and as much as 37% at 3.2-mm wavelength. Furthermore, it is worth noting that any other atmospheric target of size similar to cloud and precipitation hydrometeors can be detected by radars, which includes insects for cloud radars and birds for weather radars.

The antenna sizes of cloud and weather radars also differ widely yet both are designed to achieve the same ideal balance. Ideally, the radar observation volume should be large enough to allow for rapid sampling of the entire atmosphere but small enough to avoid encompassing inhomogeneous regions that would cause issues known as partial beam filling. The angular dimension of this volume, formally referred to as the radars' 3-dB antenna beamwidth, is defined as the angular separation between points where the transmitted intensity has fallen to half its maximum value or 3 dB below the maximum. It is proportional to the transmitted wavelength but inversely proportional to the geometrical diameter of the radar antenna. For reference, a 8.6-mm wavelength cloud radar with a 1.8-m diameter antenna can achieve an antenna 3-dB beamwidth of 0.3°, while a 10-cm wavelength weather radar with an 8.5-m diameter antenna can at best achieve an antenna 3-dB beamwidth of 0.96°. This relationship also explains why cloud radars are more lightweight and mobile than their weather radar counterparts and as such more suitable for deployments on ships, aircraft, and satellites.

14.2.2. Lidars

In parallel, advancements in laser sources and photon detectors led to the development of sophisticated, highly reliable, portable light detection, and ranging (LiDAR) systems (Weitkamp, 2006). Just like radars, lidars use coherent light pulses to provide range-resolved information about the atmosphere. The lidar sensor architecture is similar to that of a radar with the key components being a laser source (akin to the radar transmitter), a telescope for receiving backscattered light (akin to the radar antenna), and an optical detector to analyze the received backscattered signal (akin to the radar receiver).

Different lidar techniques make use of specific interactions between the emitted radiation and atmospheric constituents. Elastic-backscatter lidars receive light scattered by atoms, molecules, aerosols, and hydrometeors that has the same wavelength as the transmitted pulse. Two types of elastic scattering occur: (i) Rayleigh scattering when the particle size (e.g., atmospheric molecules) is very small compared to the lidar wavelength and (ii) Mie scattering when the particle size (e.g., aerosols and hydrometeors) is comparable or larger than the lidar wavelength. For this reason, typical elastic-backscatter lidar channels (e.g., 1,064, 532, and 355 nm) are equally rapidly attenuated by clouds with near-infrared signals being especially sensitive to aerosols and visible and ultraviolet signals being sensitive to both aerosol and molecules. Raman lidars transmit short pulses of visible and ultraviolet light and make use of the Raman effect in which light is inelastically scattered by atmospheric N_2, O_2, and H_2O molecules. Several other detection channels can be configured to measure elastically backscattered light from atmospheric aerosol. The raw signals from these detection channels can then be combined and processed to yield measurements of water vapor mixing ratio, temperature, aerosol backscatter coefficient, extinction, and depolarization ratio. (e.g., Ansmann et al., 1992; Baars et al., 2016; Balin et al., 2004; Lange et al., 2019). Differential absorption lidars (DIALs) exploit molecular absorption properties in two wavelengths by emitting one wavelength in the peak of an absorption band of the molecule of interest, such as O_3, NO_2, CH_4, NH_4, and other gases, and the second in a low-absorption region. The difference in the backscattered signal at these two wavelengths provides information about the concentration of the gas molecules along range (e.g., Ehret et al., 2008; Fix et al., 2019). Polarization lidars, capable of transmitting polarized light, may also exploit the difference in scattering response of nonspherical particles and are particularly useful in the investigation of cirrus clouds, desert dust, and volcanic ash layers (e.g., Freudenthaler et al., 2009; Gasteiger et al., 2011; Sassen, 1991). Polarization information is provided by elastic lidars such as micro pulse lidars (MPL) or by Raman lidars and high spectral resolution lidars (HSRLs). Unlike MPLs, HSRL and Raman systems measure aerosol extinction at 355 and 532 nm and aerosol backscatter and depolarization at 355, 532, and 1,064 nm (e.g., Bruneau et al., 2015; Eloranta, 2005; Hair et al., 2008; Serikov et al., 2018; Wirth et al., 2009). Recent advancements also enabled the construction of Doppler lidars that can measure the velocity of particles contained within their observation volume relative to the lidar (e.g., Lux et al., 2020; Pearson et al., 2002). Like Doppler radars, Doppler lidars can record backscattered power as a function of velocity. This capability however remains largely unexploited since most sites operating Doppler lidar simply record the mean Doppler velocity.

14.2.3. Radiometers

Radiometers are passive instruments that measure radiation emitted or scattered by all types of matter at the shortwave, infrared, and microwave wavelengths. Due to their longer wavelength (submillimeter to centimeter), microwave radiometers can detect signal through clouds and precipitation and thus can provide continuous information about water vapor and liquid in the atmospheric column. While microwave radiometers are more commonly used to estimate column-integrated quantities such as integrated water vapor and liquid water path, they may be used to profile the thermodynamic structure (temperature and water vapor) of the planetary boundary layer by combining wavelength-dependent and elevation angle-dependent emission measurements at the center and on the flanks of the oxygen and water vapor absorption bands.

In the infrared region, the most sophisticated radiometer is the atmospheric emitted radiance interferometer (AERI), a Fourier transform spectrometer operating in the spectral range 3.3–19.2 μm at a resolution of $0.5\,cm^{-1}$. The AERI has sufficient spectral resolution to discriminate among gases (e.g., CO_2 and H_2O) and suspended matter (e.g., aerosols, water droplets, and ice crystals) and thus obtains vertical profiles of tropospheric temperature and water vapor, as well as trace gas measurements and downwelling infrared spectral signatures of clouds and aerosols.

In the shortwave and near-IR region, the Aerosol Robotic Network (AERONET), which formed in the early-1990s to serve NASA and other Earth System satellite programs (Holben et al., 2018), has been operating one of the most widely used sun photometers, the CIMEL. The CIMEL sun photometer is a multichannel, automatic sun-and-sky scanning radiometer that measures the direct solar irradiance and sky radiance at discrete wavelengths in the visible and near-infrared parts of the spectrum. Sun-photometer calibration depends on lamp calibrations that may vary. Complementary to sun photometers are multi-filter rotating shadowband radiometers (MFRSR; Harrison et al., 1994) such as those deployed by both DOE and NOAA programs, which provide aerosol properties by measuring the total and diffuse global horizontal radiation (Kassianov et al., 2007). The Langley calibrated MFRSR can produce high accuracy measurement of irradiance with the proper observing conditions. The MFRSR can also be used to retrieve cloud cover and optical properties during overcast conditions (Wang & Min, 2008).

14.3. RETRIEVING GEOPHYSICAL PROPERTIES FROM REMOTE-SENSING MEASUREMENTS

Interpreting remotely sense signal in terms of geophysical properties such as those simulated by atmospheric numerical models remains challenging because they represent the amalgamed properties of populations of targets of potentially different size, shape, and phase. Adding to the challenge is the fact that measurements are related to moments of these particle size distributions that differ from those of most interest; the zeroth (total number concentration) and third (mass) moments being most desirable and the sixth and second being recorded by radars and lidar, respectively.

Over the years, a number of algorithms of varying degrees of complexity have been developed to retrieve the geophysical attributes of clouds, precipitation, aerosol, and the clear boundary layer from such measurements. All involve making assumptions about certain particles properties and/or about the atmospheric state, which introduces uncertainty.

This section focuses on covering the basis of retrieval algorithms applicable to cloud systems in general and to warm and ice/mixed-phased system in particular; readers are directed to sources listed therein for additional information about the assumptions and accuracy associated with the retrieval techniques covered.

14.3.1. Locating Clouds, Rain, Supercooled Liquid, and Ice

Characterizing cloud systems generally begins with determining their location and the type of hydrometeors they contain (Hogan et al., 2005a). While millimeter radars alone can be used to determine the location of most cloud and precipitation hydrometeors, they cannot be used to distinguish between the two. Instead lidars, which are more sensitive to small particles, are typically used to infer the location of the base of liquid clouds and the location of supercooled liquid layers. Some methods place the base of the liquid layer at the height at which the backscattered signal reaches its maximum value (Pal et al., 1992), which may be somewhat inside cloud or supercooled liquid layers. Others instead examine the vertical gradient of backscatter based on the general idea that the cloud-clear air interface generates strong negative and positive slopes in the lidar backscatter signal (Vaughan et al., 2009; Wang & Sassen, 2001); these methods however need careful tuning to reject small backscatter enhancements caused by noise and aerosol returns. Lidars are also used to detect thin hydrometeor layer as well as elevated cirrus-type layers and fair-weather cumulus over land; all of which often contain very small cloud droplets. For example, Figure 14.2a shows the

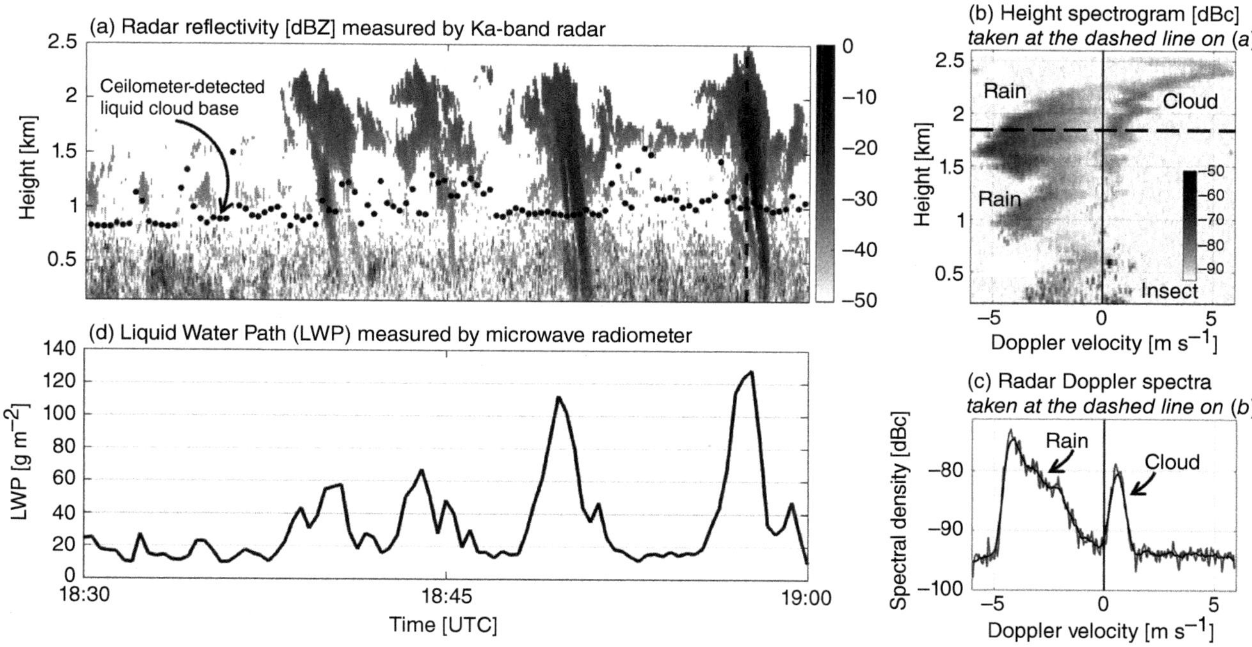

Figure 14.2 On 28 January 2013 at the Southern Great Planes (SGP) observatory, (a) the Ka-band ARM Zenith Radar (KAZR) measured the reflectivity signal of a sequence of precipitating cumuli. Estimates of cloud-base height from ceilometer lidar (black dots) allow us to visually separate the clouds (signal above the dots) from the subcloud layer (signal below the dots). Although informative, radar reflectivity remains an integrated quantity and as such it cannot be used to distinguish multiple targets present within the same radar volume. Examining a collection of radar Doppler spectra collected (b) at various heights along the atmospheric column along the position of the dashed line in (a) and (c) at a specific height in that column along the dashed line in (b) allows us to determine that both rain and cloud droplets coexisted within this cloud and that the subcloud layer was populated with both rain and insects. (d) Measurements from a microwave radiometer are used to retrieve the integrated liquid water content of these clouds.

location of the base of liquid clouds (depicted by the black dots) as determined by a zenith pointing ceilometer lidar. Figure 14.2a also shows radar reflectivity measured by a collocated zenith pointing millimeter-wavelength radar (depicted by the background shading). The radar signals above and below the ceilometer-determined cloud-base height reveal the presence of a sequence of shallow precipitating cumulus clouds. The absence of radar detections above certain ceilometer-detected cloud bases, for example around 18:43 UTC, reinforces the reality that no sensor can detect all atmospheric liquid water as radars can only be used to describe the internal structure of those clouds containing hydrometeors of size and number concentration above their detection limit (Lamer et al., 2018; Lamer & Kollias, 2015). One thing to consider though is that ground-based observatories, with their capability to sample the lowest 1–2 km of the atmosphere at high vertical resolution, are particularly well suited for the characterization of shallow convective clouds that have horizontal and vertical dimensions that are often undersampled by satellite sensors (Lamer et al., 2020a). When ceilometer lidar measurements are unavailable and radar measurements may be unreliable, microwave radiometer liquid water path retrievals can also be used to confirm the presence of liquid clouds overhead (e.g., Figure 14.2d).

Determining the phase of the hydrometeors contained within cloud systems is also critical as many quantitative retrievals rely on sphericity assumptions that are most appropriate for liquid clouds and precipitation. Single-parameter methods have relied on the height of the 0 °C isotherm or the height of the melting layer to classify radar echoes between liquid and ice phases; the melting layer is typically identifiable by its distinct radar reflectivity enhancement or "bright band" signature (Fabry & Zawadzki, 1995) and is also visible in radar polarimetric fields (Z_{DR} and especially ρ_{hv}) since those are sensitive to nonspherical hydrometeors. Improvements on these "region"-focused classifications have been made by combining temperature and radar linear depolarization ratio (LDR) measurements (e.g., Sandford et al., 2017). "Echo"-focused classifications were initially developed for weather radars; many involve fuzzy-logic algorithms that combine horizontal polarization (Z_H), differential reflectivity (Z_{DR}), differential propagation phase shift (Φ_{DP}), specific differential phase (K_{DP}), and copolar correlation coefficient (ρ_{hv}) measurements to distinguish between biological scatterers, ground clutter, light/moderate rain, heavy rain, big drops, rain/hail mixture, dry snow, wet snow, ice crystals, and graupel (e.g., Dolan & Rutledge, 2009; Dolan et al., 2013; Liu & Chandrasekar, 2000; Park et al., 2009; Ryzhkov et al., 2005; Straka et al., 2000; Vivekanandan et al., 1999); in a nutshell, these algorithms rely on Z_{DR} and K_{DP} for information about the size, aspect ratio, bulk density, and orientation of particles and on K_{DP} for particle number concentration (Kumjian, 2013). Lidar signals have also been used to determine the phase of observed echoes. Hogan et al. (2003) used regions of high lidar backscattered power as an indicator for the presence of liquid droplets. To avoid some of the misclassifications encountered when using backscattered power alone, others instead relied on a combination of lidar backscattered power and lidar linear depolarization ratio to separate lidar echoes caused by liquid droplets from those caused by ice crystals (Hu et al., 2010; Hu et al., 2007; Hu et al., 2009; Sassen, 1991; Yoshida et al., 2010). Figure 14.3 shows an example of the CloudNet target classification product, which relies on both radar reflectivity (Figure 14.3a) and lidar backscatter (Figure 14.3b) signals to identify cloud droplets, drizzle and raindrops, melting ice, supercooled liquid, ice, and aerosols (Figure 14.3c) (Hogan & O'Connor, 2004). Although the combination of radar and lidar backscatter is useful for the identification of mixed-phase conditions, their combined extent remains limited to single-layer clouds because of lidar signal attenuation. To overcome this limitation, Shupe (2007) proposed a technique that alternatively relies on Doppler spectrum width measurements to infer the presence of supercooled water and mixed-phase conditions in the upper part of multilayer systems.

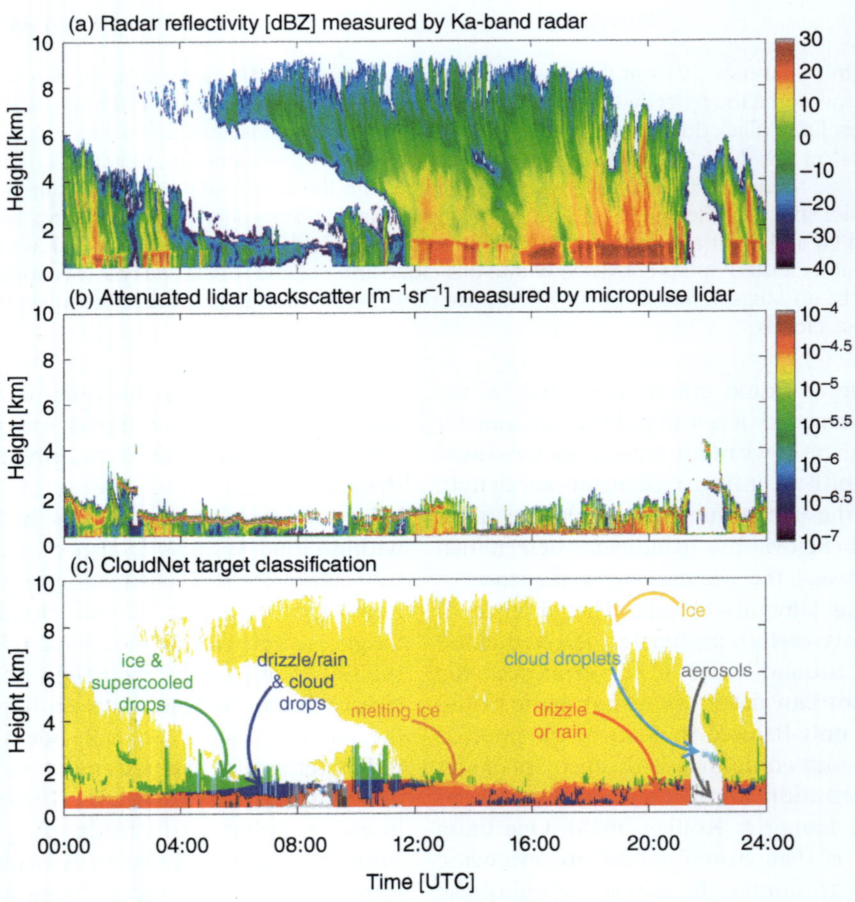

Figure 14.3 On 5 March 2020 at the Jülich Observatory for Cloud Evolution (JOYCE), a combination of (a) Ka-band radar reflectivity and (b) micropulse lidar attenuated backscatter was used as input to (c) the CloudNet target classification algorithm to identify ice (yellow), supercooled drops (green), melting ice (orange), drizzle/rain and cloud drops (royal blue), cloud droplets (light blue), drizzle or rain (red), and aerosols (light gray). Note that this classification algorithm requires other measurements (not shown).

Unfortunately, besides cloud and precipitation, millimeter-wavelength radars are also sensitive to the presence of insects that are similar in size to raindrops. Distinguishing insect from hydrometeor signals is important given that insects may penetrate clouds and bias observations (Chandra et al., 2013; Lamer & Kollias, 2015). Algorithms interpreting the spatial extent of echoes (Lamer & Kollias, 2015) in conjunction with radar reflectivity thresholds (Chandra et al., 2010) or polarimetric signatures (Luke et al., 2008) have been used to separate cloud from insects echoes near ground level (e.g., Figure 14.2a).

Finally, it is important to keep in mind that it is not uncommon for individual observation volumes to contain different particle populations that do not all dominate the return radar signal. The coexistence of multiple noise separated peak in the radar Doppler spectrum has been associated with the coexistence of multiple particle populations within the same observation volume. Moreover, the distinct properties of each population has been assed based on their particular velocity along the Doppler spectrum (Kalesse et al., 2013; Kollias et al., 2011; Tridon et al., 2013; Yu et al., 2014). Figure 14.2b shows a height spectrogram or collection of Doppler spectra collected at consecutive heights here along the dashed line presented in Figure 14.2a. The white background indicates the areas of the radar Doppler spectra that contain only receiver noise, while the darker areas indicate where atmospheric returns contribute to the observed spectral density. Clearly, the recorded radar Doppler spectra are not unimodal. Below 0.8 km, the radar Doppler spectra are composed of several narrow spectral peaks, which is a classical spectral signature of insects (Luke et al., 2008). Between 0.8 and 2.5 km, bimodal radar Doppler spectra are observed; an example of one of the spectrum collected at 1.85 km is shown in Figure 14.2c. Since we know that this cloud system is of liquid phase, we can conclude that the rightmost peak near zero velocity is related to the presence of cloud droplets since those have negligible fall speed and are effectively air motion tracers and that the left peak is associated with the presence of fast-falling raindrops. For reference, the height of the spectral peak can be used to determine which cloud or rain dominates the observed radar signals. Similar conclusions can be drawn by estimating the skewness of the entire Doppler spectra (Kollias et al., 2011; Luke & Kollias, 2013). At subzero temperatures, a thorough examination of the entire Doppler spectra can also be used to speculate about the nature of the multiple hydrometeor populations present in ice (Kalesse et al., 2013) and mixed-phased systems (Yu et al., 2014) as long as turbulence does not overwhelm the unique microphysical signature of different hydrometeor populations (i.e., there is a need to observe noise-separated spectral peaks).

14.3.2. Characterizing Warm Cloud Systems

Cloud systems composed solely of liquid cloud droplets with or without drizzle and rain are typically referred to as "warm" cloud systems. Observations of these systems are generally easier to interpret since the particles they contain are mostly spherical and our understanding of the scattering properties of liquid spherical particles is relatively complete.

When carefully calibrated using reference cold and hot loads, the measured radiance from microwave radiometers can be used to estimate the total liquid mass in the atmospheric column [i.e., liquid water path (LWP); (Crewell & Löhnert, 2003; Liljegren et al., 2001; Turner et al., 2007; Wang, 2007)]. While not range-resolved, these estimates can be used to confirm the presence of liquid in the atmospheric column and to validate more complex range-resolved liquid water content (LWC) retrievals. Microwave radiometer-based LWC retrievals have also been used to constrain methods that rely on radar reflectivity to determine the shape of the LWC profile either by assuming a linear relationships between the two (Dong & Mace, 2003; Frisch et al., 1998; Vivekanandan et al., 1997) or using optimal estimation frameworks (Löhnert et al., 2004; Löhnert et al., 2001; McFarlane et al., 2002).

In precipitating clouds, the coexistence of cloud and precipitation size droplets generally complicates quantitative retrievals. Küchler et al. (2018) proposed a hybrid technique that relies on radar Doppler spectral skewness to identify drizzle-dominated cloud portions where they assume LWC to increase linearly with height. In the absence of independent LWP estimates, other approaches to retrieve LWC include deriving fixed relationships between LWC and radar reflectivity either from aircraft in situ observations (Atlas, 1954; Fox & Illingworth, 1997; Sauvageot & Omar, 1987) or numerical models (Liao & Sassen, 1994). These techniques, like all the aforementioned, rely on observations from a single radar, make assumptions about the cloud droplet particle size distribution, and require absolute radar reflectivity calibration that is not always easy to achieve. Techniques to exploit differences in the absorption (attenuation) or scattering occurring at a different frequency, also known as dual-wavelength ratio techniques, can retrieve LWC in nonprecipitation or drizzling clouds without making a priori assumptions about the shape of the LWC profile or the droplet size distribution. Dual-wavelength ratio techniques also do not require absolutely calibrated radar reflectivity measurements but require the operation of two accurately collocated radars (Eccles & Mueller, 1971; Ellis & Vivekanandan, 2011; Hogan et al., 2005b; Hogan & O'Connor, 2004; Huang et al., 2009; Martner, 1993; Tuttle & Rinehart, 1983; Vivekanandan et al., 1999; Williams & Vivekanandan, 2007; Zhu et al., 2019).

Radar measurements have also been used to retrieve rainfall rate and the particle size distribution of raindrops. Short-term studies have capitalized on aircraft in situ measurements to relate radar reflectivity to rain rate (Comstock et al., 2004; Vali et al., 1998; VanZanten et al., 2005; Wood, 2005). However, these relationships require assumptions about the drizzle-drop size distribution that introduces appreciable uncertainty. Also, such retrievals require specific conditions where the drizzle drops are large enough to dominate the radar reflectivity signal but small enough so as not to generate non-Rayleigh signals nor significant attenuation. When larger raindrops are present, it is possible to exploit attenuation to retrieve rainfall rate (Matrosov, 2005; Matrosov et al., 2006) and to exploit non-Rayleigh signals to retrieve the raindrop size distribution (Giangrande et al., 2010; Kollias et al., 2002). Polarimetric measurements, some of which are immune to radar miscalibration and attenuation (e.g., K_{DP}), have alternatively been used to correct radar reflectivity biases and/or to directly estimate rainfall rate using direct relationships (Giangrande & Ryzhkov, 2008; Ryzhkov et al., 2005). Everything considered, retrievals based on observations collected at different wavelengths tend to be more accurate. For instance, dual-wavelength ratio estimates by radar have been used to retrieve the raindrop particle size distribution (Giangrande et al., 2012; Tridon & Battaglia, 2015; Tridon et al., 2017). Similarly, the radar reflectivity and lidar backscatter ratio (Intrieri et al., 1993; O'Connor et al., 2005) as well as the lidar color ratio (Lolli et al., 2013; Westbrook et al., 2010) have been exploited to retrieve information about the microphysical properties of drizzle in the subcloud layer.

While most retrieval algorithms are tailored toward either cloud or rain, synthetic observing system simulation experiments (OSSEs) suggest that is it possible to jointly retrieve cloud and drizzle properties using optimum estimation frameworks ingesting ground-based radar, lidar, and sun-photometer observations (Fielding et al., 2015) or microwave radiometer observations (Rusli et al., 2017); these retrievals are constructed based on the fact that active remote sensors can constrain the drizzle-drop mode while passive remote sensor can constrain the cloud droplet mode (Chiu et al., 2012), both of which coexist within drizzling clouds.

14.3.3. Characterizing Mixed-Phased and Ice Cloud and Precipitation Systems

Gaps in our understanding of the properties of ice crystals that impact their radiative properties limit the accuracy of geophysical retrievals targeting ice-containing hydrometeor systems (Comstock et al., 2007; Protat et al., 2007).

Gathering information about the shape of pristine ice crystals and aggregates, which is intimately tied to their formation process, would arguably help improve quantitative retrievals.

One of the simplest approaches to identifying an active aggregation process marked by large fast-falling hydrometeors involves estimating the relationship between radar reflectivity (a proxy for hydrometeor mass) and mean Doppler velocity (a proxy for hydrometeor fall speed) at various heights along observed profiles (Westbrook et al., 2007). A more recent study by Kalesse et al. (2016) suggests that it may be more judicious to examine radar properties along fall steaks in order to capture size-sorting effects that occur when populations of ice crystals fall (Bader et al., 1987).

Ice crystal shape discrimination also progressed as a result of polarimetric radar observations; at millimeter wavelengths, linear depolarization (LDR), the ratio of copolar and crosspolar signal has been used to identify the presence of oriented columnar ice crystals (e.g., Matrosov et al., 2001; Oue et al., 2015b; Reinking et al., 2002). In combination with differential reflectivity measurements (Z_{DR}), LDR measurements have been used to distinguish dendritic crystals from simpler hexagonal and branched crystals (Wolde & Vali, 2001). Coincident enhancements in Z_{DR} and specific differential phase (K_{DP}) have been associated with the presence dendritic crystals (Bechini et al., 2013; Kennedy & Rutledge, 2011). Also negative K_{DP} and Z_{DR} have been interpreted as indicating the presence of prolate particles with vertically oriented maximum dimensions consistent with small, slow-falling conical graupel coexisting with low concentrations of more isometric graupel (Oue et al., 2015a). Unfortunately, the polarimetric signature of crystal tends to vanish in the presence of an active aggregation process and only under few circumstances the Z_{DR} signature of aggregates and polycrystals can be decomposed to extract the intrinsic Z_{DR} signature of pristine ice crystals (e.g., Keat & Westbrook, 2017).

Combining polarimetric and Doppler spectra radar measurements largely improves our ability to identify and separate different types of ice crystals and quantify their amounts (Moisseev et al., 2015; Oue et al., 2018) or to distinguish different growth stages of rimed particles (Oue et al., 2015a). Even in the absence of polarimetric information, a thorough examination of the entire radar Doppler spectra can provide insights into the nature of active microphysical processes (e.g., rimming as showed in Kalesse et al., 2016). That being said, systematically assessing the occurrence and the importance of various microphysical processes remains challenging especially in mixed-phased clouds containing multiple hydrometeor populations of similar fall velocities. Recent work suggests that multifrequency observations could offer an avenue to distinguishing between different ice crystal shapes generated through different microphysical processes; unfortunately, collecting such observational datasets remains technically challenging since it involves the operation of several collocated radars (Kneifel et al., 2011).

In spite of outstanding uncertainties, several techniques have been developed to quantitatively retrieve the properties of ice and mixed-phased clouds; many of these techniques mirror those developed for warm clouds and their main differences reside in the assumptions they employ and in the uncertainties they carry.

Combinations of radar and in situ measurements have been used to derive empirical relationships between radar reflectivity and ice water content (IWC) (e.g., Atlas et al., 1995; Hogan et al., 2006; Kikuchi et al., 1982; Liao & Sassen, 1994; Liu & Illingworth, 2000; Sato et al., 1981). Similarly, following decomposition into its air motion and particle fall speed component, the observed radar mean Doppler velocity can be linked to particle size using empirical relationship constructed from aircraft in situ particle size distribution measurements or model simulations (e.g., Mace et al., 2002; Matrosov et al., 2002) or to constrain particle density retrievals (Mason et al., 2018). Alternatively, combinations of infrared radiance and radar reflectivity measurements have been used to retrieve ice water path, IWC, and particle size (Mace et al., 1998; Matrosov, 1999; Matrosov et al., 1992). Zhang et al. (2014) also established new relationships between radar reflectivity and ice concentrations using simulations from a 1D ice growth model. Only recently, multifrequency techniques have demonstrated potential in reducing the uncertainty in quantitative retrieval (Kneifel et al., 2015; Kulie et al., 2014; Leinonen et al., 2018; Stein et al., 2015); much of the progress has emerged from improving the scattering model used in such retrievals (Grecu et al., 2018; Kneifel et al., 2018; Leinonen & Moisseev, 2015; Tyynela & Chandrasekar, 2014).

Retrievals based on combinations of remote sensors are often preferable in mixed-phase and ice clouds especially since radars tend to lack the sensitivity required to detect the uppermost portion of clouds where ice nucleation often occurs. Not to forget that lidars tend to rapidly attenuate in the presence of thick supercooled liquid layers. Combinations of radar and lidar measurements have allowed for the retrieval of IWC and optical depth (Donovan et al., 2001; Intrieri et al., 1993; Wang & Sassen, 2002), while combinations of lidar and radiometer observations have allowed for the retrieval of visible optical depth and infrared emittance in small optical depth clouds layers (e.g., Comstock & Sassen, 2001; Yoshida et al., 2006). Going one step further, Delanoë and Hogan (2008) developed a variational scheme for retrieving ice cloud properties from combined radar, lidar, and infrared radiometer measurements.

14.3.4. Characterizing Aerosols

Because of their impact on radiation and through their interaction with clouds, aerosols can significantly affect climate. In fact, gaps in our understanding of aerosol-cloud interactions continue to contribute largely to the uncertainty of future climate predictions.

There is an understanding that an entire chain of relationships—from emissions to microphysical and chemical characteristics to optical and cloud nucleating properties—must be well understood and represented at the appropriate scales to properly represent global aerosol distributions and their critical properties with accurate regional and seasonal variability. Observations of the properties at each step in the chain, referred to as "check points" by Bates et al. (2006) and Barnard et al. (2010), are necessary to improve our understanding. Aerosol radiative forcing can be described using three first-order optical properties—optical depth, single scattering albedo, and asymmetry parameter—while in order to describe the ability of aerosols to act as cloud condensation nuclei and ice nuclei, the size and composition information must be known. Challenges with observing aerosols can be attributed to their considerable complexity and heterogeneity in space and time.

Aerosol size and composition can only be measured directly by in situ methods, but this approach is costly and complex, and for these reasons this type of observations are relatively sparse. DOE ARM deploys five aerosol-observing systems around the world which make comprehensive measurements of aerosol microphysical, chemical, and optical properties from a common air sample, but such systems designed for aerosol process studies cannot capture the high degree of variability in aerosol populations worldwide. Remote sensing offers a more spatially dense source of information on aerosol properties. For instance, sun-photometer measurements of direct solar irradiance and sky radiance measurements (Dubovik et al., 2011; Dubovik & King, 2000) and multi-filter rotating shadowband radiometer (MFRSR) measurements of total and diffuse global horizontal radiation measurements (Kassianov et al., 2007) have been used to retrieve several column-integrated optical and physical aerosol properties. Although these methods would prove more powerful combined—because sun photometry methods are more sensitive to aerosol microphysical properties aiding to the accuracy of retrievals and the MFRSR is capable of higher temporal resolution—their synergy has yet to be explored. Alternatively, lidars readily measure signals related to the vertical distribution of aerosol, yet relating measured qualities to aerosol geophysical properties remains challenging due to the aerosol complexity and heterogeneity in space and time. Multiwavelength lidar measurements of both backscatter (e.g., 1,064, 532, and 355 nm) and extinction (e.g., at 532 and 355 nm) increase the spectral information on aerosol extinction and backscatter profiles and allows the derivation of properties such as the Ångström exponent (size-related quantity) and the lidar ratio (related to particle absorption capability) in ultraviolet and visible regions. Multiwavelength elastic and Raman lidar measurements are necessary and sufficient to estimate the particle volume and surface density as well as the refractive index of particle sizes in the typical range of the accumulation mode (Burton et al., 2016).

Better sensitivity to the particle shape and the ability to retrieve the refractive indices of fine and coarse particles separately as well as to extract the vertical distribution of the fine and coarse-particle concentrations have been achieved in the framework of the European Research Infrastructure for the observation of Aerosol, Clouds and Trace Gases (ACTRIS) through the synergistic use of lidar backscatter measurements and sun-photometer measurements [Generalized Aerosol Retrieval from Radiometer and Lidar Combined (GARRLiC) data algorithm (Lopatin et al., 2013) and LIdar-Radiometer Inversion Code (LIRIC Chaikovsky et al., 2016)]. This approach can also potentially provide higher accuracy for cases of low aerosol loadings compared with the intensity-only retrieval. By combining depolarization and multiwavelength lidar techniques, specific schemes such as the polarization lidar photometer networking (POLIPHON; Mamouri & Ansmann, 2017) can deliver aerosol speciation (e.g., fine and coarse dust, maritime and anthropogenic aerosols), which is of paramount importance for aerosol model evaluation studies. For example, Figure 14.4 shows observations collected by a zenith pointing multiwavelength portable Raman lidar (PollyXT; Engelmann et al., 2016). Between 03:00 and 21:00 UTC, the attenuated backscatter coefficient measured at 1,064 nm (Figure 14.4a) suggests the presence of an aerosol layer just above 2 km with clouds forming

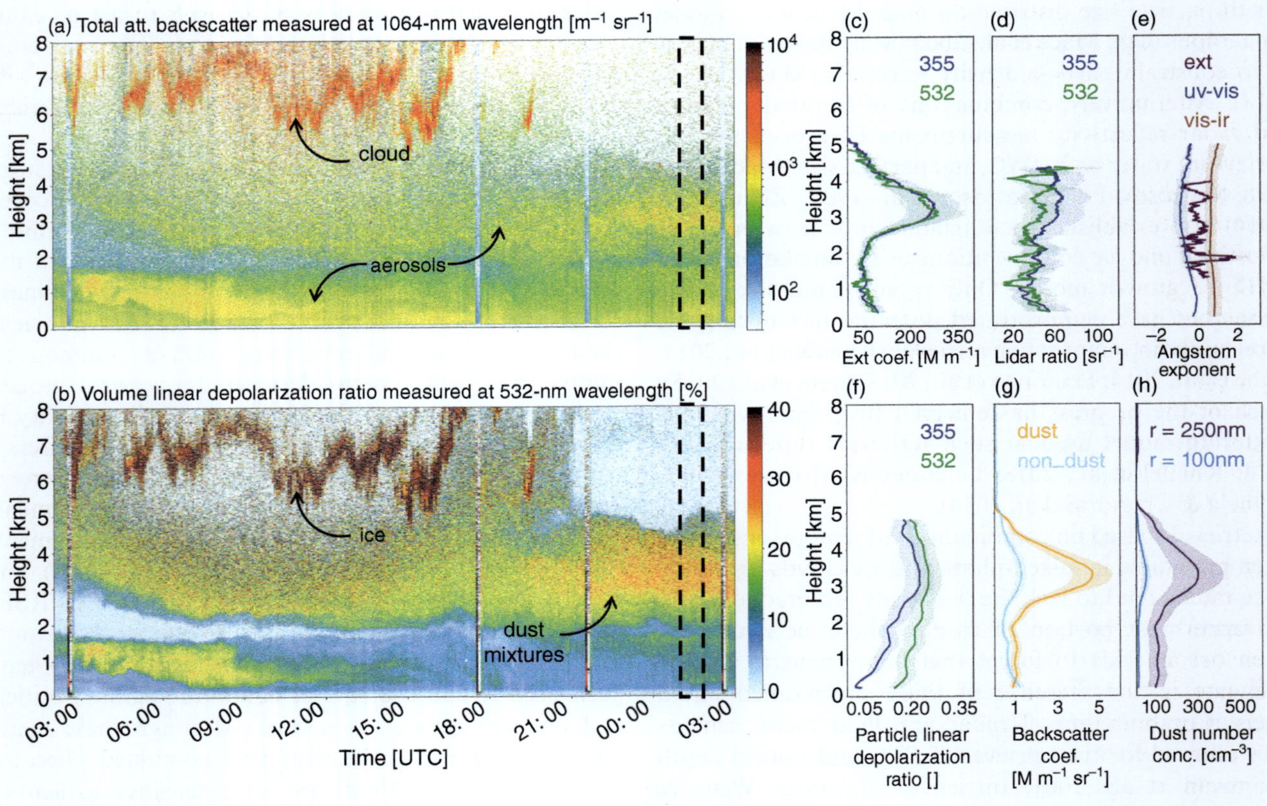

Figure 14.4 On 11 and 12 April 2018 at the PANGEA observatory in Greece a zenith pointing multiwavelength portable Raman lidar (PollyXT) measured (a) the total attenuated backscatter at 1064-nm wavelength and (b) the volume linear depolarization ratio at 532-nm wavelength. For different particle populations or from different wavelength channels (labeled within), profiles of (c) extinction coefficient, (d) lidar ratio, (e) Ångström exponent, (f) particle linear depolarization ratio, (g) backscatter coefficient, and (h) dust number concentration were estimated for the time period 1:10 to 2:30 UTC (dashed box in (a) and (b)). The shaded area in the profiles is the uncertainty of the retrieved parameters. Combinations of these measurements allow for the identification of an elevated ice cloud layer and a low-level dust layer.

above it (at 6–8 km). Measured volume linear depolarization ratio at 532 nm (δ_v; Figure 14.4b) allows for the assessment of the sphericity of these particles where values close to zero indicate the presence of spherical particles and higher values indicate the presence of nonspherical particles (e.g., dust). By averaging the profiles of measured properties for the period between 01:10 and 2:30 UTC (contained within the boxes depicted in Figure 14.4a and b), it is possible to estimate the extinction coefficient, which when integrated for the total atmospheric column provides the aerosol optical depth (Figure 14.4c) as well as the lidar ratio, which is indicative of the particles' composition (Figure 14.4d), and the Ångström exponent, which is indicative of the particles' size (Figure 14.4e). Based on their characteristic particle lidar depolarization ratio (δ_p; Figure 14.4f), ice clouds ($\delta_p \geq 40$ %; Voudouri et al., 2019) and dust mixtures can be recognized (10 % $< \delta_p <$ 30% (Groß et al., 2013; Haarig et al., 2017a). It is also possible to establish that marine aerosols dominate the boundary layer below 2 km (where $\delta_p <$ 10 %; Haarig et al., 2017b). At the same time, separating the particulate backscatter coefficient in its dust and nondust components (Figure 14.4g; Tesche et al., 2009) provides information on the mixing state of the different aerosols found in the scene and allows for the derivation of the number concentration of dust particles with radius greater than 100 and 250 nm (Figure 14.4h; Mamouri & Ansmann, 2016), which are thought to be related to the formation of cloud droplets and ice crystals (e.g., Kumar et al., 2011a; Kumar et al., 2011b). Separation of lidar extinction profiles in contributions from different aerosol types can lead to the estimation of cloud-condensation-nuclei and ice-nuclei number concentrations (e.g., Marinou et al., 2019), which can be a cornerstone for aerosol-cloud interaction studies, if combined with cloud radar techniques (e.g., Ansmann et al., 2019).

14.3.5. Characterizing the Planetary Boundary Layer

Traditionally, the thermodynamical structure of the planetary boundary layer has been characterized by in situ observations conducted via upper-air soundings measuring humidity, temperature, and wind profiles. However, apart from the high labor and material costs for regular radiosonde launches, they are often separated by 12 hours or more so that the diurnal cycle of the planetary boundary layer cannot be resolved, an aspect that is specifically crucial for atmospheric model initialization and evaluation. Until the technology evolves to the point where automatic drone can be operationally used to characterize the boundary layer thermodynamic structure, more comprehensive measurements will only be collected in limited regions including those where airplanes takeoff and land and where tall meteorological towers are deployed.

Remote sensors also offer opportunities to continuously characterize the clear boundary layer. Microwave radiometers are routinely used to retrieve path-integrated water vapor amount, typically with accuracies on the order 0.5 kg m^{-2}. Radiometers exploiting the wavelength-dependent and elevation-angle-dependent emission measurements at the center and on right flank of the 60 GHz oxygen complex can also be used to profile air temperature with accuracies of 0.5 K near the surface and 1.2 K at 1 km height because emission observations only contain ~3 degrees of freedom within the first kilometer of the planetary boundary layer (Löhnert & Maier, 2012). Figure 14.5 shows examples of retrievals estimated from a microwave radiometer. Although clouds are present, as indicated by nonzero liquid water paths (Figure 14.5b), temperature can be retrieved to characterize the vertical structure of the boundary layer (Figure 14.5c); at 0.5 km, a stable nighttime inversion is seen to vanish leaving behind a well-mixed planetary boundary layer after 6:00 UTC.

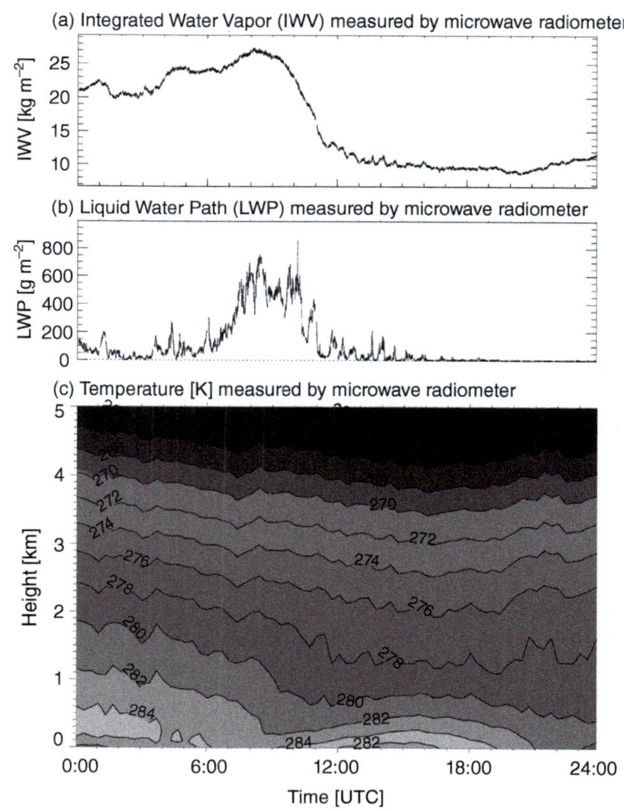

Figure 14.5 On 22 October 2018 at the Jülich Observatory for Cloud Evolution (JOYCE), microwave radiometer measurements were used to derive a time series of (a) integrated water vapor (IWV), (b) liquid water path (LWP), and (c) temperature profiles.

Observed boundary layer warming between 9:00 and 11:00 UTC is likely related to the significant (10 kg m^{-2}) integrated water vapor drying occurring around the same period (Figure 14.5a). Infrared spectrometers have also been shown to reliably retrieve boundary layer air temperature as well as humidity profiles (Turner & Löhnert, 2014). However, the degrees of freedom for infrared spectrometers signal are higher than for zenith pointing microwave radiometer (i.e., ~4–6 for temperature and ~4–8 for humidity), and typical vertical resolutions are on the order of 50–500 m in the lower troposphere.

Unlike passive systems, active systems are generally able to detect sharp vertical gradients in air temperature and/or humidity with smaller uncertainties. However, ground-based lidars have issues providing full coverage of the boundary layer in part because of incomplete transmitter/receiver overlap close to the instrument as well as because of daytime-dependent noise characteristics and blocking by opaque liquid clouds. Despite these challenges, proof of concept is given by the fully operational ARM Raman lidar (Turner et al., 2016), the NCAR field deployable micropulse DIAL (Weckwerth et al., 2016), and the portable Raman lidar network PollyNET (Dai et al., 2018). Latest developments show that many more operational ground-based lidars may be available within the next decade. Besides lidars, differential absorption radars have recently been shown to bear potential for boundary layer water vapor profiling in the presence of hydrometeors (Roy et al., 2018).

Although technical developments in the past decades have brought forth passive and active instruments for the continuous profiling of the thermodynamic structure of the atmosphere, it must be stressed that no single instrument has the capability of fully covering temperature and humidity in a sufficiently temporally and spatially resolved manner during all weather conditions. This issue can be partly overcome through composite profiling that exploits the respective strength of various sensors to complement their respective limitations. So far, only a few developments in this direction have been carried out. The challenge is that expertise in different observation systems is required simultaneously. Combined MWR and AERI observations have shown to improve the overall performance for air temperature and humidity profiling (Löhnert et al., 2009), while Barrera-Verdejo et al. (2016) have shown the advantage of using a water vapor lidar and MWR in synergy. Ebell et al. (2017) also presented a technique to combine measurements from a cloud radar and microwave radiometer with prior information in a 1D-Var approach in order to retrieve physically consistent atmospheric profiles of temperature, absolute humidity, liquid water content, and cloud droplet effective radius.

When it comes to establishing the limits of the planetary boundary layer, complications also arise in the absence of a universal definition for planetary boundary layer top leading to several retrieval techniques based on different properties. For example, some techniques identify the top of the planetary boundary layer using thermodynamic features revealed by radiosondes (Seidel et al., 2010), multivariate linear regression of multichannel microwave radiometer brightness temperature measurements (Cimini et al., 2012), or radar-detected insect signals (Chandra et al., 2010). Other approaches follow aerosol loading arguments that can be detected by taking the gradient of measured lidar backscatter (Belegante et al., 2014; Stachlewska et al., 2012), performing a wavelet decomposition of measured lidar backscatter (Granados-Muñoz et al., 2012; Talianu et al., 2006), or by analyzing Doppler lidar observations (e.g., Schween et al., 2014; Tucker et al., 2009). It is worth nothing that none of these approaches is fundamentally wrong rather they provide different insights into the structure of the boundary layer. Thus, it is of utmost importance for studies aiming to characterize or evaluate this property of the planetary boundary layer to be transparent and consistent in terms of the definition used.

14.3.6. Dynamics in Clouds and in the Clear Boundary Layer

Unarguably, Doppler lidars are the most powerful tools for the continuous characterization of the dynamical structure of the clear boundary layer (Berg et al., 2017a, 2017b). This is illustrated in Figure 14.6 that shows observations collected by zenith pointing ground-based Doppler sensors. In the clear air portion of this scene (i.e., where the radar did not detect hydrometeors), scattering of lidar signals by aerosol reveals a dynamically active boundary layer with several updrafts (depicted by the darkest regions and up-pointing arrows in Figure 14.6b) and downdrafts with velocity magnitudes of 1–2 m s^{-1}. Time series of these measurements have been used to estimate vertical velocity variance and eddy dissipation rate (Banakh et al., 1999; O'Connor et al., 2010). Other studies instead have attempted to quantify the coherency of such subcloud layer updrafts and downdrafts and to establish their relationship to shallow cumulus clouds (e.g., Ansmann et al., 2010; Lamer & Kollias, 2015; Lareau et al., 2018). Although most sensitive to small high-number concentration particles, Doppler lidar may detect signals from drizzle droplets that have appreciable fall speed and are poor air motion tracers. For this reason, it is necessary to isolate the contribution of aerosols and clouds from that of precipitation in the lidar Doppler spectra in order to use it to retrieve air motion. In the absence of Doppler lidar observations, insect have also been used to trace air motion in the boundary layer; however, it has been argued that insects may not be perfect air

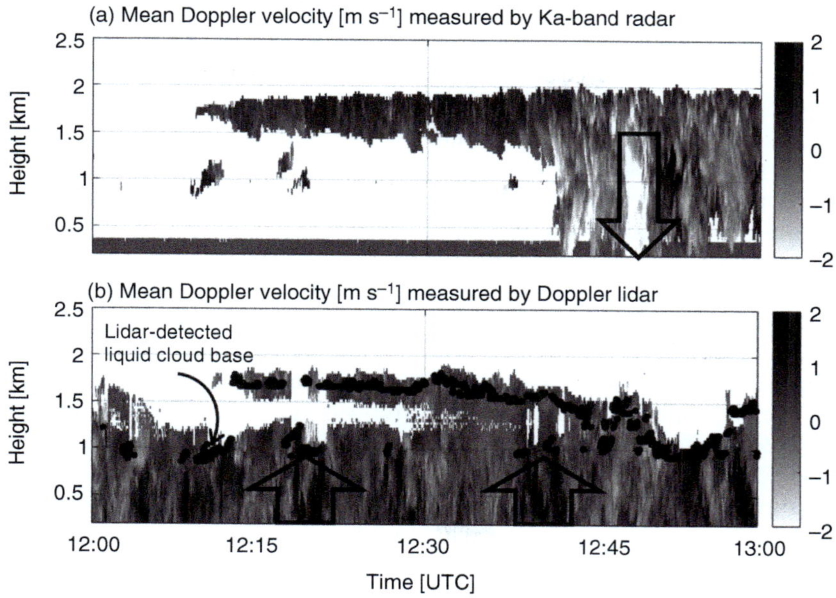

Figure 14.6 On 2 July 2018 at the Eastern North Atlantic (ENA) observatory, (a) the Ka-band ARM Zenith Radar (KAZR) measured the mean Doppler velocity of a sequence of nonprecipitating cumuli forming under a precipitating stratiform layer. (b) the Doppler lidar measured the mean Doppler velocity of aerosol and small drizzle droplets in the subcloud layer. Estimates of cloud-base height from ceilometer lidar (black dots) allow us to visually separate the location of clouds. It seems like cloud preferentially form over updrafts (in (b), up-pointing arrow indicates upward motion) and that rain clearly impacts the mean Doppler velocity measured by radar (in (a), down-pointing arrow indicates downward motion).

motion tracers because of their tendency to move against air flow to remain within a certain temperature range (Achtemeier, 1991; Chandra et al., 2010).

Above cloud-base height lidar signals quickly attenuate, and millimeter-wavelength radar signals become more useful. In nonprecipitating clouds, Doppler radars use cloud droplets as targets to trace air motion (e.g., Ghate et al., 2011; Kollias & Albrecht, 2010) and to retrieve cloud mass flux (e.g., Grinnell et al., 1996; Lamer et al., 2015) and eddy dissipation rate (Borque et al., 2016). Figure 14.6a shows the mean Doppler velocity recorded by a 35-GHz zenith pointing cloud radar. In the nonprecipitating shallow cumuli observed around 12:20 UTC, the measured mean Doppler velocity can be used to determine that these clouds contain a majority of updrafts. The absence of rain can be confirmed through a lack of relationship between observed radar reflectivity and mean Doppler velocity (Lamer et al., 2015; Orr & Kropfli, 1999). Deng and Mace (2008) went as far as to retrieve both vertical air motion and the microphysical properties of clouds along with their errors using combinations of Doppler radar moments (radar reflectivity, mean Doppler velocity, and spectrum width).

In the deeper precipitating systems such as the one presented in Figure 14.6a between 12:40 and 13:00 UTC, the cloud radar mean Doppler velocity represents a combination of air motion and the fall velocity of precipitation size particles (notice the downdraft motion associated with rain shaft around 12:50 UTC depicted by the down-pointing arrow). In the presence of larger hydrometeors, non-Rayleigh scattering signatures have been used to infer air motion (Fang et al., 2017; Kollias et al., 2002; Kollias et al., 1999). Alternatively, decomposing the Doppler spectra obtained from zenith pointing Doppler sensors can be used to retrieve both air motion and the fall speed of large hydrometeors; the offset of the slow velocity "cloud peak" from zero velocity is used to determine air motion, and, after shifting the entire Doppler spectrum to remove air motion, the mean Doppler velocity of the "precipitation peak" reveals the fall velocity of the precipitating hydrometeors (Shupe et al., 2008). Although potentially applicable to all types of sensors collecting Doppler spectra (i.e., radars and lidars), this technique has only been applied to Doppler radars to date. Note that the width of the "cloud peak" in the radar Doppler spectrum can also be used to estimate eddy dissipation rate (Kollias et al., 2001).

While it appears we have a good handle on characterizing the dynamics of shallow clouds, estimating velocity in deep convective clouds remains challenging due to a lack of adequate coverage throughout the depth of deep convective clouds, particularly at cloud top, in order to constrain traditional vertical velocity retrievals (North et al., 2017; Potvin et al., 2012). Although networks of scanning Doppler radars have been deployed to characterize these storms (Miller & Strauch, 1974), precipitation size hydrometeors cause significant attenuation at millimeter wavelengths leading to poor signal penetration and limited detection capability. Longer frequency radars, with broader field of views (but coarser resolution), are thus preferred for retrieving motions in deeper clouds. For example, radar wind profilers that operate at 50-MHz are sensitive to both turbulent clear-air (Bragg) and hydrometeor (Rayleigh) scattering processes (Giangrande et al., 2013). When signal from the two scattering peaks are well separated in the Doppler velocity spectra, then the Bragg scattering peak can be used to characterize vertical air motion and the Rayleigh scattering peak can be used to characterize hydrometeor motion (Williams, 2012).

14.4. FORWARD SIMULATORS

When measurements provide insufficient information to retrieve specific geophysical properties or when retrieval uncertainty is too large, then using forward simulators to transform model output to observables is an alternative path to creating an apples-to-apples comparison between model output and observations (Komurcu et al., 2014; Wu et al., 2014). Forward space also offers an opportunity to objectively evaluate the representation of ill-defined properties, such as water phase, by performing the same retrieval on forward-simulated model output and observation. Examples of this method are presented in Lamer et al. (2018), which showed how the Shupe (2007) multisensor water phase classifier can be used to evaluate the representation of liquid, ice, and mixed-phased clouds in general circulation model ModelE3. Similarly, Dolan et al.'s (2017) POLARRIS simulator contains a hydrometeor type ID classification for the evaluation of cloud-resolving-model-simulated mixed-phase cloud attributes. Additionally, because simulated quantities are known in forward space, forward simulators can be used to evaluate and tune existing retrievals (e.g., Lamer et al., 2018), explore new retrievals (e.g., Fielding et al., 2015; Rusli et al., 2017), or evaluate the impact of different assumptions on retrievals (e.g., Szyrmer et al., 2012). These types of studies, often referred to as observation system experiment or OSSES, can also be used to determine and optimize the design of future sensor deployments (e.g., Fielding et al., 2013).

Since the physics dictating sensor performance is largely independent of viewing angle, concepts developed to replicate spaceborne sensors are often at the origin of ground-based forward simulators. Since focus here is on active sensors readers are invited to consult Bodas-Salcedo et al. (2011) for an overview of forward simulators developed to emulate measurements from passive sensors including optical depth, albedo, and cloud fraction. Two main components form most active-sensor forward simulators, a backscattering simulator that converts simulated geophysical hydrometeor properties to sensor observable backscattering and an instrument forward model that mimics the sampling geometry and detection limitation specific to individual sensors.

Where plausible representations of hydrometeor geometrical properties are available as part of the model formulation, fundamental particle scattering calculations would be the most accurate way to transform model hydrometeor properties to observables. However, in most models such detailed hydrometeor information is highly simplified (e.g., fixed particle size distribution shapes) or not explicitly represented (e.g., orientation and realistic geometrical shapes), complicating the process of performing direct particle scattering calculations.

Output from cloud-resolving models operating single-moment, double-moment, or bin microphysical schemes can most readily be converted to radar and lidar backscattering; forward simulators like the Cloud-resolving Model Radar Simulator (CR-SIM; Oue et al., 2020) and the Passive and Active Microwave Transfer model (PAMTRA; Maahn, 2015) use T-matrix scattering lookup tables to convert simulated hydrometeor properties to ground-based radar and microwave radiometer observables assuming that simulated particles can be represented using spheroids. CR-SIM additionally reproduces the signal observed by ground-based lidars.

For one-moment or two-moment microphysical models, Chepfer et al. (2008) proposed an approach in which lidar backscattered power can be forward simulated using model output hydrometeor effective radius. Their approach, based on Mie theory, relies on the assumption that both liquid and ice cloud particles are spherical and requires additional assumptions about hydrometeor size distributions and scattering efficiencies. Using the same sphericity assumption, the CloudSat simulator part of the Cloud Feedback Model Intercomparison Project Observation Simulator Package (COSP; Bodas-Salcedo et al., 2011) and ARM Cloud Radar Simulator for general circulation model (Zhang et al., 2018) both use QuickBeam for the estimation of radar backscattered power (i.e., radar reflectivity; Haynes et al., 2007). QuickBeam additionally requires assumptions about mass-size and diameter-density relationships. Although some of these assumptions may be made to maintain consistency with the assumptions in model cloud microphysical parameterizations, some are not physically realistic (e.g., spherical ice) or complete enough for accurate backscattering calculations, and it is typically difficult to establish the sensitivity of the results to all such assumptions. The GCM-Oriented Ground-Observation forward simulator

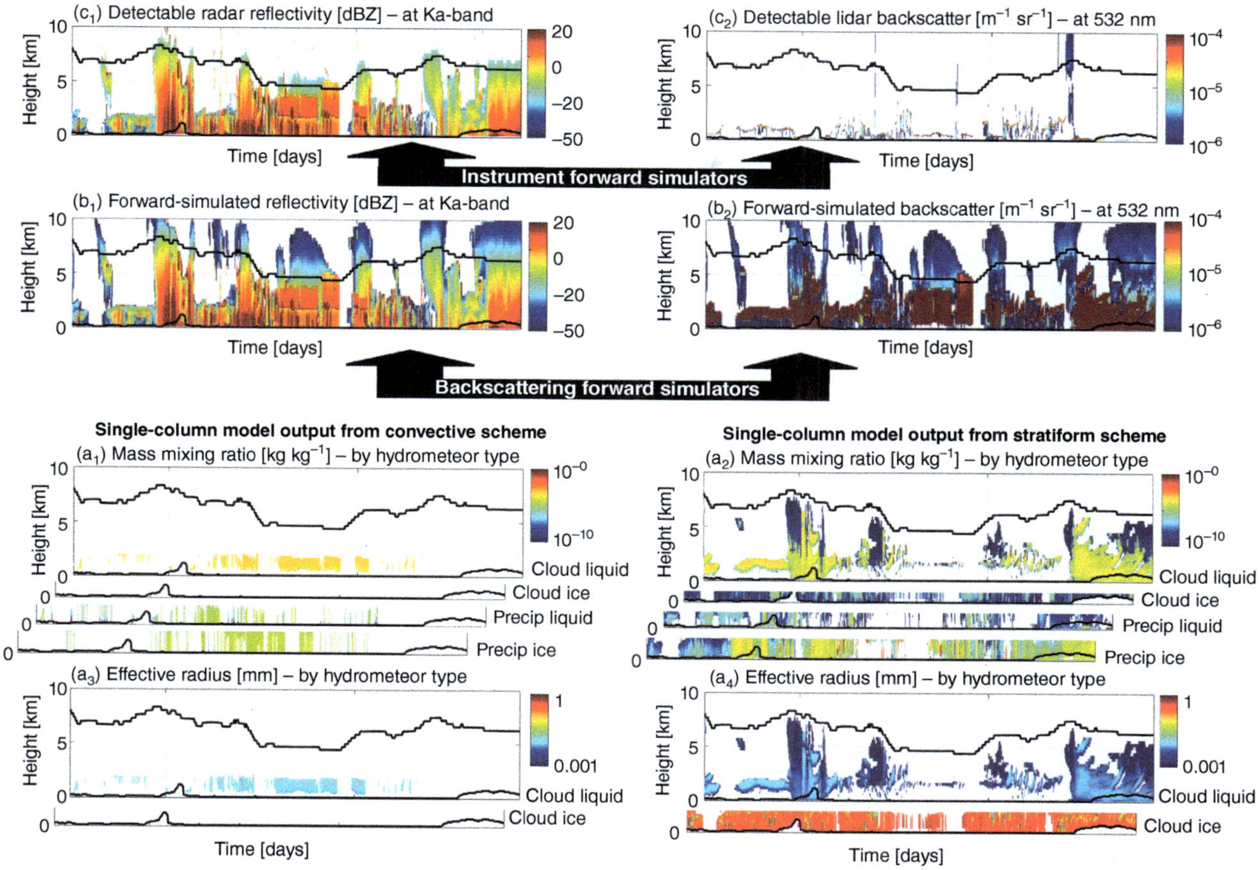

Figure 14.7 General circulation models (GCMs) commonly represent the mass mixing ratio ($a_{1\text{-}2}$) and effective radius ($a_{3\text{-}4}$) of several hydrometeor species (cloud and precipitating liquid and ice both stratiform and convective). The $(GO)^2$-SIM forward simulator can be used to emulate the backscattering signal these coexisting hydrometeor species would generate if exposed to radar (b_1) and lidar (b_2) signals; by additionally taking into account the limitation of these sensor, detectable signals can be isolated for a fairer comparison with actual observations ($c_{1\text{-}2}$).

($(GO)^2$-SIM) was recently developed to avoid having to make ad hoc assumptions about hydrometeor shapes, orientations, and compositions, which are properties that also remain poorly documented in nature. $(GO)^2$-SIM relies on empirical relationships between hydrometeor effective radius, water content and backscattering in order to convert model output to observables; Figure 14.7 shows an example of a forward simulation where $(GO)^2$-SIM was employed to convert output from a two-moment single-column model (Figure 14.7a) into radar reflectivity (Figure 14.7b_1) and lidar backscatter (Figure 14.7b_2). Because sensors cannot readily distinguish between hydrometeor populations, information about all hydrometeors forming in the atmospheric column both stratiform and convective, cloud and precipitation of both liquid and ice phases are needed in the conversion process.

Certain applications, such as water phase classification, require observables beyond backscattering. Forward simulations of Doppler quantities have been performed for cloud-resolving models (Oue et al., 2020) and two-moment large-scale models (Lamer et al., 2018). And polarimetric signatures haven been emulated with additional information about hydrometeor axis ratio, density, canting angles, and size distributions (Dolan et al. 2017).

Like the retrieval process, the forward-simulation process is not without uncertainty and very few studies attempt to quantify the uncertainty resulting from the assumptions involved. Through the use of computationally inexpensive empirical relationships, $(GO)^2$-SIM is one of the first forward simulators capable of estimating an ensemble of forward simulations in an effort to quantify part of the backscattering uncertainty (Lamer et al., 2018).

The aforementioned techniques can be used to convert any hydrometeor mixing-ratio amount, even if infinitesimal, to observables. However, in the interest of creating an objective intercomparison between simulated and observed hydrometeor properties, instrument capabilities should be reproduced in forward simulators such that only those simulated hydrometeor populations that lead to signals detectable by sensors be evaluated against observations. Literature describes how radars and lidars possess different detection capabilities because their transmitted energy, as it successively interacts with hydrometeors while progressing through the atmosphere, eventually becomes too weak to be distinguished from expected background noise (Lamer et al., 2018; Oue et al., 2020). While sensors' signal response to attenuation can be generalized, expected background noise is specific to individual sensors and depends on factors such as hardware and transmission setup that can only be reproduced if sensors are well characterized. Comparing Figure 14.7b_1 and c_1 highlights how millimeter radars are often not sensitive enough to detect the top of cloud systems containing low-liquid amounts and small droplets. Comparing Figure 14.7b_2 and c_2 highlights how near-infrared lidars are often not sensitive enough to detect low-liquid amounts like those that were simulated in the lowest kilometers and how near-infrared lidars cannot penetrate hydrometeor layers containing large amounts of liquid water that limits their use to detecting only the base of liquid clouds.

Additionally, besides being sensitive to clouds and precipitation, ground-based millimeter-wavelength radars are also sensitive to clutter such as insects, sea waves, and other man-made objects. Such signals cannot be reproduced in forward space; hence, it is important to compare forward-simulated fields to well quality-controlled observations.

Note that for comparison with scanning sensors, instrument geometry can also be reproduced in forward space (Oue et al., 2020).

14.5. USING OBSERVATIONS TO INFORM MODELS

Different approaches exist to make use of ground-based observations to improve atmospheric numerical models. For instance, observations may be used to advance our understanding of atmospheric processes or to build parametrizations for large-scale models by identifying relationships between various atmospheric properties (e.g., Hogan & Illingworth, 2003). Alternatively, ground-based observations may be used as benchmarks to evaluate the performance of atmospheric models or to construct forcing datasets to drive them. Especially for the latter two applications, interfacing observations and models requires overcoming outstanding differences between the two.

To begin one must determine if the geophysical properties of interest can be retrieved at all or with reasonable uncertainty. If that is not the case, then forward simulators that convert model output to observables offer an alternative way to use observations to evaluate model output.

Then comes the issue of scale; ignoring the fact that global cloud-resolving models (e.g., Satoh et al., 2019; Stevens et al., 2020) may well become the preferred modeling tools in the future, studies attempting to interface modern-day GCMs and high-resolution observations of the atmospheric column must overcome the scale gap between the two (e.g., Chandra et al., 2015; Lamer, 2019; Ovtchinnikov et al., 2006; Qian et al., 2012; Song et al., 2014; Wei et al., 2017; Yang et al., 2006). After years of advancements, the vertical grid resolution of GCMs has now reached the tens-to-hundred-meter scale so that individual cloud layers are now vertically resolved. In addition to removing the need to track cloud vertical fractional coverage within a model grid box, this advancement eliminates in large measure the vertical scale gap between model output and observations. An approach to eliminating the remaining vertical scale gap, one that is routinely used to produce the satellite-based GCM-Oriented CALIPSO Cloud Product (GOCCP; Chepfer et al., 2010), involves vertically averaging observed backscattering to match the vertical grid resolution of GCMs. Zhang et al. (2018) also used vertical averaging of radar reflectivity to produce 100-m vertical resolution joint frequency by altitude diagrams (CFADs) to be used as GCM evaluation targets. These approaches, however, do not treat Doppler observables. For the purpose of evaluating liquid and ice microphysics, Lamer (2019) additionally expanded on current approaches by proposing a new technique where reflectivity weights are used to degrade the vertical resolution of ground-based radar mean Doppler velocity and Doppler spectral width observables in order to emulate the scattering response of radar signals.

While it has much improved over the years, the horizontal grid resolution of modern-day GCMs remains on the order of hundreds of kilometers. Time averaging of measurements has been employed in an attempt to quantitatively relate point measurements to model grid boxes. That being said, there is no consensus on how to select the appropriate averaging time window. For instance, the well-known ARM Climate Modeling Best Estimate product (ARMBE or CMBE; Xie et al., 2010), which many studies have relied on for model evaluation (e.g., Song et al., 2014; Wei et al., 2017), employs hourly averaging while other studies suggest that longer time averaging are preferable to capture the large domain each GCM grid box represents (e.g., 3-h employed by Chandra et al. (2015), Ovtchinnikov et al. (2006), and Yang et al. (2006) or monthly employed by Qian et al. (2012)). Jakob et al. (2004) even suggest that simulated properties can be interpreted as probabilistic at an observation point such that comparisons between model output and point measurements can be performed statistically without any time averaging of observations to match the model resolution. Alternatively, cloud and precipitation vertical overlap assumptions in models permit the generation of subcolumns within a GCM column of grid boxes that capture subgrid scale variability in hydrometeor horizontal and vertical distributions. Generating subcolumns from GCM output is a multistep process. The process involves distributing both hydrometeor horizontal fractions and hydrometeor mass mixing ratio amounts in such a way as to obey all model assumptions on hydrometeor overlap within a grid box and vertical overlap from one layer to the next. The complexity of this process increases as the number of hydrometeor species and the number of model assumptions on hydrometeor overlap become more complex (e.g., Di Michele et al., 2012; Hillman et al., 2018; Klein & Jakob, 1999; Lamer, 2019; Räisänen et al., 2004; Webb et al., 2001; Zhang et al., 2010).

Another way to evaluate coarse resolution models using ground-based observations involves employing higher-resolution models as an interface. The process of evaluating high-resolution models with ground-based observations is much more straightforward (e.g., Luo et al., 2008a; Luo et al., 2008b; Varble et al., 2011; Xu et al., 2002), and benchmarked high-resolution models can provide invaluable information to coarser resolution models [e.g., WRF (Angevine et al., 2018; Jakob et al., 2004) or GCM (Endo et al., 2015; Lin et al., 2015; Vogelmann et al., 2015)], that is, because high-resolution models are more comprehensive than observations and can be used to create idealized situations that are helpful to extract a focused features from a cloud field. Multimodel efforts have stemmed from organized initiatives such as that of the Fast Physics System Testbed and Research (FASTER) project (Liu, 2019) and of the Global Energy and the Water Cycle Experiment (GEWEX) Cloud System Study (GCSS; Randall et al., 2003).

Independent of the approach, the scope of model evaluation effort must also be determined. Studies relying on single-column models or on high-resolution models generally focus on case studies, which require appropriate forcing for reproducing observed conditions (e.g., Vogelmann et al., 2015; Zhang & Lin, 1997). Recognizing the importance of collecting information beyond the cloud and precipitation of interest, the ARM program implemented routine monitoring of surface, top of the atmosphere, and column properties for the generation of forcing dataset as part of the LES ARM Symbiotic Simulation and Observation (LASSO) project (Gustafson, Jr. et al., 2020). On the other hand, care must be taken when statistically summarizing large datasets so as not to mask compensating model biases; something which is not simple because of the number of processes that affect cloud and precipitation growth, lifetime, and decay. Moreover, it is desirable that such simplified statistical summary maintains sufficient information to diagnose the cause of identified biases. The GCM-Oriented CALIPSO Cloud Product (GOCCP) provides vertically resolved contour frequency by altitude diagrams (CFADs) that characterize the global vertical distribution of lidar scattering ratio. This product also provides information about the fractions of liquid and ice clouds (without mixed-phase capability) observed within three atmospheric regions—low, middle, and high—separated by fixed pressure levels (Cesana & Chepfer, 2013; Chepfer et al., 2010). Zhang et al. (2018) are in the process of releasing routine regional vertically resolved model evaluation targets constructed using long-term ground-based radar reflectivity observations that, although not cloud-type specific, provide information comparable to GOCCP on the vertical distribution of hydrometeors. Lamer (2019) exposed cloud-type, location, and season-specific biases by creating a vertically resolved statistical summary of water phase, including ice, liquid, and mixed phase, within the context of hydrometeor vertical layering (HVL) regimes; HVL regimes are defined following a technique comparable to the one used by the International Satellite Cloud Climatology Project (ISCCP; Rossow & Schiffer, 1991) and similar to the cloud vertical structure (CVS) approach proposed by Rémillard and Tselioudis (2015) and Tselioudis et al. (2013).

14.6. THE NEXT GENERATION OF GROUND-BASED OBSERVATORIES

Developments in the modeling community and the emergence of projects producing routine simulations of the conditions forming around observatories pushed the notion that ground-based observatories need to evolve to characterize the atmosphere over domains rather than columns (Xie et al., 2006). For example, in 2015, the ARM program established the LES ARM Symbiotic Simulation and Observations (LASSO) project that targets shallow clouds at the Southern Great Plains. During its early phase, the LASSO project aimed to develop the different components (including computations resources, data bundles, large-scale forcing, etc.) needed for routine high-resolution LES simulations. One of the emerging challenges was the issue of the representativeness of one-point or *n*-point observations for the entire LES model domain. That is, although collecting observation of the same atmospheric column over long periods of time has long been believed sufficient to characterize a domain, recent work has shown that the properties of aerosols, clouds, and precipitation advected overhead zenith pointing sensors may be unrepresentative of the properties of all the aerosol, cloud, and precipitation elements forming in large domains. The "soda straw" view of the atmospheric region is especially limited when it comes to characterizing inhomogeneous clouds, precipitation (Lamer et al., 2019), and aerosol regimes (Schutgens et al., 2017), and comes with the challenging task of disentangling life cycle from domain inhomogeneity effects.

To address domain representativeness concerns, the ARM program initially installed intermediate, extended, and boundary facilities around their main SGP site (Sisterson et al., 2016). Improvement in the characterization of aerosol inhomogeneities should emerge from similar efforts involving the deployment of dense sun photometer networks over tens to hundreds of kilometers [NASA's AERONET DRAGON networks—Distributed Regional Aerosol Gridded Observation Networks (Holben et al., 2018)]. ARM's more recent investment in unmanned aerial vehicle (UAV; Boer et al., 2018; Stephens et al., 2000) could also open new doors to more systematic in situ characterization of cloud and aerosol particles. The addition scanning cloud radars showed promising results to improve some aspects of the characterization of cloud and precipitation inhomogeneity at the scales of 10–20 km (Ewald et al., 2015; Fielding et al., 2013, 2014; Lamer et al., 2019; Lamer et al., 2014; Oue et al., 2016). Figure 14.8 shows that although they are not able to cover a domain over 100 km like spaceborne sensors can (Figure 14.8c), scanning millimeter-wavelength radar observations can provide mesoscale context (Figure 14.8b) to the observations collected by zenith pointing cloud radars (Figure 14.8a).

Scanning radars and lidars may also be used to improve our understanding of cloud life cycle (Borque et al., 2014; Kollias et al., 2020) although hardware limitations continue to hamper our ability to quickly sample large observational domains. Phased-array radar systems show great promise in characterizing finer spatial and temporal scales via near-instantaneous sampling of the atmosphere with flexible beam forming, multifunctionality, and low operational and maintenance costs and without mechanical inertia limitations (Kollias et al., 2022). The possibility of deploying networks of scanning millimeter-wavelength radars (Mechem et al., 2015) and truck-based observatories (e.g., Lamer et al., 2022; Wagner et al., 2019; Wurman et al., 2021) was also put forth as a solution to improving the characterization of heterogenous atmospheric properties.

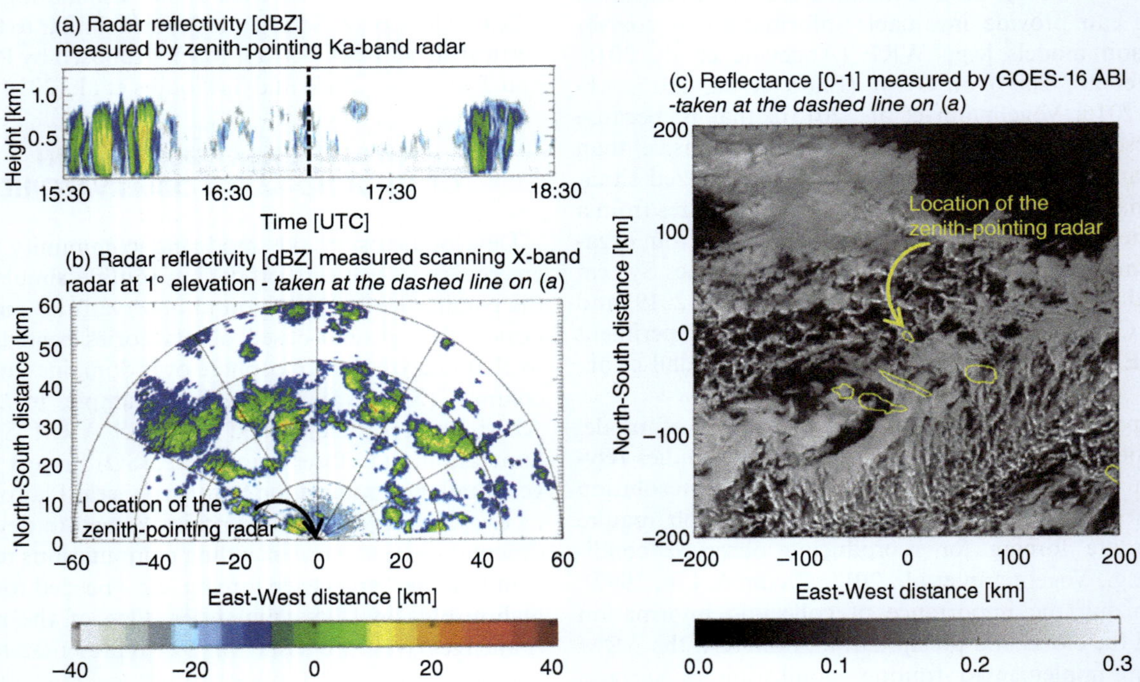

Figure 14.8 On 16 January 2018 at the Eastern North Atlantic (ENA) observatory, (a) the Ka-band ARM Zenith Radar (KAZR) measured the reflectivity signal of a sequence of cloud cells. (b) A 1° elevation plan point indicator (PPI) scan collected at 17:01 UTC along the position of the dashed line in (a) reveals that several other cloud cells formed in the region around the observatory (the arrow points to the observatory). (c) Reflectance measured by the advanced baseline imager (ABI) onboard the GOES-16 satellite provides additional regional context on the density of the cloud field (yellow outline shows the location of the Azores Islands and the arrow points to the observatory).

Some additional outstanding issues that require 3D mapping of clouds and precipitation are as follows: (i) 3D radiative transfer issues, including calculation of radiative flux profiles, (ii) cloud-aerosol interactions, (iii) convective clouds dynamics, and (iv) evaluation of satellite retrievals of cloud system properties; addressing some of these issues may require different observing paradigms (Romps & Öktem, 2018).

In a nutshell, the development of the breadth and accuracy of information from remote sensing can be achieved by following this model for spatially dense, distributed networks of active and passive remote sensors that are also collocated with in situ measurements that can provide specific information on the properties that dictate the interaction between aerosols, clouds, precipitation, and the boundary layer (Kollias et al., 2019). The distributed network can be either surfaced-based, drone-based or a mixture (de Boer et al., 2019). The information from in situ measurements can be used to improve the information content from remote sensing, which will always be more widespread in its deployment. The continuous improvement of observing capabilities specific to surface observatories is an important step toward making meaningful progress in the understanding and representation of fast physics processes in numerical models. These efforts should be developed in tandem with efforts in model computational capabilities and in the development of novel retrieval approaches that target processes understanding.

REFERENCES

Achtemeier, G. L. (1991). The use of insects as tracers for "clear-air" boundary-layer studies by Doppler radar. *Journal of Atmospheric and Oceanic Technology*, *8*(6), 746–765.

Angevine, W. M., Olson, J., Kenyon, J., Gustafson, W. I., Jr, Endo, S., Suselj, K., & Turner, D. D. (2018). Shallow cumulus in WRF parameterizations evaluated against LASSO large-eddy simulations. *Monthly Weather Review*, *146*(12), 4303–4322.

Ansmann, A., Fruntke, J., & Engelmann, R. (2010). Updraft and downdraft characterization with Doppler lidar: cloud-free versus cumuli-topped mixed layer. *Atmospheric Chemistry and Physics*, *10*(16), 7845–7858.

Ansmann, A., Mamouri, R., Bühl, J., Seifert, P., Engelmann, R., Hofer, J., et al. (2019). Ice-nucleating particle versus ice crystal number concentration in altocumulus and cirrus layers embedded in Saharan dust: a closure study. *Atmospheric Chemistry and Physics*, *19*(23), 15087–15115.

Ansmann, A., Wandinger, U., Riebesell, M., Weitkamp, C., & Michaelis, W. (1992). Independent measurement of extinction and backscatter profiles in cirrus clouds by using a combined Raman elastic-backscatter lidar. *Applied Optics*, *31*(33), 7113–7131.

Atlas, D. (1954). The estimation of cloud parameters by radar. *Journal of Meteorology*, *11*(4), 309–317.

Atlas, D., Matrosov, S. Y., Heymsfield, A. J., Chou, M.-D., & Wolff, D. B. (1995). Radar and radiation properties of ice clouds. *Journal of Applied Meteorology*, *34*(11), 2329–2345.

Baars, H., Kanitz, T., Engelmann, R., Althausen, D., Heese, B., Komppula, M., et al. (2016). An overview of the first decade of PollyNET: An emerging network of automated Raman-polarization lidars for continuous aerosol profiling. *Atmospheric Chemistry and Physics, 16*, 5111–5137.

Bader, M. J., Clough, S. A., & Cox, G. P. (1987). Aircraft and dual polarization radar observations of hydrometeors in light stratiform precipitation. *Quarterly Journal of the Royal Meteorological Society*, *113*(476), 491–515.

Bailey, M. P., & Hallett, J. (2009). A comprehensive habit diagram for atmospheric ice crystals: Confirmation from the laboratory, AIRS II, and other field studies. *Journal of the Atmospheric Sciences*, *66*(9), 2888–2899.

Balin, I., Serikov, I., Bobrovnikov, S., Simeonov, V., Calpini, B., Arshinov, Y., & Van Den Bergh, H. (2004). Simultaneous measurement of atmospheric temperature, humidity, and aerosol extinction and backscatter coefficients by a combined vibrational–pure-rotational Raman lidar. *Applied Physics B*, *79*(6), 775–782.

Banakh, V. A., Smalikho, I. N., Köpp, F., & Werner, C. (1999). Measurements of turbulent energy dissipation rate with a CW Doppler lidar in the atmospheric boundary layer. *Journal of Atmospheric and Oceanic Technology*, *16*(8), 1044–1061.

Barnard, J. C., Fast, J. D., Paredes-Miranda, G., Arnott, W., & Laskin, A. (2010). Evaluation of the WRF-Chem "Aerosol Chemical to Aerosol Optical Properties" Module using data from the MILAGRO campaign. *Atmospheric Chemistry and Physics*, *10*(15), 7325–7340.

Barrera-Verdejo, M., Crewell, S., Löhnert, U., Orlandi, E., & Girolamo, P. D. (2016). Ground-based lidar and microwave radiometry synergy for high vertical resolution absolute humidity profiling. *Atmospheric Measurement Techniques*, *9*(8), 4013–4028.

Bates, T., Anderson, T., Baynard, T., Bond, T., Boucher, O., Carmichael, G., et al. (2006). Aerosol direct radiative effects over the northwest Atlantic, northwest Pacific, and North Indian Oceans: estimates based on in-situ chemical and optical measurements and chemical transport modeling. *Atmospheric Chemistry and Physics*, *6*(6), 1657–1732.

Battaglia, A., Kollias, P., Dhillon, R., Roy, R. J., Tanelli, S., Lamer, K., et al. (2020). Space-borne cloud and precipitation radars: status, challenges and ways forward, *Reviews of Geophysics, 58*(3), e2019RG000686.

Bechini, R., Baldini, L., & Chandrasekar, V. (2013). Polarimetric radar observations in the ice region of precipitating clouds at C-band and X-band radar frequencies. *Journal of Applied Meteorology and Climatology*, *52*(5), 1147–1169. doi:10.1175/jamc-d-12-055.1.

Belegante, L., Nicolae, D., Nemuc, A., Talianu, C., & Derognat, C. (2014). Retrieval of the boundary layer height from active and passive remote sensors. Comparison with a NWP model. *Acta Geophysica*, *62*(2), 276–289.

Berg, L. K., Newsom, R. K., & Turner, D. D. (2017a). Year-long vertical velocity statistics derived from Doppler lidar data for the continental convective boundary layer. *Journal of Applied Meteorology Climatology*, *56*(9), 2441–2454.

Berg, L. K., Newsom, R. K., & Turner, D. D. (2017b). Year-long vertical velocity statistics derived from Doppler lidar data for the continental convective boundary layer. *Journal of Applied Meteorology and Climatology*, *56*(9), 2441–2454.

Bodas-Salcedo, A., Webb, M. J., Bony, S., Chepfer, H., Dufresne, J.-L., Klein, S. A., et al. (2011). COSP: Satellite simulation software for model assessment. *Bulletin of the American Meteorological Society*, *92*(8), 1023–1043. doi:10.1175/2011bams2856.1.

Boer, G. D., Ivey, M., Schmid, B., Lawrence, D., Dexheimer, D., Mei, F., et al. (2018). A Bird's-eye view: Development of an operational ARM unmanned aerial capability for Atmospheric Research in Arctic Alaska. *Bulletin of the American Meteorological Society*, *99*(6), 1197–1212. doi:10.1175/bams-d-17-0156.1.

Borque, P., Kollias, P., & Giangrande, S. (2014). First observations of tracking clouds using scanning ARM cloud radars. *Journal of Applied Meteorology Climatology*, *53*(12), 2732–2746.

Borque, P., Luke, E., & Kollias, P. (2016). On the unified estimation of turbulence eddy dissipation rate using Doppler cloud radars and lidars. *Journal of Geophysical Research: Atmospheres*, *121*(10), 5972–5989.

Bruneau, D., Pelon, J., Blouzon, F., Spatazza, J., Genau, P., Buchholtz, G., et al. (2015). 355-nm high spectral resolution airborne lidar LNG: system description and first results. *Applied Optics*, *54*(29), 8776–8785.

Burns, D., Kollias, P., Tatarevic, A., Battaglia, A., & Tanelli, S. (2016). The performance of the EarthCARE Cloud Profiling Radar in marine stratiform clouds. *Journal of Geophysical Research: Atmospheres*, *121*(24), 14,525–14,537.

Burton, S. P., Chemyakin, E., Liu, X., Knobelspiesse, K., Stamnes, S., Sawamura, P., et al. (2016). Information content and sensitivity of the 3β+ 2α lidar measurement system for aerosol microphysical retrievals. *Atmospheric Measurement Techniques*, *9*(11), 5555–5574.

Cesana, G., & Chepfer, H. (2013). Evaluation of the cloud thermodynamic phase in a climate model using CALIPSO-GOCCP. *Journal of Geophysical Research: Atmospheres*, *118*(14), 7922–7937.

Chaikovsky, A., Dubovik, O., Holben, B., Bril, A., Goloub, P., Tanré, D., et al. (2016). Lidar-Radiometer Inversion Code (LIRIC) for the retrieval of vertical aerosol properties from combined lidar/radiometer data: development and distribution in EARLINET. *Atmospheric Measurement Techniques* *9*(3), 1181–1205.

Chandra, A. S., Kollias, P., & Albrecht, B. (2013). Multiyear summertime observations of daytime fair-weather cumuli at the ARM southern great plains facility. *Journal of Climate*, *26*(24), 10031–10050. doi:10.1175/jcli-d-12-00223.1.

Chandra, A. S., Kollias, P., Giangrande, S. E., & Klein, S. A. (2010). Long-term observations of the convective boundary layer using insect radar returns at the SGP ARM climate research facility. *Journal of Climate*, *23*(21), 5699–5714.

Chandra, A. S., Zhang, C., Klein, S. A., & Ma, H.-Y. (2015). Low-cloud characteristics over the tropical western Pacific from ARM observations and CAM5 simulations. *Journal of Geophysical Research: Atmospheres*, *120*(17), 8953–8970. doi:10.1002/2015jd023369.

Chepfer, H., Bony, S., Winker, D., Cesana, G., Dufresne, J. L., Minnis, P., et al. (2010). The GCM-Oriented CALIPSO Cloud Product (CALIPSO-GOCCP). *Journal of Geophysical Research: Atmospheres*, *115*(D4). doi:10.1029/2009JD012251.

Chepfer, H., Bony, S., Winker, D., Chiriaco, M., Dufresne, J. L., & Sèze, G. (2008). Use of CALIPSO lidar observations to evaluate the cloudiness simulated by a climate model. *Geophysical Research Letters*, *35*(15).

Chiu, J. C., Marshak, A., Huang, C. H., Várnai, T., Hogan, R. J., Giles, D. M., et al. (2012). Cloud droplet size and liquid water path retrievals from zenith radiance measurements: examples from the Atmospheric Radiation Measurement Program and the Aerosol Robotic Network. *Atmospheric Chemistry and Physics*, *12*(21), 10313–10329. doi:10.5194/acp-12-10313-2012.

Cimini, D., De Angelis, F., Dupont, J.-C., Pal, S., & Haeffelin, M. (2012). Mixing layer height retrievals by multichannel microwave radiometer observations. *Atmospheric Measurement Techniques, 6*(11), 2941–2951.

Comstock, J. M., d'Entremont, R., DeSlover, D., Mace, G. G., Matrosov, S. Y., McFarlane, S. A., et al. (2007). An intercomparison of microphysical retrieval algorithms for upper-tropospheric ice clouds. *Bulletin of the American Meteorological Society*, *88*(2), 191–204.

Comstock, J. M., & Sassen, K. (2001). Retrieval of cirrus cloud radiative and backscattering properties using combined lidar and infrared radiometer (LIRAD) measurements, *Journal of Atmospheric and Oceanic Technology*, *18*(10), 1658–1673.

Comstock, K. K., Wood, R., Yuter, S. E., & Bretherton, C. S. (2004). Reflectivity and rain rate in and below drizzling stratocumulus. *Quarterly Journal of the Royal Meteorological Society*, *130*(603), 2891–2918.

Crewell, S., & Löhnert, U. (2003). Accuracy of cloud liquid water path from ground-based microwave radiometry 2. Sensor accuracy and synergy. *Radio Science*, *38*(3), 7-1–7-3.

Dai, G., Althausen, D., Hofer, J., Engelmann, R., Seifert, P., Bühl, J., et al. (2018). Calibration of Raman lidar water vapor profiles by means of AERONET photometer observations and GDAS meteorological data. *Atmospheric Measurement Techniques, 11*(5), 2735–2748.

Damiani, R., Vali, G., & Haimov, S. (2006). The structure of thermals in cumulus from airborne dual-Doppler radar observations. *Journal of the Atmospheric Sciences*, *63*(5), 1432–1450.

de Boer, G., Argrow, B., Cassano, J., Cione, J., Frew, E., Lawrence, D., et al. (2019). Advancing unmanned aerial capabilities for atmospheric research. *Bulletin of the American Meteorological Society*, *100*(3), ES105–ES108.

Delanoë, J., & Hogan, R. J. (2008). A variational scheme for retrieving ice cloud properties from combined radar, lidar, and infrared radiometer. *Journal of Geophysical Research: Atmospheres*, *113*(D7).

Deng, M., & Mace, G. G. (2008). Cirrus microphysical properties and air motion statistics using cloud radar Doppler moments. Part II: Climatology. *Journal of Applied Meteorology and Climatology*, *47*(12), 3221–3235.

Di Michele, S., Ahlgrimm, M., Forbes, R., Kulie, M., Bennartz, R., Janisková, M., & Bauer, P. (2012). Interpreting an evaluation of the ECMWF global model with CloudSat observations: Ambiguities due to radar reflectivity forward operator uncertainties. *Quarterly Journal of the Royal Meteorological Society*, *138*(669), 2047–2065.

Dolan, B., & Rutledge, S. A. (2009). A theory-based hydrometeor identification algorithm for X-band polarimetric radars. *Journal of Atmospheric and Oceanic Technology*, *26*(10), 2071–2088. doi:10.1175/2009jtecha1208.1.

Dolan, B., Rutledge, S. A., Barnum, J. I., Matsui, T., Tao, W.-K., & Iguchi, T. (2017). Investigation of hydrometeor classification uncertainties through the POLARRIS polarimetric radar simulator. Paper presented at AGU Fall Meeting Abstracts.

Dolan, B., Rutledge, S. A., Lim, S., Chandrasekar, V., & Thurai, M. (2013). A robust C-band hydrometeor identification algorithm and application to a long-term polarimetric radar dataset. *Journal of Applied Meteorology and Climatology*, *52*(9), 2162–2186. doi:10.1175/jamc-d-12-0275.1.

Dong, X., & Mace, G. G. (2003). Profiles of low-level stratus cloud microphysics deduced from ground-based measurements. *Journal of Atmospheric Oceanic Technology*, *20*(1), 42–53.

Donovan, D., Van Lammeren, A., Hogan, R., Russchenberg, H., Apituley, A., Francis, P., et al. (2001). Cloud effective particle size and water content profile retrievals using combined lidar and radar observations: 2. Comparison with IR radiometer and in situ measurements of ice clouds. *Journal of Geophysical Research: Atmospheres*, *106*(D21), 27449–27464.

Dubovik, O., Herman, M., Holdak, A., Lapyonok, T., Taure, D., Deuzé, J., et al. (2011). Statistically optimized inversion algorithm for enhanced retrieval of aerosol properties from spectral multi-angle polarimetric satellite observations. *Atmospheric Measurement Techniques, 4*(5), 975–1018.

Dubovik, O., & King, M. D. (2000). A flexible inversion algorithm for retrieval of aerosol optical properties from Sun and sky radiance measurements. *Journal of Geophysical Research: Atmospheres*, *105*(D16), 20673–20696.

Ebell, K., Löhnert, U., Päschke, E., Orlandi, E., Schween, J. H., & Crewell, S. (2017). A 1-D variational retrieval of temperature, humidity, and liquid cloud properties: Performance under idealized and real conditions. *Journal of Geophysical Research: Atmospheres*, *122*(3), 1746–1766. doi:10.1002/2016JD025945.

Eccles, P. J., & Mueller, E. A. (1971). X-band attenuation and liquid water content estimation by a dual-wavelength radar. *Journal of Applied Meteorology*, *10*(6), 1252–1259.

Ehret, G., Kiemle, C., Wirth, M., Amediek, A., Fix, A., & Houweling, S. (2008). Space-borne remote sensing of CO_2, CH_4, and N_2O by integrated path differential absorption lidar: a sensitivity analysis. *Applied Physics B*, *90*(3–4), 593–608.

Ellis, S. M., & Vivekanandan, J. (2011). Liquid water content estimates using simultaneous S and Ka band radar measurements. *Radio Science*, *46*(2), 1–15.

Eloranta, E. (2005). High spectral resolution lidar. In C. Weitkamp (Ed.), *Lidar: Range-Resolved Optical Remote Sensing of the Atmosphere* (pp. 143–163). Springer Series in Optical Sciences.

Endo, S., Fridlind, A. M., Lin, W., Vogelmann, A. M., Toto, T., Ackerman, A. S., et al. (2015). RACORO continental boundary layer cloud investigations: 2. Large-eddy simulations of cumulus clouds and evaluation with in situ and ground-based observations. *Journal of Geophysical Research: Atmospheres*, *120*(12), 5993–6014.

Evans, K. F., Wang, J. R., Racette, P. E., Heymsfield, G., & Li, L. (2005). Ice cloud retrievals and analysis with the compact scanning submillimeter imaging radiometer and the cloud radar system during CRYSTAL FACE. *Journal of Applied Meteorology*, *44*(6), 839–859.

Ewald, F., Winkler, C., & Zinner, T. (2015). Reconstruction of cloud geometry using a scanning cloud radar. *Atmospheric Measurement Techniques*, *8*(6), 2491–2508.

Fabry, F., & Zawadzki, I. (1995). Long-term radar observations of the melting layer of precipitation and their interpretation. *Journal of the Atmospheric Sciences*, *52*(7), 838–851.

Fang, M., Albrecht, B., Jung, E., Kollias, P., Jonsson, H., & PopStefanija, I. (2017). Retrieval of vertical air motion in precipitating clouds using Mie scattering and comparison with in situ measurements. *Journal of Applied Meteorology and Climatology*, *56*(3), 537–553.

Feng, Z., Hagos, S., Rowe, A. K., Burleyson, C. D., Martini, M. N., & de Szoeke, S. P. (2015). Mechanisms of convective cloud organization by cold pools over tropical warm ocean during the AMIE/DYNAMO field campaign. *Journal of Advances in Modeling Earth Systems*, *7*(2), 357–381.

Fielding, M. D., Chiu, J., Hogan, R., Feingold, G., Eloranta, E., O'Connor, E., & Cadeddu, M. (2015). Joint retrievals of cloud and drizzle in marine boundary layer clouds using ground-based radar, lidar and zenith radiances. *Atmospheric Measurement Techniques*, *8*, 2663–2683.

Fielding, M. D., Chiu, J. C., Hogan, R. J., & Feingold, G. (2013). 3D cloud reconstructions: Evaluation of scanning radar scan strategy with a view to surface shortwave radiation closure. *Journal of Geophysical Research: Atmospheres*, *118*(16), 9153–9167.

Fielding, M. D., Chiu, J. C., Hogan, R. J., & Feingold, G. (2014). A novel ensemble method for retrieving properties of warm cloud in 3-D using ground-based scanning radar and zenith radiances. *Journal of Geophysical Research: Atmospheres*, *119*(18), 10,912–10,930.

Fix, A., Steinebach, F., Wirth, M., Schäfler, A., & Ehret, G. (2019). Development and application of an airborne differential absorption lidar for the simultaneous measurement of ozone and water vapor profiles in the tropopause region. *Applied Optics*, *58*(22), 5892–5900.

Fox, N. I., & Illingworth, A. J. (1997). The retrieval of stratocumulus cloud properties by ground-based cloud radar. *Journal of Applied Meteorology*, *36*(5), 485–492.

Freudenthaler, V., Esselborn, M., Wiegner, M., Heese, B., Tesche, M., Ansmann, A., et al. (2009). Depolarization ratio profiling at several wavelengths in pure Saharan dust during SAMUM 2006. *Tellus B: Chemical and Physical Meteorology*, *61*(1), 165–179.

Frisch, A., Feingold, G., Fairall, C., Uttal, T., & Snider, J. (1998). On cloud radar and microwave radiometer measurements of stratus cloud liquid water profiles. *Journal of Geophysical Research: Atmospheres*, *103*(D18), 23195–23197.

Gasteiger, J., Wiegner, M., Gross, S., Freudenthaler, V., Toledano, C., Tesche, M., & Kandler, K. (2011). Modelling lidar-relevant optical properties of complex mineral dust aerosols. *Tellus B: Chemical and Physical Meteorology*, *63*(4), 725–741.

Ghate, V. P., Miller, M. A., & DiPretore, L. (2011). Vertical velocity structure of marine boundary layer trade wind cumulus clouds. *Journal of Geophysical Research: Atmospheres (1984–2012)*, *116*(D16).

Giangrande, S. E., Collis, S., Straka, J., Protat, A., Williams, C., & Krueger, S. (2013). A summary of convective-core vertical velocity properties using ARM UHF wind profilers in Oklahoma. *Journal of Applied Meteorology Climatology*, *52*(10), 2278–2295.

Giangrande, S. E., Luke, E. P., & Kollias, P. (2010). Automated retrievals of precipitation parameters using non-Rayleigh scattering at 95 GHz. *Journal of Atmospheric and Oceanic Technology*, *27*(9), 1490–1503.

Giangrande, S. E., Luke, E. P., & Kollias, P. (2012). Characterization of vertical velocity and drop size distribution parameters in widespread precipitation at ARM facilities. *Journal of Applied Meteorology and Climatology*, *51*(2), 380–391.

Giangrande, S. E., & Ryzhkov, A. V. (2008). Estimation of rainfall based on the results of polarimetric echo classification. *Journal of Applied Meteorology and Climatology*, *47*(9), 2445–2462.

Granados-Muñoz, M., Navas-Guzmán, F., Bravo-Aranda, J., Guerrero-Rascado, J., Lyamani, H., Fernández-Gálvez, J., & Alados-Arboledas, L. (2012). Automatic determination of the planetary boundary layer height using lidar: One-year analysis over southeastern Spain. *Journal of Geophysical Research: Atmospheres*, *117*(D18).

Grecu, M., Tian, L., Heymsfield, G. M., Tokay, A., Olson, W. S., Heymsfield, A. J., & Bansemer, A. (2018). Nonparametric methodology to estimate precipitating ice from multiple-frequency radar reflectivity observations. *Journal of Applied Meteorology and Climatology*, *57*(11), 2605–2622. doi:10.1175/jamc-d-18-0036.1.

Grinnell, S. A., Bretherton, C. S., Fraser, A. M., & Stevens, D. E. (1996). Vertical mass flux calculations in Hawaiian trade cumulus clouds from dual-Doppler radar, *Journal of the atmospheric sciences*, *53*(13), 1870–1886.

Groß, S., Esselborn, M., Weinzierl, B., Wirth, M., Fix, A., & Petzold, A. (2013). Aerosol classification by airborne high spectral resolution lidar observations. *Atmospheric Chemistry and Physics*, *13*(5), 2487.

Gustafson, W. I., Jr, Vogelmann, A. M., Li, Z., Cheng, X., Dumas, K. K., Endo, S., et al. (2020). The Large-Eddy Simulation (LES) Atmospheric Radiation Measurement (ARM) Symbiotic Simulation and Observation (LASSO) activity for continental shallow convection. *Bulletin of the American Meteorological Society, 101*(4), E462–E479.

Haarig, M., Ansmann, A., Althausen, D., Klepel, A., Groß, S., Freudenthaler, V., et al. (2017a). Triple-wavelength depolarization-ratio profiling of Saharan dust over Barbados during SALTRACE in 2013 and 2014. *Atmospheric Chemistry and Physics*, *17*(17), 10767.

Haarig, M., Ansmann, A., Gasteiger, J., Kandler, K., Althausen, D., Baars, H. M., et al. (2017b). Dry versus wet marine particle optical properties: RH dependence of depolarization ratio, backscatter, and extinction from multiwavelength lidar measurements during SALTRACE. *Atmospheric Chemistry and Physics*, *17*(23), 14199.

Hair, J. W., Hostetler, C. A., Cook, A. L., Harper, D. B., Ferrare, R. A., Mack, T. L., et al. (2008). Airborne high spectral resolution lidar for profiling aerosol optical properties. *Applied Optics*, *47*(36), 6734–6752.

Harrison, L., Michalsky, J., & Berndt, J. (1994). Automated multifilter rotating shadow-band radiometer: An instrument for optical depth and radiation measurements. *Applied Optics*, *33*(22), 5118–5125.

Haynes, J. M., Luo, Z., Stephens, G., Marchand, R., & Bodas-Salcedo, A. (2007). A multipurpose radar simulation package: QuickBeam. *Bulletin of the American Meteorological Society*, *88*(11), 1723–1727.

Heymsfield, A. J., & Miloshevich, L. M. (1993). Homogeneous ice nucleation and supercooled liquid water in orographic wave clouds. *Journal of the Atmospheric Sciences*, *50*(15), 2335–2353.

Hillman, B. R., Marchand, R. T., & Ackerman, T. P. (2018). Sensitivities of simulated satellite views of clouds to subgrid-scale overlap and condensate heterogeneity. *Journal of Geophysical Research: Atmospheres*, *123*(14), 7506–7529.

Hirsikko, A., O'Connor, E., Komppula, M., Korhonen, K., Pfuller, A., Giannakaki, E., et al. (2014). Observing wind, aerosol particles, clouds and precipitation: Finland's new ground-based remote-sensing network. *Atmospheric Measurement Techniques*, *7*, 1351–1375.

Hobbs, P. V., Rangno, A. L., Shupe, M., & Uttal, T. (2001). Airborne studies of cloud structures over the Arctic Ocean and comparisons with retrievals from ship-based remote sensing measurements. *Journal of Geophysical Research: Atmospheres*, *106*(D14), 15029–15044.

Hogan, R. J., Gaussiat, N., & Illingworth, A. J. (2005a). Stratocumulus liquid water content from dual-wavelength radar. *Journal of Atmospheric and Oceanic Technology*, *22*(8), 1207–1218.

Hogan, R. J., Gaussiat, N., & Illingworth, A. J. (2005b). Stratocumulus liquid water content from dual-wavelength radar. *Journal of Atmospheric Oceanic Technology*, *22*(8), 1207–1218.

Hogan, R. J., Illingworth, A., O'connor, E., & Baptista, J. (2003). Characteristics of mixed-phase clouds. II: A climatology from ground-based lidar. *Quarterly Journal of the Royal Meteorological Society*, *129*(592), 2117–2134.

Hogan, R. J., & Illingworth, A. J. (2003). Parameterizing ice cloud inhomogeneity and the overlap of inhomogeneities using cloud radar data. *Journal of the Atmospheric Sciences*, *60*(5), 756–767.

Hogan, R. J., Mittermaier, M. P., & Illingworth, A. J. (2006). The retrieval of ice water content from radar reflectivity factor and temperature and its use in evaluating a mesoscale model. *Journal of Applied Meteorology and Climatology*, *45*(2), 301–317.

Hogan, R. J., & O'Connor, E. (2004). Facilitating cloud radar and lidar algorithms: The Cloudnet Instrument Synergy/Target Categorization product. Cloudnet documentation.

Holben, B. N., Kim, J., Sano, I., Mukai, S., Eck, T. F., Giles, D. M., et al. (2018). An overview of mesoscale aerosol processes, comparisons, and validation studies from DRAGON networks. *Atmospheric Chemistry and Physics*, *18*(2), 655–671.

Hu, Y., Rodier, S., Xu, K. M., Sun, W., Huang, J., Lin, B., et al. (2010). Occurrence, liquid water content, and fraction of supercooled water clouds from combined CALIOP/IIR/MODIS measurements. *Journal of Geophysical Research: Atmospheres*, *115*(D4).

Hu, Y., Vaughan, M., McClain, C., Behrenfeld, M., Maring, H., Anderson, D., et al. (2007). Global statistics of liquid water content and effective number concentration of water clouds over ocean derived from combined CALIPSO and MODIS measurements. *Atmospheric Chemistry Physics*, *7*(12), 3353–3359.

Hu, Y., Winker, D., Vaughan, M., Lin, B., Omar, A., Trepte, C., et al. (2009). CALIPSO/CALIOP cloud phase discrimination algorithm. *Journal of Atmospheric and Oceanic Technology*, *26*(11), 2293–2309.

Huang, D., Johnson, K., Liu, Y., & Wiscombe, W. (2009). High resolution retrieval of liquid water vertical distributions using collocated Ka-band and W-band cloud radars. *Geophysical Research Letters, 36*(24).

Huang, Y., Siems, S. T., Manton, M. J., Hande, L. B., & Haynes, J. M. (2012a). The structure of low-altitude clouds over the Southern Ocean as seen by CloudSat. *Journal of Climate*, *25*(7), 2535–2546.

Huang, Y., Siems, S. T., Manton, M. J., Protat, A., & Delanoë, J. (2012b). A study on the low-altitude clouds over the Southern Ocean using the DARDAR-MASK. *Journal of Geophysical Research: Atmospheres*, *117*(D18).

Illingworth, A., Hogan, R., O'connor, E., Bouniol, D., Brooks, M., Delanoë, J., et al. (2007). Cloudnet: Continuous evaluation of cloud profiles in seven operational models using ground-based observations. *Bulletin of the American Meteorological Society*, *88*(6), 883–898.

Intrieri, J. M., Stephens, G. L., Eberhard, W. L., & Uttal, T. (1993). A method for determining cirrus cloud particle sizes using lidar and radar backscatter technique. *Journal of Applied Meteorology*, *32*(6), 1074–1082.

Jakob, C., Pincus, R., Hannay, C., & Xu, K. M. (2004). Use of cloud radar observations for model evaluation: A probabilistic approach. *Journal of Geophysical Research: Atmospheres*, *109*(D3).

Jensen, M. P., Petersen, W. A., Bansemer, A., Bharadwaj, N., Carey, L., Cecil, D., et al. (2016). The midlatitude continental convective clouds experiment (MC3E). *Bulletin of the American Meteorological Society*, *97*(9), 1667–1686.

Kalesse, H., Kollias, P., & Szyrmer, W. (2013). On using the relationship between Doppler velocity and radar reflectivity to identify microphysical processes in midlatitudinal ice clouds. *Journal of Geophysical Research: Atmospheres*, *118*(21), 12,168–12,179. doi:doi:10.1002/2013JD020386.

Kalesse, H., Szyrmer, W., Kneifel, S., Kollias, P., & Luke, E. (2016). Fingerprints of a riming event on cloud radar Doppler spectra: observations and modeling. *Atmospheric Chemistry and Physics*, *16*(5), 2997–3012.

Kalthoff, N., Adler, B., Wieser, A., Kohler, M., Träumner, K., Handwerker, J., et al. (2013). KITcube—A mobile observation platform for convection studies deployed during HyMeX. *Meteorologische Zeitschrift*, *22*(6), 633–647.

Karlsson, K.-G., & Devasthale, A. (2018). Inter-comparison and evaluation of the four longest satellite-derived cloud climate data records: CLARA-A2, ESA Cloud CCI V3, *ISCCP-HGM, and PATMOS-x, Remote Sensing*, *10*(10), 1567.

Kassianov, E. I., Flynn, C. J., Ackerman, T. P., & Barnard, J. C. (2007). Aerosol single-scattering albedo and asymmetry parameter from MFRSR observations during the ARM Aerosol IOP 2003. *Atmospheric Chemistry and Physics*, *7*(12), 3341–3351.

Keat, W. J., & Westbrook, C. D. (2017). Revealing layers of pristine oriented crystals embedded within deep ice clouds using differential reflectivity and the copolar correlation coefficient. *Journal of Geophysical Research: Atmospheres*, *122*(21), 11,737–11,759. doi:10.1002/2017jd026754.

Kennedy, P. C., & Rutledge, S. A. (2011). S-band dual-polarization radar observations of winter storms. *Journal of Applied Meteorology and Climatology*, *50*(4), 844–858.

Kikuchi, K., Tsuboya, S., Sato, N., Asuma, Y., Takeda, T., & Fujiyoshi, Y. (1982). Observation of wintertime clouds and precipitation in the Arctic Canada (POLEX-North). *Journal of the Meteorological Society of Japan. Ser. II, 60(6)*, 1215–1226.

Klein, S. A., & Jakob, C. (1999). Validation and sensitivities of frontal clouds simulated by the ECMWF model. *Monthly Weather Review*, *127*(10), 2514–2531.

Kneifel, S., Dias Neto, J., Ori, D., Moisseev, D., Tyynela, J., Adams, I. S., et al. (2018). Summer snowfall workshop: Scattering properties of realistic frozen hydrometeors from simulations and observations, as well as defining a new standard for scattering databases. *Bulletin of the American Meteorological Society*, *99*(3), ES55–ES58.

Kneifel, S., Kulie, M., & Bennartz, R. (2011). A triple-frequency approach to retrieve microphysical snowfall parameters. *Journal of Geophysical Research: Atmospheres*, *116*(D11).

Kneifel, S., Lerber, A., Tiira, J., Moisseev, D., Kollias, P., & Leinonen, J. (2015). Observed relations between snowfall microphysics and triple-frequency radar measurements. *Journal of Geophysical Research: Atmospheres*, *120*(12), 6034–6055.

Kollias, P., & Albrecht, B. (2010). Vertical velocity statistics in fair-weather cumuli at the ARM TWP Nauru Climate Research Facility. *Journal of Climate*, *23*(24), 6590–6604.

Kollias, P., Albrecht, B., Lhermitte, R., & Savtchenko, A. (2001). Radar observations of updrafts, downdrafts, and turbulence in fair-weather cumuli. *Journal of the atmospheric sciences*, *58*(13), 1750–1766.

Kollias, P., Albrecht, B., & Marks, F., Jr., (2002). Why Mie? Accurate observations of vertical air velocities and raindrops using a cloud radar. *Bulletin of the American Meteorological Society*, *83*(10), 1471–1484.

Kollias, P., Bharadwaj, N., Clothiaux, E., Lamer, K., Oue, M., Hardin, J., et al. (2019). The ARM Radar Network: At the leading-edge of cloud and precipitation observations. *Bulletin of the American Meteorological Society, 101*(5), E588–E607.

Kollias, P., Clothiaux, E., Miller, M., Albrecht, B. A., Stephens, G., & Ackerman, T. (2007). Millimeter-wavelength radars: New frontier in atmospheric cloud and precipitation research.

Bulletin of the American Meteorological Society, *88*(10), 1608–1624.

Kollias, P., Lhermitte, R., & Albrecht, B. A. (1999). Vertical air motion and raindrop size distributions in convective systems using a 94 GHz radar. *Geophysical Research Letters*, *26*(20), 3109–3112.

Kollias, P., Luke E., Oue M., & Lamer K. (2020). Agile adaptive radar sampling of fast-evolving atmospheric phenomena guided by satellite imagery and surface cameras. *Geophysical Research Letters*, *47*(14).

Kollias, P., Palmer, R., Bodine, D., Adachi, T., Bluestein, H., Cho, J. Y. N., et al. (2022). Science applications of phased array radars. *Bulletin of the American Meteorological Society, 103*(10), E2370–E2390.

Kollias, P., Szyrmer, W., Rémillard, J., & Luke, E. (2011). Cloud radar Doppler spectra in drizzling stratiform clouds: 2. Observations and microphysical modeling of drizzle evolution, *Journal of Geophysical Research: Atmospheres*, *116*(D13).

Komurcu, M., Storelvmo, T., Tan, I., Lohmann, U., Yun, Y., Penner, J. E., et al. (2014). Intercomparison of the cloud water phase among global climate models. *Journal of Geophysical Research: Atmospheres*, *119*(6), 3372–3400.

Küchler, N., Kneifel, S., Kollias, P., & Löhnert, U. (2018). Revisiting liquid water content retrievals in warm stratified clouds: The modified Frisch. *Geophysical Research Letters*, *45*(17), 9323–9330.

Kulie, M. S., Hiley, M. J., Bennartz, R., Kneifel, S., & Tanelli, S. (2014). Triple-frequency radar reflectivity signatures of snow: Observations and comparisons with theoretical ice particle scattering models. *Journal of Applied Meteorology and Climatology*, *53*(4), 1080–1098.

Kumar, P., Sokolik, I., & Nenes, A. (2011a). Cloud condensation nuclei activity and droplet activation kinetics of wet processed regional dust samples and minerals. *Atmospheric Chemistry & Physics*, *11*(16), 8661–8676.

Kumar, P., Sokolik, I., & Nenes, A. (2011b). Measurements of cloud condensation nuclei activity and droplet activation kinetics of wet processed regional dust samples and minerals, *Atmospheric Chemistry & Physics Discussions*, *11*(4).

Kumjian, M. R. (2013). Principles and applications of dual-polarization weather radar. Part I: Description of the polarimetric radar variables. *Journal of Operational Meteorology*, *1*(20), 243–264.

Lamer, K. (2019). Relative occurrence of liquid water, ice and mixed-phased conditions within cloud and precipitation regimes: Long term ground-based observations for GCM model evaluation (143 pp). The Pennsylvania State University.

Lamer, K., Fridlind, A. M., Ackerman, A. S., Kollias, P., Clothiaux, E. E., & Kelley, M. (2018). (GO)2-SIM: A GCM-oriented ground-observation forward-simulator framework for objective evaluation of cloud and precipitation phase. *Geoscientific Model Development*, *11*(10), 4195–4214.

Lamer, K., & Kollias, P. (2015). Observations of fair-weather cumuli over land: Dynamical factors controlling cloud size and cover. *Geophysical Research Letters*, *42*(20), 8693–8701.

Lamer, K., Kollias, P., Battaglia, A., & Preval, S. (2020a). Mind the gap—Part 1: Accurately locating warm marine boundary layer clouds and precipitation using spaceborne radars. *Atmospheric Measurement Techniques*, *13*(5), 2363–2379. doi:10.5194/amt-13-2363-2020.

Lamer, K., Kollias, P., Battaglia, A., & Preval, S. (2020b). Mind-the-gap part I: Accurately locating warm marine boundary layer clouds and precipitation using spaceborne radars. *Atmospheric Measurement Techniques Discussions, 2020*, 1–29. doi:10.5194/amt-2019-473.

Lamer, K., Kollias, P., & Nuijens, L. (2015). Observations of the variability of shallow trade wind cumulus cloudiness and mass flux. *Journal of Geophysical Research: Atmospheres*, *120*(12), 6161–6178.

Lamer, K., Luke E. P., Walsh B. Jr, Andrade S., Mages Z., Zhu Z., et al. (2022) Going mobile to address emerging climate equity needs in the heterogeneous urban environment. *Bulletin of the American Meteorological Society, 103*(9), E2069–E2080.

Lamer, K., Puigdomènech Treserras, B., Zhu, Z., Isom, B., Bharadwaj, N., & Kollias, P. (2019). Characterization of shallow oceanic precipitation using profiling and scanning radar observations at the Eastern North Atlantic ARM observatory. *Atmospheric Measurement Techniques*, *12*(9), 4931–4947.

Lamer, K., Tatarevic, A., Jo, I., & Kollias, P. (2014). Evaluation of gridded scanning ARM cloud radar reflectivity observations and vertical doppler velocity retrievals. *Atmospheric Measurement Techniques*, *7*(4), 1089–1103.

Lange, D., Behrendt, A., & Wulfmeyer, V. (2019). Compact operational tropospheric water vapor and temperature Raman lidar with turbulence resolution. *Geophysical Research Letters*, *46*(24), 14844–14853.

Lareau, N. P., Zhang, Y., & Klein, S. A. (2018). Observed boundary layer controls on shallow cumulus at the ARM Southern Great Plains site. *Journal of the Atmospheric Sciences*, *75*(7), 2235–2255.

Lawson, R., Pilson, B., Baker, B., Mo, Q., Jensen, E., Pfister, L., & Bui, P. (2008). Aircraft measurements of microphysical properties of subvisible cirrus in the tropical tropopause layer. *Atmospheric Chemistry and Physics*, *8*(6), 1609–1620.

Leinonen, J., Lebsock, M. D., Tanelli, S., Sy, O. O., Dolan, B., Chase, R. J., et al. (2018). Retrieval of snowflake microphysical properties from multifrequency radar observations. *Atmospheric Measurement Techniques*, *11*(10), 5471.

Leinonen, J., & Moisseev, D. (2015). What do triple-frequency radar signatures reveal about aggregate snowflakes?. *Journal of Geophysical Research: Atmospheres*, *120*(1), 229–239.

Liao, L., & Sassen, K. (1994). Investigation of relationships between Ka-band radar reflectivity and ice and liquid water contents. *Atmospheric Research*, *34*(1–4), 231–248.

Liljegren, J. C., Clothiaux, E. E., Mace, G. G., Kato, S., & Dong, X. (2001). A new retrieval for cloud liquid water path using a ground-based microwave radiometer and measurements of cloud temperature. *Journal of Geophysical Research: Atmospheres*, *106*(D13), 14485–14500.

Lin, W., Liu, Y., Vogelmann, A. M., Fridlind, A., Endo, S., Song, H., et al. (2015). RACORO continental boundary layer cloud investigations: 3. Separation of parameterization biases single-column model CAM5 simulations of shallow cumulus. *Journal of Geophysical Research: Atmospheres*, *120*(12), 6015–6033.

Liu, C.-L., & Illingworth, A. J. (2000). Toward more accurate retrievals of ice water content from radar measurements of clouds. *Journal of Applied Meteorology*, *39*(7), 1130–1146.

Liu, H., & Chandrasekar, V. (2000). Classification of hydrometeors based on polarimetric radar measurements: Development of fuzzy logic and neuro-fuzzy systems, and in situ verification. *Journal of Atmospheric and Oceanic Technology*, *17*(2), 140–164. doi:10.1175/1520-0426(2000)017<0140:Cohbop>2.0.Co;2.

Liu, Y. (2019). Introduction to the special section on fast physics in climate models: parameterization, evaluation, and observation. *Journal of Geophysical Research: Atmospheres*, *124*(15), 8631–8644. doi:10.1029/2019jd030422.

Löhnert, U., Crewell, S., & Simmer, C. (2004). An integrated approach toward retrieving physically consistent profiles of temperature, humidity, and cloud liquid water. *Journal of Applied Meteorology*, *43*(9), 1295–1307.

Löhnert, U., Crewell, S., Simmer, C., & Macke, A. (2001). Profiling cloud liquid water by combining active and passive microwave measurements with cloud model statistics. *Journal of Atmospheric Oceanic Technology*, *18*(8), 1354–1366.

Löhnert, U., & Maier, O. (2012). Operational profiling of temperature using ground-based microwave radiometry at Payerne: Prospects and challenges. *Atmospheric Measurement Techniques*, *5*(5), 1121–1134.

Löhnert, U., Schween, J., Acquistapace, C., Ebell, K., Maahn, M., Barrera-Verdejo, M., et al. (2015). JOYCE: Jülich observatory for cloud evolution. *Bulletin of the American Meteorological Society*, *96*(7), 1157–1174.

Löhnert, U., Turner, D., & Crewell, S. (2009). Ground-based temperature and humidity profiling using spectral infrared and microwave observations. Part I: Simulated retrieval performance in clear-sky conditions. *Journal of Applied Meteorology and Climatology*, *48*(5), 1017–1032.

Lolli, S., Welton, E. J., & Campbell, J. R. (2013). Evaluating light rain drop size estimates from multiwavelength micropulse lidar network profiling. *Journal of Atmospheric and Oceanic Technology*, *30*(12), 2798–2807.

Lopatin, A., Dubovik, O., Chaikovsky, A., Goloub, P., Lapyonok, T., Tanré, D., & Litvinov, P. (2013). Enhancement of aerosol characterization using synergy of lidar and sun-photometer coincident observations: The GARRLiC algorithm. *Atmospheric Measurement Techniques*, *6*(8), 2065–2088.

Luke, E. P., & Kollias, P. (2013). Separating cloud and drizzle radar moments during precipitation onset using Doppler spectra. *Journal of Atmospheric and Oceanic Technology*, *30*(8), 1656–1671.

Luke, E. P., Kollias, P., Johnson, K. L., & Clothiaux, E. E. (2008). A technique for the automatic detection of insect clutter in cloud radar returns. *Journal of Atmospheric and Oceanic Technology*, *25*(9), 1498–1513.

Luo, Y., Xu, K.-M., Morrison, H., & McFarquhar, G. (2008a). Arctic mixed-phase clouds simulated by a cloud-resolving model: Comparison with ARM observations and sensitivity to microphysics parameterizations. *Journal of the Atmospheric Sciences*, *65*(4), 1285–1303.

Luo, Y., Xu, K. M., Morrison, H., McFarquhar, G. M., Wang, Z., & Zhang, G. (2008b). Multi-layer arctic mixed-phase clouds simulated by a cloud-resolving model: Comparison with ARM observations and sensitivity experiments. *Journal of Geophysical Research: Atmospheres*, *113*(D12).

Lux, O., Lemmerz, C., Weiler, F., Marksteiner, U., Witschas, B., Rahm, S., et al. (2020). Intercomparison of wind observations from ESA's satellite mission Aeolus and the ALADIN Airborne Demonstrator. *Atmospheric Measurement Techniques and Discussions*, *2020*, 1–32. doi:10.5194/amt-2019-431.

Maahn, M. (2015). Exploiting vertically pointing Doppler radar for advancing snow and ice cloud observations. Universität zu Köln.

Mace, G. G., Ackerman, T. P., Minnis, P., & Young, D. F. (1998). Cirrus layer microphysical properties derived from surface-based millimeter radar and infrared interferometer data. *Journal of Geophysical Research: Atmospheres*, *103*(D18), 23207–23216.

Mace, G. G., A. J. Heymsfield, & Poellot, M. R. (2002). On retrieving the microphysical properties of cirrus clouds using the moments of the millimeter-wavelength Doppler spectrum. *Journal of Geophysical Research: Atmospheres*, *107*(D24), AAC 22-21–AAC 22-26.

Mamouri, R.-E., & Ansmann, A. (2016). Potential of polarization lidar to provide profiles of CCN-and INP-relevant aerosol parameters. *Atmospheric Chemistry and Physics*, *16*(9), 5905–5931.

Mamouri, R.-E., & Ansmann, A. (2017). Potential of polarization/Raman lidar to separate fine dust, coarse dust, maritime, and anthropogenic aerosol profiles. *Atmospheric Measurement Techniques*, *10*(9), 3403–3427.

Marinou, E., Tesche, M., Nenes, A., Ansmann, A., Schrod, J., Mamali, D., et al. (2019). Retrieval of ice-nucleating particle concentrations from lidar observations and comparison with UAV in situ measurements. *Atmospheric Chemistry and Physics*, *19*(17), 11315–11342.

Martner, B. E. (1993). *Cloud liquid water content measurement tests using dual-wavelength radar*, US Dept. of Commerce, National Oceanic and Atmospheric Administration.

Mason, S., Chiu, C., Hogan, R., Moisseev, D., & Kneifel, S. (2018). Retrievals of riming and snow density from vertically pointing Doppler radars. *Journal of Geophysical Research: Atmospheres*, *123*(24), 13807–13834.

Mather, J. H., & Voyles, J. W. (2013). The ARM Climate Research Facility: A review of structure and capabilities, *Bulletin of the American Meteorological Society*, *94*(3), 377–392.

Matrosov, S. Y. (1999). Retrievals of vertical profiles of ice cloud microphysics from radar and IR measurements using tuned regressions between reflectivity and cloud parameters. *Journal of Geophysical Research: Atmospheres*, *104*(D14), 16741–16753.

Matrosov, S. Y. (2005). Attenuation-based estimates of rainfall rates aloft with vertically pointing Ka-band radars. *Journal of Atmospheric and Oceanic Technology*, *22*(1), 43–54.

Matrosov, S. Y., Korolev, A. V., & Heymsfield, A. J. (2002). Profiling cloud ice mass and particle characteristic size from Doppler radar measurements. *Journal of Atmospheric and Oceanic Technology*, *19*(7), 1003–1018.

Matrosov, S. Y., May, P. T., & Shupe, M. D. (2006). Rainfall profiling using atmospheric radiation measurement program vertically pointing 8-mm wavelength radars. *Journal of Atmospheric and Oceanic Technology*, *23*(11), 1478–1491.

Matrosov, S. Y., Reinking, R. F., Kropfli, R. A., Martner, B. E., & Bartram, B. (2001). On the use of radar depolarization ratios for estimating shapes of ice hydrometeors in winter clouds. *Journal of Applied Meteorology*, *40*(3), 479–490.

Matrosov, S. Y., Uttal, T., Snider, J. B., & Kropfli, R. A. (1992). Estimation of ice cloud parameters from ground-based infrared radiometer and radar measurements. *Journal of Geophysical Research: Atmospheres*, *97*(D11), 11567–11574. doi:10.1029/92JD00968.

McFarlane, S. A., Evans, K. F., & Ackerman, A. S. (2002). A Bayesian algorithm for the retrieval of liquid water cloud properties from microwave radiometer and millimeter radar data. *Journal of Geophysical Research: Atmospheres*, *107*(D16).

McFarquhar, G., Wood, R., Bretherton, C., Alexander, S., Jakob, C., Marchand, R., et al. (2014). The southern ocean clouds, radiation, aerosol transport experimental study (SOCRATES): An observational campaign for determining role of clouds, aerosols and radiation in climate system. Paper presented at AGU Fall Meeting Abstracts.

Mechem, D. B., Giangrande, S. E., Wittman, C. S., Borque, P., Toto, T., & Kollias, P. (2015). Insights from modeling and observational evaluation of a precipitating continental cumulus event observed during the MC3E field campaign. *Journal of Geophysical Research: Atmospheres*, *120*(5), 1980–1995.

Miller, L., & Strauch, R. (1974). A dual Doppler radar method for the determination of wind velocities within precipitating weather systems. *Remote Sensing of Environment*, *3*(4), 219–235.

Miller, M., Nitschke, K., Ackerman, T., Ferrell, W., Hickmon, N., & Ivey, M. (2016). The ARM mobile facilities. *Meteorological Monographs*, *57*, 9.1–9.15.

Moisseev, D. N., Lautaportti, S., Tyynela, J., & Lim, S. (2015). Dual-polarization radar signatures in snowstorms: Role of snowflake aggregation. *Journal of Geophysical Research: Atmospheres*, *120*(24), 12644–12655.

Moran, K. P., Martner, B. E., Post, M., Kropfli, R. A., Welsh, D. C., & Widener, K. B. (1998). An unattended cloud-profiling radar for use in climate research. *Bulletin of the American Meteorological Society*, *79*(3), 443–456.

North, K. W., Oue, M., Kollias, P., Giangrande, S. E., Collis, S. M., & Potvin, C. K. (2017). Vertical air motion retrievals in deep convective clouds using the ARM scanning radar network in Oklahoma during MC3E. Atmospheric Measurement Techniques (Online), 10(BNL-114147-2017-JA).

O'Connor, E. J., Hogan, R. J., & Illingworth, A. J. (2005). Retrieving stratocumulus drizzle parameters using Doppler radar and lidar. *Journal of Applied Meteorology*, *44*(1), 14–27.

O'Connor, E. J., Illingworth, A. J., Brooks, I. M., Westbrook, C. D., Hogan, R. J., Davies, F., & Brooks, B. J. (2010). A method for estimating the turbulent kinetic energy dissipation rate from a vertically pointing Doppler lidar, and independent evaluation from balloon-borne in situ measurements. *Journal of Atmospheric Oceanic Technology*, *27*(10), 1652–1664.

Orr, B. W., & Kropfli, R. A. (1999). A method for estimating particle fall velocities from vertically pointing Doppler radar. *Journal of Atmospheric and Oceanic Technology*, *16*(1), 29–37.

Oue, M., Kollias, P., North, K. W., Tatarevic, A., Endo, S., Vogelmann, A. M., & Gustafson, W. I. (2016). Estimation of cloud fraction profile in shallow convection using a scanning cloud radar. *Geophysical Research Letters*, *43*(20), 10,998–11,006.

Oue, M., Kollias, P., Ryzhkov, A., & Luke, E. P. (2018). Toward exploring the synergy between cloud radar polarimetry and Doppler spectral analysis in deep cold precipitating systems in the Arctic. *Journal of Geophysical Research: Atmospheres*, *123*(5), 2797–2815.

Oue, M., Kumjian, M. R., Lu, Y., Jiang, Z., Clothiaux, E. E., Verlinde, J., & Aydin, K. (2015a). X-band polarimetric and Ka-band Doppler spectral radar observations of a graupel-producing Arctic mixed-phase cloud. *Journal of Applied Meteorology and Climatology*, *54*(6), 1335–1351.

Oue, M., Kumjian, M. R., Lu, Y., Verlinde, J., Aydin, K., & Clothiaux, E. E. (2015b). Linear depolarization ratios of columnar ice crystals in a deep precipitating system over the Arctic observed by Zenith-Pointing Ka-band Doppler radar. *Journal of Applied Meteorology and Climatology*, *54*(5), 1060–1068. doi:10.1175/jamc-d-15-0012.1.

Oue, M., Tatarevic, A., Kollias, P., Wang, D., Yu, K., & Vogelmann, A. (2020). The Cloud-resolving model Radar SIMulator (CR-SIM) Version 3.3: description and applications of a virtual observatory. *Geoscientific Model Development (Print)*, *13*(BNL-213732-2020-JAAM).

Ovtchinnikov, M., Ackerman, T., Marchand, R., & Khairoutdinov, M. (2006). Evaluation of the multiscale modeling framework using data from the atmospheric radiation measurement program. *Journal of Climate*, *19*(9), 1716–1729.

Pal, S. R., Steinbrecht, W., & Carswell, A. I. (1992). Automated method for lidar determination of cloud-base height and vertical extent. *Applied Optics*, *31*(10), 1488–1494.

Park, H. S., Ryzhkov, A. V., Zrnić, D. S., & Kim, K.-E. (2009). The Hydrometeor classification algorithm for the polarimetric WSR-88D: Description and application to an MCS, *Weather and Forecasting*, *24*(3), 730–748. doi:10.1175/2008waf2222205.1.

Pearson, G. N., Roberts, P. J., Eacock, J. R., & Harris, M. (2002). Analysis of the performance of a coherent pulsed fiber lidar for aerosol backscatter applications. *Applied Optics*, *41*(30), 6442–6450.

Potvin, C. K., Shapiro, A., & Xue, M. (2012). Impact of a vertical vorticity constraint in variational dual-Doppler wind analysis: Tests with real and simulated supercell data, *Journal of Atmospheric and Oceanic Technology*, *29*(1), 32–49.

Protat, A., Delanoë, J., Bouniol, D., Heymsfield, A., Bansemer, A., & Brown, P. (2007). Evaluation of ice water content retrievals from cloud radar reflectivity and temperature using a large airborne in situ microphysical database. *Journal of Applied Meteorology and Climatology*, *46*(5), 557–572.

Qian, Y., Long, C. N., Wang, H., Comstock, J. M., McFarlane, S. A., & Xie, S. (2012). Evaluation of cloud fraction and its radiative effect simulated by IPCC AR4 global models against ARM surface observations. *Atmospheric Chemistry and Physics*, *12*(4), 1785–1810.

Räisänen, P., Barker, H. W., Khairoutdinov, M. F., Li, J., & Randall, D. A. (2004). Stochastic generation of subgrid-scale cloudy columns for large-scale models. *Quarterly Journal of the Royal Meteorological Society*, *130*(601), 2047–2067.

Randall, D., Krueger, S., Bretherton, C., Curry, J., Duynkerke, P., Moncrieff, M., et al. (2003). Confronting models with data: The GEWEX cloud systems study. *Bulletin of the American Meteorological Society*, *84*(4), 455–469.

Rangno, A. L., & Hobbs, P. V. (2001). Ice particles in stratiform clouds in the Arctic and possible mechanisms for the production of high ice concentrations. *Journal of Geophysical Research: Atmospheres*, *106*(D14), 15065–15075.

Rauber, R. M., Stevens, B., Ochs, H. T., III, Knight, C., Albrecht, B. A., Blyth, A., et al. (2007). Rain in shallow cumulus over the ocean: The RICO campaign. *Bulletin of the American Meteorological Society*, *88*(12), 1912–1928.

Reinking, R. F., Matrosov, S. Y., Kropfli, R. A., & Bartram, B. W. (2002). Evaluation of a 45 slant quasi-linear radar polarization state for distinguishing drizzle droplets, pristine ice crystals, and less regular ice particles. *Journal of Atmospheric and Oceanic Technology*, *19*(3), 296–321.

Rémillard, J., & Tselioudis, G. (2015). Cloud regime variability over the Azores and its application to climate model evaluation. *Journal of Climate*, *28*(24), 9707–9720.

Rinehart, R. (2004). *Radar for Meteorologists* (482 pp). Boulder, CO: Rinehart Publ.

Romps, D. M., & Öktem, R. (2018). Observing clouds in 4D with multiview stereophotogrammetry. *Bulletin of the American Meteorological Society*, *99*(12), 2575–2586.

Rossow, W. B., & Schiffer, R. A. (1991). ISCCP cloud data products. *Bulletin of the American Meteorological Society*, *72*(1), 2–20.

Roy, R. J., Lebsock, M., Millán, L., Dengler, R., Rodriguez Monje, R., Siles, J. V., & Cooper, K. B. (2018). Boundary-layer water vapor profiling using differential absorption radar, *Atmospheric Measurement Techniques*, *11*(12), 6511–6523.

Rusli, S. P., Donovan, D. P., & Russchenberg, H. W. (2017). Simultaneous and synergistic profiling of cloud and drizzle properties using ground-based observations. *Atmospheric Measurement Techniques*, *10*(12), 4777–4803.

Ryzhkov, A. V., Schuur, T. J., Burgess, D. W., Heinselman, P. L., Giangrande, S. E., & Zrnic, D. S. (2005). The Joint Polarization Experiment: Polarimetric rainfall measurements and hydrometeor classification. *Bulletin of the American Meteorological Society*, *86*(6), 809–824.

Sandford, C., Illingworth, A., & Thompson, R. (2017). The potential use of the linear depolarization ratio to distinguish between convective and stratiform rainfall to improve radar rain-rate estimates. *Journal of Applied Meteorology and Climatology*, *56*(11), 2927–2940.

Sassen, K. (1991). The polarization lidar technique for cloud research: A review and current assessment, *Bulletin of the American Meteorological Society*, *72*(12), 1848–1866.

Sato, N., Kikuchi, K., Barnard, S. C., & Hogan, A. W. (1981). Some characteristic properties of ice crystal precipitation in the summer season at South Pole Station, Antarctica. *Journal of the Meteorological Society of Japan. Ser. II, 59(5)*, 772–780.

Satoh, M., Stevens, B., Judt, F., Khairoutdinov, M., Lin, S.-J., Putman, W. M., & Düben, P. (2019). Global cloud-resolving models. *Current Climate Change Reports*, *5*(3), 172–184.

Sauvageot, H., & Omar, J. (1987). Radar reflectivity of cumulus clouds. *Journal of Atmospheric and Oceanic Technology*, *4*(2), 264–272.

Schutgens, N., Tsyro, S., Gryspeerdt, E., Goto, D., Weigum, N., Schulz, M., & Stier, P. (2017). On the spatio-temporal representativeness of observations. *Atmospheric Chemistry and Physics, 17*(16), 9761–9780.

Schween, J., Hirsikko, A., Löhnert, U., & Crewell, S. (2014). Mixing-layer height retrieval with ceilometer and Doppler lidar: from case studies to long-term assessment. *Atmospheric Measurement Techniques*, *7*(11), 3685–3704.

Seidel, D. J., Ao, C. O., & Li, K. (2010). Estimating climatological planetary boundary layer heights from radiosonde observations: Comparison of methods and uncertainty analysis. *Journal of Geophysical Research: Atmospheres*, *115*(D16).

Serikov, I., Linné, H., Brügmanu, B., Worbes, L., Nicolae, D., Belegante, L., & Amiridis, V. (2018). ESA Airborne 3+ 2+ 2 HSRL for Aladin/Atlid CAL/VAL. Paper presented at IGARSS 2018-2018 IEEE International Geoscience and Remote Sensing Symposium, *IEEE*.

Shupe, M. D. (2007). A ground-based multisensor cloud phase classifier. *Geophysical Research Letters*, *34*(22).

Shupe, M. D., Kollias, P., Poellot, M., & Eloranta, E. (2008). On deriving vertical air motions from cloud radar Doppler spectra. *Journal of Atmospheric and Oceanic Technology*, *25*(4), 547–557.

Sisterson, D., Peppler, R., Cress, T., Lamb, P., & Turner, D. (2016). The ARM southern great plains (SGP) site. *Meteorological Monographs*, *57*, 6.1–6.14.

Song, H., Lin, W., Lin, Y., Wolf, A. B., Donner, L. J., Del Genio, A. D., et al. (2014). Evaluation of cloud fraction simulated by seven SCMs against the ARM observations at the SGP site. *Journal of Climate*, *27*(17), 6698–6719.

Stachlewska, I. S., Piądłowski, M., Migacz, S., Szkop, A., Zielińska, A. J., & Swaczyna, P. L. (2012). Ceilometer observations of the boundary layer over Warsaw, *Poland. Acta Geophysica, 60*(5), 1386–1412.

Stein, T. H., Westbrook, C. D., & Nicol, J. (2015). Fractal geometry of aggregate snowflakes revealed by triple-wavelength radar measurements. *Geophysical Research Letters*, *42*(1), 176–183.

Stephens, G. L., Ellingson, R., Vitko, J., Jr., Bolton, W., Tooman, T., Valero, F., et al. (2000). The department of energy's atmospheric radiation measurement (ARM) unmanned aerospace vehicle (UAV) program. *Bulletin of the American Meteorological Society*, *81*(12), 2915–2938.

Stephens, G. L., Winker, D., Pelon, J., Trepte, C., Vane, D., Yuhas, C., et al. (2018). CloudSat and CALIPSO within the A-Train: Ten years of actively observing the Earth system. *Bulletin of the American Meteorological Society*, *99*(3), 569–581.

Stevens, B., Acquistapace, C., Hansen, A., Heinze, R., Klinger, C., Klocke, D., et al. (2020). The added value of large-eddy and storm-resolving models for simulating clouds and precipitation. *Journal of the Meteorological Society of Japan. Ser. II, 98(2)*, 395–435.

Stevens, B., Farrell, D., Hirsch, L., Jansen, F., Nuijens, L., Serikov, I., et al. (2016). The Barbados Cloud Observatory: Anchoring investigations of clouds and circulation on the edge of the ITCZ. *Bulletin of the American Meteorological Society*, *97*(5), 787–801.

Stokes, G. M., & Schwartz, S. E. (1994). The Atmospheric Radiation Measurement (ARM) Program: Programmatic background and design of the cloud and radiation test bed. *75*(7), 1201–1222.

Straka, J. M., Zrnić, D. S., & Ryzhkov, A. V. (2000). Bulk hydrometeor classification and quantification using polarimetric radar data: synthesis of relations. *Journal of Applied Meteorology*, *39*(8), 1341–1372. doi:10.1175/1520-0450(2000)039<1341:Bhcaqu>2.0.Co;2.

Szyrmer, W., Tatarevic, A., & Kollias, P. (2012). Ice clouds microphysical retrieval using 94-GHz Doppler radar observations: Basic relations within the retrieval framework. *Journal of Geophysical Research: Atmospheres*, *117*(D14).

Talianu, C., Nicolae, D., Ciuciu, J., Ciobanu, M., & Babin, V. (2006). Planetary boundary layer height detection from LIDAR measurements. *Journal of Optoelectronics Advanced Materials*, *8*(1), 243.

Tesche, M., Ansmann, A., Müller, D., Althausen, D., Engelmann, R., Freudenthaler, V., & Groß, S. (2009). Vertically resolved separation of dust and smoke over Cape Verde using multiwavelength Raman and polarization lidars during Saharan Mineral Dust Experiment 2008. *Journal of Geophysical Research: Atmospheres*, *114*(D13).

Tridon, F., & Battaglia, A. (2015). Dual-frequency radar Doppler spectral retrieval of rain drop size distributions and entangled dynamics variables. *Journal of Geophysical Research: Atmospheres*, *120*(11), 5585–5601.

Tridon, F., Battaglia, A., & Kollias, P. (2013). Disentangling Mie and attenuation effects in rain using a Ka-W dual-wavelength Doppler spectral ratio technique. *Geophysical Research Letters*, *40*(20), 5548–5552.

Tridon, F., Battaglia, A., Luke, E., & Kollias, P. (2017). Rain retrieval from dual-frequency radar Doppler spectra: Validation and potential for a midlatitude precipitating case-study. *Quarterly Journal of the Royal Meteorological Society*, *143*(704), 1364–1380.

Tselioudis, G., Rossow, W., Zhang, Y., & Konsta, D. (2013). Global weather states and their properties from passive and active satellite cloud retrievals. *Journal of Climate*, *26*(19), 7734–7746.

Tucker, S. C., Senff, C. J., Weickmann, A. M., Brewer, W. A., Banta, R. M., Sandberg, S. P., et al. (2009). Doppler lidar estimation of mixing height using turbulence, shear, and aerosol profiles. *Journal of Atmospheric Oceanic Technology*, *26*(4), 673–688.

Turner, D. D., Clough, S. A., Liljegren, J. C., Clothiaux, E. E., Cady-Pereira, K. E., & Gaustad, K. L. (2007). Retrieving liquid water path and precipitable water vapor from the Atmospheric Radiation Measurement (ARM) microwave radiometers. *IEEE Transactions on Geoscience and Remote Sensing*, *45*(11), 3680–3690.

Turner, D. D., Goldsmith, J., & Ferrare, R. (2016). Development and applications of the ARM Raman lidar. *Meteorological Monographs*, *57*, 18.11–18.15.

Turner, D. D., & Löhnert, U. (2014). Information content and uncertainties in thermodynamic profiles and liquid cloud properties retrieved from the ground-based Atmospheric Emitted Radiance Interferometer (AERI). *Journal of Applied Meteorology and Climatology*, *53*(3), 752–771.

Tuttle, J. D., & Rinehart, R. E. (1983). Attenuation correction in dual-wavelength analyses. *Journal of Climate Applied Meteorology*, *22*(11), 1914–1921.

Tyynela, J., & Chandrasekar, V. (2014). Characterizing falling snow using multifrequency dual-polarization measurements. *Journal of Geophysical Research: Atmospheres, 119*(13). 8268–8283.

Uttal, T., Curry, J. A., McPhee, M. G., Perovich, D. K., Moritz, R. E., Maslanik, J. A., et al. (2002). Surface heat budget of the Arctic Ocean. *Bulletin of the American Meteorological Society*, *83*(2), 255–276.

Vali, G., Kelly, R., Pazmany, A., & McIntosh, R. (1995). Airborne radar and in-situ observations of a shallow stratus with drizzle. *Atmospheric Research*, *38*(1–4), 361–380.

Vali, G., Kelly, R. D., French, J., Haimov, S., Leon, D., McIntosh, R. E., & Pazmany, A. (1998). Finescale structure and microphysics of coastal stratus. *Journal of the Atmospheric Sciences*, *55*(24), 3540–3564.

VanZanten, M., Stevens, B., Vali, G., & Lenschow, D. (2005). Observations of drizzle in nocturnal marine stratocumulus. *Journal of the Atmospheric Sciences*, *62*(1), 88–106.

Varble, A., Fridlind, A. M., Zipser, E. J., Ackerman, A. S., Chaboureau, J. P., Fan, J., et al. (2011). Evaluation of cloud-resolving model intercomparison simulations using TWP-ICE observations: Precipitation and cloud structure. *Journal of Geophysical Research: Atmospheres*, *116*(D12).

Vaughan, M. A., Powell, K. A., Winker, D. M., Hostetler, C. A., Kuehn, R. E., Hunt, W. H., et al. (2009). Fully automated detection of cloud and aerosol layers in the CALIPSO lidar measurements. *Journal of Atmospheric and Oceanic Technology*, *26*(10), 2034–2050.

Vivekanandan, J., Li, L., Tsang, L., & Chan, C. (1997). Microwave radiometric technique to retrieve vapor, liquid and ice. II. Joint studies of radiometer and radar in winter clouds. *IEEE Transactions on Geoscience Remote Sensing*, *35*(2), 237–247.

Vivekanandan, J., Zrnic, D., Ellis, S., Oye, R., Ryzhkov, A., & Straka, J. (1999). Cloud microphysics retrieval using S-band dual-polarization radar measurements. *Bulletin of the American Meteorological Society*, *80*(3), 381–388.

Vogelmann, A. M., Fridlind, A. M., Toto, T., Endo, S., Lin, W., Wang, J., et al. (2015). RACORO continental boundary layer cloud investigations: 1. Case study development and ensemble large-scale forcings. *Journal of Geophysical Research: Atmospheres*, *120*(12), 5962–5992.

Voudouri, K. A., Giannakaki, E., Komppula, M., & Balis, D. (2019). Variability of cirrus cloud properties using a PollyXT Raman Lidar over high and tropical latitudes. *Atmospheric Chemistry and Physics Discussions.* https://doi.org/10.5194/acp-2019-565, in review.

Wagner, T. J., Klein P. M., & Turner D. D. (2019). A new generation of ground-based mobile platforms for active and passive profiling of the boundary layer. *Bulletin of the American Meteorological Society, 100*(1), 137–153.

Wang, T., & Min, Q. (2008). Retrieving optical depths of optically thin and mixed-phase clouds from MFRSR measurements. *Journal of Geophysical Research: Atmospheres, 113*(D19).

Wang, Z. (2007). A refined two-channel microwave radiometer liquid water path retrieval for cold regions by using multiple-sensor measurements. *IEEE Geoscience and Remote Sensing Letters, 4*(4), 591–595.

Wang, Z., French, J., Vali, G., Wechsler, P., Haimov, S., Rodi, A., et al. (2012). Single aircraft integration of remote sensing and in situ sampling for the study of cloud microphysics and dynamics, *Bulletin of the American Meteorological Society, 93*(5), 653–668.

Wang, Z., & Sassen, K. (2001). Cloud type and macrophysical property retrieval using multiple remote sensors. *Journal of Applied Meteorology, 40*(10), 1665–1682.

Wang, Z., & Sassen, K. (2002). Cirrus cloud microphysical property retrieval using lidar and radar measurements. Part I: Algorithm description and comparison with in situ data. *Journal of Applied Meteorology, 41*(3), 218–229.

Webb, M., Senior, C., Bony, S., & Morcrette, J.-J. (2001). Combining ERBE and ISCCP data to assess clouds in the Hadley Centre, ECMWF and LMD atmospheric climate models. *Climate Dynamics, 17*(12), 905–922.

Weckwerth, T. M., Weber, K. J., Turner, D. D., & Spuler, S. M. (2016). Validation of a water vapor micropulse differential absorption lidar (DIAL). *Journal of Atmospheric and Oceanic Technology, 33*(11), 2353–2372.

Wei, N., Zhou, L., & Dai, Y. (2017). Evaluation of simulated climatological diurnal temperature range in CMIP5 models from the perspective of planetary boundary layer turbulent mixing. *Climate Dynamics, 49*(1–2), 1–22.

Weitkamp, C. (2006). *Lidar: Range-resolved optical remote sensing of the atmosphere*. Springer Science & Business.

Westbrook, C., Hogan, R., Illingworth, A., & O'Connor, E. (2007). Theory and observations of ice particle evolution in cirrus using Doppler radar: Evidence for aggregation. *Geophysical Research Letters, 34*(2).

Westbrook, C., Hogan, R., O'Connor, E., & Illingworth, A. (2010). Estimating drizzle drop size and precipitation rate using two-colour lidar measurements. *Atmospheric Measurement Techniques, 3*(3), 671–681.

Williams, C. R. (2012). Vertical air motion retrieved from dual-frequency profiler observations. *Journal of Atmospheric and Oceanic Technology, 29*(10), 1471–1480.

Williams, J. K., & Vivekanandan, J. (2007). Sources of error in dual-wavelength radar remote sensing of cloud liquid water content. *Journal of Atmospheric Oceanic Technology, 24*(8), 1317–1336.

Winker, D. M., Pelon, J. R., & McCormick, M. P. (2003). *CALIPSO mission: Spaceborne lidar for observation of aerosols and clouds. Paper presented at Lidar Remote Sensing for Industry and Environment Monitoring III*, International Society for Optics and Photonics.

Wirth, M., Fix, A., Mahnke, P., Schwarzer, H., Schrandt, F., & Ehret, G. (2009). The airborne multi-wavelength water vapor differential absorption lidar WALES: System design and performance. *Applied Physics B, 96*(1), 201.

Wolde, M., & Vali, G. (2001). Polarimetric signatures from ice crystals observed at 95 GHz in winter clouds. Part I: Dependence on crystal form. *Journal of the Atmospheric Sciences, 58*(8), 828–841.

Wood, R. (2005). Drizzle in stratiform boundary layer clouds. Part II: Microphysical aspects. *Journal of the Atmospheric Sciences, 62*(9), 3034–3050.

Wu, W., Liu, Y., Jensen, M. P., Toto, T., Foster, M. J., & Long, C. N. (2014). A comparison of multiscale variations of decade-long cloud fractions from six different platforms over the Southern Great Plains in the United States. *Journal of Geophysical Research: Atmospheres, 119*(6), 3438–3459.

Wurman, J., Kosiba K., Pereira B., Robinson P., Frambach A., Gilliland A., et al. (2021). The Flexible Array of Radars and Mesonets (FARM). *Bulletin of the American Meteorological Society, 102*(8). E1499–E1525.

Xie, S., Klein, S. A., Zhang, M., Yio, J. J., Cederwall, R. T., & McCoy, R. (2006). Developing large-scale forcing data for single-column and cloud-resolving models from the Mixed-Phase Arctic Cloud Experiment. *Journal of Geophysical Research: Atmospheres, 111*(D19).

Xie, S., McCoy, R. B., Klein, S. A., Cederwall, R. T., Wiscombe, W. J., Jensen, M. P., et al. (2010). Clouds and more: ARM climate modeling best estimate data. *Bulletin of the American Meteorological Society, 91*(1), 13–20, doi:10.1175/2009bams2891.1.

Xu, K. M., Cederwall, R. T., Donner, L. J., Grabowski, W. W., Guichard, F., Johnson, D. E., et al. (2002). An intercomparison of cloud-resolving models with the Atmospheric Radiation Measurement summer 1997 Intensive Observation Period data. *Quarterly Journal of the Royal Meteorological Society, 128*(580), 593–624.

Yang, F., Pan, H.-L., Krueger, S. K., Moorthi, S., & Lord, S. J. (2006). Evaluation of the NCEP Global Forecast System at the ARM SGP site. *Monthly weather review, 134*(12), 3668–3690.

Yoshida, R., Okamoto, H., Hagihara, Y., & Ishimoto, H. (2010). Global analysis of cloud phase and ice crystal orientation from Cloud-Aerosol Lidar and Infrared Pathfinder Satellite Observation (CALIPSO) data using attenuated backscattering and depolarization ratio. *Journal of Geophysical Research: Atmospheres, 115*(D4).

Yoshida, Y., Asano, S., & Iwanami, K. (2006). Retrieval of microphysical properties of water, ice, and mixed-phase clouds using a triple-wavelength radar and microwave radiometer. *Journal of the Meteorological Society of Japan. Ser. II, 84(6)*, 1005–1031.

Yu, G., Verlinde, J., Clothiaux, E. E., & Chen, Y.-S. (2014). Mixed-phase cloud phase partitioning using millimeter wavelength cloud radar Doppler velocity spectra. *Journal of Geophysical Research: Atmospheres, 119*(12), 7556–7576. doi:10.1002/2013JD021182.

Zhang, D., Wang, Z., Heymsfield, A., Fan, J., & Luo, T. (2014). Ice concentration retrieval in stratiform mixed-phase clouds using cloud radar reflectivity measurements and 1D ice growth model simulations. *Journal of the Atmospheric Sciences, 71*(10), 3613–3635.

Zhang, M., & Lin, J. (1997). Constrained variational analysis of sounding data based on column-integrated budgets of mass, heat, moisture, and momentum: Approach and application to ARM measurements, *Journal of the Atmospheric Sciences*, *54*(11), 1503–1524.

Zhang, Y., Klein, S., Boyle, J., & Mace, G. (2010). Evaluation of tropical cloud and precipitation statistics of Community Atmosphere Model version 3 using CloudSat and CALIPSO data. *Journal of Geophysical Research: Atmospheres*, *115*(D12).

Zhang, Y., Xie, S., Klein, S. A., Marchand, R., Kollias, P., Clothiaux, E. E., et al. (2018). The ARM cloud radar simulator for global climate models: A new tool for bridging field data and climate models. *Bulletin of the American Meteorological Society, 99*(1), 21–26.

Zhou, X., Kollias, P., & Lewis, E. R. (2015). Clouds, precipitation, and marine boundary layer structure during the MAGIC field campaign. *Journal of Climate*, *28*(6), 2420–2442.

Zhu, Z., Lamer, K., Kollias, P., & Clothiaux, E. E. (2019). The vertical structure of liquid water content in shallow clouds as retrieved from dual-wavelength radar observations. *Journal of Geophysical Research: Atmospheres*, *124*(24), 14184–14197.

15

Satellite and Airborne Remote Sensing of Clouds and Aerosols

Alexander Marshak[1] and Anthony B. Davis[2]

ABSTRACT

We survey overhead remote sensing of cloud and aerosol properties, from either aircraft or satellites. Standoff distances range from NASA's P-3B aircraft at only ≈3.5 km above cloud top to DSCOVR at Lagrange-1, ≈1,500,000 km toward the Sun. Although not a remote-sensing technique per se, deployment of in situ radiometric sensors that collect photons rather than cloud particles is also briefly discussed. We run the spectral gamut from ultraviolet to microwaves, for both active and passive modalities. Emphasis is on the physics of sensors and retrieval algorithms. Moreover, since neither are perfect, we devote a section to uncertainty quantification. Pros and cons of high and low spatial resolutions are presented in connection with the need for fine or course spectral sampling and/or global coverage versus targeted operation. Angular sampling is also considered, particularly, in connection with polarization. We touch on the need for "research" algorithms applied to selected observations, often overflying a field campaign. These specialty algorithms require some level of human supervision and thus complement their operational counterparts, which must be globally applicable and automated with as few mishaps as possible. Finally, we do not shy away from describing a few "out-of-the-box" methods that open the path for future missions.

15.1. INTRODUCTION AND OUTLINE

Interactions between clouds and aerosols are among the largest sources of uncertainty in estimating human impacts on climate. As stated in the Intergovernmental Panel on Climate Change (IPCC) 5th Assessment Report (IPCC, 2013) Chapter 7, "Clouds and aerosols continue to contribute the largest uncertainty to estimates and interpretations of the Earth's changing energy budget [...] the quantification [...] of aerosol–cloud interactions continue[s] to be a challenge."

[1]*NASA Goddard Space Flight Center, Greenbelt, MD, USA*
[2]*NASA Jet Propulsion Laboratory, California Institute of Technology, Pasadena, CA, USA*

To address this challenge, a large contingent of atmospheric scientists have pursued research using a wide variety of approaches. Since about 2005, these efforts included the analysis of satellite observations with the goal of examining the relationship between aerosol properties and regional cloud coverage (e.g., Ignatov et al., 2005; Loeb & Manalo-Smith, 2005; Loeb & Schuster, 2008; Zhang et al., 2005). Later, scientists also examined how aerosol properties vary with the distance to the nearest clouds using observations from satellites, aircraft, and ground (Koren et al., 2007; Redemann et al., 2009; Su et al., 2008; Tackett & Di Girolamo, 2009). These remote-sensing studies were complimented by studies that used in situ observations to gain insights into some processes contributing to near-cloud aerosol changes such as swelling or chemical cloud processing (e.g., Hudson et al., 2015; Rauber et al., 2013; Twohy et al., 2009). In the meantime, other studies

Fast Processes in Large-Scale Atmospheric Models: Progress, Challenges, and Opportunities, Geophysical Monograph 282, First Edition.
Edited by Yangang Liu and Pavlos Kollias.

DOI: 10.1002/9781119529019.ch15

used time series of observations to characterize how aerosol populations change after clouds formed in the area (Eck et al., 2012, 2014; Várnai et al., 2017) or used process modeling to better understand intercloud regions (Bar-Or et al., 2012; Koren et al., 2009), in some cases, combining models with observations (e.g., Chand et al., 2012; Jeong & Li, 2010). Based on results from such studies, the IPCC 5th Assessment Report (IPCC, 2013) summarized the issue as follows: "[...] aerosol measured in the vicinity of clouds is significantly different than it would be were the cloud field, and its proximate cause (high humidity), not present. The latter results from humidification effects on aerosol optical properties, contamination by undetectable cloud fragments, and the remote effects of radiation scattered by cloud edges on aerosol retrieval."

In order to address aerosols, clouds, and their interactions, with remote-sensing assets, we must first understand remote-sensing methodology for both aerosols and clouds from both satellite and aircraft platforms. In this chapter, we therefore emphasize the physics-based algorithms used to translate remotely measured radiances into inherent cloud and aerosol properties; implementations with specific sensors are also listed, but will not be exhaustive, partly because the list is constantly evolving. If some acronyms are left undefined in the text, the reader will find a list of definitions at the end of the chapter. Also, we balance a comprehensive (but far from complete) description of operational remote-sensing products with a few examples of experimental approaches. For the theoretical basis of radiative transfer (RT) and radiometry, which is the physics that underlies all of the remote-sensing techniques described herein, we refer the reader to Chapter 2 of this volume by Y. Gu and K.-N. Liou.

A preliminary question in cloud and aerosol remote sensing is to identify the clouds in the signal, as opposed to an aerosol/surface scene. Indeed, it is intuitively clear that methodologies for the extraction of physical properties from remotely observed radiances will be qualitatively different for clouds and aerosols if only because the later are generally—but not always—optically thin enough to see the underlying surface. This first order of business is usually described as "cloud masking." Although it sounds as something easy, it is in fact a daunting task to do robustly in a fully automated way. Pitfalls are plentiful and, to appreciate that, one only needs to think about detecting clouds from above over snow and ice, or distinguishing an optically thick aerosol plume from a cloud. It is certainly not just a binary matter of brightness/opacity. There is color, height, spatial variability, texture, and so on to consider as well. Ultimately, we need to ask: What is a cloud? Isn't a subvisual cirrus not just a lofted aerosol layer that happens to be composed of solid water? We will just touch on the science—and art—of cloud detection in the next section. It would take a chapter of its own to cover in any depth. It is indeed a perennial problem in satellite remote sensing, from its very dawn (e.g., Goodman & Henderson-Sellers, 1988) to the most recent contributions of machine learning (e.g., Mahajan & Fataniya, 2020). We also leave the related question of cloud fraction estimation out of the present scope even though it is clearly paramount in climate modeling. We however discuss, as it comes up, the related notion of partial cloud cover *inside* a pixel or sensor "footprint" since it impacts directly the retrieval, most notably in the sense of error quantification, when these specifications of inherent spatial resolution are relatively large.

This chapter is structured as follows. In the following section, we describe satellite remote sensing of clouds using passive methods. In section 15.3, we do the same for aerosols. In section 15.4, we focus on the retrieval of vertical profiles for clouds and aerosols from space-based sensors, which includes but is not limited to active (radar and lidar) methods since there is sensitivity to cloud/aerosol layer height in steady radiation fields at wavelengths in the visible (VIS)/near-IR (O_2 absorption bands), as well as in the ultraviolet (UV). In section 15.5, we discuss the opportunities that open up in cloud and aerosol remote sensing when the sensors' spatial resolution is at least an order of magnitude better that the hundreds on meters achieved routinely from space; this includes but is not limited to airborne instruments. Section 15.6 is a brief discussion of vast topic of retrieval uncertainty quantification and how it relates to retrieved product validation, the very core of remote sensing; the examples used draw on the present authors' experience in assessing the impacts of forward model fidelity failures. Finally, in section 15.7, we summarize and look into the future of cloud and/or aerosol remote sensing.

15.2. PASSIVE SENSING OF COLUMN-INTEGRATED CLOUD PROPERTIES

15.2.1. Nakajima-King Bispectral Approach

Twomey and Cocks (1989) first suggested a simultaneous retrieval of the cloud optical thickness (COT), τ, and effective particle size, r_e, from the reflected solar radiation measurements. However, the procedure that has been used in cloud remote sensing for almost three decades with few substantial changes was first proposed and developed by Nakajima and King (1990). The physical principle behind this method is simple. The reflected solar radiation is measured at two wavelengths outside of all gaseous absorption regions: one in the visible where condensed water does not absorb, and one in the shortwave-IR (SWIR) spectral region where it does. This provides us with τ and r_e. The first measurement is in the VIS or near-IR (VNIR) region (e.g., 0.64 or 0.86 μm), where water absorption is negligible and therefore cloud reflection increases monotonically with τ. In the other measurement, at SWIR wavelengths (e.g., 1.6, 2.1, or 3.7 μm), where water droplets are moderately absorptive, cloud reflectance decreases with increasing r_e for optically thick clouds since, at the individual particle scale, absorption goes as its volume. The single-scattering coalbedo directly scales with effective radius (e.g., Twomey & Bohren, 1980). This technique is widely known as the "Nakajima-King" retrieval. Figure 15.1 shows how the method is implemented with look-up tables (LUTs), that is, an array of VNIR and SWIR cloud reflectances for an array of (τ, r_e) values at given solar and viewing zenith angles (SZAs and VZAs). The algorithm's refinements for different collections from the MODerate resolution Imaging Spectroradiometer (MODIS) on Terra and Aqua are discussed in detail by Platnick et al. (2003, 2017).

Pairing a VNIR channel with different SWIR channels yields somewhat different values of r_e. The question boils down to where the light comes from, i.e., what is its vertical "weighting function." In this bispectral technique, the COT τ is nearly independent of wavelength, hence so is the vertically averaged extinction coefficient $\sigma = \tau/H$, where H is cloud thickness. In contrast, the different SWIR wavelengths are associated with different

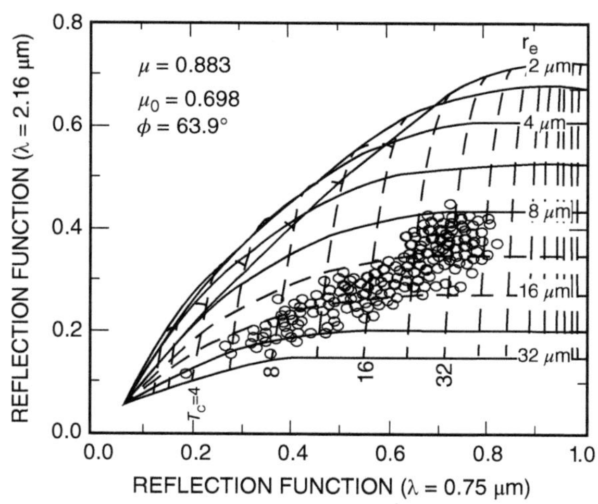

Figure 15.1 Look-up tables (LUTs) obtained from RT calculations show the relationship between the reflection functions at two wavelengths at VNIR (0.75 μm) and SWIR (2.16 μm) for various values of cloud optical thickness τ and effective radius r_e. SZA = 45.7° and VZA = 28°. Data from measurements over marine stratocumulus clouds (10 July 1987) are shown as open circles. *Source:* Reproduced with permission from Nakajima and King (1990)/American Meteorological Society.

single-scattering albedos, ω_0. Consequently, the multiply scattered SWIR light measured in the reflected radiance field has probed a layer of a specific thickness known as the diffusion length scale (e.g., Davis & Marshak, 2010),

$$L_d = \sigma^{-1}/[3(1-\omega_0)(1-\omega_0 g)]^{1/2}, \quad (15.1)$$

where g is the scattering phase function's asymmetry factor, namely, the mean value of the cosine of the scattering angle. For the present purposes, we can use $g \approx 0.85$ in warm liquid clouds, while for ice clouds, $g \approx 0.75$ (e.g., van Diedenhoven, 2014a, 2014b). Either way, we see that σL_d, the optical depth of the diffusion layer, diverges as ω_0 approaches unity. Therefore, in the VNIR, where $\omega_0 = 1$, the whole cloud is probed, hence its sensitivity to COT. In contrast, SWIR channels (Platnick, 2000), with $\omega_0 < 1$ in equation 15.1, probe ever-more-shallow layers of the cloud as σL_d becomes smaller with respect to $\tau = \sigma H$ (assumed here to be ~10 or more, for the diffusion approximation to be valid).

Implementation of the bispectral technique globally under all cloud conditions first requires selection of cloudy pixels vs. clear ones (a.k.a. cloud masking) as well as the determination of the thermodynamic phase of clouds (ice or water) since ice crystals and water droplets scatter and absorb differently. Cloud mask, in general, used to discriminate between cloudy and clear pixels depends very much on other available channels. It is usually a procedure that consists of many tests that include different spectral bands (that differ for day and night scenes): channels in the thermal IR (Ackerman et al., 1998; Frey et al., 2008), in absorption features of water vapor (Hutchison et al., 1995) or oxygen (e.g., Hollstein et al., 2015; Koelemeijer et al., 2001), or in the UV (e.g., Burrows et al., 1999; Torres et al., 1998). Cloud-masking techniques can also be different over different surfaces like such as, ocean, desert, or vegetation. To determine cloud thermodynamic phase, MODIS uses a bispectral IR algorithm that is based on relatively subtle differences in the optical properties between water droplets and ice crystals.

As previously mentioned, MODIS cloud products from both Terra and Aqua use the Nakajima-King retrieval (King et al., 2003, 2013; Platnick et al., 2017). Several other sensors do as well, e.g., VIIRS/Suomi-NPP, AHI/Himawari-8+ (Japan's geostationary series), and ABI/GOES-16+. Upcoming missions (PACE, JPSS series, 3MI, EarthCARE, among others) will follow suit.

15.2.2. Bréon-Goloub Polarimetric Technique

Hansen and Travis (1974) showed that that single scattering provides the main contribution to the polarized radiance and that the contribution of multiple scattering is fairly small. As a result, for water clouds polarized

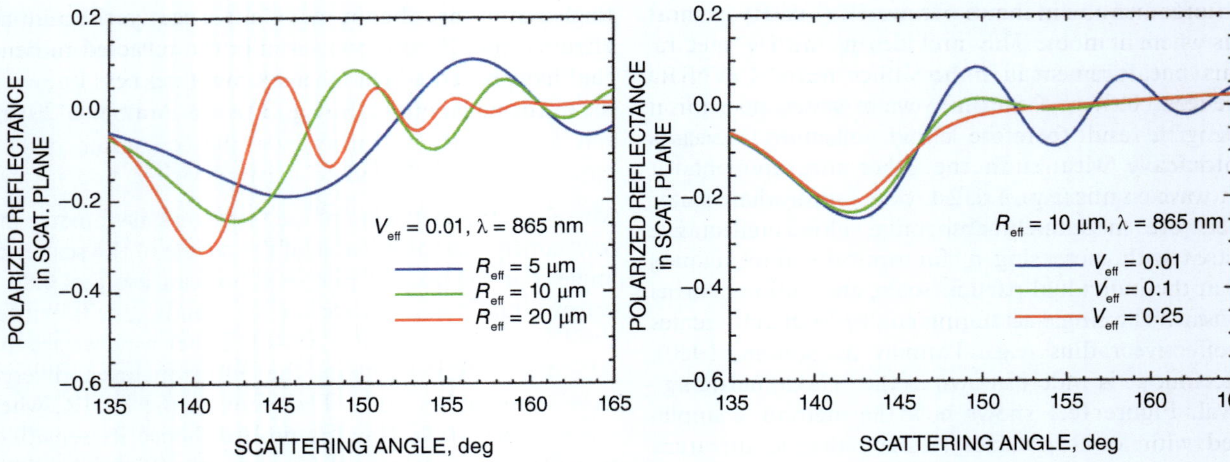

Figure 15.2 Sensitivity of polarized reflectance in the rainbow angular range to effective radius (left) and variance (right) of the cloud droplet size distribution. The data are simulated for 0.865 μm wavelength in the single-scattering approximation. Here, R_{eff} is r_e and V_{eff} is v_e. *Source:* Reproduced with permission from Alexandrov et al. (2012b)/ Elsevier.

reflectance is directly linked to the polarized phase function that has local maxima and minima at scattering angles between 140° and 170° (Bréon & Goloub, 1998). The phase matrix can be computed using Mie theory for a grid of effective radius (r_e) and variance (v_e) values. The position of the maxima and minima are different for different droplet size even if r_e varies by just a fraction of a micron. This is an indication of the high sensitivity of polarized reflectance to changes in r_e. The sharpness of the local extremes strongly depends on the dispersion of the cloud droplet size distribution, as quantified by v_e. See Figure 15.2, where we note that scattering angles must be sampled quasi continuously to capture the rainbow and supernumerary bows; we therefore speak of "hyperangular" sampling. Miller et al. (2018) determined that the angular sampling needs to be on the order of 2 degrees. According to Alexandrov et al. (2012a, 2012b), the mean accuracy of the retrieved r_e is better than 0.1 μm.

In addition, the retrievals of droplet size from polarized observations of the reflected light in this "rainbow" region allow many of the uncertainties associated with 3D effects and unknown aerosol loadings to be avoided. This advantage results from the shape of the rainbow being dominated by single scattering of light by cloud droplets, even though its amplitude can be affected by aerosols and the geometric structure of clouds (Alexandrov et al., 2012b, 2015; Emde et al., 2010; Shang et al., 2015). The dominance of single scattering as a determinant of the structure of the rainbow simplifies both forward modeling and inversions. The retrievals are just as accurate over land, or ocean (no surface albedo issues), and are valid independent of the cloud optical depth down to about unity.

This approach to cloud microphysical property retrievals has been successfully implemented for POLDER-3/ PARASOL (Tanré et al., 2011) but only for extended marine stratocumulus (Sc) decks where each pixel contributes a specific view angle and a broad range is thus covered; moreover, illumination geometry has to be favorable to imaging the rainbow. It will be used more systematically with sensors with hyperangular polarimetric sampling in future missions (e.g., PACE; see Werdell et al., 2019). In the meantime, and for the foreseeable future, the technique is used with airborne polarimetric sensors (cf. section 15.5) since they can always choose their illumination/viewing geometry. In contrast with the limited angular sampling of POLDER, airborne instruments generally have sufficient angular sampling, and so will their future space-borne counterparts.

15.2.3. Microwave-Based Methods

Methods described so far rely on reflected sunlight, hence are only applicable during the daytime half of the satellite's orbit. In contrast, microwave methods rely on thermal emission and gaseous or particle absorption properties. Microwaves have frequencies from ≈1 to ≈300 GHz, corresponding to a wavelength range of ≈30 cm to ≈1 mm (Ulaby et al., 1981). At the low end of this frequency range, the Earth's atmosphere is transparent, but it becomes more opaque with increasing frequency. Over the range of microwave frequencies, the radiative emission and absorption of the atmosphere are sensitive to geophysical parameters such as moisture, rainfall, and temperature.

The same physical laws of RT govern visible, NIR, SWIR, thermal IR (TIR), and microwave interactions with atmospheric gases and particulates, and ultimately the measured radiometric signals from remote sensors. However, the properties of geophysically important materials create important differences. For example, clouds are a fundamental difference because the opacity of most clouds is high in the VIS and across the IR spectral regions but fairly small in the microwave. The main differences between the microwave regions and shorter wavelengths result from the highly reflective surfaces in the microwave. Ocean surfaces reflect on the order of 50%, but the ocean's microwave reflectivity strongly depends on frequency, polarization, and view angle. This high surface reflectivity means that the radiance observed from space will respond strongly to the introduction of an absorber into the atmosphere at a temperature colder than at the surface. This has fundamental importance for the retrieval of constituents from the atmosphere. Even over land surfaces, the observed radiance in microwave channels depends more on the reflectivity than on the temperature.

Microwave sensors have been flown on aircraft and spacecraft and have become standard instruments for Earth observations used in operational numerical weather prediction (NWP). Extensive Earth-observation databases have been collected from various types of sensors at different frequencies.

Evans and Stephens (1995a) provide a theoretical assessment of the effects of ice crystal microphysics on microwave RT at selected frequencies. They focused first on the microwave single-scattering properties of cirrus clouds and horizontally oriented pristine ice crystals. Then, Evans and Stephens (1995b) explored the high-frequency microwave sensing of the ice water path (IWP) of cirrus clouds from high-altitude aircraft and satellite.

The high altitudes of cirrus clouds, combined with their relatively small optical depths, endows them with radiative properties rather distinct from other types of clouds (Liou, 2002). The cold temperatures of cirrus clouds mean they emit much less infrared radiation into space than clear skies, thus providing a large greenhouse-like effect, namely, a trapping of microwave radiation below the cirrus layer (e.g., Stephens & Webster, 1981).

Microwave remote sensing offers a number of potential advantages for measuring the ice water path of clouds when compared to existing visible and infrared techniques. Visible methods require a number of assumptions about ice particle shape, size distribution, and cloud spatial homogeneity to go from radiance to optical depth to IWP. In addition, TIR techniques require accurate knowledge of cloud temperature. In contrast, microwave radiation interacts with ice particles primarily through scattering, so emission and cloud temperature are relatively unimportant. Since cirrus clouds are above the absorbing part of the atmosphere, they simply modulate the upwelling microwave radiation from below. Furthermore, microwave RT occurs in the linear (single scattering) regime so the signal is directly proportional to optical depth and cloud spatial inhomogeneity effects are less important. Since they determine the relation between optical depth and IWP, the effects of particle shape and size distribution are also important for microwave remote sensing of cirrus. Moreover, particle optical properties are more amenable to calculation than for solar wavelengths because the particle sizes are at most comparable to and generally much smaller than the microwave wavelength. Microwave methods are also complementary to visible and IR methods in which microwave radiation is sensitive to larger ice crystals and thicker cirrus layers, whereas visible/IR radiation is more sensitive to smaller particles and cirrus clouds of lower IWP (e.g., Liou, 2002).

Special Sensor Microwave Imager (SSM/I) radiometers (Hollinger et al., 1990) have been continuously flown on Defense Meteorological Satellite Program (DMSP)

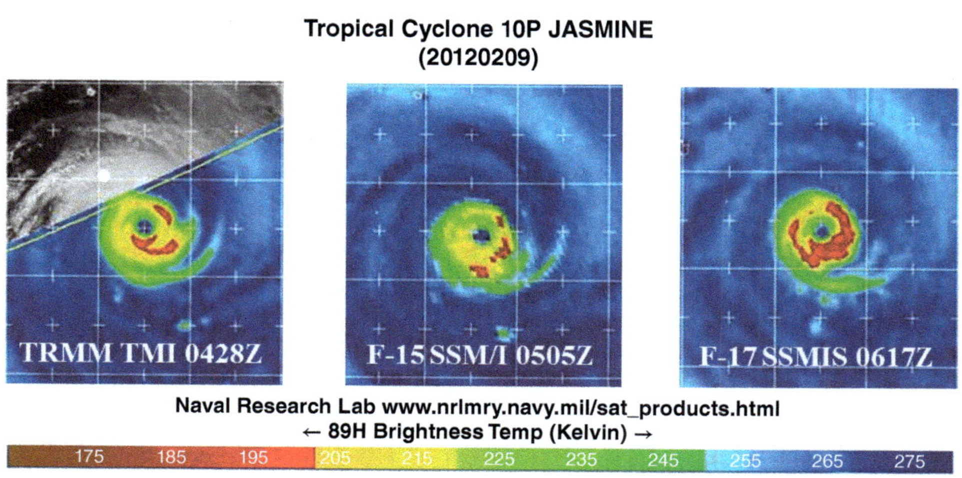

Figure 15.3 Comparison of radiance images, expressed in brightness temperature (BT), between Tropical Rainfall Measuring Mission (TRMM) microwave imager (TMI), Special Sensor Microwave/Imager (SSM/I), and Special Sensor Microwave Imager Sounder (SSMIS) for tropical cyclone Jasmine on 9 February 2012. *Source:* Song Yang et al. (2014)/MDPI/CC BY 3.0.

spacecraft since the early-1980s. Figure 15.3 illustrates a tropical cyclone Jasmine observed with different microwave imagers. There is also the Advanced Microwave Sensor Radiometer (AMSR) operating on several spacecraft. Space agencies worldwide continue to invest in microwave technology development and deployment (e.g., EarthCARE will have multichannel microwave capability).

15.3. PASSIVE SENSING OF COLUMN-INTEGRATED AEROSOL PROPERTIES

The spectral solar radiance detected by a satellite sensor from a cloud-free atmosphere overlying a reflecting surface can be written as a sum of three terms (e.g., King et al., 1999):

$$I_{\mathrm{TOA}} = I_{\mathrm{ATM}} + I_{\mathrm{SURF}} + I_{\mathrm{SURF/ATM}}. \tag{15.2}$$

They are respectively the atmospheric reflection function (a.k.a. "airlight"), the contribution of surface-reflected light, and the atmosphere-surface interaction function term resulting from multiple surface-atmosphere scatterings and reflections. Light can be scattered by the atmosphere to the sensor I_{ATM} through air molecules (Rayleigh scattering) and backscattering by aerosol particles. Absorption by aerosol particles is generally spectrally and angularly dependent. How to retrieve aerosol properties from the spectral reflectance obtained by a satellite radiometer?

We first focus on multispectral methods developed with MODIS in mind and then multiangle/multispectral counterparts developed for the Multi-angle Imaging Spectro-Radiometer (MISR, Diner et al., 1998). We finish with some comments on how polarization (in addition to multiangle) is starting to revolutionize aerosol remote sensing from space, as well as from suborbital platforms (i.e., airborne and ground-based).

15.3.1. Over Ocean

When the surface is dark, such as the ocean in the VNIR region, the atmospheric reflection function dominates. This is a function of the aerosol optical thickness (AOT), τ_a, single-scattering albedo (SSA), ω_0, and scattering phase function, which is in essence the probability distribution function of radiant energy as a function of scattering angle. The reflection function increases almost linearly as τ_a increases and is the most sensitive for low reflectance surfaces. This is the physics behind the remote sensing of aerosol optical thickness over dark surface using single channel reflectance (e.g., Stowe et al., 1997; Tanré et al., 1997). Two-channel and multichannel reflectances allow us to retrieve not only aerosol optical thickness but also some information on aerosol size distribution, at least a partition between fine (or "accumulation") mode aerosols and their coarse-mode counterparts. We note that uncertainties in aerosol retrievals over coastal regions, where the ocean is not dark enough, could be much higher than over deep ocean (Wang et al., 2017).

15.3.2. Over Land

Dark Target

In order to estimate the aerosol optical thickness over land, we need to account for surface contributions and aerosol-surface interactions. Some soils and green vegetation are dark in both the red (0.6–0.7 μm) and the blue (0.4–0.5 μm) spectral regions. To account for aerosol-surface interaction, the so-called "dark target" aerosol retrieval algorithm uses the reflectance from the spectral region with minimal aerosol contamination. With the exception of dust, the aerosol optical thickness typically decreases with wavelength, and it is many times smaller in the SWIR (2–4 μm) region than in the VIS (0.47 and 0.66 μm) for MODIS. Therefore, the effect of aerosol can be neglected in the SWIR region. Vegetated surfaces are thus expected to be relatively dark in the visible channels (0.47 and 0.66 μm) and is determined using MODIS's 2.1 μm channel.

The dark target algorithm retrieves aerosol properties simultaneously in the two visible channels and the 2.1 μm channel; it assumes that the 2.1 μm channel contains information about coarse-mode aerosols and the surface reflectance. The surface reflectance in the visible is taken to be a function of the surface reflectance at 2.1 μm (Kaufman et al., 1997). For updates on this approach, see Remer et al. (2013) and Levy et al. (2013).

Deep Blue

Due to the brightness of many land-surface types, aerosol retrieval over land is more challenging than over water. In particular, the surface contributions over desert dominate the top-of-atmosphere (TOA) reflectance acquired by the satellite sensors at red and NIR wavelengths. This makes it difficult to separate the contribution of aerosols and the surface.

The Deep Blue algorithm narrows these gaps. To infer the properties of aerosols, it utilizes blue wavelength measurements where the surface reflectance over land is much lower than for longer wavelength channels. Rayleigh scattering enters into the equation but, compared to aerosol scattering, is a well-known constant. For more details, we refer to Hsu et al. (2004, 2006) and Sayer et al. (2016).

MAIAC and GRASP

There are several unresolved issues with the Dark Target and Deep Blue algorithms that require additional information. The Dark Target algorithm makes spectral assumptions about surface reflectance and assumes a Lambertian surface model. The Deep Blue algorithm uses an ancillary global database of surface reflectance composed of monthly measurements.

The more recently developed Multi-Angle Implementation of Atmospheric Correction (MAIAC) algorithm simultaneously retrieves AOT and the surface's bidirectional reflection distribution factor (BRDF) using a time series of imager observations (Lyapustin et al., 2011a, 2011b). Specifically, MAIAC accumulates 3–16 days of calibrated and geo-located measurements. With different view angles, it helps to retrieve the surface BRDF. Gridding high-resolution data at 1 km and, assuming that surface reflectance does not change much for the accumulated time interval and that AOT changes little up to spatial scales of ≈25 km, it accumulates a sufficient number of measurements to retrieve all unknown parameters that include aerosol and surface reflectance properties (Lyapustin et al., 2012a, 2012b).

The recently released updated version of the MAIAC algorithm uses a dynamic time series analysis to separate the atmosphere and surface contribution and retrieve AOT and the surface's BRDF with improved accuracy (Lyapustin et al., 2018). Specifically, operational MAIAC Collection 6 algorithm now implements a sliding window approach accumulating up to 16 days of gridded MODIS measurements in the "Queue." The Queue dataset and minimum reflectance method are used to dynamically characterize the surface spectral reflectance ratios, required for the aerosol retrievals at 1 km resolution. Accumulation of MODIS imagery at different view angles in the Queue helps surface BRDF retrieval. A separate Queue memory is used to store a wealth of information on the surface properties for each 1 km grid cell from the clear-sky retrievals. Such detailed surface characterization helps MAIAC achieve high accuracy of its product suite, from cloud, cloud shadow and snow detection to high-resolution (1 km) AOT globally over land, and to spectral surface reflectance properties (for global MAIAC AOT validation; see Figure 15.4.

The General Retrieval of Aerosol and Surface Properties (GRASP) is an open-source code (Dubovik et al., 2014) built with similar concepts as MAIAC, namely, exploitation of temporal records and retrieval constraints based on distinct spatial correlation statistics for aerosols and surfaces. GRASP is based on several generalization principles with the idea to develop a scientifically rigorous and versatile algorithm (Dubovik et al., 2021). At the core of GRASP is an optimal estimation algorithm with a plethora of choices for the minimization technique as well as the regularization term. GRASP has significantly extended capabilities and areas of applicability; it has already been applied to several different satellite missions, databases, and sensor networks (e.g., Sayer et al., 2018; Schuster et al., 2019).

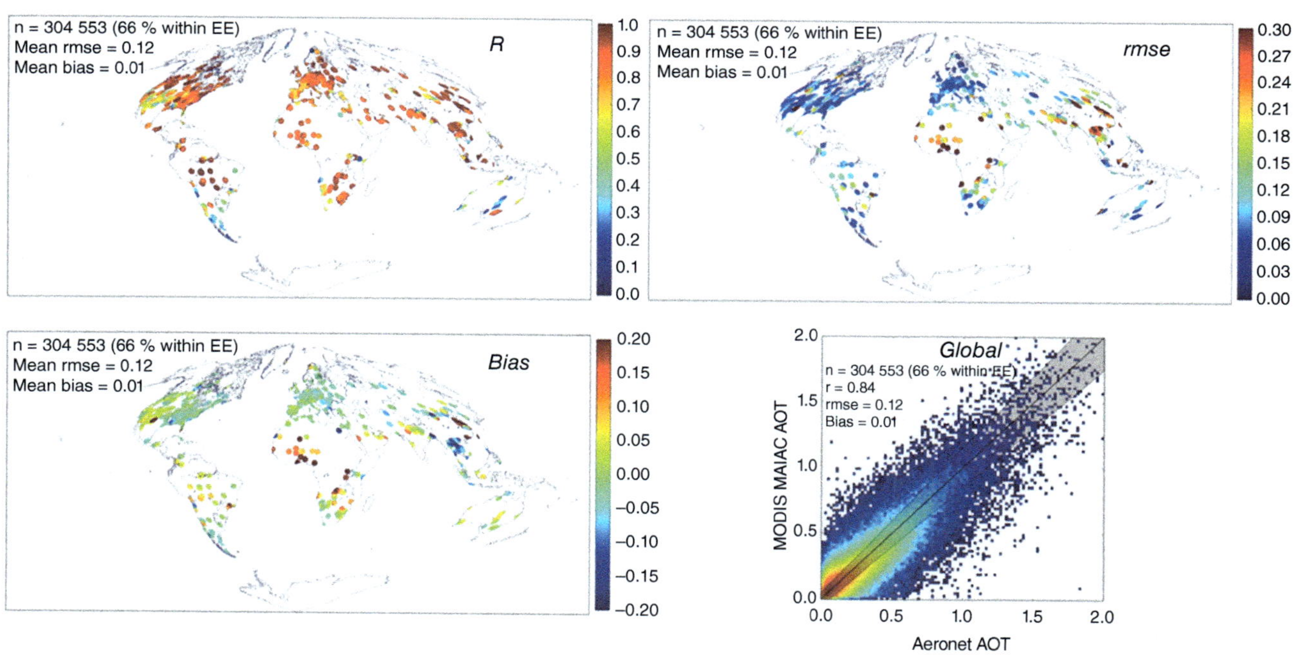

Figure 15.4 Results of global MAIAC AOT validation against AERONET (Holben et al., 1998, 2001) from the MODIS Terra and Aqua 2000–2016 record. Here "EE" is an expected error ±0.05 ± 0.1AOT. *Source:* Reproduced with permission from Lyapustin et al. (2018)/Elsevier/Public Domain CC BY 4.0.

15.3.3. MISR Aerosol Retrieval Algorithms

Although oceans cover the majority of the Earth's surface, land-surface areas are the source of most aerosols, and essentially all of the anthropogenic production. Radiances measured at TOA is a mixture of the three components shown in equation (15.2). A successful retrieval process must be able to separate and explicitly describe these radiance components, resulting in a determination of both the aerosol properties and the surface reflectance characteristics. For those instruments that observe a given scene at multiple VNIR and SWIR wavelengths but only at a single view angle (e.g., MODIS), the aerosol retrieval process at these wavelengths always requires the absolute surface reflectance or its spectral variation at the different wavelengths to be prescribed (e.g., Hsu et al., 2006; Kaufman et al., 2002). However, imaging instruments capable of viewing a scene at multiple wavelengths and multiple viewing angles over a sufficiently short time period so that atmospheric conditions can be assumed constant is much more flexible.

Flowerdew and Haigh (1995) suggested that the ratio of the surface reflectance at the two viewing angles should be nearly spectrally invariant because the surface scattering elements are much larger than the wavelengths of the scattered light. Aerosol retrieval algorithms based on this idea yield good comparisons with aerosol results from ground-based sun photometers, predominantly those of the Aerosol Robotic Network (AERONET, Holben et al., 1998, 2001). The success of this technique prompted its adaptation to the multiangle observations of MISR (Diner et al., 2005).

The operational MISR aerosol retrieval algorithm is based on a principal component analysis of the multispectral/multiangle MISR data from which empirical orthogonal functions (EOFs) for a given view angle were obtained over a $17 \times 17\,km^2$ domain (16×16 pixels) to describe the directional reflectance properties of the surface (Martonchik et al., 1998, 2009).

Both strongly wavelength-dependent Rayleigh scattering and large AOT contribute to the angular variability and strength of the diffuse radiance. The primary underlying physical assumptions for the MISR aerosol retrieval over land are as follows (Garay et al., 2017):

- Aerosols are horizontally homogeneous in the retrieval region, recently reduced in the operational product from 17×17 to $4.4 \times 4.4\,km^2$ (see Figure 15.5).
- A predefined set of aerosol microphysical models stored in LUTs is applied globally to retrievals over both land and water.
- Over land, there is sufficient surface contrast in the retrieval region so that the TOA radiances can be represented by EOFs across the multiangle imagery.

For either surface type, cost functions (i.e., "χ^2" sums of squared observation-minus-model differences) are estimated to determine how well TOA radiances are reproduced using the individual predefined aerosol models and an associated green-band AOT.

Following Flowerdew and Haigh (1995), the angular shape of the surface reflectance is assumed to be spectrally invariant, and this is used to further filter out models and AOTs for which the retrieval results do not conform to this assumption as being unlikely candidates for selection (Diner et al., 2005).

It is important to note that algorithms described so far—and most of them in the remainder—are deemed "operational," meaning that they are applied to all pixels/scenes that are determined to be cloud free in the case of aerosol retrievals and fully cloudy in the case of their counterparts for clouds, either way with a minimum number of detectable failures. There are always confidence

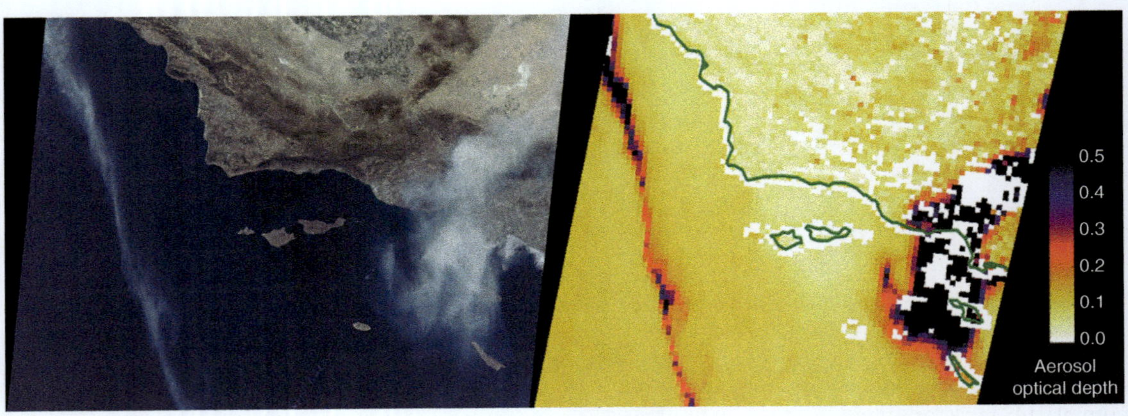

Figure 15.5 Left panel: MISR true color image, full swath, and nadir view. Right panel: MISR's 4.4 km AOT product for the same image. Note the seamless transition from land to ocean achieved with multiangle observations. *Source:* Michael J. Garay et al. (2017)/European Geosciences Union/CC BY 3.0.

estimates for successful retrieval that are determined from a variety of tests such as poor convergence or reaching the limit of an LUT. This is technically challenging but a strict requirement for building a global satellite climatology; see, e.g., Kahn and Gaitley (2015) in the case of MISR.

However, another class of retrieval scheme is extremely useful in specific science-driven investigations. This is the "research-type" algorithm that calls for careful selection of the data to be processed—maybe for a field campaign of interest—and careful supervision and tuning of retrieval parameters. For instance, the MISR aerosol research algorithm (Limbacher & Kahn, 2014, 2017) uses a much larger number of preselected aerosol types in its LUTs, and its thresholds for convergence can be tightened as needed. In the same vein, an alternative approach of the MISR aerosol retrievals over dark water is discussed by Witek et al. (2018). Other kinds of research algorithms are also described.

15.3.4. Capabilities from Multiangle Polarimetry

In short, MODIS' aerosol retrievals capitalize on its broad coverage of the solar spectrum, while MISR brings multiangle imaging to the table. Following MISR's lead, MAIAC exploits spatial variabilities (or the lack thereof) and adds temporal counterparts. One attribute of Electro-Magnetic (EM) radiation has so far not been considered, namely, its state of polarization.

CNES's POLarization and Directionality of the Earth's Reflectances (POLDER) sensor on the Polarization and Anisotropy of Reflectances for Atmospheric Science coupled with Observations from a Lidar (PARASOL) satellite (Tanré et al., 2011) pioneered the measurement of polarization from space, and in fact from the A-train. We highlighted earlier the cloud microphysical application for reasonably uniform Sc decks where each pixel contributes a different angle, leading to the required "hyperangular" sampling; cf. section 2.2. POLDER always provides multiangular sampling for a given ground position, with typically a dozen-or-so view angles, generally not including nadir, nor as widely spread as for MISR's nine angles. That is, however, more than enough for a polarimetric multiangle aerosol retrieval (Hasekamp & Landgraf, 2007). Notwithstanding, POLDER aerosol retrievals were critically appraised by Mishchenko et al. (2004) and Knobelspiesse et al. (2012) from the standpoint of climate science requirements, and found to fall short in terms of precision. This has motivated recent developments in both instrumentation and aerosol retrieval algorithms that we now survey.

Figure 15.6, from Xu et al. (2017), demonstrates the superiority of polarization inclusion in AOT and aerosol SSA retrievals (which is more challenging) but also having high scientific value in terms of aerosol speciation. Airborne Multi-angle Spectro-Polarimetic Imager (AirMSPI) data are used here in "step-and-stare" acquisition mode that leads to standard multiangle imaging. These data were collected when NASA's ER-2 was overflying the California Central Valley, during the POLarimeter Definition EXperiment (PODEX) campaign in early 2013 (Knobelspiesse et al., 2019), where JPL's AirMSPI (Diner et al., 2013) was collocated with NASA-GISS's nonimaging but hyperangular Research Scanning Polarimeter (RSP; Cairns et al., 2003) and SRON's Spectro-polarimeter for Planetary EXploration (SPEX; Hasekamp et al., 2019). All of these airborne polarimeters performed well. Since then, UMBC's Airborne Hyper-Angular Radio-Polarimeter

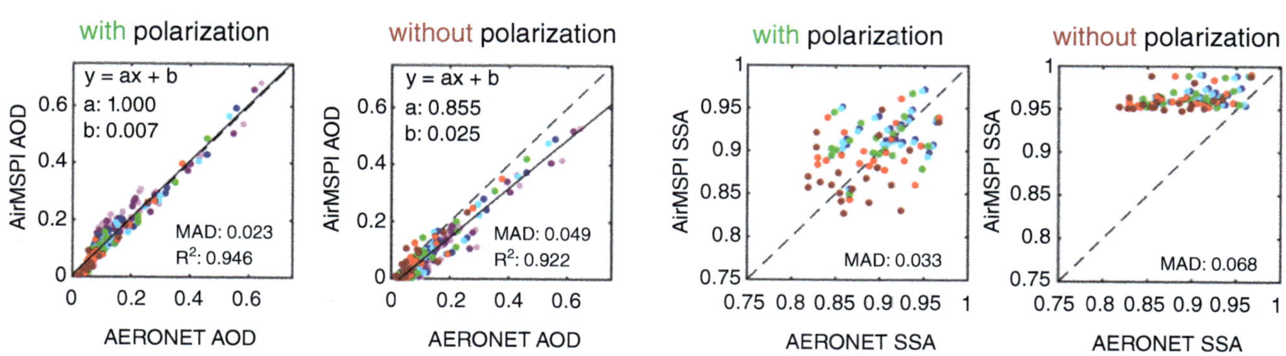

Figure 15.6 Left two panels: Comparison of AOT retrieved from AirMSPI data captured over Fresno (Ca) and from the local AERONET station; colors correspond to eight different (jointly retrieved) spectral channels. Right two panels: Same as two left ones, but for the SSA, an inherent measure of aerosol absorption. AirMSPI has eight spectral channels, seven of which have color-coded AOT and SSA in the scatter plots: 355 nm (pink), 380 nm (purple), 445 nm (dark blue), 470 nm (light blue), 555 nm (green), 660 nm (red), and 865 nm (brown); 935 nm channel is not shown, having very small AOT. Channels at 470, 660, and 865 nm are polarimetric. For this kind of retrieval, AirMSPI's "step-and-stare" acquisition mode is used, in this case, for 0° (nadir), ±29°, ±48°, ±59°, and ±66°, flying in the principal plane. *Source:* Reproduced with permission from Xu et al. (2017)/John Wiley & Sons.

(AirHARP, Martins et al., 2014) has matured to a comparable level.

In view of this demonstration of the added value of polarization in aerosol remote sensing, it will come as no surprise that CNES's POLDER-3/PARASOL was followed by NASA's Glory mission (basically abovementioned RSP in space) that, unfortunately, failed at launch in 2011 (Mishchenko et al., 2007). This mishap however sets the stage for NASA's Multi-Angle Imager for Aerosols (MAIA) investigation (Diner et al., 2018), carrying an evolved version of MSPI, and Plankton, Aerosol, Cloud, ocean Ecosystem (PACE) mission (Werdell et al., 2019), with HARP-2 and SPEXone onboard. Moreover, EUMETSAT is also preparing to launch the Multi-angle/Multi-spectral/Multi-polarimetric Imager (3MI) satellite (Marbach et al., 2015) that will carry an evolved version of POLDER. Finally (at the time of writing), we call out JAXA's Global Change Observation Mission-Climate 1 (GCOM-C1) mission (Tanaka et al., 2009), which is also centered on multiangle polarimetric imaging with the Second-Generation GLobal Imager (SGLI) sensor (Hasekamp & Landgraf, 2007). In short, aerosol remote sensing from space will be forever changed with the advent of multiangle polarimetric capability (Kokhanovsky et al., 2015).

15.4. VERTICAL STRUCTURE OF CLOUDS AND AEROSOLS

We start with active methods where the technology is designed specifically to provide height-resolved cloud and aerosol properties. We then describe three passive—hence cheaper—methods for obtaining at least some of the same information. This can be at a degraded resolution in altitude, but with far better spatial sampling. Alternatively, only a vertical boundary is determined (e.g., cloud-top height (CTH) and/or cloud-base height). Which type of probing is best depends on the application.

15.4.1. Active Remote Sensing for Clouds and Aerosols

Survey of Operational and Experimental Ground-Based Active Systems

Chapter 14 of this volume by Lamer, Kollias, Amiridis, Marinou, Loehnert, Schnitt, and McComiskey covers ground-based remote sensing, with many relevant references. Therein it is established that collocated, simultaneous ground-based measurements from lidar (light detection and ranging) and millimeter-wave cloud radar (MMCR) instruments enable accurate and comprehensive assessment of aerosol and cloud overhead. This is one of the cornerstones of the US DOE's Atmospheric Radiation Measurement (ARM) program that maintains both fixed and mobile climate observation facilities worldwide (Mather & Voyles, 2013; Stokes & Schwartz, 1994). From the standpoint of aerosol and cloud remote sensing from overhead assets covered in this chapter, ARM and similar resources have been invaluable sources of validation data.

In general, laser devices are excellent for detecting essentially all clouds that are visible from the ground. The radar fails to detect some of these clouds, especially if the clouds are composed of small hydrometeors, are thin, or are located at high altitudes. The laser instruments are required for these situations. The radar estimation of cloud-base height is often obscured or complicated by the presence of large precipitation and virga particles. Cloud radars also readily detect nonhydrometeor particulates, such as insects and airborne pieces of vegetation, which are commonly suspended in the atmospheric boundary layer and constitute a clutter that may be mistaken for echoes from stratus clouds. The ceilometer and lidar are usually insensitive to these large, but sparse, precipitation, virga, and insect targets, and therefore provide more accurate cloud-base measurements in these circumstances.

The great strength of ground-based radar is its ability to penetrate clouds and reveal multiple layers aloft. Although ceilometers and micropulse lidars (MPLs; Spinhirne, 1993; Welton & Campbell, 2002) are exquisite sensors for cloud-base height, and excellent aerosol profilers, they are unable to provide any information about higher cloud layers when lower liquid-water layers, such as stratus or fog, are present because the laser beams are readily extinguished by the cloud particles and are unable to penetrate much beyond the base of the lowest layer. In these situations, which include periods of drizzle, light rain, and most types of snowfall, the radar measurements are essential. However, when these hydrometeors are present, they are all that the cloud radar "sees." Due to their far smaller size, the cloud particles that matter for radiation are all but masked.

In spite of these limitations, the past decade-or-so has seen three significant technological developments: Doppler radar; depolarization radar; and scanning radar, using either mechanical systems or phased arrays. Doppler radar reveals the dispersion of velocity along the beam, which is vertical for a zenith-pointing antenna, and this information enables the separation of different sizes of hydrometeors. Depolarization radar transmits pulses in a specific state of polarization and receives in that one and others, which enables the detection of nonspherical hydrometeors overhead. Scanning radar enables the sampling not just of the overhead profile of hydrometeors but of their distribution in a whole volume of 3D space.

The great strength of ground-based lidars is that, in the absence of clouds, they will return a profile of aerosol backscatter from near ground to great heights. However, "backscatter" is a coefficient defined as the product of the extinction coefficient and the phase function value at 180° scattering angle, as delivered by ceilometers and MPLs, which are so-called "elastic" backscatter lidars. Both properties combined into backscatter are of more interest than their product. Two lidar technologies were developed that can separate them: Raman lidar and high-spectral resolution lidar (HSRL). The former technology exploits the purely molecular Raman shift in wavelength, while HSRL uses sophisticated spectral filtering techniques to separate the thermal agitation of molecules from the smaller velocities of the aerosols. Either way, the molecular component is a known quantity in cross section, extinction, and phase function. From there, aerosol-only extinction can be determined at every height.

All of the above active sensing technologies are currently being developed, or at least are under consideration, for future deployment in space.

We note that, not unlike the Bréon-Goloub algorithm from section 15.2.2 for passive cloud microphysics determination from polarized reflectance, the lidar and radar equations predict their respective signals using a single-scattering approximation. This strongly contrasts with the Nakajima-King VNIR+SWIR method for probing clouds. In the TIR, where sources are distributed throughout the atmosphere, one can neglect scattering altogether. In microwave remote sensing, only direct and once-scattered radiation matter in principle. Yet, multiple scattering happens and has to be corrected for in one way or another (e.g., Tsang et al., 2007).

Multiple field-of-view (MFOV) lidar radically departs from other active techniques, and several passive approaches, in which multiply scattered laser light is its signal, and it is predicted in the small-angle (paraxial) approximation for time-dependent RT. Therefore, only scattering in the narrow forward peak of cloud (Roy et al., 1999) and large aerosol particles (Hutt et al., 1994) is of interest in MFOV lidar, before and after one scattering in the near-backscattering direction. Consequently, the receiver's FOV is wider than the divergence of the transmitted beam, as in all lidar systems, but in this case wider enough to allow multiply scattered light deliberately into the detector and track it as a function of the increasing FOV. In MFOV lidar signals, there is direct sensitivity to the shape of the particle size distribution (PSD).

Backscatter, HSRL, and MFOV lidars have been used in mechanical 3D volume-scanning systems, just like typical weather radars. Some ground-based operational MMCRs scan the sky in 2D (typically at right angles from the prevailing wind direction) or in 3D. A pair of ground-based passive microwave sensors have been trained to scan the same plane from separate points to demonstrate a cloud tomography (Huang et al., 2008a, 2008b).

Lastly, we note the slow emergence of a class of Multiple Scattering Cloud Lidar (MuSCL) systems, where the receiver's FOV is as wide as necessary to capture as many orders of scattering as possible, even if they start and end far from the forward peak and backward direction. MuSCL signal models use the opposite limit of RT theory as single scattering, namely, the diffusion approximation (Davis et al., 1999a, 1999b). It has been shown that MuSCL signals have sensitivity to cloud geometrical thickness H even for very opaque layers ($\tau \gg 1$) in the highly scattered laser light that has permeated the whole cloud. The ground-based implementation of MuSCL is the Wide-Angle Imaging Lidar (WAIL). It was deployed at the ARM SGP site (Polonsky et al., 2005) where it demonstrated a joint nighttime retrieval of optical and geometrical cloud thickness, τ and H, for a solid stratus layer when τ was $\approx$30, using an FOV of 57° (later increased to 88°). Originally proposed as an antidote to ground-based MMCR when its cloud signal is swamped by drizzle (or insects), WAIL's main shortcoming was that the wide FOV requirement precluded the simple filtering solutions that would have enabled daytime operation.

In the remainder of this subsection, we briefly cover what changes in lidar or MMCR-like systems when transmitting down from space rather than up from the ground. We also gaze into the future of these technologies in space.

Space-Based Radars for Clouds

We refer the reader to Chapter 14 of this volume by Lamer et al. for a description of how cloud radar works, from the ground up. There is no fundamental difference with operating one from space, apart from using radiation-hardened components. So far, there has been only one space-based cloud radar: Cloud-Profiling Radar (CPR), which is the main payload on CloudSat (Stephens et al., 2002). CPR was developed at JPL, which has extensive heritage from Earth-observing and planetary radar systems working at longer wavelengths, with contributions from Canadian Space Agency, also with extensive radar heritage. CPR operates at 94 GHz (in the W band), hence at a wavelength of 3.2 mm, at which all cloud particles are essentially Rayleigh scatterers. Its resolution is 0.5 km in the vertical, 1.4 km across track, and 1.7 km along track. Detection threshold is <-29 dBZ, and its dynamic range is 70 dB. Like its ground-based counterparts, CPR strongly responds to the presence of drizzle, and this can mask cloud base per se. Figure 15.7 illustrates CloudSat data with a transect through a typhoon, including the eye of the massive storm.

CloudSat and CALISPO (described next) were launched together in 2006 into the "A-train," the

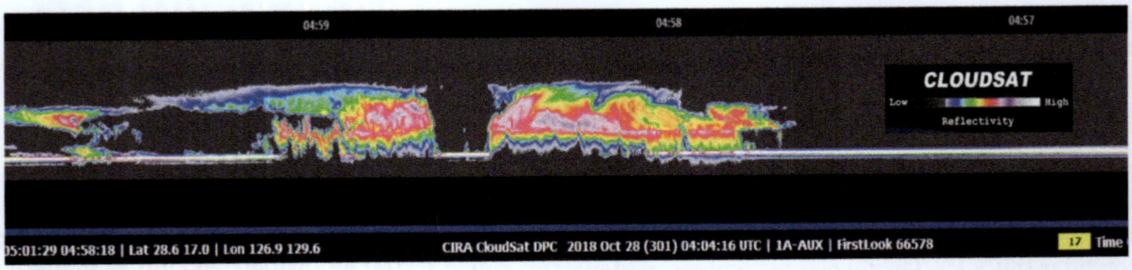

Figure 15.7 CloudSat transect of radar reflectivity through Typhoon Yutu-West Pacific, 28 October 2018. *Source:* Department of Atmospheric Science Colorado State University/https://cloudsat.atmos.colostate.edu/last accessed March 18, 2023.

sun-synchronous (1:30 equator crossing time) constellation at 705 km altitude, which contained at the time the abovementioned Aqua and PARASOL satellites, both launched in 2004. PARASOL left the A-train in 2010, and the mission was formally ended in 2013. CloudSat, then CALIPSO, were moved to a lower orbit in 2018, to manage their remaining fuel reserves for safe deorbiting in the future.

CloudSat, still operating from its lower, slowly precessing orbit, is a highly successful mission in its own right that has blazed the path for future cloud radars in space with enhanced capabilities: Doppler (for fall-speed determination), polarization (for ice particle habit characterization), and scanning (hence 3D rather than 2D cloud structure). The next satellite to bring cloud radar to space will be ESA's Earth Clouds, Aerosols and Radiation Explorer (EarthCARE) mission (Illingworth et al., 2015), which will carry a Dopplerized CPR.

Space-Based Lidars for Aerosols and Clouds

The Cloud-Aerosol Lidar and Infrared Pathfinder Satellite Observation (CALIPSO) satellite provides new insight into the vertical profile of clouds and aerosols (Winker et al., 2010). It was launched in April 2006 together with CloudSat into the A-train constellation. Though CALIPSO also has passive infrared and visible imagers, we focus here on its active remote-sensing instrument Cloud-Aerosol LIdar with Orthogonal Polarization (CALIOP), a depolarization backscatter system that provides high-resolution vertical profiles of clouds and aerosols unavailable from other Earth-observing satellites.

The main CALIOP cloud and aerosol products (height, thickness, and optical depth) have a horizontal resolution of 5 km and vertical resolution of 30 and 60 m. While cloud extinction and backscattering coefficients are at 5 km horizontal resolution, aerosol extinction and backscattering counterparts are at 40 km due to the need for along-track averaging. We note, however, that clouds can be profiled by CALIOP only up to COTs of ≈ 3 due to the rapid extinction of the direct and returned beams. Moreover, corrections need to be made for multiple scattering. For more opaque clouds, CALIOP remains an exquisite detector for cloud-top height and is often used as a benchmark for other CTH retrievals (Hagihara et al., 2013).

With their global coverage, space-based lidars offer excellent opportunities for studying cloud-aerosol interaction. With respect to the transition zone between cloudy and clear air, the active CALIOP observations complement passive measurements because lidar measurements are not affected by 3D cloud radiative effects (e.g., Qiu et al., 2000; Várnai & Marshak, 2009; Wen et al., 2007; Zhang et al., 2005). Tackett and Di Girolamo (2009) pioneered the use of the spaceborne CALIOP lidar for studying the transition zone. Their study of aerosols over the Caribbean Sea showed that aerosol properties change rapidly near clouds both because aerosols swell in the humid environment and because the collision and coalescence of aerosol-containing cloud droplets and their subsequent evaporation produce fewer but larger aerosol particles.

While the Tackett and Di Girolamo (2009) study focused on areas up to 3–4 km away from clouds in the Caribbean, a large global dataset of CALIOP measurements can reveal even the subtle changes that occur at the more distant portions of the cloud-aerosol "transition zone." A statistical analysis of a monthlong global dataset of CALIOP lidar observations over oceans shows that clouds are surrounded by a transition zone of enhanced particle backscatter and size that is ubiquitous over all oceans (Várnai & Marshak, 2011). This zone can be as wide as 15 km.

The width and ubiquity of the transition zone near clouds imply that studies of aerosol-cloud interactions and aerosol direct radiative effects need to quantitatively account using both active and passive observation for aerosol changes near clouds. These changes can cause systematic biases toward smaller AOT and radiative forcing calculated from these optical thicknesses, if areas near clouds are not included in order to avoid aerosol retrievals ambiguities near clouds (e.g., 3D radiative effects and cloud contamination). On the other hand, including aerosol products near clouds despite these ambiguities may overestimate aerosol radiative forcing. As a result, there is an urgent need for developing methods that can assess and account for remote-sensing challenges and thus allow for including the transition zone into the study.

CALIPSO was lowered out of the A-train in September 2018 but continues at the time of writing to collect science data from a slowly precessing orbit. It is a highly successful mission with an impeccable technical performance record. It carried the third backscatter lidar system into space and the first with depolarization capability. It was preceded only by the Ice, Cloud, and land Elevation Satellite (ICESat) mission (Schutz et al., 2005), launched in 2003, and by the Lidar In space Technology Experiment (LITE), a key technology demonstration on the Space Shuttle Discovery (STS-64) in September 1994 (Winker et al., 1996). CALIPSO was followed by the Cloud-Aerosol Transport System (CATS) instrument (McGill et al., 2015), launched in January 2015, on the International Space Station (ISS), which is another depolarization backscatter lidar. In addition, ICESat-2 (Abdalati et al., 2010) was launched in September 2018 with an unprecedented system of six lasers firing in sequence so that a single photon-counting receiver can be used. EarthCARE will carry the first HSRL and depolarization system into space. Many more lidars of all ilk will follow.

Apart from survival in the harsh environment of space, the main challenge of building space-based lidars coming from ground-based experience is the extraordinary demand for transmitter/receiver alignment. The receiver FOV must be as small as possible to minimize background noise from reflected sunlight and, at the same time, must overlap with the volume illuminated by transmitted beam at hundreds of kilometers down range.

Finally, we note that the early LITE mission in 1994 did not attempt this ultraprecise alignment feat in optical and micromechanical engineering. Simply put, it was a research lidar system mounted in the Space Shuttle Discovery's payload bay. It was composed of a large (1 m diameter) receiver telescope with an FOV that subtended $\approx$1 km at the surface while the beam from the potent Q-switched Nd:YAG laser transmitted its fundamental (1064 nm), doubled (532 nm), and tripled (355 nm) frequencies into a much smaller footprint. Thus, when LITE probed optically thick clouds, it inadvertently became the first MuSCL system in space (Davis, 2008; Miller & Stephens, 1999). A handful of LITE returns from marine Sc were not saturated, and their analysis using an analytical multiple-scattering lidar equation accounting for all orders of scattering lead to reasonable values for the jointly retrieved COT and cloud physical thickness (Polonsky & Davis, 2005).

15.4.2. Cloud-Top Height from Stereoscopy

The most obvious use of multiangle imaging from space does not call for a physics-based forward model to use in an inverse problem for inherent geophysical properties. It involves only geometry. Inasmuch as one can identify a given "feature" tracked between two images captured from two different positions in space (e.g., using two of MISR's views), one can then triangulate the feature and determine its position in 3D space, in particular, its height above the surface. Since satellite positions and pointing orientations are known to very high accuracy, the twice identified feature's position can be estimated to an accuracy determined primarily by the pixel scale in the imagery. In practical applications, we need an automated way of detecting an image feature and then finding it again in another view of the same scene. This is a classic problem in computer vision, with a multitude of applications, that has been addressed with many different approaches.

MISR is a pathfinder for the problem of CTH estimation based on feature detection and tracking from space. MISR's operational CTH product (Moroney et al., 2002) has been compared in detail with those from active systems described in the above, and with other passive methods described in the following, both on a per cloud basis (Genkova et al., 2007) and on a statistical basis (e.g., Wu et al., 2009).

MISR's CTH estimates are by far the most accurate among the passive methodologies due to its simplicity with no need for assumptions about atmospheric structure and composition (Sinclair et al., 2017). We note that these methods do not require any radiometric calibration, only precise registration. However, to get there, a major hurdle needs to be cleared. Indeed, MISR is not at the two stereographic vantage points at the same time, it takes $\approx$90 s for the Terra satellite to move along its orbit from MISR's view of the same cloud at 26.1° fore to 26.1° aft. (These near-nadir views facilitate feature tracking and also ensure a reasonable stereo baseline compared to altitude.) In that time interval, however, a cloud may have been advected easily by 1 or 2 of MISR's (270 m) pixels in any direction: along track, across track, or a combination of both.

Across-track motion is straightforward to detect, and therefore not an aggravating issue in CTH determination. In sharp contrast, along-track cloud motion is inherently entangled with the stereoscopy and will therefore bias the CTH estimate if it is not known independently. Since precise independent knowledge of along-track wind everywhere all the time is not an option, another solution was developed where CTH and along-track wind are jointly retrieved. It is based on the realization that MISR pixels are projected not on a tangent plane to the Earth but on

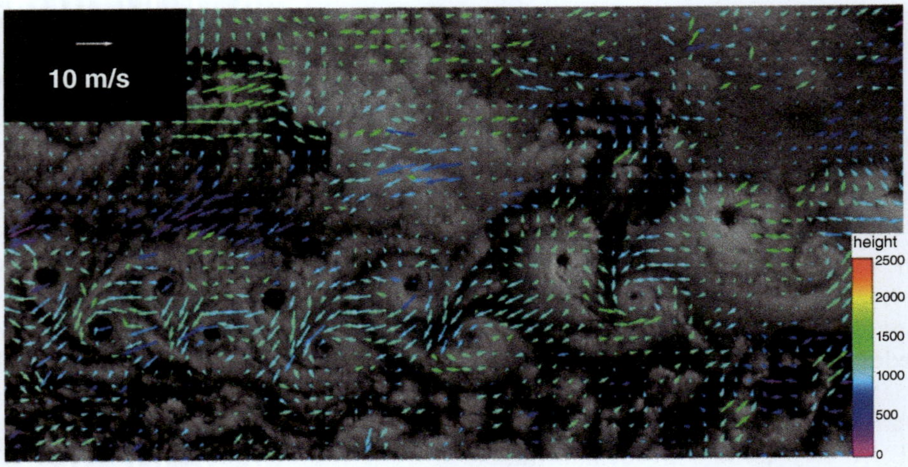

Figure 15.8 MISR red image (672 nm) overlaid with retrieved vector wind, color-coded for retrieved height. The data show a string of von Karman vortices observed off the coast of Iceland. *Source:* Courtesy of Kevin Mueller (JPL).

a standard ellipsoidal representation of the Earth's shape. The Earth's finite radius can then be used to untangle the along-track cloud motion and CTH.

MISR's basic cloud products are therefore unbiased CTH estimates and height-resolved horizontal wind vectors, after aggregation from feature/pixel space to a rectangular latitude-longitude grid, including a median filtering (to eliminate outliers) and averaging. Figure 15.8 shows a dynamically interesting composite image of the MISR red channel, horizontal wind vectors, color-coded by height.

Other sensors than MISR with stereoscopic capability include the Advanced Spaceborne Thermal Emission and Reflection Radiometer (ASTER), a JAXA imager also on the Terra platform (Yamaguchi et al., 2001), the Along-Track Scanning Radiometer (ATSR) on the ERS-1 platform (Prata & Turner, 1997), the US DoE/NNSA's Multispectral Thermal Imager (MTI; Weber et al., 1999), and ATSR-2/ERS-2 (Muller et al., 2007); in all cases, just two views are collected, but that is sufficient. Also, pairs of geostationary satellites have been used (Wylie et al., 1998).

Although this section is about *cloud*-top heights, the stereographic technique will apply seamlessly to *aerosol* plumes that only need to be are opaque enough to show "features," that is, machine-recognizable patterns in the spatial variability (i.e., strong enough gradients) in the aerosol and associated radiance field. This phenomenon is commonly observed in the case of smoke emanating from wild fires, volcano eruptions, outbreaks of lofted dust, as well as particulate pollution events. However, these scientifically interesting aerosol plumes can easily be confused with clouds by operational data-processing pipelines.

This is therefore an excellent opportunity for above-mentioned "research" algorithms to be developed and applied. A prime example is the MISR INteractive eXplorer (MINX, Nelson et al., 2013) where MISR imagery can by rapidly browsed by an image analyst in search for, say, wildfire smoke plumes. Once a plume is identified, a "region-of-interest" (ROI) is hand-drawn around it in the nadir MISR image, and the apparent wind vector is drawn by hand as well along the length of the plume. The plume height can then be stereographically computed along the line where the wind is drawn or inside the ROI. The MINX tool has been extensively used to build a climatology of smoke plume injection heights (Mims et al., 2009).

Finally, we note that CTH is not the only cloud property that can be recovered from stereoscopy, which is predicated on image feature detection and tracking across two or more views, especially using customized research algorithms. How about cloud base (e.g., Böhm et al., 2019) and cloud sides? Seiz and Davies (2006) used MISR data for a massive convective cloud to reconstruct in detail a full 2D transect of its outer shape. Cornet and Davies (2008) then used this outer shape determination to estimate the opacity of the same large cloud by adjusting its optical thickness so that (Monte Carlo) predictions of outgoing radiances approximately match the observed counterparts. Going back to outer cloud-shape determination, but extending from a 2D transect to the whole 3D cloud, Lee et al. (2018) used the intersection multiangle cloud masks backprojected from the surface back to space to find the "space-carved" volume occupied by the same large convective cloud, among others.

These pioneering advances are paving the way for satellite cloud remote sensing to leave the Nakajima-King algorithm for truly stratiform cloud layers and develop new fully 3D tomographic reconstruction methods for vertically developed convective clouds.

15.4.3. Cloud-Top and Aerosol-Layer Pressure Heights from TIR Radiometry

The physical basis of atmospheric probing in the TIR region, clouds and aerosols included, is predicated on the fact that the atmosphere is strongly stratified in terms of both temperature and opacity, hence emissivity. However, to unravel the dominant layer for the origin of the TIR radiance observed from space, one needs at least two channels where the absorbing gas is the same and is well mixed in the atmospheric column. This leads to the popular "CO_2-slicing" method (Chahine, 1974; Wielicki and Coakley, 1981). We note that since a well-mixed absorbing gas is involved, the outcome is not height but pressure level for the top of the clouds. The latter can be converted to the former with knowledge of the thermal profile of the atmosphere, which is readily obtained from a reanalysis, but this is a source of uncertainty in the final CTH retrieval.

Cloud-Top Pressure Determination

In 1980, NASA developed the Thermal Infrared Multispectral Scanner (TIMS) that acquires TIR energy in six bands from 8.2 to 12.2 μm. Successful studies using TIMS resulted in the development of the 15-channel Airborne TerrestriaL Applications Sensor (ATLAS) to monitor vegetation and soil conditions. Landsat Thematic Mapper (TM-4 and TM-5) sensors were launched in 1982 and 1984, respectively, and collected $120 \times 120\,m^2$ TIR data (10.4–12.5 μm) along with two bands of mid-IR data. However, such fine spatial resolution is only available for tasked sensors that are not designed for the global coverage needed to investigate clouds for the purposes of climate science.

Historically, the NOAA Geostationary Operational Environmental Satellite (GOES) collects TIR data at a spatial resolution of $8 \times 8\,km^2$, primarily for global weather prediction. At present (GOES-R series), TIR pixel scales are down to $2 \times 2\,km^2$. Historically, full-disk images of the Earth are obtained every 30 min both day and night by the TIR sensor. At present (GOES-R series), a full-disk scan is obtained at intervals of 10 min. Turning to LEO polar rather than GEO platforms, NOAA's Advanced Very-High-Resolution Radiometer (AVHRR) collects TIR data for local area coverage at $1.1 \times 1.1\,km^2$ and for global coverage at $4 \times 4\,km^2$. AVHRR sensors have recently been superseded in the NOAA Joint Polar Satellite System (JPSS) by VIIRS, which was demonstrated on the Suomi-NPP platform. Across the TIR, VIIRS achieves $\approx 0.75 \times 0.75\,km^2$ resolution at nadir and $\approx 1/2$ of that pixel size in its 11.5 μm channel.

The operational NOAA assets are at the core of the popular International Satellite Cloud Climatology Program (ISCCP) database (Schiffer & Rossow, 1983), and the long-term (climate change) trends detectable in the ISCCP database are the main reason that historical spatio-temporal samplings are still of considerable interest. Of course, other space agencies have assets in geostationary orbit with TIR capability due to their very high value in NWP; Spinning Enhanced Visible and Infrared Imager (SEVIRI) on EUMETSAT's Meteosat-Second Generation series and JMA's Himawari series come to mind.

Turing to MODIS sensors, we have dwelled so far on cloud microphysical and macrophysical property retrievals based its VNIR and SWIR (hence daytime only) channels. However, MODIS also has several TIR channels from which estimates of cloud-top pressure (CTP) as well as temperature, height, effective emissivity, phase (ice vs. water), and cloud fraction are derived for both daytime and nighttime swaths at a scale of $5 \times 5\,km^2$ (Platnick et al., 2003, 2017). Figure 15.9 illustrates the principle of CTP determination using the CO_2-slicing method as implemented for MODIS/Aqua channels.

In contrast with MODIS' multispectral sampling of the TIR spectrum, the Atmospheric InfraRed Sounder (AIRS) instrument on Aqua is a hyperspectral imaging sensor that is primarily targeting atmospheric composition (i.e., trace gases). It shares with MODIS a wide swath that guarantees near-global coverage daily, but with pixels that are tens of kilometers in scale. Like MODIS, it has sensitivity to clouds and AIRS delivers operational cloud products that include fractional cloud cover, as well as cloud-top pressure and temperature (Fetzer et al., 2003; Guillaume et al., 2019). There are also a number of AIRS research products for clouds, including cloud thermodynamic phase (liquid or ice), ice cloud optical thickness, and ice cloud effective diameter (e.g., Kahn et al., 2003, 2004). Cloud products from fused AIRS/AMSU data have been successfully developed (Molnar & Susskind, 2005) capitalizing on their colocation on the Aqua platform. The counterparts of AIRS and AMSU on Suomi-NPP are the Cross-track Infrared Sounder (CrIS) and the Advanced Technology Microwave Sounder (ATMS), which track clouds alone and jointly in much the same way.

In the same vein, observations from AIRS and the Microwave Limb Sounder (MLS), on Aura (in the A-Train along with Aqua), have been combined to investigate clouds in the free troposphere (Kahn et al., 2007). Another composition-centric hyperspectral sensor is the Tropospheric Emissions Spectrometer (TES), also on Aura. TES has higher spectral resolution than AIRS, but less spectral coverage, also much higher spatial resolution but for a narrower swath. TES has similar cloud products as AIRS since, unlike for molecular species, sensitivity to clouds is not highly variable in wavelength space (Eldering et al., 2008). There are, of course, other spaceborne

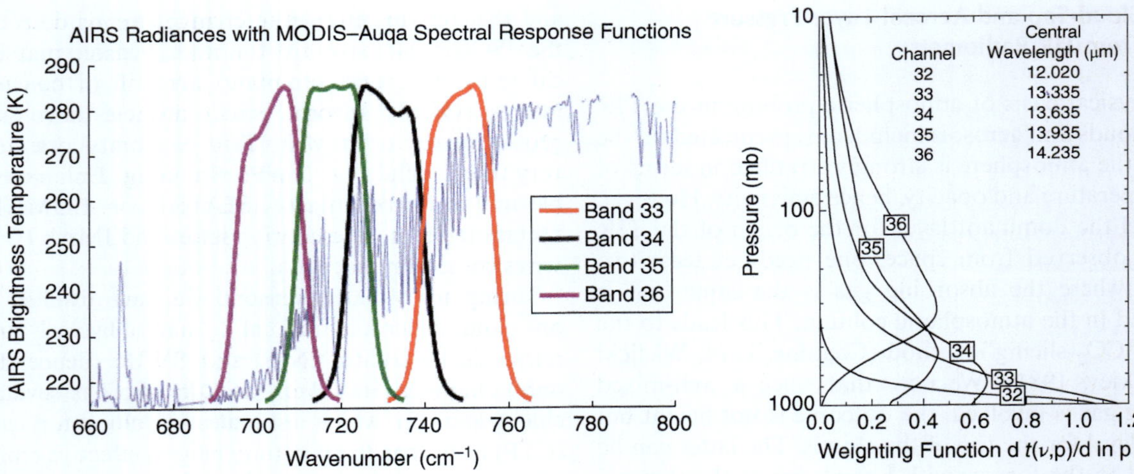

Figure 15.9 Left panel: The spectral responses of MODIS/Aqua's channels 33–36 superimposed on a scan of AIRS radiances for a model atmosphere expressed in BT [K] as a function of wavenumber $\nu = 10^4/\lambda$ in (cm^{-1}) when λ in (μm) (cf. inset table in right panel). *Source:* Courtesy of Mark Smalley, JPL. We note the deliberate sampling of the short-λ flank of the strong CO_2 absorption feature. The colder the BT, the higher the signal comes from in the atmosphere because the increasing optical thickness of the well-mixed gas inhibits penetration. Right panel: The "weighting functions" of MODIS channels 32–36, illustrating how different channels sample different atmospheric layers in the presence of increased opacity (i.e., clouds). *Source:* Adapted from Frey et al. (1999).

hyperspectral TIR sensors with cloud products, such as EUMETSAT's Infrared Atmospheric Sounding Interferometer (IASI), and other are actively being planned.

Aerosol Layer Pressure Height Determination

Aerosol detection techniques using infrared wavelengths have a distinct advantage over visible techniques by providing coverage over bright surfaces and during the night. If an aerosol layer is as opaque as a cloud can be, even at the lower end of the scale where we typically find cirrus layers, then the CO_2-slicing technique will apply just as well. Moreover, due to its large particle sizes and quite high absorption, mineral dust can be more easily detected in the TIR spectral range than clouds at a given optical thickness (Shenk & Curran, 1974). Ackerman (1997) thus studied the detection of volcanic and soil-derived aerosols using TIR observations at 8.5, 11, and 12 μm.

Different methods based either on temporal comparison of BTs (Legrand et al., 2001) on brightness temperature (BT) differences (Chaboureau et al., 2007; Dunion & Velden, 2004; Schepanski et al., 2007) have been implemented successfully for dust remote sensing from TIR instruments on geostationary satellites. For polar-orbiting satellites, a brightness temperature difference approach has been applied to MODIS observations by Zhang et al. (2006) for dust storm classification. Moreover, AIRS provides a high number of spectral bands that are suitable for remote sensing of dust AOT, particle size, and height information over ocean (DeSouza-Machado et al., 2010; Peyridieu et al., 2010; Pierangelo et al., 2004, 2005).

15.4.4. Cloud-Top/Cloud-Pressure Height from Space-Based Oxygen Absorption Channels

It was an old idea (Yamamoto & Wark, 1961) to use known absorption properties of oxygen either in the O_2 A-band region between 760 and 770 nm or in the weaker B-band between approximately 680 and 690 nm to infer column oxygen amounts, and thereby providing a way of estimating the measurements of either surface or cloud-top pressure. This sensitivity to the cumulative path of the light between TOA and the first encountered scattering layer or reflecting surface makes O_2 absorption-based probing akin to lidar but uses a passive approach where the source of radiance is the Sun. The advantage of passive over active probing is that it can be done over the full swath of an imaging sensor with the appropriate spectral channels.

As an example, Figure 15.10 shows calibrated (Geogdzhayev & Marshak, 2018) Earth Polychromatic Imaging Camera (EPIC) images in these two O_2-absorbing channels and their two neighboring "reference" channels taken on 19 November 2017, 02:24:37 UTC, when the Moon happened to be between Earth and the Deep Space Climate Observatory (DSCOVR), which is located at the Lagrange-1 point, roughly 1.5×10^6 km from Earth toward the Sun. Australia is partially covered by the Moon. The pixels size of EPIC is 10 km at nadir. For lack of any O_2 absorption in the line of sight, the Moon itself is almost equally reflective in both in-band and reference channels at both A-bands and B-bands.

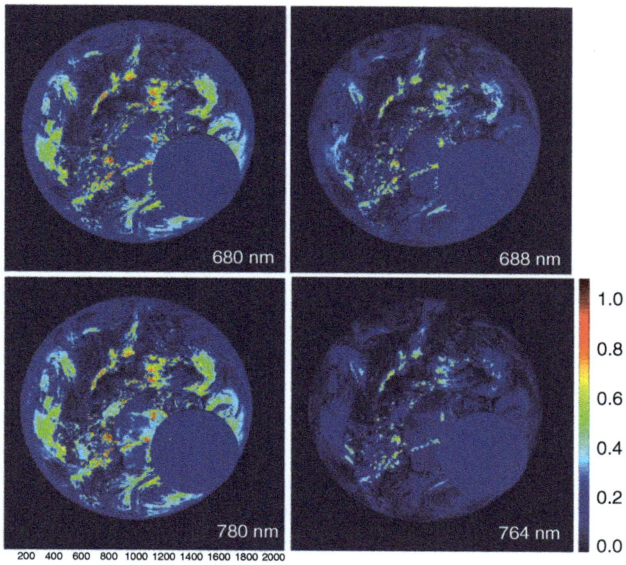

Figure 15.10 Earth-and-New-Moon EPIC images captured on 19 November 2017 at 02:24:37 UTC. Clockwise, from top left: TOA radiance $I_{TOA}(\lambda)$ spectral bands at $\lambda = 680$, 688, 764, and 780 nm in reflectance form $\pi I_{TOA}(\lambda)/F_0(\lambda)$, with $F_0(\lambda)$ being the incoming spectral solar irradiance at TOA. Left panels are nonabsorbing reference channels while right ones are in-band counterparts; B-band is on top, A-band on bottom, and the same color scale was used for all the panels. North is to the upper left, and the equator runs lower left to upper right. *Source:* Anthony B. Davis et al. (2018)/Elsevier.

The tropics are characteristically laced with relatively small but bright cloud masses typical of deep convection regimes with similar brightness in all four channels, as expected for the highest CTHs reaching the stratopause since there is little O_2 to be absorbed above them. In sharp contrast, the midlatitude cloud systems are dominated by expansive stratus layers that are not as bright in the reference purely reflecting channels as tropical systems, due to more modest COT, and are markedly darker in the absorbing channels, as expected from their lower CTHs. This is especially true for the A-band since effective O_2 absorption optical thickness is somewhat larger in the A-band EPIC filter than in its B-band counterpart.

Small amounts of atmospheric scattering can also be characterized, and thus corrected for, provided the measurements of absorption are of sufficient resolution (e.g., Davis, 2008; Ferlay et al., 2010; Heidinger & Stevens, 2000). Measurements obtained with a high-spectral resolution across the O_2 A-band can differentiate between the paths of photons that are reflected from the surface, thus experiencing the entire column of oxygen, from paths of photons that penetrate only to some level in the atmosphere before being scattered back to space (e.g., Stephens & Heidinger, 2000).

The main challenge in inferring CTH with O_2 absorption is that cloud tops are not hard reflecting surfaces. Some light is reflected back to space from the cloud-top layers, typically after a small number of scatterings. However, a significant amount of light penetrates deep into the cloud and yet emerges at cloud top, always after many scatterings. The O_2 absorption signal contains a combination of these two histories. Therefore, ideally, we want enough spectral resolution to reconstruct the distribution of path lengths in the well-mixed absorbing gas. The shortest paths tell us about CTH—no doubt about that—and the longest ones are determined by the targeted cloud's properties, primarily, its geometric thickness H. With its in-band and continuum channels, for not one but two O_2-absorption bands, EPIC can in principle retrieve both cloud-top height, CTH, and cloud geometrical thickness H (Yang et al., 2013). However, it was later shown that the effective gaseous absorption optical depths of these channels are not sufficiently independent to contain the two pieces of information (Davis et al., 2018a, 2018b). Consequently, EPIC's A-band and B-band cloud products are independent estimates of the "effective" CTH, which are known to be biased low (by a significant fraction of H) because of the confounding effect of in-cloud scattering and path cumulation. Separation of CTH and H may be possible using the CTH sensitivity in EPIC's UV channels, coming from the above-cloud Rayleigh scattering (Gao et al., 2019; Jethva et al., 2014).

POLDER also has the equivalent of in-band and continuum channels for the O_2 A-band but, in sharp contrast with EPIC (where solar and viewing angles are nearly the same), POLDER has a range of viewing angles for a targeted cloud. This is basically equivalent to having a number of spectral channels across the A-band with a range of absorption optical depths. Ferlay et al. (2010) form the ratios of in-band/continuum channel radiances—so-called "differential optical absorption spectroscopy" or "DOAS" ratios—across all view angles for a given cloudy pixel. They then proceeded to show that the dispersion of the DOAS ratios conveys information about H while their mean is controlled primarily by CTH. From there, a joint (CTH, H) retrieval was developed that compares well with collocated estimates from other A-train satellites, namely, CloudSat and CALIPSO.

A nonexhaustive but already diverse list of present and past space-based sensors with A-band capability, beyond EPIC/DSCOVR and POLDER/PARASOL, is as follows: SCIAMACHY/ENVISAT (Kokhanovsky et al., 2004), MERIS/ENVISAT (Lindstrot et al., 2006); GOME-1/-2 (Koelemeijer et al., 2001); TROPOMI on board of the Copernicus Sentinel-5 Precursor satellite (Veefkind et al., 2012); and OCO-2 in the A-train (Crisp et al., 2004) and the recently launched OCO-3/ISS scan the O_2 A-band at unprecedented spectral resolution in space. Richardson et al. (2017, 2018, 2019) have indeed showed that cloud geometric thickness is readily retrieved from OCO-2's spectrometers, as was predicted theoretically (Davis et al., 2009). In the future, we are looking forward to 3MI and MAIA (bispectral, multiangle), and PACE (multispectral, multiangle), among others.

Before leaving the topic of cloud remote sensing using O_2 absorption spectroscopy, we have to ask about ground-based deployment, which clearly enables very high spectral resolution. Many observational investigations have been conducted, primarily by Pfeilsticker et al. (1998), Pfielsticker (1999), Scholl et al. (2006), Min and Harrison (1999), and Min et al. (2001, 2004, 2014), and also by Portmann et al. (2001) and Daniel et al. (2003). Theoretical studies were also performed (Davis & Marshak, 2002; Davis et al., 2009) that largely explain the observations. The cloud information content in transmitted A-band radiances is different from their reflected counterparts. Again, it is all about the typical path lengths cumulated by the scattered sunlight through the well-mixed absorbing gas, from TOA to surface. For fixed H, say, a solid stratus layer, path length linearly increases with COT as well as with H. This is due to the increasing time (hence path) it takes for the sunlight to be diffusely transmitted through the cloud as COT increases and the mean free path between scatterings decreases. Observations (Min & Harrison, 1999; Min et al., 2004; Scholl et al., 2006) and theory (Davis et al., 2009) also show that ground-based DOAS ratios for the A-band are highly sensitive to the internal structure of the clouds above head, for instance, the occurrence of multiple solid or broken layers.

15.5. REMOTE SENSING OF CLOUDS FROM ABOVE AT HIGH SPATIAL RESOLUTION

15.5.1. The Question of Spatial Resolution Versus Global Coverage

The enterprise of observing the Earth from space has conflicting goals to have global coverage as frequently as possible (hence imaging large swaths) but also to have as fine spatial detail as possible (hence small pixels). Other constraints are aperture (telescope size, hence weight) and the related signal-to-noise ratio (SNR). The compromise is generally on the pixel size, and a further compromise is that not all spectral channels get the maximum spatial resolution. MODIS, NASA's workhorse in cloud remote sensing, has "moderate" resolution in its name for good reason, with pixels ranging from 0.25 to 0.5 to 1 km as wavelength increases across a huge (2330 km) swath enabling a 1–2 day repeat cycle. Unless it is tasked to collect in "local" mode, MISR works in "global" mode, collecting its red channel at 275 m resolution, and the three others at 1.1 km resolution, across its ≈400 km swath with a repeat coverage every 9 days. Some satellites with hyperspectral imaging sensors have huge pixels, tens of kilometers in size. In order to accommodate the vast amounts of key spectral data, some pixel shapes are not even square (e.g., GOME and OCO-2/3).

Weather-focused satellites in geostationary orbits (e.g., GOES series) and in polar orbits (e.g., NOAA and JPSS series) have larger pixels than MODIS and MISR, but they are definitely improving their resolution over time. Currently, AVHRR sensors on the NOAA-17 and NOAA-18 have pixels as small as 1.1 km and a 2500 km swath; historically, the "Very-High-Resolution" refers to temporal sampling (global coverage daily). In time, the Joint Polar Satellite System (JPSS) will replace the AVHRR/NOAA-n ($n \leq 20$) with VIIRS/JPSS-n ($n \geq 1$). The new Visible Infrared Imaging Radiometer Suite (VIIRS) sensor's swath is 3060 km and, depending on spectral channel, pixels are either 375 or 750 m at nadir. GOES-16 carries the Advanced Baseline Imager (ABI) with pixels from 0.5 to 2 km.

However, cloud science calls for the remote examination at much higher spatial resolution, and there are two ways of achieving this: specially tasked imagers in space and airborne imagers (or other kinds of radiometry). We briefly discuss these assets in the remainder of this section.

15.5.2. Space-Based Imaging at High Spatial Resolution

What if we could shrink pixel size to a few tens of meters, even if it means limiting the image size to a few tens of kilometers, with no hope of global coverage? This was the path followed by Cahalan and coworkers (1989, 1991, 2001) when using Landsat data to study clouds at 30 m resolution. A small price to pay is that, since Landsat is designed for landmass observations, its sensor will saturate on the brightest clouds. That research group made headway into characterizing cloud structures with fractal geometry and scale-invariant (power-law) spatial statistics. The present authors and their coworkers uncovered the phenomenology of radiative smoothing—which is basically Nature's version of pixel cross-talk—using Landsat data (Davis et al., 1997; Marshak et al., 1995).

The routine tasking of satellites like USGS's Landsat imager series is normally optimized to avoid clouds, but they can be programed to collect cloud data too, or the clouds can be present inadvertently and that data can benefit cloud science for free. Di Girolamo and coworkers (Dey et al., 2008; Zhao & Di Girolamo, 2006) tasked NASA's ASTER imager, with 15 m pixels in VNIR, 30 m in SWIR, and 90 m in TIR. These authors thus scrutinized

Figure 15.11 True-color channel combination of an MTI image of Los Alamos (35.875°N, 106.324°W), NM, USA, collected on 22 September 2000 at 19:05 UTC in the presence of broken clouds at 5 m resolution. VZA is 57.4° off-nadir in the aft direction, i.e., from a position to the North. SZA is 54°, about 175° away from the viewing direction in azimuth. For ease of visualization, it is corrected for pixel aspect ratio and orientation to mimic a vista from an airplane window. *Source:* Reproduced with permission from Anthony Davis 2002/SPIE.

clouds, in particular, during the Rain in Cumulus over the Ocean (RICO) field campaign (Rauber et al., 2007). Among other things, they investigated the impact of spatial resolution on cloud detection, hence cloud fraction estimation. Seiz et al. (2006) used ASTER's nadir and monospectral backward view to reconstruct photogrammetrically the outer shape of two large convective clouds.

The 5-m imagery of the US DOE's little known MTI satellite (Weber et al., 1999) inspired a method for estimating an effective optical thickness of cumulus clouds by contrasting radiances from their illuminated and shaded sides (Davis, 2002). The three clouds labeled as "Small," "Medium," and "Big" in Figure 15.11 were analyzed using that simple 3D RT technique, yielding effective COTs in the ranges 19–33, 33–52, and 33–56, respectively.

These are just a few examples of what can be done with very high spatial resolution imagery of clouds from space-based sensors. There are others (e.g., unresolved cloud contamination of aerosol products from moderate resolution sensors), and there are certainly many more research opportunities to seize with these exquisite observational tools.

15.5.3. Airborne Sensors for Imaging, Profiling, and More

The abovementioned high-resolution satellite imagers invariably need to be tasked to make the desired observations from the right place in their orbits. Space agencies tend to view airborne platforms as a stepping stone for technology heading into space. At the same time, cloud-centric field campaign organizers tended to think of aircraft as a means of sampling clouds in situ. Some aircraft can indeed reach high altitudes or even the "edge of space" at ≈20 km above sea level in the case of NASA's ER-2. But this narrow vision is now broadening rapidly. Most space agencies operate aircraft that can be outfitted interchangeably with all manner of passive and active sensors. These sensors tend to be more sophisticated than those deployed in space at any given time. Therefore, these "suborbital" assets offer a unique opportunity for enabling breakthrough science.

Dwelling only on the unique ER-2, it has carried multispectral imagers such as the enhanced MODIS Airborne Simulator (eMAS); hyperspectral ones, such as the Airborne Visual Imaging Infrared Spectrometer-Next Generation (AVIRIS-NG); imaging multiangle spectropolarimeters, such as AirMSPI, AirHARP and SPEX; nonimaging counterparts, like RSP; and active sensors, such as the cloud-profiling lidar (CPL), as well as an HSRL, and sophisticated cloud radar systems. For a given campaign, a handful of these Earth observing sensors can be mounted in various part of the fuselage. A major advantage of using airborne remote sensing is that flight plans can be optimized for the targeted cloud phenomenon and sensor performance. A spectacular example of this is illustrated in Figure 15.12, where the ER-2's pilot (upper left) was instructed to fly over an

Figure 15.12 Upper left: NASA ER-2 pilot prepared for takeoff. *Source:* Courtesy of Brian Rheingans, JPL. Right: AirMSPI sweep image of a solid marine Sc layer, 10 km wide at the waist of the bow-tie with 20 m pixels. *Source:* Courtesy of Feng Xu, JPL, and U of Oklahoma. Lower left: Step-and-stare AirMSPI nadir image showing intricate spiderweb structure of cloud-top instabilities at 10 m resolution in polarized light. *Source:* Courtesy of David Diner, JPL.

extended marine Sc layer in the "principal plane" (defined by the direction of the Sun and the local vertical). This is ideal viewing geometry for polarimeters and, in this case, the rainbow, supernumerary bows, and glory are all revealed in the "sweep" image for degree of linear polarization (DOLP, right panel). Xu et al. (2018) detail the exploitation of these data, including characterization of aerosol load above cloud, cloud microphysical retrieval on a per pixel basis, and effective pixel-scale COT. We also illustrate in Figure 15.12 (lower left panel) the intricate "spiderweb-like" networked structure of dynamical instabilities at the top of a marine Sc layer investigated by Matheou et al. (2020), imaged here at 10 m resolution in "step-and-stare" mode.

The remote-sensing data collected from, say, the sensors on the ER-2 can of course be analyzed using the very same retrieval algorithms used for their (typically, future) counterparts in space. In fact, that is standard procedure for a technology demonstration on the path to space—a prime example is used in Figure 15.6 for the MAIA investigation (Diner et al., 2018).

However, these data can also be used creatively by capitalizing on the high spatial resolution that is completely unique to the airborne platform when other sensor attributes are needed, such as hyperangular sampling or multiangle imaging. An example of exploiting the former is a recent study by Alexandrov et al. (2016) who capitalize on the RSP's quasi continuous angular sampling in the vertical plane containing the flight track to determine the size and a 2D shape of cumulus clouds by intersecting all the beams that are tangent to the cloud from all positions along the flight track where the cloud is in view.

Levis et al. (2015, 2017, 2020) use AirMSPI data acquired in the step-and-stare mode to go a step further, by applying tomographic techniques to multiangle imaging, and thus reconstructing in 3D the observed cumulus cloud's extinction field, i.e., optical thickness of each cell in a large 3D grid covering the cloud's volume. This use of real-world data comes after validating their solution to the large ill-posed inverse problem using realistic cumulus from a large eddy simulation (LES) model. At first, Levis et al. (2015) use prescribed cloud microphysics in their reconstruction that uses only AirMSPI's red channel. Then, Levis et al. (2017) relax that assumption and simply constrain the microphysical parameters (r_e, v_e) to vary only along the vertical axis and consider all of AirMSPI's spectral channels below blue. Based on LES "truth," not only they can retrieve the microphysics reasonably well but their 3D reconstruction is improved significantly. Note that this was done without polarization (because the adopted forward model did not have it at the time) nor SWIR channels (because AirMSPI does not have them, but AirMSPI-2 will). This means that there is sensitivity to droplet size in the subtle change of appearance of cumulus clouds across the VNIR spectrum. That said, the royal route to cloud microphysics, both in 1D stratiform and 3D cumuliform clouds, is via polarization, hence "vector" 3D RT. Levis et al. (2020) embrace it in the framework of cloud tomography, demonstrating accurate r_e along with improved LWC retrieval in a large 3D grid.

Aircraft can also be used for technology demonstrations. An example relevant to clouds is the THickness from Off-beam Returns (THOR), which is the airborne counterpart of ground-based WAIL and space-based LITE, and a third instantiation of a MuSCL system. It was flown over the ARM SGP site on NASA's P-3B aircraft at 5–8 km altitude, at the same time as WAIL was deployed below the solid stratiform cloud layer. THOR's off-beam (hence multiply scattered) lidar returns were analyzed by two groups using very different methodologies (Cahalan et al., 2005; Pounder et al., 2012). They both retrieved accurate values for cloud geometrical thickness in spite of its large COT (between 20 and 30, depending on exact time and location).

Lastly, we started this discussion by acknowledging the vital role of aircraft flying inside the clouds with in situ probes sampling cloud particles by size and, in the case of cirrus, by habit. But more can be done during these in-cloud flights, which is not remote sensing but is radiometry, and can significantly boost the scientific yield of field campaigns. Two examples come to mind.

First, the in situ cloud lidar that was designed at University of Colorado (Evans et al., 2003), then built and flown by SPEC, Inc., in their LearJet (Evans et al., 2006). In situ cloud lidar uses the same physical principles as the three MuSCL systems previously described (WAIL, LITE, and THOR), but the laser pulses are fired inside the cloud on one side of the aircraft while an omnidirectional detector on the other side measures the multiply scattered laser that gets back to roughly the same location after a long random walk in the cloudy airmass. A neural-net-based retrieval scheme was trained on LES stratocumulus and time-dependent 3D RT using a custom Monte Carlo model. It was then used to determine from the in situ cloud lidar data the cloud's geometric thickness H, its COT, and the position of the aircraft inside the cloud.

Second, the Cloud Absorption Radiometer (CAR) was originally designed to fly in the nose of an aircraft flying inside clouds with high COT, along with the usual suite of in situ particle probes (King et al., 1990). CAR simply measures at VNIR and SWIR wavelengths the angular distribution of diffuse radiance inside the cloud, i.e., sunlight scattered once or, more likely, several or even very many times. Deep inside the cloud the directional pattern becomes a cosine function of the view angle relative to the zenith direction, which is the signature of the transport of solar radiation via diffusion, as opposed to the RT regime that prevails near the cloud's boundaries. (In RT, light is streaming through space and occasionally scattered.) CAR data can be used with a forward RT-based or diffusion-based model to infer COT, the aircraft's optical depth below cloud top, and the SSA values for the SWIR channels. The latter can, in turn, be converted into effective droplet radii r_e. It is interesting to compare these volume-averaged estimates with the outcomes from the in situ particle probes. This can be viewed as way of understanding the representativeness of the in situ sampling of the cloud microphysics, which is necessarily very sparse in space.

Interestingly, both THOR and CAR were creatively repurposed for applications far beyond their original active/remote or passive/in situ demonstrations in optical characterization of clouds. Specifically, THOR was assessed in simulation as a probe for sea-ice thickness determination (Várnai & Cahalan, 2007), and it was even deployed to Antarctica where, unfortunately, the P-3B mission had to be aborted. In sharp contrast, CAR is now routinely deployed to fly low-altitude circles around a point on the ground. In this mode of operation, its continuous angular sampling of radiance is used to measure the surface BRDF with high accuracy and directional detail (Melnikova & Gatebe, 2018; Gatebe & King, 2016).

15.6. RETRIEVAL UNCERTAINTY QUANTIFICATION

15.6.1. Uncertainty Quantification and/or Algorithm Validation

Uncertainty quantification (UQ) in remote sensing is closely related to the key issue of product validation, which in turn generally relies (but not always) on absolute (i.e., NIST-traceable) radiometric calibration—the notorious "Cal-Val" operations in any given mission. Indeed, sensor error will directly propagate into retrieval error through the forward model that predicts the measured signal as a function of the desired properties to retrieve and sensor parameters. If there is offset in the sensor calibration, there will generally be a bias in the retrieval, which is to be avoided at all costs. The residual random sensor noise will translate to a certain dispersion in the values of the retrieved atmospheric properties. This measurement uncertainty is inevitable, but the instrument developers keep it to a minimum (SNR to a maximum) in view of all other engineering constraints.

However, radiometric calibration and sensor SNR are far from being the only sources of retrieval uncertainty. The worse offender in cloud and aerosol remote sensing is probably forward model fidelity failure. It is stated that "All models are wrong. Some are useful," an aphorism attributed to the statistician George E.P. Box (1919–2013), and this is certainly true in atmospheric remote sensing. Without a forward model, there will be no retrieval, but we need to ask: How badly does nature violate the assumptions in the (typically physics-based) model? This kind of error is generally a lack of accuracy rather than of precision, i.e., systematic rather than random, which again is the worst kind.

The present authors have made it their mission to quantify retrieval error traceable to the assumption of spatial uniformity in passive remote sensing of cloud properties, especially in the shortwave spectrum, as well as the retrieval of aerosol properties in the vicinity of clouds. The next two subsections survey some of their work and that of others engaged in similar endeavors. The plane-parallel scattering medium assumption does not hurt aerosol retrievals as much as long as the pixel of interest is not only far enough from clouds but also free of subpixel cloud contamination (see, e.g., Zhang et al., 2005). We refer the interested reader to Sayer et al. (2020) for a recent review of UQ/validation in aerosol remote sensing from space.

Finally, going back to product validation, it is not easy to get "ground truth" for clouds and aerosols due to the huge disparity in sampling scale between an in situ sensor and a satellite pixel. Therefore, much product validation is done by comparing a new remote-sensing technique to another one that is less error prone but probably more expensive and/or with poorer sampling. In this context, remote sensing of aerosols has a huge advantage over that of clouds thanks to the AERONET (Holben et al., 2001), which has become the golden standard of aerosol satellite product validation.

15.6.2. Uncertainties in Aerosol Retrievals due to Clouds

Over the past decades, the global scientific community has recognized the critical importance of aerosol-cloud interaction in terms of climate forcing, cloud feedbacks, and precipitation. The most common connotation of the term "aerosol-cloud interaction" in these contexts is a one-way interaction: aerosols modifying clouds. However, the aerosol-cloud interaction goes both ways: clouds influence aerosols possibly just as much as aerosols influence clouds. Cloud droplets indeed provide the aqueous environment for precursor gas phase substances to form new particulate matter and to process existing particles into larger size particles. Cloudy environments provide complex fields of relative humidity, updrafts/downdrafts, and precipitation that affect the loading and physical and optical properties of the particles that cohabit these environments. Thus, aerosol properties are inherently different near clouds than far away from clouds. Yet, we find a dearth of research and publications toward understanding this aspect of aerosol-cloud processes.

The lack of attention to this matter is surprising given that we live in a very cloudy world. For example, 50% of all 1 km pixels identified as cloud-free by the MODIS passive imager occur within 5 km of low clouds; see

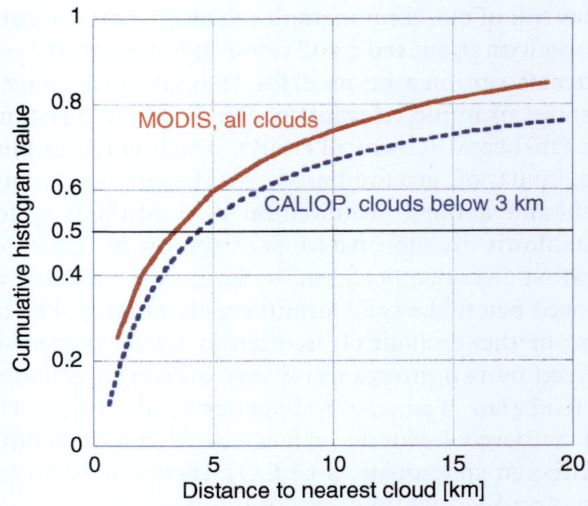

Figure 15.13 Cumulative histogram of cloud-free atmospheric columns over oceans, between 60°N and 60°S latitudes. The solid curve is for all 1 km size MODIS pixels with viewing zenith angles less than 20° from 21 June 2008; the dashed curve is for all 333 m size CALIOP columns observed at night between 15 September 2008 and 14 October 2008. The dashed curve is based on the inset in Figure 1a. *Source:* Reproduced with permission from Várnai and Marshak (2018)/MDPI/Public Domain CC BY 4.0.

Figure 15.13. Furthermore, 20% of all MODIS pixels fall "between the cracks," deemed too cloudy to be retrieved by the aerosol algorithm and not cloudy enough to be retrieved by the cloud algorithm. This means that 20% of pixels are unaccounted for in global estimates of "clear sky" and "cloudy sky" radiative effects from these types of sensors. We are thus not accounting for a significant portion of global aerosol radiative effects because the aerosol in these omitted pixels is fundamentally different in loading and in particle physical and optical properties from those retrievable pixels farther away from the cloud.

15.6.3. Cloud Retrieval Uncertainties

Among many satellite-based remote-sensing techniques for clouds, the Nakajima and King (1990) method is widely used to infer COT, denoted by τ, and cloud droplet effective radius, denoted by r_e, from satellite observation of cloud reflectance. This method uses cloud reflectance measurements from two spectral bands to simultaneously retrieve τ and r_e. As previously stated, it is implemented in practice using LUTs that contain precomputed bidirectional cloud reflectivities at VNIR and SWIR bands for various combinations of r_e and τ under different sun-satellite-viewing geometries and surface albedo values.

The Nakajima-King method makes several important assumptions about the cloud. First, it is used as an "independent pixel assumption" (IPA) that says there is no net interpixel transport of radiation. Second, within a cloudy pixel, the cloud is assumed to be horizontally homogenous. Under these assumptions, clouds are considered to be "plane-parallel." In addition to plane-parallel cloud assumptions, in the operational algorithms, clouds are often assumed to be vertically homogenous. Furthermore, the size spectrum of cloud particles is often assumed to follow certain analytical distributions, such as the single modal gamma or lognormal size distributions (e.g., Dong et al., 1997). These assumptions may be reasonable for certain types of clouds, such as stratocumulus clouds, but become problematic for others, such as broken trade wind cumuli or precipitating clouds (Di Girolamo et al., 2010; Liang & Girolamo, 2013; Painemal & Zuidema, 2011; Zhang & Platnick, 2011). As reported in numerous studies, real clouds deviate from these assumptions, and, consequently, the r_e and τ retrievals can suffer from large errors and uncertainties (e.g., Kato et al., 2006; Liang et al., 2015; Marshak et al., 2006; Miller et al., 2018; Várnai & Marshak, 2002; Zhang et al., 2012; Zhang, 2013).

It has been known for a long time that at the spatial scale of climate model grids (~hundreds of kilometers) the approximation of inhomogeneous cloud fields with plane-parallel clouds can lead to significant biases in shortwave solar radiation (Barker, 1996; Cahalan et al., 1994; Harshvardhan & Randall, 1985). Cahalan et al. (1994) described an elegant theoretical framework based on a fractal cloud model to explain the influence of small-scale horizontal variability of τ on the averaged cloud reflectance in the VNIR spectral region. It was shown that the spatially averaged reflectance is smaller than the reflectance that corresponds to the averaged COT. This inequality is well known as the "plane-parallel homogenous bias," which is a result of the nonlinear dependence of visible reflectance on τ.

To illustrate, we assume that one half of an inhomogeneous pixel is covered by a thinner cloud with $\tau_1 = 5$ and the other half by a thicker cloud with $\tau_2 = 18$, the retrieved COT based on the averaged reflectance $R^* = [R(\tau_1) + R(\tau_2)]/2$ is significantly smaller ($\tau^* = 9.8$) than the desired linear average of the subpixel $\tau = 11.5$. The impacts of plane-parallel bias on satellite-based cloud property retrievals and the implications have been investigated in a number of studies (e.g., Alexandrov et al., 2012b; Oreopoulos & Davies, 1998; Oreopoulos et al., 2007; Pincus et al., 1999; Shang et al., 2015).

We note that the variation of cloud reflectance may be a result of varying cloud properties but may also be caused by 3D radiative effects. For example, a cloudy pixel can be perfectly homogenous in terms of cloud properties, but the surrounding pixels can cast a shadow on part of this pixel leading to subpixel reflectance variations (Marshak et al., 2006). A variety of such 3D effects that cannot be explained by the 1D plane-parallel RT theory have been identified, and their impacts on cloud property retrievals have been investigated in previous studies that were surveyed, in particular, by Davis and Marshak (2010).

Recently, as the interests in aerosol-cloud interactions have grown, there is an increasing attention on the impacts of small-scale cloud variations on the satellite-based r_e retrievals (e.g., Kato et al., 2006; Liang et al., 2015; Marshak et al., 2006; Zhang & Platnick, 2011; Zhang et al., 2012). Marshak et al. (2006) pointed out that similar to the plane-parallel bias, the nonlinear dependence of the SWIR band cloud reflectance on r_e can also lead to significant biases on r_e retrievals. It must be noted that, in the framework of Marshak et al. (2006), the retrievals of r_e and τ are considered separately and assumed to be independent of one another. However, as Marshak et al. (2006) pointed out, this assumption is valid only for "large enough" τ and r_e (typically, $r_e > 5$ μm and $\tau > 10$). Zhang and Platnick (2011) showed that the subpixel variance of τ can have a significant impact on the r_e retrieval. Zhang and Platnick (2011) and Zhang et al. (2012) also found that the magnitude of the positive r_e retrieval bias caused by the subpixel variability of τ is dependent on the band chosen for the r_e retrieval. These studies showed that the same subpixel τ variability tends to induce a larger bias in retrieved r_e using the less absorptive 2.1 μm band than that using the more absorptive 3.7 μm band. This spectral dependence provides an important explanation for the fact that the MODIS operational 2.1 μm retrievals for water clouds are often significantly larger than the 3.7 μm retrievals, especially when clouds have large subpixel heterogeneity (Cho et al., 2015; Miller et al., 2018; Zhang & Platnick, 2011; Zhang et al., 2016).

Finally, we must recognize that operational cloud property retrievals, among others, always evolve in the course of the lifetime of a space-based sensor such as MODIS and, accordingly, whole-mission Level-1 databases are reprocessed into Level-2 products. A very productive development in the case of MODIS/Aqua cloud retrievals is that failure of the bispectral retrieval algorithm is flagged and documented. The IPA/LUT-based Nakajima-King algorithm can indeed fail in a number of ways that can be systematically analyzed. Cho et al. (2015) investigated in great detail MODIS bispectral cloud property (τ, r_e) retrieval failure rate for liquid water clouds over water. Figure 15.14 displays the mean failure rate when such pixels are furthermore identified as partially cloudy. We clearly see that failure rates are relatively low in regions dominated by strato-cumulus, namely, the western side of oceans in the tropics and midlatitudes. Now, low failure does not necessarily translate to accuracy as these

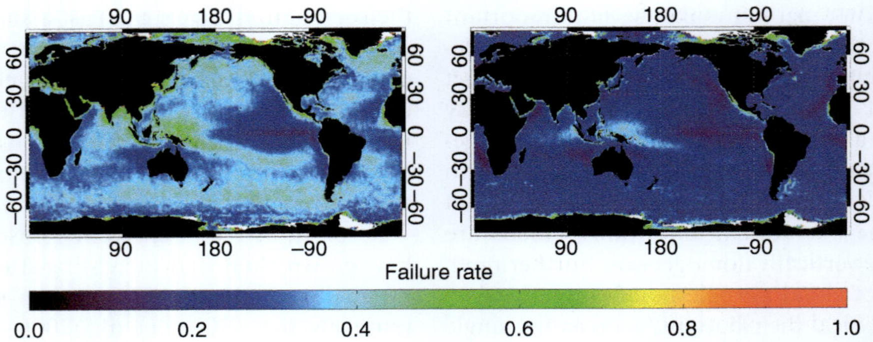

Figure 15.14 Geographic distribution of MODIS/Aqua bispectral cloud property (τ, r_e) retrieval failure rates for $1 \times 1\ \text{km}^2$ pixels over water classified as liquid phase and partly cloudy (based on 0.25 km scale VNIR pixels): (0.86, 2.1) (μm) left; (0.86, 3.7) (μm) right. *Source:* Reproduced with permission from Cho et al. (2015).

cloud types have been shown to have retrieval error due to internal 3D RT effects. In contrast, in the western side of the oceans, especially in the tropics, the bispectral algorithm fails at a significantly increased rate due to the outer and inner 3D spatial variability induced by the prevailing convective dynamics. This retrieval failure rate is very problematic in these climatically important regions and has become a major motivation for advancing the concept of 3D cloud tomography (Levis et al., 2015, 2017, 2020).

15.6.4. Information Content Analysis

The vast majority of remote-sensing algorithms are built on a (generally multivalued) physics-based forward model $\mathbf{F}(\mathbf{x})$ that predicts the various measured quantities $\mathbf{y}$ given the "state-vector" $\mathbf{x}$ of the system being observed that contains all the geophysical properties to be retrieved. Remote sensing at large is thus cast as a model fitting problem, formalized as $\mathbf{y} = \mathbf{F}(\mathbf{x}) + \boldsymbol{\varepsilon}$, where $\boldsymbol{\varepsilon}$ is measurement error vector. This is generally an overdetermined problem where the dimension m of $\mathbf{y}$ exceeds n, that of $\mathbf{x}$. The standard solution is to minimize in $\mathbf{x}$-space the cost function $\chi^2(\mathbf{x}|\mathbf{y})$, which is the weighted sum of m squared residuals $(y_j - F_j(\mathbf{x}))^2$, $1 \le j \le m$. The m weights are provided by the inverses of the variances $\sigma^2{}_j$ of the noise elements, an m-vector $\mathbf{v}_\varepsilon = (\sigma^2{}_j,\ 1 \le j \le m)^{\mathrm{T}}$, where superscript "T" means transpose.

Formally, we now have $\chi^2(\mathbf{x}|\mathbf{y}) = (\mathbf{y} - \mathbf{F}(\mathbf{x}))^{\mathrm{T}}\mathbf{S}_\varepsilon{}^{-1}(\mathbf{y} - \mathbf{F}(\mathbf{x}))$, a quadratic form where matrix $\mathbf{S}_\varepsilon = \text{diag}[\mathbf{v}_\varepsilon]$. When possible, it is even better to take $\mathbf{S}_\varepsilon$ as the full measurement error covariance matrix that will properly introduce into $\chi^2(\mathbf{x}|\mathbf{y})$ nondiagonal terms that increase it when sensor noises correlate and reduce it when they anticorrelate. The generally nonlinear data fitting problem is thus formulated as the numerical estimation of $\mathbf{x}^* = \text{argmin}_\mathbf{x}\chi^2(\mathbf{x}|\mathbf{y})$ by any convenient minimization technique such as gradient descents or the Levenberg-Marquardt algorithm, both of which call for the $m \times n$ Jacobian matrix $\mathbf{J} = \partial\mathbf{F}/\partial\mathbf{x}$ or else a gradient-free method. In the former case, uncertainties on the n retrieved properties traceable to measurement error are the diagonal terms of $\mathbf{S}_\mathbf{x} = (\mathbf{J}^{\mathrm{T}}\mathbf{S}_\varepsilon{}^{-1}\mathbf{J})^{-1}$; see, e.g., Bevington and Robinson (1992).

Rodgers (2000) brought to this formalism a Bayesian perspective that is particularly relevant to all manner of atmospheric remote sensing. In the framework of maximum likelihood estimation, that is tantamount to generalizing the cost function to include an "a priori" uncertainty term:

$$\chi^2(\mathbf{x}|\mathbf{y}) = (\mathbf{y} - \mathbf{F}(\mathbf{x}))^{\mathrm{T}}\mathbf{S}_\varepsilon^{-1}(\mathbf{y} - \mathbf{F}(\mathbf{x})) + (\mathbf{x} - \mathbf{x}_\mathbf{a})^{\mathrm{T}}\mathbf{S}_\mathbf{a}^{-1}(\mathbf{x} - \mathbf{x}_\mathbf{a}) \qquad (15.3)$$

where the second term penalizes the minimization search of $\mathbf{x}$-space if it wanders too far from the prior estimate $\mathbf{x}_\mathrm{a}$ of the unknown state vector. Although there is a somewhat controversial subjectivity to all things Bayesian, this is a natural assumption: when going into a remote-sensing situation, we generally know something already about the state vector of the system under observation. Maybe it is just a range of plausible values. The a priori uncertainty covariance matrix $\mathbf{S}_\mathrm{a}$ provides the natural weighting of the discrepancies in $(\mathbf{x} - \mathbf{x}_\mathrm{a})$. By examination of (15.3), we see that if the observations are useless ($\mathbf{S}_\varepsilon$ elements are very large), then the minimization goes to $\mathbf{x}_\mathrm{a}$ with the prior uncertainties unchanged. If, in contrast, the observations are very precise ($\mathbf{S}_\varepsilon$ elements are very small), then the minimization goes to the value of $\mathbf{x}$ that minimizes the first term in (15.3), which may be arbitrarily far from $\mathbf{x}_\mathrm{a}$. In general, the maximum-likelihood estimate of $\mathbf{x}^*$, still defined operatively as $\text{argmin}_\mathbf{x}\chi^2(\mathbf{x}|\mathbf{y})$, but now using (15.3), will have a "posteriori" uncertainty defined by the diagonal elements of $\mathbf{S}_\mathbf{x} = (\mathbf{J}^{\mathrm{T}}\mathbf{S}_\varepsilon{}^{-1}\mathbf{J} + \mathbf{S}_\mathrm{a}{}^{-1})^{-1}$. Figure 15.15 illustrates this optimal estimation method graphically in a hypothetical case where $m = n = 2$.

Possibly the most far-reaching insights from Rodgers' (2000) theory of optimal estimation come from the notion of "information content" of the observations, which may be better described as *information gain via observation*. It is quantified by the so-called "averaging kernel," the nondimensional $n \times n$ matrix $\mathbf{A} = \partial\mathbf{x}^*/\partial\mathbf{x}$ that is readily shown to be equal to $\mathbf{I}_n - \mathbf{S}_\mathbf{x}\mathbf{S}_\mathrm{a}^{-1}$, where $\mathbf{I}_n$ is the identity matrix in the n-dimensional state vector space. Drawing on the concept of entropy increase in statistical physics—and, more to the point, its negative being Shannon's information decrease—a definition of information content/gain is given by Trace [**A**]. That scalar is known as the number of "degrees of freedom" (DoF) in the observations and is necessarily $\le n$. Even more interesting are the individual diagonal elements of **A** that quantify the *partial* DoF associated with each element of **x**, which is necessarily ≤ 1. If A_{ii} $(1 \le i \le n)$ is near unity (because $S_{\mathbf{x}ii} \ll S_{\mathrm{a}ii}$), then we have gained a lot of information about it from the remote sensing. If, on the contrary, A_{ii} is very small (because $S_{\mathbf{x}ii} \approx S_{\mathrm{a}ii}$), then we have gained very little information about it.

To illustrate, we return to Figure 15.15 that is a schematic, but one that is at least qualitatively correct about a study of parametrized aerosol profiling with the O_2 A-band observations currently planned for the PACE mission described by Werdell et al. (2019). Going back to A-band DOAS considerations in section 15.4, Davis and Kalashnikova (2019) used actual sensor specifications from both the OCI spectrometer (with $m = 3$ spectral channels in the A-band) and, by default at the time, an MISR/MSPI-like multiangle imager (with $m = 9$ viewing angles). Their performance was compared in terms of partial DoFs for the retrieval of both pressure

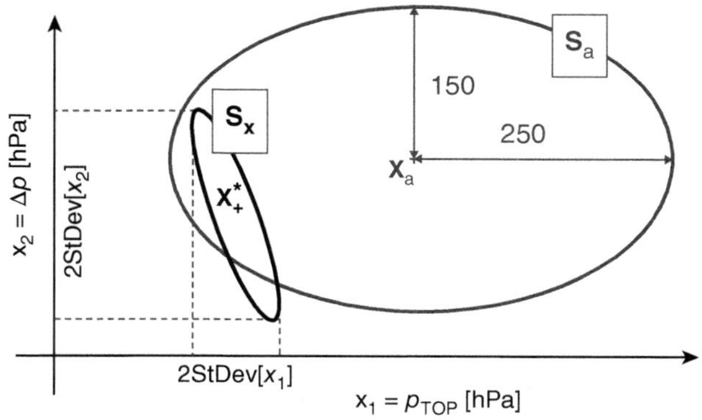

Figure 15.15 The ellipses in this schematic roughly represent the level sets where $\chi^2(\mathbf{x}|\mathbf{y}) \approx m$, i.e., where each term in (15.3) contributes equally, assuming for simplicity uniform weights from $\mathbf{S}_\varepsilon$ and $\mathbf{S}_\mathrm{a}$. Their projected dimensions along the axes scale with the prior and posterior uncertainties in the two aerosol properties in $\mathbf{x} = (p_{\mathrm{TOP}}, \Delta p)^\mathrm{T}$. *Source:* Reproduced with permission from Davis and Kalashnikova (2019)/Springer Nature.

at the top of an aerosol (p_{TOP}) along with its pressure thickness (Δp). Over the realistic ranges for $\mathbf{x} = (p_{\mathrm{TOP}}, \Delta p)^\mathrm{T}$, the DoF for p_{TOP} is high while that for Δp is small. The conclusion is therefore that, since it is desirable have the capability of estimating the (p_{TOP}, Δp) pair, it will be necessary to supplement the A-band data with other observations with sensitivity to p_{TOP}, e.g., near-UV channels. To balance the math-heavy DoF analysis, Davis and Kalashnikova (2019) conducted a physics-based counterpart that arrived at the same conclusion.

Adopting a radically different forward model, Davis et al. (2018a) applied the same DoF method as Davis and Kalashnikova (2019) to the retrieval of (p_{TOP}, Δp) for cloud- rather than aerosol layers using the two (not-so-independent) DOAS ratios from EPIC's O_2 A- and B-band channels. They came to the same conclusions about the partial DoFs and the need to use supplementary p_{TOP}-sensitive observations found in EPIC's UV channels. Davis et al. (2018b) revisited that information content/gain analysis using physics-based methods and again came to the same conclusions.

All and all, we have gained trust in the somewhat abstract formalism of Rodgers (2000) as a tool for predicting sensor performance in the context of both aerosol and cloud profiling and, by, extension, inherent property retrievals at large.

15.7. SUMMARY AND OUTLOOK

As stated in the 2017 Decadal Survey (NASEM, 2018, pp. 3–59), the "uncertainties associated with aerosols and aerosol-cloud interactions are the largest radiative forcing uncertainty. Because of this, knowledge of their characteristics and processes is critical to reducing the uncertainty of future projections of climate change." As a result, reliable remote-sensing retrieval of aerosol and cloud properties from space observations is one of the highest priorities in climate science, and supplementing the space-based observations with ones from suborbital (i.e., ground-based and airborne) platforms is equally vital.

In this chapter, the reader is served with an extensive—but by no means exhaustive—inventory of satellite and airborne remote-sensing resources. However, the emphasis is on the physics that the retrieval algorithms are founded on, in part, because these physics closely relate to how shortwave/solar and longwave/terrestrial radiation interacts with clouds and aerosols as part of the Earth's climate system, thus modulating its radiative energy budget. The other reason for drilling into the foundations of the retrieval algorithms is that they are never perfect, only reasonably good under some circumstances.

Consequently, there is a definite trend in remote sensing to navigate toward new wavelengths or new channels (and accordingly new technologies) that make the retrieval physics simpler, hence more trustworthy. But there is always a tradeoff, and an informed cost/benefit analysis is in order. There is also a trend toward bringing to bear advances in physical understanding, computer power, and inverse problem theory to better exploit measurements from proven technology. Indeed, to ensure uninterrupted monitoring of the climate system, heritage observations have to be continued into the foreseeable future. We have attempted to describe both paths forward, and we did not shy away from highlighting a few "out-of-the-box" concepts along the way.

This chapter starts with remote sensing of cloud properties; it describes two main techniques for obtaining cloud microphysics from visible and near-infrared observations. There are the Nakajima-King (1990) approach, widely used in retrievals of cloud optical depth and particle size (e.g., AVHRR, MODIS, VIIRS), and the Bréon-Goloub (1998) approach that is based on the directional signature of the polarized reflectance (applicable to the POLDER instrument under the right circumstances). One more remote-sensing technique for retrieving (mostly ice) cloud properties that uses microwave sensors (e.g., AMSR and SSM/I) is also briefly described. The chapter then switches to remote sensing of aerosol properties. Because of very different surface reflectance, aerosol retrieval techniques are very different for ocean and land. The main aerosol remote-sensing approaches used by the major satellite imagers (MODIS, VIIRS, MISR, and POLDER) are described in the chapter.

While MODIS and VIIRS are the most popular workhorses for the aerosol and cloud science communities, MISR and POLDER have blazed the path toward future aerosol and cloud observations from space that will be using both multiangle imaging and polarization. Special attention is paid to the MAIAC algorithm, now widely used for retrieval of aerosol properties over land, along with the conceptually similar GRASP technique. Although they were developed for specific sensors, MAIAC and GRASP are flexible packages that can and will be adapted to future missions.

In addition to passive remote sensing, we have also highlighted the active remote-sensing methods for aerosol and cloud profiling with an emphasis on CALIPSO and CloudSat for lidar and radar, respectively. While on the topic of locating clouds in altitude, the chapter explores oxygen A-band and B-band remote sensing of cloud and aerosol layer height, focusing mostly on the DSCOVR/EPIC instrument. Other passive methods for estimating cloud-top height are described as well, namely, stereo and TIR.

A special section is dedicated to cloud remote sensing at very high spatial resolution either from tasked imaging sensors in space, such as ASTER or the Landsat series, or from airborne platforms deployed above the clouds of interest, such as NASA's ER-2. The present authors see these assets as underutilized by the cloud science community, and there is now a burst of algorithm development that uses the unique features of airborne sensors. In view of the diversity of sensors that can be flown on even a single aircraft, there is gold to be mined for cloud-focused field campaigns in these observational tools that open the path for completely new ways of extracting cloud information by remote sensing.

The chapter closes with a too brief discussion of retrieval uncertainty for both cloud and aerosol remote sensing but offering references for further reading. Special attention is paid to the areas between clouds and aerosols; these areas are critical for understanding aerosol-cloud interaction and we call these areas the "transition zone." They are deemed too cloudy to be retrieved by the aerosol algorithm and not cloudy enough to be retrieved by the cloud one. To conclude, Charlson et al. (2007) noted that "the existence of partly cloudy regions and the fact that the clear-cloudy distinction is ambiguous and [...] the conventional expression when 'direct' and 'indirect' aerosol effects are separately calculated for the clear and cloudy regions may lead to errors."

ACKNOWLEDGMENTS

The authors wish to dedicate this chapter to their friend and mentor Warren Wiscombe who brought them in the early-1990s to the remote sensing of clouds and aerosols.

This work was performed in part at the Jet Propulsion Laboratory, California Institute of Technology, under contract with the National Aeronautics and Space Administration. The authors acknowledge financial support from Richard Eckman for DSCOVR/EPIC program involvement. A.M. thanks Ralph Kahn, Robert Levy, Tamas Varnai and Alexei Lyapustin for fruitful discussions about aerosols and their remote sensing from space. A.D. is grateful to Kevin Bowman, David Diner, Michael Garay, Brian Kahn, Matthew Lebsock, John Martonchik, Vijay Natraj, Olga Kalashnikova, Suniti Shangavi, Joao Teixeira, and Feng Xu for the same reason, for adding the airborne and polarimetric perspectives, and encouragement to look past his spectral horizon. Finally, we thank our two anonymous reviewers for their in-depth comments that helped to substantially improve this chapter. The authors acknowledge California Institute of Technology © 2020 for providing government sponsorship.

LIST OF ABBREVIATIONS AND ACRONYMS

3MI	Multiangle/Multispectral/Multipolarimetric Imager
ABI	Advanced Baseline Imager
AERONET	AErosol RObotic NETwork
AHI	Advanced Himawari Imager
AirHARP	Airborne HARP
AirMSPI	Airborne MSPI
AIRS	Atmospheric InfraRed Sounder
AMSR	Advanced Microwave Sensor Radiometer
AOT	Aerosol optical thickness
ARM	Atmospheric Radiation Measurement (DOE facility program)
ASTER	Advanced Spaceborne Thermal Emission and Reflection Radiometer
ATLAS	Airborne TerrestriaL Applications Sensor
ATMS	Advanced Technology Microwave Sounder
ATSR	Along-Track Scanning Radiometer
AVHRR	Advanced Very-High-Resolution Radiometer
AVIRIS-NG	Airborne Visual Imaging Infrared Spectrometer-Next Gen
BRDF	Bidirectional reflection distribution factor
BT	Brightness temperature
CALISPO	Cloud-Aerosol Lidar and Infrared Pathfinder Satellite Observation
CALIOP	Cloud-Aerosol LIdar with Orthogonal Polarization
CAR	Cloud Absorption Radiometer
CATS	Cloud-Aerosol Transport System
CNES	Centre National d'Études Spatiales
COT	Cloud optical thickness
CPR	Cloud-Profiling Radar
CrIS	Cross-track Infrared Sounder
CTH	Cloud-top height
CTP	Cloud-top pressure
DoE	Department of Energy
DoF	Degree of freedom
DOLP	Degree of linear polarization
DMSP	Defense Meteorological Satellite Program
DOAS	Differential optical absorption spectroscopy
DSCOVR	Deep Space Climate Observatory
EarthCARE	Earth Clouds, Aerosols, and Radiation Explorer
ENVISAT	ENVironmental SATellite
eMAS	Enhanced MODIS Airborne Simulator
EPIC	Earth Polychromatic Imaging Camera
ESA	European Space Agency
EUMETSAT	European Organisation for the Exploitation of Meteorological Satellites
FOV	Field of view
GCOM-C1	Global Change Observation Mission-Climate 1
GOES	Geostationary Operational Environmental Satellite (series)
GOME	Global Ozone Monitoring Experiment
GRASP	General Retrieval of Aerosol and Surface Properties (algorithm)
HARP	Hyper-Angular Radio-Polarimeter
HSRL	High-spectral resolution lidar
IASI	Infrared Atmospheric Sounding Interferometer
ISCCP	International Satellite Cloud Climatology Program
ICESat	Ice, Cloud, and land Elevation Satellite (mission)
IPCC	Intergovernmental Panel on Climate Change
IR	Infrared
ISS	International Space Station
IWP	Ice water path
JMA	Japan Meteorological Agency
JPSS	Joint Polar Satellite System
lidar	Light detection and ranging
LITE	Lidar In space Technology Experiment
LUT	Look-up table
LWP	Liquid water path
MAIA	Multi-Angle Imager for Aerosols
MAIAC	Multi-Angle Implementation of Atmospheric Correction (algorithm)
MFOV	Multiple field of view (lidar)
MERIS	MEdium Resolution Imaging Spectrometer
MISR	Multi-angle Imaging Spectro-Radiometer
MLS	Microwave Limb Sounder
MMCR	Millimeter-Wave Cloud Radar (specifically at DOE ARM fixed and mobile sites)
MODIS	MODerate resolution Imaging Spectroradiometer
MPL	Micropulse lidar
MSPI	Multi-angle Spectro-Polarimetic Imager
MTI	Multispectral Thermal Imager
MuSCL	Multiple Scattering Cloud Lidar

NPP	NPOESS Preparatory Project (satellite mission), a.k.a., Suomi-NPP
NNSA	National Nuclear Security Administration
NWP	Numerical Weather Prediction
PACE	Plankton, Aerosol, Cloud, ocean Ecosystem
PARASOL	Polarization and Anisotropy of Reflectances for Atmospheric Science coupled with Observations from a Lidar
PODEX	POLarimeter Definition EXperiment
POLDER	Polarization and Directionality of the Earth's Reflectances
PSD	Particle size distribution
radar	Radio detection and ranging
RICO	Rain In Cumulus over the Ocean
ROI	Region of interest
RSP	Research Scanning Polarimeter
RT	Radiative transfer
SCIAMACHY	SCanning Imaging Absorption spectroMeter for Atmospheric CHartographY
SEVIRI	Spinning Enhanced Visible and Infrared Imager
SGLI	Second-Generation GLobal Imager
SGP	Southern Great Plains (DOE/ARM site)
SNR	Signal-to-noise ratio
SPEX	Spectro-polarimeter for Planetary EXploration
SSA	Single-scattering albedo
SSM/I	Special Sensor Microwave Imager
SWIR	Shortwave IR
SZA	Solar zenith angle
TES	Tropospheric Emissions Spectrometer
THOR	THickness from Off-beam Returns
TIMS	Thermal Infrared Multispectral Scanner
TIR	Thermal IR
TM	Thematic Mapper (Landsat sensor)
TOA	Top of atmosphere
TROPOMI	TROPOspheric Monitoring Instrument
UQ	Uncertainty Quantification
UV	Ultraviolet
VIIRS	Visible Infrared Imaging Radiometer Suite
VIS	Visible
VNIR	Visible and near-IR
VZA	Viewing zenith angle
WAIL	Wide-Angle Imaging Lidar

REFERENCES

Abdalati, W., Zwally, H. J., Bindschadler, R., Csatho, B., Farrell, S. L., Fricker, H. A., et al. (2010). The ICESat-2 laser altimetry mission. *Proceedings of the IEEE*, *98*, 735–751.

Ackerman, S. A. (1997). Remote sensing aerosol using satellite infrared observations. *Journal of Geophysical Research, D102*, 17069–17079.

Ackerman, S. A., Strabala, K. I., Menzel, W. P., Frey, R. A., Moeller, C. C., & Gumley, L. E. (1998). Discriminating clear sky from clouds with MODIS, *Journal of Geophysical Research, D103*, 32141–32157.

Alexandrov, M. D., Cairns, B., Emde, C., Ackerman, A. S., Ottaviani, M., & Wasilewski, A. P. (2016). Derivation of cumulus cloud dimensions and shape from the airborne measurements by the Research Scanning Polarimeter. *Remote Sensing of Environment*, *177*, 144–152.

Alexandrov, M. D., Cairns, B., Emde, C., Ackerman, A. S., & van Diedenhoven, B. (2012b). Accuracy assessments of cloud droplet size retrievals from polarized reflectance measurements by the research scanning polarimeter. *Remote Sensing of Environment*, *125*, 92–111.

Alexandrov, M. D., Cairns, B., & Mishchenko, M. I. (2012a). Rainbow Fourier transform. *Journal of Quantitative Spectroscopy and Radiative Transfer*, *113*, 2521–2535.

Alexandrov, M. D., Cairns, B., Wasilewski, A. P., Ackerman, A. S., McGill, M. J., Yorks, J. E., & Knobelspiesse, K. D. (2015). Liquid water cloud properties during the polarimeter definition experiment (PODEX). *Remote Sensing of Environment*, *169*, 20–36.

Barker, H. (1996) A parameterization for computing grid-averaged solar fluxes for inhomogeneous marine boundary layer clouds. Part I: Methodology and homogeneous biases. *Journal of the Atmospheric Sciences*, *53*, 2289–2303.

Bar-Or, R. Z., Koren, I., Altaratz, O., & Fredj, E. (2012). Radiative properties of humidified aerosols in cloudy environment. *Atmos. Res.*, *118*, 280–294.

Bevington, P. R., & Robinson, D. K. (1992). *Data reduction and error analysis for the physical sciences*. New York (NY): WCB McGraw-Hill.

Böhm, C., Sourdeval, O., Mülmenstädt, J., Quaas, J., & Crewell, S. (2019). Cloud base height retrieval from multi-angle satellite data. *Atmospheric Measurement Techniques*, *12*, 1841–1860.

Bréon, F. M., & Goloub, P. (1998). Cloud droplet effective radius from spaceborne polarization measurements. *Geophysical Research Letters*, *25*(11), 1879–1882. doi:10.1029/98GL01221.

Burrows, J. P., Weber, M., Buchwitz, M., Rozanov, V., Ladstätter-Weißenmayer, A., Richter, A., et al. (1999). The Global Ozone Monitoring Experiment (GOME): Mission concept and first scientific results. *Journal of the Atmospheric Sciences*, *56*, 51–171.

Cahalan, R. F. (1991). Landsat observations of fractal cloud structure. In *Nonlinear Variability in Geophysics* (pp. 281–295). Dordrecht: Springer.

Cahalan, R. F., & Joseph, J. H. (1989). Fractal statistics of cloud fields. *Monthly Weather Review*, *117*, 261–272.

Cahalan, R. F., McGill, M., Kolasinski, J., Várnai, T., & Yetzer, K. (2005). THOR—Cloud thickness from offbeam lidar returns. *Journal of Atmospheric and Oceanic Technology, 22*, 605–627.

Cahalan, R. F., Oreopoulos, L., Wen, G., Marshak, A., Tsay, S. C., & DeFelice, T. (2001). Cloud characterization and clear-sky correction from Landsat-7. *Remote Sensing of Environment, 78*, 83–98.

Cahalan, R. F., Ridgway, W. L., Wiscombe, W. J., Gollmer, S., & Harshvardhan. (1994). Independent pixel and Monte Carlo estimates of stratocumulus albedo. *Journal of the Atmospheric Sciences, 51*, 3776–3790. doi:10.1175/1520-0469.

Cairns, B., Russell, E. E., LaVeigne, J. D., & Tennant, P. M. W. (2003). Research scanning polarimeter and airborne usage for remote sensing of aerosols. In J. A. Shaw & J. S. Tyo (Eds.), Polarization Science and Remote Sensing (pp. 33–44). Society of Photo-Optical Instrumentation Engineers (SPIE) Conference Series 5158.

Chaboureau, J.-P., Cammas, J.-P., Duron, J., Mascart, P. J., Sitnikov, N. M., & Voessing, H.-J. (2007). A numerical study of tropical cross-tropopause transport by convective overshoots. *Atmospheric Chemistry and Physics, 7*, 1731–1740. doi:10.5194/acp-7-1731-2007.

Chahine, M. T. (1974). Remote sounding of cloudy atmospheres. I. The single layer cloud. *Journal of the Atmospheric Sciences, 31*, 233–243.

Chand, D., Wood, R., Ghan, S., Wang, M., Ovchinnikov, M., Rasch, P. J., et al. (2012). Aerosol optical depth enhancement in partly cloudy conditions. *Journal of Geophysical Research, 117*, D17207. doi:10.1029/2012JD017894.

Charlson, R. J., Ackerman, A. S., Bender, F. A. M., Anderson, T. L., & Liu, Z. (2007). On the climate forcing consequences of the albedo continuum between cloudy and clear air. *Tellus B: Chemical and Physical Meteorology, 59*, 715–727.

Cho, H.-M., Zhang, Z., Meyer, K., Lebsock, M., Platnick, S., Ackerman, A. S., et al. (2015). Frequency and causes of failed MODIS cloud property retrievals for liquid phase clouds over global oceans A comprehensive analysis using A-Train observations. *Journal of Geophysical Research, D120*, 4132–4154. doi:10.1002/2015JD023161.

Cornet, C., & Davies, R. (2008). Use of MISR measurements to study the radiative transfer of an isolated convective cloud: Implications for cloud optical thickness retrieval. *Journal of Geophysical Research: Atmospheres, D113*, 04202. doi:10.1029/2007JD008921.

Crisp, D., Atlas, R. M., Breon, F. M., Brown, L. R., Burrows, J. P., Ciais, P., et al. (2004). The orbiting carbon observatory (OCO) mission. *Advances in Space Research, 34*, 700–709.

Daniel, J. S., Solomon, S., Miller, H. L., Langford, A. O., Portmann, R. W., & Eubank, C. S. (2003). Retrieving cloud information from passive measurements of solar radiation absorbed by molecular oxygen and O_2-O_2. *Journal of Geophysical Research: Atmospheres, 108*, 4515. doi:10.1029/2002JD002994.

Davis, A., Marshak, A., Cahalan, R. F., & Wiscombe, W. J. (1997). The Landsat scale break in stratocumulus as a three-dimensional radiative transfer effect: Implications for cloud remote sensing. *Journal of the Atmospheric Sciences, 54*, 241–260.

Davis, A. B. (2002). Cloud remote sensing with sideways looks: Theory and first results using Multispectral Thermal Imager data. In *Algorithms and Technologies for Multispectral, Hyperspectral, and Ultraspectral Imagery VIII (Proc. SPIE*, Vol. 4725, pp. 397–405). International Society for Optics and Photonics.

Davis, A. B. (2008). Multiple-scattering lidar from both sides of the clouds: Addressing internal structure. *Journal of Geophysical Research, D113*, 14S10. doi:10.1029/2007JD009666.

Davis, A. B., Cahalan, R. F., Spinhirne, D., McGill, M. J., & Love, S. P. (1999a). Off-beam lidar: An emerging technique in cloud remote sensing based on radiative Green-function theory in the diffusion domain. *Physics and Chemistry of the Earth, Part B: Hydrology, Oceans and Atmosphere, 24*, 177–185.

Davis, A. B., Cahalan, R. F., Spinehirne, D., McGill, M. J., & Love, S. P. (1999b). Off-beam lidar: An emerging technique in cloud remote sensing based on radiative Green-function theory in the diffusion domain (Erratum). *Physics and Chemistry of the Earth, Part B: Hydrology, Oceans and Atmosphere, 24*, 757–765.

Davis, A. B., & Marshak, A. (2002). Space–time characteristics of light transmitted through dense clouds: A Green's function analysis. *Journal of the Atmospheric Sciences, 59*, 2713–2727.

Davis, A. B., Polonsky, I. N., & Marshak, A. (2009). Space-time Green functions for diffusive radiation transport, in application to active and passive cloud probing. In *Light Scattering Reviews 4* (pp. 169–292). Heidelberg: Springer.

Davis, A. B., & Marshak, A. (2010). Solar radiation transport in the cloudy atmosphere: A 3D perspective on observations and climate impacts. *Reports on Progress in Physics, 73*, 026801 (70pp). doi:10.1088/0034-4885/73/2/026801.

Davis, A. B., Merlin, G., Cornet, C., Labonnote, L. C., Riédi, J., Ferlay, N., et al. (2018a). Cloud information content in EPIC/DSCOVR's oxygen A- and B-band channels: An optimal estimation approach. *Journal of Quantitative Spectroscopy and Radiative Transfer, 216*, 6–16.

Davis, A. B., Ferlay, N., Libois, Q., Marshak, A., Yang, Y., & Min, Q. (2018b). Cloud information content in EPIC/DSCOVR's oxygen A- and B-band channels: A physics-based approach. *Journal of Quantitative Spectroscopy and Radiative Transfer, 220*, 84–96.

Davis, A. B., & Kalashnikova, O. V. (2019). Aerosol layer height over water via oxygen A-band observations from space: A tutorial. In A. A. Kokhanovsky (Ed.), Springer Series in Light Scattering, vol. 3: Radiative Transfer and Light Scattering (pp. 133–166). Heidelberg (Germany):Springer. . doi:10.1007/978-3-030-03445-0.

DeSouza-Machado, S. G., Strow, L. L., Imbiriba, B., McCann, K., Hoff, R. M., Hannon, S. E., et al. (2010). Infrared retrievals of dust using AIRS: Comparisons of optical depths and heights derived for a North African dust storm to other collocated EOS A-Train and surface observations. *Journal of Geophysical Research, D115*, 15201. doi:10.1029/2009jd012842.

Dey, S., Di Girolamo, L., & Zhao, G. (2008). Scale effect on statistics of the macrophysical properties of trade wind cumuli over the tropical western Atlantic during RICO.

Journal of Geophysical Research: Atmospheres, *113*, D24214. doi:10.1029/2008JD010295.

Di Girolamo, L., Liang, L., & Platnick, S. (2010). A global view of one-dimensional solar radiative transfer through oceanic water clouds. *Geophysical Research Letters*, *37*, L18809. doi:10.1029/2010GL044094.

Diner, D. J., Beckert, J. C., Reilly, T. H., Bruegge, C. J., Conel, J. E., Kahn, R. A., et al. (1998). Multi-angle Imaging Spectro-Radiometer (MISR) instrument description and experiment overview. *IEEE Transactions on Geoscience and Remote Sensing*, *36*, 1072–1087. doi:10.1109/36.700992.

Diner, D. J., Boland, S. W., Brauer, M., Bruegge, C., Burke, K. A., Chipman, R., et al. (2018). Advances in multiangle satellite remote sensing of speciated airborne particulate matter and association with adverse health effects: From MISR to MAIA. *Journal of Applied Remote Sensing*, *12*, 042603. doi: 10.1117/1.JRS.12.042603.

Diner, D. J., Martonchik, J. V., Kahn, R. A., Pinty, B., Gobron, N., Nelson, D. L. & Holben, B. N. (2005). Using angular and spectral shape similarity constraints to improve MISR aerosol and surface retrievals over land. *Remote Sensing of Environment*, *94*, 155–171.

Diner, D. J., Xu, F., Garay, M. J., Martonchik, J. V., Rheingans, B. E., Geier, S., et al. (2013). The Airborne Multiangle SpectroPolarimetric Imager (AirMSPI): A new tool for aerosol and cloud remote sensing, *Atmospheric Measurement Techniques*, *6*, 2007–2025. doi:10.5194/amt-6-2007-2013.

Dong, X., Ackerman, T. P., Clothiaux, E. E., Pilewskie, P., & Han, Y. (1997). Microphysical and radiative properties of boundary layer stratiform clouds deduced from ground-based measurements. *Journal of Geophysical Research, D102*, 23,829–23,843.

Dubovik, O., Lapyonok, T., Litvinov, P., Herman, M., Fuertes, D., Ducos, F., et al. (2014). GRASP: A versatile algorithm for characterizing the atmosphere. *SPIE Newsroom*, *25*.

Dubovik, O., Fuertes, D., Litvinov, P., Lopatin A., Lapyonok, T., Doubovik, I., et al. (2021). A comprehensive description of multi-term LSM for applying multiple a priori constraints in problems of atmospheric remote sensing: GRASP algorithm, concept, and applications. *Frontiers in Remote Sensing*, *2*. doi:10.3389/frsen.2021.706851

Dunion, J. P., & Velden, C. S. (2004). The impact of the Saharan air layer on Atlantic tropical cyclone activity. *Bulletin of the American Meteorological Society*, *85*, 353–366.

Eck, T. F., Holben, B. N., Reid, J. S., Arola, A., Ferrare, R. A., Hostetler, C. A., et al. (2014). Observations of rapid aerosol optical depth enhancements in the vicinity of polluted cumulus clouds. *Atmospheric Chemistry and Physics*, *14*, 11,633–11,656. doi:10.5194/acp-14-11633-2014.

Eck, T. F., Holben, B. N., Reid, J. S., Giles, D. M., Rivas, M. A., Singh, R. P., et al. (2012). Fog- and cloud-induced aerosol modification observed by the Aerosol Robotic Network (AERONET). *Journal of Geophysical Research, D117*, 07206. doi:10.1029/2011JD016839.

Eldering, A., Kulawik, S. S., Worden, J., Bowman, K., & Osterman, G. (2008). Implementation of cloud retrievals for TES atmospheric retrievals: 2. Characterization of cloud top pressure and effective optical depth retrievals, *Journal of Geophysical Research*, *113*, D16S37. doi:10.1029/2007JD008858.

Emde, C., Buras, R., Mayer, B., & Blumthaler, M. (2010). The impact of aerosols on polarized sky radiance: Model development, validation, and applications. *Atmospheric Chemistry and Physics*, *10*, 383–396.

Evans, K. F., & Stephens, G. L. (1995a). Microwave radiative transfer through clouds composed of realistically shaped ice crystals. Part I. Single scattering properties. *Journal of the Atmospheric Sciences, 52*, 2041–2057.

Evans, K. F., & Stephens, G. L. (1995b). Microwave radiative transfer through clouds composed of realistically shaped ice crystals. Part II. Remote sensing of ice clouds. *Journal of the Atmospheric Sciences, 52*, 2058–2072.

Evans, K. F., Lawson, R. P., Zmarzly, P., O'Connor, D., & Wiscombe, W. J. (2003). In situ cloud sensing with multiple scattering lidar: Simulations and demonstration. *Journal of Atmospheric and Oceanic Technology*, *20*, 1505–1522.

Evans, K. F., O'Connor, D., Zmarzly, P., & Lawson, R. P. (2006). In situ cloud sensing with multiple scattering lidar: Design and validation of an airborne sensor. *Journal of Atmospheric and Oceanic Technology*, *23*, 1068–1081.

Ferlay, N., Thieuleux, F., Cornet, C., Davis, A. B., Dubuisson, P., Ducos, F., et al. (2010). Toward new inferences about cloud structures from multidirectional measurements in the oxygen A band: Middle-of-cloud pressure and cloud geometrical thickness from POLDER-3/PARASOL. *Journal of Applied Meteorology and Climatology*, *49*, 2492–2507.

Fetzer, E., McMillin, L. M., Tobin, D., Aumann, H. H., Gunson, M. R., McMillan, W. W., et al. (2003). AIRS/AMSU/HSB validation. *IEEE Transactions on Geoscience and Remote Sensing*, *41*, 418–431.

Flowerdew, R. J., & Haigh, J. D. (1995). An approximation to improve accuracy in the derivation of surface reflectances from multi-look satellite radiometers. *Geophysical Research Letters*, *22*, 1693–1696.

Frey, R., Baum, B. A., Menzel, W. P., Ackerman, S. A., Moeller, C. C., & Spinhirne, J. D. (1999). A comparison of cloud top heights computed from airborne LIDAR and MAS radiance data using CO_2-slicing. *Journal of Geophysical Research*, *104*, 24547–24555.

Frey, R. A., Ackerman, S. A., Liu, Y., Strabala, K. I., Zhang, H., Key, J., & Wang, X. (2008). Cloud detection with MODIS, Part I: Recent improvements in the MODIS cloud mask. *Journal of Atmospheric and Oceanic Technology*, *25*, 1057–1072.

Gao, M., Zhai, P. W., Yang, Y., & Hu, Y. (2019). Cloud remote sensing with EPIC/DSCOVR observations: A sensitivity study with radiative transfer simulations. *Journal of Quantitative Spectroscopy and Radiative Transfer*, *230*, 56–60.

Garay, M. J., Kalashnikova, O. V., & Bull, M. A. (2017). Development and assessment of a higher-spatial-resolution (4.4 km) MISR aerosol optical depth product using AERONET-DRAGON data. *Atmospheric Chemistry and Physics*, *17*, 5095–5106. doi:10.5194/acp-17-5095-2017.

Gatebe, C. K., & King, M. D. (2016). Airborne spectral BRDF of various surface types (ocean, vegetation, snow, desert, wetlands, cloud decks, smoke layers) for remote sensing applications. *Remote Sensing of Environment*, *179*, 131–148.

Genkova, I., Seiz, G., Zuidema, P., Zhao, G., & Di Girolamo, L. (2007). Cloud top height comparisons from ASTER, MISR, and MODIS for trade wind cumuli. *Remote Sensing of Environment*, *107*, 211–222.

Geogdzhayev, I., & Marshak, A. (2018). Calibration of the DSCOVR EPIC visible and NIR channels using MODIS Terra and Aqua data and EPIC lunar observations. *Atmospheric Measurement Techniques*, *11*, 359–368. doi:10.5194/amt-11-359-2018

Goodman, A. H., & Henderson-Sellers, A. (1988). Cloud detection and analysis: A review of recent progress. *Atmospheric Research*, *21*, 203–228.

Guillaume, A., Kahn, B. H., Fetzer, E. J., Yue, Q., Manipon, G. J., Wilson, B. D., & Hua, H. (2019). Footprint-scale cloud type mixtures and their impacts on Atmospheric Infrared Sounder cloud property retrievals. *Atmospheric Measurement Techniques*, *12*, 4361–4377. doi:10.5194/amt-12-4361-2019.

Hagihara, Y., Okamoto, H., & Luo, Z. J. (2013). Joint analysis of cloud top heights from CloudSat and CALIPSO: New insights into cloud top microphysics. *Journal of Geophysical Research*, *D119*, 4087–4106. doi:10.1002/2013JD020919.

Hansen, J. E., & Travis, L. D. (1974). Light scattering in planetary atmospheres. *Space Science Reviews*, *16*, 527–610. doi:10.1007/BF00168069.

Harshvardhan & Randall, D. (1985). Comments on "The parameterization of radiation for numerical weather prediction and climate models". *Monthly Weather Review*, *113*, 1832–1833.

Hasekamp, O. P., Fu, G., Rusli, S. P., Wu, L., Di Noia, A., van de Brugh, J., et al. (2019). Aerosol measurements by SPEXone on the NASA PACE mission: Expected retrieval capabilities. *Journal of Quantitative Spectroscopy and Radiative Transfer*, *227*, 170–184.

Hasekamp, O. P., & Landgraf, J. (2007). Retrieval of aerosol properties over land surfaces: Capabilities of multiple-viewing-angle intensity and polarization measurements. *Applied Optics*, *46*, 3332–3344. doi:10.1364/AO.46.003332.

Heidinger, K., & Stephens, G. L. (2000). Molecular line absorption in a scattering atmosphere. II: Application to remote sensing in the O_2 A-band. *Journal of the Atmospheric Sciences*, *57*, 1615–1634.

Holben, B. N., Eck, T. F., Slutsker, I., Tanré, D., Buis, J. P., Setzer, A., et al. (1998). AERONET-A federated instrument network and data archive for aerosol characterization, *Remote Sensing of Environment*, *66*, 1– 16. https://doi.org/10.1016/S0034-4257(98)00031-5

Holben, B. N., Tanré, D., Smirnov, A., Eck, T. F., Slutsker, I., Abuhassan, N., et al. (2001). An emerging ground-based aerosol climatology: Aerosol Optical Depth from AERONET. *Journal of Geophysical Research*, *D106*, 12067–12097.

Hollstein, A., Fischer, J., Carbajal Henken, C., & Preusker, R. (2015). Bayesian cloud detection for MERIS, AATSR, and their combination. *Atmospheric Measurement Techniques*, *8*, 1757–1771. doi:10.5194/amt-8-1757-2015.

Hollinger, J. P., Peirce, J. L., & Poe, G. A. (1990). SSM/I instrument evaluation. *IEEE Transactions on Geoscience and Remote Sensing*, *28*, 781–790.

Hsu, N. C., Tsay, S. C., King, M. D., & Herman, J. R. (2004). Aerosol properties over bright-reflecting source regions. *IEEE Transactions on Geoscience and Remote Sensing*, *42*, 557–569.

Hsu, N. C., Tsay, S. C., King, M. D., & Herman, J. R. (2006). Deep blue retrievals of Asian aerosol properties during ACE-Asia. *IEEE Transactions on Geoscience and Remote Sensing*, *44*, 3180–3195.

Hudson, J. G., Noble, S., & Tabor, S. (2015). Cloud supersaturations from CCN spectra Hoppel minima. *Journal of Geophysical Research*, *D120*, 3436–3452. doi:10.1002/2014JD022669.

Huang, D., Liu, Y., & Wiscombe, W. (2008a). Determination of cloud liquid water distribution using 3D cloud tomography. *Journal of Geophysical Research: Atmospheres*, *113*, 13201. doi:10.1029/2007JD009133.

Huang, D., Liu, Y., & Wiscombe, W. (2008b). Cloud tomography: Role of constraints and a new algorithm. *Journal of Geophysical Research: Atmospheres*, *113*, 23203. doi:10.1029/2008JD009952.

Hutchison, K. D., Hardy, K. R., & Gao, B.-C. (1995). Improved detection of optically thin cirrus clouds in nighttime multispectral meteorological satellite imagery using total integrated water vapor information. *Journal of Applied Meteorology*, *34*, 1161–1168.

Hutt, D. L., Bissonnette, L. R., & Durand, L. (1994). Multiple field of view lidar returns from atmospheric aerosols. *Applied Optics*, *33*, 2338–2348.

Ignatov, A., Minnis, P., Loeb, N. G., Wielicki, B., Miller, W., Sun-Mack, S., et al. (2005). Two MODIS aerosol products over ocean on the Terra and Aqua CERES SSF. *Journal of the Atmospheric Sciences*, *62*, 1008–1031.

Illingworth, A. J., Barker, H. W., Beljaars, A., Ceccaldi, M., Chepfer, H., Clerbaux, N., et al. (2015). The EarthCARE satellite: The next step forward in global measurements of clouds, aerosols, precipitation, and radiation. *Bulletin of the American Meteorological Society*, *96*, 1311–1332.

IPCC (2013). Climate Change 2013. In T. F. Stocker, D. Qin, G.-K. Plattner, M. Tignor, S. K. Allen, J. Boschung, et al. (Eds.), The Physical Science Basis. Contribution of Working Group I to the Fifth Assessment Report of the Intergovernmental Panel on Climate Change (1535 pp). Cambridge, United Kingdom and New York, NY, USA: Cambridge University Press.

Jeong, M. J., & Li, Z. (2010). Separating real and apparent effects of cloud, humidity, and dynamics on aerosol optical thickness near cloud edges. *Journal of Geophysical Research*, *D115*, 00K32. doi:10.1029/2009JD013547.

Jethva, H. T., Torres, O., Waquet, F., Chand, D., & Hu, Y. (2014). How do A-train sensors intercompare in the retrieval of above-cloud aerosol optical depth? A case study-based assessment. *Geophysical Research Letters*, *41*, 186–192. doi:10.1002/2013GL058405.

Kahn, B. H., Eldering, A., Clough, S. A., Fetzer, E. J., Fishbein, E., Gunson, M. R., et al. (2003). Near micron-sized cirrus cloud particles in high-resolution infrared spectra: An orographic case study. *Geophysical Research Letters*, *30*, 1441. doi:10.1029/2003GL016909.

Kahn, B. H., Eldering, A., Braverman, A. J., Fetzer, E. J., Jiang, J. H., Fishbein, E., & Wu, D. L. (2007). Toward the characterization of upper tropospheric clouds using Atmospheric Infrared Sounder and Microwave Limb Sounder observations. *Journal of Geophysical Research*, *112*, D05202. doi:10.1029/2006JD007336.

Kahn, B. H., Eldering, A., Ghil, M., Bordoni, S., & Clough, S. A. (2004). Sensitivity analysis of cirrus cloud properties from high-resolution infrared spectra. Part I: Methodology and synthetic cirrus. *Journal of Climate*, *17*, 4856–4870.

Kahn, R. A., & Gaitley, B. J. (2015). An analysis of global aerosol type as retrieved by MISR. *Journal of Geophysical Research: Atmospheres*, *120*, 4248–4281.

Kato, S., Hinkelman, L., & Cheng, A. (2006). Estimate of satellite-derived cloud optical thickness and effective radius errors and their effect on computed domain-averaged irradiances. *Journal of Geophysical Research*, *D111*, 17201. doi:10.1029/2005JD006668.

Kaufman, Y. J., Gobron, N., Pinty, B., Widlowski, J. L., & Verstraete, M. M. (2002). Relationship between surface reflectance in the visible and mid-IR used in MODIS aerosol algorithm—Theory. *Geophysical Research Letters*, *29*, 2116.

Kaufman, Y. J., Wald, A. E., Remer, L. A., Gao, B. C., Li, R. R., & Flynn C. L. (1997). The MODIS 2.1-μm channel—Correlation with visible reflectance for use in remote sensing of aerosol. *IEEE Transactions on Geoscience and Remote Sensing*, *35*, 1286–1298.

King, M. D., Kaufman, Y. J., Tanré, D., & Nakajima, T. (1999). Remote sensing of tropospheric aerosols from space: Past, present and future. *Bulletin of the American Meteorological Society*, *80*, 2229–2259.

King, M. D., Menzel, W. P., Kaufman, Y. J., Tanré, D., Gao, B.-C., Platnick, S., et al. (2003). Cloud and aerosol properties, precipitable water, and profiles of temperature and water vapor from MODIS. *IEEE Transactions on Geoscience and Remote Sensing*, *41*, 442–458.

King, M. D., Platnick, S. E., Menzel, W. P., Ackerman, S. A., & Hubanks, P. A. (2013). Spatial and temporal distribution of clouds observed by MODIS onboard the Terra and Aqua satellites. *IEEE Transactions on Geoscience and Remote Sensing*, *51*, 3826–3852.

King, M. D., Radke, L. F., & Hobbs, P. V. (1990). Determination of the spectral absorption of solar radiation by marine stratocumulus clouds from airborne measurements within clouds. *Journal of the Atmospheric Sciences*, *47*, 894–908.

Knobelspiesse, K., Cairns, B., Mishchenko, M., Chowdhary, J., Tsigaridis, K., van Diedenhoven, B., et al. (2012). Analysis of fine-mode aerosol retrieval capabilities by different passive remote sensing instrument designs. *Optics Express*, *20*, 21457–21484.

Knobelspiesse, K., Tan, Q., Bruegge, C., Cairns, B., Chowdhary, J., van Diedenhoven, B., et al. (2019). Intercomparison of airborne multi-angle polarimeter observations from the Polarimeter Definition Experiment. *Applied Optics*, *58*, 650–669.

Koelemeijer, R. B. A., Stammes, P., Hovenier, J. W., & de Haan, J. F. (2001). A fast method for retrieval of cloud parameters using oxygen A-band measurements from GOME. *Journal of Geophysical Research*, *106*, 3475–3490.

Kokhanovsky, A. A., Davis, A. B., Cairns, B., Dubovik, O., Hasekamp, O. P., Sano, I., et al. (2015). Space-based remote sensing of atmospheric aerosols: The multi-angle spectro-polarimetric frontier. *Earth-Science Reviews*, *145*, 85–116.

Kokhanovsky, A. A., Rozanov, V. V., von Hoyningen-Huene, W., Bovensmann, H., Burrows, J. P., & Baltink, H. K. (2004). The determination of cloud altitudes using SCIAMACHY onboard ENVISAT. *IEEE Geoscience and Remote Sensing Letters*, *1*, 211–214.

Koren, I., Feingold, G., Jiang, H., & Altaratz, O. (2009). Aerosol effects on the inter-cloud region of a small cumulus cloud field. *Geophysical Research Letters*, *36*, L14805. doi:10.1029/2009GL037424.

Koren, I., Remer, L. A., Kaufman, Y. J., Rudich, Y., & Martins, J. V. (2007). On the twilight zone between clouds and aerosols. *Geophysical Research Letters*, *34*, L08805. doi:10.1029/2007GL029253.

Lee, B., Di Girolamo, L., Zhao, G., & Zhan, Y. (2018). Three-dimensional cloud volume reconstruction from the Multi-angle Imaging Spectro-Radiometer. *Remote Sensing*, *10*, 1858.

Legrand, M., Plana-Fattori, A., & N'Doume, C. (2001). Satellite detection of dust using the IR imagery of Meteosat: 1. Infrared difference dust index. *Journal of Geophysical Research*, *D106*, 18251–18274. doi:10.1029/2000JD900749.

Levis, A., Aides, A., Schechner, Y. Y., & Davis, A. B. (2015). Airborne three-dimensional cloud tomography. *In Proceedings of the IEEE International Conference on Computer Vision 2015 (*ICCV15*)*, pp. 3379–3387.

Levis, A., Schechner, Y. Y., & Davis, A. B. (2017). Multiple-scattering microphysics tomography. In Proceedings of the IEEE Conference on Computer Vision and Pattern Recognition (CVPR17), pp. 5797–5806.

Levis, A., Schechner, Y. Y., Davis, A. B., & Loveridge, J. (2020). Multi-view polarimetric scattering cloud tomography and retrieval of droplet size. *Remote Sensing*, *12*, 2831. Special Issue on "*Remote Sensing* of Cloud and Aerosol Properties in a Three-Dimensional Atmosphere." doi:10.3390/rs12172831.

Levy, R. C., Mattoo, S., Munchak, L. A., Remer, L. A., Sayer, A. M., Patadia, F., & Hsu, N. C. (2013). The Collection 6 MODIS aerosol products over land and ocean. *Atmospheric Measurement Techniques*, *6*, 2989–3034. doi:10.5194/amt-6-2989-2013.

Liang, L. S., & Di Girolamo, L. (2013). A global analysis on the view-angle dependence of plane-parallel oceanic liquid water cloud optical thickness using data synergy from MISR and MODIS. *Journal of Geophysical Research*, *D118*, 2389–2403.

Liang, L., Di Girolamo, L., & Sun, W. (2015). Bias in MODIS cloud drop effective radius for oceanic water clouds as deduced from optical thickness variability across scattering angles. *Journal of Geophysical Research*, *D120*, 7661–7681.

Liou, K. N. (2002). *An Introduction to Atmospheric Radiation* (2nd edition). San Diego (CA), USA: Academic Press.

Limbacher, J. A., & Kahn, R. A. (2014). MISR research-aerosol-algorithm refinements for dark water retrievals. *Atmospheric Measurement Techniques*, *7*, 3989–4007.

Limbacher, J. A., & Kahn, R. A. (2017). Updated MISR dark water research aerosol retrieval algorithm–Part 1: Coupled 1.1 km ocean surface chlorophyll-A retrievals with empirical calibration corrections. *Atmospheric Measurement Techniques*, *10*, 1539–1555.

Lindstrot, R., Preusker, R., Ruhtz, T., Heese, B., Wiegner, M., Lindemann, C., & Fischer, J. (2006). Validation of MERIS cloud-top pressure using airborne lidar measurements. *Journal of Applied Meteorology and Climatology*, *45*, 1612–1621. doi: 10.1175/JAM2436.1.

Loeb, N. G., & Manalo-Smith, N. (2005). Top-of-atmosphere direct radiative effect of aerosols over global oceans from merged CERES and MODIS observations. *Journal of Climate, 18*, 3506–3526.

Loeb, N. G., & Schuster, G. L. (2008). An observational study of the relationship between cloud, aerosol and meteorology in broken low-level cloud conditions. *Journal of Geophysical Research, D113*, 14214. doi:10.1029/2007JD009763.

Lyapustin, A., Martonchik, J. V., Wang, Y., Laszlo, I., & Korkin, S. V. (2011a). Multi-Angle Implementation of Atmospheric Correction (MAIAC): 1. Radiative transfer basis and look-up tables. *Journal of Geophysical Research, D116*, 03210. doi:10.1029/2010JD014985.

Lyapustin, A., Wang, Y., Laszlo, I., Kahn, R., Korkin, S. V., Remer, L., et al. (2011b). Multi-Angle Implementation of Atmospheric Correction (MAIAC): 2. Aerosol algorithm, *Journal of Geophysical Research, D116*, 03211. doi:10.1029/2010JD014986.

Lyapustin, A. I., Wang, Y., Laszlo, I., Hilker, T., Hall, F. G., Sellers, P. J., et al. (2012b). Multi-Angle Implementation of Atmospheric Correction (MAIAC) for MODIS: 3. *Atmospheric correction. Remote Sensing of Environment*, *127*, 385–393. doi:10.1016/j.rse.2012.09.002.

Lyapustin, A. I., Wang, Y., Laszlo, I., & Korkin, S. V. (2012a). Improved cloud and snow screening in MAIAC aerosol retrievals using spectral and spatial analysis. *Atmospheric Measurement Techniques*, *5*, 843–850. doi:10.5194/amt-5-843-2012.

Lyapustin, A., Wang, Y., Korkin, S. V., & Huang, D. (2018). MODIS Collection 6 MAIAC algorithm. *Atmospheric Measurement Techniques*, *11*, 5741–5765. doi:10.5194/amt-11-5741-2018.

Mahajan, S., & Fataniya, B. (2020). Cloud detection methodologies: Variants and development—A review. *Complex & Intelligent Systems*, *6*, 251–261. doi:10.1007/s40747-019-00128-0.

Marbach, T., Riédi, J., Lacan, A., & Schlüssel, P. (2015). The 3MI mission: Multi-viewing-channel-polarisation imager of the EUMETSAT Polar System—Second Generation (EPS-SG) dedicated to aerosol and cloud monitoring. In Polarization Science and Remote Sensing VII (Proc. SPIE, Vol. 9613, p. 961310).

Marshak, A., Davis, A., Wiscombe, W. J., & Cahalan, R. F. (1995). Radiative smoothing in fractal clouds. *Journal of Geophysical Research: Atmospheres*, *100*, 26,247–26,261.

Marshak, A., Platnick, S., Várnai, T., Wen, G., & Cahalan, R. F. (2006). Impact of 3D radiative effects on satellite retrievals of cloud droplet sizes. *Journal of Geophysical Research, D111*, 09207. doi:10.1029/2005JD006686.

Martins, J. V., Nielsen, T., Fish, C., Sparr, L., Fernandez-Borda, R., Schoeberl, M., & Remer, L. A. (2014). HARP CubeSat—An innovative hyperangular imaging polarimeter for Earth science applications. In *Small Sat Pre-Conference Workshop*, Logan, Utah.

Martonchik, J. V., Diner, D. J., Kahn, R., Verstraete, M. M., Pinty, B., Gordon, H. R., & Ackerman, T. P. (1998). Techniques for the retrieval of aerosol properties over land and ocean using multiangle imaging. *IEEE Transactions on Geoscience and Remote Sensing*, *36*, 1212–1227.

Martonchik, J. V., Kahn, R. A., & Diner, D. J. (2009). Retrieval of aerosol properties over land using MISR observations. In A. A. Kokhanovsky & G. de Leeuw (Eds.), *Satellite Aerosol Remote Sensing Over Land* (pp. 267–293). Berlin: Springer.

Matheou, G., Davis, A. B., & Teixeira, J. (2020). The spiderweb structure of stratocumulus clouds, *Atmosphere*, Special Issue on "Turbulent Transport in Atmospheric Boundary Layers," *11*, 730. doi:10.3390/atmos11070730.

Mather, J. H., & Voyles, J. W. (2013). The ARM Climate Research Facility: A review of structure and capabilities. *Bulletin of the American Meteorological Society*, *94*, 377–392.

McGill, M. J., Yorks, J. E., Scott, V. S., Kupchock, A. W., & Selmer, P. A. (2015). The Cloud-Aerosol Transport System (CATS): A technology demonstration on the International Space Station. In Lidar Remote Sensing for Environmental Monitoring XV (Proc. SPIE, Vol. 9612, p. 96120A).

Melnikova, I., & Gatebe, C. K. (2018). Vertical profile of cloud optical parameters derived from airborne measurements above, inside and below clouds. *Journal of Quantitative Spectroscopy and Radiative Transfer*, *214*, 39–60.

Miller, S. D., & Stephens, G. L. (1999). Multiple scattering effects in the lidar pulse stretching problem. *Journal of Geophysical Research: Atmospheres*, *104*, 22,205–22,219.

Miller, D. J., Zhang, Z., Platnick, S., Ackerman, A. S., Werner, F., Cornet, C., & Knobelspiesse, K. (2018). Comparisons of bispectral and polarimetric retrievals of marine boundary layer cloud microphysics: Case studies using an LES satellite retrieval simulator. *Atmospheric Measurement Techniques*, *11*, 3689–3715. doi:10.5194/amt-11-3689-2018.

Mims, S. R., Kahn, R. A., Moroney, C. M., Gaitley, B. J., Nelson, D. L., & Garay, M. J. (2009). MISR stereo heights of grassland fire smoke plumes in Australia. *IEEE Transactions on Geoscience and Remote Sensing*, *48*, 25–35.

Min, Q., & Harrison, L. (1999). Joint statistics of photon pathlength and cloud optical depth. *Geophysical Research Letters*, *26*, 1425–1428.

Min, Q., Harrison, L. C., & Clothiaux, E. E. (2001). Joint statistics of photon path length and cloud optical depth: Case studies. *Journal of Geophysical Research: Atmospheres*, *106*, 7375–7385.

Min, Q. L., Harrison, L. C., Kiedron, P., Berndt, J., & Joseph, E. (2004). A high-resolution oxygen A-band and water vapor band spectrometer. *Journal of Geophysical Research: Atmospheres*, *109*, D02202. doi:10.1029/2003JD003540.

Min, Q., Yin, B., Li, S., Berndt, J., Harrison, L., Joseph, E., Duan, M., & Kiedron, P. (2014). A high-resolution oxygen A-band spectrometer (HABS) and its radiation closure. *Atmospheric Measurement Techniques*, *7*, 1711–1722.

Mishchenko, M. I., Cairns, B., Hansen, J. E., Travis, L. D., Burg, R., Kaufman, Y. J., et al. (2004). Monitoring of aerosol forcing of climate from space: Analysis of measurement requirements. *Journal of Quantitative Spectroscopy and Radiative Transfer*, *88*, 149–161. doi:10.1016/j.jqsrt.2004.03.030.

Mishchenko, M. I., Cairns, B., Kopp, G., Schueler, C. F., Fafaul, B. A., Hansen, J. E., et al. (2007). Accurate monitoring of terrestrial aerosols and total solar irradiance: Introducing the

Glory Mission. *Bulletin of the American Meteorological Society*, *88*, 677–692.

Molnar, G. I., & Susskind, J. (2005). Validation of AIRS/AMSU cloud retrievals using MODIS cloud analyses. In *Algorithms and Technologies for Multispectral, Hyperspectral, and Ultraspectral Imagery XI* (Vol. 5806, pp. 618–629). International Society for Optics and Photonics. doi:10.1117/12.603706.

Moroney, C., Davies, R., & Muller, J. P. (2002). Operational retrieval of cloud-top heights using MISR data. *IEEE Transactions on Geoscience and Remote Sensing*, *40*, 1532–1540.

Muller, J. P., Denis, M. A., Dundas, R. D., Mitchell, K. L., Naud, C., & Mannstein, H. (2007). Stereo cloud-top heights and cloud fraction retrieval from ATSR-2. *International Journal of Remote Sensing*, *28*, 1921–1938.

Nakajima, T., & King, M. D. (1990). Determinations of the optical thickness and effective particle radius of clouds from reflected solar radiation measurements. *Part I: Theory, Journal of the Atmospheric Sciences*, *47*, 1878–1893.

National Academies of Sciences, Engineering and Medicine (NASEM) (2018). *Thriving on Our Changing Planet: A Decadal Strategy for Earth Observation from Space*. Available online: https://doi.org/10.17226/24938.

Nelson, D., Garay, M., Kahn, R., & Dunst, B. (2013). Stereoscopic height and wind retrievals for aerosol plumes with the MISR INteractive eXplorer (MINX). *Remote Sensing*, *5*, 4593–4628.

Oreopoulos, L., & Davies, R. (1998). Plane parallel albedo biases from satellite observations. Part I: Dependence on resolution and other factors. *Journal of Climate*, *11*, 919–932.

Oreopoulos, L., Cahalan, R., & Platnick, S. (2007). The plane-parallel albedo bias of liquid clouds from MODIS observations. *Journal of Climate*, *20*(20), 5114–5125. doi:10.1175/JCLI4305.1.

Painemal, D., & Zuidema, P. (2011). Assessment of MODIS cloud effective radius and optical thickness retrievals over the Southeast Pacific with VOCALS-REX in situ measurements. *Journal of Geophysical Research*, *116*(D24), 24206. doi: 10.1029/2011JD016155.

Peyridieu, S., Chédin, A., Tanré, D., Capelle, V., Pierangelo, C., Lamquin, N., & Armante, R. (2010). Saharan dust infrared optical depth and altitude retrieved from AIRS: A focus over North Atlantic—Comparison to MODIS and CALIPSO. *Atmospheric Chemistry and Physics*, *10*, 1953–1967. doi:10.5194/acp-10-1953-2010.

Pfeilsticker, K. (1999). First geometrical path length probability density function derivation of the skylight from high-resolution oxygen A-band spectroscopy: 2. Derivation of the Lévy index for the skylight transmitted by midlatitude clouds. *Journal of Geophysical Research: Atmospheres*, *104*, 4101–4116.

Pfeilsticker, K., Erle, F., Funk, O., Veitel, H., & Platt, U. (1998). First geometrical pathlengths probability density function derivation of the skylight from spectroscopically highly resolving oxygen A-band observations: 1. Measurement technique, atmospheric observations and model calculations. *Journal of Geophysical Research: Atmospheres*, *103*, 11,483–11,504.

Pierangelo, C., Chédin, A., Heilliette, S., Jacquinet-Husson, N., & Armante, R. (2004). Dust altitude and infrared optical depth from AIRS. *Atmospheric Chemistry and Physics*, *4*, 1813–1822.

Pierangelo, C., Mishchenko, M., Balkanski, Y., & Chédin, A. (2005). Retrieving the effective radius of Saharan dust coarse mode from AIRS. *Geophysical Research Letters*, *32*, L20813. doi:10.1029/2005GL023425.

Pincus, R., McFarlane, S. A., & Klein, S. A. (1999). Albedo bias and the horizontal variability of clouds in subtropical marine boundary layers: Observations from ships and satellites. *Journal of Geophysical Research: Atmospheres*, *104*, 6183–6191. doi:10.1029/1998JD200125.

Platnick, S. (2000). Vertical photon transport in cloud remote sensing problems. *Journal of Geophysical Research*, *D105*, 22919–22935. doi:10.1029/2000JD900333.

Platnick, S., King, M. D., Ackerman, S. A., Baum, B. A., Riédi, J. C., & Frey, R. A. (2003). The MODIS cloud products: Algorithms and examples from Terra," *IEEE Transactions on Geoscience and Remote Sensing*, *41*(2), 459–473.

Platnick, S., Meyer, K. G., King, M. D., Wind, G., Amarasinghe, N., Marchant, B., et al. (2017). The MODIS cloud optical and microphysical products: Collection 6 updates and examples from Terra and Aqua. *IEEE Transactions on Geoscience and Remote Sensing*, *55*, 502–525. doi:10.1109/TGRS.2016.2610522.

Polonsky, I. N., & Davis, A. B. (2005). *Off-Beam Cloud Lidar: A New Diffusion Model and an Analysis of LITE Returns*. Los Alamos National Laboratory Report, LAUR-14219, Los Alamos (NM).

Polonsky, I. N., Love, S. P., & Davis, A. B. (2005). Wide-angle imaging lidar deployment at the ARM Southern Great Plains site: Intercomparison of cloud property retrievals. *Journal of Atmospheric and Oceanic Technology*, *22*, 628–648.

Portmann, R. W., Solomon, S., Sanders, R. W., Daniel, J. S., & Dutton, E. G. (2001). Cloud modulation of zenith sky oxygen photon path lengths over Boulder, Colorado: Measurement versus model. *Journal of Geophysical Research: Atmospheres*, *106*, 1139–1155.

Pounder, N. L., Hogan, R. J., Várnai, T., Battaglia, A., & Cahalan, R. F. (2012). A variational method to retrieve the extinction profile in liquid clouds using multiple-field-of-view lidar. *Journal of Applied Meteorology and Climatology*, *51*, 350–365.

Prata, A. J., & Turner, P. J. (1997). Cloud-top height determination using ATSR data. *Remote Sensing of Environment*, *59*, 1–13.

Qiu, S., Godden, G., Wang, X., & Guenther, B. (2000). Satellite-Earth remote sensor scatter effects on Earth scene radiometric accuracy. *Metrologia*, *37*, 411–414.

Rauber, R. M., Stevens, B., Ochs, H. T., III, Knight, C., Albrecht, B. A., Blyth, A. M., Fairall, C. W., et al. (2007). Rain in shallow cumulus over the ocean: The RICO campaign. *Bulletin of the American Meteorological Society*, *88*, 1912–1928.

Rauber, R. M., Zhao, G., Di Girolamo, L., & Colón-Robles, M. (2013). Aerosol size distribution, particle concentration, and optical property variability near Caribbean trade cumulus clouds: Isolating effects of vertical transport and cloud processing from humidification using aircraft measurements. *Journal of the Atmospheric Sciences*, *70*, 3063–3083.

Redemann, J., Zhang, Q., Russell, P. B., Livingston, J. M., & Remer, L. A. (2009). Case studies of aerosol remote sensing in the vicinity of clouds. *Journal of Geophysical Research, D114*, 06209. doi:10.1029/2008JD010774.

Remer, L. A., Mattoo, S., Levy, R. C., & Munchak, L. A. (2013). MODIS 3 km aerosol product: Algorithm and global perspective. *Atmospheric Measurement Techniques, 6*, 1829–1844. doi:10.5194/amt-6-1829-2013.

Richardson, M., Leinonen, J., Cronk, H. Q., McDuffie, J., Lebsock, M. D., & Stephens, G. L. (2019). Marine liquid cloud geometric thickness retrieved from OCO-2's oxygen A-band spectrometer. *Atmospheric Measurement Techniques, 12*, 1717–1737.

Richardson, M., McDuffie, J., Stephens, G. L., Cronk, H. Q., & Taylor, T. E. (2017). The OCO-2 oxygen A-band response to liquid marine cloud properties from CALIPSO and MODIS. *Journal of Geophysical Research: Atmospheres, 122*, 8255–8275.

Richardson, M., & Stephens, G. L. (2018). Information content of OCO-2 oxygen A-band channels for retrieving marine liquid cloud properties. *Atmospheric Measurement Techniques, 11*, 1515–1528.

Rodgers, C. D. (2000). *Inverse Methods for Atmospheric Sounding: Theory and Practice*. Singapore: World Scientific.

Roy, G., Bissonnette, L., Bastille, C., & Vallée, G. (1999). Retrieval of droplet-size density distribution from multiple-field-of-view cross-polarized lidar signals: Theory and experimental validation. *Applied Optics, 38*, 5202–5211.

Sayer, A. M., Goaverts, Y., Kolmonen, P., Lipponen, A., Luffarelli, M., Mielonen, T., et al. (2020). A review and framework for the evaluation of pixel-level uncertainty estimates in satellite aerosol remote sensing. *Atmospheric Measurement Techniques, 13*, 373–404.

Sayer, A. M., Hsu, N. C., Bettenhausen, C., Lee, J., Redemann, J., Schmid, B., & Shinozuka, Y. (2016). Extending "Deep Blue" aerosol retrieval coverage to cases of absorbing aerosols above clouds: Sensitivity analysis and first case studies. *Journal of Geophysical Research, D121*, 4830–4854. doi:10.1002/2015JD024729.

Sayer, A. M., Hsu, N. C., Lee, J., Kim, W., Dubovik, O., Dutcher, S., et al. (2018). Validation of SOAR VIIRS over-water aerosol retrievals and context within the global satellite aerosol data record. *Journal of Geophysical Research, D123*, 13–496. doi:10.1029/2018JD029465.

Seiz, G., & Davies, R. (2006). Reconstruction of cloud geometry from multi-view satellite images. *Remote Sensing of Environment, 100*, 143–149.

Seiz, G., Davies, R., & Grün, A. (2006). Stereo cloud-top height retrieval with ASTER and MISR. *International Journal of Remote Sensing, 27*, 1839–1853.

Schepanski, K., Tegen, I., Laurent, B., Heinold, B., & Macke, A. (2007). A new Saharan dust source activation frequency map derived from MSG-SEVIRI IR-channels. *Geophysical Research Letters, 34*, L18803. doi:10.1029/2007GL030168.

Schiffer, R. A., & Rossow, W. B. (1983). The International Satellite Cloud Climatology Project (ISCCP): The first project of the world climate research program. *Bulletin of the American Meteorological Society, 64*, 779–784.

Shenk, W. E., & Curran, R. J. (1974). The detection of dust storms over land and water with satellite visible and infrared measurements. *Monthly Weather Review, 102*, 830–837.

Scholl, T., Pfeilsticker, K., Davis, A. B., Klein Baltink, H., Crewell, S., Löhnert, U., et al. (2006). Path length distributions for solar photons under cloudy skies: Comparison of measured first and second moments with predictions from classical and anomalous diffusion theories. *Journal of Geophysical Research: Atmospheres, 111*, D12211. doi:10.1029/2004JD005707.

Schuster, G., Espinosa, W. R., Ziemba, L. D., Beyersdorf, A., Rocha-Lima, A., Anderson, B. E., et al. (2019). A laboratory experiment for the STatistical Evaluation of Aerosol Retrieval (STEAR) algorithms. *Remote Sensing, 11*, 498. doi:10.3390/rs11050498.

Schutz, B. E., Zwally, H. J., Shuman, C. A., Hancock, D., & DiMarzio, J. P. (2005). Overview of the ICESat mission. *Geophysical Research Letters, 32*, L21S01. doi:10.1029/2005GL024009.

Shang, H., Chen, L., Bréon, F. M., Letu, H., Li, S., Wang, Z., & Su, L. (2015). Impact of cloud horizontal inhomogeneity and directional sampling on the retrieval of cloud droplet size by the POLDER instrument. *Atmospheric Measurement Techniques, 8*, 4931–4945. doi:10.5194/amt-8-4931-2015.

Sinclair, K., van Diedenhoven, B., Cairns, B., Yorks, J., Wasilewski, A., & McGill, M. (2017). Remote sensing of multiple cloud layer heights using multi-angular measurements. *Atmospheric Measurement Techniques, 10*, 2361–2375. doi:10.5194/amt-10-2361-2017.

Spinhirne, J. D. (1993). Micro-pulse lidar. *IEEE Transactions on Geoscience and Remote Sensing, 31*, 48–55.

Stephens, G. L., & Heidinger, A. K. (2000). Line absorption in a scattering atmosphere. I: Theory. *Journal of the Atmospheric Sciences, 57*, 1599–1614.

Stephens, G. L., Vane, D. G., Boain, R. J., Mace, G. G., Sassen, K., Wang, Z., et al. (2002). The CloudSat mission and the A-Train: A new dimension of space-based observations of clouds and precipitation. *Bulletin of the American Meteorological Society, 83*, 1771–1790.

Stephens, G. L., & Webster, P. J. (1981). Clouds and climate: Sensitivity of simple systems. *Journal of the Atmospheric Sciences, 38*, 235–247.

Stokes, G. M., & Schwartz, S. E. (1994). The Atmospheric Radiation Measurement (ARM) Program: Programmatic background and design of the cloud and radiation test bed. *Bulletin of the American Meteorological Society, 75*, 1201–1222.

Stowe, L. L., Ignatov, A. M., & Singh, R. R. (1997). Development, validation, and potential enhancements to the second-generation operational aerosol product at the National Environmental Satellite, Data, and Information Service of the National Oceanic and Atmospheric Administration. *Journal of Geophysical Research D, 102*, 16,923–16,934.

Su, W., Schuster, G. L., Loeb, N. G., Rogers, R. R., Ferrare, R. A., Hostetler, C. A., et al. (2008). Aerosol and cloud interaction observed from high spectral resolution lidar data. *Journal of Geophysical Research, D113*, 24202. doi:10.1029/2008JD010588.

Tackett, J. L., & Di Girolamo, L. (2009). Enhanced aerosol backscatter adjacent to tropical trade wind clouds revealed by satellite-based lidar. *Geophysical Research Letters*, *36*, L14804. doi:10.1029/2009GL039264.

Tanaka, K., Okamura, Y., Amano, T., Hiramatsu, M., & Shiratama, K. (2009). Development status of the Second-Generation GLobal Imager (SGLI) on GCOM-C. In *Sensors, Systems, and Next-Generation Satellites XIII* (*Proc. SPIE*, Vol. 7474, p. 74740N).

Tanré, D., Bréon, F. M., Deuzé, J. L., Dubovik, O., Ducos, F., François, P., et al. (2011). Remote sensing of aerosols by using polarized, directional and spectral measurements within the A-Train: The PARASOL mission. *Atmospheric Measurement Techniques*, *4*, 1383–1395.

Tanré, D., Kaufman, Y. J., Herman, M., & Matoo, S. (1997). Remote sensing of aerosol properties over oceans using the MODIS/EOS spectral radiances. *Journal of Geophysical Research*, *102*, 16,971–16,988.

Torres, O., Bhartia, P., Herman, J. R., Ahmad, Z., & Gleason, J. F. (1998). Derivation of aerosol properties from satellite measurements of backscattered ultraviolet radiation. Theoretical basis. *Journal of Geophysical Research*, *103*, 17,099–17,110. 10.1029/98JD00900.

Tsang, L., Pan, J., Liang, D., Li, Z., Cline, D. W., & Tan, Y. (2007). Modeling active microwave remote sensing of snow using Dense Media Radiative Transfer (DMRT) Theory with multiple-scattering effects. *IEEE Transactions on Geoscience and Remote Sensing*, *45*(4), 990–1004.

Twomey, S., & Bohren, C. F. (1980). Simple approximations for calculations of absorption in clouds. *Journal of the Atmospheric Sciences*, *37*, 2086–2094.

Twomey, S., & Cocks, T. (1989). Remote sensing of cloud parameters from spectral reflectance in the near-infrared. *Beitraege zur Physik der Atmosphaere*, *62*, 172–179.

Twohy, C. H., Coakley, J. A., Jr., & Tahnk, W. R. (2009). Effect of changes in relative humidity on aerosol scattering near clouds. *Journal of Geophysical Research*, *D114*, 05205. doi:10.1029/2008JD010991.

Ulaby, F. T., Moore, R. K., & Fung, A. K. (1981). *Microwave remote sensing fundamentals and radiometry* (Vol. 1). Reading, MA: Addison-Wesley.

van Diedenhoven, B., Ackerman, A. S., Cairns, B. & Fridlind, A. M. (2014a). A flexible parameterization for shortwave optical properties of ice crystals. *Journal of the Atmospheric Sciences*, *71*(5), 1763–1782. doi:10.1175/JAS-D-13-0205.1.

van Diedenhoven, B., Fridlind, A. M., Cairns, B., & Ackerman, A. S. (2014b). Variation of ice crystal size, shape and asymmetry parameter in tops of tropical deep convective clouds. *Journal of Geophysical Research: Atmospheres*, *119*, 11809–11825. doi:10.1002/2014JD022385.

Várnai, T., & Cahalan, R. F. (2007). Potential for airborne offbeam lidar measurements of snow and sea ice thickness. *Journal of Geophysical Research—Oceans*, *112*, C12S90. doi:10.1029/2007Jc154091.

Várnai, T., & Marshak, A. (2002). Observations and analysis of three-dimensional radiative effects that influence MODIS cloud optical thickness retrievals. *Journal of the Atmospheric Sciences*, *59*, 1607–1618.

Várnai, T., & Marshak, A. (2009). MODIS observations of enhanced clear sky reflectance near clouds. *Geophysical Research Letters*, *36*, L06807. doi:10.1029/2008GL037089.

Várnai, T., & Marshak, A. (2011). Global CALIPSO observations of aerosol changes near clouds. *Geoscience and Remote Sensing Letters*, *8*, 19–23.

Várnai, T., & Marshak, A. (2018). Satellite observations of cloud-related variations in aerosol properties. *Atmosphere*, *9*, 430. doi:10.3390/atmos9110430.

Várnai, T., Marshak, A., & Eck, T. F. (2017). Observation-based study on aerosol optical depth and particle size in partly cloudy regions. *Journal of Geophysical Research*, *122*, 10013–10024. doi:10.1002/2017JD027028.

Veefkind, J. P., Aben, I., McMullan, K., Förster, H., de Vries, J., Otter, G., et al. (2012). TROPOMI on the ESA Sentinel-5 Precursor: A GMES mission for global observations of the atmospheric composition for climate, air quality and ozone layer applications. *Remote Sensing of Environment*, *120*, 70–83.

Wang, Y., Wang, J., Levy, R. C., Xu, X., & Reid, J. S. (2017). MODIS retrieval of aerosol optical depth over turbid coastal water. *Remote Sensing of Environment*, *9*, 595. doi:10.3390/rs9060595.

Weber, P. G., Brock, B. C., Garrett, A. J., Smith, B. W., Borel, C. C., Clodius, W. B., et al. (1999). Multispectral Thermal Imager mission overview. In Imaging Spectrometry V (Proc. SPIE, Vol. 3753, pp. 340–346).

Welton, E. J., & Campbell, J. R. (2002). Micropulse Lidar signals: Uncertainty analysis. *Journal of Atmospheric and Oceanic Technology*, *19*, 2089–2094. doi:10.1175/1520-0426(2002)019<2089:MLSUA>2.0.CO;2.

Wen, G., Marshak, A., Cahalan, R. F., Remer, L. A., & Kleidman, R. G. (2007). 3-D aerosol-cloud radiative interaction observed in collocated MODIS and ASTER images of cumulus cloud fields. *Journal of Geophysical Research*, *112*, D13204. doi:10.1029/2006JD008267.

Werdell, P. J., Behrenfeld, M. J., Bontempi, P. S., Boss, E., Cairns, B., Davis, G. T., et al. (2019). The Plankton, Aerosol, Cloud, ocean Ecosystem (PACE) mission: Status, science, advances. *Bulletin of the American Meteorological Society*, *100*, 1775–1794. doi:10.1175/BAMS-D-18-0056.1.

Wielicki, B. A., & Coakley, J. A., Jr. (1981). Cloud retrieval using infrared sounder data: Error analysis. *Journal Of Applied Meteorology*, *20*, 157–169.

Winker, D. M., Couch, R. H., & McCormick, M. P. (1996). An overview of LITE: NASA's Lidar In-space Technology Experiment. *Proceedings of the IEEE*, *84*, 164–180.

Winker, D. M., Pelon, J., Coakley, J. A., Jr., Ackerman, S. A., Charlson, R. J., Colarco, P. R., et al. (2010). The CALIPSO mission: A global 3D view of aerosols and clouds. *Bulletin of the American Meteorological Society*, *91*, 1211–1230.

Witek, M. L., Garay, M. J., Diner, D. J., Bull, M. A., & Seidel, F. C. (2018). New approach to the retrieval of AOD and its uncertainty from MISR observations over dark water. *Atmospheric Measurement Techniques*, *11*, 429–439. doi:10.5194/amt-11-429-2018.

Wu, D. L., Ackerman, S. A., Davies, R., Diner, D. J., Garay, M. J., Kahn, B. H., et al. (2009). Vertical distributions and relationships of cloud occurrence frequency as observed by MISR, AIRS, MODIS, OMI, CALIPSO, and CloudSat. *Geophysical Research Letters*, *36*, L09821. doi:10.1029/2009GL037464.

Wylie, D. P., Santek, D., & Starr, D. O. C. (1998). Cloud-top heights from GOES-8 and GOES-9 stereoscopic imagery. *Journal of Applied Meteorology*, *37*, 405–413.

Xu, F., van Harten, G., Diner, D. J., Davis, A. B., Seidel, F. C., Rheingans, B., et al. (2018). Coupled retrieval of liquid water cloud and aerosol above cloud properties using the Airborne Multi-angle Spectro-Polarimetric Imager (AirMSPI). *Journal of Geophysical Research: Atmospheres*, *123*, 3175–3204. doi:10.1002/2017JD027926.

Xu, F., van Harten, G., Diner, D. J., Kalashnikova, O. V., Seidel, F. C., Bruegge, C. J., & Dubovik, O. (2017). Coupled retrieval of aerosol properties and land surface reflection using the Airborne Multiangle Spectro-Polarimetric Imager. *Journal of Geophysical Research: Atmospheres*, *122*, 7004–7026. doi:10.1002/2017JD026776.

Yamaguchi, Y., Fujisada, H., Kahle, A. B., Tsu, H., Kato, M., Watanabe, H., et al. (2001). ASTER instrument performance, operation status, and application to Earth sciences. In *Proceedings of IEEE 2001 International Geoscience and Remote Sensing Symposium—IGARSS 2001* (Vol. 3, pp. 1215–1216).

Yamamoto, G. A., & Wark, D. Q. (1961). Discussion of the letter by R. A. Hanel, Determination of cloud altitude from satellite. *Journal of Geophysical Research*, *66*, 3596.

Yang, S., Hawkins, J. D., & Richardson, K. (2014). The improved NRL tropical cyclone monitoring system with a unified microwave brightness temperature calibration scheme. *Remote Sensing*, *6*, 4563–4581.

Yang, Y., Marshak, A., Mao, J., Lyapustin, A., & Herman, J. (2013). A method of retrieving cloud top height and cloud geometrical thickness with oxygen A and B bands for the Deep Space Climate Observatory (DSCOVR) mission: Radiative transfer simulations. *Journal of Quantitative Spectroscopy and Radiative Transfer*, *122*, 141–149.

Zhang, J., Reid, J. S., & Holben, B. N. (2005). An analysis of potential cloud artifacts in MODIS over ocean aerosol optical thickness products. *Geophysical Research Letters*, *32*, L15803. doi:10.1029/2005GL023254.

Zhang, P., Lu, N. M., Hu, X. Q., & Dong, C. H. (2006). Identification and physical retrieval of dust storm using three MODIS thermal IR channels. *Global and Planetary Change*, *52*, 197–206.

Zhang, Z. (2013). On the sensitivity of cloud effective radius retrieval based on spectral method to bi-modal droplet size distribution: A semi-analytical model. *Journal of Quantitative Spectroscopy and Radiative Transfer*, *129*, 79–88. doi:10.1016/j.jqsrt.2013.05.033.

Zhang, Z., Ackerman, A. S., Feingold, G., Platnick, S., Pincus, R., & Xue, H. (2012). Effects of cloud horizontal inhomogeneity and drizzle on remote sensing of cloud droplet effective radius: Case studies based on large-eddy simulations. *Journal of Geophysical Research*, *117*, D19208. doi:10.1029/2012JD017655.

Zhang, Z., & Platnick, S. (2011). An assessment of differences between cloud effective particle radius retrievals for marine water clouds from three MODIS spectral bands. *Journal of Geophysical Research*, *D116*, 20215. doi:10.1029/2011JD016216.

Zhang, Z., Werner, F., Cho, H. M., Wind, G., Platnick, S., Ackerman, A. S., et al. (2016). A framework based on 2-D Taylor expansion for quantifying the impacts of sub-pixel reflectance variance and covariance on cloud optical thickness and effective radius retrievals based on the bi-spectral method. *Journal of Geophysical Research*, *D024837*. doi:10.1002/2016JD024837.

Zhao, G., & Di Girolamo, L. (2006). Cloud fraction errors for trade wind cumuli from EOS-Terra instruments. *Geophysical Research Letters*, *33*, L20802. doi:10.1029/2006GL027088.

16

In Situ and Laboratory Measurements of Cloud Microphysical Properties

Kamal Kant Chandrakar[1] and Raymond A. Shaw[2]

ABSTRACT

This chapter provides a sketch of in situ measurements and laboratory experiments that have provided new insights into the microphysics of clouds. The emphasis is on processes that are small in spatial or temporal scale, in keeping with the focus of this book. Some illustrative historical examples are given, followed by an overview of what coarse-resolution atmospheric models need to know about clouds. Specifically, how do clouds influence radiative fluxes, precipitation fluxes, and convective transport, and how do those processes depend on the cloud particle size distribution? Examples are then provided to give a sense of significant advances and capabilities in current cloud research, in three areas: (1) airborne measurements of cloud properties; (2) ground-based measurements of cloud properties; and (3) laboratory experiments focused on cloud processes.

16.1. INTRODUCTION

16.1.1. The Complementary Roles of Field and Laboratory Measurements

Scientific observation is the firm foundation of reality on which all theories and modeling frameworks are based. The ideal scientific observation is more than a sanity check; it is the reference of reality to which theoretical predictions and numerical models must be compared. What do we mean by an ideal scientific observation? It is a controlled, repeatable measurement of a physical quantity or phenomenon in a well-defined system of interest with known external influences. However, in practice, especially in geophysical systems, an ideal measurement does not exist due to the absence of some of the above criteria. Therefore, these limitations must be identified and carefully considered during analysis so that they can still serve their intended and indispensable purpose. In the context of this chapter, measurements serve two primary purposes: they provide input and constraints needed for implementation of known theory (e.g., radiative transfer), and they guide the exploration of poorly understood physical processes (e.g., ice formation through nucleation and secondary processes).

Broadly, there are two significant categories of observation in atmospheric science: field and laboratory observations. Both complement each other to advance scientific understanding and to provide input for modeling frameworks, and in fact they inform each other: laboratory experiment provides understanding of specific mechanisms for idealized systems needed to interpret the complexity of the field observations, and field observations provide guidance on what ranges of parameter space and what limits are relevant for exploration in the laboratory. Thus, both approaches have limitations and strengths, which make them unique in their capabilities. For example, field measurements bring us closest to the actual representation of natural processes but are complicated due to the presence of multiple simultaneous

[1]*National Center for Atmospheric Research, Boulder, CO, USA*

[2]*Atmospheric Sciences Program, Michigan Technological University, Houghton, MI, USA*

Fast Processes in Large-Scale Atmospheric Models: Progress, Challenges, and Opportunities, Geophysical Monograph 282, First Edition.
Edited by Yangang Liu and Pavlos Kollias.

DOI: 10.1002/9781119529019.ch16

processes and feedbacks, limited data availability due to the remoteness of systems of interest (such as clouds), limited ability to perform adequate averaging due to the transient nature of many natural processes, and limitations in resources and technical capabilities for the instruments needed for field measurements. Nevertheless, field measurement provides the closest picture of reality and most relevant input data needed for developing and evaluating models. On the other hand, laboratory studies are designed to have better control of governing parameters and boundary conditions, and are sufficiently isolated and simplified to provide process level understanding. They can be designed to overcome the sampling limitations, and steady-state conditions are also easier to achieve in some cases. However, lab experiments are not able to provide a complete picture of complex atmospheric interactions, and therefore must be carefully designed if their results are to be relevant in providing individual pieces of a complex puzzle.

There are significant contributions from field and laboratory measurements in the development of atmospheric science. Undoubtedly, there are an enormous number of field studies that are constantly providing input over a range of spatial and temporal scales. Laboratory measurements have become somewhat less common, but their contributions have been profound and it is safe to say that a resurgence in laboratory experimentation relevant to "fast physics," the topic of this book, is occurring. This venue is far too constrained to allow for a thorough review of modern methods and contributions of in-situ and laboratory measurements. Our purpose, therefore, is to provide illustrative examples of measurements that have provided new insight. Our own research makes us most qualified to consider this topic from the perspective of the cloud microphysics and turbulence, so we acknowledge that bias from the outset: excellent reviews of current capabilities and advancements in aerosols, chemistry, dynamics, and radiation are available in the text by Wendisch and Brenguier (2013). Furthermore, simply because of familiarity, and without any intention of priority, we draw some of the examples from our own work. Finally, because of length constraints, the primary emphasis is on warm cloud microphysics; ice microphysics is addressed in several of the sections, but is not the focus.

16.1.2. Illustrative Historical Examples of Field and Laboratory Measurements

The early cloud chamber experiments of Gunn and Phillips (1957) are an instructive example of the insight that can be gained from controlled laboratory measurements. The experiments were performed in an enormous cloud chamber with a diameter of 18.3 m and a volume of more than 3000 m^3 (Reichelderfer, 1967). Findings for two different aerosol conditions are illustrated in Figure 16.1. The related discussion is remarkable to us because it encapsulates much of what we consider our current understanding of aerosol indirect effects of cloud microphysics. We cannot help but wonder how much sooner such indirect effects could have been thoroughly explored had this cloud chamber facility remained in operation (in passing, we have not been able to determine what factors led to the facility being decommissioned).

> It was found that the abundance of condensation nuclei profoundly influences the character of the rain-initiating processes. Hundreds of clouds artificially produced in the chamber have shown that normal expansions of originally saturated air usually produce a dense clouds of droplets having approximate radii of 7 μm or less. Such clouds are quite stable, persist in the cloud chamber for times exceeding a half hour, and ultimately evaporate as the chamber recovers from the initial cooling of the processed air. It is also observed that if the expansion rate is much less than that typically produced in nature, so that relatively few of the available nuclei are activated, some of the droplets can grow to radii of something like 15 μm. In striking contrast to the above results, it is found that the behavior of clouds formed by the expansion of air that has been carefully freed of pollution and of condensation nuclei behaves in a radically different manner. The cloud droplets so formed are notably larger than those formed from ordinary air, and the cloud

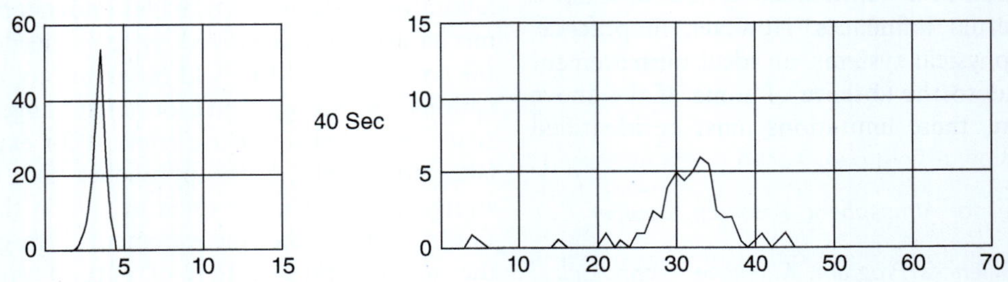

Figure 16.1 Cloud droplet size distributions measured in a large cloud chamber, illustrating what today is known as the first indirect effect or Twomey effect. The clouds were formed in "uncleaned Gulf air" on the left and "air electrostatically cleaned" on the right. Source: Figure adapted from Figure 7 of Gunn and Phillips (1957).

largely disappears by precipitation rather than evaporation. In fact, we have successfully produced a mist-like rain by expanding well cleaned and saturated air in our 60-ft cloud chamber. This is impossible when ordinary air is used. These observations are consistent with experimental data reported by Wilson nearly 60 years ago, but their significance and application to meteorological problems were not appreciated.

Gunn and Phillips (1957)

One of the most influential series of laboratory measurements came from convection tank experiments of Deardorff and colleagues, reported in a series of papers beginning with Deardorff et al. (1969) and continuing for more than 15 years. The work was idealized so as to inform the theory of mixed-layer models for the atmospheric boundary layer, such as the the stratocumulus-topped boundary model developed by Lilly (1968). It is not an overstatement that the experiments also eventually led to the first large-eddy simulations (Deardorff, 1970; Wyngaard, 2015), which has since become a ubiquitous tool in the atmospheric sciences and beyond. Perhaps symbolically, the experiments were carried out in the basement of the National Center for Atmospheric Research; according to rumor this was so as to avoid attention from administrators unimpressed by such research methods (Wyngaard, 2015). In spite of that skepticism, the measurements continue to inspire and set the standard of comparison for computational work on developing boundary layers (Van Heerwaarden & Mellado, 2016).

Some of the most insightful field measurements are those that were carefully planned in the context of clear theoretical structure and questions. The Kansas experiment that explored boundary layer turbulence was designed with simplifications and measurement strategies guided by such a theoretical framework (Wyngaard et al., 1971). One of many outcomes was to confirm the Monin-Obukhov similarity theory for the atmospheric surface layer and mixed-layer models for the region of the boundary layer above the surface layer (Kaimal & Wyngaard, 1990). Besides the numerous papers that emerged from the project itself, this well-designed experiment has continued to guide theoretical and computational research for decades since (Khanna, 1998).

Availability of a fundamentally new measurement technique can also allow a field project to contribute greatly to the community's understanding of the atmosphere. The saturation hygrometer (Gerber, 1980, 1991) is an example of such an instrument, allowing for measurement of water vapor supersaturation with unprecedented precision. Unfortunately, the development of that instrument was not continued and only a small number of profiles of supersaturation were measured in clouds. An example is shown in Figure 16.2, with vertical profiles of supersaturation and cloud liquid water content in a stratocumulus cloud, obtained from the slow ascent of a blimp carrying a package of instruments (Gerber, 1994).

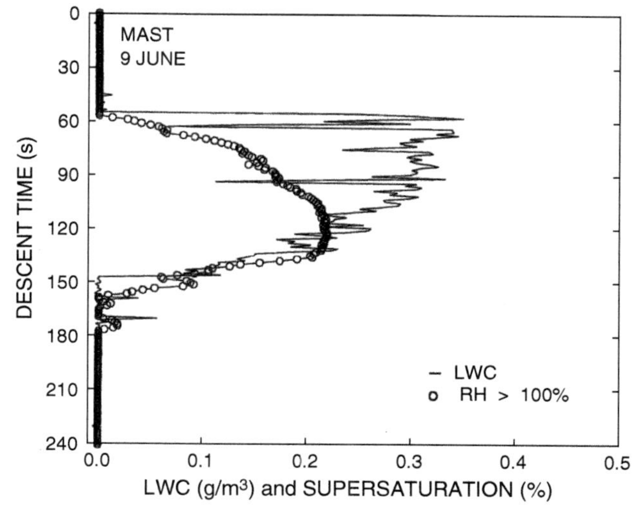

Figure 16.2 A profile of water vapor supersaturation and cloud liquid water content through a stratocumulus cloud sampled during the MACE field project. Source: Figure adapted from Gerber (1994).

Hobbs and Rangno carried out a set of extensive airborne field studies that have motivated laboratory and computational work in cloud physics for many years since the measurements themselves (Hobbs & Rangno, 1985; Rangno & Hobbs, 1988). These measurements helped inform and strengthen the idea that mechanisms must exist for ice generation beyond simple, primary ice nucleation (Beard, 1992). A key step in that work was the recognition that aircraft themselves can generate ice when they pass through clouds (Rangno & Hobbs, 1983), and the resulting adjustments in sampling strategies led to much greater confidence in the measurements. Their observations also strongly suggested that the shape of the droplet size distribution, and in particular the presence of large droplets, is relevant to secondary ice production. That is consistent with the understanding of the Hallett-Mossop mechanism (Mossop, 1978) and the field measurements that support it (Blyth & Latham, 1993; Harris-Hobbs & Cooper, 1987). Subsequently, it was determined that ice crystal shattering likely contaminated some of the concentration and size measurements (Jackson et al., 2014; Korolev et al., 2011); it is still generally accepted, however, that the discrepancy holds and mechanisms for ice multiplication are needed to account for observed ice concentrations (Korolev et al., 2017).

16.2. WHAT DO MODELS NEED TO KNOW ABOUT CLOUDS?

Large-scale models (e.g., global circulation models), as well as cloud resolving models (or large eddy simulations), are inherently limited in their ability to directly represent phenomena occurring at spatial scales below their grid sizes and temporal scales below their time steps. For example, GCMs are not able to resolve the atmospheric boundary layer or cloud-scale processes; even cloud-resolving models and large eddy simulations require some input for surface fluxes and subgrid-scale cloud and turbulence processes. Inputs needed for these unresolved processes come in the form of parameterizations from measurements of cloud, boundary-layer, and turbulent flow properties. For example, as a boundary-layer property, information about surface momentum, heat, moisture, and other scalar fluxes is needed. For cloud processes, information related to radiative, precipitation, and convective fluxes is needed. In this chapter, we limit the focus to cloud properties that are introduced briefly in the subsequent subsections.

16.2.1. Radiation (Radiative Fluxes)

Radiative fluxes through a cloud depend on properties such as cloud optical depth, single-scattering albedo, and asymmetry parameter. All these properties depend on the size distribution, shape and orientation of hydrometeors, and the partitioning of water phases (liquid versus ice). For a liquid cloud, these dependencies can be further reduced to two microphysical variables, effective radius that is a ratio between the third and second moments of a droplet size distribution and liquid water path (Hansen & Travis, 1974; Slingo, 1989; Stephens, 1978). As an input parameterization, models need the relation between amount of cloud water content and effective radius. This information can be obtained from measurements; for example, Martin et al. (1994) showed nearly a linear relation between the cube of mean volume radius and effective radius of cloud droplets for stratocumulus clouds. Therefore, a parameterization of effective radius (r_e) can be expressed as

$$r_e \equiv \frac{\int r^3 n(r) \mathrm{d}r}{\int r^2 n(r) \mathrm{d}r} \approx \left[\frac{3L}{4\pi \rho_l n_d k}\right]^{1/3}, \qquad (16.1)$$

where ρ_l is the density of liquid water, n_d is the droplet number concentration, L is the liquid water content, and k is a constant. The parameter k is a function of relative dispersion and skewness of the droplet size distribution (Pontikis and Hicks). Consequently, the cloud optical depth can be parameterized with the assumption of a constant liquid water content throughout the cloud depth:

$$\begin{aligned} \tau &\equiv \int_0^h \int_0^\infty Q_{ext}\, r^2\, n(r)\, \mathrm{d}r\, \mathrm{d}z \\ &= 3/4\pi\, \overline{Q_{ext}}\, \rho_l^{-1} r_e^{-1} L h \\ &\approx \left(\frac{3}{4\pi \rho_l}\right)^{2/3} \overline{Q_{ext}}\, L^{2/3} n_d^{1/3} k^{1/3} h. \end{aligned} \qquad (16.2)$$

Here, h and Q_{ext} are the cloud depth and extinction efficiency (bar indicating average value) respectively.

A lab experiment confirmed the above effective radius parameterization ($r_e^3 = k^{-1}\overline{r^3}$) in a controlled turbulent environment (Chandrakar et al., 2018). However, there are some conditions when this simple parameterization may not be an appropriate assumption, such as in regions dominated by entrainment or when there is a significant amount of drizzle formation. Additionally, these parameterizations also depend on aerosol population (Chandrakar et al., 2018; Gerber, 1996; Liu & Daum, 2002; Martin et al., 1994). The above picture assumes uniform cloud within a region of interest (for example, within a grid cell of a large-scale mode). This assumption might be inadequate if grid cells are large; therefore, numerical models require an additional measure of nonuniformity in a system; indeed, measurements show spatial correlations in cloud liquid water content on kilometer to millimeter scales and this must be represented in the calculation of radiative fluxes (Davis et al., 1999).

Ice clouds and mixed-phase clouds are still more complicated due to ice nucleation and growth mechanism that are still not fully understood. Moreover, their interaction with shortwave radiation involves additional factors like growth mechanisms, crystal shape, aggregation, orientation, and ice fraction (in the case of mixed-phase clouds). These demand substantive field observations and controlled laboratory experiments to understand these clouds so that useful parameterizations can be developed for models. In essence, measurements of cloud properties help guide the development of useful parameterizations and, additionally, aid in placing bounds on their applicability.

16.2.2. Precipitation (Water Flux)

Similar to the radiative flux through clouds, precipitation flux of water (liquid/ice) depends on the size distribution of hydrometeors, their type, and concentration. An estimate (or parameterization) of the rate of conversion of vapor-grown particles (like cloud droplets or small ice crystals) to precipitating particles requires information about the rate of different growth processes such as autoconversion, accretion, riming, and aggregation. In large-scale models, these processes are modeled based on the assumed size distribution of hydrometeors. For example, the autoconversion rate of cloud droplets to rain droplets can be estimated if the distribution shape (most commonly assumed as gamma distribution), critical size for conversion, and collision kernel are known (Liu & Daum, 2004):

$$P = \kappa L \int r^6 n(r)\mathrm{d}r. \tag{16.3}$$

Here, the Long collision kernel (κ) is used for the derivation (Long, 1974). In this example, the autoconversion rate depends on the zeroth (number concentration), sixth, and third moments of a droplet size distribution. It can be further simplified in terms of lower moments if a distribution shape is assumed. In case of other precipitation formation processes, additional information such as the shape of hydrometeors (ice crystal), fall speed, and density of particles is required. Therefore, observational studies are aimed at determining the size distribution of hydrometeors and other required microphysical details, as discussed above, under different cloud conditions.

16.2.3. Convection (Water Flux)

Atmospheric water vapor is one of the crucial factors contributing to climate sensitivity. Convective clouds are one of the vital sources of vertical transport of water vapor and latent heat in the free troposphere (Jensen & Del Genio, 2006; Sherwood et al., 2010). The extent of this transport depends on the entrainment and detrainment flux in these clouds (Blyth et al., 1988; Kain & Fritsch, 1990; Taylor & Baker, 1991). Moreover, these fluxes affect cloud microphysical process and precipitation formation. Measurement of thermodynamic quantities helps to quantify the profile of these fluxes as well as guide the development of an effective parameterization scheme for large-scale models.

Cumulus convection is one of the most significant sources of uncertainties in climate models (Klocke et al., 2011). In large-scale models where individual cumulus clouds are not resolved explicitly, the cumulus convection ensemble is modeled through a mass flux approach, i.e., an average rate of vertical transport of cumulus mass (Kain & Fritsch, 1993; De Rooy et al., 2013). This mass transport is coupled with the momentum, heat, and moisture transport equations in numerical weather prediction and climate models. The entrainment and detrainment rate in individual cumuli determines the amount of overall mass transport as a function of height. The amount of cloud top detrainment or mass flux profile depends on the properties of entrained air (relative humidity), the stability of the environment, and the size distribution of clouds (De Rooy & Siebesma, 2008; De Rooy et al., 2013). The turbulent entrainment and detrainment rate in all cumulus within the same environment are nearly equal. However, the total amount is thought to depend on the surface to volume ratio of a cloud. A higher instability and relative humidity of the environment promotes deep clouds, and hence the decrease in mass flux with altitude is relatively slow. Therefore, the mass flux profile can be parameterized using the critical dilution fraction that produces neutral buoyancy of a parcel (De Rooy et al., 2013). The deep convection is further complicated due to the precipitation flux and associated downdraft and cold pools. A critical part of an effective parametrization development of cumulus convection is to compare the parameterized entrainment/detrainment rate with measurements or with detailed large-eddy-simulation studies.

In boundary layer clouds, cloud-top entrainment influences cloud cover and thermodynamic properties. The layer which governs the entrainment properties is not well resolved in atmospheric models. Therefore, a parameterization is required to model the cloud top entrainment velocity. This entrainment velocity depends on wind shear, turbulence generated by radiative and evaporative cooling, and coupling to microphysics through drizzle formation and droplet settling (Mellado, 2017). For example, the entrainment velocity due to the droplet settling flux depends on the fifth moment of a droplet size distribution (de Lozar & Mellado, 2017). Measurement of the entrainment velocity is essential for verification of its parameterizations. However, accurate measurement of thermodynamics, flow, and cloud properties at the entrainment zone are often challenging due to small-scale variability in its vertical structure (Lenschow et al., 2000). Nevertheless, average flux measurements and the jump in scalar quantities across the interface are often used to infer the average entrainment velocity. For example, Lenschow et al. (1999), Gerber et al. (2013), and Faloona et al. (2005) reported measurements of the entrainment velocity at the top of the boundary layer during several different field campaigns.

16.2.4. Processes Affecting the Droplet Size Distribution

Condensation Growth

Condensation growth is the mechanism for cloud droplet growth at the initial stage after the activation of cloud condensation nuclei. A traditional picture of droplet condensation growth in a closed parcel suggests that the droplet size distribution should get narrower with time, but this is contrary to many observations (e.g., see the size distributions shown in Figure 16.3). This predicted narrowing is a result of the inverse radius dependence of the droplet radius growth rate. In most of the models, a central assumption in the growth calculation is uniform supersaturation within a parcel or model grid cell. If grid cells are large enough, significant fluctuations due to the presence of turbulent flow conditions may be filtered out. In the absence of any subgrid-scale model or parameterization, the result is a narrow droplet size distribution, spurious liquid water content, and other corresponding feedbacks. How do we know this is the case or how significant are these effects? Answers to these questions lie in exploration of the droplet size distribution primarily via in situ measurements.

The following basic calculations serve to illustrate: assume there is an adiabatic cloud parcel experiencing mean updraft velocity ($\overline{w}$) and containing a fixed concentration of activated cloud droplets. The starting mean and dispersion of the droplet size distribution are ($\overline{r}$) and (σ_r). If the growth in droplet size is larger than the assumed starting size, the difference in sizes of two droplets of starting sizes $\overline{r} - \sigma_r$ and $\overline{r} + \sigma_r$, after the parcel has been displaced Δz vertically, would be

$$\delta r(t) \approx \frac{\sqrt{2\overline{r}\sigma_r}}{\sqrt{\xi s \Delta z \overline{w}^{-1}}}. \tag{16.4}$$

Here, the droplet growth rate due to condensation is approximated as $d_t r^2 = 2\xi s$, with ξ being a thermodynamic constant and s being the supersaturation in the parcel (Yau & Rogers, 1996). The vertical displacement of the parcel is simplified as $\Delta z = \overline{w}t$. The equation demonstrates that the width of the droplet size distribution δr decreases with vertical displacement Δz.

The idea that the droplet size distribution gets narrower with time (or with vertical position) under assumed fixed supersaturation is inconsistent with observations (e.g., Martin et al. (1994); Miles et al. (2000); Pawlowska et al. (2006)). Figure 16.3 shows droplet size distributiona measured at different altitudes and for different aerosol levels (maritime versus continental) in stratocumulus clouds. The observations confirm the argument that the droplet size distribution not only does not get narrower with altitude but instead gets slightly broader (although there is a need for caution when inferring from the FSSP measurements because of their sizing uncertainties due to the nonuniform laser beam, coincidence error, and electronics response time (Baumgardner & Spowart, 1990)). An aspect of this figure related to aerosol concentration is discussed subsequently in section 16.2.4.3.

There are different proposed mechanism to explain observed broad droplet size distribution: the curvature and solute effects (and associated mechanisms such as spectral ripening and competing activation of cloud condensation nuclei of varying size, solubility, etc.) (Johnson, 1982; Korolev, 1995; Wood et al., 2002; Yang et al., 2018), turbulent fluctuations (Chandrakar et al., 2016; Cooper, 1989; Paoli & Shariff, 2009; Sardina et al., 2015; Siewert et al., 2017), entrainment and mixing (Baker et al., 1980; Lehmann et al., 2009; Yum et al., 2015), microphysical variability (Cooper, 1989; Desai et al., 2018), and internal mixing of parcels of different growth history (Hudson & Yum, 1997; Lasher-trapp et al., 2005). Liu and Hallett (1998) based on systems theory suggest that the width of the droplet size distribution also depends on the averaging length scale. If the length scale of averaging is below some minimum threshold, the average distribution shape would depend on the length scale (Liu & Hallett, 1998; Liu et al., 2002). They postulate that above this threshold, the size distribution would be independent of the length scale and would have a maximum width. The systems theory suggests a Weibull distribution shape for a maximum likelihood distribution with a variable shape factor based on the intensity of turbulence (Liu et al., 2002).

Collision Growth

Condensation growth in a typical adiabatic cloud parcel tends to be slow and insufficient to produce precipitation hydrometeors within typically observed times required for precipitation formation (Beard & Ochs III, 1993). Droplet collision-coalescence is the mechanism required to accelerate drop growth for radius approximately $r > 20$ μm. In a quiescent flow, droplet collision occurs due to the difference in relative settling velocity of large and small droplets with some efficiency of collision due to their hydrodynamic interactions. However, in the case of turbulent flow conditions, the relative velocity of droplets depends on the intensity of turbulence and relative particle response timescale. Additionally, another parameter that influences the rate of collision is the spatial distribution of droplets which also depends on turbulence intensity and particle response timescale (Shaw, 2003). The radial distribution function (or spatial covariance of droplets) is generally used to characterize the correlations (Saw et al., 2012). During a collision event, surface interactions of droplets determine the coalescence efficiency. All of the above microscale interactions are studied extensively using laboratory experiments and high-resolution direct-numerical-simulations (Grabowski & Wang, 2013). With this information, different collision kernels are proposed for estimating collision rate in atmospheric models where droplets are not resolved explicitly. However, Witte et al. (2017) based on estimation from measurements suggest that these collision kernels underestimate the collision rate in large-scale models due to the variability of the droplet distribution on scales smaller than their grid cells. Using local-volume measurements, the small-scale variability in droplet distribution can be assessed (Beals et al., 2015; Larsen et al., 2018).

A significant question that is still a topic of ongoing research is: how are the droplets large enough to start the collision-coalescence process ("precipitation embryos")

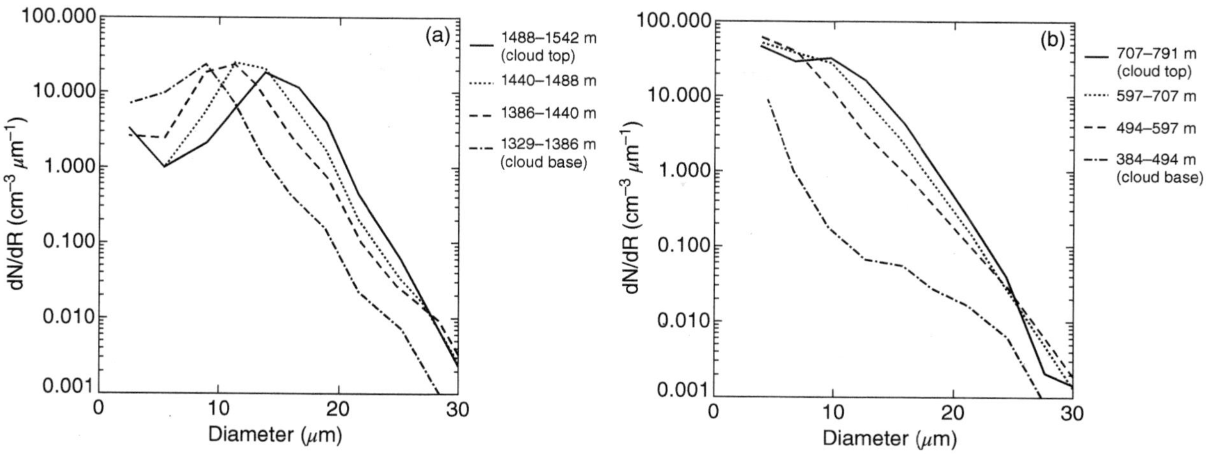

Figure 16.3 Cloud droplet size distribution measured using a Forward Scattering Spectrometer Probe (FSSP) in (a) Maritime and (b) Continental condition at different altitude level of stratocumulus cloud field. Source: The figure is from Martin et al. (1994)/American Meterological Society.

produced? The simplest calculations of drizzle onset in an adiabatic parcel rising in an updraft with constant speed show unrealistically narrow droplet size distributions and long times to drizzle formation. Measurements from the Rain in Cumulus Over the Ocean experiment (RICO), along with modeling efforts constrained by the measurements, serve to illustrate the complexity of these interacting components. Hudson et al. (2012) observed that the width of cumulus droplet size distributions decreased with increasing number concentration of CCN and that parcel model calculations were unable to represent this trend unless variability in the vertical velocity within the cumulus clouds was accounted for. While these observations are from cumulus clouds, they are related to the dispersion effect discussed further in the next section in the context of stratocumulus clouds. In contrast, Lowenstein et al. (2010) found that observations of the large-drop tail of the cloud droplet size distribution made with the 2D-S instrument could be matched by initializing a parcel model with the full aerosol size distribution observed at cloud base. Bewley and Lasher-Trapp (2011) and Cooper et al. (2013) found that variability in vertical velocity together with entrainment and secondary activation significantly contributed both to the width of the droplet size distribution and the first formation of large droplets needed to initiate collision-coalescence. Generally speaking, modeling efforts that were carefully initialized and constrained by observations, accounting for entrainment, variable updraft strength, and observed aerosol distributions near cloud base, were able to produce precipitation at a rate consistent with observations. For example, Blyth et al. (2013) found reasonable consistency between the rate of development of radar echoes from developing precipitation and the modeled cloud droplet growth initialized by the observed aerosol distribution. One hypothesis investigated in that study is that giant cloud condensation nuclei are important for producing precipitation embryos. Multiple studies concluded, however, that giant CCN did not play a significant role in precipitation development for the observed clouds (Blyth et al., 2013; Gerber & Frick, 2012; Reiche & Lasher-Trapp, 2010). Subsequent work has suggested compelling evidence that a ripening effect may play a significant role when large CCN particles are present (Jensen & Nugent, 2017).

Aerosol Effects on Cloud Droplet Number Density and Moments: First-Order Interest is Aerosol Concentration

Clouds are significant contributors in Earth's radiation budget. A change in the boundary-layer cloud coverage can alter the radiative forcing at a similar order as greenhouse effect (Slingo, 1990; Wood, 2012). Furthermore, atmospheric aerosols can directly alter cloud microphysical properties and, indirectly, cloud macroscopic properties. Cloud microphysical properties are modulated through aerosol indirect effects (first, second, and dispersion effects) (Albrecht, 1989; Liu & Daum, 2002; Pincus & Baker, 1994; Twomey, 1977, 1974). With a decrease in aerosol, the mean size of cloud droplets increases, and the distribution becomes broader. These distribution changes could cause increased precipitation and a decrease in cloud liquid water content (at least in a relatively clean cloud) and, consequently, cause a change in the radiative fluxes. Traditionally, the influence of aerosol on cloud radiative properties through a change in the mean size of cloud droplets is known as the "first indirect effect" or "Twomey effect," and on cloud lifetime and precipitation formation is known as the "second indirect effect" or "lifetime effect." The "dispersion effect" is an extension that accounts for the droplet size distribution width. In marine stratocumulus clouds, the precipitation suppression enhances the latent heating and the entrainment-induced evaporative cooling due to reduced sedimentation and higher integral radius (Hill et al., 2009; Wood, 2012). Thus, it alters the dynamics, turbulent kinetic energy, and entrainment rate. Moreover, cloud thickness positively correlates with aerosol concentration (Pincus & Baker, 1994). As a result of the above feedbacks, the entrainment rate may be enhanced, effectively forcing the system toward a thinner cloud. Although the cloud thickening increases, the liquid water and moisture flux facilitate buoyant production (Wood, 2012). In addition, the cloud-top entrainment velocity (w_e) decreases with an increase in aerosol due to a change in droplet settling flux ($w_e \propto -d^5$) (de Lozar & Mellado, 2017). Therefore, there are considerable positive and negative feedbacks governed by aerosol perturbation through microphysics. The overall effect on the cloud system depends on whether these perturbations are amplified or dampened by dynamic forcing and the aerosol regime (clean versus polluted). Finally, apart from microscale properties, the aerosol population can similarly influence mesoscale cloud structure (associated radiative fluxes) and the convection pattern. Indeed, the transformation of a cloud from a closed-cell structure to an open-cell structure is feasible merely due to the reduced aerosol concentration (Wang & Feingold, 2009).

Most of the aerosol feedbacks were previously considered a result of changes in only the mean droplet size. However, Liu and Daum (2002) pointed out that the droplet size relative dispersion changes with aerosol concentration, which influences the effective radius and cloud albedo. Yum and Hudson (2005) offer an explanation for this behavior of relative dispersion based on the assumption of an adiabatic parcel, similar to the example presented in section 16.3.2 equation (16.4). For higher aerosol concentration in a cloud parcel, supersaturation decreases due to increased competition for water vapor, which decreases the narrowing effect as a result of inverse radius dependence of droplet radius growth. Consequently, droplet size relative dispersion shows a positive correlation with aerosol concentration. Although, as pointed by Yum and Hudson (2005), a model based on this adiabatic parcel produces a lower relative dispersion than typically measured. Moreover, it cannot explain the observations of decreasing or nearly a constant trend of relative dispersion with increase in aerosol concentration (Arabas et al., 2009; Lu et al., 2007, 2008, 2012, 2013a; Miles et al., 2000; Tas et al., 2015; Zhao et al., 2006).

In stratocumulus cloud, the mean vertical velocity is relatively low; thus, supersaturation fluctuations and microphysical variability could have a significant effect on aerosol-cloud feedbacks. Recent lab experiments (e.g., Chandrakar et al., 2016), field observations (e.g., Siebert & Shaw, 2017), and some numerical simulations (e.g., Grabowski & Abade, 2017; Paoli & Shariff, 2009; Sardina et al., 2015) indicate that supersaturation fluctuations can be a crucial part of microphysical interactions. For example, turbulence induced broadening can facilitate precipitation formation by enhancing the droplet collision rate and can influence cloud radiative properties. Furthermore, supersaturation fluctuations might be a key element in the aerosol activation process, specifically, when the mean updraft velocity is low. Similar to the traditional aerosol-indirect effects due to mean properties, the concentration of aerosols can also modulate these effects of supersaturation fluctuations. Theory and laboratory results suggest that with increase in aerosol concentration, supersaturation fluctuations and associated droplet distribution broadening rate tend to decrease. Therefore, radiative and precipitation fluxes would be influenced as a result of the turbulence-induced aerosol effects (Chandrakar et al., 2018).

Entrainment and Turbulence Effects on Cloud Properties

Mixing between a cloud "parcel" and the outside environment results in liquid water contents substantially lower than the quasi adiabatic ideal. This departure is universally observed near cloud boundaries, such as at the top of stratocumulus clouds and the sides of cumulus clouds. This is particularly relevant for the stability and persistence of stratocumulus clouds that are so radiatively important to determining the climate (Malinowski et al., 2013; Mellado, 2017; Stevens, 2002; Vanzanten & Duynkerke, 2002). Assessing the entrainment rate in stratocumulus clouds remains a primary experimental challenge, and major field projects have been developed to address the question (Ghate et al., 2019; Wang et al., 2009; Yeom et al., 2017). Entrainmnent of environmental air is also crucial to determining the vertical extent and microphysical properties of cumulus clouds (Heus et al., 2009; Lu et al., 2016). Cloud microphysical properties respond to the nature of the mixing: the liquid water content is reduced, but because $L \propto n\overline{d^3}$, where n is the cloud droplet number density and d is the cloud droplet diameter, it is evident that a given reduction in L can be achieved with different combinations of n and L. For example, in the limit of homogeneous mixing, a cloud droplets is exposed to subsaturated air evaporate together, resulting in reductions of both n and $\overline{d^3}$; in the limit of extreme inhomogeneous mixing, entrained air is humidified through the complete evaporation of a subset of cloud droplets, with the remaining droplets remaining unaltered, ultimately resulting in a cloud with reduced n but constant $\overline{d^3}$ (Burnet & Brenguier, 2007; Lu et al., 2013b).

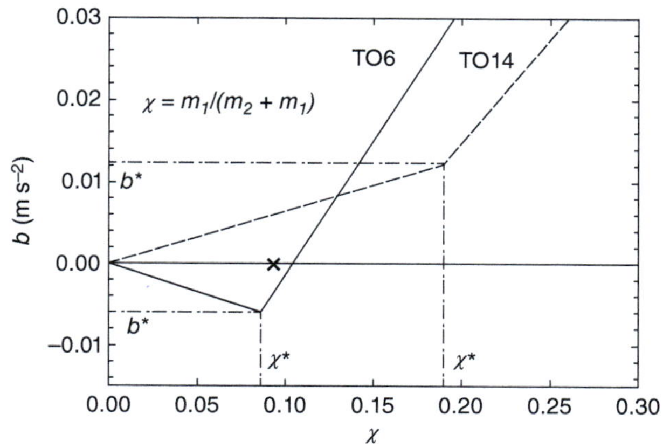

Figure 16.4 Buoyancy perturbation b of a mixture of air from stratocumulus cloud top with air from the free troposphere. The mass fraction of cloud air is given by $1 - \chi$. The lines plotted are for two flights from the POST field project (flights TO6 and TO14). The vertices denoted $\chi^\star$ and $b^\star$ are the values resulting when all cloud liquid water has evaporated and the resulting mixture has a relative humidity of 100%. The conditions observed in flight TO6 result in negative $b^\star$, indicating the possibility of buoyancy instability. Flight TO14 is an example of conditions under which buoyancy reversal is not expected. Finally, a third flight (TO12) resulted in $b^\star = 0$, with the corresponding $\chi^\star$ denoted by the ×. Source: The figure is from Gerber et al. (2016)/American Meterological Society.

The importance of cloud-top entrainment to stratocumulus clouds is illustrated in Figure 16.4. The figure depicts the culculated buoyancy of a mixture of cloud-top air with free-tropospheric air. The buoyancy is plotted versus the mass fraction χ of free-tropospheric air, and the value corresponding to complete evaporation of cloud liquid water and saturation of the resulting mixture is denoted by $\chi^\star$. Two different cases corresponding to stratocumulus clouds measured during the POST field campaign are shown to illustrate that in some conditions buoyancy reversal can occur (flight TO6). In other words, the fully saturated mixture is more dense than either the cloud-top or free-tropospheric air, with the possibility that this will lead to further entrainment and ultimately to destabilization of the cloud. The relevance of this cloud-top-mixing instability remains a focus of research (Mellado, 2017).

16.2.5. Ice Formation

Ice formation is of crucial importance to the precipitation efficiency and optical properties of clouds (Storelvmo, 2017). Ice can be formed directly through homogeneous or heterogeneous nucleation (Cantrell & Heymsfield, 2005). Multiple pathways for heterogeneous ice nucleation exist, but generally speaking it is found that ice-nucleating particles are a relatively rare subset of the atmospheric aerosol (Kanji et al., 2017). Under many circumstances, additional (secondary) mechanisms are required to account for observed ice concentrations (Field et al., 2017). Once formed, the ice phase can evolve rapidly as a result of the differential vapor pressure of liquid and solid water at temperatures below the melting point, known as the Wegener-Bergeron-Findeisen process (Storelvmo & Tan, 2015). As mentioned in the introduction, due to space constraints the focus of this chapter is on warm cloud microphysics; more complete reviews of the problem of mixed-phase and cold clouds are available elsewhere (Andronache, 2017; Korolev et al., 2017).

The complexity of the problem is illustrated by the continuing evidence in support of a correlation between the presence of large cloud drops and the ice formation process (Coopman et al., 2020; Lance et al., 2011). Early support for the correlation was discussed in section 16.1.2. The observations of Lance et al. (2011) are illustrated in Figure 16.5, where it is evident that liquid water content, large cloud drop concentrations, and ice precipitation rates differed greatly between clean and polluted cloud cases. They concisely summarize the observations as follows: "Clean clouds, identified from lower CO mixing ratios exhibit broader droplet size distributions, more than 10 times greater concentrations of precipitating ice particles, and more than 10 times greater ice water content than polluted clouds. Even though the minimum temperature in the polluted case is colder than that in the clean case, ice concentrations remain 12 orders of magnitude lower, in spite of greater dust and BC concentrations." There is still no clear mechanism to explain the seeming importance of large droplets, but recent laboratory work on ice generation by freezing drizzle-sized drops provides a compelling possibility (Lauber et al., 2018). It has even been speculated that droplet distortion or coalescence could play a role (Yang et al., 2020). Ultimately, laboratory investigation is the likely pathway for identifying and assessing specific mechanisms for secondary ice production, and the relevant literature on laboratory results is thoroughly reviewed by Korolev and Leisner (2020).

16.3. A SAMPLE OF NEW MEASUREMENT CAPABILITIES AND RECENT RESULTS FROM FIELD STUDIES AND LABORATORY EXPERIMENTS

As discussed in the previous section, the size distribution of hydrometeors is one of the key property determines the interactions of a cloud with incoming solar radiation, precipitation formation, transport of water vapor, dynamics of boundary layer, and more. Therefore, investigation of the droplet size distribution (or their moments) using

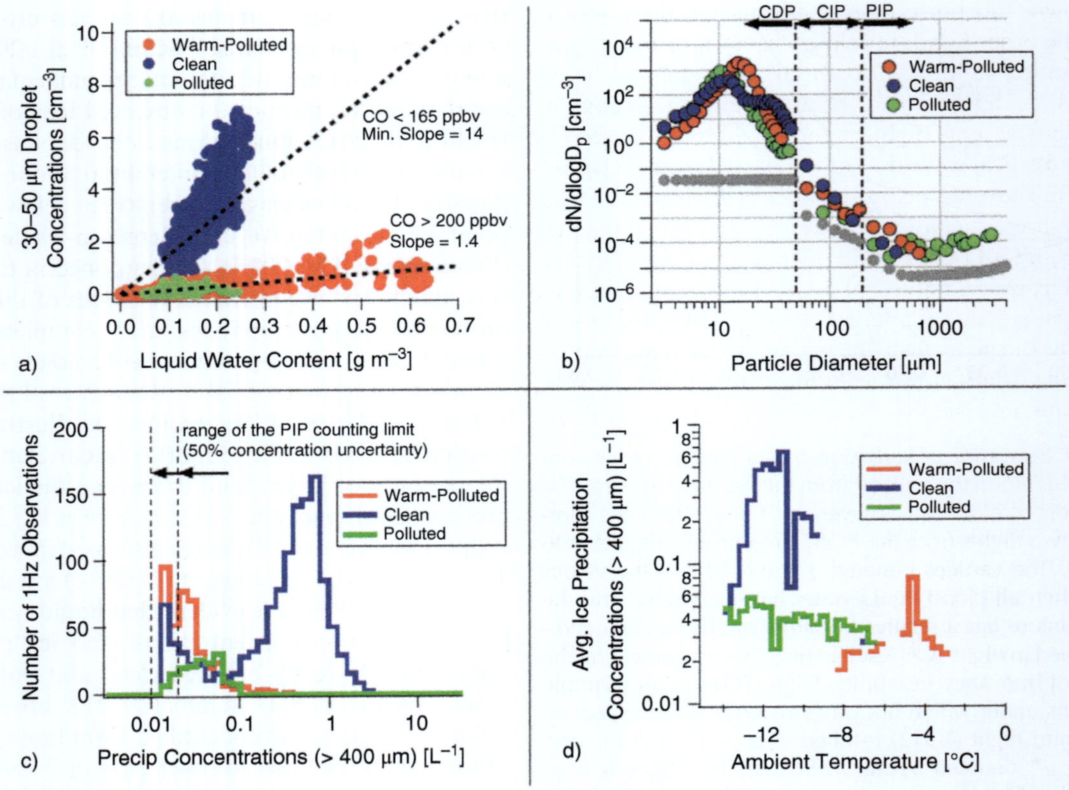

Figure 16.5 In situ measurements of cloud microphysical properties for the three cases observed during ARCPAC. (a) Number concentrations of large droplets (3050 μm) as a function of liquid water content, demonstrating the significant contribution of large droplets in the clean case. (b) Examples of cloud particle size distributions measured near cloud top. The gray markers show the minimum concentration for each instrument. (c) Histograms of ice precipitation concentrations, showing that the clean cloud had the highest frequency of high precipitation values. (d) Ice crystal concentrations versus temperature. Source: Figure is from Figure 6 of Lance et al. (2011)/Copernicus Publications/CC BY 3.0.

measurement provides some valuable input to the modeling framework. For example, droplet number concentration, second, and third moments of a droplet size distribution are essential for estimating radiative fluxes through a cloud. Similarly, droplet number concentration, third, and fifth moments of a droplet size distribution are important for autoconversion rate and cloud-top entrainment flux.

A next question would be: how do we measure droplet (or ice) size distributions and their moments, as well as other fine-scale or "fast" cloud properties? In the next section, we provide illustrative examples of recent advances and findings within the three broad areas of in situ observation: airborne measurement, ground-based measurement, and laboratory measurements.

16.3.1. Airborne Measurements of Cloud Properties

In situ field observations from airborne platforms, such as aircraft, helicopters, drones, and balloons, provide a valuable opportunity to measure cloud microphysical properties in their natural setting. Clouds are typically far from the surface and are often transient in nature, and cloud properties vary widely over different portions of the globe due to thermodynamic and aerosol conditions. Airborne measurements are ideal for reaching cloud systems ranging from marine stratocumulus, to deep convection, to high-altitude cirrus. Instrumentation suitable for airborne deployment has advanced to a very sophisticated and specialized levels and thorough reviews are available in the literature (Wendisch & Brenguier, 2013). Airborne measurements typically provide much more detailed information than ground-based remote sensing, and in fact are necessary in assessing retrieval methods. Through airborne measurements, spatial variation of cloud microphysics, thermodynamic, and turbulent flow properties can be measured in cloud as well as properties of the air above and below cloud. The cloud sampling strategy using airborne instruments depends on the purpose of the study: for example, vertical profiles of cloud properties for cloud optical depth studies may require multiple up and down "porpoise dive" patterns for the aircraft path. On the other hand, for cloud fraction or mesoscale cloud structure, a long, horizontal flight pattern may be required. A downward drifting spiral or Lagrangian spiral descents/ascents flight pattern is commonly used to observe the vertical evolution of the particle size distribution in ice/mixed-phase cloud studies (Heymsfield et al., 2002). Ever-present challenges of airborne measurements are the expense and complexity of operating an aircraft, and the high-speed nature of most measurement platforms. The emergence of uncrewed aerial vehicles (drones) for scientific purposes has the potential to significantly alter the type, frequency, duration, and range of in situ measurements available to the community (e.g., de Boer et al., 2018).

Airborne instrumentation for measuring hydrometeor size distributions and moments (such as liquid water content) are mainly of the following categories: (a) hot wire, (b) forward scattering, (c) phase Doppler, (d) shadowgraph, (e) imaging, and (f) holographic. More detailed classifications of these instrumentations and descriptions of each type can be found in Baumgardner et al. (2011) and Brenguier et al. (2013). Table 16.1 lists some of the instruments with their specifications (Baumgardner et al., 2011; Gerber et al., 1994).

Here, the first five instruments in the table can detect both cloud droplets and ice particles but are not able to differentiate their phases; furthermore, the sizing by forward-scattering instruments relies on the assumption of spherical particles. The SID-2, 2D-S, CIP, CPI, HVPS, and holographic instruments can be used for both liquid and ice phases. Hot-wire instruments are based on the change in resistance of a hot-wire junction due to the evaporation of droplets, and it provides an estimate of the liquid water content and therefore the third moment of the droplet size distribution. The second class of instruments operate by collecting light scattered by spherical droplets in the near-forward direction. They provide individual, drop-by-drop measurement of the diameter within a relatively small instrument cross section that is swept through the cloud the the aircraft. Examples of instruments in this category are the Forward Scattering Spectrometer Probe (FSSP), Cloud Droplet Probe (CDP), Cloud-Aerosol Spectrometer (CAS), and Particulate Volume Monitor (PVM-100A). The PVM-100A uses modified forward-scattered light phase functions for measuring simultaneously total liquid water content and surface area of droplets in a volume of $\approx 1.6\,\mathrm{cm}^3$ and at a data rate up to 1 kHz. The ratio of liquid water content to surface area is proportional to r_e. Since the forward scattering probes are based on the intensity of scattered light, care must be taken to maintain clean optics and regular calibration.

Phase-Doppler Interferometry (PDI) is based on detection of phase difference rather than intensity, and thereby avoids some of the challenges of forward-scattering

Table 16.1 List of airborne instruments for measuring hydrometeor size distribution and moments.

Airborne instruments for cloud particle measurement			
Instrument name	Working principle	Size range	Sampling rate
Hot-wire liquid water sensor (LWC-300)	Resistance change	L: 0.05–3 g m^{-3}	4 l s^{-1}
CAS	Forward scattering	D: 0.5–50 μm	25 cm^3 s^{-1}
Fast-CDP	Forward scattering	D: 1.5–50 μm	25 cm^3 s^{-1}
Fast-FSSP	Forward scattering	D: 1.5–50 μm	40 cm^3 s^{-1}
PVM-100A	Forward scattering	L: 0.002–10 g m^{-3} S: 5–20,000 cm^2 m^{-3}	1,600 cm^3 s^{-1}
Small ice detector (SID-2)	Scattering	D: 2–140 μm	100 cm^3 s^{-1}
Artium Flight PDI	Doppler effect	D: 0.5–2500 μm	≈ 100 cm^3 s^{-1}
2D-C	Shadowgraph	D: 25–800 μm	5 l s^{-1}
2D-S	Shadowgraph	D: 10–1260 μm	8 l s^{-1}
CIP	Imaging	D: 25–1550 μm	16 l s^{-1}
CPI	Imaging	$D > 3\ \mu m$	400 cm^3 s^{-1}
HVPS	Imaging	D: 300–19200 μm	310 l s^{-1}
PHIPS	Imaging and scattering	D: 4–800 μm	25 cm^3 s^{-1}
Holodec, HaloHolo	Holography	$D > 6\ \mu m$	Instantaneous sample volume ≈ 20 cm^3

The volume sampling rates correspond to a flight speed of 100 m s^{-1}. The table is adapted from Baumgardner et al. (2011).

instruments. The phase difference between scattered Doppler bursts measured at two angles from a moving droplet is directly proportional to the diameter of that droplet. The coefficient of proportionality is a function of geometrical properties and wavelength of the source light. As with forward-scattering instruments, phase-Doppler instruments have a relatively small sample cross section that is swept through the cloud. Due to a small instantaneous sampling volume, these instruments have very coarse temporal and corresponding spatial resolution during in-flight measurements. Moreover, they are subjected to higher statistical uncertainties as a result of low sampling rates. Thus, obtaining a statistically significant measurement of clouds having very low droplet concentration can be challenging with these instruments.

The other class of instruments provides volumetric or at least two-dimensional information about the particle distribution and size. The Shadowgraph system (e.g., 2D Stereo Probe, 2D-S) records the shadow of an illuminated particle in the focal plane of a detector. Similarly, there are other imaging probes that capture two-dimensional images of particles illuminated by light sources (e.g., Cloud Particle Imager (CPI), Cloud Imaging Probe (CIP), and High Volume Precipitation Spectrometer (HVPS)). The Particle Habit Imaging and Polar Scattering (PHIPS) probe allows simultaneous measurement of three-dimensional structure and associated microphysical and optical characteristics of cloud particles (Abdelmonem et al., 2011). Holographic instruments have the unique capability of capturing the three-dimensional information about particle position as well as particle size. These instruments capture the interference pattern created between a reference wave and the light scattered from a dilute population of particles within the measurement volume. Reconstruction of the hologram is performed digitally in modern systems. These instruments provided detailed information on cloud microphysical properties, although they have a limitation of high computation cost and lower time resolution.

Example: Cloud Droplet Size Distribution Moments.

As discussed in the previous section, the shape of the cloud droplet size distribution is relevant to radiative properties of clouds. As an example of airborne measurement with a focus on assessing the shape of droplet size distributions in stratocumulus clouds, we refer to the extensive compilation of Miles et al. (2000). They show moments of the droplet size distribution at different altitude and for different aerosol concentrations. Among the other significant aspects, they reported that the cloud droplet distribution gets broader with height contrary to predictions from parcel models. The width of the size distribution is relevant to the drizzle rate and can be significant contribution to the loss rate of cloud condensation nuclei in the marine boundary layer (Wood, 2006). That, in turn, can lead to ultraclean layers and may play a role in stratocumulus-to-cumulus transitions (O et al., 2018; Wood et al., 2018).

Glienke et al. (2017) and Witte et al. (2018) have recently demonstrated the importance of measuring the full droplet size distribution, from cloud droplet to precipitation diameters, for calculating cloud radiative properties. By carefully measuring the full size distribution with phase-Doppler interferometry (Witte et al., 2018) and digital holography (Glienke et al., 2017), it was shown that effective radius values retrieved from satellite measurements do now show the strong discrepancies previously indicated. Example size distributions from the CSET field project are shown in Figure 16.6. The three panels illustrate how the size distribution responds to different relevant powers of droplet diameter: the top

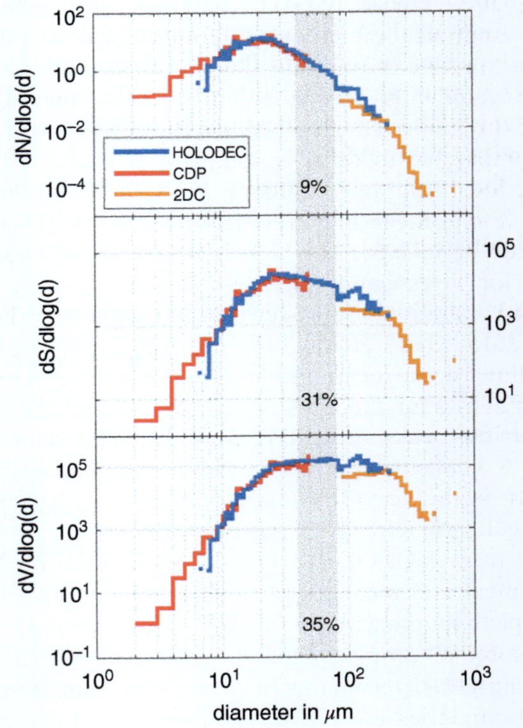

Figure 16.6 Cloud droplet size distributions, weighted by number concentration, surface area, by volume. HOLODEC data provide continuous coverage between the CDP and 2DC measurement ranges. Intriguingly, no local minimum in the size distribution is observed between cloud droplet and drizzle sizes, as is typically expected from parcel model calculations. The shaded region indicates this "autoconversion gap" and the numbers in that region indicate the number, surface area, and volume contributed by hydrometeors in this "drizzlet" size range. Source: The figure is from Glienke (2017)/John Wiley & Sons.

shows a number distribution, the middle shows a surface area (d^2) distribution, and the bottom shows a volume (d^3) distribution. The shaded region shows the range of droplet diameters that are in the "gap" between cloud droplet and drizzle sizes, with the corresponding contributions to total number, surface area, and volume. Some of the most common aircraft instruments for measuring cloud droplet size distributions miss this gap region, and therefore miss significant contributions to radiative properties. (To be precise, it should be noted that instrument measurement ranges do cover some of this gap, but the estimated concentrations therein are commonly considered to be unreliable.)

Example: Digital Holographic Measurements of Cloud Microphysics.

The HOLODEC instrument has a novel sampling strategy based on digital in-line holography, which provides an estimate of the cloud particle size distribution from a single sample volume. This is in contrast to the typical approach of counting droplets one at a time along a thin line through the cloud. Each instantaneous sample volume is approximately 20 cm^3, repeated at 3.3 Hz; the time-averaged volume sampling rate is therefore comparable to other instruments (e.g., the CDP), but the local sampling approach means assumptions of spatial homogeneity and ergodicity required when averaging can be relaxed. The "local volume" configuration allows for investigation of the influence of averaging and the variability of local-scale size distributions (Beals et al., 2015; Desai et al., 2018). The ability to look at three-dimensional locations of particles within the sample volume allows for the investigation of droplet clustering and particle breakup (Fugal & Shaw, 2009; Larsen et al., 2018). A challenge for widespread use of holography, however, is the heavy computational burden associated with the digital hologram reconstruction and analysis.

Data from HOLODEC have helped clarify whether entrainment and mixing lead to uniform evaporation of all droplets (homogeneous mixing) or total evaporation of a subset of droplets (inhomogeneous mixing). This is relevant to determining microphysical properties at cloud boundaries and is therefore important for both drizzle formation and cloud radiative properties. For example, Figure 16.7 shows mixing diagrams acquired with HOLODEC data from liquid-phase cumulus clouds, and the scatter of points along a horizontal line is indicative of extremely inhomogeneous mixing (Beals et al., 2015). Each point corresponds to a single hologram, so the spatial averaging that was previously identified as an ambiguity within mixing diagrams (Burnet & Brenguier, 2007) was bypassed (that ambiguity results for single-drop-counting instruments that often require sampling over ~tens of meters to obtain reliable droplet

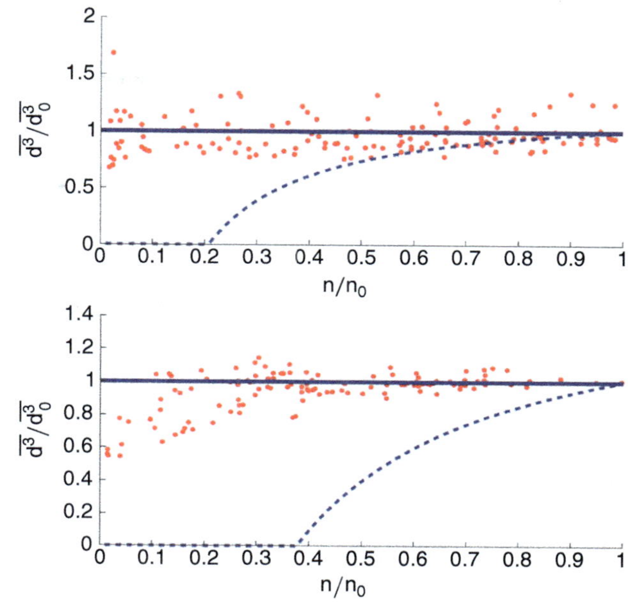

Figure 16.7 Mixing diagrams generated from HOLODEC data obtained in small, liquid-phase cumulus clouds. Each dot represents the mean-volume diameter and the cloud droplet concentration estimated from a single HOLODEC sample volume. The dashed curve represents homogeneous mixing, and the solid blue line represents extremely inhomogeneous mixing. Source: The figure is from Beals (2015)/American Association for the Advancement of Science.

size distributions). There is a need for this kind of investigation in a wider range of cloud types and environments in order to determine the conditions for homogeneous versus inhomogeneous mixing.

The ability of holographic instruments to detect spatial patterns of particles within the sample volume enables particle-shattering artifacts to be removed (Fugal & Shaw, 2009). This is an important measurement capability because of the now well-understood limitations in measuring small-particle concentrations when precipitation-sized hydrometeors are present, especially in mixed-phase or ice-phase clouds (Jackson et al., 2014; Korolev et al., 2011). Figure 16.8 shows an example of how the holographic instrument with shattering effects removed differs dramatically from small ice concentrations estimated by other instruments (e.g., for ice particles with dimensions less than 100 μm). The airborne measurements with the HaloHolo instrument were made in a cirrus cloud, where small ice concentrations are especially important in determining optical properties.

Example: High-resolution, Collocated Microphysical, and Turbulence Measurements.

An important new development is airborne platforms that allow for high-resolution and collocated

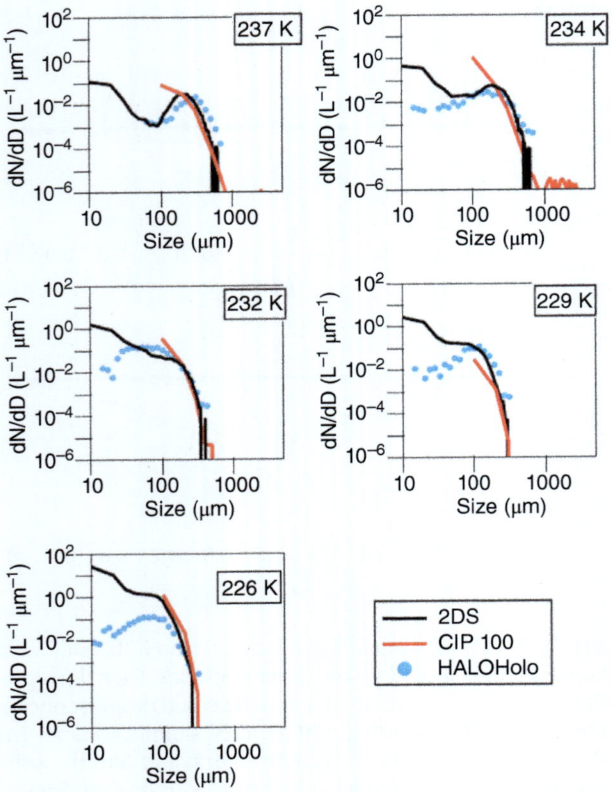

Figure 16.8 Ice crystal size distributions from a cirrus cloud as measured by the 2D-C, CIP, and HALOHolo instruments. Source: The figure is adapted from O'Shea et al. (2016).

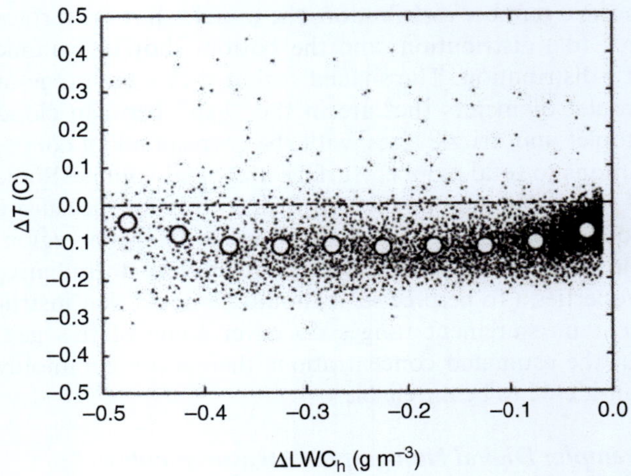

Figure 16.9 High-resolution measurements of temperature and liquid water content at stratocumulus cloud top, obtained during the POST campaign. The figure shows the codependence of differences in temperature ΔT and liquid water content ΔLWC_h between cloud entrainment holes and the cloud unaffected by entrainment. The instruments for recording T and LWC were closely collocated and operating at 50 Hz (approximately 1-m resolution), allowing fine-scale features at stratocumulus cloud top to be measured. The circles are ΔT averages over LWC_h increments to show the lack of any significant trend, which implies a weak role of evaporative cooling in the formation of cloud holes. Source: The figure is from Gerber et al. (2016)/American Meterological Society.

measurements of microphysics and thermodynamic properties. The Cloud, Aerosol, Radiation and Turbulence in the Trade Wind Regime over Barbodos (CARRIBA) project (Siebert et al., 2013) and the Physics of Stratocumulus Top (POST) project (Gerber et al., 2013; Malinowski et al., 2013) are illustrative examples. In the former, the Airborne Cloud Turbulence Observation System (ACTOS) was deployed, carried by a helicopter at relatively low flight speed and with a full suite a microphysical, thermodynamics, turbulence, and radiation instrumentation. In the latter, the CIRPAS Twin Otter was configured with instruments for high-frequency measurement of temperature, humidity, liquid water content, cloud droplet size distribution, and turbulence. The key instruments were collocated at the nose of the aircraft. It should be noted that high-frequency response is necessary to achieve high spatial resolution, but measurements of some key quantities like temperature and humidity can be degraded by wetting of the instruments.

These kinds of measurement strategies are extremely useful in understanding the entrainment-microphysics interactions at sharp cloud boundaries, such as at stratocumulus cloud top. For example, these measurements reveal the critical role played by turbulence generated in the "entrainment interfacial layer that spans the top of the cloud and into the clear air above, (Jen-La Plante et al., 2016; Katzwinkel et al., 2012; Malinowski et al., 2013). It is observed that under a wide range of conditions, this thin, turbulent layer adjusts in thickness so that it is just at the critical Richardson number needed for turbulence generation through mean or local shear (Jen-La Plante et al., 2016; Katzwinkel et al., 2012). Gerber et al. (2016) show that the entrainment interfacial layer becomes conditioned through detrainment of cloud so as to have similar thermodynamic properties as at the cloud-top; the result is that when entrainment occurs, there is little evaporative cooling and therefore cloud-top instability through buoyancy reversal is minimized. Figure 16.9 shows high-frequency temperature and liquid water content differences for cloud holes relative to cloud top. The observed ΔT is nearly independent of ΔLWC_h, meaning that reduction in liquid water content in cloud holes is occurring primarily through dilution rather than through evaporation.

Measurements through small cumulus clouds with ACTOS during the CARRIBA project are illustrated in Figure 16.10, resulting in a high-resolution view of the cloud edge region. The shaded regions show the subsiding shell that results from entrainment of relatively dry air

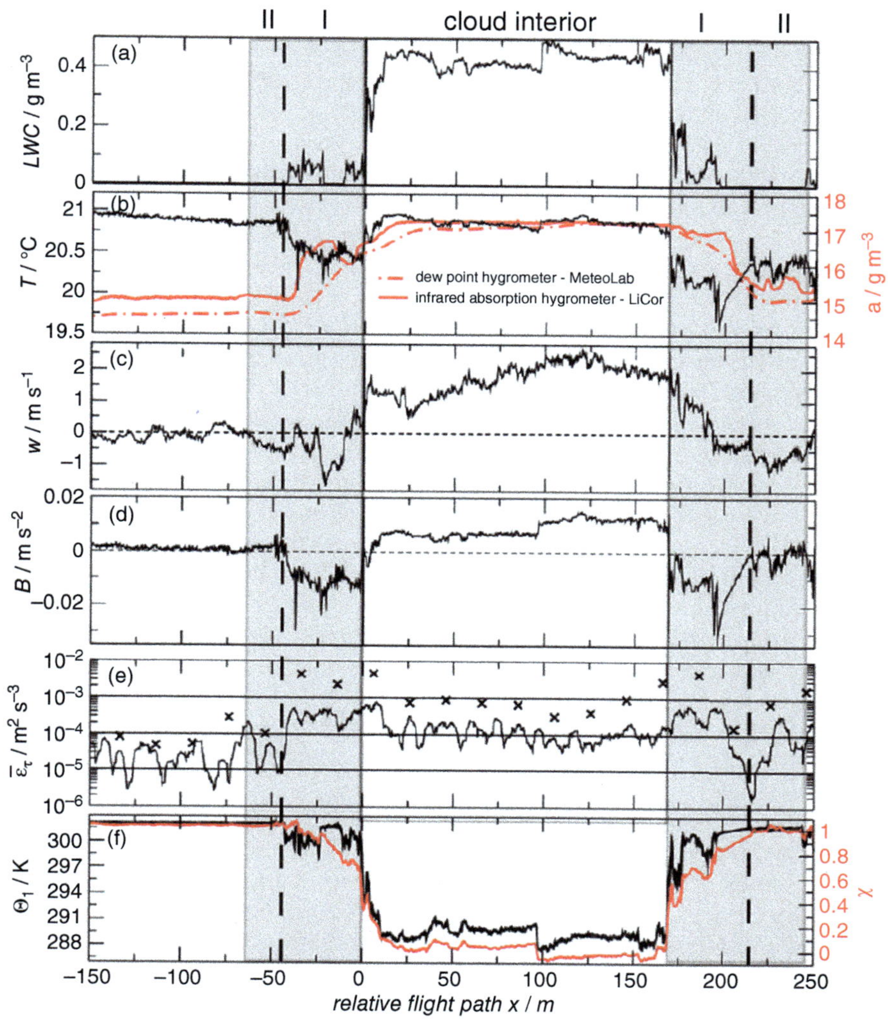

Figure 16.10 High-resolution measurements through a single cumulus cloud from ACTOS: (a) LWC, (b) temperature and absolute humidity (the two red lines correspond to two different instruments), (c) vertical wind velocity, (d) buoyancy, (e) energy dissipation rate (the line and the crosses denote two methods for calculating dissipation rate), and (f) liquid water potential temperature and mixing fraction. The two shaded regions show the subsiding shell at the cloud edge, which can be further divided into cloud-free and inner cloudy regions. Source: The figure is from Katzwinkel et al. (2014)/American Meteorological Society.

from the environment outside the cloud. The top panel shows that in the subsiding shell there are intermittent levels of cloud liquid water, clearly distinguishable from the cloud core (interior). The second panel shows the decrease in temperature and variation of absolute humidity, both of which contribute to the overall buoyancy shown in the fourth panel. There is vividly apparent that the subsiding shells are negatively buoyant relative to both the cloud and the surrounding environment, and indeed, the third panel shows a negative vertical component of air velocity within much of the shell. The fifth panel shows the turbulent energy dissipation rate, indicating a slight enhancement in the subsiding shell, and therefore suggesting that there may be a positive feedback in entrainment process (initial entrainment leading to stronger turbulence, thereby leading to additional entrainment). Finally, the bottom panel shows two commonly used variables that are conserved during mixing: the liquid water potential temperature and the water mixing fraction. The conserved variables show a clear transition between environmental and cloud conditions.

16.3.2. Ground-based Field Measurements

The third category of in situ measurements is the ground-based measurements at high altitude sites, for example, Storm Peak Laboratory at Colorado, USA (Hallar et al., 2013); Schneefernerhaus Environmental

Research Station at mount Zugspitze, Germany (Risius et al., 2015); Pico Mountain Observatory in Portugal (China et al., 2015); and the High Altitude Cloud Physics Laboratory at Mahabaleshwar, India (Leena et al., 2018). Ground-based stations have advantages of cost-effectiveness, long-term statistics, more detailed measurements over the field campaign using aircraft.

Example: Mountain-top Measurement of Mixed-phase Clouds.

Mountain stations have allowed for detailed investigation of the properties of mixed-phase and ice-phase clouds. Such measurements have helped clarify the conditions under which the differential vapor pressure effect (Wegener-Bergeron-Findeissen process) is operating (Lohmann et al., 2016). Mountain stations also allow for investigation of ice nucleation and secondary ice generation processes (e.g., Baustian et al., 2012). For example, comparison of ice crystal concentrations measured at mountain-top stations can exceed the estimated concentration of ice nucleating particles (INP) by up to 4 orders of magnitude (Lloyd et al., 2015). This has been interpreted as strong evidence in support of secondary ice generation, even outside the window of conditions needed for the Hallett-Mossop process. Subsequent work, however, has shown that blowing snow and ice particles can strongly augment observed ice crystal concentrations (Beck et al., 2018). It is therefore crucial to properly account for the various sources of ice in making surface-based measurements.

Examples of mixed-phase cloud measurements taken with a holographic instrument at the High Altitude Research Station Jungfraujoch (in the Swiss Alps at 3580 m MSL) are shown in Figure 16.11 (Schlenczek et al., 2017). The instrument was configured to have a sample volume of 305 cm^3, and the spatial and size distribution of particles within each hologram "snapshot" can be determined. This is an intriguing capability that allows the spatial partitioning of supercooled liquid and ice to be determined even down to the μm-scale. The top panel of the figure shows the size distributions of supercooled cloud droplets and of small (<50 μm) and large (>50 μm) ice crystals, illustrating the relative sparseness of the ice phase in terms of number concentration. The bottom panel shows example images of hydrometeors, also classified according to the three categories. The middle panel shows a single hologram, in this case revealing that the mixed-phase cloud is indeed mixed down to the smallest scales.

Example: Mountain-top Measurement of Turbulence-microphysics Interactions.

Ground-based measurements have the advantage that clouds can be sampled continuously and with great detail over long periods of time, of course with the caveat that ambient conditions must cooperate in producing steady clouds. The Environmental Research Station on Mount Zugspitze is climatologically prone to steady cloud conditions suitable for investigation of turbulence-microphysics interactions, and Risius et al. (2015) used a "Lumley-triangle" approach to demonstrate that in spite of ground effects and complex topography, the turbulence is nearly as isotropic as typical laboratory flows. Figure 16.12 shows the power spectral density calculated for two different clouds, one averaged over 100 min and another averaged over 40 min. Droplet size distributions and high-resolution turbulence properties were measured simultaneously (Siebert et al., 2015). Besides showing the expected $-5/3$ slope within the inertial subrange of the turbulent air, it is immediately evident that the power spectra exhibit enhanced variance at frequencies above several Hz. Possibly analogous scale breaks have been observed in high-resolution airborne measurements of liquid water content in stratocumulus clouds (Davis et al., 1999; Gerber et al., 2001). It has been suggested that greater variability in cloud thermodynamic and microphysical properties should exist at temporal scales shorter than the phase relaxation time (Lehmann et al., 2009; Mazin, 1999; Siebert & Shaw, 2017) and that this may lead to broadening of the cloud droplet size distribution (e.g., Korolev et al., 2013). Here, the phase relaxation time is a measure of the response time of cloud droplets to a perturbation in water vapor saturation ratio. This follows the qualitative idea that for eddy timescales of the same order as or less than the phase relaxation time, the supersaturation is not able to reach its quasi steady value for a given forcing. Therefore, droplets will grow in an environment different than that typically assumed in steady-state models. A frequency scale can be estimated from the inverse of the phase relaxation time, which is independently calculated from direct measurement of the droplet size distribution with a phase-Doppler interferometer. In Figure 16.12 the phase-relaxation frequency is shown by vertical dashed lines and it is evident that the estimated phase-relaxation frequency matches the scale break quite closely. Using the measured turbulent kinetic energy dissipation rate and Kolmogorov scaling for the inertial subrange, this timescale can be converted to an eddy length scale for phase relaxation, which in the two cases is found to be 1.6 and 3.2 m. A small value of this transitional length scale also implies a higher probability of the inhomogeneous mixing signatures resulting from entrainment events (Lehmann et al., 2009). Similar arguments have been found to have relevance for droplet size broadening measured in the laboratory (Chandrakar et al., 2016) and are likely also for observations in stratocumulus and small cumulus clouds (Desai et al., 2019; Gerber et al., 2001).

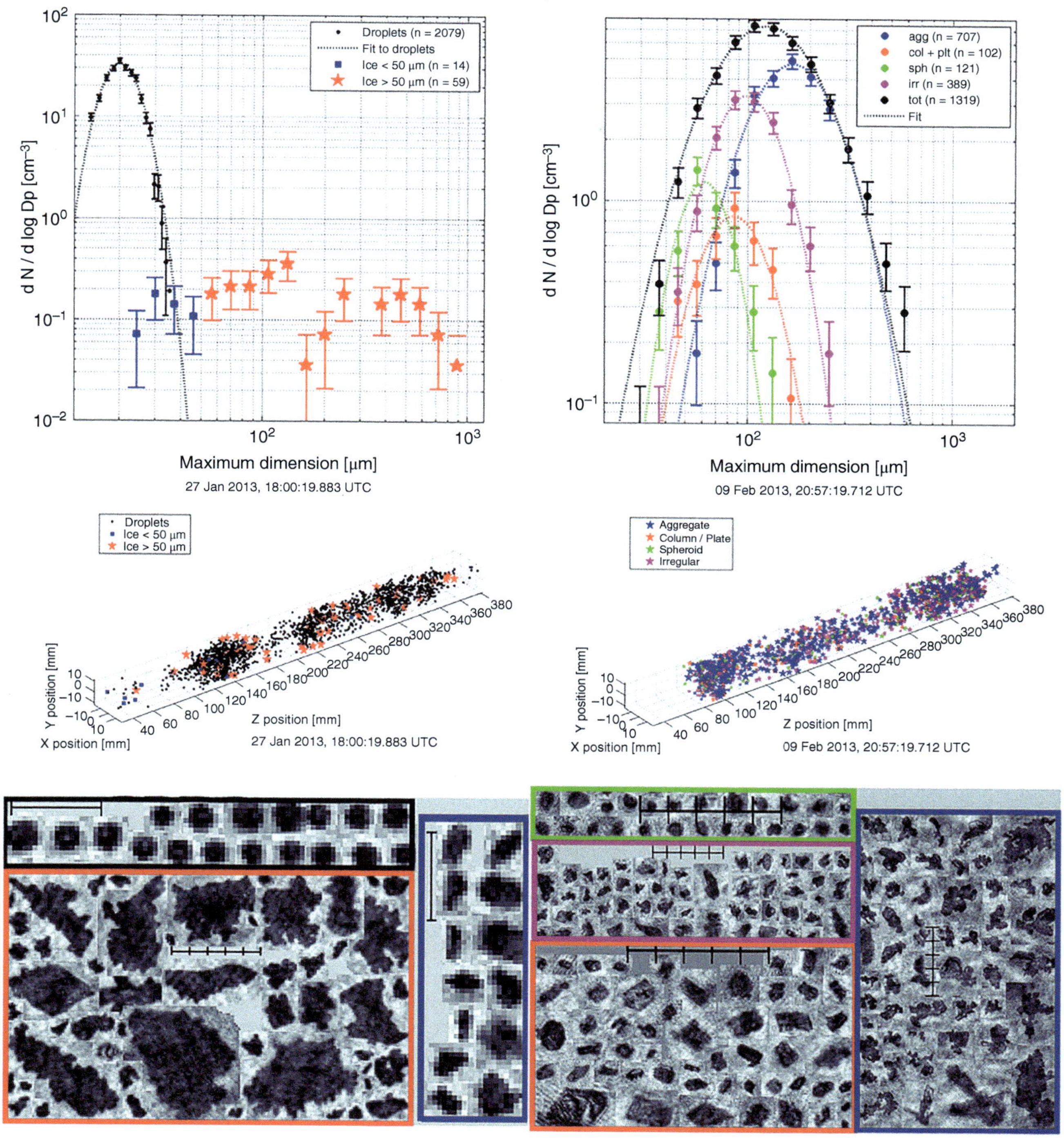

Figure 16.11 Examples of hydrometeors sampled in a mixed-phase cloud at the Jungfraujoch station using the GipfelHalo instrument. The top panel shows the size distribution with contributions from cloud droplets, small ice crystals, and large ice crystals. The middle panel shows the spatial distribution of hydrometeors within a single hologram, with particles identified according to the three categories. The bottom panel shows examples of particles in the three categories: cloud droplets (outlined by a black box), small ice crystals (outlined by a blue box), and large ice crystals (outlined by a red box). Source: The figure is Reproduced from Oliver Schlenczek et al. (2017)/American Meteorological Society.

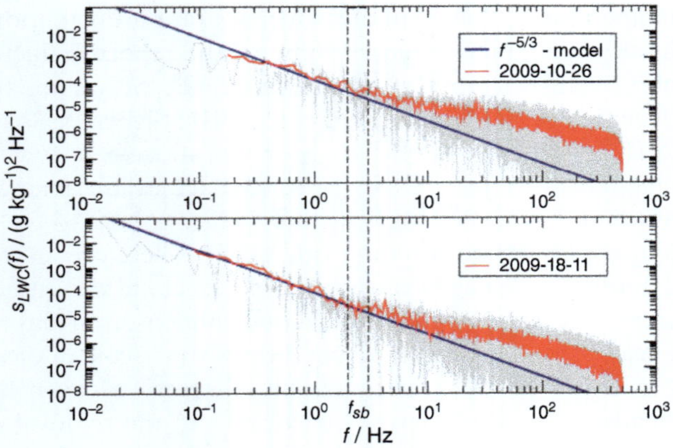

Figure 16.12 Power spectral densities of liquid water content for two cloud time series sampled on mount Zugspitze. The light gray and red lines indicate the raw and averaged spectra, respectively. The blue line show the −5/3 slope expected for inertial subrange scaling. The vertical lines denote the frequency that corresponds to the phase relaxation time estimated from phase-Doppler measurements. Source: The figure is from Siebert et al. (2015)/Copernicus Publications/CC BY 3.0.

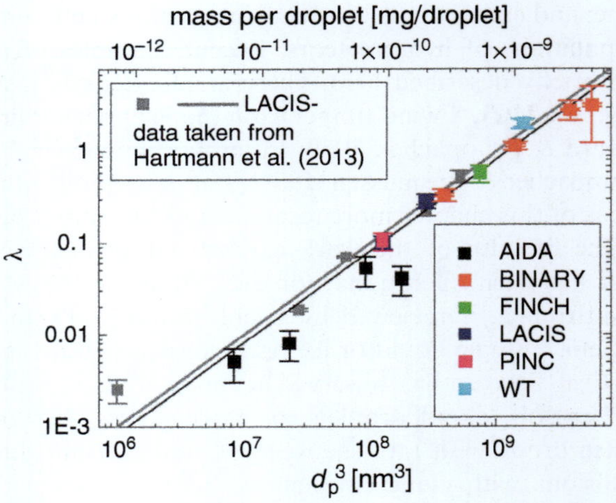

Figure 16.13 The average number of ice-nucleating macromolecules per droplet, versus Snowmax particle mass, as observed with six independent measurement techniques. Source: The figure is adapted from Wex et al. (2015).

16.3.3. Laboratory Methods and Experiments

Controlled laboratory experiments are a unique category of experimental cloud investigation, in the sense that specific mechanisms can be explored under idealized conditions. The scale of laboratory experiments ranges from bench-top investigations of single-particle processes, all the way to large-scale facilities for studying full clouds. Some of the currently existed facilities include the Aerosol Interaction and Dynamics in the Atmosphere (AIDA) chamber, the Manchester Ice Cloud Chamber (MICC), the Meteorological Research Institute (MRI) cloud chamber, the Mainz Vertical Wind Tunnel Facility, the Leipzig Aerosol Cloud Interaction Simulator (LACIS), and the Pi Chamber. Two examples of cloud processes that have been investigated in these facilities are described below.

Example: Ice Nucleation on Biological Particles.

The Leipzig Aerosol Cloud Interaction Simulator (LACIS) is a representative example of a laboratory facility that has been productive in the investigation of aerosol activation and ice nucleation studies (Niedermeier et al., 2015; Wex et al., 2007). The system operates on the principle of differential diffusion of water vapor and heat within the transition region between sections of different temperature in a long flow tube (Stratmann et al., 2004; Wex et al., 2006). Aerosol particles are carried along the centerline of the tube and are exposed to precisely controlled profiles of water vapor supersaturation. At temperatures below the ice melting point, the boundary conditions enable ice nucleation to be studied.

As an example of the results and insight that have come out of LACIS, we refer to the findings of Hartmann et al. (2013) who found that supercooled water containing biological materials freezes as a result of discrete ice-nucleating macromolecules. Figure 16.13 shows results from an intercomparison study using results from LACIS, from the Aerosol Interaction and Dynamics in the Atmosphere (AIDA) cloud chamber, and from several other ice nucleation instruments (Wex et al., 2015). The figure shows the average number of ice-nucleating macromolecules per droplet as determined from the plateau formed in the fraction of frozen droplets as temperature is decreased. This is plotted versus particle mass per droplet, and the agreement between six quite different measurement methods can be considered very encouraging.

Example: Wind-tunnel Investigations of Cloud Hydrometeor Collision Rates.

Conversion of cloud liquid water to precipitation plays a substantial role in cloud dynamics, as well as in controlling water mass fluxes in the atmosphere. This conversion crucially depends on the rate at which cloud droplets collide with each other, and the rate at which cloud droplets are collected by larger hydrometeors, both by collision and coalescence with liquid hydrometeors and by riming of ice hydrometeors. Wind tunnels in their various configurations have been very productive in cloud physics and chemistry research, especially in the study of particle collision rates (or the quantities that contribute to the collision rate, such as relative velocity of hydrometeors). Two main approaches have been taken: experiments involving a single hydrometeor held in a vertical wind tunnel for an extended period of growth time: and experiments involving interactions between full populations of hydrometeors. Recent examples of both are briefly described here. There is an extensive history (e.g., the UCLA wind tunnel from the Pruppacher group (Beard & Pruppacher, 1969; Pitter & Pruppacher, 1973; Pruppacher & Rasmussen, 1979), but in keeping with the focus of this chapter more recent results are emphasized.

The breadth of the work carried out in one facility that has been operational for over 30 years, the Mainz wind tunnel, is reviewed by Diehl et al. (2011). In this system, a single liquid or ice particle is suspended in the vertical flow and is observed to grow as it is exposed to aerosols, cloud droplets, or trace gases. Growth of water drops with radii between 70 and 170 μm due to collisions with cloud droplets with radii ranging from 1 to 7 μm has been studied, leading to empirical collision kernels (Vohl et al., 2007). That work has been extended to riming of ice spheres in a similar size range (von Blohn et al., 2009). Riming collection kernels were parameterized as power-law functions of collector momentum, showing a monotonic increase over nearly 2 decades of momentum. Turbulence generated by flow blockages, with kinetic energy dissipation rates of 0.01–0.07 W kg^{-1} led to enhancements of 10%–20% above the laminar values for collection of cloud droplets by collector drops (Vohl et al., 1999).

Recent wind tunnel studies of populations of particles have focused on turbulence processes that contribute to the collision kernel, such as inertial droplet clustering (Saw et al., 2008, 2012) and droplet relative velocity (Bewley et al., 2013; Saw et al., 2014). Although it is experimentally quite challenging, there have been successful measurements of droplet collisions in turbulent flow, leading to constraints of the various theoretical expressions that have been provided in the literature (Bordás et al., 2013).

Example: Laboratory Investigation of Turbulence-microphysics Interactions.

The Pi chamber can create conditions for cloud formation using two different principles: turbulent moist convection (Rayleigh-Bénard between differentially heated, water-saturated surfaces) and adiabatic expansion (Chang et al., 2016). The most commonly used mode, turbulent convection, is similar to the convection in atmospheric boundary-layer where the buoyancy different due to heating at the bottom and cooling at the top drives the convective mixing and turbulence. The turbulent kinetic energy dissipation rate generated through this process is around 10^{-3} m^2 s^{-3}, which is similar to observed values in stratocumulus clouds. However, the Reynolds number is lower, $Re_\lambda \approx 55$ due to the much smaller length scales. To create a supersaturated environment, the top and bottom boundaries have kept saturated with water at different temperatures. The resultant supersaturation fluctuations have similar magnitude as some of the reported fluctuation statistics from atmospheric clouds, with standard deviation ≈1%, (Chandrakar et al., 2016; Siebert & Shaw, 2017). Moreover, the mean supersaturation for a typical chamber configuration is <1%. The achievable temperature and pressure range in this facility is −50 to 50 °C and 60–1,000 hPa (Chang et al., 2016). Therefore, in this chamber, warm as well as mixed-phase cloud conditions can be created using different cloud condensation and ice nuclei. Long, steady cloud conditions (timescale of days) can be achieved by maintaining a balance between the steady injection of aerosol particles and the removal of cloud droplets due to settling. It is also possible to perform transient cloud experiments (Chandrakar et al., 2017).

An example of steady-state moist and cloudy convection in the Pi chamber is given in Figure 16.14. It shows steady-state thermodynamic properties, turbulent flow, and cloud properties during moist condition, and the transition to cloudy conditions when aerosol injection is started at approximately 9 hours. Among different scientific contributions, the Pi-chamber results have provided insight into the process of stochastic condensation. It has been found that the aerosol concentration not only affects the average size of cloud droplets but also influences the size distribution width (Chandrakar et al., 2016). Measurements suggest that supersaturation fluctuations due to the turbulent convection broaden the width of droplet size distribution, and aerosol concentration modulates this broadening process similar to the mean size growth. This phenomenon of turbulence-induced aerosol effects has been further studied with controlled, steady-state experiments in the Pi chamber and has provided insight into the role of microphysical variability (Desai et al., 2018) and the dispersion indirect effect (Chandrakar et al., 2018).

16.4. SUMMARY AND FUTURE DIRECTIONS

We close by sharing a thought from Wyngaard (1998) regarding advances in measurement versus modeling in the field of atmospheric boundary layers, but which we feel should at least be considered in the field of cloud physics: "Geophysical observations have not kept pace with the model predictions, nor could they have; modeling and observational work have cruelly different time scales." We agree about the comment on the different timescales of computational versus experimental work, and indeed the asymmetry can seem cruel at times. Nevertheless, our sense is that in the field of cloud physics the progress on in situ observations has been impressive over the recent decades. The space constraints of this chapter do not allow for a thorough review of the many impressive advances and findings, but we have attempted to provide a representative sample. Airborne measurement

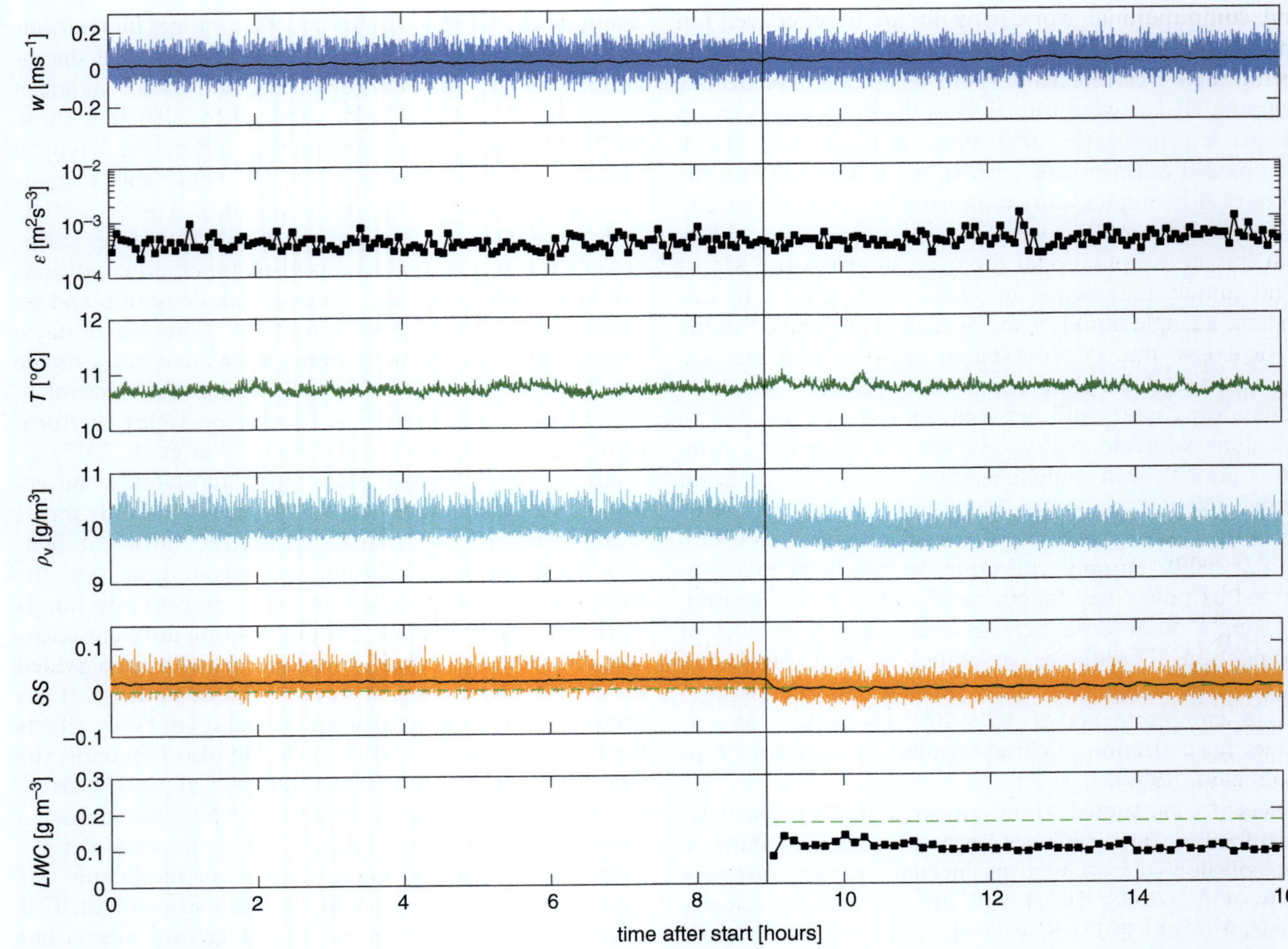

Figure 16.14 Steady-state moist and cloudy conditions: measurements of vertical velocity fluctuations (w), kinetic energy dissipation rate (ϵ), temperature (T), water vapor density (ρ_v), and supersaturation (ss) during moist convection near the center of the chamber. The vertical black line indicates the point where aerosol injection started, and the last panel shows the formation of cloud liquid water content (*LWC*) after that point. Source: The figure is from Niedermeier et al. (2018)/American Physical Society.

capabilities are stronger than ever, and the development of helicopter-borne and UAV measurements are examples of the creative energy in that area. Ground-based measurement capabilities also have continued to develop, with impressive facilities present on many continents and innovative measurement strategies emerging.

There has been a sentiment that laboratory experiments have lagged compared to field measurement capabilities. This is being corrected as of late, with several new and innovative facilities having been developed. But it is the opinion of these authors that much more can be accomplished with well-crafted laboratory investigations of specific mechanisms. We offer this prescient quote from Gunn and Phillips (1957) made in the very early days of cloud physics, referring to the measurements described in the introductory section of this chapter: "We are of the opinion that an understanding of the basic mechanics of precipitation processes can best be developed by bringing these processes into the laboratory, where physical conditions are fully known or can be controlled and measured. The results outlined above (i.e., in their 1957 paper) would have been long delayed if it had been necessary for the processes to be studied and developed against the background of complex variations commonly observed in the free atmosphere." Their view resonates with us: the laboratory is a powerful venue for scientific discovery because it allows processes to be studied in isolation, to be distilled to their most essential elements so that simple models can be developed. Laboratory experiments do not contain all of the moving and interconnected parts existing in the atmosphere—that is, in fact, the very point and therefore need to be combined with field observation and computational work in order to be effective. The approaches are complementary and we look forward to seeing them continue to develop together in the coming years.

ACKNOWLEDGMENTS

We are grateful to Patrick Chuang for helpful discussions about topics of interest for this chapter. We acknowledge support from National Science Foundation Grant AGS-1754244 and NASA Earth and Space Science Fellowship Program Grant 80NSSC17K0449.

REFERENCES

Abdelmonem, A., Schnaiter, M., Amsler, P., Hesse, E., Meyer, J., & Leisner, T. (2011). First correlated measurements of the shape and light scattering properties of cloud particles using the new Particle Habit Imaging and Polar Scattering (PHIPS) probe. *Atmospheric Measurement Techniques*, *4*(10), 2125–2142.

Albrecht, B. A. (1989). Aerosols, cloud microphysics, and fractional cloudiness. *Science*, *245*(4923), 1227–1230.

Andronache, C. (2017). *Mixed-Phase Clouds: Observations and Modeling*. Netherlands, UK, USA: Elsevier.

Arabas, S., Pawlowska, H., & Grabowski, W. (2009). Effective radius and droplet spectral width from in-situ aircraft observations in trade-wind cumuli during RICO. *Geophysical Research Letters*, *36*(11), L11803.

Baker, M., Corbin, R., & Latham, J. (1980). The influence of entrainment on the evolution of cloud droplet spectra: I. A model of inhomogeneous mixing. *Quarterly Journal of the Royal Meteorological Society*, *106*(449), 581–598.

Baumgardner, D., Brenguier, J., Bucholtz, A., Coe, H., DeMott, P., Garrett, T., et al. (2011). Airborne instruments to measure atmospheric aerosol particles, clouds and radiation: A cook"s tour of mature and emerging technology. *Atmospheric Research*, *102*(1–2), 10–29.

Baumgardner, D., & Spowart, M. (1990). Evaluation of the forward scattering spectrometer probe. Part III: Time response and laser inhomogeneity limitations. *Journal of Atmospheric and Oceanic Technology*, *7*(5), 666–672.

Baustian, K. J., Cziczo, D. J., Wise, M. E., Pratt, K. A., Kulkarni, G., Hallar, A., & Tolbert, M. A. (2012). Importance of aerosol composition, mixing state, and morphology for heterogeneous ice nucleation: A combined field and laboratory approach. *Journal of Geophysical Research: Atmospheres*, *117*(D6), D06217.

Beals, M. J., Fugal, J. P., Shaw, R. A., Lu, J., Spuler, S. M., & Stith, J. L. (2015). Holographic measurements of inhomogeneous cloud mixing at the centimeter scale. *Science*, *350*(6256), 87–90.

Beard, K. V. (1992). Ice initiation in warm-base convective clouds: An assessment of microphysical mechanisms. *Atmospheric Research*, *28*(2), 125–152.

Beard, K. V., & Ochs, H. T., III (1993). Warm-rain initiation: An overview of microphysical mechanisms. *Journal of Applied Meteorology*, *32*(4), 608–625.

Beard, K., & Pruppacher, H. (1969). A determination of the terminal velocity and drag of small water drops by means of a wind tunnel. *Journal of the Atmospheric Sciences*, *26*(5), 1066–1072.

Beck, A., Henneberger, J., Fugal, J. P., David, R. O., Lacher, L., Lohmann, U., & Möhler, O. (2018). Impact of surface and near-surface processes on ice crystal concentrations measured at mountain-top research stations. *Atmospheric Chemistry & Physics*, *18*(12), 8909–8927.

Bewley, G. P., Saw, E.-W., & Bodenschatz, E. (2013). Observation of the sling effect. *New Journal of Physics*, *15*(8), 083,051.

Bewley, J. L., & Lasher-Trapp, S. (2011). Progress on predicting the breadth of droplet size distributions observed in small cumuli. *Journal of the Atmospheric Sciences*, *68*(12), 2921–2929.

Blyth, A. M., Cooper, W. A., & Jensen, J. B. (1988). A study of the source of entrained air in Montana cumuli. *Journal of the Atmospheric Sciences*, *45*(24), 3944–3964.

Blyth, A. M., & Latham, J. (1993). Development of ice and precipitation in New Mexican summertime cumulus clouds. *Quarterly Journal of the Royal Meteorological Society*, *119*(509), 91–120.

Blyth, A. M., Lowenstein, J. H., Huang, Y., Cui, Z., Davies, S., & Carslaw, K. S. (2013). The production of warm rain in shallow maritime cumulus clouds. *Quarterly Journal of the Royal Meteorological Society*, *139*(670), 20–31.

Bordás, R., Roloff, C., Thévenin, D., & Shaw, R. (2013). Experimental determination of droplet collision rates in turbulence. *New Journal of Physics*, *15*(4), 045,010.

Brenguier, J.-L., Bachalo, W. D., Chuang, P. Y., Esposito, B. M., Fugal, J., Garrett, T., Gayet, J.-F., et al. (2013). In situ measurements of cloud and precipitation particles. In M. Wendisch and J.-L. Brenguier (Eds.), *Airborne measurements for environmental research: Methods and instruments* (chap. 5, pp. 225–301). Weinheim, Germany: Wiley-VCH.

Burnet, F., & Brenguier, J.-L. (2007). Observational study of the entrainment-mixing process in warm convective clouds. *Journal of the Atmospheric Sciences*, *64*(6), 1995–2011.

Cantrell, W., & Heymsfield, A. (2005). Production of ice in tropospheric clouds: A review. *Bulletin of the American Meteorological Society*, *86*(6), 795–808.

Chandrakar, K., Cantrell, W., Ciochetto, D., Karki, S., Kinney, G., & Shaw, R. (2017). Aerosol removal and cloud collapse accelerated by supersaturation fluctuations in turbulence. *Geophysical Research Letters*, *44*(9), 4359–4367.

Chandrakar, K., Cantrell, W., Kostinski, A., & Shaw, R. (2018). Dispersion aerosol indirect effect in turbulent clouds: Laboratory measurements of effective radius. *Geophysical Research Letters*, *45*(19), 10–738.

Chandrakar, K. K., Cantrell, W., Chang, K., Ciochetto, D., Niedermeier, D., Ovchinnikov, M., et al. (2016). Aerosol indirect effect from turbulence-induced broadening of cloud-droplet size distributions. *Proceedings of the National Academy of Sciences*, *113*(50), 14,243–14,248.

Chang, K., Bench, J., Brege, M., Cantrell, W., Chandrakar, K., Ciochetto, D., et al. (2016). A laboratory facility to study gas-aerosol-cloud interactions in a turbulent environment: The Π Chamber. *Bulletin of the American Meteorological Society*, pp. BAMS–D–15–00,203.1.

China, S., Scarnato, B., Owen, R. C., Zhang, B., Ampadu, M. T., Kumar, S., et al. (2015). Morphology and mixing state of aged soot particles at a remote marine free troposphere site: Implications for optical properties. *Geophysical Research Letters*, *42*(4), 1243–1250.

Cooper, W. A. (1989). Effects of variable droplet growth histories on droplet size distributions. Part I: Theory. *Journal of the Atmospheric Sciences*, *46*(10), 1301– 1311.

Cooper, W. A., Lasher-Trapp, S. G., & Blyth, A. M. (2013). The influence of entrainment and mixing on the initial formation of rain in a warm cumulus cloud. *Journal of the Atmospheric Sciences*, *70*(6), 1727–1743.

Coopman, Q., Riedi, J., Zeng, S., & Garrett, T. (2020). Space-based analysis of the cloud thermodynamic phase transition for varying microphysical and meteorological regimes. *Geophysical Research Letters*, *47*(6), e2020GL087,122.

Davis, A. B., Marshak, A., Gerber, H., & Wiscombe, W. J. (1999). Horizontal structure of marine boundary layer clouds from centimeter to kilometer scales. *Journal of Geophysical Research: Atmospheres*, *104*(D6), 6123–6144.

de Boer, G., Ivey, M., Schmid, B., Lawrence, D., Dexheimer, D., Mei, F., et al. (2018). A birds-eye view: Development of an operational arm unmanned aerial capability for atmospheric research in Arctic Alaska. *Bulletin of the American Meteorological Society*, *99*(6), 1197–1212.

de Lozar, A., & Mellado, J. P. (2017). Reduction of the entrainment velocity by cloud droplet sedimentation in stratocumulus. *Journal of the Atmospheric Sciences*, *74*(3), 751–765.

De Rooy, W. C., Bechtold, P., Fröhlich, K., Hohenegger, C., Jonker, H., Mironov, D., et al. (2013). Entrainment and detrainment in cumulus convection: An overview. *Quarterly Journal of the Royal Meteorological Society*, *139*(670), 1–19.

De Rooy, W. C., & Siebesma, A. P. (2008). A simple parameterization for detrainment in shallow cumulus. *Monthly Weather Review*, *136*(2), 560–576.

Deardorff, J. W. (1970). Preliminary results from numerical integrations of the unstable planetary boundary layer. *Journal of the Atmospheric Sciences*, *27*(8), 1209–1211.

Deardorff, J. W., Willis, G. E., & Lilly, D. K. (1969). Laboratory investigation of non-steady penetrative convection. *Journal of Fluid Mechanics*, *35*(1), 7–31.

Desai, N., Chandrakar, K., Chang, K., Cantrell, W., & Shaw, R. (2018). Influence of microphysical variability on stochastic condensation in a turbulent laboratory cloud. *Journal of the Atmospheric Sciences*, *75*(1), 189–201.

Desai, N., Glienke, S., Fugal, J., & Shaw, R. (2019). Search for microphysical signatures of stochastic condensation in marine boundary layer clouds using airborne digital holography. *Journal of Geophysical Research: Atmospheres*, *124*, 2739–2752.

Diehl, K., Mitra, S. K., Szakáll, M., von Blohn, N., Borrmann, S., and Pruppacher, H. R. (2011). The Mainz vertical wind tunnel facility—A review of 25 years of laboratory experiments on cloud physics and chemistry. In J. D. Pereira (Ed.), *Wind Tunnels: Aero-dynamics, models and experiments* (chap. 2, pp. 69–91). Nova Science Publishers.

Faloona, I., Lenschow, D. H., Campos, T., Stevens, B., Van Zanten, M., Blomquist, B., et al. (2005). Observations of entrainment in eastern pacific marine stratocumulus using three conserved scalars. *Journal of the Atmospheric Sciences*, *62*(9), 3268–3285.

Field, P. R., Lawson, R. P., Brown, P. R., Lloyd, G., Westbrook, C., Moisseev, D., et al. (2017). Secondary ice production: Current state of the science and recommendations for the future. *Meteorological Monographs*, *58*, 7–1.

Fugal, J., & R. Shaw (2009). Cloud particle size distributions measured with an airborne digital in-line holographic instrument. *Atmospheric Measurement Techniques*, *2*(1), 259–271.

Gerber, H. (1980). A saturation hygrometer for the measurement of relative humidity between 95 and 105%. *Journal of Applied Meteorology*, *19*(10), 1196–1208.

Gerber, H. (1991). Supersaturation and droplet spectral evolution in fog. *Journal of the Atmospheric Sciences*, *48*(24), 2569–2588.

Gerber, H. (1994). Hygrometer measurement of relative humidity in clouds. *NSF SBIR Final Report*, *9261243*, 1–19.

Gerber, H. (1996). Microphysics of marine stratocumulus clouds with two drizzle modes. *Journal of the Atmospheric Sciences*, *53*(12), 1649–1662.

Gerber, H., Arends, B., & Ackerman, A. (1994). New microphysics sensor for air- craft use. *Atmospheric Research*, *31*(4), 235–252.

Gerber, H., & Frick, G. (2012). Drizzle rates and large sea-salt nuclei in small cumulus. *Journal of Geophysical Research: Atmospheres*, *117*(D01205), doi:10.1029/2011JD016,249.

Gerber, H., Frick, G., Malinowski, S. P., Jonsson, H., Khelif, D., & Krueger, S. K. (2013). Entrainment rates and microphysics in post stratocumulus. *Journal of Geophysical Research: Atmospheres*, *118*(21), 12,094–12,109.

Gerber, H., Jensen, J., Davis, A., Marshak, A., & Wiscombe, W. (2001). Spectral density of cloud liquid water content at high frequencies. *Journal of the Atmospheric Sciences*, *58*(5), 497–503.

Gerber, H., Malinowski, S. P., & Jonsson, H. (2016). Evaporative and radiative cooling in post stratocumulus. *Journal of the Atmospheric Sciences*, *73*(10), 3877–3884.

Ghate, V. P., Mechem, D. B., Cadeddu, M. P., Eloranta, E. W., Jensen, M. P., Nordeen, M. L., & Smith, W. L., Jr. (2019). Estimates of entrainment in closed cellular marine stratocumulus clouds from the magic field campaign. *Quarterly Journal of the Royal Meteorological Society, 145*, 1589–1602.

Glienke, S., Kostinski, A., Fugal, J., Shaw, R., Borrmann, S., & Stith, J. (2017). Cloud droplets to drizzle: Contribution of transition drops to microphysical and optical properties of marine stratocumulus clouds. *Geophysical Research Letters*, *44*(15), 8002–8010.

Grabowski, W. W., & Abade, G. C. (2017). Broadening of cloud droplet spectra through eddy hopping: Turbulent adiabatic parcel simulations. *Journal of the Atmospheric Sciences*, *74*(5), 1485–1493.

Grabowski, W. W., & Wang, L.-P. (2013). Growth of cloud droplets in a turbulent environment. *Annual Review of Fluid Mechanics*, *45*, 293–324.

Gunn, R., & Phillips, B. (1957). An experimental investigation of the effect of air pollution on the initiation of rain. *Journal of Meteorology*, *14*(3), 272–280.

Hallar, A. G., Lowenthal, D. H., Clegg, S. L., Samburova, V., Taylor, N., Mazzoleni, L. R., et al. (2013). Chemical and hygroscopic properties of aerosol organics at storm peak laboratory. *Journal of Geophysical Research: Atmospheres*, *118*(10), 4767–4779.

Hansen, J. E., & Travis, L. D. (1974). Light scattering in planetary atmospheres. *Space Science Reviews*, *16*(4), 527–610.

Harris-Hobbs, R. L., & Cooper, W. A. (1987). Field evidence supporting quantitative predictions of secondary ice production rates. *Journal of the Atmospheric Sciences*, *44*(7), 1071–1082.

Hartmann, S., Augustin, S., Clauss, T., Wex, H., Šantl-Temkiv, T., Voigtländer, J., et al. (2013). Immersion freezing of ice nucleation active protein complexes. *Atmospheric Chemistry and Physics*, *13*(11), 5751–5766.

Heus, T., Pols, C. F. J., Jonker, H. J., Van den Akker, H. E., & Lenschow, D. H. (2009). Observational validation of the compensating mass flux through the shell around cumulus clouds. *Quarterly Journal of the Royal Meteorological Society*, *135*(638), 101–112.

Heymsfield, A. J., Bansemer, A., Field, P. R., Durden, S. L., Stith, J. L., Dye, J. E. et al. (2002). Observations and parameterizations of particle size distributions in deep tropical cirrus and stratiform precipitating clouds: Results from in situ observations in TRMM field campaigns, *Journal of the Atmospheric Sciences*, *59*(24), 3457–3491.

Hill, A. A., Feingold, G., & Jiang, H. (2009). The influence of entrainment and mixing assumption on aerosol–cloud interactions in marine stratocumulus. *Journal of the Atmospheric Sciences*, *66*(5), 1450–1464.

Hobbs, P. V., & Rangno, A. L. (1985). Ice particle concentrations in clouds, *Journal of the Atmospheric Sciences*, *42*(23), 2523–2549.

Hudson, J. G., Noble, S., & Jha, V. (2012). Cloud droplet spectral width relationship to ccn spectra and vertical velocity. *Journal of Geophysical Research: Atmospheres*, *117*, D11211. doi:10.1029/2012JD017,546.

Hudson, J. G., & Yum, S. S. (1997). Droplet spectral broadening in marine stratus. *Journal of the Atmospheric Sciences*, *54*(22), 2642–2654.

Jackson, R. C., McFarquhar, G. M., Stith, J., Beals, M., Shaw, R. A., Jensen, J., et al. (2014). An assessment of the impact of antishattering tips and artifact removal techniques on cloud ice size distributions measured by the 2D cloud probe, *Journal of Atmospheric and Oceanic Technology*, *31*(12), 2567–2590.

Jen-La Plante, I., Ma, Y., Nurowska, K., Gerber, H., Khelif, D., Karpinska, K., et al. (2016). Physics of stratocumulus top (post): Turbulence characteristics. *Atmospheric Chemistry and Physics*, *16*(15), 9711–9725.

Jensen, J. B., & Nugent, A. D. (2017). Condensational growth of drops formed on giant sea-salt aerosol particles. *Journal of the Atmospheric Sciences*, *74*(3), 679–697.

Jensen, M. P., & Del Genio, A. D. (2006). Factors limiting convective cloud-top height at the ARM Nauru island climate research facility. *Journal of Climate*, *19*(10), 2105–2117.

Johnson, D. B. (1982). The role of giant and ultragiant aerosol particles in warm rain initiation. *Journal of the Atmospheric Sciences*, *39*(2), 448–460.

Kaimal, J., & Wyngaard, J. (1990). The Kansas and Minnesota experiments. *Boundary-Layer Meteorology*, *50*(1–4), 31–47.

Kain, J. S., & Fritsch, J. M. (1990). A one-dimensional entraining/detraining plume model and its application in convective parameterization. *Journal of the Atmospheric Sciences*, *47*(23), 2784–2802.

Kain, J. S., & Fritsch, J. M. (1993). Convective parameterization for mesoscale models: The Kain-Fritsch scheme. In K.A. Emanuel & D.J. Raymond (Eds.), *The representation of cumulus convection in numerical models. Meteorological Monographs*. Boston, MA: American Meteorological Society. https://doi.org/10.1007/978-1-935704-13-3_16

Kanji, Z. A., Ladino, L. A., Wex, H., Boose, Y., Burkert-Kohn, M., Cziczo, D. J., & Krämer, M. (2017). Overview of ice nucleating particles. *Meteorological Monographs*, *58*, 1.1–1.33. https://doi.org/10.1175/AMSMONOGRAPHS-D-16-0006.1.

Katzwinkel, J., Siebert, H., Heus, T., & Shaw, R. A. (2014). Measurements of turbulent mixing and subsiding shells in trade wind cumuli. *Journal of the Atmospheric Sciences*, *71*(8), 2810–2822.

Katzwinkel, J., Siebert, H., & Shaw, R. (2012). Observation of a self-limiting, shear-induced turbulent inversion layer above marine stratocumulus. *Boundary-Layer Meteorology*, *145*(1), 131–143.

Khanna, S. (1998). Comparison of Kansas data with high-resolution Large-Eddy simulation fields. *Boundary-Layer Meteorology*, *88*(1), 121–144.

Klocke, D., Pincus, R., & Quaas, J. (2011). On constraining estimates of climate sensitivity with present-day observations through model weighting. *Journal of Climate*, *24*(23), 6092–6099.

Korolev, A., Emery, E., Strapp, J., Cober, S., Isaac, G., Wasey, M., & Marcotte, D. (2011). Small ice particles in tropospheric clouds: Fact or artifact? Airborne icing instrumentation evaluation experiment. *Bulletin of the American Meteorological Society*, *92*(8), 967–973.

Korolev, A., & Leisner, T. (2020). Review of experimental studies on secondary ice production. *Atmospheric Chemistry and Physics Discussions*, *20*, 11767–11797. https://doi.org/10.5194/acp–2020–537.

Korolev, A., McFarquhar, G., Field, P. R., Franklin, C., Lawson, P., Wang, Z., et al. (2017). Mixed-phase clouds: Progress and challenges, *Meteorological Monographs*, *58*, 5–1.

Korolev, A., Pinsky, M., & Khain, A. (2013). A new mechanism of droplet size distribution broadening during diffusional growth. *Journal of the Atmospheric Sciences*, *70*(7), 2051–2071.

Korolev, A. V. (1995). The influence of supersaturation fluctuations on droplet size spectra formation. *Journal of the Atmospheric Sciences*, *52*(20), 3620–3634.

Kuan-Ting, O., Wood, R., & Bretherton, C. S. (2018). Ultraclean layers and optically thin clouds in the stratocumulus-to-cumulus transition. Part II: Depletion of cloud droplets and cloud condensation nuclei through collision–coalescence. *Journal of the Atmospheric Sciences*, *75*(5), 1653–1673.

Lance, S., Shupe, M. D., Feingold, G., Brock, C. A., Cozic, J., Holloway, J. S., et al. (2011). Cloud condensation nuclei as a modulator of ice processes in arctic mixed-phase clouds. *Atmospheric Chemistry and Physics*, *11*(15), 8003–8015. doi:10.5194/acp-11-8003-2011.

Larsen, M. L., Shaw, R. A., Kostinski, A. B., & Glienke, S. (2018). Fine-scale droplet clustering in atmospheric clouds: 3d radial distribution function from airborne digital holography. *Physical Review Letters*, *121*(20), 204,501.

Lasher-trapp, S. G., Cooper, W. A., & Blyth, A. M. (2005). Broadening of droplet size distributions from entrainment and mixing in a cumulus cloud. *Quarterly Journal of the Royal Meteorological Society*, *131*(605), 195–220.

Lauber, A., Kiselev, A., Pander, T., Handmann, P., & Leisner, T. (2018). Secondary ice formation during freezing of levitated droplets. *Journal of the Atmospheric Sciences*, *75*(8), 2815–2826.

Leena, P., Anilkumar, V., Sravanthi, N., Patil, R., Chakravarty, K., Saha, S., & G. Pandithurai (2018). On the precipitation susceptibility of monsoon clouds to aerosols using high-altitude ground-based observations over Western Ghats, India. *Atmospheric Environment*, *185*, 128–136.

Lehmann, K., Siebert, H., & Shaw, R. A. (2009). Homogeneous and inhomogeneous mixing in cumulus clouds: Dependence on local turbulence structure. *Journal of the Atmospheric Sciences*, *66*(12), 3641–3659.

Lenschow, D. H., Krummel, P. B., & Siems, S. T. (1999). Measuring entrainment, divergence, and vorticity on the mesoscale from aircraft. *Journal of Atmospheric and Oceanic Technology*, *16*(10), 1384–1400.

Lenschow, D. H., Zhou, M., Zeng, X., Chen, L., & Xu, X. (2000). Measurements of fine-scale structure at the top of marine stratocumulus. *Boundary-Layer Meteorology*, *97*(2), 331–357.

Lilly, D. K. (1968). Models of cloud-topped mixed layers under a strong inversion. *Quarterly Journal of the Royal Meteorological Society*, *94*(401), 292–309.

Liu, Y., & Daum, P. H. (2002). Anthropogenic aerosols: Indirect warming effect from dispersion forcing. *Nature*, *419*(6907), 580–581.

Liu, Y., & Daum, P. H. (2004). Parameterization of the autoconversion process. Part I: Analytical formulation of the Kessler-type parameterizations. *Journal of the Atmospheric Sciences*, *61*(13), 1539–1548.

Liu, Y., Daum, P. H., & Hallett, J. (2002). A generalized systems theory for the effect of varying fluctuations on cloud droplet size distributions. *Journal of the Atmospheric Sciences*, *59*(14), 2279–2290.

Liu, Y., & Hallett, J. (1998). On size distributions of cloud droplets growing by condensation: A new conceptual model. *Journal of the Atmospheric Sciences*, *55*(4), 527–536.

Lloyd, G., Choularton, T. W., Bower, K. N., Gallagher, M. W., Connolly, P. J., Flynn, M., et al. (2015). The origins of ice crystals measured in mixed-phase clouds at the high-alpine site Jungfraujoch. *Atmospheric Chemistry and Physics*, *15*(22), 12,953–12,969.

Lohmann, U., Henneberger, J., Henneberg, O., Fugal, J., Bühl, J., & Kanji, Z. A. (2016). Persistence of orographic mixed-phase clouds. *Geophysical Research Letters*, *43*(19), 10–512.

Long, A. B. (1974). Solutions to the droplet collection equation for polynomial kernels. *Journal of the Atmospheric Sciences*, *31*(4), 1040–1052.

Lowenstein, J. H., Blyth, A. M., & Lawson, R. P. (2010). Early evolution of the largest-sized droplets in maritime cumulus clouds. *Quarterly Journal of the Royal Meteorological Society: A Journal of the Atmospheric Sciences, Applied Meteorology and Physical Oceanography*, *136*(648), 708–717.

Lu, C., Liu, Y., Niu, S., Krueger, S., & Wagner, T. (2013b). Exploring parameterization for turbulent entrainment-mixing processes in clouds. *Journal of Geophysical Research: Atmospheres*, *118*(1), 185–194.

Lu, C., Liu, Y., Niu, S., & Vogelmann, A. M. (2012). Observed impacts of vertical velocity on cloud microphysics and implications for aerosol indirect effects. *Geophysical Research Letters*, *39*(21), L21808.

Lu, C., Liu, Y., Zhang, G. J., Wu, X., Endo, S., Cao, L., et al. (2016). Improving parameterization of entrainment rate for shallow convection with aircraft measurements and Large-Eddy simulation. *Journal of the Atmospheric Sciences*, *73* (2), 761–773.

Lu, C., Niu, S., Liu, Y., & Vogelmann, A. M. (2013a). Empirical relationship between entrainment rate and microphysics in cumulus clouds. *Geophysical Research Letters*, *40*(10), 2333–2338. doi:10.1002/grl.50445.

Lu, M., Feingold, G., Jonsson, H. H., Chuang, P. Y., Gates, H., Flagan, R. C., & Seinfeld, J. H. (2008). Aerosolcloud relationships in continental shallow cumulus. *Journal of Geophysical Research: Atmospheres (1984–2012)*, *113*(D15). doi: 10.1029/2007JD009354.

Lu, M.-L., Conant, W. C., Jonsson, H. H., Varutbangkul, V., Flagan, R. C., & Seinfeld, J. H. (2007). The Marine Stratus/Stratocumulus Experiment (MASE): Aerosol-cloud relationships in marine stratocumulus. *Journal of Geophysical Research*, *112*(D10), D10209.

Malinowski, S. P., Gerber, H., Plante, J.-L., Kopec, M., Kumala, W., Nurowska, K., et al. (2013). Physics of stratocumulus top (post): Turbulent mixing across capping inversion. *Atmospheric Chemistry and Physics*, *13*(24), 12,171–12,186.

Martin, G., Johnson, D., & Spice, A. (1994). The measurement and parameterization of effective radius of droplets in warm stratocumulus clouds. *Journal of the Atmospheric Sciences*, *51*(13), 1823–1842.

Mazin, I. (1999). The effect of condensation and evaporation on turbulence in clouds. *Atmospheric Research*, *51*, 171–174.

Mellado, J. P. (2017). Cloud-top entrainment in stratocumulus clouds. *Annual Review of Fluid Mechanics*, *49*, 145–169.

Miles, N. L., Verlinde, J., & Clothiaux, E. E. (2000). Cloud droplet size distributions in low-level stratiform clouds. *Journal of the Atmospheric Sciences*, *57*(2), 295–311.

Mossop, S. (1978). The influence of drop size distribution on the production of secondary ice particles during graupel growth. *Quarterly Journal of the Royal Meteorological Society*, *104*(440), 323–330.

Niedermeier, D., Augustin-Bauditz, S., Hartmann, S., Wex, H., Ignatius, K., & Stratmann, F. (2015). Can we define an asymptotic value for the ice active surface site density for heterogeneous ice nucleation?, *Journal of Geophysical Research: Atmospheres*, *120*(10), 5036–5046.

Niedermeier, D., Chang, K., Cantrell, W., Chandrakar, K. K., Ciochetto, D., & Shaw, R. A. (2018). Observation of a link between energy dissipation rate and oscillation frequency of the large-scale circulation in dry and moist Rayleigh-Bénard turbulence. *Physical Review Fluids*, *3*(8), 083,501.

O'Shea, S., Choularton, T., Lloyd, G., Crosier, J., Bower, K., Gallagher, M., et al. (2016). Airborne observations of the microphysical structure of two contrasting cirrus clouds. *Journal of Geophysical Research: Atmospheres*, *121*(22), 13–510.

Paoli, R., & Shariff, K. (2009). Turbulent condensation of droplets: direct simulation and a stochastic model. *Journal of the Atmospheric Sciences*, *66*(3), 723–740.

Pawlowska, H., Grabowski, W. W., & Brenguier, J.-L. (2006). Observations of the width of cloud droplet spectra in stratocumulus. *Geophysical Research Letters*, *33*(19), L19810.

Pincus, R., & Baker, M. (1994). Effect of precipitation on the albedo susceptibility of clouds in the marine boundary layer. *Nature*, *372*(6503), 250–252.

Pitter, R., & Pruppacher, H. (1973). A wind tunnel investigation of freezing of small water drops falling at terminal velocity in air. *Quarterly Journal of the Royal Meteorological Society*, *99*(421), 540–550.

Pontikis, C., & Hicks, E. (1992). Contribution to the cloud droplet effective radius parameterization. *Geophysical Research Letters*, *19*, 2227–2230.

Pruppacher, H., & Rasmussen, R. (1979). A wind tunnel investigation of the rate of evaporation of large water drops falling at terminal velocity in air. *Journal of the Atmospheric Sciences*, *36*(7), 1255–1260.

Rangno, A. L., & Hobbs, P. V. (1983). Production of ice particles in clouds due to aircraft penetrations. *Journal of Climate and Applied Meteorology*, *22*(2), 214–232.

Rangno, A. L., & Hobbs, P. V. (1988). Criteria for the onset of significant concentrations of ice particles in cumulus clouds. *Atmospheric Research*, *22*(1), 1–13.

Reiche, C. H., & Lasher-Trapp, S. (2010). The minor importance of giant aerosol to precipitation development within small trade wind cumuli observed during RICO. *Atmospheric Research*, *95*(4), 386–399.

Reichelderfer, F. (1967). Ross Gunn—The scientist and the individual. *Monthly Weather Review*, *95*(12), 815–821.

Risius, S., Xu, H., Di Lorenzo, F., Xi, H., Siebert, H., Shaw, R., & Bodenschatz, E. (2015). Schneefernerhaus as a mountain research station for clouds and turbulence. *Atmospheric Measurement Techniques*, *8*(8), 3209–3218.

Sardina, G., Picano, F., Brandt, L., & Caballero, R. (2015). continuous growth of droplet size variance due to condensation in turbulent clouds. *Physical Review Letters*, *115*(18), 1–5.

Saw, E.-W., Bewley, G. P., Bodenschatz, E., Sankar Ray, S., & Bec, J. (2014). Extreme fluctuations of the relative velocities between droplets in turbulent airflow. *Physics of Fluids*, *26*(11), 111,702.

Saw, E. W., Shaw, R. A., Ayyalasomayajula, S., Chuang, P. Y., & Gylfason, A. (2008). Inertial clustering of particles in high-Reynolds-number turbulence. *Physical Review Letters*, *100*(21), 214,501.

Saw, E.-W., Shaw, R. A., Salazar, J. P., & Collins, L. R. (2012). Spatial clustering of polydisperse inertial particles in turbulence: II. Comparing simulation with experiment. *New Journal of Physics*, *14*(10), 105,031.

Schlenczek, O., Fugal, J. P., Lloyd, G., Bower, K. N., Choularton, T. W., Flynn, M. et al. (2017). Microphysical properties of ice crystal precipitation and surface-generated ice crystals in a high alpine environment in Switzerland. *Journal of Applied Meteorology and Climatology*, *56*(2), 433–453.-

Shaw, R. A. (2003). Particle-turbulence interactions in atmospheric clouds. *Annual Review of Fluid Mechanics*, *35*(1), 183–227.

Sherwood, S., Roca, R., Weckwerth, T., & Andronova, N. (2010). Tropospheric water vapor, convection, and climate. *Reviews of Geophysics*, *48*(2), RG2001.

Siebert, H., Beals, M., Bethke, J., Bierwirth, E., Conrath, T., Dieckmann, K., et al. (2013). The fine-scale structure of the trade wind cumuli over Barbados—An introduction to the Carriba project. *Atmospheric Chemistry and Physics*, *13*(19), 10,061–10,077.

Siebert, H., & Shaw, R. A. (2017). Supersaturation fluctuations during the early stage of cumulus formation. *Journal of the Atmospheric Sciences*, *74*(4), 975–988.

Siebert, H., Shaw, R., Ditas, J., Schmeissner, T., Malinowski, S., Bodenschatz, E., & Xu, H. (2015). High-resolution measurement of cloud microphysics and turbulence at a mountaintop station. *Atmospheric Measurement Techniques*, *8*(8), 3219–3228.

Siewert, C., Bec, J., & Krstulovic, G. (2017). Statistical steady state in turbulent droplet condensation. *Journal of Fluid Mechanics*, *810*, 254–280.

Slingo, A. (1989). A GCM parameterization for the shortwave radiative properties of water clouds. *Journal of the Atmospheric Sciences*, *46*(10), 1419–1427.

Slingo, A. (1990). Sensitivity of the earth's radiation budget to changes in low clouds. *Nature*, *343*, 49–51. https://doi.org/10.1038/343049a0

Stephens, G. L. (1978). Radiation profiles in extended water clouds. ii: Parameterization schemes. *Journal of the Atmospheric Sciences*, *35*(11), 2123–2132.

Stevens, B. (2002). Entrainment in stratocumulus-topped mixed layers. *Quarterly Journal of the Royal Meteorological Society*, *128*(586), 2663–2690.

Storelvmo, T. (2017). Aerosol effects on climate via mixed-phase and ice clouds. *Annual Review of Earth and Planetary Sciences*, *45*, 199–222.

Storelvmo, T., & Tan, I. (2015). The Wegener–Bergeron–Findeisen process its discovery and vital importance for weather and climate. *Meteorologische Zeitschrift*, *24*, 455–461.

Stratmann, F., Kiselev, A., Wurzler, S., Wendisch, M., Heintzenberg, J., Charlson, R., et al. (2004). Laboratory studies and numerical simulations of cloud droplet formation under realistic supersaturation conditions. *Journal of Atmospheric and Oceanic Technology*, *21*(6), 876–887.

Tas, E., Teller, A., Altaratz, O., Axisa, D., Bruintjes, R., Levin, Z., & Koren, I. (2015). The relative dispersion of cloud droplets: its robustness with respect to key cloud properties. *Atmospheric Chemistry and Physics*, *15*(4), 2009–2017.

Taylor, G. R., & Baker, M. B. (1991). Entrainment and detrainment in cumulus clouds. *Journal of the Atmospheric Sciences*, *48*(1), 112–121.

Twomey, S. (1974). Pollution and the planetary albedo. *Atmospheric Environment (1967)*, *8*(12), 1251–1256.

Twomey, S. (1977). The influence of pollution on the shortwave albedo of clouds. *Journal of the Atmospheric Sciences*, *34*(7), 1149–1152.

Van Heerwaarden, C. C., & Mellado, J. P. (2016). Growth and decay of a convective boundary layer over a surface with a constant temperature. *Journal of the Atmospheric Sciences*, *73*(5), 2165–2177.

Vanzanten, M. C., & Duynkerke, P. G. (2002). Radiative and evaporative cooling in the entrainment zone of stratocumulus—The role of longwave radiative cooling above cloud top. *Boundary-Layer Meteorology*, *102*(2), 253–280.

Vohl, O., Mitra, S., Wurzler, S., Diehl, K., & Pruppacher, H. (2007). Collision efficiencies empirically determined from laboratory investigations of collisional growth of small raindrops in a laminar flow field. *Atmospheric Research*, *85*(1), 120–125.

Vohl, O., Mitra, S., Wurzler, S., & Pruppacher, H. (1999). A wind tunnel study of the effects of turbulence on the growth of cloud drops by collision and coalescence. *Journal of the Atmospheric Sciences*, *56* (24), 4088–4099.

von Blohn, N., Diehl, K., Mitra, S. K., & Borrmann, S. (2009). Riming of graupel: Wind tunnel investigations of collection kernels and growth regimes. *Journal of the Atmospheric Sciences*, *66*(8), 2359–2366.

Wang, H., & Feingold, G. (2009). Modeling mesoscale cellular structures and drizzle in marine stratocumulus. Part I: Impact of drizzle on the formation and evolution of open cells. *Journal of the Atmospheric Sciences*, *66*, 3237–3256.

Wang, J., Daum, P. H., Yum, S. S., Liu, Y., Senum, G. I., Lu, M.-L., et al. (2009). Observations of marine stratocumulus microphysics and implications for processes controlling droplet spectra: Results from the marine stratus/stratocumulus experiment. *Journal of Geophysical Research: Atmospheres*, *114*(D18).

Wendisch, M., & Brenguier, J.-L. (2013). *Airborne measurements for environmental research: Methods and instruments*. Weinheim, Germany: Wiley-VCH.

Wex, H., Augustin-Bauditz, S., Boose, Y., Budke, C., Curtius, J., Diehl, K., et al. (2015). Intercomparing different devices for the investigation of ice nucleating particles using Snomax® as test substance, *Atmospheric Chemistry and Physics*, *15*(3), 1463–1485.

Wex, H., Hennig, T., Salma, I., Ocskay, R., Kiselev, A., Henning, S., et al. (2007). Hygroscopic growth and measured and modeled critical super-saturations of an atmospheric HULIS sample. *Geophysical Research Letters*, *34* (2), L02818.

Wex, H., Kiselev, A., Ziese, M., & Stratmann, F. (2006). Calibration of LACIS as a CCN detector and its use in measuring activation and hygroscopic growth of atmo spheric aerosol particles. *Atmospheric Chemistry and Physics*, *6*(12), 4519–4527.

Witte, M. K., Ayala, O., Wang, L.-P., Bott, A., & Chuang, P. Y. (2017). Estimating collision–coalescence rates from in situ observations of marine stratocumulus. *Quarterly Journal of the Royal Meteorological Society*, *143*(708), 2755–2763.

Witte, M. K., Yuan, T., Chuang, P. Y., Platnick, S., Meyer, K. G., Wind, G., & Jonsson, H. H. (2018). Modis retrievals of cloud effective radius in marine stratocumulus exhibit no significant bias. *Geophysical Research Letters*, *45*(19), 10–656.

Wood, R. (2006). Rate of loss of cloud droplets by coalescence in warm clouds. *Journal of Geophysical Research: Atmospheres*, *111* (D21205). doi:10.1029/2006JD007,553.

Wood, R. (2012). Stratocumulus clouds. *Monthly Weather Review*, *140* (8), 2373–2423.

Wood, R., Christopher, K.-T. O., Bretherton, S., Mohrmann, J., Albrecht, B. A., Zuidema, P., et al. (2018). Ultraclean layers and optically thin clouds in the stratocumulus-to-cumulus transition. Part I: Observations. *Journal of the Atmospheric Sciences*, *75*(5), 1631–1652.

Wood, R., Irons, S., & Jonas, P. R. (2002). How important is the spectral ripening effect in stratiform boundary layer clouds? Studies using simple trajectory analysis. *Journal of the Atmospheric Sciences*, *59*(18), 2681–2693.

Wyngaard, J., Coté, O., & Izumi, Y. (1971). Local free convection, similarity, and the budgets of shear stress and heat flux. *Journal of the Atmospheric Sciences*, *28*(7), 1171–1182.

Wyngaard, J. C. (1998). Boundary-layer modeling: history, philosophy, and sociology. In A. Holtslag & P. G. Duynkerke, *Clear and Cloudy Boundary Layers: Proceedings of the Colloquium "Clear and Cloudy Boundary Layers," Amsterdam, 26–29 August 1997* (chap. 15, pp. 325–332). Royal Netherlands Academy of Arts and Sciences.

Wyngaard, J. C. (2015). Obituary: James W. *Deardorff. Boundary-Layer Meteorology*, *157*, 343–344.

Yang, F., Cantrell, W. H., Kostinski, A. B., Shaw, R. A., & Vogelmann, A. M. (2020). Is contact nucleation caused by pressure perturbation? *Atmosphere*, *11*(1), 1.

Yang, F., Kollias, P., Shaw, R. A., & Vogelmann, A. M. (2018). Cloud droplet size distribution broadening during diffusional growth: Ripening amplified by deactivation and reactivation. *Atmospheric Chemistry and Physics*, *18*(10), 7313–7328.

Yau, M. K., & Rogers, R. R. (1996). *A short course in cloud physics*. UK: Elsevier.

Yeom, J. M., Yum, S. S., Liu, Y., & Lu, C. (2017). A study on the entrainment and mixing process in the continental stratocumulus clouds measured during the RACORO campaign. *Atmospheric Research*, *194*, 89–99.

Yum, S. S., & Hudson, J. G. (2005). Adiabatic predictions and observations of cloud droplet spectral broadness. *Atmospheric Research*, *73*(3–4), 203–223.

Yum, S. S., Wang, J., Liu, Y., Senum, G., Springston, S., McGraw, R., & Yeom, J. M. (2015). Cloud microphysical relationships and their implication on entrainment and mixing mechanism for the stratocumulus clouds measured during the vocals project. *Journal of Geophysical Research: Atmospheres*, *120*(10), 5047–5069.

Zhao, C., Tie, X., Brasseur, G., Noone, K. J., Nakajima, T., Zhang, Q., et al. (2006). Aircraft measurements of cloud droplet spectral dispersion and implications for indirect aerosol radiative forcing. *Geophysical Research Letters*, *33*(16), L16809. doi:10.1029/2006GL026653, 116809.

17

Frameworks for Testing and Evaluating Fast Physics Parameterizations in Climate and Weather Forecasting Models

Wuyin Lin[1] and Shaocheng Xie[2]

ABSTRACT

The success of fast physics parameterizations for climate and weather forecasting models hinges on robust understanding of the processes and creative means to conceptualize and represent them in the models. While the ultimate tests are their performance in climate simulations and weather prediction, multipronged modeling frameworks are needed to efficiently and effectively test the parameterizations, assess their predictive skills, and characterize the model behavior from process level to climate scale in order to confidently advance the parameterization for production use. An integrated yet complementary modeling and evaluation framework, which is also widely adopted by the modeling centers and the research community, is described and discussed. The framework promotes process-oriented evaluation while effectively bridging parameterization development with observations and modeling, a promising pathway for continually improving physical parameterizations for the models.

17.1. OVERVIEW

Climate and weather prediction models rely on parameterizations to represent fast physical processes of turbulence, convection, cloud, and radiation in order to predict sensible weather phenomena and climate variability, study climate sensitivity and feedbacks, and project climate changes. Deficiencies in the parameterizations are commonly blamed for long-standing model biases in simulating diurnal precipitation (e.g., Covey et al., 2016; Dai, 2006; Lee et al., 2007), Madden-Julian oscillation and other convectively coupled tropical waves (e.g., Ahn et al., 2017; J. Lin et al., 2006; Kim et al., 2009), the structure of intertropical convergence zone (ITCZ; e.g., Lin, 2007; Oueslati & Bellon, 2015; Qin & Lin, 2018; X. Zhang et al., 2007), to name a few. Furthermore, the large spread in climate sensitivity among climate models in response to external forcing has been mostly attributable to poor representation of marine low-level clouds in the tropics and subtropics that involve the interplay of turbulent, convective, and microphysical and radiative processes (Boucher et al., 2013; Dufresne & Bony, 2008; Neggers, 2015; Randall et al., 2007; Sherwood et al., 2014; Vial et al., 2017; M. Zhang et al., 2013). One consequence of this is a large uncertainty in the projection of global climate change, say due to increasing greenhouse gas emissions, a major scientific undertaking for almost all climate models. Continuously evaluating model performance and testing and improving physical process parameterizations therefore have been a critical part of the development process for climate and weather forecasting models (Jakob, 2010).

The ultimate tests of all atmospheric physical parameterizations for climate models are their performance in climate simulations and in the case of weather forecasting models in daily forecasts over an extended period of time. Performance metrics consisting of a set of quantitative statistical measures such as pattern correlations and

[1] *Environmental and Climate Sciences Department, Brookhaven National Laboratory, Upton, NY, USA*

[2] *Atmospheric, Earth and Energy Division, Lawrence Livermore National Laboratory, Livermore, CA, USA*

Fast Processes in Large-Scale Atmospheric Models: Progress, Challenges, and Opportunities, Geophysical Monograph 282, First Edition.
Edited by Yangang Liu and Pavlos Kollias.

DOI: 10.1002/9781119529019.ch17

standard deviations are commonly used to routinely assess model performance and track model progress (Gleckler et al., 2008). Despite the effectiveness in assessing the overall model performance, these kinds of metrics, however, are not able to tell what causes model errors. To provide more insights into model errors, particularly those that are presumed to be closely associated with the atmospheric fast physics, process studies by utilizing a hierarchy of process models and detailed process-oriented observations from major field campaigns are necessary (Jakob, 2010; Randall et al., 2003). Such a modeling and evaluation framework provides a powerful tool to test and understand critical physical processes that are required to be accurately represented in climate and weather forecasting models by directly confronting models with observations.

The modeling framework widely used for process studies includes single-column models (SCMs), cloud-resolving models (CRMs), and large-eddy simulations (LES), with the scale ranging from a few hundred kilometers to a few tens of meters. The use of process models, particularly CRM and LES models to explicitly simulate convective and turbulent processes, has long been envisioned by Global Energy and Water Cycle Experiment (GEWEX) Cloud System Study (GCSS) to facilitate the development of improved parameterizations for climate and weather prediction models (Browning et al., 1993; Randall et al., 2003). These types of models require horizontal advective tendencies of heat, water, and momentum, also known as the large-scale forcing, to be prescribed to focus on the response of model physics. In the SCM framework, the model physics are isolated and the errors in model simulations can be attributed to deficiencies of model physical parameterizations if the boundary and forcing conditions are perfect (Randall et al., 1996). The required large-scale forcing can be derived from radiosonde data collected from a well-defined sounding network (Zhang & Lin, 1997, and the references therein), which is often available from intensive observational periods (IOPs) during major field campaigns such as the Tropical Ocean Global Atmosphere (TOGA) Coupled Ocean-Atmosphere Experiment (COARE, Webster & Lukas, 1992) and a growing list of field campaigns targeting specific cloud regimes conducted by the U.S. Department of Energy (DOE) Atmospheric Radiation Measurement (ARM) program (Mather & Voyles, 2013). The forcing data can also be derived from operational analysis data produced by numerical weather prediction (NWP) centers to accompany continuous or mobile cloud property measurements at research sites (Xie et al., 2004a) in order to facilitate broader modeling studies of cloud processes or for the SCM evaluations of physical parameterizations (Christensen et al., 2018; Smalley et al., 2019). In many case studies targeting specific cloud regime or the evolution of cloud processes, the forcing data can be constructed or further adjusted to idealize representative environmental conditions (e.g., Guichard et al., 2004; Moeng et al., 1996; Zhang & Bretherton, 2008; M. Zhang et al., 2012; Zhu et al., 2005). By design, the same large-scale forcing data can also be used to drive CRM and LES models, which are run at cloud-resolving scale and turbulence-resolving scale, respectively. CRMs and LES can provide insight into important physical processes that need to be parameterized in SCMs, as well as detailed diagnostics that would be difficult or impossible to obtain directly from observations. This hierarchy of process-modeling framework has been widely used in developing, testing, and improving the representations of aerosol, clouds, convection, and precipitation processes in global climate models (GCMs) or numerical weather prediction models that are used in climate projection and weather forecast (e.g., Ahlgrimm & Forbes, 2014; Angevine et al., 2018; Bengtsson et al., 2019; Christensen, 2019; Fridlind et al., 2012; Ghonima et al., 2017; Hourdin et al., 2013; Huang et al., 2013; Klein et al., 2009; Luo et al., 2005, 2006; Neggers et al., 2009, 2012; Xie et al., 2002, 2005; Xu et al., 2002, 2005).

The model physics in SCM simulations however lack interaction with dynamics, often causing difficulties in transferring improved features obtained from those isolated process studies to full GCMs. To complement the use of SCMs and test model physical parameterizations with process-level observations in full GCMs, the idea of running climate models in "weather forecast mode" with initial condition data from NWP analyses has been facilitated through the U.S. Department of Energy's (DOE) Cloud-Associated Parameterizations Testbed (CAPT) (Ma et al., 2015; Phillips et al., 2004) and the Transpose Atmospheric Model Intercomparison Project II (Transpose-AMIP II) (Ma et al., 2014; Williams et al., 2013). Since the large-scale state of the atmosphere in the early periods of a forecast is realistic enough, the model errors in cloud-related fields may be ascribed to the parameterizations of atmospheric processes. Using the weather forecast approach can identify model errors before longer timescale feedbacks develop and facilitate the comparison to detailed process observations. This approach has been applied to many global climate models and weather forecasting models to diagnose the development of simulated biases in cloud and precipitation (e.g., Boyle & Klein, 2010; Y. Lin et al., 2012; Xie et al., 2004b, 2008; Williamson et al., 2005; Williams et al., 2013; Y. Zhang et al., 2019) as well as other systematic model biases that are at least partly attributable to deficiency in simulated clouds (e.g., Ma et al., 2018; van Weverberg et al., 2015, 2018).

Along with the use of the modeling framework, process-oriented diagnostics and metrics are also instrumental in support of process studies and providing more insights into model errors. The process-oriented diagnostics in the ARM Data-oriented Metrics and Diagnostics package (ARM-DIAGS; C. Zhang et al., 2020) and those developed by the Model Diagnostics Task Force (MDTF) under the NOAA Climate Program Office (CPO) Modeling, Analysis, Predictions and Projections (MAPP) program (Maloney et al., 2019) as well as quantitative performance metrics for clouds (e.g., Eyring et al., 2016; Hogan et al., 2009) are well suited to enhance the evaluation of model physics. With these process-oriented diagnostics and metrics, models can be evaluated in more details against the observations of physical properties associated with specific processes. In addition, new techniques are being developed to improve the comparison between models and observations. Examples include the community satellite simulator package, the Cloud Feedback Model Intercomparison Project (CFMIP) Observation Simulator Package (COSP; Bodas-Salcedo et al., 2011), which was developed to exploit satellite observations for model cloud evaluation (e.g., Kay et al., 2012; Klein et al., 2013; Y. Zhang et al, 2019). COSP contains several independent satellite simulators, which diagnose model clouds in a similar way that these instruments would view an atmosphere with physical properties (e.g., cloud cover, optical depth, water vapor mixing ratio, precipitation fluxes, or even cloud microphysical properties) produced by the models. Similar cloud diagnostic packages can also be developed for ground-based cloud radar and lidar simulators to enable the evaluation of SCM or hindcast simulations against detailed observations by vertically pointing cloud radar and lidar over field campaign sites (e.g., Y. Zhang et al., 2018), which can facilitate the investigation of the detailed vertical structures within clouds, including the skill in capturing thin clouds, and the quantification of their temporal evolution including diurnal variability.

In this chapter, we provide an overview of the commonly used modeling frameworks and diagnostic tools for evaluating and testing climate and weather forecasting models at process level. Examples on applying the hierarchy of modeling approaches and diagnostic tools to understanding and improving the models are provided. Observational data that are suitable for such studies are also discussed.

17.2. SCM, CRM, AND LES—AN INTEGRATED DEVELOPMENT AND EVALUATION FRAMEWORK

17.2.1. The Concept of SCM and Its Integration with CRM and LES

An SCM refers to a grid column of a GCM and often has the same physical parameterizations as those in its parent GCM. It involves the calculation of time evolution of the vertical distribution of temperature and advective water species, which include water vapor and cloud hydrometeors (see also Yanai et al., 1973; M. Zhang et al., 2016):

$$\frac{\partial \theta_m}{\partial t} = \left(\frac{\partial \theta_m}{\partial t}\right)_{\text{phy}} - (\vec{V} \cdot \nabla \theta)_{\text{LS}} - \omega_{\text{LS}} \frac{\partial \theta_m}{\partial p} \quad (17.1)$$

$$\frac{\partial q_m}{\partial t} = \left(\frac{\partial q_m}{\partial t}\right)_{\text{phy}} - (\vec{V} \cdot \nabla q)_{\text{LS}} - \omega_{\text{LS}} \frac{\partial q_m}{\partial p}, \quad (17.2)$$

where θ and q are potential temperature and mixing ratios for water species, respectively; subscript "m" denotes model values; "LS" stands for prescribed large-scale fields; "phy" represents physical parameterizations; and other symbols are as commonly used in atmospheric science. By specifying the large-scale advective tendencies (or the large-scale forcing) from either observations or suitable model output, SCM can be run for a selected period of interest where rich field observations are available. Model results are then compared to the relevant observations. Since the large-scale dynamics are specified, the model errors are thought to be due to the deficiencies in the model physical parameterizations. Given its simplicity and the capability to isolate the model physics, SCM has been widely used as an effective tool to test/develop and evaluate physical parameterizations used in GCMs and weather forecasting systems (Angevine et al., 2018; Betts & Miller, 1986; Christensen et al., 2018; Ghonima et al., 2017; Lord, 1982; Luo et al., 2006; Morrison et al., 2005; Randall et al., 1996; Xie & Zhang, 2000; Yang et al., 2006; Wang et al. 2020; M. Zhang et al., 2016), including parameterizations for shallow and deep moist cumulus convection, convective trigger function, boundary layer turbulence, and cloud macrophysics and microphysics. Today almost all weather and climate modeling centers have developed the companion configuration of SCM as an integrated tool to facilitate the continuing developments of physics parameterizations and their evaluations against direct observations. Examples include the SCM versions of Community Atmosphere Model (CAM) Version 6 (Gettelman et al., 2019), Energy Exascale Earth System Model (E3SM) (Bogenschutz et al., 2020), the Met Office's Unified Model (UM, Lock et al., 2000; Walters et al., 2019), European Centre for Medium-Range Weather Forecasts' (ECMWF) Integrated Forecasting System (IFS; Christensen et al., 2018; Köhler & Teixeira, 2002), and Weather Research and Forecasting (WRF) model (Angevine et al. 2018; Hacker & Angevine, 2013). These SCMs are publicly available, any of them hence can serve as the host model for new parameterization developments. It should be cautioned, however, that the uncertainty in the specified large-scale forcing and the lack of interaction between model dynamics and physics can cause challenge in interpreting the SCM results.

SCM tests are often combined with CRM simulations that have sufficient resolution to resolve fine-scale cloud dynamics and cloud systems. Since CRMs can explicitly compute clouds and convection, they have been widely used to study detailed cloud structures and their evolutions and evaluate parameterizations of convective and stratiform clouds (Christensen, 2018; Khairoutdinov & Randall, 2006; Krueger, 1988; Krueger et al., 1995; Redelsperger & Lafore, 1988; Xu & Krueger, 1991; Xu & Arakawa, 1992; Xu et al., 1992; Xu & Randall, 1996a, b). Some critical geophysical parameters such as cloud mass fluxes, size of convective cores and cold pool properties associated with organized convection, and other cloud microphysical properties, which are difficult to observe, can be obtained from CRMs and used as a proxy for observations for improving cumulus and stratiform cloud parameterizations. Therefore, CRMs can also play an important role in bridging the gap between parameterization development and observations (Randall et al., 1996). For the purpose of facilitating the SCM evaluation of cloud and convection parameterizations, CRMs are typically run over a domain size comparable to a coarse-resolution GCM grid box. While early on CRMs were mostly 2D and used resolutions of 1–4 km, with increasingly available computing power, the use of 3D CRMs has become a new norm for explicit cloud modeling with around 1-km grid resolution (Guichard & Couvreux, 2017; Krueger et al., 2016). It is noted that CRM still needs to parameterize its own subgrid-scale processes such as cloud microphysics, turbulences, and radiation. Uncertainties in representing these subgrid-scale phenomena may limit the accuracy of CRM simulations. The CRM results therefore still need to be evaluated before they can be confidently used to aid the SCM evaluations and parameterization developments.

The increasing power of computers also allows the routine use of large-eddy simulation as part of the integrated process-modeling framework for model parameterization developments (Gustafson Jr. et al., 2020; Neggers et al., 2012). LES is a methodology for the simulations of turbulent flows in which the largest, most energetically important scales of turbulence are explicitly resolved. A typical model resolution of LES is from a few meters to a hundred meters. Given the computational constraint, in the past LES has been mainly applied to investigate turbulence and cloud processes in the atmospheric boundary layer (ABL) over a relatively small domain of a few kilometers and fill in missing details from the observations for model evaluations (Deardorff, 1970; Moeng, 1984; Siebesma et al., 2003). The rapid progress in massively parallel supercomputing has promoted the application of LES for deep convections, which requires a much bigger domain (i.e., a hundred kilometers) than that used for shallow convection cases. Because this type of simulations uses billions of grid cells, the simulation has been nicknamed the "Giga-LES" (Khairoutdinov et al., 2009; Moeng et al., 2009). Entrainment rate, characteristics of turbulent kinetic energy, and interaction between turbulence and microphysics that can be diagnosed from LES simulations can be used to aid the development of parameterizations and their evaluations for boundary layer turbulence, cloud microphysics, and cumulus convection. As in CRMs, LES also needs to parameterize its own subgrid-scale processes such as subgrid scale turbulence, cloud microphysics, and radiation. For example, the LES simulation results are found to have strong sensitivity to microphysical schemes used (e.g., VanZanten et al., 2011), and simulating strongly stable boundary layer can be challenging even for LES (e.g., Zhou & Chow, 2011, and the discussions therein).

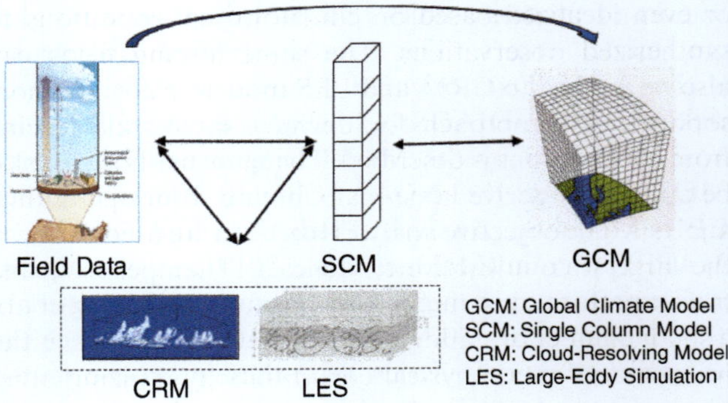

Figure 17.1 A schematic diagram for the SCM-CRM-LES modeling framework and how they link rich field observations with GCM developments.

Figure 17.1 illustrates how the SCM-CRM-LES modeling framework is used in testing and developing physical parameterizations used in GCMs and filling in the gap between field data and GCMs as envisioned by the GCSS program (Browning et al., 1993; Randall et al. 2003). Data from a well-defined surface network can provide the necessary large-scale initial and forcing conditions to drive SCM, CRM, and LES models. The results from SCM simulations can be compared to the observations as well as the results from CRM/LES which provide additional detailed turbulence and cloud information that complement the observations. The CRM/LES simulations themselves can be validated by more detailed fine-scale observations of cloud and turbulence properties derived from ground-based profilings by radars and lidars as well as aircraft measurements.

17.2.2. The Large-Scale Forcing

As mentioned in the overview section, the required large-scale forcing for driving SCM can be obtained from observations or operational analysis from NWP centers or even idealized based on climatological conditions or synthesized observations. The same forcing fields can also be applied to CRM and LES models. Here, the more sophisticated approach for deriving large-scale forcing from observations is described. The same method can also be applied to derive large-scale forcing from operational analysis. The objective analysis methods are used to derive the large-scale advective tendencies of temperature and moisture from sounding measurements over a network of a small number of stations. Due to inevitable uncertainties in sounding data, physical constraints by utilizing other observations are often enforced in the objective analysis. Earlier objective analyses (e.g., Barnes, 1964; Lin & Johnson 1996; O'Brien, 1970; Ooyama; 1987) often apply the mass constraint that adjusts horizontal winds to conserve column-integrated mass. The objectively analyzed fields typically do not conserve column-integrated heat, moisture, and momentum. The unbalanced budgets can cause serious problems in SCMs (Ghan et al., 2000). To address these problems, Zhang and Lin (1997) proposed a constrained variational analysis (CVA) approach, in which the atmospheric state variables are forced to satisfy the conservation of mass, heat, moisture, and momentum through a variational technique. These constraints can be written as

$$< \nabla \cdot \vec{V} > = -\frac{1}{g}\frac{\mathrm{d}p_s}{\mathrm{d}t} \quad (17.3)$$

$$\frac{\partial < q >}{\partial t} + < \nabla \cdot \vec{V}q > = E_S - \text{Prec} - \frac{\partial < q_l >}{\partial t} \quad (17.4)$$

$$\frac{\partial < s >}{\partial t} + < \nabla \cdot \vec{V}s > = R_{TOA} - R_{SRF} + L_v\text{Prec} + \text{SH} + L_v\frac{\partial < q_l >}{\partial t} \quad (17.5)$$

$$\frac{\partial < \vec{V} >}{\partial t} + < \nabla \cdot \vec{V}\vec{V} > + f\vec{k} \times < \vec{V} > + \nabla < \phi > = \vec{\tau}_s \quad (17.6)$$

where $\vec{V}$, s, and q are the atmospheric state variables of winds, dry static energy, and water vapor; p_s is the surface pressure; q_l is the cloud liquid water content; and ϕ denotes the geopotential height. The bracket represents vertical integration; Prec is the surface precipitation; R is the net downward radiative flux; E_s and SH are the surface evaporation and sensible heat fluxes; τ_s is the surface wind stress; and the other variables are as commonly used. These constraints ensure what comes into the column equals to what comes out of the column plus the sinks/sources within the column. Terms on the right-hand sides of equations (17.3)–(17.6) are obtained from available surface and satellite measurements.

The large-scale forcing derived from CVA is dynamically and thermodynamically consistent with the surface and top-of-the-atmosphere (TOA) flux measurements through the constraints. It has substantially improved the performance of SCMs (Ghan et al., 2000). This method has been widely applied to process the observational data collected from major field campaigns including the routine processing of the ARM data (Schumacher et al., 2007; S. Tang et al., 2016; Xie et al., 2006, 2010a, 2014). To enable the SCM research beyond IOPs with coordinated balloon soundings, Xie et al. (2004a) extended the CVA method to derive the so-called continuous forcing data by utilizing NWP operational analysis of atmospheric fields as input to the constrained variational analysis. In this method, the operational analyses are constrained by the observed surface and TOA fields. The advantage of the continuous forcing is that it can be derived for long periods and over most regions of the globe as long as surface and TOA measurements are available. The use of the observed constraints in the analysis can significantly improve the accuracy of the forcing data directly derived from the NWP products (Xie et al., 2003, 2004a).

17.2.3. Specification of the Large-Scale Forcing

As described in Randall and Cripe (1999), three approaches are often used to specify the large-scale advective forcing in SCM/CRM/LES studies: (1) revealed forcing, (2) horizontal advective forcing, and (3) relaxation forcing. The revealed forcing often refers to "3D forcing" where both horizontal and vertical advective tendencies (e.g., $-(\vec{V} \cdot \nabla\theta)_{LS} - (\omega\frac{\partial\theta}{\partial p})_{LS}$) in equation (17.1)) are prescribed. The horizontal advective forcing also refers to "2D forcing" in which the horizontal advective tendencies $-(\vec{V} \cdot \nabla\theta)_{LS}$ and the vertical velocity ω_{LS} are prescribed. It allows the vertical advective tendencies to respond to the simulated profiles of θ_m and q_m. The relaxation forcing adds a relaxation term $\frac{\theta-\theta_m}{\tau}$ to the horizontal advective tendencies to ensure that the simulated temperature and specific humidity do not drift away from observations. τ is the relaxation timescale, which is often set to 6–12 hours. Overall, the SCM results are comparable driven by the revealed forcing or the horizontal forcing while the relaxation forcing may lead to large errors in parameterized processes (e.g., precipitation) although it can keep the state variables closer to the observations (Randall & Cripe, 1999). Given its simplicity, the revealed forcing is the most used approach to specifying the large-scale forcing in SCM studies. These large-scale forcing terms can be similarly applied for CRM and LES simulations (e.g., Endo et al., 2015). To overcome the lack of feedbacks of the parameterizations on the dynamics, especially for tropical cloud regimes, the approximation of large-scale vertical velocity has been introduced using weak temperature gradient (e.g., Daleu et al., 2017; Sobel et al., 2001) or damped gravity wave approach (e.g., Daleu et al., 2015).

Surface fluxes are important driving force for the atmospheric fast physics processes. Prescribed boundary conditions and surface fluxes from large-scale forcing data set are usually used for the SCM as well as CRM/LES simulations. When the interactive flux exchange is of interest for a particular study, the model computed surface fluxes can be used, but the quality is subject to the surface flux parameterization and how the surface model is initialized. For CRM and LES simulations, the prescribed and model computed fluxes could be combined to reflect the influence of the inhomogeneity of the surface conditions.

Horizontal winds are usually prescribed for SCM simulations to maintain the large-scale dynamical conditions. This is also because the impact of physical parameterizations on large-scale winds is often considered secondary, if any. For CRM and LES simulations, fine-scale dynamics are important and the winds are computed at each grid point. The domain averaged winds can then be nudged toward the prescribed large-scale winds to maintain the large-scale dynamical environment during the course of the simulations (e.g., Endo et al., 2015).

17.2.4. Applications

Over the past 40 years, SCM has played a critical role in developing and testing physical parameterizations, in particular cumulus parameterizations, for use in GCMs. Pioneering examples include Lord (1982) for testing an implementation of the Arakawa-Schubert cumulus parameterization (Arakawa & Schubert, 1974) and Betts and Miller (1986) for testing their convective adjustment parameterizations. Recent examples are the developments and evaluations of eddy-diffusivity mass flux (EDMF) scheme (Neggers et al., 2009; Siebesma et al., 2006; Smalley et al., 2019; Šuselj et al. 2013) and cloud layer unified by binormal (CLUBB) scheme (Golaz et al., 2002; Larson & Golaz, 2005) for a unified representation of turbulence, shallow convection, and stratiform clouds, as well as other new developments or improvements for cloud microphysics, boundary layer turbulence, and shallow and deep cumulus convection (e.g., Ahlgrimm & Forbes, 2014; Angevine et al., 2018; Freitas et al., 2020; Morrison et al., 2005; Rio et al., 2010). In these studies, the results from CRMs and LES are often used to validate and understand the SCM simulations. The former is used as the primary tool for deep convection and convective cloud systems while the latter is mainly used for shallow convection and the ABL turbulence.

In fact, using the integrated SCM-CRM-LES testbed to aid the model development process is a strategy adopted by major weather prediction and climate modeling centers. One example is the framework consisting of an SCM testbed and an LES model established in the Royal Netherlands Meteorological Institute (KNMI) that allows the continuous evaluation of its newly developed parameterizations at a permanent observational site (Neggers et al., 2012). When a development is mature, the SCM framework is again often used as the first line of testing for a new parameterization before its transfer to the full model. The development of the University of Washington shallow cumulus and moist turbulence schemes for the NCAR Community Atmosphere Model (CAM) (Park & Bretherton, 2009), the porting of the CLUBB scheme for unified representation of turbulence and shallow convection to CAM (Bogenschutz et al., 2012, 2013) and to GFDL's Atmosphere Model (Guo et al., 2010, 2014), and the implementation of the EDMF scheme and other improvements to the ECMWF's Integrated Forecasting System model (Ahlgrimm & Forbes, 2014; Neggers et al., 2009) are some examples that followed this path.

The SCM-CRM-LES integrated framework has also been widely used in multimodel intercomparison projects given its simplicity in setting up model configuration with the same large-scale forcing and initial and boundary conditions since it was promoted by GCSS in the early-1990s (Browning et al., 1993). The climate modeling community including the GCSS and the ARM cloud modeling groups has conducted a series of multimodel intercomparison studies to assess the model ability to reproduce a variety of cloud types observed during major field campaigns around the globe (e.g., TOGA-COARE, Webster & Lukas, 1992; Mixed-Phase Arctic Cloud Experiment (MPACE), Verlinde et al., 2007; Rain in Cumulus over the Ocean (RICO) Field Campaign, Rauber et al., 2007; Tropical Warm Pool-International Cloud Experiment (TWP-ICE), May et al., 2008; ARM SCM IOPs, Cederwall et al., 1998; Krueger et al., 2016; M. Zhang et al., 2016). These include the studies organized or jointly organized by the GCSS working groups for marine and continental boundary layer clouds (e.g., Bechtold et al., 1996; Brown et al., 2002; Dal Gesso et al., 2015; Lenderink et al., 2004; Lock et al., 2000; Moeng et al., 1996; Zhu et al., 2005), cirrus clouds (e.g., Luo et al., 2003, 2005, 2006; Yang et al., 2012), midlatitude cloud systems (e.g., Ryan et al., 2000; Song et al., 2013, 2014; Xie et al., 2005; Xu et al., 2005), deep precipitating convective clouds in the tropics and summer continental regions (Bechtold et al., 2000; Davies et al., 2013; Fridlind et al., 2012; Ghan et al., 2000; Moncrieff et al., 1997; Petch et al., 2014; Redelsperger et al., 2000; Xie et al., 2002; Xu et al., 2002), and polar clouds (Klein et al., 2009; Morrison et al., 2009, 2011). Many of these case studies include the intercomparison of the CRM and LES simulations, and using the CRM/LES results to aid in the intercomparison and interpretation of the SCM simulations. The model intercomparison studies have proven useful to identify strengths and weaknesses of model parameterizations;

and systematic model deficiencies have been found through these studies. Examples include the large underestimation of mesoscale downdrafts in the cumulus parameterizations (Xie et al., 2002), the early onset of continental summertime convection in most SCMs that distorts the diurnal cycle of precipitation (Guichard et al., 2004; Xie et al. 2002), the large intermodel differences in the simulated cloud properties of midlatitude frontal cloud systems (Xie et al., 2005), and a wide variety of the simulated Arctic mixed-phase clouds (Klein et al., 2009). The results from these studies have promoted active researches in improving the representation of the various processes in cloud and convective parameterizations. The integrated SCM-CRM-LES model evaluation framework, as further demonstrated in W. Lin et al. (2015), can be particularly valuable in untangling the complex interactions among the parameterized processes in the model.

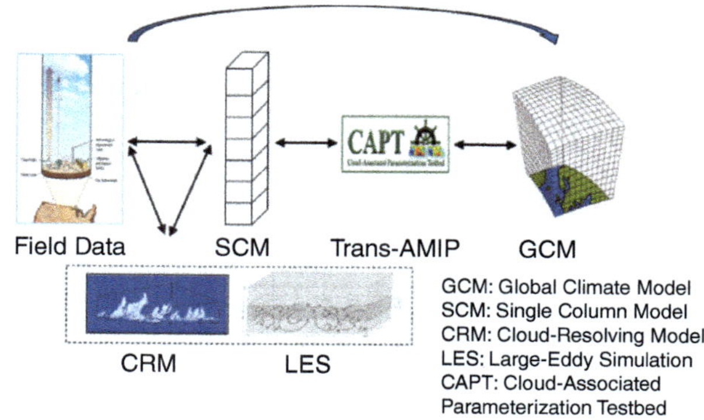

Figure 17.2 A schematic diagram that integrates Trans-AMIP into the SCM-CRM-LES modeling framework for GCM developments. CAPT is shown as a representative framework to facilitate the Trans-AMIP experiments.

17.3. NWP HINDCAST APPROACH FOR CLIMATE MODEL EVALUATION

In addition to the integrated process-modeling framework described in the previous section, there is a growing interest in the community in using a weather forecasting technique for climate model evaluation and process-level understanding of model physics and biases. This approach has been promoted by the US DOE CAPT project (Ma et al., 2015; Phillips et al., 2004) and the Working Group on Numerical Experiment (WGNE) Transpose-AMIP project (Williams et al., 2013). In this approach, a climate model is initialized with realistic conditions from NWP analysis, and its representation of atmospheric physical processes is tested in short-range weather forecasts. As there are no real-time operational constraints, retrospective forecasts (hindcasts) are performed. The rationales behind this approach are (1) the large-scale state of the atmosphere in the early periods of a forecast is realistic enough that errors may be ascribed to the parameterizations of the atmospheric physical processes; and (2) the atmospheric physical processes (e.g., moist process) are often fast (~hours) and the large-scale state changes slowly (~days). As indicated in the earlier studies (Ma et al., 2014; Xie et al., 2012; Williams & Brooks, 2008), there is a strong correspondence between the short-term and long-term systematic errors in climate models, particularly for those fields that are related to fast physics (e.g., clouds). Most systematic errors are apparent in the day 2 hindcasts. These errors steadily grow with the hindcast lead time and typically saturate after 5 days with amplitudes comparable to the climate errors. Using initialized climate models for testing physical parameterizations allows for evaluating the nature of parameterization errors before longer-timescale feedbacks develop and provides essential clue to the origin of errors in long-term climate simulations. It also facilitates process studies by directly linking model deficiencies with atmospheric processes through case studies using data collected from major field campaigns. Figure 17.2 demonstrates how the forecast approach is integrated into the process-modeling framework.

Given the fact that many climate models lack a compatible data assimilation system to initialize, obtaining a proper initial conditions (ICs) is one of the major challenges for using the hindcast approach. For models without data assimilation, ICs for the state variables (u, v, T, and q) can be created by simply mapping NWP analysis data to climate model grids. The mapping procedures routinely used in the operational NWP centers to provide ICs for their ensemble forecast systems or other applications can be utilized for this purpose. For example, CAPT has adopted the mapping procedures described in White (2001), i.e., the ECMWF "Full-pos" procedures for the Integrated Forecasting System. These procedures involve a slightly different interpolation method for each of the atmospheric state variables of temperature, winds, specific humidity, and surface pressure along with careful adjustments to account for the topographical differences between the analyses and the climate models.

Creating ICs for other atmospheric variables (e.g., aerosols and clouds) and land model involves more sophisticated procedures. Typically, a model spin-up using either "nudging" or "forecast/analysis" methods for a few months or longer before the hindcast starts is needed to produce appropriate initial conditions for the land variables such as soil moisture and temperature that are consistent with the interactive atmospheric environment. The basic idea of nudging is to continuously adjust the predicted atmospheric state variables toward a target analysis over a relaxation timescale (typically 6 hours) during the integration of the model. This is achieved by adding an extra tendency term in the prognostic state equations to decrease the distance between the predicted value and a target analysis. The nudging approach has been applied in weather forecasting for many decades (Hoke & Anthes, 1976). More details about "nudging" is described in the Appendix A of Boyle et al. (2005). The forecast/analysis method is similar to what is used in an NWP data assimilation system except that the atmospheric state variables are periodically updated by the analysis or reanalysis data at the time when analysis data are available (e.g., 6 hourly). Recently, Ma et al. (2015) proposed an improved nudging approach by nudging only the horizontal wind fields to produce more realistic initial aerosols and cloud fields. They also suggested that an offline land simulation forced by observed atmospheric forcing (e.g., precipitation, winds, temperatures, and surface fluxes) yields more realistic soil fields that are critical to land-atmosphere interactions.

The hindcast approach has been a powerful tool to understand various aspects of climate biases, facilitate model evaluation with field data, test new parameterizations, and aid the development of climate models (Barton et al., 2012; Boyle et al., 2008; Klein et al., 2006; Ma et al. 2014, 2018; Medeiros et al., 2012; Qin et al., 2018; Xie et al., 2004b, 2008, 2012; Van Weverberg et al., 2015; Williamson et al., 2005; Williams et al. 2013; Zheng et al., 2016, 2017). It complements the SCM approach by including full interactions between the model dynamics and physics as demonstrated in Figure 17.2 but without the need to develop a large-scale forcing data set. It is also an effective tool for optimizing model parameter choices during climate model tuning especially for expensive high-resolution model tuning (T. Zhang et al., 2018). This approach has been recently extended to diagnose the growth of biases in initialized coupled ocean-atmosphere systems (Ma et al., 2020; Siongco et al., 2019).

One issue with the hindcast approach is that the model errors from such a test could partly be a result of using a foreign analysis to initialize the model. To address this issue, testing alternate NWP analysis can be used to show the robustness of the model results. In addition, the hindcast approach may not be useful to study the climate errors that are related to slower processes or the result of various feedbacks between dynamical and physical processes that take longer time to impact model simulations such as the double ITCZ problem as shown in Xie et al. (2012).

17.4. EVALUATION METRICS AND DIAGNOSTICS

Evaluation of physical parameterizations is to assess their overall performance in simulations, determine whether specific processes are reasonably represented to produce the right answers for the right reasons, diagnose the bias characteristics, and identify the source of the biases and the induced compensation among the parameterized processes, with an aim to improve the understanding of the processes and their parameterizations. The metrics and diagnostics designed must align with this purpose. Here, the metrics are used to quantify the performance in simulating individual physical quantities, while the diagnostics are used to diagnose the relationship between different quantities or the relationship between different regions or times in one or more quantities. In addition, the term "diagnostics" here also includes characteristics that can be derived from synthesized quantities computed using multiple model-predicted straight variables (e.g., simulator output).

The commonly used metrics include those described in Gleckler et al. (2008) for climate simulations and described in Ma et al. (2013) for reinitialized hindcasts with a focus on cloud, aerosol, precipitation, radiative fluxes, and cloud radiative forcings. For each quantity, the basic metrics of mean climatology, seasonal cycles, spatial pattern correlation, zonal mean, and the vertical structure for vertically varying variables are effective in gauging the overall performance of the physical parameterizations in the models. These metrics, except for pattern correlation and zonal mean, can also be used to evaluate continuous SCM simulations. Taylor diagram (Taylor, 2001) is often used to summarize the performance in terms of these metrics. Various skill scores have also been proposed as effective metrics to evaluate simulated clouds against observations (Hogan et al., 2009).

To gain further physical insights, a growing body of process-oriented metrics and diagnostics have been recently developed to facilitate the evaluation of cloud and convective parameterizations at process level. The NOAA MDTF process-oriented diagnostics framework (Maloney et al., 2019 and the associated diagnostic packages cited therein) and the ARM-DIAGS package (C. Zhang et al., 2020) well represent these efforts. Among the process-oriented metrics, diurnal composites are easy to compute and can effectively characterize the cloud life cycle and prevailing precipitation time. They are important metrics for evaluating fast physics parameterization because the phenomena are ubiquitous around the globe but their simulations have long been challenging for climate models. Probability density function (PDF) of daily or sub-daily precipitation rate can describe how frequent it rains, the characteristics of the rainfall, and the skill in simulating extreme precipitation. Aside from their associated diabatic effect on large-scale circulation, all are critical for climate models in determining their fidelity in simulating hydrological cycles in the current and future climates. Emergent diagnostics based on conditional samplings have been found to be able to reveal critical knowledge about the precipitation processes both in the observations and in the models.

A popular convection onset diagnostics is built on the relationship between column water vapor (CWV) and the conditionally averaged precipitation rate (Bretherton et al., 2004; Holloway & Neelin, 2009; Schiro et al., 2016), in which a sharp increase in precipitation rate above a critical CWV value represents the onset of convection. This is a useful diagnostic to gauge the effectiveness of new convective triggering functions and their ability to alleviate the too-frequent-too-weak model precipitation problem that is primarily due to model convection being too easily triggered. A similar conditionally sampled diagnostics, as a probability density function with respect to radar reflectivity and cloud optical depth (Jing et al., 2017; Suzuki et al., 2015), separately classified according to the ranges of cloud-top particle size, is demonstrated to be skillful in revealing the microphysical conditions that promotes rain initiation in warm clouds and the transition from nonprecipitating regime to precipitating regime. Such a diagnostic can provide direct insight into the rain microphysical processes, and in the context of model evaluation expose the flaws in the model representation of the related processes. Moreover, conditionally sampled diagnostics can be even more effective for process-oriented evaluation over a broader region that experiences varying environmental conditions. Such evaluations can be further partitioned according to the cloud regimes or the dynamical regimes (e.g., Bony et al., 2004; Jakob, 2003; Lin & Zhang, 2004; Smalley et al., 2019; Tselioudis & Jakob, 2002; Webb et al., 2001), allowing the focus to remain on physical parameterizations when all weather conditions could occur over the study region during the simulation period.

In addition, novel cloud diagnostic tools have been developed to improve the comparison between the models and the observations. Examples include the use of instrument simulators to address the significant differences between how a climate model represents clouds and what the instruments observe, in terms of spatial scales and the properties of the hydrometeors that are either modeled or observed. Space-borne cloud measurements such as that compiled by the International Satellite Cloud Climatology Project (ISCCP) (Rossow & Schiffer, 1999) are the primary source of observational data for evaluating climate-model-simulated clouds because of the unmatched spatiotemporal coverage. The introduction of ISCCP simulator by Klein & Jakob (1999) and Webb et al. (2001) leads to the wide use of the ISCCP data for model cloud evaluations by converting the simulated clouds to be directly comparable to the satellite retrievals. Following the similar approach, simulators to exploit data from other space-borne sensors and imagers have been developed over time and integrated to become the community satellite simulator package (COSP) (see Bodas-Salcedo et al., 2011, and the references therein for individual simulators). COSP has since become a standard tool in the community for comparing climate model clouds with satellite observations (e.g., Kay et al., 2012; Klein et al., 2013; Konsta et al., 2012, 2016; Lin & Zhang, 2004; Nam et al., 2012; Xie et al., 2013; M. Zhang et al., 2005; Y. Zhang et al., 2019). In this regard, the ground-based radar simulators (Lamer et al., 2018; Y. Zhang et al. 2018) provide an important tool to fill in the spatial gap between GCM clouds and detailed cloud observations from field programs. In the same spirit, forward Cloud Resolving Model Radar Simulator (CR-SIM) (Oue et al., 2020) enables direct comparison with remote-sensing products for CRM simulations.

High-quality observations are needed to facilitate the evaluations using these metrics and diagnostics. Space-borne measurements of clouds (e.g., Chepfer et al., 2010; Rossow & Schiffer, 1999), precipitation (e.g., Adler et al., 2003; Huffman et al., 2001; Huffman et al., 2009; Xie & Arkin, 1997), and radiation (e.g., Kato et al., 2013; Loeb et al., 2009) are the primary observational data for evaluating global simulations. High-frequency data from field programs are best suited for evaluating the process models using the process-oriented diagnostics. Among them, the value-added products derived from field measurements at permanent or mobile research sites, such as the ARM Best Estimate (ARMBE, Xie et al., 2010b), the Active Remote Sensing of Cloud (ARSCL, Clothiaux et al., 2000) products, and the ARM Cloud Retrieval Ensemble Data (ACRED, Zhao et al., 2012), are the most commonly used data products to evaluate cloud amount, cloud vertical structure and cloud properties, precipitation, as well as surface radiative fluxes. This is particularly true for parameterization evaluation using the integrated SCM-CRM-LES framework.

17.5. CONCLUDING REMARKS AND FUTURE OUTLOOK

Rigorous testing and evaluation of atmospheric fast physics parameterizations is an integral part of the model development process. It is obvious the ultimate tests of the parameterizations are their performance in climate simulations or numerical weather predictions. But accurate representation of the fast physics processes in large-scale models is often challenging because of the complexity of the physical processes and the practical need of simplification for parameterizations. The successes of the parameterizations therefore hinge on robust understanding of the processes and creative means to conceptualize and represent them in the models. This process will take many trials and errors. Multipronged modeling frameworks are therefore needed to efficiently and effectively test the parameterizations and characterize the model behavior from process level to climate scale in order to confidently advance the parameterizations.

Single-column model that represents an independent column atmosphere of a full-fledged GCM has been used as an efficient testbed for physical parameterizations. Its wide adoption is powered by the research field measurement programs such as ARM (Mather & Voyles, 2013) and COUDNET (Illingworth et al., 2007; Neggers et al., 2012) that provide both large-scale forcing data required for SCM simulations and cloud property measurements to evaluate them. Detailed field data collected during major field campaigns and at permanent research sites, as well as the companion forcing data, enable SCM simulations and evaluations for specific cloud regimes that a parameterization may target or for long-term continuous evaluation that may reflect the overall performance of the full model climate simulations. Since the same data can also be used to drive and evaluate cloud-resolving or large-eddy-resolving process models, namely CRM and LES models, the integrated SCM-CRM-LES process-modeling framework is promoted to facilitate both the development and evaluation of the parameterizations. In such an integrated modeling framework, CRM and LES simulations can provide critical additional information, for example cloud and turbulence statistics to complement instrument measurements to aid the effort.

Placing climate model in weather forecasting mode is another framework that is being increasingly used for parameterization evaluation on local to global scale. Unlike SCM, this full model hindcast approach allows full interaction between the dynamics and the physics. It is built on the hypothesis that the model biases in cloud-related quantities are largely attributable to the parameterized physics given that the errors in large-scale atmospheric condition are limited during short-range simulations. It initializes the climate model from operational analysis or reanalysis data, along with a nudging simulation to spin up the aerosol and cloud fields as well as the land surface model that is coupled to the atmosphere model. Strong correspondences in the model biases related to fast physics between short-term hindcasts and long-term climate simulations have been documented, paving the way to using this framework to investigate and attribute the climatological biases in clouds, precipitation, and radiation with a focus on their representations in GCMs.

A set of well-established climate metrics and process-oriented diagnostics can be used to guide the evaluation of physical parameterizations. Common climate metrics including climatological mean, seasonal cycle, pattern correlation, and zonal and vertical structures can be used to assess the overall performance and track the progress in simulating cloud-related fields in global climate simulations or hindcasts. Most of these metrics can also be used in continuous evaluation with SCM. Process-oriented diagnostics are particularly valuable in characterizing the model biases and gaining insights into the underlying causes. Some of such diagnostics, including the diurnal cycle of precipitation, the shape of the probability distribution of precipitation rate, the onset of convection, and the initialization of warm rain, can provide insight into fundamental underlying processes that remain challenging to model. Evaluations using these diagnostics may very likely inform whether a new parameterization represents an improvement, and therefore they should be routinely performed. High-frequency value-added data products from permanent research sites and other IOP field campaigns can best facilitate such process-oriented evaluations.

Development, testing, and evaluation of fast physics parameterizations are a continuing endeavor. Climate models are trending toward increasingly higher resolution, and physical parameterizations are becoming more physically based, thanks to the increase in computing power and improving understanding of the fast physics processes. The parameterization evaluation framework described above will continue to be relied on though it is expected to have a gradual shift of focus away from convective parameterization. With increase in resolution, it is desirable that parameterizations to be scale aware, allowing the schemes developed and validated for the current generation models continue to be effective when grid resolution increases. The SCM framework will always be an efficient platform for initial testing and evaluation of new parameterization developments. The integrated SCM-CRM-LES framework will see even more use in the development and evaluation cycle in order for both to be more process-oriented to meet the new needs of more physically based parameterizations. The potentials of projects such as LASSO (Gustafson et al., 2020), the large-eddy simulation Atmospheric Radiation Measurement (ARM) Symbiotic Simulation and Observation, which provides unprecedented observationally constrained synthetic four-dimensional cloud process data, will further promote the concept of using the integrated framework. The hindcast framework will become more valuable. Its ability to represent the interaction between the physics and dynamics at multiple scales, particularly when grid resolution is high and cloud processes become partially explicitly resolved, can be well suited to test the scale awareness of the parameterizations. More importantly, a growing number of climate models start to implement regional refinement mesh (RRM) capability. In theory, a same model can be a seamless framework (such as the Met Office's Unified Model; Walters et al., 2019) for GCM, SCM, and CRM, with RRM to configure for CRM simulations for regions of interest. The combination of the hindcast framework with the RRM capability will be a unique platform to transform the development and evaluation of physical parameterization toward more process-oriented and physically-based as demonstrated in Tang et al. (2019).

ACKNOWLEDGMENTS

Work at LLNL was supported by the U.S. Department of Energy Atmospheric Radiation Measurement and Earth and Environmental System Modeling (EEMD) programs and performed under the auspices of the U.S. Department of Energy by Lawrence Livermore National Laboratory under contract no. DE-AC52-07NA27344. Work at BNL was supported by the U.S. DOE's EEMD programs under contract no. DE-SC0012704. The authors wish to thank the two anonymous reviewers for their constructive inputs that help to broaden the view of this chapter.

REFERENCES

Adler, R. F., Huffman, G. J., Chang, A., Ferraro, R., Xie, P., Janowiak, J., et al. (2003). The version 2 Global Precipitation Climatology Project (GPCP) monthly precipitation analysis (1979-Present). *Journal of Hydrometeorology, 4*(6), 1147–1167. https://doi.org/10.1175/1525-7541(2003)004<1147:TVGPCP>2.0.CO;2.

Ahlgrimm, M., & Forbes, R. (2014). Improving the representation of low clouds and drizzle in the ECMWF model based on ARM observations from the Azores. *Monthly Weather Review, 142*, 668–685. https://doi.org/10.1175/MWR-D-13-00153.1.

Ahn, M.-S., Kim, D., Sperber, K. R., Kang, I.-S., Maloney, R., Waliser, D., & Hendon, H. (2017). MJO simulation in CMIP5 climate models: MJO skill metrics and process-oriented diagnosis. *Climate Dynamics, 49*, 4023–4045.

Angevine, W. M., Olson, J., Kenyon, J., Gustafson, W., I., Endo, S., Suselj, K., & Turner, D. D. (2018). Shallow cumulus in WRF parameterizations evaluated against LASSO large-eddy simulations. *Monthly Weather Review, 146*, 4303–4322, https://doi.org/ 10.1175/MWR-D-18-0115.1.

Arakawa, A., & Schubert, W. H. (1974). Interaction of a cumulus cloud ensemble with the large-scale environment, Part I. *Journal of the Atmospheric Sciences, 31*, 674–701. https://doi.org/10.1175/1520-0469(1974)031<0674:IOACCE>2.0.CO;2.

Barnes, S. L. (1964). A technique for maximizing details in numerical map analysis. *Journal of Applied Meteorology, 3*, 396–409.

Barton, N. P., Klein, S. A., Boyle, J. S., & Zhang, Y. (2012) Arctic synoptic regimes: Comparing domain-wide Arctic cloud observations with CAM4 and CAM4 during similar dynamics. *Journal of Geophysical Research, 117*, D15205, doi:10.1029/2012JD017589.

Bechtold, P., Krueger, S. K., Lewellen, W. S., van Meijgaard, E., Moeng, C.-H., Randall, D. A., et al. (1996) Modeling a stratocumulus-topped PBL: Intercomparison among different one-dimensional codes and with large eddy simulation. *Bulletin of the American Meteorological Society, 77*, 2033–2042. doi:10.1175/1520-0477(1996)077,2033:MASTPI.2.0.CO;2.

Bechtold, P., Redelsperger, J.-L., Beau, I., Blackburn, M., Brinkop, S., Grandper, J.-Y., et al. (2000). A GCSS model intercomparison for a tropical squall line observed during TOGA-COARE: II: Intercomparison of single-column models and a cloud-resolving model. *Quarterly Journal of the Royal Meteorological Society, 126*, 865–888.

Bengtsson, L., Dias, J., Gehne, M., Bechtold, P., Whitaker, J., Bao, J., et al. (2019). Convectively coupled equatorial wave simulations using the ECMWF IFS and the NOAA GFS cumulus convection schemes in the NOAA GFS model. *Monthly Weather Review, 147*, 4005–4025. https://doi.org/10.1175/MWR-D-19-0195.1.

Betts, A. K., & Miller, M. J. (1986). A new convective adjustment scheme. Part II: Single column tests using GATE wave, BOMEX, ATEX, and arctic air mass data sets. *Quarterly Journal of the Royal Meteorological Society, 112*, 693–709. doi:10.1002/qj.49711247308.

Bodas-Salcedo, A., Webb, M. J., Bony, S., Chepfer, H., Dufresne, J.-L., Klein, S. A., et al. (2011). COSP: Satellite simulation software for model assessment. *Bulletin of the American Meteorological Society, 92*, 1023–1043. https://doi.org/10.1175/2011BAMS2856.1.

Bogenschutz, P. A., Gettelman, A., Morrison, H., Larson, V. E., Schanen, D. P., Meyer, N. R., & Craig, C. (2012). Unified parameterization of the planetary boundary layer and shallow convection with a higher-order turbulence closure in the Community Atmosphere Model. *Geoscientific Model Development, 5*, 1407–1423.

Bogenschutz, P. A., Gettelman, A., Morrison, H., Larson, V. E., Craig, C., & Schanen, D. P. (2013). Higher-order turbulence closure and its impact on climate simulations in the Community Atmosphere Model. *Journal of Climate, 26*, 9,655–9,676. doi: 10.1175/JCLI-D-13-00075.1.

Bogenschutz, P. A., Tang, S., Caldwell, P., Xie, S., Lin, W., & Chen, Y. (2020). The E3SM version 1 Single Column Model. *Geoscientific Model Development, 13*, 4443–4458. https://doi.org/10.5194/gmd-13-4443-2020.

Bony, S., Dufresne, J. L., Le Treut, H., Morcrette, J. J., & Senior, C. (2004). On dynamic and thermodynamic components of cloud changes. *Climate Dynamics, 22*, 71–86.

Boucher, O., Randall, D., Artaxo, P., Bretherton, C., Feingold, G., Forster, P., et al. (2013). Clouds and aerosols. In T. F. Stocker et al. (Eds.), *Climate Change 2013: The Physical Science Basis*. Contribution of Working Group I to the Fifth Assessment Report of the Intergovernmental Panel on Climate Change. Cambridge, U.K., and New York: Cambridge University Press.

Boyle, J., & Klein, S. A. (2010), Impact of model horizontal resolution on climate model forecasts of tropical precipitation and diabatic heating for the TWP-ICE period. *Journal of Geophysical Research, 115*, D23113, doi:10.1029/2010JD014262

Boyle, J., Klein, S. A., Zhang, G., Xie, S., & Wei, X. (2008). Climate model forecast experiments for TOGA-COARE.

Monthly Weather Review, 136, 808–832, doi: 10.1175/2007 MWR2145.1.

Boyle, J. S., Williamson, D. L., Cederwall, R., Fiorino, M., Hnilo, J., Olson, J. G., et al. (2005). Diagnosis of Community Atmospheric Model 2 (CAM2) in numerical weather forecast configuration at Atmospheric Radiation Measurement sites. *Journal of Geophysical Research, 110*, D15S15. doi:10.1029/2004JD005042.

Bretherton, C. S., Peters, M. E., & Back, L. E. (2004). Relationships between water vapor path and precipitation over the tropical oceans. *Journal of Climate, 17*, 1517–1528. https://doi.org/10.1175/1520-0442(2004)017<1517:RBWVPA>2.0.CO;2.

Brown, A. R., Cederwall, R. T., Chlond, A., Duynkerke, P. G., Golaz, J.-C., Khairoutdinov, M., et al. (2002). Large-eddy simulation of the diurnal cycle of shallow cumulus convection over land. *Quarterly Journal of the Royal Meteorological Society, 128*, 1075–1093. doi:10.1256/003590002320373210.

Browning, K. A., Betts, A., Jonas, P. R., Kershaw, R., Manton, M., Mason, P. J., et al. (1993). The GEWEX Cloud System Study (GCSS). *Bulletin of the American Meteorological Society, 74*, 387–400. https://doi.org/10.1175/1520-0477(1993)074<0387:TGCSS>2.0.CO;2.

Cederwall, R. T., Yio, J. J., & Krueger, S. K. (1998). The ARM SCM intercomparison study—Overview and preliminary results for case 1. Proc. Eighth Atmospheric Radiation Measurement (ARM) Science Team Meeting, Tucson, AZ, U.S. Department of Energy, 123–126. [Available online at http://www.arm.gov/publications/proceedings/conf08/extended_abs/cederwall_rt.pdf.]

Chepfer, H., Bony, S., Winker, D. M., Cesana, G., Dufresne, J.-L. Minnis, P., et al. (2010). The GCM Oriented CALIPSO Cloud Product (CALIPSO-GOCCP). *Journal of Geophysical Research, 115*, D00H16. doi: 10.1029/2009JD012251.

Christensen, H. M. (2019). Constraining stochastic parametrisation schemes using high-resolution simulations. *Quarterly Journal of the Royal Meteorological Society, 146*, 938–962. https://doi.org/10.1002/qj.3717.

Christensen, H. M., Dawson, A., & Holloway, C. E. (2018). Forcing single-column models using high-resolution model simulations. *Journal of Advances in Modeling Earth Systems, 10*, 1833–1857. https://doi.org/10.1029/2017MS001189.

Clothiaux, E. E., Ackerman, T. P., Mace, G. G., Moran, K. P., Marchand, R. T., Miller, M. A., & Martner B. E. (2000). Objective determination of cloud heights and radar reflectivities using a combination of active remote sensors at the Atmospheric Radiation Measurement Program Cloud and Radiation Test Bed (ARM CART) sites. *Journal of Applied Meteorology, 39*, 645–665.

Covey, C., Gleckler, P. J., Doutriaux, C. Williams, D. N., Dai, A., Fasullo, J., et al. (2016). Metrics for the diurnal cycle of precipitation: Toward routine benchmarks for climate models. *Journal of Climate, 29*, 4461–4471. https://doi.org/10.1175/JCLI-D-15-0664.1

Dai, A. (2006). Precipitation characteristics in eighteen coupled climate models. *Journal of Climate, 19*, 4605–4630. https://doi.org/10.1175/JCLI3884.1.

Daleu, C. L., Plant, R. S., & Woolnough, S. J. (2017). Using the weak-temperature gradient approximation to evaluate parameterizations: An example of the transition from suppressed to active convection. *Journal of Advances in Modeling Earth Systems, 9*, 2350–2367, https://doi.org/10.1002/2017MS000940.

Daleu, C. L., Plant, R. S., Woolnough, S. J., Sessions, S., Herman, A., Sobel, A., et al. (2015). Intercomparison of methods of coupling between convection and large-scale circulation: 1. Comparison over uniform surface conditions. *Journal of Advances in Modeling Earth Systems, 7*, 1576–1601. doi:10.1002/2015MS000468.

Dal Gesso, S., van der Dussen, J. J., Siebesma, A. P., de Roode, S. R., Boutle, I. A., Kamae, Y., et al. (2015). A single-column model intercomparison on the stratocumulus representation in present-day and future climate. *Journal of Advances in Modeling Earth Systems, 7*, 617–647. https://doi.org/10.1002/2014MS000377.

Davies, L., Jakob, C., Cheung, K., Del Genio, A., Hill, A., Hume, T., et al. (2013). A single Column model ensemble approach applied to the TWP-ICE experiment. *Journal of Geophysical Research, 118*, 6544–6563. doi:10.1002/jgrd.50450.

Deardorff, J. W. (1970). Preliminary results from numerical integration of the unstable planetary boundary layer. *Journal of the Atmospheric Sciences, 27*, 1209–1211.

Dufresne, J. L., & Bony, S. (2008). An assessment of the primary sources of spread of global warming estimates from coupled atmosphere-ocean models. *Journal of Climate, 21*, 5135–5144.

Endo, S., Fridlind, A., Lin, W., Vogelmann, A. M., Toto, T., Ackerman, A., et al. (2015). RACORO continental boundary layer cloud investigations. 2: Large-eddy simulations of cumulus clouds and evaluation with in-situ and ground-based observations. *Journal of Geophysical Research, 120*, 5993–6014. doi:10.1002/2014JD022525.

Eyring, V., Righi, M., Lauer, A., evaldsson, M., Wenzel, S., Jones, C., et al. (2016). ESMValTool (v1.0)—a community diagnostic and performance metrics tool for routine evaluation of Earth system models in CMIP. *Geoscientific Model Development, 9*, 1747–1802. https://doi.org/doi:10.5194/gmd-9-1747-2016.

Freitas, S. R., Grell, G. A., & Li, H. (2020). The GF Convection Parameterization: Recent developments, extensions, and applications. *Geoscientific Model Development, 14*, 5393–5411. https://doi.org/10.5194/gmd-14-5393-2021.

Fridlind, A. M., Ackerman, A. S., Chaboureau, J.-P., Fan, J., Grabowski, W. W., Hill, A. A., et al. (2012). A comparison of TWP-ICE observational data with cloud-resolving model results. *Journal of Geophysical Research, 117*, D05204. doi: 10.1029/2011JD016595.

Ghan, S. J., Randall, D., Xu, K., Cederwall, R., Cripe, D., Hack, J., et al. (2000). A comparison of single-column model simulations of Summertime midlatitude continental convection. *Journal of Geophysical Research: Atmospheres, 105(*D2*)*, 2091–2124.

Ghonima, M., Yang, H., Kim, C., Heus, T., & Kleissl, J. (2017). Evaluation of WRF SCM simulations of stratocumulus-topped marine and coastal boundary layers and improvements to turbulence and entrainment parameterizations. *Journal of Advances in Modeling Earth Systems, 9*, 2635–2653. https://doi.org/10.1002/2017MS001092.

Gettelman, A., Truesdale, J. E., Bacmeister, J. T., Caldwell, P. M., Neale, R. B., Bogenschutz, P. A., & Simpson, I. R. (2019). The Single Column Atmosphere Model Version 6 (SCAM6): Not a scam but a tool for model evaluation and development. *Journal of Advances in Modeling Earth Systems, 11,* 1381–1401. https://doi.org/10.1029/2018MS001578.

Gleckler, P. J., Taylor, K. E., & Doutriaux, C. (2008). Performance metrics for climate models. *Journal of Geophysical Research, 113,* D06104. doi:10.1029/2007JD008972.

Golaz, J. C., Larson, V. E., & Cotton, W. R. (2002). A PDF-based model for boundary layer clouds. Part I: Method and model description. *Journal of the Atmospheric Sciences, 59*(24), 3540–3551. https://doi.org/10.1175/1520-0469(2002)059<3540:APBMFB>2.0.CO;2.

Guichard, F., & Couvreux, F. (2017). A short review of numerical cloud-resolving models. *Tellus A: Dynamic Meteorology and Oceanography, 69,* 1–36. https://doi.org/10.1080/16000870.2017.1373578.

Guichard, F., Petch, J. C., Redelsperger, J.-L., Bechtold, P., Chaboureau, J.-P., Cheinet, S., et al. (2004). Modelling the diurnal cycle of deep precipitating convection over land with cloud-resolving models and single-column models. *Quarterly Journal of the Royal Meteorological Society, 130,* 3139–3172. doi:https://doi.org/10.1256/qj.03.145.

Guo, H., Golaz, J.-C., Donner, L. J., Ginoux, P., & Hemler, R. S. (2014). Multivariate probability density functions with dynamics in the GFDL atmospheric general circulation model: Global tests. *Journal of Climate, 27,* 2087–2108. 10.1175/JCLI-D-13-00347.1.

Guo, H., Golaz, J.-C., Donner, L. J., Larson, V. E., Schanen, D. P., & Griffin, B. M. (2010). Multi-variate probability density functions with dynamics for cloud droplet activation in large-scale models: Single column tests. *Geoscientific Model Development, 3,* 475–486. doi:https://doi.org/10.5194/gmd-3-475-2010.

Gustafson, W. I., Jr., Vogelmann,A. M., Li, Z., Cheng, X., Dumas, K. K., Endo, S., et al. (2020). The Large-Eddy Simulation (LES) Atmospheric Radiation Measurement (ARM) Symbiotic Simulation and Observation (LASSO) activity for continental shallow convection. *Bulletin of the American Meteorological Society, 101,* E462–E479. https://doi.org/10.1175/BAMS-D-19-0065.1.

Hacker, J. P., & Angevine, W. M. (2013). Ensemble data assimilation to characterize surface-layer errors in numerical weather prediction models. *Monthly Weather Review, 141,* 1804–1821. https://doi.org/10.1175/MWR-D-12-00280.1.

Hogan, R. J., O'Connor, E. J., & Illingworth, A. J. (2009). Verification of cloud fraction forecasts. *Quarterly Journal of the Royal Meteorological Society, 135,* 1494–1511. https://doi.org/10.1002/qj.481.

Hoke, J. E., & Anthes, R. A. (1976). The initialization of numerical models by a dynamic-initialization technique. *Monthly Weather Review, 104,* 1551–1556. https://doi.org/10.1175/1520-0442(2004)017%3C3302:EOCATR%3E2.0.CO;2.

Holloway, C. E., & Neelin, J. D. (2009). Moisture vertical structure, column water vapor, and tropical deep convection. *Journal of the Atmospheric Sciences, 66,* 1665–1683. https://doi.org/10.1175/2008JAS2806.1.

Hourdin, F., Grandpeix, J.-Y., Rio, C., Bony, S., Jam, A., Cheruy, F., et al. (2013). LMDZ5B: The atmospheric component of the IPSL climate model with revisited parameterizations for clouds and convection. *Climate Dynamics, 40,* 2193–2222. https://doi.org/10.1007/s00382-012-1343-y.

Huang, H.-Y., Hall, A., & Teixeira (2013). Evaluations of the WRF PBL parameterizations for marine boundary layer clouds: Cumulus and stratocumulus. *Monthly Weather Review, 141,* 2265–2271. https://doi.org/10.1175/MWR-D-12-00292.1.

Huffman, G. J., Adler, R. F., Bolvin, D. T., & Gu, G. (2009). Improving the global precipitation record: GPCP version 2.1. *Geophysical Research Letters, 36,* L17808. https://doi.org/10.1029/2009GL040000.

Huffman, G. J., Adler, R. F., Morrissey, M. M., Bolvin, D. T., Curtis, S., Joyce, R., et al. (2001). Global precipitation at one-degree daily resolution from multisatellite observations. *Journal of Hydrometeorology, 2,* 36–50. https://doi.org/10.1175/1525-7541(2001)002<0036:GPAODD>2.0.CO;2.

Illingworth, A. J., Hogan, R. J., O'Connor, E. J., Bouniol, D., Brooks, M. E., Delanoë, J., et al. (2007). CLOUDNET. Continuous Evaluation of cloud profiles in seven operational models using ground-based observation. *Bulletin of the American Meteorological Society, 88,* 883–898. https://doi.org/10.1175/BAMS-88-6-883.

Jakob, C. (2003). An improved strategy for the evaluation of cloud parameterizations in GCMs. *Bulletin of the American Meteorological Society, 84,* 1387–1401. https://doi.org/10.1175/BAMS-84-10-1387.

Jakob, C. (2010). Accelerating progress in global atmospheric model development through improved parameterizations. *Bulletin of the American Meteorological Society, 91,* 869–876. https://doi.org/10.1175/2009BAMS2898.1.

Jing, X., Suzuki, K., Guo, H., Goto, D., Ogura, T., Koshiro, T., & Mülmenstädt, J. (2017). A multimodel study in warm precipitation biases in global models compared to satellite observations. *Journal of Geophysical Research: Atmospheres, 122,* 11,806–11,824. https://doi.org/10.1002/2017JD027310.

Kato, S., Loeb, N. G., Rose, F. G., Doelling, D. R., Rutan, D. A., Caldwell, T. E., et al. (2013). Surface irradiances consistent with CERES derived top-of-atmosphere shortwave and longwave irradiances. *Journal of Climate, 26,* 2719–2740. https://doi.org/10.1175/JCLI-D-12-00436.1

Kay, J. E., Hillman, B., Klein, S., Zhang, Y., Medeiros, B., Gettelman, G., et al. (2012). Exposing global cloud biases in the Community Atmosphere Model (CAM) using satellite observations and their corresponding instrument simulators. *Journal of Climate, 25,* 5190–5207. http://dx.doi.org/10.1175/JCLI-D-11-00469.1.

Khairoutdinov, M. F., Krueger, S. K., Moeng, C.-H., Bogenschutz, P. A., & Randall, D. A. (2009). Large-eddy simulation of maritime deep tropical convection. *Journal of Advances in Modeling Earth Systems, 1,* 15. https://doi.org/10.3894/JAMES.2009.1.15.

Khairoutdinov, M., & Randall, D. (2006). High-resolution simulation of shallow-to-deep convection transition over land. *Journal of the Atmospheric Sciences, 63,* 3421–3436.

Kim, D., Sperber, K., Stern, W., Waliser, D., Kang, I.-S., Maloney, E., et al. (2009). Application of MJO simulation diagnostics to climate models. *Journal of Climate, 22,* 6413–6436.

Klein, S. A., & Jakob, C. (1999). Validation and sensitivities of frontal clouds simulated by the ECMWF model. *Monthly Weather Review, 127*, 2514–2531.

Klein, S. A., Jiang, X., Boyle, J., Malyshev, S., & Xie, S. (2006). Diagnosis of the summertime warm and dry bias over the U. S. Southern Great Plains in the GFDL climate model using a weather forecasting approach. *Geophysical Research Letters, 33*, L18805. doi:10.1029/2006GL027567.

Klein, S. A., McCoy, R. B., Morrison, H., Ackerman, A. S., Avramov, A., de Boer, G., et al. (2009). Intercomparison of model simulations of mixed-phase clouds observed during the ARM Mixed-Phase Arctic Cloud Experiment. Part I: Single-layer cloud. *Quarterly Journal of the Royal Meteorological Society, 135*, 979–1002. doi:10.1002/qj.416.

Klein, S. A., Zhang, Y., Zelinka, M. D., Pincus, R., Boyle, J., & Gleckler, P. J. (2013). Are climate model simulations of clouds improving? An evaluation using the ISCCP simulator. *Journal of Geophysical Research: Atmospheres, 118*, 1329–1342. doi: 10.1002/jgrd.50141.

Köhler, M., & Teixeira, J. (2002). ECMWF Single column model—Reference and user's guide Cycle 23R4, Shinfield Park, Reading RG2 9AX, UK. *On request from* openifs-support@ecmwf.int.

Konsta, D., Chepfer, H., & Dufresne, J.-L. (2012). A process oriented characterization of tropical oceanic clouds for climate model evaluation, based on a statistical analysis of daytime A-train observations. *Climate Dynamics, 39,* 2091–2108.

Konsta, D., Dufresne, J.-L., Chepfer, H., Idelkadi, A., & Cesana, G. (2016). Use of A-train satellite observations (CALIPSO-PARASOL) to evaluate tropical cloud properties in the LMDZ5 GCM. *Climate Dynamics, 46,* 1263–1284. https://doi.org/10.1007/s00382-015-2900-y.

Krueger, S. K. (1988). Numerical simulation of tropical cumulus clouds and their interaction with the subcloud layer. *Journal of the Atmospheric Sciences, 45*, 2221–2868.

Krueger, S. K., McLean, G. T., & Fu, W. (1995). Numerical simulation of the stratus-to-cumulations transition in the subtropical marine boundary layer. Part I: Boundary-layer circulation. *Journal of the Atmospheric Sciences, 52*, 2851–2868.

Krueger, S. K., Morrison, H., & Fridlind, A. (2016). Cloud-resolving modeling: ARM and the GCSS story. *Meteorological Monographs, 57,* 25.1–25.16. https://doi.org/10.1175/AMSMONOGRAPHS-D-15-0047.1.

Lamer, K., Fridlind, A. M., Ackerman, A. S., Kollias, P., Clothiaux, E., & Kelley, M. (2018). $(GO)^2$-SIM: A GCM-oriented ground-observation forward-simulator framework for objective evaluation of cloud and precipitation phase. *Geoscientific Model Development, 11*, 4195–4214. https://doi.org/10.5194/gmd-11-4195-2018.

Larson, V. E., & Golaz, J. (2005). Using probability density functions to derive consistent closure relationships among higher-order moments. *Monthly Weather Review, 133*, 1023–1042. doi: 10.1175/MWR2902.1.

Lee, M.-I., Schubert, S. D., Suarez, M. J., Held, I. M., Lau, N.-C., Ploshay, J. J., et al. (2007). An analysis of the warm-season diurnal cycle over the continental United States and Northern Mexico in General Circulation Models. *Journal of Hydrometeorology, 8*(3), 344–366. https://doi.org/10.1175/JHM581.

Lenderink, G., Siebesma, A. P., Cheinet, S., Irons, S., Jones, C. G., Marquet, P., et al. (2004) .The diurnal cycle of shallow cumulus clouds over land: A single-column model intercomparison study. *Quarterly Journal of the Royal Meteorological Society, 130*, 3339–3364. doi:10.1256/qj.03.122.

Lin, J. (2007). The double-ITCZ problem in IPCC AR4 coupled GCMs: Ocean-Atmosphere feedback analysis, *Journal of Climate, 20,* 4497–4525. https://doi.org/10.1175/JCLI4272.1.

Lin, J., Kiladis, G. N., Mapes, B. E., Weickmann, K. M., Sperber, K. R., Lin, W., et al. (2006). Tropical intraseasonal variability in 14 IPCC AR4 climate models. Part I: Convective signals. *Journal of Climate, 19,* 2665–2690. doi:10.1175/JCLI3735.1.

Lin, W., Liu, Y., Vogelmann, A. M., Fridlind, A. M., Endo, S., Feng, S., et al. (2015). RACORO continental boundary layer cloud investigations. Part III: Separation of parameterization biases in Single-Column Model CAM5 simulations of shallow cumulus. *Journal of Geophysical Research: Atmospheres, 120,* 6015–6033. doi: 10.1002/2014JD022524.

Lin, W., & Zhang, M. (2004). Evaluation of clouds and their radiative effects simulated by the NCAR Community Atmospheric Model CAM2 against satellite observations. *Journal of Climate*, *17*, 3302–3318.

Lin, X., & Johnson, R. H. (1996). Kinematic and thermodynamic characteristics of the flow over the western Pacific warm pool during TOGA COARE. *Journal of the Atmospheric Sciences, 53*, 695–715. doi:10.1175/1520-0469(1996)053,0695:KATCOT.2.0.CO;2.

Lin, Y., Donner, L. J., Petch, J., Bechtold, P., Boyle, J., Klein, S. A., et al., (2012). TWP-ICE global atmospheric model intercomparison: Convection responsiveness and resolution impact. *Journal of Geophysical Research, 117*, D09111. doi: 10.1029/2011JD017018.

Lock, A. P., Brown, A. R., Bush, M. R., Martin, G. M., & Smith, R. N. B. (2000). A new boundary layer mixing scheme. Part I:: Scheme description and single-column model tests. *Monthly Weather Review, 128,* 3187–3199.

Loeb, N. G., Wielicki, B. A., Doelling, D. R., Smith, G. L., Keyes, D. F., Kato, S., et al. (2009). Toward optimal closure of the Earth's top-of atmosphere radiation budget. *Journal of Climate, 22,* 748–766. https://doi.org/10.1175/2008JCLI2637.1.

Lord, S. J. (1982). Interaction of a cumulus cloud ensemble with the large-scale environment. Part III: Semi-prognostic test of the Arakawa–Schubert cumulus parameterization. *J Journal of the Atmospheric Sciences, 39*, 88–103. doi:10.1175/1520-0469(1982)039,0088:IOACCE.2.0.CO;2.

Luo, Y., Krueger, S. K., Mace, G. G., & Xu, K.-M. (2003). Cirrus cloud properties from a cloud-resolving model simulation compared to cloud radar observations. *Journal of the Atmospheric Sciences, 60*, 510–525. doi:10.1175/1520-0469(2003)060,0510:CCPFAC.2.0.CO;2.

Luo, Y., Krueger, S. K., & Moorthi, S. (2005). Cloud properties simulated by a single-column model. Part I: comparison to cloud radar observations of cirrus clouds. *Journal of the*

Atmospheric Sciences, 62, 1428–1445. https://doi.org/10.1175/JAS3425.1.

Luo, Y., Krueger, S. K., & Xu, K.-M. (2006). Cloud properties simulated by a single-column model. Part II: evaluation of cumulus detrainment and ice-phase microphysics using a cloud-resolving model. *Journal of the Atmospheric Sciences, 63*, 2831–2847. https://doi.org/10.1175/JAS3785.1.

Ma, H.-Y., Chuang, C. C., Klein, S. A., Lo, M.-H., Zhang, Y., Xie, S., et al. (2015). An improved hindcast approach for evaluation and diagnosis of physical processes in global climate models. *Journal of Advances in Modeling Earth Systems, 7*, 1810–1827. doi:10.1002/2015MS000490.

Ma, H.-Y., Klein, S. A., Xie, S., Zhang, C., Tang, S., Tang, Q., et al. (2018). CAUSES: On the role of surface energy budget errors to the warm surface air temperature error over the Central United States. *Journal of Geophysical Research, 123*, 2888–2909. https://doi.org/10.1002/2017JD027194.

Ma, H.-Y., Siongco, A. C., Klein, S. A., Xie, S., Karspeck, A. R., Raeder, K., et al. (2020). On the correspondence between seasonal forecast and long-term climate errors in sea surface temperatures. *Journal of Climate, 34*, 427–446. https://doi.org/10.1175/JCLI-D-20-0338.1.

Ma, H., Xie, S., Boyle, J. S., Klein, S. A., & Zhang Y. (2013). Development of Metrics and Diagnostics for precipitation-related processes in climate model short-range hindcasts. *Journal of Climate, 26*, 1516–1534. Doi: 10.1175/JCLI-D-12-00235.1.

Ma, H.-Y., Xie, S., Klein, S. A., Williams, K. D., Boyle, J. S., Bony, S., et al. (2014). On the correspondence between mean forecast errors and climate errors in CMIP5 models. *Journal of Climate, 27*, 1781–1798. doi: 10.1175/JCLI-D-13-00474.1.

Maloney, E. D., Gettelman, A., Ming, Y., Neelin, J. D., Barrie, D., Mariotti, A., et al. (2019). Process-oriented evaluation of climate and weather forecasting models. *Bulletin of the American Meteorological Society, 100*, 1665–1686. https://doi.org/10.1175/BAMS-D-18-0042.1.

Mather, J. H., & Voyles, J. W. (2013). The ARM Climate Research Facility: A review of structure and capability. *Bulletin of the American Meteorological Society, 94*, 377–392. https://doi.org/10.1175/BAMS-D-11-00218.1.

May, P. T., Mather, J. H., Vaughan, G., & Jakob, C. (2008). Characterizing oceanic convective cloud systems: The Tropical Warm Pool International Cloud Experiment. *Bulletin of the American Meteorological Society, 154*, 153–155. doi:10.1175/BAMS-89-2-153.

Medeiros, B., Williamson, D. L., Hannay, C., & Olson, J. G. (2012). Southeast Pacific Stratocumulus in the Community Atmosphere Model. *Journal of Climate, 25*, 6175–6192. doi: 10.1175/JCLI-D-11-00503.1.

Moeng, C.-H. (1984). A large-eddy-simulation model for the study of planetary boundary-layer turbulence. *Journal of the Atmospheric Sciences, 41*(13), 2052–2062.

Moeng, C.-H., Cotton, W. R., Bretherton, C., Chlond, A., Khairoutdinov, M., Krueger, S., et al. (1996). Simulation of a stratocumulus-topped PBL: Intercomparison among different numerical codes. *Bulletin of the American Meteorological Society, 77*, 261–278. doi:10.1175/1520-0477(1996)077,0261:SOASTP.2.0.CO;2.

Moeng, C.-H., LeMone, M. A., Khairoutdinov, M. F., Krueger, S. K., Bogenschutz, P. A., & Randall, D. A. (2009). The tropical marine boundary layer under a deep convection system: A large-eddy simulation study. *Journal of Advances in Modeling Earth Systems, 1*, 16. doi: 10/3894/JAMES.2009.1.16.

Moncrieff, M. W., Krueger, S. K., Gregory, D., Redelsperger, J.-L., & Tao, W.-K. (1997). GEWEX Cloud System Study (GCSS) Working Group 4: Precipitating convective cloud systems. *Bulletin of the American Meteorological Society, 78*, 831–845. doi:10.1175/1520-0477(1997)078,0831:GCSSGW.2.0.CO;2.

Morrison, H., Curry, J. A., Shupe, M. D., & Zuidema, P. (2005). A new double-moment microphysics parameterization for application in cloud and climate models. Part II: Single-column modeling of arctic clouds. *Journal of the Atmospheric Sciences, 62*, 1678–1693. https://doi.org/10.1175/JAS3447.1.

Morrison, H., McCoy, R. B., Klein, S. A., Xie, S., Luo, Y., Avramov, A., et al. (2009). Intercomparison of model simulations of mixed-phase clouds observed during the ARM Mixed-Phase Arctic Cloud Experiment. Part II: Multi-layered cloud. *Quarterly Journal of the Royal Meteorological Society, 135*, 1003–1019. doi:10.1002/qj.415.

Morrison, H., Zuidema, P., Ackerman, A. S., Avramov, A., de Boer, G., Fan, J., et al. (2011). Intercomparison of cloud model simulations of Arctic mixed-phase boundary layer clouds observed during SHEBA/FIRE-ACE. *Journal of Advances in Modeling Earth Systems, 3*, M06003. doi:10.1029/2011MS000066.

Nam, C., Bony, S., Dufresne, J. L., & Chepfer, H. (2012). The 'too few, too bright'tropical low-cloud problem in CMIP5 models. *Geophysical Research Letters, 39*, L21801. https://doi.org/10.1029/2012GL053421.

Neggers, R. A. J., Köhler, M., & Beljaars, A. C. M. (2009). A dual mass flux framework for boundary layer convection. Part I: Transport. *Journal of the Atmospheric Sciences, 66*, 1465–1487.

Neggers, R. A. J., Siebesma, A. P., & Heus, T. (2012). Continuous single-column model evaluation at a permanent meteorological supersite. *Bulletin of the American Meteorological Society, 93*, 1389–1400. https://doi.org/10.1175/BAMS-D-11-00162.1.

Neggers, R. (2015). Attributing the behavior of low-level clouds in large-scale models to subgrid-scale parameterizations. *Journal of Advances in Modeling Earth Systems, 7*, 2029–2043. https://doi.org/10.1002/2015MS000503.

O'Brien, J. J. (1970). A Note on the vertical structure of the eddy exchange coefficient in the planetary boundary layer. *Journal of the Atmospheric Sciences, 27*, 1213–1215.

Ooyama, K. (1987). Scale-controlled objective analysis. *Monthly Weather Review, 115*, 2476–2506.

Oue, M., Tatarevic, A., Kollias, P., Wang, D., Yu, K., & Vogelmann, A. M. (2020). The Cloud-resolving model Radar SIMulator (CR-SIM) Version 3.3: description and applications of a virtual observatory. *Geoscientific Model Development, 13*, 1975–1998. https://doi.org/10.5194/gmd-13-1975-2020.

Oueslati, B., & Bellon, G. (2015). The double ITCZ bias in CMIP5 models: interaction between SST, large-scale circulation and precipitation. *Climate Dynamics, 44*, 585–607.

Park, S., & Bretherton, C. S. (2009). The University of Washington shallow convection and moist turbulence schemes and their impact on climate simulations with the Community Atmosphere Model. *Journal of Climate, 22*, 3449–3469. doi:10.1175/2008JCLI2557.1.

Petch, J., Hill, A., Davies, L., Fridlind, A., Jakob, C., Lin, Y., et al. (2014). Evaluation of intercomparisons of four different types of model simulating TWP-ICE. *Quarterly Journal of the Royal Meteorological Society, 140*, 826–837. doi:10.1002/qj.2192.

Phillips, T. J., Potter, G. L., Williamson, D. L., Cederwall, R. T., Boyle, J. S., Fiorino, M., et al. (2004). Evaluating parameterizations in general circulation models: Climate simulation meet with weather prediction. *Bulletin of the American Meteorological Society, 85*, 1903–1915. doi: 10.1175/BAMS-85-12-1903.

Qin, Y., & Lin, Y. (2018). Alleviated double ITCZ problem in the NCAR CESM1: A new cloud scheme and the working mechanisms. *Journal of Advances in Modeling Earth Systems, 10*, 2318–2332.

Qin, Y., Lin, Y., Xu, S., Ma, H.-Y., & Xie, S. (2018). A diagnostic PDF cloud scheme to improve subtropical low clouds in NCAR Community Atmosphere Model (CAM5). *Journal of Advances in Modeling Earth Systems, 10*, 320–341. doi: 10.1002/2017MS001095.

Randall, D. A., & Cripe, D. G. (1999). Alternative methods for specification of observed forcing in single-column models and cloud system models. *Journal of Geophysical Research, 104*, 24527–24545. doi:10.1029/1999JD900765.

Randall, D., Krueger, S., Bretherton, C., Curry, J., Duynkerke, P., Moncrieff, M., et al. (2003). Confronting models with data: The GEWEX cloud systems study. *Bulletin of the American Meteorological Society, 84*, 455–470. https://doi.org/10.1175/BAMS-84-4-455.

Randall, D. A., Wood, R. A., Bony, S., Colman, R., Fichefet, R., Fyfe, J., et al. (2007). Climate models and their evaluation. In S. Solomon et al. (Eds.), *Climate Change 2007: The Physical Science Basis*. Contribution of Working Group I to the Fourth Assessment Report of the Intergovernmental Panel on Climate Change (pp. 589–662). Cambridge Univ. Press: Cambridge, U.K., and New York.

Randall, D. A., Xu, K.-M., Somerville, R. J. C., & Iacobellis, S. (1996). Single-column models and cloud ensemble models as links between observations and climate models. *Journal of Climate, 9*(8), 1683–1697. https://doi.org/10.1175/1520-0442(1996)009<1683:SCMACE>2.0.CO;2.

Rauber, R. M., Stevens, B., Ochs III, H. T., Knight, C., Albrecht, B. A., Blyth, A., M., et al. (2007). Rain in shallow cumulus over the ocean–the RICO campaign. *Bulletin of the American Meteorological Society, 88*, 1912–1928. doi:10.1175/BAMS-88-12-1912.

Redelsperger, J.-L., Brown, P. R A., Guichard, F., How, C., Kawasima, M., Lang, S., et al. (2000). A GCSS model intercomparison for a tropical squall line observed during TOGA-COARE. I: Cloud-resolving models. *Q. J. R. Meteorol. Soc., 126*, 823–864.

Redelsperger, J.-L., & Lafore, J.-P. (1988). A three-dimensional simulation of a tropical squall line: convective organization and thermodynamic vertical transport. *Journal of the Atmospheric Sciences, 45*, 1334–1356.

Rio, C., Hourdin, F., Couvreux, F., & Jam, A. (2010). Resolved versus parametrized boundary-layer plumes. Part II: continuous formulations of mixing rates for mass-flux schemes. *Boundary-Layer Meteorology, 135*, 469–483.

Rossow, W. B., & Schiffer, R. A. (1999). Advances in understanding clouds from ISCCP. *Bulletin of the American Meteorological Society, 80*, 2261–2288. doi:10.1175/1520-0477(1999)080<2261:AIUCFI>2.0.CO;2

Ryan, B. F., Katzfey, J. J., Abbs, J., Jakob, C., Lohmann, U., Rockel, B., et al. (2000). Simulations of a cold front by cloud-resolving, limited-area and large-scale models and a model evaluation using in situ and satellite observations. *Monthly Weather Review, 128*, 3218–3235.

Sherwood, S. C., Bony, S., & Dufresne, J.-L. (2014). Spread in model climate sensitivity traced to atmospheric convective mixing. *Nature, 505*, 37–42. doi:10.1038/nature12829.

Siebesma, A. P., Bretherton, C. S., Brown, A., Chlond, A., Cuxart, J., Duynkerke, P. G., et al. (2003). A large eddy simulation intercomparison study of shallow cumulus convection. *Journal of the Atmospheric Sciences, 60*, 1201–1219.

Siebesma, A. P., Soares, P. M. M., & Teixeira (2006). A combined eddy-diffusivity mass-flux approach for the convective boundary layer. *Journal of the Atmospheric Sciences, 64*, 1230–1248.

Schiro, K. A., Neelin, J. D., Adams, D. K., & Lintner, B. R. (2016). Deep convection and column vapor over tropical land versus tropical ocean: A comparison between the Amazon and the tropical western Pacific. *Journal of the Atmospheric Sciences, 73*(10), 4043–4063.

Schumacher, C., Zhang, M., & Ciesielski, P. E. (2007). Heating structures of the TRMM field campaigns. *Journal of the Atmospheric Sciences, 64*, 2593–2610. https://doi.org/10.1175/JAS3938.1.

Siongco, A. C., Ma, H.-S., Klein, S. A., Xie, S., Karspeck, A. R., Raeder, K., & Anderson, J. L. (2020). A hindcast approach to diagnosing the equatorial Pacific cold tongue SST bias in CESM1. *Journal of Climate, 33*, 1437–1453. https://doi.org/10.1175/JCLI-D-19-0513.1

Smalley, M., Sušelj, K., Lebsock, M., & Teixeira, J. (2019). A novel framework for evaluating and improving parameterized subtropical marine boundary layer cloudiness. *Monthly Weather Review, 147*, 3241–3260. https://doi.org/10.1175/MWR-D-18-0394.1.

Sobel, A. H., Nilsson, J., & Polvani, L. M. (2001). The weak temperature gradient approximation and balanced tropical moisture waves. *Journal of the Atmospheric Sciences, 58*, 3650–3665. doi:10.1175/1520-0469(2001)058,3650:TWTGAA.2.0.CO;2.

Song, H, Lin, W., Lin, Y., Wolf, A., Neggers, R., Donner, L., et al. (2013). Evaluation of precipitation simulated by seven SCMs against the ARM observation at the SGP site. *Journal of Climate, 26*, 5467–5492.

Song, H, Lin, W., Lin, Y., Wolf, A., Neggers, R., Donner, L., et al. (2014). Evaluation of cloud fractions simulated by seven SCMs against the ARM observation at the SGP site. *Journal of Climate, 27*, 6698–6719.

Sušelj, K., Teixeira, J., & Chung, D. (2013). A unified model for moist convective boundary layers based on a stochastic eddy diffusivity/mass-flux parameterization. *Journal of the*

Atmospheric Sciences, 70, 1929–1953. https://doi.org/10.1175/JAS-D-12-0106.1.

Suzuki, K., Stephens, G. L., Bodas-Salcedo, A., Wang, M., Golaz, J.-C., Yokohata, T., & Tsuyoshi, K. (2015). Evaluation of the warm rain formation process in global models with satellite observations. *Journal of the Atmospheric Sciences, 72*, 3996–4014. https://doi.org/10.1175/JAS-D-14-0265.1.

Tang, Q., Klein, S. A., Xie, S., Lin, W., Golaz, J.-C., Roesler, E. L., et al. (2019). Regionally refined capability in E3SM Atmosphere Model Version 1 (EAMv1) and applications for high-resolution modelling. *Geoscientific Model Development, 12*, 2679–2706. https://doi.org/10.5194/gmd-12-2679-2019.

Tang, S., Xie, S., Zhang, Y., Zhang, M., Schumacher, C., Upton, H., et al. (2016). Large-scale vertical velocity, diabatic heating and drying profiles associated with seasonal and diurnal variations of convective systems observed in the GoAmazon2014/15 Experiment. *Atmospheric Chemistry and Physics Discussions, 16*, 1–39. doi: 10.5194/acp-16-14249-2016.

Taylor, K. E. (2001). Summarizing multiple aspects of model performance in a single diagram. *Journal of Geophysical Research: Atmospheres, 106*(D7), 7183–7192. https://doi.org/10.1029/2000JD900719.

Tselioudis, G., & Jakob, C. (2002). Evaluation of midlatitude cloud properties in a weather and a climate model: Dependence on dynamic regime and spatial resolution. *Journal of Geophysical Research, 107*, D24, 4781. doi:10.1029/2002JD002259.

Van Weverberg, K., Morcrette, C. J., Ma, H.-Y., Klein, S. A., & Petch, J. C. (2015). Using regime analysis to identify the contribution of clouds to surface temperature errors in weather and climate models. *Quarterly Journal of the Royal Meteorological Society, 141*, 3190–3206. doi: 10.1002/qj.2603.

Van Weverberg, K., Morcrette, C. J., Petch, J., Klein, S. A., Ma, H.-Y., Zhang, C., et al. (2018). CAUSES: Attribution of surface radiation biases in NWP and climate models near the U.S. Southern Great Plains. *Journal of Geophysical Research, 123*, 3612–3644. https://doi.org/10.1002/2017JD027188.

vanZanten, M., Stevens, B., Nuijens, L., Siebesma, A., Ackerman, A. S., Burnet, F., et al. (2011). Controls on precipitation and cloudiness in simulations of trade wind cumulus as. *Observed during RICO. Journal of Advances in Modeling Earth Systems, 3*, M06001. https://doi.org/10.1029/2011MS000056.

Verlinde, J., Harrigton, J. Y., McFarquhar, G. M., Yannuzzi, V. T., Avramov, A., Greenburg, S., et al. (2007). The Mixed-Phase Arctic Cloud Experiment (M-PACE). *Bulletin of the American Meteorological Society, 88*, 205–221. doi:10.1175/BAMS-88-2-205.

Vial, J., Bony, S., Stevens, B., & Vogel, R. (2017). Mechanisms and model diversity of trade-wind shallow cumulus cloud feedbacks: A review. *Surveys of Geophysics, 38*, 1331–1353. doi:10.1007/s10712-017-9418-2.

Walters, D., Baran, A. J., Boutle, I., Brooks, M., Earnshaw, P., Edwards, J., et al. (2019). The Met Office Unified Model global atmosphere 7.0/7.1 and JULES global land 7.0 configurations. *Geoscientific Model Development, 12*, 1909–1963. https://doi.org/10.5194/gmd-12-1909-2019.

Wang, Y.-C., Xie, S., Tang, S., & Lin, W. (2020). Evaluation of an improved convective triggering function: Observational evidence and SCM tests. *Journal of Geophysical Research, 125*, e2019JD031651. https://doi.org/10.1029/2019JD031651.

Webb, M., Senior, C., Bony, S., & Morcrette, J. J. (2001). Combining ERBE and ISCCP data to assess clouds in the Hadley Centre, ECMWF and LMD atmospheric climate models. *Climate Dynamics, 17*, 905–922.

Webster, P. J., & Lukas, R. (1992). TOGA COARE: The coupled ocean–atmosphere response experiment. *Bulletin of the American Meteorological Society, 73*, 1377–1416.

White, P. W. (Eds.) (2002). FULL-POS postprocessing and interpolation. IFS Documentation Part VI: Technical and Computational Procedures (CY25R1), Chapter 2. [Available at https://www.ecmwf.int/file/25439/download?token=ojphD9ZX].

Williams, K. D., Bodas-Salcedo, A., Déqué, M. , Fermepin, Medeiros, B., Watanabe, M., et al. (2013). The Transpose-AMIP II experiments and its application to the understanding of Southern Ocean cloud biases in climate models. *Journal of Climate, 26*, 3258–3274.

Williams, K. D., & Brooks, M. E. (2008). Initial tendencies of cloud regimes in the Met Office Unified Model. *Journal of Climate, 21*, 833–840.

Williamson, D. L., Boyle, J., Cederwall, R., Fiorino, M., Hnilo, J., Olson, J. G., et al. (2005). Moisture and temperature budgets at the Atmospheric Radiation Measurement Souhthern Great Plain site in forecasts with the Community Atmosphere Model (CAM2). *Journal of Geophysical Research, 110*, D15S16. doi:10.1029/2004JD005109.

Xie, P., & Arkin, P. A. (1997). Global Precipitation: A 17-year monthly analysis based on gauge observations, satellite estimates, and numerical model outputs. *Bulletin of the American Meteorological Society, 78*, 2539–2558.

Xie, S., Boyle, J. S., Klein, S. A., Liu, X., & Ghan, S. (2008). Simulations of Arctic Mixed-Phase Clouds in Forecasts with CAM3 and AM2 for M-PACE. *Journal of Geophysical Research, 113*, D04211. doi:10.1029/2007JD009225.

Xie, S., Cederwall, R. T., & Zhang, M. (2004a). Developing long-term single-column model/cloud system-resolving model forcing using numerical weather prediction products constrained by surface and top of the atmosphere observations. *Journal of Geophysical Research, 109*, D01104. doi:10.1029/2003JD004045.

Xie, S., Cederwall, R. T., Zhang, M., & Yio, J. J. (2003). Comparison of SCM and CSRM forcing data derived from the ECMWF model and from objective analysis at the ARM SGP site. *Journal of Geophysical Research, 108*(D16), 4499. doi:10.1029/2003JD003541.

Xie, S., Hume, T., Jakob, C., Klein, S. A., McCoy, R., & Zhang, M. (2010a). Observed large-scale structures and diabatic heating and drying profiles during TWP-ICE. *Journal of Climate, 23*, 57–79. doi:10.1175/2009JCLI3071.1.

Xie, S., Klein, S. A., Zhang, M., Yio, J. J., Cederall, R. T., & McCoy, R. B. (2006). Developing large-scale forcing data for single-column and cloud-resolving models from the Mixed-Phase Arctic Cloud Experiment. *Journal of Geophysical Research, 111*, D19104. doi:10.1029/2005JD006950.

Xie, S., Liu, X., Zhao, C., & Zhang, Y. (2013). Sensitivity of CAM5 simulated arctic clouds and radiation to ice nucleation. *Journal of Climate, 26*, 5981–5999. doi: 10.1175/JCLI-D-12-00517.1.

Xie, S., Ma, H.-Y., Boyle, J., Klein, S. A., & Zhang, Y. (2012). On the correspondence between short- and long- timescale systematic errors in CAM4/CAM5 for the Years of Tropical Convection. *Journal of Climate, 25*, 7937–7955. doi: http://dx.doi.org/10.1175/JCLI-D-12-00134.1.

Xie, S., McCoy, R. B., Klein, S. A., Cederwall, R. T., Wiscombe, W. J., Jensen, M. P., et al. (2010b). Clouds and more: ARM climate modeling best estimate data. *Bulletin of the American Meteorological Society, 91*, 13–20. doi:10.1175/2009BAMS2891.1.

Xie, S., Xu, K.-M., Cederwall, R. T., Bechtold, P., Del Genio, A., Klein, S. A., et al. (2002). Intercomparison and evaluation of cumulus parameterization under summertime midlatitude continental conditions. *Quarterly Journal of the Royal Meteorological Society*, *128*, 1095–1136. doi: 10.1256/003590002320373229.

Xie, S., & Zhang, M. H. (2000). Impact of the convection triggering function on single-column model simulations. *Journal of Geophysical Research, 105*, 14983–14996.

Xie, S., Zhang, M., Branson, M., Cederwall, R. T., Del Genio, A., Eitzen, Z. A., et al. (2005). Simulations of midlatitude frontal clouds by SCMs and CSRMs during the ARM March 2000 Cloud IOP. *Journal of Geophysical Research, 110*, D15S03. doi:10.1029/2004JD005119.

Xie, S., Zhang, M., Boyle, J. S., Cederwall, R. T., Potter, G. L., & Lin, W. (2004b). Impact of a revised convective triggering mechanism on Community Atmosphere Model, Version 2, simulations: Results from short-range weather forecasts. *Journal of Geophysical Research, 109*(D14), D14102. https://doi.org/10.1029/2004JD004692.

Xie, S., Zhang, Y., Giangrande, S. E., Jensen, M. P., McCoy, R., & Zhang, M. (2014). Interactions between cumulus convection and its environment as revealed by the MC3E sounding array. *Journal of Geophysical Research: Atmospheres, 119*, 11,784–11,808. https://doi.org/10.1002/2014JD022011.

Xu, K.-M., & Arakawa, A. (1992). Semi-prognostic tests of the Arakawa-Schubert cumulus parameterization using simulated data. *Journal of the Atmospheric Sciences*, *49*, 2421–2436.

Xu, K.-M., Arakawa, A., & Krueger, S, K. (1992). The macroscopic behavior of cumulus ensembles simulated by a cumulus ensemble model. *Journal of the Atmospheric Sciences, 49*, 2404–2420.

Xu, K.-M., Cederwall, R. T., Donner, L. J., Grabowski, W. W., Guichard, F., Johnson, D. E., et al. (2002). An Intercomparison of cloud-resolving models with the Atmospheric Radiation Measurement summer 1997 Intensive Observation Period data. *Quarterly Journal of the Royal Meteorological Society, 128*, 593–624.

Xu, K.-M., & Krueger, S. K. (1991). Evaluation of cloudiness parameterizations using a cumulus ensemble model. *Monthly Weather Review*, *119*, 342–367.

Xu, K.-M., & Randall, D. A. (1996a). A semiempirical cloudiness parameterization for use in climate models. *Journal of the Atmospheric Sciences, 53*, 3084–3102. doi:10.1175/1520-0469(1996)053,3084: ASCPFU.2.0.CO;2.

Xu, K.-M., & Randall, D. A. (1996b). Evaluation of statistically based cloudiness parameterizations used in climate models. *Journal of the Atmospheric Sciences, 53*, 3103–3119.

Xu, K.-M., Zhang, M., Eitzen, Z. A., Ghan, S. J., Klein, S. A., & Wu, X., (2005). Modeling springtime shallow and deep frontal clouds with CRMs and SCMs. *Journal of Geophysical Research, 110*, D15S04. doi:10.1029/2004JD005153.

Yanai, M., Esbensen, S., & Chu, J.-H. (1973). Determination of bulk properties of tropical cloud clusters from large-scale heat and moisture budgets. *Journal of the Atmospheric Sciences, 30*, 611–627.

Yang, F., Pan, H.-L., Krueger, S. K., Moorthi, S., & Lord, S. J. (2006). Evaluation of the NCEP global forecast system at the ARM SGP site. *Monthly Weather Review, 134*, 3668–3690. https://doi.org/10.1175/MWR3264.1

Yang, H., Dobbie, S., Mace, G. G., Ross, A., & Quante, M. (2012). GEWEX Cloud System Study (GCSS) cirrus cloud working group: Development of an observation-based case study for model evaluation. *Geoscientific Model Development, 5*, 829–843. doi:10.5194/gmd-5-829-2012.

Zhang, C., Xie, S., Tao, C., Tang, S., Emmenegger, T., Neelin, J. D., et al. (2020). The ARM data-oriented metrics and diagnostics package for climate models – A new tool for evaluating climate models with field data. *Bulletin of the American Meteorological Society, 101*, E1619–E1627. https://doi.org/10.1175/BAMS-D-19-0282.1.

Zhang, M., & Bretherton, C. S. (2008). Mechanism of low cloud–climate feedback in idealized single-column simulations with the Community Atmospheric Model, Version 3 (CAM3). *Journal of Climate, 21*, 4859–4878. https://doi.org/10.1175/2008JCLI2237.1.

Zhang, M., Bretherton, C. S., Blossey, P. N., Bony, S., Brient, F., & Golaz, J.-C. (2012). The CGILS experimental design to investigate low cloud feedbacks in general circulation models by using single-column and large-eddy simulation models. *Journal of Advances in Modeling Earth Systems, 4*, M12001. https://doi.org/10.1029/2012MS000182.

Zhang, M., Bretherton, C. S., Blossey, P. N., Austin, P. H., Bacmeister, J. T., Bony, S., et al. (2013). CGILS: Results from the first phase of an international project to understand the physical mechanisms of low cloud feedbacks in single column models. *Journal of Advances in Modeling Earth Systems, 5*, 826–842. https://doi.org/10.1002/2013MS000246.

Zhang, M., Somerville, R. C. J., & Xie, S. (2016). The SCM concept and creation of ARM forcing data. *Meteorological Monographs, 57*, 24.1–24.12. doi: 10.1175/AMSMONOGRAPHS-D-15-0040.1.

Zhang, M. H., & Lin, J. L. (1997). Constrained variational analysis of sounding data based on column-integrated budgets of mass, heat, moisture and momentum: approach and application to ARM measurements. *Journal of the Atmospheric Sciences, 54*, 1503–1524.

Zhang M. H., Lin, J. L., Cederwall, R. R., Yio, J., & Xie, S. (2001). Objective analysis of ARM IOP data: Method, features, and sensitivity. *Monthly Weather Review, 129*, 295–311. doi: 10.1175/1520-0493(2001)129<0295:OAOAID>2.0.CO;2.

Zhang, M. H., Lin, W., Klein, S. A., Bacmeister, J. T., Bony, S., Cederwall, R. T., et al. (2005). Comparing clouds and their seasonal variations in 10 atmospheric general circulation models with satellite measurements. *Journal of Geophysical Research, 110, D15S02*, doi:10.1029/2004JD005021.

Zhang, T., Zhang, M., Lin, W., Lin, Y., Xue, W., Yu, H., et al. (2018). Automatic tuning of the Community Atmospheric Model (CAM5) by using short-term hindcasts with an improved downhill simplex optimization method. *Geoscientific Model Development, 11*, 5189–5201. https://doi.org/10.5194/gmd-11-5189-2018.

Zhang, X., Lin, W., & Zhang, M. (2007). Toward understanding the double Intertropical Convergence Zone pathology in coupled ocean-atmosphere general circulation models. *Journal of Geophysical Research, 112*, D12102. doi:10.1029/2006JD007878.

Zhang, Y., Xie, S., Klein, S. A., Marchand, R., Kollias, P., Clothiaux, E. E., et al. (2018). The ARM cloud radar simulator for global climate models: Bridging field data and climate models. *Bulletin of the American Meteorological Society, 99*, 21–26. https://doi.org/10.1175/BAMS-D-16-0258.1.

Zhang, Y., Xie, S., Lin, W., Klein, S. A., Zelinka, M., Ma, P.-L., et al. (2019). Evaluation of clouds in version 1 of the E3SM atmosphere model with satellite simulators. *Journal of Advances in Modeling Earth Systems, 11*, 1253–1268. https://doi.org/10.1029/2018MS001562.

Zhao, C., Xie, S., Klein, S. A., Protat, A., Shupe, M. D., McFarlane, S. A., et al. (2012). Toward understanding of differences in current cloud retrievals of ARM ground-based measurements. *Journal of Geophysical Research, 117*, D10206. doi: 10.1029/2011JD016792.

Zheng, X., Klein, S. A., Ma, H.-Y., Bogenschutz, P., Gettelman, A., & Larson, V. E. (2016). Assessment of marine boundary layer cloud simulations in the Community Atmosphere Model with Cloud Layers Unified by Binormals and updated microphysics scheme based on ARM observations from the Azores. *Journal of Geophysical Research, 121*, 8472–8492. doi: 10.1002/2016JD025274.

Zheng, X., Klein, S. A., Ma, H.-Y., Caldwell, P. M., Larson, V. E., Gettelman, A., & Bogenschutz, P. (2017). A cloudy planetary boundary layer oscillation arising from the coupling of turbulence with precipitation in climate simulations. *Journal of Advances in Modeling Earth Systems, 9*, 1973–1993. doi: 10.1002/2017MS000993.

Zhou, B., & Chow, F. K. (2011). Large-eddy simulation of the stable boundary layer with explicit filtering and reconstruction turbulence Modeling. *Journal of the Atmospheric Sciences, 68*, 2142–2155. https://doi.org/10.1175/2011JAS3693.1.

Zhu, P., Bretherton, C. S., Köhler, M., Cheng, A., Chlond, A., Geng, Q., et al. (2005). Intercomparison and interpretation of single-column model simulations of a nocturnal stratocumulus-topped marine boundary layer. *Monthly Weather Review, 133*, 2741–2758. https://doi.org/10.1175/MWR2997.1.

18

Future Research Outlook: Challenges and Opportunities

Yangang Liu[1] and Pavlos Kollias[1,2]

ABSTRACT

This chapter synthesizes the recent cross-cutting advances and overarching trends across the various topics discussed in this book, highlights new challenges confronting further developments, and presents new opportunities for future research.

18.1. RECENT ADVANCES

In addition to the significant progresses of the various topics elucidated in the preceding chapters, several, worth highlighting, advancements in cross-cutting research areas are mentioned here. First, fast physics parameterizations have been gradually advancing from being primarily practical fix for individual fast processes to deep theoretical underpinning of stochastic scale interactions, and process interactions within the complex "4M-2N" system (recall Figure 3 in Chapter 1). These advancements are schematically illustrated in Figure 18.1 as three levels of development. Second, several different approaches/ideas have been proposed to unify the representation of distinct processes and their interactions. Third, the importance and necessity of top-down approaches have been increasingly recognized and ideas in other disciplines, such as statistical physics and chaos theory, have found valuable applications in atmospheric and climate sciences. Fourth, the essential role of organic integration of numerical models of different types, synthesis of measurements from different observing platforms, and effective evaluation of various parameterizations has been increasingly emphasized for further development. The growing interest of a systematic, process-oriented model-measurement synthesis further reinforces the pressing need for interdisciplinary research. Fifth, with the advancement in computer technology and growing computational power, model resolutions have been continually increasing, along with the development of unstructured or resolution-adaptive meshes. Finally, more physical processes that were ignored in early studies have started to be considered, for example, turbulent entrainment-mixing processes detailed in Chapter 4.

[1]*Department of Environmental and Climate Sciences, Brookhaven National Laboratory, Upton, NY, USA*

[2]*School of Marine and Atmospheric Sciences, Stony Brook University, Stony Brook, NY, USA*

18.2. CHALLENGES AND OPPORTUNITIES FOR FUTURE RESEARCH

The material presented in the previous chapters—especially the recent advances—reveal new challenges and opportunities for future research. Below we highlight 12 points with a focus on those relevant to more than one physical process/phenomenon.

First, for developing unified parameterizations that consider a variety of different processes together and address process interactions, new issues likely emerge that differ fundamentally from parameterizing individual processes, e.g., self-consistency between different physical

Fast Processes in Large-Scale Atmospheric Models: Progress, Challenges, and Opportunities, Geophysical Monograph 282, First Edition.
Edited by Yangang Liu and Pavlos Kollias.

DOI: 10.1002/9781119529019.ch18

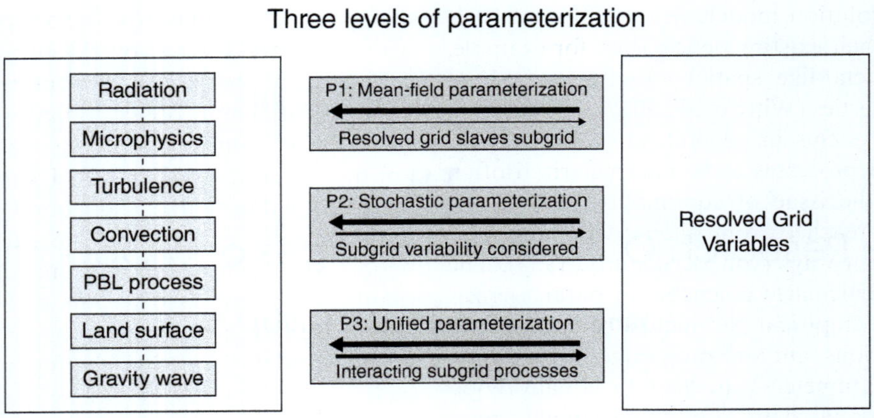

Figure 18.1 Schematic to illustrate the three levels of fast physics parameterizations. The left box lists the main subgrid processes, with the dotted vertical lines to indicate the fact that the interactions among subgrid processes have not been well represented in existing parameterizations. The three boxes in the middle show the three levels of parameterizations (P1, P2, and P3), respectively, from the top to the bottom. The thickness of the left-to-right arrow indicates the strength of considering the subgrid process couplings/interactions.

processes/variables (Liu & Daum, 2008; Park et al., 2017; Shaw and Shepherd, 2009).

Second, the feedbacks between different processes remain poorly understood and understudied, so is the numerical dynamics-physics coupling (Keyes et al., 2013; Gross et al. 2018). As the representation errors associated with each individual model component decrease, the correlated errors introduced by the component coupling will eventually dominate. Furthermore, the calling order of different process parameterizations in the conventional splitting treatment was recently found to affect not only current-climate behavior but also the model's prediction of climate sensitivity, suggesting that process ordering should be considered carefully during model development and that optimizing the order of calling different processes should be explored together with proper parameter tuning (Donahue & Caldwell, 2018).

Third, fast physics parameterizations will continue to be major sources of uncertainty in weather and climate prediction models. In particular, those associated with sub-LES scale processes fraught with significant knowledge gaps will likely become more acute as the resolutions of weather and climate models increase, including adequate representation of shallow convection, turbulent entrainment-mixing processes, and turbulence-microphysics interactions. Using process modeling studies for parameterization development is essential because it ensures that increased knowledge on processes can be incorporated into models in a physically realistic way. Particle-resolved direct numerical simulation models that can resolve the smallest turbulent eddies and follow individual cloud particles and its integration with LES models hold great potential in this area (Gao et al., 2018; Wang et al., 2011).

Fourth, as for cloud microphysics and aerosol-cloud interactions, several areas of research warrant more attention in future, including aerosol-cloud interactions in the updraft-limited regime, multimoment schemes that consider the roles of spectral shape of hydrometeor size distributions (Milbrandt & Yau, 2015a, 2015b), droplet size distributions (Chen et al., 2016, 2018; Liu et al., 2023), and wet removal of aerosols due to ice processes. Adequately coupling entrainment/detrainment and turbulent mixing processes with cloud microphysics and turbulence representation merits further investigation (Xu et al., 2022). Also moving to the forefront is the issue of aerosol-cloud continuum and transitional zone between clouds and aerosols (Charlson et al., 2007; Marshak et al., 2021), which calls for further unification of aerosol processes, cloud microphysics, and those processes expecting to significantly affect cloud edges such as entrainment-mixing and radiative effects on cloud microphysics (Zeng et al., 2022).

Fifth, most fast physics parameterizations remain essentially one dimensional in the vertical direction; such 1D representation will likely experience a shift to more realistic 3D representation for processes such as radiative transfer, gravity waves, and turbulence. This shift will pose new challenges not only to representing individual processes in question but more so to considering the process interactions (e.g., gravity-turbulence interactions) and coherent structures (Dritschel, 2013; Penner et al., 2018). Results from high-resolution 3D numerical simulations and machine learning modeling are expected to play growing roles in addressing such challenges. For this, improving high-resolution models (e.g., LES models) is crucial in view of their existing deficiencies, for example, in representing fractal-like spatial structures of cloud microphysical properties (Witte et al., 2022).

Sixth, as open systems in general, atmospheric phenomena or the fast processes to be parameterized often need to confront the issue of equifinality because the final state may be reached from different initial conditions and in different ways (von Bertalanffy, 1950). The same concept is particularly amenable to parameterizations with multiple empirical parameters constrained by uncertain observations and/or knowledge gaps. Many different sets of parameters can lead to equally good fit to the observational data, and these equally good models/parameterizations may lead to widely divergent model results under new conditions (Tang & Zhuang, 2008). The barrier of equifinality likely conspires with the curse of parameter dimensionality to present challenges to constraining models with observations, and to conducting model evaluation and parameter calibration (Mülmenstädt et al., 2020). Without a direct way of measuring key quantities in fast physics parameterizations, tuning parameters have been often estimated indirectly. Such an exercise implicitly assumes that errors from other parts of the model including the dynamical core and other parameterizations are negligible, and that the mathematical form of the parameterization to be tuned is correct. However, these assumptions are generally not valid, which results likely in the occurrence of error compensations or correct results for wrong physical reasons. Somewhat related are the use of emergent constraints to reduce the uncertainty in climate sensitivity (Caldwell et al., 2018; Eyring et al., 2019) and the use of known physical relationships to improve data-driven machine learning forecast models (Liu et al., 2021, 2022). All these potential issues together call for rigorous uncertainty quantification of observational data, modeling results, and the parameters to be calibrated. Again, machine learning models hold tremendous potential(s) in these areas.

Seventh, the climate or atmospheric system can be seen as a forced and dissipative nonequilibrium thermodynamic system (Lucarini et al., 2014; Wu & Liu, 2010) and generating entropy via a variety of irreversible processes. Subgrid-scale processes have long been known to be a critical issue when attempting to provide an accurate climate entropy budget. The consistency with the second law of thermodynamics has been understudied and needs further attention (Akmaev, 2008; Gassmann, 2018; Gassmann and Herzog, 2015; Kleidon and Lorenz, 2005; Kunz et al., 2008). An important question to address is whether the standard model equations and physical parametrizations are already compliant with the second law. It may also be useful to analyze other processes and their parametrizations from the perspective of compliance with the second law and overall entropy inventory.

Eighth, this book also demonstrates the complementarity of different observational platforms, including space-based remote sensing, surface-based remote sensing, aircraft-enhanced field campaigns, distributed observational networks, and laboratory experiments with well-controlled conditions. The next decade will bring major advancements in our measurement capabilities thanks to the rise of CubeSats and the development of data-driven dynamical observing systems (Kollias et al., 2020). Multiple lines of developing trends are anticipated, including multiangular sampling or scanning, multiwavelength, polarization, and 3D tomography for remote sensing; and high-resolution 3D holographic in situ sampling. These measurements are collected at different sampling scales and rates, and for disparate variables, and thus adequate data synthesis and integration will be needed to generate multiscale scale and 3D data sets readily available for process study and evaluation of various model types with different grid spacing and domain sizes. Such big-data tasks will be daunting and can benefit from advances in artificial intelligence and machine learning.

Ninth, another challenge we need to address to optimize atmospheric science experiments is the "valley of death" between modeling requirements and observational capabilities. Often scientists involved in the design of experiments to test theory and models have limited understanding of the real information content of measurements provided by a particular field experiment. We advocate for the development of open access, comprehensive Observing System Simulation experiment (OSSE) software packages that will generate in-silico versions of field experiments. The "digital twin" of an observational campaign will allow atmospheric scientists to evaluate proposed measurements before the experiment is conducted. It will simulate the effects of observing system design decisions (i.e., type of radar, sampling strategy, and drone-sampling strategy) on measurement data (Oue et al., 2022; Sanchez et al., 2022) and their downstream ability to constrain and improve model predictions. Developing digital twins presents a tremendous opportunity to transform the indispensable integration of measurements and models across multiple scales (Bauer et al., 2021)

Tenth, the problem of modeling weather and climate can generally be organized in terms of seven dimensions: resolution, complexity, integration length, ensemble size, data assimilation, parameter calibration, and measurements, each of which competes for computational resources (Navarra et al., 2010). The inevitable resource limitation necessitates the need to strike an optimal balance among the various tasks associated with these dimensions, for example, having better fast physics representation vs. assimilating more quality measurements or high model resolutions. The growing use of ML models is anticipated in such endeavors.

Eleventh, the continual increase of model resolutions and development of unstructured or resolution-adaptive meshes present new requirements for parameterization development. For example, the need for representing coupled subgrid variabilities and developing scale-aware parameterizations will become increasingly important. The resolution-dependent partitioning between resolved and subgrid contributions is also part of the scale-awareness solution as the weather and climate models move toward the gray zone of "terra incognita" (Malavelle et al., 2014; Schneider et al., 2007; Wyngaard, 2004).

Last but not least, despite the growing recognition of atmospheric sciences in improving prediction of weather, climate, and renewable sources of energy, the sizes of related communities are relatively small, which is particularly true for the parameterization development community (Jakob, 2010). The lack of sufficient manpower remains a serious issue to our society as a whole in view of the high benefit-to-cost ratio of the fields. New efforts to inspire and expand the next-generation workforce are desirable, including training of multidisciplinary scientists and outreach to the public.

ACKNOWLEDGMENTS

The volume editors thank all the chapter authors for their valuable contributions to this book. Dr. Leo Donner deserves special recognition; he co-convened various sessions on relevant topics at AGU Fall Meetings and helped with the conception of this book. We are indebted to Mr. Theodore Sampieri of the BNL Guest Service for his help with English presentation. The relevant research has been mainly supported by the U.S. Department of Energy's Office of Science Biological and Environmental Research program as part of the Atmospheric Systems Research (ASR) program and Earth System Modeling (ESM) program, and the Office of Energy Efficiency and Renewable Energy (EERE) under Solar Energy Technologies Office (SETO) award number 33504. Brookhaven National Laboratory is operated by Battelle for the U.S. Department of Energy under contract DE-Sc18112704.

REFERENCES

Akmaev, R. A. (2008). On the energetics of maximum-entropy temperature profiles. *Quarterly Journal of the Royal Meteorological Society*, *134*, 187–197.

Bauer, P., Stevens, B., & Hazeleger, W. (2021). A digital twin of Earth for the green transition. *Nature Climate Change*, *11*(2), 80–83. https://doi.org/10.1038/s41558-021-00986-y.

Caldwell, P. M., Zelinka, M. D., & Klein, S. A. (2018). Evaluating emergent constraints on equilibrium climate sensitivity. *Journal of Climate*, *31*, 3921–3942.

Charlson, R. J., Ackerman, A. S., Bender, F. A. M., Anderson, T .L., & Liu, Z. (2007). On the climate forcing consequences of the albedo continuum between cloudy and clear air. *Tellus B: Chemical and Physical Meteorology*, *59*, 715–727.

Chen, J., Liu, Y., Zhang, M., & Peng, Y. (2016). New understanding and quantification of the regime dependence of aerosol-cloud interaction for studying aerosol indirect effects. *Geophysical Research Letters*, *43*. doi:10.1002/2016GL067683.

Chen, J., Liu, Y., Zhang, M., & Peng, Y. (2018). Height dependency of aerosol cloud interaction regimes. *Journal of Geophysical Research: Atmospheres*, *123*, 491–506. https://doi.org/10.1002/2017JD027431.

Donahue, A. S., & Caldwell, P. M. (2018). Impact of physics parameterization ordering in a global atmosphere model. *Journal of Advances in Modeling Earth Systems*, *10*, 481–499. https://doi.org/10.1002/2017MS001067.

Dritschel, D. G. (2013). Waves and turbulence: Their cooperative role in structure formation. *Procedia IUTAM*, *8*, 85–93.

Eyring, V., & coauthors (2019). Taking climate model evaluation to the next level. *Nature Climate Change*, *9*, 102–110. https://doi.org/10.1038/s41558-018-0355-y.

Gao, Z., Liu, Y., Li, X., & Lu, C. (2018). Investigation of turbulent entrainment-mixing processes with a new particle-resolved direct numerical simulation model. *Journal of Geophysical Research: Atmospheres*, *123*, 2194–2214. https://doi.org/10.1002/2017JD027507 modeling. *Journal of Geophysical Research: Atmospheres*, *120*, 3852–3871. https://doi.org/10.1002/2014JD022180.

Gassmann, A. (2018). Entropy production due to subgrid-scale thermal fluxes with application to breaking gravity waves. *Quarterly Journal of the Royal Meteorological Society*, *144*, 499–510. doi:10.1002/qj.3221

Gassmann, A., & Herzog, H. J. (2015). How is local material entropy production represented in a numerical model? *Quarterly Journal of the Royal Meteorological Society*, *141*, 854–869.

Gross, M. S., Wan, H., Rasch, P. J., Caldwell, P. M., Williamson, D. L., Klocke, D., et al. (2018). Physics–dynamics coupling in weather, climate, and Earth system models: Challenges and recent progress. *Monthly Weather Review*, *146*, 3505–3544. doi: 10.1175/MWR-D-17-0345.1

Jakob, C. (2010). Accelerating progress in global atmospheric model development through improved parameterizations: Challenges, opportunities, and strategies. *Bulletin of the American Meteorological Society*, *91*(7), 869–876. https://doi.org/10.1175/2009BAMS2898.1

Keyes, D. E., McInnes, L. C., Woodward, C., et al. (2013). Multiphysics simulations: Challenges and opportunities. *International Journal of High Performance Computing Applications*, *27*, 4–83, https://doi.org/10.1177/1094342012468181.

Kleidon, A., & Lorenz, R. D. (Eds.). (2005). *Non-equilibrium thermodynamics and the production of entropy: Life, earth, and beyond*. Berlin: Springer.

Kollias, P., Bharadwaj, N., Clothiaux, E. E., Lamer, K., Oue, M., Hardin, J., et al. (2020). The ARM radar network: At the leading edge of cloud and precipitation observations. *Bulletin of the American Meteorological Society*, *101*, E588–E607. doi: 10.1175/bams-d-18-0288.1

Kollias, P., Luke, E., Oue, M., & Lamer, K. (2020). Agile adaptive radar sampling of fast-evolving atmospheric phenomena guided by satellite imagery and surface cameras. *Geophysical Research Letters, 45*, e2020GL088440. https://doi.org/10.1029/2020GL088440

Kunz, T., Fraedrich, K., & Kirk, E. (2008). Optimisation of simplified GCMs using circulation indices and maximum entropy production. *Climate Dynamics*, *30*, 803–813. doi 10.1007/s00382-007-0325-y.

Liu, W., Liu, Y., Zhou, X., Xie, Y., Han, Y., Yoo, S., & Sengupta, M. (2021). Use of physics to improve solar forecast: Physics-informed persistence models for simultaneously forecasting GHI, DNI, and DHI. *Solar Energy*, *215*, 252–265. https://doi.org/10.1016/j.solener.2020.12.045.

Liu, W., Liu, Y., Zhang, T., Han, Y., Zhou, X., Xie, Y., & Yoo, S. (2022). Use of physics to improve solar forecast: Part II, machine learning and model interpretability. *Solar Energy*, *244*, 362–378. https://doi.org/10.1016/j.solener.2022.08.040.

Liu, Y. (2019). Introduction to special section on fast physics in climate models: parameterization, evaluation and observation. *Journal of Geophysical Research: Atmospheres*, *124*, 8631–8644. https://doi.org/10.1029/2019JD030422.

Liu, Y., & Daum, P. H. (2008). Relationship of refractive index to mass density and self-consistency of mixing rules for multicompoent mixtures like ambient aerosols. *Journal of Aerosol Science*, *39*, 974–986.

Liu, Y., Yau, M.-K., Shima, S., Lu, C., & Chen, S. (2023). Parameterization and explicit modeling of cloud microphysics: Approaches, challenges, and future directions. *Advances in Atmospheric Sciences*, *40*, 747–790. https://doi.org/10.1007/s00376-022-2077-3

Lucarini, V., Blender, R., Herbert, C., Ragone, F., Pascale, S., & Wouters, J. (2014). Mathematical and physical ideas for climate science. *Reviews of Geophysics*, *52*, 809–859. doi:10.1002/2013RG000446.

Malavelle, F. F., Haywood, J. M., Field, P. R., Hill, A. A., Abel, S. J., Lock, A. P., et al. (2014). A method to represent subgrid-scale updraft velocity in kilometer-scale models: Implication for aerosol activation. *Journal of Geophysical Research: Atmospheres*, *119*, 4149–4173. doi:10.1002/2013JD021218.

Marshak, A., Ackerman, A., Da Silva, T., Eck, B., Holben, R., Kahn, R., et al. (2021). Aerosol properties in cloudy environments from remote sensing observations: review of current state of knowledge. *Bulletin of the American Meteorological Society*, E2176–E2107, https://doi.org/10.1175/BAMS-D-20-0225.1.

Milbrandt, J. A., & Yau, M. K. (2005a). A multimoment bulk microphysics parameterization. Part I: Analysis of the role of the spectral shape parameter. *Journal of the Atmospheric Sciences*, *62*(9), 3051–3064. https://doi.org/10.1175/jas3534.1

Milbrandt, J. A., & Yau, M. K. (2005b). A multimoment bulk microphysics parameterization. Part II: A proposed three-moment closure and scheme description. *Journal of the Atmospheric Sciences*, *62*(9), 3065–3081. https://doi.org/10.1175/jas3535.1

Mülmenstädt, J., Nam, C., Salzmann, M., Kretzschmar, J., L'Ecuyer, T. S., Lohmann, U., et al. (2020). Reducing the aerosol forcing uncertainty using observational constraints on warm rain processes. *Science Advances*, *6*, eaaz6433

Navarra, A., Kinter III, J. L., & Tribbia, J. (2010): Crucial experiments in climate science. *Bulletin of the American Meteorological Society*, *91*, 343–352.

Oue, M., Saleeby, S. M., Marinescu, P. J., Kollias, P., & van den Heever, S. C. (2022). Optimizing radar scan strategies for tracking isolated deep convection using observing system simulation experiments. *Atmospheric Measurement Techniques*, *15*, 4931–4950. https://doi.org/10.5194/amt-15-4931-2022.

Park, S., Baek, E.-H., Kim, B.-M., & Kim, S.-J. (2017). Impact of detrained cumulus on climate simulated by the Community Atmosphere Model Version 5 with a unified convection scheme. *Journal of Advances in Modeling Earth Systems*, *9*, 1399–1411, doi:10.1002/2016MS000877.

Penner, J. E., Zhou, C., Garnier, A., & Mitchell, D. L. (2018). Anthropogenic aerosol indirect effects in cirrus clouds. *Journal of Geophysical Research: Atmospheres*, *123*, 11,652–11,677. https://doi.org/10.1029/2018JD029204.

Sanchez, G. M., Lundquist, J. K., Mirocha, J. D., Arthur, R. S., Muñoz-Esparza, D., & Robey, R. (2022). Can lidars assess wind plant blockage in simple terrain? A WRF-LES study, *Journal of Renewable and Sustainable Energy, 14*, 063303, https://doi.org/10.1063/5.0103668.

Schneider, T., Teixeira, J., Bretherton, C. S., Brient, F., Pressel, K. G., Schär, C. S., & Siebesma, A. P. (2007). Climate goals and computing the future of clouds. *Nature Climate Change*, *7*, 3–5.

Shaw, T. A., & Shepherd, T. G. (2009). A theoretical framework for energy and momentum consistency in subgrid-scale parameterization for climate models. *Journal of the Atmospheric Sciences*, *66*, 3095, 3114. doi: 10.1175/2009JAS3051.1.

Tang, J., & Zhuang, Q. (2008). Equifinality in parameterization of process-based biogeochemistry models: A significant uncertainty source to the estimation of regional carbon dynamics. *Journal of Geophysical Research*, *113*, G04010. doi:10.1029/2008JG000757.

von Bertalanffy, L. (1950). The theory of open systems in physics and biology. *Science*, *111*(2872), 23–29. https://doi.org/10.1126/ science.111.2872.23.

Wang, L., Orlando, A., Parishani, H., Grabowski, W. W., Wyszogrodzki, A. A., Piotrowski, Z., et al. (2011). Towards an integrated multiscale simulation of turbulent clouds on PetaScale computers. *Journal of Physics: Conference Series*, *318*, 072021 doi:10.1088/1742-6596/318/7/072021.

Witte, M. K., Morrison, H., Davis, A. B., & Teixeir, J. (2022). Limitations of bin and bulk microphysics in reproducing the observed spatial structure of light precipitation. *Journal of the Atmospheric Sciences*, *79*, 161–178. doi: 10.1175/JAS-D-21-0134.1.

Wu, W., & Liu, Y. (2010). Radiation entropy flux and entropy production of the Earth system. *Reviews of Geophysics*, *48*, RG2003. doi:10.1029/2008RG000275.

Wyngaard, J. (2004). Toward numerical modelling in the "terra incognita." *Journal of the Atmospheric Sciences*, *61*, 1816–1826.

Xu, X., Lu, C., Liu, Y., Luo, S., Zhou, X., Endo, S., et al. (2022). Influences of entrainment-mixing parameterization on numerical simulations of cumulus and stratocumulus clouds. *Atmospheric Chemistry and Physics, 22*, 5459–5475. https://doi.org/10.5194/acp-22-5459-2022.

Zeng, X., Heymsfield, A. J., Ulanowski, Z., Neely, III, R. R., Li, X., Gong, J., & Wu, D. L. (2022). The radiative effect on cloud microphysics from the arctic to the tropics. *Bulletin of the American Meteorological Society, 103*(9), E2108–E2129. doi 10.1175/BAMS-D-21-0039.1.

INDEX

Fast Processes in Large-Scale Atmospheric Models: Progress, Challenges, and Opportunities, Geophysical Monograph 282, First Edition.
Edited by Yangang Liu and Pavlos Kollias.

DOI: 10.1002/9781119529019.bindex